Phytoplankton Whispering: An Introduction to the Physiology and Ecology of Microalgae

Patricia M. Glibert

Phytoplankton Whispering: An Introduction to the Physiology and Ecology of Microalgae

Springer

Patricia M. Glibert
University of Maryland Center for Environmental Science
Cambridge, MD, USA

ISBN 978-3-031-53896-4 ISBN 978-3-031-53897-1 (eBook)
https://doi.org/10.1007/978-3-031-53897-1

Cover Image: Satellite image of algae blooms in the Baltic Sea in July 2019. Image from the Copernicus Sentinel-2 mission, processed by the European Space Agency, reproduced under a creative commons license.

This Springer imprint is published by the registered company Springer Nature Switzerland AG
The registered company address is: Gewerbestrasse 11, 6330 Cham, Switzerland

To

Coe, Emma, Exy, Sanne, Penrhyn, Otto, and Piper

May your adventures on the water also fill you with wonder of the water itself

Preface

In a review of the history of the study of plankton productivity, oceanographers Dick Barber and Anna Hilting wrote, "1952 is a milestone in the advance of biological oceanography" (Barber and Hilting 2002). While I would love to think that they were referring to the year I was born, they used that milestone instead to mark the long struggle to develop a method for primary productivity in the sea, the ^{14}C method (Steemann Nielsen 1952). This method—and the controversies of its use—features prominently in the early chapters of this book. As methods advanced, and continue to advance, our insights also advance.

The year 1952 is also the year in which nitrogen fertilizer began to be used commercially, forever altering the waters of the world through nutrient pollution. Nutrient pollution is examined in detail in the latter chapters of this book. I entered this world on its very early trajectory of eutrophication. I was not concerned about this phenomenon early in my scientific career, during which I worried about how the oligotrophic oceans could obtain sufficient nitrogen to sustain productivity. I did not imagine then—and am still stunned by—the fact that my questions now center around how phytoplankton cope, and how phytoplankton communities change, with the excesses of nutrients.

The concept of primary productivity is at the heart of understanding how the oceans function. Understanding what it is, who carries out the process, how different species and functional groups differ, and how the processes work are the underlying themes of this book. This includes topics ranging from carbon and nutrient acquisition and how and why it varies to why it is important to food webs and the global Earth and the changes that are occurring. My insights into plankton ecology and biological oceanography have all been shaped by the era in which global productivity have been studied with advanced analytical tools. I have witnessed the power of new tools and methods in unraveling complex questions. My insights have also been shaped by profound global changes—in temperatures, in atmospheric CO_2, in nutrients, and in the phytoplankton themselves and their dynamics. Along this journey, I have learned what Colin Reynolds learned years ago; that is, we are motivated today by many of the same questions and phenomena that many have asked before us. As he said, "*Thus, the circle has turned fully and the questions asked today are the same ones which engaged the pioneers of phytoplankton ecology: what lives where and why?*" (Reynolds 1998). However, we ask these questions with new methods and with new urgency of the global changes around us.

This book is intended to take the student on the journey of discovery of the amazing microorganisms that drift in the sea. It is a book about change—changes in how we study these creatures, and how global changes are changing them. It is also a book that introduces not only the beautiful drifters of the sea, but the many, in some cases colorful, in some cases curmudgeonly, characters who have looked at, poked, prodded, measured, and modeled the lifestyle, biogeography, and importance of the phytoplankton.

This book is aimed at undergraduates and graduate students launching into phytoplankton studies—or even those who have been marveling at these creatures for a long time. It is by no means comprehensive regarding all topics and concepts related to phytoplankton. This is not a book that covers all the concepts of biological oceanography, although many concepts related to biological oceanography are introduced. This is not a book on marine ecology although many concepts related to marine ecology are introduced. This is not a how-to book, nor a what-is book; this book is geared toward conceptual relationships. The diversity of the 80,000-plus species of phytoplankton means that there is broad scope for adaptive strategies and for adaptation. This effort is intended to make sense of why compositional changes are occurring with phytoplankton considering the global changes that have occurred across the globe. Who are the winners and the losers as conditions change? By recognizing the "rules" by which these organisms live, the features of different types of phytoplankton, and how they adapt to their conditions, the future condition may be less of a mystery. The choice of emphasis on certain processes and questions is of course mine, as are the errors that have crept into this manuscript.

Why *whisper* about phytoplankton? An animal "whisperer" is one who thinks, feels, and understands the needs of the animal. The term first entered the popular lexicon following the novel—and subsequent movie—*The Horse Whisperer* (Evans 1995; Redford 1998), the story of a man with the skill and calm demeanor to tame horses. He was able to understand and feel the horse. Here, it is hoped that one comes away being a phytoplankton whisperer: someone who understands the needs and responsive behaviors of these most amazing and complex organisms.

The book begins with some history and a look at how methods shape our understanding. Light and photosynthesis are next addressed, followed by a description of the algae themselves and how they have evolved. Nutrient cycles and the physiology of nutrient acquisition are next addressed, leading to the topic of mixotrophy. The high-nutrient-low chlorophyll regions of the sea are also addressed, and the role of iron as a limiting nutrient. Next are topics related to size, stoichiometry, conceptual models of succession, loss processes, and the phycosphere. The next section of the book looks at global changes, from temperature to hypoxia, harmful algal blooms, ocean acidification, and some extreme changes that are occurring in various parts of the oceans. The last sections of the book take a closer look at modeling and controlling blooms and, finally, a look at the progress that has (or has not) been made in diversifying our community of aquatic scientists.

I am forever indebted to my partner, my husband, Todd Kana. His insight, his knowledge of physiology, and his perspective on life have kept me going. He patiently read many chapter drafts, corrected errors, and asked probing questions. Many colleagues also contributed in multiple ways, from reading and editing drafts to sharing preprints and images. I especially thank Shady Amin, JoAnn Burkholder, Dick Dugdale, Kevin Flynn, Cynthia Heil, Ming Li, Aditee Mitra, Sandy Shumway, Diane Stoecker, Frances Wilkerson, and others who will remain nameless. The fact that you shared images, comments, edits, and insights at various stages along this journey means so much. My students and postdocs (past and present) are also thanked for long contributing to the phytoplankton discovery journey that I have undertaken. I thank our phytoplankton reading group, helping me to sharpen and clarify all sorts of confusing concepts and ideas. Sophia Ahn, Yuren Chen, Amanda Muni-Morgan, Bruna Sobrinho, and Nayani Vidyarathna, thank you. I am also exceptionally grateful for the help of Jane Hawkey, whose artful hand and careful editing skills helped to bring this effort over the finish line. I also thank Kate, Austin, and Patrick for their encouragement—it means more than you can possibly ever know. Finally, I thank Éva Lőrinczi from Springer Nature for embracing this effort in their portfolio and for her patience as the manuscript was prepared. This could not have been completed without you! Thank you all.

Funding for this effort was provided by the Ecosystem Science Program of the US National Science Foundation, award number DEB 1911349, the US National Oceanographic and Atmospheric Administration, award numbers NA17NOS4780180 and NA19NOS4780183, and the University of Maryland Center for Environmental Science (UMCES). This is contribution number 6360 from UMCES and number 1095 from the NOAA Ecology of Harmful Algal Blooms Program (ECOHAB). Thank you.

References

Barber RT, Hilting AK (2002) History of the study of plankton productivity. In: Williams PJ leB, Thomas DN, Reynolds CS (eds) Plankton productivity: carbon assimilation in marine and freshwater ecosystems. Blackwell Scientific Publishing, Oxford, pp 16–43

Evans, N (1995) The horse whisperer. Delacorte Press, New York. ISBN 978-0-385-31523-4

Redford, R (director) (1998) The horse whisperer. Touchstone Pictures

Reynolds CS (1988) Functional morphology and the adaptive strategies of freshwater phytoplankton. In Sandgren CD (ed) Growth and reproductive strategies of freshwater phytoplankton. Cambridge University Press, Cambridge, pp 338–433

Steemann NE (1952) The use of radioactive carbon (^{14}C) for measuring organic production in the sea. J Cons Int Explor Mer 18:117–140

Cambridge, MD, USA Patricia M. Glibert

Contents

Historical Phytoplankton Ecologists, Early Concepts and Debates

1

1.1 Introduction

There has long been a fascination with plankton, how they differ, where and why they occur, and the processes that control them. In 1676, the Dutch scientist, Antoni van Leeuwenhoek discovered what he called "animalcules," using his then recently developed—but still very crude—light microscope. He described these creatures as "so small, in my sight, that I judged that even if 100 of these very wee animals lay stretched out one against another, they could not reach the length of a grain of coarse sand." He further described "little feet" on protists, now known to be the cilia that propel them along (Cochlan 2015). It is thought that among the earliest plankton he saw may have been the photosynthetic mixotroph *Mesodinium rubrum* (de Baar 1994), and thus there is good reason that he may have been confused as to whether he was witnessing a microscopic plant or animal.

Captain James Cook, the British explorer, although not a scientist by trade, left a legacy of descriptions of the sea from his voyages. In 1770, as he pursued his exploration along the coast of what is now Australia, Cook noted that many parts of the sea were covered with a kind of brown scum, which he examined under a microscope and concluded "it was of the vegetative kind" (Cook 1825). This is clearly a description of an algal bloom, now thought to be a bloom of the cyanobacterium *Trichodesmium*.

It was about a 100 years later when the Danish scientist Otto Friedrich Müller, as well as others, published a number of works describing some individual species of phytoplankton. The period of 1753–1882 was consequently termed the "discovery of microalgae period" (Anderson 2004). During this time, microalgae were described as infusoria, a term that included any microscopic organism that lived in freshwater and was largely treated as animals (Anderson 2004). The late 1880s and early 1900s were a period of great exploration and development of theories related to phytoplankton production.

Whether driven by efforts to understand fisheries, or just by the natural beauty of these elegant microbial creatures, there has been an abundance of questions,

P. M. Glibert, *Phytoplankton Whispering: An Introduction to the Physiology and Ecology of Microalgae*, https://doi.org/10.1007/978-3-031-53897-1_1

methods, debates, and arguments, some friendly, some less so, then and now, on what plankton are, how to best measure them, and how they function. Some of these debates are introduced here to underscore two important points. First, many of our current ideas and debates have had a long history, and second, it is the acquisition of new data, often with new approaches and methods, that allow us to resolve long standing disputes. It is also worth mentioning that not all now well-accepted concepts were well received when first introduced; this is a good lesson for students who may become discouraged if their ideas (and manuscripts) are not immediately accepted.

1.2 Victor Hensen: The Water Is Blooming!

Victor Hensen (Fig. 1.1) is credited with stating in 1887, "My original aim was to arrive at an estimate of the plankton production of the sea" (quoted in Mills 1989, p. 7). He is also credited with first use of the term *plankton* to represent the floating plants and animals of the sea. With only a handful of publications on plankton (most of them between 1890 and 1901), and no formal training in the field (he was actually trained in anatomy and studied the physiology of hearing, inspired perhaps by his father who ran an institution for the deaf), this professor of physiology at Kiel, Germany, had a profound impact on the field of plankton. Coming into the field by way of Germany's growing interest in fisheries research and funding of research expeditions in the Baltic (Mills 1989), Hensen used the opportunity to focus on the

Fig. 1.1 Christian Andreas Victor Hensen, February 10, 1835–April 05, 1924. Image reproduced from Schleswig-Holsteinische Landesbibliothek with permission

physics and chemistry of the water and its relation to commercial fisheries. He was convinced that life in the sea was nourished by the plankton (Dolan 2021). As he noted (Hensen 1906),

> The water is blooming! . . . the surface of quiet, brackish bays appears to be covered with a green layer which consists of minute spheres or threads. . .At such times the water is filled with these primitive kinds of plants.

Although most of Hensen's work was published in German (e.g., Hensen 1887, 1906), he did establish both techniques for quantifying plankton and worked not only on identifications of organisms but also their weight and chemical composition to the extent possible with then-available methods (which largely involved collecting phytoplankton by plankton nets). Hensen also proposed the notion that other than in back bays, the oceans were poor in phytoplankton and in those nutrient-poor waters, their distribution was fairly uniform in space and time. This conclusion was based on his expeditions not only in the Baltic, but also across and around the North Atlantic, during which he observed that the physical properties of water (temperature, salinity) were well mixed in surface waters due to winds and waves, and thus he concluded, the regularity of plankton should also follow. His axiom was that mixing yields regularity, and only where coasts or currents existed could this uniformity be disrupted. Although proven wrong in this respect, he also described that the oceans, including the Sargasso Sea, were poor in plankton, suggesting that lack of inorganic nutrients might be responsible. Hensen based his ideas on cells collected from nets and counts, including from nets that he specifically designed to get quantitative vertical tows (Dolan 2021). He developed a suite of mathematical approaches which lent a degree of credence to this theory. His approach—of applying mathematics to his observations—was novel, but it was not accepted by all. Indeed, his quantitative approach was "totally repugnant to many German biologists during the late nineteenth century" (Mills 1989, p. 37). As will be seen below, it was years before it was acceptable to apply mathematics to the study of these natural phenomena.

Hensen's later work moved away from direct mathematical constructs, but mathematics was never far from his reasoning. He began to reason that plankton—and other organisms—were part of the larger cycles of matter and energy. He also distinguished stock (biomass) from production (dynamics), and stated, "to estimate production we must either try to learn the speed of destruction of preexisting material or the rate of origin; of course, it would be best to know both" (quoted in Mills 1989, p. 41). He understood that there could be no more animals than the food available to them, and he characterized fish as the "apex of the sea" (Mills 1989, p. 40). As he began to understand the differences between production and total number of organisms, he incorporated the understanding of carbon (C) and nitrogen (N) flow into an early conceptual model. His model required that in order to make fish, C and N had to pass through a "cylinder of metabolism" and "the cross-section of this cylinder must then correspond with the quantity of fish existing at the moment being considered. . .Thus it would be sufficient to know this [the cross-section] and the total height of the cylinder to obtain a satisfactory picture of relative metabolism [i.e., production]. This is the task which will make precise the statistics of metabolism of the sea" (quoted in Mills, 1989, p. 41).

1.3 Ernst Haeckel and Other Hensen Antagonists

A colleague of Hensen, Ernst Haeckel, lambasted the work of Hensen, criticizing not only his conclusions (the regularity of plankton in the sea) but also his approaches (counting cells collected in nets) (Dolan 2021). Haeckel was a German zoologist and is known for his beautiful illustrations of plankton. He is best known for his iconic and exquisite drawings of the plankton collected during the Challenger expedition (Fig. 1.2).

Haeckel was flamboyant and aggressive. Based on years of his own observations in the Atlantic, the Indian Ocean, and the Mediterranean Sea, he concluded that phytoplankton were unevenly distributed in the oceans, and stated, "The fundamental fault of Hensen's plankton theory in my opinion lies in the fact that he regards a highly complicated problem of biology as a relatively simple one, that he regards its

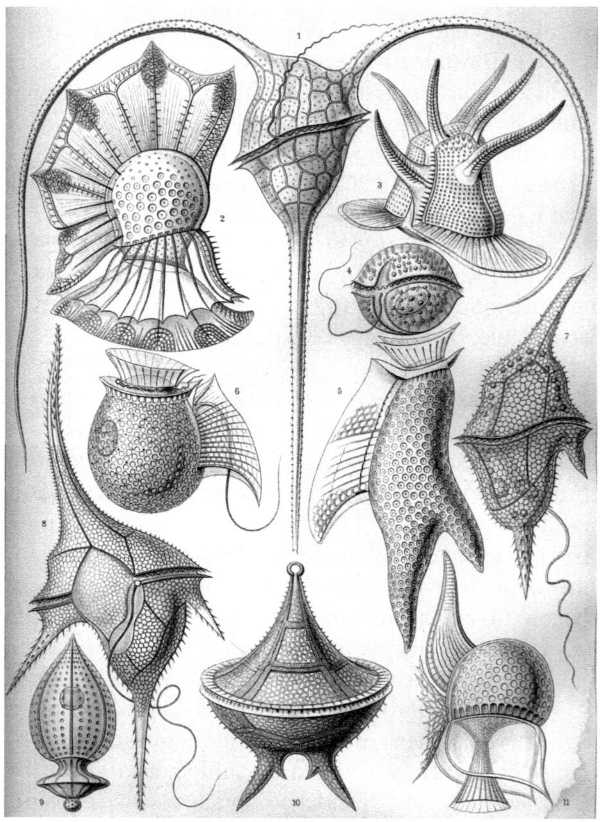

Fig. 1.2 Some of Ernst Haeckel's iconic drawings, from the 14th plate from "Kunstformen der Natur" (1904), depicting organisms classified as Peridinea. Image reproduced from https://commons.wikimedia.org/wiki/File:Haeckel_Peridinea.jpg under a creative commons license

many oscillating parts as proportionately constant bulks, and that he believes that a knowledge of these can be reached by the exact method of mathematical counting and computation" (quoted in Mills 1989, p. 29). Reflecting the thinking of the time—that biological sciences and mathematical tools should not be mixed, Haeckel further stated (quoted in Mills 1989, p. 31) that, "Mathematical treatment. . .does more harm than good, because it gives a deceptive semblance of accuracy, which in fact is not attainable." Haeckel felt that Hensen had applied defective and unsuitable collecting techniques by simply counting phytoplankton. Haeckel viewed the distribution of phytoplankton through the lens of rates of change and suggested that local conditions and temporal changes were important controlling factors. It was not just the total cell numbers that mattered. Moreover, he also believed that large predators that devour massive quantities of plankton daily exert a great influence, and thus he introduced the idea of "top-down" control (Mills 1989).

Haeckel also had considerable interest in evolution and extended some of his ideas to the evolution of humans and scientific racism. In a blunt assessment of Haeckel's concepts of human evolution, the Harvard Paleontologist Stephen Jay Gould (1977, pp. 77–78) wrote, "Haeckel's evolutionary racisms; his call to the German people for racial purity and unflinching devotion to a 'just' state; his belief that harsh, inexorable laws of evolution rules human civilization and nature alike, conferring upon races the right to dominate others. . .all contributed to the rise of Nazism."

Haeckel was not the only critic of Hensen or his methodologies. More criticisms of Hensen's work followed, but Hensen held fast to the notion that counting cells was not only suitable but provided basic information. Hans Lohmann, also working in Germany at the time, a more subtle critic of Hensen (Mills 1989) observed that small cells, which he terms "nannoplankton" passed through nets in common use, and he speculated that these small cells could be important, especially in seas which were seemingly plankton poor (Mills 1989). Lohmann's work went on to show that the sea had as much as 5–100 times as much plankton as captured in Hensen's nets. So believing in his theory of "regularity" of phytoplankton distributions, Hensen dismissed reports of irregular distributions as being the product of faulty collective methods. It would be a hundred years before the discovery of picoplankton and the full recognition of the role of these tiny cells in the ocean (see Chap. 2).

1.4 Karl Brandt, the Nitrogen Cycle and the Importance of Bacteria

One of Hensen's assistants in Kiel (and later an investigator in his own right) was Karl Brandt. His focus was the link between plankton abundance, its differential distribution at high latitudes and the tropics, and nutrient control (Mills 1989). Particularly interested in chemistry, and relating agricultural principles to the study of plankton, he focused on the chemical composition of the plankton, aiming to show whether different types of plankton could be related to the abundance of other organisms, as is the case with forage crops on land. By assessing C, N, and silica

(Si), along with lipid cellulose and protein and carbohydrate composition, he was able to show that diatoms were likely to be important food sources for planktonic animals. At the time, Hensen held the view that diatoms were likely indigestible because of their Si shells (Mills 1989).

Brandt set out to understand not only patterns in distribution but the laws underlying these distributions. He worked to apply, "the fundamental principles and approved methods of animal and vegetable physiology which have been deduced from the terrestrial world," and in particular, "it is. . .of the first importance to follow the cycle of transformation of nitrogen in nature" (quoted in Mills 1989, p. 50). Thus, he described the N cycle on land, including such processes as the oxidation of ammonia (NH_4^+) to nitrite (NO_2^-) and nitrate (NO_3^-), reduction of NO_3^- to atmospheric N_2, and the fixation of atmospheric N_2 by legumes, and he then applied these principles to the sea. He concluded that N must be the limiting nutrient in the sea.

This notion was particularly important because it required the presence of bacteria in the sea—then still a topic of considerable debate. As Mills (1989, pp. 50–51) summarized Brandt's thinking, "Bacteria like those active in the terrestrial nitrogen cycle must be present in the sea, since one would expect the nitrogen (mainly nitrate) from the land to be washed into the oceans, where over millenia the nitrate concentration of seawater could rise to poisonous levels. But the sea appeared to contain relatively small amounts of combined nitrogen compared to land. How could this be? Brandt suggested that 'according to the current state of knowledge, this inconsistency will only be made comprehensible by the hypothesis that *denitrifying bacteria in the ocean remove the excess of nitrogen compounds and that it is they that restore the prevailing equilibrium of nature's household* (Brandt 1899, p. 8).'" Years of work were devoted to the study of the N cycle, but the processes of nitrification and denitrification remained elusive and enigmatic. Sampling and experiments by Brandt, his students and associates, did confirm the presence of denitrifying bacteria. Finding nitrifiers proved a bit more problematic. In fact, decades after Brandt's death, Claude ZoBell (1946), perhaps the most influential marine microbiologist of the time (Williams and Ducklow 2019) stated that "Although, in light of circumstantial evidence, it is tacitly assumed by many oceanographers that nitrate is formed by bacteria at the bottom of the sea, and thence carried into the photosynthetic zone, there is not conclusive evidence bearing on either the mode or the place of formation of nitrate in the sea" (quoted in Mills 1989, p. 72). ZoBell's thinking on nitrifiers was consistent with his general view—and that of most Western marine microbiologists at the time—that there were very few bacteria living in the water column of the oceans (reviewed by Williams and Ducklow 2019). This view evolved from the practice at the time of isolating and growing bacteria in the lab using amended seawater or solid media (agar) and finding that, in general, few bacteria grew.

Brandt continued to pursue his theory of the importance of the N cycle in the ocean. Lack of nitrifiers in isolated samples did not deter him. He suggested that perhaps organic N compounds played an important role, and thus NH_4^+-oxidizing bacteria could be important, in solving the nitrification problem (Mills 1989). It was only in the 1980s that nitrifiers in the sea were clearly isolated, after issues of plate colony isolation were overcome with new methods and the sensitivity to light became well recognized (e.g., Olson 1981a, 1981b; Ward et al. 1982). Here, ZoBell's work was also related. ZoBell recognized the importance of organic matter and bacterial activity, but he held to the view that such activity would be found associated with particles and surfaces where organic matter concentrations were much higher than in open seawater. As Williams and Ducklow (2019) summarized, ZoBell, because he was so well regarded at the time, was rarely questioned, and thus, "his immediate influence on the field of ocean bacterial ecology was to stifle consideration of the scale of bacterial activity in relation to other components of the plankton because of the great attention directed to bacterial proliferation in bottles and on surfaces" (Williams and Ducklow 2019, p. 43).

Brandt was heavily influenced by Justus von Liebig, an agricultural chemist of the mid-to-late nineteenth century, who is largely credited with the development of the Law of the Minimum. It is important to underscore, however, that correct attribution of this concept should go to Phillipp Carl Sprengel, a German botanist, but it was Liebig who popularized the concept (Elser and Haygarth 2012). Brandt's application of this concept to the sea was encapsulated in his writing (Brandt 1902, pp. 45–46, quoted in Mills 1989, p. 105).

> Plants produce as much organic material as is possible under the general conditions of life under which they exist and according to what is possible by the inorganic nutrients at their disposal; *but thereby they are subjected to the Law of the Minimum. If one alone of the indispensable plant nutrients is present in relatively smaller amounts the production is also sparse*. . .As a result, the intensity of production depends upon the amount of that indispensable plant nutrient which is relatively most scantily replaced.

Over the years, Brandt continued to pursue the idea that the N cycle and the production of plankton were related; with each expedition or experiment he added complexity and nuance to his basic framework. He was not deterred by his detractors. He observed a striking inverse relationship between NO_3^- values and temperature after summarizing samples from various parts of the world. He also explored how NO_3^- concentrations varied with depth, testing the hypothesis that if pressure or temperature alone were responsible for this relationship, then a smooth relationship would be observed. But, he found a peak in NO_3^- at about 800 meters. This led to further pursuits of the relationships with physical factors (circulation), and the relationship between NO_3^- and NH_4^+. His data showed that the 800-meter maximum in NO_3^- corresponded with minima in NH_4^+ (although generally low everywhere), adding further evidence to his notion that NH_4^+ oxidizers played an important role (Mills 1989).

Yet, through the early twenty-first century, the disconnect between theory (bacterial cycling of N) and evidence, from the bacteriologists of the time, such as ZoBell (Williams and Ducklow 2019) remained highly contentious. Mills (1989, p. 95), quoting W.A. Herdman, a professor at Liverpool in the 1920s, highlighted this disagreement between the Brandt "camp" and the microbiologists "camp" when he stated,

> Brandt first attributed the poverty of plankton in the tropics to the destruction of nitrates in the sea as a result of the greater intensity of the metabolism of denitrifying bacteria in the warmer water; and various other writers have since then have more or less agreed that the presence of these denitrifying bacteria, by keeping down to a minimum the nitrogen concentration in the tropical waters, may account for the relative scarcity of the phytoplankton, and consequently of the zoo-plankton, that has been observed. But [H.H.] Gran, [Alexander] Nathansohn, Murray, Hjort and others have shown that such bacteria are rare or absent in the open sea, that their action must be negligible, and that Brandt's hypothesis is untenable.

Haaken Hasberg (hereafter H.H.) Gran, a Norwegian botanist, maintained that while denitrifying bacteria existed, they could not play a major role in the marine N cycle. Gran doubted there was enough organic matter in waters away from the coast to support the C source required for denitrification. Brandt countered by noting that microenvironments—such as on feces of organisms, and other organic matter could be important sources. Supporting Gran was Alexander Nathansohn, who also believed that denitrification played little role in the oceans, believing instead that NO_3^- in the oceans was low because it was rapidly consumed by algae, outcompeting the denitrifiers for this nutrient source. Another supporter of Gran was J. Gebbing, also swayed by the lack of evidence of denitrifiers and nitrifiers. He stated with confidence that, "Indeed one can assume with considerable certainty that this oxidation of ammonia to nitrite and nitrate may also take place in the sea by purely chemical means. Therefore one would be able to dispense with the hypothesis of nitrifying bacteria as essential for the cycle of nitrogen in the sea" (Gebbing 1909, pp. 191–192 as quoted in Mills 1989, p. 99). Moreover, both Nathansohn and Gebbing denied the relationship between temperature and nutrients or plankton abundance, and also held the view that dissolved N was never low enough to be limiting (Mills 1989). Even with the heavy weight of criticism, Brandt stayed true to his original hypothesis, maintaining that nitrifiers used dissolved organic nutrients rather than NH_4^+.

Brandt's critics, Nathansohn and Gebbing, responded by claiming that exhaustion of nutrients in the sea was not possible where grazers and decomposers "self-fertilized" the phytoplankton and recycled nutrients as excreta (Mills 1989). Countering the Nathansohn and Gebbing view, and drawing on then-known concepts from plant physiology, Brandt surmised that as long as there was a lower limit below which plants (including phytoplankton) could not absorb the inorganic nutrients, limitation by nutrients was possible even if there were low residual, measurable concentrations. Moreover, he held that the sea was not unique in being grazed or "self-fertilized" (Mills 1989).

Although not embracing the idea that nutrient limitation alone was controlling, Nathansohn understood that the planktonic system was in *dynamic balance* (Nathonsohn 1908, quoted in de Baar 1994). He expressed this concept—a concept well ahead of its time—by articulating his ideas as follows:

> When we now turn to the study of external, especially chemical, conditions for the plankton and its quantitative distribution, we will clearly recognize the great variety of problems that

> ensue once we consider the continual growing and dying of the plankton, and once we keep in mind that its biomass results from the balance of these two processes.
>
> It makes a great difference whether we study a nutrient solution in respect to its maximum potential of production. . .or in respect to the rate at which the production in it proceeds, as we will have to do when we want to clarify the conditions of plankton development. The first variable is, in accordance with the repeatedly mentioned Liebig's Law, always dependent on the one nutrient which happens to be in the minimum, while the latter can be affected by all possible components.

Here, Brandt was the skeptical one. He responded to Nathansohn's distinction between production and productivity by calling Nathansohn's concept, "a mere play on words, because productivity, the realized rate of growth, must depend on nutrients. Productivity is governed by growth rate, growth rate by limiting nutrients. To claim otherwise was to confuse concepts, to employ empty verbiage" (Brandt 1920 quoted in Mills 1989, p. 108).

The debates took on more complexity as the physics of the oceans began to be better understood. Nathansohn used "compelling logical grounds" to argue that vertical water circulation took the place of the N cycle and the Law of the Minimum in explaining the distribution of plankton (Nathansohn 1906 quoted in Mills 1989, p. 110). Accordingly, "if ammonia was lost from the sea to the atmosphere there would be no need for denitrification. And because ammonia was lost, and thus scarce in the sea, nitrifiers too could be dispensed with. . .large attached algae would out-compete denitrifying bacteria for nitrate. . .once the role of denitrifying bacteria had been discounted, temperature became relatively unimportant in phytoplankton abundance (Mills 1989, pp. 110–111). The appeal of the theory of control by physical processes was powerful at the time. The many expeditions of the time were yielding large amounts of supportive data from all the oceans of the world, and according to Mills (1989), Nathansohn's ideas were widely read and accepted well into the early 20th century, and "Brandt's theory was quietly set aside as idiosyncratic and out of fashion" (Mills 1989, p. 119).

1.5 Haaken Hasberg Gran and Understanding Seasonality

The debate over regularity of the plankton in the sea (Hensen's notion) gradually gave way to curiosity of the spring bloom. Franz Schütt, a botanist and plant physiologist of the late twentieth century, was fascinated by seasonally recurring blooms, describing the sequence as follows (Schütt 1892 quoted in Mills 1989, p. 125):

> One form appears, grows and vanishes yet again from the surface waters and makes way for another form, which now asserts its dominance for its own time, yet again to fade away, and this play repeats itself year after year with the same regularity as every spring the trees turn green and in autumn lose their leaves; with just such absolute certainty as the cherries bloom before the sunflowers, so Skeletonemas arrive at their yearly peak earlier than the Ceratiums.

H.H. Gran, when not debating the N cycle, also undertook analysis of the spring bloom phenomenon, especially that which occurred in the Norwegian Sea. He described the "striking characteristics" that accompanied a spring bloom, namely "it always began in shallow water; it was brought about by neritic, spore-forming diatoms; it was dominated by a succession of arctic or northern species; it reached a maximum between March and May, depending on altitude, usually just after the water reached its lowest temperature; and it disappeared one to two months after it began" (Mills 1989, p. 128). His perspective of the spring bloom, which incorporated the concept of nutrient supply and demand, and physical processes and changes in vertical distributions was one that contributed to early conceptual models of the annual plankton cycle.

Hans Lohmann's views also weighed in on the factors controlling the spring bloom. Recall he was the more subtle critic of Hensen and introduced the notion that "nannoplankton" were important. With regard to the spring bloom, he argued that the interactions of light and temperature were the important factors, with the increase in light the most significant factor leading to bloom initiation, and that the volume of summer plankton was a product of temperature and light. Thus, he was dismissive of the law of the minimum as a controlling factor (Mills 1989). This raised an interesting important question: was the seeming decline in the biomass of plankton in the summer a real phenomenon, or was it as artifact of the fact that plankton collections were being underrepresented because collections were then based on plankton nets, mostly sampled in summer when communities were dominated by small cells? Brandt, however, continued to be convinced of the importance of the regularity in the changes in nutrients in relation to the phytoplankton. By this time in the 1920s, enough nutrient data had been collected that true trends were being revealed. Brandt maintained that plankton were controlled by several nutrients, each obeying the law of the minimum, and that different species required different proportions of nutrients or had different rates of uptake, and thus a succession of species could be related directly to changes in nutrients. Bloom decline ended by nutrient depletion, and only when NO_3^- increased in late winter could the bloom renew.

By the 1930s, new insight into the factors controlling production was emerging. Gran's work again played an important role. His interest in vertical distribution of phytoplankton was growing, fueled by his appreciation of the physiological balance between photosynthesis and respiration. Thus, he maintained that there must be a depth in the water column where photosynthesis and respiration would be balanced, and below which no net production could occur. Moreover, in the Plymouth Laboratory, new insights were being gained about the levels of nutrients and their changes during spring blooms. This latter work, undertaken largely by W.R.G. Atkins brought to bear a new phosphate (PO_4^{3-}) assay that confirmed depletion of this nutrient after the spring bloom. Earlier analyses were found to be contaminated as the assay also picked up arsenate. He advocated that as waters warmed, a stable surface layer was formed, and in such a layer the phytoplankton could remove virtually all PO_4^{3-}. Collectively, these views led to the conclusions that vertical circulation positively promoted blooms, while thermal stratification negatively affected blooms by creating conditions where nutrients could be deleted.

Brandt added to this notion by pointing out that as vertical circulation occurs, it mixes phytoplankton throughout the water column, and unless diatoms have mechanisms to maintain their water column position, they would inevitably sink out.

These trailblazing phytoplankton ecologists further speculated on the role of nutrients other than N and P—a micronutrient that a century later became a major focus of research. This was iron. Gran hypothesized that (Mills 1989, p. 166),

> . . .if the productivity of coastal waters is dependent on any factor of a chemical nature acting as a minimum factor, it must be an element which in its circulation does not follow the nitrates and phosphates accumulating in solution in the deep sea and reaching the surface again by vertical circulation of any kind. If such minimum stuffs exist, they must be irreversibly go out of circulation in the sea, so that they can only be renewed from land.

Gran even demonstrated the importance of iron in nutrient addition experiments in cultures.

Collectively, Brandt, Nathansohn, and Gran brought great insight into many of the questions of phytoplankton abundance and dynamics and the underlying factors controlling them. They were hampered by limited methods, but their perspectives and their debates—sometimes contentious—laid a strong foundation for much of the work to come. What is clear is that the roles of biology (physiology), chemistry (nutrients and nutrient cycles), and physics were all being elucidated in a complex, intertwined picture. No longer were comparatively simplistic "naturalist" observations sufficient for science. As noted by W.A. Herdman (1920), a British marine biologist (quoted in Mills 1989, p. 208),

> In the best interest of biological progress the day of the naturalist who merely collects, the day of the anatomist and histologist who merely describe, is over, and the future is with the observer and the experimenter animated by a divine curiosity to enter into the life of the organism and understand how it lives and moves and has its being.

The next decades saw a shift in research on plankton dynamics, or as Mills (1989) described, "it fell into an institutional gap." This was partly due to the institutional structure in Kiel, as the number of professorial opportunities declined, and due to the disruptions of World War I, and the Kiel school itself was ultimately disbanded. As many institutions have experienced, "research groups have short lives, because scientific fashion, the availability of resources, and the relevance of scientific questions change quickly in tangible and not-so-tangible ways" (Mills 1989, pp. 185–196). The shift was also motivated by the concerning trend of variable and declining fisheries.

The ensuing decade, leading into the mid-1930s, was one that saw the major work on phytoplankton shifting from the German investigators to those at the Plymouth Laboratory in England. Among the most well-known plankton investigators of that time was H.W. Harvey, a chemist by training. He studied gases, then temperature effects on fisheries, and he compiled hydrographic data and studied relationships between nutrient changes, pH, and physical characteristics of the water column. Harvey also developed new methods for NO_3^-, and the lower levels resolved by this new method revealed the very low levels present in late summer, especially in the English Channel, leading to renewed interest in the N cycle. He, however, was perplexed by levels of NH_4^+, which showed variations unrelated to changes in NO_3^- a finding that was unexpected given the transformations known in the N cycle. He noted, "it is not clear how the plants get a sufficiency of nutrient salts during the summer months, at least in such places as the upper layers of the English Channel unless they utilize ammonium as it is being formed" (Harvey 1933 quoted in Mills 1989, p. 229). This spawned numerous experiments on NH_4^+ uptake by phytoplankton, and a better understanding of nutrient regeneration. Harvey published a paper in 1926 (cited in Mills 1989), in which he introduced (or re-introduced if one recalls the work of Nathansohn) the important concept of the balance of production and nutrient supply. He articulated that nutrient concentrations per se were less important in determining phytoplankton production than was the rate at which nutrients were supplied. He further described the various pathways of resupply, including excretion, bacterial remineralization, and efflux of nutrient-rich waters from deeper layers (Mills 1989). Harvey's book, *Biological Chemistry and Physics of Sea Water*, which he published in 1928, summarized these perspectives and his view of the ecosystem as a whole, synthesizing light, nutrient levels, uptake of nutrients, regeneration, phytoplankton production and grazing. Three decades later, still an active researcher, his thoughts were again echoing those of Nathansohn at the early turn of the century. In this work, he focused on the dynamic balance of the ecosystem and how "an equilibrium between the standing crop of plants, herbivores and carnivores is continually passing in and out of balance. . .The extent and frequency which *the balance of life* in the sea passes in and out of equilibrium increases the proportion of organic matter destroyed by non-biological and bacterial respiration" (Harvey 1955 quoted in Mills 1989, p. 249). Harvey's laboratory work ended when his laboratory was bombed in 1941.

1.6 Einer Steeman Nielsen and Light in the Sea

Light in the sea became a focus of study as new instruments to measure light became available in the mid-1930s. This history is a story in and of itself, a story in which the Danish botanist and phytoplankton ecologist, Einer Steemann Nielsen, plays a starring role, but a fuller account of his contribution is deferred to Chap. 2. He is best known for introducing the ^{14}C technique to phytoplankton research in the 1950s and revolutionizing both the methods for, and our understanding of, rates of primary productivity. Suffice it to say here that he, too, was interested in the intersection of quantitative ecology and plankton production. As Søndergaard (2002, p. 1) said in his biography of Steemann Nielsen, "Problems with his eyesight forced him to give up the many hours at the microscope and he changed, in his own words, 'to an experimental biologist', where marine primary production, plant physiology and specifically photosynthesis, light adaptation and inorganic carbon assimilation became his 'pet areas.'"

Controversies abounded as Steemann Nielsen's newly introduced ^{14}C method for primary productivity did not yield results that matched the then-widely used oxygen method and the reasons for the discrepancy. Steeman Nielsen was especially critical of the method that was common at the time, that of measuring the change in oxygen over time. It is also worth noting that "Steemann had a temper and was not very diplomatic in his critique of other people's work. . .He never strove for a diplomatic tone in his judgements" (Søndergaard (2002, p. 8). With the oxygen method, obtaining a sufficient signal for detection required long incubations, which was thought to lead to large experimental error due to the confinement of cells in small volumes. Discrepancies between methods were up to an order of magnitude. Settling the argument—and an argument it was—of how phytoplankton grow in the oceans took several more decades (see Chap. 2), although discrepancies between productivity measurements exist to this day (Marra 2019).

1.7 Gordon Riley and Colleagues and Their Quantitative Approaches

The work of H.W. Harvey and his colleagues at the Plymouth Laboratory inspired researchers in the USA, primarily at Yale University and at the Woods Hole Oceanographic Institution. One of those scientists, Gordon Riley, became a graduate student in the lab of the limnologist, G. Evelyn Hutchinson, at Yale. Together they transformed the world of aquatic population ecology into a rigorous, mathematical science. The resistance to mathematical applications, which Hensen encountered, still existed. In fact, in a monograph that Riley submitted based on his thesis research, the reviewers required that the theoretical statistical analysis be removed, as only the descriptive sections were deemed appropriate for publication! (Mills 1989). Or, as Riley said, his early papers were received with "a dull thud" (Riley 1980 interview, quoted in Mills 1989, p. 287). Riley's primary approach was to test relationships and hypotheses using multiple regression analysis. This approach was very sophisticated for the time, given that all calculations had to be done by hand. Riley and Hutchinson were not deterred. Riley described their approach (Riley 1980, quoted in Mills 1989, pp. 261–262):

> [We were]. . .a little arrogant about this. I felt that this was the way we were going to and damn it we were going to do it even if it wasn't appreciated, and we had a rather snooty attitude toward some of these old boys that were still content to just publish species lists and call it ecology.

Just as surprising is the fact that these scientists built their science *based on testing hypotheses*. Although all the researchers described so for in this chapter implicitly were pursuing new questions and new data based on questions they had, true hypothesis testing was only formally being applied in fields such as agriculture and fisheries at that time (Mills 1989). Few biologists of the time had the necessary knowledge of mathematics or statistics. World War II, 1941–1946, also ended much of the laboratory and at sea work of Riley. Yet, the war also transformed oceanography in the USA. Navy contracts through the Woods Hole Oceanographic Institution brought in many talented scientists, and mathematical modeling greatly advanced. Years later, in 1946, Riley's crowning achievement was a mathematical model of the production on George's Bank, based on a set of simple differential equations—and an explicit inclusion of physical processes (Riley 1946). While his thinking aligned in large measure with that of Harvey (whom he never met), and in many regards with that of Nathansohn and Gran, Riley was able to achieve what they had not: that of uniting physical and biological processes in a unified, rigorous framework.

Riley's approach on the one hand was "a way of thinking" (Riley 1963, quoted in Mills 1989, p. 304), and on the other hand was "a form of artistic expression" (Riley 1984, quoted in Mills 1989, p. 305), and he was later recognized as "one of the most creative minds to enter the field in the last five decades" (Mills 1989, p. 307). Yet, even in the 1960s–1970s, skeptics remained. John Strickland, one of the distinguished scientists and founder of the Food Chain Research Group at Scripps Institution of Oceanography, as late as the 1960s, "complained that models restate the obvious," and that they were data poor and premature, to which Riley responded that "models clarify thought, but that they should be combined with the detailed study of biological processes to yield the best understanding of the complexity of biological relationships" (Riley 1963, quoted in Mills 1989, p. 305).

1.8 Summary

It is the contributions of these individuals described in this chapter—and many others—who over the course of these decades from the early 1900s began to understand, and quantify phytoplankton, as well as the N cycle. They elucidated how nutrients, vertical circulation and light, and the stability of the surface waters all played a role in regulating algal growth and the annual cycle of production and how production is related to fish production. The many clashes of creativity and controversy by strong-willed (almost exclusively) men laid the foundation of what is modern-day plankton biology.

By the early 1960s, biological oceanography and plankton dynamics had become fully established disciplines. As articulated by Mills (1989, p. 313), many vexing problems still remained, such as,

> The relationship between carbon and chlorophyll in phytoplankton cells was so variable that it was difficult to find a common currency for biomass and production estimates. Very little was known about the response of most phytoplankton cells to light, so that it was unclear if a single or many photosynthesis: light intensity (PI) curves would have to be determined. What was the level of plant respiration, and how could it be considered constant in modeling or metabolic calculations? And how could one best express the uptake of nutrient by plant cells, including the complications of caused by inhibiting levels and the previous history of cells? The environmental setting of the cells also posed many unsolved problems, many of which had been overcome by unrealistic [experiments] in Riley's and Steele's models. Nutrient regeneration was such a problem; where, by what mechanisms, and at what rates

> did it take place during the seasonal plankton cycle? How could vertical mixing best be estimated and how did it affect nutrient supply to the euphotic zone? What environmental factors governed the varying sinking rates of the cells, thus controlling their nutrient and light relations? Grazing, accepted as a major controlling factor since at least the mid 1930s, still could not be measured reliably. Even primary production, the most basic of measurements, had become uncertain because of the discrepancy between estimates based on ^{14}C uptake and those on the light and dark bottle technique. How could this be overcome? Even the successful models, which deliberately evaded or ignores these problems were very simple; they seldom could be extended beyond the grazing animals, or at best beyond the primary carnivores, so that models encompassing a whole food web from phytoplankton to fish seemed beyond reach...Even the postulate that the sea was uniform enough to sample easily had to be questioned...Similarly, the reasonable postulate that the time delays occurring between, say, the growth of phytoplankton and the growth of herbivores feeding on them could be ignored in simple models was unrealistic.

To these questions, we now have to recognize that since the mid-1980s, the following world changes have occurred: 85% of the world's N fertilizer has been created, atmospheric CO_2 has increased 30%, world population has increased >40%, 50% of world's wetlands have been lost, and global temperatures have risen by more than 1 °C (see Chap. 22). Algal toxins are creeping into drinking water supplies and contaminating seafood. How are phytoplankton and the global oceans responding? What can we expect in a future world? Why are some species succeeding while others not? What are the principles and concepts—built on years of study of these delicate organisms—that inform the trajectory of change? What are the implications of change for the food web? How can we quantify and project the changes that will occur? What can we do to control the change of the future?

It should be clear by now that the world of phytoplankton has classically been viewed through the notion that energy (i.e., C) and material (i.e., nutrients) *acquisition* are the major regulators of how much biomass accumulates in a system and what types of species dominate (accounting for the fact that phytoplankton are also eaten thus modulating biomass accumulation). Indeed, phytoplankton must acquire the necessary substrates to grow and survive. However, this overarching view that *acquisition* alone is the driving force is based on the assumption that phytoplankton mostly live where these resources are scarce. Since we now live in a world with, and understand more, about nutrient pollution, eutrophication, and other environmental changes, we recognize that cells differ in their ability to regulate not only how much they *acquire*, but in their ability to *regulate* energy and material supply in excess. Cells must balance what they need with that they do not need: a delicate but *dynamic balance*. This notion may be simple, but the underlying physiology is complex.

It is here that our work now begins, but the story of the shoulders on whom we stand is not over; historical figures are introduced (or re-introduced) throughout subsequent chapters in relation to specific concepts. The next chapter introduces how changing methodologies have changed our perspectives of what organisms exist in the plankton and the processes they perform. Then, the concept of growth is introduced and how we measure and manipulate it. The next part of this book deals with light as a resource, how plankton differ in their abilities to capture light, how we measure both light and photosynthesis, and the diversity of algae, their evolution and their adaptations. Next, nutrients as a resource are considered, how nutrients are cycled, and how plankton acquire and assimilate nutrients and balance their stoichiometry. Life in the community of plankton is then considered, from the perspective of microbial interactions and frameworks by which we can tame the complexity. In the latter section of the book, the focus turns to the consequences of our changing world, including rising temperatures, hypoxia, harmful algal blooms, acidification, and examples of systems under extreme change. Lastly, the book circles back to modeling approaches and the advances that have been made in modeling and touches on some of the approaches—and the corresponding debates—about how algal blooms and other global changes can be controlled.

References

Anderson RA (2004) Biology and systematics of heterokont and haptophyte algae. Amer J Bot 91: 1508–1522

Brandt K (1899) Ueber den Stoffwechsel im Meere. Rede beim Antritt des Rektorates der Koniglichen Christian-Albrechts-Univeritat zu Kiel am 6 Marz 1988. Universitats-Buchhandlung, Kiel, p 36

Brandt K (1902) Ueber den Stoffwechsel im Meere.2. Abhandlung. Wissenschaftliche Meeresuntersuchungen, Abteilung Keil, Neue Folge 6:23–79

Brandt K (1920) Ueber den Stoffwechsel im Meere. 3. Abhaandlung. Wissenschaftliche Meeresuntersuchungen, Abteilung Keil, Neue Folge 18:246–248

Cochlan A (2015) Leeuwenhoek's 'animalcules', just as he saw them 340 years ago. New scientist 20 May, 2015. https://www.newscientist.com/article/dn27563-leeuwenhoeks-animalcules-just-as-he-saw-them-340-years-ago/#ixzz6JbiFyh9N

Cook J (1825) The voyages of captain James Cook round the world; comprehending a history of the South Sea islands, vol 1. Jaques and Wright, Paternoster Row

de Baar HJW (1994) VonLiebig's law of the minimum and plankton ecology (1899–1991). Prog Oceanogr 33:347–386

Dolan JR (2021) Pioneers of plankton research: victor Hensen (1985–1924). J Plankton Res 43: 507–510. https://doi.org/10.1093/plankt/fbab045

Elser J, Haygarth P (2012) Phosphorus past and future. Oxford University Press, New York, p 229

Gebbing J (1909) Chemische untersuchungen von meeresboden–, meerwasser–, und luft–proben der Deutschen südpolar expedition, 1901–1903, pp. 75–234 in E. von Drygalski, Deutsche südpolar expedition, 1901–1903, Band 7(2)

Gould SJ (1977) Ontogeny and phylogeny. Belknap Press of Harvard University Press, Cambridge MA, p 501

Haeckel E (1904) Kunstformen der Natur. Republished by Bibliographisches Institut, Leipzig und Wien, p 2012

Harvey HW (1926) Nitrate in the sea. J Mar Biol Assoc UK 15:183–190

Harvey HW (1928) Biological chemistry and physics of sea water. Cambridge University Press, Cambridge, p 194

Harvey HW (1933) On the rate of diatom growth. J Mar Biol Assoc UK 19:253–275

Harvey HW (1955) The chemistry and fertility of sea waters. Cambridge University Press, Cambridge, p 234

Hensen V (1887) Über die Bestimmung des Planktons oder des im Meere treibenden Materials an Pflanzen und Thieren. Kommission zur wissenenschaftlichen Untersuchung der deutschen Meere in Kiel, 1882-1886. V. Bericht, Jahrgang 12–16:1–107

Hensen V (1906) Die biologie des meres. Rede am Stifungsfest des naturwissenschaftlichen vereins in Kile. Arch Fur Hydrobiol 1:360–377

Herdman WA (1920) Oceanography and the sea-fisheries. Report of the British Association for the Advancement of Science, Cardiff 88:1–33

Marra JF (2019) Hot carbon: carbon-14 and a revolution in science. Columbia University Press, New York, p 264

Mills EL (1989) Biological oceanography: an early history, 1870–1960. Cornell University Press, Ithaca and London, p 378

Nathansohn A (1906) Über die Bedeutung vertikaler Wasserbewegung für die Produktion des Planktons. Abhandlungen der Königlichen Sächsischen Gesellschaft der Wissenschaften zu Leipzip, Mathematische-Physische Klass 29:355–441

Nathonsohn A (1908) Über die allgemeinen Produktionsbedingungen in Meere, Beiträge zur Biologie des Planktons, von H.H. Gran and Nathansohn. Internationale Revue der gestamten Hydrobiologie 1:38–72

Olson RJ (1981a) ^{15}N studies of the primary nitrite maximum. J Mar Res 39:203–226

Olson RJ (1981b) Differential photoinhibition of marine nitrifying bacteria: a possible mechanism for the formation of the primary nitrite maximum. J Mar Res 39:227–237

Riley GA (1946) Factors controlling phytoplankton populations on George's Bank. J Mar 6:54–73

Riley GA, ed (1963) Marine biology, vol. 1: First international interdisciplinary conference. Port City Press, Baltimore, p 286

Riley GA (1980) Interview, October 20, 1980, quoted in Mills EL (1989), biological oceanography an early history, 1870–1960. Cornell University Press, Ithaca and London, p 287

Riley GA (1984) Reminiscenes of and oceanographer. Unpublished mss. Oceanography Department, Dalhousie Univ, Halifax, NS, p 161

Schütt F (1892) Das Pflanzenleben der Hochsee. Ergebnisse der Plankton-Expedition der Humboldt-Stiftung, Band I.A., pp. 243–314

Søndergaard M (2002) A biography of Einer Steemann Nielsen: the man and his science. In: Williams PJLB, Thomas DN, Reynolds CS (eds) Plankton productivity: carbon assimilation in marine and freshwater ecosystems. Blackwell Scientific Publishing, Oxford, pp 1–15

Ward BB, Olson RJ, Perry MJ (1982) Microbial nitrification rates in the primary nitrite maximum off southern California. Deep-Sea Res 29:247–255

Williams PJLB, Ducklow HW (2019) The microbial loop concept: a history, 1930-1974. J Mar Res 77(2):23–81

ZoBell C (1946) Marine microbiology. A monograph on hydrobacteriology. Chronica Botanica Co., Waltham, MA

Additional Websites Accessed

https://commons.wikimedia.org/wiki/File:Haeckel_Peridinea.jpg. Accessed 6 Feb 2023

2 Methods Shape Our Understanding

2.1 The More We Learn, the More We Don't Know

Seeing is believing. How we see the microbial world and how it has changed over time has shaped what we believe. Barber and Hilting (2002, p. 17) stated, "It is difficult to study something without a concept of what is to be studied." Without the ability to see something, we have difficulty conceptualizing it or its importance. In plankton ecology, we have historically asked basically three major questions (albeit at various scales and with varying levels of detail): (1) *who* is in the plankton and at what concentration? (2) *how* (and *how fast*) and *why* do they change with time? and (3) what is the *yield* or *fate* of primary production (loss to the deep sea or fish production)? Now, we are also asking how human actions are changing these dynamics. With the right tools, we not only can ask interesting questions, but the tools also shape the kinds of questions we ask. We are no longer limited by microscopic examinations and major field campaigns; our questions can now look inward at the genomics of cells and outward toward global coverage. The more we learn, the more questions we have. Falkowski et al. (1991, p. 407) said,

> New techniques always bring new discoveries, which cannot be predicted. . .These are often exciting, precisely because they are unexpected. However, it is problem-driven science which generally appeals to most of the scientific community. In oceanography, however, problem-driven science tends to be staid primarily because oceanographers focus on questions related to "how much" rather than "how."

With modern advances in molecular, sampling, and computational techniques, we are on the verge of major advances in understanding the "how" of phytoplankton ecology.

P. M. Glibert, *Phytoplankton Whispering: An Introduction to the Physiology and Ecology of Microalgae*, https://doi.org/10.1007/978-3-031-53897-1_2

2.2 Measuring the *Who* in the Plankton

2.2.1 Advancing Tools and Surprising Findings

Floristic enumerations historically have been the mainstay of phytoplankton characterizations. Any microscopist will tell you that the ability to see small things depends on the lens through which you view them. For hundreds of years, plankton have been looked at through microscopes, and for hundreds of years, microscopists missed some of the most abundant organisms of the sea because they did not have the specialized microscopes needed to detect the tiniest cells. Though major taxonomic realignments are being made based on molecular data, there is still an important role for microscopic identification, but unfortunately, microscopist taxonomists are becoming rarer and rarer. Guides such as those of Hallegraeff et al. (2004) and Karlson et al. (2010) provide many of the details of the protocols an investigator needs in order to collect, preserve, and identify phytoplankton.

The first semi-autonomous plankton recorder was developed in the 1920s by Alister Hardy, the Continuous Plankton Recorder (CPR; Fig. 2.1). First used in the Antarctic on the Discovery cruises to sample krill, and later in the North Sea, the instrument samples the water by filtering water over moving banks of silk that are then wound on rollers (cprsurvey.org). After the voyage, the silk is unwound and examined. Still in use today with methods virtually unchanged, this technique has provided nearly a century of marine plankton observations. By late 2020, it had been towed over 7 million ocean miles by nearly 300 ships, from merchant ships to sailing ships, and it has sampled waters the world over. One of the unique aspects of these samples is that they can be archived, and thousands of samples are available for retrospective analysis. From this sampling many new species have been identified, as has the northern migration of many species in the North Atlantic due to warming (Edwards et al. 2016).

Fig. 2.1 The continuous plankton recorder. Image reproduced from www.cprsurvey.org with permission of the Marine Biological Association, UK

In the late 1970s, the tool of using fluorescent dyes to label nucleic acids was new. So was the tool of epifluorescent microscopy. These new methods were first used for quantitatively enumerating bacteria in the sea (Hobbie et al. 1977). On a cruise in the Arabian Sea, John Waterbury and Stanley Watson, from the Woods Hole Oceanographic Institution, tried out their new tool and dyes, presumably to count bacteria that should fluoresce green when excited with blue light. What they saw were samples that glowed orange, the natural fluorescence of phycoerythrin, the pigment of the picocyanobacterium, *Synechococcus* (Waterbury et al. 1979; Waterbury 2005). This discovery immediately answered a lingering question of the time: why were there seemingly so few cyanobacteria in the ocean? It turns out they were there all the time—just too small to have been seen with previous methods.

A decade later, Robert Olson of Woods Hole Oceanographic Institution and Sallie (Penny) Chisholm of MIT made an equally astounding discovery when they applied another new method (Chisholm et al. 1988). In this case, they took to sea—for the first time—a flow cytometer. In flow cytometry, cells are passed through a detector with a laser beam one cell at a time, encased in a sheath fluid. Cell characteristics are then identified based on light scattering and fluorescence properties. Fluorescence may either emanate from the natural pigments directly or from dyes or stains used to label specific constituents. Scattered light is measured at different angles to yield information on cell size and structure.

The objective of the early cruise by Olson and Chisholm—as had been Waterbury and Watson's—was to apply this tool to count bacteria in their ocean samples. What they discovered was the world of cyanobacteria even smaller than *Synechococcus*, *Prochlorococcus*. Later, another cyanobacterium, *Crososphaera*, was also discovered. These cells, *Synechococcus* and *Prochlorococcus*, which can reach the staggering concentrations of 100,000 cells per milliliter, may represent half the photosynthetic production in the central gyres of the oceans, and represent the two most abundant organisms on Earth (Waterbury 2005). It is amazing to think that the discovery of these organisms was only a few decades ago. It is also staggering to think that these discoveries came by chance—the chance to apply a new tool to the study of bacteria. What they saw changed the understanding of cyanobacteria in the oceans profoundly.

While sampling and seeing cells directly remains central to finding out *who* is there, proxies for cell enumerations have advanced, such as "marker" pigments indicating major algal groups determined using high-performance liquid chromatography (HPLC) (e.g., Jeffrey and Vesk 1997). Chlorophyll (chl) is, of course, the diagnostic photosynthetic pigment and marker of all phytoplankton, but accessory pigments, principally carotenoids and xanthophylls, vary among phytoplankton groups and some carotenoids or xanthophylls are specific to algal groupings (see Chap. 5). The application of pigment chemotaxonomy has advanced, allowing classes of microalgae to be identified based on their signature pigments. These techniques, along with in vivo and in vitro fluorometric approaches for measuring chl have increased sensitivity many-fold, allowing better resolution phytoplankton groups in low chl oligotrophic waters (Holm-Hansen et al. 1965; see Chap. 7).

2.2.2 Advancing Speed and Temporal Resolution of Observations

An imaging microscope was developed at Bigelow Laboratory for Ocean Sciences several decades ago, based on a need for a tool that could more quickly identify phytoplankton. This instrument, now known as the FlowCam®, is a microscope that can capture high-resolution images of flowing particles injected via a fluid stream. It can perform high-quality, automated image analysis based on a user-developed library of images and image processing software (e.g., Camoying and Yñiguez 2016), allowing images to be characterized, classified, and quickly enumerated. The size range of cells identified by the FlowCam and the flow cytometer are different; they are complementary instruments. Extending the FlowCam concept to include flow cytometry increased the parameters used to identify single cells. Combining the tools of flow cytometry and the FlowCam led to a new instrument, the Imaging FlowCytobot (Olson et al. 2003). With its high throughput, it can detect multiple species and resolve multiple life stages of different types of cells by generating high-resolution time series on the order of 30,000 images per hour.

As powerful as these instruments are, in their conventional operational state, they are fixed onto moored platforms—and they are expensive. Consequently, while they provide highly resolved temporal data, and may even provide highly resolved vertical data, fixed platform instruments are not able to provide spatial information. It is for this reason that these technologies are being transitioned to autonomous underwater vehicles (AUVs) and next-generation instruments are being developed for deployment on gliders or other remote vehicles. This is not an easy transition, as AUVs tend to be comparatively small in terms of payload volume, and power is much more restricted than on a moored buoy where solar power can be more readily used. With advancing miniaturization of optics, electronics, and fluidic systems, these tools will be increasingly operational. Major advances in other in situ instruments, such as nutrient analyzers, are becoming more common (e.g., Pellerin et al. 2016) and, when coupled with imaging systems, will help to provide much more substantial insights about *why* cells respond in near-real time (Glibert et al. 2005, 2018).

Satellite remote sensing—which was also "launched" in the mid-1970s—allowed a glimpse into the patterns and complexities of the global ocean, something that had not been previously possible with ship-based expeditions. Oceanographers such as Charlie Yentsch, from the Bigelow Laboratory of Ocean Sciences, and his colleague Giff Ewing, were visionaries and urged NASA to take on the challenge. As described by Barber and Hilting (2000, p. 13),

> ...NASA was looking for challenges, and the quantitative assessment of ocean surface chlorophyll and related pigments by reflected light was a challenge they took on with enthusiasm. Ironically, biological oceanographers don't even know the names of these creative NASA engineers who built the [Coastal Zone Color Scanner, launched in 1978] but that doesn't reduce our debt to them.

These images led to a new cohort of scientists who would develop and apply algorithms to decode these new data. Developing and validating algorithms for satellite remote sensing is a continual challenge.

Satellite remote sensing, specifically ocean color radiometry, provides the systematic spatial coverage that moored instruments and ship surveys are unable to achieve—and the techniques are widely used both for operational bloom monitoring and analysis of bloom phenology (Glibert et al. 2018). Ocean color and other satellite data, such as sea surface temperature and surface wind, have considerable value for indirect assessment of ecological variability and probabilistic bloom formation models (e.g., Oliveira et al. 2009). New satellites with greater sensitivity and resolution will be available in the coming years, furthering the ability to make broad taxonomic distinctions at lower phytoplankton biomass. However, there are some aspects of phytoplankton ecology that are difficult or challenging to assess. Ocean color satellites only obtain signals from the surface layers and cannot resolve sub-surface structures such as ocean-deep chl maxima. The accuracy of ocean color remote sensing products, most typically chl *a* concentrations, is diminished in coastal and estuarine environments where there is a significant presence of non-phytoplankton material (Dall'Olmo et al. 2005; Gitelson et al. 2007) that severely reduces the ability to distinguish the chl component of the spectrum of light leaving the water. Nevertheless, ocean color is now very cost-effective and an indispensable tool for surveying the oceans. As Barber and Hilting said (2000, p. 13), "Having seen the totality of the oceans, mankind can no longer maintain the concept of discrete or isolated components of the ocean."

The advent and rapid commercialization of hyperspectral optical sensors provides another major step forward toward improved spatial monitoring of phytoplankton. Hyperspectral measurements can exploit pronounced differences in the shape of phytoplankton absorption spectra relative to dissolved and suspended matter absorption spectra (particularly in the UV) currently not possible with extant multispectral instruments (Craig et al. 2006). Not only can hyperspectral measurements better tease apart the phytoplankton optical signal in optically complex waters, such measurements also provide a means to discriminate phytoplankton taxa. Drones are being used now with hyperspectral measurements of phytoplankton absorption spectra to produce spectral fingerprints of phytoplankton, including harmful taxa.

2.2.3 The 'Omics View

Just as we have advanced the tools to look outward and see the whole ocean, we have also advanced tools to look inward. Molecular tools are advancing at breakneck speed. Methods based on ribosomal RNA (rRNA) and DNA (rDNA) targeted oligonucleotides and the polymerase chain reaction (PCR) technique are becoming routine in algal monitoring. The vast amount of rapidly accumulating sequence data for a wide range of organisms is allowing a diversity of probes to be developed. These probes work by reacting a fluorescent "tag" with specific target organisms. The fluorescent tag can then be identified using epifluorescence microscopy. An example of the use of different tags and fluorescence techniques is shown in Fig. 2.2.

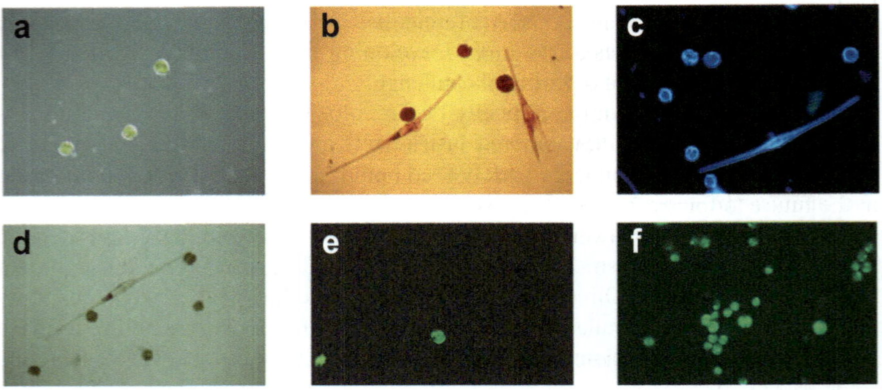

Fig. 2.2 *Alexandrium fundyense* as seen in the microscope using different techniques. *Ceratium* spp. are also clearly visible in Panels (**b**), (**c**) and (**d**). Panel (**a**): Sample was filtered and frozen with contrast enhancement; Panel (**b**): a sedimentation flask was used; Panel (**c**): The sample was filtered and stained with Calcofluor; Panel (**d**): The sample was filtered using transparent filters; Panels (**e**) and (**f**): Whole cell hybridization assay. Photos (**a**)–(**e**) from B. Karlson, and F D. Kulis. Images reproduced from Karlson et al. (2010, Microscopic and molecular methods for quantitative phytoplankton analysis) with permission of the Intergovernmental Oceanographic Commission of UNESCO

These tools are particularly attractive in monitoring for certain harmful algal taxa (see Chap. 25). Importantly, molecular data have become central to the description of new species, the rearrangement of the systematic position of species or genera, and the interpretation of evolutionary changes in phytoplankton (Saldarriaga et al. 2001; Fraga et al. 2008; see Chaps. 9 and 10).

Metabolic pools and their relation to different groups of phytoplankton are being explored through genomics, metabolomics, proteomics, and transcriptomics (Fig. 2.3, Table 2.1). 'Omics is the term used to encompass the suite of molecular and biochemical characterizations, such as genes and genomes, transcripts and transcriptomes, proteins and proteomes, and other metabolites and metabolomes (Gilbert et al. 2014). These approaches are now at the forefront of microbial ecology. Genome sequences are presently available for a wide range of phytoplankton species, and metagenomic sequencing data from oceanic samples have become available for long-term time series stations such as the Hawaii Ocean Time series (HOT) and the Bermuda Time series (BATs) (e.g., Biller et al. 2018). New integrated data management tools, such as Planet Microbe are allowing 'omics data to be contextualized with other biological and environmental data (Ponsero et al. 2020).

Recent global expeditions such as Tara (e.g., Carradec et al. 2018 and references therein) have provided phenomenal insights into diversity, biogeography, cell associations, and relationships with environmental gradients. Proteome, transcriptome, and meta transcriptome profiling are unraveling the expression of genes associated with environmental or laboratory-imposed environmental conditions. Knockout gene techniques and other tools are providing further insights about the function and control of the genetics underlying various metabolic

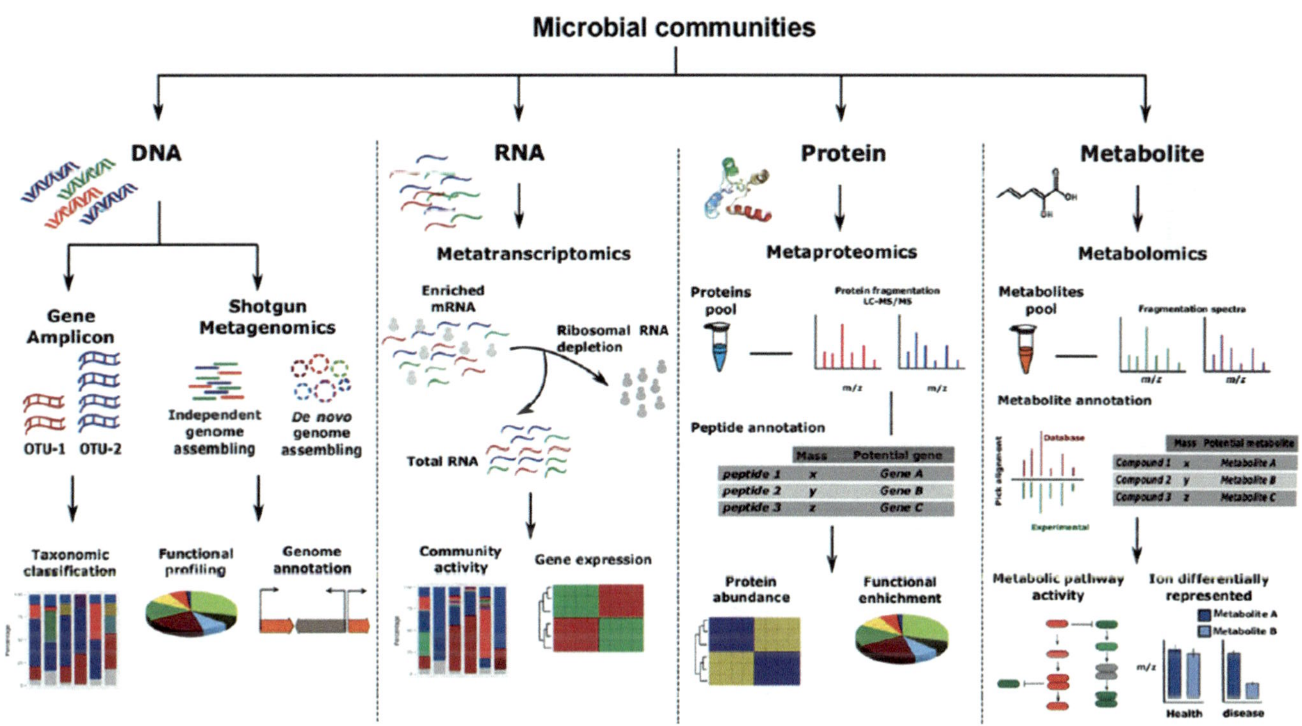

Fig. 2.3 Overview of different omics techniques. Different experimental approaches and computational strategies are applied to the study of microbial communities. During gene amplicon analysis, SSUs or ITS sequences are clustered into operational taxonomic units (OTUs) and the taxonomic identity is assigned for each OTU based on sequence homology against known sequences in a database. The resulting OTUs are used to calculate the relative abundance of each organism and quantify the population diversity between samples. In shotgun metagenomics, genomic DNA sequences can either be mapped to a reference database or used for de novo assembly of genomes. The recovered genomes can be used to assign phylogeny, calculate the relative abundance of the identified

Table 2.1 Advances and tools associated with advancing 'omics approaches (modified from Dyhrman 2016)

Term	Definition
Genome	Composed of DNA that encodes the full set of an organism's genes, the genome is the physiological blueprint of an organism
Metagenome	The complete amalgam of DNA from a microbial community in a given environment, such as an ambient water sample
Transcriptome	The complete set of an organism's transcribed genes, typically assayed on cultured isolates
Proteome	The complete set of proteins present within an organism, typically a cultured isolate. The proteome represents all transcribed and translated genes at any given point in time
Metaproteome	The complete amalgam of proteins from a given microbial community
Metabolome	All of the low molecular weight metabolites present within an organism
Lipidome	The total profile of all lipids present within an organism

processes. These tools have aided in identifying both new organisms and relationships among organisms and their oceanic provinces from poles to the tropics, leading to key insights about relationships between traits and specific environmental conditions (reviewed by Mock et al. 2016; Chust et al., 2017).

2.3 Measuring the *How Fast*

2.3.1 Early Approaches (and Debates) for Measuring Primary Production

More than hundred years ago, Victor Hensen, who was introduced in Chap. 1, was grappling with the problem of distinguishing biomass from production and is thought to have said, "to estimate production we must either try to learn the speed

Fig. 2.3 (continued) genome and assess the functional capability. In metatranscriptomics, messenger RNA (mRNA) is used to generate complementary DNA libraries that can either be mapped to reference genomes to generate gene expression profiles. These expression profiles are used to identify active pathways, genes, and organisms. In metaproteomics, mass spectrometry and fragmentation are used to reveal the amino acid sequence of peptides. The identified peptides are associated with full-length proteins by sequence homology searches against a reference database. Similar to metatranscriptomic analysis, protein expression profiles can be used to identify active pathways as well as active organisms. In metabolomics, metabolites are separated using chromatography techniques and identified and quantified using mass spectrometry. Similar to metaproteomics, the comparison between fragmentation profiles and reference databases is used to annotate the metabolic compound. Enrichment and clustering analysis can be applied to reveal patterns between sets of samples or to identify condition-dependent compounds. Image reproduced from Zuniga et al. (2017, Microb Biotechnol 10: doi: 10.1111/1751-7915) under a creative commons license of destruction of preexisting material or the rate of origin; of course it would be best to know both" (quoted in Mills 1989, p. 41). However, measuring rates and processes was limited when all that could be measured was a change in standing stock over time. Barber and Hilting (2002, p. 17) put it this way, "The central features of the primary production process - light absorption, uptake of gaseous carbon dioxide and production of gaseous oxygen - could not be sensed by humans until the necessary technology was developed in the eighteenth century… The concept of primary production waited in the wings until an accurate description of photosynthesis entered the scene late in the eighteenth century." It was many decades before the process was measured in the ocean with any accuracy. It was not until about 1850 that the nature of the photosynthetic process in the sea emerged and another half century before it was described (Barber and Hilting 2002). Although Hensen tried valiantly throughout his career to measure ocean productivity, he failed, not only because the methods were not available, but also because he depended on nets for sampling—with all their flaws—and he held the notion that growth of plankton should be measured in years (a carryover from land productivity measurements), not imagining that rates of turnover per day was possible (Barber and Hilting 2002).

H.H. Gran, also introduced in Chap. 1, was one of the early investigators addressing actual hypotheses regarding the regional, latitudinal, depth, and seasonal variations in primary production (Barber and Hilting 2002). He and colleagues in the 1910s were among the first to conduct experiments hanging light and dark bottles at different depths to measure the change in oxygen. He was interested in determining the depth at which there was just enough light for photosynthesis to balance the respiratory demand, the compensation point (Barber and Hilting 2002). Biomass measurements were also made to estimate primary production, but different ways of measuring biomass (e.g., chl or carbon (C) content) did not necessarily give the same answer, as chl is unique to primary producers, while C measurements include heterotrophic and detrital components. Also, C:chl relationships can vary even in the same organism depending on environmental conditions, particularly the light environment.

Steemann Nielsen, the shrewd, no-nonsense Danish scientist introduced in the previous chapter, began doing light and dark bottle measurements with his new ^{14}C isotope technique in the North Sea in the 1930s. It was through a bit of good fortune, good students, and good timing that he developed the ^{14}C method (Steemann Nielsen 1952; Marra 2019). It has been noted that he had "no doubt about his idea and did not bother to involve other scientists too much in the work despite the fact that he …had never worked with radioisotopes before" (Søndergaard 2002, p. 5). The method remains essentially unchanged today.

Gordon Riley of Yale also conducted productivity measurements about the same time as Steeman Nielsen, in the tropical oligotrophic North Atlantic, but Riley measured changes in O_2 concentration. In order to get a large enough signal to determine change, Riley's on-deck incubations lasted for 3 days. Riley's estimates were much higher than those of Steeman Nielsen, up to tenfold higher. Thus, a feud with Steemann Nielsen began (and which lasted years; Marra 2019). Steeman

Nielsen—with all the confidence that arrogance provides—"embarked on a small crusade to disparage the work of others" and "continued his diatribe in several publications" (Marra 2019, p. 177, 179). Steeman Nielsen's doggedness, together with the fact that his method was comparatively "easy," led to it being the commonly adopted approach for measuring rates of primary productivity in lakes, estuaries, and the oceans. It seemed like Steeman Nielsen had won the day (or the decades). However, even though the ^{14}C technique was widely used, it had its problems, and methodological questions continued to persist.

After years of comparisons of methods by many investigators, it is now thought that Riley's estimates of productivity were indeed about a factor of two to four too high; however, Steeman Nielsen's were likely a factor of about two to three too low (Marra 2019). Some of the early discrepancies may be related to where each of these men focused their sampling: Steeman Nielsen mainly in the tropics, and Riley in the more dynamic temperate ocean (Marra 2019). Barber and Hilting (2002, p. 30), summarized this episode of oceanographic history this way,

> The contrast between the methodological, but innovative experimentalist, Steemann Nielsen, and Riley, the brilliant analytical thinker whose casual lab manner was legendary, could not have been greater. . . .the productivity methods of the pre-World War II period were good enough to reveal the temporal pattern but not good enough to answer quantitative questions involving primary productivity. Steemann Nielsen recognized this better than any biological oceanographer.

Although gruff at times, Steemann Nielsen imparted considerable wisdom to his students. He is known to have said, "An impeccable experiment is more worth than a lousy experiment with lots of replicates" (Søndergaard 2002, p. 13). After 50 years of application, Marra (2002) quipped that due to the ease of application of the method and "sometimes dreadful techniques," ^{14}C measurements have not always been done with the care Steemann Nielsen would have required.

2.3.2 Advancing Isotope and Other Approaches for Measuring Primary Production

Isotope measurements are now one of the major methods for measuring primary production. Add ^{14}C as inorganic C, incubate for a period, filter the sample, and measure the ^{14}C incorporated into particulate material. With increased restrictions on the use of radioisotopes, the use of the stable isotope ^{13}C has become an alternative for measuring C assimilation (Hama et al. 1993). It yields estimates similar to those determined by ^{14}C (e.g., Cullen 2001; Pimenov et al. 2008). Primary productivity can also be measured by adding isotopically labeled water, $H_2{}^{18}O$, and following the evolved $^{18}O_2$ over time by mass spectrometry. The fundamental problem—not easily resolved by any of these methods—is the extent to which net or gross photosynthesis is being measured (gross photosynthesis is the total C fixed; net is that minus loss to respiration or other processes). The answer to this question depends, in part, on the investigator's choice of incubation period. Short incubation

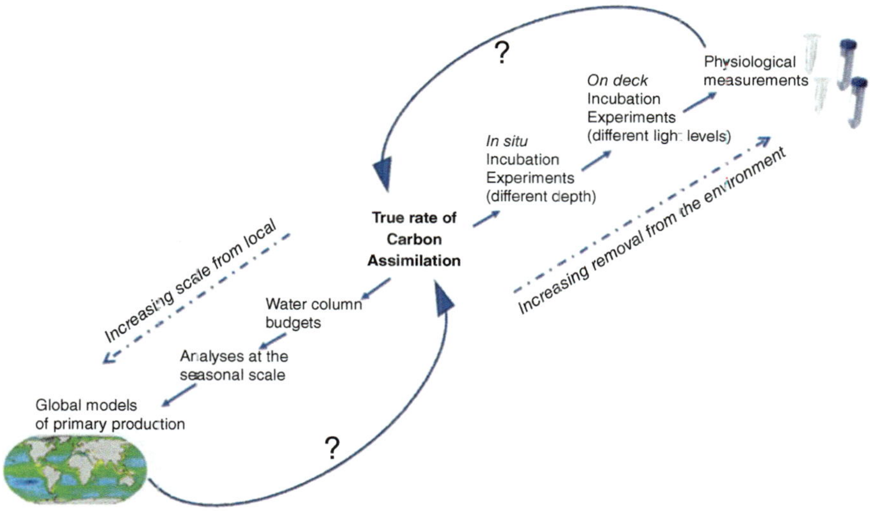

Fig. 2.4 The approaches involved in establishing the rate of carbon assimilation in aquatic systems. Experiments and physiological measurements increasingly involve removing the cells from their natural environment, while global models involve increasing space and time distance from the local scale. Image modified from Marra (2002, Phytoplankton productivity, carbon assimilation in marine and freshwater ecosystems)

should yield gross photosynthesis, and long incubations net photosynthesis, with lots of gray area in between, a function of water type and many other aspects of operator manipulations (Fig. 2.4). Chapter 7 explores these and newer methodologies for primary production.

A recent comparison of estimates of marine primary productivity derived by these and other methods confirms that the highest reported rates of primary production are, in fact, derived from ^{18}O enrichment methods (Regaudie-de-Gioux et al. 2014). An alternative ^{18}O method for measuring productivity was introduced using a modified membrane inlet mass spectrometer (MIMS) dissolved gas analyzer. The analytical method for MIMS is simple, with quick sample analysis time and it cross-calibrates well with the traditional approach (Ferrón et al. 2016). With shipboard processing possible using MIMS, many samples can be run with real-time data acquisition and sampling protocols can be modified in real time depending on results. A recent review (Mader et al. 2017) explored recent developments in the use of stable isotope ratios of O ($^{18}O/^{16}O$) and $^{17}O/^{16}O$) and how isotopic shifts have provided insights into sources and sinks of O, rates of productivity, and other processes such as gas–water exchange.

Enormous advances have been made with isotopic methods over the recent years [Further discussion of isotope methods for measuring nutrient uptake and metabolism can be found in Chaps. 12, 13, 14 and Glibert et al. 2019]. Advanced mass spectrometric methods allow resolution of metabolic rates at finer scales (Fig. 2.5). For example, cell sorting techniques such as flow cytometry coupled to mass

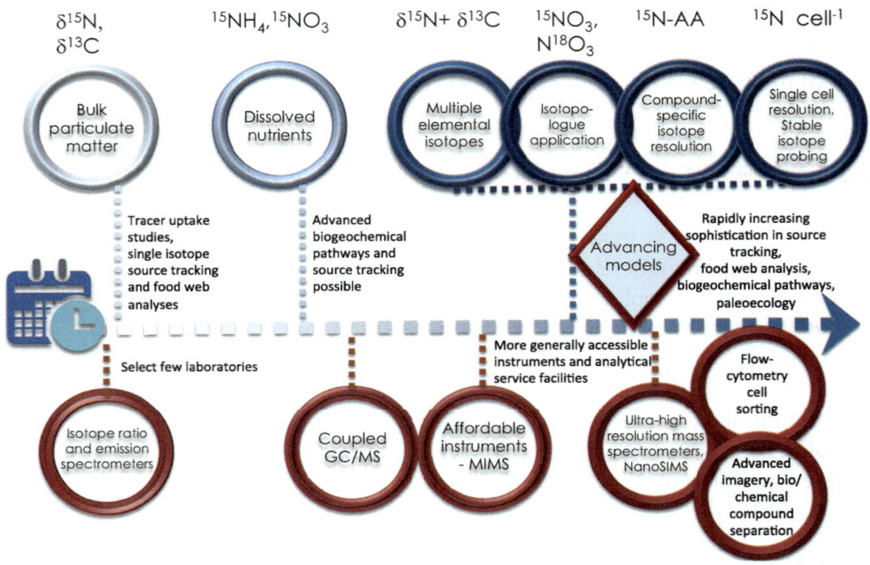

Fig. 2.5 Infographic depicting the generalized time line of isotope applications (top) and instrumental developments (bottom). Image reproduced from Glibert et al. (2019, Limnol Oceanogr 64: doi: 10.1002/lno.11087) under a creative commons license

spectrometry allow rate measurements of specific taxa or even cells within mixed communities. Recently, van Roij et al. (2017) described a laser ablation combustion isotope ratio mass spectrometer capable of measuring nanograms of particulate organic C. That technique was used to measure $\delta^{13}C$ in single dinoflagellate cysts (Sluijs et al. 2017). Nanometer-scale secondary ion mass spectrometry (nanoSIMS) provides isotope ratio imaging with a spatial resolution of <100 nm and has been used together with ^{15}N, ^{13}C or ^{2}H labelling to study dissolved inorganic C uptake, N_2 fixation, NH_4^+, and NO_3^- assimilation by phytoplankton communities (Ploug et al. 2010; Musat et al. 2012), mixotrophy (Terrado et al. 2017), symbioses (Thompson et al. 2012), and consortia of microbes in sediments (Hatzenpichler et al. 2016) on very small scales. Compound-specific isotope analysis has also rapidly advanced (Hayes et al. 1990; Boschker and Middelburg 2002; Middelburg 2014). This technique allows various extracted biomarkers, such as lipids, carbohydrates, or amino acids, to be isotopically compared (Boschker and Middelburg 2002). An additional technique gaining rapid application within microbial ecology is stable isotope probing (SIP, e.g., Radajewski et al. 2000; Neufeld et al. 2007). In this approach, an enrichment is made with an isotopic tracer, e.g., ^{13}C or ^{15}N, and after uptake by natural or cultured organisms, separation of nucleic acids is undertaken using density gradient ultracentrifugation. Nucleic acids with different densities can thus be separated and their differential incorporation of the label can be measured. Various molecular analyses of these labeled nucleic acids can be undertaken in tandem using fingerprinting, clone libraries, or metagenomics. When coupled with other highly sensitive tools such as molecular approaches, information on microbial associations and other difficult-to-resolve pathways is now possible. By linking function and taxonomic identity, the responses and interactions of some microbial taxa to ecological conditions can be resolved at very fine scales.

Measuring primary production without incubation and without isotopes is now possible using fluorescence. These measurements take advantage of the fluorescence that comes from the process of photosynthesis (see Chap. 7). In brief, a small fraction (3–10%) of light absorbed by photosynthesis is re-emitted as fluorescence and the measured variability in this signal can be related to the quantum yield of photosynthesis (i.e., the photochemistry that yields high-energy compounds). In addition, techniques for measuring the regulation of energy dissipation by photosynthesis are available to study the manner in which the photosynthetic process regulates energy flow under various physiological stress conditions.

Applying the tools of 'omics to questions of *how fast* continues to be problematic. Whether it is applications to physiological rate processes or biogeochemical processes, there remains a mismatch between the tried-and-true methods and these new tools. Genomics only provide information on the functional potential of an organism (e.g., Sunagawa et al. 2015). Transcriptomics attempt to provide relationships between gene expression (mRNA abundance) and environmental adaptation, but such relationships are variable. For example, the amplitude and timing of the mRNA pool does not align necessarily with protein expression (e.g., Waldbauer et al. 2012). However, the metabolic information derived from omics approaches is yielding new and exciting questions about the relationships between gene expression and metabolic rates, from transporters that bring nutrients across cell membranes to enzymes that metabolize them.

Strzepek et al. (2022) used the example of understanding rates of N_2 fixation in the ocean to illustrate insights that have come from omics approaches coupled with more traditional methodologies. Physiological studies (as well as the tools described above for cell detection) helped to reveal the importance of N_2 fixation in the oceans and the major players in the process. The enzyme of N_2 fixation, nitrogenase, and expression of the *nif* gene was developed as a tool for measuring N_2 fixation potential (Zehr et al. 1996; Zehr and Montoya 2007), but the presence of the gene did not correlate with rates for a number of reasons, including the diversity of types of cells that carry out this process. Nevertheless, such tools did allow high throughput measurements which allowed a greater spatial and temporal coverage of the global oceans (Fig. 2.6).

Future studies attempt to measure *how fast* (rate processes) will (or should) marry the approaches of physiology (traditional methodologies) with those of omics (Strzepek et al. 2022). Using the example of phosphate metabolism (see also Chap. 13), the co-design, or co-application of methods leads to a sum greater than its parts. Omics can help to unravel or highlight key metabolites or processes that can be measured using more standard techniques; altering growth conditions can lead to both an understanding of how rates or processes change over time, but also how that links to gene expression and transcript abundance.

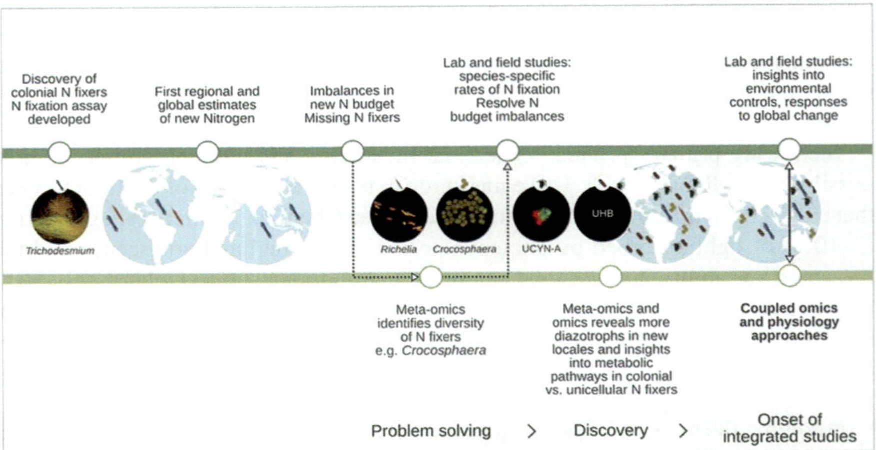

Fig. 2.6 The contributions of physiology and omics to understanding the role of diazotrophs in the ocean N cycle (based on Zehr and Capone 2020). Key events in the physiology timeline (top green line) include the estimation of N fluxes through nitrogenase activity (Dugdale and Dugdale 1962), initial estimates of global marine N_2 fixation rates (Capone et al. 2005), and the combining of lab and field measurements to understand individual diazotrophs and community contributions and constraints. Pivotal events in the omics timeline (lower green line) include problem solving (Zehr and Montoya 2007) and the discovery of diazotroph diversity including in unicellular cyanobacteria group A (UNCYN-A) and diverse uncultured heterotrophic bacteria (Martínez-Pérez et al. 2016). Image reproduced from Strzepek et al. (2022, J Plankton Res 44: doi: 10.1093/plankt/fbac026) under a creative commons license

As will be seen in the coming chapters, rate measurements are the core of understanding environmental responses (Fig. 2.7). Quantifying them is essential for mechanistic modeling. Whether the rate is photosynthesis as a function of light intensity, nutrient uptake as a function of substrate concentration, rates as a function of temperature, or grazing as a function of food concentration, all such measurements should be done with investigators' eyes wide open to potential artifacts. Linking "omics" changes with physiological rate measurements can be a helpful test of rate experiments as it can provide validation of the underlying process (es) responsible for the measured rate or flux result. Genes, of course, do not predict activity. The choices of methods are many, each with its own unique advantages and limitations—and each tailor suited to match a specific question.

2.4 Measuring *Fates* of Primary Production

Two questions stumped scientists in the mid-nineteenth century: what happens to the primary production, and how is it possible (or is it even possible) for tiny organisms to support the large animals and fish of the sea? In 1844, a Danish marine biologist, Ørsted, suggested after a cruise to the oligotrophic Sargasso Sea, that the algae were

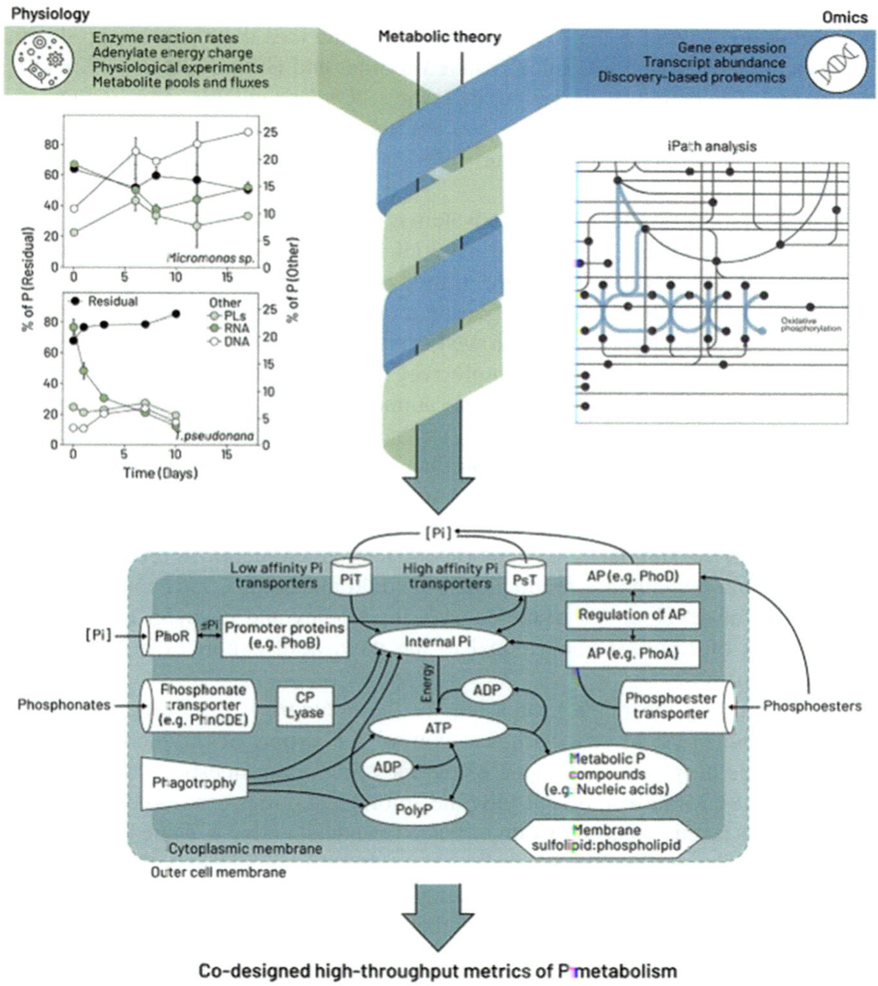

Fig. 2.7 The potential of reverse engineering physiological metrics to provide better linkages with molecular tools using the example of phosphorus (P). Findings from a physiological study (top left: redrawn from Liefer et al. 2019. PLs = phospholipids) using a cluster of long-established metrics (residual P pools/intracellular storage of inorganic P) to compare the P allocation strategies of a diatom (*Thalassiosira pseudonana*) and a prasinophyte (*Micromonas* sp.). Other physiological approaches (top left green box) include measuring enzyme reaction rates, clever experimental design, quantifying adenylate energy charge, and tracking metabolite pools and fluxes. The top right figure is a Kyoto Encyclopedia of Genes and Genomes (KEGG) map from Interactive Pathways Explorer v3 (iPath) (https://pathways.embl.de; Letunic et al. 2008; Darzi et al. 2018). iPath is a web-based tool for the visualization and analysis of cellular pathways from omics. Combining approaches—denoted by the green and blue intersecting arrows—will improve underpinning biochemical theory and identify candidate enzymes and pathways for the co-design of new physiological assays, as depicted in the bottom cartoon summarizing putative phosphorus uptake, metabolism, and storage in phytoplankton (redrawn and simplified from Lin et al. 2016). The

the primary food for oceanic food webs. Not only was the fate of primary production a question unresolved at that time, but the fate of the idea that "primary producers" supported the food web was initially nixed. Barber and Hilting (2002, p. 22), citing Wolff and Peterson (1991) put it this way, "Understandably, this iconoclastic concept of the tiny feeding the large was not accepted by most contemporary marine biologists, and Ørsted did not again refer to this idea during his long career in botany."

Primary production in the surface waters is indeed eaten, but not all of it (a topic to be revisited in Chap. 19). Some material fluxes out of the euphotic zone into deeper waters. The nutrients that are formed by this decomposing material may be transported back to the surface, supporting more production. The isotope revolution of the mid-twentieth century allowed these processes to be measured and conceptualized in an elegantly simple way. Dugdale and Goering (1967) used isotopically labelled nitrate (NO_3^-) and ammonium (NH_4^+) and distinguished the portion that was dependent on NO_3^- as "new" production and that which was dependent on NH_4^+ as "regenerated" production (see also Chap. 12). Their reasoning was that NO_3^- is transported by mixing and upwelling, and therefore provides new N nutrient to the surface waters. In contrast, NH_4^+ was largely derived from direct recycling of N via zooplankton excretion or bacterial remineralization in the surface, and thus could not enhance production, but could maintain it. This concept laid the foundation for development of the concept (by Eppley and Peterson 1979) that the upward flux of NO_3^- is about equal to the downward export of organic matter, export production.

Understanding the relationships among the ecological, physiological, biogeochemical, and physical processes that control the export of production from the surface to the deep remains as one of the most important questions of the present time (Fig. 2.8). Moreover, quantifying the "biological pump," the export of C to the deep sea, has taken on enormous importance in understanding oceanic C cycling and how it may change with climate- or how it might be manipulated anthropogenically to reduce atmospheric CO_2. While primary production can now be measured quite accurately from ships and from the sky, that which sinks out of the euphotic zone is more difficult to measure. First used in the 1970s, traps for the collection of settling material (sediment traps) have undergone many iterations (Fig. 2.9). Some now are coupled to time-lapse cameras. One insight that has come from long-term time series is that organic matter fluxes are not just a continuous "rain of detritus," but can be punctuated by high flux events. Often such flux events are linked to blooms of diatoms in surface waters. They reach the trap as cells or in fecal pellets, or other

Fig. 2.7 (continued) present cartoon is for a cyanobacterium, which in addition to the plasma membrane also contains a semipermeable outer membrane (shown as a dotted line). Abbreviations: ADP = adenosine diphosphate; ATP = adenosine triphosphate; AP = alkaline phosphatase; Pi = inorganic phosphorus; PolyP = polyphosphate; gene names as described in Lin et al. (2016). Image reproduced from Strzepek et al. (2022, J Plankton Res 44: doi: 10.1093/plankt/fbac026) under a creative commons license

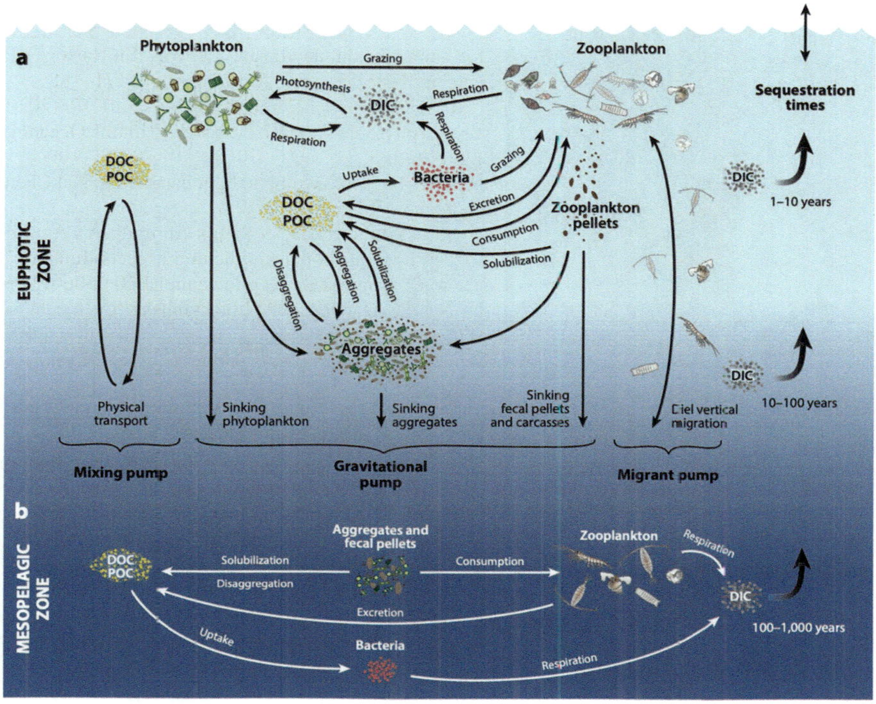

Fig. 2.8 Food web diagram illustrating the three export pathways of the biological pump, their regulating processes, and the timescales for carbon (C) sequestration. (**a**) The euphotic zone food web and the many ecological and biogeochemical processes regulate its relationship with the gravitational, migrant, and mixing pumps that transport organic C to depth. (**b**) Beneath the euphotic zone, organic C is remineralized back to DIC via food web processes in the mesopelagic zone. The depths to which that organic carbon is transported set the timescale of its sequestration. Abbreviations: DIC, dissolved inorganic C; DOC, dissolved organic C; POC, particulate organic C. Image reproduced from Seigel et al. (2023, Annu Rev Mar Sci 15: doi: 10.1146/annurev-marine-040722-115226) under a creative commons license

large aggregates. Various remineralization processes alter this material as it sinks, and it may be transported via physical processes (Siegel et al. 2016, 2023). The material in such collections is now being sequenced or fingerprinted using various molecular approaches (Metfies et al. 2017). There is work to be done to continue to refine the most appropriate methods to understand all the various fates of primary production.

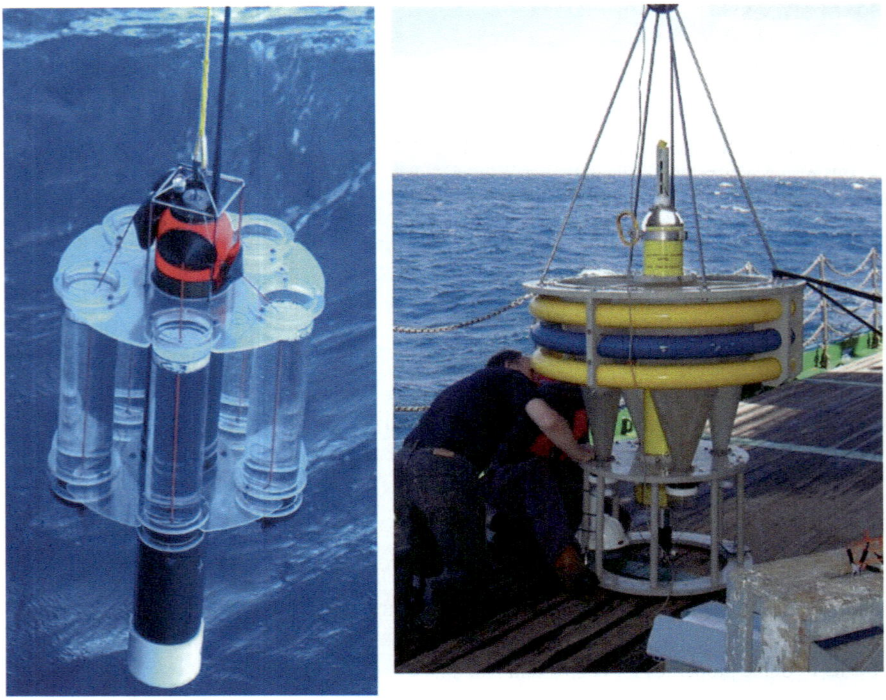

Fig. 2.9 Photos of two neutrally buoyant traps used successfully to collect sinking particles in the ocean. Images reproduced from Buesseler et al. (2007, J Mar Res 65) with permission of the author and the Woods Hole Oceanographic Institution

2.5 Summary

This chapter has barely skimmed the surface in terms of methodologies that have revolutionized the understanding of biomass and productivity of the aquatic world. New methods have punctuated new discoveries. From identifying *who* is in the plankton with improved instrumentation, to measuring *rates* and *processes* with isotopes and their *fate* in food webs the recent progress in methods and understanding is indeed monumental. Measuring productivity from space, and application of the now wide range of molecular tools have allowed us to broaden our spatial perspective as well as our understanding of the biochemical underpinnings of metabolism. New discoveries have been the catalyst for new concepts. Not all new data are readily accepted into the mainstream of ideas of the time. Critical is the need for high-quality data, as our conceptual understanding of aquatic dynamics—and our ability to model organisms and processes—is only growing with the challenges of the global changes today and how the aquatic world will change in the future.

References

Barber RT, Hilting AK (2000) Achievements in biological oceanography. In: Steele J (ed) 50 years of ocean discovery. National Academy Press, Washington, DC, pp 11–21

Barber RT, Hilting AK (2002) History of the study of plankton productivity. In: Williams PJLB, Thomas DN, Reynolds CS (eds) Plankton productivity: carbon assimilation in marine and freshwater ecosystems. Blackwell Scientific Publishing, Oxford, pp 16–43

Biller SJ, Berube PM, Dooley K, Williams M, Satinsky BM, Hackl T, Hogle SL, Coe A, Bergauer K, Bouman HA, Browing TJ, De Corte D, Hassler C, Hulston D, Jacquot JE, Maas EW, Reinthaler T, Sintes E, Yokokawa T, Chisholm SW (2018) Marine microbial genomics sampled across time and space. Sci Data 5:180176. https://doi.org/10.1038/sdata.2018.176

Boschker HTS, Middelburg JJ (2002) Stable isotopes and biomarkers in microbial ecology. FEMS Microbiol Ecol 40:85–95. https://doi.org/10.1111/j.1574-6941.2002.tb00940.x

Buesseler KO, Antia AN, Chen M, Fowler SW, Gardner WD, Gustafsson C, Harada K, Michaels AF, van der Loeff MR, Sarin M, Steinberg DK, Trull T (2007) An assessment of the use of sediment traps for estimating upper ocean particle fluxes. J Mar Res 65:345–416

Camoying MG, Yñiguez AT (2016) FlowCAM optimization: attaining good quality images for higher taxonomic classification resolution of natural phytoplankton samples. Limnol Oceanogr Methods 14:3051–3314

Capone DG, Burns JA, Montoya JP, Subramaniam A, Mahaffey C, Gunderson T, Michaels A, Carpenter E (2005) Nitrogen fixation by *Trichodesmium* spp.: an important source of new nitrogen to the tropical and subtropical North Atlantic Ocean. Global Biogeochem Cycles 19: GB2024. https://doi.org/10.1029/2004GB002331

Carradec Q, Pelletier E, Da Silva C, Alberti A, Seeleuthner Y, Blanc-Mathieu R, Lima-Mendez G et al (2018) A global ocean atlas of eukaryotic genes. Nat Commun 9:373. https://doi.org/10.1038/s41467-017-02342-1

Chisholm SW, Olson RJ, Zettler ER, Goericke R, Waterbury J, Welschmeyer NA (1988) A novel free-living prochlorophyte abundant in the ocean euphotic zone. Nature 334:340–343

Chust G, Vogt M, Benedetti F, Nakov T, Villéger S, Aubert A, Vallina SM et al (2017) *Mare incognitum*: a glimpse into future plankton diversity and ecological research. Front Mar Sci 4: 68. https://doi.org/10.3339/fmars.2017.00068

Craig SE, Lohrenz SE, Lee Z, Mahoney KL, Kirkpatrick GJ, Schofield OM, Steward RG (2006) Use of hyperspectral remote sensing reflectance for detection and assessment of the harmful alga, *Karenia brevis*. Appl Opt 45:5414–5425. https://doi.org/10.1364/ao45.005414

Cullen JJ (2001) Primary production methods. In: Steele JH, Thorpe SA, Turekian KK (eds) Encyclopedia of ocean sciences. Academic Press, San Diego, CA. https://doi.org/10.1006/rwos.2001.0172

Dall'Olmo G, Gitelson AA, Rundquist DC, Leavitt B, Barrow T, Holz JC (2005) Assessing the potential of SeaWiFS and MODIS for estimating chlorophyll concentration in turbid productive waters using red and near-infrared bands. Remote Sens Environ 96:176–187

Darzi Y, Letunic I, Bork P, Yamada T (2018) iPath3.0: interactive pathways explorer v3. Nucleic Acids Res 46:W510–W513. https://doi.org/10.1093/nar/gky299

Dugdale VA, Dugdale RC (1962) Nitrogen metabolism in lakes II. Role of nitrogen fixation in sanctuary Lake, Pennsylvania. Limnol Oceanogr 7:170–177. https://doi.org/10.4319/lo.1962.7.2.0170

Dugdale RC, Goering JJ (1967) Uptake of new and regenerated forms of nitrogen in primary productivity. Limnol Oceanogr 12:1961–1206

Dyhrman ST (2016) Nutrients and their acquisition: phosphorus physiology in microalgae. In: Borowitzka MA, Beardall J, Raven JA (eds) The physiology of microalgae. Springer, Switzerland, pp 155–183

Edwards M, Helaouet P, Alhaija RA, Batten S, Beaugrand G, Chiba S et al. (2016) Global marine ecological status report. Marine Ecological Status Report: results from the global CPR survey 2014/2015. SAHFOS technical report 11: 11–32, Plymouth UK, ISSN 1744-0750

Eppley RW, Peterson BJ (1979) Particulate organic matter flux and planktonic new production in the deep ocean. Nature 282:677–680

Falkowski PG, Demers S, Legendre L (1991) Concluding remarks: promises and limitations of individual cell and particle analysis. In: Demers S (ed) Particle analysis in oceanography, vol G27. Springer-Verlag, NATO ASI Series, Berlin, pp 403–408

Ferrón S, del Valle DA, Björkman KM, Quay PD, Church MJ, Karl DM (2016) Application of membrane inlet mass spectrometry to measure aquatic gross primary production by the ^{18}O in vitro method. Limnol Oceanogr Meth 14:610–622

Fraga S, Penna A, Bianconi I, Paz B, Zapata M (2008) *Coolia canariensis* sp. nov (Dinophyceae), a new nontoxic epiphytic benthic dinoflagellate from the Canary Islands. J Phycol 44:1060–1070

Gilbert JA, Dick GJ, Jenkins B, Heidelberg J, Allen E, Mackey KRM, Delong EF (2014) Meeting report: ocean 'omics science, technology and cyberinfrastructure: current challenges and future requirements (august 20–23, 2013). Stand Genomic Sci 9:1251–1258

Gitelson AA, Schalles JF, Hladik CM (2007) Remote chlorophyll-a retrieval in turbid, productive estuaries: Chesapeake Bay case study. Remote Sens Environ 109:464–472

Glibert PM, Seitzinger S, Heil CA, Burkholder JM, Parrow MW, Codispoti LA, Kelly V (2005) The role of eutrophication in the global proliferation of harmful algal blooms: new perspectives and new approaches. Oceanography 18(2):198–209

Glibert PM, Pitcher GC, Bernard S, Li M (2018) Advancements and continuing challenges of emerging technologies and tools for detecting harmful algal blooms, their antecedent conditions and toxins, and applications in predictive models. In: Glibert PM, Berdalet E, Burford M, Pitcher G, Zhou M (eds) Ecology and oceanography of harmful algal blooms (GEOHAB). Springer, Cham, pp 339–357

Glibert PM, Middelburg JJ, McClelland JW, Vander Zander J (2019) Stable isotope tracers: enriching our perspectives and questions on sources, fates, rates and pathways of major elements on aquatic systems. Limnol Oceanogr 64:950–981. https://doi.org/10.1002/lno.11087

Hallegraeff GM, Anderson DM, Cembella AD (eds) (2004) Manual on harmful marine microalgae. UNESCO Publishing, Paris, p 793

Hama T, Hama J, Handa N (1993) ^{13}C tracer methodology in microbial ecology with special reference to primary production processes in aquatic environments. In: Jones JG (ed) Advances in microbial ecology, vol 13. Plenum, New York, pp 39–83

Hatzenpichler R, Connon SA, Goudeau D, Malmstrom RR, Woyke T, Orphan VJ (2016) Visualizing in situ translational activity for identifying and sorting slow-growing archaeal–bacterial consortia. Proc Natl Acad Sci USA 113(28):E4069–E4078. https://doi.org/10.1073/pnas.1603757113

Hayes JM, Freeman KH, Popp BN, Hoham CH (1990) Compound-specific isotopic analyses: a novel tool for reconstruction of ancient biogeochemical processes. Organ Geochem 16(4–6): 1115–1128

Hobbie JE, Daley RJ, Jasper S (1977) Use of nuclepore filters for counting bacteria by fluorescence microscopy. Appl Environ Microbiol 33:1225–1228. https://doi.org/10.1128/aem.33.5.1225-1228.1977

Holm-Hansen O, Lorenzen CJ, Holmes RW, Strickland JDH (1965) Fluorometric determination of chlorophyll. J Cons Cons Perm Int Explor Mer 30:3–15

Jeffrey SW, Vesk M (1997) Introduction to marine phytoplankton and their pigment signatures. In: Jeffrey SW, Mantoura RFC, Wright SW (eds) Phytoplankton pigments in oceanography: guidelines to modern methods. UNESCO, Paris, p 661

Karlson B, Godhe A, Cusack C, Bresnan E (2010) Introduction to methods for quantitative phytoplankton analysis. In: Karlson B, Cusack C, Bresnan E (eds) Microscopic and molecular methods for quantitative phytoplankton analysis. Intergovernmental Oceanographic Commission of UNESCO, Paris, pp 5–12. https://cdn.ioos.noaa.gov/media/2017/12/WKNCT_Handbook_ver5May2010_2.pdf

Letunic I, Yamada T, Kanehisa M, Bork P (2008) iPath: interactive exploration of biochemical pathways and networks. Trends Biochem Sci 33:101–103

Liefer JD, Garg A, Fyfe MH, Irwin AJ, Benner I, Brown CM, Follows MJ, Omta AW, Finkel ZV (2019) The macromolecular basis of phytoplankton C:N:P under nitrogen starvation. Front Microbiol 10:763. https://doi.org/10.3389/fmicb.2019.00763

Lin S, Litaker RW, Sunda WG (2016) Phosphorus physiological ecology and molecular mechanisms in marine phytoplankton. J Phycol 52:10–36. https://doi.org/10.1111/jpy.12365

Mader M, Schmidt C, van Geldern R, Barth JAC (2017) Dissolved oxygen in water and its stable isotope effects: a review. Chem Geol 473:10–21. https://doi.org/10.1016/j.chemgeo.2017.10.003

Marra JF (2002) Approaches to the measurement of plankton production. In: Williams PJLB, Thomas DN, Reynolds CS (eds) Phytoplankton productivity, carbon assimilation in marine and freshwater ecosystems. Blackwell Science, Oxford, UK, pp 78–108

Marra JF (2019) Hot carbon: Carbon-14 and a revolution in science. Columbia University Press, New York

Martínez Pérez C, Mohr W, Löscher CR, Dekaezemacker J, Littmann S, Yilmaz P, Lehnen N, Fuchs BM, Lavik G, Schmitz RA, LaRoche J, Kuypers MMM (2016) The small unicellular diazotrophic symbiont, UCYN-A, is a key player in the marine nitrogen cycle. Nat Microbiol 1: 16163. https://doi.org/10.1038/nmicrobiol.2016.163

Metfies KE, Bauerfeind E, Wolf C, Sprong P, Frickenhaus S, Kaleschke L, Nicolaus A, Nöthig EM (2017) Protist communities in moored long-term sediment traps (Fram Strait, Arctic)–preservation with mercury chloride allows for PCR-based molecular genetic analyses. Front Mar Sci 4: 301. https://doi.org/10.3389/fmars.2017.00301

Middelburg JJ (2014) Stable isotopes dissect aquatic food webs from the top to the bottom. Biogeosciences 11:2357–2371. https://doi.org/10.5194/bg-11-2357-2014

Mills EL (1989) Biological oceanography: an early history, 1870–1960. Cornell University Press, Ithaca and London, p 378

Mock T, Daines SJ, Geider R, Collins S, Metodiev M, Millar AJ, Moulton V, Lenton TM (2016) Bridging the gap between omics and earth system science to better understand how environmental change impacts marine microbes. Glob Change Biol 22:61–75. https://doi.org/10.1111/gcb.12983

Musat N, Foster R, Vagner T, Adam B, Kuypers MMM (2012) Detecting metabolic activities in single cells, with emphasis on nanoSIMS. FEMS Microbiol Rev 36:486–511. https://doi.org/10.1111/j.1574-6976.2011.00303.x

Neufeld JD, Vohra J, Dumont MG, Lueders T, Manefield M, Friedrich MW, Murrell JC (2007) DNA stable isotope probing. Nat Protoc 2:860–866

Oliveira PB, Moita T, Silva A, Monteiro IT, Sofia Palma A (2009) Summer diatom and dinoflagellate blooms in Lisbon Bay from 2002 to 2005: pre-conditions inferred from wind and satellite data. Progr Oceanogr 83:270–277

Olson RJ, Shalapyonok AA, Sosik HM (2003) An automated submersible flow cytometer for pico- and nanophytoplankton: FlowCytobot. Deep-Sea Res I Oceanogr Res Pap 50:301–315

Ørsted AS (1844) De regionibus marinus. Elementa topagraphiae historica-naturalis freti Geresund

Pellerin BA, Stauffer BA, Young DA et al (2016) Emerging tools for continuous nutrient monitoring networks: sensors advancing science and water resources protection. J Am Water Resour Assoc 52:993–1008

Pimenov NV, Zyakum AM, Prusakova TS, Lunina ON, Ivanov MV (2008) Application of ^{13}C mineral carbon for assessment of the primary production of organic matter in aquatic environments. Microbiology 77:224–227

Ploug H, Musat N, Adam B, Moraru CL, Lavik G, Vagner T, Bergman B, Kuypers MM (2010) Carbon and nitrogen fluxes associated with the cyanobacterium *Aphanizomenon* sp. in the Baltic Sea. ISME J 4:1215–1223

Ponsero AJ, Bomhoff M, Blumberg K, Youens-Clark K, Herz NM, Wood-Charlson EM, Delong EF, Hurwitz BL (2020) Planet microbe: a platform for marine microbiology to discover and analyze interconnected 'omics and environmental data. Nucleic Acids Res 49:D792–D802. https://doi.org/10.1093/nar/gkaa637

Radajewski S, Ineson P, Parekh NR, Murrell JC (2000) Stable-isotope probing as a tool in microbial ecology. Nature 403:646–649

Regaudie-de-Gioux A, Lasternas S, Agusti S, Duarte CM (2014) Comparing marine primary production estimates through different methods and development of conversion equations. Front Mar Sci 1:19. https://doi.org/10.3389/fmars.2014.00019

Saldarriaga JF, Taylor FRJ, Keeling PJ, Cavalier-Smith T (2001) Dinoflagellate nuclear SSU rRNA phylogeny suggests multiple plastid losses and replacements. J Mol Evol 53:204–213. https://doi.org/10.1007/s002390010210

Siegel DA, Buesseler KO, Behrenfeld MJ, Benitez-Nelson CR, Boss E, Brzezinski MA, Burd A, Carlson CA, D'Asaro EA, Doney SC, Perry MJ, Stanley RHR, Steinberg DK (2016) Prediction of the export and fate of global ocean net primary production: the EXPORTS science plan. Front Mar Sci 3:22. https://doi.org/10.3389/fmars.2016.00022

Siegel DA, DeVries T, Cetiníc I, Bisson KM (2023) Quantifying the ocean's biological pump and its carbon cycle impacts on global scales. Annu Rev Mar Sci 15:329–356. https://doi.org/10.1146/annurev-marine-040722-115226

Sluijs A, van Roij L, Frieling J, Laks J, Reichart G-J (2017) Single-species dinoflagellate cyst carbon isotope ecology across the Paleocene-Eocene thermal maximum. Geology 46:79–82

Søndergaard M (2002) A biography of Einer Steemann Nielsen: the man and his science. In: Williams PJLB, Thomas DN, Reynolds CS (eds) Plankton productivity: carbon assimilation in marine and freshwater ecosystems. Blackwell Scientific Publishing, Oxford, pp 1–15

Steemann Nielsen E (1952) The use of radioactive carbon (^{14}C) for measuring organic production in the sea. J Cons Int Explor Mer 18:117–140

Strzepek RF, Nunn BL, Bach LT, Berges JA, Young EB, Boyd PW (2022) The ongoing need for rates: can physiology and omics come together to co-design the measurements needed to understand complex ocean biogeochemistry? J Plankton Res 44:485–495. https://doi.org/10.1093/plankt/fbac026

Sunagawa S, Coelho LP, Chaffron S, Kultima JR, Labadie K, Salazar G, Djahanschiri B et al (2015) Structure and function of the global ocean microbiome. Science 348:1261359

Terrado R, Pasulka AL, Lie AA-Y, Orphan VJ, Heidelberg KB, Caron DA (2017) Autotrophic and heterotrophic acquisition of carbon and nitrogen by a mixotrophic chrysophyte established through stable isotope analysis. ISME J 11:2022–2034. https://doi.org/10.1038/ismej.2017.68

Thompson A, Foster RA, Krupke A, Carter BJ, Musat N, Vaulot D, Kuypers MM, Zehr JP (2012) Unicellular cyanobacterium symbiotic with a single-celled eukaryotic alga. Science 337:1546–1550. https://doi.org/10.1126/science.1222700

van Roij L, Sluijs A, Laks JJ, Reichart G-J (2017) Stable carbon isotope analyses of nanogram quantities of particulate organic carbon (pollen) with laser ablation nano combustion gas chromatography/isotope ratio mass spectrometry. Rapid Commun Mass Spectr 31:47–58

Waldbauer JR, Rodrigue S, Coleman ML, Chisholm SW (2012) Transcriptome and proteome dynamics of a light-dark synchronized bacterial cell cycle. PLoS One 7:e43432

Waterbury JB (2005) Little things matter a lot: overlooked in the ocean until the 1970s, cyanobacteria are among Earth's most important organisms. Oceanus March 11:2005

Waterbury JB, Watson SW, Guillard RRL, Brand LE (1979) Widespread occurrence of unicellular, marine, planktonic, cyanobacterium. Nature 277:292–294

Wolff T, Peterson ME (1991) A brief biography of a.S. Ørsted, with noted on his travels in the West Indies and Central America and illustrations of collected polychaetes. Ophelia Supp 5:669–685

Zehr JP, Capone DG (2020) Changing perspectives in marine nitrogen fixation. Science 368: eaay9514. https://doi.org/10.1126/science.aay9514

Zehr JP, Montoya JP (2007) Measuring N_2 fixation in the field. In: Bothe H, Ferguson SJ, Newton WE (eds) Biology of the nitrogen cycle. Elsevier, Netherlands, pp 193–205. https://doi.org/10.1016/B978-044452857-5.50014-X

Zehr JP, Braun S, Chen YB, Mellon M (1996) Nitrogen fixation in the marine environment: relating genetic potential to nitrogenase activity. J Exp Mar Biol Ecol 203:611–673. https://doi.org/10.1016/0022-0981(96)02570-1

Zuniga C, Zaramela L, Zengler K (2017) Elucidation of complexity and prediction of interactions in microbial communities. Microb Biotechnol 10:1500–1522. https://doi.org/10.1111/1751-7915.12855

Additional Websites Accessed

www.cprsurvey.org. Accessed 22 Jan 2023

Manipulating and Measuring Growth and Biomass in Simple to Complex Systems and a Curve for All Reasons

3

3.1 Introduction: What Is Phytoplankton Growth?

What is phytoplankton growth? This is such a simple question—but one with complex answers. A hundred years ago, Alexander Nathansohn, Karl Brandt and others (introduced in Chap. 1) debated the factors contributing to the growing and dying of phytoplankton and the distinction between growth and biomass. Many of the same debates continue to this day.

It is common to think of growth as the increase in size of an organism. With phytoplankton or microalgae, this definition is not often used because the time scale of the growth of a single cell is brief (typically hours to a few days) and cell size changes are small and not often measured. For phytoplankton, growth is more commonly thought of as the growth of the population of cells, i.e., the increase in the number or biomass of the cell population or assemblage of organisms. In the laboratory, algal culture growth can be measured accurately using cell counts or the proxies for biomass such as chlorophyll *a* (chl *a*) or particulate carbon (C) or nitrogen (N). However, in nature, phytoplankton live in a world with grazers and the loss of cell numbers or biomass happens concurrent with cell population growth. Thus in natural systems, measuring cell numbers over time only measures *net* growth which is not equivalent to measuring the actual (*gross*) growth rate of the phytoplankton. The difference between net and gross growth can be large when grazing or viral lysis rates are large. Under such conditions, the phytoplankton may be physiologically growing fast but the population may not be growing at all. Thus, direct methods of measuring biomass or cell numbers do not provide information on how fast the phytoplankton are growing. As Hildebrand W. Harvey (1945) said,

> The standing crop of phytoplankton, or breeding stock, at any time is merely a 'momentary balance' between the processes of production and of consumption by animals and bacteria. The concentration of nutrient salts in the water mass at any time is likewise a momentary balance between the regeneration by animals and bacteria and their consumption by the phytoplankton.

P. M. Glibert, *Phytoplankton Whispering: An Introduction to the Physiology and Ecology of Microalgae*, https://doi.org/10.1007/978-3-031-53897-1_3

It has therefore long been recognized that the amount of biomass at any moment in time represents a balance—a *dynamic balance*—between production of new cells and removal of cells. Phytoplankton growth rate has physiological limits, but the limitation as to how much primary producer biomass can accumulate is determined by external factors—total nutrient load and rate of removal by grazing. Responses to nutrient perturbations on the organism level can be very rapid, while responses to a system's total nutrient supply are much slower; physiological processes are faster than those of system biomass (Malone et al. 1996).

At the core of understanding growth are two fundamental concepts (that apply not only to phytoplankton but to all organisms): first, understanding the rate of growth of an organism is not the same as measuring its biomass, and second, rates of biological processes (including growth) increase with increasing concentration of the limiting substrate up to a point of "saturation" where limitation is overtaken by some other factor such as another substrate or physiological limitations. These two core concepts are the foundation for understanding, quantifying, and measuring growth and other metabolic processes. This chapter uses these fundamental concepts to introduce how growth of phytoplankton is measured, how cell division occurs, and how growth can be manipulated and quantified. These have implications for understanding physiological regulation and the concept of limiting factors—at the scale of cells or ecosystems. An overarching message is that while it is necessary to understand these fundamental concepts, it is also necessary to understand how they may limit our understanding of natural complex systems.

3.2 Saturating Response Curves and the Concept of Nutrient Limitation

A saturating response (Fig. 3.1) relates a biological response to a single substrate and identifies two important phenomena—a region where the substrate influences the rate, and a region where the process is functionally at or near a maximal rate and substrate has no or little influence. A key idea is that as the substrate increases in concentration or magnitude, its influence on the process ultimately stops. Accordingly, the substrate is controlling at low concentrations and noncontrolling at high concentrations (Kana and Glibert 2016). Cells are said to be limited by a substrate (which can include nutrients or light) if their response (growth, nutrient uptake, photosynthesis) increases once the supply of that substrate increases. The responses that are determined by such a saturating response include not only growth as a function of nutrient concentration (Monod 1942, 1949), but also each component physiological process, including photosynthesis as a function of irradiance, enzyme activities (Menten and Michaelis 1913), and nutrient uptake as a function of nutrient supply (both of the media and of the cell (Droop 1968, 1973, 1979) and for grazers, ingestion as a function of food availability. Rao (2000) considered the saturating response curve so fundamental to biological processes that he termed it "a curve for all reasons."

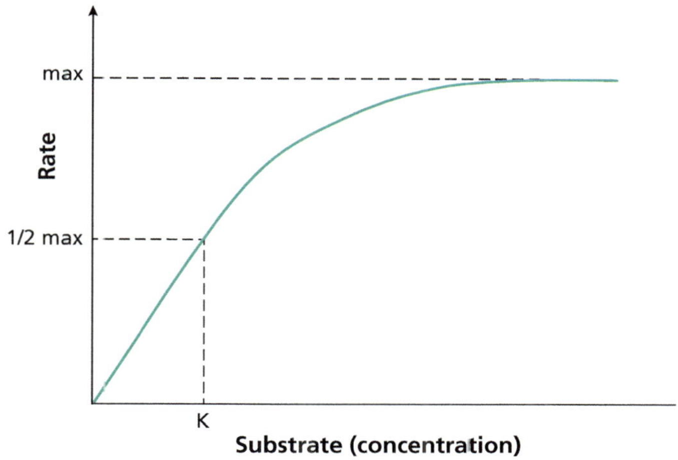

Fig. 3.1 General saturating response curve in which a physiological rate is related to the availability of a substrate

The concept of a saturating relationship in relation to "resources" is also applied at the population level. At larger scales, saturation responses have been applied to ecosystem management of nutrient loading responses; there is a region of rapid response where systems are changing and a region of slow response where systems have considerable resilience to change to either increasing nutrient loading or nutrient reductions (Glibert et al. 2010).

Mathematically, these relationships are defined typically by a maximum (saturating) rate, a half-saturation constant (K) or an index of the concentration at which the rate plateaus (I_k), and the rate of change (initial slope).

Application of the saturation curve to the concept of substrate limitation is perhaps one of the most well-known concepts in phytoplankton ecology, with the half-saturation constant as the break-point wherein substrate limitation begins to set in. What should become abundantly clear as these relationships are described in more detail is that the behavior of the saturating response is, on the one hand, constrained by organismal physiology, and on the other hand, *dynamic* depending on the physiological state of the cells (Chap. 10). At the most simple conceptual level, we can think of a cell's physiology as "running up and down" the saturation curve as substrate availability changes, i.e., as substrate concentration changes, the rate changes. Add more substrate (light, nutrients) and the rate increases. However, physiology is much more complicated. Saturating response curves depend on the physiological state of the cell and/or the manner in which the experiment is performed (e.g., Harris 1978; Goldman and Glibert 1983 and references therein), so any specific curve has some arbitrariness related to the measurement protocol. Thus, it can be argued that a cell does not just "run up and down the curve" as substrate concentration changes, but the shape of the curve fundamentally changes as its environment changes (Kana and Glibert 2016). Catalogs of curves—and the literature is replete with tables of comparisons of kinetic responses of different organisms—have little meaning unless the organisms being compared were subjected to the same measurement approach and physiological prehistory. A different approach is to understand how the cell responds to its environment, and incorporate that understanding in a more dynamic model of resource availability on physiological responses.

3.3 Concept of Balanced Growth

In balanced growth, the growth of all chemical constituents within the cell remains the same when averaged from generation to generation. Thus, whether examining C-, N-, phosphorus (P)-, or DNA-specific growth, rates should not only remain constant through time but should also result in the same value (e.g., Falkowski and Raven 2007). For natural phytoplankton in balanced growth, the proportions of major elements, C, N, and P, should not only remain constant, but they are expected to be in the proportions of the Redfield ratio, 106C:16 N:1P (Redfield 1934, 1958; Redfield et al. 1963; see Chap. 11). This condition would be expected where the environment and biomass are comparatively stable.

In changing environments, such as those undergoing a spring bloom, influenced by upwelling, or nutrient enriched due to anthropogenic causes, a condition of balanced growth is rarely observed. A decoupling of the rates of nutrient and C acquisition and that of growth rate may occur. Cells may take up more nutrients than instantaneously required for growth, leading to increased internal storage. In some cases, depending on the nutrient form, this may lead to a depression in growth rate, release of organic C, or, in the case of toxic cells, alteration of toxicity.

3.4 Cell Division and the Cell Cycle

Phytoplankton exist through cell division and the production of daughter cells. This process normally involves cellular growth, DNA replication, nuclear division, and cellular division. The cell division cycle for eukaryotic cells is generally depicted as four distinct phases: G1–S–G2–D (or M) (Chisholm 1981; Zahleder et al. 2016). The G1 (or gap) phase is that phase during which the cell becomes ready to divide. It can also be called a pre-commitment period (Zahleder et al. 2016) as the cell is not committed to proceed further through division if conditions are unfavorable. At this point, if cells do not divide they may go into a reversible cell cycle arrest (Chisholm 1981). At some point, the cell commits to division. This period can be termed the commitment point, CP, or the "point of no return" (Moberg et al. 1968). The initiation of DNA replication begins in this next period, the S phase. Another gap period (G2) follows. This is the period during which DNA replication is complete and mitosis is initiated. The D (division) phase, sometimes called M (mitosis) phase, is the period during which division actually occurs. In some cells, an additional phase occurs, the C phase, representing the extra step required for cleavage of the daughter cells, and this may require yet a third gap phase, G3 (Zahleder et al. 2016; Fig. 3.2).

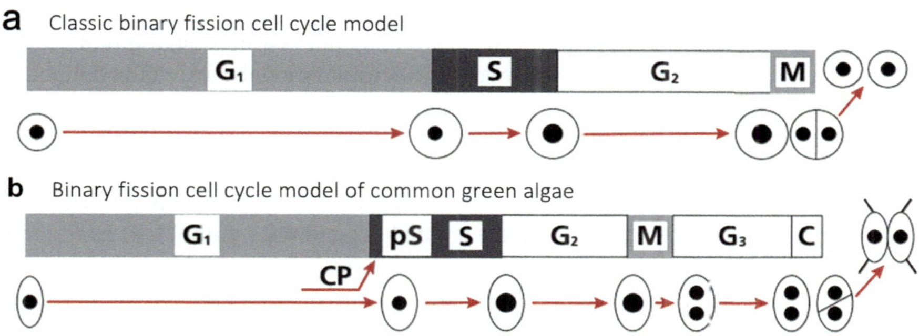

Fig. 3.2 Conceptual diagrams of different types of cell cycle phases. Panel (**a**) The classical cell cycle model, and Panel (**b**) The variation on this model undertaken by the green alga, *Desmodesmus* (reported as *Scenedesmus*) *quadricucauda* and *Clamydomonas*. The G1 phase is the phase in which the threshold size of the cell is attained. S is the phase in which replication occurs. G2 is the phase between the termination of DNA replication and the start of mitosis. M is the phase in which nuclear division occurs. In the green alga, pS is a pre-replication phase, G3 is the phase between nuclear division and cell division, and C is the phase in which cell cleavage and daughter cell formation occur. Image redrawn based on Zahleder et al. (2016, The physiology of microalgae. doi: 10.10007/978-3-319-24945-2_1)

For some cells, such as colonial algae, cells may divide into many daughter cells. In such a sequence of events, there may be multiple commitment points during a "single" cell cycle with subsequent commitment points dependent on successful completion of the first commitment cycle. Such a sequence requires additional gap phases and several rounds of DNA replication before daughter cells are ultimately produced (Zahleder et al. 2016).

When cells are grown on light/dark cycles, growth and division can become synchronized across the population, such that it occurs at the same time of day. When all cells within a population divide at the same time of day, growth is said to be "synchronous," whereas if only a portion of the population divides at the same time of day, growth is said to be "phased" (Chisholm 1981). Some species have strict, biological clock-controlled cycles that appear to be phased with the periodic supply of energy (light) to the cell. Other taxa have cycles that adjust depending on environmental conditions. How these different cell cycle strategies relate to relative success of different types of taxa in stable, predictable environments versus more dynamic environments continues to be an area of active study.

3.5 Growth Rate

One of the most important parameters of investigation in phytoplankton ecology and physiology is growth rate. How fast the cells are growing generally reflects the physiological state of the cell. Are they growing at their maximal rate? If they are growing slower, then what factor(s) contributes to the slowed growth? This is fundamental to understanding the response of phytoplankton to their environment. Measuring growth rate has its origins in population biology and the mathematics therein. For single-celled organisms, population growth usually corresponds to the doubling of cells after cell division. These are countable entities, so the first models were population number based. However, in microbial models, growth is defined by a number of proxy measures of population size, including optical measurements such as chl *a* fluorescence or optical density or biomass measurements such as cell C, cell N, or even dry weight. Conventionally, the symbol representing growth rate of a microbial population (including phytoplankton) is the Greek letter mu, μ. In classical population biology, this parameter is designated "r" and this symbol will be found in the phytoplankton literature. The term r is also called the intrinsic rate of increase (Wood et al. 2005). The two terms, r and μ, are only equivalent if and when mortality, m, is equal to zero.

When cells are growing at a constant rate, the doubling of the population happens over a particular time interval, let us say daily. Daily from day one, one cell will become 2, 4, 8, 16, 32... This increase is called exponential growth, and the rate of increase is calculated using the equation:

$$dN/dt = \mu N \tag{3.1}$$

or,

$$N_t = N_0 e^{\mu t} \tag{3.2}$$

where N refers to the size of the population at time 0 or time t and dN and dt refer to the change in cell number and time, respectively. Both μ (or r) can be solved for by rearranging Eq. 3.2:

$$\mu = (\ln (N_t/N_0)/dt \text{ or } \mu = (\ln N_t - \ln N_0)/dt \tag{3.3}$$

$$r = (\ln (N_t/N_0)/dt \text{ or } r = (\ln N_t - \ln N_0)/dt \tag{3.4}$$

Over time the rate of growth of the culture will slow and may be followed by a population crash (die-off). The slowing of growth may be due to nutrient depletion, CO_2 limitation, decrease in light availability as the cells accumulate, autoinhibition, or other factors. Unless cultures are contaminated—perhaps with a virus that causes mortality—mortality in typical culture experiments is usually low as there are no grazers, and until substrate declines or other conditions change, death rates should be low.

One of the most confusing aspects of the reporting of growth rate is the units in which it is reported. It is not unusual to see in the literature growth rate expressed as "specific growth rate" or "doubling time." While both use dimensions of time (i.e., μ in reciprocal time (e.g., h^{-1}, day^{-1})) and doubling time in time (e.g., h, day), their relationship is not interconvertible by simply taking the reciprocal of each. Instead, these conversions are made by dividing μ by the natural log of 2, or 0.69 which corresponds to $N_t/N_o = 2$. A value of 1 division d^{-1} is not equal to a specific or intrinsic growth rate, r or μ, of 1, but rather a value of 0.69 d^{-1}. A thorough review of these relationships is presented by Wood et al. (2005).

The symbol k has been recommended to distinguish it from μ, and thus,

$$k = (\log_2(N_t/N_O))/dt \tag{3.5}$$

Accordingly, doubling time, T_2, is calculated as,

$$T_2 = 0.693/\mu \tag{3.6}$$

Caution is advised in interpreting the literature as many investigations interchange the symbols μ, r, or k and reporting thereof as divisions day^{-1} or $time^{-1}$.

Growth rate (as μ or r) is central to studies of phytoplankton physiology and ecology. It is dependent on resource availability (e.g., nutrients), energy availability (e.g., irradiance), and various physical and chemical factors (e.g., temperature and salinity). The most common relationship describing the rate of growth as a function of nutrient availability is that of a saturating hyperbola, the "curve for all reasons" (Fig. 3.1). This relationship was originally defined by Monod (1942) in studies of bacterial cultures,

$$\mu = \mu_{max}(S)/(K_s + S) \tag{3.7}$$

where S is the nutrient concentration and K_s is the concentration of the limiting nutrient at which μ is half maximal. While the term μ_{max} may appear to imply some definitiveness—an absolute maximum, it is, in fact, a relative term. Maximum growth rate is a function not only of the availability of sufficient nutrients (here using limiting nutrient growth condition as an example), but it is also a function of other environmental conditions, namely temperature, light, salinity, and other environmental conditions that influence growth. Thus, μ_{max} may differ for an organism measured at 15° compared to 30°, or at high irradiance vs low irradiance. In the case of nutrient limitation, μ_{max} provides the maximum rate for metabolic activity when the substrate is provided in excess. At μ_{max}, rate limitation is conferred to a factor other than the nutrient under study. Within the experimental context, rate limitation can be assessed by simply relating μ to μ_{max}. Thus,

$$\text{Relative growth rate} = \mu/\mu_{max} \tag{3.8}$$

The concept of relative growth rate has become important in understanding general regulatory patterns across species. Growth rate, per se, corresponds to a rate of cellular metabolism and can be thought of as a "grand integrator" of metabolic processes that are affected by the limiting nutrient, a conclusion that would be drawn a priori from mass balance considerations (Shuter 1979). Relative growth rate reflects the level of downregulation of metabolism. Relative growth rate allows comparisons of physiological limitations across species, despite the fact that μ_{max} may differ between species. For example, species A and species B may have different μ_{max}, but if their relative growth rate is 0.5, then their metabolisms are both operating at 50% of capacity (under the experimental conditions). In addition, μ_{max} serves as a reference rate that corresponds to the particular physiological state

Fig. 3.3 The Droop relationship, in which growth is related to the cell quota of substrate. Note there must be a minimum, k_Q, of substrate in the cell for it to grow

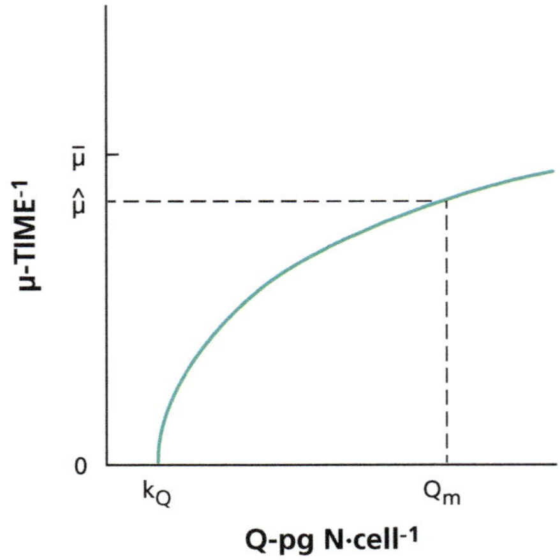

of non-limitation. Relative growth rate is thus a means to understand general patterns among species in their response to environmental variables.

An alternative relationship between substrate and growth was developed by Droop (1973). This relationship expresses growth as a function of nutrients (or other substrates, such as vitamin B12) inside the cell, rather than in the external media (Fig. 3.3). This relationship has the form:

$$\mu = \mu_{max}[1 - (k_Q/Q)] \tag{3.9}$$

in which Q is the cell quota (cellular nutrient per total cell mass), k_Q is the minimum cell quota required for growth and $\bar{\mu}$ is the unattainable value of μ at infinite Q. The symbol $\hat{\mu}$ is used to characterize μ_{max} at Q_{max} (Goldman and Glibert 1983). For those substrates making up only a small fraction of the cell mass, such as vitamin B_{12}, or PO_4^{3-}, approximates μ_{max}.

3.6 Growth in Cultures

One of the best ways to understand the difference between growth and biomass and its dynamic balance is to understand how growth is measured and manipulated in the laboratory and in systems with increasing reality. There are several methods for cultivating microalgae that give different levels of control over the growth conditions. They are generally distinguished by the level of steady-state (constant) growth conditions imposed by the culturing apparatus and the manner in which steady state is achieved. The next sections introduce the approaches and apparatus for culture growth and how they differ with respect to growth rate.

3.6.1 Batch Culture Growth

Batch cultures are the most common culture technique because they require minimal equipment to establish, can be conducted in vessels of virtually any size (from a microplate to a large vat), and can be manipulated in many ways. A batch culture consists of a vessel, liquid nutrient-enriched growth media, and cells. Growth rate and biomass vary in time in batch culture (Fig. 3.4). Often upon initial inoculation, there is often a lag phase during which cells are not dividing, or dividing at a slow rate as they acclimate to their current condition. Eventually, the cells begin to divide with a constant doubling time. This is the classical exponential growth phase where every cell division interval is a doubling of cells and biomass. Expressed graphically, the logarithm of cell counts increases linearly with time. This is an important experimental phase because it is when the cells are in balanced, steady-state, maximum growth. With time in batch culture, cells begin to deplete nutrients, and growth resources become limiting. Growth rate of the culture slows and eventually stops. At this point, the culture is in stationary phase, or the phase where there is no net increase. Note that at this phase cell death may occur and there is limited cell division to maintain a stable population. Because of the ease of growing batch cultures and the desire often to study algae that are actively growing, a commonly used method is to frequently transfer cells into fresh media to maintain the culture in exponential phase growth. This method is called "semi-continuous batch culture."

If cells are held in the stationary phase for an extended period, more cell death may occur (Fig. 3.4). Cell death may occur because as organisms grow, not only are nutrients consumed, but CO_2 availability is reduced, and extracellular products may

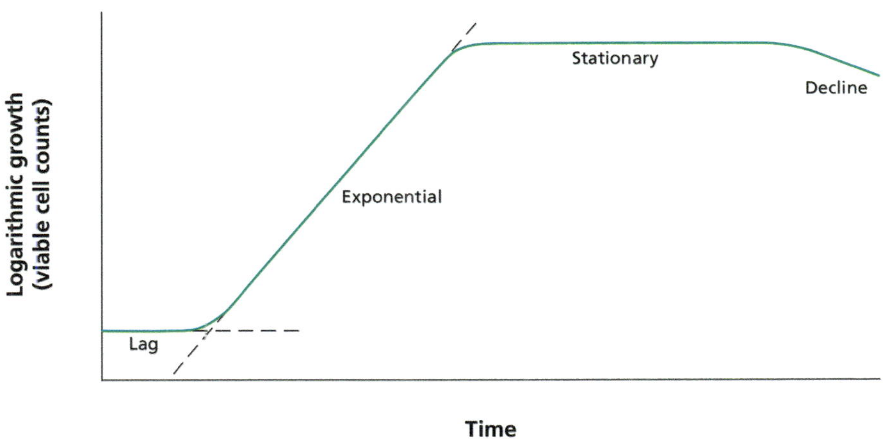

Fig. 3.4 The general progression of growth in a batch culture. Upon inoculation into fresh media, there may be an initial lag period before growth begins. Then, growth occurs at an exponential rate, eventually slowing as one or more resources become limiting. If cells are held in this limiting condition, cell death may occur

accumulate, all of which change the conditions for the organisms (culture conditions may alleviate some of these effects—for example, if cultures are bubbled or not).

Whereas batch cultures are elegant in their simplicity, even during exponential phase where growth rate is maximal, the environmental conditions undergo changes. Nutrients become depleted (even though they may initially be in excess), irradiance changes as the culture biomass becomes denser, and exudates or dead cells accumulate. The semi-continuous batch culture reduces these effects and the logical extension of repeated dilution of the culture is to provide continuous dilution. This mode of culturing is treated in the next sections.

3.6.2 Chemostat Culture Growth

3.6.2.1 Chemostat Theory

It is often desirable to maintain cells in a specific growth condition and to be confident that growth rates do not change with time as nutrients are consumed. Continuous cultures allow such conditions to be established. Continuous cultures are essentially pump-controlled versions of the aforementioned manually controlled semi-continuous cultures.

Chemostats are continuous cultures in which fresh media is dispensed (using a pump) at a regular interval with excess volume removed to a receiving or waste container (Fig. 3.5). The apparatus is designed to keep the volume of the culture constant. In practice, the media that is dispensed has all chemical nutrients in excess except for one, which is the limiting nutrient for growth. Chemostat theory assumes there is only one limiting nutrient, that a single species is in culture, and that mixing of each drop of new media is instantaneous. These continuous culture devices were independently developed by Monod (1950) and Novick and Szilard (1950) for studying the growth of bacteria and have since been well reviewed in classic syntheses (e.g., Rhee 1973).

To imagine how a chemostat works, consider a vessel with a continuous supply of media and an inoculum of cells. Initially, the small biomass of cells takes up nutrients at a lower rate than the rate of resupply of media. This is analogous to a batch culture with the cell density increasing with time where they grow faster than the rate of dilution. With time, the cells attain a density where they are consuming the limiting nutrient almost completely and the residual concentration of the limiting nutrient approaches zero (assuming the cells consume it with high efficiency). This is the critical point where the input of nutrient is all within the biomass. *The concentration of the nutrient in the media determines the amount of nutrient in biomass.* Thus, the investigator sets the cell density by manipulating the limiting nutrient concentration. At this point in the culture, if the cell density remains constant with time (something that should be measured), then the cells are growing in a *steady-state*. For the cell density to remain constant, the cells must be growing at a rate equal to the rate of dilution of the culture, which is set by the flow rate and volume of the chamber. *The dilution rate equals the growth rate of the cells.* The significance of chemostats is that the investigator has control over setting both the culture density

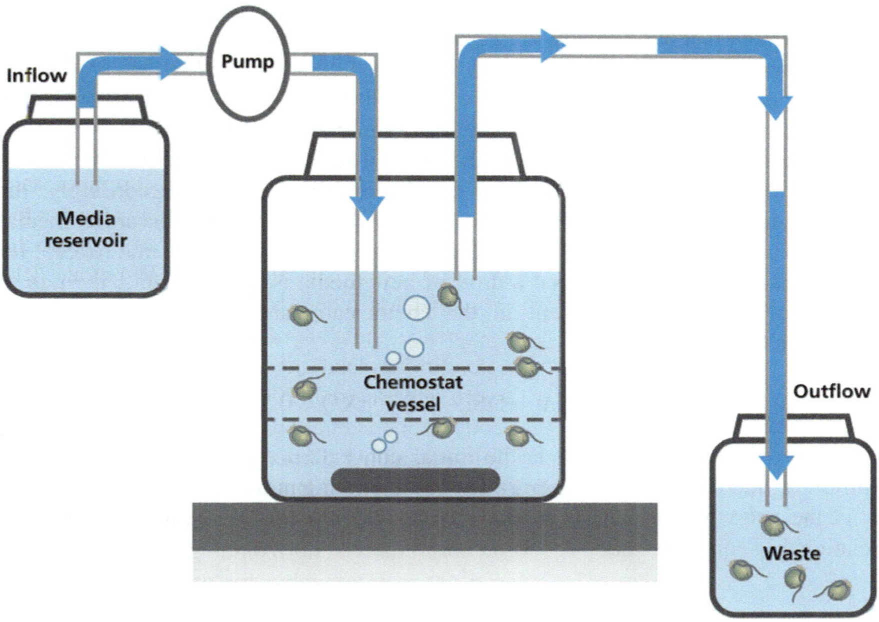

Fig. 3.5 A chemostat system. Media is provided at a fixed rate controlled by a pump. As new media is added, the equivalent amount of culture water is withdrawn to an overflow vessel. The culture may be stirred or bubbled (or both), depending on the species. In a chemostat system, the dilution rate of the media sets the growth rate, while the biomass is set by the limiting nutrient in the media. In chemostats, cultures grow at $<\mu_{max}$

and culture growth rate. These concepts have undergone considerable mathematical reasoning and some of this formalism is described next.

In chemostats there are two "control valves": the first is the rate of supply of new media and the second is the concentration of the limiting nutrients within that media. The former sets the growth rate of the cells and the latter sets the biomass that can be attained. Thus,

$$dX/dt = \mu X - DX \tag{3.10}$$

where X is the biomass concentration (as cell number or cell biomass proxy), μ is the growth rate, and D is the dilution rate. Dilution rate is the rate of delivery of the fresh media (e.g., ml per time) divided by the volume of the culture vessel (ml). For every increment of increase in cell biomass, an equal aliquot is removed to waste. Accordingly, at any given point in time,

$$\mu = D \tag{3.11}$$

The resulting dimensions of both μ and D are time^{-1}. Quite simply, by setting the rate of flow of media (typically via a peristaltic pump), the specific growth rate of a culture can be "dialed in."

In a well-functioning chemostat, in which cells have been given adequate time to acclimate, and with proper mixing, the cells should be poised to consume all of the limiting nutrient as soon as they enter the vessel, i.e., in a nutrient-limited chemostat, there is always a biological demand for the limiting nutrient. Accordingly, the concentration of limiting nutrient in the culture vessel (or in the discarded media) should be at or near detection levels and should stay constant over time (except for the instantaneous moment when a drop of new media is added). Thus, the rate of change of the limiting nutrient in the chemostat chamber can be determined according to:

$$dS/dt = DSi - DS - \mu XQ = 0 \quad (3.12)$$

where Q is the cell quota, Si is the initial concentration of the limiting nutrient (inflowing media), S is the concentration of the limiting nutrient in the culture vessel, μ is the growth rate, and X is the concentration of cells. At steady state, this relationship equals 0, and

$$X = (Si - S)/Q \quad (3.13)$$

Thus, by setting the concentration of the limiting nutrient in the inflowing media, the concentration of resulting biomass can also be "dialed in."

Since $\mu = D$, what happens when cells are growing at μ_{max}? This is the boundary where if D is set greater than μ_{max}, then the higher dilution rate will 'wash out' the cells and the culture is lost. This is an important practical constraint as it is very difficult to maintain chemostat cultures at μ_{max} without washout. Consequently, the determination of μ_{max} is often done experimentally using a batch culture approach with saturating nutrients. Similarly, under extremely low growth rates, the culture system may be unstable and cell death may increase and the concentration of the limiting nutrient will increase.

Thus, by simply turning the speed control on a pump (and giving an acclimation period), one can establish specific steady-state growth rates in the range of rates between $0 < \mu < \mu_{max}$ (in practicality this range is closer to ~$0.2 < \mu < 0.8\ \mu_{max}$). Importantly in the case of chemostats, growth rate is not a function of the limiting nutrient concentration, as the concentration remains constant in the vessel (and ideally at non-detectable levels); only the biomass is a function of the nutrient concentration in the inflowing media.

3.6.2.2 Variations on the Chemostat Theme

In traditional chemostat operations, the rate of supply of the limiting nutrient is controlled by a pump at a rate set by the experimentalist. However, there are multiple modifications to such an experimental design. If chemostats are set on a light-dark cycle rather than being provided with continuous light, cell division may be phased and instantaneous growth rate μ will not necessarily equate to D at any moment in time, but equivalency will be established if values are averaged over the light-dark cycle. Such chemostats are commonly called cyclostats. The rate of supply can also be varied in different ways such that the same daily dose is achieved but the frequency of the pulses of media injection is varied. For example, the daily dose may be delivered once per day, at discrete intervals, or in a near-continuous fashion. Olson and Chisholm (1983) carried out exactly this experiment, while also imposing a light-dark cycle and found that cell division in different species responded differently. They were able to show that phased cell division in the dinoflagellate tested cued to the light-dark cycle, while that of the diatom was cued to the nutrient pulse.

Various investigators have grown and/or modeled the behavior of multiple species within a single chemostat. For example, Baron et al. (2014) examined the stabilities of two and three competing species with a single chemostat. Other authors have taken the approach of establishing multiple species in tandem. That is, the outflow of media from one chemostat is used to feed the next chemostat in series. Sometimes called gradostats, chemostats in line may support different species. Such a setup might, for example, be useful in understanding whether exudates of one taxon positively or negatively influence another. The complexities of these culture systems are not dealt with here, but the reader should be aware that there are many versions of chemostat culture systems.

3.6.3 Turbidostat Culture Growth

Another type of continuous culture system is a turbidostat. In turbidostat cultures, cells are maintained at a constant density (turbidity) using a sensor that controls a dilution pump. In this manner cells can be maintained at μ_{max}, in contrast to chemostats in which constant growth is possible only at $<\mu_{max}$. Turbidostats are basically automated semi-continuous batch cultures, but the interval between nutrient pulses can be very short (seconds to minutes). In turbidostat growth, the system is monitored to detect small changes in biomass accumulation, typically using fluorescence or turbidity sensors (Fig. 3.6). The operator sets a tolerance range that signals the dilution pump cycle. The resulting signal is typically a saw-tooth pattern showing a rapid decline during the short pump stage and a slow growth period and an increase in turbidity. Each interval between nutrient pulses represents a time during which exponential, or μ_{max}, growth is attained. The calculated growth rates during each of these short time intervals can be plotted to show growth rates over time. Thus, growth rate in turbidostats is not a pre-set condition as in chemostats, but rather a dependent variable that is set by environmental conditions. Turbidostats are commonly used to understand growth as a function of light. One practical convenience of a turbidostat over a chemostat is that culture can be removed for analysis without interfering with the culture density control, whereas with a chemostat, any removal of volume will change the dilution rate of the culture apparatus and hence interfere with the steady-state condition.

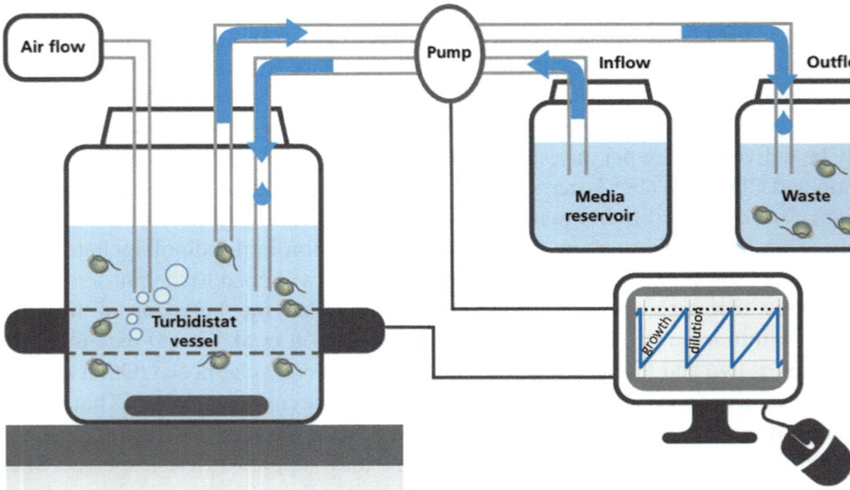

Fig. 3.6 A turbidostat system. In contrast to a chemostat system, the rate of media delivery is determined by the change in biomass of the culture. As the culture grows, sensors on the vessel record the optical signal, and when this exceeds a pre-set value, the computer signals the pump to add an amount of media to dilute the optical signal back to the desired range. In turbidostats, cultures grow at μ_{max}

3.7 Microcosms and Mesocosms

Moving from controlled laboratory culture systems, microcosms or mesocosm systems—enclosed experimental ecosystems—provide an intermediate scale of complexity compared to field studies (Fig. 3.7). Such systems allow for experiments and measurements that include multiple trophic levels, longer time periods of growth, and are of a size that is large enough not to be perturbed by sample collection. Although definitions vary by users, in general, a microcosm system contains a volume <1 m^3, while mesocosms are larger enclosed systems. The word "mesocosm" appears to have been originally proposed by Banse (1982) to describe any experimental system larger than that used on a benchtop but smaller than the natural environment. Mesocosms may take the form of large containers, in-water enclosures, "limnocorals," or indoor or outdoor tanks to which natural water may be added and responses monitored for periods of days to months (Fig. 3.8). Mesocosms may also be used in series, like chemostats, although such systems on a large scale are exceptionally difficult to maintain.

Mesocosms have become an important tool for studying natural communities and for testing responses to varying environmental changes. Such tools have been used in studying phytoplankton for decades. The Food Chain Research Group at the Scripps Institution of Oceanography in the 1960s and 1970s, under the helm of John

Fig. 3.7 With mesocosms, the scale of complexity of community interactions on longer time and larger space scales can be examined

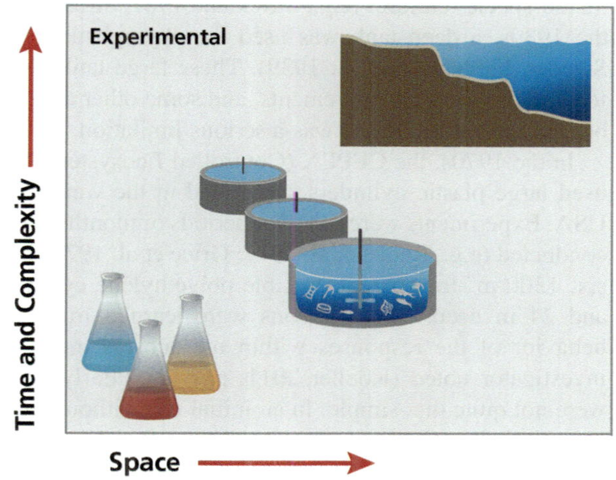

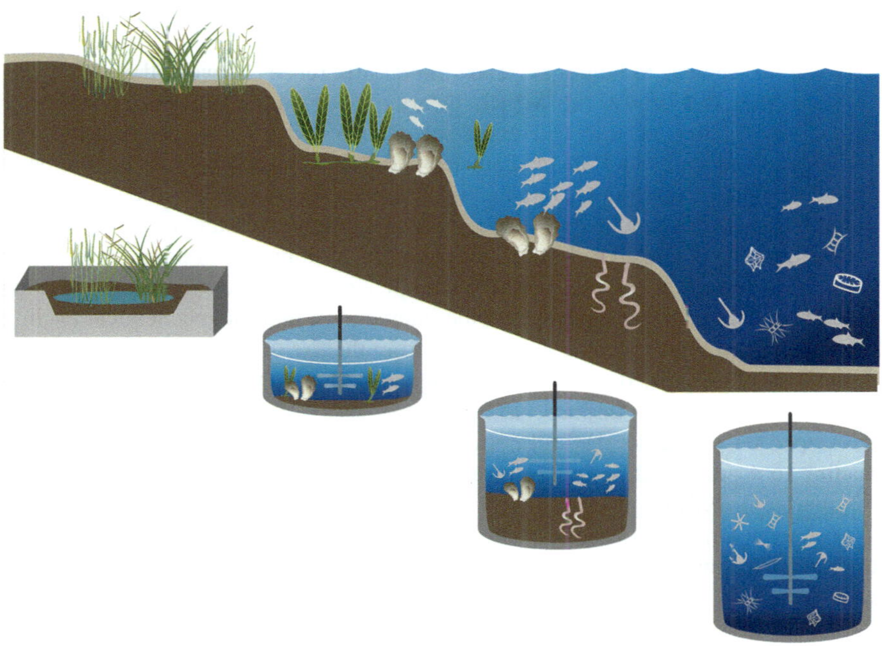

Fig. 3.8 Mesocosms can be used for experiments that address a range of questions, and which encompass different fractions of an ecosystem, from water to benthos to marshes

Strickland, used a 70,000-liter tank (3 m diameter, 10 m deep) for controlled environmental studies—which he characterized as being "between beakers and bays." With this deep tank, numerous experiments were conducted with known mixed-species cultures (e.g., Strickland 1967; Strickland et al. 1969). Even earlier, in the 1930s, a deep tank was used for phytoplankton experiments in Gothenburg, Sweden (Pettersson et al. 1939). These large tanks were useful for understanding food chains, vertical movements, and some other aspects of phytoplankton ecology, but the lack of replicates was a serious limitation.

In the 1970s, the CEPEX (Controlled Ecosystem Pollution Experiments) project used large plastic cylinders suspended in the waters of coastal Washington State, USA. Experiments were run over periods of months and many chemical studies were conducted (e.g., Reeve et al. 1977; Grice et al. 1977, 1980). These mesocosms were six, 1300 m^3 transparent, flexible polyethylene cylinders, nearly 10 m in diameter and 24 m deep. Many lessons were learned from these experiments, including behavior of the responses within mesocosms, not just pollutant effects. As one investigator noted (Koeller 2011, p. 642), "Early on, it became clear that things were not quite that simple. In an initial run without pollutants, the mesocosms went off in all directions—one turned into jellyfish, another into a monoculture of a copepod rarely seen. If unperturbed mesocosms did not replicate, it would be impossible to draw conclusions from any observed differences between controls and experimental treatments. The replication problem was quickly solved by methodological improvements, including more simultaneous capture and pumping between mesocosms to homogenize any initial differences between them."

Soon after that study, the MERL (Marine Ecosystems Research Laboratory) mesocosm facility at the University of Rhode Island was developed. With 14 large, fixed tanks, 1.8 m in diameter and 5.5 m in depth, with a volume of 13 m^3, maintained outdoors, this facility has been the site of countless studies of pollutants, nutrient enrichments, sewage effluent, and many other effects (Fig. 3.9). Experiments have ranged from short-term (weeks) to durations of longer than 1 year. These enormous "test tubes," with their replication, have yielded enormous information on organismal responses, trophic interactions, and biogeochemical cycling. One interesting unforeseen result emerged from the use of these tanks in the 1980s. Tanks were being filled with local Narragansett Bay water when a bloom of the brown tide organism, *Aureococcus anophagefferens*, developed. The brown tide developed only in tanks that did not receive specific nutrient enrichments, in turn, leading to a great deal of further research on the nutrition of this harmful species (Cosper et al. 1987).

Yet, even large mesocosm systems are subject to potential artifacts: those related to the dimensions of the mesocosm and how the system is maintained. These factors were examined in a long-term suite of studies in the MEERC (Multiscale Experimental Ecosystem Research Center) at the University of Maryland Center for Environmental Science. In that facility, replicate mesocosm of different dimensions were compared (Petersen et al. 1999, 2003, 2009). The MEERC tanks were designed so that there was a series of constant depth and a series with a constant wall surface: volume ratio. The size and shape of an enclosure can have important effects on dynamics of the ecological relationships and systems that develop over time (e.g., Berg et al. 1999; Petersen et al. 2009). In enclosures of varying dimensions, food web dynamics may be altered due to differences in wall area: volume ratios or to the

Fig. 3.9 Photo from the MERL mesocosm system at the University of Rhode Island

presence or absence of sediment, or the degree of mixing and manner of mixing (Petersen et al. 2009; Fig. 3.10). For example, Glibert and Berg (2009) enriched whole mesocosms of different sizes (Fig. 3.10) with $^{15}NH_4^+$, exchanging 10% of the mesocosm volume daily. They found in the smallest tanks tested, the isotopic enrichment in the larger plankton fraction (<202 μm) exceeded that in the small size fraction (<10 μm), whereas in the larger tanks, the isotopic enrichment in the small plankton fraction exceeded that in the large fraction. A combination of light and wall effects as well as different grazing rates contributed to these differences. Over the days of incubation, these tanks also retained nutrients, even with daily water exchange, becoming model systems for eutrophic estuaries.

One of the recurring problems in mesocosm design is how to mix large volumes of water. In natural waters, mixing and flow are critical processes that determine heat and nutrient and material fluxes. At the small scale of organisms, the motion of the fluid (shear) affects nutrient and material supplies to a cell, affects contacts between predator and prey, and can be a source of physical stress (Sanford et al. 2009). Many older mesocosms lacked mixing, while overly aggressive mixing can also be problematic. Mixing can also affect sediment resuspension or sediment-water exchange if the mesocosms have a sediment layer. More recently, the effects of different techniques for mixing have been investigated in experimental mesocosms (Stiebel et al. 2013). Those authors compared mesocosms using a mixing disk, an airlift system, and bubbling relative to mesocosms with no mixing. They found that simple lowering and raising of a disk was an effective technique.

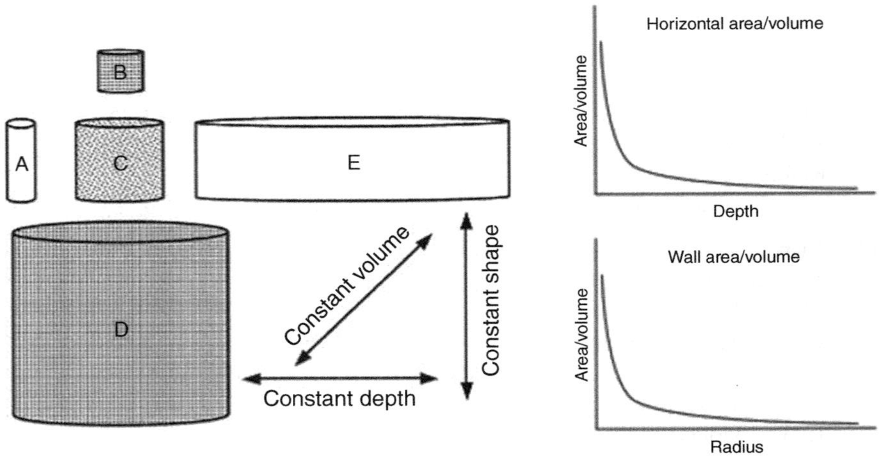

Fig. 3.10 The MEERC project investigated how biological response in mesocosms was affected by mesocosms of different sizes and shapes. Mesocosms A, C, and E had a constant depth, A and B as well as D and E had constant volumes, and B, C, and D had constant shapes

Countless recent experiments have been conducted using mesocosms of all dimensions and designs. Whether uniquely designed with independent temperature and light controls, such as the Planktotrons in Germany (Gall et al. 2017), outdoor tanks with manual hand mixing (e.g., Shangguan et al. 2017), or large in situ cylindrical bags used to study CO_2 exchange and rates of diazotrophy or nutrient enrichment (e.g., Hyun et al. 2020; Arvind et al. 2021; Böttjer-Wilson et al. (2021; Fig. 3.11), there is much that can be explored in mesocosms. It is always important to remember that such systems are intermediate in complexity—they cannot replicate field conditions, but they nevertheless can be valuable tools in studies of plankton dynamics, nutrient enrichment, pollutant impacts, and many other fundamental processes in aquatic research.

3.8 Whole-Ecosystem Studies

There is a long history of using whole lakes as experimental chambers. The Canadian Experimental Lakes Area is the most well-known of these, and over the years they were manipulated with nutrients in order to address the question of what limited their biomass (Schindler 1974, 2006). These studies were instrumental in the ultimate removal of PO_4^{3-} from laundry detergents, as the lake data showed that enriching with this nutrient led to algal blooms (Fig. 3.12). Many such studies have been continuous, or nearly continuous for decades (e.g., Carpenter and Pace 2018). In some studies, the driver of chl *a* biomass is readily apparent, in others, it has taken years of data to discern patterns. Scaling studies to whole lakes allows questions of nutrient (or other substrate) limitation to be investigated in the context of the natural

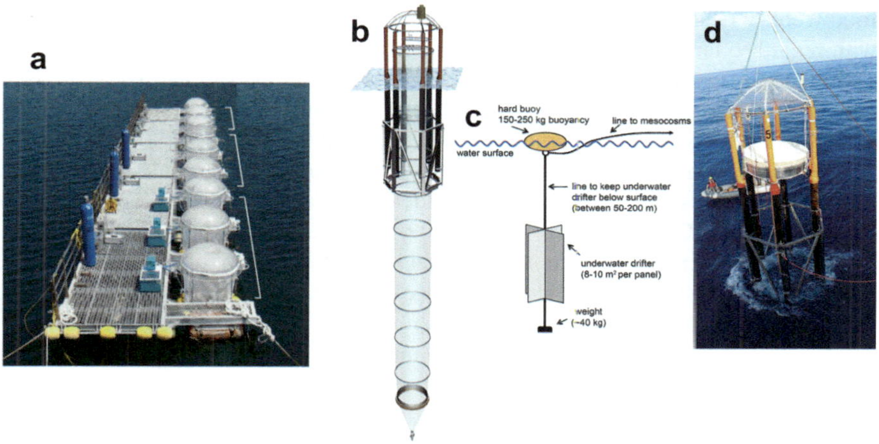

Fig. 3.11 Two views of in situ mesocosms. Panel (**a**) image of the mesocosms facility used in Korea to study the effects of ocean acidification. Panels (**b**)–(**d**) image of the mesocosm facility used to understand nutrient enrichments on oligotrophic phytoplankton near Hawai'i. Image (**a**) reproduced from Hyun et al. (2020, J Mar Sci Eng 8: 821. doi.org/10.3390/jmse8100821) and images (**b**)–(**d**) reproduced from Böttjer-Wilson et al. (2021, Aquat Microb Ecol 87: 167–187. doi.org/10.3354/ame01977), both under creative commons licenses

Fig. 3.12 This divided body of water shows the remarkable difference between mesotrophic (moderately enriched) (upper basin) and eutrophic water (lower basin). Image provided by Fisheries and Oceans Canada

food web. In Chaps. 16 and 26, whole system enrichment experiments are revisited with regard to large manipulation experiments with iron and CO_2.

In the past two decades, there have been multiple mesocosm and whole ecosystem isotope labeling studies designed to measure integrated biogeochemical pathways or flows of C and N through food webs. Examples of whole system tracer enrichment that have lasted up to months can be found in streams, rivers, lakes, estuaries, wetlands, and marshes (e.g., Middelburg et al. 2000; Hamilton et al. 2004; Pace et al. 2004; Gribsholt et al. 2005, 2009; Oakes et al. 2010, 2012). Whole-lake labeling experiments have been instrumental in identifying the role of allochthonous subsidies for lacustrine consumers (e.g., Pace et al. 2004; Carpenter et al. 2005; Cole et al. 2002).

One of the important lessons from such meta-ecosystem research is that shifts in grazing pressure can result in substantially different levels of primary productivity and biomass, even when the amount of nutrient loading is the same (Jeppesen et al. 1997; Massol et al. 2011; Carpenter et al. 2016). Adding complexity can shed light on the critical interactions that drive ecosystem structure.

3.9 Summary

The seemingly simple concept of growth by phytoplankton is not so simple. The "momentary balance" of biomass (drawing on the words of Harvey 1945) that may be measured at any given point in time is the sum of multiple physiological processes as well as varying trophic complexities in which the cell may momentarily live. From batch culture to chemostat to turbidostat or mesocosm, the choice of model system depends on the question of the investigator and the degree to which growth rate and environmental parameters are closely controlled, or, alternatively, the degree of realism desired. Different growth rates, communities, and processes develop in isolated parcels of water, regardless of their size. The saturating hyperbola is a universal relationship, applicable for many fundamental parameters as a function of substrate availability.

References

Arvind S, Bach LT, Löscher C, Paul AJ, Ojiha N, Riebesell U (2021) Impact of increasing carbon dioxide on dinitrogen and carbon fixation rates under oligotrophic conditions and simulated upwelling. Limnol Oceanogr 66:2855–2867. https://doi.org/10.1002/lno.11795

Banse K (1982) Experimental marine ecosystem enclosures in a historical perspective. In: Grice GD, Reeve MR (eds) Marine mesocosms: biological and chemical research in experimental ecosystems. Springer-Verlag, New York, pp 11–24

Baron R, Davie A, Gaines A, Grant D, Okay O, Ozsoy E (2014) Continuous cultures of phytoplankton. Int Aquat Res 6:95–111

Berg GM, Glibert PM, Chen CC (1999) Dimension effects of enclosures on ecological processes in pelagic systems. Limnol Oceangr 44:1331–1340. https://doi.org/10.4319/lo.1999.44.5.1331

Böttjer-Wilson D, White AE, Björkman KM, Church MJ, Poulos S, Shimabukuro E, Riis YM et al (2021) Effects of nutrient enrichments on oligotrophic phytoplankton communities: a

mesocosm experiment near Hawai'i, USA. Aquat Microb Ecol 87:167–187. https://doi.org/10.3354/ame01977
Carpenter SR, Pace ML (2018) Synthesis of a 33-yr series of whole-lake experiments: effects of nutrients, grazers, and precipitation-driven water color on chlorophyll. Limnol Oceanogr Lett 6: 419–427. https://doi.org/10.1002/lol2.10094
Carpenter SR, Cole JJ, Pace ML, Van de Bogert M, Bade DL, Bastviken D, Gille C et al (2005) Ecosystem subsidies: terrestrial support of aquatic food webs from ^{13}C addition to contrasting lakes. Ecology 86:2737–2750
Carpenter SR, Cole JJ, Pace ML, Wilkinson GM (2016) Response of plankton to nutrients, planktivory and terrestrial organic matter: a model analysis of whole-lake experiments. Ecol Lett 19:230–239
Chisholm SW (1981) Temporal patterns of cell division in unicellular algae. In T. Platt. (ed), Physiological bases of phytoplankton ecology. Can Bull Fish Aquat Sci 210:150–181
Cole JJ, Carpenter SR, Kitchell JF, Pace ML (2002) Pathways of organic C utilization in small lakes: Results from a whole-lake ^{13}C addition and coupled model. Limnol Oceanogr 47:1664–1675
Cosper EM, Dennison WC, Carpenter EJ, Bricelj VM, Mitchell JG, Kuenstner SH, Colflesh D, Dewey M (1987) Recurrent and persistent brown tide blooms perturb coastal marine ecosystems. Estuaries 10:282–290
Droop MR (1968) Vitamin B-12 and marine ecology: IV. The kinetics of uptake, growth and inhibition in *Monochrysis lutheri*. J Mar Biol Assoc UK 41:69–76
Droop MR (1973) Some thoughts on nutrient limitation in algae. J Phycol 9:264–272
Droop MR (1979) On the definition of X and of Q in the cell quota model. J Exp Mar Biol Ecol 39: 203
Falkowski PG, Raven JA (2007) Aquatic photosynthesis, 2nd edn. Princeton University Press, Princeton, NJ, p 484
Gall A, Uebel U, Ebensen U, Hillebrand H, Meier S, Singer G, Wacker A, Striebel M (2017) Planktotrons: a novel indoor mesocosm facility for aquatic biodiversity and food web research. Limnol Oceanogr-Methods 15:663–677. https://doi.org/10.1002/lom3.10196
Glibert PM, Berg GM (2009) Nitrogen form, fate and phytoplankton composition. In: Peterson JE, Kennedy VS, Dennison WC, Kemp WM (eds) Experimental ecosystems and scale: tools for understanding and managing coastal ecosystems. Springer, New York, pp 183–189
Glibert PM, Boyer J, Heil C, Madden C, Sturgis B, Wazniak C (2010) Blooms in lagoons: different from those of river-dominated estuaries. In: Kennish M, Paerl H (eds) Coastal lagoons: critical habitats of environmental change. Taylor and Francis, Boca Raton FL, pp 91–114
Goldman JC, Glibert PM (1983) Kinetics of inorganic nitrogen uptake. In: Carpenter EJ, Capone DG (eds) Nitrogen in the marine environment. Academic Press, New York, pp 233–274
Gribsholt B, Boschker HTS, Struyf E, Andersson M, Tramper A, De Brabandere L, van Damme S et al (2005) Nitrogen processing in a tidal freshwater marsh: a whole-ecosystem ^{15}N labeling study. Limnol Oceanogr 50:1945–1959
Gribsholt B, Veuger B, Tramper A, Middelburg JJ, Boschker HTS (2009) Long-term ^{15}N-nitrogen retention in tidal freshwater marsh sediment: elucidating the microbial contribution. Limnol Oceanogr 54:13–22. https://doi.org/10.4319/lo.2009.54.1.0013
Grice GD, Reeve MR, Koeller P, Menzel DW (1977) The use of large volume, transparent, enclosed sea-surface water columns in the study of stress on plankton ecosystems. Helgolander Wiss Meeresunters 30:118–133. https://doi.org/10.1007/BF02207830
Grice GD, Harris RP, Reeve MR, Heinbokel JF, Davis CO (1980) Large-scale enclosed water column ecosystems. An overview of food-web I, the final CEPEX experiment. J Mar Biol Ass UK 60:401–414
Hamilton SK, Tank JL, Raikow DE, Siler ER, Dorn NJ, Leonard NE (2004) The role of instream vs allochthonous N in stream food webs: modeling the results of an isotope addition experiment. J North Am Benthol Soc 23:429–448

Harris G (1978) Photosynthesis, productivity and growth: the physiological ecology of phytoplankton. Archiv für Hydrobiol IV:171

Harvey HW (1945) Recent advances in the chemistry and biology of seawater. Cambridge University Press

Hyun B, Kim J-M, Jang P-G, Jang M-C, Choi K-H, Lee K, Yang EJ (2020) The effects of ocean acidification and warming on growth of a natural community of coastal phytoplankton. J Mar Sci Eng 8:821. https://doi.org/10.3390/jmse8100821

Jeppesen E, Jensen J, Søndergaard M, Lauridsen T, Pedersen L, Jensen L (1997) Top-down control in freshwater lakes: the role of nutrient state, submerged macrophytes and water depth. Hydrobiologia 342–343:151–164

Kana TM, Glibert PM (2016) On saturating response curves from the dual perspectives of photosynthesis and nitrogen acquisition. In: Glibert PM, Kana TM (eds) Aquatic microbial ecology and biogeochemistry: a dual perspective. Springer International Publishing, Cham, Switzerland, pp 93–104

Koeller P (2011) Satellites and fisheries: a personal view. ICES J Mar Sci 68:642–643

Malone TC, Conley DJ, Glibert PM, Harding LW Jr, Sellner K (1996) Scales of nutrient limited phytoplankton productivity: the Chesapeake Bay example. Estuaries 19:371–385

Massol F, Gravel D, Mouquet N, Cadotte MW, Fukami T, Leibold MA (2011) Linking community and ecosystem dynamics through spatial ecology. Ecol Lett 14:313–323

Menten L, Michaelis MI (1913) Die kinetik der invertinwirkung. Biochem Zeitung 49:333–369

Middelburg JJ, Barranguet C, Boschker HTS, Herman PMJ, Moens T, Heip CHR (2000) The fate of intertidal microphytobenthos carbon: an *in situ* ^{13}C-labeling study. Limnol Oceanogr 45: 1224–1234

Moberg S, Knutsen G, Goksoyr J (1968) The point of no return concept in cell division. The effects of some metabolic inhibitors on synchronized *Chlorella pyrenoidosa*. Physiol Plant 21:390–400

Monod J (1942) Recherches sur la criossance des cultures bactériennes, 2nd edn. Hermann, Paris

Monod J (1949) The growth of bacterial cultures. Ann Rev Microbiol 3:371–394

Monod J (1950) La technique de culture continue, theorie et applications. Ann L'Instit Pasteur 79: 390–410

Novick A, Szilard L (1950) Description of the chemostat. Science 112:715–716

Oakes JM, Eyre BD, Middelburg JJ, Boschker HTS (2010) Composition, production, and loss of carbohydrates in subtropical shallow subtidal sandy sediments: rapid processing and long-term retention revealed by C-13-labeling. Limnol Oceanogr 55:2126–2138. https://doi.org/10.4319/lo.2010.55.5.2126

Oakes JM, Eyre BD, Middelburg JJ (2012) Transformation and fate of microphytobenthos carbon in subtropical shallow subtidal sands: a C-13-labeling study. Limnol Oceanogr 57:1846–1856. https://doi.org/10.4319/lo.2012.57.06.1846

Olson RJ, Chisholm SW (1983) Effects of photocycles and periodic ammonium supply on three marine phytoplankton species. I Cell division patterns. J Phycol 19:522–528

Pace ML, Cole JJ, Carpenter SR, Kitchell JF, Hodgson JR, Van de Bogart MC, Bade DL et al (2004) Whole-lake carbon-13 additions reveal terrestrial support of aquatic food webs. Nature 427:240–243

Petersen JE, Cornwell JC, Kemp WM (1999) Implicit scaling in the design of experimental aquatic ecosystems. Oikos 85:3. https://doi.org/10.2307/3546786

Petersen JE, Kemp WM, Bartleson R, Boynton WR, Chen C-C, Cornwell JC, Gardner RH et al (2003) Multiscale experiments in coastal ecology: improving realism and advancing theory. Bioscience 53:1181. https://doi.org/10.1641/0006-3568(2003)053[1181:MEICEI]2.0.CO;2

Petersen JE, Kennedy VS, Dennison WC, Kemp WM (eds) (2009) Enclosed experimental ecosystems and scale tools for understanding and managing coastal ecosystems. Springer Science and Media LLC, New York, p 221

Pettersson H, Gross F, Koczy F (1939) Large-scale plankton culture. Nature 144:332–333

Rao TR (2000) A curve for all reasons: the rectangular hyperbola in biology. Resonance 5:85–90

Redfield AC (1934) On the proportions of organic derivatives in sea water and their relation to the composition of plankton. James Johnstone Memorial Vol 176

Redfield AC (1958) The biological control of chemical factors in the environment. Am Sci 46:205–222

Redfield AC, Ketchum BH, Richards FA (1963) The influence of organisms on the composition of sea water. In: Hill MN, others (eds) The sea, vol 2. Interscience, New York and London, pp 26–77

Reeve MR, Gamble JC, Walter MA (1977) Experimental observations on the effects of copper on copepods and other zooplankton: controlled ecosystem pollution experiment. Bull Mar Sci 27: 92–104

Rhee G-Y (1973) A continuous culture study of phosphate uptake, growth rate and polyphosphate in *Scendesmus* sp. J Phycol 9:495–506

Rhee G-Y (1978) Effects of N:P atomic ratios and nitrate limitation on algal growth, cell composition, and nitrate uptake. Limnol Oceanogr 23:10–24

Sanford LP, Suttles SE, Porter ET (2009) Physical factors: mixing and flow. In: Petersen JE, Kennedy VS, Dennison WC, Kemp WM (eds) Enclosed experimental ecosystems and scale tools for understanding and managing coastal ecosystems. Springer Science and Media LLC, New York, pp 63–74

Schindler DW (1974) Eutrophication and recovery in experimental lakes: implications for lake management. Science 184:897–899. https://doi.org/10.1126/science.184.4113 9.897

Schindler DW (2006) Recent advances in the understanding and management of eutrophication. Limnol Oceanogr 51:356–363

Shangguan Y, Glibert PM, Alexander JA, Madden CJ, Murasko S (2017) Phytoplankton community response to changing nutrients in Florida bay: results of mesocosm studies. J Exp Mar Biol Ecol 494:38–53. https://doi.org/10.1016/j.embe.2017.05.006

Shuter B (1979) A model of physiological adaptation in unicellular algae. J Theor Biol 78:519–552

Stiebel M, Kirchmeier L, Peter H (2013) Different mixing techniques in experimental mesocosms—does mixing affect plankton biomass and community composition? Limnol Oceanogr Meth 11:176–186. https://doi.org/10.4319/lom.2013.11.176

Strickland JDH (1967) Between beakers and bays. New Sci 2:276–278

Strickland JDH, Holm-Hansen O, Eppley RW, Linn RJ (1969) The use of a deep tank in plankton ecology. I. Studies of the growth and composition of phytoplankton crops at low nutrient levels. Limnol Oceanogr 14:23–34

Wood AM, Everroad RC, Wingard LM (2005) Measuring growth rates in microalgal cultures. In: Anderson RA (ed) Algal culturing techniques. Elsevier, Amsterdam, pp 269–286

Zahleder V, Bisová K, Vitová M (2016) The cell cycle of microalgae. In: Borowitzka MA, Beardall J, Raven JA (eds) The physiology of microalgae. Springer Inter Publishing, Cham, Switzerland, pp 3–46. https://doi.org/10.10007/978-3-319-24945-2_1

Shining Light on the Complexity of Light 4

4.1 Introduction

The illuminated underwater world has fascinated scientists and naturalists since the beginning of time. In the mid-1800s, Lord Rayleigh, whose full name was John William Stutt, the 3rd Baron Rayleigh of Terling Place in England—and a Nobel Prize winner and physicist in the field of acoustics and optics—held the view that the oceans are blue due to reflection from the sky (Dickey et al. 2011). Actually, the blueness of the oceans is not due to reflection of the sky but rather the absorption of the longer visible wavelengths (red and yellow) and the scattering of the remaining blue light. The world of ocean optics is an incredibly complex and rapidly expanding field in and of itself, and only the barest elements can be described here. The treatise by Kirk (2011) remains the most comprehensive introductory manual on this topic. Here, the focus is on how light availability changes for phytoplankton, including its penetration in the water and absorption, how we measure light availability and how light sources for phytoplankton in the laboratory vary.

4.2 The Nature of Light and Its Variations

Light is electromagnetic energy and is made up of electromagnetic waves that vary in wavelength and frequency (Fig. 4.1). This energy is packaged as *photons*, and as such, it has properties like particles. The particle behavior of light is convenient for considering it in the context of photosynthesis. A photon is a "countable" entity and when light intensity is measured, the number of photons for a given time period and area can be counted. An alternative measure of light is the energy associated with electromagnetic radiation and that can be measured in joules s^{-1} m^{-2} (or W m^{-2}). Photons, however, are the preferred light measurement because they have a more direct relationship with the units of measurement of photosynthetic products as grams or moles. Every photon has a wavelength that is inversely proportional to its frequency and energy. With photosynthesis using chlorophyll (chl) pigments, a

P. M. Glibert, *Phytoplankton Whispering: An Introduction to the Physiology and Ecology of Microalgae*, https://doi.org/10.1007/978-3-031-53897-1_4

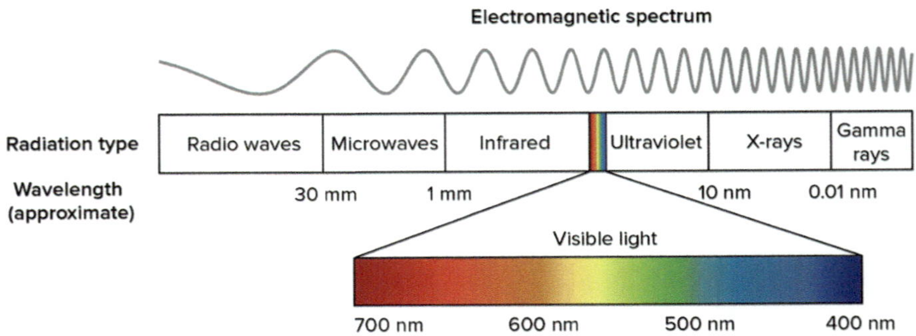

Fig. 4.1 Electromagnetic spectrum. The visible region (400–700 nm) accounts for only a small portion of the spectrum. Image reproduced from https://www.khanacademy.org/science/biology/photosynthesis-in-plants/the-light-dependent-reactions-of-photosynthesis/a/light-and-photosynthetic-pigments under a creative commons license

blue (high energy) photon has the same photochemical effect or chemical energy input as a red (low energy) photon and consequently, the biochemical energy is related to the *number* of absorbed photons rather than the energy of the absorbed photons.

The spectrum of visible light is defined by the sensitivity of the human eye to the electromagnetic spectrum. Plants and algae more-or-less share the same spectral window although humans and plants have different spectral sensitivities within the visible spectrum. Visible light is generally described by the wavelength range, as opposed to frequency, with 400 nm–700 nm being the conventional boundaries (Fig. 4.1). The peak distribution of solar radiation at the earth's surface happens to be in the visible portion of the spectrum, something that evolution took great advantage of.

Natural light is dynamic. It varies with latitude, with high latitude regions having the potential of having higher daily insolation in mid-summer than tropical regions owing to longer day lengths (Fig. 4.2). Of course, day length is much shorter in winter in high latitude regions, so annually, the higher the latitude, the lower the insolation.

Daily variations are approximately sinusoidal, but atmospheric conditions—e.g., clouds—can lead to highly variable daily total irradiance (Fig. 4.3). In addition to the well-recognized regular and irregular light variations with day and season, there are many fine-scale fluctuations in water that are the result of waves. Sosik and Johnsen (2004) describe how a diver might witness such flickering,

> In calm weather, the diver can look upward to see the entire hemisphere of the sky compressed into a circle over her head—a phenomenon called Snell's window, caused by the bending of light as it enters water. Rough weather and waves shatter this window. The waves act like lenses to focus light, creating a scintillating visual field whose brightness increases and decreases by a factor of a hundred as each wave passes by, making it impossible for eyes to adjust.

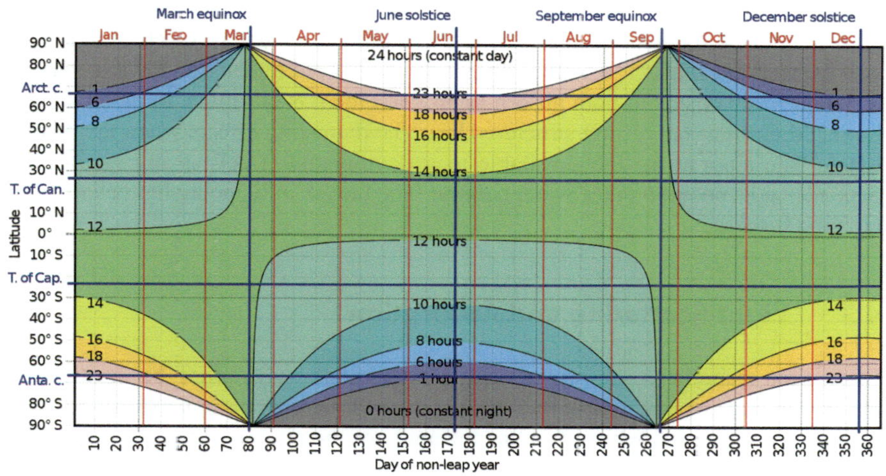

Fig. 4.2 Hours of daylight versus latitude versus day of the year. Image reproduced from Wikipedia Commons (https://en.wikipedia.org/wiki/Sunrise_equation#/media/File:Hours_of_day light_vs_latitude_vs_day_of_year_with_tropical_and_polar_circles.svg) under a creative commons license

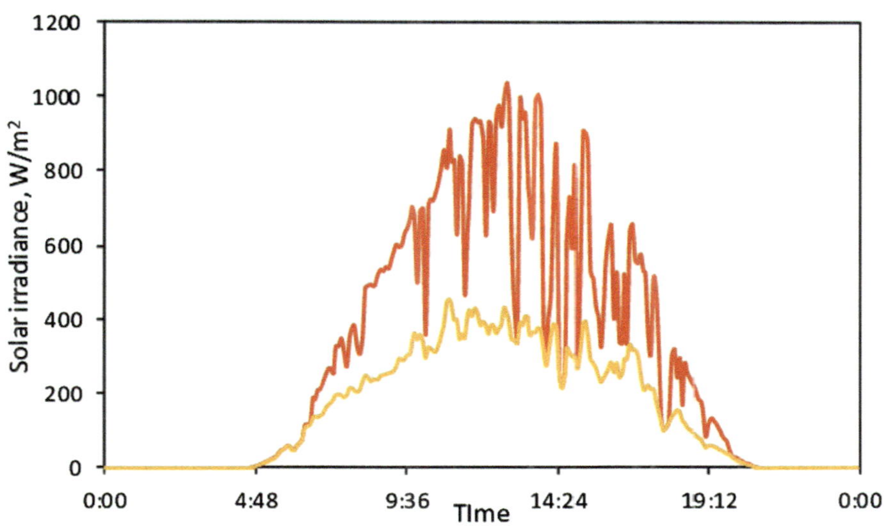

Fig. 4.3 Spectral distribution of sunlight energy for a sunny day with intermittent clouds (orange line) and a cloudy day (yellow line). Image reproduced from Jaszczur et al. (2018, doi.org/10.1051/mateconf/201824004005) under a creative commons license

Phytoplankton thus have regular cycles to which they adapt over time, while high-frequency fluctuations in clouds and flickering are much more challenging. As will be seen in Chaps. 5, 6, and 7, different types of phytoplankton have developed different mechanisms to both absorb light when there is not much to be had and to protect their light-harvesting structures from the damaging effects when there is too much. The amount of light available to a phytoplankton cell at any given point depends on its position in the water column. The factors influencing light availability are next discussed.

4.3 Light Scattering and Absorption

Once light penetrates the water's surface, photons may either be scattered or absorbed. All of the particles in water—sediment, phytoplankton, bacteria, detritus—scatter light which changes the angle of the photon. Accordingly, coastal and estuarine waters, or any water with high sediment load or high turbidity, can have high scattering functions. Phytoplankton, too, can scatter light. For example, blooms of coccolithophores scatter light from their coccoliths.

It is easy to appreciate the effect of scattering by looking at a glass of milk. Milk consists of minute scattering particles that direct incident light in all directions. That is why it looks diffusely white to our eyes and why little light is transmitted through the glass. Scattering particles redirect downward light to side and back angles, thus randomizing the direction of light. This has the effect of reducing the penetration of light in a water column and this is called "extinction."

In the absence of scattering particles, light availability rapidly declines with depth due to absorption of light (Figs. 4.4 and 4.5). Light absorption takes place by four components in water: the water itself, the photosynthetic organisms, dissolved humic material or "yellow stuff," and inorganic particulate material. Lord Rayleigh would be fascinated to know that water is not blue because it reflects the sky, but because it absorbs only very weakly in the blue and green regions of the spectrum and much more strongly in the red, orange, and yellow, the longer wavelength, regions of the spectrum. In fact, red light is absorbed in shallow depths, with about 35% of the incident red light absorbed within the top 1 m of surface waters (Kirk 2011; Fig. 4.5). Accordingly, deeper waters appear more blue while shallower waters appear more green. The absorption of light by water itself is, however, not a constant, and varies with salinity and temperature, especially in the wavelength region of 700–800 nm (Zhang et al. 2009).

Humic material absorbs light, so waters with high humic material can appear to be the color of tea. Detritus or other inorganic particulate loads, such as may be due to dust storms or erosion events delivering soil, will also contribute to light absorption. Light penetration in coastal waters, estuaries, lakes, and rivers can therefore appear more green.

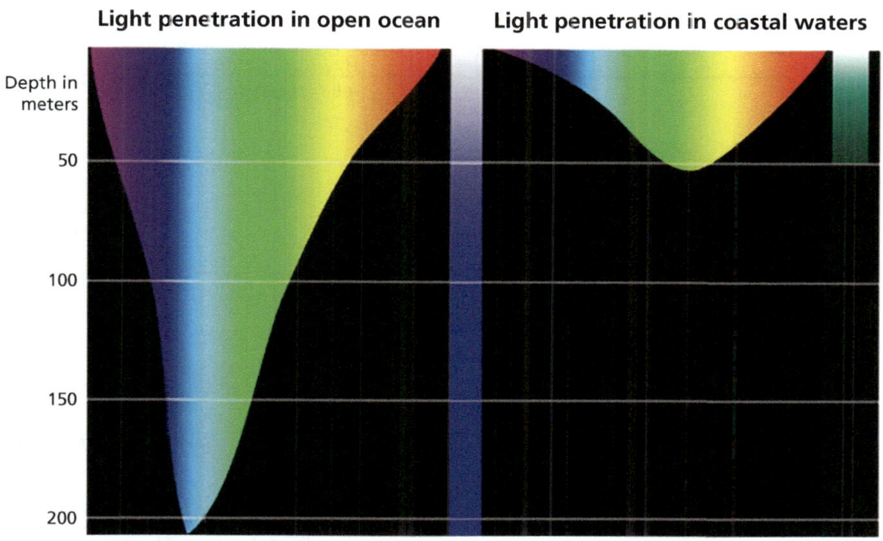

Fig. 4.4 Penetration of light in the open ocean and in coastal waters. Image from K. Carothers, National Oceanic and Atmospheric Administration (NOAA), public domain (https://oceanexplorer.noaa.gov/explorations/04deepscope/background/deeplight/media/diagram3.html)

Total absorption by wavelength, a(λ), is the sum of the absorption by constituent components (Sosik and Mitchell 1995),

$$a(\lambda) = a_w(\lambda) + a_p(\lambda) + a_s(\lambda) \tag{4.1}$$

where $a_w(\lambda)$, $a_p(\lambda)$, and $a_s(\lambda)$ are absorption due to water itself, particulate material and soluble material, and, $a_p(\lambda)$ has multiple components itself,

$$a_p(\lambda) = a_{ph}(\lambda) + a_d(\lambda) \tag{4.2}$$

where $a_{ph}(\lambda)$ is that associated with phytoplankton and $a_d(\lambda)$ is that associated with detritus. It is also useful to express $a_{ph}(\lambda)$ in relation to the pigment concentration (chl plus phaeopigments), and in doing so, this term becomes:

$$a_{ph}{}^{*}(\lambda) = \left(a_{ph}(\lambda)\right)/(\text{chl} + \text{phaeo}) \tag{4.3}$$

There is now a considerable literature comparing and contrasting $a_{ph}{}^{*}(\lambda)$ among phytoplankton (reviewed by Sosik and Mitchell 1995). High values are generally found under higher light, lower nutrient conditions, and lower temperatures. Further resolution of $a_{ph}{}^{*}(\lambda)$ according to photochemical reaction centers and photoprotective pigments provides more useful information when modeling this variability.

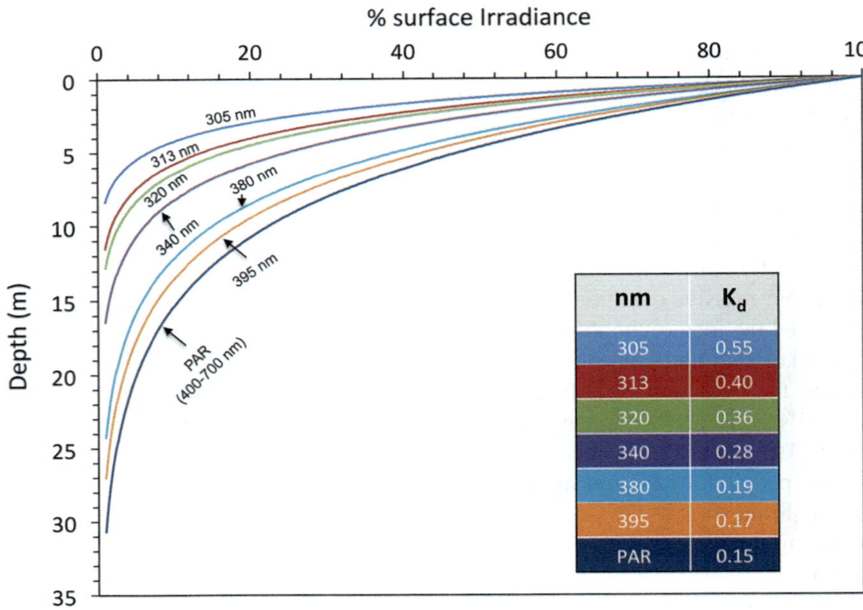

Fig. 4.5 Vertical profile of solar radiation penetration and attenuation coefficients (Kd in m^{-1}) for different wavelengths of UV-B and UV-A radiations and PAR band (400–700 nm). Values represent the average of a representative sunny day in mid-summer (January 2013) at Fildes Bay, western Antarctic Peninsula. Image reproduced from Gomez and Huovinen (2015, PLoS One 10, e0134440. doi.org/10.1371/journal.pone.0134440) under a creative commons license

Combining the effects of scattering and absorption, the total attenuation of light with depth, assuming uniform vertical distribution of particles and absorbing substances, can be expressed as an exponential declining function of depth:

$$I_Z = I_0 \, e^{-(kw+kp+kd)\,Z} \tag{4.4}$$

where:

k_w = extinction due to scatter and absorption due to water,

k_p = extinction due to scatter and absorption due to particles,

k_d = extinction due to scatter and absorption due to dissolved color. Total extinction is due to the sum of the three,

Z = depth (from surface),

I_Z = light intensity at depth Z,

I_0 = light intensity at the surface.

4.4 Light Attributes and Terminology

The world of light measurement is beleaguered with confusing terminologies and tools owing to different emphases on spectral information depending on the purpose. In the area of light perception by humans, e.g., light sources, there are a number of conventional measurements and units that do not translate well in photosynthesis research. The lumen is a standard measurement of light intensity but relates specifically to human eye perception, which is most sensitive in the yellow and least sensitive in the blue and red regions. Thus, lumens are not used for light measurements pertaining to photosynthesis. Light measurements pertinent to photosynthesis and phytoplankton include the following:

Radiance is the radiant flux of emitted or scattered *from* a unit of surface area.

Irradiance is the radiant flux incident *on* a unit area of a surface. That surface may be the frond of a giant kelp, a blade of seagrass, or a phytoplankton cell. In the oceans, most light comes from downward irradiance, but there may also be upward irradiance or from diffuse angles. *Scalar irradiance* is the measure of radiant intensity at any point, treating radiation from all directions equally (e.g., downward irradiance and upward irradiance). Consider a spherical phytoplankton cell that does not care which direction a photon is coming from.

In many scientific and commercial fields, both of these fluxes (radiance and irradiance) are often measured in units of energy or Watts, W or W m^{-2}. However as indicated above, photosynthesis physiologists prefer countable photons, often expressed in mole units:

$$1 \text{ mole photons} = 1 \text{ Avogadro's number } \left(6 \times 10^{23}\right) \text{ photons} \tag{4.5}$$

The older literature also has "Einsteins" as the unit of photon measurement with 1 Einstein equal to 1 mole of photons. Einstein is not an SI unit and contemporary literature uses moles. Solar irradiance typically has units of μmole photons m^{-2} s^{-1}, which is a flux (per time) density (per area) measurement. A bright sunny day would have an irradiance of ~1500–2000 μmole photons m^{-2} s^{-1}.

Conversions between W and photons are not straightforward and, as will become apparent by the end of this chapter, are dependent on light sources and the spectral distribution output (see Biggs 1984 for further details). A few more terms are necessary:

Photosynthetically Active Radiation (PAR) is radiation in the 400–700 nm wavelength region and can be expressed in both energy and photon units.

Photon flux density is defined as the number of photons in the range of 400–700 nm wavelength incident light per unit time per unit surface.

Photon flux fluency rate is an older (now out of favor) term introduced to specifically distinguish PAR when measured with a spherical sensor (see below).

Photosynthetically Utilizable Radiation (PUR) is a modified PAR obtained by multiplying the photon flux for each wavelength, which accounts for the spectral distribution of the light, by a dimensionless quantity proportional to phytoplankton absorption in that wavelength, and then summing the spectrum (Morel 1991;

MacIntyre and Cullen 2005; Kirk 2011). The distinction between PAR and PUR becomes important in studies of light quality on phytoplankton productivity because the usefulness of a given light to phytoplankton is not just a function of its availability, but also how well the spectrum of the available light matches the cell's ability to use it.

4.5 Absorption of Light by Phytoplankton

What is left over after absorption by water, particles, and dissolved colored material is the light available for phytoplankton. Capturing light can be a dangerous business for phytoplankton. Primary producers (all primary producers, not just phytoplankton) rely on their absorption and conversion of light energy to produce chemical energy. The diversity of pigments that is described in Chap. 5, and the mechanism of photosynthesis which is detailed in Chap. 6—the mechanism of actually making the chemical energy—yield the diversity of colors that produce the visible blooms seen in fresh or coastal waters during bloom events (Fig. 4.6). Light absorption removes some of the wavelengths of light and what is not absorbed from the spectrum is what provides the color that we see. A pigment that absorbs blue looks red to our eyes. Chlorophyll appears green because it absorbs both red and blue wavelengths strongly.

The act of light absorption by a pigment causes the energy level of the pigment to rise. The absorption event is very rapid, on the picosecond (10^{-12} s) time scale, but the excited state of the pigment is very unstable and is typically lost (discharged) within 1 nanosecond (10^{-9} s). What happens to the excited state (i.e., where does the energy go) is one of the most researched areas in photosynthesis and is described in more detail in Chap. 6.

Fig. 4.6 Examples of various colors observed in fresh and marine waters influenced by the presence of varying optically active constituents. Image reproduced from Mascarenhas and Keck (2018, doi.org/10.1007/978-3-319-93284-2_4) under a creative commons license

4.5.1 Photochemistry and Quantum Yield

In photosynthesis, an absorbed photon causes a chemical reaction typically as a redox reaction. This is termed "photochemistry" (Chap. 6). For most pigments, only a fraction of absorbed photons cause photochemistry and this fraction is referred to as *quantum yield*, denoted as ϕ. Thus, a quantum yield of 0.6 means that only 60% of the absorbed photons are capable of driving photochemistry. For photochemical reactions, the rate of reaction is determined by three parameters:

$$\text{Rate} = \phi I \sigma \tag{4.6}$$

where ϕ is quantum yield, I is irradiance, and σ is the absorption cross-section (the cross-sectional area of the pigment or pigment complex. For phytoplankton, σ can be substituted by the absorptance of the cell or colony which is the fraction of incident light absorbed; Kirk 2011). Therefore, in a photosynthetic cell,

$$\text{Photosynthesis} = \phi IA \tag{4.7}$$

where A is absorptance. Equation 4.7 belies reality as it implies that photosynthesis is always directly proportional to light intensity, i.e., it has no bounds to its rate at ever-increasing light intensity. As will be described later (Chap. 6), the quantum yield for photosynthesis is highly variable and under physiological control.

4.5.2 Absorption and Action Spectra

While quantum yields relate to the effectiveness of *absorbed* light in driving chemistry, the effectiveness of a pigment in absorbing light depends on the wavelength (or frequency) of the photon. *Absorption spectra* illustrate the differences in absorption as a function of wavelength of incident light—differences that are due to differences in pigments that can vary as a function of different growth conditions or different taxa. *Action spectra*, on the other hand, describe the wavelength dependency of the rate of photosynthesis. With phytoplankton, there are also non-photosynthetic pigments, so the absorption spectrum may differ from the action spectrum (Chap. 5).

Representative absorption spectra of several photosynthetic pigments are shown in Fig. 4.7. Chlorophyll *a*, common to all phytoplankton except *Prochlorococcus*, has dominant peaks in the blue and red portions of the visible spectrum. Green light is inefficient (low quantum yield) at exciting chl *a*. Evolution has produced pigments that "fill in the gap" of the absorption of chl, such as carotenoids and biliproteins (Fig. 4.7). The overall functioning and variability of these pigment complexes are dealt with in greater detail in Chap. 5. It is sufficient here to understand that different pigments have different spectral efficiencies, which ultimately relate to the success of different taxa in different light fields (Heggerud et al. 2023; Fig. 4.8).

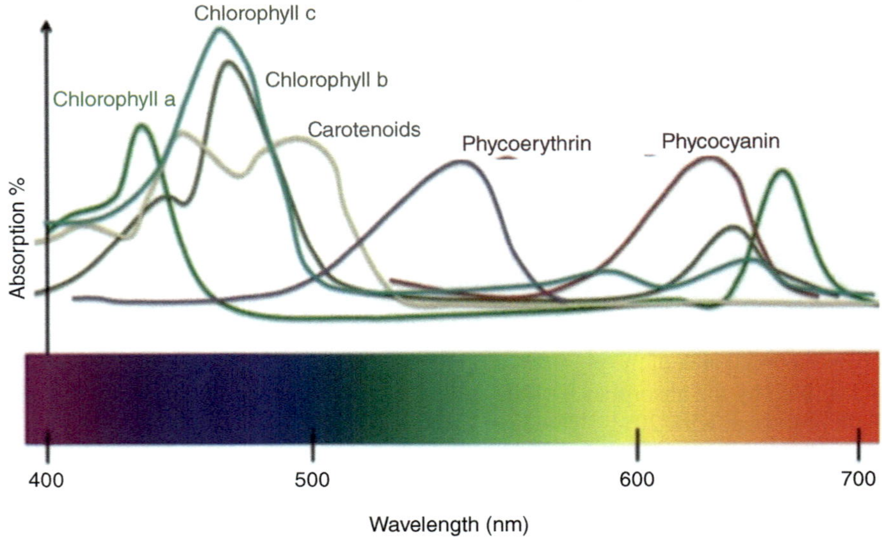

Fig. 4.7 Light absorption spectra for marine algal pigments. See Chap. 5 for more detail on absorption spectra by different types of pigments. Image modified from a figure published by Yarish et al. (2012, http://digitalcommons.uconn.edu/wracklines/72) under a creative commons license

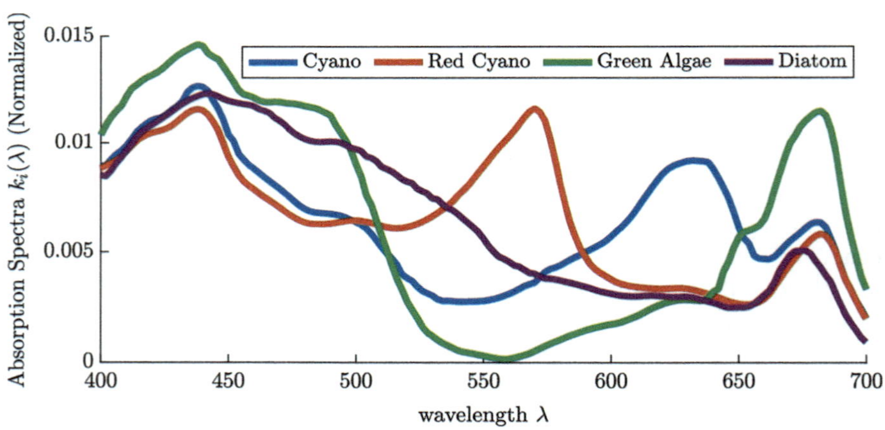

Fig. 4.8 Normalized absorption spectra for four phytoplankton species: green cyanobacteria (*Synechocystis* strain), red cyanobacteria (*Synechococcus* strain), green algae (*Chlorella* strain) and a diatom (*Nitzschia* strain). The differences in absorption spectra among species imply niche differentiation throughout the spectrum. Image reproduced from Heggerud et al. (2023, doi.org/10.1007/s00285-023-01890-z) under a creative commons license

4.6 Light Measurements and Meters

One of the more confusing aspects of the study of light is the dizzying array of light sensors, light meter options, and reporting units for light. One of the most famous tools in the measurement of light was developed in the mid-1800s: the Secchi disk (Fig. 4.9). It was the creation of Fr. Peitro Angelo Secchi, scientific advisor to the Pope, who was instructed by the Papal Navy to measure the transparency of the Mediterranean Sea (https://www.nalms.org/secchidipin/monitoring-methods/the-secchi-disk/what-is-a-secchi/). It remains a standard tool in virtually all aquatic scientists' toolboxes; it is cheap, simple, and reliable. A Secchi depth is the depth at which the disk disappears from view.

Older photosynthesis literature reported light based on brightness to the eye. Units such as foot candles, lumen, and lux, were commonly used and all were based on standardized visual responses. These units are retired and importantly, there is no way to retroactively convert these units to today's standards. Other than for strictly qualitative purposes, data reported in these units should not be used.

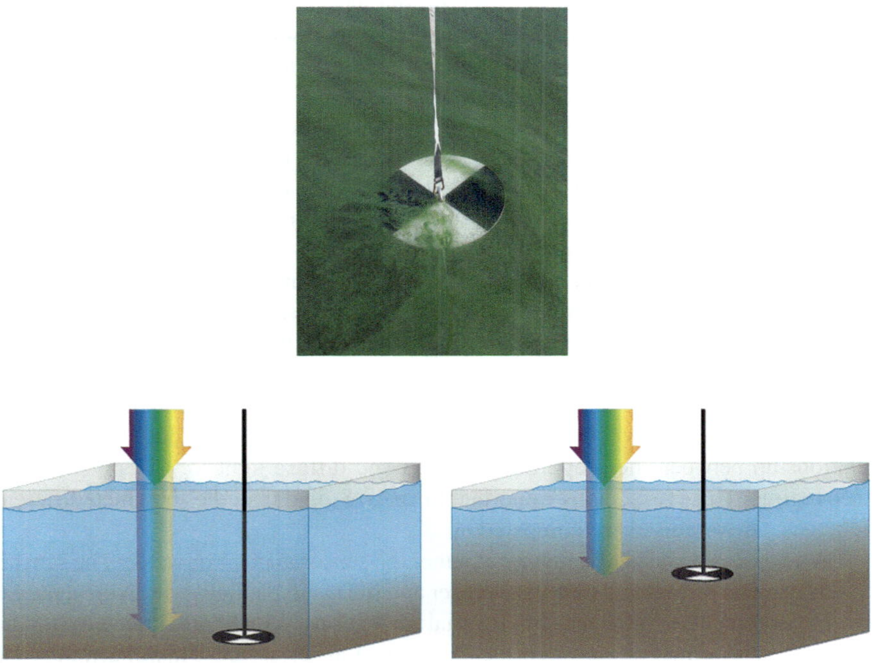

Fig. 4.9 Upper panel: A Secchi disk is a black and white disk used to measure the transparency of the water. Image from https://www.usgs.gov/media/images/secchi-disc (public domain). Lower panel: Open ocean waters allow more light penetration than coastal waters. Image reproduced from Mascarenhas and Keck (2018, doi.org/10.1007/978-3-319-93284-2_4) under a creative commons license. Secchi disk symbol from the University of Maryland Center for Environmental Science Integration and Application Network symbol library

Fig. 4.10 Common light sensors used in phytoplankton physiology. Panel (**a**) 2 π quantum sensor, Panel (**b**) 4 π underwater sensor, and Panel (**c**) 4 π sensor typically used in culture flasks. Note the difference in diameter between the spheres in Panels (**b**) and (**c**)

Preferred light measurements for photosynthesis are based on PAR, where the light sensor contains optical filters that remove wavelengths <400 nm and >700 nm and even out the sensitivity of the detector to the 400–700 nm interval. These PAR sensors are available in two different geometries, flat and spherical. A flat disk sensor, also referred to as a cosine-corrected or 2-pi (2 π) sensor, is designed to measure the irradiance across a flat surface (Fig. 4.10a). A horizontal surface receives a higher amount of PAR from the sun when directly overhead than does a vertical surface. Think about how this may be important to leaves in nature. Thus, the orientation of the sensor is important and that is determined by the application.

For phytoplankton, light comes from all directions given the near-spherical nature of the organisms. Thus, a spherical sensor is appropriate. Spherical PAR sensors are also known as 4-pi (4 π) or scalar sensors are used both for water column and laboratory measurements of light fields (Fig. 4.10b, c). Different manufacturers make spheres with different diameters, and with more or less shading on the detector side of the spherical collector.

One of the important aspects of these two types of instruments is that there is no single relationship between values obtained by a 2 pi sensor and a 4 pi sensor. As clearly explained by one of the major manufacturers of light meters (Biggs 1984, p. 3), "For a collimated beam at normal irradiances, they are equal, while for perfectly diffuse incidence [they may differ by a factor of 4]. In practical situations the ratio will be somewhere between 1 and 4." This variation is due to the unique relationship between the proportion of direct and diffuse light for any given irradiance regime. This is especially important to keep in mind if studies in which different light sensors were used are compared.

PAR sensors are also affected by whether the sensor is in air or water. The sensor, typically made of Teflon, is itself a diffuser and the available light within the material for measurement is affected by internal reflections which are affected by the differences in refraction between the Teflon and air or water. This difference is a constant and PAR meters typically have a switch for air or water to account for the different optical responses.

4.7 Light Sources/Light Quality

So much of what we do in phytoplankton ecology involves studying phytoplankton in the laboratory. We culture them for a variety of purposes. To do so, we place them under artificial lights. There is a vast literature of light studies using different lamp technologies and it is useful to have an appreciation of the spectral output of various fixtures. The spectrum of such lights does not match that of the sun, and the spectrum of different lights can be vastly different with the dominant wavelengths ranging across the spectrum (Fig. 4.11, Tosini et al. 2016). A practical aside is that the heat output of various lamps differs, with incandescent lights producing more heat energy than light energy. Spectral quality of different light sources may age differently over time as well. Most artificial lights produce spectra that look more like a series of lines than a continuous curve. If a culture shows a very different growth response when transferred to a new location or incubator with a different light source, the spectral quality may be the reason.

LED technology has rapidly advanced and these light sources are now the preferred fixtures for growing and experimenting with algae. From the action

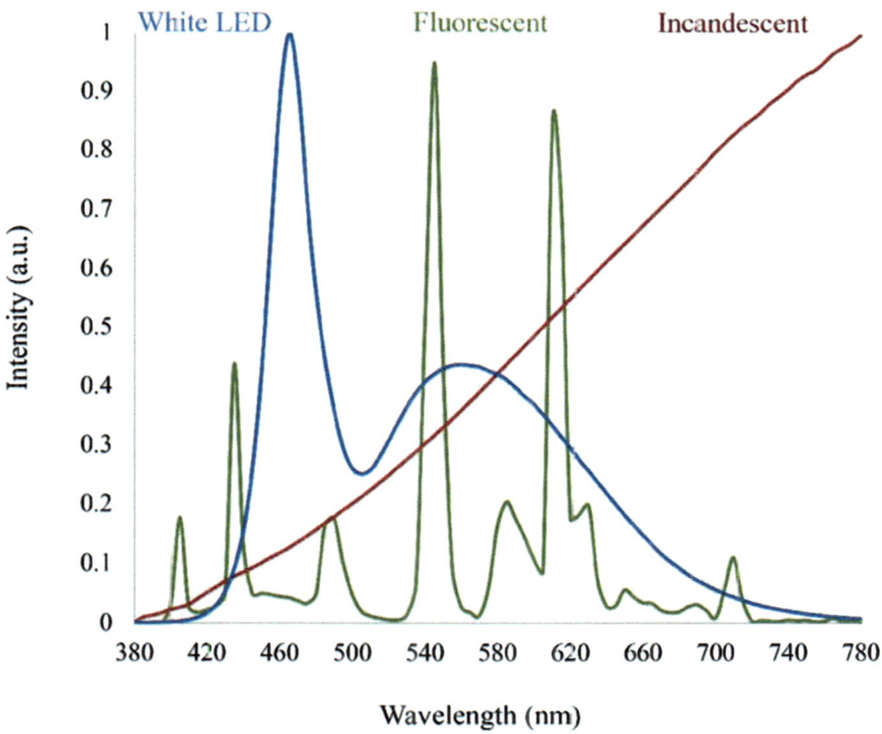

Fig. 4.11 Spectral outputs of various light sources. Image reproduced from Tosini et al. (2016, Mol Vision 22: 61–72) under a creative commons license

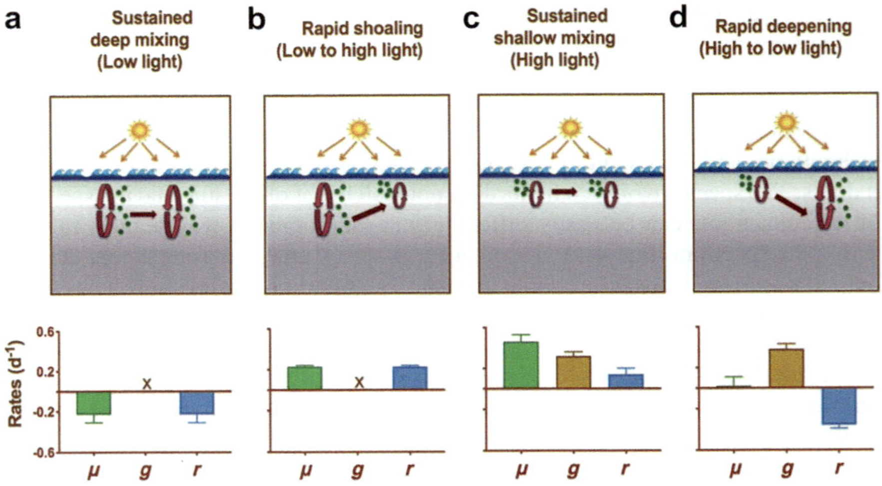

Fig. 4.12 Rates (d^{-1}) of phytoplankton growth (μ), grazing mortality (g), and biomass accumulation (r) under four mixed layer scenarios simulated using light as a proxy of (Panel **a**) sustained deep mixing, (Panel **b**) rapid shoaling, (Panel **c**) sustained shallow mixing, and (Panel **d**) rapid mixed layer deepening. Error bars represent one standard deviation of the mean of duplicate experiments. Grazing was measured but not detected in the sustained deep mixing and rapid shoaling conditions, denoted with x. Image reproduced from Morison et al. (2020, Limnol Oceanogr Letts 5: 346–353. doi.org/10.1002/lol2.10156) under a creative commons License

spectra, it follows that providing red and blue light generally provides higher growth rates for most species for a given energy input. LEDs will be the preferred tool in future studies of how metabolism changes in response to both light intensity and quality.

Recently, Bercel and Kranz (2022) examined the growth of cultures of *Prochlorococcus marinus*, *Synechococcus* sp., and *Thalassiosira weissflogii* grown under three different LED lights as well as one fluorescent growth light. They found all species displayed changes in cellular carbon (C), nitrogen (N), and chl *a* quotas as a direct response to the different light spectra. Abiusi et al. (2013) studied the effect of light quality on the prasinophyte *Tetraselmis suecica.* They found that biomass and photosynthetic efficiency were comparable under white and red light, but cells under red light were smaller and more motile. Also, under blue and green light, growth dropped significantly. Another study with the same organism, as well as a diatom assemblage, recently examined the effects of different artificial light availability at night (Diamantopoulou et al. 2021), as the use of such lights in coastal communities can be considered a form of light pollution. They found that red and green lights have the strongest effects on growth, leading to selective dominance of specific diatom species. Their results were not entirely consistent with those of Abiusi et al. (2013), raising the question of interaction of light quantity and quality as well as the timing of light supply.

Rapid changes in light also affect grazers, and therefore the coupling between growth and loss. A recent study undertaken in the North Atlantic examined how variations in light exposure, both static and varied light conditions affected both primary producer growth and grazing (Morison et al. 2020; Fig. 4.12). These results suggest that growth and grazing responded to fluctuations on different time scales, underscoring the need for further attention to the dynamics of light.

4.8 UV and Phytoplankton

In the upper euphotic zone, phytoplankton are exposed not only to visible light, but also to UV radiation. UV radiation is composed of two spectral regions: UV-A, 320–400 nm and UV-B, 280–320 nm (Kirk 2011). UV-A has far less damaging effects, and may even have some positive growth effects, on phytoplankton. UV-A has been shown to reduce growth in Antarctic diatoms (Davidson et al. 1994) and enhance photoinhibition in freshwater diatoms (Kim and Watanabe 1993), but it has also been shown to increase C fixation in coastal tropical phytoplankton (Helbling et al. 2003) and in the diatom *Pseudo-nitzschia* (Mengelt and Prézelin 2005).

Photosynthesis is highly susceptible to damage by UV-B; UV-B is responsible for a considerable fraction of photoinhibition of photosynthesis (see Chap. 6 for a more complete description of photoinhibition). UV-B affects both the light reactions of photosynthesis at photosystem II and the dark reactions, mainly the enzyme Rubisco (Vincent et al. 2000). UV-B can also block enzymatic reactions, interfere with motility and orientation, and can directly or indirectly induce the formation of reactive oxygen species (ROS), the formation of which damages DNA, lipids, and proteins and can lead to cell death (He and Häder 2002). Furthermore, the blue pigments, phycobilins, in cyanobacteria and red algae are particularly sensitive to damage by UV-B (Häder et al. 2015).

In addition to the direct effects of UV on phytoplankton photosynthesis or growth, there are numerous indirect effects. In coastal or fresh waters, where humic acid concentrations are higher, direct UV effects are low, as these substances absorb UV radiation more strongly than they do radiation in the visible spectrum (Kirk 2011). However, UV can photo-oxidize this material, releasing bound nutrients, especially NH_4^+ in marine waters (e.g., Bushaw et al. 1996) or phosphate in fresh waters (Sereda et al. 2012). This would provide a nutrient source for phytoplankton growth.

Global change is leading to increased PAR and UV in many parts of the globe, particularly polar regions experiencing thinner ice and where degradation of the ozone layer is pronounced. Global change is also leading to increasing humic and other DOM sources in surface waters, due to land use changes and practices (Chap. 22). The complexity of DOM—its type, source, and reactivity—is an important determinant of UV absorptivity.

4.9 Summary

The field of optics is complex, confusing, and incredibly important. It behooves the student of phytoplankton physiological ecology to understand terminology, light sources, light meters, and behavior of light in water. Modelers are forewarned that older literature values based on units of brightness cannot be compared with modern-day literature, nor can experiments calibrated with 2 π sensors be compared with those calibrated with 4 π sensors. Light drives the process of photosynthesis, but different functional groups are adapted to different amounts and different qualities of light. Light is changing—with complex ramifications for phytoplankton community structure. The supply of, and the quantification and characterization of, light and its response by phytoplankton should not be taken lightly. With these basics in hand, our attention turns to the light-harvesting antennas of phytoplankton and their characteristic responses to irradiance in the following chapter.

References

Abiusi F, Sampietro G, Marturano G, Biondi N, Rodolfi L, D'Ottavio M, Tredici MR (2013) Growth, photosynthetic efficiency, and biochemical composition of *Tetraselmis suecica* F&M-M33 grown with LEDs of different colors. Biotechnol Bioeng 111:956–964. https://doi.org/10.1002/bit.25014

Bercel TL, Kranz SA (2022) Effects of spectral light quality on the growth, productivity, and elemental ratios in differently pigmented marine phytoplankton species. J Appl Phycol 34:185–202

Biggs WW (1984) Principles of radiation measurement. Li-Cor. http://www.licor.com/env/pdf/light/Rad_Meas.pdf. Accessed 10 Mar 2023

Bushaw KL, Zepp RG, Tarr MA, Schulz-Jander D, Bourbonniere RA, Hodson RE, Miller WL, Bronk DA, Moran DA (1996) Photochemical release of biologically available nitrogen from aquatic dissolved organic matter. Nature 381:404–407

Davidson AT, Bramich D, Marchant HJ, McMinn A (1994) Effects of UV-B irradiation on growth and survival of Antarctic marine diatoms. Mar Biol 119:507–515

Diamantopoulou C, Christoforou E, Dominoni DM, Kaiserli E, Czyzewski J, Mirzai N, Spatharis S (2021) Wavelength-dependent effects of artificial light at night on phytoplankton growth and community structure. Proc R Soc B Biol Sci 288:20210525. https://doi.org/10.1098/rspb.2021.0525

Dickey TD, Kattawar GW, Voss KJ (2011) Shedding new light on light in the ocean. Phys Today 64:44. https://doi.org/10.10163/1.3580492

Gomez I, Huovinen P (2015) Lack of physiological depth patterns in conspecifics of endemic Antarctic brown algae: a trade-off between UV stress tolerance and shade adaptation? PLoS One 10:e0134440. https://doi.org/10.1371/journal.pone.0134440

Häder D-P, Williamson CE, Wängberg S-Å, Rautio M, Rose KC, Guo K, Helbling EW, Sinha RP, Worrest R (2015) Effects of UV radiation on aquatic ecosystems and interactions with other environmental factors. Photochem Photobiol Sci 14:108–126. https://doi.org/10.1039/C4PP90035A

He Y-Y, Häder D-P (2002) UV-B-induced formation of reactive oxygen species and oxidative damage of the cyanobacterium *anabaena* sp.: protective effects of ascorbic acid and *N*-acetyl-l-cysteine. J Photochem Photobiol B: Biol 66:115–124. https://doi.org/10.1016/S1011-1344(02)00231-2

Heggerud CM, Lam KY, Wang H (2023) Niche differentiation in the light spectrum promotes coexistence of phytoplankton species: a spatial modelling approach. J Math Biol 86:54. https://doi.org/10.1007/s00285-023-01890-z

Helbling EW, Gao K, Gonçalves RJ, Wu H, Villafañe VE (2003) Utilization of solar UV radiation by coastal phytoplankton assemblages off SE China when exposed to fast mixing. Mar Ecol Prog Ser 259:59–66. https://doi.org/10.3354/meps259059

Jaszczur M, Hassan Q, Tenata J, Styszko K, Nawrot W, Hanus R (2018) Study of dust deposition and temperature impact on solar photovoltaic module. MATEC Web Conf 240:04005. https://doi.org/10.1051/mateconf/201824004005

Kim D-S, Watanabe Y (1993) The effect of long wave ultraviolet radiation (UV-A) on the photosynthetic activity of natural population of freshwater phytoplankton. Ecol Res 8:225–234. https://doi.org/10.1007/BF02348535

Kirk JTO (2011) Light and photosynthesis in aquatic ecosystems, 3rd edn. Cambridge University Press, Cambridge, p 649

MacIntyre HL, Cullen JJ (2005) Using cultures to investigate the physiological ecology of microalgae. In: Anderson RA (ed) Algal culturing techniques. Elsevier Academic Press, Amsterdam, pp 287–326

Mascarenhas V, Keck T (2018) Marine optics and ocean color remote sensing: proceedings of the 2017 conference for YOUng MARine RESearchers in Kiel, Germany. In: YOUMARES 8-Oceans across boundaries: learning from each other. https://doi.org/10.1007/978-3-319-93284-2_4

Mengelt C, Prézelin B (2005) UVA enhancement of carbon fixation and resilience to UV inhibition in the genus *Pseudo-nitzschia* may provide a competitive advantage in high UV surface waters. Mar Ecol Prog Ser 301:81–93. https://doi.org/10.3354/meps3010081

Morel A (1991) Light and marine photosynthesis: a spectral model with geochemical and climatological implications. Prog Oceanogr 26:263–306

Morison F, Franzé G, Harvey E, Menden-Deuer S (2020) Light fluctuations are key in modulating plankton trophic dynamics and their impacts on primary production. Limnol Oceanogr Lett 5: 346–353. https://doi.org/10.1002/lol2.10156

Sereda JM, Vandergucht DM, Hudson JJ (2012) In situ UVA exposure modulates change in the uptake of radiophosphate in size-fractionated plankton assemblages following UVR exposure. Micro Ecol 63:751–760

Sosik HM, Johnsen S (2004) Shedding light on light in the ocean. Oceanus Mag 15:2004

Sosik HM, Mitchell BG (1995) Light absorption by phytoplankton, photosynthetic pigments and detritus in the California current system. Deep-Sea Res I 42:1717–1748

Tosini G, Ferguson I, Tsubota K (2016) Effects of blue light on the circadian system and eye physiology. Mol Vis 22:61–72

Vincent WF, Neale P (2000) The effects of UV radiation in the marine environment. In: deMora S, Demers S, Vernet M (eds) The effects of UV radiation in the marine environment, Cambridge environmental chemistry series 10, pp 149–176

Yarish C, Redmond S, Kim JK (2012) *Gracilaria* culture handbook for New England. Wrack Lines 72. http://digitalcommons.uconn.edu/wracklines/72. Accessed 1 Mar 2023

Zhang X, Lianbo H, He M-X (2009) Scattering by pure seawater: effect of salinity. Opt Express 17: 5698–5710

Additional Websites Accessed

https://en.wikipedia.org/wiki/Sunrise_equation#/media/File:Hours_of_daylight_vs_latitude_vs_day_of_year_with_tropical_and_polar_circles.svg. Accessed 1 Apr 2024

https://oceanexplorer.noaa.gov/explorations/04deepscope/background/deeplight/media/diagram3.html. Accessed 2 Jan 2023

https://www.khanacademy.org/science/biology/photosynthesis-in-plants/the-light-dependent-reactions-of-photosynthesis/a/light-and-photosynthetic-pigments. Accessed 6 Feb 2023
https://www.nalms.org/secchidipin/monitoring-methods/the-secchi-disk/what-is-a-secchi/. Accessed 2 Jan 2023
https://www.usgs.gov/media/images/secchi-disc. Accessed 2 Jan 2023

Capturing Light and the Diversity of Pigments

5

5.1 Introduction and Some Early History

The 1700s was an important time in the evolution of ideas as to how plants can grow. Joseph Priestley (who had discovered oxygen (O_2)) conducted a simple but elegant experiment that showed that plants reverse the respiratory process of animals (Hall and Rao 1994). His experiment involved burning a candle in an enclosed container first with a mouse (that died), and then with a mouse and a branch of mint. The mouse lived. A few years after that, Jan Ingenhousz, a Dutch physician, discovered that the O_2 put out by plants only occurred in sunlight and only the green parts of plants did this (Hall and Rao 1994).

Roughly 50 years went by, when in 1817, two French chemists, Pelletier and Caventou isolated this green substance and named it *chlorophyll* (Hall and Rao 1994). Then, 30 years after that, a German physician, Mayer, pronounced that plants transform light energy into chemical energy (reviewed by Pol 2007).

Fast forward to the early 1900s. Experiments were being conducted with varying amounts of light, and light saturation responses had been reported. In 1905, the British physiologist Blackman interpreted these curves as having two regions of importance: a "light" reaction" and a "dark" reaction (Hall and Rao 1994). In the early 1930s, two scientists from the US state of Illinois, Emerson and Arnold, tested cultures of *Chlorella* with flashes of light and measured O_2 evolution in relation to the energy of the flashes (Emerson and Arnold 1932a). They showed that the maximum rate of photosynthesis was not a function of the amount of chlorophyll (chl), but instead by processes (enzymatic reactions) that controlled the dark reactions. Yet, even with this advancing knowledge, during that time the dogma was that the primary reaction in photosynthesis was the splitting of CO_2 by light to yield carbon (C) and O_2. As will be seen in this in detail in the following chapter, this is not the case; the O_2 comes from H_2O. It was work by Ruben and Kamen in California in 1941, using newly developed isotopic methods with ^{18}O-enriched water, that showed that the isotopic composition of the O_2 evolved was the same

P. M. Glibert, *Phytoplankton Whispering: An Introduction to the Physiology and Ecology of Microalgae*, https://doi.org/10.1007/978-3-031-53897-1_5

as that of water, but not the same as that of CO_2 (Ruben and Kamen 1941). It was 20 years later when Emerson and co-workers made another fundamental discovery. Using the green alga *Chlorella* and the red alga *Porphyridium*, they showed that two superimposed light beams of different wavelengths yielded a higher quantum yield than when the two beams were separately supplied. This became known as the "Emerson enhancement effect." They used these data to propose the existence of two photosystems, with different pigment systems (Emerson et al. 1957; Emerson and Chalmers 1958). At the same time, in 1960, Hill and Bendall related the two photosystems and photosynthetic electron transport with their Z-scheme model of photosynthesis (Hill and Bendall 1960). Unraveling this complex web of reactions was complicated. Over the next several chapters, the complexity of this process will hopefully be made clear. The process starts with the capture of light by phytoplankton pigment proteins.

There are three basic reasons for understanding algal pigments. The first is that pigments play an important role in photosynthesis, so understanding how pigments harvest light and regulate the flow of energy through the processes of photosynthesis tells us about the physiological conditions of algae at any given point in time or under varying environmental conditions. The second is that different types of algae have different pigment composition and thus knowing something about pigments tells us something about community composition. The third is that there are pigments that are harvested from algae for commercial applications. Thus, if this is of interest, then the physiological regulation of pigments must be understood.

Now, we dive into these processes in detail. This chapter reviews different pigments and their properties, and the next two chapters describe the processes of photosynthesis and the complex suite of tools by which photosynthesis can be measured.

5.2 Harvesting Light for Photosynthesis

The first rule of photochemistry is that for light to promote a chemical reaction, it must be absorbed. This is the role of pigments. Pigments are colored substances, which means that they absorb only part of the visible spectrum and therefore appear colored to our eyes. Evolution has led to a vast array of pigment types and combinations that confer different light-absorbing efficiencies and spectral sensitivities that are optimum under specific natural light regimes. As introduced in Chap. 4, among algae, green, pink, red, and blue-green are the common hues. Chlorophyll (chl) is the pigment that gives plants and algal cells their green color. Different taxa have different pigmentation, and even within a single species, coloration can also change depending on growth conditions. Despite the vast array of colors, the pigments that participate in photosynthesis originate from only three molecular classes of compounds, chlorophylls, carotenoids/xanthophylls, and

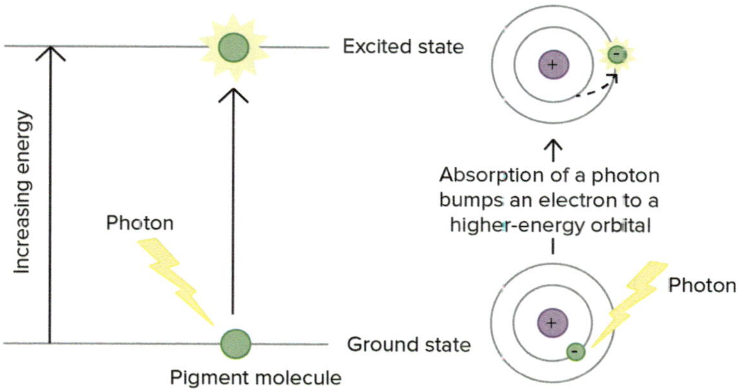

Fig. 5.1 The fate of a pigment molecule when a photon is absorbed. Image reproduced from Singer (2015, OpenStax-CNX, https://docplayer.net/41741485-Bis2a-06-3-photophosphorylation-the-light-reactions-in-photosynthesis.html) under a creative commons license

phycobilins. It is the small molecular variations and different pigment mixtures within these cells that lead to a vast array of algae colors.

The role of photosynthetic pigments is to harvest light, that is, to acquire the energy of a photon and ultimately transfer that energy into useful biochemical energy. A pigment molecule can be thought of as a simple antenna that can "absorb" a photon. In actuality, the electromagnetic energy of the photon interacts with a pigment molecule causing the molecule to shift into a higher energy state, at which point it becomes unstable (Fig. 5.1). What happens to this higher energy state of a single pigment molecule is the fascinating story of photosynthesis and the biological regulation that has evolved.

Absorbing light and transferring it to biochemical energy requires delicate balancing. Absorb too little and metabolism cannot get enough energy to function. Absorb too much and the cell can be overexcited and potentially damaged. The act of photochemistry or photosynthesis only occurs by two specialized chl molecules, termed P680 and P700 for their dominant absorption wavelength, that are able to donate a high-energy electron to an acceptor molecule. All of the other pigments that participate in photosynthesis serve as antennae pigments that absorb or regulate light energy to transfer the elevated energy state to the photochemically reactive chls (Fig. 5.2). Incredibly, a single reaction center chl may have an antenna consisting of several hundred pigment molecules. For the antenna to function efficiently with little or no energy loss, the individual pigment molecules must be spatially arranged in a relatively tight matrix to allow energy transfer among the antennae pigments. Adjacent pigments are only separated by angstroms to have pigment molecule-to-molecule energy transfer. This positioning is accomplished by embedding or covalently bonding the pigment molecules in protein structures. These pigment structures

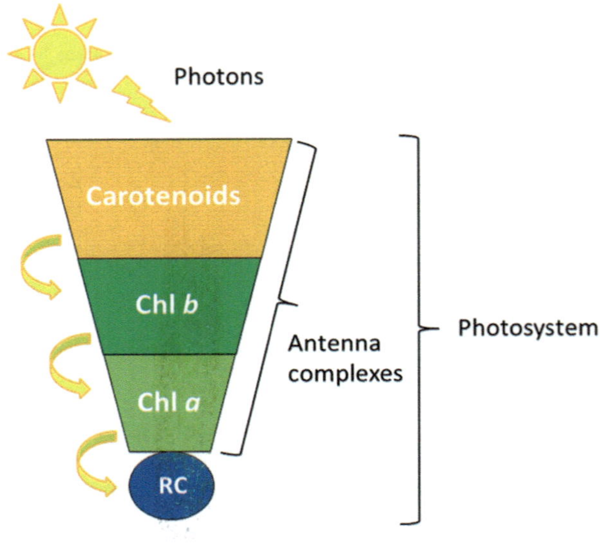

Fig. 5.2 A schematic representation of the light absorption process, showing how antenna pigments funnel light to chl a and the reaction center. Antenna complexes and the reaction center form a photosystem. Image reproduced from Guidi et al. (2016, InTechOpen. doi.org/10.5772/65594) under a creative commons license

are associated with thylakoid membranes, the arrangement of which is highly ordered, but variable depending on the type of organism (as detailed in Chap. 6). The architecture of the arrangement of pigments is, on the one hand, remarkably simple, being composed of only three major types of pigments, but at the same time, remarkably complex and fine-tuned. As Croce and van Amerongen (2014, p. 498) have summarized, "nature uses a limited number of pigments and a polymeric scaffold that acts as a smart matrix, allowing the control of the assembly in terms of affinity of the binding sites for different pigments, geometrical arrangement (correct orientation and high density without self-quenching) and tuning of the pigments' spectroscopic properties. In addition, the scaffold permits the switching on and off of light harvesting to respond to light fluctuations." The essence of light harvesting in algae is that it involves physiologically adjustable protein-pigment molecules that serve as tunable light antennae for photochemical reaction centers.

Photosynthetic pigments all contain a molecular structure with a conjugated bond system consisting of alternating double and single bonds. The wavelength of absorption depends on the number and arrangement of the conjugated bonds and generally, molecules with a greater number of conjugated bonds absorb light of longer wavelength. When looking at the chemical structure of pigments, look for the conjugated bond region. For chl (Fig. 5.3), this is found across the pyrrole structures that form a porphyrin ring. This is the "business" end of the molecule responsible for absorbing light energy. For carotenoids and xanthophylls, the conjugation extends across a hydrophobic chain (Fig. 5.4). For phycobilins, the conjugated system cuts across a linear series of four pyrrole rings (Fig. 5.5). It is necessary for these conjugated regions of the pigment to be placed and constrained at very close

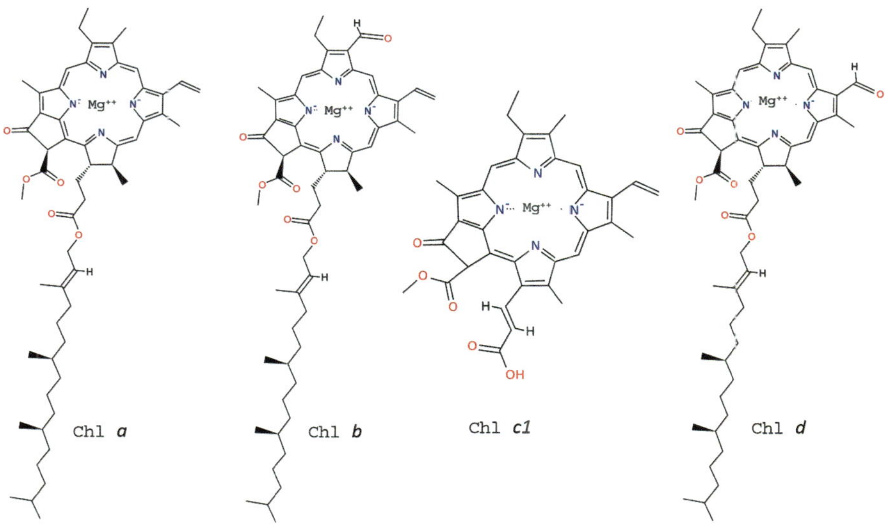

Fig. 5.3 The structure of chlorophylls *a*, *b*, *c1* and *d*. Images by D. Richfield and reproduced from https://en.wikipedia.org/wiki/Chlorophyll under a creative commons license

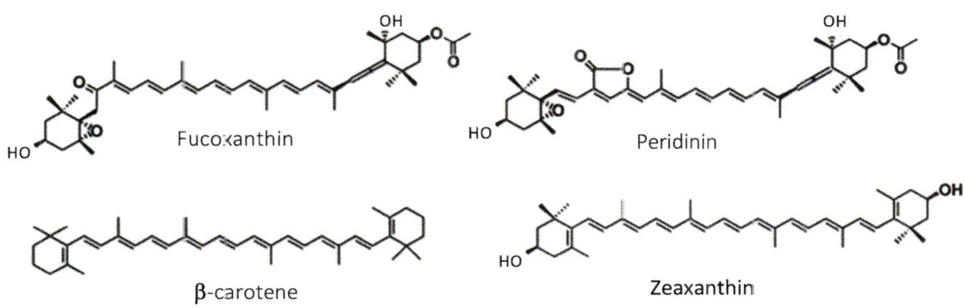

Fig. 5.4 Basic structures of several common algal carotenoids. Image modified and reproduced from Maoka (2020, J Nat Med 74: 1–16) under a creative commons license

distances to other pigment molecules in order for them to biophysically share or transfer the excited state from one antenna pigment to another and ultimately to the different photosystems, P680 or P700. The chemical structures of the pigments give clues on how this is accomplished.

The structure of these pigment–protein complexes, and the diversity of such structures in different types of photosynthetic organisms have been increasingly resolved in recent years (Hofmann et al. 1996; Lokstein et al. 2021). Typically, the antennae pigments, the carotenoids and phycobiloproteins, surround the chl *a* and are attached to it, forming a pigment complex, as shown for the dinoflagellate

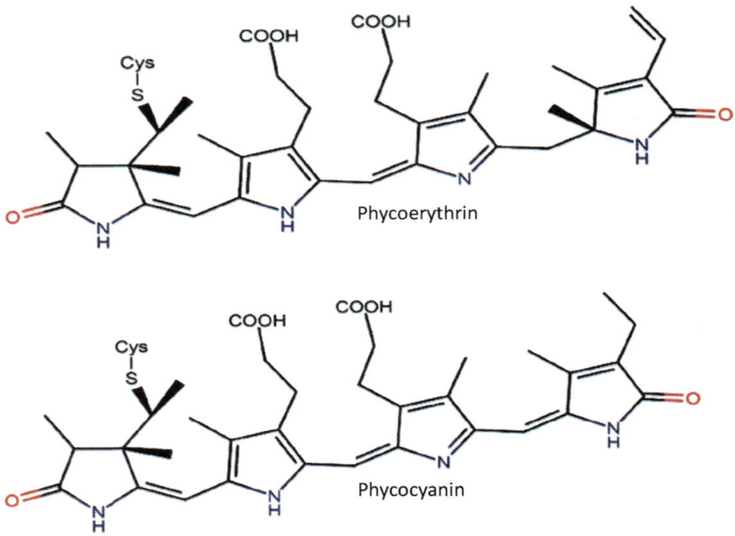

Fig. 5.5 Phycoerythrin and phycocyanin of cyanobacteria. Image reproduced from Abidizadegan et al. (2021, *Front Pharmacol* 11:618836. doi: 10.3389/fphar.2020.618836) under a creative commons license

Amphidinium carterae, which has peridinin carotenoids as its primary accessory pigment (Fig. 5.6). As will be seen below, the types and proportion of accessory pigments and chls serve as distinguishing features of different taxa.

It was once thought that the "tail" of the chl molecule was present in order to provide a hydrophobic region that would embed the chl molecules among the lipid bilayer of the photosynthetic membrane. This applied to the hydrophobic carotenoid molecules as well. There was some logic to this, as chl and carotenoids are easily extracted using acetone or methanol, and indeed, such extractions are used to quantify these pigments. However, the discovery of water-soluble pigment–protein complexes (e.g., Murata et al. 1971) changed this thinking. Protein "scaffolding" is responsible for associating and orienting pigment molecules that allow efficient energy transfer, and it is the hydrophobic regions of the protein that interact with the hydrophobic regions of the pigments. An exception to this is the phycobiliproteins of cyanobacteria and cryptophytes wherein the pigment–protein connection is via covalent bonding. Phytobiliproteins are water soluble and the pigments (phycobilins) cannot be extracted using acetone or methanol.

Each antenna is a result of multiple gene products and their associated pigments and the expression of these genes is under physiological control. This gives the organism a way of adjusting the antennae composition and therefore light absorption efficiency for each photochemical reaction center. Adding a larger antenna increases light when it is low, and decreasing the antenna size when light is too high is one

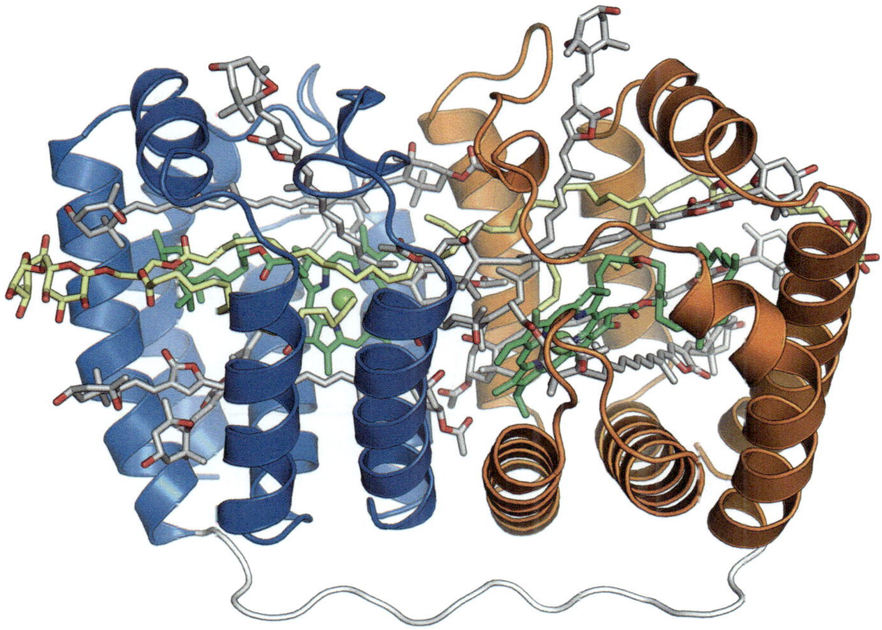

Fig. 5.6 Structure of the peridinin–chlorophyll a–protein complex in the dinoflagellate *Amphidinium*. The chlorophyll is shown in green (with the magnesium ion shown as a green sphere), peridinins in gray. The structure also has eight helix-shaped segments (blue and orange) that wrap around the structure. Image reproduced from Hofmann et al. (1996, Science 272: 1788–1791, doi.org/10.10.1126/science.272.5269.1788) and wikipedia (https://en.wikipedia.org/wiki/Peridinin-chlorophyll-protein_complex#/media/File:1ppr_peridinin_chlorophyll_protein.png) under a creative commons license

mechanism by which the cell can protect itself from overexcitation. Other protective mechanisms are described more fully in Chap. 6.

5.3 Pigment Diversity

5.3.1 Chlorophylls

The work of photosynthesis depends at the core—literally and figuratively—on chl *a*. The structure of chl *a* is a porphyrin ring, with magnesium at its center, chelated to four nitrogen (N) atoms (Fig. 5.3). The molecule also has a long phytol chain which provides a hydrophobic region that can be constrained in hydrophobic regions of antenna proteins. All photosynthetic algae and cyanobacteria contain chl *a* or a slightly modified form. Additional forms of chl (chl *b* through *f*) are distinguished by substitutions on the porphyrin side branches or length of the phytol tail (Fig. 5.3). As will be seen below, chls *a, b,* and *c* are the most common in algae. The form chl

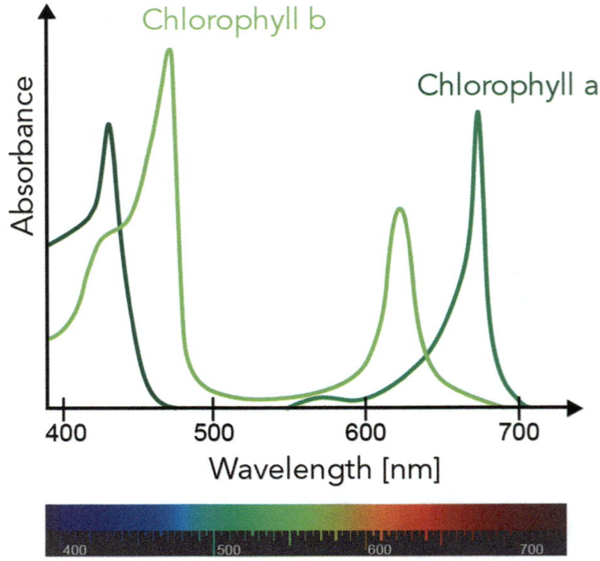

Fig. 5.7 Absorption spectra of Chlorophyll *a* and Chlorophyll *b*. Image by D. Pugliesi and reproduced from Wikipedia Commons (https://commons.wikimedia.org/wiki/File:Chlorophyll_ab_spectra-en.svg) under a creative commons license

d is mostly found in several red macroalgae, although it and chl *f* can also be found in filamentous and other cyanobacteria (Willows 2020; Larkum 2020). Chl *f* is induced under far-red light. In the marine *Prochlorococcus*, chls *a* and *b* are found in the divinyl forms, chls *a2* and *b2*.

Chlorophylls are characterized by two major absorption bands, one in blue (or blue-green) and one in red (Fig. 5.7). Thus, chl appears green. Compared to Chl *a*, the other forms of chl have the blue absorption band shifted to longer wavelengths and the red band shifted to shorter wavelengths. Consequently, chls *b-f* extend the spectral absorption window of the antennae (Fig. 5.7).

With two major absorption bands, blue and red, does it make a difference whether a blue photon or a red photon is absorbed? As described in Chap. 4, when a photon is absorbed, it can be considered as a countable event. Thus, for a molecule like chl, it does not matter whether the absorbed photon is blue or red in terms of causing a single photochemical reaction. A blue (higher energy) photon causes the same photosynthetic reaction as a red (lower energy) photon. The additional energy that is associated with the blue photon happens to be lost as heat while the balance of the energy drives photosynthesis. It would matter, however, if the cell were provided exclusively green light if it had no other pigments to absorb in that wavelength, in which case the efficiency of light absorption would be much less.

5.3.2 Carotenoids and Xanthophylls

Carotenoids and xanthophylls share the same basic structure and are only distinguished by xanthophylls containing one or more atoms of O. Carotenoids lack O in

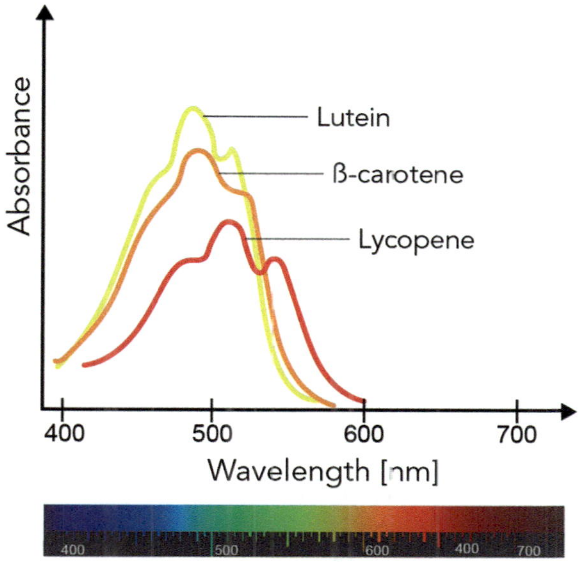

Fig. 5.8 Several typical carotenoids and their characteristic absorption patterns. Image by D. Pugliesi and reproduced from Wikipedia Commons (https://commons.wikimedia.org/wiki/File:Chlorophyll_ab_spectra-en.svg) under a creative commons license

their molecular structure. Xanthophylls are typically classed as a type of carotenoid, although they do not serve the same light-absorbing function as other carotenoids. The carotenoids are a diverse suite of pigments that differ chemically from the chls (Fig. 5.4). They have two 6-carbon rings connected by a chain of conjugated C atoms (Fig. 5.4), leading to absorption of light in the blue to green region of the spectrum (Kirk 2011; Fig. 5.8). Thus, their characteristic colors are yellow, orange, and red (Fig. 5.8). The presence of carotenoids in the photosynthetic antennae extends the efficient absorption of light into the green region.

Carotenoids serve two major roles: light harvesting and photoprotection. For light harvesting, the carotenoids serve as accessory pigments in the antennae to extend the spectral response of photosynthesis. In particular, carotenoids help close much of the green window where the chls do not absorb. The most common carotenoids in phytoplankton for light harvesting are fucoxanthin and peridinin, found in diatoms, many dinoflagellates, and chrysophytes. Their absorption properties make them excellent photoreceptors in the blue-green region of the spectrum and they are efficient at transferring absorbed light energy into the photosynthetic antennae. Absorption in the blue-green region gives those algae with high carotenoid content, an advantage as blue and green light penetrate deeper in the water column (see Chap. 4).

Several xanthophylls (e.g., violoxanthin, antheraxanthin, zeaxanthin, diadinoxanthin, and diadoxanthin) serve a different function than light absorption for promoting photosynthesis. Rather than passing light energy to chl *a*, these pigments are part of a 'protection' system that regulates the efficiency of energy transfer to the reaction center. In situations where light is in excess of what can be utilized immediately by the cell, such as transfer to very bright light or when metabolism is limited by another factor, the excess light energy can cause over-energization of, and damage to, the photosynthetic apparatus. A method of protecting the cell from this damage is to convert the excited state in the antenna to heat before the energy reaches the reaction center where it can potentially cause protein damage. Other processes that the cell may have to protect from overexcitation are discussed in Chap. 6.

The process of regulating energy flow in the antennae by xanthophylls is termed the "xanthophyll cycle." There are two forms of xanthophyll cycles, one involving diadinoxanthin and diadoxanthin interconversions and the other involving violaxanthin, antheraxanthin, and zeaxanthin interconversions (Fig. 5.9). Common to many algal classes, including diatoms, chrysophytes, chloromonads, dinoflagellates, and euglenoids is the diadinoxanthin and diatoxanthin cycle, whereas chlorophytes, chrysophytes, as well as higher plants, employ the violaxanthin, antheraxanthin, and zeaxanthin cycle. The essential mechanism is

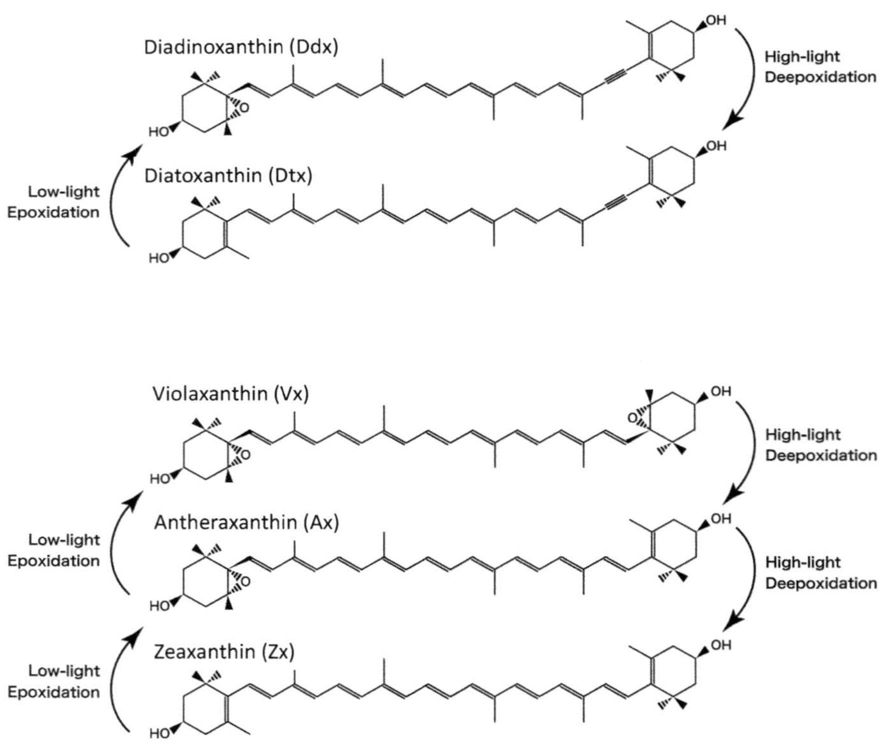

Fig. 5.9 The two xanthophyll cycles common in algae. The Diadinoxanthin cycle is commonly found in diatoms, dinoflagellates, and haptophytes, while the Violaxanthin cycle is common in green algae and higher plants. Arrows between pigments indicate conversions caused by xanthophyll cycling. Image reproduced from Tanabe et al. (2011, PLoS ONE 6(2): e14690. doi.org/10.1371/journal.pone.0014690) under a creative commons license

the same for both cycles. One form of the xanthophyll, either violaxanthin or diadinoxanthin, located in the antennae, allows the excited state to persist and provides an opportunity to carry out photosynthesis. The other forms, antheraxanthin, zeaxanthin, and diatoxanthin, are quenching forms, which rapidly cause the excited state to dissipate as heat, thereby preventing the excitation energy from propagating to the reaction center where damage could occur. The interconversion of these xanthophyll cycle pigments is enzymatically driven and regulated by the energy state of the photosynthetic apparatus (reviewed by Larkum 2020). Under low light conditions, the xanthophyll cycle pigments contain epoxides, e.g., violaxanthin and diadinoxanthin, which makes them non-quenching. In a high energy state de-epoxidases become active converting the xanthophylls to a quenching condition. The reverse happens under a shift to low light by epoxidases. It should be noted that cyanobacteria only have a limited xanthophyll cycle, but they have a high zeaxanthin content, and thus changes in the proportion of zeaxanthin function in this role. Zeaxanthin is effective in photoprotection as it dissipates energy as heat and it is a strong antioxidant and scavenges reactive oxygen species (ROS) that may form when the cell is overexcited (Schiphorst and Bassi 2020).

5.3.3 Phycobilins and Phycobiliproteins

Phycobilins are the predominant accessory pigments found only in cyanophyta, cryptophyta, glaucophyta, and rhodophyta, suggestive of their common ancestral cyanobacterial lineage (see Chap. 10). Phycobilins differ in their conjugated bond structure and therefore have different absorption spectra (Fig. 5.10). The absorption characteristics of the biliproteins differ from that of the chls and of the carotenoids, being further shifted to the region of the red (Fig. 5.10). Peak absorption ranges from about 500 nm to over 650 nm. Algae possess one or more phycobilins to exploit different regions of the visible spectrum.

Unlike algae with chl and carotenoids as accessory pigments, phycobilins are covalently bound to proteins and the combined structure is called a phycobiliprotein (Fig. 5.11). The antenna formed by phycobiliproteins is quite different from that in non-phycobiliprotein-containing algae (Fig. 5.2). These pigments occur as aggregates, or phycobilisomes, which are attached to the outer surfaces of the thylakoid membranes (Fig. 5.11). Allophycocyanin sits at the center and from it the transfer of energy to chl *a* occurs. The other biliproteins form a "shell" around allophycocyanin, first with phycocyanin and then phycoerythrin. In cyanobacteria, biliproteins are either of the red type (phycoerythrin, phycoerythrocyanin) or blue type (phycocyanin, allophycocyanin; Fig. 5.5). The arrangement of biliproteins in the cryptophyta is somewhat different than that in the cyanophyta and rhodophyta. Each cryptophyte species appears to contain only one biliprotein, not allophycocyanin. Moreover, rather than being located on the thylakoid membranes they occur as dense granular matrix within the interior of the thylakoid (Kirk 2011).

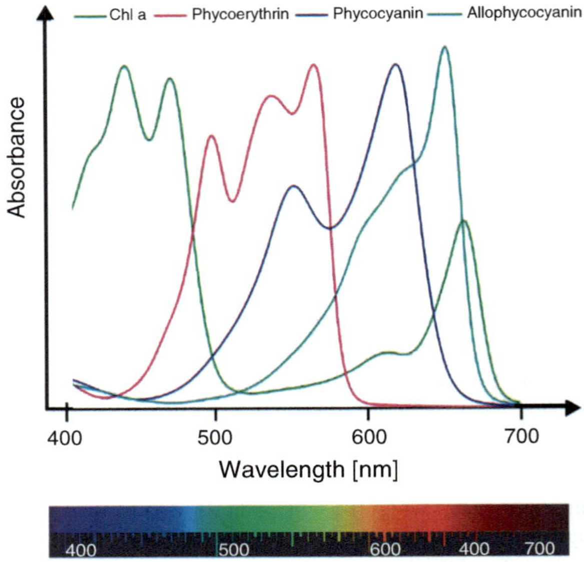

Fig. 5.10 Absorption characteristics of phycoerythrin, phycocyanin, and allophycocyanin compared to that of chlorophyll *a*. Image modified from Dagnino-Leone et al. (2022, Comp Struct Biotechnol J 20: 1506–1527) under a creative commons license

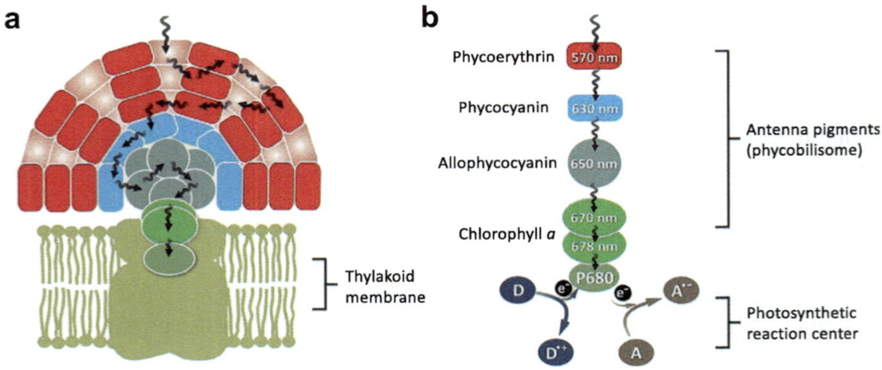

Fig. 5.11 Schematic diagram of a phycobilisome situated on the thylakoid membrane (Panel **a**) and the energy transfer steps which include charge separation (Panel **b**). Each phycobilisome contains a core consisting of allophycocyanins. From this core, several rods made up of phycocyanin and phycoerythrin extend in an outward orientation and function as light harvesting antennae. Energy absorbed by these rods is efficiently transferred to the chlorophyll pigments for their use in photosynthetic reactions. Image reproduced from Govindjee and Shevela (2011, Front Plant Sci doi.org/10.3389/fpls.2011.00028) under a creative commons license

5.4 Use of Diagnostic Accessory Pigments to Classify Algae

Although all phytoplankton face the same challenges of capturing light and efficiently transferring light energy to chemical energy, and all phytoplankton do so with basically similar architecture, their antennae and light harvesting properties differ, a result of their adaptation to the environmental conditions of different taxa. Among the algae, the chl content and type vary, with chl *a* always constituting the bulk of chl (Table 5.1, but note the divinyl forms for Prochlorophyta). The major carotenoids also differ (Table 5.2). A simplified view of pigment chemotaxonomy is illustrated in Fig. 5.12.

With the development of techniques such as high-performance liquid chromatography (HPLC), the ability to quantify accessory pigments within natural samples has greatly advanced. There have been various efforts to apply diagnostic pigment signatures to identify phytoplankton and/or to relate these pigment signatures to size structures of plankton communities. Vidussi et al. (2001) suggested that seven pigments were sufficient to classify phytoplankton size classes. These pigments were fucoxanthin, peridinin, 19′-hexanoyloxyfucoxanthin, 19′-butanoyloxyfucoxanthin, alloxanthin, total chl, and zeaxanthin. They and others (reviewed by Chase et al. 2020) have presented various approaches for weighting the pigment signature values to size classes of phytoplankton. Others have used single, dominant pigments, or their relationships to chl *a*, for example, fucoxanthin for diatoms, peridinin for peridinin-containing dinoflagellates, alloxanthin for cryptophytes, and chl *b* for chlorophytes (e.g., Glibert et al. 2004; Shangguan et al. 2017). The use of ratios works in principle because, although the total pigment complement within a cell may change due to changing conditions (see Chap. 8), the proportion of accessory pigments to chl *a* stays roughly constant within algal taxa when compared across algal taxa. Of course, use of pigment signatures is imperfect; some pigments are shared by different classes. Some taxa are mixotrophic (Chap. 15), and thus their pigments at a given point in time may be a combination of their own and that which they consumed. Chemical taxonomy software is available for calculating the relative contribution of phytoplankton groups to overall biomass based on accessory pigments (Mackey et al. 1996). It is a valuable tool, but like all such tools, appropriate calibration with other approaches is desirable.

5.5 Pigments and Their Applications

Algal pigments are used in a variety of applications, which are only expected to increase in the coming years (Patel et al. 2022; Zittelli et al. 2023). They have antioxidant, anti-carcinogenic, anti-inflammatory, anti-obesity, anti-angiogenic, and neuroprotective properties (Patel et al. 2022 and references therein). Fucoxanthin has found widespread use in pharmaceuticals as it is antimicrobial, anti-inflammatory, and has other beneficial effects (Mumu et al. 2022) A variety of algae pigment-

Table 5.1 Chlorophyll distribution among algal groups. Modified from Kirk and Tilney-Bassett (1978) and Kirk (2011)

Algal group	Chlorophyll *a*	Chlorophyll *b*	Chlorophyll *c1*	Chlorophyll *c2*	Chlorophyll *d*
Chlorophyta	+	+	–	–	–
Euglenophyta	+	+	–	–	–
Heterokontophyta	+	+	–	–	–
Chrysophycea	+	–	+	+	–
Xanthophyceae	+	–	–	–	–
Eustingmatophyceae	+	–	–	–	–
Bacillariophyceae	+	–	(+)–	+	–
Haptophyta	+	–	+/–	+	–
Pyrrophyta (Dinophyta)	+	–	–(+)	+	–
Phaeophyta	+	–	+	+	–
Cryptophyta	+	–	–	+	–
Rhodophyta	+	–	–	–	–
Cyanophyta	+	–	–	–	+ (*Acaryochloris*)
Prochlorophyta	+ (Divinyl forms)	+ (Divinyl forms)	–	–	–

Table 5.2 Major carotenoids in various algal classes. Modified from Kirk and Tilney-Bassett (1978) and Kirk (2011). Note that (+) indicates that the pigment is present in at least some members of this class

Carotenoids	*Chlorophyta*	*Xanthophyceae*	*Bacillariophyta*	*Chrysophyta*	*Euglenophyta*	*Phaeophyta*	*Pyrrophyta*	*Cryptophyta*	*Rhodophyta*	*Cyanophyta*	*Prochlorophyta*
α-Carotene	+	–	–	+	–	–	–	+	+	–	+
β-Carotene	+	+	+	+	+	+	+	–	+	+	+
Lutein	+	–	–	–	–	–	–	–	–	–	–
Zeaxanthin	+	–	–	–	–	–	–	–	+	+	+
Neoxanthin	+	–	–	–	+	–	–	–	+	–	–
Violaxanthin	+	–	–	–	–	+	–	–	–	–	–
Fucoxanthin	–	–	+	+	–	+	(+)	–	–	–	–
Diatoxanthin	–	+	+	+	+	–	–	–	–	–	–
Diadinoxanthin	–	+	+	+	+	–	+	–	–	–	–
Peridinin	–	–	–	–	–	–	+	–	–	–	–
Alloxanthin	–	–	–	–	–	–	–	+	–	–	–
Myxoxanthophyll	–	–	–	–	–	–	–	–	–	+	–

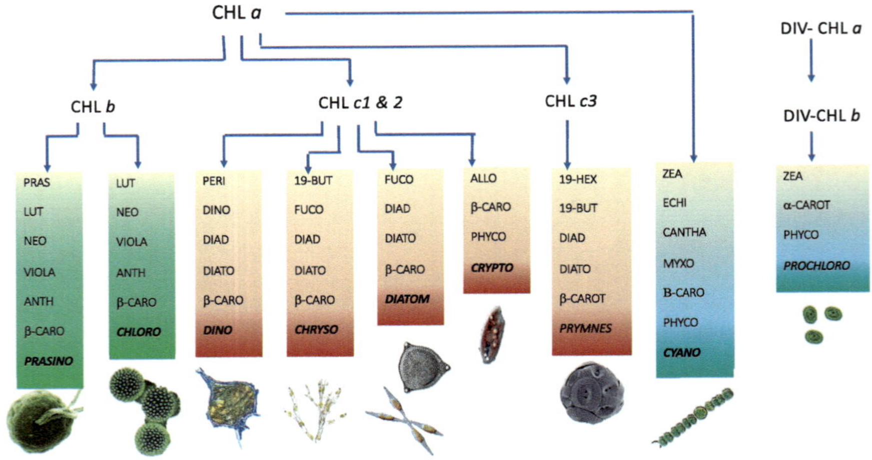

Fig. 5.12 Simplified illustration of the pigment chemotaxonomy of phytoplankton. Abbreviations: PRAS-Prasinoxanthin, LUT-Lutein, NEO-Neoxanthin, VIOLA-Violaxanthin, ANTH-Anthoxanthin, b-CAROT-b-Carotene, PERI-Peridinin, DINO-Dinoxanthin, DIAD- Diadinoxanthin, DIATO-Diatoxanthin, 19-BUT- 19′-Butanoyloxyfucoxanthin, FUCO- Fucoxanthin, ALLO-Alloxanthin, PHYCO-Phycoerythrin, 19-HEX- 19′-Hexanoyloxyfucoxanthin, ZEA-Zeaxanthin, ECHI-Echinenone, CANTHA-Canthoxanthin, MYXO-Myxoxanthin, a-CAROT- a-Carotene, Image modified based on https://www.geomar.de/en/research/fb2/fb2-bi/infrastructure/hplc-analyses. Prasinophyte image (*Pyramimonas* sp.) reproduced from Nigaku-Ike, https://en.wikipedia.org/wiki/File:Pyramimonas_sp_color.jpg; Chlorophyte (*Volvox*), reproduced from F.Fox, https://en.wikipedia.org/wiki/Volvox#/media/File:Mikrofoto.de-volvox-4.jpg; Dinoflagllate (*Protoperidinium*) reproduced from https://en.wikipedia.org/wiki/Protoperidiniaceae#/media/File:Protoperidinium_dinoflagellate.jpg; Chrysophyte (*Dinobryon*), reproduced from https://en.wikipedia.org/wiki/Chrysophyta; Diatom (upper) Haeckel's 1904, https://en.wikipedia.org/wiki/Diatom#/media/File:Diatomeas-Haeckel.jpg, (lower, *Phaeodactylum tricornutum*), reproduced from de Martino and Bowler (https://en.wikipedia.org/wiki/Diatom#/media/File:Phaeodactylum_tricornutum.png); Cryptophyte (*Rhinomonas nottbecki*) reproduced from Abidizadegan et al. Front Pharmacol. 2021; Prymnesiophyte (*Coccolithus pelagicus*) reproduced from https://en.wikipedia.org/wiki/Prymnesiophyceae; Cyanobacteria (*Anabaena sperica*) reproduced from https://en.wikipedia.org/wiki/Cyanobacterial_morphology#Movement; and *Prochlorococcus* reproduced from https://en.wikipedia.org/wiki/Cyanobacteria, all under creative commons licenses

derived natural colors are ecofriendly and used as synthetic food colorants, including dietary supplements (Chakdar and Pabbi 2017) and in cosmetics, aquaculture, textiles, and other light manufacturing fields (Zittelli et al. 2023; Fig. 5.13). Food dyes are of particular interest given health concerns over synthetic dyes. Understanding the biotic and abiotic factors regulating pigment content and quality is a focus of the industry, as is the optimization of extraction and purification methods.

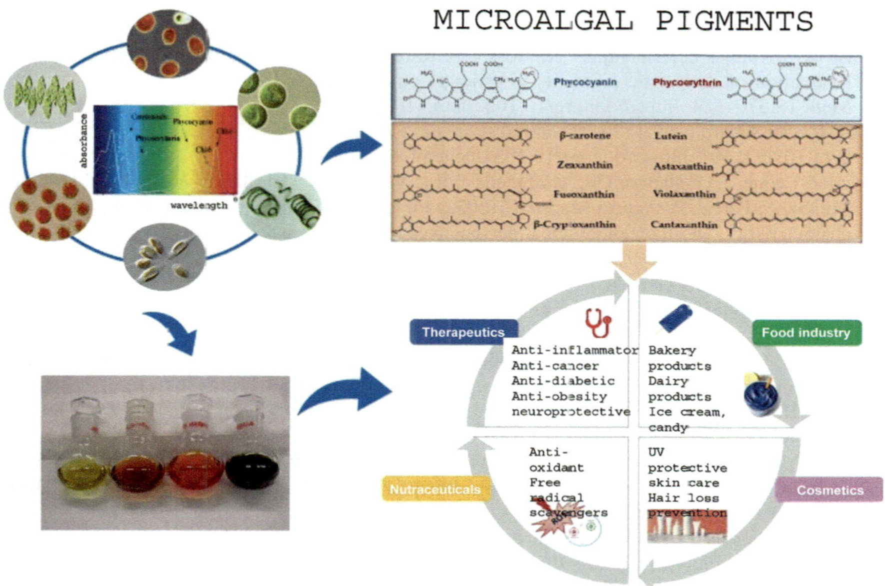

Fig. 5.13 Applications of algal pigments. They are being used, and have great potential for expanded applications as therapeutics, in the food industry, as nutrachemicals and in cosmetics. Image reproduced from Zittelli et al. (2023, Photochem Photobiol Sci 22: doi.org/10.1007/s43630-023-00407-3) under a creative commons license

5.6 Summary

With a fundamental understanding of the types of pigments in phytoplankton, it is clear that the light harvesting mechanisms differ in different types of phytoplankton. Yet, the fundamental building blocks of the apparatus of capturing light follow the same basic principles. The diversity of pigments gives us a powerful tool to classify these organisms. It gives us insight into what types of light different types of phytoplankton may favor. This basic information is also foundational in the understanding of the processes of photosynthesis. Next, the complexities of how light energy is converted to chemical energy are explored.

References

Abidizadegan M, Peltomaa E, Blomster J (2021) The potential of Cryptophyte algae in biomedical and pharmaceutical applications. Front Pharmacol 11:618836. https://doi.org/10.3389/fphar.2020.618836

Chakdar H, Pabbi S (2017) Algal pigments for human health and cosmeceuticals. In: Rastogi RP, Madamwar D, Pandey A (eds) Algal green chemistry. Elsevier, Amsterdam, pp 171–188

Chase AP, Kramer SJ, Haëntjens N, Boss ES, Karp-Boss L, Edmondson M, Graff JR (2020) Evaluation of diagnostic pigments to estimate phytoplankton size classes. Limnol Oceanogr Meth 18:570–584. https://doi.org/10.1002/lom3.10385

Croce R, van Amerongen H (2014) Natural strategies for photosynthetic light harvesting. Nat Chem Biol 10:492–501. https://doi.org/10.1038/nchembio.1555

Dagnino-Leone J, Figueroa CP, Castañeda ML, Youlton AD, Vallejos-Almirall A, Agurto-Muñoz A, Pérez JP, Agurto-Muñoz C (2022) Phycobiliproteins: structural aspects, functional characteristics, and biotechnological perspectives. Comput Struct Biotechnol J 20:1506–1527

Emerson R, Arnold W (1932a) A separation of the reactions in photosynthesis by means of intermittent light. J Gen Physiol 15:391–420

Emerson R, Arnold W (1932b) The photochemical reaction in photosynthesis. J Gen Physiol 16: 191–205

Emerson R, Chalmers RV (1958) Speculations concerning the function and phylogenetic significance of the accessory pigments of algae. Phycol Soc News Bull 11:51–56

Emerson R, Chalmers RV, Cederstrand CN (1957) Some factors influencing the long wave limit of photosynthesis. Proc Natl Acad Sci USA 43:133–143

Glibert PM, Heil CA, Hollander D, Revilla M, Hoare A, Alexander J, Murasko S (2004) Evidence for dissolved organic nitrogen and phosphorus uptake during a cyanobacterial bloom in Florida bay. Mar Ecol Prog Ser 280:73–83

Govindjee, Shevela D (2011) Adventures with cyanobacteria: a personal perspective. Front Plant Sci 2:28. https://doi.org/10.3389/fpls.2011.00028

Guidi L, Tattini M, Landi M (2016) How does the chloroplast protect against excessive light? In: Jacob-Lopes E, Zepka LQ, Queiroz MI (eds) Chlorophyll. InTechOpen. https://doi.org/10.5772/65594

Haeckel E (1904) Kunstformen der Natur. Republished by Bibliographisches Institut, Leipzig und Wien, 2012

Hall DO, Rao KK (1994) Photosynthesis, 5th edn. Cambridge University Press, Cambridge

Hill R, Bendall F (1960) Function of the two cytochrome components in chloroplasts: a working hypothesis. Nature 186:136–140

Hofmann E, Wrench PM, Sharples FP, Hiller RG, Welte W, Diederichs K (1996) Structural basis of light harvesting by carotenoids: peridinin-chlorophyll-protein from *Amphidinium carterae*. Science 272:1788–1791. https://doi.org/10.1126/science.272.5269.1788

Kirk JTO (2011) Light and photosynthesis in aquatic ecosystems, 3rd edn. Cambridge University Press, Cambridge, p 649

Kirk JTO, Tilney-Bassett RAE (1978) The plastids, 2nd edn. Elsevier, Amsterdam

Larkum AWD (2020) Light-harvesting in cyanobacteria and eukaryotic algae: an overview. In: Larkum AWD, Grossman AR, Raven JA (eds) Photosynthesis in algae: biochemical and physiological mechanisms. Springer Publ, Cham, Switzerland, pp 207–260

Lokstein H, Renger G, Götze JP (2021) Photosynthetic light-harvesting (antenna) complexes–structures and functions. Molecules 26:3378. https://doi.org/10.3390/molecules26113378

Mackey MD, Mackey DJ, Higgins HW, Wright SW (1996) CHEMTAX- a program for estimating class abundances from chemical markers: application to HPLC measurements of phytoplankton. Mar Ecol Prog Ser 144:265–285

Maoka T (2020) Carotenoids as natural functional pigments. J Nat Med 74:1–16

Mumu M, Das A, Emran TB, Mitra S, Islam F, Roy A, Karim MM et al (2022) Fucoxanthin: a promising phytochemical on diverse pharmacological targets. Front Pharmacol 13:929442. https://doi.org/10.3389/fphar.2022.929442

Murata T, Toda F, Uchino K, Yakushiji E (1971) Water-soluble chlorophyll protein of *Brassica oleracea* var. *botrys* (cauliflower). Biochim Biophys Acta 245:208–215

Patel AK, Albarico FPJB, Perumal PK, Vadrale AP, Nain CT, Chau HTB, Anwar C et al (2022) Algae as an emerging source of bioactive pigments. Bioresour Technol 351:12690. https://doi.org/10.10106/j.biotech.2022.126910

Pol D (2007) History of plant biology, history of knowledge of plant physiology. Fondation La main à la pâte: https://www.fondation-lamap.org/fr/page/11407/histoire-des-connaissances-sur-la-physiologie-des-plantes

Ruben S, Kamen MD (1941) Long-lived radioactive carbon: C14. Phys Rev 59:349–354

Schiphorst C, Bassi R (2020) Chlorophyll-xanthophyll antenna complexes: in between light harvesting and energy dissipation. In: Larkum AWD, Grossman AR, Raven JA (eds) Photosynthesis in algae: biochemical and physiological mechanisms. Springer Publ, Cham, Switzerland, pp 27–55

Shangguan Y, Glibert PM, Alexander JA, Madden CJ, Murasko S (2017) Phytoplankton community response to changing nutrients in Florida bay: results of mesocosm studies. J Exp Mar Biol Ecol 494:38–53. https://doi.org/10.1016/j.embe.2017.05.006

Singer M (2015) Bis2A 06.3 photophosphorylation: the light reactions in photosynthesis. OpenStax-CNX. https://docplayer.net/41741485-Bis2a-06-3-photophosphorylation-the-light-reactions-in-photosynthesis.html. Accessed 13 Aug 2013

Tanabe Y, Shitara T, Kashino Y, Hara Y, Kudoh S (2011) Utilizing the effective xanthophyll cycle for blooming of *Ochromonas smithii* and *O. itoi* (Chrysophyceae) on the snow surface. PLoS One 6(2):e14690. https://doi.org/10.1371/journal.pone.0014690

Vidussi F, Claustre H, Manca BB, Luchetta A, Marty J-C (2001) Phytoplankton pigment distribution in relation to upper thermocline circulation in the eastern Mediterranean Sea during winter. J Geophys Res-Atmos 106:19939–19956. https://doi.org/10.1029/1999JC000308

Willows RD (2020) In: Larkum AWD, Grossman AR, Raven JA (eds) Photosynthesis in algae: biochemical and physiological mechanisms. Springer Publ, Cham, Switzerland, pp 83–103

Zittelli GC, Lauceri R, Faraloni C, Benavides AMS, Torzillo G (2023) Valuable pigments from microalgae: phycobiliproteins, primary carotenoids, and fucoxanthin. Photochem Photobiol Sci 22:1733–1789. https://doi.org/10.1007/s43630-023-00407-3

Additional Websites Accessed

https://commons.wikimedia.org/wiki/File:Chlorophyll_ab_spectra-en.svg. Accessed 30 Aug 2023

https://en.wikipedia.org/wiki/Chrysophyta. Accessed 30 Aug 2023

https://en.wikipedia.org/wiki/Cyanobacteria. Accessed 30 Aug 2023

https://en.wikipedia.org/wiki/Cyanobacterial_morphology#Movement. Accessed 30 Aug 2023

https://en.wikipedia.org/wiki/Diatom#/media/File:Diatomeas-Haeckel.jpg. Accessed 30 Aug 2023

https://en.wikipedia.org/wiki/Diatom#/media/File:Phaeodactylum_tricornutum.png). Accessed 30 Aug 2023

https://en.wikipedia.org/wiki/File:Pyramimonas_sp_color.jpg. Accessed 30 Aug 2023

https://en.wikipedia.org/wiki/Peridinin-chlorophyll-protein_complex#/media/File:1ppr_peridinin_chlorophyll_protein.png). Accessed 30 Aug 2023

https://en.wikipedia.org/wiki/Protoperidiniaceae#/media/File:Protoperidinium_dinoflagellate.jpg. Accessed 30 Aug 2023

https://en.wikipedia.org/wiki/Prymnesiophyceae. Accessed 30 Aug 2023

https://en.wikipedia.org/wiki/Volvox#/media/File:Mikrofoto.de-volvox-4.jpg. Accessed 30 Aug 2023

6 Photosynthesis: The Complex Path from Light to Carbon Fixation

6.1 Introduction

Roughly half of the world's O_2 comes from the photosynthesis of the phytoplankton of the oceans. This reaction, however, is not so simple. The history of discoveries associated with photosynthesis were briefly touched upon in Chap. 5, but there are many early investigators who made critical discoveries. Experiments dating back to the 1600s had begun to unravel the roles of water and oxygen (O_2) in photosynthesis, as reviewed by Stirbet et al. (2020, p. 513, 514),

> Jan van Helmont (1648) showed that plant growth was mainly from the water that plants had absorbed; it was only later that Nicolas Théodore de Saussure (1804) clearly showed that water was an essential reactant of photosynthesis. Joseph Priestley (1776) showed, in elegant experiments, that plants produced 'oxygen' (then called de-phlogisticated air) needed by a mouse to live, whereas Jan Ingen-Housz (1773) convincingly established that light was necessary for photosynthesis. The role of CO_2 in photosynthesis was shown by Jean Senebier (1782), whereas the synthesis of starch was shown by Julius von Sachs (1862, 1864). However, the involvement of chlorophyll in this process has a long history. For some of the earliest concepts, we must remember to mention Pierre Joseph Pelletier and Joseph Bienaimé Caventou (1817, 1818), and René Joachim Henri Dutrochet (1837). However, Theodor Engelmann (1882) provided the first action spectrum of photosynthesis, showing that red and blue light, absorbed by chlorophyll, produce oxygen... By measuring photosynthesis as a function of light intensity, Frederick Frost Blackman (1905) suggested that photosynthesis consists of two separate phases: a light-dependent phase (i.e. so-called 'light' reactions), and a temperature-dependent biochemical phase (so-called 'dark' reactions...).

In addition to the important discovery of Blackman, another important discovery of the early 1900s was that of Emerson and Arnold (1932a, b). They conclusively separated the photochemical steps. They did so by studying how photosynthesis varied with light intensity, flash duration, and flash interval. They were able to conceptualize how the pigment antenna functioned and developed the concept of what is now known as a photosynthetic unit. In addition, Robert Hill (1937) discovered the light and dark phases of photosynthesis operated differently. Thus,

P. M. Glibert, *Phytoplankton Whispering: An Introduction to the Physiology and Ecology of Microalgae*, https://doi.org/10.1007/978-3-031-53897-1_6

by the time Gordon Riley and Steeman Nielsen made their measurements of primary productivity with O_2 and ^{14}C in the 1930s (see Chaps. 1 and 2), there was much that was understood about this important process (or processes).

The complexities of photosynthesis are hidden in the summary reaction,

$$CO_2 + 2H_2O + \text{light energy} \rightarrow (CH_2O) + H_2O + O_2 \quad (6.1)$$

This chapter examines some of these complexities, the reactions that occur in the two "phases" of photosynthesis, how the cell copes with a condition of over-reduction or over-excitation, and the various processes of respiration. This chapter is, however, only able to provide a cursory look at the processes of photosynthesis; more thorough reviews are provided by Kirk (2011), Falkowski and Raven (2007), and the compendia of Suggett et al. (2010), Larkum et al. (2020), among many others. These sources shaped the summary provided here. The next chapters (Chap. 7 and 8) examine the confusing array of methods now available to measure aquatic photosynthesis, and how to relate the different measurements of photosynthesis with respect to growth and how cells can acclimate to changing environmental conditions on short time scales.

6.2 Units Matter: Photosynthetic Units

The excitation energy—the photons—received by the algal cell are initially captured by the antenna pigments of which there are many more molecules of these pigments than of chlorophyll (chl) *a* (Chap. 5). These pigments can be envisioned as a funnel through which the light energy passes until it reaches its respective reaction center. The excitation energy moves within the reaction centers in a more or less random fashion until it reaches the specialized chl *a* molecule, whereupon it is used for electron transfer. There are two photosystems, each composed of chl *a* with its accessory pigments that funnel light to the core chl *a*. Each photosystem has its own type of chl *a*, with that for photosystem I (PSI), P_{700}, having maximal absorption of light at a wavelength of 700 nm, while that for photosystem II (PS II), P_{680}, having maximal absorption of light at a wavelength of 680 nm. For cyanobacteria, the accessory pigments are phycobilisomes that are attached to the thylakoid membrane. For eukaryotic algae, the accessory pigments are typically carotenoids or other forms of chl attached to *light harvesting complexes* (LHS) embedded in the membrane (see Chap. 5). When light energy is absorbed by antennae pigments, it is then transferred as excitation energy to its chl *a*, the process is said to occur within a *reaction center* (Figs. 6.1 and 6.2). The LHC's are composed of the PSII and PSI pigment–protein supercomplexes. When a single PSI and a PSII work together in energy conversion of the light energy with their light-harvesting pigments, it is referred to as a *photosynthetic unit* (Fig. 6.2).

Reaction centers reside in the thylakoid membranes and are associated with energy-transforming components (the electron transport chain), ATP synthesis components, and various enzymes. The PSII complex is associated with the splitting

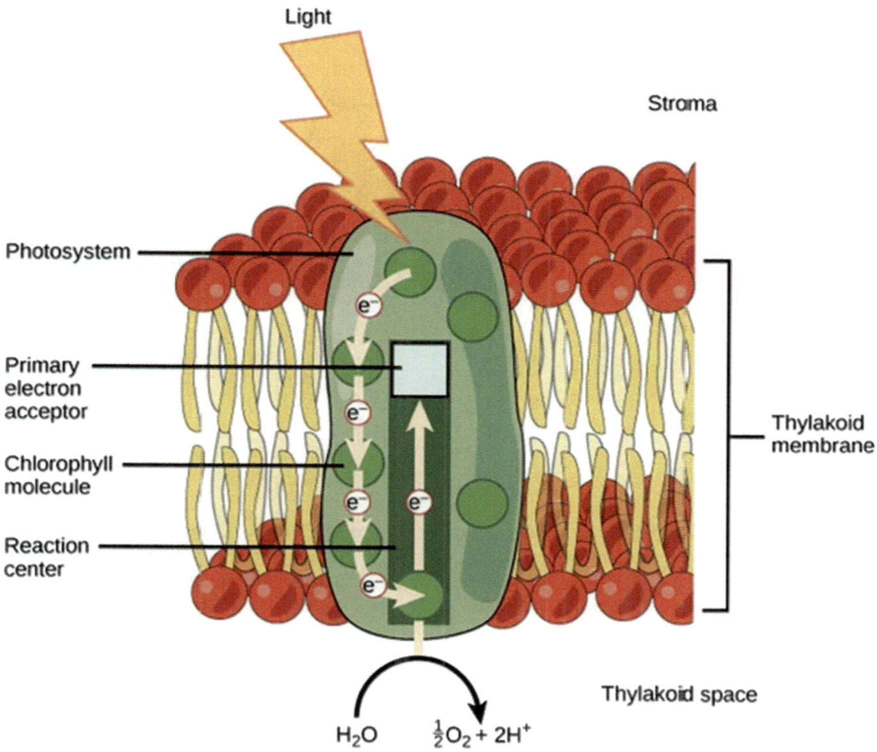

Fig. 6.1 Schematic representation of photosystem II in the thylakoid, showing the light-harvesting complex and the reaction center. Light energy is absorbed by a chlorophyll molecule and is passed along a pathway to other chlorophyll molecules. The energy culminates in a molecule of chlorophyll found in the reaction center. The energy "excites" one of its electrons enough to leave the molecule and be transferred to a nearby primary electron acceptor. A molecule of water splits to release an electron, which is needed to replace the one donated. Oxygen and hydrogen ions are also formed from the splitting of water. Image reproduced from OpenStax https://openstax.org/books/concepts-biology/pages/5-2-the-light-dependent-reactions-of-photosynthesis under a creative commons license

of H_2O, and that of PSI is associated with the reduction of $NADP^+$ to NADPH via the activity of $NADP^+$ reductase. The thylakoid membranes may be stacked or unstacked. Whereas PSI is found typically in unstacked regions of the membranes, PSII and its associated light-harvesting pigments are found most typically in stacked regions of these membranes. The internal space between membranes provides a means for generating an electrochemical proton gradient used by ATP-producing complexes. These processes and reactions are described in more detail in the next sections.

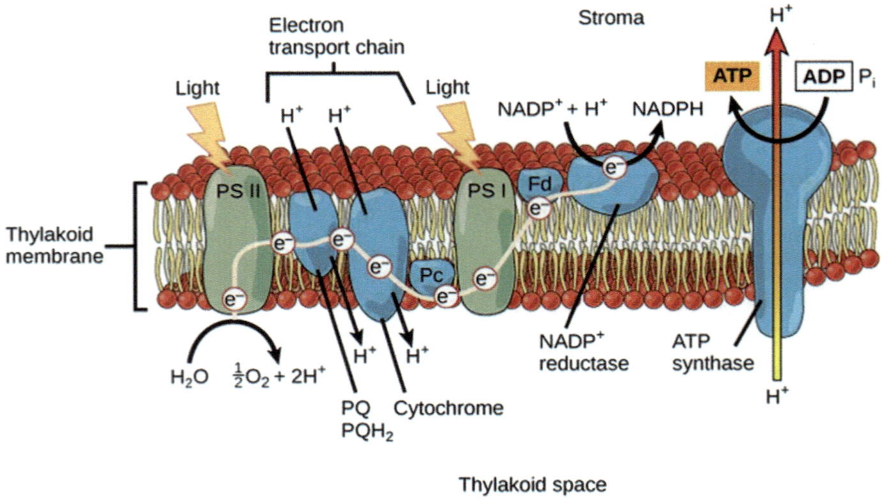

Fig. 6.2 Schematic representation of an electron transport chain. Using light and the two photosystems, electrons from the water are transferred to $NADP^+$ to form NADPH. Oxidation of water produces oxygen and protons in the lumen of the thylakoids. The functioning of the electron transfer chain—a series of oxidation–reduction processes—transfers protons to the inner side of the thylakoids (lumen) from the outer side (stroma) leading to the formation of an electrochemical gradient of protons that will allow the synthesis of ATP. Image reproduced from OpenStax https://openstax.org/books/concepts-biology/pages/5-2-the-light-dependent-reactions-of-photosynthesis under a creative commons license

6.3 Photons, Protons, and Electrons

The photosynthetic electron transport system operates through a successive series of oxidation–reduction reactions involving a number of components, the most important of which are plastoquinone (PQ), cytochrome b6/f complex, plastocyanin, and ferredoxin (Fig. 6.3). The first three couple PSII to PSI, the latter couples PSI to NADP and to downstream reactions, such as those involving the enzymes nitrite reductase, glutamine synthase, and thioredoxin reductase, and CO_2 fixation. Each component successively transfers an electron to an acceptor, and then receives its next electron.

Absorption of a photon results in P_{680} being raised to an excited electronic state (Fig. 6.3). When P_{680} reaches this excited state, it can reduce (or transfer an electron to) an appropriate acceptor molecule, quinone QA, at which point the chl *a* returns to its original state. The acceptor QA then passes its electron to the next carrier, quinone QB. For the near-instantaneous moment this process is occurring, the reaction center is referred to as being "closed" as it is unable to donate another electron until it is reduced (receives an electron) from the water-splitting reaction (see Chap. 7 for how this phenomenon is applied in fluorescence measurements). When the transfer of

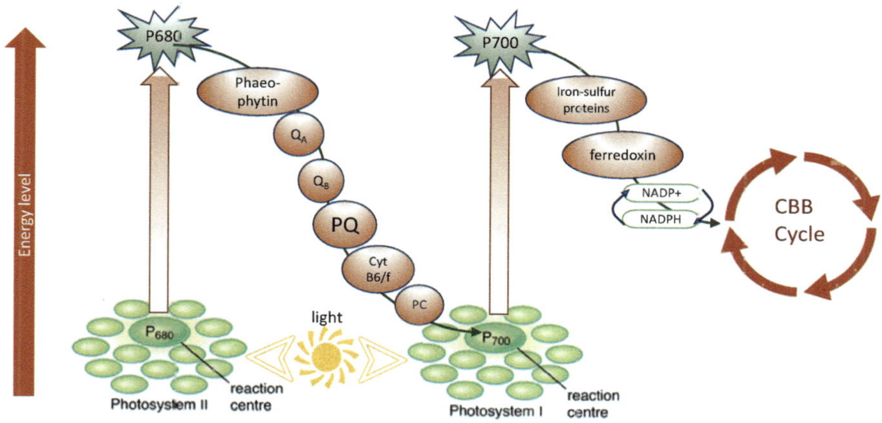

Fig. 6.3 Flow of electrons during the light reaction stage of photosynthesis: the Z-scheme. When light energy is received by reaction center P680 of PSII, it sequentially donates its electron to the quinone-cytochrome b_6/f chain, which results in proton pumping (not shown). The electron is then transferred to reaction center P700 of PSI, which in its excited state transfers the electron onto ferredoxin and NADP, forming NADPH, which is required to drive the Calvin–Benson–Basham (CBB) Cycle among other processes. See text for further description of the Z scheme

electrons eventually reduces $NADP^+$ to NADPH, an associated transmembrane difference in electrical potential and H^+ concentration gradient across the thylakoid membranes drives ATP synthesis via the enzyme ATP synthase (Figs. 6.2 and 6.3). The path of the flow of electrons along the electron transport chain is commonly called the Z-scheme of photosynthesis.

The PQ pool is considered a gatekeeper of electron carriers; it is a regulator of the acclimation state of the cells (Rochaix 2014). The PQ pool determines both short and long-term acclimation of the cell based on its redox state. It accepts 2 electrons and then takes 2 protons and then detaches from the reaction center and becomes mobile within the thylakoid membrane. There are about 5–10 PQ/electron chain. PQ also functions in more than one electron transport chain and therefore can provide alternate routes for electron transport when acceptor molecules are already reduced. The multifunctionality of PQ is further addressed below (Sect. 6.5.2).

The oxidized PSII reaction center is re-reduced by an *oxygen evolving complex* (OEC) attached to PSII. The OEC is the site of O_2 evolution from the oxidation of 2 water molecules. With one molecule of O_2, 4 protons are released to the thylakoid lumen. An additional 2 protons are released during redox transfer at cytochrome b_6/f. Associated with this proton release is the conversion of ADP to ATP, with 2 ATP molecules formed for every 2 electrons transferred or a molecule of NADP+ reduced. Thus:

$$H_2O + NADP \rightarrow \tfrac{1}{2}\, O_2 + NADPH_2 \quad (6.2)$$

$$2\ ADP + 2\ Pi \rightarrow 2\ ATP \tag{6.3}$$

Accordingly, only in PSII does the splitting of water occur and water is the ultimate source of electrons for photosynthesis (Fig. 6.2). The released plastoquinol molecule then reaches a cytochrome b6/f complex which it reduces to cytochrome b6. That, in turn, reduces cytochrome f, then plastocyanin, which releases its electron to the oxidized PSI reaction center chl *a*, molecule P_{700}.

An important aspect of electron transport is the coordinated production and translocation of H^+ to the thylakoid membrane. As the reduced PQ molecule is oxidized by the cytochrome b6/f complex, 2 hydrogen (H^+) atoms are released to the lumen. Also, as water is oxidized, and electrons removed, there is a transfer of H^+ to the intra-thylakoid space. As the light reactions of photosynthesis progress and electrons are transferred along the electron transport chain, the generation of these protons results in the thylakoid space becoming increasingly acidic, in some cases approaching a pH of 4. For each splitting of water (2 molecules), 4 protons are released, and the transport of electrons through the cytochrome b_6/f complex leads to 8 protons being moved into the thylakoid space (Croce and van Amerongren 2013). In PSI, ferredoxin functions analogously to PQ in PSII. It is the primary electron acceptor and ultimately reduces $NADP^+$ with the aid of ferredoxin-NADP reductase.

When energy is transferred to P_{680} and P_{700} chl *a* from the carotenoid pigments, such as fucoxanthin or peridinin, which primarily absorb in the 500–560 nm region, it is highly efficient. Similarly, efficient transfer to chl *a* occurs from phycoerythrin, phycocyanin, and allophycocyanin (Kirk 2011). However, even though efficient, such transfer is not 100% and some loss of energy occurs at a longer wavelength. This is the fluorescent signal. The topic of how this fluorescent signal can be used as a measure of photosynthesis is covered in the following chapter (Chap. 7).

If cells are in balanced growth or growing at steady-state (see Chap. 3), the process of light harvesting and the eventual fixation of CO_2 into organic molecules are fine-tuned and in balance with that of the assimilation of other required elements. However, under stress, whether that is light, temperature, or nutrient stress, these processes can become imbalanced and uncoupled. The cell may become over-excited or redox conditions may be altered. Kanazawa et al. (2020) refer to this as a condition of "metabolic congestion" or "indigestion." When this happens, the cell has various mechanisms to help rebalance and self-repair, but there are limits. The ability of cells to cope with stress is central to their ability to cope with new environmental conditions. Like a finely tuned supply chain, the flow of electrons from PSII to PSI and on to downstream metabolic reactions depends on the orientation and spatial configuration of all the components, but—as will be seen below—also like a supply chain, it functions at the rate of the slowest link in the chain and can be disrupted when conditions change. Cell metabolism lives and grows off the energy that reaches the last link (like the consumer in a supply chain). How cells cope with over-excitation is discussed in detail below. First, a review of the C assimilation processes.

6.4 The Calvin–Benson–Bassham Cycle: The Spinning Wheel of Metabolism, Inappropriately Termed the "Dark Reactions"

The NADPH and ATP produced in the light reactions are subsequently used to convert CO_2 to carbohydrates in the series of reactions commonly known as the "dark reactions" or "light-independent" reactions. The term "dark reactions" is, however, a misnomer, as darkness is not required; hence, "light-independent reactions" is a more appropriate term. This cycle was elucidated by Calvin, Benson, and Bassham in the 1950s, and thus it is also frequently termed the Calvin or the Calvin–Benson–Bassham (CBB) Cycle.

At the center of the discussion of these reactions is the enzyme Rubisco, ribulose bisphosphate carboxylase oxygenase (Fig. 6.4). Rubisco is considered the most abundant protein on Earth. Evolutionarily, there are various forms of Rubisco, and as will be seen below and discussed more thoroughly in Chap. 26, these different forms have different affinities for CO_2, therefore dictating the various C concentrating mechanisms (CCMs) of different groups of algae.

The first stage of the CCB cycle begins with carboxylation, via Rubisco's reaction of CO_2 and a 5-C molecule, ribulose biphosphate (RuBP). The 6-C product splits into 2, 3-C molecules, 3-phosphoglycerate (3-PG) (Fig. 6.5). Then, 3-PG is phosphorylated by ATP and reduced by NADPH to make glyceraldehyde-3-phosphate (G3P). In this phase, the $NADP^+$ and ADP are consequently formed, and they return to thylakoids to continue their role in the light reactions. Finally, a series of reactions convert G3P to various intermediates, regenerating RuBP, but requiring

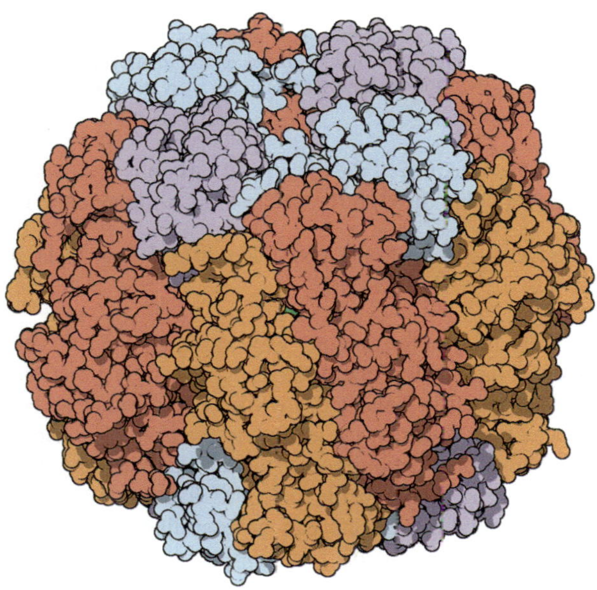

Fig. 6.4 The structure of Rubisco. Image reproduced from Goodsell D (2000) PDB-101.rcsb.org, doi: 10.2210/rcsb_pdb/mom_2000_11) under a creative commons license

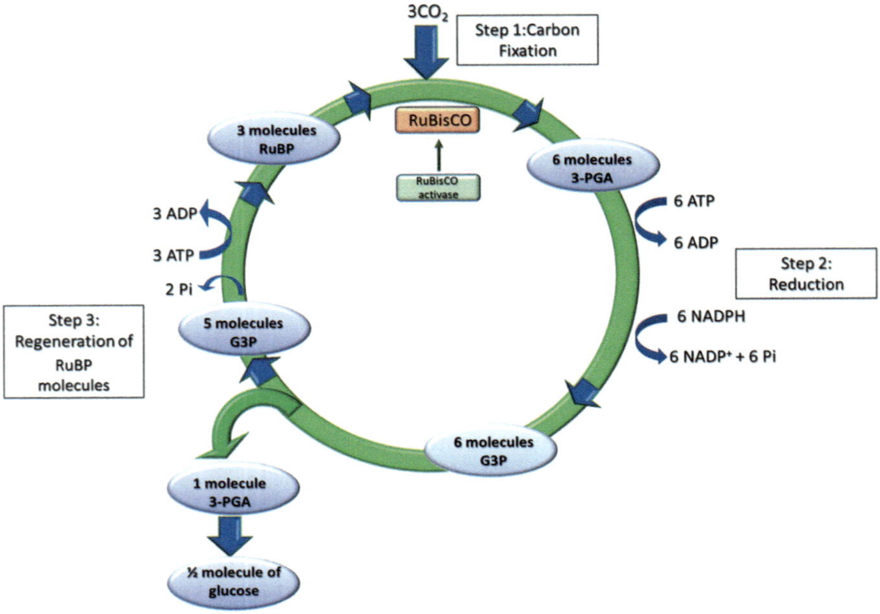

Fig. 6.5 Pathway of carbon dioxide fixation and reduction in photosynthesis, The Calvin–Benson–Bassham (CBB) Cycle reactions. The CBB Cycle has three stages. In stage 1, the enzyme RuBisCO incorporates 3 CO_2 molecules into the 5-carbon sugar ribulose-1,5-bisphosphate (RuBP) to form 6 molecules of 3-phosphoglycerate (3-PGA). In stage 2, 6 molecules of 3-PGA are converted into 6 molecules of Glyceraldehyde-3-P (G3P) by using 6 molecules of ATP and 6 molecules of NADPH as reducing power. In stage 3, RuBP is regenerated so that the cycle can continue. Stage 3 includes a complex series of reactions combining 3-, 4-, 5-, 6-, and 7-carbon sugar phosphates, which are not explicitly shown in the diagram. Image reproduced from Vecchi et al. (2020, Plants 9: 67, doi.org/10.3390/plants9010067) under a creative commons license

ATP. For one turn of the cycle, one carbon is removed to form starch, while the other 5 are recycled through the process. In summary, the reaction is,

$$6\ CO_2 + 18\ ATP + 12\ NADPH + 12H_2O \rightarrow C_6H_{12}O_6 + 18ADP + 18\ Pi + 12\ NADP + +6H^+ \quad (6.4)$$

Returning to the analogy used earlier—that of a supply chain, the light reactions and the light-independent reactions generally function in sync and are finely tuned, that is, a cell in balanced growth. Yet, various parts of the assembly process can proceed faster or slower, leading to the need for various storage processes or various mechanisms to accommodate metabolic pathways that may not be functionally in perfect synchrony. At the site of C fixation, when photosynthesis occurs at a rate faster than the products can be assimilated into new growth, starch is typically stored. This may occur, for example, as cells are shifted to different temperatures, as different enzymes have different temperature optima. In Chap. 9, the different starch storage products of the different classes of algae are introduced. In green algae, as is the case with higher plants, starch is stored in the form of starch grains in the stroma. For all other algae, the various forms of starch are stored outside the chloroplast. In cyanobacteria and prochlorophyta, the storage product is a-1-4-glucan, very similar to starch. In the cryptophytes, a starch like that of higher plants accumulates. The different starch types are a distinguishing feature of the different types of algae. Other mechanisms by which cells maintain the balance of production along the "supply chain" are part of the respiration processes of the cells.

6.5 Respiration

Most students who have taken elementary biology would be able to recite the notion that respiration and photosynthesis are complementary processes. The products of one reaction become substrates for the other. However, the metabolic interactions between these processes are far more complicated. Students of oceanography often think of the process of respiration as occurring at a constant rate regardless of the light provided. It is generally assumed to be light-independent. Thus, in measuring photosynthesis in aquatic samples (see Chap. 7), bottles in light and dark are typically incubated, and the dark values (considered to be representative of the respiration rate) are added to the rate determined in the light to yield a value of gross photosynthesis. It is now recognized that photosynthesis and respiration are not independent processes, but instead are intricately related to each other and to the processes of assimilation of N and other nutrients, through various intracellular metabolite pools and redox conditions (e.g., Nunes-Nesi et al. 2008, 2011; Araújo et al. 2014). There has been considerable interest in these relationships and especially how respiration changes under varying conditions and if these relationships can be manipulated (bioengineered) to provide higher yields of photosynthesis. As will be seen, not all respiration is light-independent.

6.5.1 Mitochondrial Respiration

A major difference between prokaryotic cyanobacteria and eukaryotic algae and plants is that spatial separation of photosynthetic and respiratory bioenergetic pathways in chloroplasts and mitochondria has evolved in eukaryotes, that is, these processes occur in separate organelles, while in cyanobacteria all bioenergetic and metabolic pathways exist in the same compartment. Cellular respiration is the breakdown of glucose ($C_6H_{12}O_6$) in the presence of O_2,

$$C_6H_{12}O_6 + 6O_2 \rightarrow 6CO_2 + 6H_2O + 36\ ATP \tag{6.4}$$

In mitochondria, its electron transport chain derives its NADH and $FADH_2$ from the citric acid cycle. There are four protein complexes in the chain (Fig. 6.6). NADH first donates electrons, returning it to NAD^+, and in the process, pumping protons from

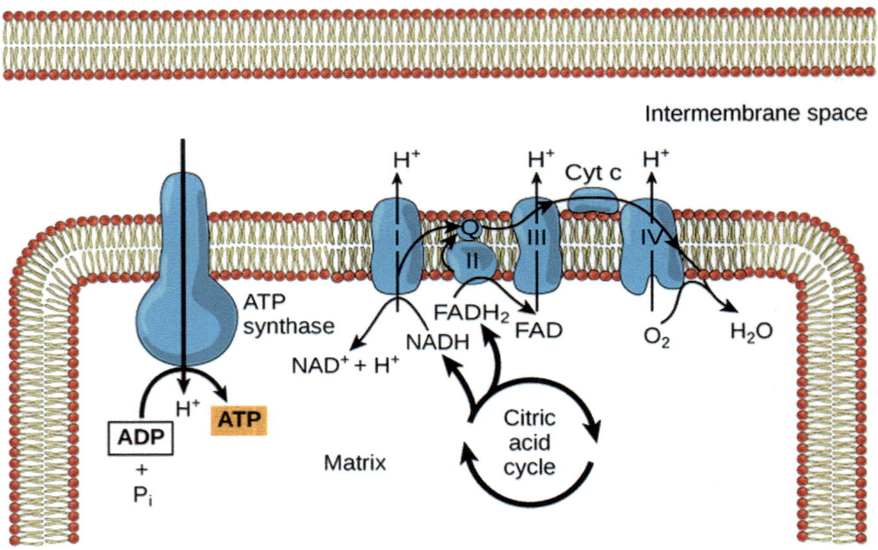

Fig. 6.6 Electron transport chain of respiration. There are four major protein complexes, each of which pumps H+ to the intermembrane space. Cytochrome c and the enzyme cytochrome c oxidase catalyze the final reaction in which electrons are passed to O_2, forming H_2O. The pH gradient formed by the electron transport chain is used by ATP synthase to form ATP. Image modified from OpenStax under a creative commons license. Figure available at: https://www.khanacademy.org/science/ap-biology/cellular-energetics/cellular-respiration-ap/a/oxidative-phosphorylation-etc

the matrix to the intermembrane space. As electrons move through the transport chain, more H^+ ions are pumped to the intermembrane space. Eventually, electrons are passed to O_2, forming H_2O. Cytochrome c and cytochrome c oxidase catalyze this final reaction of this electron transport chain. The proton gradient generates ATP. Thus, this pathway is also known as the *cytochrome oxidase pathway*.

The electron transport chain of mitochondria also has a secondary route to produce H_2O. This is the *alternate oxidase pathway*, or *alternate respiration pathway* (Fig. 6.7). In this pathway, the enzyme alternative oxidase (AOX) passes electrons to O_2 after protein complex II. Since this pathway bypasses the protein complexes III and IV, it has a lower ATP yield. The benefit of this pathway is that it can help to maintain an oxidized state of the upstream parts of the electron transport pathway, and thus is a "safety valve" reducing oxidative stress caused by over-reduced electron carriers (Maxwell et al. 1999).

6.5.2 Light-Dependent Respiration Processes

Redox regulation is a function of not only the flow of reductant through photosynthesis but also the demands for energy via metabolism. Compared to the high

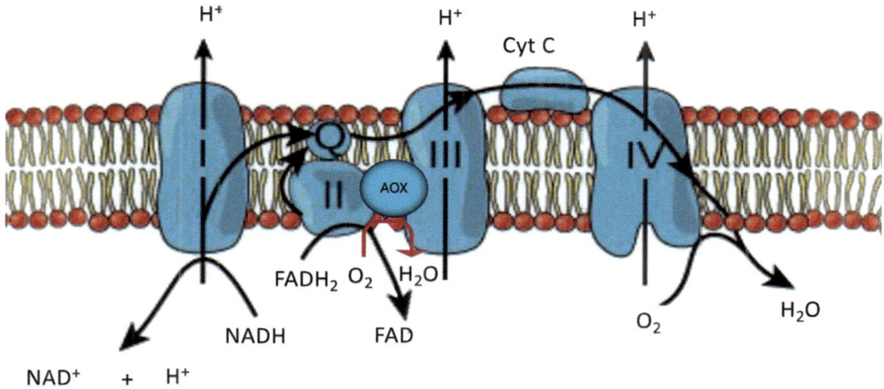

Fig. 6.7 The electron transport chain of respiration showing the alternate oxidase pathway. The AOX enzyme, bound to the inner membrane, bypasses the normal flow of electrons and passes electrons directly to O_2, forming H_2O (the red arrows in the diagram). It is expressed when other stresses may reduce the normal flow of electrons. Image modified from Open Stax reproduced under a creative commons license

potential fluxes of energy from the light reactions, the pool sizes of NADPH and those of ATP are comparatively small. The "energy pressure" can be thought of as a measure of the reductant state or availability for a cell. Cellular energy balance has been termed the "broker" of coordinated regulation between N and C interactions (Foyer et al. 2011), a delicate balance as the assimilation of N and that of C are linked in multiple biochemical pathways (e.g., Turpin and Harrison 1979; Turpin 1991; Coruzzi and Bush 2001; Wang et al. 2014). Thus, any imbalance in their production or consumption can cause a depletion, or alternatively a buildup, of metabolic intermediates (Kanazawa et al. 2020). As the ratio of ATP/NADPH is critical for the functioning of downstream reactions, any change in the production of one or the other has immediate metabolic consequences. Both ATP and NADPH must be supplied in exactly the right proportions to meet demand for the process to function seamlessly (Kramer and Evans 2011).

There are a number of processes that aid in the regulation of cellular redox and energy balance that often arise from imbalances in photochemistry and C assimilation. The first is non-photochemical quenching operating around the thylakoid pH gradient and xanthophyll cycle, as described in Chap. 5. Chlororespiration, cyclic electron flow around photosystem II, the Mehler reaction (the water-water cycle) and photorespiration are other important processes (e.g., Lavaud et al. 2003; Ruban et al. 2004; Scheibe 2004; Wilhelm et al. 2006; Fig. 6.8). These mechanisms help to dissipate excess electron energy that may result from light or other stresses, when the light reactions or electron transport chain of photosynthesis go into "overdrive" compared to the rate of assimilation of C (Kana et al. 1997; Chung et al. 2008). As all of these processes are O_2 consuming, it is of note that identifying and characterizing these processes was aided with the application of the isotope ^{18}O (see Chap. 7). However, despite the fact that O_2 consumption is common to these

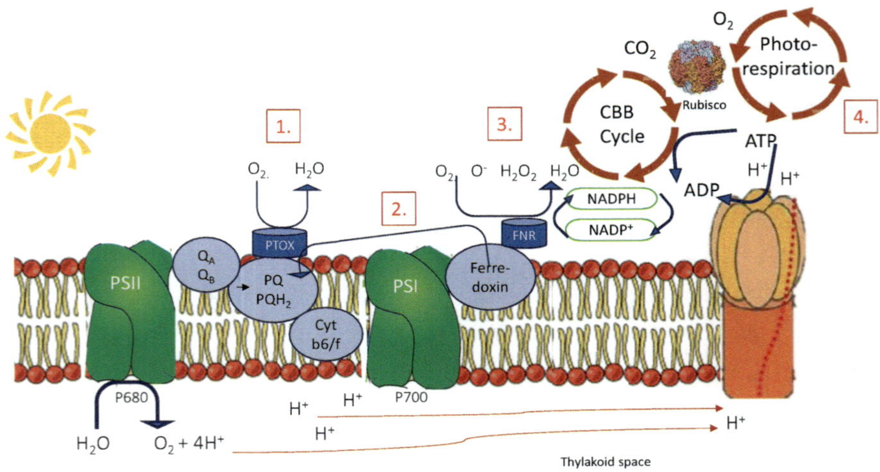

Fig. 6.8 Schematic representation of a photosynthetic unit with alternate pathways for O_2 uptake shown (identified in numbered red boxes). Pathway #1 is *chlororespiration*, the electron transport chain initiated at the site of PQ. Pathway #2 is *cyclic electron transport*, the process in which ferredoxin passes electrons back to PQ by way of several possible PQ reductases. Pathway #3 is the *Mehler reaction* (or water-water cycle) in which O_2 is converted to $O_2^{\bullet-}$ under stress at PSI. Then, this $O_2^{\bullet-}$ is dismutated to H_2O_2 and then to H_2O. Pathway #4 is *photorespiration*, the oxygenation reaction of Rubisco

'respiratory processes', there are different energetic consequences of the different pathways.

Chlororespiration is a respiratory process that involves the photosynthetic electron transport chain, analogous to that in mitochondrial respiration, but it is initiated at the site of PQ when the cell is under light or temperature stress (Fig. 6.8, identified as pathway #1). Indeed, and especially for cyanobacteria where there is no physical separation of the photosynthetic and respiratory pathways, the photosynthetic electron transport chain and the respiratory electron transport chain are intertwined in the chloroplast (Nixon and Rich 2007). The process was first described in the green alga *Chlamydomonas* (Bennoun 1982). Chlororespiration is thus a spillover process that helps to prevent the over-excitation of the PQ pool; it thus plays an important role in the photoprotection of PSII (Zolotareva and Polischuk 2021). Two enzymes are involved in this process, NADPH dehydrogenase complex and the nucleus-encoded plastoquinol oxidase (PTOX). There are several key features of this process. First, because it functions in a manner analogous to that of mitochondrial respiration, it is ATP producing and the product of this reaction is H_2O. Second, even though this process occurs within the chloroplast, it does not need light to be initiated. As light intensity increases, the rate of this process decreases as other pathways become more favorable (Peltier and Sarrey 1988; Lewitus and Kana 1995). Third, chlororespiration as a whole is an H^+ pumping process.

While not a respiratory process, *cyclic electron transport* is another energy balancing process. It is the process in which ferredoxin (PSI) passes electrons back to PQ (PSII) by way of several possible PQ reductases (Fig. 6.8, identified as pathway #2). During this process, additional protons are pumped to the intra-thylakoid space. With this, a pH gradient is established, and ATP synthesis is promoted and non-photochemical quenching (NPQ) is initiated. This process may help in maintaining the necessary balance of ATP:NADPH for the CBB Cycle. The specific pathways by which this occurs have been well studied (e.g., Kramer and Evans 2011) and how it varies among species (reviewed by Kanazawa et al. 2020). It is of note that there are multiple pathways of cyclic electron flow (Kramer and Evans 2011). Most importantly, this process has been observed to increase under conditions of high ATP demand, under environmental stressful conditions, and when C4 CCMs are active (Kohzuma et al. 2009; Finazzi et al. 2002; Kanazawa et al. 2020).

As ferredoxin is clearly iron rich, in the open ocean where Fe may be Fe-limited (Chap. 16) PSI content may be decreased (Allen et al. 2008; Schuback et al. 2015). Accordingly, an over-reduction of PSII can damage such cells, leading to increased susceptibility to photoinhibition (Barber and Andersson 1992). Cyclic electron transport may help to keep such cells from such a fate. Many algae can synthesize and substitute flavodoxin for this reaction. Flavodoxin synthesis may therefore indicate Fe deficiency.

These two processes, chlororespiration and cyclic electron transport, have recently been suggested to be complementary processes (Li et al. 2016). Whereas both processes function in photoprotection, cyclic electron transport appears to be the more important process in this function, but chlororespiration is more important at very low light. Both work together to alleviate photoinhibition during temperature stress (Raven et al. 2020).

Another light-dependent respiratory process is that of the *Mehler reaction*, or the water-water cycle (Fig. 6.8, identified as pathway #3). This is the process of reduction of O_2 at the site of ferredoxin in PSI (Raven and Beardall 1981). If photosynthetic production of ATP and NADPH is in balance with the requirements of metabolism, then the supply chain functions seamlessly. Under high light stress, however, when the electron transport chain is overexcited, some of the electrons are directed to O_2, causing $O_2^{\bullet-}$ to be generated at PSI. Then, this $O_2^{\bullet-}$ is dismutated to H_2O_2 either spontaneously or by the enzyme superoxide dismutase (SOD) in the thylakoid membrane. This is the Mehler reaction. Thus, the electron flow from H_2O at PSII to H_2O at PSI is called the water-water cycle (Asada 2006; Raven et al. 2020). Interestingly, this process generates an increased H^+ gradient across the thylakoid membrane; there is no net synthesis of NADPH. Accordingly, it may be ATP-generating (Asada 1999, 2000). Mehler activity has been confirmed across taxa. Kana (1992) showed that Mehler activity in *Synechococcus* was only important at irradiances that saturated C assimilation. Other studies have similarly demonstrated that this process is greatest in high energy (high light) environments.

The presence, or rate of activity of Mehler reactions, varies considerably with species and environmental conditions. Similarly, the ability of the cell to detoxify

ROS also varies by species and conditions. Excess production of ROS can have substantial damaging effects both on the cell itself and if it diffuses from the cell, can cause ecological damage to large swaths of the community. In the symbiotic dinoflagellate *Symbiodinium* spp., which exists in corals, Mehler activity is part of the initiation of the process of coral bleaching (Suggett et al. 2008).

Mehler activity plays an important role in the metabolism of the seemingly enigmatic cyanobacterium, *Trichodesmium*, which also fixes N_2. The enzyme involved in N_2 fixation, nitrogenase, is highly sensitive to O_2, and thus cyanobacteria that do fix N_2 have developed a number of strategies to protect this enzyme from the presence of O_2: temporal separation over the diel cycle, spatial separation of the enzyme within specialized cells (heterocysts), or restriction to microenvironments that may be depleted in O_2. *Trichodesmium* is enigmatic in that it fixes N_2 during the day when it simultaneously photosynthesizes (see also Chap. 11). Kana (1993) showed that *Trichodesmium* express a high rate of O_2 uptake in the light especially at midday. The rate of O_2 reduction could be >50% of the rate of photosynthetic O_2 evolution. Such high O_2 cycling has two benefits to *Trichodesmium*. One is that net O_2 evolution is reduced, preventing damage to nitrogenase, and another is that it is ATP generating in support of N_2 fixation. These findings were consistent with observed Mehler activity during photosynthetic induction and operation of a CCM in cyanobacteria (Radmer and Kok 1976; Miller et al. 1988).

Of the light-dependent respiratory processes, *photorespiration* is the most well-known and understood (Fig. 6.8, identified as pathway #4). Photorespiration occurs due to the dual substrate affinities of Rubisco. Rubisco reacts with both CO_2 and O_2 (Fig. 6.9); it has been characterized as "hamstrung by slow catalysis and confusion between CO_2 and O_2 as substrates, an 'abominably perplexing' puzzle" (Tcherkez

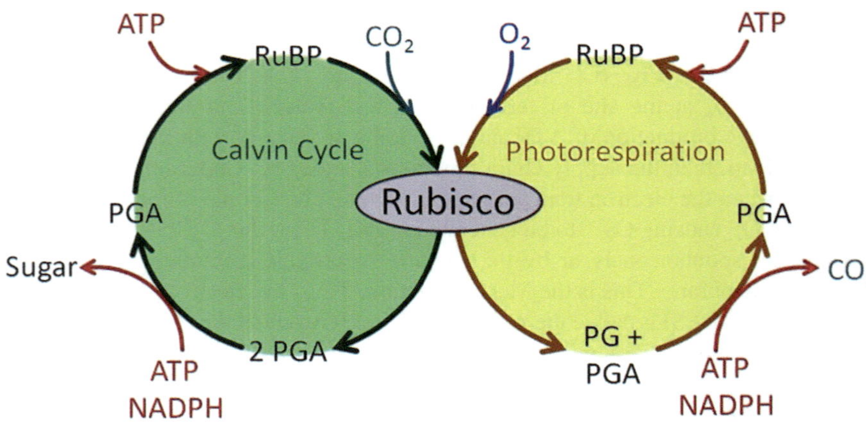

Fig. 6.9 The main reactions of Rubisco: photosynthesis and photorespiration. During photorespiration, O_2 is bound to RuBP and forms phosphoglycolate (PG) and phosphoglycerate (PGA), then a number of additional reactions, ultimately releasing CO_2. Image reproduced from OpenStax (https://openstax.org/books/biology-2e/pages/8-3-using-light-energy-to-make-organic-molecules) under a creative commons license

et al. 2006, p. 7246). The oxygenation reaction of Rubisco is termed photorespiration and it effectively reduces the amount of photosynthesis that can be produced. The rate of photorespiration is controlled by the ratio of CO_2 to O_2 at the active site of Rubisco.

Photorespiration has been considered as an unproductive, energy expensive, wasteful pathway. The affinity of Rubisco for O_2 is thought to be a hold-over from the evolution of Rubisco and the CBB Cycle, which occurred during a period when O_2 levels were still very low. Thus, at that time, having affinity for a substrate that was low in concentration would not have a major effect on C fixation. However, this dual affinity, while creating a challenge for the cell to obtain CO_2 in levels high enough to overcome the more favorable oxygenase reaction, may play an important role as an important redox-balancing path. Thus, photorespiration may be "an important pathway that makes the best of a bad situation caused by Rubisco's seemingly inevitable oxygenase activity" (Peterhensel et al. 2010, p. 10).

The photorespiratory pathway involves chloroplasts, mitochondria, and peroxisomes. Photorespiration shares many metabolic products with those of N assimilation using C skeletons synthesized in the tricarboxylic acid (TCA) cycle (Fig. 6.10; see also Glibert et al. 2016). It is linked to N metabolism via the release and re-incorporation of NH_4^+ (see Chap. 12 for a more detailed description of its role in N metabolism). Photorespiration provides no net gain in C or energy for the cell (i.e., no net growth) and it imposes other cellular costs in terms of the repair, quenching, and other functions affected by increased oxygenase activity (Raven

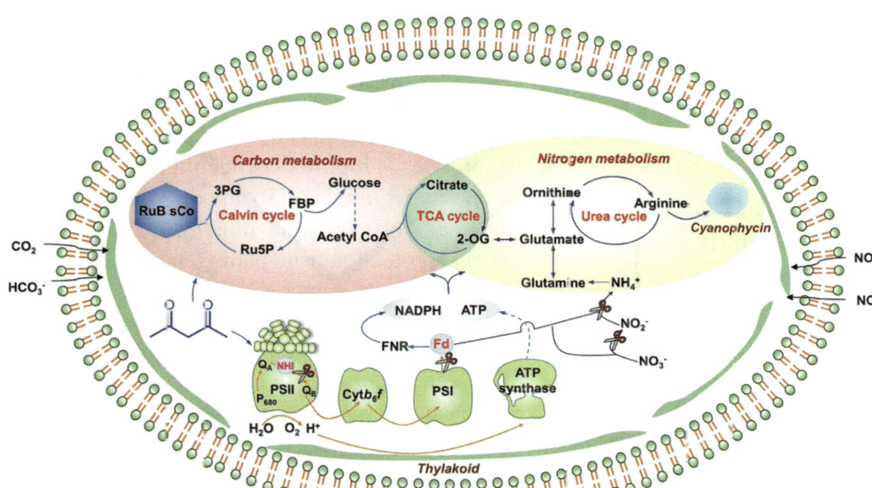

Fig. 6.10 Schematic diagram of the interaction between the assimilation of C in Calvin (Calvin–Benson–Bassham) cycle, the TCA cycle and the urea cycle with a *Microcystis* cell. Image reproduced from Yilimulati et al. (2022, Environ Sci Technol 56: 9683–9692. doi.org/10.1021/acs.estt.2c00776) with permission of the American Chemical Society

2011; Voss et al. 2013; Raven et al. 2014). Photorespiration increases under conditions of high light, high O_2, and high temperature—all factors that favor the oxygenation:carboxylation ratio of Rubisco (Kangasjärvi et al. 2012).

Due to the favorability of the oxygenase reaction, many cells have developed mechanisms to concentrate CO_2 at the site of Rubisco. These CCMs are necessary for algae because pCO_2 concentrations in natural waters, especially seawater, are too low to saturate Rubisco carboxylase by diffusion alone; they serve to increase the concentration of CO_2 at the site of Rubisco (Roberts et al. 2007a,b; Wu et al. 2014 and references therein; Chap. 26). Thus, CCMs increase the rate of carboxylation relative to oxygenation, and as a consequence, there should be less photorespiration under a given set of environmental conditions. Diatoms and most other algae are well documented to have CCMs (Raven and Beardall 2003; Armbrust et al. 2004; Granum et al. 2005; Wilhelm et al. 2006; Roberts et al. 2007a,b; Raven et al. 2008; Hopkinson et al. 2011). Basal, peridinin-containing dinoflagellates have an apparent disadvantage in the carboxylase pathway in that their Rubisco is the so-called Form II (Morse et al. 1995; Fig. 6.11). This form is novel among eukaryotic algae and was likely acquired from anaerobic proteobacteria via horizontal gene transfer. The kinetic properties of Form II Rubisco are less favorable for carboxylase activity

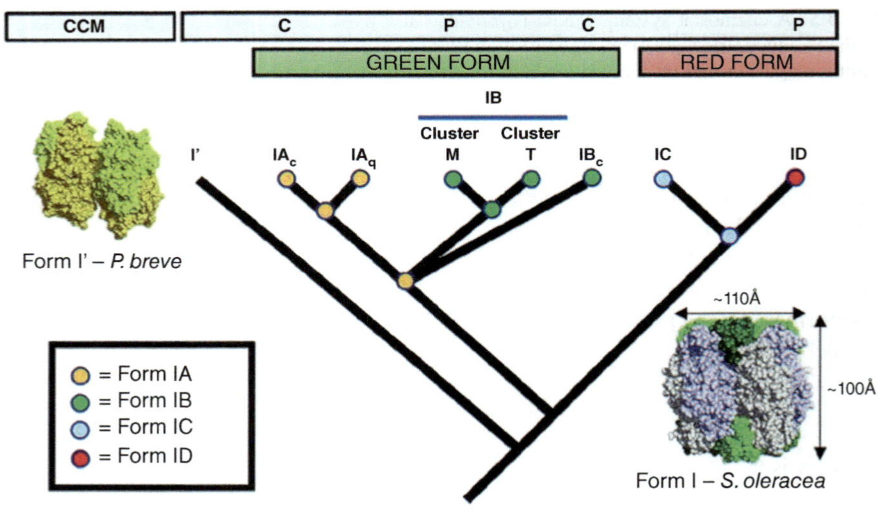

Fig. 6.11 Evolutionary origins and diversification of Form I Rubisco. Differently colored nodes represent the four broad types of Form I Rubisco, and their positions on the phylogram reflect currently hypothesized evolutionary trajectories. The biophysical carbon-concentrating mechanisms (CCMs) in which the various Form I Rubisco lineages take part are shown, with "C" and "P" signifying carboxysomal and pyrenoid-based mechanisms, respectively. Note, pyrenoid formation has been gained and lost several times in different organisms, so not all extant Form IB (M) and ID Rubisco molecules form pyrenoids. The representative crystal structures of Form I′ and I Rubisco are from *Promineofilum breve* [Protein Data Bank (PDB) code 6URA] and *Spinacia oleracea* (PDB: 1RCO), respectively. Image reproduced from Mao et al. (2022, J Exp Bot 74, doi.org/10.1093/jxb/erac309) under a creative commons license

than Forms IA and IB and Form ID Rubisco (Whitney and Andrews 1998; Tcherkez et al. 2006; Marin et al. 2007) and this low CO_2:O_2 selectivity of the Form II Rubisco should be more favorable for photorespiration at the direct expense of photosynthesis (see also Chap. 26). In fact, high rates of photorespiration have been measured or inferred at least in some dinoflagellate species (e.g., Burris 1977; Suggett et al. 2009).

When exposure to excessive light is prolonged or severe, these various overflow pathways are insufficient for full photoprotection. Too much light exceeds the capacity of the cells to self-regulate and balance the over-excitation pressure and damage to the PSII-O_2 reaction centers can occur. The PSII light-harvesting complex becomes "disorganized and damaged," particularly at the site of the D1 protein (Yamamoto 2016). This process, *photoinhibition*, is formally described as "the light-dependent, slowly reversible retardation of photosynthesis, independent of any developmental change" (Long et al. 1994; Neale 1987, quoted in Falkowski and Raven 2007, p. 246). Typically, photoinhibition takes time to develop—that is, cells can handle short-term exposures to excess light, but over time, damage develops. It also takes the cell time to repair the damage and synthesize a new copy of the D1 protein. With even higher irradiance levels, the repair mechanism of the D1 protein is prevented and the damage to PSII becomes irreparable. Photoinhibition is highly temperature sensitive, as the rate of damage and rate of repair are also dependent on the fluidity of the thylakoid membranes (Yamamoto 2016). Photoinhibition is the reason why primary production in surface waters may be less than rates measured just below the surface.

6.6 Photosynthesis as a Function of Irradiance

Photosynthesis can be described as a function of light according to a saturating hyperbolic relationship, one of the many "curves for all reasons" (Rao 2000 see Chap. 3). The curve has a light-limited region, a light-saturated region, and a light-inhibited region (Fig. 6.12). In general, the light-limited portion of the curve reflects processes and genes involved in light harvesting and electron transport, while the light-saturated portion of the curve reflects processes, including enzymes, involved in C fixation. The initial slope of the curve is denoted by the symbol α. The initial part of the curve is essentially linear, and it corresponds to the light-use efficiency by chloroplasts and photosynthetic efficiency (Φ). When normalized to biomass (e.g., Chl *a*), a superscript is typically added, α^B. As irradiance increases, the photosynthetic response begins to reach its maximum capacity (P_{max}), and beyond this point, an increase in PAR does not affect the rate of photosynthesis (light-saturation point). At saturation, or P_{max}, the rate of photon absorption exceeds the rate of electron transport to CO_2 (Falkowski and Raven 2007). The intersection of α and P_{max} (equal to P_{max}/α) is termed the *light-saturation point* and is denoted as I_k (or E_k). Thus, at irradiance values $<I_k$, the quantum yield of photosynthesis is high and the rate of electron transport capacity exceeds the rate at which photons are absorbed and delivered to PSII, but absolute photosynthetic rates are less than maximal. At values

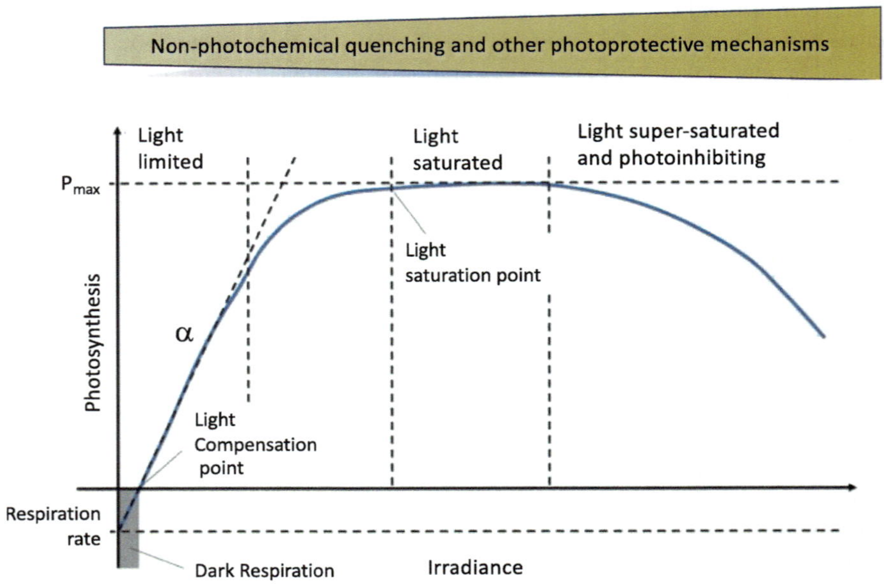

Fig. 6.12 Light response curve for photosynthesis. The light compensation point is the minimum light intensity at which the organism shows a gain of carbon fixation. The net photosynthetic rate shows a linear rise in response to increased light, in the range of light limitation. At higher light levels, saturation occurs as the efficiency of the photosynthetic mechanism is reduced due to the activation of energy quenching processes. Under excess light conditions, net photosynthesis can decline as a result of photooxidative stress. Image modified and reproduced from Benedetti et al. (2018, Microbial Cell Factories 17, doi.org/10.1186/s12934-018-1019-3) under a creative commons license

$>I_k$, the quantum yield decreases which is a necessary consequence of photosynthesis being limited by C assimilation (Falkowski and Raven 2007). As light intensities increase further, there can be a reduction in the photosynthetic rate, that is, photoinhibition may set in.

The light intensity at which photosynthesis balances respiration is termed the *compensation point* or *compensation light intensity*. The respiration rate considered here is that of dark or basal respiration.

There are several ways to parameterize the relationship between photosynthesis and irradiance. The first is analogous to Michaelis-Menten kinetics,

$$P = P_{max}[I]/(K_I + [I]) \tag{6.6}$$

where P = photosynthetic rate at a given light intensity, P_{max} is the maximum photosynthetic rate, I is the irradiance, and K_i is the half-saturation constant, the light intensity at which the photosynthetic rate is ½ maximal. This formula does not account for photoinhibition, and thus is rarely used in oceanography.

In aquatic sciences, the more common models are those of Jassby and Platt (1976) and Platt (1980). The former does not include a term for photoinhibition (β), but the latter does. These are commonly written as:

$$P = P_{\max} \tan\text{h}\ (I/I_k) \tag{6.7}$$

$$P = P_{\max}\left(1 - \mathrm{e}^{-\alpha I/Pmax}\right)\mathrm{e}^{-\beta\ I/Pmax} \tag{6.8}$$

in which α is the initial slope β_I = light intensity at the start of photoinhibition (all other terms as in Eq. 6.6). In oceanography, it is common to see *I* (which typically has units of energy, such as Watts m^{-2}) replaced with the symbol *E* (for Einsteins, with units of μmole photons m^{-2} sec^{-2}; see Chap. 4), and thus the literature may refer to these curves as *P* vs *I* or *P* vs *E* curves.

There have been extensive reviews of the different models for the photosynthesis-irradiance curve (Jassby and Platt 1976; Eileers and Peeters 1988; MacIntyre et al. 2002) and in general, they all give good estimates of P_{max}, but may vary in the estimate of α. The curvature of the initial slope and the maximal rate vary between different models (Jones et al. 2014). This may be due to different methods used (e.g., change in O_2 vs CO_2). It is therefore important to apply only one model in making comparisons between and among species or experiments; in general, the ^{14}C method tends to overestimate net photosynthesis at low irradiance levels (MacIntyre and Cullen 2005).

6.7 Respiration as a Function of Irradiance

The common concept that respiration is a constant and can be measured by measuring change in O_2 in the dark must now be reconsidered. The pathways described above make clear that there are multiple respiration pathways, and several of these pathways are light-dependent. Thus, respiration can also be characterized as a function of irradiance (Fig. 6.13). Using a combination of tracer methods and selective inhibitors, Lewitus and Kana (1995) were able to differentiate the various respiration pathways in cells growing strictly as autotrophs and those that were given an exogenous organic C as a respiratory substrate. They observed a many-fold increase in respiration rate as irradiance increased for all species studied. Their results allowed them to develop a conceptual model of phytoplankton *R* vs *I* responses. In the low light region, chlororespiration dominated, especially for those cells that provided exogenous glucose, and as light increased, Mehler activity increased, especially in cells that were energy (C) poor. The newer $^{18}O_2$ methods for measuring simultaneous photosynthesis and respiration make measurements of the relationship between photosynthesis and respiration as a function of irradiance much more accessible.

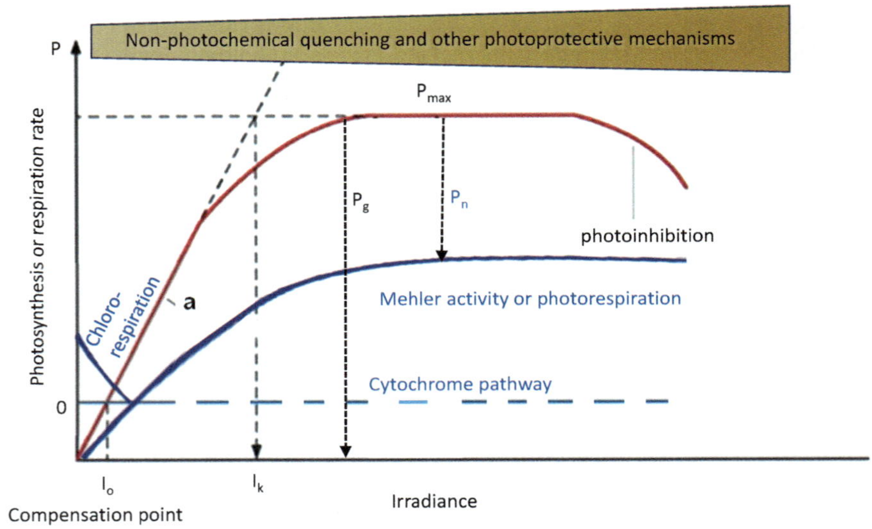

Fig. 6.13 The photosynthesis-irradiance curve (red line, as per Fig. 6.12) now with respiration vs irradiance curve superimposed (blue line) with different components of light respiration indicated. The "dark" respiration (cytochrome pathway) remains unchanged. Chlororespiration is shown to decrease as light increases. As more and more light is imposed, Mehler activity, alternative pathway respiration, and photorespiration increase. Note the difference in net photosynthesis rates if light-dependent respiration rates are high compared to the calculation of net photosynthesis rates without considering these fluxes (Fig. 6.12). The amount of light-dependent respiration depends on species and the amount of stress they may experience

6.8 Summary

Photosynthesis is a balancing act. Cells need to acquire light to drive the energy transformation through the electron transport chain, but the process is delicate. Too much light and the "supply chain" can be disrupted. The cell has multiple mechanisms to cope with stresses associated with excess light, or fluctuations in temperature. Various respiration pathways are involved in protecting cells from over-excitation, and thus respiration is both a process occurring in the dark and varies with light availability. Plastoquinone is a slow step in the transfer of energy from PSII to PSI, and it plays an outsized role in regulating energy use or energy dissipation. The processes are far more complex than conveyed in this chapter, but the message is clear: the photoprotective mechanisms and energy balancing mechanisms are interconnected and these various pathways give the cells ability to thrive under varying environmental conditions. The next chapter explores how different methods for measuring photosynthesis target different regions of the photosynthetic process, and the following chapter looks at how cells adapt their photosynthetic machinery to fluctuating environments.

References

Allen AE, LaRoche J, Maheswari U, Lommer M, Shauer N, Lopez PJ, Finazzi G et al (2008) Whole-cell response of the pennate diatom *Phaeodactylum tricornutum* in iron starvation. Proc Natl Acad Sci USA105:10438–10443

Araújo WL, Nunes-Nesi A, Fernie AR (2014) On the role of plant mitochondrial metabolism and its impact on photosynthesis in both optimal and sub-optimal growth conditions. Photosynth Res 119:141–156. https://doi.org/10.1007/s11120-013-9807-4

Armbrust EV, Berges JA, Bowler C, Green BR, Martinez D, Putnam NH, Zhou S et al (2004) The genome of the diatom *Thalassiosira pseudonana*: ecology, evolution, and metabolism. Science 306:79–86. https://doi.org/10.1126/science.1101156

Asada K (1999) The water-water cycle in chloroplasts: scavenging of active oxygens and dissipation of excess photons. Annu Rev Plant Physiol Plant Mol Biol 50:601–639

Asada K (2000) The water-water cycle as alternate photon and electron sinks. Philos Trans R Soc B 355:1419–1431

Asada K (2006) Production and scavenging of reactive oxygen species in chloroplasts and their functions. Plant Physiol 141:391–396

Barber J, Andersson B (1992) Too much of a good thing: light can be a bad thing for photosynthesis. Trends Biochem Sci 17:61–66

Benedetti M, Vecchi V, Barera S, Dall'Osto L (2018) Biomass from microalgae: the potential of domestication towards sustainable biofactories. Microb Cell Factories 17:1–18. https://doi.org/10.1186/s12934-018-1019-3

Bennoun P (1982) Evidence for a respiratory chain in the chloroplast. Proc Natl Acad Sci USA 79: 4352–4356

Blackman FF (1905) Optima and limiting factors. Ann Bot 19:281–295

Burris JE (1977) Photosynthesis, photorespiration, and dark respiration in eight species of algae. Mar Biol 39:371–379

Chung C-C, Hwang S-PL, Chang J (2008) Nitric oxide as a signaling factor to upregulate the death-specific protein in a marine diatom, *Skeletonema costatum*, during blockage of electron flow in photosynthesis. Appl Environ Microbiol 74:6521–6527. https://doi.org/10.1128/AEM.01481-08

Coruzzi G, Bush DR (2001) Nitrogen and carbon nutrient and metabolite signaling in plants. Plant Physiol 125:61–64

Croce R, van Amerongen H (2013) Light-harvesting in photosystem I. Photosynth Res 116:153–166

Eileers PHC, Peeters JCH (1988) A model for the relationship between light intensity and the rate of photosynthesis in phytoplankton. Ecol Monogr 42:199–215

Emerson R, Arnold W (1932a) A separation of the reactions in photosynthesis by means of intermittent light. J Gen Physiol 15:391–420

Emerson R, Arnold W (1932b) The photochemical reaction in photosynthesis. J Gen Physiol 16: 191–205

Falkowski PG, Raven JA (2007) Aquatic photosynthesis. Princeton University Press, Princeton NJ, p 484

Finazzi G, Rappaport F, Furia A, Fleishmann M, Rochaix JD, Zito F, Forti G (2002) Involvement of state transitions in the switch between linear and cyclic electron flow in *Chlamydomonas reinhardii*. EMBO Rep 3:280–285

Foyer CH, Noctor G, Hodges M (2011) Respiration and nitrogen assimilation: targeting mitochondria-associated metabolism as a means to enhance nitrogen use efficiency. J Exp Bot 62:1467–1482

Glibert PM, Wilkerson FP, Dugdale RC, Raven JA, Dupont C, Leavitt PR, Parker AE et al (2016) Pluses and minuses of ammonium and nitrate uptake and assimilation by phytoplankton and implications for productivity and community composition, with emphasis on nitrogen-enriched conditions. Limnol Oceanogr 61:165–197. https://doi.org/10.1002/lno.10203

Goodsell D (2000) Molecule of the month: rubisco. PDB-101.rcsb.org. https://doi.org/10.2210/rcsb_pdb/mom_2000_11

Granum E, Raven JA, Leegood RC (2005) How do marine diatoms fix 10 billion tonnes of inorganic carbon per year? Can J Bot 83:898–908

Hill R (1937) Oxygen evolution by isolated chloroplasts. Nature 139:881–882

Hopkinson BM, Dupont CL, Allen AE, Morel FMM (2011) Efficiency of the CO_2-concentrating mechanism of diatoms. Proc Natl Acad Sci USA 108:3830–3837

Jassby AD, Platt T (1976) Mathematical formulation of the relationship between photosynthesis and light for phytoplankton. Limnol Oceanogr 21:540–547

Jones C, Craig S, Ciochetto A, MacIntyre HL, Cullen JJ (2014) Curvature in models of the photosynthesis-irradiance response. J Phycol 50. https://doi.org/10.1111/jpy.12164

Kana TM (1992) Relations between photosynthetic oxygen cycling and carbon assimilation in *Synechococcus* 7805 (Cyanophyta). J Phycol 28:304–308

Kana TM (1993) Rapid oxygen cycling in *Trichodesmium theibautii*. Limnol Oceanogr 38:18–24

Kana TM, Geider RJ, Critchley C (1997) Regulation of photosynthetic pigments in micro-algae by multiple environmental factors: a dynamic balance hypothesis. New Phytol 137:629–638. https://doi.org/10.1046/j.1469-8137.00857.x

Kanazawa A, Neofotis P, Davis GA, Fisher N, Kramer DM (2020) Diversity in photoprotection and energy balancing in terrestrial and aquatic phototrophs. In: Larkum AWD, Grossman AR, Raven JA (eds) Photosynthesis in algae: biochemical and physiological mechanisms. Springer Publ, Cham, Switzerland, pp 299–327

Kangasjärvi S, Neukermans J, Li S, Aro EM, Noctor G (2012) Photosynthesis, photorespiration, and light signalling in defense responses. J Exp Bot 63:1619–1636

Kirk JTO (2011) Light and photosynthesis in aquatic ecosystems, 3rd edn. Cambridge University Press, Oxford

Kohzuma K, Cruz JA, Akashi K, Hoshiyasu S, Munekage Y, Yokota A, Kramer DM (2009) The long-term responses of the photosynthetic proton circuit to drought. Plant Cell Environ 32:209–219

Kramer DM, Evans JR (2011) The importance of energy balance in improving photosynthetic productivity. Plant Physiol 155:70–78

Larkum AWD, Grossman AR, Raven JA (eds) (2020) Photosynthesis in algae: biochemical and physiological mechanisms. Springer Nature, Cham, 514 pp

Lavaud J, Bousseau B, Etienne AL (2003) Enrichment of the light-harvesting complex in diadinoxanthin and implications for the non-photochemical fluorescence quenching in diatoms. Biochemist 442:5802–5808

Lewitus AJ, Kana TM (1995) Light respiration in six estuarine phytoplankton species: contrasts under photoautotrophic and mixotrophic growth conditions. J Phycol 31:754–761

Li Q, Yao Z-J, Mi H (2016) Alleviation of photoinhibition by co-ordination of chlororespiration and cycling flow mediated by NDH under heat stressed conditions in tobacco. Front Plant Sci 7:285. https://doi.org/10.3389/fpls.2016.00285

Long SP, Humphries S, Falkowski PG (1994) Photoinhibition of photosynthesis in nature. Annu Rev Plant Physiol 45:633–662. https://doi.org/10.1146/Annurev.Pp.45.060194.003221

MacIntyre HL, Cullen JJ (2005) Using cultures to investigate the physiological ecology of microalgae. In: Anderson RA (ed) Algal culturing techniques. Academic Press, Amsterdam, pp 287–326

MacIntyre HL, Kana TM, Anning T, Geider RJ (2002) Photoacclimation of photosynthesis irradiance response curves and photosynthetic pigments in microalgae and cyanobacteria. J Phycol 38:17–38

Mao Y, Diaz-Ramos A, Greiff GRL, Azinas S, Gunn L, McCormick AJ (2022) The small subunit of Rubisco and its potential as an engineering target. J Exp Bot 74:543–561. https://doi.org/10.1093/jxb/erac309

Marin B, Nowack EC, Glöckner G, Melkonian M (2007) The ancestor of the *Paulinella* chromatophore obtained a carboxysomal operon by horizontal gene transfer from a *Nitrococcus*-like gamma-proteobacterium. BMS Evol Biol 7:85

Maxwell DP, Wang Y, McIntosh L (1999) The alternative oxidase lowers mitochondrial reactive oxygen production in plant cells. Proc Natl Acad Sci USA 96:8271–8276. https://doi.org/10.1073/pnas.96.14.8271

Miller AG, Espie GS, Canvin DT (1988) Active transport of inorganic carbon increases the rate of O_2 photoreduction by the cyanobacterium *Synechococcus* UTEX 625. Plant Physiol 88:6–9

Morse D, Salois P, Markovic P, Hastings JW (1995) A nuclear-encoded form II RuBisCO in dinoflagellates. Science 268:1622–16244

Neale PJ (1987) Algal photoinhibition and photosynthesis in the aquatic environment. In: Kyle DJ, Osmond CB, Arntzen CJ (eds) Photoinhibition. Elsevier, New York, pp 39–65

Nixon PJ, Rich P (2007) Chlororespiratory pathways and their physiological significance. In: Wise RR, Hoober JK (eds) The structure and function of plastids. Springer, Dordrecht, The Netherlands, pp 237–251

Nunes-Nesi A, Sulpice R, Gibon Y, Fernie AR (2008) The enigmatic contribution of mitochondrial function in photosynthesis. J Exp Bot 59:1675–1684. https://doi.org/10.1093/jxb/ern002

Nunes-Nesi A, Araújo WL, Fernie AR (2011) Targeting mitochondrial metabolism and machinery as a means to enhance photosynthesis. Plant Physiol 155:101–107. https://doi.org/10.1104/pp.110.163816

Peltier G, Sarrey F (1988) The Kok effect and light-inhibition of chlororespiration in Chlamydomonas reinhardtii. FEBS Lett 228:259–262. https://doi.org/10.1016/0014-5793(88)80011-5

Peterhensel C, Horst I, Niesssen M, Blume C, Kebeish R, Kürkcüoglu S, Kreuzaler F (2010) Photorespiration. The Arabidopsis Book/American Society of Plant Biologists 8:e0130. https://doi.org/10.1199/tab.0130

Platt T, Gallegos CL, Harrison WG (1980) Photoinhibition of photosynthesis in natural assemblages of marine phytoplankton. J Mar Res 38:687–701

Radmer RJ, Kok B (1976) Photoreduction of O_2 primes and replaces CO_2 assimilation. Plant Physiol 58:336–340. https://doi.org/10.1104/pp.58.3.336

Rao TR (2000) A curve for all reasons: the rectangular hyperbola in biology. Resonance 5:85–90

Raven JA (2011) The cost of photorespiration. Physiol Plant 142:87–104

Raven JA, Beardall J (1981) Respiration and photorespiration. In Platt T (ed) Physiological bases of phytoplankton ecology. Can J Fish Aquat Sci 210:55–82

Raven JA, Beardall J (2003) Carbon acquisition mechanisms of algae: carbon dioxide diffusion and carbon dioxide concentrating mechanisms. In: AWD L, Douglas SE, Raven JA (eds) Photosynthesis in algae, Advances in photosynthesis and respiration Vol. 14. Kluwer Academic Pub, Dordrecht, The Netherlands, pp 225–244

Raven JA, Cockell CS, De La Rocha CL (2008) The evolution of inorganic carbon concentrating mechanisms in photosynthesis. Philos Trans Roy Soc B 363:2641–2650

Raven JA, Beardall J, Giordano M (2014) Energy costs of carbon dioxide concentrating mechanisms in aquatic organisms. Photosynth Res 121:111–124

Raven JA, Beardall J, Quigg A (2020) Light-driven oxygen consumption in the water-water cycles and photorespiration, and light stimulated mitochondrial respiration. In: Larkum AWD, Grossman AR, Raven JA (eds) Photosynthesis in algae: biochemical and physiological mechanisms. Springer Publ, Cham, Switzerland, pp 162–178

Roberts K, Granum E, Leegood RC, Raven JA (2007a) C_3 and C_4 pathways of photosynthetic carbon assimilation in marine diatoms are under genetic, not environmental, control. Plant Physiol 145:230–245

Roberts K, Granum E, Leegood RC, Raven JA (2007b) Carbon metabolism in diatoms. Photosynth Res 93:79–88. https://doi.org/10.1007/s11120-007-9172-2

Rochaix J-D (2014) Regulation and dynamics of the light-harvesting system. Annu Rev Plant Biol 65:287–309. https://doi.org/10.1146/annurev-arplant-050213-040226

Ruban AV, Lavaud J, Rousseau B, Gugleilmi G, Horton P, Etienne A-L (2004) The super-excess energy dissipation in diatom algae: comparative analysis with higher plants. Photosynth Res 82: 165–175

Scheibe R (2004) Malate valves to balance cellular energy supply. Physiol Plant 120:21–26

Schuback N, Schellenberg C, Duckham C, Maldonado MT, Tortell PD (2015) Interacting effects of light and iron availability on the coupling of photosynthetic electron transport and CO_2-assimilation of marine phytoplankton. PLoS One 10:e0133235

Stirbet A, Lazár D, Guo Y, Govindjee G (2020) Photosynthesis: basics, history and modelling. Ann Bot 126:511–537. https://doi.org/10.1093/aob/mcz171

Suggett DJ, Warner ME, Smith DJ, Davey P, Hennige S, Baker NR (2008) Photosynthesis and production of hydrogen peroxide by *Symbiodinium* (Pyrrhophyta) phylotypes with different thermal tolerances. J Phycol 44:948–956

Suggett DJ, MacIntyre HL, Kana TM, Geider RJ (2009) Comparing electron transport with gas exchange: parameterising exchange rates between alternative photosynthetic currencies for eukaryotic phytoplankton. Aquat Microb Ecol 56:147–162. https://doi.org/10.3354/ame01303

Suggett DJ, Prášil O, Borowitzka MA (eds) (2010) Chlorophyll *a* fluorescence in aquatic sciences: methods and applications. Springer Publ, Dordrecht

Tcherkez GGB, Farquhar GD, Andrews TJ (2006) Despite slow catalysis and confused substrate specificity, all ribulose bisphosphate carboxylases may be nearly perfectly optimized. Proc Natl Acad Sci USA 103:7246–7251

Turpin DH (1991) Effects of inorganic N availability on algal photosynthesis and carbon metabolism. J Phycol 27:14–20

Turpin DH, Harrison PJ (1979) Limiting nutrient patchiness and its role in phytoplankton ecology. J Exp Mar Biol Ecol 39:151–166

Vecchi V, Barera S, Bassi R, Dall'Osto L (2020) Potential and challenges of improving photosynthesis in algae. Plan Theory 9:67. https://doi.org/10.3390/plants9010067

Voss I, Sunil B, Scheibe R, Raghavendra AS (2013) Emerging concept for the role of photorespiration as an important part of abiotic stress response. Plant Biol 15:713–722

Wang M, Shen Q, Xu G, Guo SW (2014) New insights into the strategy for nitrogen metabolism in plant cells. Int Rev Cell Mol Biol 310:1–37

Whitney SM, Andrews TJ (1998) The CO_2/O_2 specificity of single subunit ribulose-bisphosphate carboxylase from the dinoflagellate *Amphidinium carterae*. Aust J Plant Physiol 25:131–138

Wilhelm C, Büchel J, Fisahn R, Goss R, Jakob T, LaRoche J, Lavaud J et al (2006) The regulation of carbon and nutrient assimilation in diatoms is significantly different from green algae. Protist 157:91–124

Wu Y, Jeans J, Suggett DJ, Finkel ZV, Campbell DA (2014) Large centric diatoms allocate more cellular nitrogen to photosynthesis to counter slower RUBISCO turnover rates. Front Mar Sci 1: 68. https://doi.org/10.3389/fmars.2014.00068

Yamamoto Y (2016) Quality control of photosystem II: the mechanisms for avoidance and tolerance of light and heat stresses are closely linked to membrane fluidity of the thylakoids. Front Plant Sci 7:1136. https://doi.org/10.3389/fpls.2016.01136

Yilimulati M, Zhou L, Shevela D, Zhang S (2022) Acetylacetone interferes with carbon and nitrogen metabolism of *Microcystis aeruginosa* by cutting off the electron flow to ferredoxin. Environ Sci Technol 56:9683–9692. https://doi.org/10.1021/acs.estt.2c00776

Zolotareva EK, Polischuk OV (2021) Chlororespiration as a protective stress-inducible electron transport pathway in chloroplasts. Open Agricul J 16. https://doi.org/10.2174/18743315-v16-e2208151

Additional Websites Accessed

https://openstax.org/books/biology-2e/pages/8-3-using-light-energy-to-make-organic-molecules. Accessed 15 Aug 2023

https://openstax.org/books/concepts-biology/pages/5-2-the-light-dependent-reactions-of-photosynthesis. Accessed 15 Aug 2023

https://www.khanacademy.org/science/ap-biology/cellular-energetics/cellular-respiration-ap/a/oxidative-phosphorylation-etc. Accessed 15 Aug 2023

7 Photosynthesis and the Confusing Array of Methodologies

7.1 Introduction

New methods can change the course of science. With advancing tools have come exciting discoveries (Chap. 2). At the same time, some methods—that may seem appropriate at the time—can lead to false conclusions. The debates that have been introduced in the previous chapters—for example, between Victor Hensen and Ernst Haeckel and between Gordon Riley and Einer Steeman Nielsen—were not just debates about how to interpret data, they arose from fundamental differences in how data were acquired. More details on those legendary, heated, and often bitter debates are given in Chaps. 1 and 2.

The question of how to measure primary productivity is fundamental. Riley measured changes in oxygen (O_2), using long incubation times, as that was required to get a sufficiently measurable signal. Steeman Nielsen introduced the ^{14}C (carbon-14) technique, a far more sensitive method and one that was relatively easy to apply. Yet both have specific methodological issues. John Marra (2019, pp. 177, 179, 180, 181, 196, 197) picks up the narrative,

> The simplicity of the [carbon-14] method hides significant problems . . .First, because the method is so easy, a large number of measurements can be made. Second, the method has unmatched sensitivity for even very low rates of photosynthesis in the ocean's "deserts." Third, the method measures appearance of carbon-14 in organic matter. Only a positive outcome can occur; there is no such thing as a negative incorporation of carbon-14. Together, these attributes mean that one can collect lots of data with lots of measurements, be consoled by universally positive results, and have those results stand by themselves. The extreme sensitivity of the method and the small rates that exist in most parts of the ocean mean that results from the carbon-14 method cannot be compared or validated against other kinds of estimates of fertility or productivity. Herein lies a trifecta of a problem in understanding ocean fertility.Biological oceanography went through an unhappy time in the 1950s. . . .and foreshadowed decades more of controversy, uncertainty, and misunderstanding. The 1960s saw many, many observations using carbon-14 in lakes, estuaries, and the ocean, and even in large plastic spheres submerged below the water's surface. [By the] early 1980s, [t]he carbon-14 method had become everyone's favorite "whipping boy.". . .The

P. M. Glibert, *Phytoplankton Whispering: An Introduction to the Physiology and Ecology of Microalgae*, https://doi.org/10.1007/978-3-031-53897-1_7

conflicts came from many quarters and mostly had to do with squaring carbon-14 primary productivity results with activities of the other parts of the planktonic realm. . .

Steeman Nielsen's numbers for the organic production in the ocean, at least in certain areas of the North Atlantic, have not withstood the test of time.

There is a whole other dimension to the problem regarding the fertility of the ocean. Productivity occurs at different levels and with different fates. . .Permutations and processes that might go unmeasured complicate the whole scheme, and the methods themselves—notably, oxygen changes and the carbon-14 method—actually measure different things.

The approach to measuring primary productivity has become even more complex with recently developed fluorescent methods. This chapter explores the techniques for measuring photosynthesis and why different methods may yield very different results, what each one measures, and their relationships to growth.

7.2 Classic Approaches

The classic measurements of photosynthesis involve assessing either the O_2 generated or the CO_2 assimilated. The overall equation for photosynthesis (Eq. 6.1) would suggest that both measurements should give the same answer; that is, one molecule of O_2 should be consumed for every molecule of CO_2 fixed. However, as also apparent from previous chapters, and the imperfect nature of the "supply chain" of the electron transport process (especially under variable conditions), the pathway from the "light" reactions to carbon (C) assimilation is variable. The overall stoichiometry, or photosynthetic quotient, of $O_2 : CO_2$ can range from 0.5 to 3.5 (Williams and Robertson 1991). Of course, there is an analytical error associated with measuring any changes in O_2 or CO_2 as well. The production of O_2 is typically measured chemically (Winkler titration) or via an O_2 electrode, whereas CO_2 assimilation is typically measured using the radioisotope ^{14}C, or more recently, the stable isotope ^{13}C.

The ^{13}C method, analogous to that of the radiotracer ^{14}C, is relatively straightforward, as $H^{13}CO_3$ is added as a tracer and its incorporation into particulate material (and sometimes dissolved organic material) over time is measured (Hama et al. 1993; see also Glibert et al. 2019). It yields estimates similar to those determined by ^{14}C (e.g., Cullen 2001; Pimenov et al. 2008). There are basic differences, however. With ^{14}C, an absolute amount of isotope incorporation is measured, whereas with ^{13}C, a ratio is determined (the ratio of enriched C, ^{13}C, to normal C, ^{12}C). Thus, if C sources other than the $H^{13}CO_3$ label are incorporated (e.g., organic C substrates), there can be biases leading to underestimates of productivity, but these can be corrected if the final particulate C concentrations are considered (Collos and Slawyk 1985). The ^{13}C method can be applied in situ without the concerns associated with use of radioactive isotopes and can, in combination with other methods, be used to quantify group-specific primary production via measurement of ^{13}C incorporation into biomarkers such as alkenones (Popp et al. 2006) or phospholipid-derived fatty acids (Dijkman et al. 2009; de Kluijver et al. 2010, 2013). The application of ^{13}C

tracers coupled with metabolomics studies as a means to understand functional pathways is rapidly advancing (You et al. 2014 and references therein).

7.3 Advancing Isotopic Approaches

Isotope measurements using O_2 isotopes are also advancing (Glibert et al. 2019). The stable isotope of O_2, ^{18}O, can be used as a tracer as a gas or as water. The latter is favored, as it is easier to handle a tracer as a liquid than as a gas. The measurement of productivity with ^{18}O begins with the enrichment of sample water with $H_2{}^{18}O$ (Bender et al. 1987, 1999, 2000) and the enrichment of $^{18}O_2$ over time is determined, reflecting the product of the splitting of water during photosynthesis. As classically used in oceanography, this method is much more labor intensive and technically difficult than that of ^{13}C incorporation but is thought to be a better measure of true gross primary production (PSII activity). There are uncertainties, however. Key questions are the associated cycling of O_2 in the cells via both dark respiration and light-dependent respiration (Kana 1990, 1992; Robinson et al. 2009; Regaudie-de-Gioux et al. 2014; see Chap. 6). In addition to oceanographic applications, the ^{18}O method has also been used to estimate gross primary production, community metabolism, and net metabolism in a productive agricultural stream in the Midwestern US (Tobias et al. 2007).

An alternative ^{18}O method for measuring productivity using $H_2{}^{18}O$ was introduced using a modified membrane inlet mass spectrometer (MIMS) dissolved gas analyzer (Kana et al. 1994). This method overcomes the tedious extraction process used in traditional ^{18}O mass spectrometry, giving a much simpler method, less sample handling post-incubation, with quick turn-around, ~ 5 min total, and which cross-calibrates well with the traditional approach (Ferrón et al. 2016). With shipboard processing possible, this allows many more samples to be run, more questions to be addressed as well as contingency planning at sea. For example, at the Pacific Ocean long-term time series station, where gross primary production ranged from 0.2 to 1.1 μmol $O_2L^{-1}d^{-1}$, a standard primary production depth profile with six depths and triplicate samples for each depth could be analyzed in approximately 3–4 h.

7.4 Gross Versus Net Productivity and Excess Energy: A Recap

Gross photosynthesis is the total assimilation of C or the liberation of O_2, without consideration of the fact that CO_2 is simultaneously lost, and O_2 consumed in the process of respiration (whether mitochondrial or light-dependent pathways). To recap from Chap. 6, photosynthesis is related to irradiance by a saturated hyperbolic relationship, with the initial slope of the response representing the efficiency of photon capture, and the maximal rate set by the metabolic condition of the cell, particularly the enzyme Rubisco. Photosynthesis can be inefficient; that is, not every absorbed photon contributes to photochemistry. When photosynthesis reaches its maximum, any photon capture that cannot be used directly in photosynthesis is dissipated by any of a number of reactions previously described, for example, xanthophyll cycling, chlororespiration, Mehler activity, photorespiration (Chap. 6). Very high irradiance levels can be harmful or damaging to photosynthesis and photoinhibition develops. New tools, such as rapid fluorescence measurements, allow us to probe these relationships in considerable detail.

7.5 Fluorometry

7.5.1 Basic Principles and Instruments

Rapid photosynthesis measurements are increasingly being applied in limnology, oceanography, and algal physiology. The literature with respect to these techniques can be very confusing for the novice user, as the terminology is complex, and different investigators may prefer different terms. The aim here is to provide an introduction to these techniques, to demystify terms and relationships, launching users and readers to more advanced literature. For further details, the reader is referred to the excellent reviews by Consalvey et al. 2005, Falkowski and Raven (2007), Hout and Babin (2010), Ralph et al. (2010), Suggett et al. (2010), Ruban (2017), among many others.

When a photon is absorbed by the reaction center chlorophyll *a* (chl *a*) (after having been transferred from the accessory or antennae pigments), it has one of three fates: it can carry out photochemistry, or it can be lost to the processes of fluorescence or to heat (Fig. 7.1). These processes are in competition, and thus learning about (measuring) one gives us clues about the other. A boost in one signal must come at the cost of a decrease in another; it is a zero-sum game. Thus, fluorescence and its different regulation as light availability changes give us a powerful tool to understand photosynthesis—and the heat or dissipation processes that dial down the efficiency of photosynthesis.

There are a number of attributes of fluorescence that are key to this powerful tool or its interpretation. Most of the fluorescence originates from PSII, with that from PSI making a smaller contribution (Consalvey et al. 2005). Thus, fluorescence is not equal to C fixation and the "imperfect supply chain" of electrons (see Chap. 6) may mean that while photochemistry is occurring, it is not the final step before growth. Fluorescence is always emitted at a longer wavelength than the absorbed photon—thus the fluorescence spectrum is shifted toward the red. This shift is known as the Stokes shift. Fluorescence is also an extremely fast process, occurring within pico- to nanoseconds of the absorption event. Carotenoids do not fluoresce; any photon energy not transferred to photosynthesis from carotenoids is emitted only as heat. It is these properties that have allowed the detection of productivity from space using flourescence emission (Yentsch and Phinney 1982), with increasing numbers of spectral bands included with advancing satellites.

Fluorescence intensity is a function of the light intensity, as well as the dissipation of absorbed energy via the processes of heat and photochemistry. This is usually

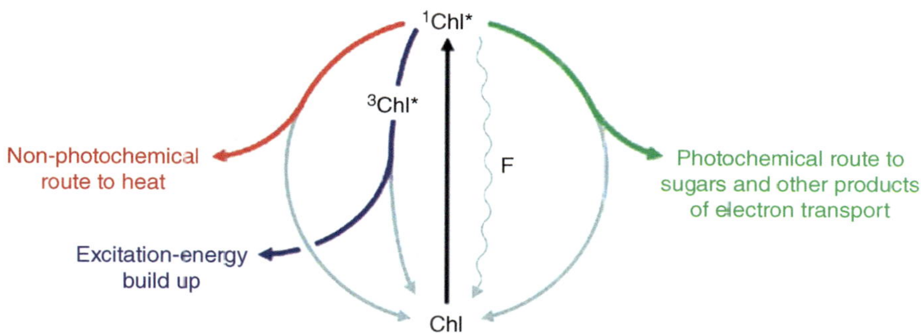

Fig. 7.1 Possible fates of excitation energy once a photon of light is absorbed by a molecule of chlorophyll resulting in conversion from the ground state (chl) to the singlet excited state (^{1}Chl*). When light is not saturating for photosynthesis, the majority of chl molecules pass their excitation energy to neighboring chl molecules until the energy reaches a reaction center chl molecule entering the photosynthetic electron transport chain (photochemical route denoted in green, returning the reaction center chl molecule to its ground state). As light levels become increasingly excessive, excess excitation energy is increasingly dissipated preemptively as thermal energy via a regulated pathway (non-photochemical route denoted in red that also returns the excited chl molecule to its ground state). When the combined levels of photochemical and regulated non-photochemical routes are insufficient to swiftly de-excite all excited singlet chl, conversion to triplet excited chl (^{3}Chl*) takes place, which represents a temporary build-up of excess excitation energy (purple route). In the presence of molecular oxygen, triplet excited chl returns to its ground state by transfer of excitation energy to oxygen, resulting in the formation of singlet excited oxygen (not shown). A small percentage of the chl molecules return to ground state directly from the singlet excited state (wavy line), emitting fluorescence (F). Although only representing a very small fraction of the excitation energy, fluorescence emission provides valuable quantitative information about the other fates (particularly the regulated non-photochemical and photochemical routes). Image reproduced from Demmig-Adams et al. (2020, Molecules 25: 5825. doi.org/10.3390/molecules25245825) under a creative commons license

studied as *fluorescence yield*, which is the probability that an absorbed photon is dissipated as fluorescence. Similarly, *photosynthetic yield* is the probability of a photon causing photochemistry. The higher the yield of photosynthesis, the less energy is available for fluorescence. Photochemistry quenches the fluorescence signal. Accordingly, the mimimim fluorescence yield occurs when photosynthesis is at its maximum. Thus, the maximum quantum yield of photosynthesis can be determined not only from the initial slope of an irradiance curve (see Chap. 6), but also by fluorescence methods, as described below.

Applications of fluorescence as a measure of photosynthesis were advanced with the development of instruments that take advantage of sensitive fluorescence measurements in conjunction with high-intensity flashes of light. These instruments, the most common of which are pulse-amplitude-modulated beam (PAM) fluorometers and fast-repetition rate fluorometers (FRRF). Each instrument type operates slightly differently (and there is variability in instrument function with different manufacturers), but they share several fundamental principles, illustrated here only for PAM (see Hout and Babin 2010 and Cosgrove and Borowitzka

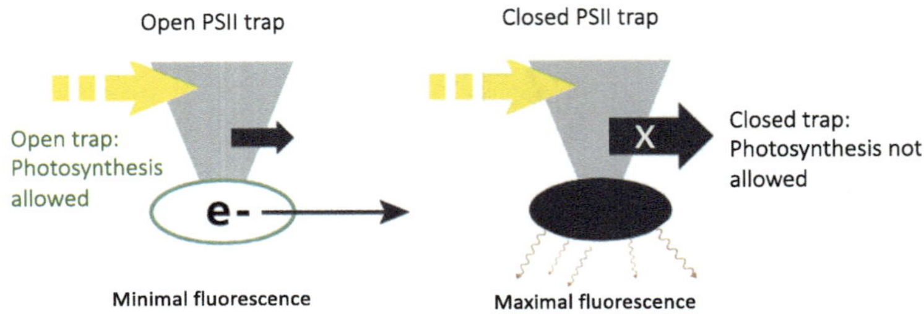

Fig. 7.2 Schematic showing an "open" trap at the site of PSII, which would yield minimal fluorescence and a "closed" trap which would yield maximal fluorescence

2011 for reviews and comparisons of the differences between PAM and the FRRF approaches). The field is advancing very fast and references to specific instruments and advantages or limitations need to be interpreted with this in mind. The versatility of instruments—and potential options for the investigator for manipulation—give the investigator a simultaneously powerful—but also bewildering and confusing array of options by which to make such measurements or to interpret measurements in the literature.

Both PAM and FRRF instruments operate on the same fundamental principles. This begins with manipulating the "readiness" of a cell to carry out photosynthesis. When a cell is in darkness, not carrying out photosynthesis, the reaction center is poised and ready to receive a photon; the 'trap' is considered to be "open" or fully relaxed (Fig. 7.2). Thus, measurements with either instrument typically start by placing the sample in darkness for a period of time, 10–30 min.

Once photochemistry starts, i.e., once light is supplied, there is a near-instantaneous period when the cell cannot accept more photons, and fluorescence occurs. This is because, for the instant that a chl *a* molecule is processing a photon, it is not available to accept another photon; the "trap" is considered closed (Fig. 7.2). At this moment, maximal fluorescence should be observed as this excess light energy has to be diverted from photochemistry. Recall that the cycling of electrons within the plastoquinone (PQ) pool is comparatively slow step (Chap. 6). The PQ pool must be "drained" or recycled back to its original state, requiring that Q_A be oxidized, before it can accept another electron. Thus, by applying a flash or pulse of high light, photosynthesis is driven to zero and the remaining processes (heat and fluorescence) can be quantified.

There are three different types of light used in PAM fluorometry. The first is the *measuring light*. This is a very minimal light, the intensity of which provides a very weak background that induces fluorescence but not any substantial amount of photosynthesis and there is no regulated heat dissipation. This induces minimal fluorescence, designated F_0. The measuring light is not provided continuously, but rather "modulated," turned off and on at the microsecond scale. Thus, "Modulated" is the M in PAM. The value of F_0 correlates with the amount of chl *a* in the sample (but see Consalvey et al. 2005 regarding cautions in interpreting F_0 as a measure of biomass). The second type of light is a *saturating pulse*. These are short pulses of very high light, the purpose of which is to close all reaction centers, ensuring they will be unable to quench fluorescence by photochemistry (Fig. 7.3). These flashes induce maximal fluorescence or F_M. The third type of light is *actinic light*, which is provided in a normal range of irradiance and manipulated to induce photosynthesis and its changes.

As noted above, if photosynthesis is not an option for the reaction center (if the 'trap" is closed; Fig. 7.2), the only other pathways for the absorbed energy are fluorescence and heat. To ensure that fluorescence yield is truly maximal, the intensity of the saturating pulse is typically many-fold (2–10 times) that of full sunlight. The difference between F_m and F_0 ($F_m - F_0$) is the variable fluorescence, or F_v, and thus the ratio F_v/F_m is termed the variable fluorescence yield and is related to the quantum yield of PS II photochemistry (Genty et al. 1992). This can be viewed as the maximal ratio of photons directed toward photosynthesis relative to photons received,

$$\text{Quantum efficiency of photosynthesis} = \phi_{\text{PSII}} = F_v/F_m = (F_m' - F_0)/F_m' \quad (7.1)$$

Healthy algae typically have an F_v/F_m of ~0.65. That is, healthy algae have about a 65% efficiency—at their maximum—in converting light energy to photochemical energy. Note that healthy leaves often have values as high as 0.85, while values of healthy cyanobacteria (and some dinoflagellates) can have F_v/F_m values substantially lower, 0.40–0.50. Other small pico-eukaryotes also have comparatively low F_v/F_m (Suggett et al. 2009). Chronic stress, such as that imposed by nutrient deprivation or non-optimal temperatures, will also lead to reductions in F_v/F_m. Consequently, F_v/F_m can be viewed as a measure of the physiological state of the cell, and comparisons of F_v/F_m of algae under different conditions, or different algae under the same conditions, can be very insightful. However, all measures of F_v/F_m are relative, and actual values may depend on the instrument used and its unique settings, as well as the physiological condition of the cells. There is no such thing as a unique F_v/F_m for a given taxa. It is also worth noting that due to the way in which FRRF and PAM instruments deliver the saturating pulses, an FRRF instrument generally produces a lower F_m signal and thus a lower F_v/F_m than a PAM instrument for the same sample.

When the saturating pulse used in measuring F_m is removed (its duration is generally <1 s), the fluorescence yield will rapidly drop, reflecting the progressive opening of "traps" or reaction centers of PSII (Fig. 7.3). This slow, progressive opening occurs because the electron transport chain takes time to oxidize the intermediate carriers. It requires an election sink such as C metabolism.

7.5.2 Measuring Rates of Photosynthesis with PAM

With this basic background, we can now look at how we can manipulate fluorescence to measure photosynthetic rates. Reaction centers will begin to close again if actinic light is applied (Fig. 7.4). When exposed to actinic light and then re-exposed

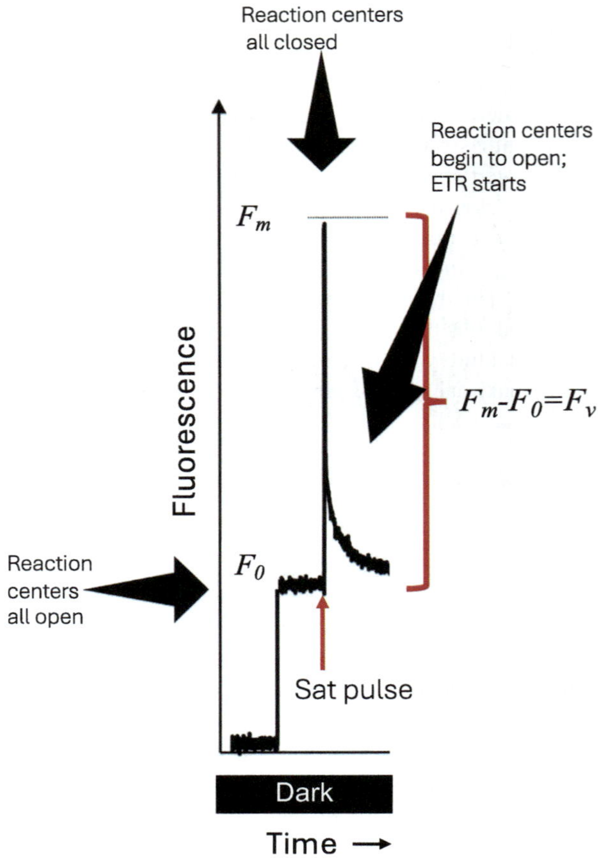

Fig. 7.3 Measurement protocol applied to measure F_o and F_m with a PAM fluorometer. F_m and F_o are the maximum and minimum fluorescence levels after dark adaptation. The pulse represents a many-fold increase above ambient light, resulting in the closure of reaction centers and maximal fluorescent signal results. The difference between F_m and F_o is F_v, and the ratio F_v/F_m is a common measure of photosynthetic efficiency. Image modified and reproduced from Ruban (2017, Phil Trans Royal Soc B 372: doi.org/10.1098/rstb.2016.0393) under a creative commons license

to saturating pulses, the kinetics of fluorescence change reflecting transient closure of PSII reaction centers, but the change in these kinetics and the exact pattern of the increase-then-decrease depends on the structural and functional properties of the cell and how the pulses are applied. It takes time for the signal to return to the original F_0 signal due to the fact that the acceptor of PSII, namely the PQ pool, remains reduced for a finite period of time. Most notably, the maximal fluorescence measured under these new conditions will be different from that obtained from a cell following darkness; the signal is reduced (Fig. 7.4). The decline in maximum fluorescence yield is the result of an increase in heat dissipation in the photosystem. The time steps and intensities of pulses provided in any given set of measurements can be manipulated and varied. Values of F_m measured after the cells have been illuminated are designed with a prime (or apostrophe) symbol, thus F_m becomes F_m' (or F'_m).

Changes in amplitude of actinic light allow rates to be measured at different light intensities. When more actinic light is applied, more photochemistry is carried out,

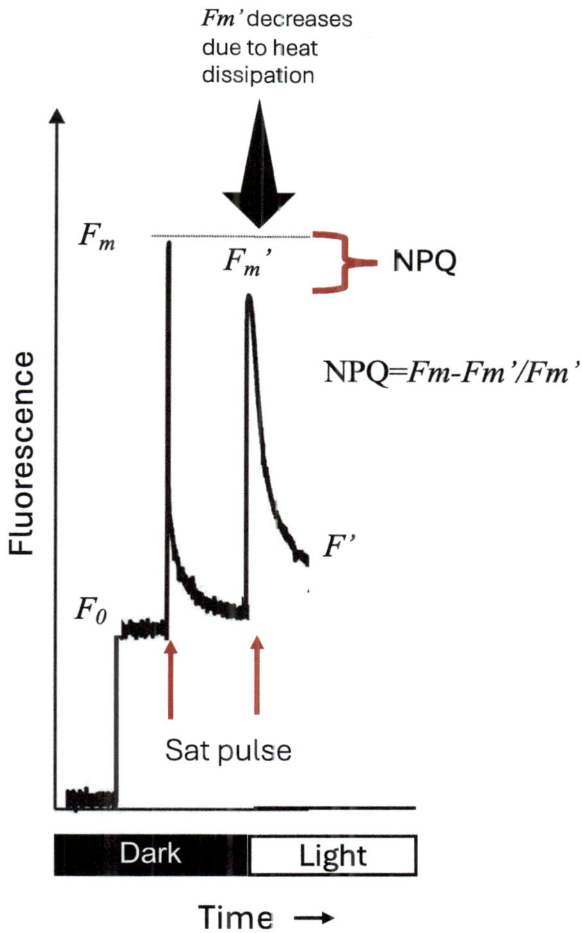

Fig. 7.4 As for Fig. 7.3, but with the addition of a pulse of actinic light after the fluorescence had returned to a value close to F_o. Note that now the fluorescence does not reach the same level as it did with the first pulse. If subsequent light pulses are given with the actinic light on, F_m continues to drop. Note that *Fm* measured under actinic light conditions is termed Fm' and the difference between F_m and F_m' divided by F_m' yields a measure of NPQ. Image modified and reproduced from Ruban (2017, Phil Trans Royal Soc B 372: doi.org/10.1098/rstb.2016.0393) under a creative commons license

and thus from the perspective of fluorescence, the fluorescent signal will drop or be quenched. If light is increased in sequential steps, the channeling of electrons to light or fluorescence is not only a function of the light applied, but also a function of how well the photosystem has recovered from the last pulse. Thus, using sequential changes in actinic light, the fluorescence at each step can be measured and related to photosynthesis response for each light intensity, that is, a photosynthesis–irradiance curve can be developed.

Before turning to how such curves are developed, a summary of common terms, for dark and light-adapted cells, and relationships between and among various measured parameters that can be derived in PAM measurements is provided (Table 7.1). Some of these relationships are discussed further below, and some are included here simply to ease a budding user in reading the literature.

Table 7.1 Fluorescence terminology and common relationships (adapted from Heinz Walz GmbH 2003, Consalvey et al. 2005, Serôdio et al. (2006), and Murchie and Lawson (2013). Note that multiple formulations are reported for some relationships and the referenced literature should be consulted for rationale of preferred forms

Experimental conditions and terms		Description
Dark-adapted	F_O	Minimum fluorescence yield; measured at very low light to ensure PSII reaction centers are in open state
Dark-adapted	F_m	Maximum fluorescence yield during the saturating pulse, thereby closing all PSII reaction centers
Dark-adapted	F_v	Variable fluorescence, or difference between F_m and F_O
Dark-adapted	F_v/F_m	Quantum efficiency of PSII
Light	F'	Fluorescence yield in light-adapted state
Light	F_s	Fluorescence yield at steady-state in light conditions
	F_t or F	Current or momentary level of fluorescent yield in light conditions, measured before application of a saturating pulse
Light	F_m'	Maximum fluorescence yield during saturing pulse in light-adapted state
Light	Fq'	Fluorescence quenched at a particular light level relative to light-adapted maximal fluorescence
Light	Fq'/F_m'	Light utilization efficiency
Light	F_O'	Minimum fluorescence yield in light-adapted state, typically determined during dark conditions following a saturating pulse
Light	F_v'	Variable fluorescence in light-adapted state
Light	NPQ	Non-photochemical quenching
Light	RLC	Rapid ETR vs irradiance curve
Light	ETR	Relative electron transport rate measured (dimensionless)
Light	ETRm	Maximum relative electron transport rate of the rapid light curve (dimensionless)
Light	I_k	Light-saturation parameter of the ETR vs irradiance curve
Relationships		
	F_v/F_m	$(F_m - F_O)/F_m$ Maximum quantum efficiency of PSII (dimensionless)
	F_v'/F_m'	$(F_m' - F_O')/F_m'$ Maximum quantum efficiency of PSII in the light (if all reaction centers were in the open state)
	ϕ_{PSII}	$(F_m' - F')/F_m'$ Operating efficiency of PSII; effective quantum efficiency (quantum yield) of PSII
	NPQ	$(F_m - F_m')/F_m'$ or $F_m/F_m' - 1$ or YNPQ/YNO

(continued)

Table 7.1 (continued)

Experimental conditions and terms		Description
		Non-photochemical quenching or heat loss (dimensionless) (also termed Stern-Volmer-type NPQ)
	qN	$(F_m - F_m')/(F_m - F_O)$ Alternate formulation of non-photochemical quenching
	qI	Irreversible component of NPQ related to photoinhibition
	qP	$(F_m' - Fs)/(F_m' - F_O)$, or Fq'/F_v' An index of the proportion of reaction centers in the open state or coefficient of photochemical quenching 1-qP represents the balance between delivery of excitation and removal of electrons, or fraction of closed reaction centers
	qL	$(Fq' - F_v')/(F_O' - F')$ An indicator of PSII redox state or fraction of open PSII reaction centers; coefficient of photochemical quenching assuming interconnected PSII antennae 1-qL represents the balance between delivery of excitation and removal of electrons, or fraction of closed reaction centers
	qE	The rapidly reversing component of NPQ
	ETR	$(Fq'/F_m') \times$ PPFD/2 $\times$ a^*
	rETR	Relative ETR where a^* is not measurable $(Fq'/F_m') \times$ PPFD/2
Yields		Y(PSII) + Y(NPQ) + Y(NO =1
	Y(II) = ϕ_{PSII}	$(F_m' - F')/F_m'$ Excitation energy/ quantum yield of PSII
	Y (NPQ) = ϕ_{NPQ}	$(F/F_m') - (F/F_m)$ or $Fs/F_m' - Fs/F_m$ Quantum yield of light-induced NPQ due to ΔpH and xanthophyll cycling (regulated quenching mechanisms)
	Y(NO) = ϕ_{NO}	Fs/F_m Sum of heat and fluorescent loss of excitation energy (non-regulated quenching mechanisms); high values suggest an inability to protect against photodamage

There are two ways to establish the response of a cell to changes in actinic light, that is, to develop a photosynthesis–irradiance curve. First, a given sample can be exposed to actinic light that is progressively stepped up in intensity, and a response measured at each increasing step (Fig. 7.5). Using a PAM instrument, the variable fluorescence, Fv', can be established at each step, and if light is applied in stepped-up pulses, the relationship between photosynthesis and light can be determined. This

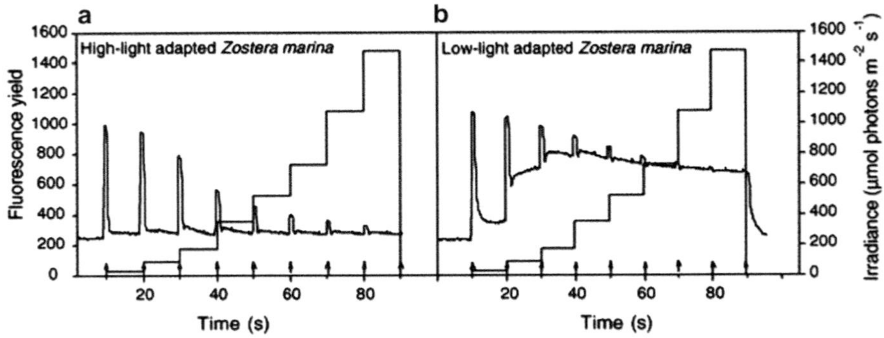

Fig. 7.5 Recording of fluorescence signal yielding rapid light curves for a sample of *Zostera*. Panel (**a**)—RLC from a high light-adapted *Zostera marina* leaf, and Panel (**b**)—from a low light-adapted *Z. marina* leaf. Time-response curve of fluorescence yield (F), showing maximum fluorescence (F_m') peaks after each saturating flash is plotted against time. The horizontal axis is seconds and the vertical axis shows fluorescence units (arbitrary). The actual irradiance during the RLC is superimposed (stepped curve). Arrows indicate the saturating pulses. Image reproduced from Ralph and Gademann (2005, Aquat Bot 82:222–237. doi.org/10.1016/j.aquabot.2005.02.006) with permission of Elsevier

approach, termed a rapid light curve (RLC; White and Critchley 1999), can typically be programmed in a PAM instrument and run automatically. Once an RLC program has been set, different algae or experimental treatments can be compared. The time interval between steps and magnitude of each step change can be varied by the user, but the end result is a hyperbolic response curve relating the rate of photosynthesis to light intensity. In conducting such an experiment, it must be borne in mind that the response will be a function not only of the light intensity but also the duration of exposure and the time interval between exposures. Thus, a curve in which light intensity is stepped up vs a curve in which light is stepped down will not be equivalent. Also, cells (or, in the example shown, *Zostera* sp.) previously acclimated to high light will have a different RLC than cells previously acclimated to low light (Fig. 7.5; Ralph and Gademann 2005). The accumulated light dose influences the response of the cell (or plant) to subsequent light exposures. All of these parameters can be manipulated and thus comparisons of responses must take these factors into account.

Using such an approach, the rate of photosynthesis, as electron transport rate (ETR) for each irradiance, can be determined according to:

$$\text{Photochemistry} = \text{irradiance} \times \text{absorption} \times \text{quantum yield} \tag{7.2}$$

And

$$\text{ETR} = \text{Y(II)} \times \text{PAR} \times 0.5 \times \text{absorptivity} \tag{7.3}$$

where ETR is expressed as mol electrons mg chl a^{-1} s^{-1}, Y(II) is the quantum yield of PSII, PAR (photosynthetically active radiation) is the irradiance as mol quanta m^{-2} s^{-1}, and absorptivity (a*) is the absorption cross-section as m^2 mg chl a^{-1}. Some investigators apply a factor of 0.5 if they assume the absorbed light is distributed equally between two photosystems. In the higher plant literature, the absorptivity is generally assumed to be 0.84 (e.g., Genty et al. 1992). If a* is not known or measurable, ETR is expressed as a relative measure, rETR.

An alternative to measuring a RLC, the fluorescence of independent replicate samples can be measured at unique actinic light levels. In this case, the resulting signal will not be influenced by the previous measurement but will be influenced by the previous state (growth, light condition, etc.) of the cell when the subsample was collected. Using either approach, making such measurements across a gradient of light intensities, a photosynthesis-irradiance response curve can be generated. As will be discussed later, these two approaches are not unlike the differing approaches used in measuring nutrient kinetics (see Chap. 12). One approach involves sequential measurement of a single sample over time, while another approach involves exposing individual samples to a different concentration at a single time. The important point is that the physiological state of the cell differs in each of these approaches. In RLC measurements, the physiology changes due to exposure to irradiance in the immediate prior light pulse.

Some of the newer PAM instruments use LED lights of different wavelengths as measuring and actinic lights. This allows deconvolution of signals (with instrument provided software) from different functional groups (brown algae, green algae, blue-green algae, and phycoerythrin-containing algae); it determines the photophysiological condition of each algal group individually. Glibert et al. (2022) were able to measure the change in signals of the brown (diatom), green and phycoerythrin-type algae in different stations of the San Francisco Bay Delta (Fig. 7.6). The composition of the different signals as detected by the PAM was confirmed microscopically.

7.5.3 Energy Loss: Non-photochemical Quenching, Photoinhibition, and Non-regulated Photochemical Quenching

Due to the fundamental relationship that the fate of a photon has only three options (i.e., to carry out photosynthesis, give off heat, or fluorescence), there is much that can be learned about energy dissipation using fluorescence.

The fraction of photosystems that are "open for business" at any given time is designated qP. Alternatively, the fraction of closed photosystems is designated 1-qP, which is also a measure of the reduction state of PQ (Table 7.1). Note that an alternate calculation for the fraction of open photosystems is designated qL, and while these terms are calculated based on some different assumptions, they represent the same fundamental concept (Table 7.1). More photosystems would be expected to be closed as more light is provided in a RLC.

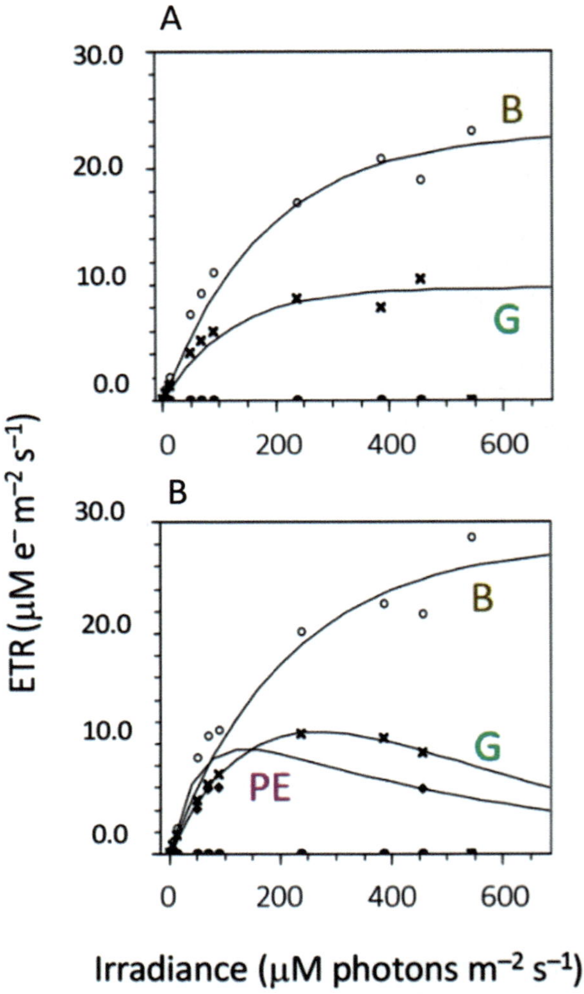

Fig. 7.6 Examples of Rapid Light Curves from PAM fluorometry measurements. Samples were collected from two sites in the San Francisco Bay Delta. The algal groups are differentiated as B (brown algae, diatoms), G (chlorophytes), and as PE (phycoerythrin-containing cells). Image reproduced from Glibert et al. (2022, Nitrogen 3: 569–591. doi.org/10.3390/nitrogen3040037) under a creative commons license

Recently it was shown that manipulations of the amount of nitrogen (N, as NO_3^- or NH_4^+) provided to the dinoflagellate *Karenia mikimotoi* can decrease the fraction of closed photosystems, suggesting that metabolic demand can play an important regulatory role in the balance of open vs closed photosystems (Ahn and Glibert 2022). The RLCs in their experiment all showed an increase in ETR at each irradiance level as the concentration of N substrate provided increased from 1 to 50 μM (Fig. 7.7a, b), and concomitantly, the fraction of closed photosystems decreased with the provision of more N at each irradiance level (Fig. 7.7c, d). The fraction of closed photosystems also increased as the irradiance increased. These

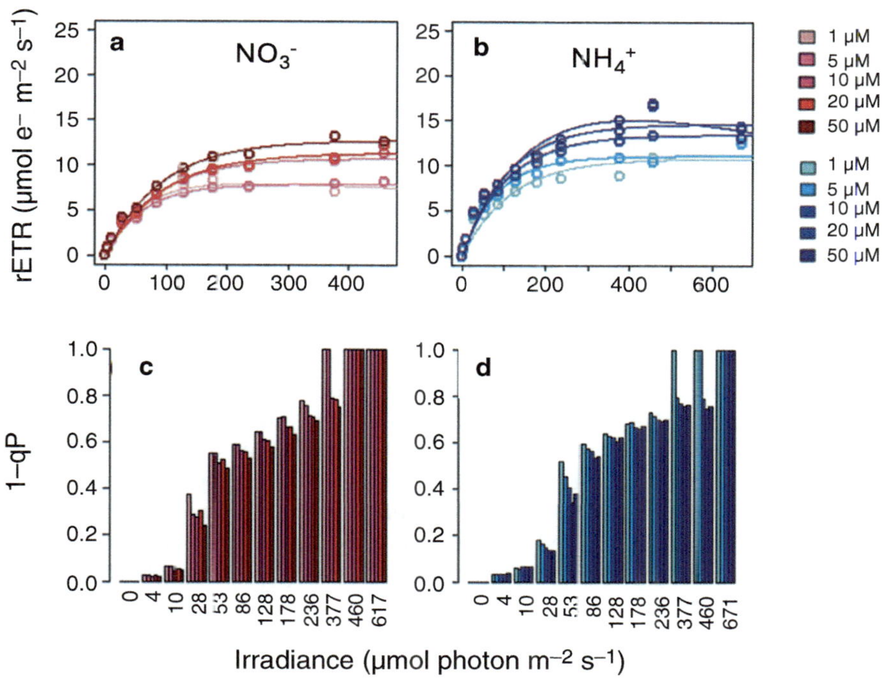

Fig. 7.7 Panels (**a**, **b**)—The rapid light responses of the dinoflagellate *Karenia mikimotoi* at 25 °C when enriched with different nitrogen forms (NO_3^- and NH_4^+) and amounts (1, 5, 10, 20, 50 μM-N). For each depicted ETR curve, the darker the color of the lines, the higher the enrichment with added N. Panels (**c**, **d**)—The extent of PSII reaction centers that were closed for the same treatments. Note that at each light intensity, the fraction of closed photosystems is generally lower with increasing N concentrations. At the highest light provided, all reaction centers were closed. Images reproduced from Ahn and Glibert (2022, Phycology 2:30–44. doi.org/10.3390/phycology2010002) under a creative commons license. Note that the original reference also provides data for samples also treated with urea as well as at different temperatures

results indicate that additional nutrients promote the rate of metabolism that drives the oxidation of PQ, and hence the opening of photosystems.

The first component of heat is that of non-photochemical quenching, NPQ. The concept of NPQ was introduced in Chap. 5 in relation to photoprotective pigments and the xanthophyll cycle. Using fluorescence, NPQ can be quantified as,

$$\mathrm{NPQ} = \left(F_m - F_m{}'\right)/F_m{}' \tag{7.4}$$

Another related parameter, qN, is also used to assess non-photochemical quenching, or heat dissipation as,

$$\mathrm{qN} = \left(F_m - F_m{}'\right)/\left(F_m - F_O\right) \tag{7.5}$$

Note that both of these values require measurements of *Fm* in the dark and in the light. Following Shreiber (2004) and Ralph and Gademann (2005), NPQ is a preferred term as it is not dependent upon $F_O{'}$ and not affected by F_O quenching (the NPQ quenching mechanism has been shown to reduce $F_O{'}$ to variable extents; Ruban 2017). It is more sensitive to energy dissipation within the antenna matrix, whereas qN is more sensitive to thylakoid membrane energy (Shreiber 2004). The NPQ component of heat can be considered the component that the cell regulates to protect its photosystems from over-reduction. The process is thought to be initiated by the increase in H^+ ions in the thylakoid, which are generated during the photosynthetic process (Chaps. 5 and 6).

The most significant mechanism of NPQ is the change in xanthophyll pigments. Xanthophylls are the carotenoids that reside in the thylakoid membranes and which undergo interconversions in response to irradiance (Chap. 5). As such, the extent to which xanthophyll cycling occurs is a function of the cell's need to downregulate photosynthesis and is therefore a measure of the energy state of the cell. There are two groups of xanthophyll cycles. The first, known to operate in diatoms, chrysophytes, and dinoflagellates includes the pigments diadinoxanthin (dd) and diatoxanthin (dt). When exposed to light, dd is converted to dt via de-epoxidation, a process that removes an epoxide group from the xanthophyll molecule. The second group of xanthophylls, more common in chlorophytes and many higher plants, includes violaxanthin, antheroxanthin, and zeaxanthin. In the light, the conversion is a two-step process, first involving a de-epoxidation of violaxanthin to antheroxanthin, and then a second reaction to convert antheroxanthin to zeaxanthin. The more de-epoxidated forms of these pigments are less efficient at transferring energy and thus are considered photoprotective pigments: they have lost (or greatly reduced) their ability to transfer electrons to chl *a*.

The downregulation of photosynthesis via xanthophyll cycling reduces photosynthesis but does not result in long-term damage (although xanthophyll repair is a slow process). When cells receive far more light than that which saturates the photosynthetic rate, and energy from which exceeds the various protective mechanisms, the cell experiences photoinhibition or photodamage, processes that require de novo synthesis for repair (see Chap. 6). Photoinhibition or damage typically involves extreme stress to the cell and involves a long period of recovery, hours to days, as it can involve the repair of the reaction center proteins. In general, the term NPQ is most commonly used for the rapid and/or moderately slowly reversible photoprotective mechanisms and does not include true photoinhibition, which is differentiated as qI. Collectively, the dissipation of energy (as heat) via these non-regulated processes is termed NO (Table 7.1).

Accordingly, and as previously emphasized, the Yield (Y) of a given photon in driving photosynthesis (YII), regulated non-photochemical quenching (YNPQ), or non-regulated non-photochemical quenching (YNO) is a zero-sum game. The three yield parameters sum to 1.0,

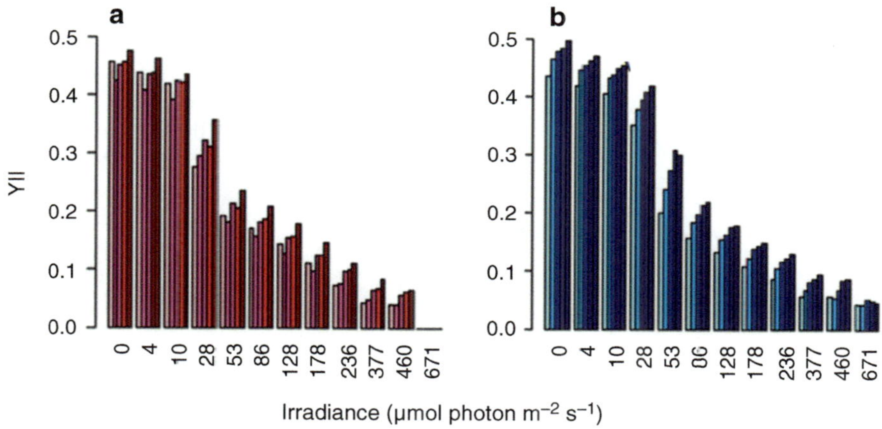

Fig. 7.8 Effective Photosystem II photochemical efficiency, YII, as a function of irradiance for *Karenia mikimotoi* at 25 °C) when enriched with different nitrogen forms (NO_3^- (Panel **a**) and NH_4^+ (Panel **b**) and amounts (1, 5, 10, 20, 50 μM-N). For each depicted ETR curve, the darker the color of the bars, the higher the enrichment level, with added N (see Fig. 7.7). Images reproduced from Ahn and Glibert (2022, Phycology 2:30–44. doi.org/10.3390/phycology2010002) under a creative commons license. Note that the original reference also provides data for samples treated with urea as well as at different temperatures

$$Y(II) + Y(NPQ) + Y(NO) = 1 \tag{7.6}$$

The higher the value of Y(NO), the greater the degree of stress and a cell's inability to downregulate photosynthesis in response to damaging levels of irradiance. In the example cited above in which samples received different N treatments prior to measuring RLC, the yields of PSII also increased with the provision of additional N (Fig. 7.8, Ahn and Glibert 2022).

7.5.4 Putting It to the Test: Methods Comparisons

Various investigators have reported differences between the various approaches for measuring photosynthesis, not only between techniques but across irradiance levels (Suggett et al. 2003). A recent comparison of estimates of marine primary productivity derived from $H_2{}^{18}O$ and other methods confirms that the highest reported rates of primary production are, in fact, derived from ^{18}O enrichment methods (Regaudie-de-Gioux et al. 2014). As examples, Masojídek et al. (2001) showed for a natural cyanobacterial bloom that ETR (based on PAM measurements) and O_2 evolution were linearly related at irradiance levels that were <800 μmol quanta m^{-2} sec^{-1}, but deviated at higher light levels. Similarly, Beer and Axelsson (2004) also found that ETR and O_2 were directly correlated at the lower end of the irradiances tested, but at the highest irradiances, ETR also declined while O_2 evolution did not. Suggett et al.

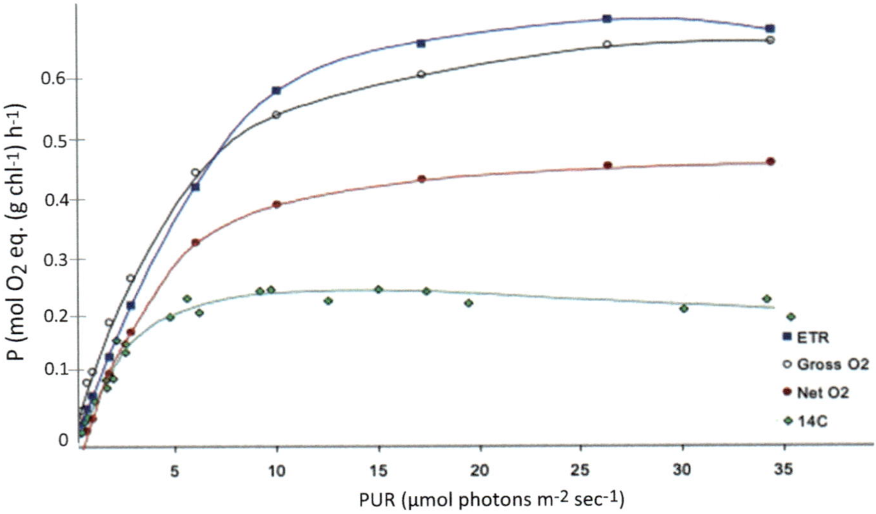

Fig. 7.9 Comparison of the rate of photosynthesis measured in a culture of the Pelagophyte *Aureococcus anophagefferens* using a range of techniques. Measurements of ETR and gross O_2 fluxes measured processes at PSII, net O_2 fluxes account for pathways of respirations, and ^{14}C measurements determine the ultimate amount of carbon assimilated. Image provided by T. Kana and H. MacIntyre based on unpublished data

(2010) have summarized the broader literature of comparisons of active fluorescence measurements and other approaches.

One example is provided here to highlight these methodological differences in common. This experiment involved the pelagophyte *Aureococcus anophagefferens* which was measured using PAM fluorometry, ^{18}O (as gross and net values), and ^{14}C incorporation. The results illustrate very clearly that not only do different methods for measuring photosynthesis yield very different values, they also illustrate the progressive drop in P_{max} from ETR and gross O_2 fluxes, which measure processes related to PSII, to net O_2 fluxes, which account for the loss in O_2 due to respiration processes (in the dark and in the light, and finally to ^{14}C incorporation (Fig. 7.9). Gross O_2 and ETR gave similar values as both measure processes associated with PSII. Lowest values were from ^{14}C as it is the process furthest along the electron transport chain; the loss from ETR to ^{14}C reflected the inefficiencies of the "supply chain" as discussed above. Note, also, that in order to make this comparison, the various irradiances from the different light sources used for the different measurements had to be converted to photosynthetically utilizable radiation (see Chap. 4).

7.6 Summary

Riley and Steeman Nielsen had their debates about the best approaches for measuring photosynthesis. They may not have fully understood why their measured values differed so significantly. With new approaches for photosynthesis, it is even more important that investigators understand what they are measuring (what specific process in the photosynthetic process does the technique target?), what time scales they are measuring (rapid responses of electron transport or long-term acclimation?), and how methods manipulations may affect results. The wisdom here of Steemann Nielson referred to in Chap. 2, is worth repeating (even though his context was ^{14}C measures of primary productivity it applies equally to all measurement approaches), "An impeccable experiment is worth more than a lousy experiment with lots of replicates" (quoted in Søndergaard 2002, p. 13).

It is also important for modelers to take note of the different methodologies, and why a value from the literature may or may not be representative of the model parameter desires (typically growth). The new and advancing tools are powerful, allowing sophisticated ecological and physiological questions to be addressed. As summarized by Suggett et al. (2010, p. 123), "Answering the question as to whether it is appropriate to use ETR_{PSII} to predict C fixation may ultimately rest with the level of accuracy and precision we are willing to accept and, perhaps importantly, how much effort we are prepared to invest in improving upon these levels."

In the following chapter, the dynamic behavior of photosynthesis is further explored. It is not sufficient to understand the instantaneous rate of the process of photosynthesis, it is also important to understand how the rate may change in a dynamic aquatic environment. All photosynthetic measurements are, quite simply, an exercise in chasing a moving target.

References

Ahn SH, Glibert PM (2022) Shining light on photosynthesis in the harmful dinoflagellate *Karenia mikimotoi*– responses to short-term changes in temperature, nitrogen form and availability. Phycology 2:30–44. https://doi.org/10.3390/phycology2010002

Beer S, Axelsson L (2004) Limitations in the use of PAM fluorometry for measuring photosynthetic rates of macroalgae at high irradiances. Eur J Phycol 39:1–7

Bender ML, Grande K, Johnson K, Marra J, Williams PJLB, Sieburth J, Pilson M et al (1987) A comparison of four methods for determining planktonic community production. Limnol Oceanogr 32:1085–1098. https://doi.org/10.4319/lo.1987.32.5.1085

Bender ML, Orchardo J, Dickson M-L, Barber R, Lindley S (1999) In vitro O_2 fluxes compared with ^{14}C production and other rate terms during the JGOFS equatorial Pacific experiment. Deep Sea Res II 46:637–654

Bender ML, Dickson M-L, Orchardo J (2000) Net and gross production in the Ross Sea as determined by incubation experiments and dissolved O_2 studies. Deep Sea Res II 47:3141–3158. https://doi.org/10.1016/S0967-0645(00)00062-X

Collos Y, Slawyk G (1985) On the compatibility of carbon uptake rates calculated from stable and radioactive isotope data: implications for the design of experimental protocols in aquatic primary productivity. J Plankton Res 7:595–603. https://doi.org/10.1093/plankt/7.5.595

Consalvey M, Perkins RG, Paterson DM, Undewood GJC (2005) PAM fluorescence: a beginners guide for benthic diatomists. Diatom Res 20(20):1–22. https://doi.org/10.1080/0269249X.2005.9705619

Cosgrove J, Borowitzka MA (2011) Chlorophyll fluorescence terminology: an introduction. In: Suggett DJ, Prasil O, Borowitzka (eds) Chlorophyll fluorescence in aquatic sciences: methods and applications. Springer Science and Business Media, pp 1–17

Cullen JJ (2001) Primary production methods. In: Steele JH, Thorpe SA, Turekian KK (eds) Encyclopedia of ocean sciences. Academic Press, San Diego, CA. https://doi.org/10.1006/rwos.2001.0172

Demmig-Adams B, Stewart JJ, López-Pozo M, Polutchko SK, Adams WW III (2020) Zeaxanthin, a molecule for photoprotection in many different environments. Molecules 25:5825. https://doi.org/10.3390/molecules25245825

Dijkman NA, Boschker HTS, Middelburg JJ, Kromkamp JC (2009) Group-specific primary production based on stable-isotope labeling of phospholipid-derived fatty acids. Limnol Oceanogr-Meth 7:612–625

Falkowski PG, Raven JA (2007) Aquatic photosynthesis. Princeton University Press, Princeton, NJ

Ferrón S, del Valle DA, Björkman KM, Quay PD, Church MJ, Karl DM (2016) Application of membrane inlet mass spectrometry to measure aquatic gross primary production by the ^{18}O in vitro method. Limnol Oceanogr-Meth 14:610–622

Genty B, Goulas Y, Dimon B, Peltier G, Briantais JM, Moya I (1992) Modulation of efficiency of primary conversion in leaves, mechanisms involved at PS2. In: Murata N (ed) Research in photosynthesis, Vol. IV: Proceedings of IXth International Congress on Photosynthesis, Nagoya, Japan. August 30–September 4, pp 603–610

Glibert PM, Middelburg JJ, McClelland JW, Vander Zander J (2019) Stable isotope tracers: enriching our perspectives and questions on sources, fates, rates and pathways of major elements on aquatic systems. Limnol Oceanogr 64:9501–9981. https://doi.org/10.1002/lno.11087

Glibert PM, Wilkerson F, Dugdale RC, Parker AE (2022) Ecosystem recovery in progress? Initial nutrient and phytoplankton response to nitrogen reduction from sewage treatment upgrade in the San Francisco Bay Delta. Nitrogen 3:569–591. https://doi.org/10.3390/nitrogen3040037

Hama T, Hama J, Handa N (1993) ^{13}C tracer methodology in microbial ecology with special reference to primary production processes in aquatic environments. In: Jones JG (ed) Advances in microbial ecology, vol 13. Plenum, New York, pp 39–83

Heinz Walz GmbH (2003) Phytoplankton analyzer, PhytoPAM and Phyto-Win software V 1.45. System components and principles of operation. https://www.walz.com/files/downloads/manuals/phyto-pam/phyto_4e.pdf. Accessed 1 Aug 2023

Hout Y, Babin M (2010) Overview of fluorescence protocols: theory, basic concepts, and practice. In: Suggett DJ, Prášil O, Borowitzka MA (eds) Chlorophyll *a* fluorescence in aquatic sciences: methods and applications. Springer Publ, Heidelberg, pp 31–74. https://doi.org/10.1007/978-90-481-9268-7_3

Kana TM (1990) Light-dependent oxygen cycling measured by an oxygen-18 isotope dilution technique. Mar Ecol Prog Ser 64:293–300

Kana TM (1992) Relationship between photosynthetic oxygen cycling and carbon assimilation in *Synechococcus* WH7803 (Cyanophyta). J Phycol 28:304–308. https://doi.org/10.1111/j.0022-3646.1992.00304.x

Kana TM, Darkangelo C, Hunt MD, Oldham JB, Bennett GE, Cornwell JC (1994) Membrane inlet mass spectrometer for rapid high-precision determination of N_2, O_2, and Ar in environmental water samples. Anal Chem 66:4166–4170. https://doi.org/10.1021/ac00095a009

de Kluijver A, Soetaert K, Schulz KG, Riebesell U, Bellerby RGJ, Middelburg JJ (2010) Phytoplankton-bacteria coupling under elevated CO_2 levels: a stable isotope labelling study. Biogeosciences 7:3783–3797. https://doi.org/10.5194/bg-7-3783-2010

de Kluijver A, Soetaert K, Czerny J, Schulz KG, Boxhammer T, Riebesell U, Middelburg JJ (2013) A ^{13}C labelling study on carbon fluxes in Arctic plankton communities under elevated CO_2 levels. Biogeosciences 10:1425–1440. https://doi.org/10.5194/bg-10-1425-2013

Marra JF (2019) Hot carbon: Carbon-14 and a revolution in science. Columbia University Press, New York

Masojídek J, Grobbelaar JU, Pechar L, Koblížek M (2001) Photosystem II electron transport rates and oxygen production in natural waterblooms of freshwater cyanobacteria during a diel cycle. J Plankton Res 23:57–66

Murchie EH, Lawson T (2013) Chlorophyll fluorescence analysis: a guide to good practice and understanding some new applications. J Exp Bot 64:3983–3998

Pimenov NV, Zyakun AM, Prusakova TS, Lunina ON, Ivanov MV (2008) Application of ^{13}C mineral carbon for assessment of the primary production of organic matter in aquatic environments. Microbiol 77:224–227

Popp BN, Bidigare RR, Deschenes B, Laws EA, Prhal FG, Tanimoto JK, Wallsgrove R (2006) A new method for estimating growth rates of alkenone-producing haptophytes. Limnol Oceanogr-Meth 4:114–129

Ralph PJ, Gademann R (2005) Rapid light curves: a powerful tool to assess photosynthetic activity. Aquat Bot 82:222–237. https://doi.org/10.1016/j.aquabot.2005.02.006

Ralph PJ, Wilhelm C, Lavaud J, Jakob T, Petrou K, Kranz SA (2010) Fluorescence as a tool to understand changes in photosynthetic electron flow regulation. In: Suggett DJ, Prášil O, Borowitzka MA (eds) Chlorophyll *a* fluorescence in aquatic sciences: methods and applications. Springer Publ, Heidelberg, pp 75–90. https://doi.org/10.1007/978-90-481-9268-7_4

Regaudie-de-Gioux A, Lasternas S, Agustí S, Duarte CM (2014) Comparing marine primary production estimates through different methods and development of conversion equations. Front Mar Sci 1:19. https://doi.org/10.3389/fmars.2014.00019

Robinson C, Tilstone GH, Rees AP, Smyth TJ, Fishwick JR, Tarran GA, Luz B et al (2009) Comparison of in vitro and in situ plankton production determinations. Aquat Microb Ecol 54: 13–34. https://doi.org/10.3354/ame01250

Ruban AV (2017) Quantifying the efficiency of photoprotection. Philos Trans Royal Soc B Biol Sci 372:20160393. https://doi.org/10.1098/rstb.2016.0393

Serôdio J, Vierira S, Cruz S, Coelho H (2006) Rapid light-response curves of chlorophyll fluorescence in microalgae: relationship to steady-state light curves and non-photochemical quenching in benthic diatom-dominated assemblages. Photosynth Res 90:29–43

Shreiber U (2004) Detection of rapid induction kinetics with a new type of high-frequency modulated chlorophyll fluorometer. Photosynth Res 9:261–271

Søndergaard M (2002) A biography of Einer Steemann Nielsen: the man and his science. In: Williams PJLB, Thomas DN, Reynolds CS (eds) Plankton productivity: carbon assimilation in marine and freshwater ecosystems. Blackwell Scientific Publishing, Oxford, pp 1–15

Suggett DJ, Oxborough K, Baker NR, MacIntyre HL, Kana TM, Geider RJ (2003) Fast repetition rate and pulse amplitude modulation chlorophyll a fluorescence measurements for assessment of photosynthetic electron transport in marine phytoplankton. Eur J Phycol 38:371–384

Suggett DJ, Moore CM, Hickman AE, Geider RJ (2009) Interpretation of fast repetition rate (FRR) fluorescence: signature of phytoplankton community structure versus physiological state. Mar Ecol Prog Ser 376:1–19

Suggett DJ, Moore CM, Geider RJ (2010) Estimating aquatic productivity from active fluorescence measurements. In: Suggett DJ, Prášil O, Borowitzka MA (eds) Chlorophyll *a* fluorescence in aquatic sciences: methods and applications. Spring Publ, Heidelberg, pp 103–128. https://doi.org/10.1007/978-90-481-9268-7_6

Tobias C, Böhlke JK, Harvey JW (2007) The oxygen-18 isotope approach for measuring aquatic metabolism in high-productivity waters. Limnol Oceanogr 52:1439–1453

White AJ, Critchley C (1999) Rapid light curves: a new fluorescence method to assess the state of the photosynthetic apparatus. Photosynth Res 59:63–72

Williams PJLB, Robertson JE (1991) Overall plankton oxygen and carbon dioxide metabolism: the problem of reconciling observations and calculations of photosynthetic quotients. J Plankton Res 13:153–169

Yentsch CS, Phinney DA (1982) The use of the attenuation of light by particulate matter for the estimate of phytoplankton chlorophyll with reference to the coastal zone color scanner. J Plankton Res 4:93–102

You L, Zhang B, Tang YJ (2014) Application of stable isotope-assisted metabolomics for cell metabolism studies. Metabolites 4:142–165. https://doi.org/10.3390/metabo4020142

8 Changing in the Short-Term: Acclimation

8.1 Introduction

The parameterization of a P vs I (or P vs E) curve, with values that correspond to the initial slope of the reaction, α, the maximal rate, P_{max}, and the light saturation value, I_k (see Chap. 6), gives an investigator a sense of definitiveness of a species in response to light. Similarly, parameterization of nutrient uptake responses to nutrient substrates gives investigators a similar sense of definitiveness with respect to nutrient availability. These values make it easy to compare species or values from different stations or points in time. So, for many years, people measured and compared and tabulated "catalogs" of curves in order to understand different species' responses to various environmental conditions. Models applied kinetic constants reported in the literature, regardless of the conditions under which measurements were made. Photosynthetic responses were classically grouped according to their "sun" (high light) vs "shade" (low light) responses. These ideas all stemmed from the perception that responses were generally fixed for a given species or condition.

Now it is well understood that cells have multiple mechanisms to modify pigmentation and enzyme activity and other processes so that they can balance energy and material flow through the cell as conditions change. As reviewed in the previous chapter (Chap. 7, see also Chap. 12), the response curve parameters are also dependent on the method of the investigator, and, for photosynthesis, which part of the photosynthetic apparatus is being targeted with a given technique. Thus, the seeming definitiveness of parameters on a response curve belies their true variability.

Phytoplankton moving through the water column, either via their inherent diel rhythms or forced by physical factors, must acclimate or will experience stress. The gradients they experience will not only be in light quantity but light quality (see Chap. 4), possibly temperature and nutrient availability (see Chap. 11). Phytoplankton generally can rapidly acclimate, but sometimes they experience damage to their photosystems, and sometimes severe damage (Chap. 6). On short time scales (hours to days), cells may be able to acclimate; on long time scales they may be able to adapt. It is thus important to recognize the distinction between short-term

P. M. Glibert, *Phytoplankton Whispering: An Introduction to the Physiology and Ecology of Microalgae*, https://doi.org/10.1007/978-3-031-53897-1_8

acclimation and longer-term adaptation. Raven and Geider (2003) suggested that the time scale for acclimation is on the order of several hours, and involves changes in protein expression or other metabolic fluxes, while adaptation occurs over many generations, and involves a changed genome through mutation, natural selection, or other processes (Falkowski and LaRoche 1991).

Photoacclimation involves the change in macromolecular composition in all aspects of the photosynthetic apparatus. Photoacclimation is the phenotypic adjustments that a cell makes in response to changes in environmental conditions—in the short term. Thus, in the short term, as irradiance changes, not only can chlorophyll (chl) *a* per cell change, but so too do the antennae pigments. Similarly, as light increases, the pigment content per cell decreases, while photosynthetic rates—and growth rates—may increase.

When cells are exposed to new conditions, sometimes gradually, and over many generations, they can adapt (or not). Mutations are the basis for adaptation. Adaptation involves the acquisition of new metabolic traits. Such experiments are of particular interest now, in this era of climate change, and many such experiments are being undertaken. In the context of microbial biotechnology, there are also efforts underway to engineer traits to maximize photosynthetic efficiency, temperature tolerance, CO_2 assimilation, lipid production, and the like.

Here, photoacclimation will be explored with the aim toward understanding how the cell knows when and how to regulate, the factors that affect regulation and the consequences for cell stoichiometry and ultimately for growth. The regulatory mechanisms that function in photoacclimation will be compared to acclimation strategies for cells under varying nutrient conditions.

8.2 Stepping Back: A Reminder that Units (Including Photosynthetic Units) Matter

A *P* vs *I* (or *P* vs *E*) curve can be parameterized in a number of ways. It can be measured as Electron Transport rates (ETR), oxygen (O_2) production, and/or Carbon (C) assimilated (Chap. 7). It can also be normalized to cell biomass, which can be expressed as chl *a*, cell numbers, cell C, or any other biomass metric. The early literature is replete with studies addressing whether cells regulate via adjusting their photosynthetic unit size or number. If a photosynthetic unit increases in size, it is due to the increase in optical cross-section, that is, an increase in accessory pigments, allowing for a greater "funnel" of pigments to transfer light to chl *a*. If a cell increases its number of photosynthetic units, the ratio of accessory pigments to chl *a* stays the same but the total amount of absorbed light increases. These differing relationships have consequences for how a P vs I curve "looks" if a cell is grown under conditions of high light (reduced numbers of units or their size) or low light (increased numbers of units or their size; Fig. 8.1). Decades ago, it was the general view that cells generally were pre-disposed to have either one strategy or another (either the ability to alter photosynthetic unit size or number). This was borne out of

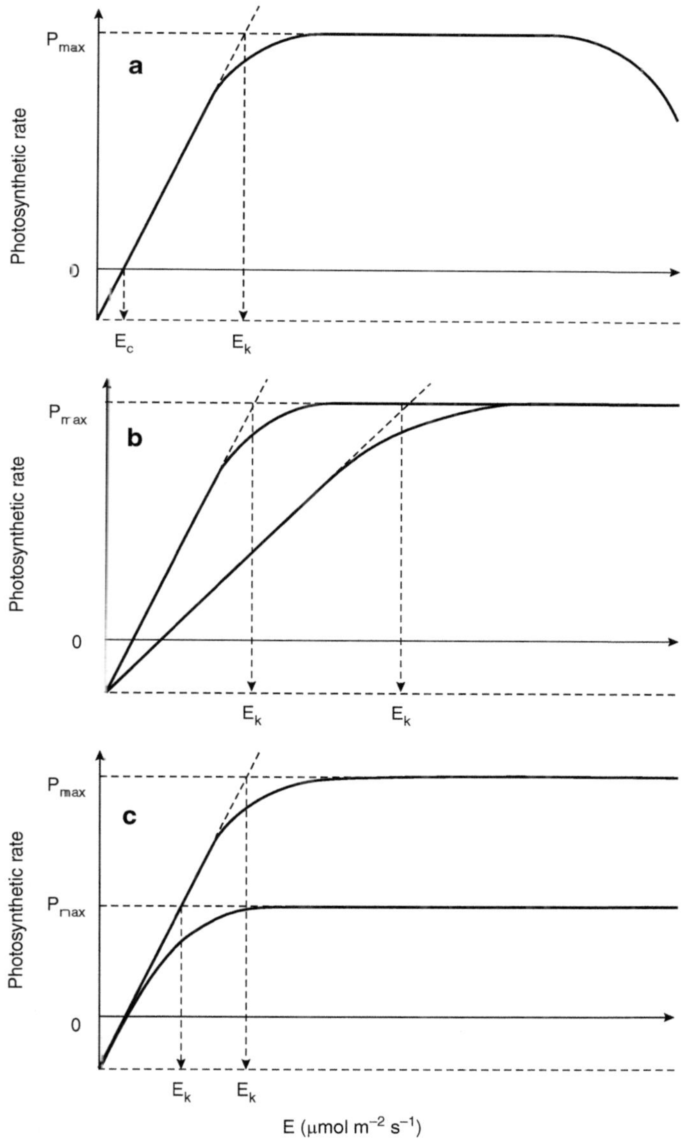

Fig. 8.1 Photosynthesis—irradiance response curves: E_c, irradiance compensation point; E_k, saturating irradiance; and P_{max}: maximum photosynthetic rate. Panel (**a**) typical plot; Panel (**b**) comparison of two curves with different slopes: keeping constant the number of photosynthetic units, but increasing the functional absorption cross-section, the slope increases; and Panel (**c**) comparison of two curves with different maximum photosynthetic rate: increasing the number of photosynthetic units, P_{max} increases. Image redrawn from https://biocyclopedia.com/index/algae/working_with_light/photosynthesis_irradiance_response_curve_p_versus_e_curve.php

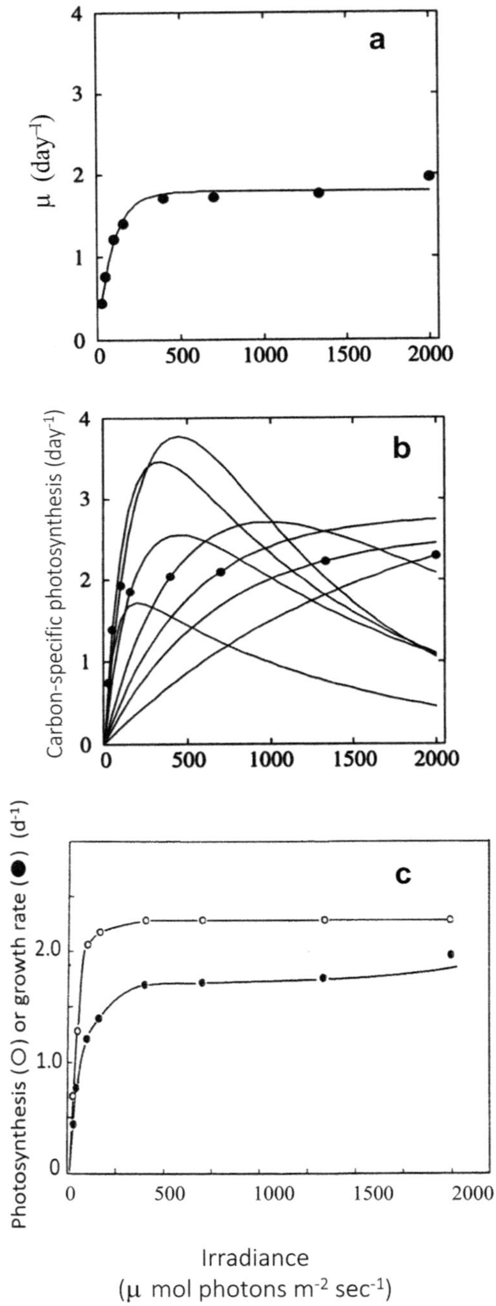

Fig. 8.2 Panel (**a**) The effect of growth irradiance on growth rate. Panel (**b**) In situ carbon uptake rates across an irradiance range for each pre-adapted culture of *Synechococcus* WH7803 shown in

the typical comparative study in which cultures (or natural samples) were placed in two light conditions, high and low.

Thirty-plus years ago, the then-newly-discovered, picocyanobacterium *Synechococcus* (see Chap. 2) was considered a low-light adapted organism. It was found at depth in the sea, and when it was found at the surface, it was severely bleached in color, leading investigators to surmise that it was stressed. Kana and Glibert (1987a, b) carried out a suite of experiments in which they grew this species in the laboratory and slowly adapted the cells to higher and higher light intensities. They measured photosynthetic rates (using ^{14}C methodologies), growth rates, as well as pigmentation and other cell biomass parameters. Their goal was to explore how the cells adapted over a gradient of light intensities, not just a "high" and "low" treatment. Growth rate increased with light intensity, following a saturation response, up to the highest light level tested, 2000 μE m^{-2} sec^{-1} (essentially what the cells would see at the surface under full sun; Fig. 8.2a). They observed that the photosynthesis response for each acclimated growth rate was different, with photosynthesis exceeding that required to match growth at each light intensity. However, this was most pronounced for cells acclimated to low light, and the maximum rates of photosynthesis (P_{max}) at low light far exceeded that which was required to balance growth (Fig. 8.2b). On the other hand, for cells acclimated to high light, P_{max} was better matched to that which was needed for growth (Fig. 8.2b). In other words, at light limitation for growth, there was an excess capacity for of photosynthesis that was not utilized under the growth conditions. Although the total chl *a* did decline from low to high-light grown cells, the accessory phycobilins declined far more. Across the entire gradient, the cell produced just enough pigment to yield a photosynthetic rate while operating a high quantum yield; there was no excess pigment wastefully produced that would operate at a low quantum yield. The consequence is that the cell's capacity for photosynthesis is maximized to produce a balance point that satisfies the maximal growth rate, μ_{max}. The cell demonstrated all of the previously described strategies (photosynthetic unit numbers and size) in a single organism when observed over a growth irradiance range that encompassed growth-limiting to growth-saturating intensities. Thus, it was not an either/or response. Comparisons of other species yield similar relationships: that is, excess capacity for photosynthesis at low light compared to high light (Fig. 8.3). This implied that there must be a single mechanism for photoacclimation rather than multiple or differing strategies. This mechanism was subsequently shown to be the reduction state of the plastiquinone (PQ) pool. How PQ serves this role and the implications are next discussed.

Fig. 8.2 (continued) Panel (**a**). The points on each curve in Panel (**b**) indicate the rate of photosynthesis at the growth irradiance. Panel (**c**) A comparison of the photosynthesis rates at each growth irradiance (open circles) with growth rates at the same irradiances (closed circles). Note that across the gradient, the photosynthesis rate (d^{-1}) exceeded the growth rate, implying the excess "capacity" for photosynthesis relative to that which was required to just match growth rate. Images redrawn from Kana and Glibert (1987b, Deep-Sea Res 34: 497–516) with permission of Elsevier

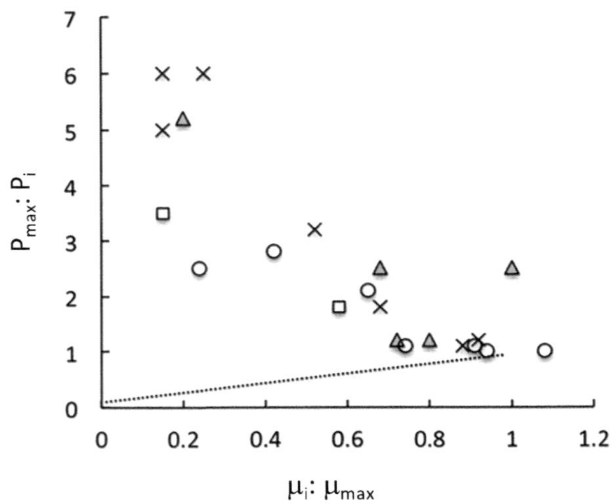

Fig. 8.3 Comparison of the relationships between P_{max}:P_i (P_i is the in situ photosynthetic rate at a given light intensity) and μ_i:μ_{max} (m_i is the in situ growth rate at a given light intensity). Four species are compared: circle, *Synechococcus* WH7803; triangle, *Microcystis aeruginosa*; crosses, *Phaeodactylum tricornutum*; and squares, *Alexandrium tamarensis*. The fine dashed line represents the equivalency of the rates. Figure redrawn from, and original data sources given in, Kana and Glibert (1987b, Deep-Sea Res 34: 497–516). Image reproduced from Kana and Glibert (2016, doi.org/10.1007/978-3-319-30259-1–8) with permission of Springer Nature

8.3 Plastiquinone and Its Role as a "Light Meter"

The central role that PQ plays in regulating photosynthesis has been well described in previous chapters (Chaps. 6 and 7). Studies on *Dunaliella* sp. (Chlorophyta) have shown that the redox state of the PQ pool controls the transcription of the *cab* gene that directs the production of the antenna pigment–protein (Escoubas et al. 1995; Maxwell et al. 1995). Maxwell et al. (1995) have further pointed out that an increase in light intensity or a decrease in temperature can lead to a decline in *cab* mRNA, light harvesting, and chl *a*. Thus, PQ becomes a "light sensor," via the degree of excitation upon it, and it regulates the electron flow between PSII and PSI (see also Chap. 6). It is also a "redox sensor" and "connects photosynthesis with metabolism, light acclimation, and stress tolerance" (Havaux 2020, p. 1252). There have been many recent advances in understanding the regulation of photosynthesis upon changes in environmental cues. The supercomplex of PSII and its light-harvesting pigment–proteins is functionally connected to that of PSI and state transitions occur in order to balance the excitation levels of the two photosystems when conditions change. From work on cyanobacteria and green algae to higher plants, the key role of redox sensing/signaling associated with PQ has been confirmed (reviewed by Hüner et al. 2012).

One way to view this regulation is as a proportion of the flux of light energy into the cell and the rate of use of reductant leading to growth, energy "in" vs energy "out" (e.g., Kana et al. 1997). An interesting feature emerging from this notion of a ratio of "in" vs "out" is that it applies equally well across irradiance, temperature, nutrient gradients, and diverse species (Geider et al. 1996, 1997, 1998; Kana et al. 1997). Thus, photoacclimation of pigments is not strictly a function of light intensity, it also requires knowledge of how the rate of light absorption is related to the rate of assimilation. The ratio of absorbed light to capacity for photosynthesis defines the *dynamic balance* regulation at any point along the irradiance response curve. Accordingly, pigment responses scale to growth rate, the "grand integrator." Viewing this concept in relation to a *P* vs *I* curve, it is clear that a cell does not "ride up and down" a *P* vs *I* curve only as irradiance changes, but instead the entire shape of the curve at any given point in time can change as a consequence of the nature of the nutrient acquisition machinery and the rate constants of a suite of assimilating enzymes (Kana and Glibert 2016).

One implication of this dynamic balance dance that PQ plays as irradiance and other factors change, is that it operates as much through control of pigment losses as it does through pigment synthesis (Kana et al. 1997; Fig. 8.4). The number of photosynthetic units adjusts to a level where energy in and energy out are in balance. Pigment downregulation can be achieved by converting functional PSII units into non-functional ones. This may involve cleavage of the D1 protein from this complex, a switch-off that can be reversed. It follows, then, that at light limitation, the dynamic balance is dominated by the synthesis of new photosynthetic units, while at P_{max}, the opposite is the case; it is dominated by pigment degradation.

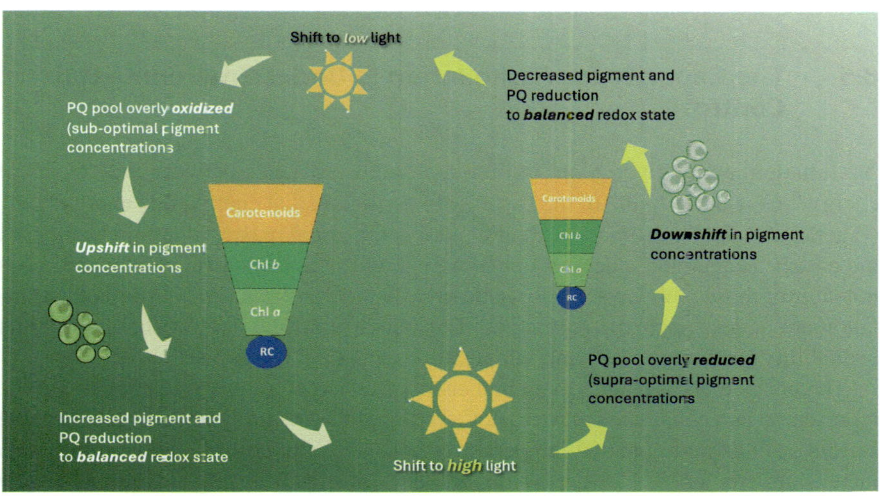

Fig. 8.4 The dynamic balance of photoacclimation as a cell is shifted to low light or high light. Low light will result in an upshift in pigmentation, while high light will result in a downshift in pigmentation. After the appropriate shifts, the cell reaches a balanced redox state

There is a continual push and pull of PQ regulation as a cell moves through different irradiance regimes. Shift a cell to lower light and the PQ pool oxidizes and sends a signal to initiate a shift up in pigment–protein synthesis. When this is accomplished, the PQ pool is able to increase reduction and becomes poised at an acclimated state (Fig. 8.4). Now shift the same cell to higher light and the PQ pool becomes overly reduced, signaling a downshift in light-harvesting pigment synthesis, ultimately leading to a new stable, acclimated state at the higher light (Fig. 8.4).

In all, the slow step, the bottleneck in electron transport through PSII, PQ is truly the ringmaster, directing the extent of up- or downregulation.

8.4 Units Matter: Interpretations of Acclimation Depend on How Photosynthesis Is Normalized

The relationship between photosynthesis and irradiance and the interpretation of acclimation depends on how the data are normalized, as different components of the photosynthetic apparatus change differently with irradiance supply (Kana and Glibert 1987a, b; MacIntyre et al. 2002). Rates of photosynthesis are most commonly reported per chl *a*, but they can also be reported per C, per N, or per concentrations of accessory pigments. Carbon-specific rates can be directly compared to rates of growth (Fig. 8.2). Normalizations based on chl *a* or based on cell numbers will not account for variations in cell volume or mass (MacIntyre et al. 2002). Maximum pigment: biomass ratios can differ by an order of magnitude, and cell volumes/cell sizes up to three orders of magnitude. Thus, comparisons of photosynthetic responses between or among species and conditions are insufficient unless all are scaled similarly.

8.5 The Capacity for Assimilation Is Under Environmental Control

Acclimation at any given point in time represents a balance between the "push" of energy "in" and the "pull" of the capacity to assimilate that energy, i.e., energy "out" (Fig. 8.5). In photoacclimation, the PQ pool is the grand master directing changes at the energy acquisition side and funneling energy to assimilatory pathways. It does not control those pathways. Those pathways, from C assimilation to nutrient assimilation are under enzymatic regulation. Thus, environmental control such as nutrient availability (including iron) and temperature regulate those processes.

The balance between energy in and energy out across a light gradient can also be represented as a change in chl *a* to cell C. With light as the limiting substrate (holding factors such as nutrients and temperatures constant), the chl *a*: C will decline with increasing irradiance. Now, increase nutrient (especially nitrogen, N) availability and drive increased growth rates (as in a nutrient-limited chemostat, Chap. 3), the chl *a*:C ratio will increase directly with growth rate. Recall that chl *a* is N rich: it is a porphyrin ring chelated to four N atoms (Chap. 5). Lastly, add a change in

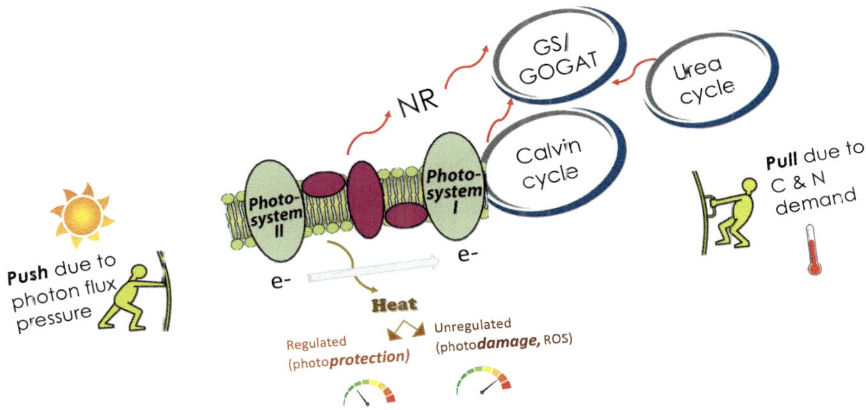

Fig. 8.5 Conceptual figure representing the push and pull of energy and metabolic reactions (carbon assimilation via the Calvin cycle, nitrogen assimilation via the GS/GOGAT pathway and urea metabolism via the urea cycle (see Chap. 12 for more details on these pathways). Image reproduced from Ahn and Glibert (2022, Phycology 2: 30–44, doi.org/10.3390/phycology2010002) under a creative commons license

temperature. Broadly speaking, metabolism increases with increasing temperatures (see Chap. 23), and this is especially true for the enzyme Rubisco, which has a comparatively high temperature optimum. Proportionately more C can be assimilated as temperatures warm, but this may be modulated depending on the nutrient condition of the cells. More details of the model and parameterizations of these relationships can be found in Geider et al. (1996, 1997), and Kana et al. (1997). More recently, Flynn (2001) has developed a mechanistic model that describes the growth of phytoplankton as functions of NH_4^+, NO_3^-, light, iron, silicon, phosphorus (P), and temperature.

8.6 Is There Comparable Dynamic Balance Control in Nutrient Assimilation?

Nutrient uptake typically follows a saturating response with respect to nutrient availability. And, like the years of efforts spent measuring and cataloging *P* vs *I* curves, the literature is replete with comparisons of "fixed" kinetic constants, comparing half-saturation values and maximal uptake rates for species and conditions. Like *P* vs *I* curves, it is now well known that saturating response curves depend on the physiological state of the cell, and the manner in which an experiment is conducted (see Chap. 12 for discussion with respect to N measurement methodologies).

Decades ago, McCarthy and Goldman (1979) first demonstrated that when cells that were grown in N-limited chemostats across a gradient of growth rates were pulsed with saturating concentrations of NH_4^+, the maximum capacity for uptake

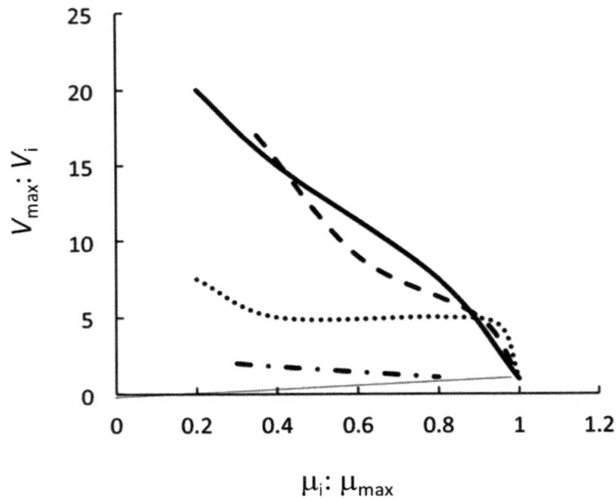

Fig. 8.6 Comparison of the relationships between V_{max}:V_i (where V_i is the rate of nutrient assimilation that just balanced growth) and μ_i:μ_{max}. Four species are compared: solid line, *Phaeodactylum tricornutum*; dashed line, *Thalassiosira weissflogii*; dotted line, *Chaetoceros simplex;* and dot-dash line, *Dunaliella teriolecta.* The fine dashed line represents the equivalency of the rates. Figure redrawn from and original data derived from Goldman and Glibert (1982). Image reproduced from Kana and Glibert (2016, doi.org/10.1007/978-3-319-30259-1–8) with permission of Springer Nature

(V_{max}) exceeded that necessary to just "match" the N required for growth. The excess capacity was greater at low relative growth rates and at low substrate concentration levels. Follow-up experiments with multiple species confirmed this response (Goldman and Glibert 1982; Fig. 8.6). Growth rate (μ_i:μ_{max}) again emerged as the "grand integrator" of metabolism.

In terms of N acquisition and metabolism, there are many parallels with the dynamic balance concept of "energy in" vs "energy out." However, in this case, one needs to think about "materials in" vs "materials out." The balance between carrier proteins and the enzymes for nutrient metabolism are analogous to the balance of pigments and the enzymes for C assimilation (Kana and Glibert 2016). In nutrient assimilation, the carriers at the cell surface are analogous to the light-harvesting apparatus (see Chaps. 12, 13, and 14, for details on transporters and assimilation in N, P, and Si metabolism). In terms of nutrient acquisition, however, there are nuances not perfectly analogous with light acquisition, as different types of transporters function differently as nutrient content and form change (Kana and Glibert 2016; Chap 12). Nevertheless, the general concept holds.

The analogy can be taken one step further. If the "light meter" for photosynthesis is the PQ pool, is there a "nutrient meter"? In terms of N, one can envision this being the NH_4^+ assimilation enzyme pool. All N forms are ultimately reduced to NH_4^+ once taken up by the cell (Chap. 12). Assimilation subsequently occurs via a series of reactions involving the enzymes glutamine (Gn) synthetase (GS) and glutamate

(Glu) synthase (GOGAT). The availability of Gln and the Gln/Glu ratio governs the reducing capacity in the cell (Kana and Glibert 2016). In all, cells have "tuning knobs," the PQ redox state, and the relative levels of Gln:Glu that serve as signals to upregulate or downregulate acquisition strategies to meet the needs of the cells when resources are low, and to switch off acquisition when resources are sufficient. Approaches to incorporate a dynamically fluctuating balance between acquisition and assimilation in nutrient kinetics have been advanced with new approaches (e.g., Smith et al. 2009; Bonachela et al. 2012), and these are described in more detail in Chap. 12.

8.7 Summary

Both light acquisition and CO_2 assimilation and nutrient acquisition and assimilation have strong biological regulation between the demand side (getting what is needed when the supply is low) and the assimilation of the resource (the maximal rate always set by the biochemical reactions and their constants) for each set of environmental conditions. The implications are many. First, resource stress can develop before availability of the resource declines below a conventionally defined half-saturation value. Excess uptake capacity is built into the cells' resource acquisition strategy when resources are low. Such regulation is fine-tuned and balanced at a steady state. However, natural communities are rarely growing at steady state under single limiting resources and fixed concentrations. Conceptualizing these relationships between physiological processes and growth as dynamic rather than as fixed entities and kinetic relationships has further implications for understanding the variability in cell properties and ecosystem metabolism. Thus, *P* vs *I* curves and nutrient kinetic curves are nothing more than a momentary consequence of the acquisition machinery and the assimilation machinery and their rate constants for the given environmental conditions at any point in time. Nevertheless, the ability to perform a dynamic dance can vary across taxa. The next two chapters examine taxonomic differences and how evolutionary relationships help us understand how responses of different taxa are related.

References

Ahn SH, Glibert PM (2022) Shining light on photosynthesis in the harmful dinoflagellate *Karenia mikimotoi*–responses to short-term changes in temperature, nitrogen form and availability. Phycology 2:30–44. https://doi.org/10.3390/phycology2010002

Bonachela JA, Raghib M, Levin SA (2012) Dynamic model of flexible phytoplankton nutrient uptake. Proc Natl Acad Sci 108(51):20633–20638. https://doi.org/10.1073/pnas.1118012108

Escoubas JM, Lomas M, La Roche J, Falkowski PG (1995) Light-intensity regulation of CAB gene transcription is signaled by the redox state of the plastoquinone pool. Proc Natl Acad Sci USA 92:10237–10241

Falkowski PG, LaRoche J (1991) Acclimation to spectral irradiance in algae. J Phycol 27:8–14. https://doi.org/10.1111/j.0022-3646.1991.00008.x

Flynn KJ (2001) A mechanistic model for describing dynamic multi-nutrient light, temperature interactions in phytoplankton. J Plankton Res 23:977–997. https://doi.org/10.1093/plankt/23.9.977

Geider RJ, MacIntyre HL, Kana TM (1996) A dynamic model of photoadaptation in phytoplankton. Limnol Oceanogr 41:1–15

Geider RJ, MacIntyre HL, Kana TM (1997) Dynamic model of phytoplankton growth and acclimation: responses of the balanced growth rate and chlorophyll a:carbon ratio to light, nutrient-limitation and temperature. Mar Ecol Prog Ser 148:187–200

Geider RJ, MacIntyre HL, Kana TM (1998) A dynamic regulatory model of phytoplankton acclimation to light, nutrients and temperature. Limnol Oceanogr 43:679–694

Goldman JC, Glibert PM (1982) Comparative rapid ammonium uptake by four species of marine phytoplankton. Limnol Oceanogr 27:814–827

Haveaux M (2020) Plastiquinone in and beyond photosynthesis. Trends Plant Sci 25:1252–1265. https://doi.org/10.1016/j.tplants.2020.06.011

Hüner NPA, Bode R, Dahal K, Hollis L, Rosso D, Krol M, Ivanov AG (2012) Chloroplast redox imbalance governs phenotypic plasticity: the “grand design of photosynthesis” revisited. Front Plant Sci 3:255. https://doi.org/10.3389/fpls.2012.00255

Kana TM, Glibert PM (1987a) Effect of irradiances up to 2000 μE m^{-2} sec^{-1} on marine *Synechococcus* WH7803: I. Growth, pigmentation, and cell composition. Deep-Sea Res 34: 479–495

Kana TM, Glibert PM (1987b) Effect of irradiances up to 2000 μE m^{-2} sec^{-1} on marine *Synechococcus* WH7803: II. Photosynthetic responses and mechanisms. Deep-Sea Res 34: 497–516

Kana TM, Glibert PG (2016) On saturating response curves from the dual perspectives of photosynthesis and nitrogen acquisition. In: Glibert PM, Kana TM (eds) Aquatic microbial ecology and biogeochemistry: a dual perspective. Springer International Publ, Cham, Switzerland, pp 93–104. https://doi.org/10.1007/978-3-319-30259-1

Kana TM, Geider RJ, Critchley C (1997) Photosynthetic pigment regulation in microalgae by multiple environmental factors: a dynamic balance hypothesis. New Phytol 137:629–638

MacIntyre HL, Kana TM, Anning T, Geider RJ (2002) Photoacclimation of photosynthesis irradiance response curves and photosynthetic pigments in microalgae and cyanobacteria. J Phycol 38:17–38

Maxwell DP, Laudenbach DA, Huner HPA (1995) Redox regulation of the light harvesting complex II and cab mRNA abundance in *Dunaliella salina*. Physiol Psychol 109:787–795

McCarthy JJ, Goldman JC (1979) Nitrogenous nutrition of marine phytoplankton in nutrient-depleted waters. Science 203:670–672

Raven JA, Geider RD (2003) Adaptation, acclimation and regulation in algal photosynthesis. In: Larkum AWD, Douglas SE, Raven JA (eds) Photosynthesis in algae. Kluwer, Dordrecht, The Netherlands

Smith S, Yamanaka Y, Pahlow M, Oschlies A (2009) Optimal uptake kinetics: physiological acclimation explains the patterns of nitrate uptake by phytoplankton in the ocean. Mar Ecol Prog Ser 384:1–12

Additional Websites Accessed

https://biocyclopedia.com/index/algae/working_with_light/photosynthesis_irradiance_response_curve_p_versus_e_curve.php. Accessed 12 Jan 2023

The Algae and Their (Imperfect!) Classifications

9

9.1 Introduction: Defining Species and Classifications

Ever since Aristotle, the classification of plants and animals has been a fascination (South and Whittick 1987). Scientific classifications are highly technical and precise, and at the same time very messy. Taxonomists frequently recategorize species based on new knowledge—especially so now that more and more molecular data are being amassed. We aim to speak precisely with our colleagues, and we try to teach distinct definitions. Yet, we are also sloppy in terminology, based on historical or colloquial designations. Take, for example, the Joshua Tree—the iconic "tree" of the desert of the southwestern USA. A national park is named after it (the Joshua Tree National Park). These trees were named after the Biblical figure Joshua because Mormon settlers felt they were guided by their outstretched arms (https://www.nwf.org/Educational-Resources/Wildlife-Guide/Plants-and-Fungi/Joshua-Tree). Yet, Joshua trees are not trees at all—they are succulents belonging to the genus *Yucca*. Even knowing that, we continue to call them Joshua "trees."

It was earlier introduced that the terms "phytoplankton" and "algae" are used throughout this book as general terms. In this volume, the focus is on planktonic microalgae, although benthic species, multicellular algae, and symbiotic algae are contrasted where appropriate. More precisely, we would refer to the collective aggregate of these organisms as microbial photoautotrophs, but this precision is not always practical. New terminology will be introduced in later chapters, and while the importance of being accurate in usage is important, the context always matters. A general term may be the most descriptive for a given context even when we, as precise, data-driven scientists, know it is not the most accurate.

The Swedish botanist Linnaeus (1754) first applied the term "algae" to a mix of algae and bryophytes (mosses and liverworts, primitive vascular plants). Fritsch (1935) proposed a broad classification of the algae and defined them as, "Unless purely artificial limits are drawn, the designation of an alga must involve all holophytic organisms (as well as their colourless derivatives) that fail to reach the level of differentiation of archegonite plants" (reproduced in South and Whittick

P. M. Glibert, *Phytoplankton Whispering: An Introduction to the Physiology and Ecology of Microalgae*, https://doi.org/10.1007/978-3-031-53897-1_9

1987, p. 5). In other words, they are "oxygen-producing, photosynthetic, non-vascular organisms with unprotected reproductive structures" (South and Whittick 1987, p. 5), wherein photosynthesis refers to oxygenic photosynthesis conducted with the same biochemistry as that of higher plants. Van den Hoek et al. (1995, p. 9) recognized the unnatural "grouping" of the algae and simplified their definition as "algae are very diverse photosynthetic plants that have neither roots nor leafy shoots, and which also lack vascular tissues; certain algae are not able to photosynthesize but are nevertheless classified as algae because of their close resemblance to photosynthetic forms." The definition by Graham et al. (2009) states that "algae are (with numerous exceptions) aquatic organisms that (with frequent exceptions) are photosynthetic oxygenic autotrophs that are (except for the kelps) typically smaller and less structurally complex than land plants." In other words, the term "algae" has, at best, a very loose definition! There is—and never has been—a definition of algae that fits all of the many organisms considered in the term.

The term "phytoplankton" comes from the Greek words, *phyto*, meaning plant and *planktos* for wandering. Thus, this definition covers both prokaryotes and eukaryotes—all the photosynthetic wanderers of water. But, just as in use of the term algae, the term phytoplankton is often loosely used to include some taxa that are exclusively heterotrophic. Such dinoflagellates are lumped together with phytoplankton by many ecologists, while others consider the heterotrophs separately as zooplankton. About half of the known dinoflagellate species are also mixotrophic (see Chap. 15). In this book—and for lack of better options—the definitions of the terms algae and phytoplankton will follow Van den Hoek et al. (1995).

The classification of algae—historically based on pigments, food reserve materials, and morphology under light microscopy (LM)—has long been in a state of debate and flux. Harvey (1836, not the same Harvey introduced in Chap. 1) first recognized four major divisions (phyla) of algae based on *predominant color* (brown, red, green, and the golden-brown diatoms). From the 1970s on, scanning electron microscopy and transmission electron microscopy (SEM and TEM, respectively) greatly advanced understanding about the fine structure of algal cells and, from there, dramatically altered how some taxonomic groups were described (Moestrup 2006). Beginning in the 1990s, the "molecular revolution" has reshaped our thinking about systematics. The field of algal taxonomy, faced with such upheaval, is thus sometimes viewed "as an essentially retrogressive and outmoded science unable to cope with the current biodiversity crisis" (De Clerck et al. 2013). An excellent database, mostly current although "running to keep up" with constant changes, is the web resource AlgaeBase (www.algaebase.org; Guiry and Guiry 2023). It is frustrating, but not surprising, that algae undergo frequent name changes. This creates challenges for students and others researching literature on a particular species; it may not be obvious that they were known by a different name in the past. Some examples of name changes among common taxa are given in Table 9.1, and many more synonyms are given by Tomas (1997).

Part of the difficulty, including within the "molecular takeover" of algal taxonomy, is that defining species is an ever-challenging and ever-evolving problem. The original species concept, the *biological species concept*, defines species as those taxa

Table 9.1 Representative examples of microalgal species names in the older literature and the currently accepted/recommended name

Currently accepted name	Older name/synonym
Dolichospermum sp.	*Anabaena* sp.
Planktothricoides raciborskii	*Plantothrix raciborskii*
Prymnesium polylepis	*Chrysochromulina polylepis*
Karlodinium veneficum	*Karlodinium micrum* *Gymnodinium veneficum* *Gymnodinium galatheanum* *Gyrodinium galatheanum*
Karenia brevis	*Gymnodinium breve* *Ptychodiscus brevis*
Noctiluca scintillans	*Noctiluca miliaris*
Prorocentrum cordatum	*Prorocentrum minimum* *Exuviella marie-lebouriae* *Prorocentrum traingulatum* *Exuviaella minima*

Modified from Tomas (1997) and Borowitzka (2016). See also the World Register of Marine Species (www.marinespecies.org)

that interbreed and produce viable offspring (Graham et al. 2009). This concept is, at best, a major challenge considering cyanobacteria, the prokaryotes of the algal realm, that do not engage in sexual reproduction. Difficulties extend to the eukaryote algae: breeding experiments are difficult; many eukaryote algae only produce gametes under select environmental conditions and many do not—or rarely—undertake sexual reproduction.

In algal taxonomy, two concepts are especially prevalent, genospecies and morphospecies (e.g., Manhart 1992). In the *genospecies concept*, a species is the smallest group of organisms that exhibits at least one molecular character. In contrast, the *morphological species concept* relies on morphological differences, that is, structural characteristics that are relatively easy to distinguish under LM or SEM. The morphological information, combined with dominant pigments and food reserves, has been widely used to differentiate species and describe algal diversity. Yet, another approach is to distinguish species based on phylogeny, *the phylogenic species concept*, which differentiates species based on at least one distinctive and unifying structural, biochemical or molecular characteristics (Graham et al. 2009). New algal species at present are commonly defined using a combination of morphological and molecular characters.

Biology further complicates these complexities. As one example, cryptic species are defined as two or more taxa erroneously grouped under a single name because they are indistinguishable morphologically, yet are incapable of interbreeding and have considerable genetic disparity (Korshunova et al. 2019). "Cryptic" can also refer to species that are easily overlooked due to small size, nondescript appearance, or other features. Another complication is the well-known phenomenon of strain differences, wherein algal populations within the same species differ so strikingly in various fundamental characteristics (life history stages, homo/heterozygosity, resting stages, morphological features, toxin production, etc.) that they are more similar to *other species* than to other populations within the same species (Burkholder and Glibert 2006 and references therein, Borowitzka 2016).

While taxonomy is simply defined as the science of classification of organisms, systematics is also strongly considered now in discerning taxa from phylum to species levels. Systematics is the study of organismal diversity and evolutionary relationships (phylogenetics) among organisms. Phylogenetic studies describe the stages in the evolutionary history of groups of organisms, and provides a foundational basis for their classification. The evolutionary history is commonly represented as a "phylogenetic tree" that indicates relatedness of species and higher groupings.

Molecular sequences (DNA or RNA) are now heavily relied upon to differentiate major classes of algae; in some cases, new phyla as well; and the degree of relatedness within and among groups. They are also being used to define—and identify—species and gene expression in natural environments. Species level identifications are necessary for many applications, as for example, the fact that expression of toxicity by the potentially harmful diatom genus *Pseudo-nitzschia* differs depending on species. This becomes important in activities such as predictive models, where individual species niches differ and a model built on characteristics of one species may not be transferable to a different species (Chap. 28). While the aggregation of traits may be useful for characterizing how and why certain types of taxa occur where they do (Chap. 18), it is important to keep in mind differences between species and between clones of the same species, as it may be a slight difference in species may explain an unusual occurrence, an unusual laboratory outcome, or a model prediction that does not fit the presumed calibration data set.

The purpose of this chapter is to introduce the major phyla of phytoplankton—not with the intent of covering all of their features, but rather to give a "flavor" of the diversity, distribution and evolutionary relationships. Thus, this chapter is lengthy yet is only able to cover most major algal groups (phyla) only briefly. Consider this a highly condensed "moving target" taxonomic overview. By getting a sense of the relationships between—and differences among—major taxonomic groups, the hope is that it may be easier to understand how physiological responses of these different groups relate. The organization described herein is based on classic works (South and Whittick 1987; Bold and Wynne 1978; Margulis et al. 1990; Van den Hoek et al. 1995; Barsanti and Gualtieri 2014; Graham et al. 2009, 2016), updated to include more recent information and changes contributed by Graham et al. (2016, and references therein) and many others. Emphasis is placed on the pigments and cell structures that relate to the physiological processes emphasized in this book; other sources should be consulted for more details on microstructures, life histories, and taxonomy and phylogenetics.

9.2 General Classifications and Key Features of Algae

Recognizing that many differences of opinion remain regarding the details of classification—differences of opinion that will not be expanded on here, a generally accepted classification recognizes 11 major phyla (pl.; sing., phylum; word ending,

-phyta) or divisions of algae (Table 9.2). Phyla and classes differ in molecular profiles, cellular fine structure (flagella, chloroplasts, nuclei, other organelles, wall presence/absence/structure, etc.), pigmentation, main food reserves, and other biochemical characteristics (Table 9.3). Not only their diversity, but also their regulation by environmental factors such as light and nutrients, are critically important in understanding phytoplankton success.

A distinguishing morphological feature of different major algal groups is the fine structure of their chloroplasts. The term "plastid" is derived from a Greek word meaning formed or moldable (Solymosi 2012). Due to the different (primary, secondary, tertiary or serial) endosymbiotic events by which the type of plastid known as chloroplasts have been acquired by host organisms (see Chap. 10), chloroplasts either have a double membrane "envelope" or 3–4 (or more) outer membranes.

9.3 Dominant Algal Groups

9.3.1 Cyanobacteria

Cyanobacteria have been abundant on Earth for at least 1.5 billion (perhaps up to 3 billion) years, being the most ancient photosynthetic, oxygen-producing organisms. They are aptly described (Lang and Walsby 1982) as,

> often inobtrusive inhabitants of nearly every habitat from polar pond to thermal pool, the spray-soaked littoral fringe of the sea to the sunbaked rock varnish of the desert, and from the plankton to the benthos in freshwater lakes and hypersaline pools.

As Graham et al. (2009) noted, there is ongoing, decades-long debate as to whether these organisms should be classified as "blue bacteria" emphasizing their prokaryote cell structure, or emphasizing their ecological role as primary producers like eukaryotic algae (e.g., Lewin 1976; Stanier et al. 1978; Golubic 1979). Cyanobacteria reproduce by simple binary fission. Similarly as for most other algal groups, cyanobacterial systematics are in a "fluid state of affairs" (Graham et al. 2009). There presently are about 2700 described cyanobacterial species (Nabout et al. 2013). Planktonic cyanobacteria are the seasonal numerical dominants of many aquatic ecosystems; they are notorious high-biomass, toxic "bloom" formers in eutrophic lakes and rivers worldwide, as well as the most widely distributed picoplankton in marine systems (see Chap. 25).

The cyanobacteria ("blue bacteria"), also commonly known as blue-green algae, are prokaryotes, and therefore, like eubacteria, they have no organelles—no nuclei, mitochondria, chloroplasts, etc. (Fig. 9.1). Their simple morphology includes coccoid unicells, groups of such cells in mucilaginous colonies, or strings of cells connected together in filaments (trichomes). Although they mostly have small cells (length ~2 μm), they can range in size from the minute *Prochlorococcus* of the oligotrophic oceans (0.5–0.8 μm; Chisholm et al. 1992) to some tropical

Table 9.2 Classification and common general distribution of algae, following the International Code of Botanical Nomenclature for eukaryotes (from Graham et al. 2016 and references therein, with number of species from Hopes and Mock 2015, unless otherwise indicated). Bolding/italics indicates predominance in estuarine/marine (M) versus freshwater (F) aquatic habitats, true of some groups that "cross the salinity gradient"

Phylum (P) or Division (D)[a]	Class(es)[b]	Common name	Aquatic habitat	Described species[c]
D. Cyanobacteria (historic, P. Cyanophyta)	Chroobacteria, Hormogoneae	Cyanobacteria, blue-green algae	***F***, M	~2700
P. Glaucophyta	Glaucophyceae	Glaucophytes	F	~20
P. Rhodophyta	Bangiophyceae, Compsopogonophyceae, Cyanidiophyceae, Florideophyceae, Porphyridiophyceae, Rhodellophyceae, Stylonematophyceae	Red algae (nearly all are macroalgae)	***M***, F	~6500
P. Dinophyta	Desmophyceae, Dinophyceae[c]	Dinoflagellates	***M***, F	~3300
P. Haptophyta	Pavlophyceae, Coccolithophyceae	Haptophytes (some, coccolithophorids)	***M***, F	~600
P. Heterokontophyta (mostly photosynthetic stramenopiles)	Aurearenophyceae, Bacillariophyceae, Boldiophyceae, Chrysomeridophyceae[d], Chrysoparadoxophyceae[d], Chrysophyceae, Dictyochophyceae, Eustigmatophyceae, Pelagophyceae, Phaeophyceae, Phaeosacciophyceae[d], Phaeothamniophyceae, Pinguiophyceae, Raphidophyceae, Schizocladiophyceae, Synchromophyceae, Synurophyceae, Xanthophyceae[c]	Mostly golden-brown algae (xanthophyceans—yellow-green algae)	***M***, F	~16,000 (diatoms, ~50–80%)
P. Chlorophyta	Prasinophytes ("still a ragbag"; class(es) not assigned)[c], Trebouxiophyceae, Ulvophyceae, Chlorophyceae	Green algae	***F***, M	~4500
P. Streptophyta	Charophyceae[c], Chlorokybophyceae, Coleochaetophyceae, Klebsormidiophyceae, Mesostigmatophyceae (including some prasinophytes!)	Green algae	***F***, M	~3500

P. Euglenophyta	Euglenophyceae	Euglenophytes, euglenoids	***F***, M	~1325
P. Chlorarachniophyta	Chlorarachniophyceae	Chlorarachniophytes	M	20[c]
P. Cryptophyta	Cryptophyceae	Cryptophytes, cryptomonads	M, F	~200[c]

[a] The order given for the major groupings roughly follows what is known about their evolutionary "age," except that glaucophytes are listed adjacent to cyanobacteria because of the clear, close relationship of their chloroplasts to extant coccoid cyanobacterial cells. Division is used under the International Code of Bacteriological Nomenclature. Phylum or division is used for eukaryote algae; this table follows preference for phylum as in Graham et al. (2016)

[b] Under the International Code of Botanical Nomenclature, the class ending is -phyceae. Cyanobacteria are under the International Code of Bacteriological Nomenclature

[c] Species number for dinoflagellates from Barsanti and Gualtieri (2006). Species number for glaucophytes is based on 14 species from Barsanti and Gualtieri (2006) together with at least 1 species in each of 6 additional genera Seckbach (1994)

Cyanobacteria ("blue bacteria")—number of species from Nabout et al. (2013), with an estimated ~4700 remaining to be described. Cyanobacterial taxonomy is, as politely stated by Graham et al. (2009), "a fluid state of affairs." This division now includes the "apple green" prochlorophytes, which previously were within a separate division

Red algae—At least three major taxonomic schemes have been published; this is the simplest

Dinoflagellates—only the two major photosynthetic ("algae-like") classes are included here (but note: these classes also include heterotrophic members). Excluded are at least two classes of parasitic dinoflagellates (some photosynthetic, generally poorly known) and two classes of other heterotrophic and photosynthetic taxa. Indication of the ongoing status of dinoflagellate taxonomy is captured by Graham et al. (2009): "Because the overall phylogeny of dinoflagellates is not yet well understood...the genera [in updated group descriptions] have not been clustered into classes, orders, or families"

Heterokontophytes—an amazingly huge, diverse grouping of algae; species range from tiny unicells (diameter ~2 μm) to giant kelps up to 60 m long with primitive vascular tissue (Graham et al. 2009). Classes listed are taken from Graf et al. (2020) and references therein. This phylum is based especially on molecular sequence data and flagellum fine structure (especially the distinctive straw-like "hairs" or mastigonemes found on one of the two usually heterokont flagella in swimming cells of many taxa. In species number, classes vary from the Bacillariophyceae (with ~8000 species or more) to the Aurearenophyceae, Schizocladiophyceae, and Synchromophyceae (thus far each with one species)

Streptophyte algae—classes as in Mikhailyuk et al. (2018) and references therein

Cryptophytes—taxonomy is in *major* disarray (Novarino 2012; Hoef-Emden and Archibald 2017). As examples, the genera *Cryptomonas* and *Campylomonas* are linked in the same life cycle, and the genus formerly known as *Chilomonas* is now realized to consist of nonphotosynthetic species of *Cryptomonas*. Molecular science is slowly providing a firmer foundation for the taxonomy of this small but ecologically important group

[d] The class name Chrysomeridophyceae should be used rather than "Chrysomerophyceae" (Graf et al. 2020). The small marine classes Chrysoparadoxophyceae and Phaeosacciophyceae were erected by Wetherbee et al. (2019) and Graf et al. (2020), respectively

Table 9.3 Some basic features of algal phyla (here, including Division Cyanobacteria), compiled from van den Hoek et al. (1995), Graham et al. (2016), and references therein unless otherwise indicated. Note that molecular information (18S rRNA and rbcL sequences, gene phylogenies) is omitted here

Division or phylum	Diagnostic pigments	Chloroplast and/or thylakoid features	Food reserves	Cell covering(s)	Flagella	Other notable features
Cyanobacteria	Chlorophyll (chl) *a* only. Phycobilins phycocyanin (PC), allo-PC, phycoerythrin (PE). Carotenoids β-carotene, myxoxanthin, zeaxanthin Note: the "apple-green" subgroup (prochlorophytes) has chlorophylls *a* and *b*, lacks phycobilins, and has more α-carotene than β-carotene (Griffiths 2006)	No chloroplasts Most: Thylakoids singly occurring (unstacked) + phycobilisomes, free in cytoplasm Apple-green subgroup: Stacked thylakoids lacking phycobilisomes, free in cytoplasm	Cyanophycean starch (cyanophycin—an arginine and asparagine polymer as an α-1,4-glucan)	Wall consisting of peptidoglycans; often with an outer mucilaginous sheath	–	Carboxysomes enhance carbon acquisition; phycobili-somes; these are the only prokaryotes with the same photosynthetic biochemistry as higher plants; many can "fix" nitrogen gas, including heterocyte cells in some filamentous species
Glaucophyta	Chl *a* only. Phycobilins PC, allo-PC Carotenoids β-carotene, zeaxanthin, β-cryptoxanthin	Bright blue-green chloroplasts (2 to many per cell); each with a thin vestigial peptidoglycan wall and inner membrane, and in	Mainly glucose from photosynthesis, in chloroplast; and glaucophyte starch (an α-1,4-glucan) in cytoplasm	Wall ± depending on the species and life history stage; if present, wall consists of peptidoglycan ± cellulose. Also covered by a	2, unequal subapically inserted in a shallow depression; each with 2 rows of delicate "hairs"	Chloroplasts are extremely similar to coccoid cyanobacteria cells in fine structure, pigments, etc.

		a special vacuole; thylakoids singly occurring; no girdling lamella		superficial layer of flat vesicles subtended by microtubules		
Rhodophyta	Chl *a* only Phycobilins PE, PC, allo-PC. Carotenoids α and β-carotene, lutein, zeaxanthin, antheraxanthin, violaxanthin	1 to many chloroplasts, each with a simple envelope (2 outer membranes); thylakoids singly occurring (unstacked); no girdling lamella	Floridean starch (α-1,4-glucan substance similar to the amylopectin of higher plants) and protein crystals, both stored outside the chloroplast	Extracellular matrix (ECM, a loose cellulosic microfibrillar network with amorphous gel-like hydrophilic sulfated galactan polymers and mucilage; ± calcite/aragonite crystals and/or Mg, Fe, Zn). Often with an outer multilayered proteinaceous "cuticle"	–	Primitive reds—biphasic life history with shell-boring stage. Advanced reds—unique triphasic life history, often with a protective sterile jacket of cells around sexually produced offspring. Phycocolloids (agars, carageenans)
Dinophyta	Usually chl *a* and *c* but, due to symbiotic events, some species have chl *a* and *b* Carotenoids β-carotene; peridinin, fucoxanthin, diadinoxanthin, dinoxanthin, gyroxanthin	Phototrophs: 2 to many chloroplasts per cell, each with at least 3 (4) membranes. Thylakoids in stacks of 3 (2–4); no girdling lamella	True (higher plant) starch (amylose + amylopectin, α-1,4 + α-1,6 linkages) outside the chloroplast; and lipids (including some unique sterols)	Vegetative cell outer covering is the amphiesma, a peripheral layer of membrane-bound vesicles = outer continuous membrane + thecal vesicles (± cellulose plates) ± pellicle + (innermost) cell	Vegetative cells—usually 2, with mastigonemes ("hairs"): Transverse flagellum has a striated strand and is helically structured (causes cell to whirl and move forward). Longitudinal	Nucleus with permanently condensed chromosomes; unusual organelles such as mucocysts, projectiles (trichocysts, nematocysts, taeniocysts), oscelloid (ocellus—eye-like

(continued)

Table 9.3 (continued)

Division or phylum	Diagnostic pigments	Chloroplast and/or thylakoid features	Food reserves	Cell covering(s)	Flagella	Other notable features
				membrane A few also with outer organic scales	flagellum swings in a narrow arc and acts as a rudder (steers)	organelle), peduncle, etc.
Haptophyta	Chl *a* and *c*. Carotenoids α- and β-carotene; fucoxanthin and derivatives	Usually 1–2 chloroplasts; 4 membranes around each; thylakoids in stacks of 3; no girdling lamella	Chrysolaminarin (type of β-1,3-glucan formed outside the chloroplast); sometimes also lipids. (Note: *Pavlova* has paramylon, another type of β-1,3-glucan as its major food reserve)	Lacking; or fibrillous organic scales embedded in mucilage (1-several layers); or cellulose scales; or $CaCO_3$ scales (coccoliths); or, rarely, outer siliceous scales	Usually 2 (4) flagella without "hairs" (except in *Pavlova*); no flagellar swelling; unique flagellar basal body structure	Mostly biflagellate unicells, or colonies with flagellate stages; also can be amoeboid, coccoid, palmelloid, or filamentous. Unique haptonema organelle and distinct Golgi
Heterokontophyta	Chl *a* and *c*. Carotenoids α-, β-, and ε-carotenes; fucoxanthin (predominant in most classes), violaxanthin	1 to several chloroplasts per cell; mostly with 4 outer membranes (including 1 that is connected to the nuclear envelope). Thylakoids in 3s; girdling lamella (e) present in most species	Chrysolaminarin (type of β-1,3-glucan), outside the chloroplast. Browns (pheophyceans) may also have mannitol, sucrose, and/or glycerol; and lipids inside chloroplasts	Absent; or organic or siliceous scales; or siliceous wall pieces; may have a lorica. Browns—wall of 3 components in at least 2 layers: outer mucilage and sometimes a cuticle of mostly alginic acid and	Most species have stage(s) with 2 heterokont, subapically inserted: Forward—long pleuronematic flagellum (+ tripartite, stiff "hairs" or mastigonemes in 2 rows). Backward	Huge, diverse group ranging from unicells to complex thallus with true tissues, meristems, and primitive phloem. Phycocolloids (algin, alginic acid)

				alginates; cellulose provides the main structural support; + amorphous alginic acid and fucoidan, fucans in the intercellular matrix	shorter (usually) acronematic flagellum with basal swelling (senses light, + stigma). + transitional helix	
Chlorophyta	Chl *a* and *b*. Carotenoids α-, β-, and γ-carotenes; lutein, prasinoxanthin, or siphonein or siphonoxanthin	1 to many chloroplasts per cell; envelope (2 membranes); thylakoids stacked in 2–6–20 (grana may be present); no girdling lamella	True (higher plant) starch (amylose + amylopectin, α-1,4 + α-1,6 linkages) inside the chloroplast	Most with a cell covering: ± scales (sometimes in layers), or ± a cell wall that may consist of cellulose, algaenans, or other materials	Usually 2 flagella (1–16) per cell. Flagellar scales and "hairs" uncommon; rhizoplast usually present	Photorespiratory enzymes—most with glycolate dehydrogenase
Streptophyta	Chl *a* and *b*. Carotenoids α-, β-, and γ-carotenes; lutein, prasinoxanthin	1 to many chloroplasts per cell; envelope (2 membranes); thylakoids stacked in 2–6–20 (grana may be present); no girdling lamella	True (higher plant) starch (amylose + amylopectin, α-1,4 + α-1,6 linkages) inside the chloroplast	Most with a cell wall mainly of cellulose; scales usually present	Many without flagella. If present, flagellar scales common, no "hairs," and usually no rhizoplast	Photorespiratory enzymes—glycolate oxidase and catalase in peroxisome. Include closest extant ancestors of higher plants. Some with a protective sterile jacket of cells surrounding sexually produced offspring

(continued)

Table 9.3 (continued)

Division or phylum	Diagnostic pigments	Chloroplast and/or thylakoid features	Food reserves	Cell covering(s)	Flagella	Other notable features
Euglenophyta	Chl *a* and *b* Carotenoids α-, β-, and γ-carotenes; neoxanthin, diadinoxanthin	3 chloroplast membranes; thylakoids in stacks of 3 or more; no girdling lamella	Paramylon (type of β-1,3-glucan), lipids	No cell wall; but a protein-rich pellicle (periplast) lies beneath the cell membrane in the cytoplasm	2 flagella (1-several) but usually only 1 emergent flagellum with a paraflagellar rod	Nucleus with permanently condensed chromosomes; metaboly ("cell squirting") movement. Some have a cytostome used in prey capture
Chlorarachniophyta	Chl *a* and *b*. Carotenoids lutein, neoxanthin, violaxanthin	1 or many bilobed chloroplasts per cell, each with 4 outer membranes and a prominent pyrenoid. Thylakoids loosely stacked in 3s; no girdling lamella	A type of β-1,3-glucan stored in cytoplasmic vesicles that are often tightly associated with pyrenoids	Flagellate cells have 1 flagellum that wraps around the cell when swimming	Cell wall (single- or multilayered) in some stages; others without a wall	Solitary amoeboid, coccoid, and/or flagellate cells. Nucleomorph (remnant nucleus) in the periplastidal compartment
Cryptophyta	Chl *a* and *c*. Phycobilins—either PC or PE. Carotenoids α-, β-, and ε-carotene; alloxanthin	1–2 chloroplasts per cell; each surrounded by 4 membranes. The 3rd membrane can be continuous with the outer nuclear envelope. Thylakoids stacked in 2s (sometimes more); no girdling lamella	True (higher plant) starch (amylose + amylopectin, α-1,4 + α-1,6 linkages) grains between the outer 2 and inner 2 chloroplast membranes (periplastidal compartment)	No cell wall, but have a periplast (pellicle); the cell membrane is sandwiched between protein layers that form the periplast	2 slightly unequal flagella with fine "hairs"; sometimes + organic rosette scales. Unique rhizostyle flagellar apparatus	Nucleomorph in the periplastidal compartment. Other unique organelles: ejectisomes, Maupas bodies

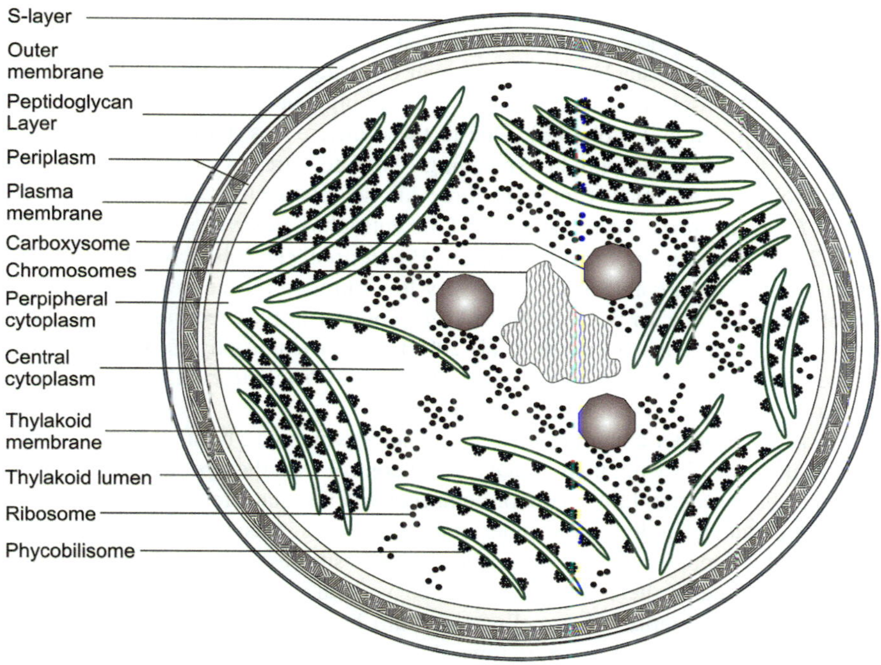

Fig. 9.1 Schematic detailing the ultrastructure of the cyanobacterium *Synechocystis* sp. PCC 6803 showing various subcellular components. Image reproduced from Mills et al. (2020, Biosci Rep 40(4): BSR20193325. doi: 10.1042/bsr20193325) under a creative commons license

cyanobacteria filaments with diameters at ~100 mm (Schulz-Vogt et al. 2007). The cell wall in cyanobacteria consists of peptidoglycans, and is multilayered (Griffiths 2006), differing substantially from the cell walls of eukaryote algae which are mainly lipid bilayers with proteins on either side. Most cyanobacteria have an array of internal thylakoid membranes which incorporate their photosynthetic pigments. They contain chl *a*, but in most taxa phycobilin pigments predominate (see Chap. 5). Cyanobacteria also have abundant carotenoids that help protect the cells from harmful UV light. For example, zeaxanthin is associated with the cellular membrane rather than the thylakoids, suggesting that it serves a light-shielding function rather than as an accessory pigment in light harvesting for chl *a* (Kana et al. 1988; Griffiths 2006).

Reflected in an historic name for this group ("Myxophyta," slime plants), cyanobacteria cells commonly are surrounded by a thick, mucilaginous slimy, protective sheath. While sexual reproduction is absent in this important group, conjugation can result in some genetic recombination (Doolittle 1979; Gale et al. 2019).

A small subgroup within this division are presently called "apple-green" or "green" cyanobacteria; it is difficult to make sense of such names, since they literally mean "apple-green blue bacteria." The fine structure of the apple-green subgroup is distinct from that of most cyanobacteria in that the thylakoids are stacked rather than singly occurring; the stacked thylakoids can be structured in folds, similarly as in green algal chloroplasts (Griffiths 2006); and the pigment complement includes chl *b* but no phycobilins (Table 9.3). Due to these differences, they are in a separate division, the Prochlorophyta (meaning "before green plants"). Although prochlorophytes are mostly marine, three known filamentous species within the genus *Prochlorothrix* have been found in eutrophic freshwater lakes (Pinevich et al. 2012). Among the Prochlorophytes, *Prochlorococcus* also contains a divinyl form of chl. They are able to thrive in very low-nitrogen (N) environments such as the oligotrophic oceans because the chl *a/b* protein complex requires much less N than phycobiliproteins (Graham et al. 2009). All Prochlorophytes have abundant carotenoids, dominated by β-carotene, except for *Prochlorococcus* which has a higher proportion of α-carotene (Griffiths 2006).

Some cyanobacteria can produce gas vesicles which can be important in buoyancy regulation, allowing them to have some control over their position in the water column. Gas vesicles have rigid proteinaceous walls that are fully permeable to gases, but with a hydrophobic inner surface that prevents water influx. In quiet waters with a dependable pressure gradient, ideally such as a stratified water column, gas vacuoles enable the cells to control their vertical position in the water column (Fig. 9.2). For example, when cells are in dim light, they tend to form more gas vesicles, causing them to move up toward the water surface. In contrast, photosynthesis in high light results in production of sugars that increase the internal cell turgor pressure, causing gas vesicles to shrink. The cells sink into deeper, more nutrient-rich waters. Cyanobacteria buoyancy regulation involves slow changes in external pressure that afford cells sufficient time to adjust their internal pressure (Reynolds and Walsby 1975). If conditions change too rapidly from low to high light, such as during or following a storm event, when calm re-establishes, the cyanobacterial cells are still "over-buoyant." Another reason for over-buoyancy is that senescent cells cannot increase their internal turgor pressure enough to sink, or photosynthesis becomes limited by the rate of CO_2 diffusion through calm water (Graham et al. 2009). Alternatively, based on laboratory studies of the cyanobacterium *Microcystis*, it has been hypothesized that low N and low light in combination may not limit carbohydrate production of cells in deeper waters, so that they are not able to increase their gas vesical production enough to rise toward the water surface (Fig. 9.3).

Cyanobacteria gain some protection from damage by high-UV surface light through increased carotenoid production and their mucilaginous sheaths. But if over-buoyant cells are trapped by the strong force of surface tension at the water surface, they become stressed. In turn, their decomposing high surface biomass can block oxygen from reaching aquatic life below, creating hazardous conditions for beneficial aquatic life that are exacerbated if the dying cyanobacteria release toxins.

Many cyanobacteria are capable of fixing nitrogen (N_2) gas. Filamentous N_2 fixers (diazotrophs) often accomplish this process within specialized cells known as heterocytes (formerly heterocysts; Graham et al. 2009). N_2 fixation is inhibited by

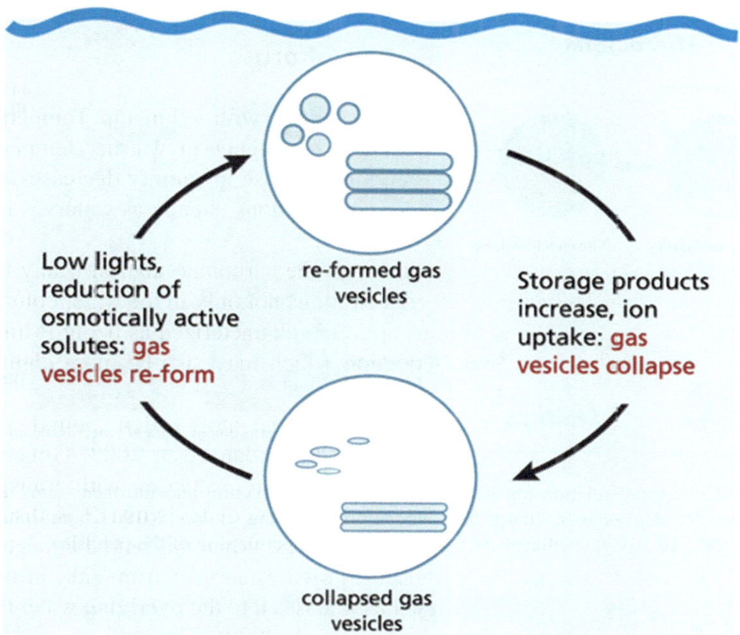

Fig. 9.2 Gas vesicles and the buoyancy of cyanobacteria having gas vesicles. Under low-light conditions, gas vesicles form, causing the cells/colonies to rise upward in the water column. At the surface, enhanced photosynthetic production may give rise to higher turgor pressures and the accumulation of dense storage particles, causing the cells to sink. As cells at lower depths utilize the stored nutrients, gas vesicles may re-form, reinitiating the cycle. Image modified and redrawn based on Graham et al. (2016, Algae. LJLM)

oxygen. The thick-walled heterocytes block oxygen from reaching the key enzyme involved in the process, nitrogenase. More information about N_2 fixation and oxygen protective mechanisms, including in some cyanobacteria that lack heterocytes, is given in Chaps. 6 and 11.

9.3.2 Glaucophytes

This small but structurally diverse freshwater group consists of only 14 described species of unicellular flagellates, coccoid unicells, or planktonic and benthic colonies (Fig. 9.4), occasionally abundant in soft freshwaters (bogs, acid swamps). Cells are readily discerned under LM by their bright blue-green chloroplasts in unpigmented cytoplasm. Glaucophytes are evolutionarily important because they have provided key insights about algal lineages (Graham et al. 2016, and references therein). Their chloroplasts are extremely similar to unicellular coccoid cyanobacterial cells; in fact, their plastids are often referred to as "cyanelles." The singly occurring thylakoids have phycobilisomes (see Chap. 5 for detailed descriptions of these structures), and

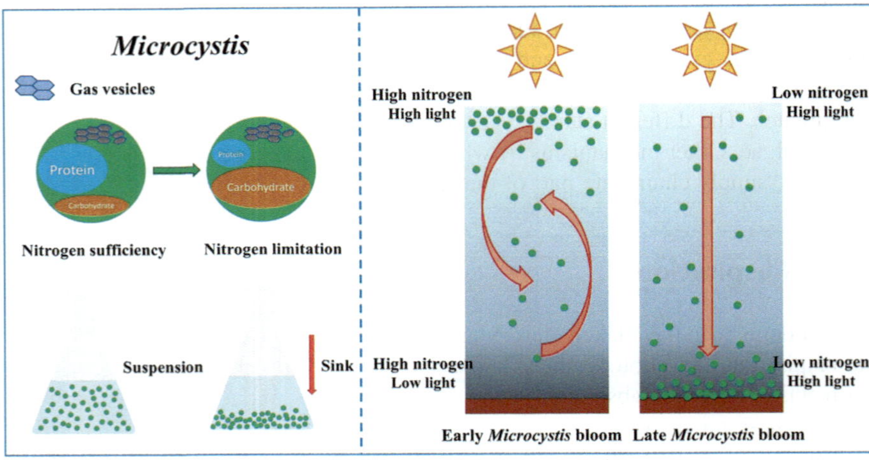

Fig. 9.3 Conceptual relationship between gas vesicle production and nitrogen status and light availability for *Microcystis*. Image reproduced from Huang et al. (2019, Chemosphere 221: 665–671. doi: 10.1016/j.chemosphere.2019.01.080) with permission of the publisher

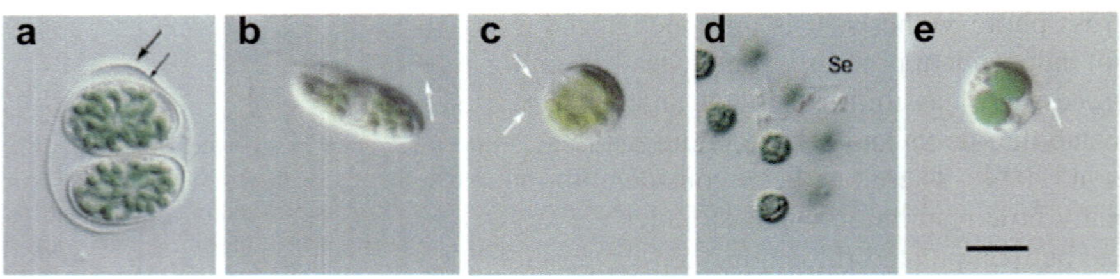

Fig. 9.4 Glaucophyte cells in light microscope with differential interference optics. Panel (**a**) Autospores of *Glaucocystis nostochinearum* (strain BBH, UNB culture collection). Offspring cell with outer wall (thin arrow) surrounded by a parent wall (thick arrow). The individual plastids are in a star-like cluster. Panel (**b**) Zoospore of *Cyanoptyche gloeocystis* (SAG 4.97) motile at image capture (arrow points to one of the flagella). Panel (**c**) A rounded spore of *Cyanoptyche gloeocystis* (SAG 4.97) with thin cellular extensions (arrows) having the appearance of flagella. Panel (**d**) Cluster of *Gloeochaete wittrockiana* cells (SAG 46.84) bearing setae (Se). Panel (**e**) *Cyanophora cuspidata* (SAG 45.841) with putative flagella (arrow). Scale: 10, 20, and 6.7 μm (**a**–**c**, **d**, and **e**, respectively). Image reproduced from Jackson et al. (2015, Acta Soc Bot Pol 84: 149–165. doi: 10.5586/asbp.2015.020) under a creative commons license

blue phycobilins pigments, phycocyanin (PC) and allo-PC, predominate (Jackson et al. 2015). Glaucophyte chloroplasts are sometimes considered intermediate between cyanobacteria and the chloroplasts of other algae and higher plants (van den Hoek et al. 1995).

Nine genera (cell length ~6–35 μm) have been described (Seckbach 1994); the four that are currently available in culture include *Cyanophora* (solitary biflagellate cells), *Cyanoptyche* (single or palmelloid cells, or zoospores), *Glaucocystis* (cells solitary or in colonies formed by retention of ovoid offspring cells within a persistent parent cell wall; flagella uncertain), and *Gloeochaete* (cells solitary or in a group of 2 or 4, with up to 2 setae [bristles or hairs] per cell) (Jackson et al. 2015). Sexual reproduction is unknown.

9.3.3 Red Algae

Rhodophytes are mostly considered important ecologically (along with their enormous importance economically) because this group also includes common macroalgae or seaweeds. Actually, rhodophytes range morphologically from unicells to filamentous and more complex multicellular forms. The life histories of many species include tough, cryptic perennating crustose stages that make them superior competitors for suitable space (Sheath and Vis 2015). Red algae are an evolutionarily old group, with a fossil record dating back ~750 million to 1.2 billion years (Graham et al. 2016 and references therein). They are almost entirely marine, with only a small number of species found in freshwaters (Sheath and Vis 2015). There are about 6000 described species in this relatively large group; even so, it is thought that their diversity is underestimated due to cryptic species (Maggs et al. 2007; Guiry 2012). None have flagella; instead, they have amoeboid male gametes. Their chloroplasts are most closely related to the cyanobacteria. Red algae have phycobiliproteins phycoerythrin (PE) and PC arranged in phycobilisomes (like cyanobacteria and glaucophytes), and their thylakoid membranes lie singly within the chloroplasts (Barsanti and Gualtieri 2006). The only chl present is chl *a*. Other accessory pigments include the carotenoids α-carotene, β-carotene, lutein and zeaxanthin, and some other pigments associated with the xanthophyll cycle have been found in some species. The triphasic life histories of advanced red algae (Class Florideophyceae) are the most complex life histories known among plants and plant-like organisms. Red algae have many mechanisms, reflected in their life histories, that allow them to thrive in habitats ranging from small streams, with unidirectional flow and scouring, to rocky intertidal and subtidal coastal waters with bidirectional flow.

The genus *Porphyridium* is one of only a few red algal phytoplankters (see Miyagishima and Tanaka 2021), widely distributed in seawater, freshwater, and moist soils (Fig. 9.5). Optimal conditions for its growth were reported at salinity 34 and pH 8 (Lu et al. 2020). *Porphyridium* also recently was discovered to be an endosymbiont within large tropical marine foraminiferan protozoa (Prazeres and Renema 2019).

9.3.4 Dinoflagellates

Dinoflagellates are dominant primary producers of the tropical/subtropical oceans, and they are also abundant in the late spring/summer plankton of temperate/subarctic seas (e.g., Fig. 9.6). They range in diameter from 0.01 to 2 mm, and are

Fig. 9.5 A culture of the unicellular alga *Porphyridium*, showing coccoid cells each with a large stellate chloroplast, and predominant red phycoerythrin pigment (scale bar, 10 μm). Image by T. Darienko (https://sagdb.uni-goettingen.de/detailedList.php?str_number=1380-1c), reproduced under a creative commons license

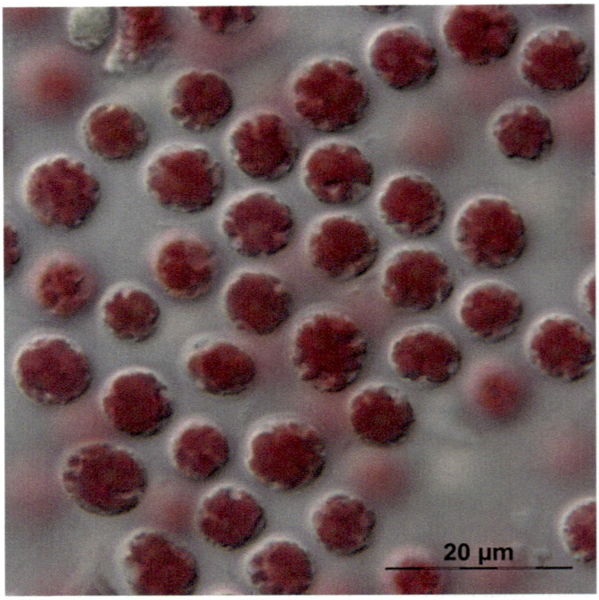

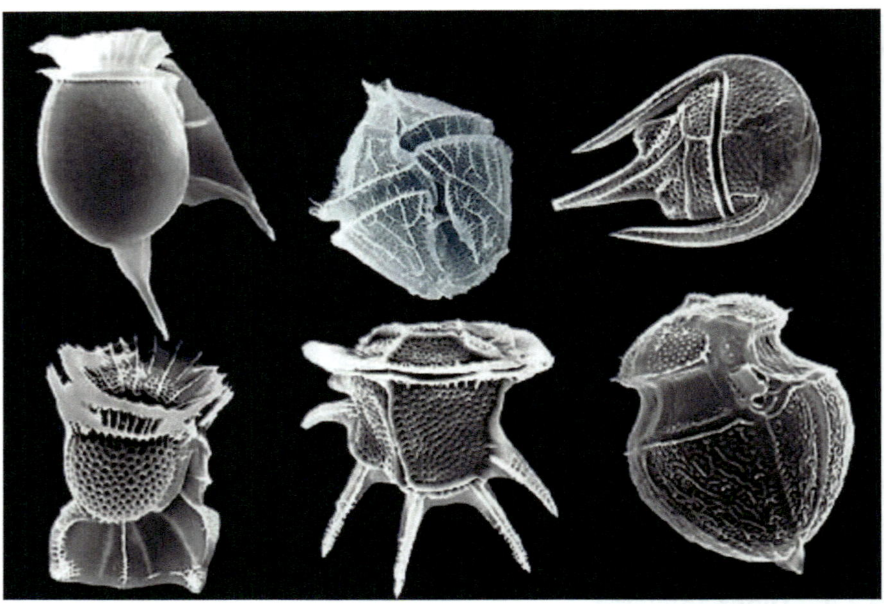

Fig. 9.6 The various morphologies of dinoflagellates. Image from fickleandfreckled and reproduced from https://commons.wikimedia.org/wiki/File:Dinoflagellates.jpg, under a creative commons license

morphologically diverse as flagellate cells (solitary or colonial), coccoid cells, amoebae, filamentous forms, or palmelloid aggregates (Spector 1984). Fossil cysts date them back at least to the Silurian (~450 million years ago), and they are thought to have been the dominant phytoplankton in Paleozoic seas (Graham et al. 2016 and references therein).

Dinoflagellates have been described as "a very aberrant protist group" cytologically (Moestrup and Daugbjerg 2007) that originated from a fundamentally heterotrophic, phagotrophic lineage (particle-feeding; Graham et al. 2016 and references therein, and see Chap. 10). About half of the known extant species are mixotrophs (Chap. 15), and the rest are heterotrophs, so this group is confusingly claimed taxonomically by both botanists and zoologists. Dinoflagellates play remarkably varied, key roles in ocean ecosystems as primary producers, predators, parasites, symbionts, and causative agents of toxic blooms that can cause disease and death for beneficial aquatic life and human consumers of seafood (see Chap. 25). Heterotrophy and symbioses are extremely well developed as well, including the endosymbiotic zooxanthellae that drive coral reef energetics. Although they are thought to be predominantly planktonic, benthic species are increasingly being recognized as ecologically important (Hoppenrath et al. 2014).

The chloroplasts of phototrophic dinoflagellates have arisen by at least eight independent symbioses, ranging from stable long-term relationships to temporary kleptoplastidic events. A diverse array of predominant pigments among these species reflects evolutionary endosymbioses across various algal divisions. Thus, dinoflagellates are thought to have acquired photosynthesis multiple times from different algal prey that became symbionts. Host cells subsequently lost the ability to photosynthesize and relied on phagotrophy to acquire more chloroplasts from algal prey (see Chap. 10).

Although this group generally has been considered to have a simple life history dominated by haploid stages, that information is based on phototrophic species. Most heterotrophic species (from free-living to parasitic) are very poorly known; some of them have been described to have complex life histories with active free-living haploid and diploid stages. Many of their transformations have been described in detailed observations, but have been difficult to photo-document because they happen sporadically and rapidly; triggering environmental and endogenous controlling factors are mostly unknown; some stages rapidly increase in size relative to the stages that produced them; others suddenly sink from near the water surface to the bottom of the slide or petri dish. In addition, many dinoflagellates have not yet been possible to culture except under nonaxenic conditions, apparently requiring unknown organic substances (Skelton et al. 2008). Others have been possible to culture only with certain prey sources (e.g., Park et al. 2006); or some stages have not been expressed in (highly artificial) culture conditions (Burkholder and Marshall 2012, and references therein).

Sexual reproduction is known for ~40 of the ~3000 species, mostly phototrophic plankters. A difficulty in recognizing sexual reproduction is that gametes can be identical in appearance to asexual (vegetative) cells. Nutrient deficiency—especially low N—has induced sexual reproduction for some species, but not others under certain circumstances. For many species, the swimming zygote stage (planozygote) is short-lived and undergoes meiosis quickly, whereas dormant nonmotile hypnozygotes can remain viable for many years (Graham et al. 2009). Temporary cysts during stressful conditions, resting cysts, and other cysts typically are part of the life history (Burkholder 1992; Dale et al. 1999; Parrow and Burkholder 2003). As van den Hoek et al. (1995) noted, "...for the great majority of dinophyte resting spores, which are very diverse in form, we do not know whether they are sexual or asexual. Both types...are often referred to simply as 'cysts.'"

Hypnozygotes and resting cysts may function to enhance survival during adverse condition. Based on relatively few species, dormancy can last for days to weeks to months; germination of phototrophs appears to require light and preferred temperature conditions. Cyst "seed beds" have—but also have not—served as inocula for blooms when conditions became favorable for germination (Taylor 1987). Population (strain) differences within the same dinoflagellate species, even within subclones from the same clone that were grown identically, can be so pronounced—for a wide array of features such as gamete numbers, cyst production, median dormancy period, resting cyst germination, progeny viability, nutrition, and toxicity—that clonal populations from a given species have been shown to differ much more than clonal populations from two distinct species (Burkholder and Glibert 2006, and references therein). These often-stark intraspecific differences appear to at least partly reflect the environmental history of clones, and/or high potential for cryptic species within "a" species that remain to be detected, separated, and profiled.

Some phototrophic and heterotrophic dinoflagellates can be bioluminescent (Fig. 9.7). The phototrophic ones are the only photosynthetic organisms to be able to emit light. The function of bioluminescence is thought to be as a deterrent to predators (Marcinko et al. 2013). For the cells to emit light, there must be a mechanical stimulus (e.g., waves breaking or physical disturbance of otherwise calm waters). For example, shear due to proximity or contact with zooplankton is known to stimulate dinoflagellate bioluminescence. Then each cell emits a short flash of light produced by the reaction of the enzyme luciferin with the substrate luciferase (a structure chemically similar to chl). Light production occurs within scintillons, the flashing units of the cell. The luciferin/luciferase reaction occurs when the scintillons are exposed to reduced pH, originating in a cellular vacuole. Bioluminescence typically occurs only at night and is subject to an endogenous rhythm (Hastings 1983, 2007).

Various other behaviors in dinoflagellates also involve a diel rhythm. Many make daily migrations to deeper waters at night and return to the well-lit surface waters during the day. There is an apparent cellular "memory" for such behavior. For example, vertical excursions by a *Ceratium* species through the water column can continue to occur for days after the cells are placed into continual darkness (van den Hoek et al. 1995 and references therein). Over a diel period, dinoflagellates can move 20–40 m up and down through the water column (Gaines and Taylor 1985).

Heterotrophic dinoflagellates derive nutrition via phagotrophy (Chap. 15). *Noctiluca* is a heterotrophic genus that lacks plastids but can harbor endosymbionts

Fig. 9.7 Bioluminescence from *Noctiluca scintillans* at the Yacht Port of Zeebrugge, Belgium. Photo by Hans Hillewaert reproduced from https://commons.wikimedia.org/wiki/File:Noctiluca_scintillans.jpg under a creative commons license

or consume other algae, including other dinoflagellates. Prey can range from bacteria and other algae to ciliates and other particles. Feeding generally involves suction of the prey cellular contents through a specialized extended organelle for feeding (van den Hoek et al. 1995). Using structures such as a peduncle, feeding tube, cytoplasmic net, tentacle, etc., heterotrophic dinoflagellates can feed on organisms much larger than themselves. Multiple heterotrophic dinoflagellates commonly engage in swarming behavior to surround, attack, and share consumption of the same prey item.

Dinoflagellates have a unique nucleus (dinokaryon) with permanently condensed chromosomes, even during interphase. These nuclei are characterized by very large genomes (range in size from 1.5 to 112 Gbp [giga-base pairs], or 0.5- to 40-fold larger than the human haploid genome) (Murray et al. 2016). In addition to extremely large amounts of DNA, the nucleus contains notably high concentrations of bivalent cations (Ca^+ and Mg^{2+}) and transition metals (Cu, Zn, Fe and Ni) (Gornik et al. 2019). Dinoflagellates have other bizarre, amazing organelles as well, such as "ballistic nematocysts" (Gavelis et al. 2017); an ocelloid, "the most elaborate photoreceptor known in a unicellular organism"; and a piston, a fast contractile appendage unknown in any other organism (Gómez 2017). The piston is responsible for locomotion via successive extensions and contractions, and with it the cell can attain speeds up to 50 mm s^{-1}. Linear velocities of dinoflagellate cells without this organelle range from 75 to 500 μm s^{-1}. The piston appears to act as a tactile organelle that scans the surrounding waters for prey such as copepod eggs, which the cell then engulfs.

Some dinoflagellates are maintained as endosymbionts in other organisms such as reef-building corals many planktonic foraminiferans, and radiolarians, and in such mode, they are known as zooxanthellae. When within the hosts, many distinctive cellular features are changed, such that the endosymbionts may only be recognizable by their dinokaryon. The zooxanthellae provide the hosts with metabolites such as glycerol, glucose, and organic acids, while the N from the host's NH_3 or urea and phosphorus (P) in phosphoglycerate and nucleic acids can be transferred to the dinophyte cells.

9.3.5 Haptophytes

Haptophytes are named for a unique organelle, the haptonema, which emerges between the two similar flagella of most flagellate cells, and can be used to assist in attachment and capture of bacterial prey (Graham et al. 2016). Life history stages may also be amoeboid, coccoid, palmelloid, or filamentous. Haptophytes are mostly golden-brown marine pico- and nanoplankton, with relatively few freshwater representatives. They have chl *a* and *c*, and their predominant pigment generally is fucoxanthin. The chloroplasts are surrounded by four membranes, thylakoids are stacked in threes, and the food reserve is usually chrysolaminarin (Table 9.2).

Cells may lack a covering, or may have one to several layers of organic scales embedded in mucilage, and/or large, ornate calcium carbonate ($CaCO_3$) scales called coccoliths (Fig. 9.8). Siliceous scales also (uncommonly) may occur. The process of calcification in coccolith-forming haptophytes is highly dependent on photosynthesis, which provides the (considerable) energy required (ATP) and the sink for CO_2 in the calcification reaction (Graham et al. 2009). Possible functions of coccoliths include helping to focus light into the cells, protection from pathogens and predators, buoyancy regulation, and acquisition of nutrients such as P (Young et al. 2009). Coccolith features (size, design, etc.) historically were used to discern species, but it is now realized that within the same species, diploid cells and haploid cells may produce different types of coccoliths, or haploid cells may lack coccoliths. The haploid and diploid stages now known for some haptophytes had been described as separate species! The (messy) taxonomy is being sorted out with heavy reliance on molecular sequence data in combination with culture studies.

This relatively small group (~600 species) of tiny cells has enormous ecological importance. Widespread across the oceans are picohaptophytes, only recently detected and described as having "massive, unique diversity...dramatically diverse and ecologically dominant" in chl *a*-normalized biomass (Liu et al. 2009). Their ecology and evolutionary success, thought to have been based on mixotrophy, may have significantly affected the oceans' carbon (C) dynamics. Sedimented coccoliths are major contributors to ocean floor limestone, and the huge deposits that have

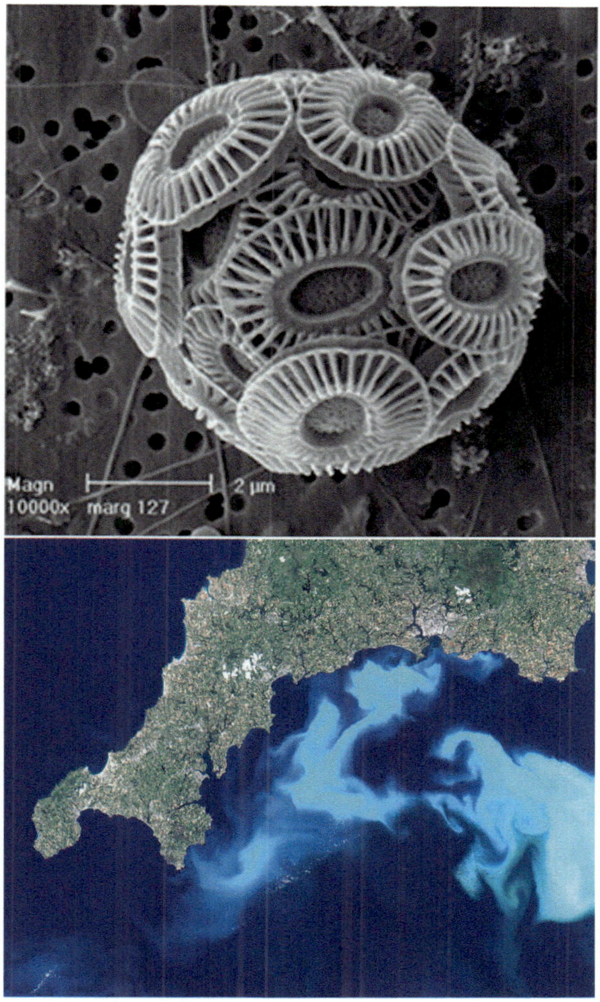

Fig. 9.8 Top: Scanning electron micrograph of *Emiliania huxleyi*. Image from Neukermans and Fournier (2018, Front Mar Sci 5: doi: 10.3389/fmars.2018.00146) and reproduced from https://commons.wikimedia.org/wiki/File:Emiliania_huxleyi_coccosphere_and_coccolith.jpg. Bottom: Landsat image of a 1999 *E. huxleyi* bloom in the English Channel. Image from S. Groom (USGS) reproduced from https://commons.wikimedia.org/wiki/File:Cwall99_lg.jpg. Both images reproduced under creative commons licenses

accumulated over time are the largest long-term sink of C on Earth. Dimethylsulfide (DMS), a volatile sulfur-containing substance produced by major bloom-forming haptophytes (below) during osmoregulation, is released to the atmosphere and transformed photochemically into sulfuric acid which can act as nuclei in cloud formation. Thus, haptophytes strongly influence biogeochemistry and climate via sequestering coccoliths' C in deep ocean deposits and by producing volatile organic substance like DMS (Graham et al. 2016).

Massive haptophyte blooms occur in the oceans. As examples, large blooms of the coccolithophorid (coccolith-producing), *Emiliania huxleyi*, can be seen from space due to reflection from the coccoliths (Fig. 9.8), and they contribute the calcareous oozes after their death. This haptophyte species has been "found in nearly every sample of ocean water and sediments from the Late Quaternary [~85,000 years ago] to the present" (Graham et al. 2016, pp. 10–17). Dense blooms (up to 1×10^5 cells mL^{-1}) can extend more than 50,000 km^2 and are most frequent in nutrient-rich temperate and subpolar waters.

Another major haptophyte bloom former in temperate and polar seas, *Phaeocystis pouchetti*, is thought to be a "species aggregate" (Medlin and Zingone 2007). Various species of *Phaeocystis* spp. produce copious mucilage (Fig. 9.9) that may act as energy/nutrients reserves and increase buoyancy. Storms and wave action

Fig. 9.9 Upper panel: Live colony of *Phaeocystis globosa*. Image from Alexandra, reproduced from https://planktonnet.awi.de/index.php?contenttype=image_details&itemid=68308#content. Lower panel: Foam produced by a *Phaeocystis* bloom. Image reproduced from https://commons.wikimedia.org/wiki/Category:Phaeocystis#/media/File:Ecomare_-_schuimalg_strand_(7037-schuimalg-phaeocystis-ogb).jpg. Both images reproduced under creative commons licenses

often beat the blooms into foams, sometimes more than 1 m thick, that clog fishing nets and suffocate fish (Lancelot et al. 1998). *Phaeocystis* spp. are especially stimulated by low N:P ratios in raw sewage and animal wastes (Riegman et al. 1992). Other harmful haptophytes include some populations of *Prymnesium parvum* and *Prymnesium* (formerly *Chrysochromulina*) *polylepis*, which can produce an array of toxins that adversely affect beneficial aquatic life (see review in Burkholder et al. 2018).

9.3.6 Heterokontophytes

This phylum, with 18 classes presently recognized, contains the most remarkably diverse array of algae across aquatic ecosystems, yet they share some key molecular, fine structural, and biochemical characteristics (Tables 9.2 and 9.3). The phylum name means "different flagella," referring to the fact that they generally have one flagellum that is usually directed forward and another backward in movement. In some species, these are only present in the gamete stage. The two flagella also are dissimilar in other features as well. The chloroplasts of heterokontophytes generally contain chl *a* and *c* (c_1, c_2), usually masked by golden-brown fucoxanthin which is an important accessory pigment that helps capture green light for chl *a* in photosynthesis. The main food reserve of the majority of huge group is chrysolaminarin, a type of β-1-3-linked glucan. Here, the most ecologically important classes (subgroups) among the 18 presently recognized are emphasized, along with brief descriptions of the others. Classes consisting mostly of microalgae are described first, ending with brown macroalgae (Phaeophyceae) which are mostly benthic.

9.3.6.1 Diatoms

Diatoms (Class Bacillariophyceae) are arguably the best known of the heterokontophytes. People have been mesmerized by the beauty of their ornate siliceous coverings since they first became the focus of research and artistic efforts in the nineteenth century—to such an extent that some wealthy individuals spent much of their lives arranging cells into "diatom art" on microscope slides, about 10 years per slide (Hasle and Syertsen 1996). Diatoms are mostly unicells but can also be colonial or filamentous microalgae, even a few filamentous macroalgae. Among "lumper" versus "splitter" taxonomists, estimates range from ~8000 to more than 100,000 species. This subgroup is extremely important both ecologically and economically. They typically are golden-brown from predominant fucoxanthin, as are many other heterokontophytes, and they occur in two basic forms: Planktonic species generally are centric (round), radially symmetrical in "face" or valve view, whereas benthic species are mostly boat-like pennate cells that are bilaterally symmetrical in valve view.

The diatom cell wall consists of two pieces (frustules or valves), likened in structure to a petri dish: The larger piece, called the epivalve, extends over the lower piece or hypovalve. Organic, somewhat elastic structures (girdle bands, etc.) wrap around the overlapping pieces and hold them together. The side view of the cell is called the girdle view; as for a petri dish, in girdle view both centric and pennate cells have a rectangular shape. Diatoms are unique because they require major amounts of silica (Si) as a macronutrient, needed to construct their cell walls. The frustules consist of polymerized silicic acid ($Si(OH)_4$; literally, glass) along with membranes of proteins, polysaccharides, and lipids. The degree of silicification varies among taxa; planktonic diatoms tend to be more weakly silicified, whereas benthic taxa appear to be strongly silicified and, therefore, heavier cells (Harris 1986; see Chap. 14 for discussion of additional factors that affect silicification). Concentrations of Si are generally abundant in natural waters, so it is an energetically inexpensive wall material (Falkowski and Raven 2007). It is a relatively heavy material and can strongly affect planktonic cells ballast regulation. Many species have long spine-like extensions that can assist the cells in staying afloat and may deter some potential grazers. Whereas centric diatoms are incapable of movement except for their flagellate sperm, pennate diatoms that have a slit (raphe) in one or both frustules can move quickly by secreting mucilage through that opening.

Diatoms communicate with the external environment via membrane-covered holes in their glass walls. The various pores (with daunting terminology) are arranged in intricate patterns, difficult to discern under LM on living cells (or on cells that were alive when preservative was added) because an outer membrane obscures the wall details. That membrane is rapidly decomposed from dead cells so that their ornamentation is clearly visible under LM and SEM (Fig. 9.10).

As a consequence of having cell walls that cannot stretch, the same diatom population in the same location on the same date usually varies substantially in cell size. This happens because during asexual (vegetative) cell division, the larger frustule (epivalve, "petri dish top") of the parent cell becomes the epivalve of one offspring cell, and the other, smaller frustule of the parent (hypovalve, "petri dish bottom") becomes the bigger valve of the second offspring cell. Through repeated vegetative reproduction over time, a portion of the population becomes smaller and smaller until there is a "point of no return" beyond which the cells cannot survive further size reduction.

Diatoms have several mechanisms at that point to regain maximum cell size. One mechanism they use is to make auxospores through sexual reproduction (Figs. 9.11 and 9.12). Changes in temperature, day length and nutrient availability may trigger this process (Graham et al. 2016). Within a population, sexual reproduction can sometimes be synchronous (completed by large numbers of cells) under favorable environmental conditions. Centric and pennate species differ in certain major features of sexual reproduction: The only flagellates found in diatoms are the sperm cells of centric species. Each sperm has one flagellum, modified in structure from that of the classic eukaryote flagellar structure. Centric diatoms are oogamous, with motile sperm and stationary "egg" (oogonial) cells. In contrast, pennate diatoms have "+" and "–" amoeboid gametes that move toward one another and fuse within a mass of mucilage secreted by the parent cells.

Although diatoms are mostly beneficial, a small estuarine/marine group of species within the genera *Pseudo-nitzschia* (26 species), *Nitzschia* (2 species), and *Halamphora* (1 species) has some populations that can produce the potent

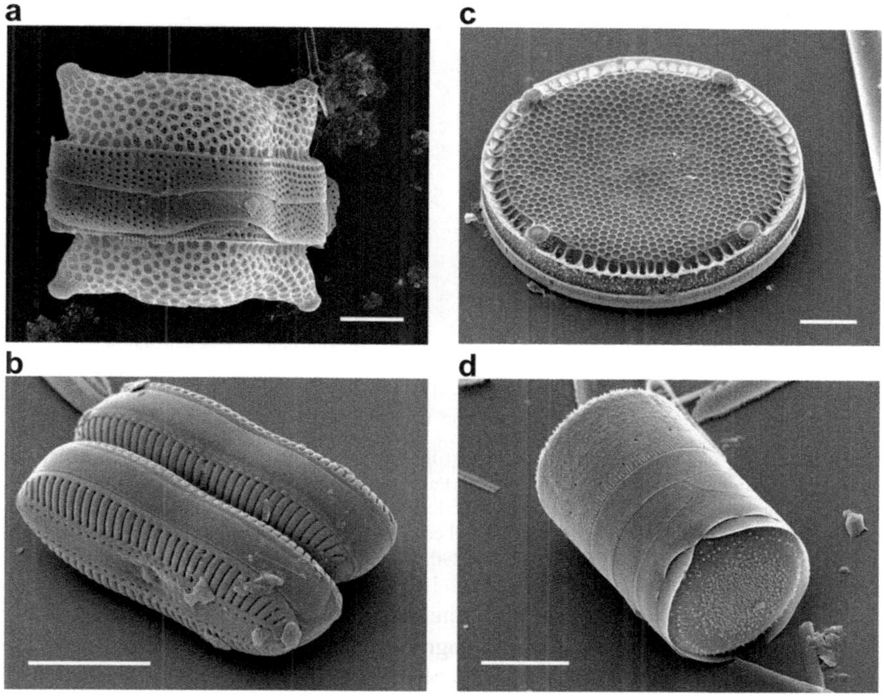

Fig. 9.10 Scanning electron micrographs of diatoms. Panel (**a**) *Biddulphia reticulata*. The whole shell or frustule of a centric diatom showing valves and girdle bands (size bar = 10 μm). Panel (**b**) *Diploneis sp.* This picture shows two whole pennate diatom frustules in which raphes or slits, valves, and girdle bands can be seen (size bar = 10 μm). Panel (**c**) Eupodiscus radiatus. View of a single valve of a centric diatom (size bar = 20 μm). Panel (**d**) Melosira varians. The frustule of a centric diatom, showing both valves and some girdle bands (size bar = 10 μm). Images reproduced from Bradbury (2004, PLoS Biol 2(10): e306. doi: 10.1371/journal.pbio.0020306) under a creative commons license

neurotoxin domoic acid (Caruana and Amzil 2018). This toxin bioaccumulates and can cause adverse impacts across trophic levels, from other phytoplankton to marine mammals (Chap. 25).

9.3.6.2 Small Subgroups, Mostly Estuarine-Marine

Eleven "minor" classes of heterokontophytes, nearly all golden-brown in color from predominance of fucoxanthin, are known thus far mostly—or entirely—from marine environments. Although these classes are small in species number, some members have major ecological, economic, and/or evolutionary "puzzle piece" importance as described in Chap. 10. The molecular phylogenetic information about these subgroups reflects the enormous diversity of heterokontophytes, including various taxa that were only recently detected and described.

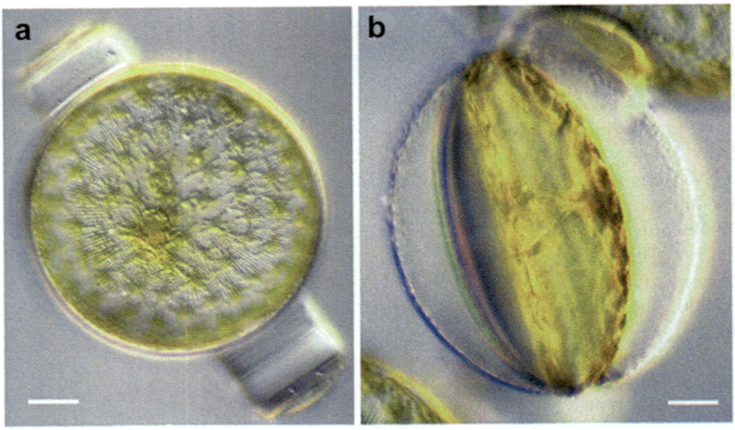

Fig. 9.11 Panel (**a**) Initial cell in valve view, still enclosed within the auxospore and containing many chloroplasts and a nucleus near its center. Panel (**b**) Girdle view of a fully formed initial cell enclosed within the auxospore, showing the spherical auxospore wall and the rugby ball-like profile of the initial cell, formed as a result of two lateral contractions of the protoplast. Images reproduced from Idei et al. (2012, PLoS One 7(8): e41890. doi: 10.1371/journal.pone.0041890) under a creative commons license

The best known among these subgroups are the dictyochophyceans, the pelagophyceans, and the raphidophyceans. Class Dictyochophyceae, silicoflagellates and pedinellids, consists of unicells that may be uninucleate flagellates or multinucleate amoebae, found mostly in estuarine and marine waters. Unlike most heterokontophytes, each cell has only one flagellum. At least one life history stage has a distinctive, highly perforated siliceous "exoskeleton" (outside the cell membrane) composed of hollow tubes of silica and surrounded by a mucilaginous sheath. The cytoplasm is also "netlike," highly reticulate (perforated, vacuolated). Silicoflagellates were much more abundant and diverse back in the Miocene (~20 million years ago), and there is only one extant genus, *Dictyocha*. Silicoflagellates are most abundant in colder marine waters. Pedinellids are unicellular marine flagellates covered with organic scales. The cells are radially symmetrical in end view.

Class Pelagophyceae was first differentiated when molecular data revealed a novel organism that was not closely related to other heterokontophyte classes (Andersen et al. 1993). There presently are ~20 genera and these species were described using DNA sequences (Wetherbee et al. 2022). Planktonic pelagophyceans are coccoid or unicellular flagellate picoplankton (diameter 1.5–3 μm) or small nanoplankton (~5–7 μm) with a novel perforated thecal covering (Wetherbee et al. 2022). Their habitats range from open ocean (e.g., the widespread *Pelagococcus, Pelagomonas*) to (mostly) coastal. Two species, *Aureococcus anophagefferens* (U.S. East Coast) and *Aureoumbra lagunensis* (Laguna Madre, Texas, USA), became well known beginning in the 1980s for forming dense,

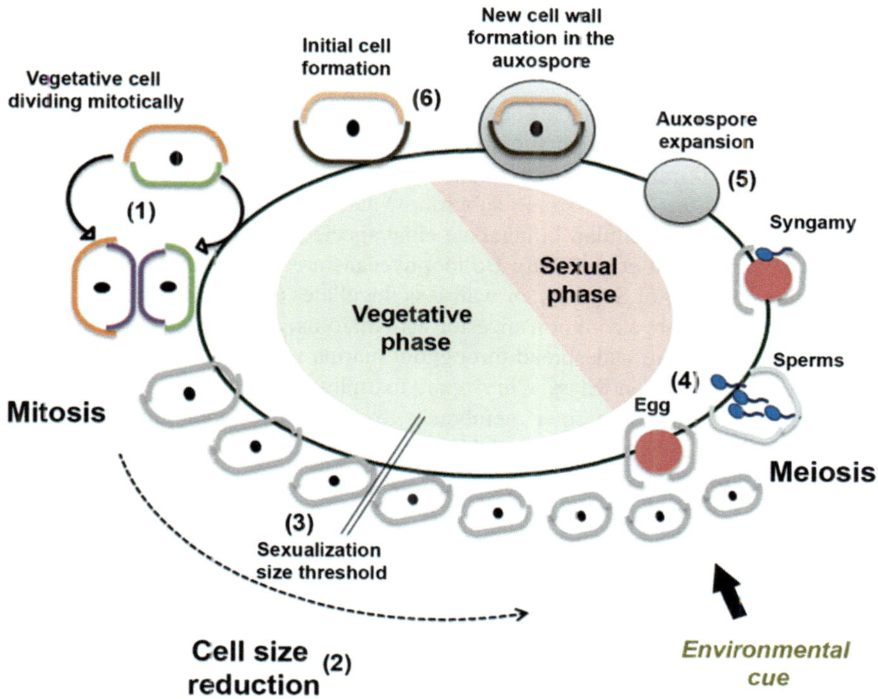

Fig. 9.12 Schematic drawing of the life cycle of a centric diatom. When a vegetative cell divides mitotically (1), each daughter cell inherits one of the two valves (thecae) of the rigid silica cell wall (orange and green lines) and builds a new smaller thecae (purple lines). This leads to a progressive reduction of the cell size of the cell population (2). Above a species-specific sexualization size threshold (SST) (3), diatom cells are incapable of performing sexual reproduction. Below the SST, if a proper trigger is present (such as a salinity shock), meiosis is induced and cells can either produce one egg or four sperms (4). Gamete conjugation (syngamy) leads to the formation of a zygote that becomes an auxospore, a soft stage which can expand (5). A new cell wall is built inside the auxospore, a soft stage which can expand (5). A new cell wall is built inside the auxospore, which eventually becomes an initial cell of the maximum cell size (6). Image reproduced from Ferrante et al. (2019, Genes 10(7): 494. doi: 10.3390/genes10070494) under a creative commons license

ecosystem-disruptive "brown tides" with many adverse impacts across trophic levels (Graham et al. 2009 and references therein; see Chap. 25).

Raphidophyceans (~30 species) are most widespread and abundant among estuarine/marine plankton, can be locally abundant in freshwaters (Graham et al. 2016 and references therein). Oddly, the estuarine/marine taxa are golden-brown due to predominance of fucoxanthin, whereas the freshwater genera are green and have minimal or no fucoxanthin. Most are relatively large flagellates (length ~30–100 μm) without a cell wall or scales; some have been shown to form benthic palmelloid masses or cysts when stressed. They may have unusual organelles such as mucocysts and trichocysts. Raphidophyceans differ from most heterokontophytes in some fine structure (e.g., no connection between a chloroplast membrane and the nuclear envelope, no transitional helix of microtubules at the base of the flagellum, no photoreceptor apparatus, etc.). About 25% (7) of the known species have populations capable of producing toxins including reactive oxygen species (ROSs such as hydrogen peroxide and superoxide anions, together with fatty acids) and/or two brevetoxins (Graham et al. 2016 and references therein; Chap. 25).

Among the other eight small subgroups, Class Aurearenophyceae thus far contains only one unicellular, biflagellate algal species that was found along sandy beaches in Japan (Kai et al. 2008). Bolidophyceans are picoplankton consisting of cells with a covering of Si plates, or wall-less flagellates (Kuwata et al. 2018). They are considered to share a common ancestor with diatoms. Although apparently minor in abundance, they are widespread throughout marine waters from tropics to poles. The species thus far in Class Chrysoparadoxophyceae ("golden paradox") has chloroplasts with only two outer membranes, one of which is continuous with the nuclear envelope. The tubular cells of this odd species, found in a high intertidal pool in Australia, grow attached to sand grains by a complex adhesive plug (Wetherbee et al. 2019). Offspring are initially amoeboid and may produce zoospores. Class Chrysomeridophyceae is in taxonomic flux (Graf et al. 2020 and references therein); the cryptic members of this subgroup include coccoid or flagellate unicells, colonies, and filaments. Unlike filamentous brown algae, filamentous forms within this class have no alginates in their cell walls.

Phaeosacciophyceans (seven species) are benthic (except for zoospore flagellates) and may be unicellular, colonial, filamentous, palmelloid, or thalloid forms (thallus, primitive "plant body" lacking vascular tissue) (Graf et al. 2020). Pinguiophyceans (Kawachi et al. 2002; pingue—fat or grease) can be planktonic or benthic, and taxa may have wall-less biflagellate cells, coccoid cells, and/or loricate cells. They are mostly known for high production of unsaturated fatty acids, especially eicosapentaenoic acid. The one species described within Class Schizocladiophyceae has branched filaments that produce biflagellate zoospores (Kawai et al. 2003). It is barely visible to the human eye (up to 0.5 mm in height) and was detected via germling growth from subtidal coral debris in culture in the Mediterranean Sea (Rizouli et al. 2020). It is a close relative of brown algae but, unlike browns, its cell wall lacks cellulose and plasmodesmata (Kawai et al. 2003). Yet another one-species class, the Synchromophyceae, consists of amoebae collected from among basaltic rocks in a tide pool with sublittoral environmental conditions. The new class was designated based on unique phylogenetic and biochemical data, and conspicuous chloroplast complexes that represent multiplastidic red algal secondary endosymbionts (Horn et al. 2007).

9.3.6.3 Small Subgroups, Mostly Freshwater

Five minor classes of heterokontophytes mainly or entirely occur in freshwaters. Plankton of Class Chrysophyceae (~1200 species) generally are most abundant in softwater, cold, oligotrophic waters, although they can also be common in estuarine/marine habitats. They generally are biflagellate unicells or colonies, but may also be coccoid, filamentous, palmelloid, plasmodial, and amoeboid. Chrysophyceans have

Fig. 9.13 Light micrograph of the mixotroph *Dinobryon*, a colonial chrysophycean with each cell contained within a lorica. Photo by F. Fox. Image reproduced from https://en.wikipedia.org/wiki/Dinobryon#/media/File:Mikrofoto.de-Dinobryon_divergens.jpg under a creative commons license

mostly classical heterokontophyte traits. Their cell covering ranges from absent (cell membrane only) to organic scales, or a lorica. In addition, their cysts (stomatocysts or statospores) may have a siliceous wall. The chrysophycean *Dinobryon* (Fig. 9.13) is famous ecologically because Bird and Kalff's (1986) discovery of its voracious consumption of bacteria contributed to the understanding of the importance of phagotrophic mixotrophy among phytoplankton from nutrient-poor to nutrient-rich habitats (Burkholder et al. 2008; Chap. 15). Mixotrophy is now known to be widespread among many chrysophyceans.

Class Synurophyceae (~270 species; https://eol.org/pages/3600), thus far exclusively freshwater, is closely related to chrysophyceans and shares many traits including their golden-brown color. Synurophyceans tend to be most abundant and diverse in neutral to slightly acidic waters low in nutrients, although some species thrive in eutrophic alkaline lakes. Notably, they lack heterokont flagella—instead, their flagella have a parallel orientation, directed forward. Another unusual feature is the lack of connection between a chloroplast membrane and the nuclear envelope. Like chrysophyceans, the cell covering varies; it may be absent, or may consist of organic scales, or a lorica, and stomatocysts can be produced. Synurophyceans additionally may have a covering of ornate siliceous scales but, unlike diatoms, they do not require Si (Graham et al. 2016).

The enigmatic Class Eustigmatophyceae (~50 species; Amaral et al. 2020) consists of coccoid algae (diameter 2–32 μm), solitary or in loose colonies, sometimes with asexual reproduction also including flagellate cells (zoospores). They mostly occur in freshwaters and soils. The class was named for a large orange-red stigma (function unknown) in the cytoplasm of flagellate cells. As Amaral et al. (2020) noted, "The formal classification of the class, developed decades ago, does not fit the diversity and phylogeny of the group as presently known and is in urgent need of revision." Members of this class are greenish-yellow in color because they lack fucoxanthin; they also have little or no chl *c* and accumulate unique lipids as their main food reserve. Although most species are not marine, the marine species within two genera (*Nannochloropsis, Microchloropsis*) are best studied, due to their importance as food for some finfish and shellfish in aquaculture, and to their potential for various uses in biotechnology (e.g., Schädler et al. 2021).

The last two classes of these five are close relatives of brown algae (pheophyceans, below) and most are small macroalgae (sensu Lapointe et al. 2018). Members of the Class Xanthophyceae, like eustigmatophyceans, are mostly found in freshwaters or soils (Graham et al. 2016). This subgroup has ~600 species and includes few phytoplankters. They are yellow-green, sometimes dark green, in color because they lack fucoxanthin, and they also lack chl *c*. Morphs include flagellates (rare except for zoospores), coccoid unicells, mucilaginous colonies (palmelloid forms), amoeba-like forms, filaments, and multinucleate siphonous tubular structures that resemble filaments but lack cross-walls. Cell walls are comprised of cellulose, sometimes with some Si. The food reserves consist mostly of lipids, but xanthophyceans can also store chrysolaminarin and mannitol. Class Phaeothamniophyceae consists of ~25 mostly freshwater benthic microalgae and macroalgae (length up to ~1 cm) (Bailey et al. 1998; Graf et al. 2020). They differ from brown algae in having walls of cellulose and certain fine structural features.

9.3.6.4 Brown Algae

The most advanced heterokontophytes, the phaeophyceans, are nearly all marine and range from microscopic free-living unicellular zoospores to giant kelps. Their dominant pigment is often golden-brown fucoxanthin, but in higher light chl *a* and *c* can predominate, resulting in greenish color. Brown algae are a large group (~1500–2000 species), best known for the famous "seaweeds" (e.g., *Fucus*) of mid-/lower intertidal and subtidal marine rocky coasts mostly of colder latitudes. A few brown genera are pelagic such as some species of *Sargassum* that form spectacular high-biomass blooms across hundreds of square kilometers when fueled by higher temperatures and nutrient pollution (Wang et al. 2019). A few species are strictly freshwater, or include both freshwater and marine representatives (McCauley and Wehr 2007; Graham et al. 2016). The life histories of many browns commonly have both haploid and diploid free-living stages in alternation of generations that may look nearly identical in appearance or may markedly differ. In kelps, for example, the large 2N sporophyte releases spores that undergo meiosis, settle on rocks (often adjacent to the sporophyte), and form a microscopic 1N gametophyte that produces sexual gametes.

9.3.7 Green Algae

Green algae have so many representatives, and are so diverse ecologically and evolutionarily (especially regarding their ancestry to land plants), that they recently were split into two phyla (Table 9.2) (Graham et al. 2016 and references therein; Mikhailyuk et al. 2018). Phylum Chlorophyta contains nearly all of the estuarine/marine species, although it mostly consists of freshwater taxa. Several subgroupings, including two well-known subgroups, Classes Zygnematophyceae and Charophyceae, were moved to the same phylum that contains higher vascular plants (embryophytes), the Streptophyta. Both classes are nearly all freshwater inhabitants. Thus, the Chlorophyta are emphasized here.

Chlorophyceans are ecologically important and diverse especially in freshwater plankton, where many are beneficial, high-quality food resources for primary consumers. Greens generally have chl *a* and *b* as predominant pigments. They are nearly all freshwater inhabitants, but some marine members, mostly macroalgae, are highly ecologically and economically significant as well. Most planktonic chlorophytes have a relatively simple life history with predominant haploid stages and the zygote as the only diploid stage. Flagellate cells are part of the life history of most chlorophytes. They provide clues, especially the microtubule fine structural arrangement at the base of the flagella, that are used to infer evolutionary lineages (van den Hoek et al. 1995; Liliaert et al. 2012). A cell wall is usually present, variably comprised of fibrous glycoproteins and various sugars, sometimes with cellulose. The wall can include sulfur or Si, and it is often cuticular, chemically resistant, and physically resistant. The chloroplast has two outer membranes (referred to as a simple "envelope"), and DNA in this organelle is aggregated into nucleoids (van den Hoek et al. 1995). The main food reserve, "true" or higher plant starch (grains of amylose + amylopectin, α-1,4 and α-1,6 linkages) is stored inside the chloroplast(s). Thus, the chloroplast fine structure closely resembles that of higher plants. In addition to the predominant chl, green algae have various carotenoids and xanthophylls such as ß-carotene, lutein, violoxanthin, neoxanthin, antheraxanthin, and zeaxanthin.

Regarding chlorophyte ecology, prasinophytes are sometimes important components of marine nano- and picoplankton. The prasinophyte *Tetraselmis* is an important feeder alga in molluscan aquaculture, and the marine flatworm (turbellarians) *Convoluta roscoffensis* contains *Tetraselmis convolutae* endosymbionts (Lee 1980). Trebouxiophyceans such as *Chlorella* are ubiquitous in freshwaters and also form symbiotic interactions as zoochlorellae, for example, with freshwater *Hydra viridis*. *Chlorella* additionally is a "work horse" of biochemistry and physiological research (see Chap. 12)—but caution is warranted, as the genus *Chlorella* (~2500 species) actually belongs to at least twelve independent lineages of both trebouxiophyceans and chlorophyceans (Darienko and Pröschold 2019, and references therein). Accordingly, within the past two decades, some species within this genus have been reclassified into at least seven different genera based on molecular sequences (Graham et al. 2016; Darienko and Pröschold 2019). Similarly, many of the ~1800 species included within the well-known genus

Chlamydomonas are in the process of being reclassified to other genera (Graham et al. 2016). Ulvophycean marine macroalgae such as *Ulva* can sometimes become abundant in the plankton for short periods due to sudden releases of their asexual *Chlamydomonas*-like zoospores (e.g., Niesenbaum 1988). Ulvophyceans include some of the most notorious floating filamentous macroalgal responders to nutrient pollution in freshwaters, such as *Cladophora* and *Pithophora* (note that *Cladophora* is also a known responder to nutrient pollution in estuarine and marine coastal waters; see Lapointe et al. 2018).

9.3.8 Euglenoids

Euglenoids were among the microscopic motile "animalcules" referred to by van Leeuwenhoek (1674, Chap. 1; cited in Dobell 1932). Although mostly known from plankton studies, increasingly euglenoids are suspected to be primarily benthic in habit. As Graham et al. (2016, p. 162) noted, they exhibit "extraordinary variety of nutritional types. . . .[they are found in] farm ponds, marshes, swamps, fens, bogs, oceanside beaches, mud flats, epizoic on cladoceran zooplankton, in hindguts of dipteran larvae, and in tadpole rectums" as just some of the habitats frequented by euglenoids.

This group of ~1300 described species consists mostly of solitary flagellates in fresh and estuarine habitats, with a few also inhabiting marine waters (Graham et al. 2016). The cells have two flagella but usually only one is emergent; in sudden stressful conditions, the cells can rapidly lose their emergent flagellum and form benthic mucilaginous palmelloid masses, where they can continue cell production (Zimba et al. 2010). Some euglenoids have been documented to form resting cysts as well (Olli 1996), which may be a widespread phenomenon in response to unfavorable conditions. Flagellate cells use the emergent flagellum for swimming, but can also engage in a unique, bizarre "squirting" movement called metaboly (Digumarti et al. 2018) (Fig. 9.14). Metaboly consists of large-amplitude, highly concerted deformations of the entire cell, mediated by the highly plastic pellicle. Euglenoid fine structure includes notable features such as the complex pellicle covering beneath the cell membrane in the cytoplasm, a nucleus with permanently condensed chromosomes as in dinoflagellates, and chloroplasts with three outer membranes. The predominant pigments are chl *a* and *b* as in green algae.

About two-thirds of euglenoids are heterotrophs that depend on osmotrophy or phagocytized bacteria for food. Based on molecular evidence, phototrophic euglenoids evolved from phagotrophic ancestors. They form a clade with lineages of phagotrophic euglenoids that originated through a secondary endosymbiosis with chlorophytes (Chap. 10). The phototrophic species (euglenophytes) are all auxotrophic (vitamin-requiring) and most, if not all, are mixotrophic as osmotrophs and/or bacteriovores. The exciting recent discovery of a novel marine euglenoid, *Rapaza viridis* ("green grasper"), represents the first example of euglenoid mixotrophy involving phagotrophy of eukaryote prey (Yamaguchi et al. 2012). This obligate mixotroph uses a novel type of cytostome to consume a specific strain of the small

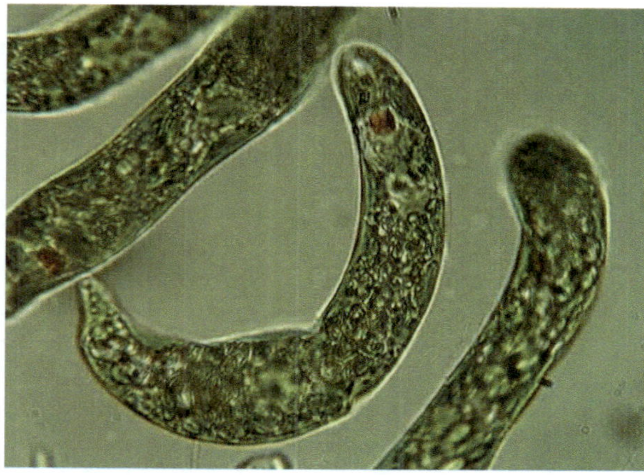

Fig. 9.14 *Euglena* sp. Image reproduced from https://commons.wikimedia.org/wiki/File%3AEuglena_sp..jpg under a creative commons license

green microalga *Tetraselmis* as prey, while maintaining a functional chloroplast that is homologous with that of other phototrophic euglenoids. An unusual combination of fine structural features in *R. viridis*, together with its molecular phylogenetic position between heterotrophic and phototrophic euglenoids, conforms to the transitional stage expected to have occurred during early evolution of phototrophic euglenoids from heterotrophic ancestors.

Although euglenoids historically have been considered mostly as plankton, that generalization may be in error. As Lackey (1968) wrote, "There may be no truly planktonic euglenoids"; they appear to be "occupants of interfaces" (Walne and Kivic 1990) such as the sediment-water interface. Euglenophytes have been major contributors to our knowledge about phototaxis in algae, given their response to light via direction of movement and swimming velocity; they also are well known for vertical migration in surficial sediments, a circadian rhythm that is also influenced by tides in estuarine/marine species (Kingston 1999). At least seven freshwater or brackish species recently were discovered to be capable of producing a neurotoxin lethal to fish, as well (Zimba et al. 2010, 2017). It is suspected that fish kills from this euglenophycin toxin may be common but have been overlooked because the culprit cells leave the water column and move to a benthic, cryptic habit where they are not detected (Zimba et al. 2017)

9.3.9 Chlorarachniophytes

This small—but evolutionarily fascinating—group occurs in temperate and tropical marine waters on mud, in tide pools, or among seaweeds, sand grains, or pelagic phytoplankton. Chlorarachniophytes are named for their green chloroplast color, from chl *a* and *b*, and the spider-like shapes of their filose amoeboid cells (Fig. 9.15). The life history also includes coccoid forms. Chlorarachniophytes are a key group for understanding how chloroplasts have been retained in many different groups of

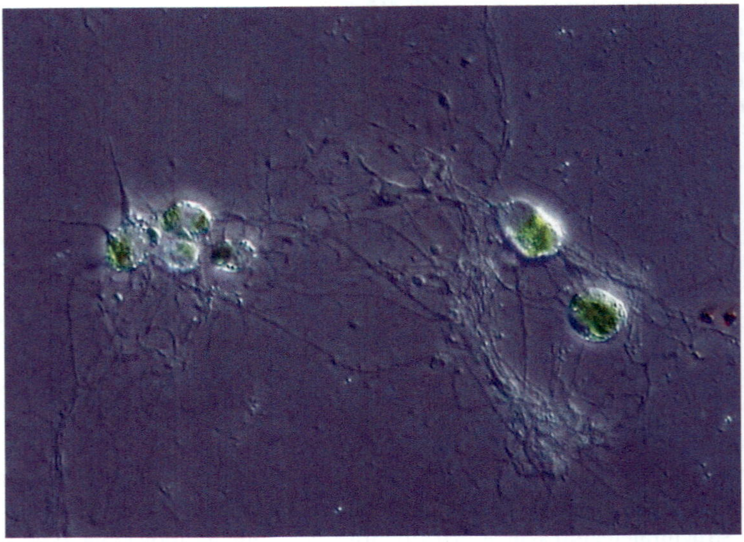

Fig. 9.15 Amoeboid cells of *Chlorarachnion reptans* extending many filose pseudopodia. Image from Amani-Oshima, Kagoshima Pref., Japan. Image reproduced from https://en.wikipedia.org/wiki/Chlorarachniophyte#/media/File:Chlorarachnion_reptans.jpg under a creative commons license

eukaryotes; they are considered to be "evolution in progress" wherein various eukaryote algae have resulted from heterotrophic microbes incorporating smaller eukaryote algal prey and retaining chloroplasts and, sometimes, other organelles as endosymbionts (Graham et al. 2016; Ponce-Toledo et al. 2018; see Chap. 10). These organisms acquired photosynthesis via a host heterotroph cell engulfing a green alga and retaining its chloroplast. Chlorarachniophyte chloroplasts are thought to have been acquired by the secondary endosymbiosis of a green alga, and the plastids also harbor a relict nucleus (nucleomorph) of the endosymbiont (Suzuki et al. 2017). Chloroarachniphytes (e.g., *Bigelowiellanatans*, *Gymnochlora stellata*) have been found to be mixotrophs; they consume bacterial cells and digest them in food vacuoles (Ishida et al. 1996; Moestrup and Sengco 2001).

9.3.10 Cryptophytes

Cryptomonads, mostly phototrophs (cryptophytes) along with a small number of heterotrophic species, are small biflagellate unicells (length ~5–50 μm found in diverse freshwater, brackish, and marine habitats (Graham et al. 2009, 2016). They are characterized by a distinct cellular asymmetry and the presence of ejectisomes (explosive organelles) at the ventral side of the cell. Cells are often described as being dorso-ventrally flattened, with swimming that resembles a "flat tire" appearance. The two flagella are inserted with a shift to the right side of in the vestibulum of the cell invagination. The asymmetric insertion of the flagella causes an unbalanced rotation around the longitudinal cell axis when swimming. The larger the cells and the more dorsally or ventrally bent, the more the cell sways or wobbles.

Cryptophytes are extremely complex from molecular and fine structural perspectives (Hoef-Emden and Archibald 2010). This small, enigmatic group is considered of central importance in understanding the diversification of eukaryote algae (Graham et al. 2009) and the evolution of chloroplasts (Hoef-Emden 2008). These organisms are mostly mixotrophs, also with some heterotrophs. They acquired photosynthesis through secondary (eukaryote-eukaryote) endosymbiosis by engulfing and "adopting" a red algal endosymbiont during their evolutionary past (Gould 2008).

As a result, cryptophytes have a four-membrane-bound chloroplast, and the outer two membranes can be continuous with the outer nuclear envelope. Between the inner two and outer two chloroplast membranes is a periplastidial compartment containing starch grains, 80S ribosomes, and the remnant, highly reduced nucleus (nucleomorph) of what was the endosymbiont. Their food reserve is true starch (amylose + amylopectin) as in green algae, dinoflagellates, and higher plants, stored within the chloroplasts. In addition, lipid droplets can occur in the cytoplasm. As another unique trait, the chloroplasts contain one, and only one, highly modified phycobilin, either PC or PE (Hill and Rowan 1989). Light energy passes from phycobilin to chl c_2 to chl a (Falkowski and Raven 2007). The flagellar apparatus is also unique, especially in the root arrangement. A rhizostyle (longitudinal microtubular root containing a cluster of 6–10 microtubules) runs from near the basal body of the longer, dorsal flagellum to the cell posterior. Light sensing in cryptophytes is accomplished by two distinct rhodopsins that serve as photoreceptors (Sineshchekov et al. 2005). The cells change their position in response to light through alternate shading of the rhodopsins by a stigma composed of lipid droplets containing carotenoids, and then illumination of the photoreceptor rhodopsins as the cells move.

Large ejectosomes occur along the cryptophyte cell invagination at the ventral side (i.e., the "gullet" region), and small ejectosomes are scattered around the rest of the cell periphery. When discharged, two coiled ribbons in each ejectisome unfurl and form a narrow, kinked barb. Ejectosomes discharge explosively as the coil suddenly unrolls, causing the cell to abruptly move backward away from an irritant. This mechanism may be an avoidance reaction to predators. As "nature in progress," however, more serious disturbance leads to massive discharge of ejectosomes that can cause the cell itself to disintegrate (Graham et al. 2009).

Cryptophytes are generally regarded as among the most inconspicuous of eukaryote algae, not only because many of them are small (length ~5–10 μm) and nondescript under LM, but also because their high proportions of polyunsaturated fatty acids make then favored prey for a wide array of herbivores, so they tend to occur in low numbers despite sometimes-rapid growth. In addition, they are often

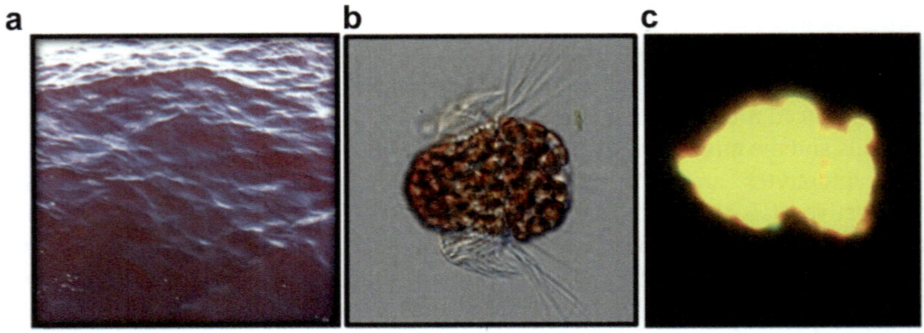

Fig. 9.16 Panel (**a**) Photograph of red water in western Long Island Sound during a *Mesodinium rubrum* bloom. Photo courtesy of K. Howard-Strobel (University of Connecticut, Groton, CT). Panel (**b**) Micrograph of *M. rubrum* with enslaved cryptophyte chloroplasts. Micrograph courtesy of National Oceanic and Atmospheric Administration Phytoplankton Monitoring Network. Panel (**c**) Fluorescent microscopy under green light excitation showing characteristic yellow fluorescence (565–570 nm) of *M. rubrum* due to high amounts of phycoerythrin pigment. Image reproduced from Dierssen et al. (2015, Proc Nat Acad Sci 112: 14783–14787. doi: 10.1073/pnas.1512538112) under a creative commons license

most abundant in cold or deep waters, not frequently sampled. Moreover, the cells tend to burst when preserved, or when suddenly brought from depth to the surface in sampling devices, so that they effectively "disappear" (Graham et al. 2016 and references therein). Some cryptophytes vertically migrate up to ~5 m day^{-1}, moving from P-depleted surface waters during the day to P-rich deeper waters at night (Arvola et al. 1991).

"Red water blooms" of the "*Mesodinium rubrum* species complex" can span 160 km^2 (100 square miles) in nutrient-enriched estuarine and marine coastal waters (and, sometimes, freshwaters) as a regular feature in temperate and tropical regions (Graham et al. 2009 and references therein; Fig. 9.16). These blooms consist of holotrich protozoan ciliates with obligate cryptophyte endosymbionts in a mutualistic symbiosis. Six clades of different protozoan ciliate species have been found so far (Johnson et al. 2016). The ciliates are obligate mixotrophs; they rely on light and ingestion of red-pigmented cryptophytes for sustained growth. Ingestion rates are usually low; ingestion of only 1 free-living cryptophyte cell per day (~1% of the carbon required for growth) can sustain 1 doubling per day, indicating that the ciliates can grow almost entirely dependent on cryptophyte photosynthesis. The symbionts occupy more than one-third of the host cell, and each retains a nucleus, mitochondria, chloroplasts with nucleomorphs, and endoplasmic reticulum. The chloroplasts can synthesize chl and undergo replication under regulation by the ciliates (Johnson et al. 2006), allowing the ciliates to function as phototrophs. The *Mesodinium*/cryptophytes compete exceptionally well with [other] phototrophic phytoplankton: their swimming speeds (2.0–7.2 m h^{-1}) are about tenfold faster than the maximum swimming speed of dinoflagellates; and they have extremely high photosynthetic rates (1000–2000 mg C m^{-3} h^{-1}), the highest ever observed for marine phytoplankton (Graham et al. 2009 and references therein). This symbiosis undoubtedly is of major importance, as yet unquantified, to the aquatic ecosystems where the massive red-water blooms occur.

9.4 The Confusing "Fluorescent Picoplankton"

No longer are microscopes—even a combination of LM, SEM, and TEM—sufficient to identify most phytoplankton to species. While this is now undoubtedly the case for many "larger" cells (> 5 μm), this problem is exacerbated for picoplankton. Biological sensors and a wide array of other new tools and approaches are being used to identify and quantify algal cells. Such identifications and characterizations are often derived from cytograms that are grouped by experts or by machine learning protocols with categorizations that unfortunately vary widely between instrument types and user groups (Thyssen et al. 2022). Flow cytometry and instruments such as the Imaging FlowCytoBot (Olson and Sosik 2007; see also Chap. 2) are coming into more routine use, but taxonomic descriptions of the microbial groups—despite ongoing, major efforts—have not kept pace. Different user groups have applied different size cut-offs to discriminate pico- and nanoplankton groups, and have used a wide range of descriptive names.

A new vocabulary and decision tree was recently proposed to distinguish the most common of these tiny cells. Based on this effort, 13 groups of phototrophic, mixotrophic or heterotrophic microorganisms were classified in order to better homogenize datasets (Table 9.4), together with a conceptual diagram of relative position in cytograms of scatter and fluorescent properties (Fig. 9.17) and a decision tree for separating red and orange fluorescing particles when using a blue laser (Fig. 9.18). Such an effort represents a major step forward in comparing global production of these tiny but critically important organisms.

9.5 Summary

The taxonomy and classification of algae is an ever-evolving challenge. However imperfect, classifications give us a framework around which common physiological traits (see Chap. 18) may be identified. Pigmentation, storage products, cell wall characteristics, and structures such as flagella, chloroplast organization, and other cellular structures are useful in defining major taxonomic groups of species. Molecular data are now providing new tools by which relationships between species can be more definitively discerned. Fluorescent properties facilitate identification of many tiny prokaryotes and eukaryotes. With this abbreviated review of the general features of different algal taxa, we now turn to how they are evolutionarily related.

Table 9.4 Comparison of terms identifying the main groups considered in flow cytometry that appear in the literature, and the common nomenclature accessible in the database https://vocab.nerc.ac.uk/collection/Fo2/current/

Identifier	Common vocabulary (alternative label)	Short description (preferred label)	Example of terms that appear in the literature
SDN: F02:: F0200002	RedPicoProk	Red-only fluorescing prokaryote picophytoplankton	Prochlorophytes, *Proclorococcus*, very small red fluorescing bodies, PRO, Prochl
SDN: F02:: F0200013	RedRedPico	Red and Red-only fluorescing prokaryote picophytoplankton, rich in pigment phycocyanin (PC)	PC-rich *Synechococcus*, PC-CYAN, PC SYN
SDN: F02:: F0200003	OraPicoProk	Orange fluorescing prokaryote picophytoplankton, rich in pigment phycoerythrin (PE)	Syn, *Synechococcus*, cyanobacteria, PE-type *Synechococcus*, *Crocospaera*, PE-SYN
SDN: F02:: F0200004	RedPico	Red-only fluorescing picophytoplankton	Pocoeuk, redpico, Peuk, picophytoplankton, PicoEuk1
SDN: F02:: F0200005	RedNano	Red-only fluorescing nanophytoplankton	Nanoeuk, rednano, Neuk, nanophytoplankton
SDN: F02:: F0200014	OraPico	Orange and red-fluorescing picophytoplankton, rich in pigment phycoerythrin (PE)	Pico cryptophytes
SDN: F02:: F0200006	OraNano	Orange and red-fluorescing nanophytoplankton, rich in pigment phycoerythrin (PE)	*Cryptomonas*, Cryptophytes, HighOrgnano, Nanocyanobacteria, CRYPTO
SDN: F02:: F0200015	RedRedNano	Red and red-fluorescing nanophytoplankton, rich in pigment phycocyanin (PC)	Nanocyanobacteria
SDN: F02:: F0200007	HsNano	Red-fluorescing nanophytoplankton, with relatively high sideward light scattering properties	Coccolithophorids, HighSWSnano, COCCO
SDN: F02:: F0200008	RedMicro	Red-only fluorescing microphytoplankton	Microphytoplankton
SDN: F02:: F0200016	OraMicro	Orange fluorescing microphytoplankton, rich in pigment phycoerythrin (PE)	Microphytoplankton
SDN: F02:: F0200010	HetHNA	Heterotrophic prokaryotes with relatively high nucleic acid	HNA
SDN: F02:: F0200011	HetLNA	Heterotrophic prokaryotes with relatively low nucleic acid	LNA

Reproduced from Thyssen et al. (2022, Front Mar Sci 9: 975877. doi: 10.3389/fmars.2022.975877) under a creative commons license

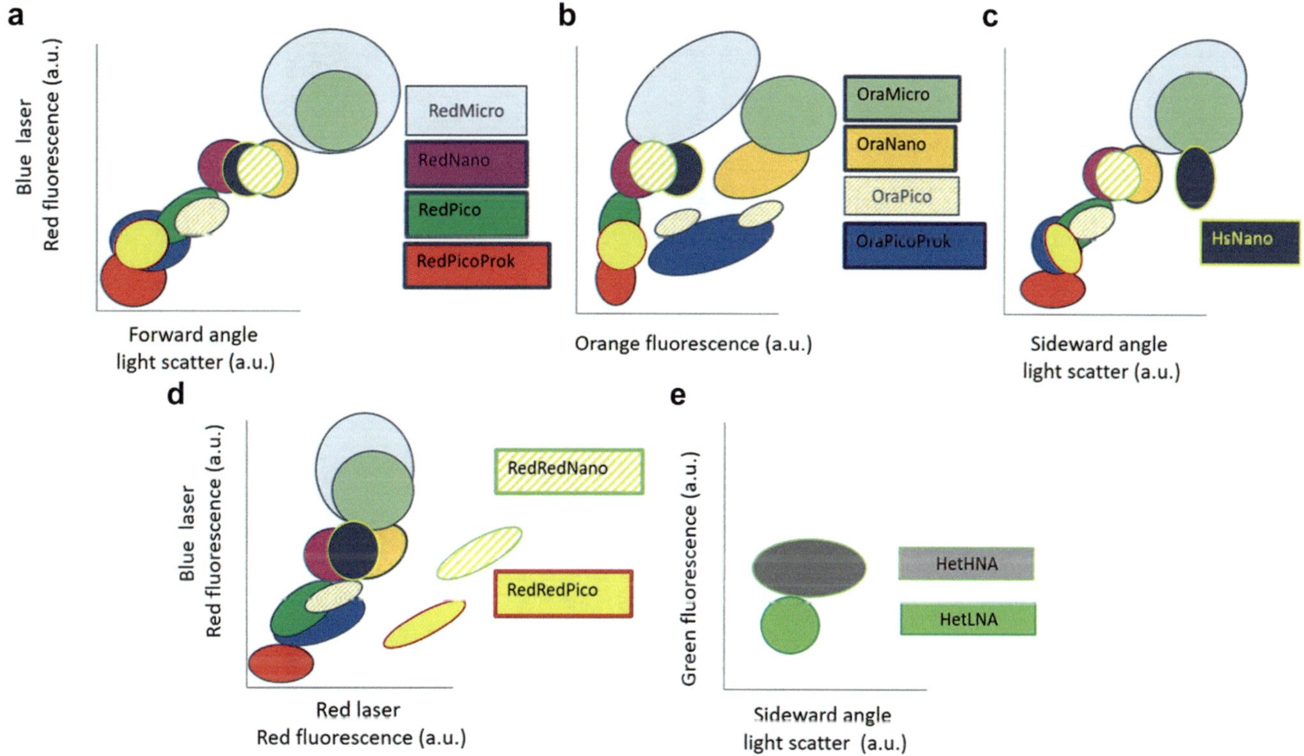

Fig. 9.17 Conceptual cytograms depicting the relative position of the most commonly observed groups in marine waters. Axes are in log vs. log scale. Panel (**a**) Relationship between forward light scatter (FSC), a proxy for cell size, and chl *a* FLR from a blue laser. Panel (**b**) Phytoplankton group separation based on phycoerythrin orange fluorescence and chl *a* fluorescence. Panel (**c**) Relationship between sideward angle light scatter (SSC), a proxy for cell surface roughness, and chl *a* FLR from a blue laser. Panel (**d**) Relationship between chl *a* and phycocyanin FLR from a red laser, and chl *a* FLR from a blue laser. Panel (**e**) HetHNA and HetLNA are evidenced after DNA staining with a green emission dye such as SybrGreen I® and are separated on a green fluorescence vs (SSC) with a blue laser. Image reproduced from Thyssen et al. (2022, Front Mar Sci 9: 975877. doi: 10.3389/fmars.2022.975877) under a creative commons license

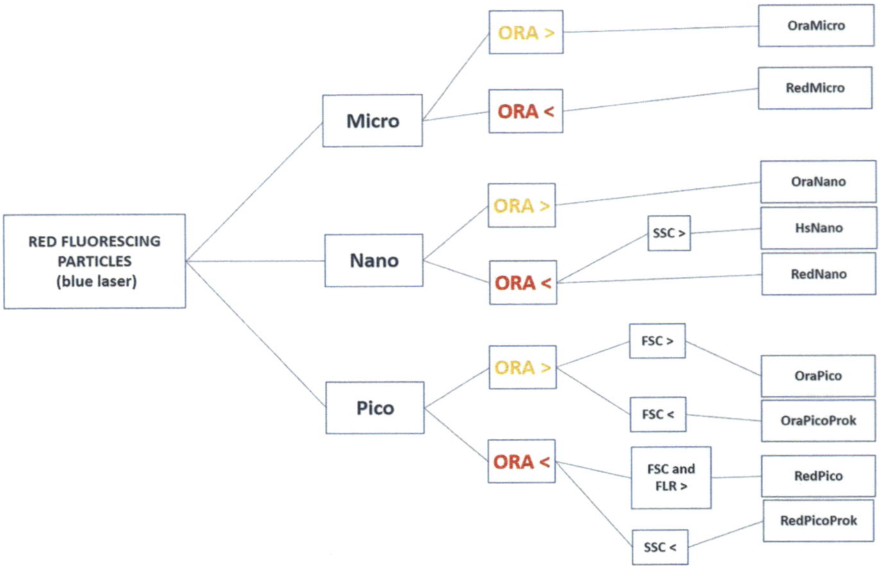

Fig. 9.18 Decision tree for helping in the group's discrimination on cytograms for unstained natural red fluorescing particles excited with a blue laser. Red fluorescing particles corresponding to the phytoplankton groups are selected above a threshold that corresponds to the lowest limit of detection of the FCM. The lowest limit of detection is defined when the optical noise and the signature of the detritus in the sheath signal are well identified and do not affect the smallest resolved phytoplankton group the FCM can count (at best, the RedPicoProk). FSC, forward scatter; FLR, red fluorescence; FLO, orange fluorescence; SSC, sideward scatter. The groups RedRedPico and RedRedNano (discriminated with a red laser), as well as the HetHNA and HetLNA, are not presented here. Image reproduced from Thyssen et al. (2022, Front Mar Sci 9: 975877. doi: 10.3389/fmars.2022.975877) under a creative commons license

References

Amaral R, Fawley KP, Němcová Y, Ševčíková T, Lukešová A, Fawley MW, Santos LMA, Eliáš M (2020) Towards modern classification of eustigmatophytes, including the description of Neomonodaceae, fam. nov. and three new genera. J Phycol 56:630–648. https://doi.org/10.1111/jpy.12980

Andersen RA, Saunders GW, Paskind MP, Sexton JP (1993) Ultrastructure and 18S $_{R}$RNA gene sequence for *Pelagomonas calceolata* gen. et sp. nov. and the description of a new algal class, the Pelagophyceae classis nov. J Phycol 29:701–715. https://doi.org/10.1111/j.0022-3646.1993.00701.x

Arvola L, Ojala A, Barbosa F, Heaney SI (1991) Migration behaviour of three cryptophytes in relation to environmental gradients: an experimental approach. Br Phycol J 26:361–373

Bailey JC, Bidigare RR, Christensen SJ, Andersen RA (1998) Phaeothamniophyceae classis nova: a new lineage of chromophytes based upon photosynthetic pigments, rbcL sequence analysis and ultrastructure. Protist 149:245–263

Barsanti L, Gualtieri P (2006) Algae anatomy, biochemistry and biotechnology. CRC Press, Boca Raton, FL

Barsanti L, Gualtieri P (2014) Algae: anatomy, biochemistry, and biotechnology, 2nd edn. Taylor & Francis, Boca Raton, p 362

Bird DF, Kalff J (1986) Bacterial grazing by planktonic lake algae. Science 231:493–495. https://doi.org/10.1126/science.231.4737.493

Bold HC, Wynne MJ (1978) Introduction to the algae—structure and reproduction. Prentice-Hall, Englewood Cliffs, NJ

Borowitzka MA (2016) Systematics, taxonomy and species names: do they matter? In: Borowitzka MA, Beardall J, Raven JA (eds) The physiology of microalgae, Developments in applied phycology, vol 6. Springer, Cham, pp 655–681. https://doi.org/10.1007/978-3-319-24945-2_24

Bradbury J (2004) Nature's nanotechnologists: unveiling the secrets of diatoms. PLoS Biol 2(10): e306. https://doi.org/10.1371/journal.pbio.0020306

Burkholder JM (1992) Phytoplankton and episodic suspended sediment loading: phosphate partitioning and mechanisms for survival. Limnol Oceanogr 37:974–988

Burkholder JM, Glibert PM (2006) Intraspecific variability: an important consideration in forming generalizations about toxigenic algal species. Afr J Mar Sci 28:177–180

Burkholder JM, Marshall HG (2012) Toxigenic *Pfiesteria* species—updates on biology, ecology, toxins, and impacts. Harmful Algae 14:196–230

Burkholder JM, Glibert PM, Skelton HM (2008) Mixotrophy, a major mode of nutrition for harmful algal species in eutrophic waters. Harmful Algae 8:77–93

Burkholder JM, Shumway SE, Glibert PM (2018) Food web and ecosystem impacts of harmful algae. In: Shumway SE, Burkholder JM, Morton SL (eds) Harmful algal blooms and their management: a compendium desk reference. Wiley, Hoboken, NJ, pp 243–336

Caruana AMN, Amzil Z (2018) Microalgae and toxins. In: Levine I, Fleurence J (eds) Microalgae in health and disease prevention. Academic, New York, pp 263–305

Chisholm SW, Frankel SL, Goericke R, Olson RJ, Palenik B, Waterbury JB, West-Johnsrud L, Zettler ER (1992) *Prochlorococcus marinus* nov. gen. nov. sp.: an oxyphototrophic marine prokaryote containing divinyl chlorophyll *a* and *b*. Arch Microb 157:297–300

Dale B, Thorsen TA, Fjellså A (1999) Dinoflagellate cysts as indicators of cultural eutrophication in the Oslofjord, Norway. Estuar Coastal Shelf Sci 48:371–382

Darienko T, Pröschold T (2019) The genus *Jaagichlorella* Reisigl (Trebouxiophyceae, Chlorophyta) and its close relatives: an evolutionary puzzle. Phytotaxa 388:47–68. https://doi.org/10.11646/phytotaxa.388.1.2

De Clerck O, Guiry MD, Leliaert F, Samyn Y, Verbruggen H (2013) Algal taxonomy: a road to nowhere? J Phycol 49:215–225. https://doi.org/10.1111/jpy.12020

Dierssen H, McManus GB, Chlus A, Qiu D, Gao B-C, Lin S (2015) Space station image captures a red tide ciliate bloom at high spectral and spatial resolution. Proc Natl Acad Sci USA 112: 14783–14787. https://doi.org/10.1073/pnas.1512538112

Digumarti M, Conn A, Rossiter J (2018) EuMoBot: replicating euglenoid movement in a soft robot. J R Soc Interface 15. https://doi.org/10.1098/rsif.2018.0301

Dobell C (1932) Antony van Leeuwenhoek and his "little animals". Dover, New York

Doolittle WF (1979) The cyanobacterial genome, its expression, and the control of that expression. Adv Microb Physiol 20:1–102

Falkowski P, Raven JA (2007) Aquatic photosynthesis, 2nd edn. Princeton University Press, Princeton, NJ. 484 pp

Ferrante MI, Entrambasaguas L, Johansson M, Töpel M, Kremp A, Montresor M, Godhe A (2019) Exploring molecular signs of sex in the marine diatom *Skeletonema marinoi*. Genes 10(7):494. https://doi.org/10.3390/genes10070494

Fritsch FE (1935) Structure and reproduction of the algae, vol I. Cambridge University Press, Cambridge

Gaines G, Taylor FJR (1985) Form and function of the dinoflagellate transverse flagellum. J Eukaryot Microbiol 32:290–296. https://doi.org/10.1111/j.1550-7408.1985.tb03053.x

Gale GA, Schiavon Osorio AA, Puzorjov A, Wang B, McCormick AJ (2019) Genetic modification of cyanobacteria by conjugation using the CyanoGate Modular Cloning Toolkit. J Vis Exp 152. https://doi.org/10.3791/60451

Gavelis GS, Wakeman KC, Tillmann U, Ripken C, Mitarai S, Őzbek S, Holstein T, Keeling PJ, Leander BS (2017) Microbial arms race: ballistic "nematocysts" in dinoflagellates represent a new extreme in organelle complexity. Sci Adv 3(3). https://doi.org/10.1126/sciadv.1602552

Golubic S (1979) Cyanobacteria (blue-green algae) under the bacteriological code? An ecological objection. Taxon 28:387–389

Gómez F (2017) The function of the ocelloid and piston in the dinoflagellate *Erythropsidinium* (Gymnodiniales, Dinophyceae). J Phycol 53:629–641

Gornik SG, Hu I, Lassadi I, Waller RF (2019) The biochemistry and evolution of the dinoflagellate nucleus. Microorganisms 7:8. https://doi.org/10.3390/microorganisms7080245

Gould SB (2008) Ariadne's thread: guiding a protein across five membranes in cryptophytes. J Phycol 44:23–26

Graf L, Yang EC, Han KY, Küpper FC, Benes KM, Oyadomari JK, Herbert RJH, Vergruggen H, Wetherbee R, Andersen RA, Yoon HS (2020) Multigene phylogeny, morphological observation and re-examination of the literature lead to the description of the Phaeosacciophyceae Classis Nova and four new species of Heterokontophyta SI clade. Protist 171(6):125781. https://doi.org/10.1016/j.protis.2020.125781

Graham LE, Graham JM, Wilcox LW (2009) Algae, 2nd edn. Pearson Benjamin Cummings, San Francisco

Graham LE, Graham JM, Wilcox LW, Cook ME (2016) Algae, 3rd edn. LJLM Press, Madison

Griffiths DJ (2006) Chlorophyll *b*-containing oxygenic photosynthetic prokaryotes: oxychlorobacteria (Prochlorophytes). Bot Rev 72:330–366

Guiry MD (2012) How many species of algae are there? J Phycol 48:1057–1063

Guiry MD, Guiry GM (2023) AlgaeBase. World-wide electronic publication, National University of Ireland, Galway. https://www.algaebase.org. Accessed 11 Mar 2023

Harris GP (1986) Phytoplankton ecology: structure, function and fluctuation. Chapman and Hall, London

Harvey WH (1836) Algae. In: Mackay JT (ed) Flora hibernica comprising the flowering plants ferns characeae musci hepaticae lichenes and algae of Ireland arranged according to the natural system with a synopsis of the genera according to the Linnaean system. William Curry Jun, Dublin, Ireland, pp 157–254

Hasle GR, Syertsen EE (1996) Marine diatoms. In: Tomas CR (ed) Identifying marine phytoplankton. Academic, San Diego, pp 44–45

Hastings JW (1983) Chemistry and control of luminescence in marine organisms. Bull Mar Sci 33: 818–828

Hastings JW (2007) The Gonyaulax clock at 50: translational control of circadian expression. Cold Spring Harbor Symp Quant Biol 72:141–144

Hill DRA, Rowan KS (1989) The biliproteins of the Cryptophyceae. Phycologia 28:455–463

Hoef-Emden K (2008) Molecular phylogeny of phycocyanin-containing cryptophytes: evolution of biliproteins and geographical distribution. J Phycol 44:985–993

Hoef-Emden K, Archibald JM (2010) Tree of life cryptomonads. Available at http://tolweb.org/Cryptomonads/2396. Accessed 10 Mar 2023

Hoef-Emden K, Archibald JM (2017) Cryptophyta (Cryptomonads). In: Archibald J, Simpson A, Slamovits C (eds) Handbook of the protists, 2nd edn. Springer, Cham, pp 851–891. https://doi.org/10.1007/978-3-319-28149-0_35

Hopes A, Mock T (2015) Evolution of microalgae and their adaptations in different marine ecosystems. eLS. https://doi.org/10.1002/9780470015902.a0023744

Hoppenrath M, Murray SA, Chomérat N, Horiguchi T (2014) Marine benthic dinoflagellate—unveiling their worldwide biodiversity. Kleine Senckenberg-Reihe 54. Schweizerbart'sche Verlagsbuchhandlung (Nägele U. Obermiller), Stuttgart, Germany

Horn S, Ehler K, Fritsch G, Gil-Rodríguez MC, Wilhelm C, Schnetter R (2007) *Synchroma grande* spec. nov. (Synchromophyceae class. nov., Heterokontophyta): an amoeboid marine alga with unique plastid complexes. Protist 158:277–293. https://doi.org/10.1016/j.protis.2007.02.004

Huang Y, Hong J, Liang C, Li G, Shen L, Zhang H, Hu X, Chen X (2019) Nitrogen limitation affects the sinking property of *Microcystis* by promoting carbohydrate accumulation. Chemosphere 221:665–671. https://doi.org/10.1016/j.chemosphere.2019.01.080

Idei M, Osada K, Sato S, Toyoda K, Nagumo T, Mann DG (2012) Gametogenesis and Auxospore Development in *Actinocyclus* (Bacillariophyta). PLoS One 7(8):e41890. https://doi.org/10.1371/journal.pone.0041890

Ishida K, Nakayama T, Hara Y (1996) Taxonomic studies on the Chlorarachniophyta. II. Generic delimitation of the chlorarachniophytes and description of *Gymnochlora stellata* gen. et sp. nov. and *Lotharella* gen. nov. Phycol Res 44:37–45. https://doi.org/10.1111/j.1440-1835.1996.tb00036.x

Jackson C, Clayden S, Reyes-Prieto A (2015) The Glaucophyta: the blue-green plants in a nutshell. Acta Soc Bot Pol 84:149–165. https://doi.org/10.5586/asbp.2015.020

Johnson MD, Tengs T, Oldach D, Stoecker DK (2006) Sequestration, performance, and functional control of cryptophyte plastids in the ciliate *Myrionecta rubra* (Ciliophora). J Phycol 42:1235–1246

Johnson MD, Beaudoin DJ, Laza-Martinez A, Dyhrman ST, Fensin E, Lin S, Merculief A, Nagai S, Pompeu M, Setälä O, Stoecker DK (2016) The genetic diversity of *Mesodinium* and associated cryptophytes. Front Microbiol 7. https://doi.org/10.3389/fmicb.2016.02017

Kai A, Yoshii Y, Nakayama T, Inouye I (2008) Aurearenophyceae *classis nova*, a new class of Heterokontophyta based on a new marine unicellular alga *Aurearena cruciata* gen. et sp. nov. inhabiting sandy beaches. Protist 159:435–457. https://doi.org/10.1016/j.protis.2007.12.003

Kana TM, Glibert PM, Goericke R, Welschmeyer NA (1988) Zeaxanthin and B-carotene in *Synechococcus* WH7803 respond differently to irradiance. Limnol Oceanogr 33:1623–1627

Kawachi M, Inouye I, Honda D, O'Kelly CJ, Bailey JC, Bidigare RR, Andersen RA (2002) The Pinguiophyceae *classis nova*, a new class of photosynthetic stramenopiles whose members produce large amounts of omega-3 fatty acids. Phycol Res 50:31–47. https://doi.org/10.1046/j.1440-1835.2002.00260.x

Kawai H, Maeba S, Sasaki H, Okuda K, Henry EC (2003) *Schizocladia ischiensis*: a new filamentous marine chromophytes belonging to a new class, Schizocladiophyceae. Protist 154:211–228. https://doi.org/10.1078/143446103322166518

Kingston MB (1999) Effect of light on vertical migration and photosynthesis of *Euglena proxima* (Euglenophyta). J Phycol 35:245–253

Korshunova T, Picton B, Furfaro G, Mariottini P, Pontes M, Prkić J, Fletcher K, Malmberg K, Lundin K, Martynov A (2019) Multilevel fine-scale diversity challenges the 'cryptic species' concept. Sci Rep 9:6732. https://doi.org/10.1038/S41598-019-42297-5

Kuwata A, Yamada K, Ichinomiya M, Yoshikawa S, Tragin M, Vaulot D, dos Santos AL (2018) Bolidophyceae, a sister picoplanktonic group of diatoms—a review. Front Mar Sci 5. https://doi.org/10.3389/fmars.2018.00370

Lackey JB (1968) Ecology of euglena. In: Buetow DE (ed) The biology of euglena. Academic, New York, pp 27–44

Lancelot C, Kellar DM, Rosseau V, Smith WO, Mathot S (1998) Autoecology of the marine haptophyte *Phaeocystis* sp. In: Anderson DA, Cembella A, Hallegraeff GM (eds) Physiological ecology of harmful algae. NATO ASI series G: Ecological sciences, vol 41. Springer, Berlin, pp 209–224

Lang NJ, Walsby AE (1982) Cyanophyceae: introduction and bibliography. In: Rosowski JR, Parker BC (eds) Selected Papers in phycology II. Allen Press, Lawrence, KS, pp 647–655

Lapointe BE, Burkholder JM, Van Alstyne KL (2018) Harmful macroalgal blooms in a changing world: causes, impacts, and management. In: Shumway SE, Burkholder JM, Morton SL (eds) Harmful algal blooms and their management: a compendium desk reference. Wiley, Hoboken, NJ, pp 515–560

Lee RE (1980) Phycology. Cambridge University Press, New York
Lewin RA (1976) Naming the blue-greens. Nature 259:360
Liliaert F, Smith DR, Moreau H, Herron MD, Verbruggen H, Delwiche CF, de Clerck O (2012) Phylogeny and molecular evolution of the green algae. Crit Rev Plant Sci 31:1–46
Linnaeus C (1754) Genera plantarum, 5th edn. H.R. Engelmann, Berlin
Liu H, Probert I, Uitz J, Claustre H, Aris-Brosou S, Frada M, Not F, de Vargas C (2009) Extreme diversity in noncalcifying haptophytes explains a major pigment paradox in open oceans. Proc Natl Acad Sci USA 106:12803–12808
Lu X, Nan F, Feng J, Lv J, Liu Q, Liu X, Xie S (2020) Effects of different environmental factors on the growth and bioactive substance accumulation of *Porphyridium purpureum*. Int J Environ Res Public Health 17:2221. https://doi.org/10.3390/ijerph17072221.0
Maggs CA, Verbruggen H, de Clerck O (2007) Molecular systematics of red algae: building future structures on firm foundations. In: Brodie J, Lewis J (eds) Unraveling the algae—the past, present, and future of algal systematics. CRC, Boca Raton, pp 103–121
Manhart JR (1992) Molecular data and species concepts in the algae. J Phycol 28:730–737
Marcinko CLJ, Painter SC, Martin AP, Allen JT (2013) A review of the measurement and modelling of dinoflagellate bioluminescence. Prog Oceanogr 109:117–129
Margulis L, Corliss JO, Melkonian M, Chapman DJ (1990) Handbook of Protoctista: the structure, cultivation, habitat and life histories of the eukaryotic microorganisms and their descendants exclusive of animals, plants and fungi: a guide to the algae, ciliates, foraminifera, sporozoan, water molds, slime molds and the other protoctists. Jones and Bartlett, Boston
McCauley LAR, Wehr JD (2007) Taxonomic reappraisal of the freshwater brown algae *Bodanella*, *Ectocarpus*, *Heribaudiella*, and *Pleurocladia* (Phaeophyceae) on the basis of *rbc*L sequences and morphological characters. Phycologia 46:429–439
Medlin L, Zingone A (2007) A taxonomic review of the genus *Phaeocystis*. Biogeochem 83:3–18. https://doi.org/10.1007/s10533-007-9087-1
Mikhailyuk T, Lukešová A, Glaser K, Holzinger A, Obwegeser S, Nyporko S, Friedl T, Karsten U (2018) New taxa of streptophyte algae (Streptophyta) from terrestrial habitats revealed using an integrative approach. Protist 169:406–431
Mills LA, McCormick AJ, Lee-Smith DJ (2020) Current knowledge and recent advances in understanding metabolism if the model cyanobacterium *Synechocystis* sp. PCC 6803. Biosci Rep 40(4):BSR20193325. https://doi.org/10.1042/bsr20193325
Miyagishima S-Y, Tanaka K (2021) The unicellular red alga *Cyanidioschyzon merolae*—the simplest model of a photosynthetic eukaryote. Plant Cell Physiol 62:926–941. https://doi.org/10.1093/pcp/pcab052
Moestrup Ø (2006) Algal taxonomy: historical overview. eLS. https://doi.org/10.1002/9780470015902.a0000328.pub2
Moestrup Ø, Daugbjerg N (2007) On dinoflagellate phylogeny and classification. In: Brodie J, Lewis J (eds) Unraveling the algae—the past, present, and future of algal systematics, The systematics association special volume series, vol 75. CRC, Boca Raton, pp 215–230
Moestrup Ø, Sengco M (2001) Ultrastructural studies on *Bigelowiella natans*, gen. et sp. nov., a chlorarachniophyte flagellate. J Phycol 37:624–646
Murray SA, Suggett DJ, Doblin MA, Kohli GS, Seymour JR, Fabris M, Ralph PJ (2016) Unravelling the functional genetics of dinoflagellates: a review of approaches and opportunities. Perspect Phycol 3(1):37–52. https://doi.org/10.1127/pip/2016/0039
Nabout JC, da Silva RB, Carneiro FM, Sant'Anna CL (2013) How many species of cyanobacteria are there? Using a discovery curve to predict the species number. Biodivers Conserv 22:2907–2918. https://doi.org/10.1007/s10531-013-0561-x
Neukermans G, Fournier G (2018) Optical modeling of spectral backscattering and remote sensing reflectance from *Emiliania huxleyi* blooms. Front Mar Sci 5. https://doi.org/10.3389/fmars.2018.00146
Niesenbaum R (1988) The ecology of sporulation by the macroalga *Ulva lactuca* L. (Chlorophyceae). Aquat Bot 32:155–166

Novarino G (2012) Cryptomonad taxonomy in the 21st century: the first 200 years. In: Phycological reports. Institute of Botany, Polish Academy of Sciences, Krakow, pp 19–52

Olli K (1996) Resting cyst formation of *Eutreptiella gymnastica* (Euglenophyceae) in the northern coastal Baltic Sea. J Phycol 32(5):35–542

Olson RJ, Sosik HM (2007) A submersible imaging-in-flow instrument to analyze nano- and microplankton: imaging FlowCytobot. Limnol Oceanogr Methods 5:195–203. https://doi.org/10.4319/lom.2007.5.195

Park MG, Kim S, Kim HS, Myung G, Kang YG, Yih W (2006) First successful culture of the marine dinoflagellate *Dinophysis acuminata*. Aquat Microb Ecol 45:101–106

Parrow MW, Burkholder JM (2003) Reproduction and sexuality in *Pfiesteria shumwayae* (Dinophyceae). J Phycol 39:697–711

Pinevich A, Velichko N, Ivanikova N (2012) Cyanobacteria of the genus *Prochlorothrix*. Front Microbiol 3. https://doi.org/10.3389/fmicb.2012.00173

Ponce-Toledo RI, Moreira D, López-García P, Deschamps P (2018) Secondary plastids of euglenids and chlorarachniophytes function with a mix of genes of red and green algal ancestry. Mol Biol Evol 35:2198–2204. https://doi.org/10.1093/molbev/msy121

Prazeres M, Renema W (2019) Evolutionary significance of the microbial assemblages of large benthic Foraminifera. Biol Rev 94:828–848. https://doi.org/10.1111/brv.12482

Reynolds CS, Walsby AE (1975) Water-blooms. Biol Rev 50:437–481. https://doi.org/10.1111/j.1469-185X.1975.tb01060.x

Riegman R, Noordeloos AAM, Cadée GC (1992) *Phaeocystis* blooms and eutrophication of the continental coastal zones of the North Sea. Mar Biol 112:479–484

Rizouli A, Küpper FC, Louizidou P, Mogg AOM, Azzopardi E, Sayer MDJ, Kawai H, Hanyuda T, Peters AF (2020) The minute alga *Schizocladia ischiensis* (Schizocladiophyceae, Ochrophyta) isolated by germling emergence from 24 m depth off Rhodes (Greece). Diversity 12:102. https://doi.org/10.3390/d12030102

Schädler T, Thurn AL, Brück T, Weuster-Botz D (2021) Continuous production of lipids with *Microchloropsis salina* in open thin-layer cascade photobioreactors on a pilot scale. Energies 14(2):500. https://doi.org/10.3390/en14020500

Schulz-Vogt HN, Angert ER, García-Pichel F (2007) Giant bacteria. eLS. Wiley, Chichester. https://doi.org/10.1002/9780470015902.a0020371

Seckbach J (1994) The natural history of *Cyanidium* (Geitler 1933: past and present perspectives). In: Seckbach J (ed) Evolutionary pathways and enigmatic algae: Cyanidium caldarium (Rhodophyta) and related cells. Kluwer Academic, Boston, pp 99–112

Sheath RG, Vis M (2015) Red algae. In: Wehr JD, Sheath RG, Kociolek JP (eds) Freshwater algae of North America, 2nd edn. Academic, San Diego, pp 237–264. https://doi.org/10.1016/B978-0-12-385876-4.00005-0

Sineshchekov OA, Govorunova EG, Jung K-H, Zauner S, Maier U-G, Spudich JL (2005) Rhodopsin-mediated photoreception in cryptophyte flagellates. Biophys J 89:4310–4319

Skelton HM, Burkholder JM, Parrow MW (2008) Axenic cultivation of the heterotrophic dinoflagellate *Pfiesteria shumwayae* and observations of feeding behavior. J Phycol 44:1614–1624

Solymosi K (2012) Plastid structure, diversification and interconversions I. Algae. Curr Chem Biol 6:167–186. https://doi.org/10.2174/2212796811206030002

South GR, Whittick A (1987) Introduction to phycology. Blackwell Scientific, Boston

Spector DL (ed) (1984) Dinoflagellates. Academic, New York

Stanier RY, Sistrom WR, Hansen TA, Whitton BA, Castenholz RW, Pfennig N, Gorlenko VN, Kondratieva EN, Eimhjellen KE, Whittenbury R, Gherna RL, Truper HC (1978) Proposal to place the nomenclature of the cyanobacteria (blue-green algae) under the rules of the International Code of Nomenclature of Bacteria. Int J Bacteriol 28:335–336

Suzuki S, Ishida K-I, Hirakawa Y (2017) Diurnal transcriptional regulation of endosymbiotically derived genes in the chlorarachniophyte *Bigelowiella natans*. Genome Biol Evol 8:2672–2682. https://doi.org/10.1093/gbe/evw188

Taylor FJR (ed) (1987) The biology of dinoflagellates. Blackwell Scientific, Oxford, UK

Thyssen M, Grégori G, Créach V, Lahbib S, Dugenne M, Aardema HM, Artigas L-F, Huang B et al (2022) Interoperable vocabulary for marine microbial flow cytometry. Front Mar Sci 9:975877. https://doi.org/10.3389/fmars.2022.975877

Tomas CR (ed) (1997) Identifying marine phytoplankton. Academic, San Diego, p 858

van den Hoek C, Mann DG, Jahns HM (1995) Algae—an introduction to phycology. Cambridge University Press, Cambridge, UK

Walne PL, Kivic PA (1990) Phylum Euglenida. In: Margulis L, Corliss JO, Melkonian M, Chapman DJ (eds) Handbook of protoctista. Jones and Bartlett, Boston, pp 270–287

Wang M, Hu C, Barnes BB, Mitchum G, Lapointe B, Montoya JP (2019) The great Atlantic Sargassum belt. Science 365:83–87. https://doi.org/10.1126/science.aaw7912

Wetherbee R, Jackson CJ, Repetti SI, Clementson LA, Costa JF, van de Meene A, Crawford S, Verbruggen H (2019) The golden paradox– a new heterokont lineage with chloroplasts surrounded by two membranes. J Phycol 55:257–278. https://doi.org/10.1111/jpy.12822

Wetherbee R, Bringloe TT, van de Meene A, Andersen RA, Verbruggen H (2022) Structure and formation of the perforated theca defining the Pelagophyceae (Heterokonta), and three new genera that substantiate the diverse nature of the class. J Phycol 59:126–151. https://doi.org/10.1111/jpy.13294

Yamaguchi A, Yubuki N, Leander BS (2012) Morphostatis in a novel eukaryote illuminates the evolutionary transition from phagotrophy to phototrophy: description of *Rapaza viridis* n. gen. et sp. (Euglenozoa, Euglenida). BMC Evol Biol 12:29. https://doi.org/10.1186/1471-2148-12-29

Young JR, Andruleit H, Probert I (2009) Coccolith function and morphogenesis: insights from appendage-bearing coccolithophores of the family Syracosphaeraceae (Haptophyta). J Phycol 45:213–226. https://doi.org/10.1111/j.1529-8817.2008.00643

Zimba PV, Moeller PD, Beauchesne K, Lane HE, Triemer RE (2010) Identification of euglenophycin—a toxin found in certain euglenoids. Toxicon 55:100–104

Zimba PV, Huang I-S, Gutierrez D, Shin W, Bennett MS, Triemer RE (2017) Euglenophycin is produced by at least six species of euglenoid algae and six of seven strains of *Euglena sanguinea*. Harmful Algae 63:79–84

Additional Websites Accessed

https://commons.wikimedia.org/wiki/Category:Phaeocystis#/media/File:Ecomare_-_schuimalg_strand_(7037-schuimalg-phaeocystis-ogb).jpg. Accessed 9 Mar 2023

https://commons.wikimedia.org/wiki/File%3AEuglena_sp..jpg. Accessed 12 Mar 2023

https://commons.wikimedia.org/wiki/File:Cwall99_lg.jpg. Accessed 8 Mar 2023

https://commons.wikimedia.org/wiki/File:Dinoflagellates.jpg. Accessed 9 Mar 2023

https://commons.wikimedia.org/wiki/File:Emiliania_huxleyi_coccosphere_and_coccolith.jpg. Accessed 10 Mar 2023

https://commons.wikimedia.org/wiki/File:Noctiluca_scintillans.jpg. Accessed 9 Mar 2023

https://en.wikipedia.org/wiki/Chlorarachniophyte#/media/File:Chlorarachnion_reptans.jpg. Accessed 9 Mar 2023

https://en.wikipedia.org/wiki/Dinobryon#/media/File:Mikrofoto.de-Dinobryon_divergens.jpg. Accessed 12 Mar 2023

https://eol.org/pages/3600. Accessed 11 Mar 2023

https://planktonnet.awi.de/index.php?contenttype=image_details&itemid=68308#content. Accessed 10 Mar 2023

https://sagdb.uni-goettingen.de/detailedList.php?str_number=1380-1c. Accessed 6 Feb 2023

https://vocab.nerc.ac.uk/collection/Fo2/current/. Accessed 23 Jan 2023

https://www.nwf.org/Educational-Resources/Wildlife-Guide/Plants-and-Fungi/Joshua-Tree. Accessed 6 Sept 2022

www.algaebase.org. Accessed 11 Mar 2023

www.marinespecies.org. Accessed 11 Mar 2023

Endosymbiosis and Evolution 10

10.1 Introduction: Endosymbiotic Proponents and Skeptics

Paul Falkowski, a renowned phytoplankton physiologist, together with his colleagues, once posed the question (Falkowski et al. 2004a), "Why is the land green and the oceans red?" He was not referring to the occurrences of red tides, but instead to the fact that phytoplankton in the oceans are represented to a greater extent by those species that contain "red" plastids, and chlorophyll *c*, while land plants are almost exclusively dominated by plants that are green. The answer to his question of why the land is green and the ocean red lies in the most important event in the evolution of photosynthesis: the acquisition of photosynthetic prokaryotes by eukaryotes. It is now generally accepted that the acquisition of photosynthesis by eukaryotes occurred through endosymbiosis (*symbiosis* in Latin means "living together"), the engulfment of one organism by another, resulting eventually in both plastids and mitochondria (Martin et al. 2015). The evolutionary scheme most widely accepted involves a series of endosymbiotic events from which the diversity of species has radiated.

The notion that eukaryote evolution is linked to prokaryote engulfment has been highly controversial over time, and many aspects still remain controversial. More than 100 years ago, the theory of endosymbiosis (or, as originally termed, symbiogenesis) was introduced by the Russian biologist Constantin Mereschkowski (1905; summarized by Martin 2005; Martin et al. 2015). Mereschkowski (1905) said (as translated into English and as summarized by Martin et al. (2015, p. 2), "(i) plastids are unquestionably reduced cyanobacteria that early in evolution entered into a symbiosis with a heterotrophic host, (ii) the host that acquired plastids was itself the product of an earlier symbiosis between a larger, heterotrophic, amoeboid host cell and a smaller 'micrococcal' endosymbiont that gave rise to the nucleus, and (iii) the autotrophy of plants is an inheritance, *in toto*, from cyanobacteria." Mereschkowski had the basic concept of endosymbiosis correct (Martin et al. 2015).

However, while there was a period of about 20 years during which the theory was accepted (early 1900s)—largely by botanists, the idea was subsequently condemned

P. M. Glibert, *Phytoplankton Whispering: An Introduction to the Physiology and Ecology of Microalgae*, https://doi.org/10.1007/978-3-031-53897-1_10

by E.B. Wilson, a leading cell biologist of the time (Wilson 1928, p. 56). He referred to Mereschkowski's ideas as "fantasy" and "flights of imagination" (cited by Martin et al. 2015). As a result of his criticism, "the notion [was banned] from the realm of serious scientific endeavor or, as he [Wilson] put it, from 'polite biological society' for several decades" (Martin 2007). The theory that emerged through much of the 1970s was that eukaryotes descended directed from cyanobacteria and that "algae [were] the most primitive eukaryotes, from which the remaining eukaryotes arose thorough loss of autotrophy (and plastids) and a transition to the heterotrophic lifestyle" (Martin 2007, p. 56).

Enter Lynn Margulis. She reintroduced the endosymbiosis theory and provided biochemical evidence in its favor (Margulis 1970; she also published under her married name, Sagan 1967). Although elements of her theory were controversial, she ultimately showed that (Martin 2007, p. 57):

> Endosymbiotic theory predicted the sequences of genes encoded in organelle DNA to be more similar to their homologues among free-living prokaryotes than to nuclear-encoded counterparts, whereas competing alternatives did not. That strong line of evidence eventually brought virtually all resistance against endosymbiotic theory for mitochondria (and chloroplasts) to an end...

It is worth noting—and indicative of the times—that her ideas were not immediately accepted, nor did her success come easily and she was considered somewhat of a radical. Her initial paper was reportedly rejected by 15 different journals (Knoll 2012). As noted in one of her obituaries (Lake 2011), "She had to stand up for her ideas and accept repeated rejections... Remarkably, she developed her theories without training in molecular biology – RNA sequencing was just being developed, and DNA sequencing was years away. Her discovery of the endosymbiotic origins of animals and plants was ahead of the field. She got there on her own terms, and her main insights have been proven correct."

10.2 An Oxygenated World

Earth was anaerobic for its first 2 billion years or so (Green 2007). Cyanobacteria are singularly responsible for oxygenating the atmosphere and transforming a once-reducing environment into an oxidizing one (Holland 2006). Only after cyanobacteria evolved the ability to extract electrons from water, using two photosystems in series, did O_2 evolve. The world changed. The production of O_2 first infiltrated the atmosphere, and slowly the shallow waters. This meant that all every living organism faced a new world, with its rising "pollutant," O_2 (Green 2007). Those organisms that could not adapt to the changing world were poisoned by it. As Martin et al. (2015, p. 8) underscored, "[e]ndosymbiosis and organelle origins are not about digestion. Microbial symbiosis, the process that gave rise to bioenergetic organelles, is about chemistry."

Rising O_2 levels led to the evolution of eukaryotes via one or more mechanisms. The evolution of photosynthesis is at the center of understanding the history and diversity of life as we know it. Using light energy in the presence of O_2 is a "dangerous business" (Green 2007 p. 37), especially for those with delicately balanced photosystems. and once O_2 came on the scene, it provided two challenges. First, if an excited chlorophyll molecule cannot immediately transfer its electron to a reaction center, singlet oxygen, $^1O_2$, may be formed. This is a reactive oxygen species (ROS). In addition, if O_2 acts as an electron acceptor, as electrons flow from PSII to PSI, superoxide, O_2^-, is formed. This reacts with water to produce H_2O_2, another form of ROS. When O_2 levels rose, the evolution of larger and more communication-related transmembrane proteins allowed for subsequent compartmentalization. Alternatively, or in addition, rising O_2 levels may have promoted cellular compartmentalization in order to protect the metabolic activities of the plasma membrane from rising levels of ROS in the cellular environment (Gross and Bhattacharya 2010).

The mechanisms by which cyanobacteria obtained two distinct photosystems are a topic of considerable debate (Cardona 2019). There is evidence that the two photosystems of photosynthesis (which are covered more extensively in Chap. 6), arose originally from different events. Oxygenic photosynthesis originated in cyanobacteria when an anoxygenic photosystem evolved into a water-splitting photosystem (Cardona et al. 2015). How these photosystems fused into a finely tuned machine remains unclear. Were these triggered by a gene duplication event or via horizontal gene transfer from lineages of anoxygenic phototrophs before the great oxidation event (Cardona 2019)? Some evidence suggests that such a merger likely occurred only once, giving rise to ancestral prokaryotes (Martin 2007). The two photosystems are fundamentally and functionally different, yet intricately coupled. Both photosystems have two different polypeptides embedded within a nonphospholipid membrane. Based on structural evidence and amino acid sequences, one hypothesis is that photosystem II originally derived from a purple bacteria, while photosystem I derived from green sulfur bacteria (Falkowski and Knoll 2007 and references therein). Thus, this line of reasoning suggests that bacteria containing photosystems in which O_2 is not produced, i.e., purple bacteria, evolved first, and only later did cyanobacteria evolve, giving rise eventually to the chlorophyll in algae and plants. This is based on the notion that anoxygenic gave rise to oxygenic photosynthesis, but this is an unproven assumption (Cardona 2017, 2019; Oliver et al. 2023) and has been called into question with new molecular evolution studies of photosystems.

10.3 Genome Inheritance

Photosynthetic eukaryotes emerged from a heterotrophic eukaryote which engulfed a free-living photosynthetic cyanobacterium. All plastids appear to have evolved from a cyanobacterial endosymbiont (Delwiche 1999). The ancestral eukaryote was likely a facultative anaerobe (Martin et al. 2015). Such a eukaryote must already

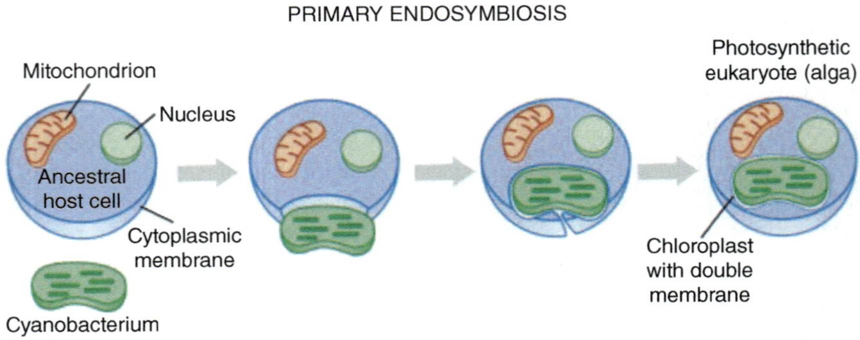

Fig. 10.1 Schematic of a primary endosymbiosis event in which a free-living bacterium is engulfed phagotrophically by a eukaryotic cell already containing a mitochondrion, which then is transformed into plastid. Primary endosymbiosis yields organelles with two membranes. In this example, the original independent cyanobacterium has a cytoplasmic membrane, which is retained, and an outer membrane, which is lost during symbiosis. When the two cells associate, the host-cell cytoplasmic membrane surrounds the cyanobacterium, which is therefore left surrounded by two membranes. Image reproduced from Clark and Pazdernik (2013, doi: 10.1016/C2009-0-01986-2) with permission of Elsevier

have possessed a mitochondrion which had transferred genes from the endosymbiont to the host's chromosomes, as "only cells that possessed mitochondria had the bioenergetic means to attain eukaryotic cell complexity, which is why there are no true intermediates in the prokaryote-to-eukaryote transition" (Martin et al. 2015, p. 1). The endosymbiotic origin of mitochondria is considered to be different from that of plastids, having been likely an α-proteobacterium (purple non-sulfur bacteria), and the host possibly a methanogen (an archaeon not a eukaryote), although this has been debated (Martin et al. 2015; Archibald 2015a, b). As this proteobacterium evolved into a mitochondrion, its initial function was to detoxify O_2 by reducing it to water. The rise of O_2 did provide a better terminal electron acceptor for respiratory chains, and diversification of heterotrophic prokaryotes benefitted thereof (Green 2007). How the integration into a mitochondrion of the host occurred, physiologically and genetically, is the subject of considerable study, not reviewed here (see Roger et al. 2017). Suffice it to say here that gene duplication, gene loss, among other processes contributed to this complex evolution.

Primary endosymbiosis is the process of incorporation of an ingested cyanobacterial cell within a eukaryotic host (Fig. 10.1). Such endosymbionts evolved into plastids with two envelope membranes; their membranes protected them from digestion. In order for an endosymbiont to become an organelle, there must be evidence that at least some genes were transferred to the host nucleus and host proteins imported into the endosymbionts. Endosymbionts eventually came under the control of the host, and as more and more genes were transferred to the host nucleus, they took on the role of an organelle.

There are two hypotheses about the number of primary endosymbiotic events that may have occurred. One hypothesis is based on a single successful event: that is, a heterotrophic eukaryote once successfully engulfed a cyanobacterium. This hypothesis suggests that this event occurred about 1.5–2 billion years ago (Hackett et al. 2007). First, a cyanobacterium became a symbiont of the eukaryote, then it was reduced to its metabolic core, serving a permanent role as a plastid. This required that metabolically there was an eventual "establishment of a reliable connection between the host cell and the ancestral plastid," as unregulated flux of materials would not be metabolically fit for purpose. Gene transfer from the endosymbiont to the host had to have occurred (Allen et al. 2005; Hackett et al. 2007). This means that the genes of the endosymbiont were transferred to the host nucleus, rendering the endosymbiont unable to survive without the host. With this hypothesis, such an event is thought to have given rise to three lineages with plastids of primary lineage, the Glaucophyta, the Rhodophyta and the Viridiplantae (green algae and land plants). These three groups are collectively termed the Archaeplastida, or the Plantae (Archibald 2009).

Although there is strong support for a single endosymbiotic event as the origin of plastids, alternate scenarios are emerging (Larkum et al. 2007; Gruber and Medlin 2023). An alternative hypothesis posits that multiple primary endosymbiotic events occurred, with different plastid-less hosts acquiring different endosymbionts (reviewed by Graham et al. 2016; Larkum 2020). Sato (2019, 2020) studied various chloroplast enzymes and compared them with their cyanobacterial analogs and found various origins for the enzymes, leading to the conclusion that enzymes and chloroplasts were acquired at different times, suggesting multiple gene transfers. As stated by Keeling et al. (2015),"The mitochondria and plastids we see today may, accordingly, have only been the luckiest of a longstanding series of doomed endosymbionts who were saved by transfer of genes to the nucleus." There are more "recent" cyanobacterium-eukaryote endosymbioses, such as that of *Paulinella chromatophora* and the diatom *Rhopalodia gibba* (Prechtl et al. 2004; Nowack 2014), that are informative with respect to the process of cellular events from endosymbiont to plastid and are actively being investigated.

Secondary endosymbiosis is the process by which a eukaryotic cell with its primary endosymbionts themselves becomes endosymbionts in a new eukaryotic host cell (Fig. 10.2). Algae other than Archaeplastida, including haptophytes, cryptophytes, dinoflagellates, heterokonts, and other photosynthetic eukaryotes were hosts and appear to have had their plastids originate from acquisition of primary plastids during secondary endosymbiosis. Secondary endosymbionts, have three or four membranes as plastids. All secondary plastids appear to have originated from endosymbionts of the two primary plastid lineages, rhodophytes and chlorophytes (Oborník 2019).

Tertiary endosymbiosis repeats the process, with a eukaryotic cell engulfing a eukaryotic endosymbiont with a secondary plastid. Tertiary plastids appear to have arisen multiple times (Delwiche 1999; Fig. 10.3).

Although these processes of inheritance seem straightforward, the reality is more complex, with the result representing a "tangled web across a very large fraction of

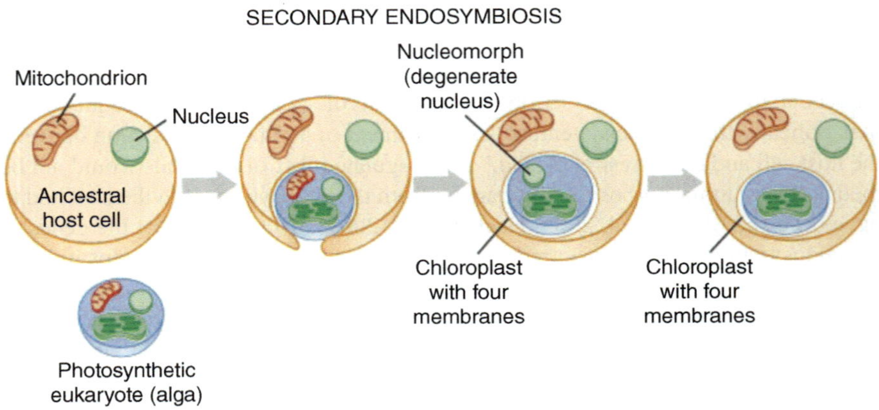

Fig. 10.2 Secondary endosymbiosis consists of a eukaryotic alga being engulfed by another eukaryote, forming a chloroplast with three or four membranes. The alga already has a chloroplast with two membranes as well as a nucleus and other organelles. Since the host cell only needs the energy from the chloroplast, the other captured organelles degenerate and eventually disappear. However, the membranes often remain and the chloroplast is left with four membranes, rather than two. Image reproduced from Clark and Pazdernik (2013, doi: 10.1016/C2009-0-01986-2) with permission of Elsevier

the eukaryotic tree" (Archibald 2009, p. R81). The mosaic of algal species has subsequently been shaped by a number of major evolutionary forces, not only endosymbiosis, but also mutation, selection, genetic drift and gene transfer. Larkum et al. (2007) compared how hosts incorporated new genetic material into their cells to "bargain hunt[ing] at a trash and treasure outlet." Gruber and Merlin (2023) suggest that DNA from the endosymbiont may have integrated into hosts via rupture of the symbiont and incorporation of random pieces into DNA nuclei. In all, numerous "super-groups" of eukaryotes have evolved in this manner, based on their plastid composition, of which the Archaeplastida is just one such group (Fig. 10.3). These groups also include the Rhizaria, the Stramenopiles (also called ochrophytes) and Alveolates (collectively referred to as the SAR groups), and the Excavates (including the euglenophytes). Alveolates (dinoflagellates, apicomplexans, ciliates) and the stramenopiles (diatoms, golden algae, brown algae, oomycetes) are commonly grouped as the "Chromalveolates," due to an assumed common ancestor. Subsequently, these taxa were presumed to have undergone serial endosymbioses. However, a reclassification of "Chromalveolates" has occurred as new data has come to light, questioning whether they indeed have a common ancestor (Burki et al. 2020). Furthermore, molecular clock studies suggest earlier branching of different algal groups which would be at odds with endosymbiotic acquisition of a red algal plastid in common ancestral host (Strassert et al. 2021; Gruber and Merlin 2023). The classic groupings have now been reshuffled due to the increased availability of molecular sequence data and a number of newly discovered, but rare, protists, resulting in a new evolutionary tree of life (eToL; Burki et al. 2020). In

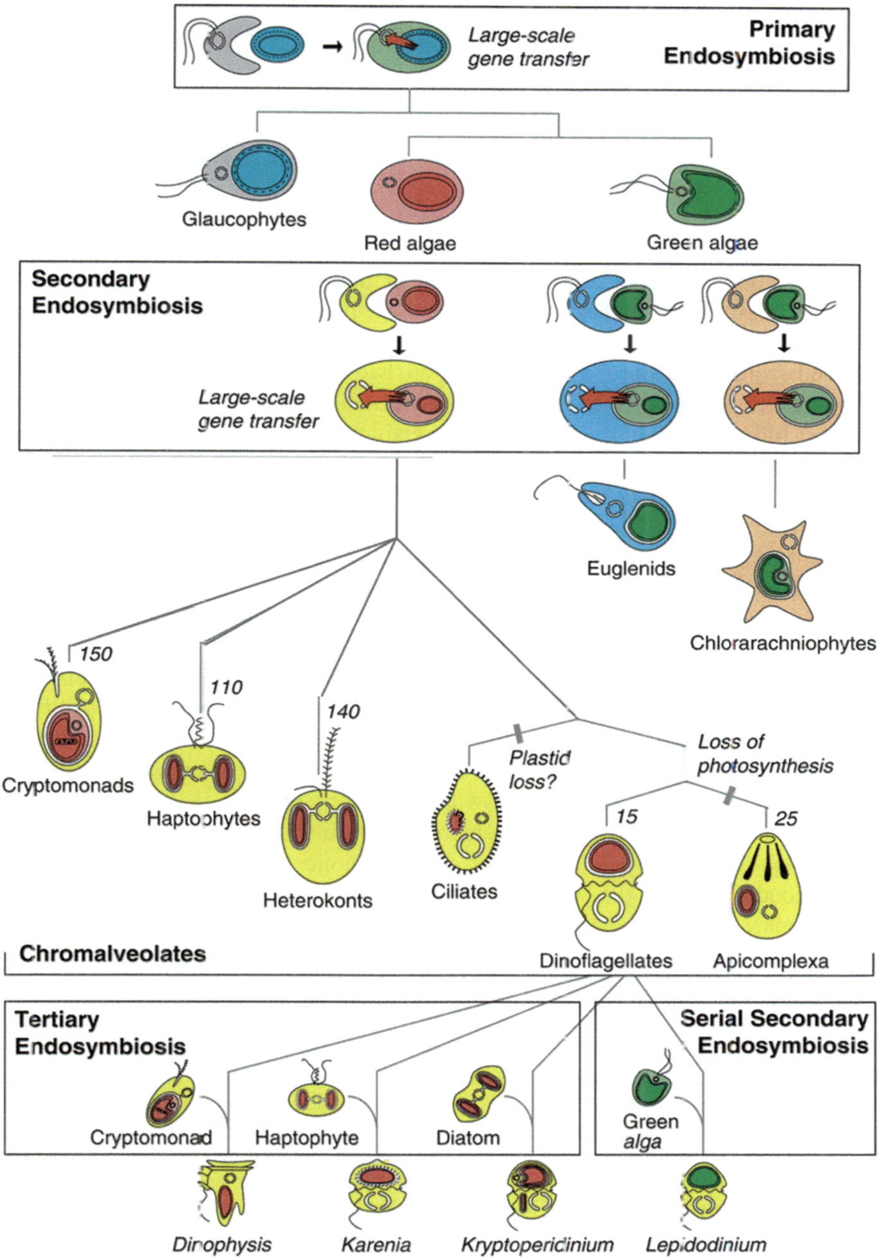

Fig. 10.3 Conceptual diagram of algal evolution and the origin and spread of plastids by endosymbiosis. At the top is the single origin of plastids by primary endosymbiosis between a cyanobacterium and a eukaryotic host. This endosymbiont was reduced and integrated, and part of this process involved the transfer of hundreds of genes from the cyanobacterium/plastid to the eukaryotic host nucleus (red arrow). Glaucophytes, red and green algae all descended from this fully integrated partnership. Next, plastids spread to other eukaryotic groups by secondary

developing historical context, the term "Chromalveolates" will be used below, recognizing that it is imprecise and that the chromalveolate hypothesis is currently being challenged.

10.4 Why the Land Is Green and the Ocean Is Red

The diversification of algal lineages in the ocean, and the diversification of plant lineages on land, took separate paths. The acquisition of plastids is considered critical in both the diversity of algae and in the early evolution of land plants (Graham et al. 2009). On land, photosynthesis by plants is derived from endosymbiosis of a green alga; every plant (virtually speaking) is green. The "green" lineage evolved from the Chlorophyta, resulting in two groups, the Euglenoids and the "Chlorachniophytes" via secondary endosymbiotic events. It did, however, take about 2 billion years of evolution and diversification before colonization on land was possible (Larkum et al. 2020).

In the oceans, a much more diverse pattern, emerged. The major eukaryotic phytoplankton in the oceans all have "red" plastids except for the euglenophytes and chlorachniophytes (Gruber and Merlin 2023). The "red" lineage evolved, resulting in four groups, the cryptophytes, haptophytes, heterokontophytes and dinoflagellates. Within the dinoflagellates, tertiary endosymbiotic events, differentially involving a cryptomond, a haptophyte, a diatom and a serial secondary endosymbiotic event involving a green alga, explain the radiation of different types of dinoflagellates (Fig. 10.3). The idea that an ancient green endosymbiont was replaced via repeated secondary and tertiary endosymbioses of a red alga-derived chloroplast is consistent with changes in the selective pressures acting on algae during a protracted period of low atmospheric CO_2.

It may also be that trace element composition differs for the green algae compared to the algae that have had secondary endosymbiotic events, leading to differing redox conditions and this may contribute to why the radiation of diversity of photosynthesis is so different on land and sea (Quigg et al. 2003). Specifically, red algae and the "Chromalveolates" (see note on reclassification above) tend to contain

Fig. 10.3 (continued) endosymbiosis (middle). Green algae were most likely involved in two independent events, giving rise to euglenids (turquoise) and chlorarachniophytes (orange). A single endosymbiosis involving a red alga probably gave rise to the chromalveolates (yellow); this group is supported by several molecular characters and gene trees (plotted on the figure; but see text regarding the current status of this group). Plastids have apparently been lost in ciliates and Cryptosporidium (and perhaps other lineages) and photosynthesis has been lost in apicomplexa and many other individual lineages Numbers indicate the approximate number of protein genes in the plastid genomes of the various lineages with secondary red plastids. Finally, dinoflagellates have substituted their ancestral plastid several times, most notably by tertiary endosymbioses involving other chromalveolates (a cryptomonad, a haptophyte, and a diatom) and by serial secondary endosymbiosis involving a green alga. Image reproduced from Keeling et al. (2004, Proc Natl Acad Sci USA 112:10101–10103) with permission of the American Association for the Advancement of Science much lower levels of iron, copper, and zinc than green algae, which might give them greater resource efficiency. Other evidence for differential evolutionary processes resulting from endosymbiotic events of the red and green lineage comes from the different types of the enzyme Rubisco in different lineages. Red algal plastids have Rubisco of the Form 1D, which is known only to occur in β-proteobacteria. In contrast, in green algae and land plants, the Rubisco is of the Form 1B and is known to occur in cyanobacteria, green algae, and land plants (Graham et al. 2009).

10.5 The Many Shades of Red

Algal lineages with red algal-derived plastids dominate the oceans. They have chlorophyll *c* and xanthophyll pigments. The nonphotosynthetic lineages are thought to have descended from photosynthetic ancestors and secondarily lost their photosynthetic capacity, but may retain some evidence of their photosynthetic past, such as nuclear genes acquired from an ancient chloroplast (Dorrell and Smith 2011).

Why has the red algal-derived lineage been so successful in the sea? The "portable plastid" hypothesis (Grzebyk et al. 2003; Falkowski et al. 2004b), albeit controversial (Keeling et al. 2004), proposes that algae with red-derived chloroplasts have an advantage over algae with green-derived chloroplasts because they have larger genomes than green algal chloroplasts, that red algae have been incorporated more often than green algae by secondary endosymbiosis, and that gene transfer to the host is quite rare. Specifically, red algal plastids have been thought to retain a number of genes involved in regulating photosynthesis and metabolism and may depend less on their host's proteins (Graham et al. 2016). Accordingly, red plastids should be able to better adjust within plastid-less host cells, leading to their wide diversity. Additional sequencing and analysis of genomes of those species with red-derived plastids will help to define these lineages and mechanisms of plastid acquisition in more detail.

Although an enlarged chloroplast genome would be selectively advantageous, e.g., due to more efficient redox regulation of gene expression (Allen 2003), the chloroplast genomes of diatoms and dinoflagellates are significantly more reduced than those of red algae (Bowler et al. 2008; Howe et al. 2008). Such an "endosymbiotic mosaic" of red, green, and chromalveolate-derived genes could have provided the genetic diversity to allow extant "chromalveolates" to dominate post-Permian algal communities (Dorrell and Smith 2011).

Tertiary endosymbiosis has been particularly important in the diversity of dinoflagellates. Some dinoflagellates evolved from endosymbiosis events involving a green alga. The most primitive of the dinoflagellates are the gymnodinoids, which have dinophyte chloroplasts, chlorophyll *c*2, peridinin, and a triple membrane envelope, and feed via a peduncle (Van den Hoek et al. 1995). Peridinin plastids are thought to have evolved from a secondary endosymbiotic event involving a red alga. Others are thought to have arisen from endosymbiotic events involving a cryptomonad, a diatom or a haptophyte (Palmer and Delwiche 1996). Some species subsequently lost their dinophyte chloroplasts, becoming either exclusively heterotrophic (phagotrophic). In addition to the chloroplast evidence, it is also thought that over evolutionary time, there was reduction in the thecal plates of dinoflagellates.

10.6 Endosymbiosis and Mixotrophy: The Transition from Food to Function

Plastids derived from ingestion appear to represent early stages of endosymbiosis. Following this line of reasoning, a link between comparatively poor inorganic nutrient acquisition abilities, mixotrophic nutrition and the possibility that early endosymbiosis may have been a response to resource limitation in the primitive ocean has been hypothesized (see also Chap. 15). If the ocean is indeed severely nutrient limited, mixotrophic capability would have conferred competitive advantages, and over time, this stimulated algal specialization and diversification (Anbar and Knoll 2002; Yoon et al. 2004). There is relatively little doubt that plastids and mitochondria resulted from endosymbiosis of an early eukaryotic cell. However, there remains much uncertainty about how many endosymbiotic events occurred, who ate whom and how early the incorporation of an endosymbiont may have occurred. The distinction between early endosymbionts and permanent plastids is, inevitably, blurry. The time scale of endosymbiont gene transfer to the host, and the many other metabolic processes involved in making a food stuff permanently incorporated in the cell, and why this all occurred remain questions without full answers.

Many taxa that have secondary or tertiary-acquired plastids are mixotrophic (see Chap. 15). Yet, extant red algae or glaucophytes with their primary plastids, as well as diatoms, have lost or substantially reduced their phagotrophic ability (Raven et al. 2009). Some capacity for phagotrophy has been retained in all major eukaryotic phytoplankton lineages except for diatoms (Raven 2013). There is some evidence that diatoms may originally had a green endosymbiotic plastid that was replaced with a red one (Gruber and Medline 2023). Loss of plastid function may have occurred once the environmental conditions changed or when not required for the cell's survival any longer (Williams and Keeling 2003; Mansour and Anestis 2021). Could it be that those with secondary or tertiary plastids are evolutionarily younger and therefore have still maintained their phagotrophic ability, while those taxa that are evolutionarily older may have lost this ability once their plastids were fully established (Mansour and Anestis 2021)?

Kleptochloroplasts, the transient or nonpermanent "stolen" plastids acquired by mixotrophs during grazing (see Chap. 15), may last within the host cell for periods that range from days to substantially longer, depending on the host taxa and conditions for growth. These short-term plastids may provide clues to the pathways to permanent plastid retention. They may be "in effect new plastids *en route* to becoming permanent plastids, a 'work in progress'" (Mansour and Anestis 2021; Fig. 10.4). Nonconstitutive mixotrophs (those that acquire their plastids by grazing) may represent a transitional stage to constitutive mixotrophs (those that have their

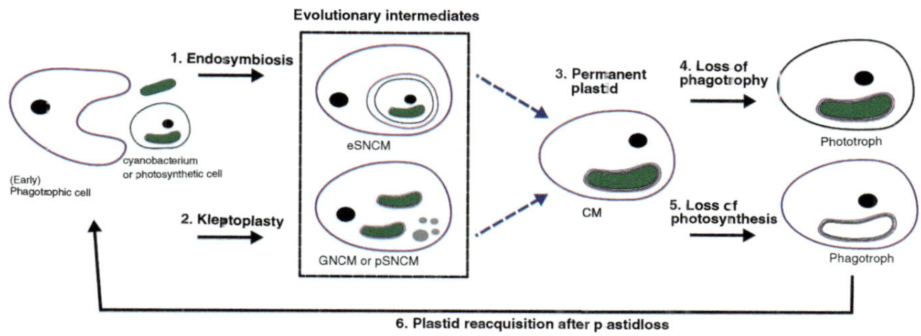

Fig. 10.4 Conceptual diagram of protist plastid evolution and the role of mixotrophy. The cell nucleus is shown as a black circle, kleptokaryons as gray circles, and photosynthetic plastids in green. For the initial primary plastid acquisition, the plastid came from a cyanobacterium (green oval), the acquired primary plastid would be bound by only two membranes as depicted in the photosynthetic cell on the left. While complex permanent plastids gained by secondary or tertiary endosymbiosis are bound by three or four membranes, respectively as illustrated by the number of lines around the plastid. Blue dashed lines are the hypothetical route toward permanent plastid retention. (1) Uptake of a photosynthetic alga and internalization of the entire organism as a symbiont (in a symbiosome); (2) Kleptoplasty shows the phagotrophic uptake of photosynthetic algae (prey), keeping the chloroplasts temporarily functional; (3) The establishment of permanent plastids; (4) Secondary loss of phagotrophy; (5) Loss of photosynthesis, though only in rare cases is the plastid completely lost; and (6) Plastid reacquisition after loss of phototrophy (known in dinoflagellates). Image reproduced from Mansour and Anestis (2021, Front Mar Sci 8:666160) under a creative commons license. See Chap. 15 (Sect. 15.9) for further details about the types and strategies of different types of mixotrophs (CM, constitutive mixotroph (those with innate ability to photosynthesize but also feed); eSNCM, endosymbiotic specialist nonconstitutive mixotroph (those that harbor photosynthetic endosymbionts); GNCM, general non constitutive mixotrophs (generalist nonconstitutive mixotrophs using chloroplasts from a range of species; pSCM, plastidic specialist nonconstitutive mixotrophs (those that acquire plastids from specific species)

own plastids but also feed; Gast et al. 2007; Mansour and Anestis 2021; see Chap. 15). Some modeling studies have lent support to the notion that environmental conditions may contribute, or have contributed, to the specialization of mixotrophic feeding (Troost et al. 2005a, b). Mixotrophic plankton are expected to become more prevalent in future oceans (e.g., Wilken et al. 2013; Glibert 2020). Experimental evolutionary studies will be critical in providing the needed insight into the past and future of mixotrophs and how their development relates to their endosymbiotic lineage.

10.7 The Algal Tree of Life: From Kingdoms to Supergroups

There has been a long history of systematizing organisms. Aristotle proposed two kingdoms, plants and animals. Haeckel, whose contributions to plankton ecology are discussed throughout this volume (see Chaps. 1 and 19, for example), concluded that bacteria, protozoans, algae, and fungi belonged in a separate kingdom, the Protista

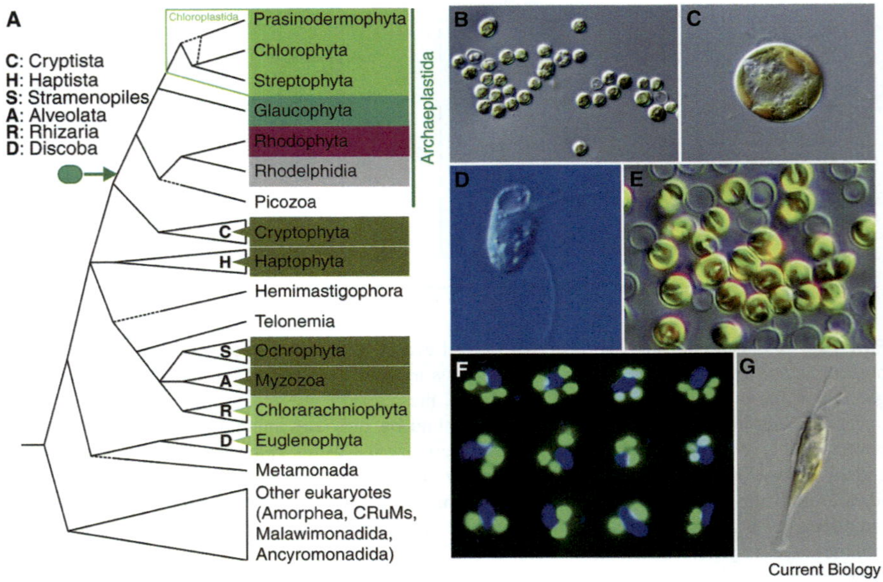

Fig. 10.5 Phylogenetic diversity of algae. Image reproduced from Eliáš (2021, Curr Biol 31: R733–R735, doi: 10.1016/j.cub.2021.04.025) with permission of Elsevier

(reviewed by Margulis and Chapman 2009). More recently, Whittaker developed a five kingdom system, Monera, prokaryotes; Protista, eukaryotic, mostly unicellular organisms; Plantae, autotrophic multicellular eukaryotes; Fungi, osmotrophic multicellular eukaryotes; and Animalia, phagotrophic multicellular eukaryotes. Algae were divided between the first three kingdoms in the Whittaker system (reviewed by Gololobova and Belyakova 2022). With new tools, new data and improved knowledge of cell biology and genetics, new classifications systems have been developed. This gave rise to the concept of supergroups rather than kingdoms. Simpson and Roger (2002) proposed that several supergroups be isolated in eukaryotes, including Opisthokonta, Plantae, Chromalveolata, Cercozoa, Amoebozoa, and Excavata, and they placed various groups of algae in four of the supergroups (Gololobova and Belyakova 2022).

Recent trees (Burki et al. 2020; Eliáš 2021; Gololobova and Belyakova 2022; Fig. 10.5) have reshuffled relationships, based on recent phylogenomic analyses. These new trees have many more branches, have elevated some supergroups, and united others. One thing is certain: Phylogenetic reconstructions are rapidly changing as new data are amassed. These trees, like all others before, should be considered a work in progress.

10.8 Summary

The mosaic of algal species has been shaped by many forces, but evolutionarily, the acquisition of prokaryotic photosynthetic machinery by eukaryotes in the sea is perhaps the most important. Subsequent endosymbiotic engulfment along with mutations, selections, genetic drift, and environmental factors have led to the radiation of species with their unique, but sometimes related, distinguishing characteristics. Much is being learned about genetic diversity and relationships between and among groups, with new genomic data lending insight to eukaryotic evolution, gene gains, and losses. What is clear is that there is still much to be learned about evolutionary relationships, and new methods and approaches will be critical to advancing our understanding.

References

Allen JF (2003) The function of genomes in bioenergetic organelles. Philos Trans R Soc Lond B Biol Sci 358:19–37

Allen JF, Puthiyaveetil S, Ström J, Allen CA (2005) Energy transduction anchors genes in organelles. BioEssays 27:426–435. https://doi.org/10.1002/bies.20194

Anbar AD, Knoll AH (2002) Proterozoic ocean chemistry and evolution: a bioinorganic bridge? Science 297:1137–1142

Archibald JM (2009) The puzzle of plastid evolution. Curr Biol 19:R81–R88

Archibald JM (2015a) Genomic perspectives on the birth and spread of plastids. Proc Natl Acad Sci USA 112:10147–10153. https://doi.org/10.1073/pnas.1421374112

Archibald JM (2015b) Endosymbiosis and eukaryotic cell evolution. Curr Biol 25:R911–R921

Bowler C, Allen AE, Badger JH, Grinwood J, Jabbari K, Kuo A, Maheswari U et al (2008) The *Phaeodactylum* genome reveals the evolutionary history of diatom genomes. Nature 456:239–244

Burki F, Roger AJ, Brown MW, Simpson AGB (2020) The new tree of eukaryotes. Trends Ecol Evol 35:43–55. https://doi.org/10.1016/j.tree.2019.08.008

Cardona T (2017) Photosystem II is a chimera of reaction centers. J Mol Evol 84:149–151

Cardona T (2019) Thinking twice about the evolution of photosynthesis. Open Biol 9:180246. https://doi.org/10.1098/rsob.180246

Cardona T, Murray JW, Rutherford AW (2015) Origin and evolution of water oxidation before the last common ancestor of the Cyanobacteria. Mol Biol Evol 32:1310–1328. https://doi.org/10.1093/molbev/msv024

Clark DP, Pazdernik NJ (2013) Molecular biology, 2nd edn. Elsevier, Amsterdam. https://doi.org/10.1016/C2009-0-01986-2

Delwiche CF (1999) Tracing the thread of plastid diversity through the tapestry of life. Am Nat 154:S164–S177

Dorrell RG, Smith AG (2011) Do red and green make brown?: Perspectives on plastid acquisitions within Chromalveolates. Eukaryot Cell 10:856–868. https://doi.org/10.1128/EC.00326-10

Eliáš M (2021) Protist diversity: novel groups enrich the algal tree of life. Curr Biol 31:R733–R735. https://doi.org/10.1016/j.cub.2021.04.025

Falkowski PG, Knoll AH (2007) An introduction to primary producers in the sea: who they are, what they do, and when they evolved. In: Falkowski PG, Knoll AH (eds) Evolution of primary producers in the sea. Elsevier Academic, China, pp 1–6. https://doi.org/10.1016/B978-012370518-1/50002-3

Falkowski PG, Katz ME, Knoll AH, Quigg A, Raven JA, Schofield O, Taylor FJR (2004a) The evolution of modern eukaryotic phytoplankton. Science 305:354–360. https://doi.org/10.1126/science.1095964

Falkowski PG, Schofield O, Katz ME, de Schootbrugge V, Knoll AH (2004b) Why is the land green and the ocean red? In: Thierstein HR, Young JR (eds) Coccolithophores. Springer, Berlin. https://doi.org/10.1007/978-3-662-06278-4_16
Gast RJ, Moran DM, Dennett MR, Caron DA (2007) Kleptoplasty in an Antarctic dinoflagellate: caught in evolutionary transition? Environ Microbiol 9:39–45
Glibert PM (2020) Harmful algae at the complex nexus of eutrophication and climate change. Harmful Algae 91:101583
Gololobova MA, Belyakova GA (2022) Position of algae on the tree of life. Doklady Biol Sci 507: 312–326. Translated by Tkacheva T
Graham LE, Graham JM, Wilcox LW (2009) Algae, 2nd edn. Pearson Benjamin Cummings, San Francisco
Graham LE, Graham JM, Wilcox LW, Cook ME (2016) Algae, 3rd edn. LJLM, Madison
Green BR (2007) Evolution of light-harvesting antennas in an oxygen world. In: Falkowski PG, Knoll AH (eds) Evolution of primary producers in the sea. Elsevier Academic, China, pp 38–54. https://doi.org/10.1016/B978-012370518-1/50005-9
Gross J, Bhattacharya D (2010) Uniting sex and eukaryotic origins in an emerging oxygenic world. Biol Direct 5:53. https://doi.org/10.1186/1745-6150-55-53
Grzebyk D, Schofield O, Vetriani C, Falkowski PG (2003) The mesozoic radiation of eukaryotic algae: the portable plastid hypothesis. J Phycol 39:259–267
Gruber A, Medlin LK (2023) Complex plastids and the evolution of the marine phytoplankton. J Mar Sci Engine 11:1903. https://doi.org/10.3390/jmse11101903
Hackett JD, Yoon HS, Li S, Reyes-Prieto A, Rümmele SE, Bhattacharya D (2007) Phylogenomic analysis supports the monophyly of cryptophytes and haptophytes and the association of Rhizaria with Chromalveolates. Mol Biol Evol 24:1702–1713
Holland HD (2006) The oxygenation of the atmosphere and oceans. Philos Trans Roy Soc B 361: 903–915
Howe CJ, Nisbet RER, Barbrook AC (2008) The remarkable chloroplast genome of dinoflagellates. J Exp Bot 59:1035–1045
Keeling PJ, Archibald JM, Fast NM, Palmer JD (2004) Comment on "The evolution of modern eukaryotic phytoplankton". Science 5705:2191
Keeling PJ, McCutcheon JP, Doolittle WF (2015) Symbiosis becoming permanent: survival of the luckiest. Proc Natl Acad Sci USA 112:10101–10103. https://doi.org/10.1073/pnas.1513346112
Knoll AH (2012) Lynn Margulis, 1938–2011. Proc Natl Acad Sci USA 109:1022. https://doi.org/10.1073/pnas.1120472109
Lake JA (2011) Lynn Margulis (193–2011). Nature 480:458
Larkum AWD (2020) Light-harvesting in cyanobacteria and eukaryotic algae: an overview. In: Larkum AWD, Grossman AR, Raven JA (eds) Photosynthesis in algae: biochemical and physiological mechanisms. Springer, Cham, pp 207–260. https://doi.org/10.1007/978-3-030-33397-3_10
Larkum AWD, Lockhart PJ, Howe CJ (2007) Shopping for plastids. Trends Plant Sci 12:189–195
Larkum AWD, Grossman AR, Raven JA (2020) Preface: Brief introduction to the algae. In: Larkum AWD, Grossman AR, Raven JA (eds) Photosynthesis in algae: biochemical and physiological mechanisms. Springer, Cham, pp xxv–xxvi
Mansour JS, Anestis K (2021) Eco-evolutionary perspectives on mixoplankton. Front Mar Sci 8: 666160. https://doi.org/10.3389/fmars.2021.666160
Margulis L (1970) Origin of eukaryotic cells. Yale University Press, New Haven, CT
Margulis L, Chapman MJ (2009) Kingdoms and domains: an illustrated guide to the phyla of life on Earth. Academic, London. 711 pp
Martin W (2005) Archaebacteria (Archaea) and the origin of the eukaryotic nucleus. Curr Opin Microbiol 8:630–637
Martin WF (2007) Eukaryote and mitochondrial origins: two sides if the same coin and too much ado about oxygen. In: Falkowski P, Knoll AH (eds) Primary producers of the sea. Elsevier Academic, New York, pp 53–73

Martin WF, Garg S, Zimorski V (2015) Endosymbiotic theories do eukaryotic origin. Philos Trans Roy Soc B 370:20140330. https://doi.org/10.1098/rstb.2014.0330
Mereschkowski C (1905) Über Natur und Ursprung der Chromatophoren im Pflanzenreiche. Biol Centralbl 25:593–604
Nowack EC (2014) *Paulinella chromatophora* – rethinking the transition from endosymbiont to organelle. Acta Soc Botan Polon 83:387–397
Oborník M (2019) Endosymbiotic evolution of algae, secondary heterotrophy and parasitism. Biomolecules 9:266. https://doi.org/10.3390/biom9070266
Oliver T, Kim TD, Trinugroho JP, Cordón-Preciado V, Wijayatilake N, Bhatia A, Rutherford AW et al (2023) The evolution and evolvability of photosystem II. Annu Rev Plant Biol 74:225–257. https://doi.org/10.1146/annurev-arplant-070522-062509
Palmer JD, Delwiche CF (1996) Second-hand chloroplasts and the case for the disappearing nucleus. Proc Natl Acad Sci USA 93:7432–7435. https://doi.org/10.1073/pnas.93.15.7432
Prechtl J, Kneip C, Lockhart P, Wenderoth K, Maier UG (2004) Intracellular spheroid bodies of *Rhopalodia gibba* have nitrogen-fixing apparatus of cyanobacterial origin. Mol Biol Evol 2: 1477–1481. https://doi.org/10.1093/molbev/msh086
Quigg A, Finkel ZV, Irwin AJ, Rosenthal Y, Ho T-Y, Reinfelder JR, Scofield O et al (2003) The evolutionary inheritance of elemental stoichiometry in marine phytoplankton. Nature 425:291–294
Raven JA (2013) Cells inside cells: symbiosis and continuing phagotrophy. Curr Biol 23:R530–R531
Raven JA, Beardall J, Flynn KJ, Maberly SC (2009) Phagotrophy in the origins of photosynthesis in eukaryotes and as a complementary mode of nutrition in phototrophs: relation to Darwin's insectivorous plants. J Exp Bot 60:3975–3987. https://doi.org/10.1093/jxb/erp282
Roger AJ, Muñoz-Gómez SA, Kamikawa R (2017) The origin and diversification of mitochondria. Curr Biol 27:PR1177–PR1192. https://doi.org/10.1016/j.cub.2017.09.015
Sagan L (1967) On the origin of mitosing cells. J Theor Biol 14:225–274. https://doi.org/10.1016/0022-5193(67)90079-3
Sato N (2019) Endosymbiotic theories of organelles revisited: retrospects and prospects. Springer, Singapore
Sato N (2020) Complex origins of chloroplast membranes with photosynthetic machineries: multiple transfers of genes from divergent organisms at different times or a single endosymbiotic event? J Plant Res 133:15–33
Simpson AG, Roger AJ (2002) Eukaryotic evolution: getting to the root of the problem. Curr Biol 12:R691–R693
Strassert JFH, Irisarri I, Williams TA, Burke FA (2021) A molecular timescale for eukaryotic evolution with implications for the origin of red algal-derived plastids. Nat Commun 12:1879
Troost TA, Kooi BW, Kooijman SALM (2005a) Ecological specialization of mixotrophic plankton in a mixed water column. Am Nat 166:E45–E61
Troost TA, Kooi BW, Kooijman SALM (2005b) When do mixotrophs specialize? Adaptive dynamics theory applied to a dynamic energy budget model. Math Biosci 193:159–182
Van den Hoek C, Mann DG, Jahns HM (1995) Algae: An introduction to phycology. Cambridge University Press, Cambridge
Wilken S, Huisman J, Naus-Wiezer S, Van Donk E (2013) Mixotrophic organisms become more heterotrophic with rising temperature. Ecol Lett 16:225–233
Williams B, Keeling APPJ (2003) Cryptic organelles in parasitic protists and fungi. Adv Parasitol 54:9–67
Wilson EB (1928) The cell in development and heredity, third revised. Macmillan, New York. Reprinted (1987), Garland, New York
Yoon HS, Hackett YD, Ciniglia C, Pinto G, Bhattacharya D (2004) A molecular timeline for the origin of photosynthetic eukaryotes. Mol Biol Evol 21:809–818

11 Nutrient Supplies, Cycles, and the Oceanic Dynamic Balance

11.1 Introduction: Early Insights

Hildebrand W. Harvey, who was introduced in Chap. 1, spent years studying nutrients and their relationships to physical characteristics of the water column in the 1920s and 1930s. He wrote (Harvey 1926) about the,

> remarkable fact that plant growth should be able to strip sea water of both nitrates and phosphates, and that in the English Channel the store of these nutrient salts, formed during autumn and winter, should be used up about the same time.

It was only a few years later that Alfred C. Redfield documented the relative consistency of nitrogen (N) and phosphorus (P) in water and phytoplankton biomass (Redfield 1934).

Redfield (Fig. 11.1), variously described as urbane, courtly, gracious, and demanding, initially studied the coloration of toads for his dissertation, and then respiratory functions of blood in marine invertebrates (Revelle 1995). He became a more-or-less chemical oceanographer when he was recruited to be part of the small staff that constituted the early Woods Hole Oceanographic Institution. He became well known for his discovery of the consistency of the proportions of biologically active elements in seawater and phytoplankton. Redfield did not fully understand the reasons why this was the case. He suggested that bacterial N fixation might increase the ratio of NO_3^- to PO_4^{3-}, and that denitrifying bacteria limited the increase of NO_3^- in the ocean more than phytoplankton uptake. Understanding the balance between N_2 fixation and denitrification has motivated many oceanographers both before and since Redfield. Before reviewing these and other processes, it is important to understand what Redfield showed.

Redfield (1934) described the relationships between nutrients and between nutrients and oxygen and CO_2 based on data collected from several cruises in the Atlantic:

P. M. Glibert, *Phytoplankton Whispering: An Introduction to the Physiology and Ecology of Microalgae*, https://doi.org/10.1007/978-3-031-53897-1_11

Fig. 11.1 Alfred C. Redfield. Photo from the Woods Hole Oceanographic Institution

The straight line drawn through the points describing the nitrate-phosphate correlation has the slope demanded if, for every three grams of nitrogen added to or subtracted from the water, one gram of PO_4 (or ¾ g of P_2O_5) also made its appearance or disappearance (Fig. 11.2a). This corresponds to a ratio of 20 atoms of nitrogen to 1 atom of phosphorus. On the assumption that these changes are due solely to the decomposition or synthesis of organic matter, these ratios may be taken to reflect the proportions of nitrogen and phosphorus in the plankton community as a whole.

The correlation between the gain in concentration of nitrate and the loss in concentration of oxygen, as shown in Fig. 11.2b, is almost equally good for the water samples collected from depths less than 1000 m in the Sargasso Sea and 400 m in the Gulf Stream, depths which mark the transition between the intermediate layer of low oxygen content and the deeper layer of cold water. . . .the variation in nitrate concentration to oxygen utilization is in the proportion of 1 millimol of nitrate to 6 millimol of oxygen.

The correlation between the concentrations of nitrate and carbonate shown in Fig. 11.2c is less precise owing to the greater variation in the carbonate measurements. . . Nevertheless, taking the data as a whole, it is clear that carbonate content is correlated with nitrate content. The envelopes drawn around the points. . .have a slope corresponding to a ratio of 1 millimol of nitrate to 7 of carbonate. . . .it follows that the ratio of carbonate to oxygen is 7:6. This ratio is close to that characterizing photosynthetic processes and the decomposition of the principal classes of organic compounds.

Twenty-five years later, Redfield (1958) further developed and expanded his theory on the relationships between biogeochemical cycles and the strong elemental ratios previously described. He wrote:

In considering the influence of the biogeochemical cycle on the environment it should be remembered that the cycle can, and apparently for the most part does, run its course without adding more to the environment at one phase than it withdraws from another. Photosynthesis cannot increase the oxygen content of the environment unless the products of photosynthesis are in some way withdrawn from reoxidation.

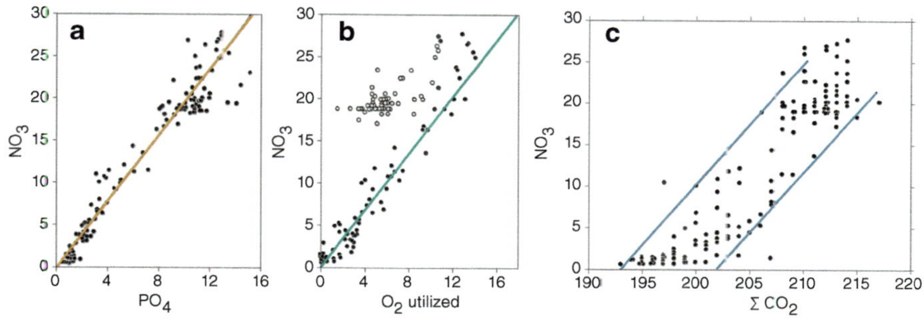

Fig. 11.2 Panel (**a**) Correlation between concentrations of nitrate and phosphate in the waters of western Atlantic Ocean. The line represents the ratio of ΔN:ΔP = 20:1 mg atoms. For these nutrients, a milligram atom is equivalent to a μM. Panel (**b**) Correlation between concentrations of nitrate and amount of oxygen utilized in waters of western Atlantic Ocean. The line represents a ratio of oxygen used: ΔN = 6.1. Points representing stations from a depth below 400 m in the Gulf Stream and below 1000 m in the Sargasso Sea are represented by open circles. Panel (**c**) Correlations between concentration of nitrate and total carbonate in waters of the western Atlantic Ocean. The slope of the envelope corresponds to a ratio of ΔC:ΔN = 7.1 mg atoms. Images redrawn from Redfield (1934, James Johnstone Memorial 176)

These biogeochemical cycles can now be explored in detail—something Redfield was unable to do, hampered as he was by limited data and methodologies. He wrote (Redfield 1958):

> What is lacking is direct evidence that the postulated processes do, in fact, act in a regulated way so as to bring the atomic ratios of phosphate nitrate and oxygen in the environment into conformity with the requirements of the biogeochemical cycle.

This chapter takes an overview of nutrient availability and cycling pathways; the following chapters explore the acquisition and assimilation of each nutrient by phytoplankton in more detail.

11.2 Nutrient Availability: The Broad Scale View

Nutrient concentrations, loads, and forms are constantly changing on local scales, but on oceanic scales, they display remarkably consistent patterns. Phytoplankton consume nutrients, so concentrations are lower in surface relative to deeper waters, where nutrients are remineralized. However, due to the differences in the relative ages of deepwater masses in the different oceans of the world, inorganic nutrients are more enriched in the deep waters of the Pacific than the Atlantic (Fig. 11.3). Intermediate concentrations with depth are observed in the Southern Ocean.

All the major inorganic nutrients are also more enriched in surface waters near the poles, but the equatorial waters of the Pacific also show surface enrichments of all

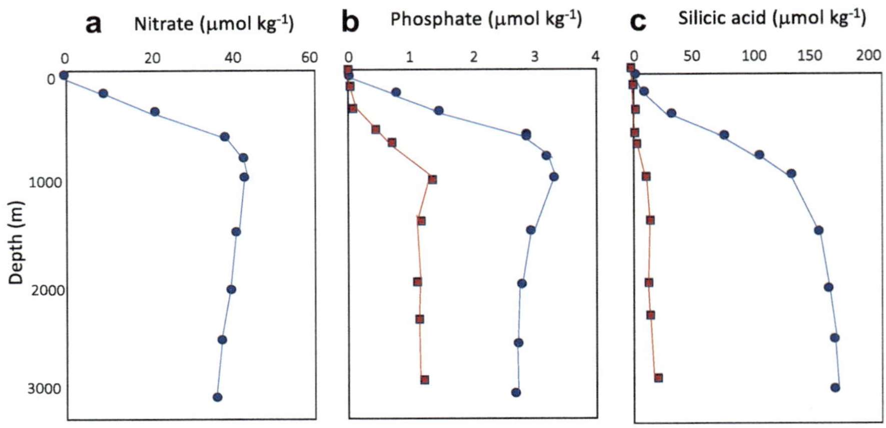

Fig. 11.3 Depth profiles of nitrate (Panel **a**), phosphate (Panel **b**), and silicic acid (Panel **c**) for the central North Pacific (blue circles) and North Atlantic (red squares). Image reproduced from Sunda (2012, Front Microbiol 3: doi: 10.3389/fmicb.2012.00204) under a creative commons license

the major nutrients (Fig. 11.4). The equatorial waters, however, are traditionally considered unproductive, and have low chlorophyll *a* (chl *a*) accumulations. They have thus been termed high nutrient-low-chlorophyll (HNLC) regions. Southern Ocean waters have also been given this moniker, as they too often have high nutrients but low chl *a* accumulations. These regions are now known to be limited by iron (Fe; Chap. 16).

While inorganic nutrients are typically low in oceanic surface waters due to phytoplankton uptake, dissolved organic N (DON) and dissolved organic P (DOP) can be present at significant concentrations in the open ocean, including surface waters. In fact, DON can be the largest pool of N in some waters, including freshwaters. This fact leads to the general conclusion that these forms are comparatively less bioavailable, although many phytoplankton use various forms of these organic nutrients (see Chaps. 12 and 13). Classically, DON was characterized as composed of truly refractory compounds, those that persist in the environment for months to years, and a semi-labile fraction, including proteins, dissolved combined amino acids and polysaccharides (reviewed by Bronk et al. 2007). There are also some highly labile compounds, including urea, dissolved free amino acids and nucleic acids which may be important N substrates for phytoplankton uptake with turnover in time scales of minutes to days (Antia et al. 1991; Berman and Bronk 2003). The pathways of cycling of these compounds have historically been difficult to study, but some approaches for measuring specific pathways and organic substrates are described in Chaps. 12 and 13.

The importance of DOP in phytoplankton nutrition is becoming better understood. Polyphosphates (polyP), an important DOP form, can be common in marine and freshwaters (Chap. 13). Most organic P compounds are ester linkages with an O atom between the P and the C, with the "R" groups representing any one of many C

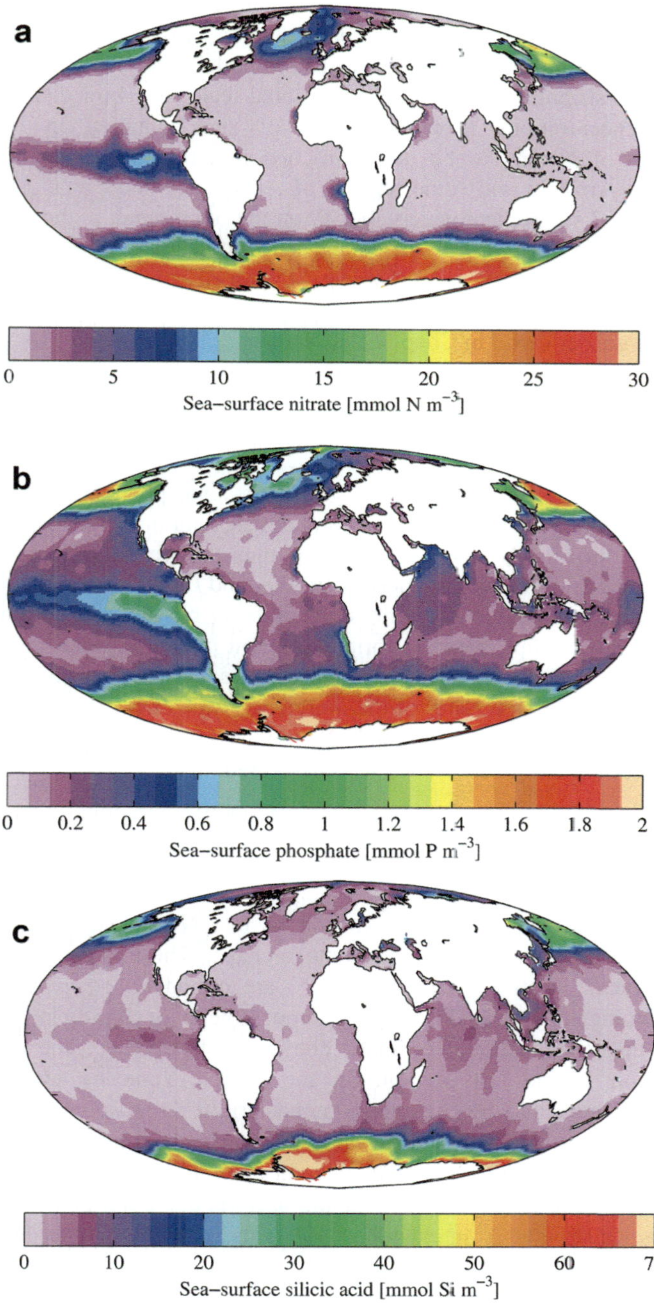

Fig. 11.4 Global annual distributions of nitrate (Panel **a**), phosphate (Panel **b**), and silicic acid (Panel **c**) at the surface based on the World Ocean Atlas data of 2009. Images were reproduced from Wikipedia (https://en.wikipedia.org/wiki/World_Ocean_Atlas) under creative commons licenses

groups including lipids, sugars, proteins, and nucleic acids. Monophosphate esters and phosphonates are probably the best studied, the latter associated with phosphoproteins and phospholipids (Bianchi-Smiraglia et al. 2015).

Freshwater distributions of N and P on a global scale also show distinct patterns. Based on concentrations from thousands of sites sampled in freshwaters, concentrations of N tend to be most enriched in temperate regions, especially Europe, North America and some regions of Australasia and South America and some tropical and subtropical waters. For P, on the other hand, higher enrichments are observed in the dry areas of Africa and the Middle East, some areas of North America and areas above the Arctic circle (McDowell et al. 2020; Fig. 11.5). The changes that are occurring due to anthropogenic loads and altered climate are considered in greater detail in Chap. 22.

11.3 Nitrogen Cycling

Elemental nitrogen (N_2) makes up about 80% of the atmosphere and is the dominant form of N. It is simultaneously the most common element in the atmosphere and the least available form of N to organisms. It is thus thought that N limits primary production in much of the world's oceans (Ryther and Dunstan 1971). Compared to many of the other biologically required elements, N has the most complex cycle (Fig. 11.6a), with multiple reactive forms, different oxidation states and reactions, each of which is catalyzed by different organisms. Redox reactions drive the N cycle (Fig. 11.6b). The oxidation states of the inorganic forms of N range from a valence of −3 to +5. The most oxidized form is NO_3^- oxidized form, while NH_4^+ is the most reduced. The dominant processes are next described.

11.3.1 Biological Nitrogen Fixation

Biological N_2 fixation is the conversion of N_2 into NH_4^+ and is carried out by cyanobacteria, some eubacteria, and some archaea. These organisms, with the capability to break the strong, triple bond of N_2 are termed *diazotrophs*. The amount of N_2 fixed biologically on land is roughly comparable to, or slightly larger than, that produced in the ocean, although estimates continue to be updated. The vast majority of N_2 fixation in the sea is done in the pelagic zone by cyanobacterium *Trichodesmium* spp. and by the picocyanobacterium *Crocosphaera* (Zehr et al. 2001). *Trichodesmium* spp. can form large floating mats consisting of clumps of filaments ("puffs" or "tufts"), and they commonly form oceanic surface slicks (Fig. 11.7). There are also cyanobacteria and bacteria that contribute to N_2 fixation in coral beds, on seagrasses, in cyanobacterial mats, and among other benthic sites. These are regionally important. Some N_2-fixing bacteria, such as *Richelia intracellularis,* are symbiotic with certain diatom taxa, and such symbioses can be regionally or locally significant as well. In freshwaters, estuaries, and coastal seas,

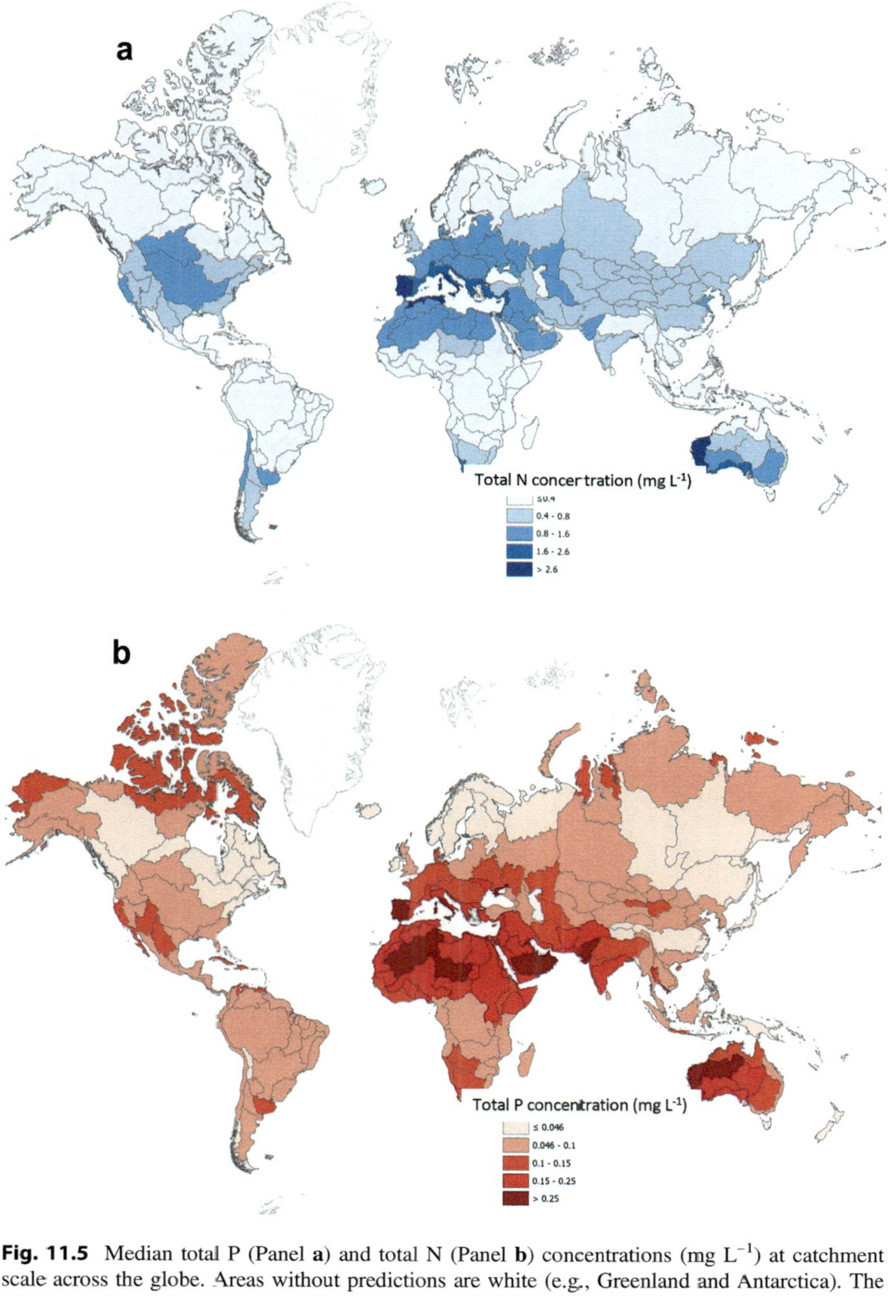

Fig. 11.5 Median total P (Panel **a**) and total N (Panel **b**) concentrations (mg L^{-1}) at catchment scale across the globe. Areas without predictions are white (e.g., Greenland and Antarctica). The relative median values for each of the nutrients are depicted as five different concentration ranges. Image reproduced from McDowell et al. (2020, Sci Rep 10, 3568) under a creative commons license

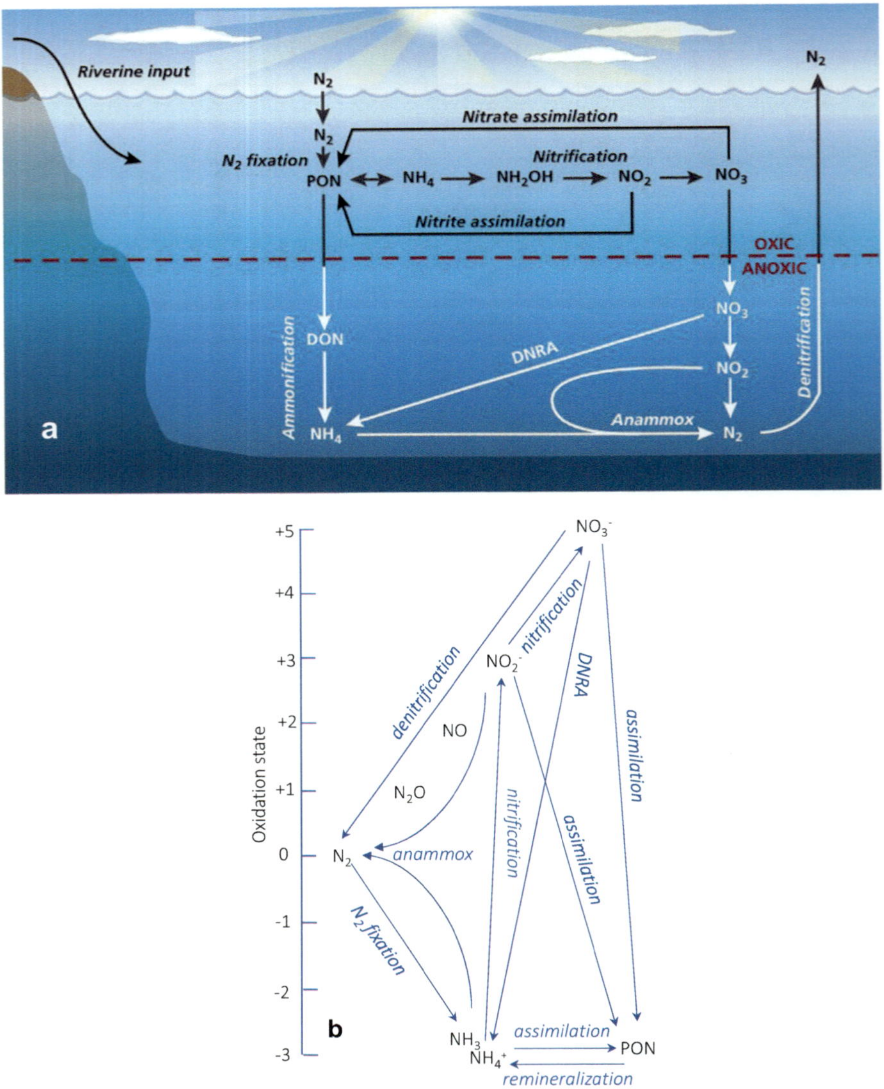

Fig. 11.6 Panel (**a**) The nitrogen cycle, depicting where in the water column the dominant processes occur. Panel (**b**) The processes of the nitrogen cycle and chemical forms of nitrogen relative to their oxidation state. Images modified based on Arrigo (2005, Nature 437: doi: 10.1038/nature04158) and Hutchins et al. (2009, Oceanography 22: doi: 10.5670/oceanog.2009.103)

other cyanobacteria with N_2-fixing ability can take on increasing importance, such as *Aphanizomenon, Dolichospermum* (formerly *Anabaena*), and *Nodularia*.

Fixing N_2 requires considerable energy, as well as a source of electrons for reduction. The biological process can be written as,

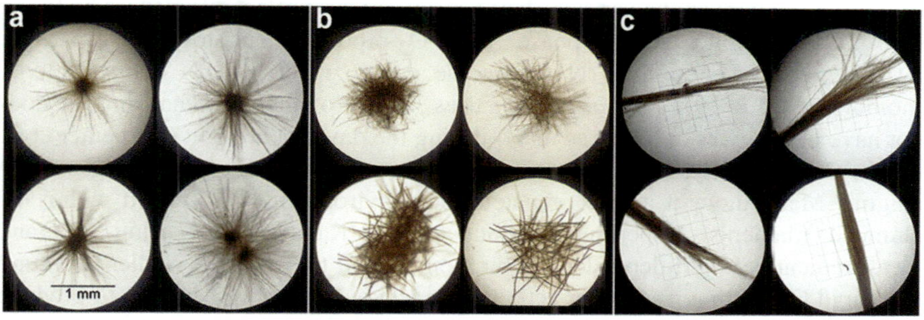

Fig. 11.7 The morphological classes of *Trichodesmium*. Panel (**a**): radial puffs, Panel (**b**): nonradial puffs, and Panel (**c**): tufts. Images reproduced from Gradoville et al. (2017, Front Microbiol 8:1122) under a creative commons license

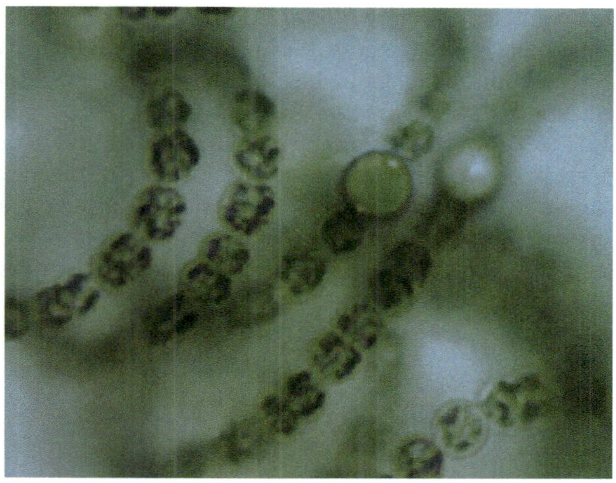

Fig. 11.8 *Dolichospermum crassum*, showing the specialized heterocyst. Image reproduced from https://www.wikidata.org/wiki/Q35949474#/media/File:Dolichospermum_crassum_akinete.jpg reproduced under a creative commons license

$$N_2 + 8\,H^+ + 8\,e^- + 16\,ATP \rightarrow 2\,NH_3 + 16\,ADP + 16\,P_i + H_2 \qquad (11.1)$$

The reaction is carried out by the enzyme nitrogenase, a complex of two proteins, each containing molybdenum and Fe. For active N_2 fixers, nitrogenase can make up as much as 10% of the total cell protein (Dingler et al. 1988). As this process is energy requiring, it also requires PO_4^{3-}. This requirement has made N_2 fixation a central "player" in the debate over which nutrient is most important to manage in freshwaters; that discussion is developed in Chap. 29.

The nitrogenase enzyme is very sensitive to the presence of oxygen (O_2). Thus, N_2 fixation occurs in anaerobic zones. In the aerobic waters of oceans and lakes, however, cyanobacteria have other mechanisms to protect this enzyme from O_2 damage. Some cyanobacteria form specialized cells called heterocytes (heterocysts; Fig. 11.8), which carry out N_2 fixation but do not carry out photosynthesis, or only carry out cyclic photophosphorylation (photosystem I), which does not produce O_2.

Various other cyanobacteria, such as *Crocosphaera*, protect nitrogenase by fixing N_2 only at night, out of phase with O_2-producing photosynthesis. These cyanobacteria take up fixed N during the day to meet their N demands (e.g., Inomura et al. 2019). *Trichodesmium* uses neither of these mechanisms: it does not have heterocytes and it fixes its N_2 during the day, not at night (Carpenter et al. 1992). Consequently, how it protects its nitrogenase from O_2 was a longstanding enigma. Many thought a deoxygenated microenvironment was created within the clumped filaments. Experiments conducted by Kana (1992), using ^{18}O and measurements of light-dependent respiration, helped to solve this riddle. He found that rapid O_2 cycling by Mehler activity (see Chap. 6) was high enough to produce zones of low O_2 within the cells themselves, allowing N_2 fixation and photosynthesis to simultaneously occur. Rapid O_2 cycling provides another benefit—a mechanism for generating excess ATP through pseudocyclic photophosphorylation. Milligan et al. (2007) later confirmed these findings. There is also some evidence, from applications of high-resolution secondary ion mass spectrometry (NanoSIMS), that temporal separation may also be involved. For example, Finzi-Hart et al. (2009) found that photosynthesis was proportionately higher earlier in the day, whereas $^{15}N_2$ accumulated later in the day. Both of these mechanisms may be involved in N_2 fixation by *Trichodesmium*.

The Tara Oceans project has amassed an enormous gene catalog from pan-global surveys. By mining these data, including >1300 metagenomes and >2,000,000 high-throughput images, the diversity, distribution and abundance of various diazotrophs have been synthesized (Pierella Karlusich et al. 2021). The data indicate that larger diazotrophs (>10 μm) dominate the tropical belts, while unicellular cyanobacterial and noncyanobacterial diazotrophs are globally distributed in surface and mesopelagic layers (Fig. 11.9).

11.3.2 Denitrification

Denitrification is the process by which fixed N_2 is returned to the atmosphere. The overall reaction is:

$$4HNO_3^- + 5\,CH_2O \rightarrow 5CO_2 + 7H_2O + 2N_2 \uparrow \tag{11.2}$$

in which CH_2O represents organic matter. However, denitrification is actually a summed series of reactions (Fig. 11.6), each of which is involves different enzymes and different organisms,

$$NO_3^- \rightarrow NO_2^- \rightarrow NO \rightarrow N_2O \rightarrow N_2 \uparrow \tag{11.3}$$

Denitrification is performed by a diverse suite of organisms, largely thriving under anaerobic or low-O_2 conditions with NO_3^- as an electron acceptor. The first step involves NO_3^- reductase, the next nitrite (NO_2^-) reductase, then nitric oxide (NO) reductase and finally nitrous oxide (N_2O) reductase. If the latter enzyme is not present, N_2O will be released to the atmosphere instead of N_2. Different steps in the

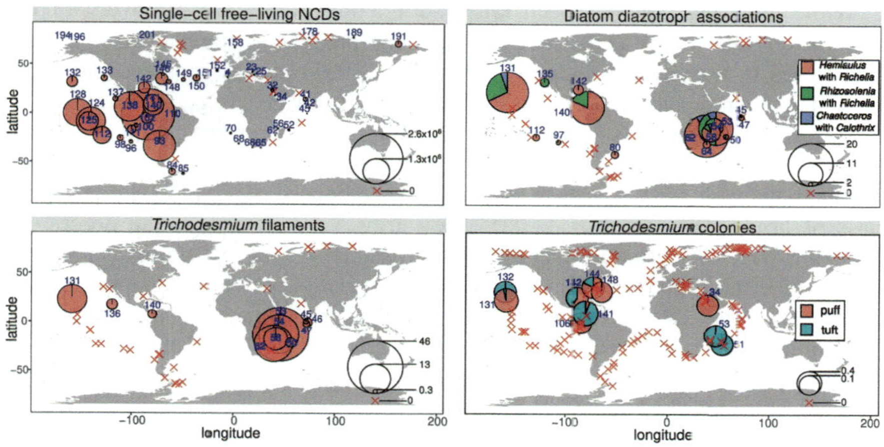

Fig. 11.9 Biogeography in surface waters. Bubble size varies according to the corresponding diazotroph concentration (individuals L^{-1}), while red crosses indicate their absence. Station labels with detection of diazotrophs are indicated in blue. NCD is noncyanobacterial diazotrophs. Images reproduced from Pierella Karlusich et al. (2021, Nat Comm 12, 4160) under a creative commons license

denitrification process are sensitive to O_2. The reduction step from NO_3^- to NO_2^- is less sensitive than the NO_2^- or N_2O reduction steps, for example (Bonin and Raymond 1990). Anoxia does not need to be widely present for denitrification to occur, as there can be anoxic or hypoxic microsites that may also be sites of denitrification.

Denitrification is a major pathway of loss of fixed N from estuaries and freshwaters, and thus, it is a positive process in the removal of N that contributes to eutrophication. Many factors regulate denitrification, such as temperature, salinity, sediment–water NH_4^+ exchange, sediment O_2 status, and rates of input of N to estuaries or freshwater—especially the availability of NO_3^- (Kemp et al. 1990; Seitzinger et al. 1991; Bianchi 2006). Nixon et al. (1996) showed that across a range of rivers and estuaries, the total N denitrified was directly proportional to the residence time of the water body; water with more time to react with sediments had greater overall denitrification activity.

11.3.3 Ammonium Oxidation (Anammox)

Another mechanism by which N_2 may be lost to the atmosphere is the reaction of NO_2^- with NH_4^+ to yield N_2 (Fig. 11.6):

$$NO_2^- + NH_4^+ \rightarrow N_2 \uparrow + 2H_2O \quad (11.4)$$

While denitrification has been known for a long time, the discovery of the anammox (ANaerobic AMMonium OXidation) process is comparatively recent. Discovered in wastewater sludge (Mudler et al. 1995), it was soon recognized to be of major significance in the oceans and in some freshwaters (e.g., Zhu et al. 2015).

Discovery of the organisms responsible for this process was difficult. These bacteria are "peculiar in several ways" (Ward 2003). They use hydrazine as an intermediate, which they encase in a membrane composed of a previously unknown class of lipids (Damsté et al. 2002). They grow very slowly, as the process of oxidizing NH_4^+ is not at all energetically favorable. Using ^{15}N tracer techniques, Dalsgaard et al. (2003) confirmed that in stations off Costa Rica, anammox rates contributed from ~20 to 35% of the N_2 flux. However, this process only occurred in O_2 minimum zones. Devol (2003) extrapolated from the available evidence to suggest that 70% of the N lost in the ocean is a result of denitrification and that 30% is lost through anammox.

11.3.4 Nitrification

The oxidation of NH_4^+ to NO_2^- and ultimately to NO_3^- is the process of nitrification (Fig. 11.6)

$$NH_4^+ + 3/2\,(O_2) \rightarrow NO_2^- + H_2O + 2H^+ \tag{11.5}$$

$$NO_2^- + \tfrac{1}{2}\,(O_2) \rightarrow NO_3^- \tag{11.6}$$

It is because of these reactions that the deep ocean has high concentrations of NO_3^-. The first reaction is carried out by NH_4^+ oxidizers, which grow very slowly, and the second by NO_2^- oxidizers that are somewhat faster growers. The first step is catalyzed by the enzyme ammonia monooxygenase (AMO), and the amoA gene has been used as a marker for presence of such microbial oxidizers in the environment. The second step is catalyzed by the enzyme NO_2^- oxidoreductase (NXR), which is closely related to NO_3^- reductase (Pajares and Ramos 2019). Ammonia oxidizing bacteria (AOB) and ammonia oxidizing archaea (AOA) both carry out the first reaction, but in the oceans, AOA are much more abundant than AOB (Wang et al. 2017; Shafiee et al. 2021) and appear to have far greater affinity for NH_4^+ than their bacterial counterparts. AOA can also produce N_2O (Santoro et al. 2011).

Nitrifiers are chemoautotrophs, as they gain their energy by chemical oxidation, but they rely on external sources of organic matter. As they depend on O_2, they are sensitive to the absence of dissolved O_2. Other factors affecting the distribution of nitrifiers are the substrates NH_4^+ and NO_2^-, as well as pH, temperature and salinity. Light may be an indirect control, as photoinhibition of nitrifiers is commonly observed (Pajares and Ramos 2019), making the base of the euphotic zone an ideal location for nitrifiers. The higher affinity for NH_4^+ compared with AOB gives AOA the opportunity to occupy greater depths where the NH_4^+ supply is lower, but photoinhibitory effects are substantially reduced. Indeed, AOA maxima

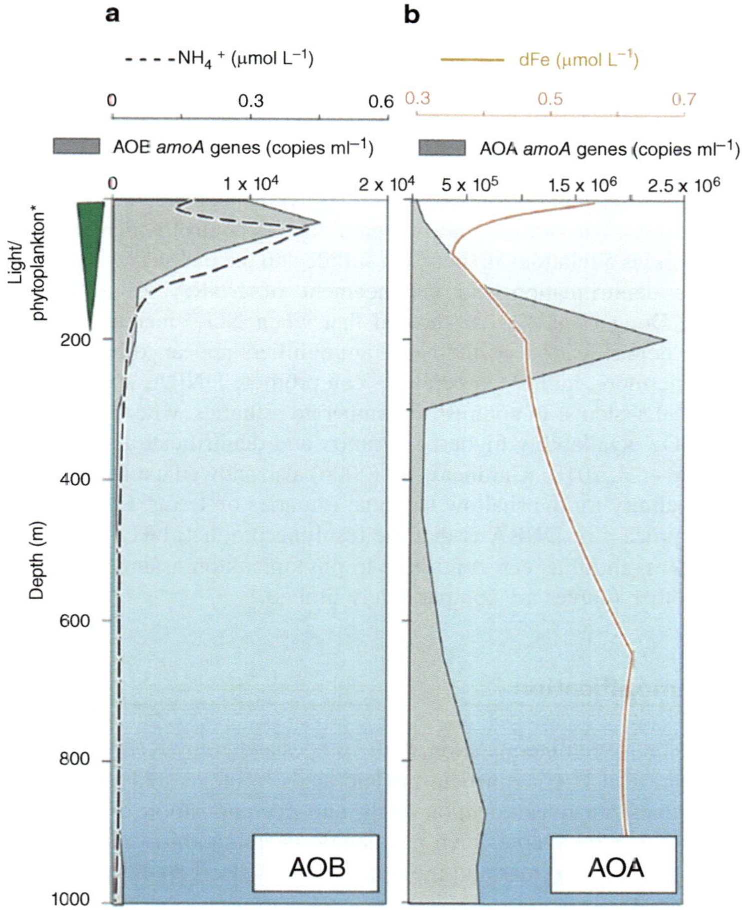

Fig. 11.10 Schematic showing hypothesized spatial niches of AOB and AOA. Published amoA gene abundance depth profiles for (panel **a**) AOB and (panel **b**) AOA are plotted against NH_4^+ depth profiles (black dashed line, μmol L^{-1}) and dissolved Fe (dFe, orange solid line, nmol L^{-1}). Green wedge labeled light/phytoplankton* reflects the relative decrease in light, phytoplankton, and marine bacteria with increasing depth. Image reproduced from Shafiee et al. (2021, ISME Comm 1: 1) under a creative commons license

are commonly at the base of the photic zone deeper than AOB maxima (Fig. 11.10). Vertical niche separation of the competing consumers of NH_4^+, phytoplankton and nitrifiers, was recently confirmed in the western North Pacific Ocean. There, Xu et al. (2019) found that light stimulated the uptake of NH_4^+ (phytoplankton), while the opposite was observed for oxidation kinetics (nitrifiers). It has also recently been found that AOA and AOB have different physiologies with respect to trace metals, which may also have implications for niche separation in the oceans (Shafiee et al. 2021).

11.3.5 Dissimilatory Nitrate Reduction to Ammonia (DNRA)

While NO_3^- reduction most commonly leads to N_2 when the process is carried to completion, in some cases, the process can instead lead to NH_4^+. This process is NO_3^- reduction to NH_4^+, DNRA (Fig. 11.6). In tidal flats or shallow estuaries, rates of DNRA can be as high as rates of denitrification (e.g., Tobias et al. 2001). Under hypoxic or anaerobic conditions, where NO_3^- is available, denitrification and DNRA compete for NO_3^-, but various environmental factors control which process will be favorable. There is a relationship between sulfide and the pathway of N cycling, with inhibition of denitrification and enhancement of DNRA as sulfide increases (Fig. 11.11). Dong et al. (2010) showed that when NO_3^- increased, DNRA was favored over denitrification, as the NO_3^- ammonifiers appear to have higher NO_3^- affinity. Furthermore, higher temperatures can promote DNRA, such as may be the case in tropical systems. In contrast, in temperate estuaries, where temperatures are lower and NO_3^- availability higher, anammox and denitrification are favored over DNRA (Dong et al. 2010). Gardner et al. (2006) also showed a relationship between DNRA and salinity for the shallow lagoonal estuaries of Texas. Clearly, the importance of the process of DNRA is that the resulting product, NH_4^+, can be retained within a system, and thus can contribute to phytoplankton assimilation even when inputs from other sources are comparatively limited.

11.3.6 Ammonification

Ammonification, or remineralization, is the process whereby NH_4^+ is released from organic matter. The N contained in nucleic acids, proteins and amino-sugars and other compounds are released upon death and decomposition of organisms. The process may either be simple, such as a single deamination reaction, or complex, involving hydrolytic enzymes and multiple reactions (Herbert 1999):

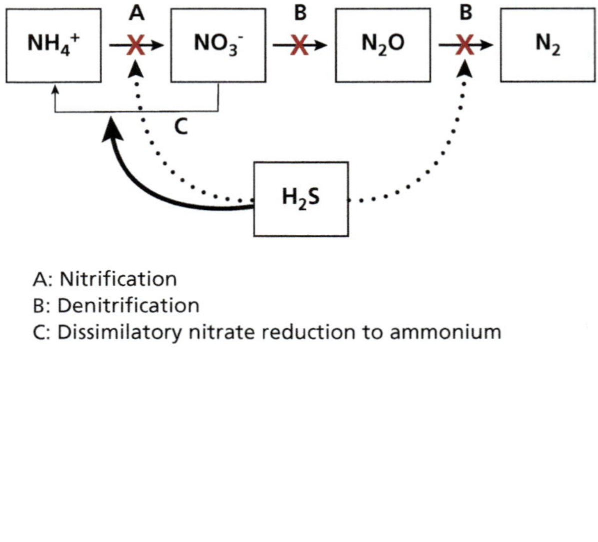

Fig. 11.11 Relationship between sulfide and N cycling in sediments. Solid arrows indicate positive effects, and dashed lines indicate negative effects. Image modified and redrawn based on An and Gardner (2002, Mar Ecol Prog Ser 237: 41–50)

$$\text{Protein} \xrightarrow{\text{proteinases}} \text{Peptides} \xrightarrow{\text{peptidases}} \text{Amino acids} \qquad (11.7)$$

$$\text{Amino acids} \xrightarrow{\text{deamination}} \text{Organic acids} + NH_4^+ \qquad (11.8)$$

The regeneration of N also occurs from excretion and degradation of fecal material. In oligotrophic waters, the rate of regeneration of N can be rapid enough to provide virtually all of the N required for phytoplankton growth (Glibert et al. 1982). Regeneration of N is also amplified in aggregates such as marine snow.

11.3.7 Assimilation of N

Although multiple factors regulate the assimilation of N in surface waters, the relative concentration of each N substrate is a dominant factor in growth and in community composition. The NH_4^+ inventory (supply) is generally many orders of magnitude smaller than that of NO_3^-, but because NH_4^+ is typically used preferentially over NO_3^-, its cycling and rate of regeneration are especially critical to phytoplankton nutrition and to the oceanic N cycle. The complexities of N assimilation and metabolism, the preferences for different forms of N by different phytoplankton functional groups, interactions between different forms of N and effects on phytoplankton and relationships to various environmental factors are fully developed in Chap. 12.

11.4 Phosphorus Cycling

Whereas redox processes dominate the N cycle and N speciation is highly dynamic, in both fresh and seawater, nearly all dissolved phosphorus (P) is present in only one major form, inorganic phosphate, PO_4^{3-}. It has an oxidation state of +5. At normal pH, phosphoric acid, H_3PO_4, is essentially all dissociated. PO_4^{3-} is often equated to soluble reactive P, SRP, which is the P fraction that forms a phosphomolybdate complex under acid conditions (Bianchi 2006). Differences between SRP and PO_4^{3-} are more common in freshwaters, so it is important to be aware that while the two terms are commonly used interchangeably, they technically differ. The cycling of P, rather than dominated by redox reactions, is dominated by adsorption–desorption processes. Redox processes do not control the cycling of P directly but can do so indirectly: Biological reactions of microbes continually alter the redox potential of the sediments, which therefore control the flux and form of P.

The dominant mechanism by which P is recycled is particle-colloid interactions. Phosphorus is "sticky"; thus, much P is associated with oxyhydroxides, calcium carbonate, humic, or other colloidal fractions (Stumm and Morgan 1981). Much P is bound in sediments. The adsorption takes place by the binding of PO_4^{3-} to positively charged edges in the clays or other particles and by substitutions with cations such as Ca^{2+}, Al^{3+}, and Fe^{3+} (Stumm and Morgan 1981). The cycling of P (Fig. 11.12),

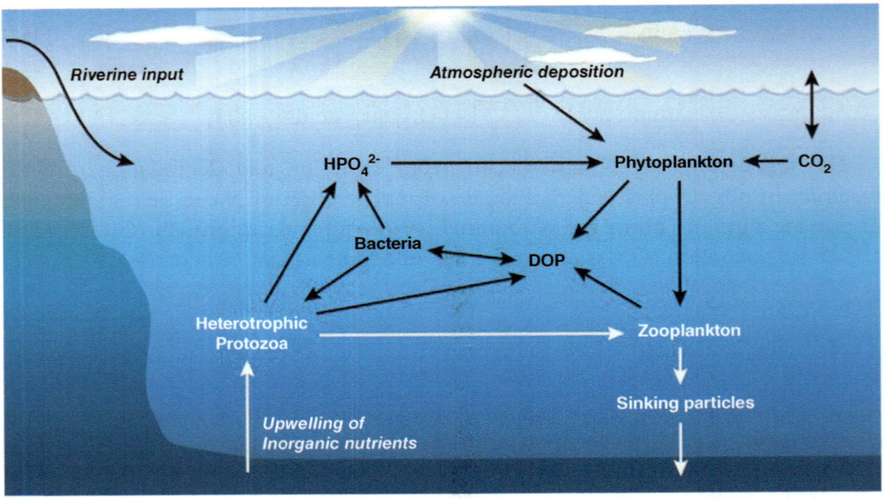

Fig. 11.12 The marine phosphorus cycle

which is mainly driven by the processes of uptake, adsorption, and release, is usually very fast. Accordingly, dissolved concentrations of P typically remain very low in freshwaters, estuaries, and the oceans.

11.4.1 Sources of Phosphorus: Between a Rock and a Hard Place

There is no significant atmospheric source of P, and no analog of N_2 fixation; P cannot be manufactured like N (via the Haber Bosch process, see Chap. 22). It is not found in nature as a free element, and it is not readily biologically available. The largest source is rock. Phosphorus is also found in guano reserves, and thus in the eighteenth century, guano was a prized commodity that was exploited from coastal Peru, and from various South Pacific islands. Other sources included crushed bones which were used by both British and American farmers in the same time period. In the Great Plains of the USA, as European settlers spread in the late 1800s, buffalo bones were gathered and sold by the wagonload (Bennett 2020). However, these supplies dwindled, and it became apparent that mining P rock was the only solution to supply this necessary nutrient (Fig. 11.13). Thus, globally, phosphate rock weathering is the primary source of this nutrient to oceanic waters over time.

With mining as the only means to obtain P for agricultural and other anthropogenic uses, there are clearly many concerns. First, there is a finite supply and there is concern that demand will outstrip supply in the coming years. There is no substitute for P in food production. Only five countries hold 88% of the world's P reserves, with the largest reserves in Morocco and the Western Sahara (Jasinski 2010). As for the USA, its reserves are only expected to last a few more decades (Cordell 2010). In 2019, just four states, Florida, North Carolina, Idaho, and Utah, still had active PO_4^{3-}

Fig. 11.13 Phosphorus mining operation at Societe Nouvelle des Phosphates de Togo. Image from Alexandra Pugachevsky and reproduced from Wikimedia (https://commons.wikimedia.org/wiki/File:Togo_phosphates_mining.jpg) under a creative commons license

mining operations. Mining of P also comes with a host of associated environmental consequences, from the scouring of the earth to the waste that it leaves behind on land and adjacent waters. In 2021, for example, Florida had a large "spill" (discharge) of PO_4^{3-} waste with its associated NH_4^+, derived from an abandoned P mining operation into Tampa Bay, reversing decades of efforts to reduce nutrient loads to this estuary. This discharge over a 1 week period was equivalent to about 1 year's worth of P and N loading to this Bay. In turn, this discharge was related to a sustained bloom of the toxic dinoflagellate *Karenia brevis* on the West Florida Shelf in the months following the discharge (Chen et al. 2023).

Future prospects for having adequate PO_4^{3-} and avoiding the "phosphorus time bomb" (when P supplies become inadequate to meet society's needs) include recovery from wastewater treatment plants (Cordell et al. 2011) and animal manures. Fortunately, there appear to be at least several decades before world reserves of PO_4^{3-} are depleted, so there is still time to find solutions before the predictions of dire consequences on world food supply come to be (Bennett 2020).

11.4.2 Stickiness and Inorganic Phosphorus

Various factors regulate the exchange processes of P with sediments. Temperature is one factor; adsorption of P on to sediments is higher in warmer temperatures. Adsorption of P is also related to salinity; an increase in salinity decreases adsorption. Accordingly, water-column PO_4^{3-} concentrations change as waters warm and as freshwater and saltwater mix.

In calcareous sediments, P binds tightly to the carbonate matrix. Many tropical carbonate systems have very low P concentrations not only in the water column, but also in the pore waters. These systems are often characterized as being P-limited, a designation that is enhanced by N_2 fixation which may also be more common in these waters.

In noncalcareous freshwater sediments, P is most often bound to iron oxyhydroxides (FeOOH, Compton et al. 2000; Jordan et al. 2008; Glibert et al. 2011). The FeOOH-bound P may be delivered to estuaries with transport of suspended solids or may become adsorbed to particulates when P is discharged from other sources, such as from sewage discharges. When this bound P meets saline or sulfate-rich water, either from transport down-estuary or from salt intrusion to sediments, formation of Fe sulfide minerals releases P to the overlying water (Caraco et al. 1989; Jordan et al. 2008; Lehtoranta et al. 2009). The sulfides produced in saline sediments preferentially bind with the Fe, releasing P and precipitating Fe(II); this has been termed the "Fe conveyer belt" (Jordan et al. 2008), according to:

$$\mathrm{FeOOH(s)} \cdot \mathrm{H_2PO_4^-}\ \mathrm{(ads)} + \mathrm{H_2S} \rightarrow \mathrm{FeS\ (s)} + \mathrm{H_2PO_4^-}\ \mathrm{(aq)} \tag{11.9}$$

where (s) refers to solid phase, (ads) refers to adsorbed P, and (aq) refers to aqueous P. In freshwaters, FeOOH-bound P may be released under anoxic conditions, but the cycle of Fe binding of P begins anew when O_2 is encountered again (Carignan and Flett 1981).

Salinity has an opposite effect on N fluxes, compared to P fluxes. Whereas P flux rates are higher in saltwater, rates of NH_4^+ flux are higher in freshwater (Jordan et al. 2008 and references therein), and this can accentuate the discrepancy between N:P ratios along the salinity gradient. Consistent with this idea, a study was conducted on benthic fluxes of nutrients in the upper San Francisco Bay Delta in late summer and rates of PO_4^{3-} efflux ranged from insignificant to ~0.06 μmol m^{-2} day^{-1}, considerably lower than rates previously found for the more saline reaches of the estuary and considerably lower than those of NH_4^+ (Kuwabara et al. 2009). Comparisons of the $NH_4^+ : PO_4^{3-}$ ratio along the salinity gradient of four sub-estuaries of Chesapeake Bay (Patuxent, Potomac, Choptank and Bush Rivers) revealed a common switch from molar ratios >16 to <16 as salinity increased, with a major breakpoint in the salinity range of 1–4 (Hartzell et al. 2010).

As the pH increases, which can occur during intensive algal blooms, the fundamental physical–chemical relationships related to P adsorption–desorption change. Enhancement of sediment P release under elevated water-column pH conditions has been observed in eutrophic lakes (i.e., Andersen 1975; Drake and Heaney 1987;

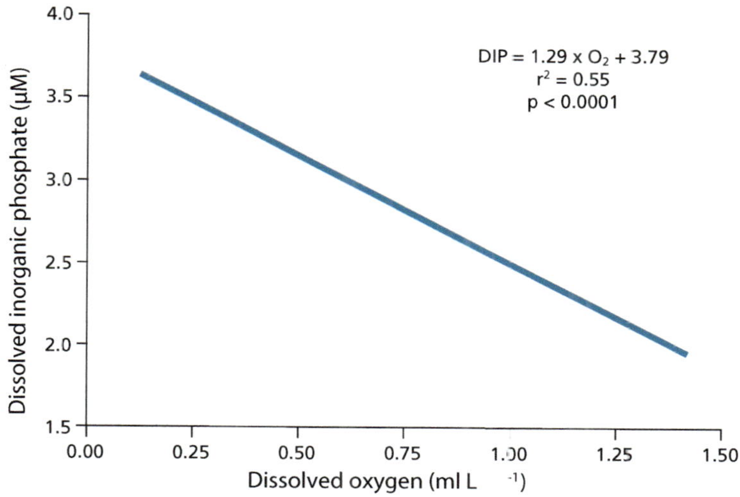

Fig. 11.14 Negative correlations between dissolved inorganic phosphorus (μM) and oxygen (mL L^{-1}) concentration in bottom waters of the Baltic Sea. Image redrawn and simplified from Conley et al. (2002, Environ Sci Technol 36: doi: 10.1021/es025763w)

Jensen and Andersen 1992; Xie et al. 2003) and tidal freshwater/oligohaline estuaries (Seitzinger 1990). For example, PO_4^{3-} flux from the sediment in the Potomac Estuary increased from < 5 μmol m^{-2} h^{-1} to nearly 30 μmol m^{-2} h^{-1} in < 24 h when the pH increased from 7.8 to 9.5. Organisms that can tolerate high pH and high NH_4^+ thus form the base of the food web in such habitats. In turn, their metabolism affects the availability of benthic PO_4^{3-}. As will be developed in Chap. 25, harmful taxa such as cyanobacteria that form high-biomass blooms can tolerate these conditions and, in turn, enhance conditions that accelerate the release of P which they rapidly consume.

The release of P is also affected by O_2 availability. Numerous examples suggest that P is released from sediments with declines in O_2 (e.g., Carlton and Wetzel 1988; Conley et al. 2002; Fig. 11.14). The Baltic Sea data highlight this relationship well. In anaerobic sediments, the reduction of Fe from ferric to ferrous iron makes the ferric oxyhydroxides unavailable to adsorb P (Stumm and Morgan 1981). Thus, diffusion of P out of anoxic/anaerobic sediments can provide a substantial flux of P into overlying water columns.

PO_4^{3-} is also released as excreta and in fecal material. While substantial amounts of P can be released by animals, the rate of excretion or release depends on many factors including the physiological state of the animals and the elemental composition of the food on which they feed (e.g., Sterner and Elser 2002; see also Chap. 18).

The uptake of P by phytoplankton and bacteria is extremely fast. As with N uptake and assimilation, the metabolism of P is discussed more fully in its own chapter, Chap. 13.

11.5 Feedbacks, Redfield Ratios, and the Grand Dynamic Balance

In the absence of anthropogenic nutrient pollution, stabilizing feedbacks help to keep the balance of N and P in the oceans. These feedbacks can be envisioned as the push and pull of N_2 fixation and denitrification. If N becomes depleted relative to that which is required to sustain productivity (i.e., a low N:P ratio), the process of N_2 fixation may increase; the ecological niche for N_2 fixers may expand. Then, as the inventory of fixed N increases, the process of N_2 fixation becomes disadvantaged, and non-N_2-fixing primary producers can again proliferate.

In contrast, a higher reservoir of fixed N (particularly in the form of NO_3^-) may lead to increased rates of denitrification. First, primary production may increase with the availability of this nutrient. Then, organisms may sink from the euphotic zone, leading to consumption of O_2 as this material is remineralized and decomposed. This reduction in O_2 alters the supply of reductant and increases the favorability of loss of N_2 via denitrification. Across the global oceans the rates of N_2 fixation and denitrification are roughly comparable (Zhang et al. 2020). These feedbacks are also dependent on other factors, namely sufficient availability of P so that it is not the limiting nutrient for primary production.

Redfield (Fig. 11.1) described the near constancy (16:1) of N:P in the oceans. Deviations from this ratio have been used to infer rates of N_2 fixation and denitrification (again, anthropogenic nutrient pollution aside). The term N* has been used to quantify the excess or deficit of NO_3^- relative to PO_4^{3-}:

$$N^* = [NO_3^-] - 16 \times [PO_4^{3-}] + 2.9 \qquad (11.10)$$

where 2.9 is a constant in mmol kg^{-1}, and it is a value which yields a mean global N* value of zero (Gruber and Sarmiento 1997; Zhang et al. 2020 and references therein). Higher N* values (up to +10) are found in the North Atlantic tropical gyre, while lower N* values (<−15) are found in the Arabian Sea, Eastern Tropical North and South Pacific. Thus, the Atlantic has relatively low rates of denitrification but higher rates of N_2 fixation. As described in later chapters, anthropogenic pressures are putting stresses on the grand dynamic balance of these nutrient cycles.

11.6 Silica Cycle

Among phytoplankton and protozoa, silica (Si) is primarily used by diatoms, silicoflagellates, and radiolarians. This nutrient is mostly added to the oceans by rivers, groundwater, hydrothermal vents and aeolian dust deposition (Fig. 11.15). It is largely recycled and/or dissolved, with only a small fraction permanently lost to the deep sea. The majority of Si is in the form of silicic acid, $Si(OH_4)$ (see Chap. 14). Grazing results in remineralization of other components of organic matter faster than the Si components, leading to particles becoming more enriched in Si with depth. Biogenic silica in sediments can be quickly dissolved, however.

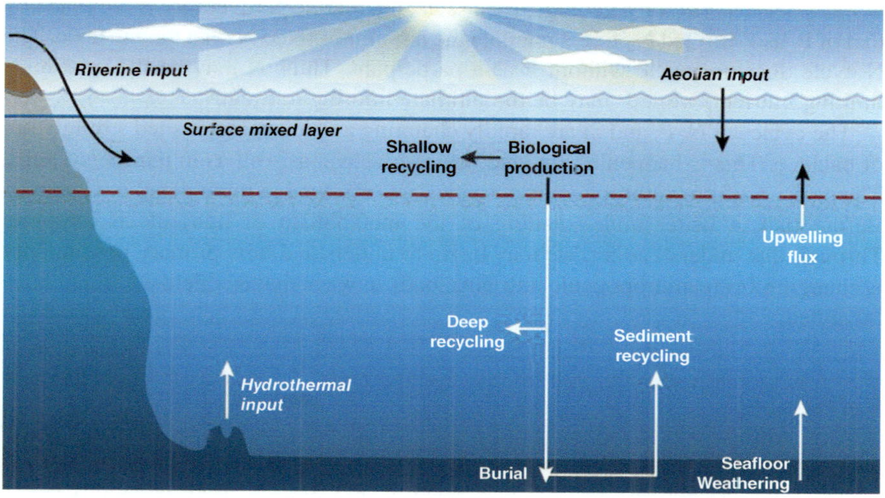

Fig. 11.15 The marine silica cycle

As further developed in Chap. 22, the global Si cycle is changing due to the large increase in dams that trap sediments and diatoms that are produced upstream. It is also of note that marine diatoms tend to contain less Si than freshwater diatoms (Conley et al. 1989; Bianchi 2006), potentially due to salinity differences or long-term evolutionary adaptations to lower Si environments.

Global change is likely to bring a shift away from diatoms in the global ocean (e.g., Glibert 2020; Dutkiewicz et al. 2021). If such a scenario turns to reality, it is expected to lead to a reduction in the biological flux of Si, which may be self-reinforcing with less Si available to be upwelled for future diatom generations (see Chap. 14). Moreover, global warming is likely to increase stratification of the oceans, further depriving the near surface waters of fluxes from the deeper water (Tréguer et al. 2021).

11.7 Summary

Redfield was able to show the relationships between concentrations of various nutrients and with dissolved oxygen and CO_2. He and many who came later understood that elemental nutrients could limit primary production in the ocean, and that the relative proportion between nutrients and between processes whereby new sources of nutrient are made available (e.g., N_2 fixation, nutrient regeneration) and processes by which they are removed (uptake, denitrification) must be in a dynamic balance. The geochemical supply of each nutrient must meet the biological demand or nutrient limitation occurs.

The perspective of limiting nutrients is also shaped by the scale on which the question is asked. On the one hand, N is clearly limiting for many parts of the world's oceans. Yet, on a geologic time scale, the marine N inventory is regulated by that of P, as P-limited N_2 fixers regulate the amount of fixed N on a longer scale. The N cycle does not cycle without P as a participant. Thus, N may be the proximate limiting nutrient, while P may be the ultimate limiting nutrient.

The cycles of N, P, and Si are rapidly changing and are being pushed strongly out of balance. This is happening on land, in freshwaters, as well as in marine systems. The pressures from anthropogenic activities are reviewed in more detail in Chap. 22. Before that, a better understanding of the metabolism of nutrients is required. Through that understanding, we may begin to understand how primary productivity is changing from anthropogenic nutrients loads in ways that Redfield could not have imagined.

References

An SM, Gardner W (2002) Dissimilatory nitrate reduction to ammonium (DNRA) as a nitrogen link, versus denitrification as a sink in a shallow estuary (Laguna Madre/Baffin Bay, Texas). Mar Ecol Prog Ser 237:41–50. https://doi.org/10.3354/meps237041

Andersen JM (1975) Influence of pH on release of phosphorus from lake sediments. Arch Hydrobiol 76:411–419

Antia NJ, Harrison PJ, Oliveira L (1991) The role of dissolved organic nitrogen in phytoplankton nutrition, cell biology and ecology. Phycologia 30:1–89

Arrigo KR (2005) Marine microorganisms and global nutrient cycles. Nature 437:349–355. https://doi.org/10.1038/nature04158

Bennett C (2020) Phosphorus time bomb for agriculture? Myth and reality. SCOOP, August 25. https://www.thedailyscoop.com/news/retail-industry/phosphorus-time-bomb-agriculture-myth-and-reality

Berman T, Bronk DA (2003) Dissolved organic nitrogen: A dynamic participant in aquatic ecosystems. Aquat Microb Ecol 31:279–305

Bianchi TS (2006) Biogeochemistry of estuaries. Oxford University Press, Oxford, UK

Bianchi-Smiraglia A, Wawrzyniak JA, Bagati A, Marvin EK, Ackroyd J, Moparthy S, Bshara W et al (2015) Pharmacological targeting of guanosine monophosphate synthase suppresses melanoma cell invasion and tumorigenicity. Cell Death Differ 22:1858–1864. https://doi.org/10.1038/cdd.2015.47

Bonin P, Raymond N (1990) Effects of oxygen on denitrification in marine sediments. Hydrobiologia 207:115–122

Bronk DA, See JH, Bradley P, Killberg L (2007) DON as a source of bioavailable nitrogen for phytoplankton. Biogeosciences 4:283–296

Caraco NF, Cole JJ, Likens GE (1989) Evidence for sulphate-controlled phosphorus release from sediments of aquatic systems. Nature 341:316–318. https://doi.org/10.1038/341316a0

Carignan R, Flett RJ (1981) Postdepositional mobility of phosphorus in lake sediments. Limnol Oceanogr 26:361–366. https://doi.org/10.4319/lo.1981.26.2.0361

Carlton RG, Wetzel RG (1988) Phosphorus flux from lake sediments: Effect of epipelic algal oxygen production. Limnol Oceanogr 33:562–571. https://doi.org/10.4319/lo.1988.33.4.0562

Carpenter EJ, Bergman B, Dawson R, Siddiqui PJA, Soderback E, Capone DG (1992) Glutamine synthetase and nitrogen cycling in colonies of the marine cyanobacteria *Trichodesmium* spp. Appl Microb Mirobiol 58:3122–3129

Chen Y, Li M, Glibert PM, Heil CA (2023) Mu*r*Ky waters: modeling the succession r to K strategists (diatoms to dinoflagellates) following a nutrient spill from a mining facility in Florida. Limnol Oceanogr 68(10):2288–2304

Compton J, Mallinson D, Glenn CR, Filippelli G, Föllmi K, Shields G, Zanin Y (2000) Variations in the global phosphorus cycle. In: Glenn CR, Prévôt-Lucas L, Lucas J (eds) Marine authigenesis: from global to microbial. SEPM Special Publication No. 66. Society for Sedimentary Geology, pp 21–33

Conley DJ, Kilham SS, Theriot E (1989) Differences in silica content between marine and freshwater diatoms. Limnol Oceanogr 34:205–213

Conley DJ, Humborg C, Rahm L, Savchuk OP, Wulff F (2002) Hypoxia in the Baltic Sea and basin-scale changes in phosphorus biogeochemistry. Environ Sci Technol 36:5315–5320. https://doi.org/10.1021/es025763w

Cordell D (2010) The story of phosphorus: sustainability implications of global phosphorus scarcity for food security. Linköping Studies in Arts and Science No. 509. Dept. of Water and Environmental Studies, Linköping University, Linköping, Sweden. http://urn.kb.se/resolve?urn:nbn:se:liu:diva-53430

Cordell D, Rosemarin A, Schröder JJ, Smit AL (2011) Towards global phosphorus security: A systems framework for phosphorus recovery and reuse options. Chemosphere 84:747–785. https://doi.org/10.1016/j.chemosphere.2011.02.032

Dalsgaard T, Canfield DE, Petersen J, Thamdrup B, Acuña-González J (2003) N_2 production by the anammox reaction in the anoxic water column of Gulfo Dulce, Costa Rica. Nature 422:606–608. https://doi.org/10.1038/nature01526

Damsté JSS, Strous M, Rijpstra WIC, Hopmans EC, Geenevasen JAJ, van Duin ACT, van Niftrik LA, Jetten MSM (2002) Linearly concatenated cyclobutane lipids from a dense bacterial membrane. Nature 419:708–712. https://doi.org/10.1038/nature01128

Devol AH (2003) Solution to a marine mystery. Nature 422:575–576. https://doi.org/10.1038/422575a

Dingler C, Kuhla J, Wassink H, Oelze J (1988) levels and activities of nitrogenase proteins in *Azotobacter vinelandii* grown at different dissolved oxygen concentrations. J Bacteriol 170: 2148–2152

Dong LF, Sobey MN, Smith CJ, Rusmana I, Phillips W, Scott A, Osborn AM, Nedwell DB (2010) Dissimilatory reduction of nitrate to ammonium, not denitrification or anammox, dominates benthic nitrate reduction in tropical estuaries. Limnol Oceanogr 56:279–291. https://doi.org/10.4319/lo.2011.56.1.0279

Drake JC, Heaney SI (1987) Occurrence of phosphorus and its potential remobilization in the littoral sediments of a productive English lake. Freshw Biol 17:513–523. https://doi.org/10.1111/j.1365-2427.1987.tb01072.x

Dutkiewicz S, Boyd PW, Riebesell U (2021) Exploring biogeochemical and ecological redundancy in phytoplankton communities in the global ocean. Glob Change Biol 27:1196–1213

Finzi-Hart JA, Pett-Ridge J, Weber PK, Capone DG (2009) Fixation and fate of C and N in the cyanobacterium *Trichodesmium* using nanometer-scale secondary ion mass spectrometry. Proc Nat Acad Sci USA 106:6345–6350. https://doi.org/10.1073/pnas.0810547106

Gardner WS, McCarthy MJ, An S, Soobolev D, Sell KS, Brock D (2006) Nitrogen fixation and dissimilatory nitrate reduction to ammonium (DNRA) support nitrogen dynamics in Texas estuaries. Limnol Oceanogr 51:558–568

Glibert PM (2020) Harmful algal at the complex nexus of eutrophication and climate change. Harmful Algae 91:101583. https://doi.org/10.1016/j.hal.2019.03.001

Glibert PM, Lipschultz F, McCarthy JJ, Altabet MA (1982) Isotope dilution models of uptake and remineralization of ammonium by marine plankton. Limnol Oceanogr 27:639–650

Glibert PM, Fullerton D, Burkholder JM, Cornwell J, Kana TM (2011) Ecological stoichiometry, biogeochemical cycling, invasive species and aquatic food webs: San Francisco Estuary and comparative systems. Rev Fish Sci 19:358–417

Gradoville MR, Crump BC, Letelier RM, Church MJ, White AE (2017) Microbiome of *Trichodesmium* colonies from the North Pacific subtropical gyre. Front Microbiol 8:1122. https://doi.org/10.3389/fmicb.2017.01122

Gruber N, Sarmiento JL (1997) Global patterns of marine nitrogen fixation and denitrification. Glob Biogeochem Cycles 11:235–266
Hartzell JL, Jordan TE, Cornwell JC (2010) Phosphorus burial in sediments along the salinity gradient of the Patuxent River, a subestuary of the Chesapeake Bay (USA). Estuaries Coast 33: 92–106
Harvey HW (1926) Nitrate in the sea. J Mar Biol Assoc UK 14:71–88. https://doi.org/10.1017/S0025315400007141
Herbert RA (1999) Nitrogen cycling in coastal marine ecosystems. FEMS Microbiol Rev 23:563–590. https://doi.org/10.1111/j.1574-6976.1999.tb00414.x
Hutchins DA, Mulholland MR, Fu F-X (2009) Nutrient cycles and marine microbes in a CO_2-enriched ocean. Oceanography 22(4):128–145. https://doi.org/10.5670/oceanog.2009.103
Inomura K, Masuda T, Gauglitz JM (2019) Active nitrogen fixation by *Crocosphaera* expands their niche despite the presence of ammonium – a case study. Sci Rep 9:15064
Jasinski SM (2010) Phosphate rock: Mineral commodity summaries, 8 November. U.S. Geological Survey. http://minerals.usgs.gov/minerals/pubs/commodity/phosphate_rock/
Jensen HS, Andersen FØ (1992) Importance of temperature, nitrate, and pH for phosphate release from aerobic sediments of four shallow, eutrophic lakes. Limnol Oceanogr 37:577–589. https://doi.org/10.4319/lo.1992.37.3.0577
Jordan TE, Cornwell JC, Boynton WR, Anderson JT (2008) Changes in phosphorus biogeochemistry along an estuarine salinity gradient: The iron conveyer belt. Limnol Oceanogr 53:172–184
Kana TM (1992) Rapid oxygen cycling in *Trichodesmium thiebautii*. Limnol Oceanogr 38:18–24
Kemp WM, Sampou P, Caffrey J, Mayer M (1990) Ammonium recycling versus denitrification in Chesapeake Bay sediments. Limnol Oceanogr 35:1545–1563. https://doi.org/10.4319/lo.1990.35.7.1545
Kuwabara JS, Topping BR, Parchaso F, Engelstad AC, Greene VE (2009) Benthic flux of nutrients and trace metals in the Northern Component of San Francisco Bay, California. U.S. Geological Survey Open-File Report 2009-1286, p 14. U.S. Geological Survey, Reston, VA
Lehtoranta J, Ekholm P, Pitkänen H (2009) Coastal eutrophication thresholds: A matter of sediment microbial processes. AMBIO 38:303–308. https://doi.org/10.1579/09-A-656.1
McDowell RW, Noble A, Pletnyakov P et al (2020) Global mapping of freshwater nutrient enrichment and periphyton growth potential. Sci Rep 10:3568. https://doi.org/10.1038/s41598-020-60279-w
Milligan AJ, Berman-Frank I, Gerchman Y, Dismukes GC, Falkowski PG (2007) Light-dependent oxygen consumption in nitrogen-fixing cyanobacteria plays a key role in nitrogenase protection. J Phycol 43:845–852
Mudler A, van de Graaf AA, Robertson LA, Kuene JG (1995) Anaerobic ammonium oxidation discovered in a denitrifying fluidized bed reactor. FEMS Microbiol Ecol 16:177–183. https://doi.org/10.1016/0168-6496(94)00081-7
Nixon SW, Ammerman JW, Atkinson LP, Berounsky VM, Billen G, Boicourt WC, Boynton WR et al (1996) The fate of nitrogen and phosphorus at the land-sea margin of the North Atlantic Ocean. Biogeochemistry 35:141–180
Pajares S, Ramos R (2019) Processes and microorganisms involved in the marine nitrogen cycle: knowledge and gaps. Front Mar Sci 6:739. https://doi.org/10.3389/fmars.2019.00739
Pierella Karlusich JJ, Pelletier E, Lombard F, Carsique M, Dvorak E, Colin S, Picheral M et al (2021) Global distribution patterns of marine nitrogen-fixers by imaging and molecular methods. Nat Comm 12:4160. https://doi.org/10.1038/s41467-021-24299-y
Redfield AC (1934) On the proportions of organic derivatives in sea water and their relation to the composition of plankton. James Johnstone Memorial 176
Redfield AC (1958) The biological control of chemical factors in the environment. Am Sci 46:205–222
Revelle R (1995) Alfred C. Redfield, November 15, 1890–March 17, 1983. In: Biographical memoirs: vol 67. The National Academies Press, Washington, DC, pp 315–329. https://nap.nationalacademies.org/read/4894/chapter/16#315

Ryther J, Dunstan W (1971) Nitrogen, phosphorus, and eutrophication in the coastal marine environment. Science 171:1008–1013. https://doi.org/10.1126/science.171.3975.1008

Santoro AE, Buchwald C, McIlvin MR, Casciotti KL (2011) Isotopic signature of N_2O produced by marine ammonia-oxidizing archaea. Science 33:1282–1285. https://doi.org/10.1126/science.1208239

Seitzinger SP (1990) Denitrification in aquatic sediments. In: Revsbech NP, Sørensen J (eds) Denitrification in soil and sediment, Federation of European microbiological societies symposium series, vol 56. Springer, Boston, pp 301–322. https://doi.org/10.1007/978-1-4757-9969_19

Seitzinger S, Gardner WS, Spratt AL (1991) The effect of salinity on ammonium sorption in aquatic sediments: implications for benthic nutrient recycling. Estuaries 14:167–174

Shafiee RT, Diver PJ, Snow JT, Zhang Q, Rickaby REM (2021) Marine ammonia-oxidizing archaea and bacteria occupy distinct iron and copper niches. ISME Commun 1:1. https://doi.org/10.1038/s43705-021-00001-7

Sterner RW, Elser JJ (2002) Ecological stoichiometry: the biology of elements from molecules to the biosphere. Princeton University Press, Princeton. 439 pp

Stumm W, Morgan JJ (1981) Aquatic chemistry: an introduction emphasizing chemical equilibria in natural waters, 2nd edn. Wiley, New York. 583 pp

Sunda WG (2012) Feedback interactions between trace metal nutrients and phytoplankton in the ocean. Front Microbiol 3. https://doi.org/10.3389/fmicb.2012.00204

Tobias CR, Anderson IC, Canuel AC, Macko SA (2001) Nitrogen cycling through a fringing marsh-aquifer ecotone. Mar Ecol Prog Ser 210:25–39

Tréguer PJ, Sutton JN, Brzezinski M, Charette MA, Devries T, Dutkiewicz S, Ehlert C et al (2021) The biogeochemical cycle of silicon in the modern ocean. Biogeosciences 18:1269–1289. https://doi.org/10.5194/bg-18-1269-2021

Wang J, Kan J, Zhang X, Xia Z, Zhang X, Qian G, Miao Y, Leng X, Sun J (2017) Archaea dominate the ammonia-oxidizing community in deep-sea sediments of the eastern Indian Ocean – from the equator to the Bay of Bengal. Front Microbiol 8:415. https://doi.org/10.3389/fmicb.2017.00415

Ward BB (2003) Significance of anaerobic ammonium oxidation in the ocean. Trends Microbiol 11: 408–410

Xie LQ, Xie P, Tang HJ (2003) Enhancement of dissolved phosphorus release from sediment to lake water by *Microcystis* blooms – an enclosure experiment in a hyper-eutrophic, subtropical Chinese lake. Environ Poll 122:391–399

Xu MN, Li X, Shi D, Zhang Y, Dai M, Huang T, Glibert PM, Kao S-J (2019) Coupled effect of substrate and light on assimilation and oxidation of regenerated nitrogen in euphotic ocean. Limnol Oceanogr 64:1270–1283. https://doi.org/10.1002/lno.11114

Zehr JP, Waterbury JB, Turner PJ, Montoya JP, Omoregie E, Steward GF, Hansen A, Karl DM (2001) Unicellular cyanobacteria fix N_2 in the subtropical North Pacific Ocean. Nature 412:635–638. https://doi.org/10.1038/35088063

Zhang X, Ward BB, Sigman DM (2020) Global nitrogen cycle: critical enzymes, organisms, and processes for nitrogen budgets and dynamics. Chem Rev 120:5308–5351. https://doi.org/10.1021/acs.chemrev.9b00613

Zhu G, Xia C, Shanyun W, Zhou L, Liu L, Zhao S (2015) Occurrence, activity and contribution of anammox in some freshwater extreme environments. Environ Microbiol 7:961–969. https://doi.org/10.1111/1758-2229.12341

Additional Websites Accessed

https://ccdatahub.ipsl.fr/world-ocean-atlas-2013-woa13-2/. Accessed 23 July 2023
https://commons.wikimedia.org/wiki/File:Togo_phosphates_mining.jpg. Accessed 31 July 2023
https://en.wikipedia.org/wiki/World_Ocean_Atlas. Accessed 30 June 2023
https://www.wikidata.org/wiki/Q35949474#/media/File:Dolichospermum_crassum_akinete.jpg. Accessed 15 Mar 2023

Nitrogen Uptake, Assimilation, and Metabolism: Making Amino Acids and Protein

12

12.1 Introduction

The green alga, *Chlorella*, is the hero of many of the early studies of nitrogen (N) metabolism. Early growth studies with this fairly-easy-to-grow taxon led to incredible insight into growth rates, utilization of inorganic and N forms, and the activity of associated enzymes (reviewed by Syrett 1981). In fact, by the 1940s and 1950s, many fundamental concepts related to N metabolism were well known. For example, it was already recognized that during batch growth, NH_4^+ is generally taken up by algae first, and only after its near depletion is NO_3^- taken up (Ludwig 1938; Harvey 1953). This single concept has both metabolic and ecological implications.

Our current understanding of the rates and processes by which the N cycle turns and how primary productivity varies in space, time, and with each form of this complex nutrient, would be quite limited without the application of its stable isotope, ^{15}N. The rapid progress and advancements in methods in stable isotope application were described in Chap. 2, but it is worth taking a step back and understanding the early applications of ^{15}N, and most importantly, who was responsible for pioneering this technique. That credit goes to Richard C. (Dick) Dugdale (Fig. 12.1, although he graciously gives John Ryther (Woods Hole Oceanographic Institution) the credit for bringing this tool specifically into biological oceanography.

Dugdale recently described (Dugdale 2018, p. 5–6) how his early ^{15}N analyses for measuring rates of N_2 fixation in lake waters were conducted, and how the application of the tool evolved from soil studies,

> Dr. Robert Burris had pioneered the use of ^{15}N as a tracer for nitrogen fixation studies in soil in 1943 and had a Consolidated Nier 21-201 isotope ratio mass spectrometer in his laboratory in the Department of Biochemistry in the UW School of Agriculture, which he generously made available .... The method we developed required first getting rid of the low-enrichment natural nitrogen gas in the lake water sample, which we accomplished by bubbling the water with a mixture of 20% O_2 and 80% He and then adding back N_2 enriched in ^{15}N. After incubating the sample in the light, we collected the algae by filtration and

P. M. Glibert, *Phytoplankton Whispering: An Introduction to the Physiology and Ecology of Microalgae*, https://doi.org/10.1007/978-3-031-53897-1_12

Fig. 12.1 Dick Dugdale

oxidized the nitrogen to ammonium in a Kjeldahl apparatus for 24 h. Then, on a vacuum frame, we oxidized the ammonium with a hypochlorite solution, compressed the N_2 into a glass vial with a Toepler (mercury) pump, and sealed the vial off with a gas torch. We could then transport the sample to the Burris laboratory's mass spectrometer. The whole procedure was labor intensive, and each sample required hours of preparation before mass spectrometry. The first measurements of nitrogen fixation in lakes were made in Lake Pymatuning, Pennsylvania (at the University of Pittsburgh field station, where I was an instructor), and published in *Science* (Dugdale et al. 1959). The details of the method were described by Neess et al. (1962).

Several years later, and after multiple cruises around Bermuda, and multiple permutations of ^{15}N applications, Dugdale and his former postdoc, then turned colleague, John Goering, described how primary production differs when it is based on the dominant oxidized form of N, nitrate (NO_3^-) compared to the dominant form of chemically reduced nitrogen, ammonium (NH_4^+). This distinction, now well known as the *new and regenerated nitrogen* paradigm, was described by them (Dugdale and Goering 1967, p. 205) and recognizes the difference between production resulting from those reduced N forms, primarily NH_4^+ and urea, that are regenerated in situ (from zooplankton excretion or bacterial remineralization in the water column or sediment) and production resulting from the use of oxidized N forms, primarily NO_3^-, resulting from allochthonous ("new") inputs to a system. Their conceptual model related rates of uptake of different forms of N to primary production (Fig. 12.2; see Sect. 12.2.2 below for an explanation of the calculation of ρ; for now, consider it the rate of uptake).

The same year as the now classic new and regenerated production paper was published, Dugdale published another insightful analysis (Dugdale 1967). In that study, he analyzed laboratory culture data and showed that nutrient uptake follows the same equations as many enzyme reactions and that had been used to describe phytoplankton growth: a saturating hyperbola. Dugdale's contributions—and those of many others—and the evolution of our understanding of concepts related to nitrogen uptake and metabolism are examined in detail in this chapter. Phosphorus (P), silica (Si), and iron (Fe) metabolism are considered in the following chapters. First, though, a return to methodology.

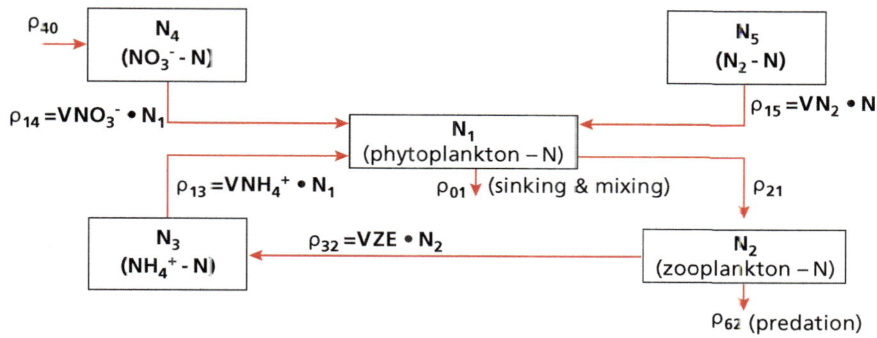

Fig. 12.2 Circulation of inorganic nitrogen through the euphotic zone. Under quasi-steady-state conditions, ammonia can circulate indefinitely if the phytoplankton population incurs no losses whatever, that is $r_{21} = r_{32} = r_{13}$. The primary production system is real, so losses are incurred primarily through sinking and mixing, r_{01}, and by production by zooplankton r_{62}. The sum of these losses, $\rho_{01} + \rho_{62}$, must at least be balanced by nitrate uptake, r_{14}, or by nitrogen fixation, r_{15}, or by any other possible source of nonregenerated nitrogen. Thus, for the phytoplankton cell in the sea, ammonia is an important source serving to maintain the cell in a healthy state and providing much of the nitrogen used in reproduction when nitrate levels are low, but nitrate and nitrogen fixation are the most important parameters with respect to nitrogen limitation of primary productivity. Image redrawn based on Dugdale and Goering (1967), Limnol Oceanogr 12:196–206

12.2 Measuring Nutrient Uptake

12.2.1 Depletion vs. Isotope Approaches

The most used method for measuring nutrient (any nutrient) uptake—by far—is the simplest. Add a pulse of nutrient and measure its change over time (i.e., the depletion method; Fig. 12.3a). This method can be applied in culture studies but can prove problematic in studies of natural waters. This is because the rate of regeneration of a nutrient, especially if one is measuring NH_4^+, can resupply that nutrient, in some cases at or near the same rate that it is consumed. This would be the case in oligotrophic waters, and the logical conclusion stemming from such a suite of measurements is that there is little or no change, therefore little or no uptake. This could be a wrong conclusion. In depletion studies, it also needs to be recognized that for each measurement taken, the physiological state of the cells reflects its condition during the prior time interval. Thus, the rate of uptake is expected to decline not only as the pulse of nutrients becomes used, but also as the cells become satiated with nutrient. That is, the cell quota, Q (see Chap. 3 and discussion of the Droop model for growth), increases with time as nutrients are consumed.

In an isotope study, one adds a known quantity of the isotopically enriched substrate, and after a period of incubation, measures its incorporation into particulate matter. If one is interested in measuring how cells respond to different concentrations of a nutrient pulse, then different aliquots of the culture (or natural sample) are

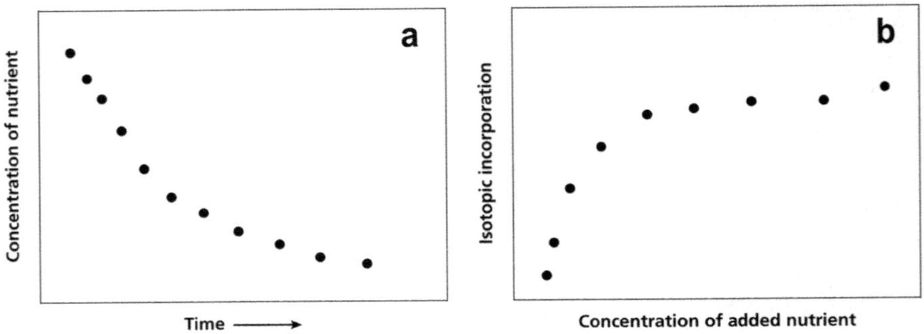

Fig. 12.3 Conceptual figure illustrating the typical response of the change in nutrient over time in a nutrient depletion experiment (Panel **a**) and the rate of isotope incorporation as a function of concentration of added nutrient in an isotope experiment (Panel **b**). In the former, the cell quota changes with time, but in the latter, all measurements are made with cells in the same initial physiological status

enriched with different substrate levels, with each treated independently and incubated for a fixed period of time (Fig. 12.3b). Thus, different treatments are not influenced by the physiological state of the adapting cells; all cells should be at the same physiological state when the incubations are started. It is also possible to do a combination of these methods; that is, by measuring the time course incorporation of an isotopically labeled substrate. Measurements of uptake are only equivalent to growth rates (μ) when cells are growing at μ_{max}. Rates of uptake can far exceed the rate necessary to sustain growth over short periods of time if the cells are nutrient limited and then resupplied the limiting nutrient in short pulses (see also Chap. 3; Fig. 12.4). The capacity for the cell to expand its cell quota depends on the extent to which the cell quota has been depleted. In reading the literature, it is important to be aware of these differences in methodologies as the resulting rates would not be expected to be comparable (Harrison et al. 1989).

Nitrogen does not have a convenient radioactive isotope, and thus, virtually all isotopic analyses of N use stable isotopes (there is a short-lived radioisotope ^{13}N which has been used in a very few studies of N metabolism; see Furhman et al. 1988). In contrast, there is no stable isotope of P, but radioactive isotopes are available. One of the important points of distinction with regard to stable versus radioactive isotopes is that the former involves measurements of ratios, namely ratios of lighter to heavier isotopes (or alternatively heavier to lighter isotopes), ratios of sample to standard, and ratios of these ratios. In radioactive isotopic studies, an absolute amount of incorporation is measured. The next chapter introduces some of methodological considerations in applying radioactive tracers for P studies.

12.2.2 Isotope Terminology and Advancing Methods

The terminology and calculation of rates of nutrient when stable isotopes are used are confusing because the vocabulary associated with enrichment studies is different

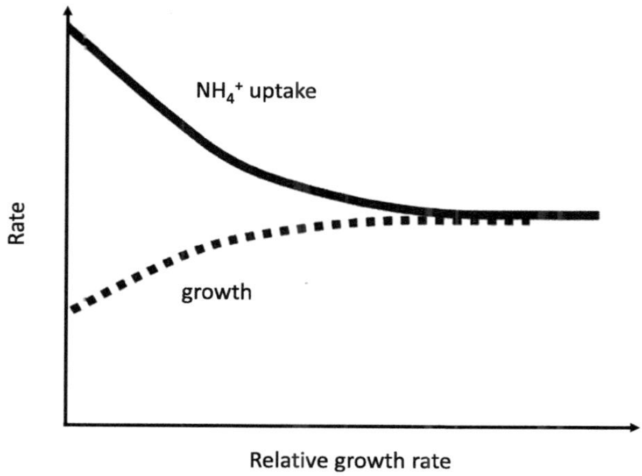

Fig. 12.4 Relationship between specific uptake of nutrient (here, NH_4^+) and specific growth rate. When cells are growing at less than μ_{max}, the capacity to expand the cell quota for the limiting nutrient is great, but is reduced as cells approach μ_{max}. Image reproduced from Glibert et al. (2016, Limnol Oceanogr 61: 165–197, doi: 10.1002/lno.10203) under a creative commons license

from that of the natural abundance stable isotope literature (when ^{15}N is used not as an enriched or added tracer but interpreted based on natural abundance changes or differences; Glibert et al. 2019). The velocity, or rate, of uptake or transformation of an isotopic enrichment into a product (V, also termed specific uptake rate, with units of reciprocal time, i.e., h^{-1}), is determined as the excess atom fraction of the target or product pool divided by that of the initial nutrient enrichment times the duration of the experiment (Sheppard 1962; Dugdale and Goering 1967). Thus:

$$V = \text{at\%excess of product}/[\text{at\%enrichment of substrate} \times \text{time}] \quad (12.1)$$

Atom fraction excess (or in % terms, at % excess, APE) is that isotope fraction in a sample in excess of the measured standard (typically taken to be atmospheric N_2). Atom % enrichment is the amount of enrichment added in the experiment (e.g., amount of $^{15}NO_3$) relative to the total NO_3 (amount added + ambient). The term atom fraction refers to the absolute measure of the number of atoms of the isotope relative to 100 atoms of the element, i.e., the atom fraction of heavy isotope relative to heavy plus light (e.g., mass 29 relative to mass 28 in studies of N, as the molecular weight of N is 14 and unlabeled N_2 would be 14 + 14 and labeled N_2 would be 14 + 15). When expressed as a percent, this value is known as atom percent (at %, or ^{15}AP, in which the superscript represents the heavy isotope—in this example ^{15}N). One can think of V as the N uptake into biomass per unit N of biomass per time (with the N's crossing out). Therefore, V has the units of reciprocal time.

To calculate an absolute flux (denoted as ρ), such as a rate of uptake or transport of a particular N compound (i.e., $\mu M\ N\ h^{-1}$), the specific uptake rate (V) is multiplied by the concentration of that element in the target pool (or other biomass parameter). For example, for N uptake,

$$\rho = V \times \text{PN} \tag{12.2}$$

where PN is the particulate N concentration of the sample, ideally measured at the end, rather than beginning of an incubation (PN will change over time but the amount of change will depend on the duration of the experiment). In nitrification studies where the added isotope is NH_4^+, the velocity of the rate is calculated the same way, but the target pool is the concentration of NO_2^-.

Stable isotope analysis requires a mass spectrometer (although some early studies used emission spectrometers). The techniques of mass spectrometry have been rapidly advancing, allowing for smaller and smaller samples to be analyzed, as well as for specific compounds to be analyzed, as for example, Nanometer-scale Secondary Ion Mass Spectrometry (nanoSIMS; see Chap. 2 and also described further below). An additional method gaining rapid application within microbial ecology is that of Stable Isotope Probing (SIP, e.g., Radajewski et al. 2000; Neufeld et al. 2007). For SIP applications, an enrichment is made with an isotopic compound, e.g., ^{13}C or ^{15}N, and after uptake by natural or cultured organisms, separation of nucleic acids is undertaken using a density gradient ultracentrifugation. Nucleic acids with different densities can thus be separated and their differential incorporation of the label can be measured. This tool, when combined with molecular analyses can lead to insights into responses and interactions of microorganisms and ecological conditions at very fine scales. Using SIP, Morando and Capone (2018) showed, for the Southern California Bight that diatoms dominated the assimilation of NO_3^- and urea and prokaryotes dominated the assimilation of NH_4^+, a finding consistent with their respective physiology, as described below.

With isotopes, one can also observe severe departures from a saturating hyperbolic response due to propagating errors from measurements in ambient substrate concentrations (especially at the very low levels necessary to achieve limiting levels), or in rates of isotopic enrichment (Eppley et al. 1977). These errors lead to calculated rates of uptake that are far too high at low substrate concentrations. The propagation of such errors is due to the fact that substrate concentration is in the denominator of the uptake calculation (see Eq. 12.1), and if underestimated, will lead to a highly inflated rate (Fig. 12.5). It is worth underscoring that measuring extremely low concentrations of ambient nutrients (especially NH_4^+) in oligotrophic waters remains an analytical challenge (but new techniques are promising).

Another approach being applied in resolving many of the complex pathways of NO_3^- assimilation and denitrification is to compare isotope assimilation of multiple isotopes within the same compound, or isotopologues (see also Glibert et al. 2019). For example, by resolving the isotope composition of ^{18}O and ^{15}N in N_2O, it was shown that the N_2O produced by ammonia-oxidizing archaea (AOA) incorporates O from O_2, not H_2O, and the ^{15}N and ^{18}O produced by AOA is more enriched than that produced by ammonia-oxidizing bacteria (AOB) (Santoro et al. 2011). Moreover, dual isotope measurements of both NO_2^- and NO_3^- have revealed NO_3^- regeneration in the euphotic zone and in O_2 deficient zones (e.g., Sigman et al. 2005), and thus, nitrification is now recognized to play a significant role in the oceanic N cycle (e.g., Casciotti and Buchwald 2012). Recently, triple isotopes (^{15}N, ^{18}O, and ^{17}O of NO_3

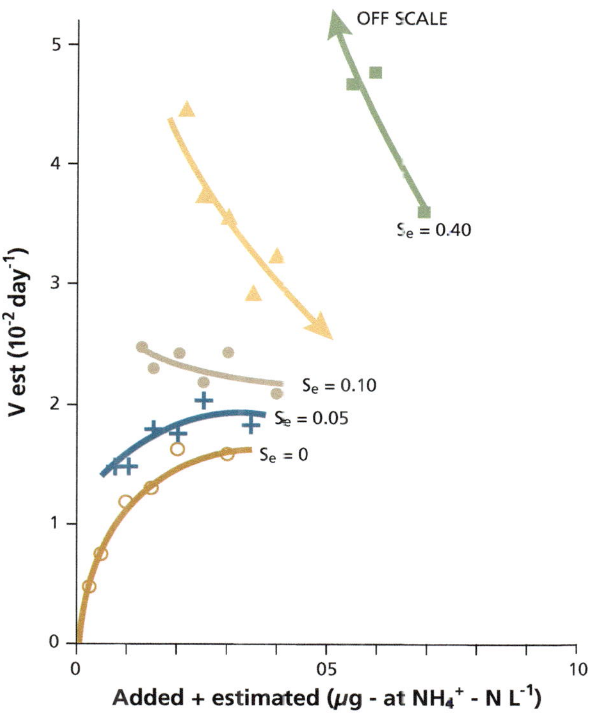

Fig. 12.5 Calculated effect of erroneously high estimates of ambient concentration (Se) on rates of nutrient uptake determined by stable isotope techniques. Image redrawn and modified from Eppley et al. (1977, Mar Biol 39:111–120)

have been applied to the study of rates of nitrification and the assimilation of NO_3 in streams (Rose et al. 2015) and lakes (Tsunogai et al. 2018).

12.3 Uptake and Incorporation

12.3.1 Transport

The first step in nutrient metabolism (for all nutrients) is the transport of nutrient across the cell membrane. In a nutrient depletion or an isotopic enrichment study, if the time course is very short, the resulting rate largely reflects transport only (uptake) and not assimilation of these nutrients into macromolecules.

In a typical nutrient-poor environment, it generally requires a transporter protein to facilitate the entry of nutrients into the cell. Net diffusion is generally limited to very high concentration of substrate (>many tens of μM). Each nutrient form has its own unique transporters, and within each nutrient form, there may be unique transporters that function under very low nutrient conditions and those that operate under high nutrient conditions (Fig. 12.6; Glibert et al. 2016). Thus, cells may have both high-affinity transporters (HATs, operating at low concentrations of substrate) and low-affinity transporters (LATs, operating at relatively high concentrations of

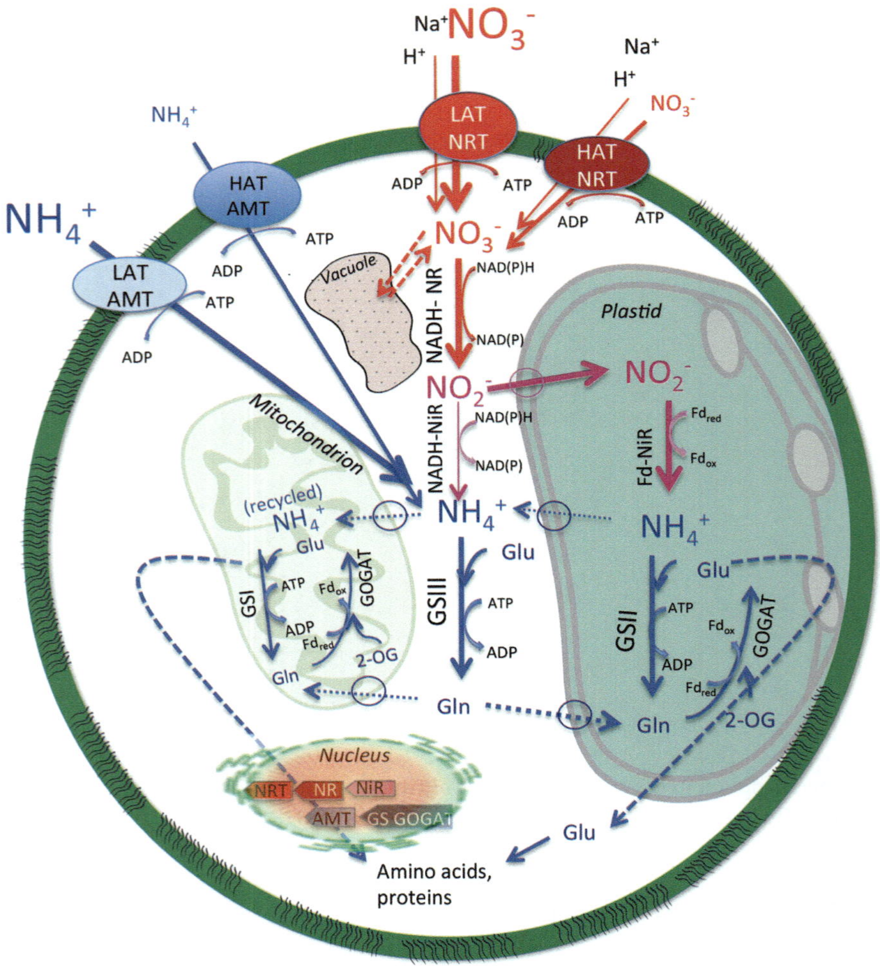

Fig. 12.6 Simplified conceptual relationship of a generic diatom cell and the major metabolic pathways of uptake and assimilation of NO_3^- and NH_4^+. Note that assimilative NO_3^- reduction typically occurs in the cytosol, while NO_2^- reduction occurs in the chloroplast. NH_4^+ assimilation can occur in the cytosol, in the chloroplast and in the mitochondrion. The nucleus controls enzyme and transporter production. See text for details. Image reproduced from Glibert et al. (2016, Limnol Oceanogr 61: 165–197, doi: 10.1002/lno.10203) under a creative commons license

substrate) at the cell surface for each substrate. HATs are saturable and are typically expressed under N-limiting conditions, while LATs may be nonsaturable and generally expressed only when the substrate is abundant—but see some differences below (Howitt and Udvardi 2000; Rogato et al. 2015 and references therein).

Transport of both NH_4^+ and NO_3^- across the cell membrane is performed by NH_4^+ and NO_3^- transporters, AMTs, and NRTs, respectively, that perform proton

(H^+)-coupled transport (e.g., Navarro et al. 1996; Galván and Fernandez 2001; Rogato et al. 2015; Glibert et al. 2016 and references therein), although there is evidence of Na^+ rather than H^+ symport with NO_3^- in marine diatoms (Rees et al. 1980; Boyd and Gradmann 1999). Diffusive influx of NH_4^+ depends on an electrogenic pump (H^+ or Na^+) to maintain the inside-negative electrical potential difference (Raven 1980). For urea, genomic sequencing has revealed that marine unicellular photosynthetic eukaryotes possess the gene DUR3, which encodes a high-affinity urea active transporter that is part of the larger Na^+: solute symporter family (SSSF) of transport proteins (Wang et al. 2008) that use the energy from a Na^+ gradient across the cell membrane to transport the solute into the cell (Jung 2002).

With time of exposure to a nutrient source, phytoplankton alter the number of nutrient transporters, as well as the internal enzymes that assimilate these nutrients once encountered (Smith et al. 2009; Fig. 12.7). Over time, cells can upregulate (shift up) or downregulate (shift down) transporters and enzymes as needed. There are important differences in the regulation of NRTs and AMTs in phytoplankton, as well as in virtually all plants (Rogato et al. 2015). In general, NO_3^- transporters are induced by the *presence* of their substrate (NO_3^-), whereas NH_4^+ transporters are induced by the *absence* or deficiency of their substrates, and repressed by increased availability of their substrate, NH_4^+ (Clarkson and Luttge 1991; Navarro et al. 1996; Crawford and Glass 1998; Daniel-Vedele et al. 1998). Thus, increasing concentrations of NO_3^- yield more NRTs, whereas increasing concentrations of NH_4^+ yield fewer AMTs. Most marine eukaryotic phytoplankton appear to be able to take up NO_3^-, and most of the genomes sequenced to date contain genes encoding for NO_3^- transporters. However, some strains of the picocyanobacterium *Prochlorococcus* have been shown to be unable to grow on—or take up NO_3^-—implying that it does not have the genetic capacity to synthesize these transporters

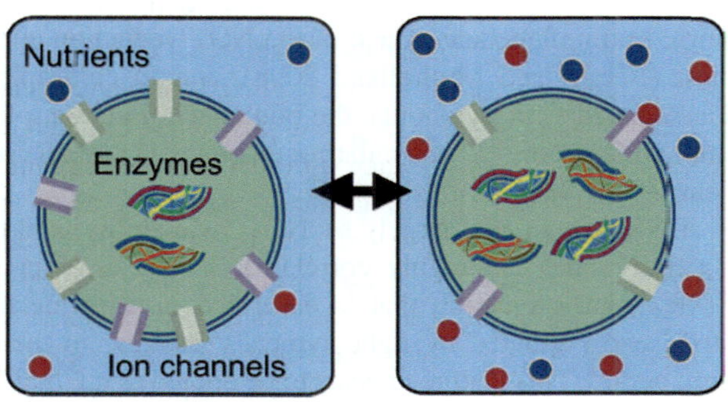

Fig. 12.7 Conceptual diagram of the relationship between acclimation to low substrate concentration (left) and high substrate concentration (right). Note that not only does the substrate concentration change, but so too does the assimilation machinery (enzymes). The important point is that cells transition between these states and are not fixed at one or the other. Image reproduced from Smith et al. (2009, Mar Ecol Prog Ser 384: 1–12) with permission of Elsevier

(e.g., Moore et al. 2002). Urea transport is also upregulated with urea availability. For example, Berg et al. (2008) found that the expression of the urea active transporter gene (AaDUR3) in *Aureococcus anophagefferens* was 20- to 50-fold higher in cultures grown on urea compared with those grown on NO_3^- and NH_4^+, respectively.

12.3.2 Enzymes

Once nutrients are inside the cell, enzymes take over. Enzymes mediate assimilation, as well as dissimilation and internal N cycling (Fig. 12.6).

12.3.2.1 Nitrate Reductase

One of the best characterized enzymes related to N metabolism is that of nitrate reductase (NR) which has been studied for decades, especially in higher plants and from early studies of *Chlorella* (e.g., Solomonson 1979; see also Syrett 1981). It is a complex structure containing molybdenum, heme, and flavin adenine dinucleotide, and it uses NADH, NADPH, or both as reductants. There are several types of NR; they are structurally different, they vary with different algal functional types, and may be regulated quite differently (e.g., Berges 1997; Morozkina and Zvyagilskaya 2007). Gene sequences of this enzyme are similar in higher plants, green algae and diatoms (Allen et al. 2005). In contrast, NR in dinoflagellates and prokaryotes may be different, and are ferredoxin-dependent and contain iron–sulfur clusters (Flores et al. 2005).

Traditionally it was thought that NR is localized in the cytosol of the cell; however, there is considerable evidence that a form of NR also exists in the plasmalemma and other cellular membranes (Jones and Morel 1988), particularly in diatoms, chlorophytes, and cyanobacteria (Jones and Morel 1988; Tischner et al. 1989; Stöhr et al. 1993; Berges 1997; Dagenais-Bellefeuille and Morse 2013 and references therein). In dinoflagellates, most of the NO_3^- reduction appears to occur in the chloroplast (Berges and Mulholland 2008), and in chlorophytes, there is substantial NR associated with pyrenoids, the bodies within chloroplasts that have high concentrations of carbon (C)-assimilating enzymes (Lopez-Ruiz et al. 1985; Fischer and Klein 1988; Chap. 26).

The activity of NR is regulated largely by NO_3^- availability, and thus transport rates and NR activity rates are highly correlated. NR is also regulated by the availability of NH_4^+, an interaction that is described in more detail below (see Sect. 12.5.2). It is also regulated by light, typically peaking in the beginning or middle of the light period. This feature appears to be related to internal C pools of the cells and not to a circadian rhythm per se, so N starved calls are capable of taking up NO_3^- in darkness (Cochlan et al. 1991). The activity of NR is also sensitive to temperature, but the temperature optima of NR may differ by phytoplankton taxa (Lomas and Glibert 1999a, b). It is thought that features of the N terminus of the NR protein (which is variable in diatoms) may play a role in this temperature sensitivity (Berges et al. 2002; Allen et al. 2005).

12.3.2.2 Nitrite Reductase

Closely linked to NR is the enzyme mediating the reduction of NO_2^- to NH_4^+, nitrite reductase (NiR). Once NO_3^- is reduced in the cytosol by NR, the resulting NO_2^- is rapidly transported into the chloroplast (typically via localized transporters) where it is further reduced to NH_4^+ by NiR (Galván et al. 2002. Fig. 12.6). NiR is a smaller molecule than NR. There may be different forms of NiR. The ferredoxin-dependent NiR is localized in the chloroplast, but, based on a putative targeting sequence, it is hypothesized that there is also a cytosolic form of NiR that is NADPH-dependent, at least in diatoms (Armbrust et al. 2004; Allen et al. 2006). In fact, the localization of NiR in the chloroplast has long been known in higher plants (Syrett 1981).

The activity of NiR can exceed that of NR, indicating that NR is the rate-limiting step and that this tight coupling may be the mechanism by which toxic amounts of NO_2^- cannot accumulate in the cells. NiR, like NR, can also be affected by C limitation. NiR contains more Fe than NR, and thus can become more readily limited under Fe limitation (Milligan and Harrison 2000).

12.3.2.3 Ammonium Assimilation Enzymes

Assimilation of NH_4^+, either derived from direct uptake or from reduction of NO_2^-, occurs via a series of reactions involving (for most algal species) the enzymes glutamine (Gln) synthetase (GS) and glutamate (Glu) synthase (GOGAT; also known as glutamine-2-oxoglutarate amidotransferase, Fig. 12.6). This pathway yields Glu, the product of Gln and oxoglutarate (2-OG) (Scanlan and Post 2008). Both Glu and Gln are essential for amino acid metabolism and cellular N regulation, as they are both N acceptors and N donors (Dagenais-Bellefeuille and Morse 2013). Most cyanobacteria assimilate N through the GS-GOGAT pathway, but some also have Glu dehydrogenase (GDH) which may present an advantage for those species in that NH_4^+ assimilation through GDH is not ATP-requiring (Muro-Pastor et al. 2005). *Prochlorococcus* and *Trichodesmium theibautii*, for example, appear to have the genes that encode for GDH but most cyanobacteria do not have these genes (Muro-Pastor et al. 2005).

As with NR and NiR, there are several forms GS, one localized to the chloroplast (form GSII), and one or two forms localized in the cytosol or the mitochondrion (GS I or III), at least in diatoms (Robertson and Alberte 1996; Takabayashi et al. 2005; Siaut et al. 2007). The plastid form is ferredoxin-dependent, while the mitochondrial form is hypothesized to be NADPH-dependent (Alipanah et al. 2015 and references therein). The chloroplastidic form is inducible by external NH_4^+ availability, but the cytosolic form is more constitutively expressed. This localization is important in the metabolism of different N forms because, for many plants and at least in diatoms, the gene encoding for the chloroplastidic GSII, as well as total GS activity, appears to be upregulated in cells assimilating NO_3^- but not in cells assimilating NH_4^+. These findings support the premise that chloroplastidic N metabolism is the important pathway of reduction of oxidized forms of N (Takabayashi et al. 2005). In fact, GSII in the diatom *Skeletonema costatum* has been shown to be a genetic marker for NO_3^- assimilation (Allen et al. 2006), while GSIII, the mitochondrial form, is expressed more in the assimilation of NH_4^+ derived from deamination and hydrolysis of organic N (Siaut et al. 2007). At least in higher plants, the GSI form is thought be a stress marker, as is catalyzes an N-recovery pathway (Pageau et al. 2006).

GS is critical to the assimilation of N, and its regulation is tightly coupled with N availability. Moreover, different pools of the GS enzyme may be regulated differently. GOGAT is a complex iron–sulfur flavoprotein, but, as with GS, there are multiple types with distinct genes (Suzuki and Knaff 2005).

GDH may play a different role than GS-GOGAT in some species having this enzyme, such as green algae or macroalgae. It may play a catabolic role, becoming important in assimilation when intracellular concentrations are high (e.g., Melo-Oliveira et al. 1996); its role in nonassimilatory pathways deserves more attention (Berges and Mulholland 2008).

12.3.2.4 Urea Catabolism

Urea is broken down into NH_4^+ and inorganic C by either urease, the more common enzyme, or by UALase (Antia et al. 1991; Solomon et al. 2010). The cell then uses the NH_4^+ produced in protein synthesis. The urease enzyme occurs as two small (gamma and beta) and one large (alpha) subunits, encoded by the ureA, ureB, and ureC genes, respectively (Mobley et al. 1995). Other genes (ureD, ureE, ureF, and ureG) aid in the production of active urease by supporting the assembly of the nickel metallocenter of the enzyme (Mobley et al. 1995).

The urease protein sequences of red-lineage phytoplankton cluster together, while those of the green lineage cluster with those of higher plants (Solomon et al. 2010; see also Chap. 10). The UALase pathway is more common in a variety of bacteria and in at least some green algae (Hausinger 2004).

Urease activity appears to be strongly regulated by the availability of different sources of N in the cell. However, there appears to be considerable variation in how this regulation occurs in different taxa. In some organisms, lower urease activity is found in NH_4^+-grown cultures than in NO_3^-- and urea-grown cultures, as in the case of the dinoflagellates *Prorocentrum cordatum* (reported as *P. minimum*) and *Karlodinium veneficum* (Fan et al. 2003a; Solomon and Glibert 2008). In these organisms, as in many cyanobacteria, when insufficient NH_4^+ is available to meet N demand, urease activity may increase in response to limitation of inorganic N. In contrast, urease activity in the dinoflagellate *Alexandrium fundyense* was shown to be the highest in N-starved and urea-grown cultures, lower in NH_4^+ and not detected in a NO_3^--grown culture (Dyhrman and Anderson 2003), while in *Heterocapsa triquetra*, urease activity was shown to be higher when grown on NH_4^+ than on NO_3^- or urea (Solomon and Glibert 2008).

In diatoms, and likely at least in dinoflagellates as well, there is a urea cycle (Armbrust et al. 2004; Allen et al. 2011, Weyman et al. 2015; Fig. 12.8). The long-known function of the urea cycle in animals is to excrete excess N produced by amino acid catabolism; thus, it has long been considered a waste pathway. However, in diatoms (and other algae), the urea cycle appears to play a role in exchange of nutrients between the mitochondria and the cytoplasm, and potentially the plastid (Bender et al. 2012) and may help to regulate NH_4^+ metabolism (Armbrust et al.

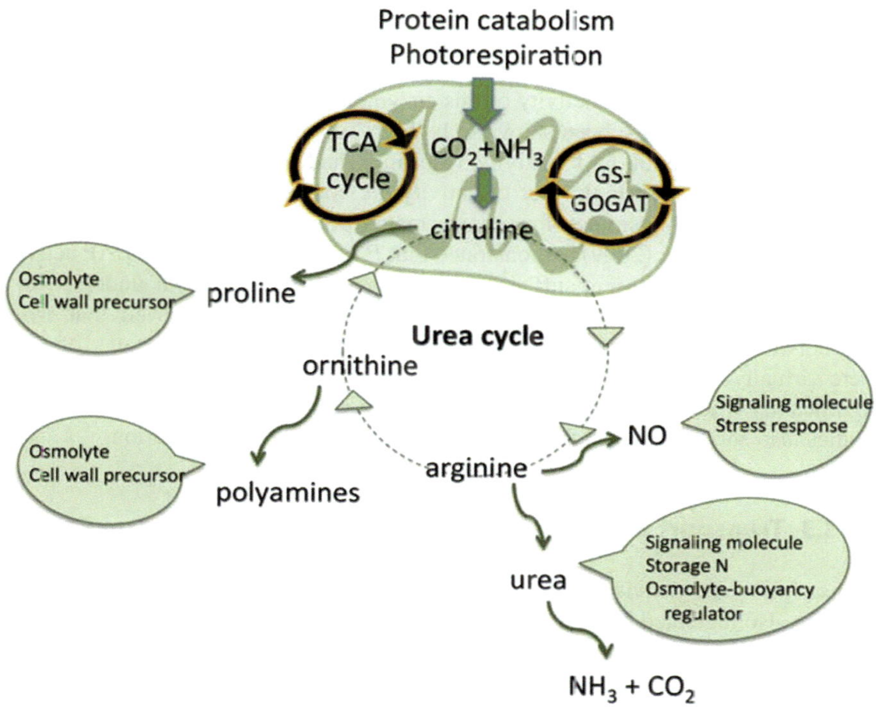

Fig. 12.8 Conceptual schematic of the mitochondrial urea cycle in a generic diatom cell and the potential fates of urea cycle intermediates. Image reproduced from Glibert et al. (2016, Limnol Oceanogr 61: 165–197, doi: 10.1002/lno.10203) under a creative commons license

2004; Allen et al. 2011). Because of this cycle, marine diatoms, in contrast to chlorophytes, also have acquired a mitochondrial urea transporter and, in fact, based on bioinformatics, a complete mitochondrial GS-GOGAT cycle has been hypothesized (Allen et al. 2011; Fig. 12.8).

12.3.2.5 Extracellular Enzymes

Not all enzyme activity occurs within the cell; some occurs on the cell surface (Chróst 1991). This is particularly true for enzymes associated with use of dissolved organic N (DON) or dissolved organic matter (DOM). There are a variety of such enzymes, but due to the complexity of the DOM pool as a whole, there is still much that is yet to be learned about such enzymes. Nevertheless, it is important to understand that activity of such enzymes may be necessary for cells to hydrolyze proteins and peptides and large biomolecules that cannot be transported into the cell. One example is extracellular amino acid oxidases that facilitate the production of inorganic N from amino acids (e.g., Palenik and Morel 1990; Pantoja and Lee 1994).

Other metabolic enzymes involved in N metabolism of dinoflagellates such as *Prorocentrum cordatum* (reported as *P. minimum*), including leucine aminopeptidase (LAP; Stoecker and Gustafson 2003; Salerno and Stoecker 2009), may also be significant (Glibert et al. 2012). LAP is a protease that catalyzes production of small peptides and amino acids from proteins and peptides and appears to have a preference for N-terminal leucine and similar amino acids (Salerno and Stoecker 2009). Field measurements of the activity of this enzyme in *P. cordatum* bloom conditions in Chesapeake Bay have been shown to be high, roughly tenfold higher than those estimated for bacterial LAP activity (Martinez and Azam 1993; Stoecker and Gustafson 2003). Such high rates of activity are consistent with an organism that has the nutritional physiology for organic matter processing. In laboratory studies, Salerno and Stoecker (2009) demonstrated that *P. cordatum* has higher LAP activity when grown under nutrient-sufficient conditions than when grown under N or P limitation (Salerno and Stoecker 2009). These authors also showed that when nutrient-sufficient cultures were further supplemented with NO_3^-, rates of LAP were virtually unchanged compared to unamended controls, whereas rates dropped significantly when NH_4^+ or urea was added (Salerno and Stoecker 2009); their field studies also showed that LAP activity declined with NH_4^+ concentration.

12.3.3 Transport and Enzyme Regulation by Environmental Factors

Transcription and activity of enzymes involved in NO_3^- and NH_4^+ metabolism for both are also regulated by various external cues, as well as cellular fluxes of various C and N substrates (Takabayashi et al. 2005 and references therein). Rates and abundance of NR typically vary on a diel basis (e.g., Packard et al. 1971; Berges et al. 1995; Brown et al. 2009) although some diatoms may continue to assimilate NO_3^- in darkness (depending on their nutrient status), while GS-GOGAT activity is generally maintained throughout the day (Clark et al. 2002). Based on a study of the diurnal expression of the genes encoding N assimilation, Brown et al. (2009) found that NR abundances and activity are not under circadian control, regulated instead by changes in the metabolic pools of NO_3^- and NH_4^+ that, in turn, regulate the N assimilating enzymes. For urea, there also appears to be an effect of irradiance, as urea uptake occurs via active membrane transporters that require energy. The energy for urea uptake may be obtained from photophosphorylation in phytoplankton during the day (Rees and Syrett 1979; Beckers et al. 2004) and from oxidative phosphorylation during the night (Cimbleris and Cáceres 1991). In Chesapeake Bay plankton, a diel pattern in urea uptake was observed with the highest rates of urea uptake in mid-afternoon and the lowest rates during the night (Bronk et al. 1998). However, urea uptake can take place during the dark in dinoflagellate-dominated assemblages (Kudela and Cochlan 2000; Fan and Glibert 2005; Sinclair et al. 2009).

Temperature also affects enzymes associated with NO_3^- and NH_4^+ metabolism differently. NR activity has an inverse relationship with temperature (generally between 12 °C and 25 °C; Kristiansen 1983; Gao et al. 1983; Lomas and Glibert 1999a, b), while GS-GOGAT activity is positively related to temperature across the same range (e.g., Clayton and Ahmed 1986; Fig. 12.9). The assimilation of NO_3^- would thus be expected to be higher at lower temperatures (10–15 °C), and indeed, over the temperature range of 5–25 °C, NO_3^- uptake by both diatom-dominated and dinoflagellate-dominated natural communities show an inverse relationship with

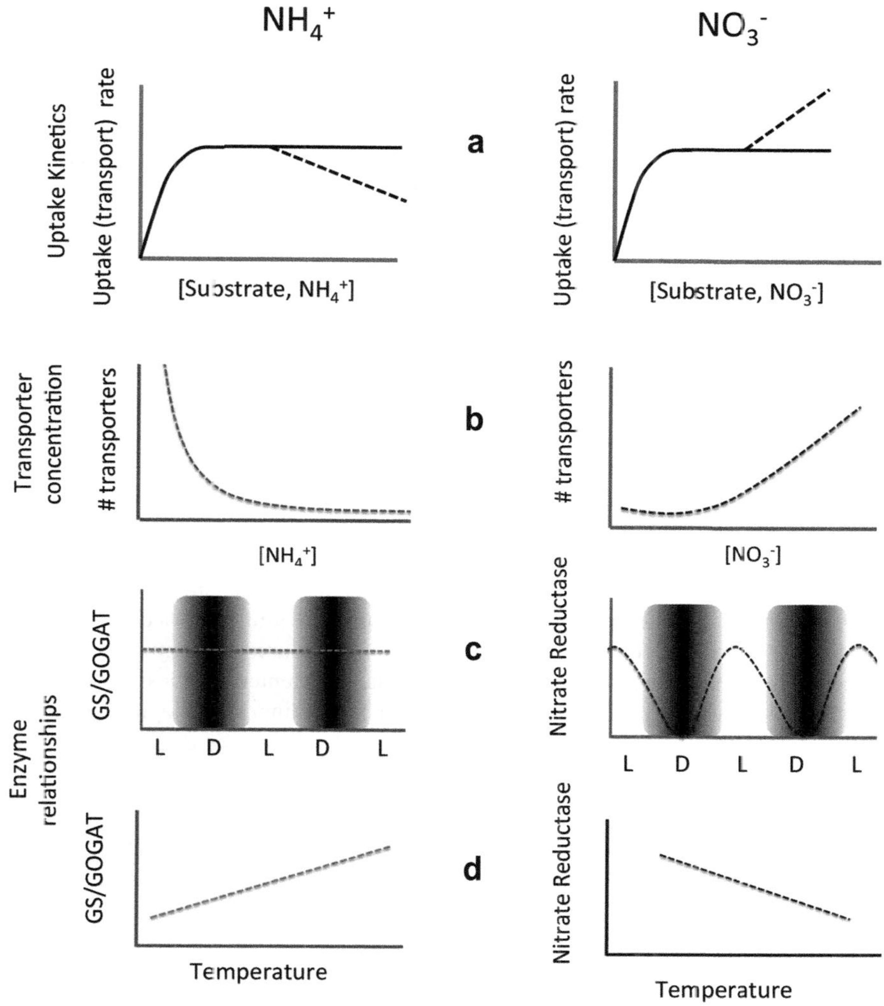

Fig. 12.9 Comparison of the regulation of N transport and their assimilative enzymes for "generic" phytoplankton cells growing exclusively on NH_4^+ (left-hand panels) or NO_3^- (right-hand panels). (Panel **a**) Generic relationships of NH_4^+ and NO_3^- uptake kinetics. Note the potential for biphasic uptake of NO_3^-. (Panel **b**) Generalized relationship of the numbers of transporters in relation to the availability of substrate for that transporter. (Panel **c**) Generalized relationship between enzyme activity of GS-GOGAT and NR as a function of time of day (Panel **d**) Generalized relationships of activity of GS-GOGAT and NR as a function of temperature. Image reproduced from Glibert et al. (2016, Limnol Oceanogr 61: 165–197, doi: 10.1002/lno.10203) under a creative commons license

temperature while NH_4^+ uptake shows a positive relationship (Lomas and Glibert 1999a, b, Fan et al. 2003b; Fig. 12.9). For urea, there appears to be a direct effect of temperature on rates of urea transport and enzyme activity, but there may be species differences. Urea uptake increased as a function of temperature in the diatom *Melosira italica*, with a Q_{10} of 1.94 (Cimbleris and Cáceres 1991). However, urea uptake has been shown to remain constant over a narrower temperature range of 10–25 °C for *P. cordatum* (reported as *P. minimum*) in the Chesapeake Bay and Neuse Estuary (Fan et al. 2003b).

Taken together, the regulation of NH_4^+, NO_3^- and urea (and other organic N sources) uptake and assimilation is quite different with respect to environmental cues, a feature that can influence phytoplankton assemblage dynamics in N-enriched systems, a feature that contributes to the difference of new and regenerated production.

12.4 Nutrient Kinetic Relationships

12.4.1 Classic Kinetic Constructs

The rate of phytoplankton nutrient uptake generally follows a saturating hyperbola (with exceptions as developed below; see also Chap. 3). It thus follows the "curve for all reasons" (Rao 2000; Chap. 3). As the process of assimilation is enzyme-mediated, the relationship theoretically should follow Michaelis–Menten kinetics in which the substrate (S) reversibly binds to the enzyme (E), yielding and enzyme–substrate (ES) complex, that ultimately yields product (P) and releases the enzyme:

$$S + E \rightleftarrows \mathrm{ES} \rightarrow E + P \tag{12.3}$$

Thus, the uptake velocity (V, in units of reciprocal time) and the absolute uptake of nutrient (ρ, in units of mass per time) can both be described by a hyperbolic function,

$$V = V_{\mathrm{max}}\ (S/(K_{\mathrm{s}} + S) \tag{12.4}$$

and

$$\rho = \rho_{\mathrm{max}}\ (S/K_{\mathrm{s}} + S) \tag{12.5}$$

In each of these relationships, the curve is defined by a maximum value (V_{max} or ρ_{max}) and a half saturation constant (K_{s}). The first publications of the kinetics of nutrient uptake by phytoplankton came from nearly coincident efforts by Dugdale (1967) and by Eppley and Coatsworth (1968).

12.4.2 Complexity in the Uptake Kinetic Response

There is a perception of some "definitiveness" that emerges from parameterization of uptake as a function of substrate concentration. The seeming "definitiveness" of these relationships has led to many comparisons of nutrient kinetics of one organism compared to another, with the idea that this may yield insight into their comparative success under different nutrient regimes. And, the similarity in the shapes of these curves and those that are defined for growth (using the Monod or Droop expressions, as described in Chap. 3) have led to the common misconception that these relationships are equitable or interchangeable. However, rather than singular entities associated with individual taxa, kinetic responses are dynamic and responsive and such dynamic behavior is the norm rather than the exception (Kana and Glibert 2016; see Chap. 8). Any measure of this sort depends on the method used to establish the relationship (as described above) and the physiological state of the cell at the moment the measurements were taken. Thus, uptake relationships are not equitable to growth relationships; uptake does not equal growth, except when cells are at steady state. Moreover, depending on the length of time an investigator used to assess these rates, the resulting kinetic properties may not only reflect transport processes, but the summed responses of carrier transport proteins, and various enzymes for assimilation, each of which can be induced, repressed, or otherwise regulated by the cellular physiological state and the environmental conditions of the assay. Single measurements and fixed kinetic values are simply a momentary image of a response that is variable.

Nevertheless, comparisons can be drawn within a single study or between cells adapted to different conditions if other experimental conditions are held constant. Cells may exhibit a high affinity (low K_s), and low affinity (high K_s) or may even show extremely low affinity with no apparent saturation (Fig. 12.10).

A further complexity is that a saturating hyperbolic relationship is not always expressed or is not the most appropriate form of the response curve. Relationships between nutrient uptake and substrate concentration may be linear, biphasic, or may involve an induction period (Fig. 12.11). Biphasic curves may show a saturation response across an initial substrate gradient, but then show a secondary increase as concentrations increase further. This may reflect enzyme regulation across the initial gradient and diffusion as concentrations reach very high levels (Fig. 12.11). Biphasic kinetics are much more commonly reported for NO_3^- uptake than for NH_4^+ uptake. For example, Lomas and Glibert (2000) showed, for six different phytoplankton species, that uptake followed a pattern that showed a saturation response at low concentrations (Fig. 12.12 inset panels), but with increasing availability of substrate, the rate of uptake increased higher than that which was defined if only the lower range of substrate values were examined (Fig. 12.12 primary panels).

For NH_4^+, at very high concentrations, the response curve may also display inhibition, analogous to the photoinhibitory response in the photosynthesis-irradiance relationship (see Chap. 6). Such an inhibition is less often seen with uptake of NO_3^-. In many cyanobacteria, the transcriptional activator of N assimilation genes, NtcA, and therefore of AMT1 expression, is negatively controlled by

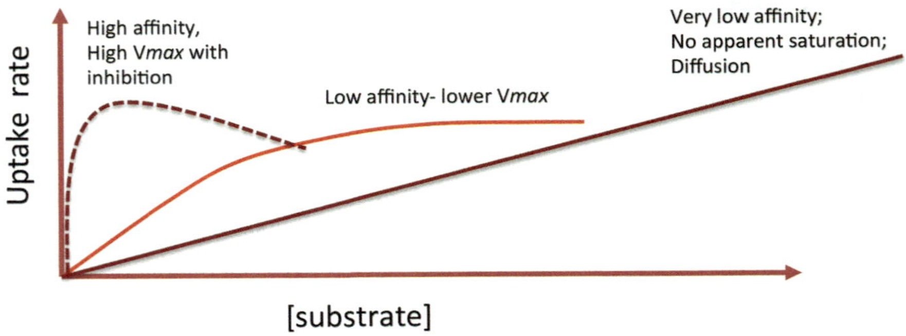

Fig. 12.10 Depiction of the range of kinetic responses that an organism can display in response to variable substrate supply. The response may be (Panel **a**) highly efficient with respect to acquisition at the low end of nutrient supply or may show inhibition of response as the substrate increases; (Panel **b**) the response may show classic saturating substrate responses; or, (Panel **c**) the response may not have characteristics of saturation and may display a linear relationship. In reality, an organism may transit through all of these response relationships depending on substrate, duration of exposure, prehistory and physiological status among other factors. Image reproduced from Glibert et al. (2018, doi: 10.1007/978-3-319-70069-4_6) with permission of the publisher

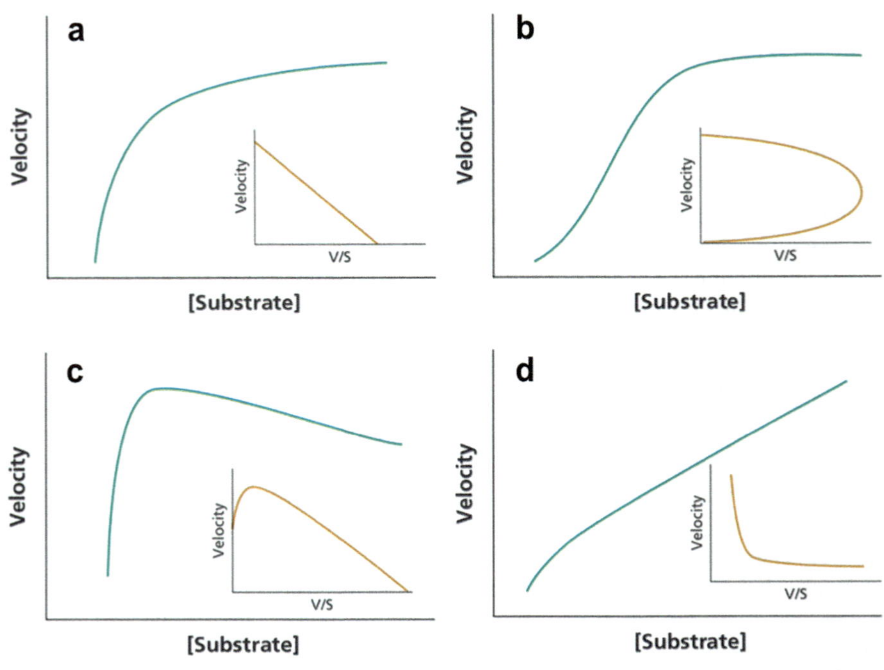

Fig. 12.11 Uptake kinetic curves may take a variety of forms, displaying saturating hyperbola (Panel **a**), an induction period (Panel **b**), inhibition at substrate levels above saturation (Panel **c**), or linear responses (Panel **d**)

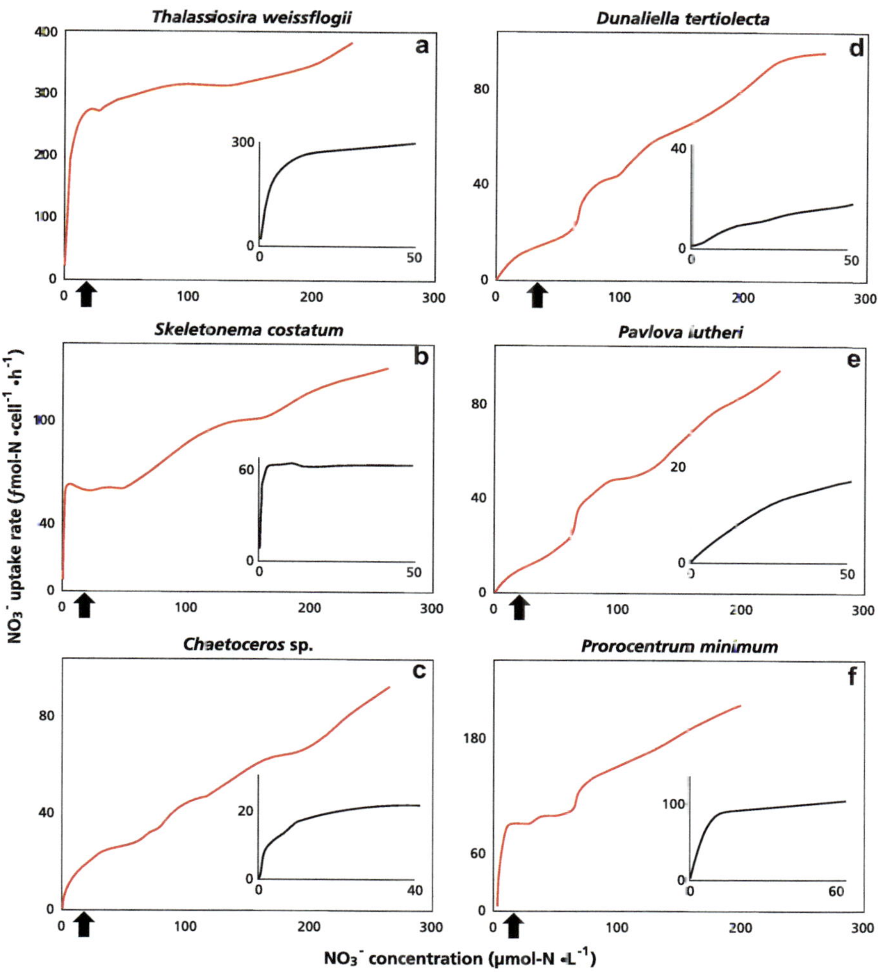

Fig. 12.12 Nitrate uptake kinetics for species indicated. The arrows on the *X* axis represent the ambient concentration of NO_3^- at the time cells were harvested. The insets are an expanded view of the low concentration range. Note that all displayed biphasic responses, with saturation evident at substrate concentrations <60 μM, but linear responses at higher substrate levels. Image redrawn based on Lomas and Glibert (2000, J Phycol 36:903–913)

NH_4^+ and the metabolite 2-OG (Lindell and Post 2001; Coruzzi and Bush 2001; Muro-Pastor et al. 2001, 2005). When this metabolite accumulates, it represses further N assimilation of NH_4^+ (Post et al. 2012). It is of note that NH_4^+ generally does not accumulate in the cell to the same extent as NO_3^- (e.g., Dortch 1982).

Here, too, methodologies come into play in interpreting the kinetic response to a substrate gradient. As described above, if one is measuring such responses based on substrate depletion, the physiology of the cells varies with time. As cells consume nutrient, the cell quota, Q, increases. Thus, the cell reduces its maximum uptake rate over time. A high rate or "spike" in uptake may be initially measured (assuming cells were limited in the nutrient under consideration initially) once the pulse is given (e.g., Fig. 12.3a). Then, the cell has a period during which uptake becomes regulated by internal cell processes. Finally, the rate declines rapidly as nutrients are consumed and become limiting once again.

12.4.3 New Kinetic Constructs: Optimum Uptake Model

On the one hand, a cell does indeed increase its rate of uptake as substrate concentrations increase, and this response can be defined from the saturating response. On the other hand, the entire curve can change (as previously shown for photosynthetic responses; see Chap. 8. The end result is that the shape of a kinetic curve is highly variable, with structure that may not actually fit a saturating hyperbola or may change with time and physiological state or the nutrient under consideration (Glibert et al. 2013; Kana and Glibert 2016).

Thus, rather than assuming an invariant relationship, the dynamic nature of uptake responses (and similarly, photosynthetic responses of C assimilation as a function of light as a substrate) can be thought of as a balance between control from external nutrient supply (nutrients, light) and that of the internal cellular nutrient content, a physiological trade-off (and ability to regulate) between the ability to acquire nutrients (or light) through transport and the ability to assimilate these nutrients into new biomass (Kana et al. 1997; Smith et al. 2009; Glibert et al. 2013). Phytoplankton not only alter the number of nutrient transporters, but the activity of their enzymes also changes (Smith et al. 2009; Pahlow and Oschlies 2013). As noted above, Dugdale et al. (1981) laid the foundation for the understanding of a time-varying uptake response by defining a period when the uptake rate is externally controlled and a period when the uptake rate is internally controlled. He suggested, based on time courses, that the kinetic relationship should be based on the last segment of the time course (as nutrients become limiting) and the internally controlled period, but not based on the initial spike. He showed these same types of relationships with N and with uptake of Si (see Chap. 14). Over time, the cell is adjusting, acclimating, and its enzyme activity becomes either up or downregulated. McCarthy and Goldman (1979) laid the foundation for understanding short-term saturation-dependent responses. While the terminology used by Dugdale et al. (1981) differs from terminology used in current modeling (and recognizing there is considerable confusion and variability in the terminology in use), the concept is the same: there is a dynamic regulation of uptake across the entire gradient of concentrations, and cellular physiology is continually adjusting accordingly. In contemporary terms of nutrient kinetics, this is known as an optimal uptake (OU) model (e.g., Smith et al. 2009). Such a model assumes a physiological trade-off between the efficiency of nutrient encounter at the cell surface (A_S) and the maximum rate at which a nutrient can be assimilated (V_{max}). Accordingly, rather than a Michaelis–Menten model, and instead of a half-saturation constant for uptake,

K_s, the dynamic regulatory approach to kinetic modeling applies the ratio of V_{max} to A_s (e.g., Aksnes and Egge 1991; Smith et al. 2009). In such a dynamic model, the uptake rate V_{aff} relates to substrate, S, availability as follows:

$$V_{aff} = [(V_{max}S)/((V_{max}/A_s) + S)] \quad (12.6)$$

The use of a ratio instead of fixed half saturation constant is a way of incorporating "self-regulating" feed-backs that the cells use to upregulate the needs of the cells when resources are low, or downregulate or switch off acquisition when resources are sufficient; it is a "tuning knob" (Kana and Glibert 2016). While the ratio may be a tuning knob, the underlying physiology may be complex and variable. As an example, and as described above, NO_3^- transporters are induced by the presence of their substrate (NO_3^-), whereas NH_4^+ transporters are induced by the absence or deficiency of their substrates, or repressed by increased availability of their substrate, NH_4^+ (Glibert et al. 2016). Bonachela et al. (2013) have similarly proposed a dynamic formulation of nutrient uptake in which a model cell allows for dynamic regulation of cell transport proteins, leading to flexibility in the maximal uptake rate as well as the limiting portion of the curve.

The implications of a dynamically varying, rather than fixed kinetic, model are important. On the one hand, nutrient stress can develop before nutrient availability declines below a conventionally-defined half saturation value (and bearing in mind how poorly this value is typically known), while on the other hand, regulation of nutrient uptake does not cease when availability of nutrient reaches values defined as "saturating" (Glibert et al. 2013). A dynamically-regulating kinetic framework recognizes that changes in the proportions of dissolved nutrients in the environment have profound effects on cells, and thus can cascade through the food web even when the availability of these elements are not in limiting proportions, with the potential of transforming ecosystems to new stable states. Across a range of substrate availability, the balance between uptake and assimilation also affects other cell properties, such as the relative proportions of ribosomes, enzyme activities, gene regulation, cellular pigmentation complement, and ultimately the cell elemental composition (Glibert et al. 2013, 2016). Thus, properties such as chlorophyll *a*: C ratios (Chl:C), C:N, N:P or other elemental properties change in response to the varying uptake rates as substrate availability changes. As the different species within a community dynamically regulate their response as nutrient availability changes, the consequence is that community composition ultimately may change (Chap. 18).

12.5 Nutrient Interactions

12.5.1 Nitrogen Forms: Defining Preference

Except in culture studies, rarely are phytoplankton in an environment in which only one form of N is available. More typically the cell is exposed to both oxidized and reduced forms of N. One of central tenets of classic phytoplankton ecology, born out of studies focused on growth in oligotrophic, nutrient-impoverished waters, is the preference for NH_4^+ in conditions for which different combined sources of N are available.

Preferential uptake or use is defined variously in the literature, generally involving a comparison of (1) rates of draw-down of one substrate relative to another, (2) uptake affinity, (3) maximal or *in situ* rate of uptake, or (4) an index of relative preference (RPI) for different N forms (e.g., McCarthy et al. 1977).

The RPI is a common index for assessing the use of one form of N relative to another (McCarthy et al. 1977). It is a version of Ivlev's (1961) electivity index,

$$\mathrm{RPI}_{\mathrm{Ni}} = [(V_{\mathrm{Ni}}/\Sigma V_{\mathrm{N}})/(S_{\mathrm{Ni}}/\Sigma S_{\mathrm{N}})] \tag{12.7}$$

in which Ni is a particular form of nitrogen, V is the specific rate of uptake of that particular nutrient form (Ni) and of total nutrient (N), and S is the concentration of that particular nutrient form (Ni) and of total nutrient (N). The relationship compares the rate of that particular N form to total N and relative to its proportion in the substrate pool. Values of unity suggest uptake of that nutrient form in proportion to its availability in the environment, values >1 indicate preference and values <1 indicate use less than its relative proportion. Not a perfect metric for sure, due to analytical and other reasons, but when viewed with its caveats such an index can be useful for comparisons. The general finding of most reports in which such metrics have been used is the same: typically NH_4^+ is preferentially used, especially when N is limiting.

The preference for NH_4^+ is considered to be due, at least in part, to lower energy requirements for the cell, and that NH_4^+ is more easily transported across the cell membrane than NO_3^- under balanced growth and N limited conditions (Raven et al. 1992). Evidence for NH_4^+ preference is grounded in the classical physiological literature, and, importantly, in studies where N was the limiting nutrient. Preferential use of NH_4^+ was originally documented in batch culture experiments in the 1930s–1950s (reviewed in Syrett 1981), and field observations of this phenomenon have been made since at least the 1960s (MacIsaac and Dugdale 1969, 1972; McCarthy et al. 1975, 1977). Delayed uptake of NO_3^- relative to that of NH_4^+ has been observed in enclosure studies in which 10 μM NH_4^+ resulted in cessation of NO_3^- uptake by diatom-dominated assemblages, while subsequent depletion of NH_4^+ concentrations to <4 μM allowed resumption of NO_3^- uptake and diatom growth (Wilkerson et al. 2006; Parker et al. 2012). Preferential uptake of NH_4^+ over NO_3^- at NH_4^+ concentrations exceeding a few μM has also been documented through the declining proportion of NO_3^- uptake in relation to total N uptake as the concentration of NH_4^+ in the water column increases. For example, McCarthy et al. (1975) illustrated that the uptake of oxidized forms of N in Chesapeake Bay never exceeded more than a few percent of the total N ration when NH_4^+ concentrations exceeded 1–2 μM. Berman et al. (1984) reported a similar finding in Lake Kinneret.

It has thus been commonly argued that phytoplankton growth on NH_4^+ should be higher than that on NO_3^-, or at a least growth on both substrates should be equal. Raven et al. (1992, p. 20) summarized the logic of this argument, "If the use of the resource needing more manipulation [e.g. NO_3^-]..., in order to achieve the same product formation [moles] product per second), then the cell doubling time will be significantly increased since more [moles] of the product of the resource manipulation will be required to double cell mass..." Consistent with this hypothesis, several studies have shown that some phytoplankton species grown on NH_4^+ or urea have higher growth rates than on NO_3^- (e.g., Herndon and Cochlan 2007; Solomon et al. 2010 and references therein), although this is far from a universal observation. The physiology of the interactions of NO_3^- and NH_4^+ within the cell is next examined.

12.5.2 Nitrogen Forms: Metabolic Interactions

Different forms of N interact inside the cell. The seeming favorability for NH_4^+ by phytoplankton is actually a function of both the preferential use of NH_4^+ and its favorable energetics, and the repressive effect (often referred to as inhibition) of NH_4^+ on NO_3^- uptake and assimilation (Dortch 1990 and references therein; reviewed by Glibert et al. 2016). Note that while NH_4^+ may repress the uptake of NO_3^-, the reverse generally does not occur. That is, NO_3^- has minimal effect on the rate of uptake of NH_4^+ (Flynn et al. 1997; Fig. 12.13).

Repression of NO_3^- uptake or assimilation by NH_4^+ has been well studied for many decades (e.g., Morris and Syrett 1963; Dortch 1990 and references therein; Lomas and Glibert 1999a, b). The repression of NO_3^- uptake by NH_4^+ occurs at NH_4^+ concentrations as low as a few μM (e.g., Eppley et al. 1969; Lund 1987; Dortch and Conway 1984; Cochlan and Harrison 1991; L'Helguen et al. 2008 among others). From work in the subarctic Pacific, Wheeler and Kokkinakis (1990) even suggested that concentrations of NH_4^+ between 0.1 and 0.3 μM caused complete repression of NO_3^- assimilation (see Chap. 16 for implications of this study with respect to regions of the ocean that are considered high nutrient-low chlorophyll), and L'Helguen et al. (2008) reported that similar concentrations of NH_4^+ caused repression of NO_3^- uptake in the oligotrophic Atlantic. In the San

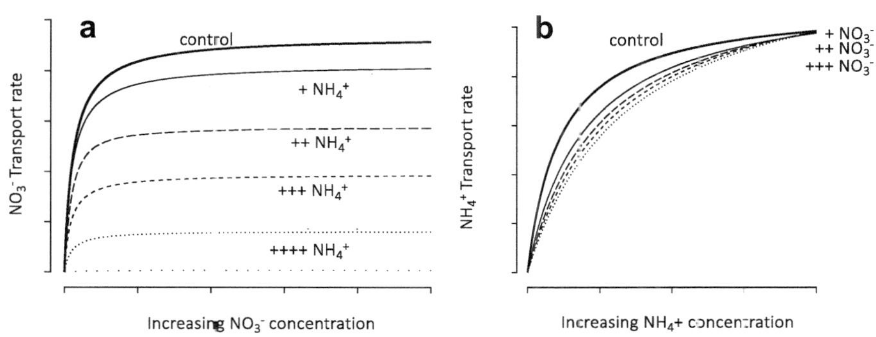

Fig. 12.13 Panel **a**: effects of increasing external ammonium concentrations (shown as + signs) on rates of nitrate transport, and panel **b**: effects of increasing external nitrate concentrations on rates of ammonium transport. Image modified and simplified from Flynn et al. (1997 Philos Trans Roy Soc B 352:1625–1645)

Francisco Bay Delta, much higher concentrations of NH_4^+, 4–10 μM, have been associated with repression of NO_3^- uptake by NH_4^+ based on direct measurements (e.g., Dugdale et al. 2007; Glibert et al. 2014a, b), and similar concentrations were found to repress NO_3^- uptake in laboratory cultures of diatoms and dinoflagellates (Lomas et al. 2000). The extent and threshold concentrations of repression by NH_4^+ on NO_3^- metabolism have been shown to depend on the algal species present, their physiological status (Dortch and Conway 1984; Dortch et al. 1991; Maguer et al. 2007) and the environmental conditions to which they have been exposed (e.g., Harrison et al. 1996; Lomas and Glibert 1999a, b; Xu et al. 2012). Cells growing on highly elevated NO_3^- concentrations, as in the case of a nutrient-rich environment, may require considerably more NH_4^+ to repress cellular NO_3^- activity than is the case for a cell with a low cellular NO_3^- content, as in oligotrophic environments.

NH_4^+ affects NO_3^- metabolism by downregulation of transport of NO_3^- across the cell membrane and repression of NR abundance and activity (Vergera et al. 1998, reviewed by Glibert et al. 2016). In most algae, the regulation is via the size of the Gln pool (e.g., Flynn et al. 1994), although in cyanobacteria and some other taxa, the metabolite 2-oxoglutarate (2-OG) also serves this regulatory function (Muro-Pastor et al. 2001, 2005; Post et al. 2012). The availability of Gln and the Gln/Glu ratio govern the NO_3^- reducing capacity in the cell; when Gln levels are low, and when NO_3^- is available, NR is upregulated. Alternatively, when Gln levels are high, NR activity levels are reduced (Flynn et al. 1994; Campbell 1999). Thus, the degree of phytoplankton uptake and growth repression and suppression conditions by a given amount of NH_4^+ varies with cell status and the environmental conditions (see, for example, Thompson et al. 1989; Yin et al. 1998; Dortch et al. 1991 and references therein). Any environmental factor that affects the availability of substrate, the nutritional state of the cell, or the rate of enzyme activity will affect the rate at which NO_3^- or NH_4^+ is transported and assimilated (Dortch 1990). In addition to the potential environmental controls, N uptake may also depend on the presence of multiple forms of NR that alter the effects of NH_4^+ on NO_3^- reduction. At least for the diatom *Thalassiosira weisflogii*, the membrane-bound form of the enzyme is seemingly not repressible by NH_4^+, whereas the cytosolic form is repressible (Jones and Morel 1988). As noted by Berges (1997), it may be of significance that the NR of dinoflagellates appears to be localized in the chloroplast, as lack of repression of NR by NH_4^+ in some dinoflagellates has been found. Repression of NR and NiR by NH_4^+ in picocyanobacteria has been shown to be quite variable, being near complete in *Synechococcus elongatus*, but being comparatively insensitive in *Synechocystis* (Kobayashi et al. 2005).

In addition to NH_4^+ repression of NO_3^- uptake and assimilation, NH_4^+ may differentially affect oxidation and reduction of the chloroplastidic metabolic pathways, leading to enhanced photorespiration or increased reactive oxygen species (ROS). Importantly, suppression of growth can occur without direct or lethal NH_4^+ toxicity. The wide array of redox sensitive pathways, proteins, and enzymes can affect processes such as photosynthesis, biosynthesis, antioxidant activity, and signaling pathway translation among other metabolic activities (Rosenwasser et al. 2014; Glibert et al. 2016).

Yoshiyama and Sharp (2006) summarized decades of data from the Delaware Bay and observed that the primary productivity rate per unit chl *a* declined exponentially with increasing NH_4^+ concentration (most of the change occurring at <10 μM NH_4^+) and classified these systems as High-Nutrient, Low-Growth (HNLG). In the San Francisco Bay Delta, it has been suggested that a similar phenomenon of growth suppression is responsible for the lack of spring blooms ever since NH_4^+ loading from sewage effluent increased several decades ago (Wilkerson et al. 2006; Dugdale et al. 2007) and observational and experimental evidence are confirmatory (e.g., Wilkerson et al. 2006; Parker et al. 2012). In the higher plant literature, this is known as the "NH_4^+ syndrome" (Gerendás et al. 1997; Britto and Kronzucker 2013). At very elevated concentrations, normally exceeding several hundred μM, NH_4^+ can be toxic for growth (e.g., Britto and Kronzucker 2002 and references therein), a condition from which the cell does not easily recover. The environmental relevance of direct toxicity by NH_4^+ in estuaries and freshwaters is limited, however, to those sites receiving such excessive loads of this N form. In most estuarine and freshwaters, reports of growth suppression by NH_4^+ have been mostly associated with increased NH_4^- at levels not normally considered to be toxic for phytoplankton growth, i.e., at levels in the tens of μM range. Collectively, the observations of preferential use at the low end of the substrate availability spectrum, together with repression and/or toxicity at the high end of the substrate spectrum, have led to NH_4^+ being characterized as a "paradoxical" nutrient (Britto and Kronzucker 2002; Dugdale et al. 2012; Fig. 12.14).

Interestingly, in higher plants it is well documented that one of the means by which NO_3^- repression by NH_4^+ can be alleviated is through the addition of NO_3^- (Goyal et al. 1982; Below and Gentry 1987; Britto and Kronzucker 2002, among others), implying that NH_4^+ repression of NO_3^- uptake and reduction is not necessarily or always absolute. It is also known that coprovision of both NO_3^- and NH_4^+ can induce synergistic growth compared to growth on either substrate alone (e.g., Weissman 1964; Britto and Kronzucker 2002). In fact, total N uptake in higher plants can be up to 75% higher when the two substrates are coprovided relative to when either substrate is provided alone (Kronzucker et al. 1999).

Importantly, sensitivity to NH_4^+ stress varies widely in both higher plants and in the eukaryotic phytoplankton.

12.5.3 Nitrogen Forms: Exceptions to the Rule of NH_4^+ Preference

The preference for NH_4^+ is not universal. In contrast to reports of repression of NO_3^- uptake by NH_4^+, a substantial body of literature has shown that in cool, nutrient-rich environments, large diatoms use a disproportionate fraction of total N as NO_3^- even when NH_4^+ is available at levels in excess of 10 μM (e.g., Maestrini et al. 1982; Probyn and Painting 1985; Lomas and Glibert 1999a, b, reviewed by Glibert et al. 2016). Diatoms appear to be NO_3^- opportunists. For example, in river-dominated estuaries and upwelling systems, the occurrence of many rapidly growing diatom

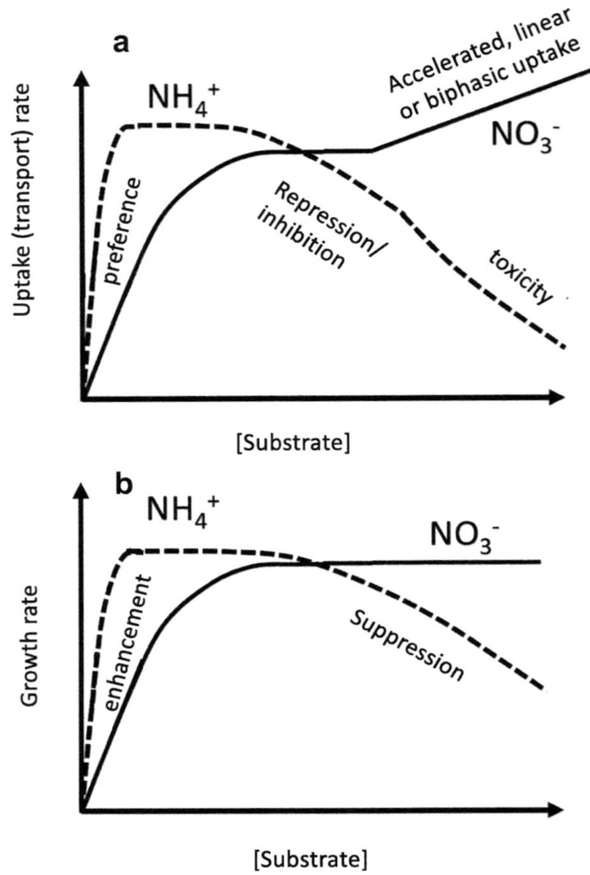

Fig. 12.14 Panel (**a**) Conceptual relationships with respect to uptake or transport kinetics of NH_4^+ (dashed line) relative to NO_3^- (solid line) illustrating the tendency toward preference, repression/inhibition of uptake of NH_4^+ uptake and toxicity in contrast to the tendency for accelerated, linear or biphasic uptake of NO_3^-. Note that at low concentrations of substrate, uptake of NH_4^+ may display not only a lower K_s, but may also have a higher V_{max}. The result is the paradoxical behavior of NH_4^+ kinetics relative to those of NO_3^-. Panel (**b**) Comparable relationship except for growth, illustrating the potential for growth enhancement and growth suppression with NH_4^+ as the growth substrate in comparison to NO_3^- as the growth substrate. Image reproduced from Glibert et al. (2016, Limnol Oceanogr 61: 165–197, doi: 10.1002/lno.10203) under a creative commons license

species has been highly correlated with the large and/or frequent additions of NO_3^- (e.g., Goldman 1993; Lomas and Glibert 1999a). Diatoms are the dominant protist in NO_3^--rich water columns during spring blooms. Marine pelagic ecosystems with predominantly NO_3^- sources are often dominated by diatoms (e.g., Kudela and Dugdale 2000; Wilkerson et al. 2000) and typically have short, efficient food webs at the base of major natural fisheries (e.g., coastal Peru) and high rates of export of organic matter from the photic zone (e.g., Eppley and Peterson 1979). Interestingly, the brown intertidal macroalgae, *Fucus* sp. and *Laminaria* sp. also appear to have very high rates of NO_3^- assimilation, especially in winter, suggesting they too may be NO_3^- opportunists (e.g., Young et al. 2007; see also Chap. 10 for evolutionary relationships between these groups).

When and why would these taxa use NO_3^- instead of the more energetically favorable NH_4^+? The answer appears to lie in the regulation of cellular energy and redox balance. Balancing energy-generating and energy-using processes with those of N- and C-acquisition and N- and C-assimilation is delicate. The assimilation of N and that of C are linked in multiple biochemical pathways and thus C and N metabolites have various "cross-talk" in the cell and mechanisms to regulate the flux of metabolites into the cell and cellular redox status (e.g., Turpin and Harrison 1979; Turpin 1991; Coruzzi and Bush 2001; Wang et al. 2014). Redox regulation is a function of not only the flow of reductant through photosynthesis, but also the demands for energy via metabolism.

Some of the well-documented processes which aid in the regulation of cellular redox and energy balance that often arise from imbalances in photochemistry and C assimilation include nonphotochemical quenching operating around the thylakoid pH gradient and xanthophyll cycle, cyclic electron flow around photosystem II, export of excess NAD(P)H through the malate shuttle, and the Mehler reaction (the water-water cycle; e.g., Lavaud et al. 2003; Ruban et al. 2004; Scheibe 2004; Wilhelm et al. 2006; Fig. 12.15, see also Chaps. 6 and 7). These mechanisms help to dissipate excess electron energy that may result from light stress, where electron

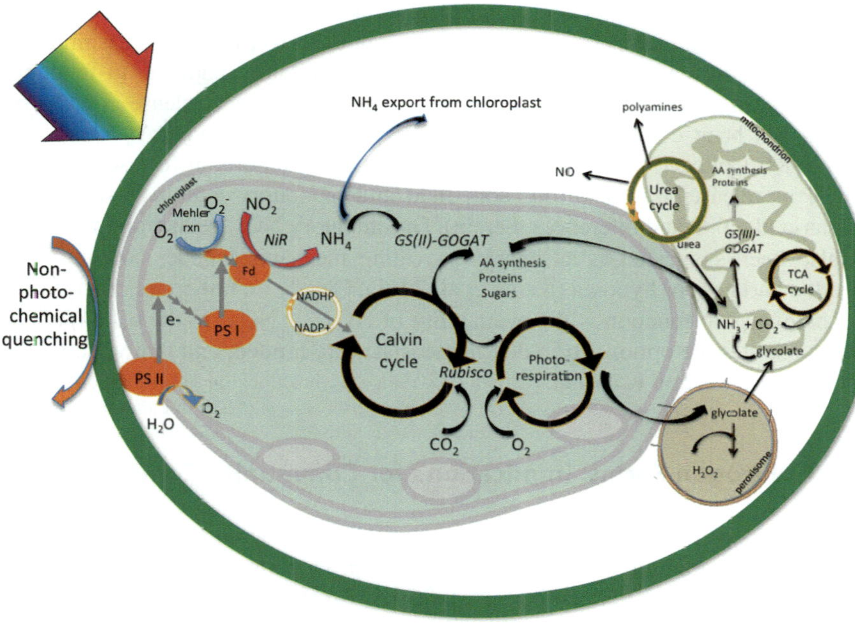

Fig. 12.15 Conceptual schematic illustrating, for a generic algal cell, the electron transport of photosynthesis, the coupling to the Calvin cycle and N assimilation pathways, and the various mechanisms for energy and excess reductant dissipation. The dissipatory mechanisms shown include nonphotochemical quenching, Mehler activity, dissimilatory NO_3^-/NO_2^- reduction to NH_4^+ and photorespiration. Also shown are the relationships between the Calvin cycle and N assimilation pathways, photorespiration, and the urea cycle in the chloroplast, peroxisome and the mitochondrion. Image reproduced from Glibert et al. (2016, Limnol Oceanogr 61: 165–197, doi: 10.1002/lno.10203) under a creative commons license

transport through the light reactions and associated thermochemical reactions in the thylakoids in photosynthesis is in excess of C assimilation capacity, i.e., when the light reactions or electron transport chain of photosynthesis go into "overdrive" (Kana et al. 1997; Chung et al. 2008). These mechanisms are expressed differentially in different phytoplankton groups. These mechanisms are also related to differential use of oxidized vs. reduced forms of N (reviewed by Glibert et al. 2016).

An important pathway regulating overall cellular energy balance in diatoms is the reduction of NO_3^- and NO_2^- via NR and NiR in a nonassimilatory mode that complements such reduction in N assimilation (e.g., Lomas and Glibert 1999a, b, Parker and Armbrust 2005; Kamp et al. 2011; Rosenwasser et al. 2014; Fig. 12.15). The reduction of NO_2^- to NH_4^+ in the chloroplast uses the reducing power of the ferredoxin system, and it can serve as a sink for excess reductant, derived from the splitting of water, that may develop when photochemistry exceeds assimilatory capacity (e.g., conditions of high light and cool temperatures). Such reactions can protect the chloroplast electron transport chain from over-reduction. Cool temperatures may enhance a condition of excess reductant because the biophysical light reactions of photosynthesis are relatively temperature-insensitive, but the biochemical reactions (e.g., Calvin Cycle reactions and nonphotochemical reactions in the thylakoid) are temperature-sensitive leading to slower rates of C assimilation than of the light reactions. In order for such a dissimilatory pathway to function, release of N in a more reduced state should be observed. In fact, release of NO_2^- by diatoms has been commonly observed during NO_3^- uptake (e.g., Anderson and Roels 1981; Collos 1982), and there are also numerous reports of release of NH_4^+, as well as release of DON from both field and laboratory cultures using NO_3^- (Lomas et al. 2000 and references therein). Dissimilatory NO_3^- reduction in diatoms has also been suggested as an energy-providing mechanism under conditions of extended darkness or hypoxia (Kamp et al. 2011). Clearly an important criterion for such pathways to function is the availability of NO_3^- or NO_2^- in the cell. Without these substrates the options for redox homeostasis are limited, and that via dissimilatory NO_3^- reduction is absent.

12.5.4 Nitrogen Forms: Implications for Community Composition

The different patterns of uptake of NO_3^- and NH_4^+ are embodied in the classic oceanographic paradigm of new and regenerated production (Dugdale and Goering 1967; Figs. 12.2 and 12.16; Glibert et al. 2016). Greater flows of organic material through the microbial loop generally occur when systems are more enriched with chemically reduced N forms, NH_4^+ and urea, and the resulting communities are often dominated by mixotrophic dinoflagellates or (pico)cyanobacteria as well as bacteria (Eppley and Peterson 1979; Legendre and Rassouzadegan 1995; Berg et al. 1997, 2003; LaRoche et al. 1997; Glibert 1998; see also Williams and Ducklow 2019). Shifts in N form and respective rates of uptake from NO_3^- to NH_4^+ have been shown in numerous systems to be associated with community shifts away from plankton communities dominated by diatoms to those dominated by flagellates,

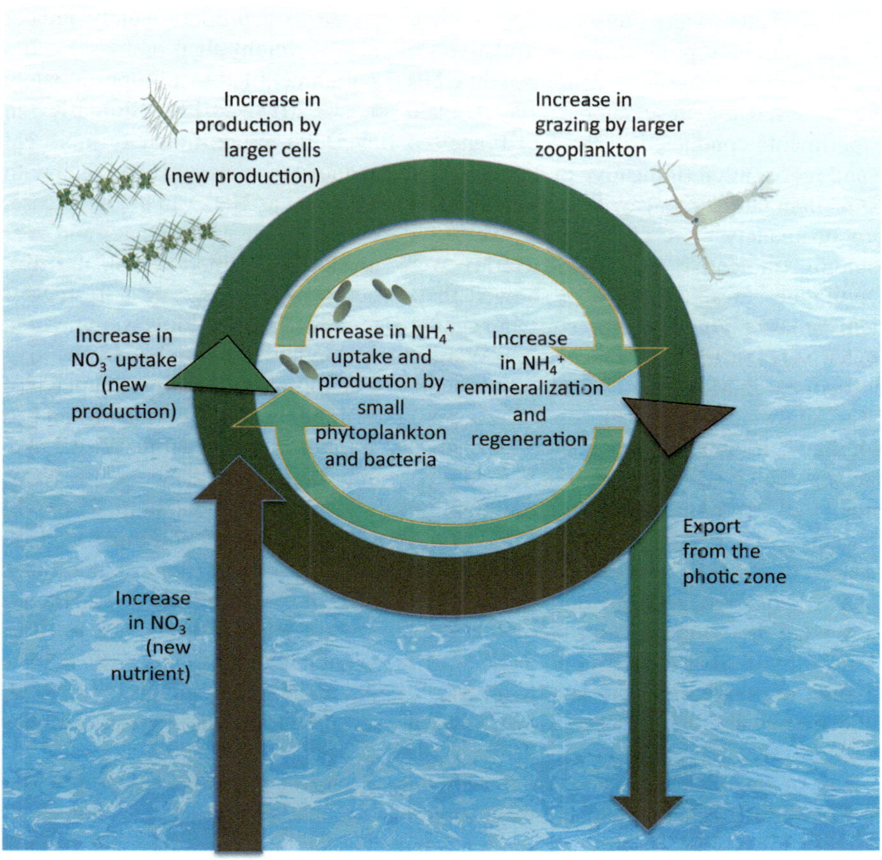

Fig. 12.16 Conceptual food web relationships resulting from changes in the ratios of NH_4^+/NO_3^-; regenerated vs. new production (sensu Dugdale and Goering 1967). In this conceptual framework, an increase in NO_3^- (via upwelling or other source) increases production by larger phytoplankton cells, typically diatoms, in turn supporting larger grazers. The recycling of NH_4^+ from excretion and remineralization pathways (inner cycle) supports growth of smaller cells, typically nondiatomaceous, as well as bacteria. Image reproduced from Glibert et al. (2016, Limnol Oceanogr 61: 165–197, doi: 10.1002/lno.10203) under a creative commons license

cyanobacteria, and bacteria, in turn, resulting in a shift in composition of higher food webs (e.g., Legendre and Rassouzadegan 1995; Glibert 1998; Berg et al. 2003; Heil et al. 2007).

Laboratory culture and large-scale nutrient manipulation experiments have provided strong evidence that not only do different functional groups predominantly rely on different forms and proportions of nutrients, but different communities develop when enriched with different forms of N even when provided at the same concentration. In mesocosm experiments conducted in hypereutrophic Wascana

Lake, Saskatchewan, Canada, NO_3^- enrichment led to a proportionately greater increase in chlorophyll *a* (chl *a*) (relative to total wet-weight algal biomass) and a greater initial response by diatoms, while NH_4^+ enrichment led to a proportionately greater increase in cyanobacteria (Donald et al. 2011, 2013). Similarly, in experiments conducted in the San Francisco Bay Delta, proportionately more chl *a* and fucoxanthin (indicative of diatoms) were produced in enclosures enriched with NO_3^- than in treatments with the same total N enrichment as NH_4^+. In the latter case, proportionately more chl *b* (indicative of chlorophytes, i.e., green algae) and zeaxanthin (indicative of cyanobacteria) were produced (Glibert et al. 2014b). Domingues et al. (2011) also showed that enrichment by NH_4^+ in a freshwater tidal estuary favored chlorophytes and cyanobacteria, whereas diatoms were favored under NO_3^- enrichment. Fawcett and Ward (2011) conducted a simulated upwelling experiment in Monterey Bay, CA, and found that an initial acceleration of uptake of NO_3^- allowed diatoms, those of the largest size fraction, to grow quickly and to dominate. In similarly conducted experiments, enriching waters from West Lake, Florida, with NH_4^+ resulted in increased picocyanobacterial abundance, while NO_3^- plus PO_4^{3-} enrichment resulted in diatom proliferation (Shangguan et al. 2017). In the equatorial Pacific, where neither NO_3^- nor $Si(OH)_4$ were limiting based on measured substrate concentrations (Barber et al. 1996), diatoms contributed more than 90% of the new production as measured by rates of $^{15}NO_3^-$ uptake, even though they were the minority contributor to the total phytoplankton community (<10%) (McCarthy et al. 1996). The abundance of the picoplankter *Prochlorococcus* was ubiquitous, negatively correlated with NO_3^-, being maintained through recycled N, reflecting what has since been shown to be its lack of NO_3^- uptake capacity (Moore et al. 2002).

Use of NanoSIMS has shed even more light on the dichotomous use of different N forms by different members of a phytoplankton assemblage. Berthelot et al. (2019) showed that photosynthetic picocyanobacterial and picoeukaryotes differed in their N use (Fig. 12.17), while Olofsson et al. (2019) were able to visualize NO_3^- and NH_4^+ fluxes to diatoms and dinoflagellates at a single cell level in a mixed field community (Fig. 12.18).

There is thus a dichotomy in use of oxidized vs. reduced N substrates and in sensitivity or tolerance to excess NH_4^+ by different phytoplankton functional groups as is the case in terrestrial plants (Britto and Kronzucker 2013; Podgórska and Szal 2015 and references therein). Phytoplankton community structure is mirrored by a suite of interconnected ratios of transporters, enzymes, regulator proteins, and synthesis and dissipatory pathways inside the cell. The extent to which various constituents and pathways are expressed by different species may impart advantages depending on the availability of substrate, from limiting to supersaturating in conjunction with other regulating environmental factors, especially temperature and light. In turn, these different phytoplankton groups may ultimately support different food webs (e.g., Eppley and Peterson 1979; Legendre and Rassouzadegan 1995; Glibert 1998).

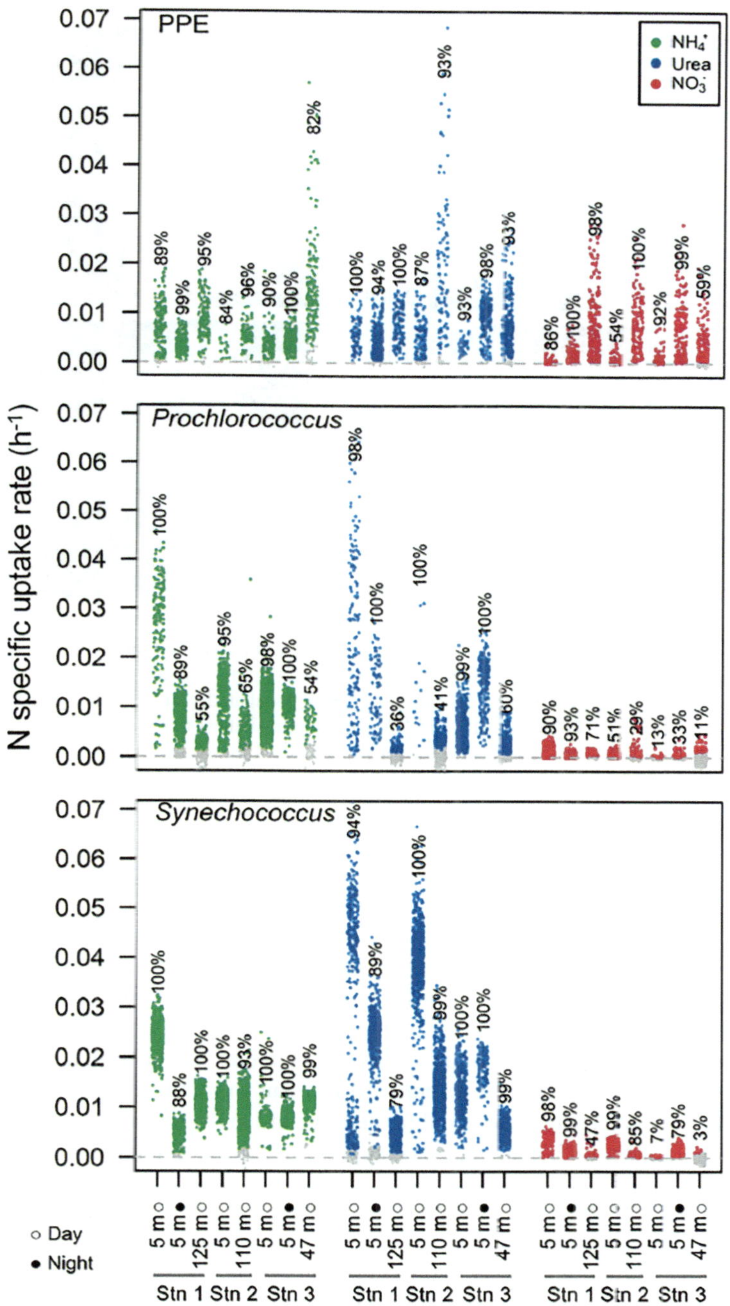

Fig. 12.17 NH_4^+, urea, and NO_3^--specific uptake rates (h^{-1}) of photosynthetic picoeukaryotes (PPE), *Prochlorococcus*, and *Synechococcus* at each sampled station and depth. Each dot represents

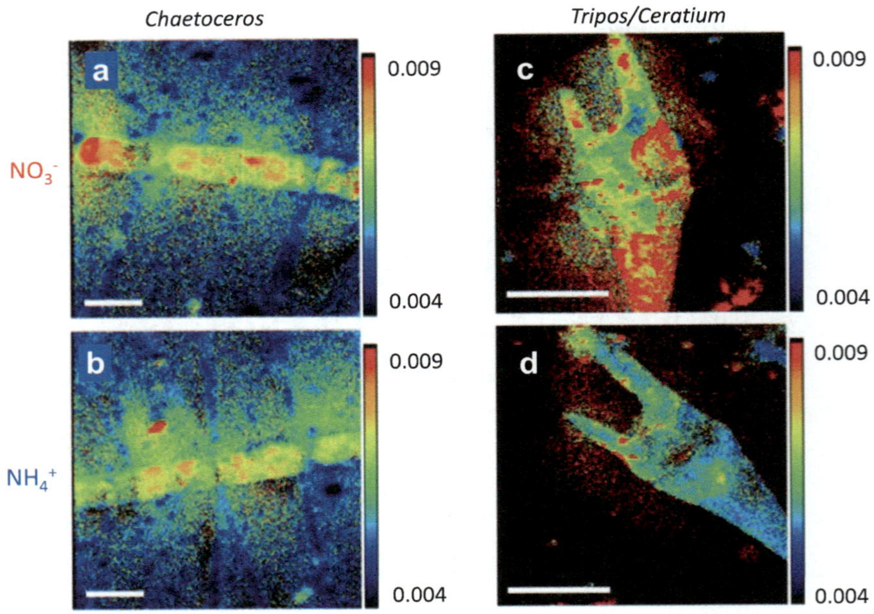

Fig. 12.18 SIMS images of stable isotopes measured within phytoplankton. Panels **a, b**: $^{15}N/^{14}N$ (dimensionless scale) measured within *Chaetoceros* spp. >8 μm; scale bar 20 μm, and Panels **c, d**: and *Tripos/Ceratium* spp. scale bar 50 μm. Cells were incubated with ^{13}C bicarbonate and ^{15}N nitrate (Panels **a, c**) or ^{13}C-bicarbonate and ^{15}N ammonium (Panels **b, d**). Image reproduced from Olofsson et al. (2019, Sci Rep 9: 1424, doi: 10.1038/s41598-018-38059-4) under a creative commons license

12.6 Summary

Our understanding of the nitrogen metabolism of phytoplankton has been rapidly evolving. While classic studies laid the foundation for many critical pathways and processes, it is with application of isotopes and new approaches that our knowledge has greatly expanded. While emphasis in the literature has been placed on understanding how cells upregulate responses to obtain the requisite nutrients in least supply, there is emerging understanding of how nutrient forms interact and how cells can downregulate to balance excess supply. It is the fundamental differences

Fig. 12.17 (continued) an analyzed cell. Colored and grey dots denote cells with detected and undetected activities, respectively. The percentages are the proportions of detected active cells in each assay. Image reproduced from Berthelot et al. (2019, ISME J 13: 651–662, doi: 10.1038/s41396-018-0285-8) under a creative commons license between functional groups or species—in active transport mechanisms, cell nitrogen assimilation genes, and pathways that provide the mechanisms for ecological competition and for the underpinnings of the concept of "new" and "regenerated" production (Dugdale and Goering 1967). These relationships not only hold at the limiting end of the substrate spectrum but also at the substrate saturated to supersaturated end. There are limits to both up- and downregulation and these vary by taxa—and by N form. It thus makes a difference which form of N is supplied, whether in proportions that are limiting or in excess. Differences in productivity and ultimately species composition should result when the form of N changes; NH_4^+ at environmentally relevant concentrations, for increasingly N-enriched systems (see Chap. 22), has profound effects on metabolism and growth, and ultimately species composition. It is no longer sufficient to characterize only the limiting nutrient if one purports to understand biotic responses. Next, P and Si metabolism are considered.

References

Aksnes DL, Egge JK (1991) A theoretical model for nutrient uptake in phytoplankton. Mar Ecol Prog Ser 70:65–72

Alipanah L, Rohloff J, Winge P, Bones AM, Brembe T (2015) Whole-cell response to nitrogen deprivation in the diatom *Phaeodactylum tricornutum*. J Exp Bot 66:6281–6296. https://doi.org/10.1093/jxb/erv340

Allen AE, Ward BB, Song B (2005) Characterization of diatom (Bacillariophyceae) genes and their detection in marine phytoplankton. J Phycol 41:95–104. https://doi.org/10.1111/j.1529-8817.2005.04090.x

Allen AE, Vardi A, Bowler C (2006) An ecological and evolutionary context for integrated nitrogen metabolism and related signaling pathways in marine diatoms. Curr Opin Plant Biol 9:264–273

Allen AE, Dupont CL, Oborník M, Horák A, Nunes-Nesi A, McCrow JP, Zheng H et al (2011) Evolution and metabolic significance of the urea cycle in photosynthetic diatoms. Nature 473: 203–207. https://doi.org/10.1038/nature10074

Anderson SM, Roels OA (1981) Effects of light intensity on nitrate and nitrite uptake and excretion by *Chaetoceros curvisetus*. Mar Biol 62:257–261

Antia NJ, Harrison PJ, Oliveira L (1991) The role of dissolved organic nitrogen in phytoplankton nutrition, cell biology and ecology. Phycologia 30:1–89

Armbrust EV, Berges JA, Bowler C, Green BR, Martinez D, Putnam NH, Zhou S et al (2004) The genome of the diatom *Thalassiosira pseudonana*: ecology, evolution and metabolism. Science 306:5693. https://doi.org/10.1126/science.1101156

Barber RT, Sanderson MP, Lindley ST, Chai F, Newton J, Trees CC, Foley DG et al (1996) Primary productivity and its regulation in the equatorial Pacific during and following the 1991-1992 El Niño. Deep Sea Res II 43:933–969. https://doi.org/10.1016/0967-0645(96)00035-5

Beckers G, Bendt AK, Kramer R, Burkovski A (2004) Molecular identification of the urea uptake system and transcriptional analysis of urea transporter- and urease-encoding genes in *Corynebacterium glutamicum*. J Bacteriol 186:7645–7652

Below FE, Gentry LE (1987) Effect of mixed N nutrition on nutrient accumulation, partitioning, and productivity of corn. J Fert Issues 4:79–85

Bender SJ, Parker MS, Armbrust EV (2012) Coupled effects of light and nitrogen source on the urea cycle and nitrogen metabolism over a diel cycle in the marine diatom *Thalassiosira pseudonana*. Protist 163:232–251. https://doi.org/10.1016/j.protis.2011.07.008

Berg GM, Glibert PM, Lomas MW, Burford M (1997) Organic nitrogen uptake and growth by the Chrysophyte *Aureococcus anophagefferens* during a brown tide event. Mar Biol 129:377–387

Berg GM, Balode M, Purina I, Bekere S, Béchemin C, Maestrini SY (2003) Plankton community composition in relation to availability and uptake of oxidized and reduced nitrogen. Aquat Microb Ecol 30:263–274

Berg GM, Shrager J, Glockner G, Arrigo KR, Grossman AR (2008) Understanding nitrogen limitation in *Aureococcus anophagefferens* (Pelagophyceae) through cDNA and qRT-PCR analysis. J Phycol 44:1235–1249

Berges JA (1997) Minireview: Algal nitrate reductases. Eur J Phycol 32:3–8

Berges JA, Mulholland MR (2008) Enzymes and nitrogen cycling. In: Capone DG, Bronk DA, Mulholland MR, Carpenter EJ (eds) Nitrogen in the marine environment. Elsevier, Burlington, MA, pp 1385–1445

Berges JA, Cochlan WP, Harrison PJ (1995) Laboratory and field responses of algal nitrate reductase to diel periodicity in irradiance, nitrate exhaustion, and the presence of ammonium. Mar Ecol Prog Ser 124:259–269

Berges JA, Varela D, Harrison P (2002) Effects of temperature on growth rate, cell composition and nitrogen metabolism in the marine diatom *Thalassiosira pseudonana* (Bacillariophyceae). Mar Ecol Prog Ser 225:139–146

Berman T, Sherr BF, Sherr E, Wynne D, McCarthy JJ (1984) The characteristics of ammonium and nitrate uptake by phytoplankton in Lake Kinneret. Limnol Oceanogr 29:287–297. https://doi.org/10.4319/lo.1984.29.0287

Berthelot H, Duhamel S, L'Helguen S, Maguer J-F, Wang S, Cetinić I, Cassar N (2019) NanoSIMS single cell analyses reveal the contrasting nitrogen sources for small phytoplankton. ISME J 13: 651–662. https://doi.org/10.1038/s41396-018-0285-8

Bonachela J, Allison SD, Martiny AC, Levin SA (2013) A model for variable phytoplankton stoichiometry based on cell protein regulation. Biogeosciences 10:4341–4356. https://doi.org/10.5194/bg-10-4341-2013

Boyd CM, Gradmann D (1999) Electrophysiology of the marine diatom *Coscinodiscus waislesii* III. Uptake of nitrate and ammonium. J Exp Bot 50:461–467

Britto DT, Kronzucker HJ (2002) NH_4^+ toxicity in higher plants: a critical review. J Plant Physiol 159:567–584

Britto DT, Kronzucker HJ (2013) Ecological significance and complexity of N-source preference in plants. Ann Bot 112:957–963

Bronk DA, Glibert PM, Malone TC, Banahan S, Sahlsten E (1998) Inorganic and organic nitrogen cycling in Chesapeake Bay: autotrophic versus heterotrophic processes and relationships to carbon flux. Aquat Microb Ecol 15:177–189

Brown KL, Twing KI, Robertson DL (2009) Unraveling the regulation of nitrogen assimilation in the marine diatom *Thalassiosira pseudonana* (Bacillariophyceae): Diurnal variations in transcript levels for five genes involved in nitrogen assimilation. J Phycol 45:413–426

Campbell WH (1999) Nitrate reductase structure, function and regulation: Bridging the gap between biochemistry and physiology. Annu Rev Physiol Plant Mol Biol 50:277–303

Casciotti KL, Buchwald C (2012) Insights on the marine microbial nitrogen cycle from isotopic approaches to nitrification. Front Microbiol 3:356. https://doi.org/10.3389/fmicb.2012.00356

Chróst RJ (1991) Environmental control of the synthesis and activity of aquatic microbial ectoenzymes. In: Chróst RJ (ed) Microbial enzymes in aquatic environments. Springer, Berlin, pp 29–59

Chung C-C, Huang S-PL, Chang J (2008) Nitric oxide as a signaling factor to up-regulate the death-specific protein in a marine diatom, *Skeletonema costatum*, during blockage of electron flow in photosynthesis. App Environ Microbiol 74:6521–7527

Cimbleris ACP, Cáceres O (1991) Kinetics of urea uptake by *Melosira italica* (Ehr.) Kütz at different luminosity conditions. Hydrobiologia 220:211–216

Clark DR, Flynn KJ, Owens NJP (2002) The large capacity for dark nitrate-assimilation in diatoms may overcome nitrate limitation for growth. J Phycol 155:101–108

Clarkson NM, Luttge U (1991) Mineral nutrition: inducible and repressible nutrient transport systems. Prog Bot 52:61–83

Clayton JR Jr, Ahmed SI (1986) Detection of glutamine synthase (GOGAT) activity in phytoplankton: evaluation of cofactors and assay optimization. Mar Ecol Prog Ser 32:115–122

Cochlan WP, Harrison PJ (1991) Inhibition of nitrate uptake by ammonium and urea in the eukaryotic picoflagellate *Micromonas pusilla* (Butcher) Manton et Parke. J Exp Mar Biol Ecol 153:143–152

Cochlan WP, Harrison PJ, Denman KL (1991) Diel periodicity of nitrogen uptake by marine phytoplankton in nitrate-rich environments. Limnol Oceanogr 36:1689–1700. https://doi.org/10.4319/lo.1991.36.8.1689

Collos Y (1982) Transient situations in nitrate assimilation by marine diatoms. 2. Changes in nitrate and nitrite following a nitrate perturbation. Limnol Oceanogr 27:528–535

Coruzzi G, Bush DR (2001) Nitrogen and carbon nutrient and metabolite signaling in plants. Plant Physiol 125:61–64

Crawford NN, Glass ADN (1998) Molecular and physiological aspects of nitrate uptake in plants. Trends Plant Sci 3:389–395

Dagenais-Bellefeuille S, Morse D (2013) Putting the N in dinoflagellates. Front Microbiol 4:369. https://doi.org/10.3389/micb.2013.00369

Daniel-Vedele F, Filleur S, Caboche M (1998) Nitrate transport: a key step in nitrate assimilation. Curr Opin Plant Biol 1:235–239

Domingues RB, Barbosa AB, Sommer U, Galvão HM (2011) Ammonium, nitrate and phytoplankton interactions in a freshwater tidal estuarine zone: potential effects of cultural eutrophication. Aquat Sci 73:331–343

Donald DB, Bogard MJ, Finlay K, Leavitt PR (2011) Comparative effects of urea, ammonium and nitrate on phytoplankton dominance, community composition and toxicity in a hypereutrophic freshwater. Limnol Oceanogr 56:2161–2175

Donald DB, Bogard MJ, Finlay K, Bunting L, Leavitt PR (2013) Phytoplankton-specific response to enrichment of phosphorus-rich surface waters with ammonium, nitrate, and urea. PLoS One 8 e53277. https://doi.org/10.1371/journal.pone.0053277

Dortch Q (1982) Effect of growth conditions on accumulation of internal nitrate, ammonium, amino acids and protein in three marine diatoms. J Exp Mar Biol Ecol 61:243–264

Dortch Q (1990) The interaction between ammonium and nitrate uptake in phytoplankton. Mar Ecol Prog Ser 61:183–201

Dortch Q, Conway HL (1984) Interactions between nitrate and ammonium uptake: variation with growth rate, nitrogen source and species. Mar Biol 79:151–164

Dortch Q, Thompson PA, Harrison PJ (1991) Short-term interaction between nitrate and ammonium uptake in *Thalassiosira pseudonana*: effect of preconditioning nitrogen source and growth rate. Mar Biol 110:183–193

Dugdale RC (1967) Nutrient limitation in the sea: dynamics, identification and significance. Limnol Oceanogr 12:685–695

Dugdale RC (2018) A biogeochemical oceanographer at sea: My life with nitrogen and a nod to silica. Annu Rev Mar Sci 10:1–18. https://doi.org/10.1146/annurev-marine-121916-063655

Dugdale RC, Goering JJ (1967) Uptake of new and regenerated forms of nitrogen in primary productivity. Limnol Oceanogr 12:196–206

Dugdale RC, Dugdale V, Neess J, Goering JJ (1959) Nitrogen fixation in lakes. Science 130:859–860

Dugdale RC, Jones BH, MacIsaac JJ (1981) Adaptation of nutrient assimilation. In Platt T (ed) Physiological bases of phytoplankton ecology. Can Bull Fish Aquat Sci 210:234–250

Dugdale RC, Wilkerson FP, Hogue VE, Marchi A (2007) The role of ammonium and nitrate in spring bloom development in San Francisco Bay. Est Coast Shelf Sci 73:17–29

Dugdale RC, Wilkerson FP, Parker AE, Marchi A, Taberski K (2012) River flow and ammonium discharge determine spring phytoplankton blooms in an urbanized estuary. Est Coast Shelf Sci 115:187–199. https://doi.org/10.1016/j.ecss.2012.08.025

Dyhrman ST, Anderson DM (2003) Urease activity in cultures and field populations of the toxic dinoflagellate *Alexandrium*. Limnol Oceanogr 48:647–655

Eppley RW, Coatsworth JL (1968) Uptake of nitrate and nitrite by *Ditylum brightwellii* – kinetics and mechanisms. J Phycol 4:151–156

Eppley RW, Peterson BJ (1979) Particulate organic flux and planktonic new production in the deep ocean. Nature 282:677–680

Eppley RW, Coatsworth JL, Solórzano L (1969) Studies of nitrate reductase in marine phytoplankton. Limnol Oceanogr 14:194–205

Eppley RW, Sharp JH, Renger EH, Perry MJ, Harrison WG (1977) Nitrogen assimilation by phytoplankton and other microorganisms in the surface waters of the central North Pacific Ocean. Mar Biol 39:111–120

Fan C, Glibert PM (2005) Effects of light on carbon and nitrogen uptake during a *Prorocentrum minimum* bloom. Harmful Algae 4:629–642

Fan C, Glibert PM, Alexander J, Lomas MW (2003a) Characterization of urease activity in three marine phytoplankton species, *Aureococcus anophagefferens, Prorocentrum minimum*, and *Thalassiosira weissflogii*. Mar Biol 142:949–958

Fan CL, Glibert PM, Burkholder JM (2003b) Characterization of the affinity for nitrogen, uptake kinetics, and environmental relationships for *Prorocentrum minimum* in natural blooms and laboratory cultures. Harmful Algae 2:283–299

Fawcett SE, Ward BB (2011) Phytoplankton succession and nitrogen utilization during the development of an upwelling bloom. Mar Ecol Prog Ser 428:13–31. https://doi.org/10.3354/meps09070

Fischer P, Klein U (1988) Localization of nitrogen-assimilating enzymes in the chloroplast of *Chlamydomonas reinhardtii*. Plant Physiol 88:947–952

Flores E, Frías JE, Rubio LM, Herrero A (2005) Photosynthetic nitrate assimilation in cyanobacteria. Photosynth Res 83:117–133. https://doi.org/10.1007/s11120-004-5830-9

Flynn KJ, Jones KJ, Raine R, Richard J, Flynn K (1994) Use of intracellular amino acids as an indicator of the physiological status of natural dinoflagellate populations. Mar Ecol Prog Ser 103:175–186

Flynn KJ, Fasham MJR, Hipkin CR (1997) Modelling the interactions between ammonium and nitrate uptake in marine phytoplankton. Philos Trans Roy Soc B 352:1625–1645

Furhman JA, Horrigan SG, Capone DG (1988) Use of ^{13}N as tracer for bacterial and algal uptake of ammonium from sea-water. Mar Ecol Prog Ser 45:271–278. https://doi.org/10.3354/meps045271

Galván A, Fernandez E (2001) Eukaryotic nitrate and nitrite transporters. Cell Mol Life Sci 58:225–233

Galván A, Rexaach J, Marischal V, Fernandez E (2002) Nitrite transport to the chloroplast in *Clamydomonas reinhardtii*: molecular evidence for a regulated process. J Exp Bot 53:845–853

Gao Y, Smith GJ, Alberte RS (1983) Nitrate reductase from the marine diatom *Skeletonema costatum.* Biochemical and immunological characterization. Plant Physiol 103:1437–1445

Gerendás J, Zhu Z, Bendixon R, Ratcliffe RG, Sattelmacher B (1997) Physiological and biochemical processes related to ammonium toxicity in higher plants. Z Pflanzener Bodenkd 160:239–251

Glibert PM (1998) Interactions of top-down and bottom-up control in planktonic nitrogen cycling. Hydrobiologia 363:1–12

Glibert PM, Burkholder JM, Kana TM (2012) Recent advances in understanding of relationships between nutrient availability, forms and stoichiometry and the biogeographical distribution, ecophysiology, and food web effects of pelagic and benthic *Prorocentrum* spp. Harmful Algae 14:231–259

Glibert PM, Kana TM, Brown K (2013) From limitation to excess: consequences of substrate excess and stoichiometry for phytoplankton physiology, trophodynamics and biogeochemistry, and implications for modeling. J Mar Sys 125:14–28

Glibert PM, Dugdale RC, Wilkerson F, Parker AE, Alexander J, Antell E et al (2014a) Major- but rare-spring blooms in 2014 in San Francisco Bay Delta, California, a result of the long-term drought, increased residence time, and altered nutrient loads and forms. J Exp Mar Biol Ecol 460:8–18. https://doi.org/10.1016/j.jembe.2014.06.001

Glibert PM, Wilkerson F, Dugdale RC, Parker AE, Alexander JA, Blaser S, Murasko S (2014b) Microbial communities from San Francisco Bay Delta respond differently to oxidized and reduced nitrogen substrates – even under conditions that would otherwise suggest nitrogen sufficiency. Front Mar Sci 1:article 17. https://doi.org/10.3389/fmars.2014.00017

Glibert PM, Wilkerson FP, Dugdale RC, Raven JA, Dupont C, Leavitt PR, Parker AE et al (2016) Pluses and minuses of ammonium and nitrate uptake and assimilation by phytoplankton and implications for productivity and community composition, with emphasis on nitrogen-enriched conditions. Limnol Oceanogr 61:165–197. https://doi.org/10 1002/lno.10203

Glibert PM, Heil CA, Wilkerson FP, Dugdale RW (2018) Nutrients and harmful algal blooms: dynamic kinetics and flexible nutrition. In: Glibert PM, Berdalet E, Burford MA, Pitcher GC, Zhou MJ (eds) Global ecology and oceanography of harmful algal blooms. Springer, Cham, pp 93–112. https://doi.org/10.1007/978-3-319-70069-4_6

Glibert PM, Middelburg JJ, McClelland JW, Vander Zander MJ (2019) Stable isotope tracers: enriching our perspectives and questions on sources, fates, rates and pathways of major elements on aquatic systems. Limnol Oceanogr 64:950–981. https://doi.org/10.1002/lno.11087

Goldman JC (1993) Potential role of large oceanic diatoms in new primary production. Deep Sea Res 40:159–168

Goyal SS, Lorenz OA, Huffaker RC (1982) Inhibitory effects of ammoniacal nitrogen on growth of radish plants. I. Characterization of toxic effects of NH_4^+on growth and its alleviation by NO_3^-. J Am Soc Hortic Sci 107:125–129

Harrison PJ, Parslow JS, Conway HL (1989) Determination of nutrient uptake kinetic parameters: a comparison of methods. Mar Ecol Prog Ser 52:301–312

Harrison WG, Harris LR, Irwin BD (1996) The kinetics of nitrogen utilization in the oceanic mixed layer: nitrate and ammonium interactions at nanomolar concentrations. Limnol Oceanogr 41: 13–35

Harvey HW (1953) Synthesis of organic nitrogen and chlorophyll by *Nitzschia closterium*. J Mar Biol Assoc UK 31:477–487

Hausinger RP (2004) Metabolic versatility of prokaryotes for urea decomposition. J Bacteriol 186: 2520–2522

Heil CA, Revilla M, Glibert PM, Murasko S (2007) Nutrient quality drives phytoplankton community composition on the West Florida Shelf. Limnol Oceanogr 52:1067–1078

Herndon J, Cochlan WP (2007) Nitrogen utilization by the raphidophyte *Heterosigma akashiwo*: Growth and uptake kinetics in laboratory cultures. Harmful Algae 6:260–270

Howitt SM, Udvardi MK (2000) Structure, function and regulation of ammonium transporters in plants. Biochim Biophys Acta 1465:152–170

Ivlev VS (1961) Experimental ecology of the feeding of fishes. Yale University Press, New Haven, CT. 302 pp

Jones GJ, Morel FMM (1988) Plasmalemma redox activity in the diatom *Thalassiosira*. Plant Physiol 87:143–147

Jung H (2002) The sodium/substrate symporter family: structural and functional features. FEBS Lett 529:73–77

Kamp A, de Beer D, Nitsch JL, Lavik G, Stief P (2011) Diatoms respire nitrate to survive dark and anoxic conditions. Proc Natl Acad Sci USA 108:5649–5654. https://doi.org/10.1073/pnas.1015744108

Kana TM, Glibert PM (2016) On saturating response curves from the dual perspectives of photosynthesis and nitrogen acquisition. In: Glibert PM, Kana TM (eds) Aquatic microbial ecology and biogeochemistry: A dual perspective. Springer, Cham, pp 93–104

Kana TM, Geider RJ, Critchley C (1997) Photosynthetic pigment regulation in microalgae by multiple environmental factors: A dynamic balance hypothesis. New Phytol 137:629–638. https://doi.org/10.1046/j.1469-8137.1997.00857.x

Kobayashi M, Takatani N, Tanigawa M, Omata T (2005) Posttranslational regulation by nitrite of the nitrate assimilation operon in the cyanobacterium *Synechococcus* sp. strain PCC 7942 and *Plectonema boryanum*. J Bacteriol 178:5822–5825

Kristiansen S (1983) The temperature optimum of the nitrate reductase assay for marine phytoplankton. Limnol Oceanogr 28:776–780

Kronzucker HJ, Glass ADM, Siddiqi MY (1999) Inhibition of nitrate uptake by ammonium in barley. Analysis of component fluxes. Plant Physiol 120:283–292

Kudela RM, Cochlan WP (2000) Nitrogen and carbon uptake kinetics and the influence of irradiance for a red tide bloom off southern California. Aquat Microb Ecol 21:31–47

Kudela RM, Dugdale RC (2000) Nutrient regulation of phytoplankton productivity in Monterey Bay, California. Deep Sea Res 47:1023–1053

LaRoche J, Nuzzi R, Waters R et al (1997) Brown tide blooms in Long Island's coastal waters linked to interannual variability on groundwater flow. Glob Change Biol 3:101–114

Lavaud J, Bousseau B, Etienne A-L (2003) Enrichment of the light-harvesting complex in diadinoxanthin and implications for the non-photochemical fluorescence quenching in diatoms. Biochemistry 442:5802–5808

Legendre L, Rassouzadegan F (1995) Plankton and nutrient dynamics in marine waters. Ophelia 41: 153–172

L'Helguen S, Maguer J-F, Caradec J (2008) Inhibition kinetics of nitrate uptake by ammonium in size-fractionated oceanic phytoplankton communities: implications for new production and *f*-ratio estimates. J Plankton Res 10:1179–1188

Lindell D, Post AF (2001) Ecological aspects of *ntcA* gene expression and its use as an indicator of the nitrogen status of marine *Synechococcus* spp. Appl Environ Microbiol 67:3340–3349

Lomas MW, Glibert PM (1999a) Temperature regulation of nitrate uptake: a novel hypothesis about nitrate uptake and reduction in cool-water diatoms. Limnol Oceanogr 44:556–572

Lomas MW, Glibert PM (1999b) Interactions between NH_4^+ and NO_3^- uptake and assimilation: comparison of diatoms and dinoflagellates at several growth temperatures. Mar Biol 133:541–551

Lomas MW, Glibert PM (2000) Comparisons of nitrate uptake, storage and reduction in marine diatoms and flagellates. J Phycol 36:903–913

Lomas MW, Rumbley CJ, Glibert PM (2000) Ammonium release by nitrogen sufficient diatoms in response to rapid increases in irradiance. J Plankton Res 22:2351–2366

Lopez-Ruiz A, Verbelen JP, Roldan JM, Diez J (1985) Nitrate reductase of green algae is localized in the pyrenoid. Plant Physiol 79:1006–1010

Ludwig CA (1938) The availability of different forms of nitrogen to a green alga (*Chlorella*). Am J Bot 25:448–458

Lund BA (1987) Mutual interference of ammonium, nitrate and urea on uptake of ^{15}N sources by the marine diatom *Skeletonema costatum* (Grev.) Cleve. J Exp Mar Biol Ecol 113:167–180

MacIsaac JJ, Dugdale RC (1969) The kinetics of nitrate and ammonia uptake by natural populations of marine phytoplankton. Deep Sea Res 16:45–57

MacIsaac JJ, Dugdale RC (1972) Interactions of light and inorganic nitrogen in controlling nitrogen uptake in the sea. Deep Sea Res 19:209–232

Maestrini SY, Robert JM, Truquet I (1982) Simultaneous uptake of ammonium and nitrate by oyster pond algae. Mar Biol Lett 3:143–153

Maguer J-F, L'Helguen S, Madec C, Labry C, Le Corre P (2007) Nitrogen uptake and assimilation kinetics in *Alexandrium minutum* (Dinophyceae): effect of N-limited growth rate on nitrate and ammonium interactions. J Phycol 43:295–303

Martinez J, Azam F (1993) Aminopeptidase activity in marine chroococcoid cyanobacteria. Appl Environ Microbiol 59:3701–3707

McCarthy JJ, Goldman JC (1979) Nitrogenous nutrition of marine phytoplankton in nutrient-depleted waters. Science 203:670–672

McCarthy JJ, Taylor WR, Taft JL (1975) The dynamics of nitrogen and phosphorus cycling in the open waters of the Chesapeake Bay. In: Church TM (ed) Marine chemistry in the coastal environment. American Chemical Society, Washington, DC, pp 664–691

McCarthy JJ, Taylor WR, Taft JL (1977) Nitrogenous nutrition of the plankton in the Chesapeake Bay. I. Nutrient availability and phytoplankton preferences. Limnol Oceanogr 22:996–1011

McCarthy JJ, Garside C, Nevins JL, Barber RT (1996) New production along 140°W in the equatorial Pacific during and following the 1992 El Niño event. Deep Sea Res 43:1065–1093

Melo-Oliveira R, Oliveira IC, Coruzzi GM (1996) *Arabidopsis* mutant analysis and gene regulation define a nonredundant role for glutamate dehydrogenase in nitrogen assimilation. Proc Natl Acad Sci USA 93:4718–4723. https://doi.org/10.1073/pnas.93.10.4718

Milligan AJ, Harrison PJ (2000) Effects of non-steady-state iron limitation on nitrogen assimilatory enzymes in the marine diatom *Thalassiosira weissflogii* (Bacillariophyceae). J Phycol 36:78–86. https://doi.org/10.1046/j.1529-8817.2000.99013x

Mobley HLT, Island MD, Hausinger RP (1995) Molecular biology of microbial ureases. Microbiol Rev 59:451–480

Moore LR, Post AF, Rocap G, Chisholm SW (2002) Utilization of different nitrogen sources by the marine cyanobacteria *Prochlorochoccus* and *Synechococcus*. Limnol Oceanogr 47:989–996

Morando M, Capone DG (2018) Direct utilization of organic nitrogen by phytoplankton and its role in nitrogen cycling within the Southern California Bight. Front Microbiol 9:2118. https://doi.org/10.3389/fmicb.2018.02118

Morozkina EV, Zvyagilskaya RA (2007) Nitrate reductases: Structure, functions and effect of stress factors. Biochemistry (Moscow) 72:1151–1160

Morris I, Syrett PJ (1963) The development of nitrate reductase in *Chlorella* and its repression by ammonium. Archiv Fur Mikrobiol 47:32–41

Muro-Pastor MI, Reyes JC, Florencio FJ (2001) Cyanobacteria perceive nitrogen status by sensing intracellular 2-oxoglutarate levels. J Biol Chem 276:38320–38328

Muro-Pastor MI, Reyes JC, Florencio FJ (2005) Ammonium assimilation in cyanobacteria. Photosynth Res 83:135–150

Navarro MT, Prieto R, Fernandez E, Galván A (1996) Constitutive expression of nitrate reductase changes the regulation of nitrate and nitrite transporters in *Clamydomonas reinhardii*. Plant J 9: 819–827

Neess J, Dugdale RC, Dugdale VA, Goering JJ (1962) Nitrogen metabolism in lakes. I. Measurement of nitrogen fixation with ^{15}N. Limnol Oceanogr 7:163–169

Neufeld JD, Vohra J, Dumont MG, Lueders T, Manefield M, Friedrich MW, Murrell JC (2007) DNA stable isotope probing. Nat Protoc 2:860–866

Olofsson M, Robertson EK, Edler L, Arneborg L, Whitehouse MJ, Ploug H (2019) Nitrate and ammonium fluxes to diatoms and dinoflagellates at a single cell level in mixed field communities in the sea. Sci Rep 9:1424. https://doi.org/10.1038/s41598-018-38059-4

Packard T, Blasco D, MacIsaac JJ, Dugdale RC (1971) Variations of nitrate reductase activity in marine phytoplankton. Inv Pesq 35:209–219

Pageau K, Reisdorf-Cren M, Morot-Gaudry J-F, Masclaux-Daubresse C (2006) The two senescence-related markers *GS1* (cytosolic glutamine synthetase) and *GDH* (glutamate dehydrogenase), involved in nitrogen mobilisation are differentially regulated during pathogen attack, by stress hormones and reactive oxygen species in *Nicotiana tabacum* L. J Exp Bot 57:547–557

Pahlow M, Oschlies A (2013) Optimal allocation backs Droop's cell-quota model. Mar Ecol Prog Ser 473:1–5

Palenik B, Morel FMM (1990) Comparison of cell-surface L-amino acid oxidases from several marine phytoplankton. Mar Ecol Prog Ser 59:195–201

Pantoja S, Lee C (1994) Cell-surface oxidation of amino acids in sea water. Limnol Oceangr 39: 1718–1725

Parker AE, Dugdale RC, Wilkerson FP (2012) Elevated ammonium concentrations from wastewater discharge depress primary productivity in the Sacramento River and the Northern San Francisco Estuary. Mar Pollut Bull 64:574–586

Parker MS, Armbrust EV (2005) Synergistic effects of light, temperature and nitrogen source on transcription of genes for carbon and nitrogen metabolism in the centric diatom *Thalassiosira pseudonana* (Bacillariophyceae). J Phycol 41:1142–1153

Podgórska A, Szal B (2015) The role of reactive oxygen species under ammonium nutrition. In: Gupta KJ, Igamberdiev AU (eds) Reactive oxygen and nitrogen species signaling and communication in plants. Springer, Heidelberg, pp 133–154

Post AF, Rihtman B, Wang Q (2012) Decoupling of ammonium regulation and *ntcA* transcription in the diazotrophic marine cyanobacterium *Trichodesmium* sp. IMS101. ISME J 6:629–637

Probyn TA, Painting SJ (1985) Nitrogen uptake by size fractionated phytoplankton populations in Antarctic surface waters. Limnol Oceanogr 30:1327–1332

Radajewski S, Ineson P, Parekh NR, Murrell JC (2000) Stable-isotope probing as a tool in microbial ecology. Nature 403:646–649

Rao TR (2000) A curve for all reasons: the rectangular hyperbola in biology. Resonance 5:85–90

Raven JA (1980) Nutrient transport in microalgae. Adv Microbiol Physiol 21:47–226

Raven JA, Wollenweber B, Handley LL (1992) A comparison of ammonium and nitrate as nitrogen sources for photolithotrophs. New Phytol 121:19–32

Rees TAV, Syrett PJ (1979) Uptake of urea by the diatom, *Phaeodactylum*. New Phytol 82:169–178

Rees TAV, Cresswell RC, Syrett PJ (1980) Sodium-dependent uptake of nitrate and urea by a marine diatom. Biochim Biophys Acta 596:141–144

Robertson DL, Alberte RS (1996) Isolation and characterization of glutamine synthetase from the marine diatom *Skeletonema costatum*. Plant Physiol 111:1169–1175

Rogato A, Amato A, Iudicone D, Chiurazzi M, Ferrante MI, d'Alcalà MR (2015) The diatom molecular toolkit to handle nitrogen uptake. Mar Genomics 1:95–108. https://doi.org/10.1016/j.margen.2015.05.018

Rose LA, Elliott EM, Adams MB (2015) Triple nitrate isotopes indicate differing nitrate source contributions to streams across a nitrogen saturation gradient. Ecosystems. https://doi.org/10.1007/s10021-015-9891-8

Rosenwasser S, van Creveld SG, Schatz D, Malitsky S, Tzfadia O, Aharoni A, Levin Y et al (2014) Mapping the diatom redox-sensitive proteome provides insight into response to nitrogen stress in the marine environment. Proc Natl Acad Sci USA 111:2740–2745. https://doi.org/10.1073/pnas.1319773111

Ruban AV, Lavaud J, Rousseau B, Guglielmi G, Horton P, Etienne A-L (2004) The super-excess energy dissipation in diatom algae: comparative analysis with higher plants. Photosynth Res 82:165–175. https://doi.org/10.1007/s11120-004-1456-1

Salerno M, Stoecker DK (2009) Ectocellular glucosidase and peptidase activity of the mixotrophic dinoflagellate *Prorocentrum minimum* (Dinphyceae). J Phycol 45:34–45. https://doi.org/10.1111/j.1529-8817.2008.00622.x

Santoro AE, Buchwald C, McIlvin MR, Casciotti KL (2011) Isotopic signature of N_2O produced by marine ammonia-oxidizing archea. Science 333:1282–1285

Scanlan DJ, Post AF (2008) Aspects of marine cyanobacterial nitrogen physiology and connection to the nitrogen cycle. In: Capone DG, Bronk DA, Mulholland MR, Carpenter EJ (eds) Nitrogen in the marine environment. Elsevier, Burlington, MA, pp 1073–1096

Scheibe R (2004) Malate valves to balance cellular energy supply. Physiol Plant 120:21–26

Shangguan Y, Glibert PM, Alexander JA, Madden CJ, Murasko S (2017) Phytoplankton community response to changing nutrients in Florida Bay: results of mesocosm studies. J Exp Mar Biol Ecol 494:38–53. https://doi.org/10.1016/j.embe.2017.05.006

Sheppard CW (1962) Basic principles of the tracer method. Introduction to mathematical tracer kinetics. Wiley, New York. 282 pp

Siaut M, Heijde M, Mangogna M, Montsant A, Coesel S, Allen A, Manfredonia A et al (2007) Molecular toolbox for studying diatom biology in *Phaeodactylum tricornutum*. Gene 406:23–35

Sigman DM, Granger J, DiFiore PJ, Lehmann MM, Ho R, Cane G, van Geen A (2005) Coupled nitrogen and oxygen isotope measurements of nitrate along the eastern North Pacific margin. Global Biogeochem Cycles 19:GB4022. https://doi.org/10.1029/2005GB002458

Sinclair G, Kamykowski D, Glibert PM (2009) Growth, uptake, and assimilation of ammonium, nitrate, and urea, by three strains of *Karenia brevis* grown under low light. Harmful Algae 8: 770–780

Smith SL, Yamanaka Y, Pahlow M, Oschlies A (2009) Optimal uptake kinetics: physiological acclimation explains the patterns of nitrate uptake by phytoplankton in the ocean. Mar Ecol Prog Ser 384:1–12

Solomon CM, Glibert PM (2008) Urease activity in five phytoplankton species. Aquat Microb Ecol 52:149–157

Solomon CM, Collier JL, Berg GM, Glibert PM (2010) Role of urea in microbial metabolism in aquatic systems: a biochemical and molecular review. Aquat Microb Ecol 59:67–88

Solomonson LP (1979) Structure of *Chlorella* nitrate reductase. In: Hewitt EJ, Cutting CV (eds) Nitrogen assimilation of plants. Academic, New York, pp 91–114

Stoecker DK, Gustafson DE (2003) Cell-surface proteolytic activity of photosynthetic dinoflagellates. Aquat Microb Ecol 30:175–183

Stöhr C, Tischner R, Ward MR (1993) Characterization of the plasma-membrane-bound nitrate reductase in Chlorella saccharophila (Krüger) Nadson. Planta 191:79–85

Suzuki A, Knaff DB (2005) Glutamate synthase: structural, mechanistic and regulatory properties, and role in the amino acid metabolism. Photosynth Res 85:191–217. https://doi.org/10.1007/s11120-004-3478-0

Syrett PJ (1981) Nitrogen metabolism of microalgae. In: Platt T (ed) Physiological bases of phytoplankton ecology. Can Bull Fish Aquat Sci 210:182–210

Takabayashi M, Wilkerson FP, Robertson DL (2005) Response of glutamine synthetase gene transcription and enzyme activity to external nitrogen sources in the diatom *Skeletonema costatum* (Bacillariophyceae). J Phycol 41:84–94

Thompson PA, Levasseur ME, Harrison PJ (1989) Light-limited growth on ammonium vs. nitrate: what is the advantage for marine phytoplankton? Limnol Oceanogr 34:1014–1024

Tischner R, Ward MR, Huffaker RC (1989) Evidence for a plasma-membrane-bound nitrate reductase involved in nitrate uptake of *Chlorella sorokiniana*. Planta 178:19–24

Tsunogai U, Miyauchi T, Ohyama T, Komatsu DD, Ito M, Nakagawa F (2018) Quantifying nitrate dynamics in a mesotrophic lake using triple oxygen isotopes as tracers. Limnol Oceanogr 63: S458–S476. https://doi.org/10.1002/lno.10775

Turpin DH (1991) Effects of inorganic N availability on algal photosynthesis and carbon metabolism. J Phycol 27:14–20

Turpin DH, Harrison PJ (1979) Limiting nutrient patchiness and its role in phytoplankton ecology. J Exp Mar Biol Ecol 39:151–166

Vergara J, Berges J, Falkowski P (1998) Diel periodicity of nitrate reductase activity and protein levels in the marine diatom *Thalassiosira weissflogii* (Bacillariophyceae). J Phycol 34:952–961

Wang M, Shen Q, Xu G, Guo SW (2014) Chapter one – New insights into the strategy for nitrogen metabolism in plant cells. Int Rev Cell Mol Biol 310:1–37

Wang WH, Kohler B, Cao FQ, Liu LH (2008) Molecular and physiological aspects of urea transport in higher plants. Plant Sci 175:467–477

Weissman GS (1964) Effect of ammonium and nitrate nutrition on protein level and exudate composition. Plant Physiol 39:947–952

Weyman PD, Beeri K, Lefebvre SC, Rivera J, McCarthy JK, Heuberger AL, Peers G et al (2015) Inactivation of *Phaeodactylum tricornutum* urease gene using transcription activator-like effector nuclease-based targeted mutagenesis. Plant Biotechnol J 13:460–470

Wheeler PA, Kokkinakis SA (1990) Ammonium recycling limits nitrate use in the oceanic subarctic Pacific. Limnol Oceanogr 35:1267–1278
Wilhelm C, Büchel J, Fisahn R, Goss R, Jakob T, LaRoche J, Lavaud J et al (2006) The regulation of carbon and nutrient assimilation in diatoms is significantly different from green algae. Protist 157:91–124
Wilkerson FP, Dugdale RC, Kudela RM, Chavez FP (2000) Biomass and productivity in Monterey Bay, California: contribution of the large phytoplankton. Deep Sea Res 47:1003–1022
Wilkerson FP, Dugdale RC, Hogue VE, Marchi A (2006) Phytoplankton blooms and nitrogen productivity in the San Francisco Bay. Estuaries Coast 29:401–416
Williams PJLB, Ducklow HW (2019) The microbial loop concept: A history, 1930–1974. J Mar Res 77(Suppl):23–81
Xu J, Glibert PM, Liu H, Yin K, Yuan X, Chen M, Harrison PJ (2012) Nitrogen sources and rates of phytoplankton uptake in different regions of Hong Kong waters in summer. Estuaries Coast 35: 559–571
Yin K, Harrison PJ, Dortch Q (1998) Lack of ammonium inhibition of nitrate uptake for a diatom grown under low light conditions. J Exp Mar Biol Ecol 228:151–165
Yoshiyama K, Sharp JH (2006) Phytoplankton response to nutrient enrichment in an urbanized estuary: Apparent inhibition of primary production by over eutrophication. Limnol Oceanogr 51:424–434
Young EB, Dring MJ, Savidge G, Birkett DA, Berges JA (2007) Seasonal variations in nitrate reductase activity and internal N pools in intertidal brown algae are correlated with ambient nitrate concentrations. Plant Cell Environ 30:764–774. https://doi.org/10.1111/j.1365-3040.2007.01666.x

Phosphorus Uptake, Assimilation, and Metabolism: Making Energy and Genetic Material

13

13.1 Introduction

To make elemental phosphorus (P), one only has to follow the recipe described by the German alchemist, Hennig Brandt (in the 1600s, and not to be confused with Karl Brandt, introduced in Chap. 2; reprinted from Ashley et al. 2011, p. 737), and as captured in the painting by Joseph Wright from 1771 (Fig. 13.1):

- Boil urine to reduce it to a thick syrup;
- Heat until a red oil distills up from it, and draw that off;
- Allow the remainder to cool, where it consists of a black spongy upper part and a salty lower part;
- Discard the salt, mix the red oil back into the black material;
- Heat that mixture strongly for 16 h;
- First white fumes come off, then an oil, then phosphorus;
- The phosphorus may be passed into cold water to solidify.

This element is central to all life, or as described by science fiction writer Isaac Asimov (1974, cited in Ashley et al. 2011, p. 737), "Life can multiply until all the phosphorus has gone and then there is an inexorable halt which nothing can prevent." The name "phosphorus" was given by ancient Greek and Roman astronomers and is based on the fact that elemental phosphorus (P) glows in the dark (Ashley et al. 2011). It derived from the Greek "phôs" meaning "light," and "phoros" meaning "bearer," the same name as the "morning star" (Venus, which is not a star at all) that appears before dawn (Vallentyne, 1974). As Ashley et al. (2011, p. 737) describe, much of the fascination with—and desire to produce—P was both medicinal and military:

> White phosphorus was a dangerous element that could now be produced in relatively high quantities. As an elemental form of phosphorus, white phosphorus is highly reactive and hence not found in nature. It is flammable when exposed to air, can spontaneously combust

P. M. Glibert, *Phytoplankton Whispering: An Introduction to the Physiology and Ecology of Microalgae*, https://doi.org/10.1007/978-3-031-53897-1_13

Fig. 13.1 A painting by Joseph Wright from 1771 entitled *The Alchymist, in Search of the Philosopher's Stone, Discovers Phosphorus, and prays for the successful Conclusion of his operation, as was the custom of the Ancient Chymical Astrologers* and refers to the discovery of phosphorus by Brandt (https://en.wikipedia.org/wiki/The_Alchemist_Discovering_Phosphorus)

> and is a deadly poison in low doses (Emsley, 2000). Phosphorus became known as the 'Devil's element', due to its life-destroying properties when used in military applications (such as artillery shells, tracers, grenades, smoke cartridges, and fire bombings) and in organophosphate biocides. The most potent of these was the nerve gas VX, which is lethal at 0.1 mg per kg of body weight when applied to exposed skin (Emsley, 2000). By the 20th century it was a common 'element of war' and Emsley notes the tragic irony that 2000 tonnes of "burning phosphorus" was used in 'Operation Gomorrah' to bomb Hamburg and create a horrific firestorm during one summer week in World War II.

[It is thus not recommended to try the above recipe at home; it could be deadly.]

Phosphorus plays may roles in the cell, from production of chemical energy (ATP) and the production of nicotinamide adenine dinucleotidephosphate (NADPH) during photosynthesis and respiration, which are required for carbon (C) fixation and cell metabolism (Falkowski and Raven 2007; Chap. 6) to phospholipids and the synthesis of proteins via RNA (Geider and La Roche 2002). Van Mooy and Devol (2008) showed that RNA synthesis was indeed the largest sink for phosphate (PO_4^{3-}) and made up roughly half of the PO_4^{3-} uptake of a *Prochlorococcus*-dominated community of the North Pacific. In a typical eukaryotic cell, the RNA pool is approximately 85% ribosomal RNA, while in contrast the ATP pool in a typical cell is much smaller.

Thus, unlike nitrogen (N) which makes up the backbone of amino acids and proteins, P facilitates biosynthetic processes by supplying the energy and genetic material to do so. Phosphorus holds the key to cellular metabolism. As apparent from the above history, P is also key to the structure of bones in animals. Thus, it has been said that N makes biomass while P makes growth happen.

This chapter examines P metabolism and contrasts it with that of N. As with the discussion of N, the focus here is on inorganic and organic sources of P and its metabolism; the topic of nutrient acquisition through mixotrophic grazing is addressed in detail in Chap. 15.

13.2 Notes on Methodology

13.2.1 Measuring Concentrations

The colorimetric technique for P is the most commonly used in both marine and freshwaters. Yet, this method does not exclusively measure inorganic PO_4^{3-}; there may be some amount of organic PO_4^{3-} that is detected also by these methods. Thus, the term soluble reactive PO_4^{3-} (SRP) is used to correctly distinguish between that which is measured and true dissolved inorganic PO_4^{3-} (DIP). A modification of the standard colorimetric technique for inorganic P, the Magnesium-Induced Co-precipitation method (MAGIC; Karl and Tien 1992) allows nanomolar levels to be detected.

There is a diverse suite of organic P compounds in ocean and freshwaters, but they can be distilled to the phosphoesters, which account for about 75% of organic P in seawater, and phosphonates which account for the rest (Kolowith et al. 2001). Whereas inorganic PO_4^{3-} is typically depleted in surface ocean waters, concentrations of dissolved organic PO_4^{3-} (DOP) may be more enriched (Paytan and McLaughlin 2007). These organics are the result of grazing and the rupture of cells from that process as well as viral lysis and bacterial decomposition (Young and Ingall 2010).

13.2.2 Measuring Rates and Genes

Chemical depletion methods and radioisotopes have been used to measure PO_4^{3-} uptake. Phosphorus has a plethora of isotopes, at least 23 (Dyhrman 2016). Both ^{32}P (half-life 14.3 days) and ^{33}P (half-life 25.3 days) have been commonly used as tracers. The longer half-life and fewer hazards associated with the use of ^{33}P make it the preferable isotope for routine use. Moreover, only inorganic PO_4^{3-} and nucleotides like ATP are commercially available as radioisotopes. Uptake rates are therefore calculated as the change in radioactivity of the particulate matter over time,

$$\text{Uptake rate} = \rho = (\text{DPM}_t - \text{DPM}_0)/R \times \text{time} \tag{13.1}$$

where ρ is in units of mass per time, DPM_t and DPM_0 are the radioactivity of the particulate matter at time t and 0, respectively, R is the specific activity of the PO_4^{3-} solution, and time is the duration of the experiment (Harrison 1983).

One of the difficulties with application of either chemical depletion or isotope methods is that the uptake of P is typically so fast that it is hard to capture an initial rate of reaction (e.g., Lean and White 1983). Time course measurements are necessary to capture these dynamics, and a commonly applied approach is to plot the activity of the incorporated radioisotope over time and to use the initial linear portion of the curve as a reflection of transport or uptake (e.g., Burmaster 1978). Nevertheless, when uptake rates are rapid, even this approach is not always satisfactory. Moreover, the calculated uptake velocity of P with very low additions of labeled substrate suffers from the same analytical error as that of N: an overestimate of the rate (Lean and Wright 1983).

Methods for measuring rates of enzymes of P metabolism have long been available. An enzyme-linked fluorescent (ELF) assay for determining rates of the enzyme alkaline phosphatase has been successfully applied across a wide range of taxa. In this technique, the colorless, soluble substrate is added to the sample, but upon reacting with the phosphomonoester, a fluorescent and insoluble product is formed at the site of enzyme activity (Dyhrman 2016). The fluorescence can subsequently be visualized microscopically with ultraviolet excitation. First used with cyanobacteria, dinoflagellates and cryptophytes (González-Gil et al. 1998; Dyhrman and Palenik 1999; Dyhrman et al. 2002), and later with a range of phytoplankton species isolated from the Sargasso Sea, this method allows for single-cell assays as opposed to the more standard bulk enzyme assay (Lomas et al. 2004).

Lack of commercially available isotopes of any DOP compound (other than ATP) has hindered classical analyses of kinetics of these compounds. Similarly, the only fluorescent substrate for P enzymes is the one described above; no such tool is available for phosphonates, for example (Dyhrman 2016).

Single-cell genomics has also advanced our understanding of P nutrition. Understanding the core set of genes and their variability has led to a greatly improved understanding of numbers of PO_4^{3-} transporters and enzymes across species and strains (e.g., Read et al. 2013). Furthermore, high-throughput RNA sequencing has been used to survey responses to P stress and differences or variability in regulation patterns of genes across and between taxa.

13.3 Phosphorus Transport and Surface Adsorption

As discussed in the previous chapter for N (Chap. 12), transporter proteins are required for P substrates to enter the cell (Fig. 13.2). And, as with the case for N, there are both low-affinity and high-affinity P transporters, PiT and PsT, respectively, of which the high-affinity PHO transporter is most common (Lin et al. 2015). The low-affinity transporters function at high ambient P concentrations, while the high-affinity transporters are upregulated when P is scarce. In eukaryotes, the

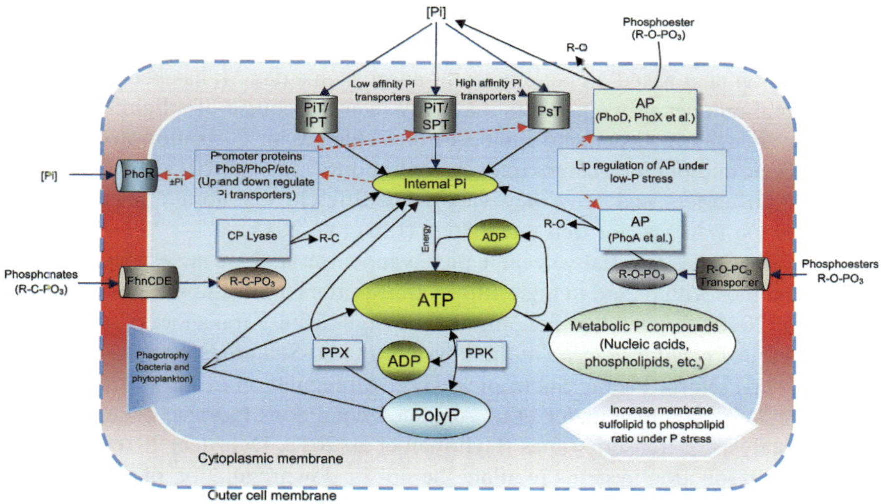

Fig. 13.2 Molecular mechanisms of phosphorus uptake, metabolism, and storage. This diagram represents a composition of known bacterial and eukaryotic phosphate acquisition and metabolic pathways which are likely to be present in some form in most phytoplankton species. Black arrows indicate pathways by which P is transported or metabolized. Red dashed arrows indicate regulatory systems that up- or downregulate specific proteins or protein complexes involved in the transport or metabolism of P. Cylinders indicate transporter protein complexes, ovals—specific phosphate pools, squares—phosphate-processing enzymes or regulatory proteins, and polygons—either phagotrophy or increases in the sulfolipid content in the membrane in response to low-P stress. Pi uptake involves low-affinity transporters (PiT) and high-affinity transporters (PsT). PiT includes subtypes of inorganic phosphate transporter (IPT) and ion-phosphate cotransporter or symporter (SPT). DOP uptake must involve transporters for phosphomonoesters (as yet unidentified) and phosphonates (PhnCDE in cyanobacteria). Alkaline phosphatase (AP) localized on the cell surface (or in the periplasm of cyanobacteria) hydrolyzes external phosphomonoesters to Pi followed by intracellular uptake of the released Pi. Inside the cell, phosphomonoesters are hydrolyzed by intracellular APs (or acid phosphatases) to simultaneously release Pi and R-O compounds while phosphonates are cleaved by phosphonate lyase (in cyanobacteria) to release Pi and various R-C compounds. Intracellular Pi is assimilated via the photosynthetic and respiratory formation of ATP, which is utilized as the major energy source for cell metabolism and as a phosphorus and energy source for the formation of P-containing cellular molecules (e.g., phospholipids, DNA, RNA, and nucleotides). In cyanobacteria, ATP is converted by the enzyme polyphosphate kinase (PPK) into polyphosphate (polyP) with simultaneous release of ADP. PolyP serves as both a major cellular phosphate and an energy storage pool and can be converted back into ATP (with reconsumption of ADP) by PPK. A PPK homolog remains to be identified in eukaryotic phytoplankton, and here the formation of polyP is catalyzed by other enzymes (e.g., Vtc 1–4). PolyP can also be hydrolyzed to Pi by exophosphatase (PPX) in cyanobacteria and eukaryotic algae. Phagotrophy of bacteria and other phytoplankton as well as potential endocytosis of surface adsorbed P represent further mechanisms for acquiring P in eukaryotic algae. Under P-limitation of growth, many species can also increase the membrane sulfolipid to compensate for the reduction in phospholipid content. The present diagram is for a cyanobacterium, which in addition to the plasmamembrane, also contains a semipermeable outer membrane (shown here as a dotted line). By contrast, eukaryotic cells do not have an outer cell membrane, but instead usually possess an outer cell wall or cell plates (e.g., coccoliths in many prymnesiophytes). Image reproduced from Lin et al. (2015, J Phycol 52: 1–36, doi: 10.1111/jpy.12365) with permission of John Wiley and Sons

low-affinity transporters are symporters, that is, they transport P and sodium or sulfate along with it. Such transporters have been documented in a diverse suite of phytoplankton taxa based on genome and transcriptome data. It has been suggested that the total number of transporters may vary over the course of a light:dark cycle reflecting changes in P demand linked to diel changes in C-fixation and the cell division cycle and/or diel changes in P concentrations (Lin et al. 2015). The diversity of these transporters and how they vary with different taxonomic groups of the phytoplankton is as yet not well understood (Lin et al. 2015).

In cyanobacteria, especially oceanic picocyanobacteria, there are especially interesting dynamics with regard to high-affinity transporters. For example, Coleman and Chisholm (2010) showed that genomes of the high-affinity transporters were over-represented in *Prochlorococcus* in the low-P Sargasso Sea relative to the comparatively higher-P North Pacific Subtropical Gyre. Upregulation of PO_4^{3-} transporters of many eukaryotic algae under PO_4^{3-}-depleted conditions has also been observed through analysis of transcriptomes (Dyhrman et al. 2012; Dyhrman 2016).

Signal transduction systems regulate the up- or downregulation of transporters. One component of the system senses the concentration of PO_4^{3-} at the cell surface or in the cytosol and the other regulates the transporters and other metabolic processes such as alkaline phosphatase (Dick et al. 2011).

Phosphorus does not always enter the cell: It can also adsorb to the surface of phytoplankton cells (Sañudo-Wilhelmy et al. 2004). Phosphorus, as described previously (Chap. 11), is sticky. This process appears to be associated with oxides (such as Mn and Fe oxides) that precipitate on the cell surface and that, in turn, adsorb the PO_4^{3-}. In the adsorbed state, this P is unavailable to the cells, but phytoplankton may possess mechanisms to reductively dissolve the Mn and Fe oxidases, thereby releasing the PO_4^{3-} for subsequent uptake (Sunda 2012). It has been speculated that this external surface-adsorbed P may provide a storage pool for when cells subsequent become P stressed (e.g., Sañudo-Wilhelmy et al. 2004; Fu et al. 2005), but this needs further investigation.

13.4 Phosphorus Metabolism: Enzymes

Until fairly recently, any description of the enzymes associated with P use would have focused only on alkaline phosphatase, the most well studied of all the P enzymes. Now, it is recognized that the list of the common enzymes for polyphosphate (PolyP) utilization and assimilation is long (Dyhrman 2016), as it is increasingly recognized that PolyP compounds can play key roles in PO_4^{3-} nutrition in both fresh and marine waters. The most common polyphosphates are phosphoesters and phosphonates. There is little evidence that phosphite can be used as a sole P source (Dyhrman 2016).

The alkaline phosphatases are, in fact, a diverse suite of enzymes, encoded by many different genes, often with many per genome (Dyhrman 2016 and references therein). In fact, at least four different alkaline phosphates have been identified in the diatom *Thalassiosira pseudonana* alone, and all are upregulated under P deficiency

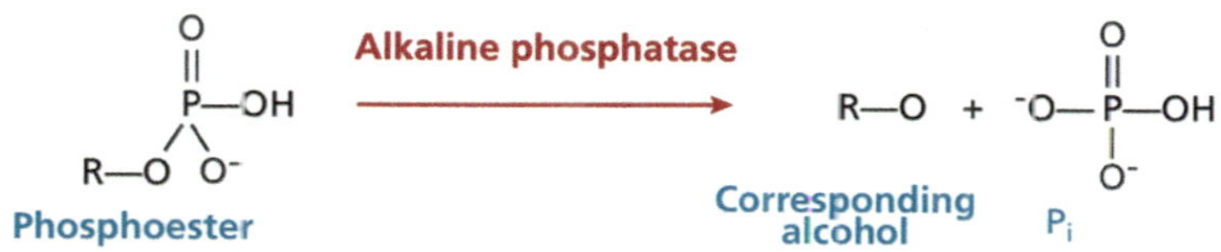

Fig. 13.3 The reaction of alkaline phosphatase

(Dyhrman et al. 2012). Alkaline phosphatases are mostly upregulated when P is in low supply, but this is not always the case. This enzyme catalyzes the reaction of phosphoesters to inorganic PO_4^{3-} (Fig. 13.3).

Three different families of genes related to alkaline phosphatase (*phoA, phoX, phoD*) are located in different parts of the cell, including the cytoplasm, the outer cell membrane and extracellular (Lin et al. 2015). Although often thought to be mostly extracellularly located, the greatest abundance of alkaline phosphatase genes is cytoplasmic. This implies that either there are phosphoester transporters or that intracellular alkaline PO_4^{3-} plays a different function inside the cell, such as cell signaling (Luo et al. 2009; Dick et al. 2011). Among the eukaryotic phytoplankton, the *phoA* genes are more similar than they are to bacteria alkaline phosphatase types, and those of the dinoflagellates, diatoms, and haptophytes are more similar to one another than to those of the chlorophytes (Lin et al. 2012, 2015).

The activity of alkaline phosphatase (as evident from its name) increases under alkaline conditions (pH > 8), which was the normal pH of oceanic waters under pre-ocean acidification (OA) conditions (see Chap. 26; Lin et al. 2015). Thus, as oceans acidify, there may be consequences for the P metabolism of phytoplankton. The activity of alkaline phosphatase has been shown to be higher among dinoflagellates relative to diatoms (e.g., Nicholson et al. 2006; Ou et al. 2006), but there is also a wide diversity of expression of this enzyme among species.

Other phosphodiesterases break the phosphodiester bonds in compounds such as DNA, RNA, and lipids (Dyhrman 2016). These, too, may be cell-surface associated, as Flynn et al. (1986) partially purified such an enzyme from the diatom *Phaeodactylum tricornutum*. Some eukaryotic algae may grow on phosphodiesters as a sole P source (e.g., Yamaguchi et al. 2014).

Nucleotides may also be used as P sources (e.g., Ammerman and Azam 1991), and nucleotidases, as well as their regulation with P availability are increasingly shown in transcriptomes (Dyhrman 2016). By applying flow cytometric sorting approaches, the utilization of ATP by the picocyanobacteria, *Trichodesmium,* and even picoeukaryotes has been shown, especially under conditions of very low inorganic PO_4^{3-} (e.g., Orchard et al. 2009, 2010; Duhamel et al. 2012).

Phosphonates may be used through a variety of enzyme systems, and thus, this source of P—long thought to be unavailable for phytoplankton—has become a topic of considerable interest (Palenik et al. 2003; Dyhrman et al. 2006; Dyhrman 2016). Although heterotrophic bacteria were known to use this source of P, its use by cyanobacteria was more recently discovered (Dyhrman et al. 2006).

13.5 Phosphate (PolyP) Storage and Release

If P is taken up in excess of the needs for balanced growth, the cell may store it, the process that has sometimes been termed luxury uptake. Such a condition may arise, for example, if a cell is N limited but not colimited by P. Unlike N (and most often in diatoms) that can be stored in inorganic form in vacuoles, the storage of PO_4^{3-} involves forming a PolyP molecule which consists of linear chains ranging from several to hundreds of PO_4^{3-} residues linked by high-energy phosphoanhydride bonds (Kornberg et al. 1999). For example, Diaz et al. (2008) found PolyP concentrations to represent about 7% of the PO_4^{3-} in a nutrient-replete coastal diatom community dominated by *Skeletonema*. Interestingly, Raven and Knoll (2010) hypothesize that because PolyP takes up less volume in the cell than does PO_4^{3-}, this storage mechanism becomes more important for small cells (such as picocyanobacteria) relative to larger cells. Moreover, the storage of PolyP also alters buoyancy of cells, resulting in them being more likely to sink (Raven and Knoll 2010). Polyphosphate has also been linked to a stationary phase adaptation, an energy storage compound, metal chelator, a buffer against alkaline conditions, among many other functions (Kornberg 1995; Kornberg et al. 1999; Dyhrman 2016). When used for energy storage, it undergoes an enzymatic reaction with ADP (Kornberg et al. 1999), such that:

$$\text{PolyP}(n) + \text{ADP} \leftrightarrow \text{PolyP}(n\text{–}1) + \text{ATP} \tag{13.2}$$

where n is the number of PO_4^{3-} residues in the PolyP chain. Forming PolyP from ATP, and reversing that reaction, involves the enzyme PolyP kinase (PPK) in heterotrophic bacteria and cyanobacteria, but a similar enzyme has yet to be found in eukaryotic phytoplankton (Kornberg et al. 1999). Another set of enzymes, exophosphatase (PPX) and endophosphatase, may also catalyze the hydrolysis of PolyP to inorganic PO_4^{3-} in bacteria and eukaryotes (Kornberg et al. 1999).

Recently, the dynamics of polyP storage and release were shown to play important roles in regulating primary productivity in eutrophic Lake Ontario (Li et al. 2019). Although this lake is eutrophic, it experiences seasonal periods of P limitation and P stress. Under initial P deficiency P stress, polyP becomes a P reserve and the enrichment of polyp in picoplankton acts as a deficiency response. As P stress becomes more acute, alkaline phosphatase is activated and polyP is recycled, making P available for uptake by other members of the phytoplankton community (Fig. 13.4; Li et al. 2019). With such a mechanism operating, as well as those abiotic recycling mechanisms previously described (Chap. 11), high productivity may be able to be maintained even under conditions that are seemingly P poor.

13.6 Phospholipid Substitution

When P is particularly low, as in oligotrophic waters, in addition to upregulating high-affinity transporters and enzymes for PolyP acquisition, many cyanobacteria—and at least some eukaryotic phytoplankton—employ yet another mechanism to live

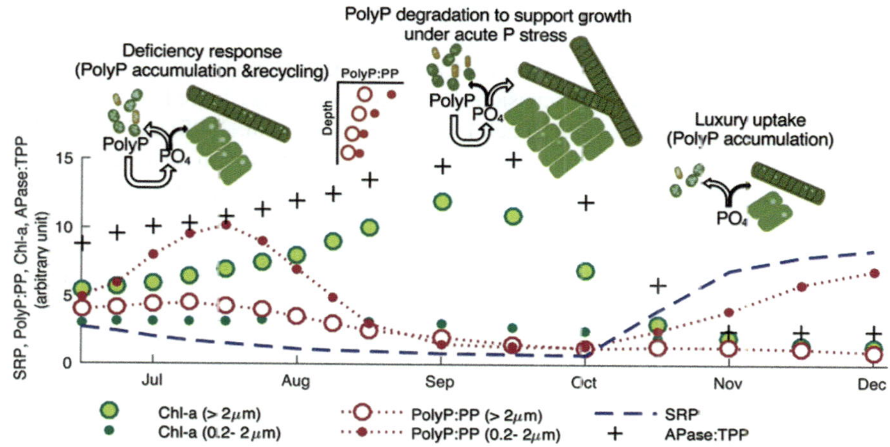

Fig. 13.4 Illustrations of polyP metabolisms and their roles in phosphorus biogeochemical cycling (data points are conceptual) in a eutrophic lake experiencing seasonal P limitation. In early-mid summer, SRP concentrations decrease due to increasing P uptake, leading to P stress (high alkaline phosphatase) and polyP enrichment in picoplankton as a P deficiency response. This enables efficient P recycling due to preferential degradation of polyP, keeping bioavailable P in the water column to support primary productivity. During the late summer peak of primary production, acute P stress (elevated alkaline phosphatase) leads to polyP degradation that provides P for blooming algae. High SRP concentration in winter triggers P luxury uptake, an important P storage mechanism that may be beneficial for algal growth in subsequent spring and summer. All values in the schematic plots are in arbitrary units. Image reproduced from Li et al. (2019, Sci Rep 9: 19563, doi: 10.1038/s41598-019-56042-5) under a creative commons license

on low P. They may replace phospholipids with sulfolipids (Van Mooy et al. 2006, 2009). Originally documented in *Synechococcus* and *Trichodesmium*, this strategy has now been documented in some diatoms, coccolithophores (Van Mooy et al. 2009), the pelagophyte *Aureococcus anophagefferns* (Wurch et al. 2011) and even some green algae (Merchant and Helmann 2012). The substitution appears to happen rapidly under P stress, and also can be rapidly reversed if the stress is alleviated (Martin et al. 2011). Although lipid substitution may help to alleviate P stress, the lipid that substitutes may be N containing (betaine lipids), and this, in turn, creates an additional demand for N—the so-called N penalty (Van Mooy et al. 2009). Both the diversity of this response and its molecular underpinnings are yet to be fully elucidated.

13.7 Kinetic Relationships

The kinetics of inorganic PO_4^{3-} uptake follow those previously described for N. That is, they fit a saturating hyperbola and data can be fit to equations previously described (Chap. 12). Classic studies have shown that there is typically a daily cycle of inorganic PO_4^{3-} uptake, increasing during the day and decreasing at night, as would be expected for the energy requirements and higher growth demand for PO_4^{3-} from light-driven carbon fixation (Chisholm and Stross 1976).

Not surprisingly (based on discussions in previous chapters), there is variability in uptake kinetics of P as a function of cell P physiology (Fu et al. 2005) and within the same taxa collected from different regions of the oceans (Orchard et al. 2010). Moreover, there are also examples of multiphase uptake kinetics for P (Chisholm and Stross 1976). Perry (1976) measured P uptake rates for the diatom *Thalassiosira pseudonana* grown in chemostats and showed that uptake rates were ten time higher for PO_4^{3-}-limited cultures than those that were PO_4^{3-} replete (N limited). Thus, responses to P as a limiting nutrient are like that of N.

Using ^{33}P as inorganic P in conjunction with ^{33}P labeled ATP, and sorting samples using a flow cytometer, the differential uptake rates of populations of *Synechococcus* and *Prochlorococcus* in the Sargasso Sea were assessed (Casey et al. 2009). More recently, the kinetics of PO_4^{3-} uptake by these taxa as well as others of different cell sizes were measured, again sorting the different populations by flow cytometry (Lomas et al. 2014). The whole community and each of the component populations displayed a saturating response with respect to added PO_4^{3-}, but the responses differed, and not necessarily by cell size. Relationships varied by station with kinetic parameters varying with ambient PO_4^{3-} concentration, suggesting regulation of transport proteins. Small cells had the lowest K_s and the lowest V_{max} (Fig. 13.5). However, when normalized to cell quotas, the affinity then ranked *Prochlorococcus* > *Synechococcus* > eukaryotic phytoplankton > *Trichodesmium*. Their data further support the notion that kinetic parameters are indeed not fixed entities and instead display considerable plasticity.

13.8 Metabolism and Implications for Phytoplankton Community Composition

As seen for N and the dichotomy of phytoplankton communities that typically develop under new vs. regenerated N sources, a similar dichotomy of phytoplankton communities develops under inorganic and organic P availability, although there is still much to be learned about preferences by different taxa (Lin et al. 2015). In general, diatoms have a greater capacity to use inorganic P forms, while dinoflagellates have a higher affinity and capacity to utilize organic P (and N) forms (Sunda et al. 2006; Burkholder et al. 2008). They, for example, have higher alkaline PO_4^{3-} activity during blooms (Nicholson et al. 2006; Ou et al. 2006), and they appear to have higher PO_4^{3-} storage capacity for polyP (Diaz et al. 2008). Cyanobacteria—and perhaps to some degree other eukaryotic phytoplankton—can also effectively substitute their phospholipids, thus downregulating their cellular P quota.

An understanding of the potentially large contribution that organic P sources can make to the nutrition of oligotrophic waters has reshaped our understanding of limiting nutrients. The oligotrophic waters of the oceans have traditionally been

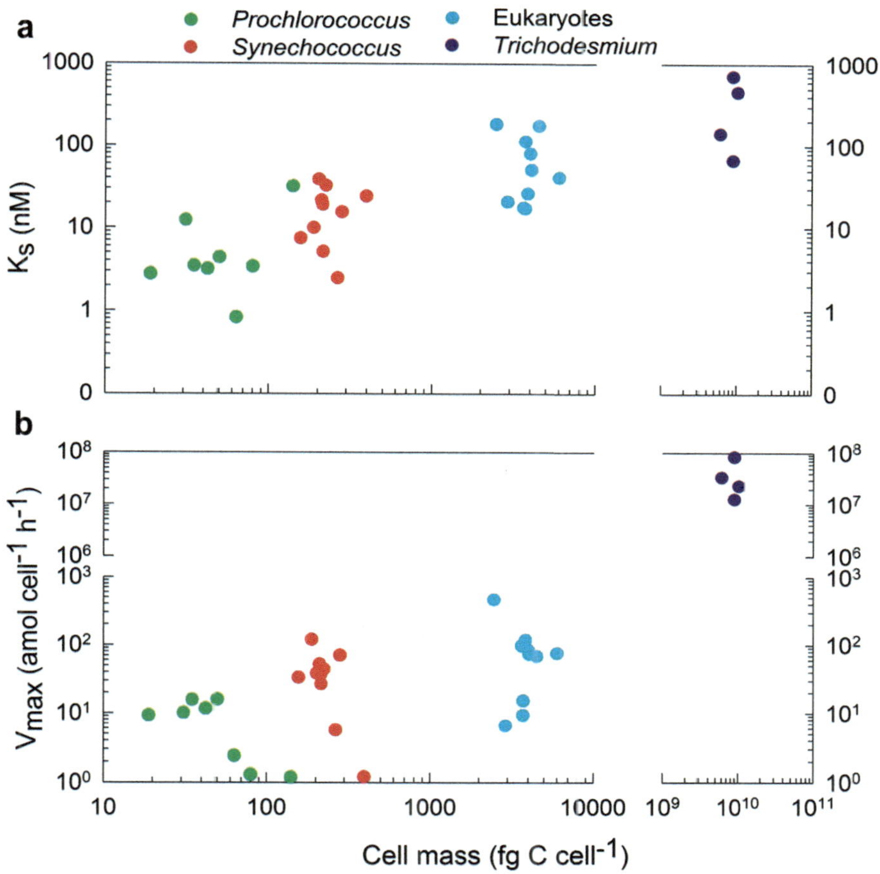

Fig. 13.5 Relationship between K_s (Panel **a**), V_{max} (Panel **b**), and cell mass across phytoplankton groups. Due to difficulties of accurately estimating cell volume, cellular carbon biomass was used as a proxy for cell size. Image reproduced from Lomas et al. (2014, Proc Nat Acad Sci (USA) 111: 17540–17545) under a creative commons license

considered N limited, but if one assumes the induction of alkaline phosphatase activity that has been well documented in such waters (e.g., Lomas et al. 2010), then the hypothesis that oceanic primary production may be limited by inorganic P (e.g., Karl et al. 2001) is further supported.

13.9 Summary

The study of PO_4^{3-} metabolism differs considerably from that of N. Fundamental to cell energy and genetic material, it holds the key to metabolism. It has traditionally been studied using different tools than those used in N studies, as it does not have a convenient stable isotope for tracer studies. While PO_4^{3-} has traditionally been considered the limiting nutrient in freshwaters and N in marine waters, the understanding of PO_4^{3-} regulation of marine communities is rapidly advancing. So too is our understanding of the important role(s) of DOP compounds. In all, PO_4^{3-} is on the one hand a major contributor to the eutrophication of aquatic systems, and there is an associated need to control its use, but it is, on the other hand, at risk of becoming globally scarce (Chap. 11).

References

Ammerman J, Azam F (1991) Bacterial 5′-nucleotidase activity in estuarine and coastal marine waters – role in phosphorus regeneration. Limnol Oceanogr 36:1437–1447. https://doi.org/10.4319/lo.1991.36.7.1427

Ashley K, Cordell D, Mavinic D (2011) A brief history of phosphorus: from the philosopher's stone to nutrient recovery and reuse. Chemosphere 84:737–746. https://doi.org/10.1016/j.chemosphere.2011.03.001

Asimov I (1974) Asimov on chemistry. Doubleday, New York

Burkholder JM, Glibert PM, Skelton HM (2008) Mixotrophy, a major mode of nutrition for harmful algal species in eutrophic waters. Harmful Algae 8:77–93. https://doi.org/10.1016/j.hal.2008.08.010

Burmaster DE (1978) Steady and unsteady continuous culture of *Monochrysis lutheri*. Ph.D. thesis, MIT, Cambridge, MA

Casey JR, Lomas MW, Michelou VK, Dyhrman ST, Orchard ED, Ammerman JW, Sylvan JB (2009) Phytoplankton taxon-specific orthophosphate (P_i) and ATP utilization in the western subtropical North Atlantic. Aquat Microb Ecol 58:31–34. https://doi.org/10.3354/ame01348

Chisholm SW, Stross RG (1976) Phosphate uptake kinetics in *Euglena gracilis* (Z) (Euglenophyceae) grown on light/dark cycles. I. Synchronized batch cultures. J Phycol 12:210–217. https://doi.org/10.1111/j.1529-8817.1976.tb00505.x

Coleman ML, Chisholm SW (2010) Ecosystem specific selection pressures revealed through comparative population genomics. Proc Natl Acad Sci USA 107:18634–18539. https://doi.org/10.1073/pnas.100948107

Diaz J, Ingall E, Benitez-Nelson C, Peterson D, de Jonge MD, McNulty I, Brandes JA (2008) Marine polyphosphate: a key player in geologic phosphorus sequestration. Science 320:652–655. https://doi.org/10.1126/science.1151751

Dick CF, Dos-Santos ALA, Meyer-Fernandes JR (2011) Inorganic phosphate as an important regulator of phosphatases. Enzyme Res 2011:103980. https://doi.org/10.4061/2011/103980

Duhamel S, Björkman KM, Karl DM (2012) Light dependence of phosphorus uptake by microorganisms in the subtropical North and South Pacific Ocean. Aquat Microb Ecol 67:225–238. https://doi.org/10.3354/ame01593

Dyhrman ST (2016) Nutrients and their acquisition: phosphorus physiology in microalgae. In: Borowitzka M, Beardall J, Raven J (eds) The physiology of microalgae. Developments in applied phycology, vol 6. Springer, Cham. https://doi.org/10.1007/978-3-319-24945-2_8

Dyhrman S, Palenik B (1999) Phosphate stress in cultures and field populations of the dinoflagellate *Prorocentrum minimum* detected by a single-cell alkaline phosphatase assay. Appl Environ Microbiol 65:3205–3212. https://doi.org/10.1128/aem.65.7.3205-3212.1999

Dyhrman S, Webb E, Anderson DM, Moffett JW, Waterbury JB (2002) Cell-specific detection of phosphorus stress in *Trichodesmium* from the western North Atlantic. Limnol Oceanogr 47:1832–1836. https://doi.org/10.4319/lo.2002.47.6.1832

Dyhrman ST, Chappell P, Haley ST, Moffett JW, Orchard ED, Waterbury JB, Webb EA (2006) Phosphonate utilization by the globally important marine diazotroph *Trichodesmium*. Nature 439:68–71. https://doi.org/10.1038/nature04203

Dyhrman ST, Jenkins BD, Rynearson TA, Saito MA, Mercier ML, Alexander H, Whitney LP et al (2012) The transcriptome and proteome of the diatom *Thalassiosira pseudonana* reveal a diverse phosphorus stress response. PLoS One 7:e33768. https://doi.org/10.1371/journal.pone.0033768

Emsley J (2000) The 13th element: the sordid tale of murder, fire, and phosphorus. Wiley, New York, p 327

Falkowski PG, Raven JA (2007) Aquatic photosynthesis, 2nd edn. Blackwell Science, London, UK

Flynn KJ, Oepik H, Syrett PJ (1986) Localization of the alkaline phosphatase and 5′-nucleotidase activities of the diatom *Phaeodactylum tricornutum*. J Gen Microbiol 132:289–298. https://doi.org/10.1099/00221287-132-2-289

Fu FX, Zhang Y, Leblanc K, Sañudo-Wilhelmy SA, Hutchins DA (2005) The biological and biogeochemical consequences of phosphate scavenging onto phytoplankton cell surfaces. Limnol Oceanogr 50:1459–1472. https://doi.org/10.4319/lo.2005.50.5.1459

Geider RJ, La Roche J (2002) Redfield revisited: variability of C: N: P in marine microalgae and its biochemical basis. Eur J Phycol 37:1–17. https://doi.org/10.1017/S0967026201003456

González-Gil S, Keafer B, Jovine R, Anderson DM (1998) Detection and quantification of alkaline phosphatase in single cells of phosphorus limited marine phytoplankton. Mar Ecol Prog Ser 164:21–35. https://doi.org/10.3354/meps164021

Harrison WG (1983) Uptake and recycling of soluble reactive phosphorus by marine microplankton. Mar Ecol Prog Ser 10:127–135. https://www.jstor.org/stable/24815185

Karl DM, Tien G (1992) MAGIC: a sensitive and precise method for measuring dissolved phosphorus in aquatic environments. Limnol Oceanogr 37 105–116. https://doi.org/10.4319/LO.1992.37.1.0105

Karl DM, Bjorkman K, Dore J, Fujieki L, Hebel D, Houlihan T, Letelier R, Tupas L (2001) Ecological nitrogen to phosphorus stoichiometry at station ALOHA. Deep Sea Res II 48:1529–1566. https://doi.org/10.1016/S0967-0645(00)00152-1

Kolowith LC, Ingall ED, Benner R (2001) Composition and cycling of marine organic phosphorus. Limnol Oceanogr 46:309–320. https://doi.org/10.4319/lo.2001.46.2.0309

Kornberg A (1995) Inorganic polyphosphate: toward making a forgotten polymer unforgettable. J Bacteriol 177:491–496. https://doi.org/10.1128/jb.177.3.491-496.1995

Kornberg A, Rao NN, Ault-Riché D (1999) Inorganic polyphosphate: a molecule of many functions. Prog Mol Subcell Biol 23:1–18. https://doi.org/10.1007/978-3-642-58444-2_1

Lean DRS, White E (1983) Chemical and radio tracer measurements of phosphorus uptake by lake plankton. Can J Fish Aquat Sci 40:147–155. https://doi.org/10.1139/f83-024

Li J, Plouchart D, Zastepa A, Dittrich M (2019) Picoplankton accumulate and recycle polyphosphate to support high primary productivity in coastal Lake Ontario. Sci Rep 9: 19563. https://doi.org/10.1038/s41598-019-56042-5

Lin S, Litaker RW, Sunda WG (2015) Phosphorus physiological ecology and molecular mechanisms in marine phytoplankton. J Phycol 52:1–36. https://doi.org/10.1111/jpy.12365

Lin X, Zhang H, Cui Y, Lin S (2012) High sequence variability, diverse subcellular localizations, and ecological implications of alkaline phosphatase in dinoflagellates and other eukaryotic phytoplankton. Front Microbiol 3:235. https://doi.org/10.3389/fmicb.2012.00235

Lomas MW, Swain A, Shelton R, Ammerman JW (2004) Taxonomic variability of phosphorus stress in Sargasso Sea phytoplankton. Limnol Oceanogr 49:2303–2310. https://doi.org/10.4319/lo.2004.49.6.2303

Lomas MW, Burke AL, Lomas DA, Bell DW, Shen C, Dyhrman ST, Ammerman JW (2010) Sargasso Sea phosphorus biogeochemistry: an important role for dissolved organic phosphorus (DOP). Biogeosciences 7:695–710. https://doi.org/10.5194/bg-7-695-2010

Lomas MW, Bonachela JA, Levin SA, Martiny AC (2014) Impact of ocean phytoplankton diversity on phosphate uptake. Proc Nat Acad Sci USA 111:17540–17545. https://doi.org/10.1073/pnas.1420760111

Luo H, Benner R, Long RA, Hu J (2009) Subcellular localization of marine bacterial alkaline phosphatases. Proc Natl Acad Sci USA 106:21219–21223. https://doi.org/10.1073/pnas.0907586106

Martin P, Van Mooy BAS, Heithoff A, Dyhrman SY (2011) Phosphorus supply drives rapid turnover of membrane phospholipids in the diatom *Thalassiosira pseudonana*. ISME J 5: 1057–1060. https://doi.org/10.1038/ismej.2010.192

Merchant SS, Helmann JD (2012) Elemental economy: microbial strategies for optimizing growth in the face of nutrient limitation. Adv Microb Phys 60:91–210. https://doi.org/10.1016/B978-0-12-398264-3.00002-4

Nicholson D, Dyhrman S, Chavez F, Paytan A (2006) Alkaline phosphatase activity in the phytoplankton communities of Monterey Bay and San Francisco Bay. Limnol Oceanogr 51: 874–883. https://doi.org/10.4319/lo.2006.51.2.0874

Orchard ED, Webb EA, Dyhrman ST (2009) Molecular analysis of the phosphorus starvation response in *Trichodesmium* spp. Environ Microbiol 11:2400–2411

Orchard ED, Ammerman JW, Lomas MW, Dyhrman ST (2010) Dissolved inorganic and organic phosphorus uptake in *Trichodesmium* and the microbial community: the importance of phosphorus ester in the Sargasso Sea. Limnol Oceanogr 55:1390–1399. https://doi.org/10.4319/lo.2010.55.3.1390

Ou L, Huang B, Lin L, Hong H, Zhang F, Chen Z (2006) Phosphorus stress of phytoplankton in the Taiwan Strait determined by bulk and single-cell alkaline phosphatase activity assays. Mar Ecol Prog Ser 327:95–106. https://doi.org/10.3354/meps327095

Palenik B, Brahamsha B, Larimer FW, Land M, Hauser L, Chain P, Lamerdin J, Regala W, Allen EE, McCarren J, Paulsen I, Dufresne A, Partensky F, Webb EA, Waterbury J (2003) The genome of a motile marine *Synechococcus*. Nature 424:1037–1042. https://doi.org/10.1038/nature01943

Paytan A, McLaughlin K (2007) The oceanic phosphorus cycle. Chem Rev 107:563–576. https://doi.org/10.1021/cr0503613

Perry MJ (1976) Phosphate utilization by an oceanic diatom in phosphorus-limited chemostat culture and in the oligotrophic waters of the central North Pacific. Limnol Oceanogr 21:88–107. https://doi.org/10.4319/lo.1976.21.1.0088

Raven JA, Knoll AH (2010) Non-skeletal biomineralization by eukaryotes: matters of moment and gravity. Geomicrobiol J 27:572–584. https://doi.org/10.1080/01490451003702990

Read BA, Kegel J, Klute MJ, Kuo A, Lefebvre SC, Maumus F, Mayer C et al, *Emiliania huxleyi* Annotation Consortium; Dacks JB, Delwiche CF, Dyhrman ST, Glöckner G, John U, Richards T, Worden AZ, Zhang X, Grigoriev IV (2013) Pan genome of the phytoplankton *Emiliania* underpins its global distribution. Nature 499:209–213. https://doi.org/10.1038/nature12221

Sañudo-Wilhelmy SA, Tovar-Sanchez A, Fu FX, Capone DG, Carpenter EJ, Hutchins DA (2004) The impact of surface-adsorbed phosphorus on phytoplankton Redfield stoichiometry. Nature 432:897–901. https://doi.org/10.1038/nature03125

Sunda WG (2012) Feedback interactions between trace metal nutrients and phytoplankton in the ocean. Front Microbiol 3:204. https://doi.org/10.3389/fmicb.2012.00204

Sunda WG, Granéli E, Gobler CJ (2006) Positive feedback and the development and persistence of ecosystem disruptive algal blooms. J Phycol 42:963–974. https://doi.org/10.1111/j.1529-8817.2006.00261.x

Vallentyne JR (1974) The algal bowl – lakes and man. Fish Res Board Can Misc Spec Publ 22, Ottawa, ON, Canada

Van Mooy BAS, Devol AH (2008) Assessing nutrient limitation of *Prochlorococcus* in the North Pacific subtropical gyre by using an RNA capture method. Limnol Oceanogr 53:78–88. https://doi.org/10.4319/lo.2008.53.1.0078

Van Mooy BAS, Rocap G, Fredricks HF, Evans CT, Devol AH (2006) Sulfolipids dramatically decrease phosphorus demand by picocyanobacteria in oligotrophic marine environments. Proc Natl Acad Sci USA 103:8607–8612. https://doi.org/10.1073/pnas.0600540103

Van Mooy BAS, Fredricks HF, Pedler BE, Dyhrman ST, Karl DM, Koblížek M, Lomas MW, Mincer TJ, Moore LR, Moutin T, Rappé MS, Webb EA (2009) Phytoplankton in the ocean use non-phosphorus lipids in response to phosphorus scarcity. Nature 458:69–72. https://doi.org/10.1038/nature07659. PMID: 19182781

Wurch LL, Bertrand EM, Saito MA, Van Mooy BAS, Dyhrman ST (2011) Proteome changes driven by phosphorus deficiency and recovery in the brown tide-forming alga *Aureococcus anophagefferens*. PLoS One 6:e28949. https://doi.org/10.1371/journal.pone.0028949

Yamaguchi H, Arisaka H, Otsuka N, Tomaru Y (2014) Utilization of phosphate diesters by phosphodiesterase-producing marine diatoms. J Plankton Res 36:281–285. https://doi.org/10.1093/plankt/fbt091

Young CL, Ingall ED (2010) Marine dissolved organic phosphorus composition: insights from samples recovered using combined electrodialysis/reverse osmosis. Aquat Geochem 16:563–574. https://doi.org/10.1007/s10498-009-9087-y

Additional Website Accessed

https://en.wikipedia.org/wiki/The_Alchemist_Discovering_Phosphorus. Accessed 3 Jan 2023

14 Silica Uptake, Assimilation, and Metabolism: Making Cell Walls

14.1 Introduction

If *Chlorella* was the hero of early studies on nitrogen (N) metabolism, there is no doubt that diatoms are the hero of early studies on silica (Si) metabolism. Hildebrand et al. (2018) summarized some of the early observations on diatoms and Si, attributing much to anonymous sources,

> It started with a microscopic observation by an anonymous English country nobleman in 1703, who observed an object that looked like a chain of regular parallelograms and debated whether it was just crystals of salt, or a plant (Anonymous 1702). The viewer decided that it was a plant because the parallelograms didn't separate upon agitation, nor did they vary in appearance when dried or subjected to warm water (in an attempt to dissolve the "salt"). Unknowingly, the viewer's confusion captured the essence of diatoms—mineral utilizing plants. It is not clear when it was determined that diatom cell walls are made of silica, but in 1939 a seminal reference characterized the material as silicic acid in a "subcolloidal" state (Rogall 1939).

Applying the terminology of Simpson and Volcani (1981), as recommended by Finkel (2016), Si refers to the element of silicon and to forms of siliceous compounds that are unknown, and silicic acid ($Si(OH)_4$) is the dominant unionized aqueous form at pH 7–8, while amorphous hydrated polymerized Si is referred to as opal or silica. Virtually all dissolved Si is in the form of $Si(OH)_4$, or silicate. Particulate Si is the sum of biogenic silica (BSi) and lithogenic silica (LSi). Currently, there is considerable interest in silicification as the process of polymerization has important implications in the work of nanotechnology. Here, metabolism and silicification are addressed. First, however, a brief look at how the oceanic global Si inventory is changing (see also Chap. 22).

P. M. Glibert, *Phytoplankton Whispering: An Introduction to the Physiology and Ecology of Microalgae*, https://doi.org/10.1007/978-3-031-53897-1_14

14.2 Where Has All the Silica Gone?

Diatoms depend on Si to grow and to make their cell walls. If global phosphorus (P) supply is a concern for the future (see Chaps. 13 and 22), then the future of Si supply is even more so. Dam and reservoir construction and other in-river consumptive uses, as well as channelization of flow, have greatly modified the timing, magnitude, form, and stoichiometry of all nutrient delivery—but especially that of Si—to the coast. Thousands of dams throughout the world have altered the delivery of Si (Grill et al. 2015; Zarfl et al. 2015; see also Chap. 22). Motivated by the production of energy from hydropower, "the world went on a dam-building spree after 1945, peaking in the 1960s and 1970s, by which time most of the good sites in rich countries had been taken" (McNeill and Engelke 2014, p. 33). For example, the retention of Si upstream following the construction of the Three Gorges Dam in China has been considered to be an important factor leading to altered nutrient proportions favoring non-Si-requiring HABs in the East China Sea (e.g., Ding et al. 2019). Even systems not impacted directly by large dams are experiencing reductions in Si. For example, the Baltic Sea has had a documented trend of reduced $Si(OH)_4$ concentrations from the 1970s into the 1990s (Wulff et al. 1990; Humborg et al. 2000). Increasing ratios of N:Si and P:Si are leading to shifts away from diatoms (e.g., Smayda 1990; Turner and Rabalais 1994).

Diatoms, themselves, are responsible for the sequestration of a large fraction of Si, locking it in sediments where comparatively slow dissolution processes are required to return it to dissolved forms where it can be again used biologically. Due to their heavy cell walls, diatoms sink. They do so directly, and when consumed by grazers, their heavy cell walls become part of sinking fecal material. Heavily silicified cells sink faster than lightly silicified cells, sequestering both C and biogenic Si (BSi) in the sediment, the so-called "Si pump" (Dugdale et al. 1995). Moreover, fecal pellets from copepods that do graze on heavily silicified cells are denser and sink faster from the water column, increasing Si export from the photic zone and carbon (C) export (Dugdale et al. 1995). The dissolution rate of biogenic Si (BSi) from diatoms that have been incorporated in fecal pellets appears to be reduced by factors of 2–3 relative to diatoms that are not aggregated (Claquin et al. 2006). By locking up BSi in sedimented diatoms and fecal pellets for those that are grazed, the biogeochemical remineralization of DSi back to the water is stalled, leading to seasonal consequences, including an acceleration of seasonal Si limitation and an increased window of opportunity for non-siliceous species, including HAB taxa, to proliferate.

Thus, this Si gets stripped from the productive waters of the sea and only slowly returns. The rate of Si incorporation into biomass is several orders of magnitude faster than the rate of regeneration of the Si pool (Barsanti and Gualtieri 2006). Living diatoms are further protected from dissolution by a protective organic coating but which can be degraded by bacterial activity (Bidle and Azam 1999; Kristiansen and Hoell 2002). Unlike the regeneration pathways for N and P, which are roughly in balance with the rates of uptake of these nutrients in the water column, the recycling of Si fundamentally differs. While rates of dissolution are slow, such rates may be becoming even slower with ocean acidification (Taucher et al. 2022; see Chap. 26). A slowing of dissolution rates slows diatom production, as they experience more Si limitation. It has been suggested that a decline in diatoms by 13–26% due to ocean acidification (OA) by the year 2022 may be expected. It is important to point out that this effect was due only to OA effects on Si dissolution rates, not biological production per se. In contrast, the siliceous skeletons of silicoflagellates are much more readily dissolved back into solution (Barsanti and Gualtieri 2006). There is clearly much to be concerned about, and much to be learned, about the global implications of changes in diatom production.

14.3 Silica Content Across Taxa

Although diatoms are the most well-known of the Si-containing algae, in fact, numerous algal groups have Si or siliceous structures (Table 14.1). Diatoms, however, are the only groups with an absolute requirement for Si. Silica is found in cyanobacteria, where, at least for natural populations of *Synechococcus*, ratios of Si: P are similar to those of many diatoms (Baines et al. 2012). While some of this may be external to the cell wall (e.g., bound to organic ligands), it is also possible that some may be located within the periplasmic spaces associated with the cell wall (Baines et al. 2012; Finkel 2016). In green algae, also, Si has been associated with the outer cell wall, and there is considerable debate as to whether this is a required element or if Si deficiency negatively impacts their growth (Finkel 2016 and references therein). One idea is that Si accumulation on the cell wall surface may improve resistance to grazing or parasites.

The Chrysophyta and (in freshwaters) the Synurophyta, as well as selected species within the Haptophyta, produce exogenous Si scales (Finkel 2016). Such scales can be complex and ornamented, but if these cells are grown without Si, they do not produce scales and can exist "naked." Under such conditions, if Si is resupplied, growth will be slowed until scales are reestablished (Sandgren et al. 1996). Silicoflagellates have amorphous, tubular, or hollow rods of siliceous material (Lipps 1970).

Some coccolithophores, but not all, also have a clear requirement for Si in calcification and may, indeed, be limited by Si (Durak et al. 2016). There is much that is not understood regarding the physiology of Si uptake and incorporation in these organisms. They have the capacity for active Si transport. Ratcliffe et al. (2023) have suggested that Si supports the calcification process but does not form a structural component of the coccolith itself.

14.4 Silica Content and Role in Diatoms

Within the diatoms, the siliceous walls surround the plasmalemma and the valves fit together, forming a frustule. Unlike other major nutrients (N and P), Si transport and silicification in diatoms are tightly linked to the cell cycle (see Chap. 3) and therefore

Table 14.1 Microalgae and related groups that accumulate silica

Phylum, class, order (common name)	Example genera	Form of Si	Marine or fresh
Cyanophyta (cyanobacteria)	*Synechococcus, Microcystis, Arthrospira*	Amorphous opal? Not localized	Marine and fresh
Chlorophyta, Chlorophyceae (green algae)	*Tetraselmis Pediastrum Hydrodictyon*	Si in cell wall	Marine and fresh
Dinophyta, Dinophyceae (dinoflagellates)	*Actinicus*	Internal skeletal	Marine and fresh
	Ceratium, Peridinium	Si in the outer cyst wall	
	Eodinia, Jusella	Fossil genera-silicified cyst or theca	
Haptophyta, Coccolithophyceae (coccolithophore)	*Prymnesium nephrolepis*	External scales	Marine
Ochrophyta (chrysophysea)	*Chrysosphaerella Paraphysomonas Spiniferomonas*	Silicified cysts or external scale in the motile stage	Marine and fresh
Ochrophyta, Synurophyceae	*Synura Mallomonas*	External scales in the motile stage	Fresh
Ochrophyta, Dictyochophyceae (silicoflagellates)	*Dictyocha*	Siliceous skeleton of hollow rods	Marine
Ochrophyta, Phaeophyceae (brown algae)	*Ectocarpus Macrocystis Pelagophycus*	Not localized	Marine
Ochrophyta, Bolidophyceae, Parmales	*Triparma Tetrasperma Pentalamina*	External Si cell wall of interlocking plates	Marine
Ochrophyta, Bacillariophyceae, Coscinodiscophyceae, Fragilariophyceae (diatoms)	*Coscinodiscus Chaetoceros Nitzschia Thalassiosira, etc.*	External 2-part ornamented skeleton, termed a frustule	Marine and fresh
Ochrophyta, Xanthophyceae	*Acanthochloris bacillifera*	Si in cell wall and/or cysts	Fresh
Protozoa incertae sedis, Ebriophyceae (Ebridians)	*Ebridia Hermesinium* (*Synechococcus*-like symbionts)	Internal skeleton of rugose or spiny rods	Marine

Modified from Finkel (2016), based on Guiry and Guiry (2015, https://www.algaebase.org)

to growth (Martin-Jézéquel et al. 2000). With each division, the daughter cell forms inside the parent frustule, and thus, cells become smaller over time. When a minimum cell size is reached, sexual reproduction is triggered, restoring cells to a larger size. The formation of new siliceous valves appears to occur just before cell division and accordingly has been associated with the G2 and M periods of cell division; an increase in the phase length of these periods leads to more Si uptake and deposition (Martin-Jézéquel et al. 2000; Claquin et al. 2002). There is considerable evidence that the precursor stages for silicification, namely the upregulation of encoded proteins, occur in advance of when they are needed for cell wall synthesis and "stockpiled" for later use (Hildebrand et al. 2018). Without enough Si, the cell cycle becomes arrested at either the G1/S or G2/M phases (Brzezinski et al. 1990). The cellular energy for silicification comes from aerobic respiration and apparently not from photosynthetic energy (Martin-Jézéquel et al. 2000). Thus, uptake of $Si(OH)_4$ can occur both day and night.

14.5 Silica Transport and Assimilation

Silicic acid ($Si(OH)_4$) can freely diffuse into the cell, but there may also be active uptake that involves Si transporters (SITs; Hildebrand et al. 1997). These transporters are upregulated during Si limitation and downregulated when Si is in sufficient supply. The upregulation typically occurs during the M phase of cell division. The SITs from diatoms are very N-rich; they have ~550 amino acids with ten trans-membrane alpha helices and a carboxyl-terminal (Finkel 2016). The genes for SITs within species generally are more closely related than those from different species (Thamatrakoln and Hildebrand 2005).

Building this complex Si structure is a "massive event," and it occurs inside the cell, requiring large numbers of genes and their proteins (Hildebrand et al. 2018). These Si structures are formed within membrane-bound structures called the Si deposition vesicles (SDVs, Drum and Pankratz 1964; Hildebrand et al. 2018). Although individual SDVs have not been isolated and studied directly, it is thought that these originate from Golgi/endoplasmic reticulum networks (Li and Volcani 1984). In the Chrysophyteae and the Synurophceae, siliceous cysts are also thought to be associated with SDVs (Sandgren 1989). Precipitation of nano-sized spheres of Si can occur within minutes in these structures, if the right pH conditions exist (Kröger et al. 1999). The morphology of the precipitated Si can change as pH and salinity change. There have been some reports that salinity may also be related to Si content, with cells growing at lower salinities having more Si per biovolume than those at high salinities (Brzezinski 1985; Conley et al. 1989; Vreiling et al. 1999). Long-chain polyamines are important components of silicification, and thus, as described below, sufficient N is required for silicification. Carbohydrates may be involved as well, but their roles are less clear (Hildebrand et al. 2018). Chitin is also involved. Once formed, the Si structures are then exported outside the cell (Ogden 1979; Finkel 2016).

14.6 Why Silicify?

Silicification is expensive—it has a high materials and energy cost. There must be a benefit. Physically, the diatom frustule alters the scattering and absorption of light (Furhmann et al. 2004), and the frustule facilitates the enzymatic conversion of bicarbonate to CO_2 at the cell surface by serving as a pH buffer, thereby mediating C limitation (Barsanti and Gualtieri 2006). The presence of a frustule also alters the diffusion of nutrients into the cell and may externally bind cadmium, thus reducing the bioavailability of potentially toxic metals (Santos et al. 2013). Furthermore, being covered in a Si protective cell wall may protect the cell from grazing (or digestion) or viral attack (Finkel 2016). Lastly, adding a Si cell wall makes the cell heavier, which on the one hand may hasten its sinking from well-lit surface waters, but may provide the cell access to nutrient-rich waters deeper in the water column (Finkel 2016).

14.7 Kinetics and Isotopes

Silica has both stable and radioisotope tracers that have both been applied in measuring rates of Si uptake. For Si, there has been substantial application of isotopes in understanding global biogeochemistry and rates of Si incorporation, and a recent review (Sutton et al. 2018) has compiled much of this literature for both marine and freshwater systems. Using the stable isotopic tracer, ^{29}Si, Goering et al. (1973) first demonstrated a saturating response for $Si(OH)_4$ uptake. Note that there are two stable isotopes of Si, ^{29}Si and ^{30}Si, and the more commonly applied stable tracer is now ^{30}Si (Nelson and Goering 1977). There has been considerable use of the stable isotope ^{30}Si as an enrichment tracer in marine studies (Nelson and Goering 1977), but far less than isotopic measurements of N incorporation, and far less in freshwater than marine systems. The principles of isotope application are similar to those applied to N, but there are more methodological constraints with respect to sample handling, and thus, comparatively few laboratories are equipped for such measurements. More recently, the radioisotope ^{32}Si was used to measure rates in the Southern Ocean (Tréguer et al. (1991), but this isotope had a comparatively low specific activity. Since then, ^{32}Si with a much higher specific activity has become available, allowing $Si(OH)_4$ uptake rates to be measured in waters of very low ambient $Si(OH)_4$ concentrations (Brzezinski et al. 1997; Kristiansen and Hoell 2002).

Krause et al. (2018) recently reported kinetic relationships for Si uptake in the European Arctic and broader high-latitude North Atlantic during the spring bloom in May 2016 using ^{32}Si isotopes (Fig. 14.1a). That study also reported that rates of NO_3^- + NO_2^- drawdown were more than twice those of Si uptake (Fig. 14.1b), suggesting that N stocks would be depleted before Si stocks in these waters and that 0.5 μM is the lower limit of Si necessary to support diatom growth. These authors also reported that a regression of the relationship of V (Si specific uptake rate) to V_{max} as a function of ambient Si concentration has an intercept of 2.3, suggesting that in their studied waters at concentrations below this value, uptake is limiting and diatoms may slow their growth in order to take up enough Si to produce new cells (Fig. 14.1c). While diatoms can vary their Si content by as much as a factor of 4 (e.g., Martin-Jézéquel et al. 2000), changes in silicification affect growth rates.

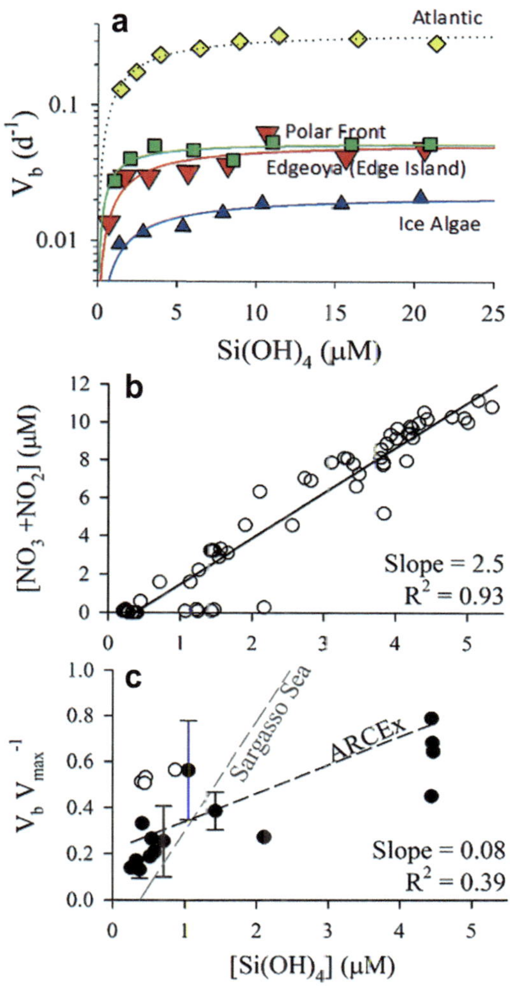

Fig. 14.1 Panel (**a**) Examples of Si uptake relationships for various stations indicated, as measured with ^{32}Si uptake on a cruise to the European Arctic and broader high-latitude North Atlantic in May 2016. Panel (**b**) Relationship between NO_3^- plus NO_2^- drawdown as a function of dissolved Si concentration. Panel (**c**) The ratio of V_b (Si specific uptake rate) at ambient [Si(OH)$_4$] to V_{max} versus dissolved Si(OH)$_4$ for samples measured on the Arctic cruise (ARCEx) and, for comparison, the same ratio for the Sargasso Sea subtropical gyre. Images reproduced from Krause et al. (2018, Biogeosci 15: 6503–6517) under a creative commons license

Another approach for measuring Si uptake rates is the use of an analog of Si, ^{68}Ge (OH)$_4$) as a radiotracer for Si(OH)$_4$. This approach was applied by Thamatrakoln and Hildebrand (2008) using various cultures of algal species. These investigators found that for cells that were Si-replete, nonsaturable kinetics were common, but became saturable over time as Si became depleted. Sigmoidal kinetic relationships (see Chap. 11) were also observed in *Thalassiosira pseudonana* at low ambient Si concentrations. They proposed the following model for the uptake of Si (Fig. 14.2). When cells are starved of Si for short time periods, the capacity for uptake is high if the substrate is replenished, initially via nonsaturable (surge)

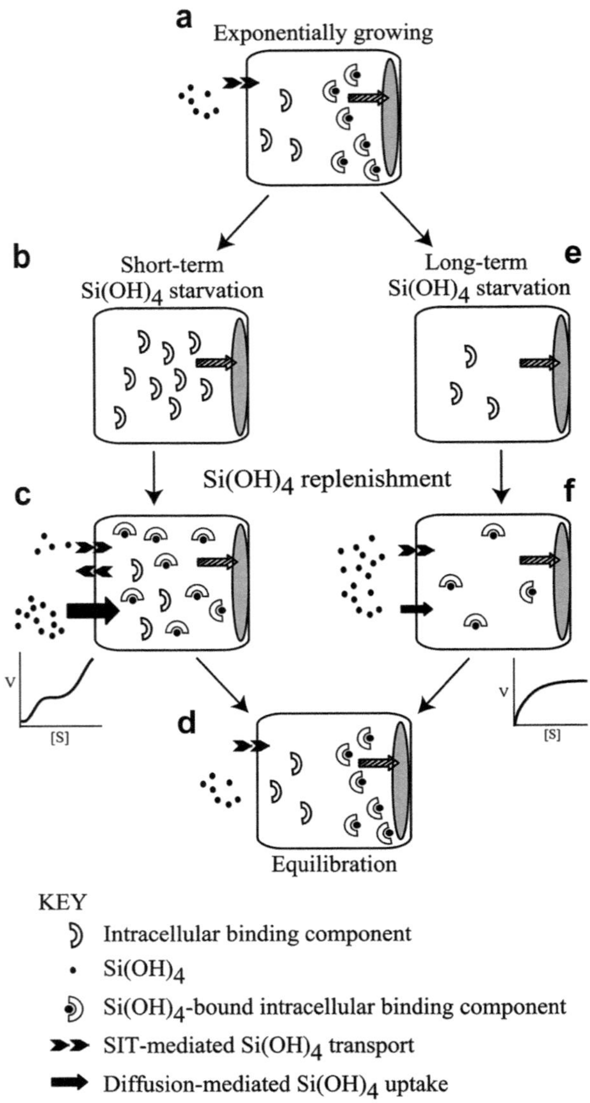

Fig. 14.2 Proposed model of Si uptake in diatoms. In each image, a diatom cell is represented as a rectangular box and cell wall Si (i.e., the Si deposition vesicle, SDV) is represented as a gray elongated oval. Black dots represent $Si(OH)_4$. Horseshoe-shaped structures represent intracellular Si-binding components that data suggest are present but have yet to be identified. Arrows denote the direction of transport with magnitude indicated by their thickness. Black arrows denote diffusion-mediated uptake, and broken arrows represent SIT-mediated transport. Hatched arrows show the movement of $Si(OH)_4$ into the SDV. Stylized graphs of uptake kinetics are shown as inset panels (**c** and **e**). Panel (**a**) In exponentially growing cells, equilibration is achieved between uptake rate, intracellular pools, and cell wall Si incorporation. Uptake is internally controlled. Panel (**b**) After a brief (5–10 min) incubation in Si-free medium, levels of intracellular-binding component have delivered $Si(OH)_4$ to the SDV and are predominantly in the uncomplexed state. Panel (**c**) Upon Si $(OH)_4$ replenishment, cells are able to accommodate surge uptake mediated by diffusion at high Si

uptake. If long-term Si starvation occurs, and then cells are replenished with substrate, the capacity for rapid uptake will decrease. This leads to immediate saturable kinetic responses. Such a model may relate to natural responses. That is, in coastal waters where cells could be exposed to high levels of Si after short-term depletion, rapid or surge uptake is possible, but kinetic responses would depend on the extent to which the cells were depleted. In most waters, internal control, via SITs and SDVs, would be more common.

14.8 Further Implications of Changing Nutrient Stoichiometry and Silicification for Community Composition

What should be clear from the above description of metabolism is that silicification is a nutrient-intensive process. Thus, any changes in nutrient availability—in amount and form—may have consequences for silicifying species. Although a shift away from silicifying species may be well appreciated as a result of decreasing dissolved Si concentrations in coastal waters, largely due to sediment trapping and elemental transformations following the construction of dams worldwide (e.g., Vörösmarty et al. 2003; Beusen et al. 2009), less appreciated is how the composition of diatoms may change with changes in nutrients from eutrophication. Under conditions of reduced Si, more lightly silicified diatoms might be expected, especially if Si levels are low enough to be limiting (Conley et al. 1993). Yet, an unusual pattern of more heavily silicified diatoms has emerged in some systems under conditions of elevated N:Si—especially in winter. The Yellow Sea is one such example (Liu and Glibert 2018).

This eutrophied water body has witnessed two- to fivefold increases in dissolved inorganic N (DIN) and P (DIP) concentrations between 1970 and 2000 and that trajectory of change is continuing (Liu et al. 2013; Strokal et al. 2014). Also associated with this eutrophication, the frequency, timing, and composition of diatoms in the Yellow Sea have changed. *Paralia sulcata*, a tychopelagic diatom species, is now common and often dominant, during the winter months (Liu et al.

Fig. 14.2 (continued) $(OH)_4$, and non-saturable uptake kinetics are observed. Biphasic curves are seen because, at low $Si(OH)_4$ concentrations, SITs are still capable of mediating uptake. In *T. pseudonana*, this results in sigmoidal kinetics. SIT-mediated efflux aids in equilibration. Panel (**d**) After time (approximately 1 h in *T. pseudonana*), equilibration is reestablished between the level of binding component and the rate of Si incorporation, and uptake becomes internally controlled. Panel (**e**) During long-term (24 h) Si starvation, the level of binding component becomes reduced. Panel (**f**) Upon $Si(OH)_4$ replenishment, intracellular capacity is low, and cells are not able to accommodate surge uptake; thus, Michaelis-Menten type saturation is observed. Image reproduced from Thamatrakoln and Hildebrand (2008, Plant Physiol 146: 1397–1407) under a creative commons license

2008, 2013; Liu and Glibert 2018). This species is particularly interesting because it is heavily silicified, with valves that have been variously described as "robust" or "sturdy" (McQuoid and Nordberg 2003; Gebühr et al. 2009; Denys 1999). Given the trend toward lower Si availability in the water column, cells that accumulate proportionately more Si would not be expected. Rather, the expected shift under conditions of eutrophication and increasing Si limitation is a trend toward the proliferation of more lightly silicified diatoms (Conley et al. 1993). Gebühr et al. (2009), in observing an increase in this species in Helgoland Roads, North Sea, described *P. sulcata* as "a typical winter alga adapted to low light conditions under colder water temperatures but high concentrations of nutrients." The recent emergence in *P. sulcata* in the Yellow Sea is associated with times when water temperatures are <5 °C, whereas non-*P. sulcata* diatoms are more abundant over a broader temperature range (5.5–22.5 °C), and non-diatoms ($\geq$50% of phytoplankton assemblage) are dominant at temperature >17 °C (Liu and Glibert 2018). Diatom (both non-*P. sulcata* and *P. sulcata*) dominance appears to be associated with a median NO_3^- water column concentration significantly higher than that associated with the period of non-diatoms, but *P. sulcata* was found across the broadest range in NH_4^+ concentrations. It may be that in the winter, the cellular energy balance of these diatoms may be stressed. This would lead to an increase in photorespiration, which, in turn, has significant effects on the production of silicified cell walls in diatoms (Liu and Glibert 2018). Photorespiration increases the flow of glycolate through the peroxisome and then to the mitochondria where its metabolism is normally linked to a putatively, fully functioning urea cycle (Armbrust et al. 2004; Allen et al. 2006, 2011; see Chap. 6). The urea cycle plays important roles in exchange of nutrients between organelles, and one of its key intermediates is polyamines (see Fig. 12.8), the suite of molecules attached to polypeptides of silifins, the Si-precipitating proteins (Allen et al. 2006; Kröger et al. 2000). Importantly, when polyamine synthesis is overexpressed, more cell wall precursors are made, ultimately leading to production of thickened Si frustules in smaller cells (Nunn et al. 2013; Fig. 14.3). Importantly, heavily silicified cells are more likely to sink faster, sequestering both carbon (C) and biogenic Si (BSi) in the sediment in the "Si pump" (Dugdale et al. 1995).

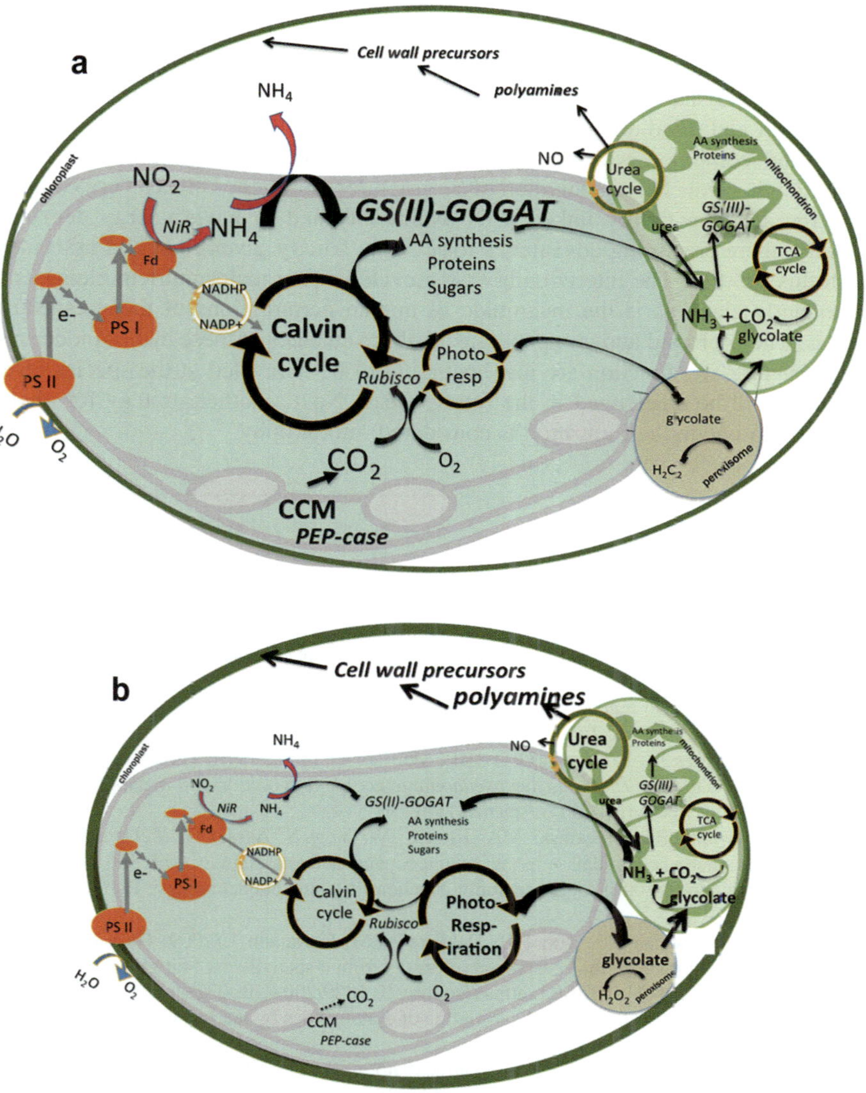

Fig. 14.3 Conceptual schematics of a typical diatom (panel **a**) and a smaller, more heavily silicified diatom (panel **b**), showing the electron transport of photosynthesis, the coupling to the Calvin cycle and photorespiration, and the urea cycle and its link to cell wall precursor formation. Note that in the heavily silicified case, photorespiration is higher, as is the export of glycolate to the peroxisome, and urea cycle cell wall precursor formation. Figure modified and reproduced from Glibert et al. (2016) based on Creative Commons license. Image reproduced from Liu and Glibert (2018, Mar Ecol Progr Ser 604: 51–63. doi: 10.3354.meps12747) reproduced under a creative commons license

14.9 Summary

Silica is fundamental to the growth of diatoms but is widely found across many algal groups that may have a diverse and complex suite of Si structures. The balance of Si production and dissolution differs from that of N or P, with dissolution rates very slow, and combined with changing Si export from rivers, the potential for Si limitation is great. The assimilation of Si structures within the cell is a very complex process, and one tightly linked to N availability and to P via energy demands. Clearly, there are strong advantages for cells to silicify given the costs associated with this process. The intertwining of the cycles of the major nutrient elements is abundantly clear, as is the magnitude of human perturbations of their respective cycles. The profound and complex effects of the changes in these cycles on ecosystem structure and function are just beginning to be unraveled at the species level. Diatoms will be challenged in the future ocean. Next, another strategy for phytoplankton to obtain their nutrition is considered: mixotrophy.

References

Allen AE, Vardi A, Bowler C (2006) An ecological and evolutionary context for integrated nitrogen metabolism and related signaling pathways in marine diatoms. Curr Opin Plant Biol 9:264–273

Allen AE, Dupont CL, Obornik M, Horák A, Nunes-Nesi A, McCrow JP, Zheng H et al (2011) Evolution and metabolic significance of the urea cycle in photosynthetic diatoms. Nature 473: 203–207

Anonymous (1702) Two letters from a gentleman in the country, relating to Mr. Leeuwenhoek's Letter in Transaction, no. 283. Philos Trans R Soc Lond B 23:1494–1501

Armbrust EV, Berges JA, Bowler C, Green BR, Martinez D, Putnam NH, Zhou S et al (2004) The genome of the diatom *Thalassiosira pseudonana*: ecology, evolution and metabolism. Science 306:5693. https://doi.org/10.1126/science.1101156

Baines SB, Twining BS, Brzezinski MA, Krause JW, Vogt S, Assael D, McDaniel H (2012) Significant silicon accumulation by marine picocyanobacterial. Nat Geosci 5:886–891

Barsanti L, Gualtieri P (2006) Algae: anatomy, biochemistry, and biotechnology. Taylor & Francis, Boca Raton

Beusen AHW, Bouwman AF, Dürr HH, Dekkers ALM, Hartmann J (2009) Global patterns of dissolved silica export to the coastal zone: Results from a spatially explicit global model. Glob Biogeochem Cycles 23:GB0A02. https://doi.org/10.1029/2008GB003281

Bidle KD, Azam F (1999) Accelerated dissolution of diatom silica by marine bacterial assemblages. Nature 397:508–512

Brzezinski MA (1985) The Si:C:N ratio of marine diatoms: interspecific variability and the effect of some environmental variables. J Phycol 21:347–357

Brzezinski MA, Olson RJ, Chisholm SW (1990) Silicon availability and cell-cycle progression in marine diatoms. Mar Ecol Prog Ser 67:83–96. https://doi.org/10.3354/meps067083

Brzezinski MA, Phillips DR, Chavez FP, Friederich GE, Dugdale RC (1997) Silica production in the Monterey, California, upwelling system. Limnol Oceanogr 42:1694–1705

Claquin P, Martin-Jézéquel V, Kromkamp JC, Veldhuis MJW, Kraay GW (2002) Uncoupling of silicon compared to carbon and nitrogen metabolism, and role of the cell cycle, in continuous cultures of *Thalassiosira pseudonana* (Bacillariophyceae) under light, nitrogen and phosphorus control. J Phycol 38:922–930

Claquin P, Leynaert A, Sferratore A, Garnier J, Ragueneau O (2006) Physiological ecology of diatoms along the river-sea continuum. In: Ittekkot V, Unger D, Humborg C, An NT (eds) The

silicon cycle, human perturbations and impacts on aquatic systems. Island Press, Washington, DC, pp 121–138
Conley DJ, Kilham SS, Theriot E (1989) Differences in silica content between marine and freshwater diatoms. Limnol Oceanogr 34:205–213
Conley DJ, Schelske CL, Stoermer EF (1993) Modification of silica biogeochemistry with eutrophication in aquatic systems. Mar Ecol Progr Ser 101:79–192
Denys L (1999) A diatom and radiocarbon perspective of the paleoenvironmental history and stratigraphy of Holocene deposits between Oostende and Nieuwpoort (western coastal plain, Belgium). Geol Belgica 2(3–4):111–140
Ding S, Chen P, Liu S, Zhang G, Zhang J, Dan SF (2019) Nutrient dynamics in the Changjiang and retention effect in the Three Gorges Dam. J Hydrol 574:96–109
Drum RW, Pankratz SH (1964) Post mitotic fine structure of *Gomphonema parvulum*. J Ultrastruct Res 10:217–223
Dugdale RC, Wilkerson FP, Minas HJ (1995) The role of a silicate pump in driving new production. Deep Sea Res 42:697–719
Durak GM, Taylor AR, Walker CE, Probert I, de Vargas C, Audic S et al (2015) A role for diatom-like silicon transporters in calcifying coccolithophores. Nat Commun 7:10543
Finkel ZV (2016) Silicification in the microalgae. In: Borowitzka MA, Beardall J, Raven JA (eds) The physiology of microalgae, developments in applied phycology, vol 6. Springer, Dordrecht, pp 289–300. https://doi.org/10.1007/978-3-319-24945-2_13
Furhmann T, Landwehr S, El Rharbi-Kucki M, Sumper M (2004) Diatoms as living photonic crystals. Appl Phys B 78:257–260
Gebühr C, Wiltshire KH, Aberle N, van Beusekom JEE, Gerdts G (2009) Influence of nutrients, temperature, light and salinity on the occurrence of *Paralia sulcata* at Helgoland Roads, North Sea. Aquat Biol 7:185–197
Glibert PM, Wilkerson FP, Dugdale RC, Raven JA, Dupont CL, Leavitt PR, Parker AE et al (2016) Pluses and minuses of ammonium and nitrate uptake and assimilation by phytoplankton and implications for productivity and community composition, with emphasis on nitrogen-enriched conditions. Limnol Oceanogr 61:165–197. https://doi.org/10.1002/lno.102033
Goering JJ, Nelson DM, Carter JA (1973) Silicic acid uptake by natural populations of marine phytoplankton. Deep Sea Res 20:777–789
Grill G, Lehner B, Lumsdon AE, Macdonal GK, Zarfl C, Liermann CR (2015) An index-based framework for assessing patterns and trends in river fragmentation and flow regulation by global dams at multiple scales. Environ Res Lett 10:015001. https://doi.org/10.1088/1748-9326/10/1/015001
Guiry MD, Guiry GM (2015) AlgaeBase. National University of Ireland. https://www.algaebase.org. Accessed 4 Oct 2022
Hildebrand M, Volcani BE, Gassmann W, Schroeder JI (1997) A gene family of silicon transporters. Nature 385:688–689
Hildebrand M, Lerch SJL, Shrestha RP (2018) Understanding diatom cell wall silicification—moving forward. Front Mar Sci 5. https://doi.org/10.3389/fmars.2018.00125
Humborg C, Conley DJ, Rahm L, Wulff F, Cociasu A, Ittekkot V (2000) Silicon retention in river basins: Far-reaching effects on biogeochemistry and aquatic food webs in coastal marine environments. Ambio 29:45–50
Krause JW, Duarte CM, Marquez IA, Assmy P, Fernéndez-Méndez M, Wiedmann I, Wassmann P et al (2018) Biogenic silica production and diatom dynamics in the Svalbard region during spring. Biogeosciences 15:6503–6517. https://doi.org/10.5194/bg-15-6503-2018
Kristiansen S, Hoell EE (2002) The importance of silicon for marine production. Hydrobiologia 484:21–31
Kröger N, Deutzmann R, Sumper M (1999) Polycationic peptides from diatom biosilica that direct silica nanosphere formation. Science 286:1129–1132. https://doi.org/10.1126/science.286.5442.1129

Kröger N, Deutzmann R, Bergsdorf C, Sumper M (2000) Species-specific polyamines from diatoms control silica morphology. Proc Natl Acad Sci USA 97:14133–14138. https://doi.org/10.1073/pnas.260496497

Li C-W, Volcani B (1984) Aspects of silicification in wall morphogenesis of diatoms. Philos Trans Roy Soc Lond B 304:519–528

Lipps JH (1970) Ecology and evolution of silicoflagellates. In: Yochelson (ed) Proceedings of the North American Paleontological Convention, vol 2. The Paleontological Society special publication, Chicago, pp 965–993

Liu D, Glibert PM (2018) Physiological linkage of nitrogen enrichment to enhanced silicification of diatoms in winter and implications for Si cycling and harmful algal blooms. Mar Ecol Progr Ser 604:51–63. https://doi.org/10.3354/meps12747

Liu D, Sun J, Zhang J, Liu G (2008) Response of the diatom flora in Jiaozhou Bay, China to environmental changes during the last century. Mar Micropaleontol 66:279–290

Liu D, Shen DX, Di B, Shi Y, Keesing JK, Wang Y, Wang Y (2013) Palaeoecological analysis of phytoplankton regime shifts in response to coastal eutrophication. Mar Ecol Progr Ser 457:1–14

Martin-Jézéquel V, Hildebrand M, Brzezinski MA (2000) Silicon metabolism in diatoms: implications for growth. J Phycol 36:821–840

McNeill JR, Engelke P (2014) The Great Acceleration: An environmental history of the Anthropocene since 1945. Harvard University Press, Cambridge, MA. 288 pp

McQuoid MR, Nordberg K (2003) The diatom *Paralia sulcata* as an environmental indicator species in coastal sediments. Estuar Coast Shelf Sci 56:339–354

Nelson DM, Goering JJ (1977) A stable isotope tracer method to measure silicate uptake by marine phytoplankton. Anal Biochem 78:139–147

Nunn BL, Faux JF, Hippmann AA, Maldonado MT, Harvey HR, Goodlett DR, Boyd PW, Strzepek RF (2013) Diatom proteomics reveals unique acclimation strategies to mitigate Fe limitation. PLoS One 8(10):e75653

Ogden C (1979) An ultrastructural study of division in Euglypha (Protozoa: Rhizopoda). Protistologica 15:541–556

Ratcliffe S, Meyer EM, Walker CE, Knight M, McNair HM, Matson PG, Iglesias-Rodriguez D et al (2023) Characterization of the molecular mechanisms of silicon uptake in coccolithophores. Environ Microbiol 25:315–330. https://doi.org/10.1111/1462-2920.16280

Rogall E (1939) Uber den feinbau der kieselmembran der diatomeen. Planta 29:279–281

Sandgren CD (1989) SEM investigations of statospore (stomatocyst) development in diverse members of the Chrysophyceae and Synurophyceae. Nova Hedwigia Beih 95:45–69

Sandgren CD, Hall SA, Barlow SB (1996) Siliceous scale production in Chrysophyte and Synurophyte algae. I. Effects of silica-limited growth on cell silica content, scale morphology, and the construction of the scale layer of *Synura petersenii*. J Phycol 32:675–692

Santos J, Almeid SF, Figueira E (2013) Cadmium chelation by frustulins: a novel metal tolerance mechanism in *Nitzschia palea* (Kützing) W. Smith. Ecotoxicology 22:166–173

Simpson TL, Volcani BE (1981) Introduction. In: Simpson TL, Volcani BE (eds) Silicon and siliceous structures in biological systems. Springer, New York, p 587

Smayda TJ (1990) Novel and nuisance phytoplankton blooms in the sea: evidence for a global epidemic. In: Granéli E, Sundstrom B, Edler L, Anderson DM (eds) Toxic marine phytoplankton. Elsevier, New York, pp 29–40

Strokal M, Yang H, Zhang Y, Kroeze C, Li L, Luan S, Wang H et al (2014) Increasing eutrophication in the coastal seas of China from 1970 to 2050. Mar Poll Bull 85:123–140

Sutton JN, André L, Cardinal D, Conley DJ, de Souza GF, Dean J, Dodd J et al (2018) A review of the stable isotope biogeochemistry of the global silicon cycle and its associated trace elements. Front Earth Sci 5:112. https://doi.org/10.3389/feart.2017.00112

Taucher J, Bach LT, Prowe AEF, Boxhammer T, Kvale K, Riebesell U (2022) Enhanced silica export in a future ocean triggers global diatom decline. Nature 605. https://doi.org/10.1038/s41586-022-04687-0

Thamatrakoln K, Hildebrand M (2005) Approaches for functional characterization of diatoms silicic acid transporters. J Nanosci Nanotechnol 5:158–166. https://doi.org/10.1166/jnn.2005.014

Thamatrakoln K, Hildebrand M (2008) Silicon uptake in diatoms revisited: a model for saturable and nonsaturable uptake kinetics and the role of silicon transporters. Plant Physiol 146:1397–1407. https://doi.org/10.1104/pp.107.107094

Tréguer P, Lindner L, Van Bennekom AJ, Leynaert A, Panouse M, Jacques G (1991) Production of biogenic silica in the Weddell-Scotia Seas measured with ^{32}Si. Limnol Oceanogr 36:1217–1227

Turner RE, Rabalais NN (1994) Coastal eutrophication near the Mississippi river delta. Nature 368: 619–621

Vörösmarty CJ, Meybeck M, Fekete B, Sharma K, Green P, Syvitski JPM (2003) Anthropogenic sediment retention: Major global impact from registered river impoundments. Glob Planet Change 39:169–190

Vreiling EG, Gieskes W, Beelen TP (1999) Silicon deposition in the diatoms: control by the pH inside the silicon deposition vesicle. J Phycol 35:548–559

Wulff F, Stigebrandt A, Rahm L (1990) Nutrient dynamics of the Baltic Sea. Ambio 19:126–133

Zarfl C, Lumsdon AE, Berlekamp J, Tydecks L, Tockner K (2015) A global boom in hydropower dam construction. Aquat Sci 77:161–170. https://doi.org/10.1007/s00027-014-0377-0

Mixotrophy: Microbial Multitasking 15

15.1 Introduction

Our understanding of "phytoplankton" growth and productivity and their relationship with "zooplankton" has historically been based on our perception that they, like their terrestrial counterparts, fall into the logical and seemingly natural divisions of plants and animals. Accordingly, decades of work, and hundreds, if not thousands, of papers and books—much like the previous chapters of this book—have been published on phytoplankton describing their photosynthetic characteristics and their dependence on inorganic nutrient substrates and how these processes are modulated by environmental conditions. What is comparatively recent (all things considered) is the recognition that most phytoplankton can consume particles, and many zooplankton can consume algae and become photosynthetic. It was once thought that organisms that could combine phototrophy and phagotrophy were just curiosities. They were worthy of reports as novelties or rarities, in much the same way as Darwin (1875) described the curious botanical, terrestrial mixotrophs that could consume insects. This thought-to-be weird or curious behavior by microorganisms was described in the early 1900s (Pascher 1917; Biecheler 1936), but little thought was given to the magnitude or ecological significance of this nutritional strategy at the time (see reviews in Jacobson 1999; Stoecker 1999). Now, it is recognized that most protist plankton communities are largely comprised of mixotrophs (Flynn et al. 2013; Mitra et al. 2014a, b), and this lifestyle fundamentally changes our understanding of the flow of energy and materials in the food web and has implications for productivity, ecosystem stability, and biodiversity. Mixotrophy represents a paradigm shift in the understanding of food web dynamics (Glibert and Mitra 2022), and mixotrophy has important implications for the understanding of the evolutionary processes that have shaped the diversity of plankton protists (see Chap. 10). Mixotrophy is now considered so important in the plankton communities that it has been proclaimed as one of the recent "revolutions in science/discoveries that could change everything" (https://static.scientificamerican.com/sciam/assets/File/OSQ318_Revolutions.pdf).

P. M. Glibert, *Phytoplankton Whispering: An Introduction to the Physiology and Ecology of Microalgae*, https://doi.org/10.1007/978-3-031-53897-1_15

The community of protists with the capability to eat and to photosynthesize is now termed mixoplankton (Flynn et al. 2019). Using this contemporary definition, "mixoplankton" can be considered to be those organisms traditionally considered algae that have been shown to eat, as well as those organisms traditionally considered pure heterotrophs that have been shown to consume algal prey and that use the photosynthetic machinery of the ingested prey. Now it is recognized that there are distinct mixotrophic strategies, but even within specific types of mixotrophs, there may be a continuum of responses (e.g., Sanders 1991; Sanders et al. 1992; Jones 1994; Stoecker 1998; Flynn et al. 2013; Stoecker et al. 2017; Selosse et al. 2017). For example, some dinoflagellates eat aggressively and some more reluctantly. This chapter explores the processes of mixotrophy, advantages and traits of mixoplankton, mechanisms of feeding and implications for conceptualizing food web interactions, and current classifications of mixoplankton and their global distribution.

15.2 Differentiating Osmo- from Mixotrophy

In its broadest definition, mixotrophy is the ability of an organism to combine autotrophy and heterotrophy, and thus for an organism to obtain its requisite nutrition from both inorganic and organic sources, dissolved or particulate. When mixotrophy only involves the uptake of dissolved organic compounds together with inorganic compounds, the process is normally termed osmotrophy. In recent years, the term mixotrophy has taken on a more focused definition when referring to plankton: the combination of photosynthesis (phototrophy) and phagotrophy and the ingestion of particulate material. One caveat, however: The use of the term mixotrophy as the combination of phagotrophy and phototrophy is more common in the marine literature. Many freshwater plankton ecologists continue to use osmotrophy and mixotrophy interchangeably (Table 15.1).

The importance of osmotrophy has been recognized for many decades (reviewed by Syrett 1981; Glibert and Legrand 2006), with phytoplankton well documented to be able to grow on amino acids or to take up other complex organic compounds. Carbohydrates (glucose), amino acids, and humic and fulvic acids are just a sampling of the diversity of compounds that can be taken up by algae. Organic compounds containing nitrogen (N) are now well recognized to be actively taken up, and there are many reviews covering the diversity of compounds and conditions under which such uptake pathways may be favored (e.g., Antia et al. 1991; Berman and Bronk 2003; Bronk et al. 2007). The phenomenon of mixotrophy as osmotrophy by plankton autotrophs has been recognized for decades and was recognized as widespread among some flagellate groups (Granéli and Carlsson 1998). Yet, even with this definition, based on results from the occasional laboratory experiment, the early understanding was that bacteria (once the significance of their abundance was fully established) were far superior to microalgae in competing for organic substrates due to their transport systems that algae presumably lacked (e.g., Wright and Hobbie 1965). Thus, it was generally concluded that osmotrophy was marginal in its

Table 15.1 Common terms related to autotrophy and mixotrophy and their accepted usage

Autotrophy/autotroph	A mode of nutrition involving the synthesis of complex organic substances using photosynthesis (phototrophy) or chemosynthesis/organisms that undertake this nutritional mode. Typically associated with the use of inorganic nutrients
Bacterivory/bacterivore	A mode of nutrition involving the consumption of bacteria/organisms that undertake this nutritional mode
Osmotrophy/osmotroph	A mode of heterotrophy involving the uptake and consumption of DOM, also referred to as osmo-hetero-trophy/organisms that undertake this nutritional mode
Heterotrophy/heterotroph	A mode of nutrition involving the uptake of organic substances or food via grazing/organisms that undertake this nutritional mode
Mixotrophy/mixotroph	Nutrition involving both autotrophy and heterotrophy/organisms that undertake this nutritional mode
Mixoplankton	Plankton protists capable of obtaining nourishment via photoautotrophy and phagoheterotrophy and osmoheterotrophy
Phagotrophy/phagotroph	A mode of heterotrophy involving the engulfment of particles (often whole organisms) into a phagocytic vacuole in which digestion occurs, also referred to as phago-hetero-trophy/organisms that undertake this nutritional mode
Phototrophy/phototroph	A mode of autotrophy involving the fixation of CO_2 using energy derived from light, also referred to as photoautotrophy/organisms that undertake this nutritional mode
Kleptoplastidy/kleptochloroplast	A mode of mixotrophy involving the consumption of cells with photosynthetic machinery—and in which the chloroplasts remain protected from digestion, and functional for periods of days to weeks/the retained chloroplast
Karyoklepty/kleptokaryon	A mode of mixotrophy involving the consumption of cells with photosynthetic machinery—and in which the chloroplasts and nucleus remain protected from digestion, and functional for periods of days to weeks/the retained nucleus
Constitutive mixotrophy/mixotrophs/constitutive mixoplankton	Protist plankton with an inherent capacity for phototrophy that can also exhibit phagotrophy/organisms that undertake this nutritional mode
Non-constitutive mixotrophy/mixotrophs/non-constitutive mixoplankton	Protist plankton that acquire the capability for phototrophy from consumption (via phagotrophy) of phototrophic prey/organisms that undertake this nutritional mode

Modified from Glibert and Mitra (2022, Limnol Oceanogr 67: 585–597. doi: 10.1002/lno.12018) under a creative commons license importance for phytoplankton (reviewed by Syrett 1981), and so mixotrophy was marginalized as an important nutritional strategy and as a concept.

15.3 Early Mixotrophy Feeding Observations: Recognizing Bacterivory

Two concurrent advances in the 1980s and 1990s advanced the recognition of mixoplankton. As with many discoveries, method advancements and conceptual discoveries went hand in hand (Chap. 2). First, the importance of mixotrophy to harmful algae became recognized, about the 1980s (Thingstad et al. 1996). At the simplest level, there are advantages to eating your competitor, and phytoflagellates that consume other algae or bacteria may gain this advantage (Thingstad et al. 1996). This strategy was suggested as one mechanism by which harmful species may gain an advantage over their competitors. Second, fluorescent tools became more readily available, and probes, including fluorescently labeled bacteria, or fluorescently labeled microbeads, were being applied. Experiments on phagotrophy of fluorescently labeled bacteria (FLB) by Nygaard and Tobiesen (1993) suggested that bacterivory could be an important source of phosphorus (P) for the haptophyte *Prymnesium polylepis* (reported as *Chrysochromulina polylepis*) during blooms in coastal Norwegian waters. Ingestion of P-rich bacteria by *P. polylepis* (6 cells h^{-1}) in the natural phytoplankton assemblage from surface waters was estimated to have comprised about 60% of the total bacterial grazing. Supporting laboratory experiments on several harmful species fed FLB or radiolabeled bacteria (RLB, ^{14}C-amino acids) indicated that bacterivory increased under P limitation and provided significantly more P than was needed to maintain equilibrium population growth rates ($k = 0.3$ day^{-1}). In experiments with P-limited continuous cultures of *Heterocapsa akashiwo* and *Karlodinium veneficum* (reported as *Gyrodinium galatheanum*), rates of consumption of up to 113 and 48 bacteria $cell^{-1}$ h^{-1}, respectively, and up to 706 bacteria $cell^{-1}$ h^{-1} by *Alexandrium fundyense* (reported as *A. tamarense*) were determined (reviewed by Burkholder et al. 2008). These efforts largely focused on who ate whom and at what rate. Less emphasis was placed on understanding the factors controlling feeding or the significance of mixotrophic feeding on the ecological functioning of the system.

These early experiments and measurements of feeding assumed that bacterivory or mixotrophy provided a nutrient source when nutrients were otherwise limiting and that feeding would be a function of the available food. Yet, mixotrophy within harmful algae was beginning to be reported by taxa that were more common in eutrophic waters, where nutrients presumably would be available in inorganic form (Burkholder et al. 2008). It became clear during the period of these studies that many harmful algae that thrive in eutrophic habitats are mixotrophs that respond both directly to nutrient inputs and indirectly through high abundance of bacterial and algal prey that are stimulated by the elevated nutrients. Mixotrophy may be promoted by low light and/or nutrient deficiency (Legrand et al. 1998; Granéli et al. 1999; Stoecker 1999; Stoecker et al. 2006), conditions that commonly occur in eutrophic systems. Low light can develop after precipitation events, wind mixing, or other disturbance, especially in shallow waters, and eutrophic habitats offer a wealth of dissolved organic substrates from high organic loads and substances excreted, secreted, or lysed by bacteria, algae, and other aquatic organisms (e.g., Bjørnsen 1988; Bronk et al. 2007; Verity et al. 2007), as well as abundant "particulate" bacterial, algal, and other prey whose growth is stimulated by the elevated nutrients (e.g., Stoecker et al. 1997; Lewitus et al. 1999; Berg et al. 2003; Adolf et al. 2008). Even in eutrophic waters, however, nutrient deficiency can develop. It can develop seasonally, or following algal blooms, and there may be conditions when one nutrient (e.g., N) may be exceptionally high while another (e.g., P) may be very low (Hecky and Kilham 1988; Cloern and Dufford 2005; Glibert and Burkholder 2011). Experiments in Chesapeake Bay showed that *Prorocentrum cordatum* (reported as *P. minimum*) ingested cryptophytes (counted as naturally orange-fluorescent inclusions, OFIs; Stoecker et al. 1997). Those studies showed that mixotrophy was higher in summer and that the frequency of OFIs was positively correlated with cryptophyte densities, but OFIs were not necessarily abundant in all populations of *P. cordatum* when cryptophyte densities were high. Ingestion rates were also highest in the afternoon and evening and lowest in the morning, and additions of $NO_3^- + PO_4^{3-}$ inhibited feeding.

15.4 Mixotrophy Synergism

The next major leap in understanding of mixotrophy came with the recognition that autotrophy and phagotrophy may be synergistic in terms of nutrition, not simply additive (Flynn and Mitra 2009; Mitra and Flynn 2010; Flynn et al. 2013; Mitra et al. 2014a, b). The synergy in the use of dissolved substrates and material gained from grazing comes from the provision of additional nutrients (N, P, Fe) from feeding to support primary production, together with a supply of C that can supplement photosynthesis. That is, even though mixotrophs may feed to acquire one nutrient, or C, they acquire other nutrients just by virtue of feeding. Furthermore, having an additional source of organic C supplements that which the cells gains from photosynthesis. Thus, by being a mixotroph the cell may be better able to balance C lost through respiration (Putt 1990; Stoecker and Michaels 1991). Consequently, mixotrophy can lead to a significant increase in growth rate. For example, when growing as a phototroph, the dinoflagellate *Margalefidinium* (reported as *Cochlodinium*) *polykrikoides* has a growth rate of 0.17 divisions day^{-1} (Jeong et al. 2004). Yet, when growing as a mixotroph, consuming cryptophytes, the division rate of *M. polykrikoides* doubles to 0.34 divisions day^{-1} (Jeong et al. 2004). Toxigenic *Karenia brevis* has been found to graze on the cyanobacterium *Synechococcus* sp. (Jeong et al. 2005a; Glibert et al. 2009). In laboratory experiments, Jeong et al. (2005a) estimated that up to 5 cells h^{-1} of *Synechococcus* were grazed by mixotrophic *K. brevis,* while Glibert et al. (2009) found that from ~1 to 80 *Synechococcus* cells h^{-1} were grazed by *K. brevis*, depending on the predator:prey ratio. The growth rate of *K. brevis* increased as the supply of *Synechococcus* increased, indicating that natural variability in this food source may affect growth rates of this HAB species in nature. In contrast, *Prymnesium parvum* (reported as *Prymnesium patelliferum*) and *Chrysochromulina* spp., which ingested prey under light- and nutrient-sufficient conditions, maintained similar growth rates with versus without phagotrophy (Pintner and Provasoli 1958; Larsen et al. 1993; Granéli and Carlsson 1998).

Collectively, these studies showed that most phytoflagellates (heterokontophytes, haptophytes, and cryptophytes) are capable of consuming bacteria and small eukaryotic cells (Unrein et al. 2007; Zubkov and Tarran 2008; Stukel et al. 2011). Dinoflagellates exhibit a wide range in capabilities, from those that express phagotrophy only upon extreme nutrient starvation to exclusive phagotrophic forms. Even the latter organisms may gain at least transitory inorganic C-fixing capacity from ingestion of phototrophs (Raven et al. 2009; Jeong et al. 2010; Hansen 2011) or other mixotrophs (Minnhagen et al. 2011). In fact, the bulk of eukaryotic planktonic protists commonly classified as phototrophs is actually mixotrophic protists (Flynn et al. 2013; Mitra et al. 2016).

15.5 Feeding Mechanisms

There is much yet to be learned about the physiological mechanisms, costs, and benefits of mixotrophy (Fig. 15.1). Although mixotrophs that have been traditionally considered heterotrophs presumably have feeding mechanisms, there are many questions related to how they acquire food and how such mechanisms may relate to food selectivity.

There are varied feeding mechanisms of protists. Regardless of how the cell captures its prey, ingestion appears to be triggered by receptors that trigger the ingestion into membrane-bound vacuoles. There, the nutrients and C are made available via biosynthetic pathways, while the indigestible material is egested (Wilken et al. 2019). Prey may be ingested via direct engulfment. For dinoflagellates, the opening for direct engulfment may vary. Most engulf their prey through a sulcus (e.g., Skovgaard 1996; Jeong et al. 2005b), but some may take in their prey via the apical horn or through sutures on the sides of their bodies (Jeong et al. 2005b). Jeong et al. (2010) reported having witnessed the dinoflagellate *Prorocentrum micans* engulfing cryptophyte prey at three different locations on the cell simultaneously. Moreover, the relative size of these openings can be a critical factor in limiting the size of the prey cell that can be consumed.

For a dinoflagellate to engulf a larger cell than that which can slip through a small opening, pallium or peduncle feeding mechanisms may be involved. Via this mechanism, a dinoflagellate can consume a cell many times its own size (e.g., Jacobson and Anderson 1986), and if several dinoflagellates are co-located, they can simultaneously feed on the same prey cell using this approach (Jeong 1994).

Protist predators may also generate feeding currents in order to concentrate the prey cells. These currents may be generated using the flagella. This is an important

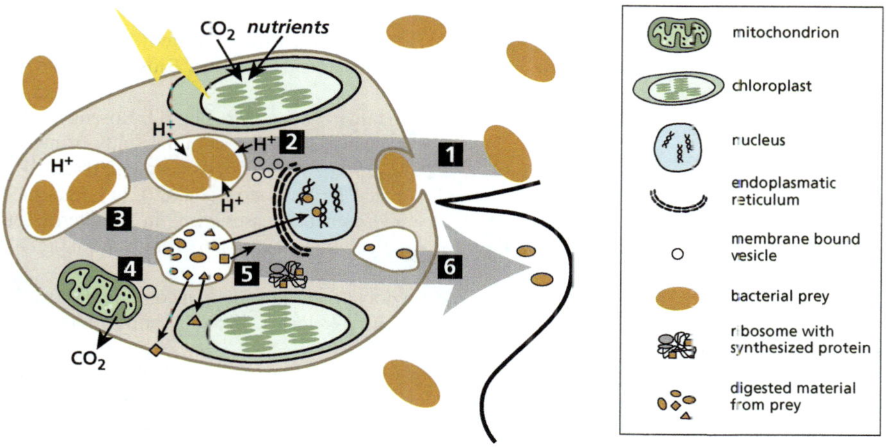

Fig. 15.1 Phagocytotic process in a mixotrophic flagellate, including detection and ingestion of a prey item (e.g., bacterium and orange) (1), modification of the food vacuole properties (2), creating the acidic conditions required for lysosomal enzyme activities and prey digestion (3), resorption of molecules from digested prey into the cytosol (4) or their incorporation in macromolecule biosynthesis, (5) and egestion of remaining material. Illustrated cellular structures are plastids with thylakoids (green), nucleus (blue), neighbored by ER (gray), and mitochondrion (purple). Image redrawn based on Wilken et al. (2019, Phil Trans Roy Soc B 374: doi: 10.1098/rstb.2019.0090)

strategy by which dinoflagellates may graze on bacteria or other tiny cells that they would otherwise have difficulty detecting or capturing (Jeong et al. 2010). These mechanisms of feeding are also not either-or; a single dinoflagellate may employ one or more of these mechanisms depending on prey availability or other factors. Thus, a meal for a dinoflagellate may vary from heterotrophic bacteria to picocyanobacteria and nanoflagellates to diatoms, other dinoflagellates and even ciliates, small larvae, eggs, or other morsels that may be in the aqueous environment (Jeong et al. 2010).

It has recently been shown that the dinoflagellate *Prorocentrum* c.f. *balticum* uses mucoid prey capture, a "mucosphere" that attracts and immobilizes prey (Larsson et al. 2022). The prey become entrapped, yielding a concentrated food source for the mixotroph. Recent evidence for this dinoflagellate suggests that it makes a new mucosphere on a daily basis (Fig. 15.2). Given that the mucosphere is composed of C-rich acidic polysaccharides, photosynthesis is required for its production. Moreover, these mucus materials are negatively buoyant and thus contribute to particulate organic C (POC) and to vertical flux of C from the euphotic zone when discarded when the mixotrophic predator has satiated its food intake.

There are many physiological implications for a small cell, otherwise photosynthetic, to engulf prey. One of the most basic is that the cell volume is finite—and small—and the ingested material may displace volume that might otherwise be occupied by photosynthetic machinery. This would be especially true for low-light-adapted mixotrophs, where the photosynthetic machinery would be upregulated, consuming more of the cell's volume. However, if rates of

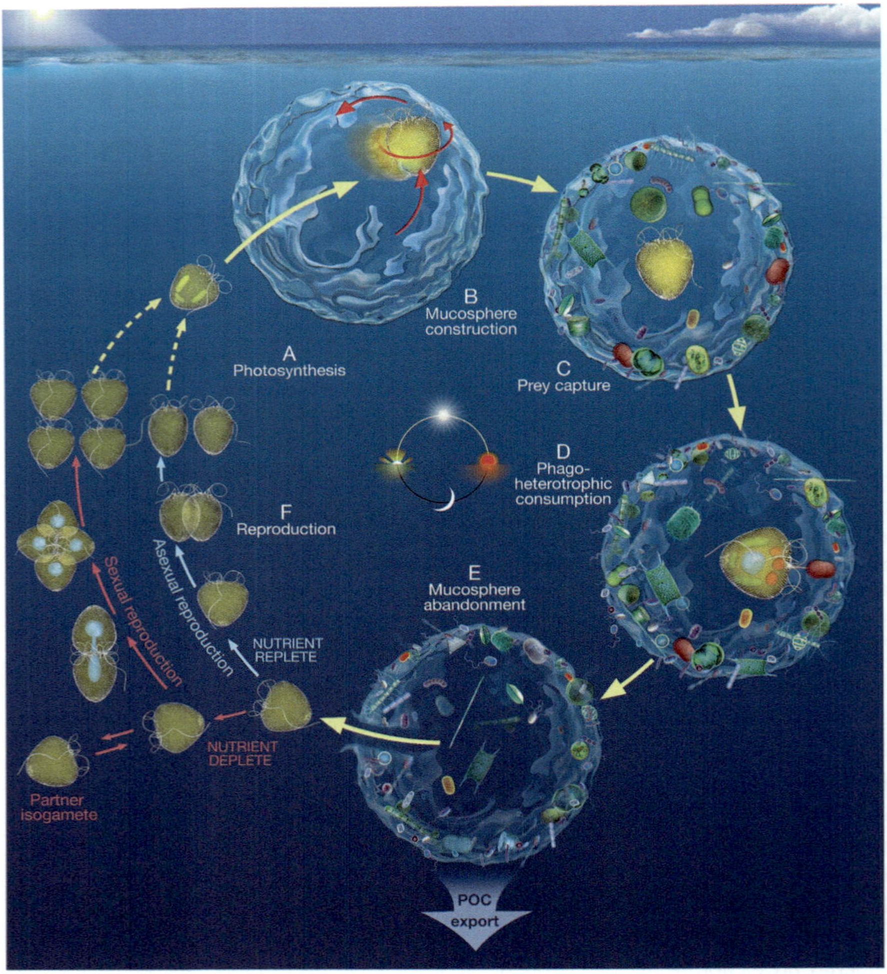

Fig. 15.2 Summary of *Prorocentrum* cf. *balticum* behaviors. An infographic showing the daily cycles of A—photosynthesis, B—mucosphere production, C—prokaryotic and eukaryotic prey attraction and capture, D—peduncular phago-heterotrophic feeding, E—mucosphere abandonment and carbon export potential, and F—asexual or sexual reproduction. The solid yellow arrows represent the daily cycle of the *P.* cf. *balticum* cell including entering the mucosphere during construction in (B) and exiting the mucosphere in (E). The light blue arrows show the *P.* cf. *balticum* cell through the asexual reproduction stage and the pink arrows represent the sexual reproduction stage. The broken yellow arrows signify one of the succeeding *P.* cf. *balticum* cells from either of the reproduction stages re-entering the daily cycle. Image reproduced from Larsson et al. (2022, Nat Commun 13: 1301. doi:10.1038/s41467-022-28867-8) under a creative commons license

photosynthesis are high in the grazing cell, then internal feedbacks might depress the rate of digestion, if the goal of taking in prey is to supplement C acquisition (e.g., Flynn and Mitra 2009). Respiration rates also change. For example, Hansen et al.

(2000) found that respiration rates were elevated twofold in a photosynthetic cell of dinoflagellate *Fragilidium subglobosum* compared to the same cell actively grazing—more of the C was retained.

For the dinoflagellate *Karlodinium veneficum* (reported as *K. micrum*), the cell seems to need a critical minimum level of photosynthesis to drive feeding. Whether this is specific fatty acids or other metabolites, the rate of ingestion and digestion is controlled, and feeding is limited to the light period (Adolf et al. 2007).

15.6 Acquiring Chloroplasts

A second general type of mixotrophs, those that are traditionally considered to be heterotrophs but which become photosynthetic via ingested prey, also has a complex physiology. It can be argued that adding photosynthesis—and all of its requisite machinery—to an organism that is primarily heterotrophic is an evolutionary expensive proposition (e.g., Raven 1997). By comparison, from an energetic perspective, adding heterotrophy to an organism that is otherwise photosynthetic is far less expensive. Despite the elegance of photosynthesis, and the extent to which a cell can regulate the rates by which it occurs (within the constraints of light and nutrient availability), it can be suggested that acquiring C via grazing is far more efficient than acquiring C via photosynthesis (Flynn and Mitra 2009). With mixotrophs that are primarily heterotrophic feeding on cells that are photosynthetic, food is taken in but the photosynthetic machinery—the chloroplasts—may be protected from digestion for periods of days to weeks (Lewitus et al. 1999). These chloroplasts, termed kleptochloroplasts, as they are stolen from their original host, retain functionality, but the extent to which their functionality is retained is highly variable (Schnepf and Elbrächter 1992). The grazer also can retain nutrients that may be regenerated in the cell from digestion, re-assimilating them rather than releasing them back to the environment. Flynn and Mitra (2009) further suggest that photosynthesis via kleptochloroplasts is more efficient than that by native chloroplasts because the C fixed by them is not used in repair or synthesis of new chloroplasts, but only used in enhancing core C biomass of the mixotroph. However, there may be costs associated with kleptochloroplasty, for example, the risk of additional oxidative stress that comes from the added chloroplasts (e.g., McManus et al. 2012). This would necessitate mitigation of reactive oxygen species (ROS), which would be produced during the photosynthesis from these chloroplasts. In fact, it has been suggested that the inactivation of chloroplast antioxidant enzymes may contribute to the loss of kleptochloroplast activity over time—and accordingly the need to replenish these chloroplasts—in the mixotroph ciliate, *Strombidium rassoulzadegani* (McManus et al. 2012).

It has also been observed that kleptochloroplasts, depending on their origin, may become enlarged after ingestion due to generation of additional thylakoid membranes (e.g., Onuma and Horiguchi 2013). Once the kleptochloroplast begins to fail, it becomes digested and its core elements are either used or voided. The integrity of ingested chloroplasts and their lifetime of service within the mixotroph may be temperature-dependent as the digestive processes may be enhanced at higher temperatures (e.g., Li et al. 2001). In some cases, dividing mixotrophs may apportion their kleptochloroplasts to the daughter cells, but functionality may be affected. For example, *Mesodinium rubrum* divides its prey chloroplasts along with its cell divisions (Johnson et al. 2006). An understanding of this process is most relevant to an understanding of the evolutionary acquisition of permanent chloroplasts. Lin et al. (2018) demonstrated that, in warmer waters (>25 °C), *K. veneficum* appeared to be more mixotrophic and showed higher biomass, reflecting temperature regulation of nutritional mode.

There are two general types of kleptochloroplasty. In the first type, the host mixotroph cell steals the plastids, but digests or rejects the other parts of the ingested cell. The second type not only retains the plastids but also retains other functionality or organelles, such as mitochondria or nuclei (Johnson et al. 2007). In the latter, as shown by the mixotrophic, kleptochloroplastidic ciliate, *Mesodinium rubrum,* the prey nucleus retains its transcriptional activity. This type of kleptochloroplasty has been termed "karyoklepty," and the symbiont nucleus has been termed a "kleptokaryon" (Johnson et al. 2007). These prey nuclei, like prey chloroplasts, may become enlarged over time during prey starvation; they may fuse (Kim et al. 2016, 2017; Nam et al. 2016).

15.7 Evolutionary Importance of Mixotrophy

The evolutionary diversity of eukaryotic plankton is thought to have involved multiple plastic acquisition events (Delwiche 1999; Falkowski et al. 2004; Keeling 2010; Caron 2016; see Chap. 10). Indeed, without phagocytosis, there would be no diversity of life as we know it (e.g., Keeling 2013; Mansour and Anestis 2021). Thus, phago-mixotrophic lineages are widespread across the tree of life (Millette et al. 2023; Fig. 15.3). While there is considerable debate about the actual number of plastic acquisition and loss events and how this has shaped the tree of life, there appears to be little doubt that the capability for heterotrophy within the "phytoplankton" may be analogous to the mechanisms that shaped ultimate diversity; the first mixotrophic eukaryote set the trajectory for phytoplankton evolution. The process of stealing a nucleus along with a chloroplast may be an evolutionary precursor to a fully functioning tertiary plastid (Johnson 2011). Furthermore, it is equally important to understand how plastids may have been lost evolutionarily and how that may have led to pure phototrophy (Fig. 15.3). Indeed, some red algae or glaucophytes harboring primary plastids, as well as diatoms, have lost their phagotrophic ability (Raven et al. 2009). As we learn more about gene transfer and the time scale of adaptation, as well as the metabolic and physiological trade-offs, the study of mixotrophy may yield great insight into evolutionary processes (Millette et al. 2023).

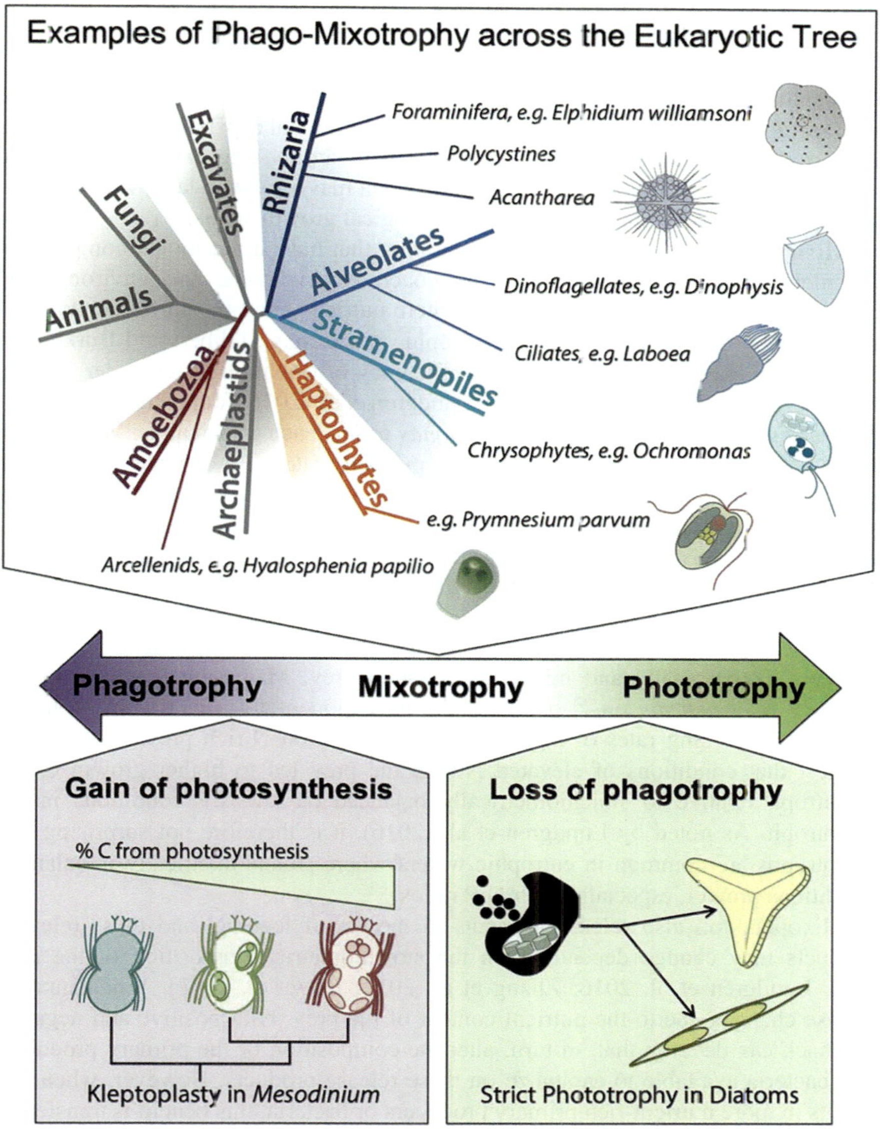

Fig. 15.3 Mixoplankton span an evolutionary and physiological spectrum. Top: Phago-mixotrophy is found across numerous lineages of the eukaryotic tree of life (some examples shown). Middle: These mixoplankton span a gradient of reliance on photosynthesis, from almost completely phagotrophic to almost completely photoautotrophic. Bottom: As such, mixoplankton also present an opportunity to study the evolutionary transition from phagotrophy to phototrophy, particularly by examining two key evolutionary transitions: (1) the evolutionary gain of phototrophy (here, exemplified by kleptoplasty in the *Mesodinium* genus, in which lineages become increasingly photosynthetic as they become more evolutionarily derived; Johnson et al. 2016) and (2) the evolutionary loss of phagotrophy (here, exemplified by the diatoms, which lost phagotrophy prior to undergoing an extensive radiation). Illustration by C. S Raphael and H. Moeller. Image reproduced from Millette et al. (2023, doi: 10.1093/plankt/fbad020) under a creative commons license

15.8 Placing Mixotrophy in the Context of Food Web Understanding

Conceptualizing and comprehending the importance of mixotrophy are much more than recognizing which plankton have this dual nutritional capability. Mixotrophy is not an either-or nutritional trait; protists have varying degrees of mixotrophic capability, yielding a spectrum of capabilities that may vary with taxa, with environmental conditions, or nutritional and physiological growth status.

Mixoplankton may selectively feed on prey that have a nutritional composition complementary to themselves—an approach to overcome the environmental challenges of living in an environment where nutrients may be out of the stoichiometric proportion needed by the mixotrophic grazer (e.g., Glibert and Burkholder 2011). This may serve to stabilize the efficiency of trophic transfer under varying nutrient regimes (Moorthi et al. 2017). Lundgren et al. (2016) found that in studies of *Prymnesium parvum* grazing on *Rhodomonas salina*, each grown under varying N:P ratios and mixed in varying combinations, mortality rates of prey were higher when prey were N-rich, regardless of the nutritional states of *P. parvum*. Furthermore, the growth of the mixotroph was higher when prey were provided in high N: P ratio conditions. Experiments by Lin et al. (2017) showed that growth rates of initially exponential- and stationary-phase *Karlodinium veneficum* cells were enhanced in the presence of prey (*Rhodomonas* sp.) with reciprocal nutrient conditions. Feeding rates (measured as prey death rates) were highest for low-N:P *K. veneficum* initially growing exponentially and mixed with N-rich prey. Maximum feeding rates of low-N:P *K. veneficum* on N-rich prey during exponential growth were ~fourfold higher than feeding rates of high N:P *K. veneficum* on N-rich prey. Both studies suggest that conditions of elevated N:P in the prey led to higher growth of the mixotroph relative to stoichiometrically balanced or low-N:P conditions in the mixotroph. As noted by Lundgren et al. (2016), it is therefore not surprising that mixotrophs are common in eutrophic waters where stoichiometrically imbalanced conditions prevail, especially high N:P ratios.

Mixoplankton also release nutrients in the act of feeding, and these released products may change depending on the stoichiometric composition of the prey (e.g., Lundgren et al. 2016; Zhang et al. 2017; Lin et al. 2017). When nutrient release changes due to the nutrient content of the prey, both positive and negative feedback can develop that, in turn, alter the composition of the primary producers and bacteria available to capitalize on these release products. However, when this results in more nutrient-rich primary producers or bacteria, this benefit is transferred to the grazer, including when the grazer is a mixoplankter, but, with mixoplankton, both the grazer and prey may be mixotrophic, as in the case of the multi-link mixotroph consortia of bacteria-fed on by a cryptophyte-fed on the ciliate *Mesodinium,* which is fed on by the dinoflagellate *Dinophysis*. Ultimately, how the resource stoichiometric proportions compare with those of the mixoplankter has propagating effects through the microbial loop, with implications for food web structure and functioning. These advantages afforded to mixotrophs may outweigh their comparatively poor growth rates in otherwise non-limiting nutrient conditions.

There are other ecological advantages to being mixotrophic—advantages that also change the functioning of the food web (Glibert and Mitra 2022). For those mixoplankton that form HABs, the relationship between mixotrophic nutrition and toxicity of many HAB taxa may also be synergistic. The toxin may harm or stun the prey, in turn making the prey easier to capture (Skovgaard and Hansen 2003; Tillmann 2003). Even for those species that are not directly toxic, mixoplanktonic activity and allelopathy may be synergistic: The mixoplankton gain, while the competitors affected by allelopathic compounds do not (e.g., John et al. 2015). All of these interactions lay themselves open to other interactions and stressors, for example, virus attack, and all of these interactions and stressors may affect in one way of the other the functioning of the microbial loop.

Productivity and/or growth may increase due to the dual channels by which the organism gains its C or nutrients. Essential elements such as N, P, or C are typically rich in microbial prey and thus mixotrophy can provide a supplemental supply when there is an elemental imbalance in dissolved (water column) nutrient substrates (Granéli et al. 1999; Vadstein 2000; Li et al. 2001; Stibor and Sommer 2003; Stoecker et al. 2006). In eutrophic environments, although nutrients may be proportionately more available than in oligotrophic environments, nutrients are often out of stoichiometric balance (Burkholder et al. 2008). In the North Sea, for example, as increasing N relative to P proportions developed in response to disproportionate nutrient reductions, mixotrophic dinoflagellates increased, and many of these were harmful species (Burson et al. 2016).

Compared with a classic "food chain," in which true phytoplankton form the base, and in which they are consumed by progressively larger grazers (Fig. 15.4a) or even one that considers the activities of other microbes (Fig. 15.4b), the inclusion of mixoplankton yields a flow of energy and materials that is fundamentally different (Fig. 15.4c). Jeong et al. (2010) described mixoplankton-dominated food webs as resembling "hubs" (in the sense of an airport hub) for energy and nutrient flow through the microbial consortium. A single species may sit at the center of the hub, serving as both predator and prey of a wide range of species. The "hub" serves to concentrate nutrients and to transfer them to higher trophic levels.

Accepting that mixotrophic plankton sit at the base of most food webs, it follows that a food web structure with "strict" phototrophic configuration represents an optimal strategy only in spring blooms in temperate waters, or in well-illuminated waters that are highly eutrophic, systems typically dominated by diatoms in what is considered immature *r*-selecting conditions (Fig. 15.5, Glibert and Mitra 2022). As the nutrient conditions change (mature) due to algal growth, and with the emergence of a prey community, the assemblage transitions to a "*K*-selecting" environment, and mixotrophy becomes selectively advantageous (Fig. 15.5; see Chap. 17 for more detail on *r* vs. *K* strategies). This more mature environment is dominated by heterotrophs that have phototrophic capabilities and phototrophs that have heterotrophic capabilities; the diversity of mixotrophic strategies is great (Mitra and Flynn 2010; Flynn et al. 2013). It is in more mature ecosystems, which often are also warmer during summer, where harmful algal blooms (HABs, see Chap. 25) often develop. The trophic and allometric upgrading in mixotrophy provides a route to

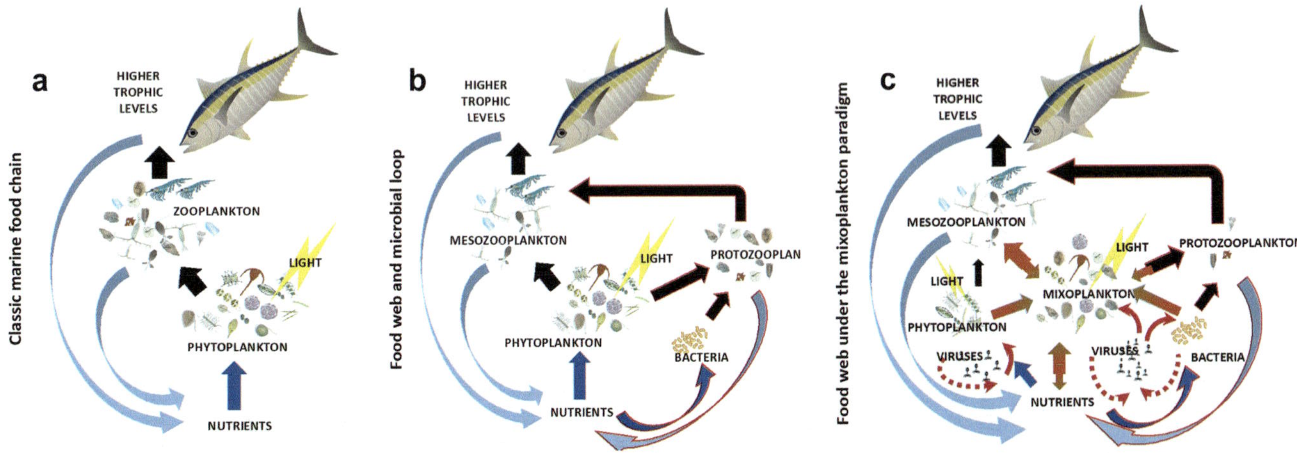

Fig. 15.4 Panel (**a**) The classic food chain in which dissolved inorganic and organic nutrients are taken up by phytoplankton, which in turn are grazed by zooplankton, and they, in turn, support higher trophic levels. Nutrient recycling occurs via the grazing processes of zooplankton and higher trophic levels. Panel (**b**) The classic food chain, now envisioned as a food web with the introduction of the microbial loop (arrows with red outline). In addition to the transfer of energy and materials depicted in panel **a**, dissolved nutrients and organic matter are taken up by bacteria that are, in turn, consumed by protozooplankton such as flagellates and ciliates, and they, in turn, also support higher trophic levels and contribute to nutrient recycling. Panel (**c**) The microbial food web, including the microbial loop and the viral shunt, under the mixoplankton paradigm (orange arrows). The phytoplankton now include diatoms and cyanobacteria. The mixoplankton community comprises constitutive mixoplankton (CM), which can graze a wide range of particles, from bacteria to other phytoplankton, and non-constitutive mixoplankton (NCM), which in addition to grazing can also photosynthesize through acquired phototrophy. All these organisms form the microbiome, and their interactions alter the flow of dissolved and particulate organic matter, as well as the nutritional quality of material transferred through the food web. Organism icons are from the University of Maryland Center for Environmental Science Integration and Application Network

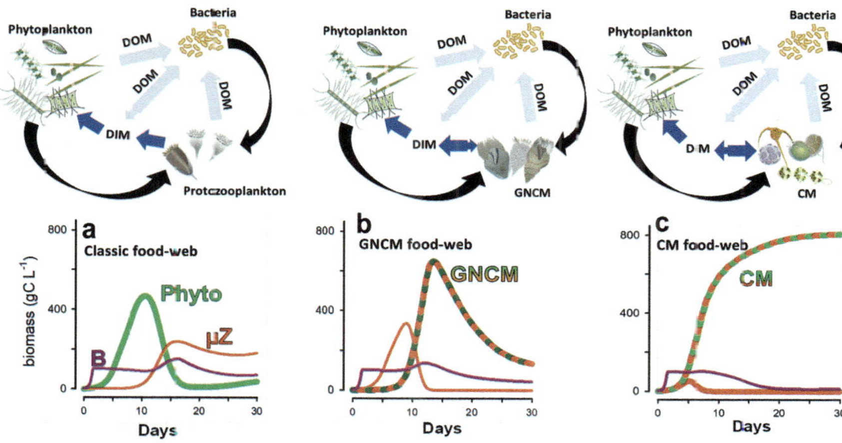

Fig. 15.5 Three contrasting food web structures (upper panels) and the corresponding temporal patterns (lower panels) of the development of biomass (gC L^{-1}) in the simulated communities over 30 days. Panel (**a**) The classic food web paradigm in which phytoplankton (Phyto) and protozooplankton (µZ) and bacteria (B) are the dominant protist plankton functional types; panel (**b**) As for Panel **a** except that the µZ functional type is replaced by the generalized non-constitutive mixoplankton (GNCM; e.g., plastidic ciliates); and panel (**c**) As for panel **a** except that the µZ functional type is replaced by the constitutive mixoplankton (CM; e.g., dinoflagellates). The release and uptake of dissolved inorganic and organic matter (DIM and DOM, respectively) are indicated. Note the higher biomass of the GNCM and CM relative to phytoplankton and the sustained duration of the CM over the time period shown. Images modified from Mitra et al. (2016) and reproduced from Glibert and Mitra (2022, Limnol Oceanogr 67: 585–597, doi: 10.1002/lno.12018) under a creative commons license

introduce important biochemicals (notably fatty acids) into what would otherwise be microbial loop organisms. Through mixotrophs, the microbial loop becomes an effective link and not just a sink to the "classic food web," based on stoichiometric and allometric principles.

Mixotrophy can be important in the maintenance of blooms, allowing them to be sustained for longer periods than would be the case if the cells depended only on dissolved nutrient availability (e.g., Glibert et al. 2009). Mixotrophy also can allow growth to be sustained or even accelerated during periods of apparent water column nutrient deficiency or imbalance (or actual nutrient deficiency, for non-mixotrophic competitors) when only inorganic forms of dissolved nutrients are considered. Numerous dinoflagellates have increased growth rates when feeding in comparison with growth on dissolved inorganic nutrients (e.g., Jeong et al. 2004; Adolf et al. 2008; Glibert et al. 2009; see also review by Burkholder et al. 2008). Thus, mixotrophy can provide a physiological mechanism for the acquisition of a nutrient(s) that may not be available in dissolved form and therefore not available to competitors without alternate nutrition strategies. Mixotrophy can provide a substantial growth advantage.

15.9 Mixotroph Categories and Terminology

The wealth of emerging data on mixotrophy has led to various efforts to categorize mixoplankton along the spectrum of their reliance on this nutritional strategy and to consider explicitly how true phytoplankton, mixoplankton, and true microzooplankton can be distinguished. Mixotrophs vary from species with poor efficiency as phototrophs but high efficiency as phagotrophs, to obligate phototrophs with minimal heterotrophy (Caron et al. 1990, 1993; Stoecker 1998). Some species only supplement their nutrition with mixotrophy, while others are capable of using mixotrophy to grow in complete darkness (Granéli et al. 1999). Some initiate mixotrophy only if high prey densities are present (Sanders et al. 1990); for others, light or nutrient limitation triggers mixotrophy (e.g., Caron et al. 1993; Jones et al. 1993, 1995). Classification becomes useful in unraveling relationships between and among groups, including evolutional relationships, but simplification of the diversity is necessary when mixotrophs are included in biogeochemical or food web dynamical models.

Jones (1997) proposed four functional groups of mixotrophic flagellates. In that scheme, the first group represented those protists primarily dependent on heterotrophy, but which supplemented growth by phototrophy. The second group in Jones' scheme is those protists that are primarily phototrophic and which feed under light limitation. The third group, like the second, is primarily phototrophic, but which feed to gain essential elements or growth substrates. Finally, the last group in this scheme are those that are primarily phototrophic, but which can ingest prey but only at low rates and only under severe conditions of light limitation or nutrient limitation.

Stoecker (1998) expanded this classification scheme and included a wider group of protists, accounting for ciliates, sarcodines, and protists with algal symbionts, in addition to flagellates in her classification scheme (Fig. 15.6). Some of the groups defined by Jones (1997) aligned with groups defined by Stoecker, but the classification was more expansive.

Using a functional (i.e., trait-based) approach, placing emphasis on photo- and phago-trophy as classifying processes, Mitra et al. (2016, 2023a) suggested a new categorization scheme for protists, which included the classic autotrophic phytoplankton at one extreme, classic phagotrophs at the other extreme, and four variations of mixotrophs in between (Figs. 15.7 and 15.8). First are the constitutive mixotrophs (CM), with an innate ability to photosynthesis, but which also feed. The constitutive mixotrophs include HAB species such as small-celled toxic taxa (e.g., the dinoflagellate *Karlodinium* and the haptophyte *Prymnesium*), as well the large dinoflagellate *Ceratium*, which is generally not considered a HAB taxa except under high biomass conditions. Another category is the non-constitutive mixotrophs (NCM), which only acquire the ability to photosynthesize from acquired prey. This group lacks any ability to synthesize, repair, or control the photosynthetic machinery (Flynn and Hansen 2013); they only have chloroplasts when they eat them. Among the non-constitutive mixotrophs, some (the plastidic ciliates) are

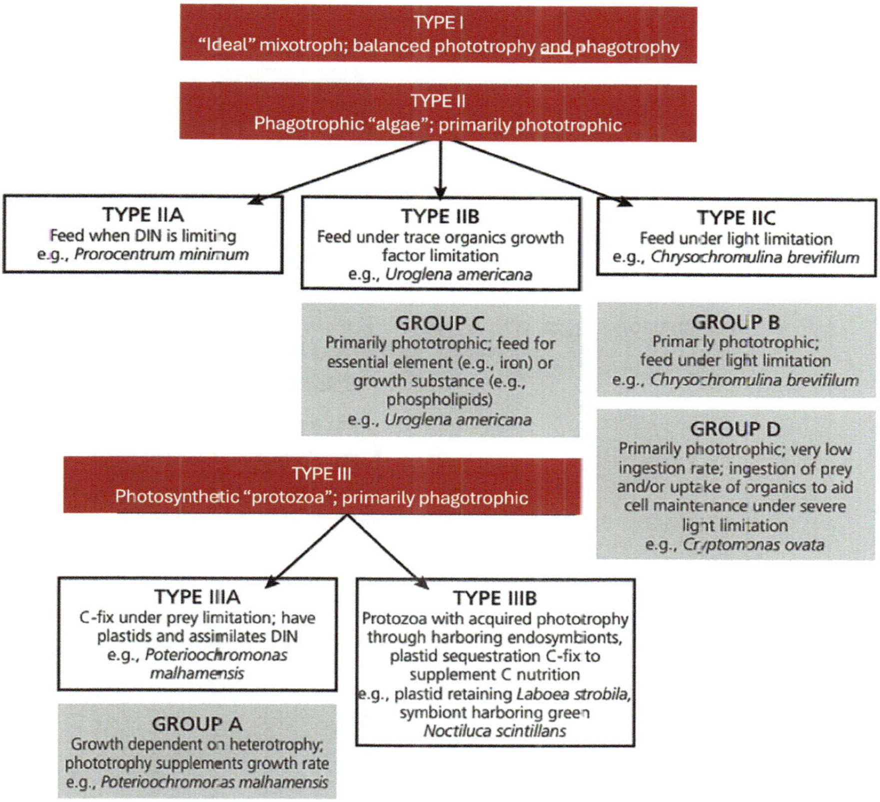

Fig. 15.6 Traditional classification of mixotrophic protists according to Stoecker (1998, open boxes) and Jones (1997, gray boxes). "Groups" proposed by Jones (1997) have been aligned against "Types" proposed by Stoecker (1998). Image modified from Mitra et al. (2016, Protist 167: 106–120, doi: 10.1016/j.protis.2016.01.003) and reproduced under a creative commons license

generalists (GNCM) and can steal chloroplasts from a range of prey types, while others are specialists (SNCM) making use of plastids acquired from very few prey sources (these include the red-tide ciliate *Mesodinium*, and the harmful dinoflagellate *Dinophysis*, which actually obtains its plastids "third-hand" from *Mesodinium*—Hansen et al. 2013 and references therein). The specialists themselves are further categorized into two functional groups—the plastidic SNCM (pSNCM) that acquires plastids from specific species (e.g., *Mesodinium rubrum and Dinophysis acuta*) and the endosymbiotic SNCM (eSNCM) that harbors photosynthetic endosymbionts (e.g., different rhizarians).

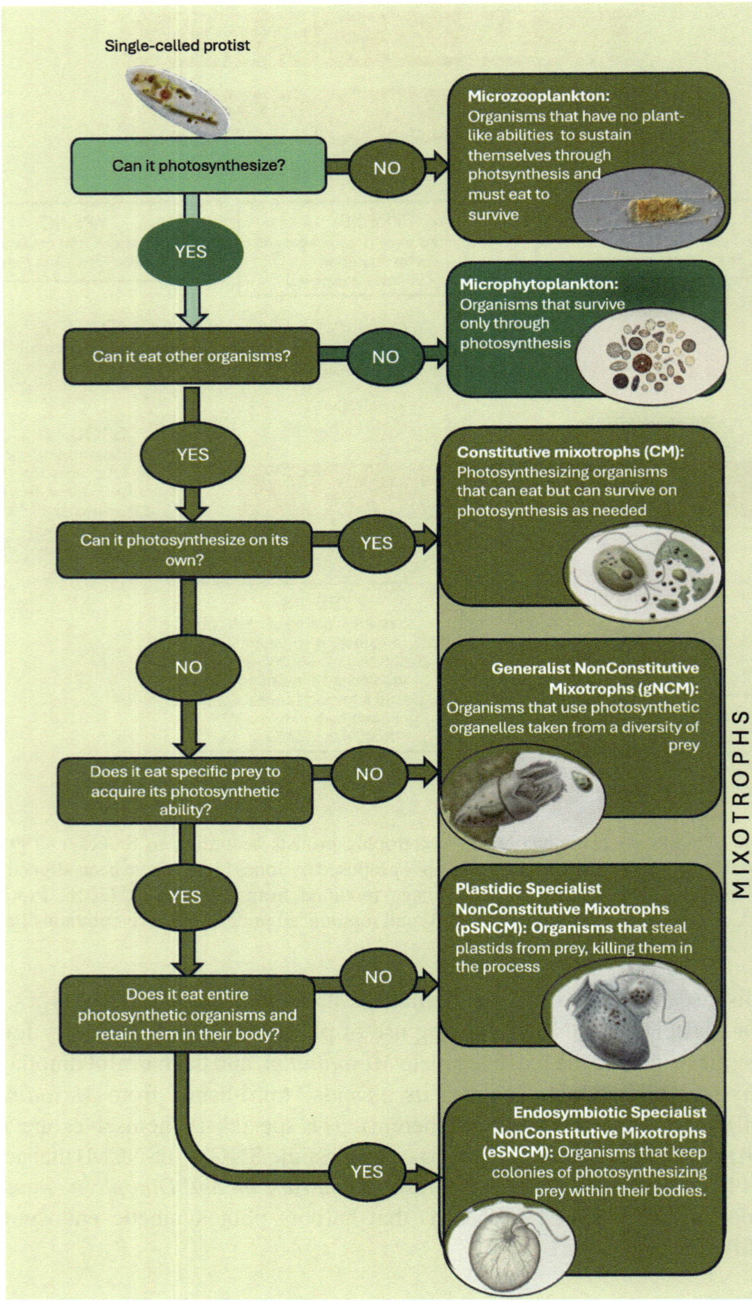

Fig. 15.7 Functional group classification for marine microbial plankton. Image redrawn based on based on Mitra et al. (2016, Protist 167:106–120. doi: 10.1016/j.protis.2016.01.003) and Mitra (2018, Sci Amer 318: 26–33. doi: 10.1038/scientificamerican0418-26) with permission of Scientific American

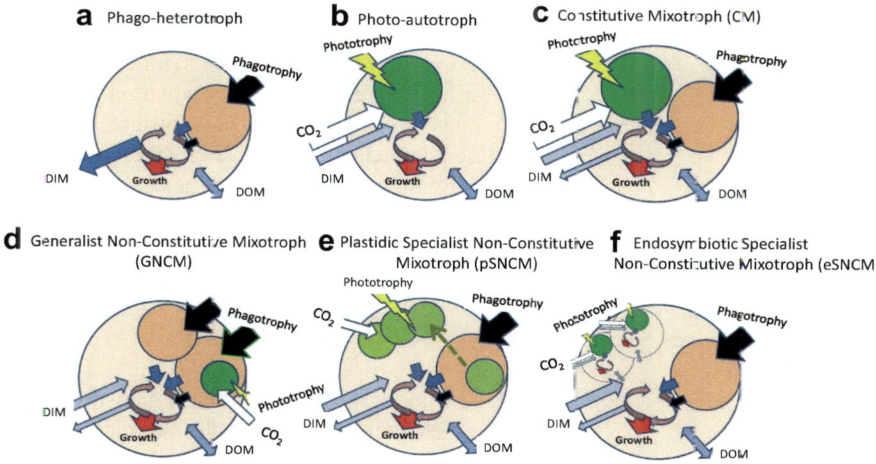

Fig. 15.8 Schematic illustrating the different levels in complexity among different types of protists. Panel (**a**) Phagotrophic (no phototrophy); Panel (**b**) Phototrophic (no phagotrophy); Panel (**c**) Constitutive mixotroph, with innate capacity for phototrophy; Panel (**d**) Generalist non-constitutive mixotroph acquiring photosystems from different phototrophic prey; Panel (**e**) Specialist non-constitutive mixotroph acquiring plastids from a specific prey type; Panel (**f**) Specialist non-constitutive mixotroph acquiring photosystems from endosymbionts. DIM, dissolved inorganic material (ammonium, phosphate, etc.). DOM, dissolved organic material. Image reproduced from Mitra et al. (2016, Protist 167: 106–120, doi: 10.1016/j.protis.2016.01.003) under a creative commons license

15.10 Biogeography

Globally, the understanding of the spatial and temporal distribution is hampered by the availability of data. There have been efforts to characterize biogeography of mixotrophs using microscopic (Barton et al. 2013; Leles et al. 2017, 2019), 'omics-based approaches (Faure et al. 2019), and modeling (Follows et al. 2007; Ward and Follows 2016; Edwards 2019).

Based on a bibliographic survey of published literature, Leles et al. (2017) developed a global distribution of non-constitutive mixotrophs and their subgroups. Spatially, the GNCMs were mainly associated with the Temperate Seas, Polar Seas, and Mediterranean Sea biomes, and eSNCMs were primarily associated with the Oligotrophic Gyres, Coastal Upwelling, and Equatorial Upwelling biomes (Fig. 15.9). The eSNCM distribution was observed to follow two distinctly diverse patterns; one group was closely related to the Equatorial Upwelling biome, while a second cosmopolitan group occurred in the intersection between most biomes (Fig. 15.9). The Coastal Sea biome was not associated with any single functional group; all the NCM functional groups occurred at least in one biogeographic province within this biome. This distribution suggests that the contribution of

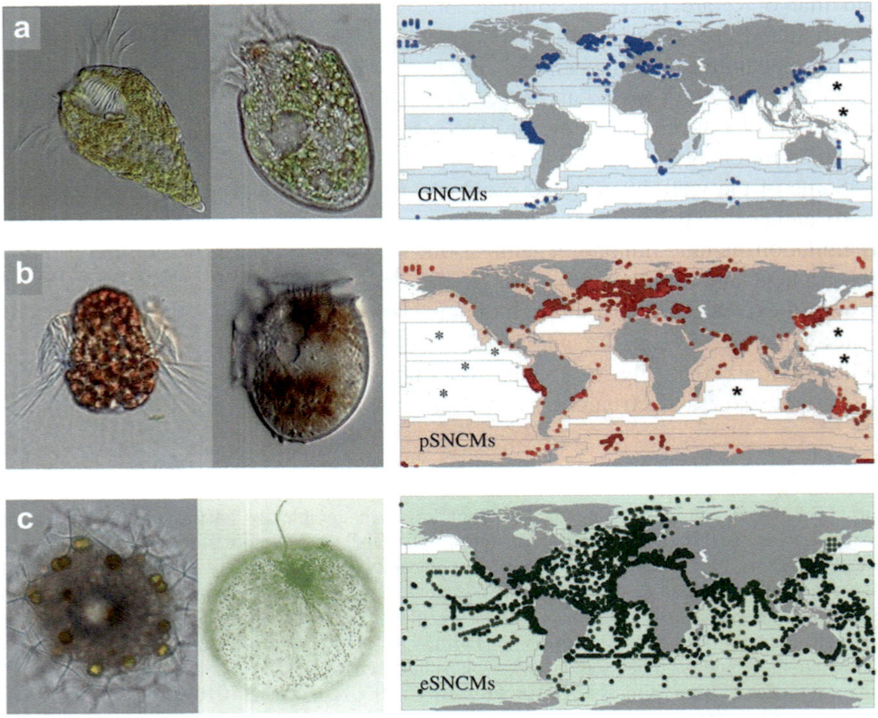

Fig. 15.9 Global distribution of protists with acquired phototrophy (non-constitutive mixotrophs, NCMs). Functional groups identify protists, which acquire plastids from a variety of prey (generalist NCMs, GNCMs; blue, Panel **a**), from specific prey (plastidic specialist NCMs, pSNCMs, red, Panel **b**), or specific autotrophic prey as symbionts (endosymbiotic SNCMs, green, Panel **c**). Images nest to each map provide protist genus examples within each functional group. From left to right (size as length): Panel **a**, GNCMs *Loboea* (100 μm) and *Strombidium* (50 μm); Panel **b**, pSNCMs, *Mesodinium* (60 μm) and *Dinophysis* (40 μm); Panel **c**, eSNCMs, *Sphaerozoum* (200 μm) and *Noctiluca* (500 μm). On maps, symbols correspond to the exact location where mixotrophic species/taxa were found; the grid indicates biogeographic provinces. Colored areas indicate the presence of NCMs and white areas their absence. Provinces marked with asterisks indicate that studies conducted in these areas did not record the presence of mixotrophic species; unmarked white provinces indicate a lack of field studies providing information on acquired phototrophy among microzooplankton. Image reproduced from Leles et al. (2017, Proc R Soc B 284: 20170664, doi: 10.1098/rspb.2017.0664) with permission of The Royal Society (UK)

mixotrophic ciliates increases toward more productive and less variable systems, while the opposite pattern is expected for mixotrophic *Rhizaria* (GNCMs tend to dominate after the phytoplankton bloom, particularly in summer, under stratified water column conditions, while pSNCMs are more commonly encountered during spring, in nutrient-replete conditions).

Using metabarcoding (amplicon sequencing), mixotrophic functional groups were also shown to have contrasting biogeography (Faure et al. 2019). Their results

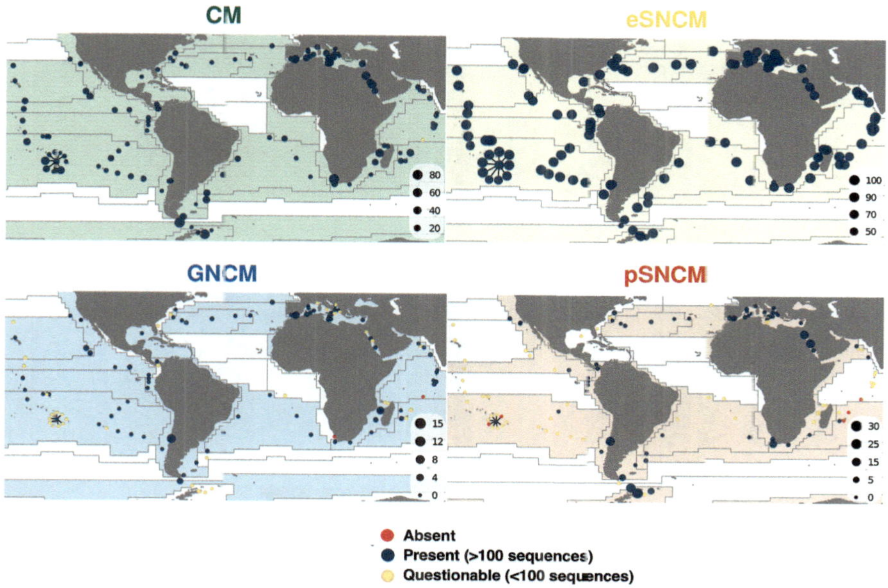

Fig. 15.10 Global distribution of mixotypes from metabarcoding data. Maps showing for each station the proportion of sequences (in %) belonging to each mixotype over the total number of mixotrophic sequences. Stations in which no sequence was found were marked as absent, and ones with less than 100 sequences were marked as questionable. Each biogeographical province is colored in the background if more than 100 sequences are detected in at least one of its stations. Image reproduced from Faure et al. (2019, ISME J 13: 1072–1083, doi: 10.1038/s41396-018-0340-5) with permission of the International Society for Microbial Ecology

highlighted the importance of certain constitutive mixotrophs, and their results, like those of Leles et al. (2017), underscored that mixotrophs are everywhere (Fig. 15.10), but—and at least in part due to the global distribution of metabarcoding data—the distributions for some types of mixotrophs differed from those reported by Leles et al. (2017). Both studies showed the global distribution of GNCMs, and pSNCMs appeared more seasonally and spatially restricted. Most CMs were found in eutrophic and shallow habitats, a finding not unexpected given that many CMS are also HABs.

Another approach for assessing the regional biogeography of mixotrophs was undertaken by Schneider et al. (2020). They used a 24-year data set from the North Sea and applied several classifications based on either known mixotrophic properties and/or statistical analyses to assess the seasonality of mixotrophy and its relationship with environmental factors. Here, showing only one of their classifications, in which only documented mixoplankton and motile, phototrophic taxa associated with the environmental envelope of documented mixoplankton were labeled as mixoplankton, a clear spatial and seasonal gradient can be seen (Fig. 15.11).

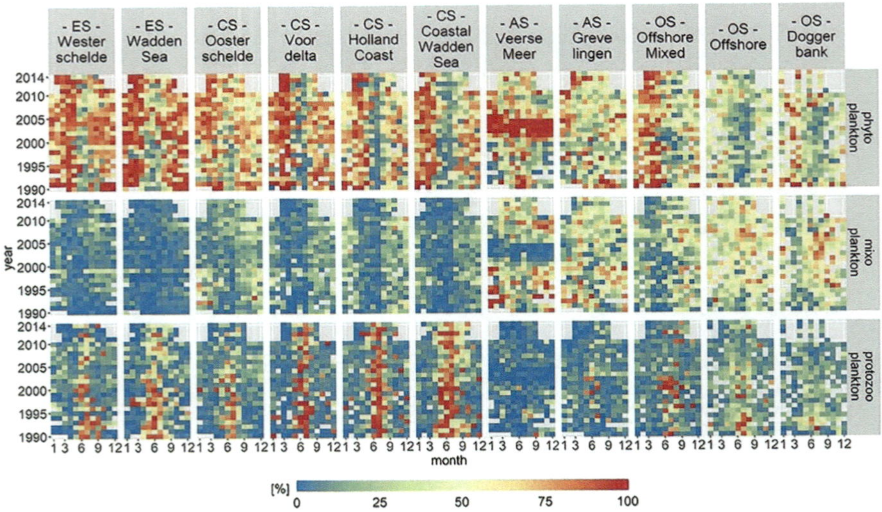

Fig. 15.11 Monthly fractions of the total biomass in which only documented mixoplankton and motile, phototrophic taxa associated with the environmental envelope of documented mixoplankton were labeled as mixoplankton. The location classes are grouped into estuary (ES), coastal (CS), anthropogenically modified (AS), and offshore systems (OS). The mixoplankton fraction displays a clear spatial and seasonal gradient, with increased orientation toward the late summer/autumn and in the offshore stations. Image reproduced from Schneider et al. (2020, Front Mar Sci 7: doi: 10.3389/fmars.2020.586915) under a creative commons license

Mixoplankton increased in the late summer months, with year-to-year variations and variations between stations.

A recent summary has provided a comprehensive database of mixoplankton based on knowledge of their identity, allometry, physiology, and tropic interactions (Mitra et al. 2023a). This database expands the datasets of Leles (2017, 2019) and Faure et al. (2019) by hundreds of species. Notable findings of this compilation for those mixotrophs identified as CM, or "phytoplankton," are that ~9% were reported as HAB taxa. Further analysis of this broad group of microbes shows that, due to the repeated occurrences of gains and losses of functional traits (e.g., Keeling et al. 2014; Bremer et al. 2022; Fig. 15.12; Chap. 10), there were likely many different evolutionary paths that resulted in mixotrophy at different times (Mansour and Anestis 2021; Bremer et al. 2022). Accordingly, it may not be appropriate to categorize advantages or disadvantages of a mixotrophic life strategy relative to that of a pure phytoplankton or pure zooplankton based on traits and their trade-offs (Mitra et al. 2023b). Mixoplankton in general thrive in more mature systems ("*K*-selected"), less so in the high-nutrient, turbulent environment favoring diatoms (Margalef 1978; Glibert 2016, Chap. 17). More mature environments are where high growth rates are not necessary, but where niche specialization, and the ability to adapt to changing environments becomes much more favorable.

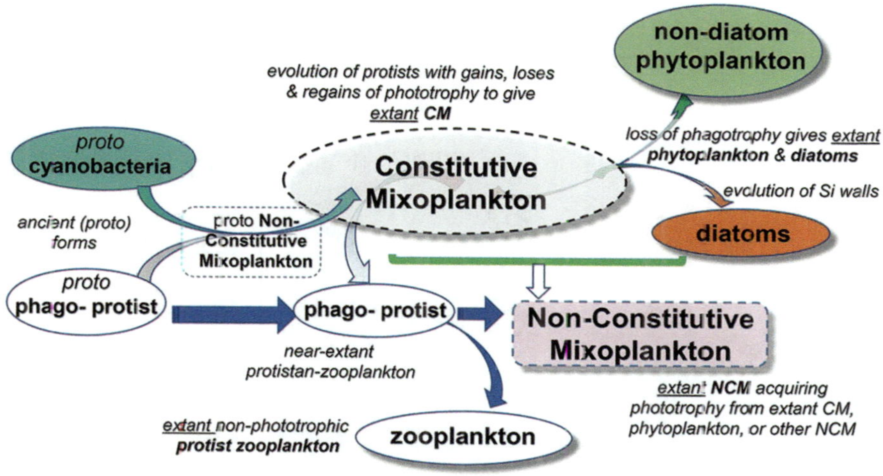

Fig. 15.12 An interpretation of the evolution of protist functional types. Multiple gains and losses of phototrophy have occurred within what we now term constitutive mixoplankton. Some of these have resulted in extant zooplankton and extant phytoplankton groups. Image reproduced from Mitra et al. (2023b, Euro. J. Eukary. Microb. doi: 10.1080/09670262.2023.2216259) under a creative commons license

15.11 Mixotrophs and the Biological Pump

Trends in numerous factors suggest that the ubiquitous mixoplankton (Leles et al. 2017, 2019; Faure et al. 2019) could become dominant in the globally changing world—where waters may be more CO_2-enriched, more eutrophic, and generally warmer (Flynn et al. 2015; Glibert 2020; see Chaps. 22–26). In contrast to the "pure" protist phytoplankton—the diatoms, mixoplankton tend to become proportionately more abundant as systems seasonally mature (become nutrient-poor) rather than in immature (nutrient-rich) ecosystems (e.g., Mitra et al. 2014a). Moreover, in a future world, short food "chains" and simple microbial "webs" may be disproportionately stressed as temperatures rise, increasing thermal stratification and reducing water column nutrient exchange, as oceanic waters become more acidic, and as land-based and atmospheric nutrient pollution shifts in amounts, forms, and proportions. These conditions strengthen the importance not only of mixoplankton but also of the microbial C pump (Fig. 15.13). Any enhancement of the microbial C pump will further enhance mixoplankton dominance by enhanced bacterial activity and abundance, in turn, increasing food availability for those microbial multitaskers capable of eating them. If organisms evolve their growth rate potential to match the flows of resources in their environment (Flynn and Skibinski 2020), then mixoplankton

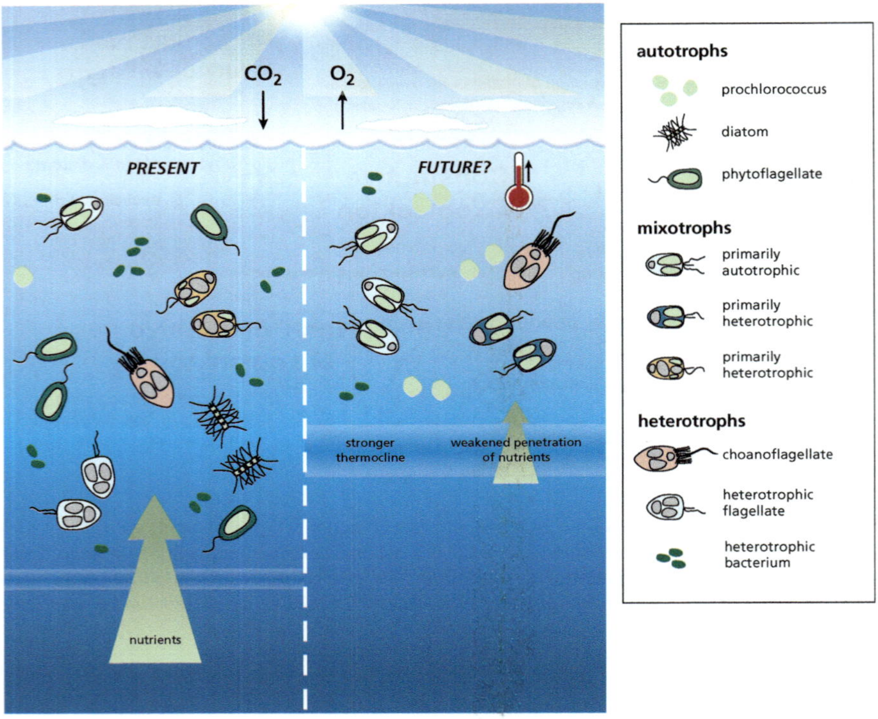

Fig. 15.13 Schematic of the predicted changes in community composition in future oceans. The future ocean is expected to have stronger stratification and shallower mixed layer, together with warmer, more oligotrophic waters subject to higher average light intensity. Protists are shown with photosynthetic potential (plastids in green) and/or with phagocytotic potential (food vacuoles containing bacterial prey). Future conditions are expected to favor strong competitors for nutrients such as *Prochlorococcus,* but effects on mixotrophs remain unclear and might be species-specific. Some mixotrophs might shift toward increasing the importance of phagotrophy (in pink), while others might shift toward increasing the importance of photosynthesis (in yellow). Overall, mixotrophs are expected to be favored under future conditions compared to specialist autotrophic or heterotrophic competitors. Image redrawn and modified based on Wilken et al. (2019, Phil Trans Roy Soc B 374: doi: 10.1098/rstb.2019.0090)

would not necessarily have to have high growth rates to succeed in a changing world. Predictions based on these considerations suggest a decline in pure phytoplankton (diatoms) relative to mixotrophs (e.g., Glibert 2020), but the species-specific or type-specific mixotrophs that may be favored remain unclear (e.g., Wilken et al. 2019). Understanding, quantifying, and predicting how plankton are interacting with changing climate and nutrients will continue to be challenging. Not only may diatoms be disadvantaged in a future ocean, but mixoplankton with harmful properties may emerge even more successful (Flynn et al. 2015; Glibert 2020).

15.12 Summary

The "weird" microbes that can eat and photosynthesize simultaneously are no longer considered mere curiosities; they are the norm. They are global. They are present in eutrophic to oligotrophic waters. They are likely favored as climate changes. These diverse microbes are complex and understanding and including them as the complicated organisms they are is absolutely necessary to understand productivity. They also give us important clues as to microbial evolution. Yet, there is much to be learned about this physiological lifestyle, and methods are advancing allowing us to do so.

References

Adolf JE, Bachvaroff TR, Krupatkina DN, Place AR (2007) Karlotoxin mediates grazing by *Oxyrrhis marina* on strains of *Karlodinium veneficum*. Harmful Algae 6:400–412

Adolf JE, Bachvaroff T, Place AR (2008) Can cryptophyte abundance trigger toxic *Karlodinium veneficum* blooms in eutrophic estuaries? Harmful Algae 8:119–128

Antia NJ, Harrison PJ, Oliveria L (1991) The role of dissolved organic nitrogen in phytoplankton nutrition, cell biology and ecology. Phycologia 30:1–89

Barton AD, Pershing AJ, Litchman E, Record NR, Edwards KF, Finkel ZV, Kiørboe T et al (2013) The biogeography of marine plankton traits. Ecol Lett 16:522–534. https://doi.org/10.1111/ele.12063

Berg GM, Balode M, Purina I, Bekere S, Béchemin C, Maestrini SY (2003) Plankton community composition in relation to availability and uptake of oxidized and reduced nitrogen. Aquat Microb Ecol 30:263–274

Berman T, Bronk DA (2003) Dissolved organic nitrogen: a dynamic participant in aquatic ecosystems. Aquat Microb Ecol 31:273–305

Biecheler B (1936) Observation de la capture et de la digestion des protes chez un peridinén vert. C R Séances Soc Biol 12:1173–1175

Bjørnsen PK (1988) Phytoplankton exudation of organic matter: why do healthy cells do it? Limnol Oceanogr 33:151–154

Bremer N, Tria FD, Skejo J, Garg SG, Martin WF (2022) Ancestral state reconstructions trace mitochondria but not phagocytosis to the last eukaryotic common ancestor. Genet Biol Evol 14: evac079. https://doi.org/10.1093/gbe/evac079

Bronk DA, See JH, Bradley P, Killberg L (2007) DON as a source of bioavailable nitrogen. Biogeosciences 4:283–296. https://doi.org/10.5194/bg-4-283-2007

Burkholder JM, Glibert PM, Skelton H (2008) Mixotrophy, a major mode of nutrition for harmful algal species in eutrophic waters. Harmful Algae 8:77–93. https://doi.org/10.1016/j.hal.2008.08.010

Burson A, Stomp M, Akil L, Brussaard C, Huisman J (2016) Unbalanced reduction of nutrient loads has created an offshore gradient from phosphorus to nitrogen limitation in the North Sea. Limnol Oceanogr 61:869–888. https://doi.org/10.1002/lno.10257

Caron DA (2016) Mixotrophy stirs up our understanding of marine food webs. Proc Natl Acad Sci USA 113:2806–2808

Caron DA, Porter KG, Sanders RW (1990) Carbon, nitrogen and phosphorus budgets for the mixotrophic phytoflagellate *Poterioochromonas malhamensis* (Chrysophyceae) during bacterial ingestion. Limnol Oceanogr 35:433–443

Caron DA, Sanders RW, Lim EL (1993) Light-dependent phagotrophy in the freshwater mixotrophic chrysophyte *Dinobryon cylindricum*. Microb Ecol 25:93–111

Cloern JE, Dufford R (2005) Phytoplankton community ecology: principles applied in San Francisco Bay. Mar Ecol Prog Ser 285:11–28

Darwin CR (1875) Insectivorous plants. D. Appleton, New York

Delwiche CF (1999) Tracing the thread of plastid diversity through the tapestry of life. Am Nat 154: S164–S177

Edwards KF (2019) Mixotrophy in nanoflagellates across environmental gradients in the ocean. Proc Natl Acad Sci USA 116:6211–6220

Falkowski PG, Katz ME, Knoll AH, Quiqq A, Raven JA, Schofield O, Taylor FJR (2004) The evolution of modern eukaryotic phytoplankton. Science 305:354–360. https://doi.org/10.1126/science.1095964

Faure E, Not F, Benoiston A-S, Labadie K, Bittner L, Ayata S-D (2019) Mixotrophic protists display contrasted biogeographies in the global ocean. ISME J 13:1072–1083. https://doi.org/10.1038/s41396-018-0340-5

Flynn KJ, Hansen PJ (2013) Cutting the canopy to defeat the "selfish gene"; conflicting selection pressures for the integration of phototrophy in mixotrophic protists. Protist 164:811–823

Flynn KJ, Mitra A (2009) Building the "perfect beast": Modelling mixotrophic plankton. J Plankton Res 31:965–992. https://doi.org/10.1093/plankt/fbp044

Flynn KJ, Skibinski DOF (2020) Exploring evolution of maximum growth rates in plankton. J Plankton Res 42:497–513. https://doi.org/10.1093/plankt/fbaa038

Flynn KJ, Stoecker DK, Mitra A, Raven JA, Glibert PM, Hansen PJ, Granéli E, Burkholder JM (2013) Misuse of the phytoplankton-zooplankton dichotomy: The need to assign organisms as mixotrophs within plankton functional types. J Plankton Res 35:3–11. https://doi.org/10.1093/plankt/fbs062

Flynn KJ, Clark DR, Mitra A, Fabian H, Hansen PJ, Glibert PM, Wheeler GL et al (2015) Ocean acidification with (de)eutrophication will alter future phytoplankton growth and succession. Proc Roy Soc 282:20142604

Flynn KJ, Mitra A, Anestis K, Anschütz AA, Calbet A, Ferreira GD, Gypens N et al (2019) Mixotrophic protists and a new paradigm for marine ecology: Where does plankton research go now? J Plankton Res 41:375–391. https://doi.org/10.1093/plankt/fbz026

Follows MJ, Dutkiewicz S, Grant S, Chisholm SW (2007) Emergent biogeography of microbial communities in a model. Ocean Sci 315:1843–1846

Glibert PM (2016) Margalef revisited: a new phytoplankton mandala incorporating twelve dimensions, including nutritional physiology. Harmful Algae 55:25–30. https://doi.org/10.1016/j.hal.2016.01.008

Glibert PM (2020) Harmful algae at the complex nexus of eutrophication and climate change. Harmful Algae 91:10153. https://doi.org/10.1016/j.hal.2019.03.001

Glibert PM, Burkholder JM (2011) Eutrophication and HABs: Strategies for nutrient uptake and growth outside the Redfield comfort zone. Chin J Oceanol Limnol 29:724–738. https://doi.org/10.1007/s00343-011-0502-z

Glibert PM, Legrand C (2006) The diverse nutrient strategies of harmful algae: focus on osmotrophy. In: Granéli E, Turner J (eds) Ecology of harmful algae. Springer, Heidelberg, pp 163–175

Glibert PM, Mitra A (2022) From webs, loops, shunts and pumps to microbial multi-tasking: Evolving paradigms of marine microbial ecology, global mixoplankton importance and implications for a future ocean. Limnol Oceanogr 67:585–597. https://doi.org/10.1002/lno.12018

Glibert PM, Burkholder JM, Kana TM, Alexander JA, Schiller C, Skelton H (2009) Grazing by *Karenia brevis* on *Synechococcus* enhances their growth rate and may help to sustain blooms. Aquat Microb Ecol 55:17–30. https://doi.org/10.3354/ame01279

Granéli E, Carlsson P (1998) The ecological significance of phagotrophy in photosynthetic flagellates. In: Anderson DM, Cembella AD, Hallegraeff GM (eds) Physiological ecology of harmful algal blooms. NATO ASI series, vol G41. Springer, Berlin, pp 539–557

Granéli E, Carlsson P, Legrand C (1999) The role of C, N and P in dissolved and particulate organic matter as a nutrient source for phytoplankton growth, including toxic species. Aquat Ecol 33: 17–27

Hansen PJ (2011) The role of photosynthesis and food uptake for the growth of marine mixotrophic dinoflagellates. J Eukaryot Microbiol 58:203–214

Hansen PJ, Skovgaard A, Glud RN, Stoecker DK (2000) Physiology of the mixotrophic dinoflagellate *Fragilidium subglobosum.* II. Effects of time scale and prey concentration on photosynthetic performance. Mar Ecol Prog Ser 201:137–146

Hansen PJ, Nielsen LT, Johnson M, Berge T, Flynn KJ (2013) Acquired phototrophy in *Mesodinium* and *Dinophysis*—a review of cellular organization, prey selectivity, nutrient uptake and bioenergetics. Harmful Algae 28:126–139

Hecky R, Kilham P (1988) Nutrient limitation of phytoplankton in freshwater and marine environments: A review of recent evidence on the effects of enrichment. Limnol Oceanogr 33:796–822. https://doi.org/10.4319/lo.1988.33.4part2.0796

Jacobson DM (1999) A brief history of dinoflagellate feeding research. J Eukaryot Microbiol 46: 376–381

Jacobson DM, Anderson DM (1986) Thecate heterotrophic dinoflagellates: feeding behavior and mechanisms. J Phycol 22:358–369

Jeong HJ (1994) Predation by the heterotrophic dinoflagellate *Protoperidinium* cf. *divergens* on copepod eggs and early naupliar stages. Mar Ecol Prog Ser 114:203–208

Jeong HJ, Yoo YD, Kim JS, Kim TH, Kim JH, Kang NS, Yih W (2004) Mixotrophy in the phototrophic harmful alga *Cochlodinium polykrikoides* (Dinophyceae): prey species, the effects of prey concentration, and grazing impact. J Eukaryot Microbiol 51:563–569

Jeong HJ, Park JY, Nho JH, Park MO, Ha JH, Seong KA, Jeng C et al (2005a) Feeding by red-tide dinoflagellates on the cyanobacterium *Synechococcus*. Aquat Microb Ecol 41:1331–2143

Jeong HJ, Yoo YD, Seong KA, Kim JH, Park JY, Kim S, Lee SH et al (2005b) Feeding by the mixotrophic red-tide dinoflagellate *Gonyaulax polygramma*: mechanisms, prey species, effects of prey concentration, and grazing impact. Aquat Microb Ecol 38:249–257

Jeong HJ, Yoo YD, Kim JS, Seong KA, Kang NS, Kim TH (2010) Growth, feeding and ecological roles of the mixotrophic and heterotrophic dinoflagellates in marine planktonic food webs. Ocean Sci J 45:65–91. https://doi.org/10.1007/s12601-010-0007-2

John U, Tillmann U, Hülskötter J, Alpermann TJ, Wohlrab S, Van de Waal DB (2015) Intraspecific facilitation by allelochemical mediated grazing protection with a toxigenic dinoflagellate population. Proc R Soc B 282:20141268. https://doi.org/10.1018/rspb.2014.1268.2015

Johnson MD (2011) Acquired phototrophy in ciliates: A review of cellular interactions and structural adaptations. J Eukaryot Microbiol 58:185–195

Johnson MD, Tengs T, Oldach D, Stoecker DK (2006) Sequestration, performance, and functional control of cryptophyte plastids in the ciliate *Myrionecta rubra* (Ciliophora). J Phycol 42:1235–1246. https://doi.org/10.1111/j.1529-8817.2006.00275.x

Johnson MD, Oldach D, Delwiche CF, Stoecker DK (2007) Retention of transcriptionally active cryptophyte nuclei by the ciliate *Myrionecta rubra*. Nature 445:426–428

Johnson MD, Beaudoin DJ, Laza-Martinez A, Dyhrman ST, Fensin E, Lin S, Merculief A, Nagai S et al (2016) The genetic diversity of Mesodinium and associated cryptophytes. Front Microbiol 7. https://doi.org/10.3389/fmicb.2016.02017

Jones HLJ (1997) A classification of mixotrophic protists based on their behaviour. Freshw Biol 37: 35–43

Jones HLJ, Leadbeater BSC, Green JC (1993) Mixotrophy in marine species of *Chrysochromulina* (Prymnesiophyceae): ingestion and digestion of a small green flagellate. J Mar Biol Assoc UK 73:283–296

Jones HLJ, Durjun P, Leadbeater BSC, Green JC (1995) The relationship between photoacclimation and phagotrophy with respect to chlorophyll a, carbon and nitrogen content, and cell size of *Chrysochromulina brevifilum* (Prymnesiophyceae). Phycologia 34:128–134

Jones RI (1994) Mixotrophy in planktonic protists as a spectrum of nutritional strategies. Mar Microb Food Webs 8:87–96

Keeling PJ (2010) The endosymbiotic origin, diversification and fate of plastids. Philos Trans R Soc B 365:729–748. https://doi.org/10.1098/rstb.2009.0103

Keeling PJ (2013) The number, speed, and impact of plastid endosymbioses in eukaryotic evolution. Annu Rev Plant Biol 64:583–607. https://doi.org/10.1146/annurev-arplant-050312-120144

Keeling PJ, Burki F, Wilco H, Allam B, Allen EE, Amaral-Zettler LA, Armbrust EV et al (2014) The Marine Microbial Eukaryote Transcriptome Sequencing Project (MMETSP): illuminating the functional diversity of eukaryotic life in the oceans through transcriptome sequencing. PLoS Biol 12:e.1001889. https://doi.org/10.1371/journal.pbio.1001889

Kim GH, Han JH, Kim B, Han JW, Nam SW, Shin W, Park JW et al (2016) Cryptophyte gene regulation in the kleptoplastidic, karyokleptic ciliate *Mesodinium rubrum*. Harmful Algae 52: 23–33. https://doi.org/10.1016/j.hal.2015.12.004

Kim M, Drumm K, Daugbjerg N, Hansen PJ (2017) Dynamics of sequestered cryptophyte nuclei in *Mesodinium rubrum* during starvation and refeeding. Front Microbiol 8. https://doi.org/10.3389/fmicb.2017.00423

Larsen A, Eikrem W, Paasche E (1993) Growth and toxicity in *Prymnesium patelliferum* (Prymnesiophyceae) isolated from Norwegian waters. Can J Bot 71:1357–1362

Larsson ME, Bramucci AR, Collins S, Hallegraeff G, Kahike T, Raina J-B, Seymour JR et al (2022) Mucospheres produced by a mixotrophic protist impact ocean carbon cycling. Nat Commun 13: 1301. https://doi.org/10.1038/s41467-022-28867-8

Legrand C, Granéli E, Carlsson P (1998) Induced phagotrophy in the photosynthetic dinoflagellate *Heterocapsa triquetra*. Aquat Microb Ecol 15:65–75

Leles SG, Mitra A, Flynn KJ, Stoecker DK, Hansen PJ, Calbet A, McManus GB et al (2017) Oceanic protists with different forms of acquired phototrophy display contrasting biogeographies and abundance. Proc R Soc B 284:20170664. https://doi.org/10.1098/rspb.2017.0664

Leles SG, Mitra A, Flynn K, Tillmann U, Stoecker DK, Jeong HJ, Burkholder JM et al (2019) Sampling bias misrepresents the biogeographic significance of constitutive mixotrophs across global oceans. Glob Ecol Biogeogr 28:418–428. https://doi.org/10.1111/geb.12853

Lewitus AJ, Willis BM, Hayes KC, Burkholder JM, Glasgow HB, Glibert PM, Burke MK (1999) Mixotrophy and nitrogen uptake by *Pfiesteria piscicida* (Dinophyceae). J Phycol 35:1430–1437

Li A, Stoecker DK, Coats DW (2001) Use of the 'food vacuole content' method to estimate grazing by the mixotrophic dinoflagellate *Gyrodinium galatheanum* on cryptophytes. J Plankton Res 23: 303–318

Lin C-H, Accoroni S, Glibert PM (2017) Mixotrophy in the dinoflagellate *Karlodinium veneficum* under variable nitrogen:phosphorus stoichiometry: Feeding response and effects on larvae of the eastern oyster (Crassostrea virginica). Aquat Microb Ecol 79:101–114. https://doi.org/10.3354/ameo01823

Lin C-H, Flynn KJ, Glibert PM, Mitra A (2018) Modeling effects of variable nutrient stoichiometry and temperature on mixotrophy in the harmful dinoflagellate *Karlodinium veneficum*. Front Mar Sci 5. https://doi.org/10.3389/fmars.2018.00320

Lundgren V, Glibert PM, Granéli E, Vidyarathna NK, Fiori E, Ou L, Flynn KJ et al (2016) Metabolic and physiological changes in *Prymnesium parvum* when grown under, and grazing on, prey of variable nitrogen:phosphorus stoichiometry. Harmful Algae 55:1–12

Mansour JS, Anestis K (2021) Eco-evolutionary perspectives on mixoplankton. Front Mar Sci 8: 666160. https://doi.org/10.3389/fmars.2021.666160

Margalef R (1978) Life-forms of phytoplankton as survival alternatives in an unstable environment. Oceanol Acta 1:493–509

McManus GB, Schoener D, Haberlandt K (2012) Chloroplast symbiosis in a marine ciliate: ecophysiology and the risks and rewards of hosting foreign organelles. Front Microbiol 3: 321. https://doi.org/10.3389/fmicb.2012.00321

Millette NC, Gast RJ, Luo J, Moeller H, Stamieszkin K, Andersen KH, Brownlee E et al (2023) Mixotrophs and mixotrophy: Future research priorities. J Plankton Res. https://doi.org/10.1093/plankt/fbad020

Minnhagen S, Kim M, Salomon PS, Yih W, Granéli E, Park MG (2011) Active uptake of kleptoplastids by *Dinophysis caudata* from its ciliate prey *Myrionecta rubra*. Aquat Microb Ecol 62:99–108

Mitra A (2018) Tiny creatures, part plant and part animal, may control the fate of the planet. Sci Am 318:26–33. https://doi.org/10.1038/scientificamerican0418-26

Mitra A, Flynn KJ (2010) Modelling mixotrophy in harmful algal blooms: More or less the sum of the parts? J Mar Syst 83:158–169. https://doi.org/10.1016/j.jmarsys.2010.04.006

Mitra A, Flynn KJ, Burkholder JM, Berge T, Calbet A, Raven JA, Granéli E et al (2014a) The role of mixotrophic protists in the biological carbon pump. Biogeoscience 11:995–1005. https://doi.org/10.5194/bg-11-995-2014

Mitra A, Castellani C, Gentleman WC, Jónasdóttir SH, Flynn KJ, Bode A et al (2014b) Bridging the gap between marine biogeochemical and fisheries sciences; configuring the zooplankton link. Prog Oceanogr 129:176–199

Mitra A, Flynn KJ, Tillmann U, Raven JA, Caron D, Stoecker DK, Not F et al (2016) Defining planktonic protist functional groups on mechanisms for energy and nutrient acquisition; incorporation of diverse mixotrophic strategies. Protist 167:106–120. https://doi.org/10.1016/j.protis.2016.01.003

Mitra A, Caron DA, Faure E, Flynn KJ, Leles SG, Hansen PJ, McManus GB et al (2023a) The mixoplankton database—diversity of photo-phago-trophic plankton in form, function and distribution across the global ocean. J Eukaryot Microbiol. https://doi.org/10.1111/jeu.12972

Mitra A, Flynn KJ, Stoecker DK, Raven JA (2023b) Trait trade-offs in phagotrophic microalgae: the mixoplankton conundrum. Eur J Phycol. https://doi.org/10.1080/09670262.2023.2216259

Moorthi S, Ptacnik R, Sanders RW, Fischer R, Busch M, Hillebrand H (2017) The functional role of planktonic mixotrophs in altering seston stoichiometry. Aquat Microb Ecol 79:235–245. https://doi.org/10.3354/ame01832

Nam SW, Park JW, Yih W, Park MG, Shin W (2016) The fate of cryptophyte cell organelles in the ciliate *Mesodinium* cf. *rubrum* subjected to starvation. Harmful Algae 59:19–30. https://doi.org/10.1016/j.hal.2016.09.002

Nygaard K, Tobiesen A (1993) Bacterivory in algae: a survival strategy during nutrient limitation. Limnol Oceanogr 38:273–279

Onuma R, Horiguchi T (2013) Morphological transition in kleptochloroplasts after ingestion in the dinoflagellates *Amphidinium poecilochroum* and *Gymnodinium aeruginosum* (Dinophyceae). Protist 164:622–642

Pascher J (1917) Von der merkwürdigen Bewegungsweise einiger Flagellaten. Biologisches Zentralblatt 37:421–442

Pintner IJ, Provasoli L (1958) Artificial cultivation of a red-pigmented marine blue-green alga, *Phormidium persicinum*. J Gen Microbiol 18:190–197. https://doi.org/10.1099/0221287-18-1-190

Putt M (1990) Abundance, chlorophyll content and photosynthesis rates of ciliates in the Nordic Seas during summer. Deep Sea Res 37:1713–1731. https://doi.org/10.1016/0198-0149(90)90073-5

Raven JA (1997) Phagotrophy in phototrophs. Limnol Oceanogr 42:198–205

Raven JA, Beardall J, Flynn KJ, Maberly SC (2009) Phagotrophy in the origins of photosynthesis in eukaryotes and as a complementary mode of nutrition in phototrophs: relation to Darwin's insectivorous plants. J Exp Bot 60:3975–3987

Sanders RW (1991) Mixotrophic protists in marine and fresh-water ecosystems. J Protozool 38:76–81

Sanders RW, Porter KG, Caron DA (1990) Relationship between phototrophy and phagotrophy in the mixotrophic chrysophyte *Poterioochromonas malhamensis*. Microb Ecol 19:97–109

Sanders RW, Patterson DJ, Larsen J (1992) Trophic strategies among heterotrophic flagellates. In: Patterson DJ, Larsen J (eds) The biology of free-living heterotrophic flagellates, vol 45. Clarendon Press, Oxford, pp 21–38

Schneider LK, Flynn KJ, Herman PMJ, Troost TA, Stolte W (2020) Exploring the trophic spectrum: placing mixoplankton into marine protist communities of the southern North Sea. Front Mar Sci 7. https://doi.org/10.3389/fmars.2020.586915

Schnepf E, Elbrächter M (1992) Nutritional strategies in dinoflagellates. Eur J Protistol 28:3–24
Selosse M-A, Charpin M, Not F, Jeyasingh P (2017) Mixotrophy everywhere on land and in water: the grand écart hypothesis. Ecol Lett 20:246–263
Skovgaard A (1996) Engulfment of *Ceratium* spp. (Dinophyceae) by the thecate photosynthetic dinoflagellate *Fragilidium subglobosum*. Phycologia 35:490–499
Skovgaard A, Hansen PJ (2003) Food uptake in the harmful alga *Prymnesium parvum* mediated by excreted toxins. Limnol Oceanogr 48:1161–1166
Stibor H, Sommer U (2003) Mixotrophy of a photosynthetic flagellate viewed from an optimal foraging perspective. Protist 154:91–98
Stoecker DK (1998) Conceptual models of mixotrophy in planktonic protists and some ecological and evolutionary implications. Eur J Protistol 34:281–290
Stoecker DK (1999) Mixotrophy among dinoflagellates. J Eukaryot Microbiol 46:397–401
Stoecker DK, Michaels AE (1991) Respiration, photosynthesis and carbon metabolism in planktonic ciliates. Mar Biol 108:441–447
Stoecker DK, Li A, Coats DW, Gustafson DE, Nannen MK (1997) Mixotrophy in the dinoflagellate *Prorocentrum minimum*. Mar Ecol Prog Ser 15:1–12
Stoecker DK, Tillmann U, Granéli E (2006) Phagotrophy in harmful algae. In: Granéli E, Turner J (eds) Ecology of harmful algae. Springer, Heidelberg, pp 177–187
Stoecker DK, Hansen PJ, Caron DA, Mitra A (2017) Mixotrophy in the marine plankton. Ann Rev Mar Sci 9:311–335. https://doi.org/10.1146/annurev-marine-010816-060617
Stukel MR, Landry MR, Selph KE (2011) Nanoplankton mixotrophy in the eastern equatorial Pacific. Deep Sea Res II 58:378–386
Syrett PJ (1981) Nitrogen metabolism in microalgae. In: Platt T (ed) Physiological bases of phytoplankton ecology, Can Bull Fish Aquat Sci 210. Dept. of Fisheries and Oceans, Ottawa, pp 182–201
Thingstad TF, Havskum H, Garde K, Riemann B (1996) On the strategy of 'eating your competitor'. A mathematical analysis of algal mixotrophy. Ecology 77:39–49
Tillmann U (2003) Kill and eat your predator: a winning strategy of the planktonic flagellate *Prymnesium parvum*. Aquat Microb Ecol 32:73–84
Unrein F, Gasol JM, Not F, Forn I, Massana R (2007) Mixotrophic haptophytes are key bacterial grazers in oligotrophic coastal waters. ISME J 8:164–176. https://doi.org/10.1038/ismej.2013.132
Vadstein O (2000) Heterotrophic, planktonic bacteria and cycling of phosphorus: phosphorus requirements, competitive ability and food web interactions. Adv Microb Ecol 16:115–168
Verity PG, Brussaard CP, Nejstgaard JC, van Leeuwe MA, Lancelot C, Medlin LK (2007) Current understanding of *Phaeocystis* ecology and biogeochemistry, and perspectives for future research. Biogeochemistry 83:311–330. https://doi.org/10.1007/s10533-007-9090-6
Ward BA, Follows MJ (2016) Marine mixotrophy increases trophic transfer efficiency, mean organism size, and vertical carbon flux. Proc Natl Acad Sci USA 113:2958–2963
Wilken S, Yung CCM, Hamilton M, Hoadley K, Nzongo J, Eckmann C, Corro-Luque M et al (2019) The need to account for cell biology in characterizing predatory mixotrophs in aquatic environments. Phil Trans Roy Soc B 374:201900090. https://doi.org/10.1098/rstb.2019.0090
Wright RT, Hobbie JE (1965) The uptake of organic solutes in lake water. Limnol Oceanogr 10:22–28
Zhang S, Liu H, Glibert PM, Guo C, Ke Y (2017) Elemental stoichiometry and nutrient excretion of *Noctiluca scintillans* in response to prey of different quality. Sci Rep 7:7622. https://doi.org/10.1038/s41598-017-05991-w
Zubkov MV, Tarran GA (2008) High bacterivory by the smallest phytoplankton in the North Atlantic Ocean. Nature 455:224–227. https://doi.org/10.1038/nature07236

Additional Website Accessed

https://static.scientificamerican.com/sciam/assets/File/OSQ318_Revolutions.pdf. Accessed 31 Mar 2023

Iron and the Anemic High-Nutrient-Low-Chlorophyll Oceanic Regions

16

16.1 Introduction: The Iron Hypothesis

The story of iron (Fe) and the understanding of its role in the oceans is rich in mystery, intrigue, controversy, characters, and, more recently, commercialization and profiteering.

Since the 1930s, scientists have been curious about the fact that in certain regions of the oceans, nitrogen (N) and phosphorus (P) are not depleted, but chlorophyll *a* (chl *a*) does not accumulate. These regions—formally identified as the high-nutrient-low chlorophyll (HNLC) regions (Martin et al. 1989)—are found only in the extremes: the Equatorial Pacific, the Subarctic Pacific, and much of the Southern Ocean (Fig. 16.1). What could be keeping growth from occurring in these regions? Why only these regions? Why are these nutrients not being consumed?

Fast forward several decades to about the 1970–1980s. It was around this time that trace metal chemists developed, applied, and demonstrated that the normal protocols used in sampling from ships could contaminate samples (e.g., Fitzwater et al. 1982). Trace metals in ships, including Fe, are a major contamination source. Hydrowires and sampling equipment are also typically built with metal components. The development of ultra-clean sampling techniques was challenging and laborious, though necessary to test the potential role of metal limitation. Ultra-clean techniques involved soaking all bottles for multiple days to weeks sequentially in detergent, HCL (distilled multiple times), and deionized water. Sampling of the water had to be done far from the zone of influence of the mother ship by launching small boats and using plastic and Teflon sampling devices (no metal wires or hinges). It was only with these techniques was the true impact of the metal contaminants seen. The data comparing conventional and "clean" sampling techniques showed that Fe was significantly lower in concentration in cleanly sampled waters than previously thought. This led to the hypothesis that Fe was the magic ingredient, the limiting micronutrient, that was missing from the Southern Ocean and other HNLC regions.

P. M. Glibert, *Phytoplankton Whispering: An Introduction to the Physiology and Ecology of Microalgae*, https://doi.org/10.1007/978-3-031-53897-1_16

Fig. 16.1 Global distribution of surface chlorophyll levels in May 2023. Observations are based on the MODIS sensor on NASA's Aqua satellite. Land is dark gray, and places where MODIS could not collect data (reasons include sea ice, polar darkness, or clouds) are light gray. Image reproduced from https://earthobservatory.nasa.gov/global-maps/MY1DMM_CHLORA/MYD28M

John Martin, an oceanographer of Moss Landing Institute (Fig. 16.2), encapsulated the Fe hypothesis when he introduced his ideas rather nonchalantly (Martin 1990, quoted by Chisholm and Morel 1991) and

> ... more or less facetiously at a Journal Club lecture at Woods Hole Oceanographic Institution in July 1988. [He] estimated that, with 300,000 tons of Fe, the Southern Ocean phytoplankton could bloom and remove two billion tons of carbon dioxide. Putting on [his] best Dr. Strangelove accent, [he] suggested that with half a ship load of Fe ... [he] could give you an ice age. After [this lecture they] all had beer on the lawn...

As noted by Chisholm and Morel (1991) who were present for the lecture (and the beer), "No one thought he was in the least bit serious about the fertilization idea; it was viewed as a joke by those who reacted at all."

It was several years later when Martin re-introduced the idea in a more serious tone in a *Nature* paper, where he wrote, "oceanic [Fe] fertilization aimed at the enhancement of phytoplankton production may turn out to be the most feasible method of stimulating the active removal of greenhouse gas CO_2 from the atmosphere, if the need arise" (Martin et al. 1990). He based this on his set of "bottle enrichments" he had done in the Subarctic Pacific and Southern Ocean, which had resulted in blooms in his bottles. Indeed, these bottle experiments showed that by adding Fe to incubation bottles, chl *a* increased.

When this idea was first proposed, it was mocked and ridiculed, if it was given any attention at all by the scientific community. Scientists balked. Dick Barber, an oceanographer from Duke Marine Laboratory, said that John Martin's quip was "enough to make 500 oceanographers grab their calculators to prove that he couldn't be right" (Trujillo 2011). Lou Codispoti, best known for his work on N cycling, offered his take on the idea, "Since growth of plants leads to a net uptake of

Fig. 16.2 John Martin. Drawing by Roger Kammerer. Image reproduced from Weier (2001, https://earthobservatory.nasa.gov/features/Martin)

atmospheric carbon dioxide, who needs to worry about the greenhouse effect. All we need to do is convert old cars into soluble [Fe] and dump them in the Southern Ocean" (Codispoti 2001, p. 95). Others freely criticized the use of incubation bottles and the distorted view that can emerge from their use due to bottle effects and altered grazer communities.

However, and in no small part because the idea was so seemingly outlandish (or, as others have said, "wild, maverick and simple" (Trujillo 2011)), the idea soon took hold. The media pounced. John Martin began appearing on TV talk shows. Fully aware of his critics and appreciative of the concern of polluting the oceans, he commented, "I agree that the ideal would be to have the average American get out of his car; have the Chinese not develop their coal resources; have the Brazilians not cut down the rain forest. However, we don't live in an ideal world" (cited in Trujillo 2011). Some coined Fe fertilization the "Geritol" solution to global warming (e.g., Weier 2001). (Geritol is the trademark name of a dietary Fe supplement.)

With the fear of the potential for a global catastrophe from a rogue experiment gone wrong, both the National Research Council and the American Society of Limnology and Oceanography (now the Association for the Sciences of Limnology and Oceanography, ASLO) held major symposia and reviewed efforts to evaluate the idea. ASLO published a special issue (Limnol Oceanogr Vol. 36, no. 8) devoted to anything related to Fe and the HNLC paradox and a summary report (https://www.whoi.edu/fileserver.do?id=56328&pt=10&p=39295) and adopted a resolution regarding experiments on Fe fertilization (Box 16.1). An editorial in *Science* (Roberts 1991) summarized the overall outcome:

ASLO trounces the idea of fertilizing the oceans with iron as a greenhouse fix, as expected. But, in an unexpected twist, the society endorses a small-scale experiment in which iron would be added to the open ocean. The idea isn't to engineer the oceans, the [ASLO] report stresses, but to test a hypothesis that might answer one of the long-standing puzzles in biological oceanography.

Box 16.1 ASLO Resolution on Iron Fertilization

The participants of symposium on "What controls phytoplankton production in nutrient-rich areas of the open ocean?" by the American Society of Limnology and Oceanography (now the Association for the Sciences of Limnology and Oceanography) adopted the following resolution related to ocean fertilization and atmospheric carbon dioxide (summarized in Chisholm and Morel 1991).

Recent research suggests that primary production in some nutrient-rich areas of the open sea may be limited by iron deficiency. This suggestion has stimulated discussion concerning the feasibility of fertilizing the Southern Ocean with iron as a means of reducing carbon dioxide concentration in the atmosphere:

- Whereas major manipulation of the Southern Ocean by iron fertilization is a scientifically uncertain mitigation measure to reduce rising carbon dioxide levels in the atmosphere.
- Whereas even if fully implemented and successful, this measure would likely at best postpone the impending climate change by a few years if not combined with a significant reduction in carbon dioxide emission.
- Therefore, it resolved that the American Society of Limnology and Oceanography urges all governments to regard the role of iron in marine productivity as an area for further research and not to consider iron fertilization as a policy option that significantly changes the need to reduce emissions of carbon dioxide.

For scientists, the questions did not go away. They pushed the need for more experimentation, but the appeal of a seemingly simple solution to climate change would not die. In this chapter, the focus is on the science, framed in the controversy of geo-engineering by Fe.

16.2 The HNLC Conundrum

High-nutrient, low-chl *a* regions of the oceans were long known and documented, and there has been no lack of hypotheses explaining these regions. Some hypotheses were location-specific, and others were more general. In polar waters, the mean light availability in deep-mixed layers was viewed as a major impediment to productivity

(Frost 1987; Banse 1990). High rates of grazing were also considered to be sufficient to keep biomass in check and to keep biomass low enough that nutrient depletion did not occur (Frost 1987; Banse 1990; Cullen 1991). Sommer and Stabel (1986) raised the issue of temperature limitation of growth rates in the Southern Ocean. There was the question of the role of Fe (Martin and Fitzwater 1988; Martin et al. 1989). Walsh's (1976, p. 1) words preceding this debate as to the controlling factor encapsulated the forthcoming disputes in a general context,

> Simple consideration of single control factors of marine productivity such as light, nutrients, mixing, chelation, or grazing can lead to rather naïve and perhaps inaccurate conceptualizations of the dynamics of nonlinear aquatic ecosystems…

The role of grazing was viewed with such interest that a major interdisciplinary, multi-investigator program was funded by the US NSF to explore its role (Project SUbarctic Pacific Ecosystem Research, SUPER). One of this program's major findings coincided with the observation emerging worldwide during the 1980s that microzooplankton controlled the bulk of the biomass. The stock of copepods was not large enough to control growth—and phytoplankton were not observed in sufficient quantities in their guts to sustain their own growth (Miller et al. 1991a, b). The period of the SUPER project was also the period during which the microbial loop came into prominence as an oceanographic concept (Azam et al. 1983). The importance of the Subarctic Pacific was that the region was dominated by small cells; thus, their control by micrograzers was perhaps not too surprising (Miller et al. 1991a, b). Protozoa, then, were determined to be the principal diet of larger copepods.

An important consequence of this type of trophic web is that small grazers make small fecal material. They are much more efficient at recycling nutrients, especially at recycling NH_4^+, which would be readily utilized by the small phytoplankton. As previously discussed (Chap. 12, see also Glibert et al. 2016), NH_4^+ can repress the uptake of NO_3^-. Wheeler and Kokkinakis (1990) demonstrated this effect for the Subarctic Pacific (Fig. 16.3). In their experiments, the percent inhibition (repression) of NO_3^- uptake varied with sampling date and station but ranged overall from 10 to 100%. Thus, it was hypothesized that available NO_3^- was not used because NH_4^+ was rapidly and efficiently recycled and then assimilated preferentially by the small phytoplankton cells.

Squaring this effect with the observations from bottle experiments that Fe leads to more chl *a* requires appreciation of the type of chl *a* that Fe promotes. Such additions change the phytoplankton composition, leading to larger cells, especially larger diatoms. Our understanding of the role of NO_3^- in the nutrition of diatoms (see Chap. 12) is now far improved relative to the time of the SUPER program and the major Fe debate. Yet, it was known even then that adding NO_3^- made proportionately more diatoms than other types of cells (e.g., Goldman 1983). The reduction of NO_3^- to NH_4^+ via NO_3^- reductase (NR) is an Fe-demanding process. Thus, the ability of large cells to fully access and reduce NO_3^- may indeed be limited by Fe. The "normal" small phytoplankton grow well on regenerated NH_4^+, which represses NO_3^- uptake, and Fe limitation reduces the ability of large diatoms to

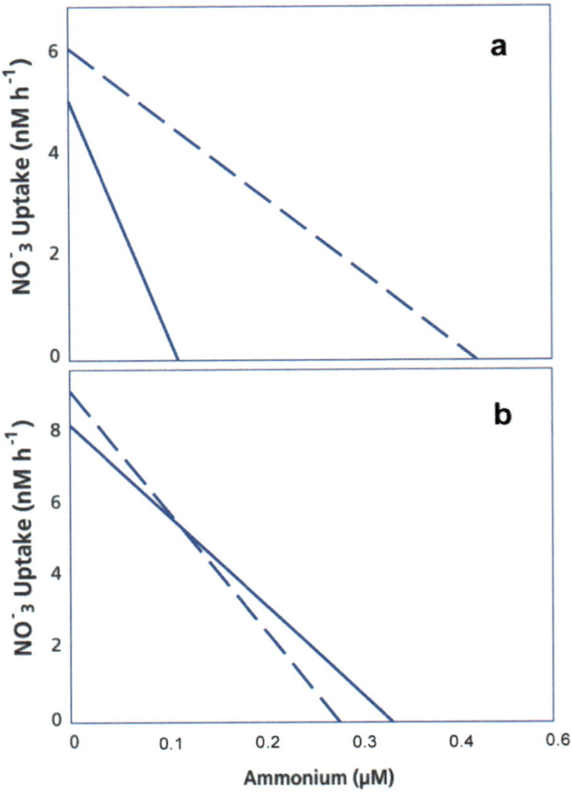

Fig. 16.3 Relationships between uptake rates of NO_3^- and concentrations of NH_4^+ measured at 6 h during deck incubation experiments. Panel (**a**) June (solid line) and September (dashed line) 1987. Panel (**b**) May (solid line) and August (dashed line) 1988 sampled during the SUPER program. Images simplified and redrawn based on Wheeler and Kokkinakis (1990, Limnol Oceanogr 35:1267–1278)

utilize the available NO_3^-. As summarized by Morel et al. (1991), HNCL regions are under Fe stress and are efficiently grazed in a tightly coupled microbial loop. Such systems are typically dominated by small cells (an adaptation to reduce their Fe requirement), depending mainly on NH_4^+, and only upregulate their use of NO_3^- when Fe is supplied. Frost (1991, p. 1627) explained the phenomenon further,

> if we incorporate a possible effect of Fe deficiency on phytoplankton species composition, surface waters of the Subarctic Pacific may remain nutrient-replete because grazing control of phytoplankton stock, coupled with preferential utilization of NH_4^+ by phytoplankton, yet the grazing control may be exerted on an assemblage exerted by Fe limitation. A large part of the indigenous phytoplankton assemblage—that responsible for the bulk of the phytoplankton production—appears not to be Fe limited.... However, when a natural phytoplankton assemblage is exposed to Fe enrichment, a rare component of the phytoplankton assemblage, normally contributing little to total phytoplankton production, is released from Fe limitation. This component seems to be large-sized species, chiefly diatoms...that cannot be grazed efficiently by the indigenous grazer assemblage.

The kinetics, physiology, and physical and chemical controls are clearly more complex than this short summary. Nevertheless, even this simple summary demonstrates the complexity of the system and also underscores themes that have reoccurred throughout this book: Limitation of growth and biomass may be caused by different factors, and limitation of biomass and species composition may also be caused by different factors. It is curious that although NH_4^+ repression was incorporated in all the early hypotheses of HNLC regions (e.g., Chisholm and Morel 1991; Frost 1991; Morel et al. 1991), this idea soon disappeared from view, with the simpler-to-understand ideas of light and grazing limitation perpetuated as the competing ideas to that of Fe limitation.

16.3 Stepping Back and a Review of Fe Physiology

16.3.1 Fe Availability and Cellular Requirements

Iron exists in the oceans primarily as soluble Fe(II) with smaller fractions as soluble Fe(III). It is also present in insoluble chemical forms, complex organic forms, and colloid materials. The speciation of Fe is dependent on O_2 concentrations, pH, and other environmental factors, including light (Sunda 2001). It is primarily delivered to surface waters via atmospheric deposition as Fe dust particles. There are also groundwater sources, and in the deep ocean, hydrothermal vents are a significant source. Glaciers, also, may contribute Fe via icebergs and their melting. In freshwaters, Fe can also be contributed via human activities, such as mining, agricultural runoff, or other activities (Marchetti and Maldonado 2016).

The role of Fe in cells primarily lies in oxygenic photosynthesis and N assimilation. There are many other cellular processes that also rely on Fe, including, but not limited to, respiration, vitamin synthesis, dehydrogenase and oxygenase reactions, and detoxification of reactive oxygen species (Marchetti and Maldonado 2016). The photosynthetic reaction centers (see Chaps. 5 and 6) contain Fe, as does the cytochrome b_6f complex, and, of course, ferredoxin. Reduced chl *a* content occurs when cells are under Fe deficiency. (This occurs not only due to a downregulation of photosynthetic units but also due to a lower abundance of Fe-containing enzymes involved in chl *a* synthesis; Raven 1990; Behrenfeld and Milligan 2013.)

There is a requirement for Fe in NO_3^- assimilation (NR), NO_2^- assimilation (nitrite reductase, NiR), and fixation of N_2 (see Chap. 11). For example, Milligan and Harrison (2000) showed in the diatom *Thalassiosira weissflogii* (now renamed as *Conticribra weissflogii*) that under Fe-limiting conditions, NiR became rate-limiting, and thus, the cell excreted NO_2^-.

As an enzyme, NR has a temperature window of optimum activity. The HNLC regions of the world—the warm Equatorial Pacific and the cold polar regions—fall outside of this optimum window for most diatoms. Indeed, different NRs in different species may have somewhat different temperature (thermal optimum) curves, but they are not so dissimilar as to be totally skewed in the directionality of extreme heat or cold. Speculating—but entirely in keeping with the contemporary understanding of the regulation of this enzyme—it is possible that the addition of Fe results in the production of more NR per cell, allowing the cell to compensate for its lower activity at non-optimal temperatures. More NR can also help to overcome the potential of NH_4^+ repression. As the presence of NH_4^+ hastens the decay of existing NR, this would be overcome by the enhanced production of more NR. Interest in the role of Fe in photosynthetic yields (Fe would regulate the ferredoxin pool) resulted in measurements of *Fv/Fm* throughout the Fe experiments described above. No comparable and consistent measurements are available for NR. It would not take a major Fe enrichment experiment to do so.

Within those cells capable of fixing N_2, the enzyme nitrogenase requires substantial quantities of Fe. Thus, for these cells (primarily cyanobacteria), there may be regions of the ocean where the process of N_2 fixation is limited by Fe, but not necessarily the growth rate of non-N_2-fixing cells that can acquire an alternate (e.g., dissolved inorganic or organic) N source (Moore et al. 2009; Marchetti and Maldonado 2016). Moreover, cyanobacteria, especially those that fix N_2, have a much higher Fe cell quota than many other groups of microalgae (Marchetti and Maldonado 2016). Cells that may become limited by Fe often reduce their cell size, by as much as 50% (e.g., Morel et al. 1991; Sunda and Huntsman 1995).

16.3.2 Fe Acquisition Mechanisms and Kinetics

As with other nutrients, Fe is transported across the cell membrane via an active transport process, and there is a high-affinity system and a low-affinity system (or systems; Marchetti and Maldonado 2016). The low-affinity transport system is mainly a function of the inorganic Fe(III) concentration, while the high-affinity system is a function, primarily, of organically bound Fe (Maldonado and Price 2001). Fe(II) is an obligate intermediate in both uptake systems (Marchetti and Maldonado 2016). There are multiple mechanisms for acquiring Fe, depending on the species, Fe availability and form, and much has recently been learned from molecular approaches (Marchetti and Maldonado 2016; Sutak et al. 2020; Fig. 16.4).

To acquire Fe, many phytoplankton employ siderophores. These are small, high-affinity chelating compounds that are secreted, and then, the chelated Fe compound is transported into the cell via specific transporters (Fig. 16.5a, b). The term derives from the Greek ("iron carriers"). Although mostly known in bacteria (including cyanobacteria) and fungi, it has been suggested that the marine diatoms *Thalassiosira pseudonana* (Armbrust et al. 2004) and *Phaeodactylum tricornutum* (Kustka et al. 2007) have a yeast-like reducing system, but many other phytoplankton do not have siderophores (Sutak et al. 2012).

Uptake kinetics of Fe follow a saturating hyperbolic relationship (Fig. 16.5c), although biphasic uptake kinetics have been observed, most notably for the Southern Ocean prymnesiophyte, *Phaeocystis antarctica* (Strzepek et al. 2011). Growth rates in relation to the cellular Fe quota, Q_{Fe}, can be related by the Droop equation (Droop 1970, see Chap. 3). Moreover, following pulsed additions of Fe, microalgae can take up and accumulate Fe quickly, a strategy analogous to luxury uptake of N or P. Storage is mainly in the form of the Fe-containing protein, ferritin, of which there are multiple types (Marchetti and Maldonato 2016).

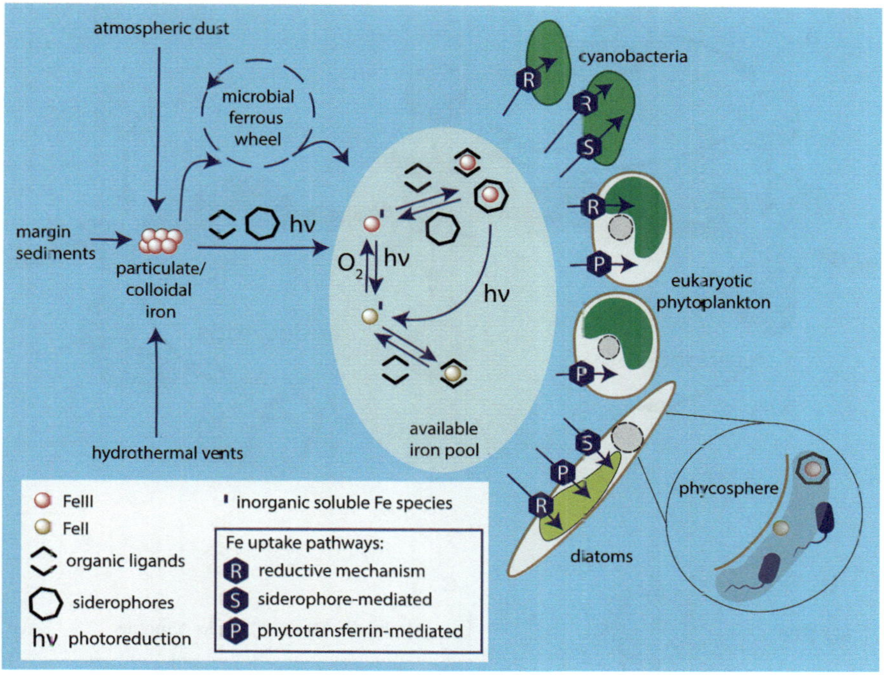

Fig. 16.4 Iron sources available to phytoplankton and associated uptake pathways. It appears that the main strategy of iron acquisition by cyanobacteria is reductive iron uptake combined with siderophore-mediated pathways. For eukaryotic phytoplankton, iron uptake is mediated by phytotransferrin, while some species may use additional mechanisms. The phycosphere (Chap. 21) is the unique environment in close proximity to the cell surface, in which iron availability may be increased due to bacterial metabolism. Image reproduced from Sutak et al. (2020, Front Microbiol 11: doi.org/10.3389/fmicb.2020.566691) under a creative commons license

The Tara Oceans program confirmed the prevalence of Fe as a limiting nutrient in vast regions of the oceans (Marchetti 2019; Caputi et al. 2019). Using 'omics data, copy number, and expression profiles of Fe-responsive genes in diatoms and other dominant phytoplankton groups, they found high expression of diatom Fe-starvation-induced protein genes in Fe-poor regions of the ocean. Moreover, they found that *Pseudo-nitzschia* had an exceptional ability to grow under low Fe as it had a minimal Fe requirement. Furthermore, *Pseudo-nitzschia* was found to use ferritin for long-term Fe storage, but at a community level there were no clear patterns in diatom ferritin gene abundance or expression as a function of Fe levels (Caputi et al. 2019).

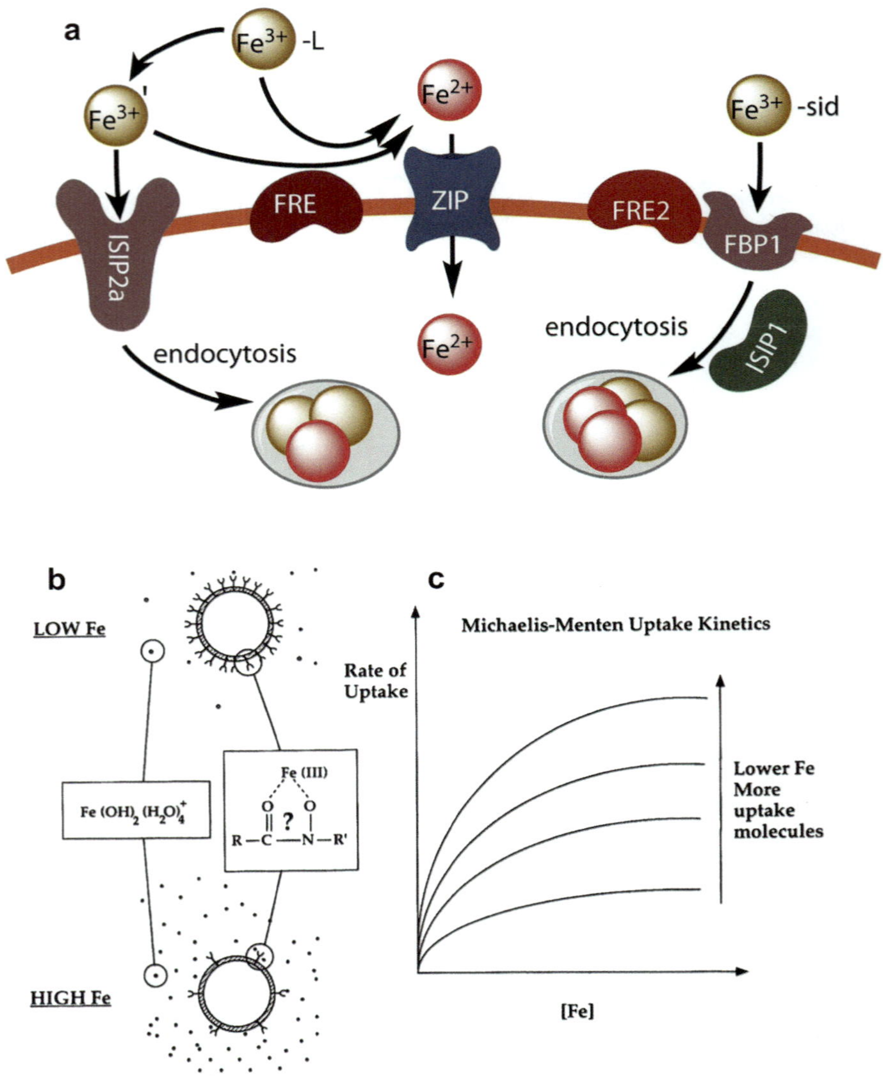

Fig. 16.5 Panel (**a**) Iron uptake by the diatom *Phaeodactylum tricornutum.* Unchelated ferric ions can be transported by endocytosis after binding to the phytotransferrin ISIP2a, or ferric chelates can be dissociated by reduction (FRE) and the resulting ferrous ions can be taken up by divalent meal transporters (ZIP). Hydroxamate siderophores are taken up by endocytosis after binding to the FBP1 protein, and iron is released by reduction (FRE2) possibly in endocytosis vesicles. Panel (**b**) Schematic representation of phytoplankton and their cell surface Fe transport ligands in seawater containing high and low concentrations of dissolved Fe. The Fe transport ligands are hypothesized to resemble hydroxamate siderophores (R-CO-NOH-R′). Panel (**c**) Increased Fe transport ligand density in phytoplankton grown in low-Fe seawater increases the maximum rate of Fe uptake by the cells. Image (**a**) reproduced from Sutak et al. (2020, Front Microbiol 11: doi.org/10.3389/fmicb.2020.566691) and images (**b**, **c**) reproduced from Morel et al. (1991, Oceanography 4(2): 56–61. doi.org/10.5670/oceanog.1991.03), both under creative commons licenses

16.4 The First Iron Tests: The New Iron Age

The first "tanker of iron" enrichment experiment was undertaken just months after John Martin passed away in 1993 at the young age of 56 (John Martin suffered from polio in young adulthood and cancer as an adult). That test involved barrels and barrels of Fe granules dumped into the waters 250 km southwest of the Galapagos Islands and tracked for weeks by some of the preeminent oceanographers and trace metal chemists. This was the first Iron Enrichment Experiment (IronEx) mission.

Zigzagging around a buoy in a calm patch of water, over 20 barrels of Fe salt crystals were first mixed with seawater and acidified to dissolve the Fe and then dumped into the ocean (Fig. 16.6). Ultimately, it resulted in a 5 square mile of Fe-fertilized water. Pharmaceutical-grade Fe(II) sulfate was the enrichment Fe source due to its low cost and solubility in seawater among other factors. The target concentration for the initial experiment was 4 nM, basically twice the concentration that had previously been shown in bottle experiments to produce enhanced growth (Coale et al. 1998). Sulfur hexafluoride was added as a tracer to identify the patch.

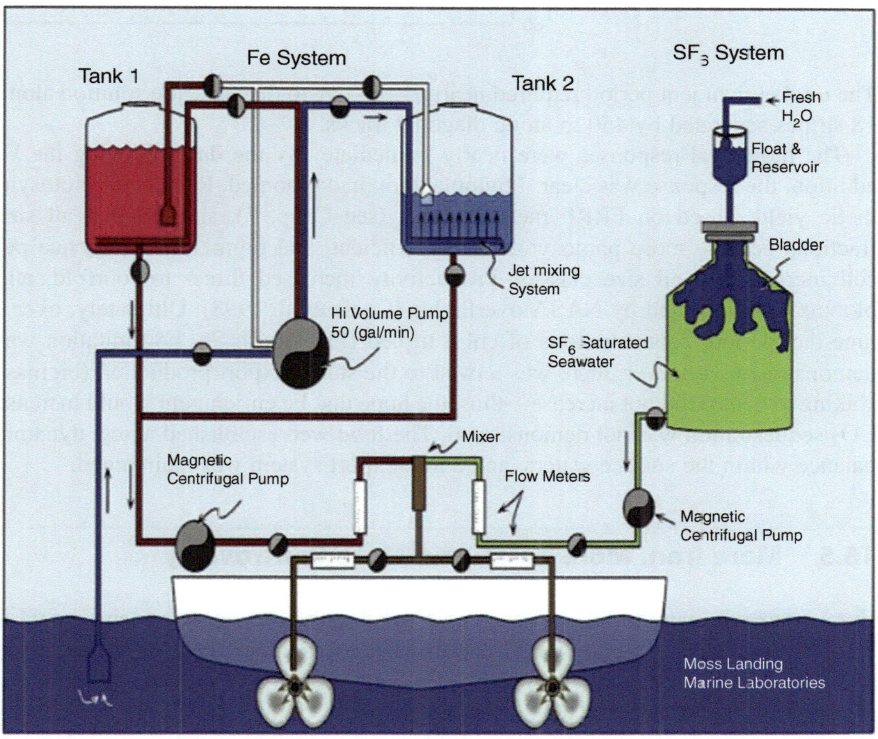

Fig. 16.6 Schematic diagram of the Fe/SF6 delivery system used in the IronEx I project. Image redrawn based on Coale et al. (1998, Deep Sea Res 45: 919–945) with permission of Elsevier

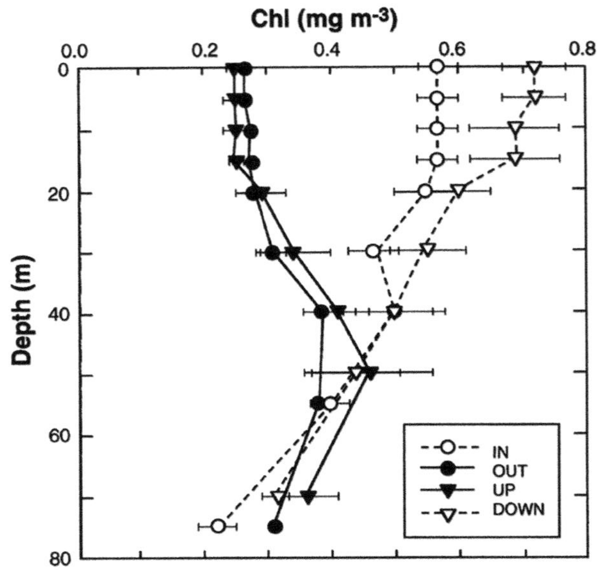

Fig. 16.7 Vertical profiles of chlorophyll concentrations both inside and outside of the IronEx I patch and stations upstream (east) and downstream (west) of the Galapagos Islands. Image redrawn based on Coale et al. (1998, Deep Sea Res 45: 919–945) with permission of Elsevier

The total enrichment period required nearly 1 full day to dispense the solution along 18 stripes separated by 450 m along diagonal tracks.

The biological responses were nearly immediate. By the day following the Fe addition, the response was clear: Phytoplankton had bloomed. Results of photosynthetic yield, based on FRRF measurements (see Chap. 7), showed that all size fractions became more photosynthetically efficient, and pigment fluorescence per cell increased in all size classes. Productivity increased three- to fourfold, and blooms were detected by NASA overflights (Coale et al. 1998). Ultimately, over a nine-day period, concentrations of chl *a* tripled (Fig. 16.7). So, Fe limitation was demonstrated. However, there was a twist to the story. Export production (biomass sinking to depth) did not increase—thus, the hope that Fe enrichment would increase CO_2 sequestration was not demonstrated. The food web established a new dynamic balance within the surface waters, and a low-export system was maintained.

16.5 More Iron, More Tests, and More Controversy

From IronEx I, many technical issues associated with open ocean enrichment were learned. Two years later, a second IronEx mission was undertaken. This time, using twice the Fe and newly understood ways of keeping the Fe on the surface, phytoplankton increased 85 times and drew down hundreds of tons of CO_2 (Trujillo 2011). Soon after the enrichment and algal bloom, vertically migrating zooplankton responded to the patch and consumed the newly fixed carbon (C) (Coale et al. 1998). So, in this test, the first hint of a major shift in community composition of the phytoplankton was observed—and the first hint that the shift may be favorable

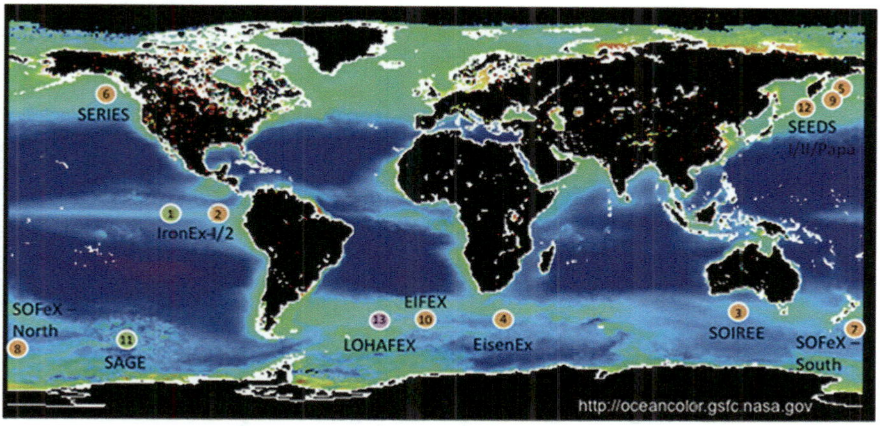

Fig. 16.8 Since 1993, these 12 small-scale open ocean experiments (*red dots*) have shown that iron additions do indeed result in phytoplankton blooms, thereby drawing carbon dioxide out of the atmosphere and into the ocean. Image reproduced from Morrissey and Bowler (2012, Front Microbiol 3: 43, doi.org/10.3389/fmicb.2012.00043) under a creative commons license

for the production of taxa that may be harmful algal species (reviewed by de Baar et al. 2005; Boyd et al. 2007; Yoon et al. 2018). In this experiment, the diatom *Pseudo-nitzschia* was stimulated, and within 5 days, it became the dominant phytoplankter. These cells were eventually grazed by ciliates and other protists. Thus, the diatoms enjoyed a several-day window in which they were able to bloom without grazing pressure.

Since the first test in 1993, scientists have conducted more than a dozen major Fe enrichment experiments (Fig. 16.8, Table 16.1). One of the concerns with each of the early experiments was that the patch was not large enough, the addition of Fe was not large enough, or the duration of the study was not long enough. Thus, such experiments expanded from 10 days in Iron Ex I with 450 kg Fe in 1993 to 40 days and 10 tons of $FeSO_4$ by the LOHAFEX study in 2009 (Table 16.1).

An overarching conclusion of all the experiments conducted during this period was that—in general—in HNLC regions, Fe enrichments did enhance photosynthesis (as *Fv/Fm*), chl *a*, and primary productivity (Fig. 16.9), primarily by increasing diatom production, confirming the hypothesis of Fe limitation. However, the anticipated increase in C flux/C export was not observed. In other words, results from all enrichment experiments determined that large-scale ocean fertilization could—at best—only sequester a modest amount of C relative to human emissions (Strong et al. 2009). In addition to the production of *Pseudo-nitzschia*, in several of the Fe enrichment experiments some other large chain-forming pennate diatoms emerged. The species *Fragilariopsis kerguelensis* was successful in SOIREE, while during SEEDS, *Chaetoceros debilis* replaced *Pseudo-nitzschia turgidula* by the sixth day (de Baar et al. 2005). Other large taxa also emerged in some cases, including *Rhizosolenia* sp., *Thalassiothrix*, *Thrichotoxon*, *Asteromphalus*, and

Table 16.1 Summary of the major iron enrichment experiments

Experiment	Year	Location	Duration (days)	Magnitude and type of addition	Hypotheses tested	General conclusions
IronEx I	1993	Eastern Equatorial Pacific Ocean	10	450 kg Fe 64 km^2	Fe addition will increase phytoplankton productivity by relieving Fe limitation	Increase in primary productivity, but fairly modest
IronEx II	1995	Eastern Equatorial Pacific Ocean	17	450 kg Fe 72 km^2	Fe addition will increase phytoplankton productivity by relieving Fe limitation	Larger bloom than iron Ex-I; taxonomic shift toward diatoms; no observed C export
SOIREE	1999	Southern Ocean-Australia; South of Antarctic Polar Front	13	1740 kg Fe 50 km^2	Fe addition will increase phytoplankton productivity by relieving Fe limitation in the Southern Ocean; fate of C	Diatom dominated bloom induced; no measurable C export
EisenEx[a]	2000	Southern Ocean-Africa; Antarctic Polar Front	21	4 tons $FeSO_4$ 38.5 km^2	In glacial periods, Fe dust contributed to lower CO_2 concentrations	Diatom dominated bloom induced; no measurable C export compared to outside patch
SEEDS-I	2001	Subarctic Pacific-Northwest	13	350 kg Fe 80 km^2	IFe addition will increase phytoplankton productivity by relieving Fe limitation in Subarctic Pacific; fate of C	Largest changes in biogeochemical parameters of experiments to date; large increase on oceanic diatoms; no C export detected
SoFex-N SOFex-S	2002	Southern Ocean-New Zealand—north and south of Antarctic Polar Front	30	North: 1712 kg Fe 225 km^2; South: 1260 kg Fe 225 km^2	Compare iron-induced blooms in regions with high and low silicate in the Southern Ocean	North: increase in diatom blooms South: enhanced export fluxes but similar to natural blooms
SERIES	2002	Subarctic Pacific-Gulf of Alaska	25	490 kg Fe 77 km^2	Fate of carbon fixed in Fe-induced bloom; efficiency of C export	Observation throughout all phases of the phytoplankton bloom; high bacterial remineralization and mesozooplankton grazing; no significant C export

EiFex[a]	2004	Southern Ocean-Atlantic	35	7 tons $FeSO_4$ 150 km^2	Fate of Fe on phytoplankton community structure; efficiency of C export	Observation throughout all phases of the phytoplankton bloom; C export documented due to the formation of aggregates
FeeP	2004	Subtropical Northeast Atlantic	21	5 tons $FeSO_4$ 20 tons PO_4^{3-} 25 km^2	Interactions of Fe and P on bloom regulation in North Atlantic	Increased production in picophytoplankton; Increased N_2 fixation
SAGE	2004	Southern Ocean-New Zealand	15	5.4 tons $FeSO_4$ 100 km^2	Effect of Fe on sea-air gas exchange; Production of DMS during Fe-induced blooms	Increase in phytoplankton production—but no shift to diatoms; no measurable DMS produced; no significant CO_2 drawdown
SEEDS II	2004	Subarctic Pacific-Northwest	26	491 kg Fe 64 km^2	Understand factors regulating the termination of Fe-induced blooms	No diatom response; increase in zooplankton (copepod) grazing
LOHAFEX[a]	2009	Southern Ocean-Atlantic	40	10 tons $FeSO_4$ 300 km^2	Ecological effects and fate of sinking C	No shift to diatom community; no C export detected; high grazing pressure

[a] Eisen means iron in German; Loha means iron in Hindi (Yoon et al. 2018); EIFEX refers to the European Iron Fertilization Experiment
Modified from de Baar et al. (2005), Boyd et al. (2007), Strong et al. (2009) and Yoon et al. (2018)

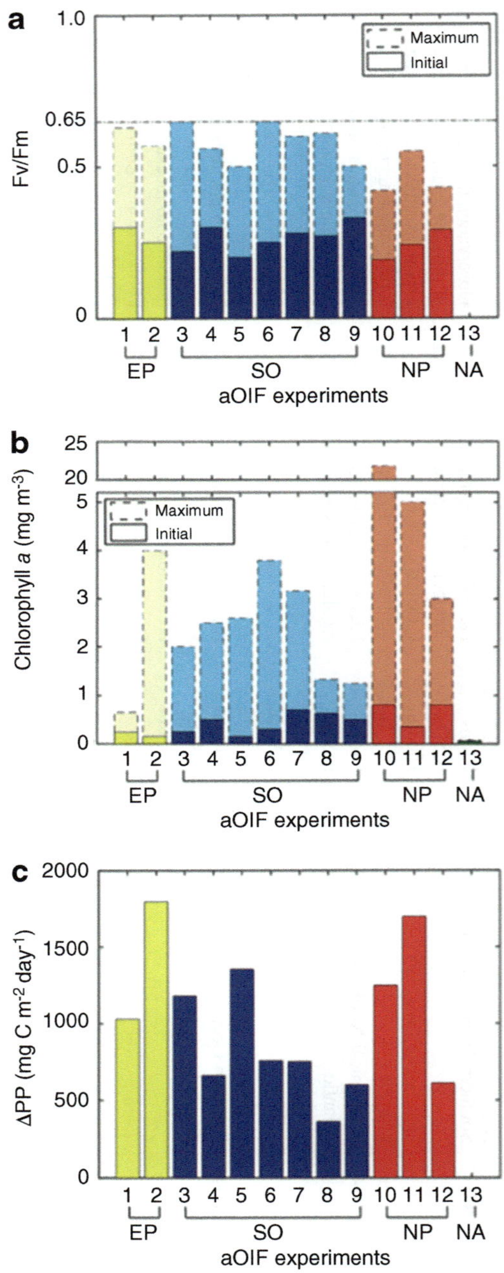

Fig. 16.9 Change in Fv/Fm Panel (**a**), chlorophyll *a* Panel (**b**), and primary production Panel (**c**) in iron enrichment experiments. Experiments summarized were conducted in the Equatorial Pacific (EP—Iron Ex I and Iron Ex II-1,2), the Southern Ocean (SOIREE-3, EisenEx-4, SoFex-N-5, SoFex-S-6, EIFEX-7, SAGE-8, and LOHAFEX-9), the North Pacific (SEEDS 1–10, SERIES-11, and SEEDS 2–12), and the North Atlantic (FeeP-13). Images reproduced from Yoon et al. (2018, Biogeosci 15: 5847–5889) under a creative commons license

Actinocyclus sp. (de Baar et al. 2005). An interesting relationship was derived from cultures of several of such species with respect to added Fe: The required ambient Fe to achieve a relative growth rate of 50% increased with diatom size (reviewed by de Baar et al. 2005 based on Timmermans et al. 2004). The response by diatoms seen in IronEx II and most experiments, however, was not reproduced in the SEEDS II experiment (Boyd et al. 2005), which showed no significant diatom bloom, but rather a bloom of pico- and nanophytoplankton, such as *Synechococcus*, picoeukaryotes, and cryptophytes. This was attributed to a heavy grazing pressure on diatoms (Landry et al. 2000).

16.6 From Iron Hypothesis to Paradigm (With Lingering Questions)

Scientifically, the major lingering question about the use of Fe fertilization as a C sequestration strategy is: How does the method of Fe supply influence the biogeochemistry of large expanses of ocean and how is net C export related to Fe fertilization on large scales? All of the Fe experiments to date are—in the grand scheme of oceanic production—small in scale both in time and space and there are no current plans going forward for expanded-scale experiments, but some are voicing their desire to develop a new suite of such experiments.

The voices of concern of side effects have been loud. Mentioned above was the observation that *Pseudo-nitzschia* was a common diatom produced in fertilized regions, with its potential for increases in domoic acid (DA; see Chap. 25). *Pseudo-nitzschia* was, in fact, the common diatom observed in Iron ExII, SOIREE, EisenEx, SOFEX-N and SoFEX-S, and SERIES (reviewed by Yoon et al. 2018). Discernable changes in DA were found in Iron Ex II and SoFex-S (and it is interesting to note that it took 12 years for the Iron Ex II samples and 4 years for the SoFex samples to be analyzed; Yoon et al. 2018). While no large-scale hypoxia has been observed in response to Fe-enriched blooms, it is worth emphasizing the short duration of many of these studies. Early modeling studies suggest that anoxic/hypoxic may develop under long-term, sustained Fe enrichment conditions (Fuhrman and Capone 1991). One of the greater concerns is the potential for the production of other greenhouse-producing gases under Fe enrichment. Fuhrman and Capone (1991) raised a number of these concerns in the ASLO symposium discussed above. Some of their predictions were realized.

Production of dimethyl sulfide (DMS) was observed in some, but not in all of the Fe fertilization experiments. DMS is a potential precursor of sulfate aerosols that contribute to cloud formation and thus may be an important influence on climate (e.g., Charlson et al. 1987). Concentrations of DMS are partially controlled by the cycling in the microbial web, as well as the species produced during Fe enrichment. The largest DMS production observed was a 6.8-fold increase in SOIREE in the patch (Turner et al. 2004) and a 4.8-fold increase in SOFex-N. However, SEEDS 1 and 2 in the western Subarctic Pacific did not show any enhancement in DMS, and in SERIES, there was an initial increase of >tenfold in DMS, but an overall decline by the end of the experiment (summarized by Yoon et al. 2018).

The potential changes in other climate gases, such as N_2O, are perhaps of greatest concern. With Fe enrichment comes an enhancement in the rates of N cycling and a shift in the relative rates of different pathways (see Chap. 11). N_2O is a potent greenhouse gas, far more so than CO_2, as it has a longer lifetime and is more effective in radiative heating of the atmosphere (Shine et al. 1990). With enhanced denitrification, N_2O, as an intermediate product, can be considerable, especially if there is the development of low O_2 waters. While substantial depletion in O_2 has not been observed in fertilized regions, depleted O_2 conditions can be expected with increased Fe additions and increased patch sizes. In the Equatorial Pacific, it has been suggested that Fe-fertilized N_2O enhancement may have offset ~40% of any potential benefits of CO_2 sequestration (Sarmiento and Orr 1991; Yoon et al. 2018). However, excess N_2O was not found in the EIFEX data. The overall concerns of N_2O cannot be dismissed despite the variable results from fertilization experiments. As summarized by Lou Codispoti (2001, pp. 94, 95, 97), albeit with a bit of tongue in cheek,

> Denitrification, by creating a "need" for compensatory nitrogen fixation, increases the amount of iron limitation in the ocean because nitrogen fixation requires approximately 100 times more iron than does uptake of ammonia and nitrate…Hmmm, could fertilizing with iron be like [being loaned] money from a loan shark? In other words, could adding iron make the ocean more iron limited. Small changes in the exact nature of dissolved oxygen distribution can produce big changes vis-à-vis denitrification and nitrous oxide production…If we are humble in addressing the uncertainties, we *might* be able to intentionally manipulate this cycle on larger scales to help mitigate our unintentional experiments. Certainly, it is worthwhile to put more thought into how we impinge upon the oceanic nitrogen cycle…

16.7 The International Community Response

The appeal of Fe fertilization was not diminished by the underwhelming results related to C sequestration following initial phytoplankton blooms upon enrichment. Indeed, there was great interest in capitalizing and profiting from such efforts. Various companies have formed to utilize ocean fertilization as a way to sell C credits (and to advertise the potential for enhanced fish production in the sea; see review by Strong et al. 2009). One of the initial entrepreneurs in such endeavors, Michael Markels, in an interview with a popular magazine journalist stated, "given 200 boats, 8.1 millions tons of iron, and say, 11 percent of the world's ocean" he could "zero out global warming. Next question?" (Graeber 2000 *Wired* Magazine; quoted in Strong et al. 2009). Numerous companies seemed to emerge and dissolve with regularity, due to lack of funds or a broader international response. In 2012, the Haida Salmon Restoration Corporation added 90 tons of Fe into the Pacific Ocean near the islands of Haida Gwaii with the aim of increasing salmon and the hope of using carbon credit offsets to cover the cost (https://en.wikipedia.org/wiki/Iron_fertilization). The results were questionable. No company has yet to gain traction in this method of C sequestration.

In the 1990s and early 2000s, when the first research-based Fe enrichment experiments were being conducted, no legal concerns were raised, even though skeptics raised environmental concerns. As commercial enterprises emerged, it was clear that international laws were weak or merky at best with regard to commercial ocean fertilization. The 1972 London Convention on Marine Pollution and Ocean Dumping, the International Maritime Organization that oversees the dumping of wastes at sea, and the 1959 Antarctic Treaty, which covers much of the Southern Ocean, had some relevance, but had no teeth at that time for regulating ocean fertilization. The London Convention resolved in November 2007 that fertilization schemes were not scientifically justified and "...urged governments to exercise utmost caution when considering such proposals" (https://ccrlaw.org/wp-content/uploads/2021/10/London-Convention-Statement-of-Concern-on-OIF-2007.pdf). In 2008, a resolution was passed at the London Convention/London Protocol Convention prohibiting ocean fertilization activities until such time as specific guidance could be developed to justify legitimate scientific research (Yoon et al. 2018). It stated, that "...ocean fertilization activities other than legitimate scientific research should not be allowed...and [should] not currently qualify for any exemption from the definition of dumping." The Convention on Biological Diversity, in May 2008, resolved that ocean fertilization should not take place (exceptions for small-scale legitimate scientific research) until there is a better assessment of the potential risks and should not be used for the selling of carbon offsets. They placed a moratorium on all such activities. It is of note that the USA is not a signatory to these conventions. In 2009, when the LOHAFEX expedition was just getting underway, the German government halted it on the basis that it violated the 2008 United Nations Convention on Biological Diversity. After further review and several independent environmental assessments, the expedition was allowed to go forth. In 2013, the London Convention adopted rules for evaluating how such studies can or should be done (Tollefson 2017).

16.8 Natural Iron Enrichment and Why the Atlantic Is Not an HNLC

A number of Fe enrichment experiments were conducted in the Atlantic, regions not characterized as HNLC regions. Additionally (and not listed in Table 16.1), several smaller-scale, privately funded, commercial endeavors were conducted in the Gulf of Mexico. Those experiments did not produce C that was exported, nor did they produce large diatoms, for the most part. The Gulf of Mexico, in fact, was an interesting choice for a demonstration of Fe fertilization, as it receives considerable deposits of aeolian Fe dust seasonally. There, Fe dust deposits have been documented to stimulate N_2 fixers, primarily *Trichodesmium*, and those blooms that are stimulated are thought to be part of the initiation cycle of blooms of the HAB dinoflagellate *Karenia brevis* (Walsh et al. 2006). The FeeP project documented increased N_2 fixation rates (Table 16.1). A very recent set of experiments has shown that in the oligotrophic waters of subtropical gyres,

Trichodesmium responds by upregulating Fe stress biomarker genes with decreasing Fe availability and upregulating genes involved in P acquisition with diverse P availability, and thus, the genes involved in N_2 fixation were mostly upregulated at the intersection of moderate Fe and P availability (Cerdan-Garcia et al. 2022). Clearly, the responses by diatoms, N_2-fixing cyanobacteria, or picoplankton depend on the suite of environmental actors from temperature to Fe availability to availability of other macronutrients, N, P, and Si.

16.9 Summary

Ocean fertilization captured the imagination of scientists and entrepreneurs from its first mention. Over several decades, the idea that Fe fertilization would solve the global warming crisis grew from an off-hand remark to an irresistibly simple idea that needed to be tested. The more-than-dozen experiments that have been conducted have shown that the primary production in vast areas—in the temperature extremes of the ocean—is Fe-limited. Iron stimulates production in these regions largely in large diatoms, which overtake the indigenous picoplankton. Grazing plays an important role, in keeping the production normally in a dynamic balance. The release products of grazing, primarily NH_4^+, also serve to regulate the ability of diatoms to access this form of N, as they have a greater dependence on NO_3^- (depending on temperature). With Fe, these diatoms may be able to produce more NR, allowing them to grow.

What has not been shown is that these blooms result in long-term sequestration of C to the deep sea. Concerns over the unintended consequences of hypoxia, DMS production, N_2O, and other negative consequences are substantial. As huge "mesocosm" studies, the Fe enrichment experiments have been insightful and have answered questions about the regulation of primary production in the oceans. It is also clear that such exercises are not the solution to global warming and pose substantial risks. The words of Chisholm and Morel (1991, p. v), written at the start of the Fe debate, hold equally true today, "perhaps the most significant outcome of this whole episode is that it has forced us to struggle hypothetically with the philosophical dimensions of massive intervention in the earth's biogeochemical cycles."

References

Armbrust EV, Berges JA, Bowler C, Green BR, Martinez D, Putnam NH, Zhou S et al (2004) The genome of the diatom *Thalassiosira pseudonana*: ecology, evolution and metabolism. Science 306:5693. https://doi.org/10.1126/science.1101156

Azam F, Fenchel T, Field JG, Gray JS, Meyer-Reil LA, Thingstad F (1983) The ecological role of water-column microbes in the sea. Mar Ecol Progr Ser 10:257–263. https://doi.org/10.3354/meps010257

Banse K (1990) Does iron really limit phytoplankton production in the offshore subarctic Pacific? Limnol Oceanogr 35:772–775

Behrenfeld MJ, Milligan AJ (2013) Photophysiological expressions of iron stress in phytoplankton. Annu Rev Mar Sci 5:217–246. https://doi.org/10.1146/annurev-marine-121211-172356

Boyd PW, Strzepek R, Takeda G, Jackson G, Wong CS, McKay RM, Law C et al (2005) The evolution and termination of an iron-induced mesoscale bloom in the northeast subarctic Pacific. Limnol Oceanogr 50:1872–1886. https://doi.org/10.4319/lo.2005.50.6.1872

Boyd PW, Jickells T, Law CS, Blain S, Boyle EA, Beusseler KO, Coale KH et al (2007) Mesoscale iron enrichment experiments 1993-2005: synthesis and future directions. Science 315(5812): 612–617

Caputi L, Carradec Q, Eveillard D, Kirilovsky A, Pelletier E, Pierella Karlusich JJ, Vieira FRJ et al (2019) Community-level responses to iron availability in open ocean planktonic ecosystems. Glob Biogeochem Cycles 33:391. https://doi.org/10.1029/2018GB006022

Cerdan-Garcia E, Baylay A, Polyviou D, Woodward EMS, Wrightson L, Mahaffey C, Lohan MC et al (2022) Transcriptional responses of *Trichodesmium* to natural inverse gradients of Fe and P availability. ISME J 16(4):1055–1064. https://doi.org/10.1038/s41396-021-01151-1

Charlson RJ, Lovelock JE, Andreae MO, Warren SG (1987) Oceanic phytoplankton, atmospheric sulphur, cloud albedo, and climate. Nature 326:655–661

Chisholm SW, Morel FMM (1991) Preface to: what controls phytoplankton production in nutrient-rich areas of the open sea? Limnol Oceanogr 36(8):i–v

Coale KH, Johnson KS, Fitzwater SE, Bain SPG, Stainton TP, Coley TP (1998) IronEx-I, an *in situ* iron-enrichment experiment: experimental design, implementation and results. Deep Sea Res II 45:919–945

Codispoti L (2001) The oceanic nitrogen cycle: a double-edged agent of change. In: Bendell-Young L, Gallaugher P (eds) Waters in peril. Klewer Academic Pub, Norwell, pp 73–101

Cullen JJ (1991) Hypotheses to explain high-nutrient conditions in the open sea. Limnol Oceanogr 36:1578–1599

de Baar HJW, Boyd PW, Coale KH, Landry MR, Tsuda A, Assmy P, Bakker DCE et al (2005) Synthesis of iron fertilization experiments: from the iron age in the age of enlightenment. J Geophys Res 110:C09S16. https://doi.org/10.1029/2004JC002601

Droop MR (1970) Vitamin B12 and marine ecology. V. Continuous culture as an approach to nutritional kinetics. Helgolander Wiss Meeresunters 20:629–636. https://doi.org/10.1007/BF01609935

Fitzwater SE, Knauer GA, Martin JH (1982) Metal contamination and its effects on primary production measurement. Limnol Oceanogr 27:544–551

Frost BW (1987) Grazing control of phytoplankton stock in the open subarctic Pacific Ocean: a model assessing the role of mesoplankton, particularly the large calanoid copepods *Neocalanus* spp. Mar Ecol Prog Ser 39:49–68

Frost BW (1991) The role of grazing in nutrient-rich areas of the open sea. Limnol Oceanogr 36: 1616–1630

Fuhrman JA, Capone DG (1991) Possible biogeochemical consequences of ocean fertilization. Limnol Oceanogr 36(8):1951–1959

Glibert PM, Wilkerson FP, Dugdale RC, Raven JA, Dupont C, Leavitt PR, Parker AE et al (2016) Pluses and minuses of ammonium and nitrate uptake and assimilation by phytoplankton and implications for productivity and community composition, with emphasis on nitrogen-enriched conditions. Limnol Oceanogr 61: 165–197. https://doi.org/10.1002/lno.10203

Goldman JC (1983) Potential role of large oceanic diatoms in new primary production. Deep-Sea Res I 40:159–168

Graeber C (2000) Dumping iron: Ecohacker Michael Markels claims he has a megafix for global warming: supercharge the growth of ocean plankton with vitamin Fe and let a zillion CO_2 scrubbers bloom. Wired Magazine. Available online at http://www.wired.com/wired/archive/8.11/ecohacking...pr.html. Accessed 1 June 2023

Kustka AB, Allen AE, Morel FMM (2007) Sequence analysis and transcriptional regulation of iron acquisition genes in two marine diatoms. J Phycol 43:715–729

Landry MR, Ondrusek ME, Tanner SJ, Brown SL, Constantinou J, Bidigare RR, Coale KH, Fitzwater S (2000) Biological response to iron fertilization in the eastern equatorial Pacific (IronEx II). I. Microplankton community abundances and biomass. Mar Ecol Prog Ser 201:27–42. https://doi.org/10.3354/meps201027

Maldonado MT, Price NM (2001) Reduction and transport of organically bound iron by *Thalassiosira oceanica* (Bacillariophyceae). J Phycol 37:298–309

Marchetti A (2019) A global perspective on iron and plankton through the *Tara* Oceans lens. Glob Biogeochem Cycles 33:239–242. https://doi.org/10.1019/2019GB006181

Marchetti A, Maldonado MT (2016) Iron. In: Borowitzka MA, Beardell J, Raven J (eds) The physiology of microalgae. Springer Publishing, pp 233–279

Martin JH (1990) A new iron age, or a ferric fantasy. JGOFS Newslett 1(4):5–11

Martin JH, Fitzwater SE (1988) Iron deficiency limits phytoplankton growth in the north-east Pacific subarctic. Nature 331:341–343

Martin JH, Gordon RM, Fitzwater SE, Broenkow WW (1989) VERTEX: phytoplankton/iron studies in the Gulf of Alaska. Deep-Sea Res I 35(6):649–680

Martin JH, Gordon RM, Fitzwater SE (1990) Iron in Antarctic water. Nature 345:156–158

Miller CB, Frost BW, Booth B, Wheeler PA, Landry MR, Welschmeyer N (1991a) Ecological processes in the subarctic Pacific: iron-limitation cannot be the whole story. Oceanography 4: 71–78

Miller CB, Frost BW, Wheeler PA, Landry MR, Welschmeyer N, Powell TM (1991b) Ecological dynamics in the subarctic Pacific, a possibly iron-limited ecosystem. Limnol Oceanogr 36: 1600–1615

Milligan AJ, Harrison PJ (2000) Effects of non-steady state iron limitation on nitrogen assimilatory enzymes in the marine diatom *Thalassiosira weissflogii* (Bacillariophyceae). J Phycol 36:78–86

Moore CM, Mills M, Achterberg E, Geider RJ, LaRoche J, Lucas MI, McDonagh EL et al (2009) Large-scale distribution of Atlantic nitrogen fixation controlled by iron availability. Nat Geosci 2:867–871. https://doi.org/10.1038/ngeo667

Morel FMM, Rueter JG, Price NM (1991) Iron nutrition of phytoplankton and its possible importance in the ecology of ocean regions with high nutrient and low biomass. Oceanography 4(2):56–61. https://doi.org/10.5670/oceanog.1991.03

Morrissey J, Bowler C (2012) Iron utilization in marine cyanobacteria and eukaryotic algae. Front Microbiol 3. https://doi.org/10.3389/fmicb.2012.00043

Raven JA (1990) Predictions of Mn and Fe use efficiencies of phototrophic growth as a function of light availability for growth and of C assimilation pathway. New Phytol 116:1–18

Roberts L (1991) Report nixes "Geritol" fix for global warming. Science 253:1490–1491. https://doi.org/10.1126/science.253.5027.1490

Sarmiento JL, Orr JC (1991) Three-dimensional simulations of the impact of Southern Ocean nutrient depletion on atmospheric CO_2. Limnol Oceanogr 36:1928–1950

Shine KP, Derwent RG, Wuebbles DJ, Morcrette J-J (1990) Radiative forcing of climate. In Houghton JT, Jenkins GJ, Ephraums JJ (eds) Climate change: the IPCC scientific assessment. Cambridge University Press, Cambridge

Sommer U, Stabel H-H (1986) Near surface nutrient and phytoplankton distributions in the Drake Passage during early December. Polar Biol 6:107–110

Strong AL, Cullen JJ, Chisholm SW (2009) Ocean fertilization: science, policy, and commerce. Oceanography 33(3):236–261

Strzepek K, McCluskey A, Boehlert B, Jacobsen M, Fant C (2011) Climate variability and change: a basin scale indicator approach to understanding the risk to water resources development and management. Water papers. World Bank, Washington, DC. http://documents.worldbank.org/curated/en/2011/09/15897484/climate-variability-change-basin-scale-indicator-approach-understanding-risk-water-resources-development-management. Accessed 3 June 2023

Sunda WG (2001) Bioavailability and bioaccumulation of iron in the sea. In: Turner DR, Hunter KA (eds) The biogeochemistry of iron in seawater. Wiley, Chichester, pp 41–84

Sunda WG, Huntsman SA (1995) Iron uptake and growth limitation in oceanic and coastal phytoplankton. Mar Chem 50:189–206. https://doi.org/10.1016/0304-4203(95)00035-P

Sutak R, Botebol H, Blaiseau PL, Leger T, Bouget FY, Camadro JM et al (2012) A comparative study of iron uptake mechanisms in marine microalgae: iron binding at the cell surface is a critical step. Plant Physiol 160:2271–2284. https://doi.org/10.1104/pp.112.204156

Sutak R, Camadro J-M, Lesuisse E (2020) Iron uptake mechanisms in marine phytoplankton. Front Microbiol 11. https://doi.org/10.3389/fmicb.2020.566691

Timmermans KR, van der Wagt B, de Baar HJW (2004) Growth rates, half-saturation constants, and silicate, nitrate, and phosphate depletion in relation to iron availability of four large, open-ocean diatoms from the Southern Ocean. Limnol Oceanogr 49:2141–2151. https://doi.org/10.4319/lo.2004.49.6.2141

Tollefson J (2017) Iron-Dumping Ocean experiments sparks controversy. Nature 545:393–394. https://doi.org/10.1038/545393a

Trujillo A (2011) The iron hypothesis. http://www.homepages.ed.ac.uk/shs/Climatechange/Carbon%20sequestration/Martin%20iron.htm. Accessed 15 June 2022

Turner SM, Harvey MJ, Law CS, Nightingale PD, Liss PS (2004) Iron-induced changes in oceanic sulfur biogeochemistry. Geophys Res Letts 31(14). https://doi.org/10.1029/2004GL020296

Walsh JJ (1976) Herbivory as a factor in patterns of nutrient utilization in the sea. Limnol Oceanogr 21:1–13

Walsh JJ, Jolliff JK, Darrow BP, Lenes JM, Milroy SP, Remsen A, Dieterle DA et al (2006) Red tides in the Gulf of Mexico: where, when, and why? J Geophys Res 111:C11003. https://doi.org/10.1029/2004JC002813

Weier J (2001) John Martin (1935–1993). https://earthobservatory.nasa.gov/features/Martin. Accessed 3 July 2023

Wheeler PA, Kokkinakis SA (1990) Ammonium recycling limits nitrate use in the oceanic subarctic Pacific. Limnol Oceanogr 35:1267–1278

Yoon J-E, Yoo K-C, Macdonald AM, Yoon H-I, Park K-T, Yang EJ, Kim H-C et al (2018) Reviews and syntheses: ocean iron fertilization experiments – past, present, and future looking to a future Korean iron fertilization experiment in the Southern Ocean (KIFES) project. Biogeosciences 15:5847–5889. https://doi.org/10.5194/bg-15-5847-2018

Additional Websites Accessed

https://cdrlaw.org/wp-content/uploads/2021/10/London-Convention-Statement-of-Concern-on-OIF-2007.pdf. London Convention: Convention on the Prevention of Marine Pollution by Dumping of Wastes and Other Matter 1972, 1972. Accessed 1 July 2023

https://earthobservatory.nasa.gov/global-maps/MY1DMM_CHLORA/MYD28M. Chlorophyll and sea surface temperature. Accessed 15 July 2023

https://en.wikipedia.org/wiki/Iron_fertilization. Iron fertilization. Accessed July 12, 2023

https://www.whoi.edu/fileserver.do?id=56328&pt=10&p=39295. The Scientific and Policy Uncertainties Surrounding the Use of Ocean Fertilization to Transfer Atmospheric Carbon Dioxide to the Oceans. A summary statement drafted by participants in a workshop sponsored by the: American Society of Limnology and Oceanography (ASLO) (www.aslo.org). April 25, 2001, Washington D.C. Accessed July 1, 2023

Diversity and Succession: Metrics, Models, Mandalas, and Intaglios

17

17.1 Introduction

Per Teodor Cleve, with his expertise in chemistry, geology and oceanography in the late nineteenth century, was one of the first to develop the concept of "plankton types" and the relationship between different types of distinct water masses and current systems. He grouped phytoplankton types into thermal "units" having abundance centered around their thermal preferences, which associated them with distinct water masses and seasons (Cleve 1887, 1900). Cleve developed the view that different plankton types reflected the water masses in which they were found, and he used these distributions to understand the distribution of ocean currents, and how the planktonic organisms could influence fisheries and climate. He wanted to understand the basic current systems of the world's oceans (reviewed in Bjørklund et al. 2014). Earlier in that century, as previously described in Chap. 1, Victor Hensen was convinced that the distribution of phytoplankton in the sea was fairly uniform in time and space and that within specific zones of the sea corresponding to different latitudes, the regularity and evenness of distribution and abundance were maintained (Mills 1989, Chap. 1). Hensen developed an evenness measure that was the difference between the geometric mean and the median values of paired samples. He interpreted deviations from "regularity" as due to measurement errors or hydrographic variations. Ernst Haeckel, and numerous others, criticized this view (Mills 1989). Over time, the changing composition and succession of phytoplankton became well recognized. Henry Bryant Bigelow (a faculty member of Harvard's Museum of Comparative Zoology and a founding director of the Woods Hole Oceanographic Institution) summarized the emerging understanding of the spring bloom (Bigelow 1926),

> Perhaps no phenomenon in the natural economy of the Gulf [of Maine] so arrests attention (certainly none is as spectacular) as the sudden appearance of enormous numbers of diatoms in early spring and their equally sudden disappearance from most of this area after a brief flowering period.

P. M Glibert, *Phytoplankton Whispering: An Introduction to the Physiology and Ecology of Microalgae*, https://doi.org/10.1007/978-3-031-53897-1_17

Thus, the recognition that phytoplankton change with the season and with the currents in which they reside resolved the "regularity" debate. In order to understand changes in diversity over time, it required the development of improved indices (relative to Hensen's evenness measure) to measure diversity and conceptual models of *why* consistent taxa occur when they do. Here, the evolution of ideas related to phytoplankton diversity and succession is explored.

Diversity, or richness, has classically reflected the number of species in a particular sample or community; however, as tools and knowledge of phytoplankton have developed, so too have our characterizations of phytoplankton diversity. In addition to phylogenetic (or taxonomic) diversity (Chap. 9), communities can be characterized based on function and morphological or physiological traits (see Chap. 18). Genetic and molecular tools are also advancing the ability to characterize biodiversity (Johnson and Martini 2015). Now, many approaches for assessing biodiversity use DNA sequencing approaches and data. With the costs of sequencing rapidly falling, such approaches are becoming more available. These molecular approaches identify genomic variation within populations of low overall diversity as measured conventionally (e.g., Zehr et al. 2008). Broadly speaking, there is exceptional diversity within the phytoplankton. This chapter explores some of the ways in which diversity and succession have been characterized.

17.2 Hutchinson's Paradox of the Plankton

G. Evelyn Hutchinson from Yale University (see Chap. 1) was concerned about which species bloom and how many species can co-exist in a body of water. Based on the principle of competitive exclusion (Hardin 1960), he puzzled about the paradox of the plankton (Hutchinson 1961, p. 137) and said, "we should expect that one species alone would outcompete all the others so that in a final equilibrium situation the assemblage would reduce to a population of a single species." His basic premise, originally focused on lakes—but equally applicable to oceans—was that limiting nutrients are well mixed and equally available to the phytoplankton. If different phytoplankton taxa have similar nutrient requirements and are competing for the same nutrients (a major assumption), then the competitive exclusion principle should limit the number of species that are present at any given time. Interestingly, in systems that trend toward high biomass blooms, including many harmful algal blooms (HABs) where a single species dominates, the process of competitive exclusion may actually apply. Hutchinson proposed that differential predation may permit some diversification, and nonequilibrium conditions, caused by patches or non-even mixing, could also contribute to the development of diversified phytoplankton (Hutchinson 1961). Since his original papers, many theories have been proposed to describe how diversity can be sustained, including chaotic oscillations of species abundances (Huisman and Weissing 1999), differential abilities of different taxa for nutrient acquisition (e.g., Roy and Chattopadhyay 2007), localized competition (Levin 1992), and others. More recently, Rynearson and Menden-Deuer (2016) suggested that by focusing on individuals and their cellular interactions that occur on time scales of seconds to hours, there can be profound impacts on ecosystem function, including species abundance and production. Such a reductionist notion is based on observed strain differences. This suggests that there is no one mechanism leading to species occurrence within assemblages, but several mechanisms that co-occur to varying extents. There is continuing interest in studying how species diversity can be maintained in the context of "grand patterns" among species distributions.

17.3 Succession and Sequence

The Norwegian plankton ecologist, H.H. Gran (see Chap. 1), and his former student, Trygre Braarud (Gran and Braarud 1935), distinguished between succession and sequential changes. They defined succession as the change in species composition within a water mass, resulting from changing physical (light, temperature), chemical (nutrients, toxins), and biological (grazing, competition) factors within a water mass. [Interestingly, Cleve and H.H. Gran had disagreements about whether diatoms may produce resting spores in unfavorable conditions and whether that can re-emerge when favorable conditions again develop. On that point, Gran was proven correct (Bjørklund et al. 2014)].

Sequence is a change in species composition resulting from a change in water mass type (as interpreted by Smayda 1980). Accordingly, in true succession, autochthonous populations are involved, but in a sequence, there can be seeding of new or introduced species via the movement or displacement of water masses (Smayda 1980). In reality—at least for marine systems—neither true succession nor sequence occurs, but instead, a continuum of responses is responsible for the observed progression of species. Ponds and quiescent lakes should generally be less subject to sequence changes as water mass movements and exchanges are less.

17.4 Measuring Diversity

Diverse communities are usually thought to be desirable. A loss of biodiversity can negatively impact ecosystem functioning (Cardinale et al. 2012; Tilman et al. 2014). As systems lose diversity, they may become less stable and less able to tolerate environmental change (e.g., Cozzoli et al. 2017). A community with different species may be better able to buffer changes and environmental stresses (Ives and Carpenter 2007). Systems with low diversity may be more stressed if perturbed. Using experimental mesocosms, Striebel et al. (2012) showed that phytoplankton diversity increased zooplankton diversity.

There are a number of metrics or indices that have been developed to quantify diversity and other measures of community composition based on taxa (Table 17.1). The simplest among the indices of diversity is the Taxonomic Richness Index, or the total number of species in a given sample. One of the most common diversity indices was developed by Claude Shannon (1948), building from a desire to quantify

Table 17.1 Various measures of diversity, evenness, and dominance. Modified from the synthesis of Francé et al. (2021) under a creative commons license

Index	Relationship	Explanation
Taxonomic richness	R'	In which R' is the number of taxa in a given sample
Shannon–Weiner Diversity Index	$H' = -\Sigma^{R}_{i=1} p_i \times \ln p_i$	In which p_i is the proportion of individuals in the taxon i, as estimated based on the number of individuals in the taxon of interest (n_i) relative to the total number of individuals in the community (N)
Margalef's Diversity Index	$M' = (R'\text{-}1)/(\log N)$	
Simpson's Diversity Index	$S' = \Sigma^{R}_{i=1} p_i^2$	
Pielou's Evenness Index	$E' = H'/\log R'$	
Sheldon's Evenness Index	$Sh' = e^{H'}/R'$	
Berger-Parker Dominance Index	$BP' = n_1/N$	In which n_1 is the abundance of the most abundant species
McNaughton's Dominance Index	$McN' = (n_1 + n_2)/N$	In which n_2 is the abundance of the second most abundant species

entropy in strings of text. His index has since become well-known throughout ecology as a measure of diversity. Now recognized as the Shannon–Weiner index, it accounts for both diversity and to a lesser extent evenness (Table 17.1, Francé et al. 2021). Another metric published about the same time (1940s–1950s) was developed by Simpson (1949). His version of a diversity index measures the probability that any two individuals drawn at random will belong to the same taxon (Table 17.1, Francé et al. 2021). Ramon Margalef (Fig. 17.1), a Catalan and Spanish ecologist, was particularly interested in how communities were structured, the patterns of ecological succession, and the abiotic determinants contributing especially to harmful algal bloom (HAB) taxa. His diversity index (Table 17.1; Margalef 1958; Francé et al. 2021) effectively describes the relationship between the taxa detected and the total individuals counted. His concepts of succession are further developed below.

Evenness is a measure of the equitable distribution of taxa in a sample, while dominance measures the extent to which a community is dominated by particular taxa (Francé et al. 2021). Evenness measures build on diversity indices. Pielou's Evenness Index (Table 17.1) is the ratio of the Shannon–Weiner Diversity index and its maximum possible value (Pielou 1975; Francé et al. 2021). Sheldon's Evenness Index takes a slightly different approach than Pielou, but also builds off of the Shannon–Weiner Index. A simpler approach was recommended by Berger and Parker (1970), and their index thus takes the ratio of the first most abundant species relative to the total number (Table 17.1, Francé et al. 2021). Using a similar approach, but accounting for the top two most abundant taxa, is McNaughton's Dominance Index (Table 17.1, Francé et al. 2021).

Fig. 17.1 Ramon Margalef. Image reproduced from Wikipedia (https://en.wikipedia.org/wiki/Ramon_Margalef) under a creative commons license

All of these metrics allow quantification of diversity in a sample, at a given point in time or space. When these metrics are applied, the taxa enumerated are limited to the functional groups of interest (e.g., all phytoplankton); they do not consider other microbes, such as bacteria, nor are they necessarily suitable for application to non-constitutive mixotrophs (Chap. 15). Incorporation of picoplankton is also problematic, wherein species enumeration is difficult (although that is changing with new approaches such as flow cytometry; Chap. 2). Even diversity indices based on DNA sequencing approaches have biases. For example, primers may be too specific to capture all the diversity within a sample, or organisms may contain multiple copies of a given gene, complicating interpretations (Johnson and Martini 2015).

Various investigators have compared the performance of each of these indices. Some are more sensitive than others to sample size, and some success has been had in using such biodiversity indices as part of assessments of pollution impacts (e.g., Cozzoli et al. 2017; Francé et al. 2021). In spite of all difficulties, a number of conceptual models of succession and sequence have been developed. The classic and next-generation models are described below.

17.5 Sverdrup and the Critical Depth Hypothesis

The Norwegian scientist, Harald Ulrik Sverdrup, conceptualized the formation of spring blooms based on the quantitative relationships between light and mixed layer depth (Sverdrup 1953). Building on concepts from Gran and Braarud (1935; Chap. 1), Sverdrup understood the importance of water column stability for

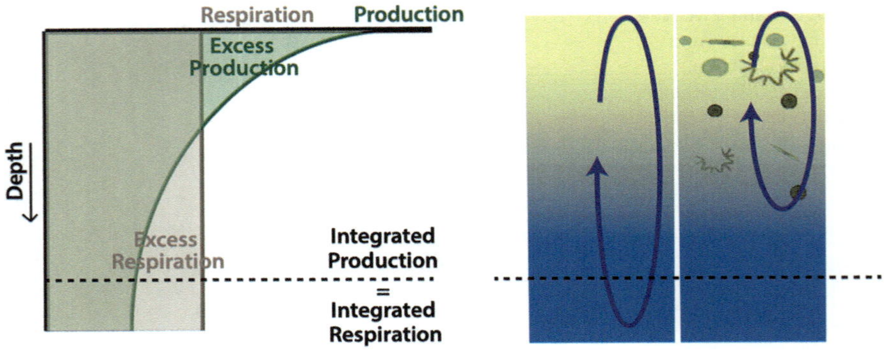

Fig. 17.2 Schematic depiction of Sverdrup's critical depth hypothesis and the other hypotheses (c, d, e) discussed in this paper. Left panel: Production is governed solely by light and thus declines exponentially with depth, while loss/respiration is constant with depth. The "critical depth" is the bottom of the layer within which the integrated production equals integrated respiration (dashed line). Right panel: When mixing is deeper than the critical depth, there is net loss, so a bloom is not expected (left ellipse); if the mixing depth is shallower than the critical depth, there is excess production and a bloom can occur (right ellipse). Image reproduced from Fisher et al. (2014, Oceanogr 27(1): 222–235. doi.org/10.5670/oceanog.2014.26), under a creative commons license, as modified from Sverdrup (1953)

phytoplankton growth. Sverdrup (1953) proposed that the deep-mixed layers in winter limit production because phytoplankton have an unfavorable light environment. He defined the *critical depth* as the depth at which vertically integrated phytoplankton growth equals vertically integrated phytoplankton losses (respiration among other losses). Thus, when the mixed layer is less than the critical depth, a bloom occurs because growth can be larger than the integrated losses. Sverdrup thus concluded that blooms initiate with stratification creating mixed layers with adequate light (Fischer et al. 2014, Fig. 17.2). The Sverdrup model is based first on the conservation of mass in a water column layer, and it identifies the major factors that change when blooms form (Sathyendranath et al. 2015).

This relatively simple but elegant concept ("a thing of beauty" according to Sathyendranath et al. 2015) has been applied frequently and, of course, challenged on many fronts, as many blooms do not necessarily fit this simple thermal stratification model. The concept of critical depth was new at the time and brought together biological-physical interactions in a rigorous framework. One particular criticism is that the loss processes are combined into a single term, encompassing respiration, grazing, and vertical export (Fischer et al. 2014). Another criticism is that there are indeed blooms that form in the absence of stratification. Without evaluating or critiquing those criticisms here and recognizing that with new tools and observations, ideas evolve, the important point is that the Sverdrup concept brought important insights about the regulation of phytoplankton production and the factors controlling into a fairly simple, testable conceptual and mathematical model.

17.6 Margalef's Mandala

Ramon Margalef was a pioneer in understanding phytoplankton succession. In his early works (Margalef 1958, 1967a, b, reviewed by Smayda 1980), he suggested four stages to a successional cycle. In Stage I, typical of turbulent and nutrient-rich waters (such as upwelling conditions), small diatom cells typically dominate. Diatoms have a high surface/volume ratio and high potential growth rates. Stage II has a more diverse suite of larger-celled diatoms and dinoflagellates, many of which are motile, and in Stage III, larger diatoms, dinoflagellates, and coccolithophores increase in abundance. In the later stage of blooms, Stage IV, when nutrients have been depleted and stratification has been established, diatoms tend to sink out, and larger species with lower growth rates predominate. During this later stage, N_2-fixing cyanobacteria and heterotrophic dinoflagellates become more common.

Margalef's successional model builds on the notion that there is a succession of species from those that are "*r*"-selected to those that are "*K*"-selected (Fig. 17.3). In ecological theory, *r* species are those that are fast-growing, typically with shorter life spans, whereas *K*-selected species may have a slower rate of growth but tend to be longer-lived. Furthermore, *r*-selected species may be considered colonizers, while *K*-selected species favor strong competitors. The *r* versus *K* model is built on the "universal" hyperbolic response curve (the "curve for all reasons"; Rao 2000; Chap. 3), with *r* being those which display a rapid initial slope in terms of rate of growth, while those that are *K* strategists may not reach the same overall biomass but can be sustained longer. It is of note, as previously discussed in Chap. 15, that this basic conceptual model underpins the notion that food webs based on pure "phytoplankton" have a different trajectory than those in which mixoplankton may dominate (Mitra et al. 2014; Glibert and Mitra 2022).

Margalef used the idea of progression from *r* to *K* species and differences in phylogenic morphotypes to develop a sequence of succession from diatoms to red-tide-forming dinoflagellates (Margalef et al. 1979). He depicted this sequence in a figure classically referred to as the Margalef *Mandala* (Fig. 17.4). A mandala is a conceptual and spiritual symbol of the cosmic truths of the universe (www.newworldencyclopedia.org), a representation of recurring order. The term mandala has recently been used to represent a chart or diagram that captures such concepts and normally uses shapes such as squares and circles in a geometric pattern. Margalef's *Mandala* simply and elegantly depicts the succession of species based on only two parameters, nutrients and turbulence. His mandala appropriately recognizes the diatoms as the *r* strategists, and the dinoflagellates as *K* strategies, with the former giving way to the latter as the season progresses (Fig. 17.4). In more recent iterations of this mandala, both Katz et al. (2004) and Kudela (2010) placed coccolithophores in the center of the trajectory from *r* to *K* strategists.

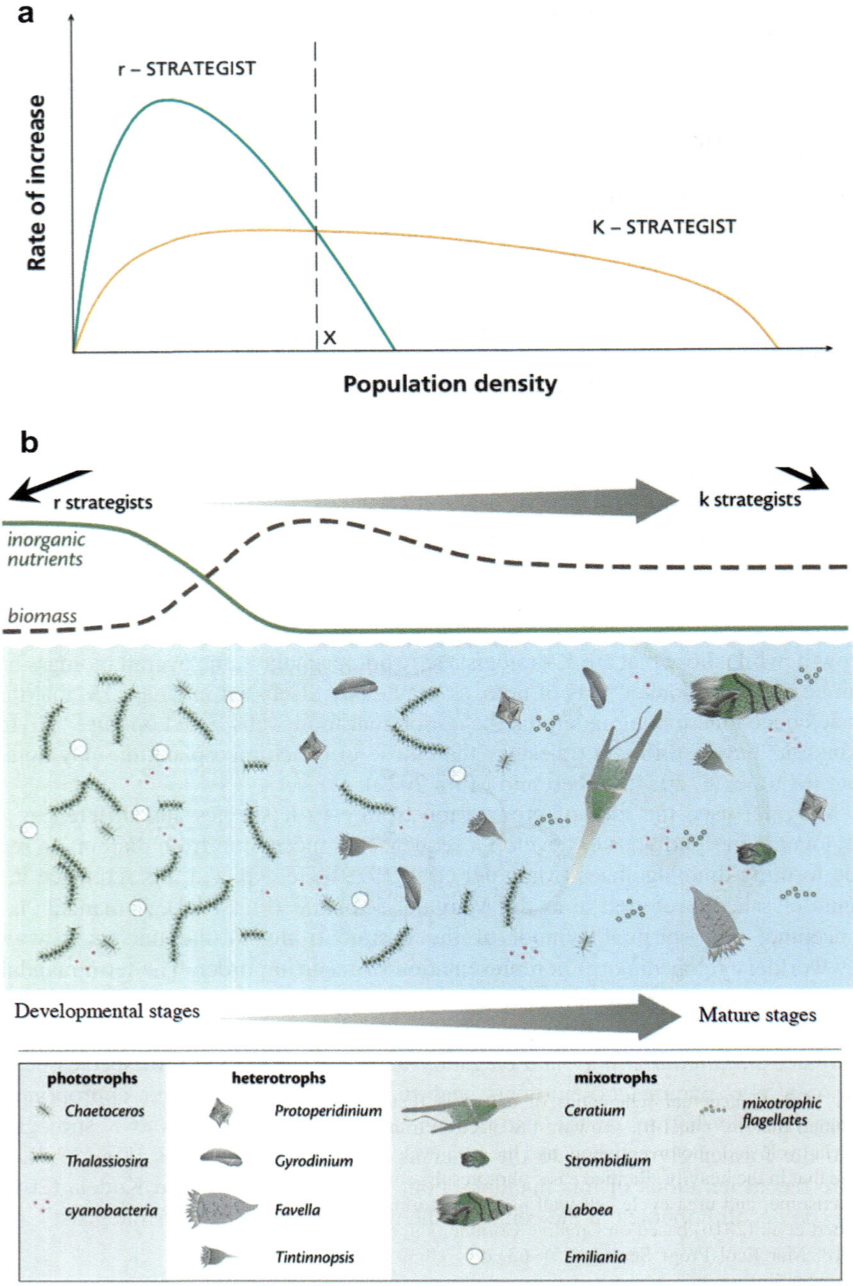

Fig. 17.3 Panel (**a**) Depiction of the rate of growth versus population density for *r* strategists and *K* strategists. Panel (**b**) Conceptual diagram of the succession of plankton community from *r* strategists, dominated by diatoms, to more mature stages during which *K* strategists dominate, where more dinoflagellates and mixoplankton become dominant. Panel (**b**) Reproduced from Mitra et al. (2014, Biogeosciences 11:995–1005. doi: 10.5194/bg-11-995-2014) under a creative commons license

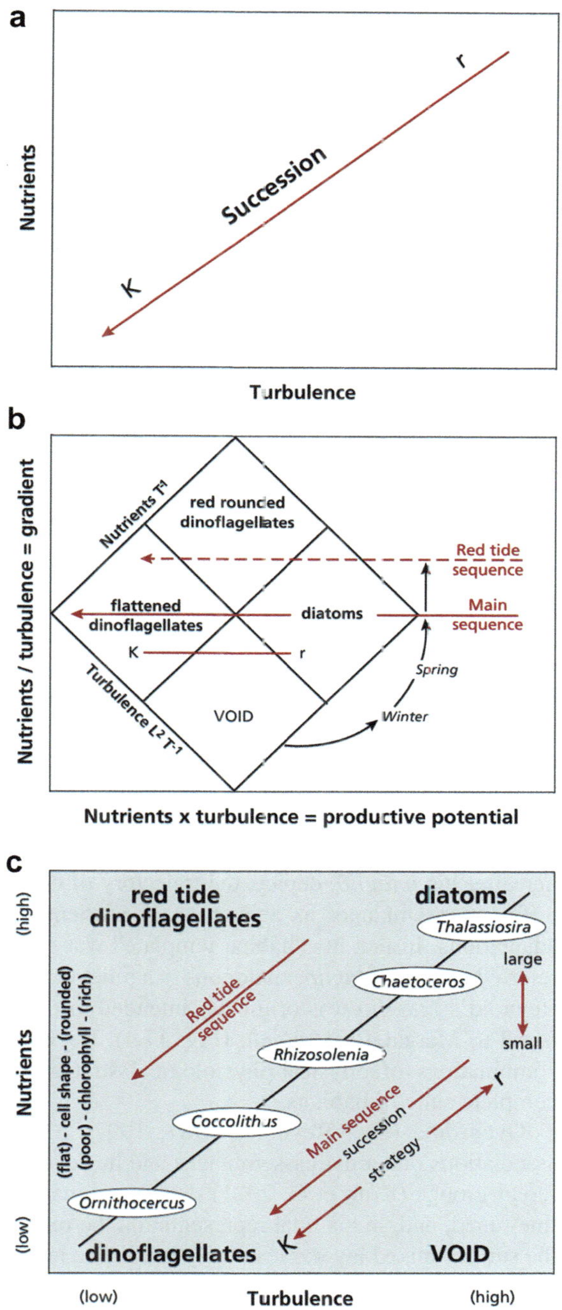

Fig. 17.4 Panel (**a**) Relationship between turbulence and nutrients and the trajectory of *r* vs. K species, according to Margalef. Panel (**b**) Margalef's mandala. Building on the concepts shown in Panel (**a**), the Mandala illustrates both a generalized winter-spring bloom sequence and a "red-tide" sequence. Panel (**c**) another view of the Mandala showing the succession of key taxa. Panel (**a**) redrawn based on Margalef et al. (1979); Panel (**b**) modified and redrawn based on Smayda and Reynolds (2001 J Plankt Res 23: 447–461. doi.org/10.1093/plankt/23.5.447) with permission of Oxford University Press. Panel (**c**) modified and redrawn based on Margalef (1978, Oceanol Acta 1:493–509)

17.7 Reynolds' Intaglio

Colin Reynolds simplified Margalef's *Mandala* and redrew the diagonal succession of the *r* to *K* strategists vertically. Then, he identified three strategies of life forms, rather than the two that underpinned Margalef's idea. Reynold's concept was built on Grime's (1977) concepts from the field of ecology describing the trade-offs faced by all organisms. All resources cannot be devoted to all processes simultaneously, or exclusively to one or another, but the range of intermediate possibilities may dictate who thrives under certain conditions. Grime envisioned this as a three-way trade-off, a triangle. Those that are rapid growers maximizing resource acquisition are considered competitive, *C*. Those that can maintain metabolism even in the presence of different stresses are stress tolerance, *S*. Finally, those that complete a life cycle and reproduction even in otherwise stressful environments are ruderal, *R*. This theory assumes that in high disturbance and stressful environments, competition will be relatively low. Thus,

C strategists are comparable to *r* strategists, which are small and fast-growing.
S strategists are similar to *K*-selected, which are larger, slower-growing, low surface/volume ratios, and nutrient-stress tolerant.
R strategists are disturbance tolerant, with a high capacity for light harvesting and nutrient acquisition via high surface/volume ratios.

In describing the main hypotheses of his approach, Reynolds (1998, pp. 12, 16) stated, "it is proposed that algal attributes are matched to the opportunities provided by the environment and that there is an encouraging fit of phytoplankton species to the range of habitats described by the trophic spectrum" and that "at any given times and under given conditions, some species of groups of species tend to increase in biomass more strongly than others." His primary conceptual diagram (Fig. 17.5), hereafter his *Intaglio*, depicts the trajectory of change, or self-assembly process for different assemblages as a function of nutrients, mixing, and their physiological adaptations. In that his "habitat template" was not strictly two-dimensional, it was referred to as an *intaglio*, analogous to a relief etching (Smayda and Reynolds 2001). Reynold's *Intaglio* was originally intended to depict a freshwater sequence. Compared to Margalef's *Mandala* (Fig. 17.4), Reynold's *Intaglio* has a fuller suite of combinations of physical-physiological strategies that can be occupied, a more complete suite of habitats.

Over time (e.g., Reynolds 1987, 1997, 2000), Reynolds began referring to associations rather than assemblages and he developed a habitat template for 31 different groups (Kruk et al. 2021). His functional groupings were continually being fine-tuned, and, in his final representation, he relied on the following axes: depth of the surface mixed layer; mean daily irradiance level experiences; water temperature; concentration of CO_2; proportion of the water processed each day by the herbivorous zooplankton; and nutrients in the water required to saturate the rate of growth, including N, P, and Si (Kruk et al. 2021).

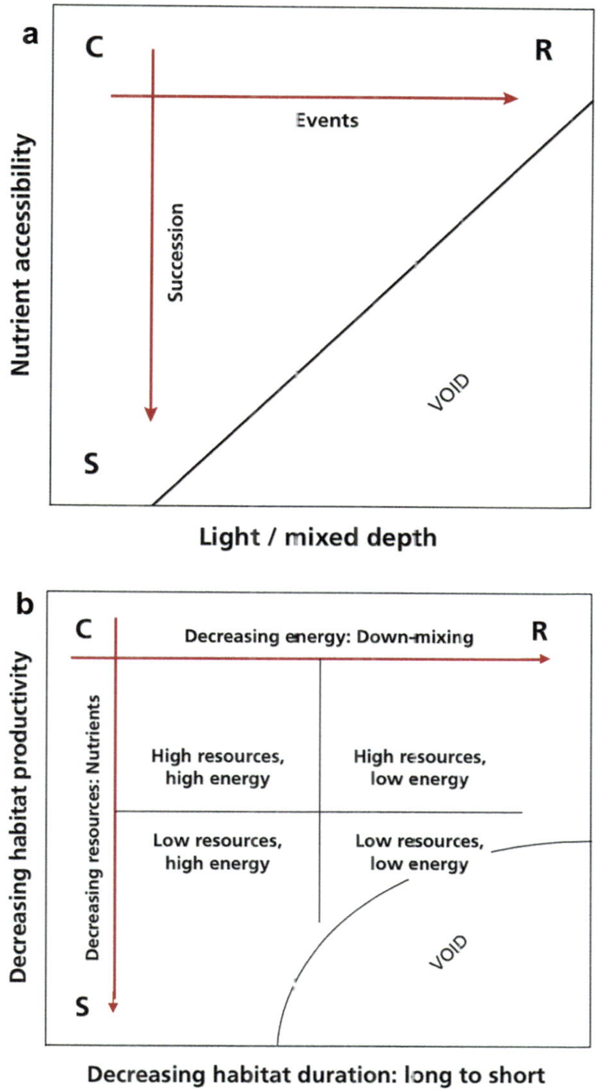

Fig. 17.5 Reynolds *Intaglios*. Panel (**a**) Intaglio version that allows the selection of species within a wide ecological space, except for habitats where nutrients and light are both continuously deficient (the "void"). Panel (**b**) revised Intaglio modified to show additional gradients in habitat productivity and duration. Panel (**a**) redrawn based on Reynold's (1987, Symp Br Ecol Soc 27:297–325) and Panel (**b**) redrawn based on Kruk et al.'s (2021, Hydrobiologia 848:113–129. doi: 10.1007/s10750-020-04340-9). Interpretation of Reynold's Intaglio and modified to include gradients in habitat productivity and duration

17.8 Smayda, Dinoflagellate Life Forms, and Community Assembly

Smayda and Reynolds (2001) extended the *intaglio* concept to the sea and examined how it relates to seasonality particularly that of dinoflagellates. By tracing the seasonally changing *Intaglio* coordinates (Fig. 17.6a), the selective bias moves

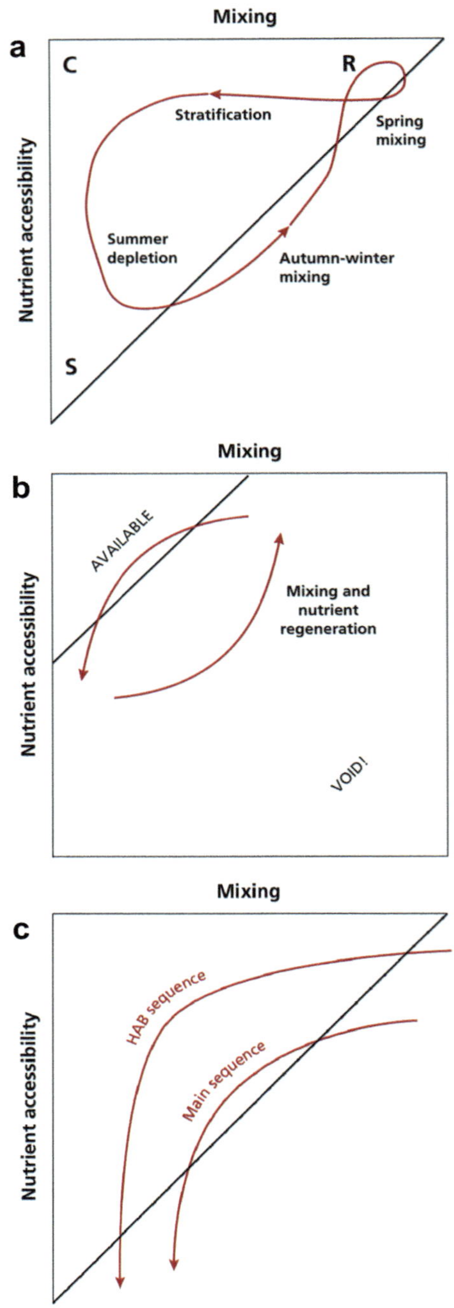

Fig. 17.6 Smayda and Reynold's *Intaglios*. (**a**) Ideal year-long trace of the selective trajectory imposed by (temperate) seasonal habitat variability and associated zones of occurrence of C, S, and R species. (**b**) Habitat suitable for phytoplankton growth in the sea is diminished (the "void") mostly because it is too nutrient- and light-deficient to support phytoplankton growth. The region depicted in the upper left-hand corner and designated as "available" shows the considerable restriction in habitat available with conditions suitable for phytoplankton exploitation. (**c**) Expansion of the top left-hand corner of (**b**) showing that the main phytoplankton sequence (literally) skirts the edge of the "available" habitat, but, with more nutrients available, the trajectories pass deeper within the mixing-nutrient matrix selecting more strongly for C-strategist algae and HAB species. Images redrawn based on Smayda and Reynolds (2001 J Plankt Res 23: 447–461. doi.org/10.1093/plankt/23.5.447) with permission of Oxford University Press

- **Type I (= *Gymnodinioids*)**
  Gymnodinium spp., *Gyrodinium instriatum, Katodinium rotundatum*
- **Type II (= *Peridinians/Prorocentroids*)**
  Heterocapsa triquetra, Scrippsiella trochoidea, Prorocentrum micans, Prorocentrum minimum
- **Type III (= *Ceratians*)**
  Ceratium tripos, Ceratium fusus, Ceratium lineatum
- **Type IV (= *Frontal Zone Taxa*)**
  Gymnodinium mikimotoi, Alexandrium tamarense
- **Type V (= *Upwelling Relaxation Taxa*)**
  Gymnodinium catenatum, Lingulodinium (Gonyaulax) polyedra
- **Type VI (= *Coastal Current Entrained Taxa*)**
  Gymnodinium breve, Ceratium spp., *Pyrodinium bahamense* var. *compressum*
- **Type VII (= *Dinophysoids*)**
  Dynophysis acuta, Dinophysis acuminata
- **Type VIII (= *Tropical Oceanic Flora*)**
  Amphisolenia, Histioneis, Ornithocercus, Certium spp.
- **Type IX (= *Tropical Shade Flora*)**
  Pyrocystis noctiluca, Pyrocystis pyriformis

Fig. 17.7 Dinoflagellate bloom and vegetation life-form types, and representative species, found along an onshore-offshore gradient of decreasing nutrients, reduced mixing, and deepened euphotic zone. Image redrawn based on Smayda and Reynolds (2001 J Plankt Res 23: 447–461. doi.org/10.1093/plankt/23.5.447) with permission of Oxford University Press. Note that species names are as listed in the original diagram, but *Gymnodinium mikimotoi* is now known as *Karenia mikimotoi* and *Gymnodinium breve* is now named *Karenia brevis*

from *R* species to *C* species, then to *S* species, and, with loss of vertical structure, back to *R* species. The original seasonal progression depicted how freshwater phytoplankton may change. However, in the oceans, the diagram is depicted with less available habitat, with the "void" region being an area of nutrient impoverishment and low light; this region is greater in the sea than in typical freshwater systems (Fig. 17.6b). Finally, zooming in on the available habitat, the window of opportunity for growth of *C*-type species, those with invasive life history strategies and faster-growing, is higher with more nutrient availability (Fig. 17.6c).

Building on the revised intaglios, Smayda and Reynolds (2001) attempted to categorize dinoflagellate bloom taxa and to determine how they ordinate along an onshore-offshore gradient of decreasing nutrients, reduced mixing, and deepened mixed euphotic zones (Figs. 17.7 and 17.8). Their efforts revealed a strong pattern. That is, in shallow, comparatively nutrient-rich habitats with reduced water exchange, the dominant dinoflagellates tend to be small to intermediate-sized gymnodinoids. Moving to habitat where nutrient availability is reduced, the peridinoids (such as *Heterocapsa triquetra* and *Scrippsiella trochoidea*) and the peridiniums (*Prorocentrum minimum*, now *P. cordatum*, and *P. micans*) become more common. The third type is those experiencing increased nutrient loading that develops in late summer in temperate or boreal coastal habitats (*Ceratium* spp.). Moving offshore to regions associated with frontal zones, dinoflagellates such as *Karenia mikimotoi* (reported by Smayda and Reynolds as *Gymnodinium mikimotoi*) and *Alexandrium tamarense* were adapted for entrainment within coastal currents. The final types include those that tolerate high shear/stress.

17.9 Glibert's Mandala

Both Margalef and Reynolds incorporated nutrient availability in their mandalas and intaglios, but they did not differentiate between nutrients. Glibert (2016) recently focused on the complexity of the nutrient dimension, underscoring that a single nutrient scale, as Margalef represented it (largely based on upwelling as a nutrient source), cannot capture the dynamics of nutrients in a bloom sequence. Given the changes in nutrient forms and stoichiometry that are occurring with eutrophication and other anthropogenic changes (Chap. 22), nutrients as a whole do not necessarily change in tandem. Ratios of nutrients are changing. Different forms of the same nutrient (e.g., NO_3^- vs. NH_4^+) also have different metabolic effects on different taxa (Chap. 12).

Glibert's *mandala* places these differences in conceptual space against other environmental characteristics and maps 12 effect or response traits (Fig. 17.9): (1) relative preference for chemically reduced vs. chemically oxidized forms of N; (2) relative availability of inorganic N and P; (3) adaptation to high vs low light or the tendency to be purely autotrophic vs mixotrophic; (4) cell motility; (5) environmental turbulence; (6) relative pigmentation quality, from higher relative proportion of carotenoids (brown) to a higher relative proportion of phycobiliproteins and chlorophylls (blue-green or green); (7) temperature; (8) cell size; (9) relative growth rate; (10) relative production of bioactive compounds such as toxins or reactive oxygen species (ROS); (11) ecological strategy, as *r* vs. *K*; and (12) fate of the production in terms of grazing. Viewing these axes from the inner square out, it places key importance on nutritional strategies and modes of motion. The next set of axes superimposes physical constraints or variability, namely turbulence and temperature, acting in opposite directions; high turbulence is associated with low temperatures and vice versa. Pigmentation is a response axis (Chap. 5). Other emergent properties are shown on the largest box, including cell size, growth

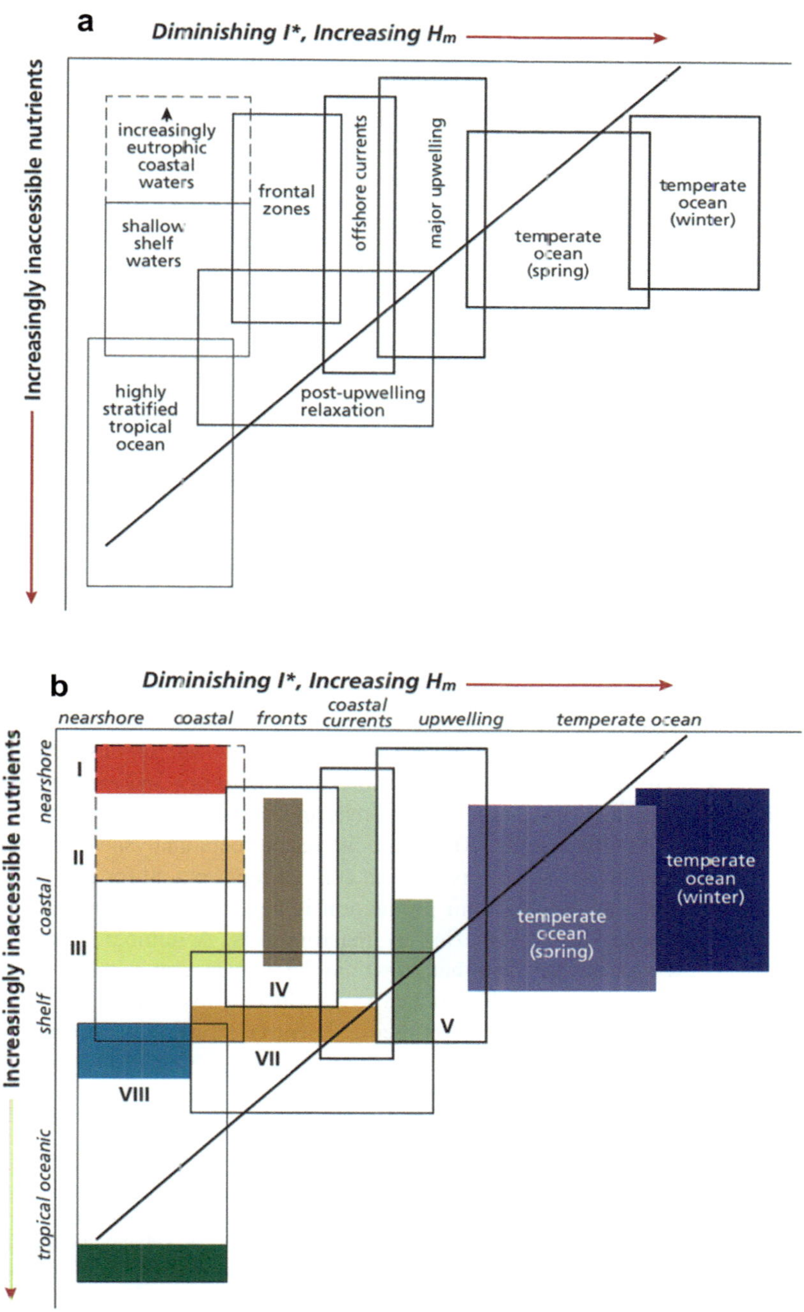

Fig. 17.8 Panel (**a**) Schematic matrix of pelagic marine habitats along an onshore-offshore gradient separating deep-mixed and well-stratified, but nutrient-deficient systems. I^* refers to the irradiance level received by cells within the water column; H_m represents the depth of mixed layer.

rates, potential to produce toxin, and overall ecological strategy. The overarching circle represents the fate of the production in the food web and the relative strength of macro- vs micro-grazing control. The fate of production via these varying grazers is the key distinction in new vs regenerated production (sensu Dugdale and Goering 1967; see Chap. 12).

Key among the notions captured in this new mandala are the relationships between and among traits (see Chap. 18). While the mandala serves to highlight the differences and trade-offs between traits, it can also be seen that, in general, some traits or associated environmental conditions tend to track each other. Notable among these is the relationship between the availability of dissolved inorganic nutrients (axis 2), growth rate (axis 9), and ecological strategy (axis 11). Such a relationship is in keeping with principles of ecological stoichiometry and the growth rate hypothesis (e.g., Elser et al. 1996; Sterner and Elser 2002; Chap. 18) suggesting that to sustain high growth rates, proportionately more P is required (lower N:P ratio), as it is a key element of ribosomes, the organelle of growth. High growth rates are a central tenet of *r* selection. In contrast, higher N:P ratios would be associated with slower growth rates and a maintenance condition of the population, i.e., a *K* strategy.

Turning to how specific phytoplankton functional types map onto this multidimensional space, it can be seen that the major groups of marine species differentiate themselves reasonably well. Diatoms fall in the quadrant of comparatively high availability of NO_3^- and relatively low N:P ratios (Fig. 17.9). They also map to high light availability and/or autotrophic nutrition and cells with a sinking mode of motility. They have proportionately more "brown" pigments, typically are larger cells, have high growth rates, generally are not toxic (*Pseudo-nitzschia* spp. being an exception), and are considered *r* strategies. Such a positioning is in keeping with the N nutritional physiology of diatoms (e.g., Lomas and Glibert 1999a, b, 2000; Parker and Armbrust 2005; Glibert et al. 2016), and the recent meta-analysis of Hillebrand et al. (2013) in which the comparatively high cellular requirement for P by diatoms was elegantly demonstrated. Diatom production is generally associated with new production and higher rates of grazing by macrograzers. A number of studies in recent years have also documented that growth in the N form may be related to toxin production for those cells that do produce toxin. Specifically, when grown under similar conditions but with only N form being the variable, more toxin per cell has been shown to be produced by cells grown on NH_4^+ than NO_3^- or under

◄

Fig. 17.8 (continued) Overlap of types within the habitat template schema does not always imply their contiguity. The diagonal approximates the main successional sequence depicted in Margalef et al. (1979). Panel (**b**) Predominant dinoflagellate life-form types associated with the turbulence-nutrient matrix (from Panel **a**) along an onshore-offshore continuum characterizing pelagic habitats. Type I = gymnodinioids; type II = peridinioids and prorocentroids; type III = ceratians; type IV = frontal zone species; type V = upwelling relaxation taxa; type VI = coastal current entrained taxa; type VII = dinophysoids; type VIII = tropical oceanic flora; type IV = tropical shade flora. Images redrawn based on Smayda and Reynolds (2001 J Plankt Res 23: 447–461. doi.org/10.1093/plankt/23.5.447) with permission of Oxford University Press

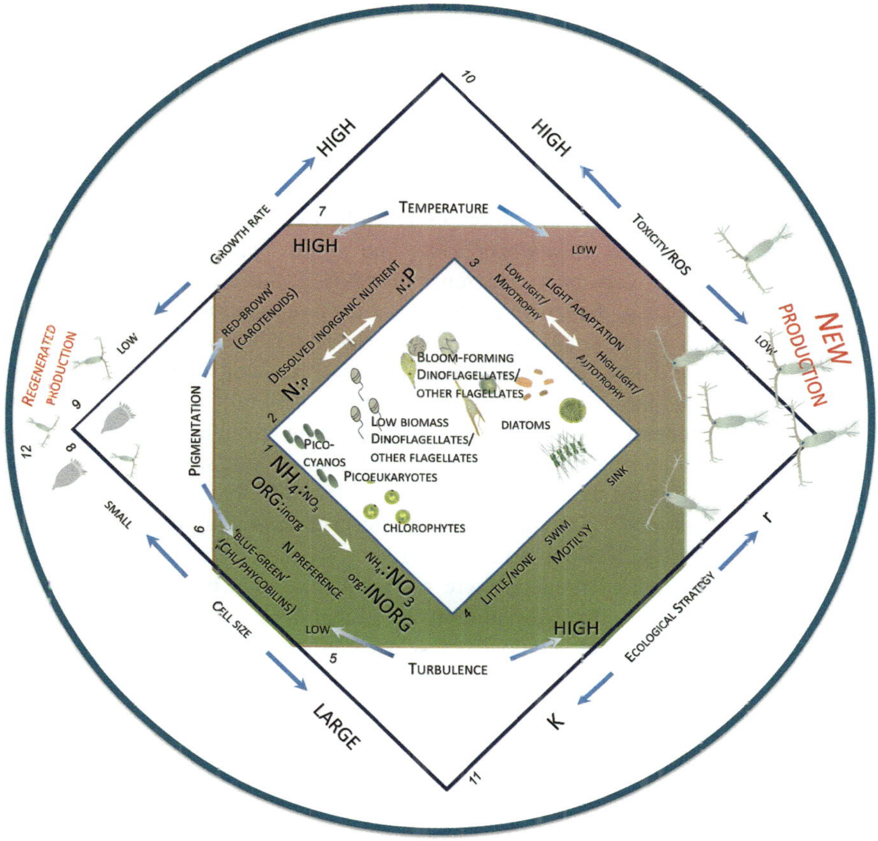

Fig. 17.9 Glibert's *Mandala* in which phytoplankton functional types are depicted along 12 axes (shown by the small numbers in the corner of each axis). Note that all scales are relative and no dimensions are implied. All responses within relative space are representative and not meant to imply that all species or individuals within a given response surface will respond similarly. Image reproduced from Glibert (2016, Harmful Algae 55: 25–30. doi.org/10.1016/j.hal.2016.01.008) with permission of Elsevier

imbalanced N:P conditions (e.g., Leong et al. 2004; Hattenrath et al. 2010; Van de Waal et al. 2014). Antioxidant activity similarly appears to be related to cells under imbalanced growth conditions of elevated irradiance and growth on NH_4^+ (e.g., Podgórska and Szal 2015), although such data for phytoplankton are comparatively few.

At the opposite quadrant, dinoflagellates and other flagellates also map to features that are well established with respect to their physiology. They map to the region in which the reduced forms of N are favored over oxidized N, they can thrive in imbalanced N:P regions, often sustained in high N:P environment, they have a comparatively high tendency to be mixotrophic to greater or lesser degree, they prefer conditions of low turbulence and higher temperatures, have slower growth rates, and can be toxic, and they are generally considered *K* strategists (Fig. 17.9). As with diatoms, there is a range of cell size within the flagellates, the larger ones being more NO_3^- tolerant. On the Glibert *mandala*, two groups of HAB dinoflagellates can be distinguished, the high biomass formers, such as *Prorocentrum* spp., and the toxic, but typically low biomass formers, such as *Alexandrium* spp. The bloom-formers are associated with a somewhat lower N:P ratio, especially during the period of the establishment phase of the bloom (e.g., Glibert et al. 2012).

Picoplankton, such as *Prochlorococcus* and some *Synechococcus*, sit at the corner represented by very high preferences for NH_4^+ (indeed, some picoplankton cannot access NO_3^-; e.g., Moore et al. 2002), may thrive in reduced light conditions, have slow growth rates, small size, live under low turbulence and often warm temperatures, have "blue-green" pigmentation, and are considered *K* strategists (Fig. 17.9). Picoplankton and other small flagellates generally are associated with higher rates of regenerated production, enhanced cycling through the microbial loop, and less efficient transfer of production to higher trophic levels. Chlorophytes may be mapped to the right of the picocyanobacteria (Fig. 17.9), having a comparatively greater preference for NO_3^- than picocyanobacteria or picoeukaryotes, but they also have a higher N:P (e.g., Hillebrand et al. 2013).

As with all conceptual models, there are limitations that even 12 axes cannot resolve. Moreover, not all phytoplankton functional types are represented in this mandala even with 12 axes. Over the past decades, there have been many other attempts to advance the application of Margalef's *Mandala* and to improve either the representation of different phytoplankton taxa or different features on the mandala. As examples, Cullen et al. (2002, 2007) provided a revised view that includes the picoplankton, and Balch (2004) suggested a revision that included coccolithophores and further suggested that day length might be an important factor to consider. Allen and Polimene (2011) suggested that the addition of a gradient of reactive oxygen activity provided an important link between physiology and ecology. Wyatt (2014) noted that a simple turbulence scale fails to capture the dynamics in the vertical and fails to incorporate fine-scale structures such as thin layers. He also stated that many adaptive strategies also cannot be incorporated into a simple trait diagram. He said (Wyatt 2014, p. 39), "If we add a few simple shapes (say spheres, discs, rods, needles), some spines, threads or trichocysts, motility, gas vesicles, colony formation, whether cells are rigid or flexible, EPS formation, bioluminescence, functional properties (specific production rates, nutrient uptake rates, half saturation coefficients, internal stores, pigment suites), some demographic features, the spectrum of responses obviously gets out of hand very rapidly…."

17.10 Summary

With the rejection of Hensen's notion of "regularity" in phytoplankton composition in the sea, efforts to conceptualize and quantify how phytoplankton change spatially and temporally have taken many different approaches. Yet, from Cleve to Sverdrup to Margalef to Reynolds and Smayda, the foundational ideas are all similar. All conceptual models involve to some extent a notion of the importance of turbulence (or water column stratification) and access to resources, light and nutrients, and/or thermal regulation. The more recent mandalas and intaglios consider life forms and life history strategies. Criticisms of these approaches are many, as they are indeed simplifications of all of the complexities of physics and nutrients and other factors. Efforts abound to improve these conceptual models with more factors and more complexity. Yet, even the early models are remarkable in the explanatory power that they have. An important lesson can be learned from the tradition of mandala-making: While the balance and interaction of the universe are fascinating, it is also in flux. Monks often destroy their intricately beautiful mandalas once completed, underscoring that nothing is permanent. The obvious lesson here is that the environmental forces that regulate species biodiversity indeed are constantly changing. Our knowledge of traits and ecosystem functioning is also continually advancing. Finally, it should not be forgotten, as Smayda and Reynolds (2001) note, one of the greatest advantages for growth for any taxa is being at the right place at the right time.

References

Allen JI, Polimene L (2011) Linking physiology to ecology: towards a new generation of plankton models. J Plankton Res 33:989–997. https://doi.org/10.1093/plankt/fbr032

Balch WM (2004) Re-evaluation of the physiological ecology of coccolithophores. In: Thiersten HR, Young JR (eds) Coccolithophores–from molecular processes to global impact. Springer, Heidelberg, pp 165–190

Berger WH, Parker FL (1970) Diversity of planktonic foraminifera in deep-sea sediments. Science 168(3937):1345–1347. https://doi.org/10.1126/science.168.3937.1345

Bigelow HB (1926) Plankton of the offshore waters of the Gulf of Maine. Bull US Bur Fish 40(968):465

Bjørklund KR, Itaki T, Dolven JK (2014) Per Theodor Cleve: a short résumé and his radiolarian results from the Swedish Expedition to Spitsbergen in 1898. J Micropaleontol 33:59–93. https://doi.org/10.1144/jmpaleo2012-024

Cardinale BJ, Duffy JE, Gonzalez A, Hooper DU, Perrings C, Venail P (2012) Biodiversity loss and its impacts on humanity. Nature 486:59–67. https://doi.org/10.1038/nature11148

Cleve PT (1887) A treatise on the of the Atlantic and its tributaries and on the periodic changes of the plankton of the Skagerak. Upsala

Cleve PT (1900) The seasonal distribution of Atlantic plankton organisms. D.F. Bonniers Boktrykkeri Aktiebolag, Göteborg, pp 3–368

Cozzoli F, Stanca E, Selmeczy GB, Francé J, Varkitzi I, Basset A (2017) Sensitivity of phytoplankton metrics to sample-size: a case study on a large transitional water dataset (WISER). Ecol Indic 82:558–573. https://doi.org/10.1016/j.ecolind.2017.07.022

Cullen JJ, Franks PJS, Karl DM, Longhurst A (2002) Physical influences on marine ecosystem dynamics. In: Robinson AR, McCarthy JJ, Rothschild BJ (eds) The sea: biological-physical interactions in the ocean. Wiley, Cambridge, MA, pp 297–335

Cullen JJ, Doolittle WF, Levin SA, Li WKW (2007) Patterns and prediction in microbial oceanography. Oceanography 20:34–46

Dugdale RC, Goering JJ (1967) Uptake of new and regenerated forms of nitrogen in primary productivity. Limnol Oceanogr 12:196–206

Elser JJ, Dobberfuhl D, MacKay NA, Schampel JH (1996) Organism size, life history, and N:P stoichiometry: towards a unified view of cellular and ecosystem processes. Bioscience 46:674–684

Fischer AD, Moberg EA, Alexander H, Brownlee EF, Hunter-Cevera KR, Pitz KJ, Rosengard SZ, Sosik HM (2014) Sixty years of Sverdrup: a retrospective of progress in the study of phytoplankton blooms. Oceanography 27(1):222–235. https://doi.org/10.5670/oceanog.2014.26

Francé J, Varkitzi I, Stanca E, Cozzoli F, Skejić S, Ungaro N, Vascotto I et al (2021) Large-scale testing of phytoplankton diversity indices for environmental assessment in Mediterranean sub-regions (Adriatic, Ionian, and Aegean Seas). Ecol Indic 126:107630. https://doi.org/10.1016/j.ecolind.2021.107630

Glibert PM (2016) Margalef revisited: a new phytoplankton mandala incorporating twelve dimensions including nutritional physiology. Harmful Algae 55:25–30. https://doi.org/10.1016/j.hal.2016.01.008

Glibert PM, Mitra A (2022) From webs, loops, shunts and pumps to microbial multi-tasking: evolving paradigms of marine microbial ecology, global mixoplankton importance and implications for a future ocean. Limnol Oceanogr 67:585–597. https://doi.org/10.1002/lno.12018

Glibert PM, Burkholder JM, Kana TM (2012) Recent insights about relationships between nutrient availability, forms, and stoichiometry, and the distribution, ecophysiology, and food web effects of pelagic and benthic *Prorocentrum* species. Harmful Algae 14:231–259

Glibert PM, Wilkerson FP, Dugdale RC, Raven JA, Dupont C, Leavitt PR, Parker AE et al (2016) Pluses and minuses of ammonium and nitrate uptake and assimilation by phytoplankton and implications for productivity and community composition, with emphasis on nitrogen-enriched conditions. Limnol Oceanogr 61:165–197. https://doi.org/10.1002/lno.10203

Gran HH, Braarud T (1935) A quantitative study of the phytoplankton in the Bay of Fundy and the Gulf of Maine. J Biol Board Can 1:279–467

Grime JP (1977) Evidence for the existence of three primary strategies in plants and its relevance to ecological and evolutionary theory. Am Nat 111:1169–1194

Hardin G (1960) The competitive exclusion principle. Science 131:1292–1297

Hattenrath TK, Anderson DM, Gobler CJ (2010) The influence of anthropogenic nitrogen loading and meteorological conditions on the dynamics and toxicity of *Alexandrium fundyense* blooms in a New York (USA) estuary. Harmful Algae 9:402–412

Hillebrand H, Steinert G, Boersma M, Malzahn A, Meunier CL, Plum C, Ptacnik R (2013) Goldman revisited: faster-growing phytoplankton has lower N:P and lower stoichiometric flexibility. Limnol Oceanogr 58:2076–2088

Huisman J, Weissing FJ (1999) Biodiversity of plankton by species oscillations and chaos. Nature 402:407–410

Hutchinson GE (1961) The paradox of the plankton. Am Nat 95:137–145

Ives AR, Carpenter SR (2007) Stability and diversity of ecosystems. Science 317:58–62. https://doi.org/10.1126/science.1133258

Johnson ZI, Martini AC (2015) Techniques for quantifying phytoplankton biodiversity. Annu Rev Mar Sci 2015:299–324. https://doi.org/10.1146/annurev-marine-0108814-015902

Katz ME, Finkel ZV, Grzebyk D, Knoll AH, Falkowski PG (2004) Evolutionary trajectories and biogeochemical impacts of marine eukaryotic phytoplankton. Ann Rev Ecol Evol System 35:523–556

Kruk C, Devercelli M, Huszar VL (2021) Reynolds functional groups: a trait-based pathways from patterns to predictions. Hydrobiologia 848:113–129. https://doi.org/10.1007/s10750-020-04340-9

Kudela RM (2010) Does horizontal mixing explain phytoplankton dynamics. Proc Natl Acad US 107:18235–18236. https://doi.org/10.1073/Pnas.1013175107

Leong SCY, Murata A, Nagashima Y, Taguchi S (2004) Variability in toxicity of the dinoflagellate *Alexandrium tamarense* in response to different nitrogen sources and concentrations. Toxicon 43:407–415

Levin LS (1992) The problem of pattern and scale in ecology: the Robert H. MacArthur Award lecture. Ecology 73:1843–1967

Lomas MW, Glibert PM (1999a) Temperature regulation of nitrate uptake: a novel hypothesis about nitrate uptake and reduction in cool-water diatoms. Limnol Oceanogr 44:556–572

Lomas MW, Glibert PM (1999b) Interactions between NH^+ and NO^- uptake and assimilation: comparison of diatoms and dinoflagellates at several growth temperatures. Mar Biol 133:541–551

Lomas MW, Glibert PM (2000) Comparisons of nitrate uptake, storage and reduction in marine diatoms and flagellates. J Phycol 36:903–913

Margalef R (1958) Temporal succession and spatial heterogeneity in phytoplankton. In: Buzzati-Traverso AA (ed) Perspectives in marine biology. University of Chicago Press, Berkeley, pp 329–349

Margalef R (1967a) The food web in the pelagic environment. Helgol Wiss Meeresunters 15:548–559. https://doi.org/10.1007/BF01618650

Margalef R (1967b) Succession in marine populations. Adv Frontiers Pl Sci New Delhi 2:137–188

Margalef R (1978) Life-forms of phytoplankton as survival alternatives in an unstable environment. Oceanol Acta 1:493–509

Margalef R, Estrada M, Blasco D (1979) Functional morphology of organisms involved in red tides, as adapted to decaying turbulence. In: Taylor D, Seliger H (eds) Toxic dinoflagellate blooms. Elsevier, New York, pp 89–94

Mills EL (1989) Biological oceanography: an early history, 1870–1960. Cornell University Press, New York, p 264

Mitra et al (2014) The role of mixotrophic protists in the biological carbon pump. Biogeosciences 11:995–1005. https://doi.org/10.5194/bg-11-995-2014

Moore LR, Post AF, Rocap G, Chisholm SW (2002) Utilization of different nitrogen sources by the marine cyanobacteria *Prochlorococcus* and *Synechococcus*. Limnol Oceanogr 47:989–996. https://doi.org/10.4319/lo.2002.47.4.0989

Parker MS, Armbrust EV (2005) Synergistic effects of light, temperature and nitrogen source on transcription of genes for carbon and nitrogen metabolism in the centric diatom *Thalassiosira pseudonana* (Bacillariophyceae). J Phycol 41:1142–1153

Pielou EC (1975) Ecological diversity. Wiley, New York, p 165

Podgórska A, Szal B (2015) The role of reactive oxygen species under ammonium nutrition. In: Gupta K, Igamberdiev A (eds) Reactive oxygen and nitrogen species signaling and communication in plants. Signaling and communication in plants 23. Springer Nature. https://doi.org/10.1007/978-3-319-10079-1_7

Rao TR (2000) A curve for all reasons: the rectangular hyperbola in biology. Resonance 5:85–90

Reynolds CS (1987) Community organization in the freshwater plankton. Symp Br Ecol Soc 27: 297–325

Reynolds CS (1997) Successional development, energetics and diversity in planktonic communities. In: Abe T, Levin SA, Higashi M (eds) Biodiversity. Springer, New York, NY. https://doi.org/10.1007/978-1-4612-1906-4_11

Reynolds CS (1998) What factors influence the species composition of phytoplankton in lakes of different trophic status? Hydrobiologia 369:11–26. https://doi.org/10.1023/A:1017062213207

Reynolds CS (2000) Phytoplankton designer–or how to predict compositional responses to trophic-state change. Hydrobiologia 424:123–132

Roy S, Chattopadhyay J (2007) Toward a resolution of 'the paradox of the plankton': a brief overview of the proposed mechanisms. Ecol Complex 4:26–33. https://doi.org/10.1016/j.ecocom.2007.02.016

Rynearson T, Menden-Deuer S (2016) Drivers that structure biodiversity in the plankton. In: Glibert PM, Kana TK (eds) Aquatic microbial ecology and biogeochemistry: a dual perspective. Springer, pp 13–24. https://doi.org/10.1007/978-3-319-30259-1_2

Sathyendranath S, Ji R, Browman HI (2015) Revisiting Sverdrup's critical depth hypothesis. ICES J Mar Sci 72:1892–1986. https://doi.org/10.1093/icesjms/fsv110

Shannon CE (1948) A mathematical theory of communication. Bell Syst Tech J 27:379–423
Simpson EH (1949) Measurement of diversity. Nature 163(4148):688. https://doi.org/10.1038/163688a0
Smayda TJ (1980) Phytoplankton species succession. In: Morris I (ed) The physiological ecology of phytoplankton. University of Chicago Press, Berkeley, pp 493–570
Smayda TJ, Reynolds CS (2001) Community assembly in marine phytoplankton: application of recent models to harmful algal blooms. J Plankton Res 23:447–461. https://doi.org/10.1093/plankt/23.5.447
Sterner RW, Elser JJ (2002) Ecological stoichiometry: the biology of elements from molecules to the biosphere. Princeton University Press, Princeton, NJ, 439 pp
Striebel M, Singer G, Stibor H, Andersen T (2012) "Trophic overyielding": phytoplankton diversity promotes zooplankton productivity. Ecology 93:2719–2727. https://doi.org/10.1890/12-0003.1
Sverdrup H (1953) On conditions for the vernal blooming of phytoplankton. J Cons Int Explor Mer 18:287–295. https://doi.org/10.1093/icesjms/18.3.287
Tilman D, Isbell F, Cowles JM (2014) Biodiversity and ecosystem functioning. Annu Rev Ecol Evol Syst 45:471–493. https://doi.org/10.1146/annurev-ecolsys-120213-091917
Van de Waal DB, Smith VH, Declerck SAJ, Stam ECM, Elser JJ (2014) Stoichiometric regulation of phytoplankton toxins. Ecol Lett 17:736–742
Wyatt T (2014) Margalef's mandala and phytoplankton bloom strategies. Deep Sea Res II 101:32–49
Zehr JP, Bench SR, Carter BJ, Hewson I, Niazi F, Shi T, Tripp HJ, Affourtit JP (2008) Globally distributed uncultivated oceanic N_2-fixing cyanobacteria lack oxygenic photosystem II. Science 322:1110–1112. https://doi.org/10.1126/scienceV.1165340

Additional Websites Accessed

https://en.wikipedia.org/wiki/Ramon_Margalef. Accessed April 2, 2023
www.newworldencyclopedia.org. Accessed April 2, 2023

18 Taming Diversity Through the Traits of Size and Stoichiometry

18.1 Introduction

With knowledge of the diversity of species, their differential responses to light, nutrients, temperatures, and other factors, how do we conceptualize not only their patterns in distribution and succession, as described in the previous chapter, but the laws underlying their distribution? Karl Brandt and colleagues had such questions in the early 1900s (Chap. 1), and global changes in nutrient supply and climate place a greater need to understand the *why*—not just the *who*—as systems change. What general principles help to determine the *why*, and can we begin to think in terms of trade-offs that may help us to categorize successful strategies rather than focusing on species and taxonomic differences as the only way to categorize species? Can we go beyond quantifying diversity in terms of cell abundance and/or taxa, and can we think in terms of diversity of form and function instead? How do traits and diversity relate?

One of the most common ways of categorizing diversity exclusive of taxonomy is by size. The use of cell size as a way of categorizing diversity stems from the common use of different mesh size to separate size fractions. As Chisholm (1992, p. 213) once noted,

> [P]hytoplankton ecologists fall out into two groups: Those who delight in finding the patterns in nature that can be explained by size, and those who delight in finding exceptions to the established size-dependent rules. . . .The mechanisms underlying the size-dependent patterns have undoubtedly steered the general course of phytoplankton evolution, but the organisms that do not abide by the rules reveal the wonderful diversity of ways in which cells have managed to disobey the "laws" scripted for them. The simplicity of the general relationships serves as a stable backdrop against which the exceptions can shine. By understanding the forces that have driven these exceptions, we can begin to understand the ecology that has shaped past and present planktonic ecosystems.

Cell size is considered a "master trait" intricately intertwined with other traits (Naselli-Flores and Padisák 2007; Litchman et al. 2010; Naselli-Flores and Barone

P. M. Glibert, *Phytoplankton Whispering: An Introduction to the Physiology and Ecology of Microalgae*, https://doi.org/10.1007/978-3-031-53897-1_18

2011; Acevedo-Trejos et al. 2015; Sommer et al. 2017) and therefore a good candidate for understanding and predicting how phytoplankton change as environmental conditions change. Hecky and Kilham (1974, p. 365) captured the importance of size by saying, "Phytoplankton are usually looked at as complex solutions with particular biochemical properties rather than as organisms acted on by natural selection. If large and small phytoplankters were heuristically looked upon as elephants and mice and basic biological questions concerning form and function were asked, we are confident that the significance of size in the planktonic environment would be more easily understood."

Cell size is just one of many traits that are now used to characterize, quantify, and model phytoplankton responses and diversity. By thinking in terms of different strategies for resource acquisition, growth, reproduction, and survival, the natural distribution of phytoplankton not only begins to make sense but also becomes more predictable. Elemental composition is another universal trait that affects all physiological processes, from nutrient acquisition to growth and reproduction. In order to succeed, one needs to grow. Growth is related to fitness, or physiological state, and this, in turn, is related to the cell's ability to acquire and maintain the necessary energy and material for growth. Thus, traits, whether they be elemental acquisition and metabolism, cell size, or life history traits or other, help to provide the *why* and *how* different species respond as they do. With traits, rather than applying individual, classical concepts such as the law of the minimum, predator-prey dynamics, or trophic interactions, the goal is to connect interactions of organisms in relation to energy and elemental flow. Traits encompass any morphological, physiological, or phenological feature measurable at the individual level, from the cell to the whole-organism level, without reference to the environment of any other level of organization (Violle et al. 2007; Martini et al. 2021). With current and future conditions altering multiple aspects of the environment in which a cell lives (temperature, pH, nutrients), a major question is whether different traits work in tandem and whether they dampen or amplify the impacts of multiple perturbations. Can we connect key physiological traits with relevant ecological variables? This chapter cannot cover all traits but is instead focused on some key traits—cell size and stoichiometry—and the ecological theory supporting them. The purpose here is to introduce a way of thinking. Once these traits are understood, ecosystem structure and how it varies begin to make sense.

18.2 Diversity of Traits: Overview

Different cellular traits are rooted in the eco-physiological processes that have been described in the chapters thus far, and by understanding limits to adaptive changes, and trade-offs between traits, the consequences for different types of phytoplankton communities can be envisioned and quantified (and subsequently modeled; Chap. 28). Martini et al. (2021) summarized the main definitions of traits applied in aquatic studies (Table 18.1) and provided a typology of key functional traits that transcend taxonomy (Fig. 18.1). Importantly, the ability to characterize functional

Table 18.1 Main definitions of aquatic organism traits

Term	Recommended definitions	References	Examples
Trait	Any morphological, physiological, or phenological feature measurable at the individual level, from the cell to the whole-organism level, without reference to the environmental or any other level of organization	Violle et al. (2007)	Cell size Cell morphology
Functional trait	Any trait that impacts fitness indirectly via its effects on growth, reproduction, and survival	Violle et al. (2007)	Mixotrophy
Realized trait	Any trait that is actually measured in situ or in the laboratory	Reu et al. (2011)	Growth rate
Potential trait	A trait described from the literature, usually at the species levels, and ideally covering a large variety of environmental conditions	Reu et al. (2011)	Temperature niche
Life history trait	Traits referring to life history	Litchman and Klausmeier et al. (2008)	Type of reproduction (sexual vs. asexual); ability to from resting cysts
Morphological trait	Traits related to the morphology of organisms	Litchman and Klausmeier et al. (2008)	Cell size, cell shape
Physiological trait	Traits related to the physiology of organisms	Litchman and Klausmeier et al. (2008)	Nutrient acquisition; response to light
Behavioral trait	Traits related to the behavior of organisms	Litchman and Klausmeier et al. (2008)	Motility

Table modified from Martini et al. (2021, Limnol Oceanogr 66: 965–994. doi.org/10.1002/lno.11655) under a creative commons license

traits can help to understand emergent properties of ecosystems without identifying species taxonomically (Violle et al. 2007; Martini et al. 2021). Traits may also be easier to measure and quantify than species, thus characterizing diversity can be accomplished without taxonomic expertise. Within increasing availability of 'omics-based tools, many functional traits linked to metabolic pathways can be characterized. For example, genes associated with photosynthesis, nutrient uptake, or presence of various metabolites can be used to code for specific metabolic or physiological traits (Martini et al. 2021).

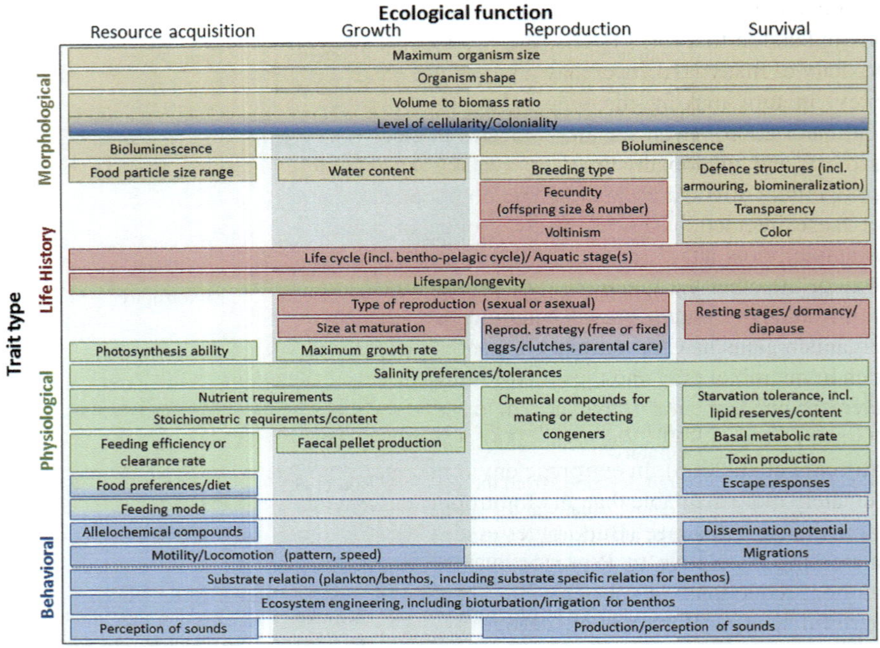

Fig. 18.1 Unified typology of aquatic functional traits that could be used in multicompartmental studies. This typology focuses on the key functional traits that transcend taxonomic peculiarities of the different aquatic ecosystems. Traits are classified by type and ecological function (as in Litchman and Klausmeier 2008), and most of them are quantitative. The dashed lines are a representation for similar traits crossing multiple ecological functions that are not close. A mental map providing a network visualization of this figure is available online, with each trait node linking toward associated research articles (https://doi.org/10.5281/zenodo.3635898). Image reproduced from Martini et al. (2021, Limnol Oceanogr 66: 965–994. doi.org/10.1002/lno.11655) under a creative commons license

Among ecosystem functions, resource acquisition (Table 18.1)—or resource use efficiency—is critically important, ultimately affecting nutrient cycling, growth, and trophic transfer (Filstrup et al. 2014; Ptacnik et al. 2008). A community with higher functional diversity should have greater resource niche partitioning and therefore greater resource use efficiency (Ye et al. 2013, 2019). Functional diversity may, in fact, have a stronger effect on resource use efficiency than species diversity (Ye et al. 2019). As with species diversity, evenness, and richness (see Chap. 17), many indices have been developed to capture functional diversity. Online tools are also now available to readily calculate such indices (Casanoves et al. 2010). Among such indices are functional richness, evenness, and diversion measures (e.g., Villéger et al. 2008), as well as functional dispersion (Laliberté and Legendre 2010).

18.3 Body Size, Allometry, and Metabolic Scaling

18.3.1 The Diversity of Algal Cell Sizes and Shapes

The biomass of phytoplankton cells varies by up to nine orders of magnitude (Malone 1980; Finkel et al. 2010). As Finkel et al. (2010) have shown, the comparison in size of a *Prochlorococcus* cell and that of a *Trichodesmium* colony is analogous to the comparison of a fish and the city of Manhattan (New York; Fig. 18.2).

Size influences growth, photosynthesis, nutrient uptake, and sinking rates, as well as grazing rates, excretion, and reproduction (Kleiber 1932, 1961; Malone 1980; Schmidt-Nielsen 1984; Finkel et al. 2010; Hirst et al. 2014). These relationships in turn have profound consequences for biogeochemical cycling, biodiversity, food web structure, and evolution (Quigg et al. 2003; Bragg and Hyder 2004; Finkel et al. 2005, 2007, 2010).

Although the various factors regulating cell size are not fully understood, it is well accepted that small cells dominate the oligotrophic oceans, whereas larger cells dominate in more productive waters, supporting shorter food chains and greater flux of organic matter to depths. In a classic illustration of this, Raimbault et al. (1988) showed that the total amount of chlorophyll (chl) *a* in different size fractions of phytoplankton from the Mediterranean Sea had distinct upper limits (Fig. 18.3). The total chl *a* in the respective fractions was 0.5 μg l^{-1} for the <1 μm fraction, 1 μg l^{-1} for the <3 μm fraction, and 2 μg l^{-1} for the <10 μm fraction. This implies that once a threshold is reached for a small size class, more chl *a* can only be added if the cells are larger. Most often, this "extra" chl *a* comes in the form of larger diatoms. Smaller cells have increased surface:volume (S/V) ratios, making them more competitive at resource acquisition in low-nutrient environments (Chisholm 1992; Kiørboe 1993; Raven 1998), whereas larger cells, typically diatoms, have the ability to sink to obtain more nutrients (Stolte et al. 1994).

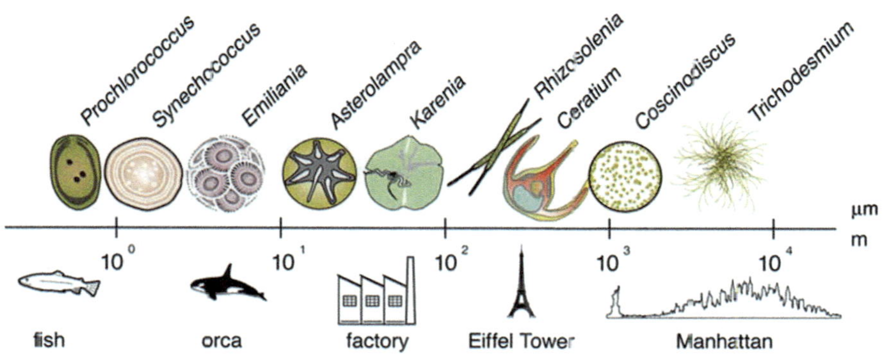

Fig. 18.2 A comparison of the size range (maximum linear dimension) of phytoplankton relative to common objects or places. Image reproduced from Finkel et al. (2010, J Plankt Res 32:119–137 doi.org/10.1093/plankt/fbp098) with permission of Oxford University Press

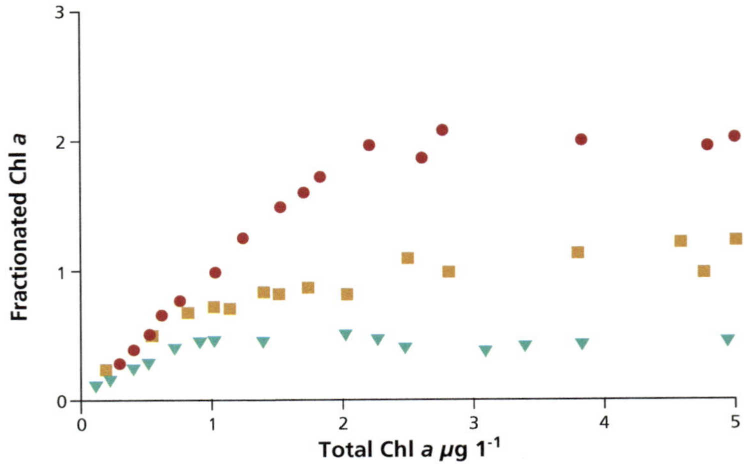

Fig. 18.3 Relationships between total chlorophyll *a* and the maximum amount of chlorophyll *a* in a <1 µm (triangles), <3 µm (squares), and <10 µm (circles) fraction in the Mediterranean Sea. Image redrawn based on Raimbault et al. (1988, Microb Food Webs 3:1–7)

The higher variation in nutrient supplies, including seasonal pulsing, in temperate regions relative to tropical regions, favors larger cells as documented in situ and reproduced in models (e.g., Acevedo-Trejos et al. 2015). Moreover, the Acevedo-Trejos model showed that in temperate regions, greater size diversity is more common. Based on the notion that grazing "pushes" a community to larger cell sizes, while nutrient uptake "pulls" a community to smaller cell sizes, the temperate region, with its greater seasonality, drives the community to greater size and functional diversity. Grazers play a role in phytoplankton cell size as they graze on preferred size ranges (often small cells, leaving the larger cells behind; see also Chap. 20).

High investments in one trait may come at the expense of other traits. While small cells have higher S/V ratios that can support a faster rate of nutrient exchange per unit cell volume than larger cells, there are limits to small cell advantages—that is, being too small places constraints on different aspects of metabolism. The essential components of a free-living cell have to be contained within the cell, including proteins, nucleic acids, ribosomes, and photosynthetic machinery, and there is simply a limit to how much can be packed into a very tiny cell. Small cells may therefore trade-off different capabilities to optimize their light acquisition. The genomes are highly streamlined, limiting their ability to adapt to change. Small cells are diffusion-limited and have limited ability to store nutrients. A cell of radius 5 µm growing at 1 day^{-1} would become diffusion-limited at N concentrations <100 nM, but a cell with a radius of 0.35 µm at the same growth rate would only become diffusion-limited if the N concentration were <5 nM (Chisholm 1992). Anderson et al. (2015) suggested that size-dependent investment in light acquisition is smallest for small cells, larger cells invest more in nutrient acquisition

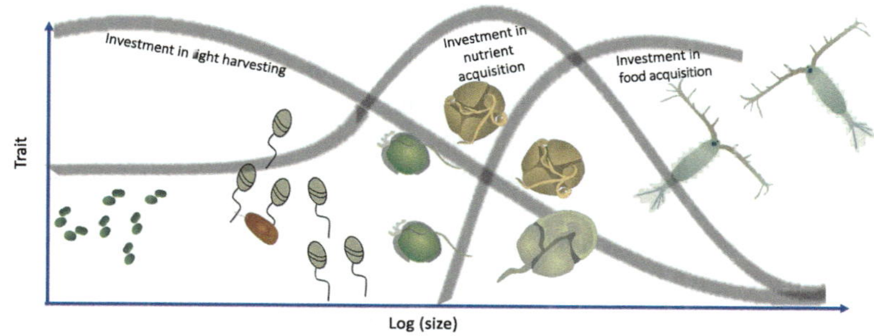

Fig. 18.4 Conceptual relationships between investments in light harvesting, nutrient acquisition, or prey capture traits as a function of cell size. Image modified and redrawn based on Anderson et al. (2015, J Plankton Res 37:862. doi: 10.1093/plankt/fbv054). Organism icons are from the University of Maryland Center for Environmental Science Integration and Application Network symbol library

mechanisms, including mixotrophy, and increasingly larger organisms invest more in prey capture. Being larger increases the yield of prey capture for heterotrophs (copepods). Expressed another way, the smallest phytoplankton are typically light-limited, medium-sized phytoplankton are nutrient-limited, and large heterotrophs are prey-limited (Fig. 18.4).

Both cell volume and surface area have long been considered the major factors for nutrient uptake rates, transformation, and allocation of energy and materials (Okie 2013). With their high S/V ratio, small cells are more likely to be spherical, whereas large cells are more ornate and irregular in shape (Sheldon and Parsons 1967). Shape is important, as a spherical shape is not optimum for nutrient acquisition (Munk and Riley 1952). Karp-Boss and Boss (2016) argued that while surface area is important for both nutrient and light acquisition, S/V per se may not be the best predictor of phytoplankton fitness. They showed that for the same unit volume and investment in surface area, an elongated cell may have an advantage over a squat or flattened cell. Nutrient competition appears to be selected for elongated shapes, while light may be neutral for small cells and selective for prolate or oblate spheroid cells for larger cells (Karp-Boss and Boss 2016). It is of note that non-diatom cells tend to be more spherical, while diatoms have a greater diversity in shape. Overall, cell volume relates to nutrient uptake rates of small phytoplankton, whereas surface area becomes important for larger cells (Dao 2013). Of course, many selective pressures bear on the shape of phytoplankton. Swimming, buoyancy, and other metabolic traits do not fit neatly into partitions established according to size.

18.3.2 Allometry: Relating Size to Metabolic Traits

Allometry is the study of cell size and its relationship with other organismal characteristics. In general, a diverse suite of traits scale with body size according to the relationship,

$$R = aM^b \tag{18.1}$$

where R is a metabolic rate process (in units of reciprocal time, i.e., time^{-1}), a is a scaling constant, M is a measure of body biomass, and b is the scaling exponent. Max Kleiber (1932, 1961) found that under favorable conditions for growth, many metabolic rates for all types of organisms scale according to three-fourths. That is, metabolism becomes slower in larger organisms by a b equal to 0.75 (Finkel et al. 2004). Since b appears so consistent across various biological processes, it was termed a natural "law" of biology (Peters 1983; Platt and Silvert 1981).

Metabolic rates are also influenced by temperature, and thus, combining the effects of body mass and temperature, the relationship (assuming nutrients are not limiting) becomes

$$R = aM^b e^{-\mathrm{Ea/kT}} \tag{18.2}$$

where E_a is the activation energy for metabolism (assumed to be −0.65 eV for respiration), T is temperature in absolute units (kelvin), and k is a constant. Thus, this relationship predicts that not only do larger organisms have higher metabolic rates, but high temperatures also promote faster rates of metabolism. This relationship is the core of the *Metabolic Theory of Ecology* (Gillooly et al. 2001; Brown et al. 2004). Moreover, this theory suggests that organisms that are *r*-selected are generally smaller and have higher body temperatures, whereas organisms that are *K*-selected are cooler and larger (see Chap. 17 for descriptions of *r*- vs. *K*-selected organisms).

There are some rates and traits of phytoplankton that scale with a power law of cell size (Finkel et al. (2010; Table 18.2). Nevertheless, within the microbial world, the scaling exponent can be variable (Finkel et al. 2010; Marañón et al. 2013). Values of 1 for many protists have been reported, as well as values of up to 2 for bacteria (Delong et al. 2010).

Kremer et al. (2017) discussed several reasons why phytoplankton may deviate from the ¾ scaling relationships above. They may have lower activation energy (closer to 0.32 eV; López-Urrutia et al. 2006), and there are extreme differences between and among taxa in terms of temperature tolerances. In addition, cell size varies between and among species and may change with environmental conditions, and cell shape may also change the relationship. Marañón et al. (2013) showed that growth rate and metabolic rates increase with cell size in prokaryotes, are relatively size-independent in protists, and decrease with body size in metazoans, a pattern arising from the trade-offs between different size-related constraints of small cells, especially related to nutrient uptake and assimilation. Thus, size and nutrient content are related in these small cells. There is also evidence that phytoplankton growth rates are highest at intermediate cell sizes across a range of temperatures (Fernandez-González and Marañón 2021). Marañón et al. (2014) summarized field data from across polar, template, and tropical regions and compared the role of temperature and nutrient supply in controlling large-scale patterns of phytoplankton metabolic rates. They concluded that resource supply was the major driver. Dutkiewicz et al. (2020),

Table 18.2 Various physiological and metabolic processes that vary with a power law of cell size. Table adapted from Finkel et al. (2010 J Plankt Res 32:119–137 doi.org/10.1093/plankt/fbp098)

Process or trait	Size dependence
Metabolic rate (time^{-1})	Metabolic rates often scale with volume with a scaling exponent of −0.25
Nutrient uptake kinetics (mol N (cell time)$^{-1}$)	Maximum nutrient uptake rate is a function of surface area and enzymes and transporter concentrations and tends to have a size scaling exponent of −2/3
Nutrient diffusion (mol N (cell time)$^{-1}$)	Nutrient diffusion rate varies as 4π RD ΔC (D, molecular diffusivity; R, cell radius, and ΔC, the difference in nutrient concentration between the cell surface and bulk medium)
Light absorption (m^2 mg chl a^{-1})	Light absorption per unit of chlorophyll is a function of cell size, and intracellular pigment concentration varies under light limitation with a scaling exponent b of −0.08
Cell nutrient or chlorophyll content (mol nutrient or chl a cell^{-1})	In phytoplankton, C, N, and P tend to have b ranging from 0.7 to 0.9, with some variability between taxonomic groups and elements The size scaling of cellular chlorophyll varies with irradiance: $b = 2/3$ under light limitation but tends to 3/4 under light-saturating conditions
Sinking velocity	Sinking velocity increases with the size ($b = 2/3$) according to Stoke's law
Population abundance (cells L^{-1})	The numerical abundance of organisms in an area or volume can often, but not always, be approximated as a power law relationship of cell size with a scaling exponent $b = -5/3$ to $-2/3$, often -1. The intercept and exponent vary with environmental conditions
Trophic interactions	The body size of the consumer is often correlated with the size of prey consumed; larger predators eat larger prey and a larger range of prey sizes
Diversity	Organism diversity within taxa is often a log-normal distribution of body size; maximum diversity occurs at a size somewhat smaller than the median

in a modeling study, concluded that phytoplankton diversity, and hence size spectrum, is a complex function of four major driving forces: rate of supply of limiting resource, imbalance in resource ratios relative to the competitor requirements, size-dependent grazing control, and transport processes. The significant metabolic rates between and among phytoplankton taxa are convincing that phytoplankton as a whole do not follow Kleiber's power rule when cells are growing exponentially under optimal conditions (López-Sandoval et al. 2014; Marañón et al. 2013).

There are multiple lines of evidence suggesting cell size is a good predictor of a number of metabolic processes. López-Sandoval et al. (2014), in a culture study, showed substantial departure for the ¾ scaling value. They found not only significant taxon-specific differences in metabolic rates, but also that intermediate-sized cells can attain faster metabolic rates. Hillebrand et al. (2021), in a meta-analysis, also found allometric relationships between cell sizes and photosynthesis, respiration, and resource affinities, but relationships were much weaker when relationships were scaled by C- or volume-specific rates or growth. Cell size is intricately related to diversity and biogeography, and nutrient content and the plasticity thereof become central in understanding size-dependent relationships. These are next discussed.

18.4 Resource Acquisition and Elemental Composition

18.4.1 Ecological Stoichiometry and the Redfield Ratio

Alfred Redfield (Chap. 11) pioneered the notion that elements in organisms and in solution vary in systematic ways. Redfield determined that the atomic ratios of carbon (C), nitrogen (N), and phosphorus (P) in marine plankton paralleled the ratios of these same elements in water. Although Redfield is best known for characterizing elemental ratios, it was Lotka and Dublin (1925, cited in Sterner and Elser 2002, p. 8) who wrote that, "we may employ the term Stoichiometry to denote that branch of the science which concerns itself with the material transformations, with the relationships between the masses of the components." The term stoichiometry comes from the Greek "stoicheoin" for "element," and therefore, the word implies measuring elements (Sterner and Elser 2002). *Ecological stoichiometry* has become a unifying principle for understanding community self-organization.

Ecological stoichiometry builds on classical concepts of Liebig's law of the minimum relating to nutrient limitation (von Liebig 1855—but see Chap. 1 and appropriate attribution to Sprengel 1828), Lotka's (Lotka and Dublin 1925) understanding of the dynamics of predators and prey, Lindeman's understanding of trophic dynamics (Lindeman 1942), and Redfield's (1934) concept of balanced proportions of elements in the ocean. It brings these concepts together by recognizing that different organisms both within and between trophic groups have fundamentally different elemental requirements, that food web structure is a function not only of food quantity but also of food quality, and that these interactions result in a complex suite of feedbacks that shape community composition. Thus, ecological stoichiometry is basically a comparison of nutrient ratios in solution or food and in consumer biomass. Organism composition is clearly influenced by size and metabolic rates—and the ability of different organisms to metabolically regulate their elemental composition (Glibert et al. 2011; Glibert 2012).

The general Ecological Stoichiometry framework was formalized by Sterner and Elser (2002) and provides "an integrated framework for merging perspectives across individual, population, community, and ecosystem levels." A key concept of the ecological stoichiometric framework is that of "stoichiometric homeostasis" or the degree to which organisms maintain and regulate their chemical composition. The stoichiometric framework recognizes that changes in the proportions of dissolved nutrients in the environment have profound effects on food webs *even when the availability of these elements are not in limiting proportions*, with the potential of transforming ecosystems to new stable states. Just as phytoplankton continually regulate their capability for nutrient assimilation along a continuum of substrate availability (according to their kinetics) and other environmental factors, grazers—together with primary producers—operate in a dynamic balance with respect to nutrient composition and availability (e.g., Glibert 1998; Boersma and Elser 2006; Meunier et al. 2014). Within the ecological stoichiometry framework, *Biological Stoichiometry* specifically relates to the study of the balance of energy and elements in living systems, integrating ecological dynamics with cellular and genetic traits as a means to characterize and understand diverse organisms in diverse habitats (Elser et al. 2000). *Nutritional Geometry* is a contrasting framework that focuses on micro- and macronutrients (amino acids, carbohydrates, and lipids), rather than elements per se, to understand nutrient acquisition and metabolic regulation.

At any given time, a cell is regulating not only the acquisition of different forms of the same nutrient (e.g., NH_4^+ and NO_3^-), but it is also balancing the uptake and metabolism of other nutrients (e.g., N and P and trace metals), along with cell energy demands, reductant, and cell redox state. Codified by Redfield (1958), phytoplankton have often been reported to have an elemental composition of 16:1 in terms of molar N:P (= ~7 on a wt:wt basis; Redfield 1934; Falkowski 2000; Arrigo 2005). However, wide ranges in this ratio have been reported (e.g., Klausmeier et al. 2008; Finkel et al. 2010; Hillebrand et al. 2013; Fig. 18.5a) and there is a growing appreciation that different taxa have different elemental compositions under ideal growth conditions (Geider and LaRoche 2002; Ho et al. 2003; Klausmeier et al. 2004) and that there is an evolutionary basis for elemental stoichiometry in phytoplankton (Quigg et al. 2003). Importantly, there is accumulating evidence that a species' elemental stoichiometry varies with environmental conditions (Leonardos and Geider 2004a, b; Cullen and Sherrell 2005; Finkel et al. 2006, 2007; Glibert et al. 2013). Phytoplankton stoichiometry is a function of taxon-specific differences, on the one hand, and their flexibility in stoichiometry through luxury uptake and storage, on the other. In a meta-analysis of both fresh and marine studies of phytoplankton stoichiometry, Hillebrand et al. (2013) confirmed that phytoplankton N:P varied with growth rate, increasing under N limitation, and decreasing under P limitation, but, in addition to having a higher P content, fast-growing phytoplankton were more constrained in their stoichiometry. Hillebrand et al. (2013) also found that the weighted molar averages for optimal N:P (defined as the ratio of minimal cell quota for N and P when available ratios exactly match uptake and growth demands and there is no luxury uptake or storage) were lowest for diatoms (14.9) and increased for dinoflagellates (15.1) and even more for cyanobacteria (25.8) and chlorophytes (27.0; Fig. 18.5b). Their findings lend support to the idea that not only do values of phytoplankton N:P become more restricted and are lower with increasing growth rate, but that at maximum growth rate, values of N:P converge to an optimal ratio (or a more narrowly defined range) that is different for species and phylogenetic groups.

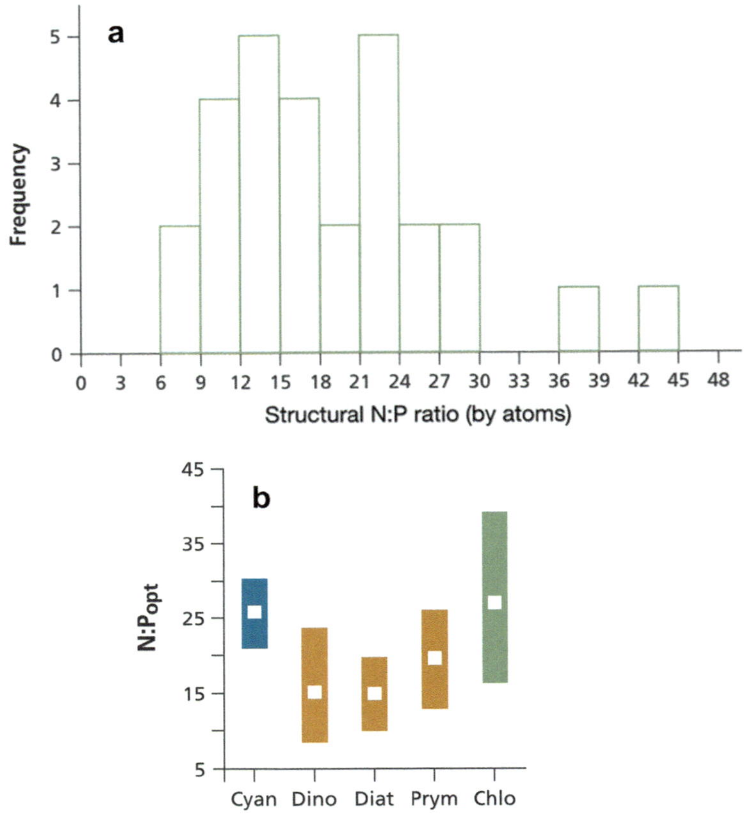

Fig. 18.5 Panel (**a**) Range in the structural N:P ratios of 29 freshwater and marine species (not including picoplankton and not including an outlier at N:P of 135) during exponential growth. Image Panel A redrawn based on Klausmeier et al. (2004). Panel (**b**) Average optimal N:P ratio (+95% confidence intervals) for different phytoplankton groups (Cyan: cyanobacteria; Dino: dinoflagellates; Diat: diatoms; Prym: prymnesiophytes; Chlo: chlorophytes). Panel (**b**) redrawn from Hillebrand et al. (2013, Limnol Oceanogr 58:2076–2088. doi: 10.4319/lo.2013.58.6.2076) excluding the data for polycultures

Meunier et al. (2017) specifically related cell size to elemental stoichiometry by making the case that physiological limits of minimum (structural) and maximum (storage capacity) cell quotas, as well as affinity for particular nutrients, determine the N:P stoichiometry of phytoplankton (Fig. 18.6). As further detailed below, phytoplankton with high growth rates will be P-rich and typically small-sized cells, while P-limited cells select for better P competitors with higher N:P ratios. Thus, N:P (and form of these nutrients) becomes a major factor determining autotrophic community composition.

In a nutshell, the concept of ecological stoichiometry provides a framework for understanding that while the total nutrient load of a system may set the amount of

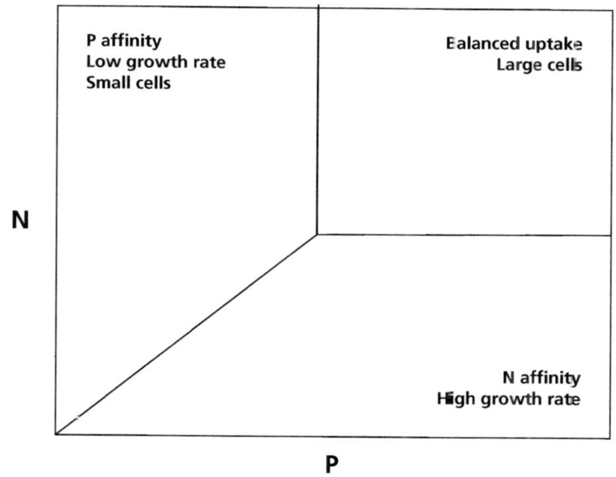

Fig. 18.6 Influence of N and P availability on trait dominance in autotrophs. Low N and P supplies will favor species with high affinities. Relatively high N availability when P is limiting may selectively favor the synthesis of N-rich secondary metabolites (e.g., toxins) and favor small cells and lower growth rates. Relatively high P availability will be selected for higher growth rates. Both N and P availability may favor larger cells and high maximum growth rates and cells with sufficient pigments per cell. Image modified from Meunier et al. (2017, Front Environ Sci 5:18. doi: 10.3389/fenvs.2017.00018)

biomass that can be supported, it also sets the trajectory of the composition, from autotrophs to heterotrophs and throughout the food web (Glibert et al. 2011). Ecological stoichiometry suggests that food webs are not merely a summation of a series of rate processes (and kinetic curves), or food quantity (as C, for example), but are an outcome of both the quantity and quality of the substrate (or food) provided and the balance of nutrients therein.

This notion—that ratios are fundamentally linked to body stoichiometry, and therefore shape ecological interactions such as competition, predation, and ultimately dominance within communities—is an idea that contradicts one of the concepts proposed by Reynolds (see also Chap. 17). He (Reynolds 1998) argued that "ratios explain very little of the distribution of species with respect to trophy." However, at that time there was little recognition that nutrient forms—as well as ratios—matter, that kinetic properties are not strictly fixed entities (see Chap. 8, Kana and Glibert 2016), that cells must regulate their nutrient stoichiometry not only by acquiring nutrients that are limiting but also by regulating the potentially repressive effects of nutrients that are in excess, and that anthropogenic changes are driving many ecosystems into nutrient imbalance (see Chap. 22, Glibert et al. 2013, 2016). Ideas evolve. As Hillebrand et al. (2014, pp. 468) more recently summarized, "understanding the consequences of eutrophication on biodiversity requires ratio-based thinking to unravel the underlying mechanisms."

The sections below dive into stoichiometric regulation in more detail at various levels of organization.

18.4.2 Stoichiometry at the Subcellular Level: The Growth Rate Hypothesis

Elemental differences in biomass are found at all levels of organismal structure across trophic levels, from the subcellular to the macrocellular structural components (Sterner and Elser 2002). At the subcellular level, biomolecules and organelles vary in their N:P content (Sterner and Elser 2002, Fig. 18.7).

Fluxes of energy and materials are linked based on kinetics and elemental compositions of subcellular structures, and biomass is comprised of metabolic structures, which can exhibit distinct allometries and composition (Allen and Gillooly 2009). In particular, ribosomes are high in P relative to N; they are "the most P rich and lowest N:P organelles in cells" (Sterner and Elser 2002, p. 73, Fig. 18.7). For cells lacking storage mechanisms for P (vacuoles in algae, bones in fish), elemental C:N:P becomes dependent on the demand for P by ribosomes (Elser et al. 2000). Ribosomes are required for growth, and an increase in ribosomes is required for a cell to have an increase in growth rate. Thus, the *Growth Rate Hypothesis* (GRH) says that variations in an organism's stoichiometry (as C:P or N:P) are a function of the growth-dependent allocation of P to ribosomal RNA.

This concept has been well illustrated for phytoplankton (Geider and LaRoche 2002; Sterner and Elser 2002; Quigg et al. 2003; Finkel et al. 2010). Fast-growing cells, such as diatoms, have a lower N:P ratio than their more slowly growing counterparts. They have proportionately more allocation of resources to "assembly machinery" (rRNA; high P) than to "acquisition machinery" (protein; high N) (Elser et al. 2003; Klausmeier et al. 2004; Fig. 18.8). In contrast, phytoplankton species that can sustain their metabolism when resources are low—i.e., more slowly growing cells—have a higher proportion of pigments and proteins with proportionately higher N:P ratio (Sterner and Elser 2002; Elser et al. 2000; Elser 2006; Arrigo 2005; Finkel et al. 2010; Fig. 18.8). Slowly growing cells are also generally, but not always, larger in size (e.g., Malone 1980; Kagaami and Urabe 2001; Finkel et al. 2010). Stated another way, if the GRH is correct, small cells should have a lower N:P due to their higher P demand and content (Sterner and Elser 2002). However, slowly growing, nutrient-stressed phytoplankton may also be capable of short-term "luxury" uptake of the limiting nutrient in excess of growth, leading to highly variable N:P ratios under transient conditions (Flynn 2002); thus, the change in N:P ratio with algal growth rate is not necessarily a linear function (Ågren 2004). More formally, the assumptions of the GRH are that (1) growth-dependent variation in P allocation to RNA is proportionally large enough to drive changes in organismal P content; (2) the number of ribosomes in a cell controls the overall rate of protein synthesis; (3) ribosomes translate proteins at or near their maximum capacity; and (4) a fixed fraction of synthesized protein accumulates in the cell, contributing to growth (as summarized by Isanta-Navarro et al. 2022).

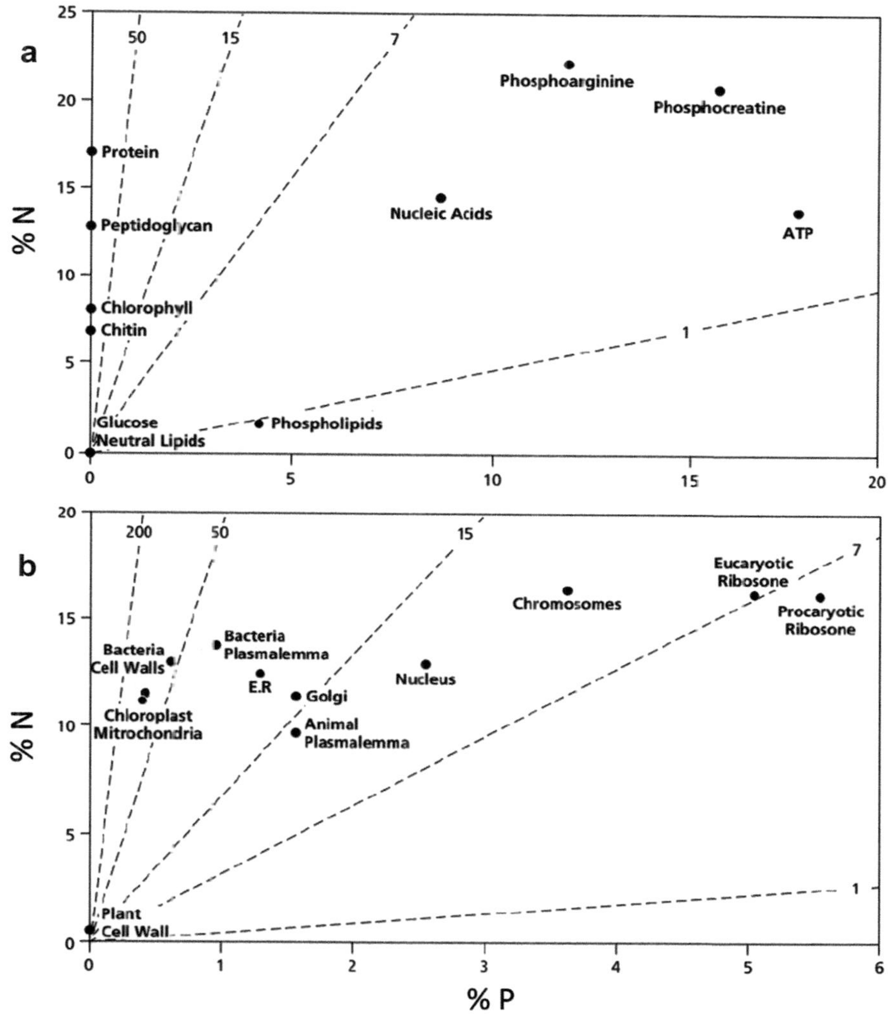

Fig. 18.7 Elemental composition (as percent N and P) of important biomolecules Panel (**a**) and of major cellular structures Panel (**b**). Images redrawn from Sterner and Elser (2002, Ecological stoichiometry: the biology of the elements from the molecules to the biosphere)

Within this framework, the GRH proposes that an organism's C:N:P and growth rate are coupled because elevated growth rate depends on increased production and maintenance of P-rich ribosomes (Elser et al. 1996, 2000; Fig. 18.9). This biochemical investment therefore affects body stoichiometry and sets constraints on growth, resource competition, and, in the case of animals, trophic efficiency, and nutrient recycling. More recent work has shown that maximum growth rate, as a key life

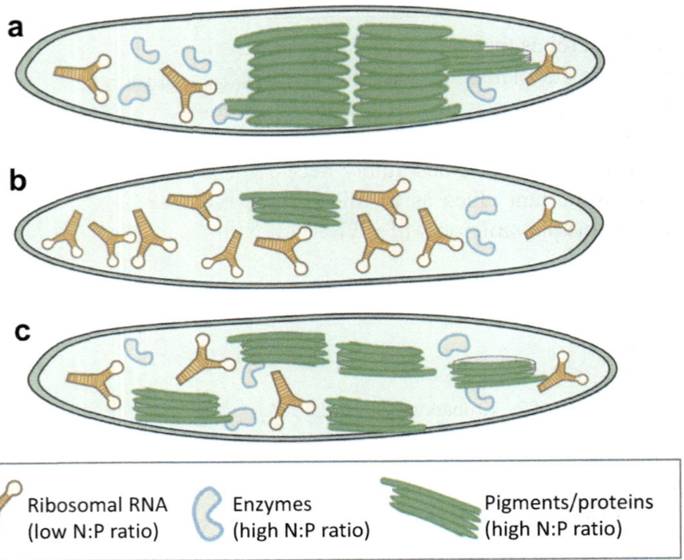

Fig. 18.8 Conceptual diagram of the allocation of N and P between acquisition (enzymes and pigments) and growth machinery (ribosomal RNA) in cells with different life strategies. Panel (**a**) The "survivalist" which would be expected to have a high N:P ratio (>30) as it has large amounts of acquisition machinery; Panel (**b**) The "bloomer" which would be expected to have a low N:P ratio as it has a high proportion of growth machinery; and Panel (**c**) The "generalist" which has a balanced N:P ratio and a balance of growth and acquisition machinery. Image modified and redrawn from Arrigo (2005, Nature 437:349–355. doi: 10.1038/nature04159)

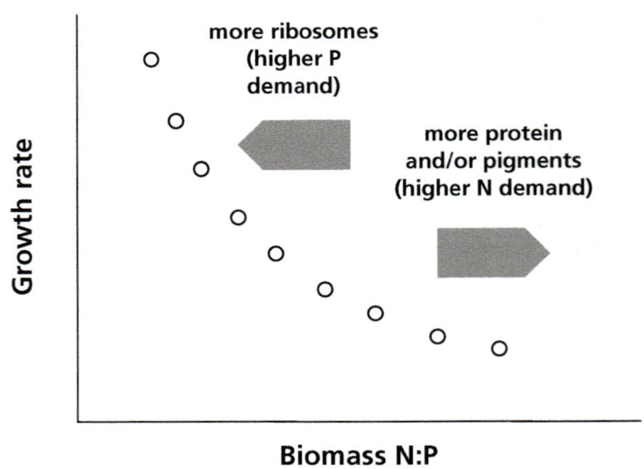

Fig. 18.9 Generalized relationships between the intrinsic growth rate of organisms and the N:P ratio in the biomass of those organisms. Image redrawn from Glibert et al. (2011, Rev Fish Sci 19:358–417)

history parameter, and ribosome production capacity have a genomic basis in the multiplicity of ribosomal RNA operons (rRNA gene copy number) (Weider et al. 2005; Stoddard et al. 2014; Roller et al. 2016). This has led to the observation that bacteria with fewer copies of rRNA genes tend to be more competitive in oligotrophic or low-nutrient conditions due to their high efficiency in resource use, whereas

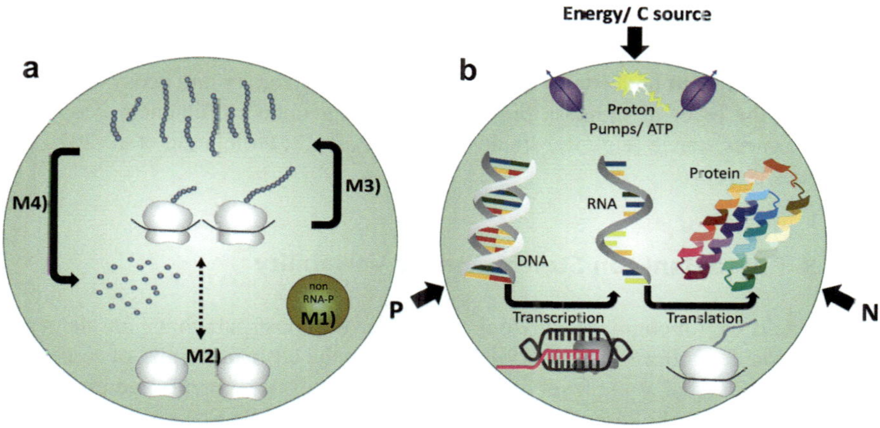

Fig. 18.10 Panel (**a**) Schematic depiction of four mechanisms that can influence the coupling of growth, RNA, and P proposed by the GRH. On the organismal scale, the first mechanism (M1) involves contributions from pools of non-RNA P, including storage of P in molecules such as polyphosphates. Mechanisms M2, M3, and M4 constitute changes on the molecular scale that individually or collectively affect the net protein production rate per ribosome. Mechanism (M2) involves change in the fraction of inactive ribosomes among all ribosomes. Mechanism (M3) entails differences in ribosome translation elongation rate and mechanism (M4) highlights protein degradation or protein turnover rate. Panel (**b**) Effects of C, N, and P limitation on cellular functions. The figure shows the relationships between environmental resource supplies and cellular functions that influence growth rate and could result in deviations from the GRH under different types of resource limitation. Image reproduced from Isanta-Navarro et al. (2022, Ecol Letts 25: 2324–2339. doi.org/10.1111/ele.14096) under a creative commons license

bacteria in eutrophic environments with higher rRNA gene copy numbers tend to respond more rapidly and are adapted to high-nutrient supply and episodic availability (Klappenbach et al. 2000; Lauro et al. 2009; Nemergut et al. 2016; Roller et al. 2016).

The GRH was recently reconsidered by Isanta-Navarro et al. (2022) in an analysis across diverse taxa including 59 species of: zooplankton ($n = 15$ species), phytoplankton (11), aquatic invertebrates (10), terrestrial invertebrates (7), plants (8), fungi (7), bacteria (3), and human cancer cells (1). Studies consisted mostly of manipulative experiments involving P (9), P co-limitation by other elements/macromolecules (8), or water stress (2). They found strong support for the relationship between RNA and P predicted by the GRH. However, support for the coupling between growth rate and P content was mixed. In all, the GRH was supported in about half of the studies analyzed. They suggest that the next-generation stoichiometric model will come from combining these more standard measurements with those of key cellular processes of ribosome activity and protein production and turnover rates (Fig. 18.10).

Other ecological theories are related to, or under the umbrella of, ecological stoichiometry. For example, the *Resource Ratio Theory* predicts that as the rates of different essential elements change, the community structure will change because of competition between algae with different optimal nutrient ratios (Tilman 1977, 1982, 1985; Smayda 1990, 1997). In addition, the concept of *Biodiversity and Ecosystem Functioning* aims to understand linkages between biodiversity and ecosystem functioning at the level of biomass production, nutrient uptake efficiency, ecosystem stability, or resistance (e.g., Cardinale et al. 2009, 2012; Hillebrand et al. 2014 and references therein).

18.4.3 Phytoplankton Stoichiometric Variability

As Redfield well recognized, phytoplankton (for the most part) have an elemental composition that reflects their surroundings. One of the basic principles of ecological stoichiometry is that at the level of the organism, fluxes of elements and energy are determined by metabolic rates (Allen and Gillooly 2009). Photoautotrophs, including phytoplankton, have relatively weak homeostasis, or as Elser et al. (2000, p. 542) expressed, "wildly variable C:N:P composition." That is, as the relative availability of the elemental resources in the environment varies, their biomass chemical composition varies (Fig. 18.11a). This has been elegantly demonstrated for many phytoplankton in culture. For example, it has been shown that the medium N:P ratio and the cellular N:P ratio of the chlorophyte *Scenedesmus* are very similar when grown over a range of N:P ratios in culture (e.g., Rhee 1978; Fig. 18.11b, c). Due to the ability of many phytoplankton to take up nutrients in excess of their growth demands (e.g., Glibert and Goldman 1981; Terry et al. 1985), they often follow the "you are what you eat" model (Sterner and Elser 2002, p.16), within reasonable limits. This is true of cells that are storage- or affinity-adapted (e.g., Meunier et al. 2017). Extreme cellular N:P ratios are observed in cultured cells that have experienced limitation of either N or P (Rhee 1978; Goldman et al. 1979; Geider and LaRoche 2002) reflecting a non-homeostatic "luxury consumption" response, which is characteristic of microalgae (Goldman and Glibert 1983)—and a condition of variable and unbalanced growth. Significantly, under nutrient-replete growth conditions (i.e., saturating ambient concentrations), variation in the N:P ratio of microalgae can be on the order of three- to fourfold (5–19 mol N:mol P; Geider and LaRoche 2002; Finkel et al. 2010). While it may be difficult in field data to differentiate the extent to which this variation is related to changing algal species (or taxonomic groups), culture experiments have clearly demonstrated plasticity in the N:P ratio as a function of growth condition (e.g., Quigg et al. 2003; Leonardos and Geider 2004a, b; Finkel et al. 2010). Field comparisons of N:P stoichiometry in dissolved substrates have been compared favorably with that of particulate matter in some regions (e.g., Glibert et al. 2006), but a range of relationships between dissolved and particulate matter can be observed. The salient point is that microalgae, especially non-size-limited picoplankton, are generally not homeostatic with respect to cellular N and P.

Viewed from the perspective of kinetic relations, phytoplankton nutrient acquisition can be categorized as being (1) velocity-adapted (*r*-adapted, with high maximum uptake and growth rates), (2) storage-adapted (high rates of nutrient uptake but

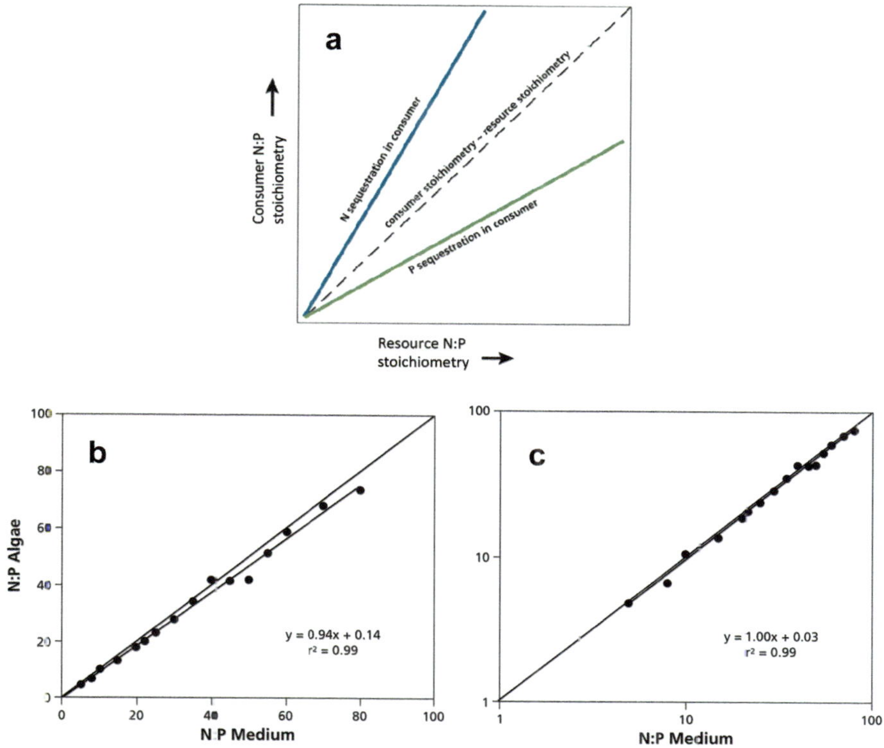

Fig. 18.11 Panel (**a**) Schematic relationship between resource N:P and consumer (i.e., phytoplankton) N:P stoichiometry. In general, phytoplankton stoichiometry reflects that of the dissolved resource. Image redrawn from Glibert et al. (2011). Panel (**b**, **c**) Example of phytoplankton N:P as a function of N:P in the growth media for the green alga, *Scenedesmus*. Data based on Rhee (1978) plotted on a linear scale Panel (**b**) and on a log scale Panel (**c**). Images (**b**, **c**) redrawn from Sterner and Elser (2002, Ecological stoichiometry: the biology of the elements from the molecules to the biosphere)

low maximum growth rates), and (3) affinity-adapted (*K*-adapted with ability to take up nutrients rapidly but lower overall rates of growth). Stoichiometric homeostasis will vary with these different strategies (Sommer 1984; Meunier et al. 2017). In all, trade-offs between nutrient acquisition capability, growth rate, homeostasis regulation, and cell size imply that P-rich conditions select for faster-growing taxa with lower N:P composition, while P-limited conditions select for slower-growing taxa, but have higher P acquisition capability and higher N:P ratios (Meunier et al. 2017, Fig. 18.8). Picoplankton, with their limited storage capability, may thrive under P-limited conditions and should have a high N:P biomass composition. It also implies that organisms are continually balancing the trade-off between homeostasis and maximal growth rate, and should a nutrient perturbation occur, homeostatic strength should decrease with variability in resource supply.

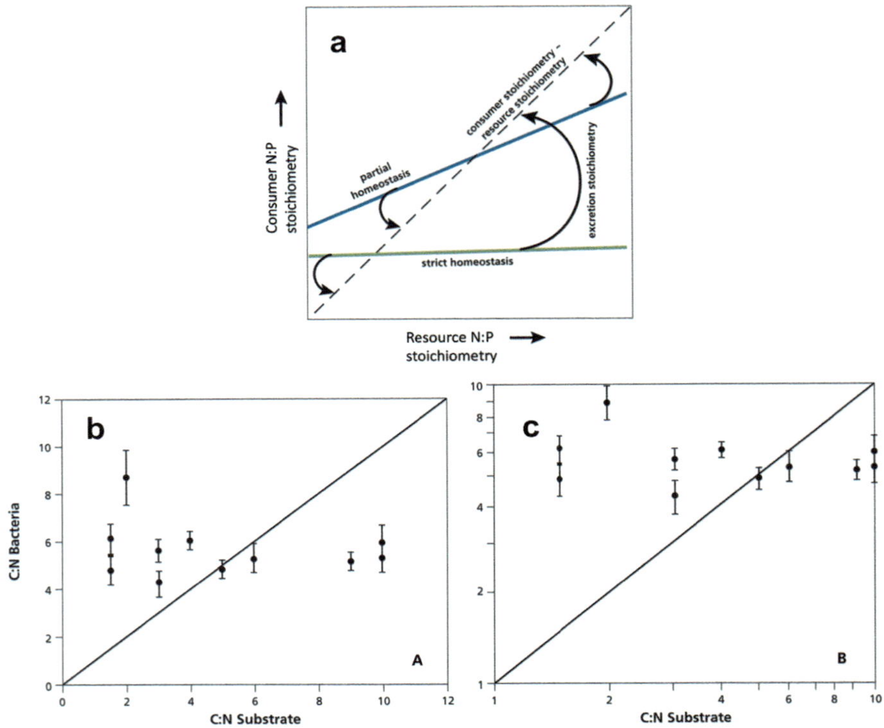

Fig. 18.12 Panel (**a**) Schematic relationship between resource (i.e., food) N:P and consumer N:P stoichiometry. In general, heterotrophs/consumers either partially or strictly regulate their stoichiometry regardless of the stoichiometry of that of resource. Image redrawn from Glibert et al. (2011). Panel (**b**, **c**) Example of bacterial stoichiometric homeostasis as a function of the C:N of the growth substrate, showing strict homeostasis. Data based on Goldman et al. (1987) plotted on a linear Panel (**b**) and on a log scale Panel (**c**). Images (**b**, **c**) redrawn from Sterner and Elser (2002, Ecological stoichiometry: the biology of the elements from the molecules to the biosphere)

18.4.4 Heterotroph Stoichiometry

Unlike phytoplankton, which have considerable plasticity in their C:N:P stoichiometry, heterotrophs—from bacteria to fish—are generally more constrained in stoichiometry, and there may be a mismatch between the stoichiometry of grazers and their food (Goldman et al. 1987; Sterner and Elser 2002; McIntyre and Flecker 2010; Fig. 18.12). This holds true for higher trophic levels even when consuming foods with large differences in C:N:P (e.g., Andersen and Hessen 1991; Sterner and Hessen 1994).

Stoichiometry of grazers affects various life stages differently (Moe et al. 2005, p. 31): "[a]n organism's requirements for different elements may vary throughout its life cycle, and thus certain life stages may be more sensitive than others to variation in the stoichiometry of its resource." As a specific example, there is a greater need for C, N, and P for developing copepod juveniles, but at a later stage, while C is still needed for metabolism, more P must be allocated to eggs; thus, P-poor food can disproportionately affect egg production while not affecting survival (Færøvig and Hessen 2003; Laspoumaderes et al. 2010).

Homeostasis comes at a cost, typically as a reduction in the rates of growth or reproduction (Boersma 2000; Boersma and Kreutzer 2002; Boersma et al. 2008). For example, an animal may respond to the nutritional demand for one nutrient (e.g., C) and in so doing obtain an excess of another nutrient (e.g., P) requiring an investment of energy to release this excess. This may impair many metabolic functions with a consequence of reduced growth (Boersma and Elser 2006; Peace et al. 2014, see Chap. 20).

18.4.5 Resource Ratio Control of Community Composition

The net effect of the stoichiometric regulation of consumers, together with different nutrient requirements of phytoplankton, is that not only are grazers affected by food quality, but they, in turn, affect food quality by altering the composition of nutrients available to them, and this process of stoichiometric interaction is seen at all levels of the food web, leading to a complex regulation of stoichiometry and species competition. Theory predicts that systems that undergo a shift in the availability of nutrient resources (bottom-up control) should experience shifts in dominant species at all levels of the food web such that consumers that successfully sequester the nutrient in least supply relative to their needs should dominate. This implies that an inverse relationship will be seen between the stoichiometry of dissolved nutrients and that available in food, or between the nutrients in prey and the one in biomass of grazers higher in the food web. Thus, interactions between alterations in food web structure cause the grazing communities to recycle nutrients differently (Sterner 1990; Sterner et al. 1992).

Grazers may (1) reflect the stoichiometry of their prey (within reasonable limits), (2) be somewhat more restrictive in their stoichiometry than their prey, or (3) be highly constrained in their stoichiometry (Sterner and Elser 2002). Although phytoplankton do have metabolic dissipative pathways to regulate energy and nutrients in excess of the nutritional demands (e.g., xanthophyll cycling, Mehler activity, dissimilatory NO_3^- reduction; see Chaps. 6 and 12), grazers have especially well-developed pathways of release and excretion to eliminate nutrients acquired in excess (Sterner and Elser 2002). Grazers are thus affected by food quality, and they, in turn, affect food quality by altering the composition of nutrients available to them through multiple processes. Grazers can sustain a mismatch in the stoichiometry of their food because they can retain what they need and release what they do not need. For example, in principle, grazers that are stoichiometrically balanced with strict stoichiometry feeding on phytoplankton that are N-rich will excrete proportionately more N than those grazing on phytoplankton that are more balanced in their N:P or N:C ratio (Fig. 18.13). By excreting more N, a condition of excess N is, in turn, maintained or even amplified for the phytoplankton (Sterner and Elser 2002). Consumer-driven changes in stoichiometry can thus maintain, and may even amplify, nutrient changes for producers (e.g., Schindler and Eby 1997; Loladze

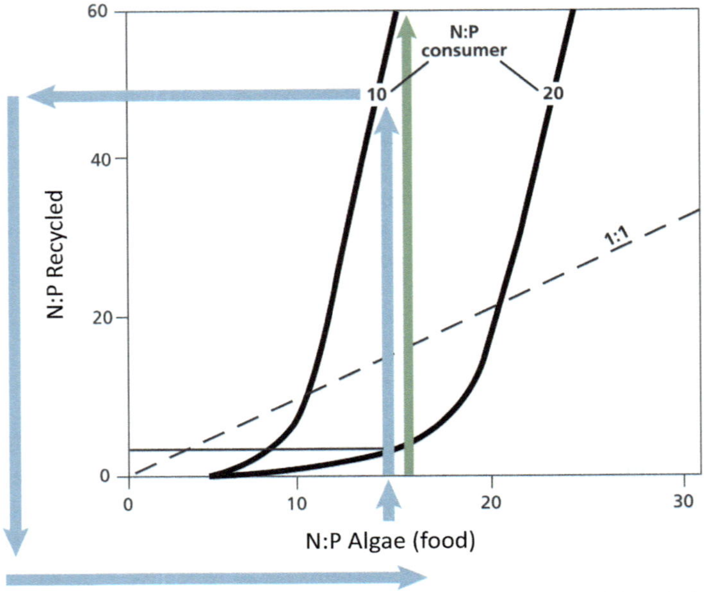

Fig. 18.13 Illustration of how N:P of food impacts N:P of recycled/excreted products, which, in turn, further alters the N:P of food, assuming weak homeostasis of algae. Consider algae initially at an N:P ratio comparable to the Redfield ratio (short blue arrow on the x-axis). In turn, a consumer with a lower N:P will be favored. If the N:P of the consumer is 10, then the excreted products will have an N:P > 40, which then will lead to a higher N:P for algae to consume, and thus, as the cycle of consumption and recycling progresses, the algae become enriched in N:P (green arrow) and so do the products. Image redrawn and modified from Sterner (1990, Am Nat 136:209–229)

et al. 2000). Imbalances in the N:P of excretion products present further cellular challenges for producers to regulate uptake of both the limiting and saturating nutrient cellular levels. Ecological stoichiometry principles would predict that the dominant predator, if its biomass N:P ratio is tightly constrained, can maintain a biomass-to-nutrient ratio that is inversely related to the nutrient ratio of its food, and homeostasis from nutrient recycling will drive the nutrient balance of the system to be self-sustaining (Fig. 18.13).

An example of the differences in elemental composition of the consumer community relative to the availability of the limiting nutrients for autotrophs is that of the differences in the common marine copepod *Eurytemora*, and the common freshwater grazer, *Daphnia*. Under the common N-limiting conditions of marine waters, copepods dominate; they have high body N:P and recycle proportionately more P. In contrast, in freshwaters where P is more likely to be limiting, *Daphia* is the more common grazer, but it has a low body N:P and recycles proportionately more N. In turn, in marine waters, this provides a source of P for phytoplankton, while in freshwaters, this recycling pathway provides a source of N (Cease and Elser 2013; Fig. 18.14).

It must be noted, however, that stoichiometric regulation of trophic interactions can be modulated when "good food goes bad" (sensu Mitra and Flynn 2005). Many

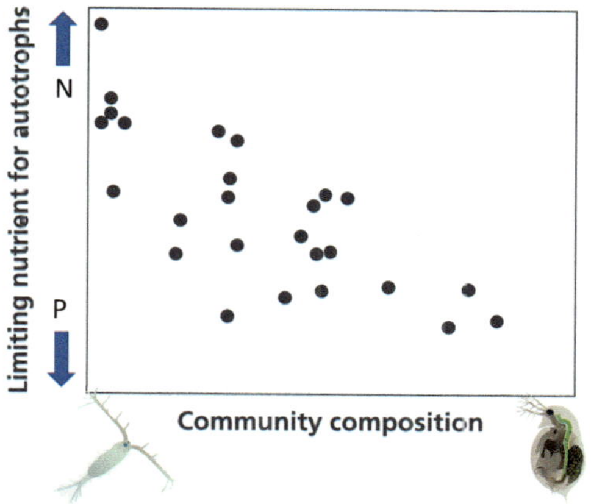

Fig. 18.14 Example of the effect of the differences in elemental composition of the consumer community on the availability of nutrients for autotrophs. Under N-limited conditions, where copepods typically dominate (e.g., marine environments), the grazers have a high body N content and recycle proportionately more P, giving the phytoplankton access to that nutrient. In contrast, under P-limited conditions, which are more common in freshwaters where *Daphna* are more likely to be the consumers, they have a high body P content and recycle proportionately more N, giving the phytoplankton greater access to that nutrient. Image modified and redrawn from Cease and Elser (2013, Nat Educ Knowl 4(5):3). Icons of the copepod and of *Daphnia* are from the University of Maryland Center for Environmental Science Integration and Application Network symbol library

attributes of food can be altered chemically and physiologically, leading to trophic interactions that would not be anticipated strictly on the basis of elemental stoichiometry. For example, the production of allelopathic compounds or toxins, or even physical aggregation of cells, can alter trophic transfer.

18.4.6 The Resource Ratio Theory

Based on Tilman (1982), every population has an intrinsic characteristic with respect to any two essential resources known as its optimum [consumption] proportion (OP), which can be graphically represented as its optimum consumption line (OCL). The optimal nutrient of a species is that internal (cellular) concentration of nutrients just meets physiological growth requirements. As growth rates are a function of resource levels, the higher the resource, the higher the growth rate, which, if mortality is assumed to be negligible, results in higher biomass (Fig. 18.15a). However, as the organism grows, it consumes resources, ultimately reducing its growth rate and growth will be held at the minimum level of resources to maintain growth (Fig. 18.15b). With two species competing, the species growing at the lowest resource level should drive the other species out of the system (Fig. 18.15c).

Now, superimpose multiple resources that are both variables. Both resources can define the growth rate response, but they may do so differently. Different cellular N:

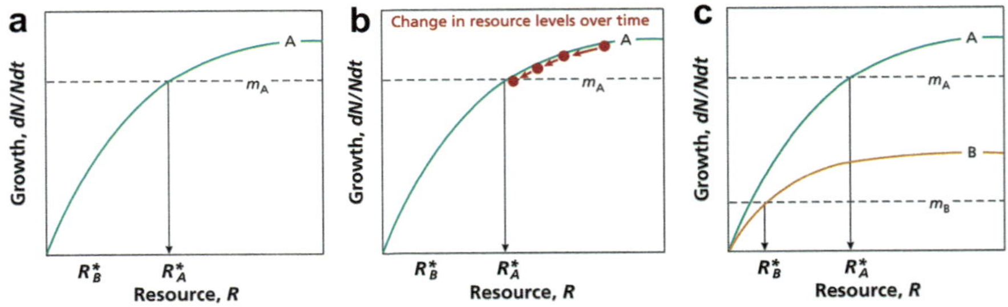

Fig. 18.15 Panel (**a**) Growth rate as a function of resource availability. The mortality rate (m) is independent of the resource levels). The value R* is the resource level at which growth and mortality are in balance. Panel (**b**) As resources are drawn down (due to organismal uptake), the relationship between growth and mortality may change; Panel (**c**) When more than one species is present, each with their own relationship between growth and resource availability, and different respective values of R*, as resources are drawn down, the relative proportions of each species with respect to growth and mortality changes, leading to altered dominance. Images redrawn from Tilman (1982, Monogr Pop Biol. 17. doi: 10.1515/9780691209654)

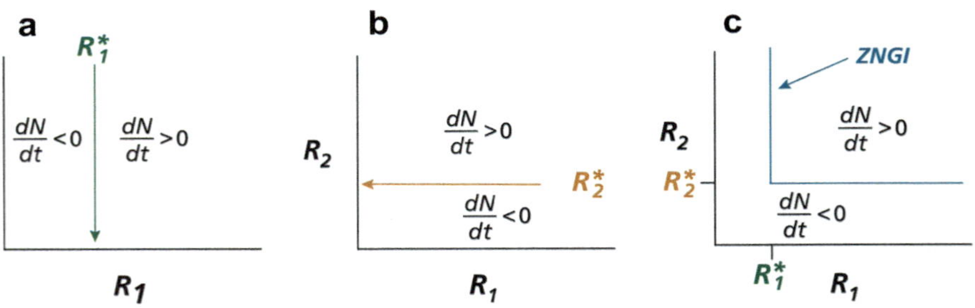

Fig. 18.16 Panel (**a, b**) Two different resources and their effects on competitive outcomes of two species. In both cases, the population declines in size with lower resources and increases with higher resource levels. Panel (**c**) When both resources are considered together, a line is defined as the Zero Net Growth Isocline, above which growth is balanced by mortality. Images redrawn from Tilman (1982, Monogr Pop Biol. 17. doi: 10.1515/9780691209654)

P ratios can develop depending on the different resource supply ratios, the degree of cellular regulation, and the taxa under consideration. Each resource divides the ecosystem into those species that succeed and those that do not (Fig. 18.16a, b). When considered together, the line that defines the combined effects of both R* values for the different resources is known as the Zero Net Growth Isocline, or the resource conditions that result in zero net growth because growth becomes balanced by mortality (Fig. 18.16c).

By applying the Droop equation (see Chap. 3), R^* can be determined as

$$R^* = [m\mu_m{}'Q_{\min}\mathrm{Kv}]/[(\mu_m{}'-m)V_{\max}-m\mu_m{}'Q_{\min}] \tag{18.3}$$

in which m is a mortality rate imposed on the population, Kv is the half-saturation constant for uptake, $V_{\max}$ is the maximum uptake rate, $Q_{\min}$ is the minimum cell quota, and $\mu_m{}'$ is the maximum growth rate at infinite Q (Grover 1989; Sterner and

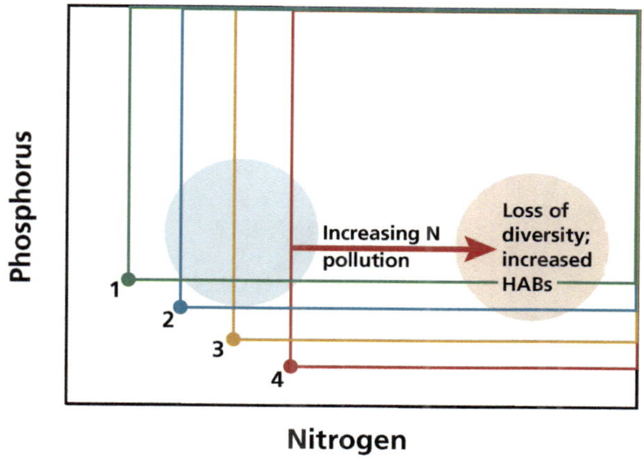

Fig. 18.17 Conceptual relationship between taxa success as nutrient proportions change. Species 1–4 each have a different optimal N/P ratio (the intersection of the right-angled lines). If the blue circular region represents historical levels of N and P supply, increasing N pollution (red arrow) might move the set of supply points to the red circular region. Under the new condition of decreased N limitation, only species 4 would persist because it is the best competitor for limiting P. Image redrawn and modified from Cleland and Harpole (2010, Ann N Y Acad Sci 1195:46–61. doi: 10.1111/j.1749-6632.2010.05458.x)

Elser 2002). Thus, a decreased cell quota reduces R^* in two ways, as it reduces the numerator and increases the denominator of the equation. As different functional groups and even species have different needs for N and P, for example, as the balance of these nutrients changes in the environment, the success of different species should lead to new dominants (Fig. 18.17). This concept, formalized as the *Resource Ratio Theory* (Tilman 1977, 1982, 1985; Smayda 1990, 1997), predicts that as the ratios of different essential elements change, the community structure may change due to competition between algae with different optimal nutrient ratios.

18.5 Testing the Concepts: Global Relationships Between Diversity and Traits

It is generally accepted that species richness and diversity increase with productivity, but only to a point. When blooms become large, they are more likely to be comprised of minimal diversity (e.g., monospecific) due to the fact they either escape grazing due to unpalatability, sheer biomass, or mismatches in timing. Functional trade-offs are important in the community composition across environmental gradients. Vallina et al. (2014) developed a global model of community composition and diversity and showed that as productivity increases, there is a dominance of slow-growing nutrient specialists, which are under selective grazing control, while the negative slope at high productivity is a function of fast-growing nutrient opportunists and competitive exclusion (Fig. 18.18).

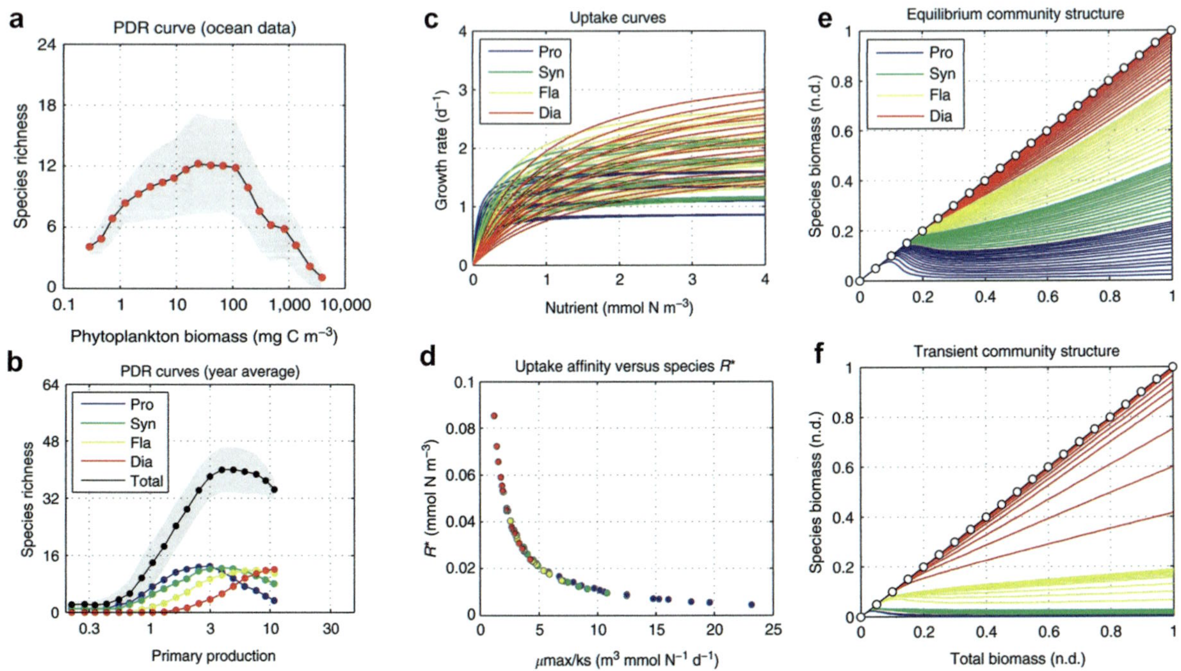

Fig. 18.18 Panel (**a**) Global productivity-diversity relationship (PDR) curve using equally spaced log10 bins of biomass. Units: phytoplankton biomass (mg C m^{-3}) and diversity (# species) (contributing >1% to total biomass). Panel (**b**) Species richness as a function of primary production using annually averaged data based on global maps of plankton diversity. Units: primary production (mmol C m^{-3} day^{-1}); phytoplankton diversity (# species) (contributing >1% total biomass). Color legend: *Prochlorococcus* (blue line), *Synechococcus* (green line), flagellates (yellow line), and diatoms (red line). Note the transient blooms of fast-growing taxa. Panel (**c**) Species nutrient uptake curves as a function of dissolved inorganic nitrogen (μmol N m^{-3}). Panel (**d**) Species subsistence nutrient concentration at equilibrium (R^*) versus their nutrient uptake affinity (*μ*max/ks). Panel (**e**) Steady-state equilibrium of species biomass as a function of total biomass; Panel (**f**) transient nonequilibrium. Species biomasses are cumulative on the *y*-axis. Units: phytoplankton (normalized) biomass no dimensions (n.d.). Images reproduced from Vallina et al. (2014, Nat Comm 5: 4299. doi.org/10.1038/ncomms5299) under a creative commons license

18.6 Summary

Phytoplankton metabolism, while not strictly following the three-fourths power rule, cell size and elemental stoichiometry, often follow predictably in response to abiotic conditions. The key traits of cell size and stoichiometry are intricately linked. Both are intricately linked to biodiversity. Stoichiometry can either diminish the chances of success of key species or be a critical aspect contributing to large shifts in primary producer species and major shifts in ecosystem nutrient cycling. An ecological stoichiometric approach is based on the transfer of elements, C, N, and P, as well as other elements through the food web, rather than on the rate of production of organic C. Improving the empirical basis for modeling variable stoichiometry and its effects through the food web requires recognition that phytoplankton stoichiometry is not a fixed proportion and that it varies as a function of both nutrient limitation and nutrient saturation. In ecological stoichiometric terms, *r* strategists generally have a low N:P ratio and have a higher proportion of growth machinery (ribosomal RNA; high P), whereas *K* strategists have a higher N:P ratio and have more resource acquisition machinery (enzymes and proteins; high N; Elser et al. 2000; Klausmeier et al. 2004; Arrigo 2005; Elser 2006; Finkel et al. 2010). Organisms that either feed on or preferentially sequester P from P-sufficient diets will have a higher P body content and can themselves sustain higher growth rates (more RNA), or may have more P-containing structural components (Elser et al. 1996; Sterner and Elser 2002). The extent to which primary producers vary in their nutritional content—across the entire spectrum of substrate availability—is considered to be a major food quality—and food web structure regulator. Understanding these relationships has important consequences for understanding how phytoplankton growth, size, and community composition may change in a future world.

Biodiversity should be a consequence of stoichiometry, and populations should self-stabilize as a result of stoichiometric constraints. In particular, Elser et al. (2010) articulated that they are closely linked on three levels.

- At the subcellular level, the fundamental "machines" of energy and material capture and transformation (e.g., chloroplasts, mitochondria, and ribosomes) generally have characteristic chemical composition.
- At the level of whole organisms, the relationships among size, composition, and function probably depend on the relative allocation to structural vs. metabolically active tissues and their respective chemical compositions. Small cells are constrained in their structure and metabolic allocation.
- At the cellular, organismal, and even the ecosystem level, material and energy perspectives are complementary in that the uptake and transformation of materials and energy both require metabolic energy, together with material substrates and the molecular machinery of life.

Size matters. Ratios matter. Traits matter.

References

Acevedo-Trejos E, Brandt G, Bruggeman J, Merico A (2015) Mechanisms shaping size structure functional diversity of phytoplankton communities in the ocean. Sci Rep 5:8918. https://doi.org/10.1038/srep08918

Ågren GI (2004) The C:N:P stoichiometry of autotrophs—theory and observations. Ecol Lett 7:185–191

Allen AP, Gillooly J (2009) Towards an integration of ecological stoichiometry and the metabolic theory of ecology to better understand nutrient cycling. Ecol Lett 12:369–384

Andersen T, Hessen DO (1991) Carbon, nitrogen and phosphorus content in freshwater zooplankton. Limnol Oceanogr 36:807–814. https://doi.org/10.4319/lo.1991.36.4.0807

Anderson KH, Asknes DL, Berge T, Fiksen Ø, Visser A (2015) Modelling emergent trophic strategies in plankton. J Plankton Res 37:862. https://doi.org/10.1093/plankt/fbv054

Arrigo KR (2005) Marine microorganisms and global nutrient cycles. Nature 437:349–355. https://doi.org/10.1038/nature04159

Boersma M (2000) The nutritional quality of P-limited algae for *Daphnia*. Limnol Oceanogr 45:1157–1161

Boersma M, Elser JJ (2006) Too much of a good thing: on stoichiometrically balanced diets and maximal growth. Ecology 87:1325–1330

Boersma M, Kruetzer C (2002) Life at the edge: is food quality really of minor importance at low quantities? Ecology 83:2552–2561

Boersma M, Aberle N, Hantzsche FM, Shoo KL, Wiltshire KH, Malzahn AM (2008) Nutritional limitation travels up the food chain. Internat Rev Hydrobiol 93:479–488

Bragg JG, Hyder CL (2004) Nitrogen versus carbon use in prokaryotic genomes and proteomes. Proc R Soc Lond B 271:S374–S377

Brown JH, Gillooly JF, Allen AP, Savage VM, West GB (2004) Toward a metabolic theory of ecology. Ecology 85:1771–1789. https://doi.org/10.1890/03-9000

Cardinale BJ, Bennett DM, Nelson CE, Gross K (2009) Does productivity drive diversity or vice versa? A test of the multivariate productivity-diversity hypothesis in streams. Ecology 90:1227–1241. https://doi.org/10.1890/08-1038.1

Cardinale BJ, Duffy JE, Gonzoles A, Hooper DU, Perrings C, Venail P, Narwani A et al (2012) Biodiversity loss and its impact on humanity. Nature 486:59–67. https://doi.org/10.1038/nature11148

Casanoves F, Pla L, Di Rienzo JA, Díaz S (2010) FDiversity: a software package for the integrated analysis of functional diversity. Methods Ecol Evol 2:233–237. https://doi.org/10.1111/j.2041-210X.2010.00082.x

Cease AJ, Elser JJ (2013) Biological Stoichiometry. Nat Educ Knowl 4(5):3

Chisholm SW (1992) Phytoplankton size. In: Falkowski PG, Woodhead AD (eds) Primary productivity and biogeochemical cycles in the sea. Plenum Press, New York, pp 213–237

Cleland EE, Harpole WS (2010) Nitrogen enrichment and plant communities. Ann N Y Acad Sci 1195:46–61. https://doi.org/10.1111/j.1749-6632.2010.05458.x

Cullen JT, Sherrell RM (2005) Effects of dissolved carbon dioxide, zinc, and manganese on the cadmium to phosphorus ratio in natural phytoplankton assemblages. Limnol Oceanogr 50:1193–1204

Dao MH (2013) Reassessment of the cell surface area limitation to nutrient uptake in phytoplankton. Mar Ecol Prog Ser 489:87–92. https://doi.org/10.3354/meps10434

DeLong JP, Okie JG, Moses ME, Sibly RM, Brown JH (2010) Shifts in metabolic scaling, production, and efficiency across major evolutionary transitions of life. Proc Natl Acad Sci USA 107:12941–12945

Dutkiewicz S, Cermeno P, Jahn O, Follows MJ, Hickman AE, Taniguchi DAA, Ward BA (2020) Dimensions of marine phytoplankton diversity. Biogeosciences 17:609–634. https://doi.org/10.5194/bg-17-609-2020

Elser JJ (2006) Biological stoichiometry: a chemical bridge between ecosystem ecology and evolutionary biology. Am Nat 168:S25–S35
Elser JJ, Dobberfuhl DR, Mackay NA, Schampel JH (1996) Organism size, life history, and N:P stoichiometry. Bioscience 46:674–684
Elser JJ, Sterner RW, Gorokhova E, Fagan WF, Markow TA, Cotner JB, Harrison JF et al (2000) Biological stoichiometry from genes to ecosystems. Ecol Lett 3:540–550. https://doi.org/10.1111/j.1461-0248.2000.00185.x
Elser JJ, Acharya K, Kyle M, Cotner J, Makino W, Markow T, Watts T et al (2003) Growth rate-stoichiometry couplings in diverse biota. Ecol Lett 6:936–943. https://doi.org/10.1046/J.1461-0248.2003.00518.X
Elser JJ, Fagan WF, Kerkoff AJ, Swenson NG, Enquist BJ (2010) Biological stoichiometry of plant production: metabolism, scaling and ecological response to global change. New Phytol 186:593–608. https://doi.org/10.1111/j.1469-8137.2010.03214.x
Færøvig PJ, Hessen DO (2003) Allocation strategies in crustacean stoichiometry: the potential role of phosphorus in the limitation of reproduction. Fresh Biol 48:1782–1792
Falkowski PG (2000) Rationalizing elemental ratios in unicellular algae. J Phycol 36:3–6
Fernandez-González C, Marañón E (2021) Effect of temperature on the unimodal size scaling of phytoplankton growth. Sci Rep 11:953. https://doi.org/10.1038/s41598-020-79616-0
Filstrup CT, Hillebrand H, Heathcote AJ, Harpole WS, Downing JA (2014) Cyanobacteria dominance influences resource use efficiency and community turnover in phytoplankton and zooplankton communities. Ecol Lett 17:464–474. https://doi.org/10.1111/ele.12246
Finkel ZV, Irwin AJ, Schofield OME (2004) Resource limitation alters the ¾ size scaling of metabolic rates in phytoplankton. Mar Ecol Prog Ser 273:269–279
Finkel ZV, Katz M, Wright J, Schofield OME, Falkowski PG (2005) Climatically driven macroevolutionary patterns in the size of marine diatoms over the Cenozoic. Proc Natl Acad Sci USA 102:8927–8932
Finkel ZV, Quigg AS, Raven JA, Reinfelder JR, Schofield OME, Falkowski PG (2006) Irradiance and the elemental stoichiometry of marine phytoplankton. Limnol Oceanogr 51:2690–2701
Finkel ZV, Sebbo J, Feist-Burkhardt S, Irwin AJ, Katz ME, Schofield OME, Young JR, Falkowski PG (2007) A universal driver of macroevolutionary change in the size of marine phytoplankton over the Cenozoic. Proc Natl Acad Sci USA 104(20416):20420
Finkel ZV, Beardall J, Flynn KJ, Quigg A, Rees TAV, Raven JA (2010) Phytoplankton in a changing world: cell size and elemental stoichiometry. J Plankton Res 32:119–137. https://doi.org/10.1093/plankt/fbp098
Flynn KJ (2002) How critical is the critical N:P ratio? J Phycol 38:961–970
Geider RJ, LaRoche J (2002) Redfield revisited: Variability of C:N:P in marine microalgae and its biochemical basis. Eur J Phycol 37:1–17
Gillooly JF, Brown JH, West GB, Savage VM, Charnov EL (2001) Effects of size and temperature on metabolic rate. Science 293:2248–2251. https://doi.org/10.1126/science.1061967
Glibert PM (1998) Interactions of top-down and bottom-up control in planktonic nitrogen cycling. Hydrobiologia 363:1–12
Glibert PM (2012) Ecological stoichiometry and its implications for aquatic ecosystem sustainability. Curr Opin Environ Sustain 4:272–277
Glibert PM, Goldman JC (1981) Rapid ammonium uptake by marine phytoplankton. Mar Biol Lett 2:25–31
Glibert PM, Heil CA, O'Neil JM, Dennison WC, O'Donohue MJH (2006) Nitrogen, phosphorus, silica and carbon in Moreton Bay, Queensland, Australia: differential limitation of phytoplankton biomass and production. Est Coasts 29:107–119
Glibert PM, Fullerton D, Burkholder JM, Cornwell JC, Kana TM (2011) Ecological stoichiometry, biogeochemical cycling, invasive species and aquatic food webs: San Francisco estuary and comparative systems. Rev Fish Sci 19:358–417
Glibert PM, Kana TM, Brown K (2013) From limitation to excess: consequences of substrate excess and stoichiometry for phytoplankton physiology, trophodynamics and biogeochemistry,

and implications for modeling. J Mar Syst 125:14–28. https://doi.org/10.1016/j.jmarsys.2012.10.004

Glibert PM, Wilkerson FP, Dugdale RC, Raven JA, Dupont C, Leavitt PR, Parker AE et al (2016) Pluses and minuses of ammonium and nitrate uptake and assimilation by phytoplankton and implications for productivity and community composition, with emphasis on nitrogen-enriched conditions. Limnol Oceanogr 61:165–197. https://doi.org/10.1002/lno.102033

Goldman JC, Glibert PM (1983) Kinetics of inorganic nitrogen uptake. In: Carpenter EJ, Capone DG (eds) Nitrogen in the marine environment. Academic Press, New York, pp 233–274

Goldman JC, McCarthy JJ, Peavey DG (1979) Growth rate influence on the chemical composition of phytoplankton in oceanic water. Nature 279:210–215

Goldman JC, Caron DA, Dennett MR (1987) Regulation of gross growth efficiency and ammonium regeneration in bacteria by substrate C:N ratio. Limnol Oceanogr 32:1239–1252

Grover JP (1989) Influence of cell shape and size on algal competitive ability. J Phycol 25:402–405

Hecky RE, Kilham P (1974) Environmental control of phytoplankton cell size. Limnol Oceanogr 19:361–366. https://doi.org/10.4319/lo.1974.19.2.0361

Hillebrand H, Steinert G, Boersma M, Malzahn A, Meunier C, Plum C, Ptacnik R (2013) Goldman revisited: faster-growing phytoplankton has lower N:P and lower stoichiometric flexibility. Limnol Oceanogr 58:2076–2088. https://doi.org/10.4319/lo.2013.58.6.2076

Hillebrand H, Cowles JM, Lewandowska A, Van de Waal DB, Plum C (2014) Think ratio! A stoichiometric view on biodiversity–ecosystem functioning research. Basic Appl Ecol 15:465–474. https://doi.org/10.1016/j.baae.2014.06.003

Hillebrand H, Acevedo-Trejos E, Moorthi SD, Ryabov A, Stiebel M, Thomas PK, Schneider M-L (2021) Cell size as driver and sentinel of phytoplankton community structure and functioning. Funct Ecol 36:276–293. https://doi.org/10.1111/1365-2435.13986

Hirst AG, Glazier DS, Atkinson D (2014) Body shape shifting during growth permits tests that distinguish between competing geometric theories of metabolic scaling. Ecol Lett 17:1274–1281. https://doi.org/10.1111/ele.12334

Ho T-Y, Quigg A, Finkel ZV, Mulligan A, Wyman K, Falkowski PG, Morel FMM (2003) Elemental composition of some marine phytoplankton. J Phycol 39:1145–1159

Isanta-Navarro J, Prater C, Peoples LM, Loladze I, Phan T, Jeyasingh PD, Church MJ et al (2022) Revisiting the growth rate hypothesis: towards a holistic stoichiometric understanding of growth. Ecol Lett 25:2324–2339. https://doi.org/10.1111/ele.14096

Kagaami M, Urabe J (2001) Phytoplankton growth rate as a function of cell size: an experimental test in Lake Biwa. Limnology 2:111–117

Kana TM, Glibert PM (2016) On saturating response curves from the dual perspectives of photosynthesis and nitrogen metabolism. In: Glibert PM, Kana TM (eds) Aquatic microbial ecology and biogeochemistry: a dual perspective. Springer, Cham, pp 93–104. https://doi.org/10.1007/978-3-319-30259-1_8

Karp-Boss L, Boss E (2016) The elongated, the squat and the spherical: selective pressures for phytoplankton shape. In: Glibert PM, Kana TM (eds) Aquatic microbial ecology and biogeochemistry: a dual perspective. Springer, Cham, pp 25–34. https://doi.org/10.1007/978-3-319-30259-1_3

Kleiber M (1961) The fire of life: an introduction to animal energetics. Wiley, New York, p 454

Kiørboe T (1993) Turbulence, phytoplankton cell size, and the structure of pelagic food webs. Adv Mar Biol 29:1–72

Klappenbach JA, Dunbar JM, Schmidt TM (2000) rRNA operon copy number reflects ecological strategies of bacteria. Appl Environ Microbiol 66:1328–1333. https://doi.org/10.1128/aem.66.4.1328-1333.2000

Klausmeier CA, Litchman E, Daufresne T, Levin SA (2004) Optimal N:P stoichiometry of phytoplankton. Nature 429:171–174. https://doi.org/10.1038/nature02454

Klausmeier CA, Litchman E, Daufresne T, Levin S (2008) Phytoplankton stoichiometry. Ecol Res 23:479–485. https://doi.org/10.1007/s11284-008-0470-8

Kleiber M (1932) Body size and metabolism. Hilgardia 6:315–353. https://doi.org/10.3733/hilg.v06n11p315

Kremer CT, Thomas MK, Litchman E (2017) Temperature- and size-scaling of phytoplankton population growth rates: reconciling the Eppley curve and the metabolic theory of ecology. Limnol Oceanogr 62:1658–1670. https://doi.org/10.1002/lno.10523

Laliberté E, Legendre P (2010) A distance-based framework for measuring functional diversity from multiple traits. Ecology 91:299–305. https://doi.org/10.1890/08-2244.1

Laspoumaderes C, Modenutti B, Balseiro E (2010) Herbivory versus omnivory: linking homeostasis and elemental imbalance in copepod development. J Plankton Res 32:1573–1582

Lauro FM, McDougald D, Thomas T, Williams TJ, Egan S, Rice S et al (2009) The genomic basis of trophic strategy in marine bacteria. Proc Natl Acad Sci USA 106:15527–15533. https://doi.org/10.1073/pnas.0903507106

Leonardos N, Geider RJ (2004a) Responses of elemental and biochemical composition of *Chaetoceros muelleri* to growth under varying light and nitrate:phosphate supply ratios and their influence on critical N:P. Limnol Oceanogr 49:2105–2114

Leonardos N, Geider RJ (2004b) Effects of nitrate:phosphate supply ratio and irradiance on the C:N:P stoichiometry of *Chaetoceros muelleri*. Eur J Phycol 39:173–180

Lindeman RL (1942) The trophic-dynamic aspect of ecology. Ecology 23:399–418

Litchman E, Klausmeier CA (2008) Trait-based community ecology of phytoplankton. Ann Rev Ecol Evol Syst 39:615–639. https://doi.org/10.1146/annurev.ecolsys.39.110707.173549

Litchman E, de Tezanos PP, Klausmeier CA, Thomas MK, Yoshiyama K (2010) Linking traits to species diversity and community structure in phytoplankton. Hydrobiologia 653:15–28. https://doi.org/10.1007/s10750-101-0341-5

Loladze I, Kuang Y, Elser JJ (2000) Stoichiometry inproducer-grazer systems: linking energy flow with element cycling. Bull Math Biol 62:1137–1116. https://doi.org/10.1007/s10750-007-0714-6

López-Sandoval DC, Rodríguez-Ramos T, Cermeño P, Sobrino C, Marañón E (2014) Photosynthesis and respiration in marine phytoplankton: relationship with cell size, taxonomic affiliation, and growth phase. J Exp Mar Biol Ecol 457:151–159

López-Urrutia Á, San Martin E, Harris RP, Irigoien X (2006) Scaling the metabolic balance of the oceans. Proc Natl Acad Sci USA 103:8739–8744. https://doi.org/10.1073/pnas.0601137103

Lotka AJ, Dublin L (1925) On the true rate of natural increase. J Am Stat Assoc 20:305–339

Malone TC (1980) Algal size. In: Morris I (ed) The physiological ecology of phytoplankton. University of California Press, Berkeley, CA, pp 433–464

Marañón E, Cermeño P, López-Sandoval R-RT, Sobrino C, Huete-Ortega M, Blanco JM et al (2013) Unimodal size scaling of phytoplankton growth and the size dependence of nutrient uptake and use. Ecol Lett 16:371–379

Marañón E, Cermeño P, Huete-Ortega M, López-Sandoval DC, Mouriño-Carballido B, Rodríguez-Ramos T (2014) Resource supply overrides temperature as a controlling factor of marine phytoplankton growth. PLoS One 9(6):e99312. https://doi.org/10.1371/journal.pone.0099312

Martini S, Larras F, Boyé A, Faure E, Aberle N, Archambault P, Bacouillard L et al (2021) Functional trait-bases approaches as a common framework for aquatic ecologists. Limnol Oceanogr 66:965–994. https://doi.org/10.1002/lno.11655

McIntyre PB, Flecker A (2010) Ecological stoichiometry as an integrative framework in stream fish ecology. Amer Fish Soc Symp 73:539–558

Meunier CL, Malzahn AM, Boersma M (2014) A new approach to homeostatic regulation: towards a unified view of physiological and ecological concepts. PLoS One 9:e107737. https://doi.org/10.1371/journal.pone.0107737

Meunier CL, Boersma M, El-Sabaawi R, Halvorson HM, Herstoff EM, Van de Waal DB, Vogt RJ, Litchman E (2017) From elements to function: toward unifying ecological stoichiometry and trait-based ecology. Front Environ Sci 5:18. https://doi.org/10.3389/fenvs.2017.00018

Mitra A, Flynn KJ (2005) Predator-prey interactions: is "ecological stoichiometry" sufficient when good food goes bad? J Plankton Res 27:393–399

Moe SJ, Stelzer RS, Forman MR, Harpole WS, Daufresne T, Yoshida T (2005) Recent advances in ecological stoichiometry: insights for population and community ecology. Oikos 109:29–39
Munk WH, Riley GA (1952) Absorption of nutrients by aquatic plants. J Mar Res 11:215–240
Naselli-Flores L, Barone R (2011) Fight on plankton! Or, phytoplankton shape and size as adaptive tools to get ahead in the struggle for life. Crytogamie Algologie 32:157–204. https://doi.org/10.7872/crya.v32.iss2.2011.157
Naselli-Flores L, Padisák AM (2007) Shape and size in phytoplankton ecology: do they matter? Hydrobiologia 578:157–161. https://doi.org/10.1007/s10750-006-2815-z
Nemergut DR, Knelman JE, Ferrenberg S, Bilinski T, Melbourne B, Jiang L et al (2016) Decreases in average bacterial community rRNA operon copy number during succession. ISME J 10: 1147–1156. https://doi.org/10.1038/ismej.2015.191
Okie JG (2013) General models for the spectra of surface area scaling strategies of cells and organisms: fractality, geometric dissimilitude, and internalization. Am Nat 181:421–439. https://doi.org/10.1086/669150
Peace A, Wang H, Kuang Y (2014) Dynamics of a producer-grazer model incorporating the effects of excess food nutrient content on grazer's growth. Bull Math Biol 76:2175–2197
Peters RH (1983) The ecological implications of body size. Cambridge Univ Press, Cambridge. https://doi.org/10.1017/CBO9780511608551
Platt T, Silvert W (1981) Ecology, physiology, allometry and dimensionality. J Theor Biol 93:885–860
Ptacnik R, Solimini AG, Andersen T, Tamminen T, Brettun P, Lepistö WE et al (2008) Diversity predicts stability and resource use efficiency in natural phytoplankton communities. Proc Natl Acad Sci USA 105:5134–5138. https://doi.org/10.1073/pnas.0708328105.2008
Quigg A, Finkel ZV, Irwin AJ, Rosenthal Y, Ho T-Y, Reinfelder JR, Schofield O, Morel FMM, Falkowski PG (2003) The evolutionary inheritance of elemental stoichiometry in marine phytoplankton. Nature 425:291–294
Raimbault P, Rodier M, Taupier-Letage I (1988) Size fraction of phytoplankton in the Ligurian Sea and the Algerian Basin (Mediterranean Sea): size distribution versus total concentration. Mar Microb Food Webs 3:1–7
Raven JA (1998) The twelfth Tansley lecture. Small is beautiful: the picophytoplankton. Funct Ecol 12:503–513. https://doi.org/10.1046/j.1365-2435.1998.00233.x
Redfield AC (1934) On the proportions of organic derivatives in sea water and their relation to the composition of plankton. In: James Johnstone memorial volume. University of Liverpool Press, Liverpool, pp 176–192
Redfield AC (1958) The biological control of chemical factors in the environment. Am Sci 46:205–221
Reu B, Proulx R, Bohn K, Dyke JG, Kleidon A, Pavlick R, Schmidtlein S (2011) The role of climate and plant functional trade-offs in shaping global biome and biodiversity patterns. Glob Ecol Biogeogr 20:570–581. https://doi.org/10.1111/j.1466-8238.2010.00621.x
Reynolds CS (1998) What factors influence the species composition of phytoplankton in lakes of different trophic status? Hydrobiologia 369(370):11–26
Rhee G-Y (1978) Effects of N:P atomic ratios and nitrate limitation on algal growth, cell composition and nitrate uptake. Limnol Oceanogr 23:10–25
Roller BRK, Stoddard SF, Schmidt TM (2016) Exploiting rRNA operon copy number to investigate bacterial reproductive strategies. Nat Microbiol 1:16160. https://doi.org/10.1038/nmicrobiol.2016.160
Schindler DW, Eby LA (1997) Stoichiometry of fishes and their prey: implications for nutrient recycling. Ecology 78:1816–1831
Schmidt-Nielsen K (1984) Scaling: why is animal size so important? Cambridge University Press, Cambridge, 241 pp. https://doi.org/10.1017/CBO9781139167826
Sheldon RW, Parsons TR (1967) A continuous size spectrum for particulate matter in the sea. J Fish Res Bd Can 24:909–915

Smayda TJ (1990) Novel and nuisance phytoplankton blooms in the sea: evidence for a global epidemic. In: Granéli E, Sundstrom B, Edler L, Anderson DM (eds) Toxic marine phytoplankton. Elsevier, New York, pp 29–40

Smayda TJ (1997) Harmful algal blooms: their ecophysiological and general relevance to phytoplankton blooms in the sea. Limnol Oceanogr 42:1137–1153. https://doi.org/10.4319/lo.1997.42.5_part_2.1137

Sommer U (1984) The paradox of the plankton: fluctuations of phosphorus availability maintain diversity of phytoplankton in flow-through cultures. Limnol Oceanogr 29:633–636. https://doi.org/10.4319/lo.1984.29.0633

Sommer U, Charalampous E, Genitsaris S, Moustaka-Gouni M (2017) Benefits, costs and taxonomic distribution of marine phytoplankton body size. J Plankt Res 39:494–508. https://doi.org/10.1093/jplant/fbw071

Sterner RW (1990) N:P resupply by herbivores: zooplankton and the algal competitive arena. Am Nat 136:209–229

Sterner RW, Elser JJ (2002) Ecological stoichiometry: the biology of the elements from the molecules to the biosphere. Princeton University Press, Princeton, p 464

Sterner RW, Hessen DO (1994) Algal nutrient limitation and the nutrition of aquatic herbivores. Ann Rev Ecol Syst 25:1–29. https://doi.org/10.1146/annurev.es.25.11019.000245

Sterner RW, Elser JJ, Hessen DO (1992) Stoichiometric relationships among producers, consumers and nutrient cycling in pelagic ecosystems. Biogeochem 17:49–67. https://doi.org/10.1007/BF00002759

Stoddard SF, Smith BJ, Hein R, Roller BRK, Schmidt TM (2014) *rrnDB* improved tools for interpreting rRNA gene abundance in bacteria and archaea and a new foundation for future development. Nucleic Acids Res 43:D593–D598. https://doi org/10.1093/nar/gku1201

Stolte W, McCollin T, Noordeloos AAM, Riegman R (1994) Effect of nitrogen source on the size distribution within marine phytoplankton populations. J Exp Mar Biol Ecol 184:83–97. https://doi.org/10.1016/0022-0981(94)90167-8

Terry KL, Laws EA, Burns DJ (1985) Growth rate variation in the N:P requirement of phytoplankton. J Phycol 21:323–329

Tilman D (1977) Resource competition between planktonic algae: an experimental and theoretical approach. Ecology 58:338–348

Tilman D (1982) Resource competition and community structure. Monogr. Pop. Biol. 17. Princeton University Press, Princeton, NJ, 296 p. https://doi.org/10.1515/9780691209654

Tilman D (1985) The resource-ratio hypothesis of plant succession. Am Nat 125:827–852

Vallina SM, Follows MJ, Dutkiewicz S, Montoya JM, Cermeno P, Loreau M (2014) Global relationship between phytoplankton diversity and productivity in the ocean. Nat Commun 5: 4299. https://doi.org/10.1038/ncomms5299

Villéger S, Mason NWH, Mouillot D (2008) New multidimensional functional diversity indices for a multifaceted framework in functional ecology. Ecology 89:2290–2301. https://doi.org/10.1890/07-1206.1

Violle C, Navas M-L, Vile D, Kazakou E, Fortunel C, Hummel I, Garnier E (2007) Let the concept of trait be functional! Oikos 116:882–892. https://doi.org/10.1111/j.0030-1299.2007.15559.x

Von den Sprengel C (1828) Substanzen der ackerkrume und des Uuntergrundes (About the substances in the plow layer and the subsoil). J für Technische und Ökonomische Chemie 2: 423–474

Von Liebig J (1855) Principles of agricultural chemistry with special reference to the late researches made in England. Reprinted In: Pomeroy LR (ed) Cycles of essential elements (benchmark papers in ecology), vol. 1. Dowden, Hutchinson and Ross Inc., Stroudsburg, PA, pp 11–28

Weider LJ, Elser JJ, Crease TJ, Mateos M, Cotner JB, Markow TA (2005) The functional significance of ribosomal (r)DNA variation: impacts on the evolutionary ecology of organisms. Ann Rev Ecol Syst 36:219–242. https://doi.org/10.2307/30033803

Ye L, Chang C-Y, García-Comas C, Gong G-C, Hsieh C-H (2013) Increasing zooplankton size diversity enhances the strength of top-down control on phytoplankton through diet niche partitioning. J Animal Ecol 82:1052–1061. https://doi.org/10.1111/1365-2656.12067
Ye L, Chang C-W, Matsuzaki S-IS, Takamura N, Widdicombe CE, Hsieh C-h (2019) Functional diversity promotes phytoplankton resource use efficiency. J Ecol 107:2353–2363. https://doi.org/10.1111/1365-2745.13192

Additional Websites Accessed

https://doi.org/10.5281/zenodo.3635898. Accessed March 23, 2023

19 Loss Processes: The Tiny Feeding the Large

19.1 Introduction

In the 1800s, Victor Hensen (Chap. 1) was convinced that phytoplankton were regularly distributed in the sea and even proclaimed that "Great evenness is ... irrefutably established by the [plankton] counts and transects" (Hensen 1892, quoted by Mills 1989, p. 27). Ernst Haeckel, the bombastic critic of Hensen, attacked this notion, based on fundamental principles and techniques. Haeckel won the day on that argument. These investigators who often did not see eye to eye also appear to have been on opposite sides of the emerging discussion in the late nineteenth century as to how food webs are nourished. Hensen also promoted the idea that had been introduced a century earlier by two Danish scientists, Hooker and Ørsted—and later their Norwegian colleague, Sars, that small ocean plants nourish the animals of the sea (Gran 1912; Barber and Hilting 2002). Two very different ecosystems led Hooker and Ørsted to reach the conclusions they did; the former studied diatom blooms in the Antarctic, while the latter studied the oligotrophic waters of the Sargasso Sea. In fact, Ørsted's idea that the tiny fed the large in a nutrient—and phytoplankton—impoverished sea was considered quite preposterous at the time. The rejection of the concept that oceanic phytoplankton supported the food web of the sea by other contemporary scientists led Ørsted to never again make reference to the idea (Wolff and Peterson 1991; Barber and Hilting 2002). Years later, Haeckel also thought "it was preposterous that to believe ...that tiny, invisible drifting plants could provide enough food to support vast schools of fish, birds and whales" (Barber and Hilting 2002, p. 22). Haeckel thought that the food supplied by rivers and benthic algae was far more important than the plankton. Here, Hensen was correct and Haeckel was not, but with one caveat. Hensen could not understand how diatoms could be digestible, and therefore, he thought they could not be a good food source.

H.H. Gran, also introduced in Chap. 1, who made major contributions to the study of the importance of nutrients in supporting primary production, strongly believed in the diatom-copepod food link as a dominant process. He suggested that divergent cyclonic oceanic systems, with water motion that supplies subsurface nutrients to the

P. M. Glibert, *Phytoplankton Whispering: An Introduction to the Physiology and Ecology of Microalgae*, https://doi.org/10.1007/978-3-031-53897-1_19

photic zone, are richer in production than convergent, anti-cyclonic gyres that dominate the tropics (Barber and Hilting 2002). Supporting Gran was his colleague Alexander Nathansohn, who recognized the balance of production and loss. He articulated the concept of *dynamic balance* and that "biomass results from the balance of production and loss processes" (Nathansohn 1908, see Chap. 1).

Box 1: John Ryther and Photosynthesis and Fish Production of the Sea
Ryther (1969) provided the first estimates of fish production in the sea based on phytoplankton production for three domains: open oceans, coastal oceans, and upwelling areas. He based his estimates on a number of comparatively simple assumptions. First, the larger the plant (phytoplankton) cells at the beginning of the food chain, the fewer the trophic levels necessary to convert it to a "useful form." He also assumed that in upwelling regions, food chains are the shortest, due not only to the larger size of the phytoplankton but also due to the fact that many form colonies of filaments, making them more readily eaten by fishes. Another assumption was that the metabolic costs of animals with respect to food availability worked in opposite ways; the less the food available, the more work the animal had to do, while the greater the quantity, the animal had to digest, assimilate, and store the food. The final calculations revealed that the open sea, representing 90% of the ocean, is essentially a biological desert; upwelling regions produce about half the world's fish supply and coastal regions the other half. This is in spite of the fact that the open ocean overall produces one to two orders of magnitude more total productivity due to differences in area.

Primary production and fish production in three ocean provinces. From Ryther (1969)

Province	Percent of the ocean	Area (km^2 × 10^6)	Total primary production (10^9 tons carbon/year	Trophic levels	Efficiency (%)	Fish production (tons (fresh wt.)
Oceanic	90	326	16.3	5	10	16 × 10^5
Coastal	9.9	36	3.6	3	15	12 × 10^7
Upwelling	0.1	0.36	0.1	1.5	20	12 × 10^7

Years later, John Ryther, an "intellectual descendent of Gran" (Barber and Hilting 2002, p. 38), applied many of Gran's ways of thinking to estimate fish production in the sea (Ryther 1969). Also, building on then-available measurements of primary production, Ryther examined a range of factors, from water color, temperature, biomass of higher trophic levels, and wind and seasonality to estimate regional fish production (Barber and Hilting 2002; Box 1). He simplified and partitioned the oceans into three major domains: oceanic, coastal, and upwelling, and he calculated the fish yields that primary production can support in each of these regions, making some assumptions of food web transfers and food web efficiency. Thus, he made the first estimate of fish production based on primary productivity, achieving the goal that Hensen had long reached for but was never able to attain (Barber and Hilting 2002).

This chapter examines the fate of phytoplankton in terms of zooplankton grazing. Grazers influence nutrient and energy transfer to higher trophic levels, they are key mediators of the biological pump (e.g., Ducklow et al. 2001), and they represent the major fate of primary production (Banse 2013; Steinberg and Landry 2017; Fig. 19.1). The following chapter examines food web structure in more detail, including mixotrophic food webs, along with the evolution of ideas and methodologies by which food webs and trophic position may be studied. Microbial interactions of the phycosphere, viral mortality, and programmed cell death are described in Chap. 21.

19.2 Phytoplankton Mortality

Phytoplankton mortality is far less understood than phytoplankton growth. Phytoplankton die for various reasons.

Reynolds (1984) formalized the inclusion of loss processes in an overall growth equation,

$$\delta N/\delta t = e^{(k' - \Sigma KL)} \tag{19.1}$$

where N is the biomass, k' is growth, and K_L is the sum of loss rates, which are in turn specified as,

$$\Sigma K_L = k_w + k_s + k_d + k_g \tag{19.2}$$

where k_w is hydraulic washout (perhaps equivalent to horizontal advection in a marine setting; recall that Reynolds focused mainly on freshwater ecosystems—Chap. 17, k_s is sedimentation, k_d is death, and k_g is grazing. The rates of k_g and k_s are measurable, and those of k_w are calculable, but those of k_d are very difficult to define, let alone measure (Franklin et al. 2006). This chapter focuses on k_g and k_s.

19.2.1 Loss Through Grazing

19.2.1.1 Zooplankton Feeding

Phytoplankton are grazed by many different consumers in the sea, from meso- and microzooplankton to filter feeders such as bivalves. Zooplankton such as copepods are generally responsible for about 10% of the loss of primary production (Calbet 2001) but grazing by herbivorous protists (microzooplankton) may remove as much as 65% of global primary production (Calbet and Landry 2004). The relative impact of macro- and micro-zooplankton on phytoplankton production is, of course,

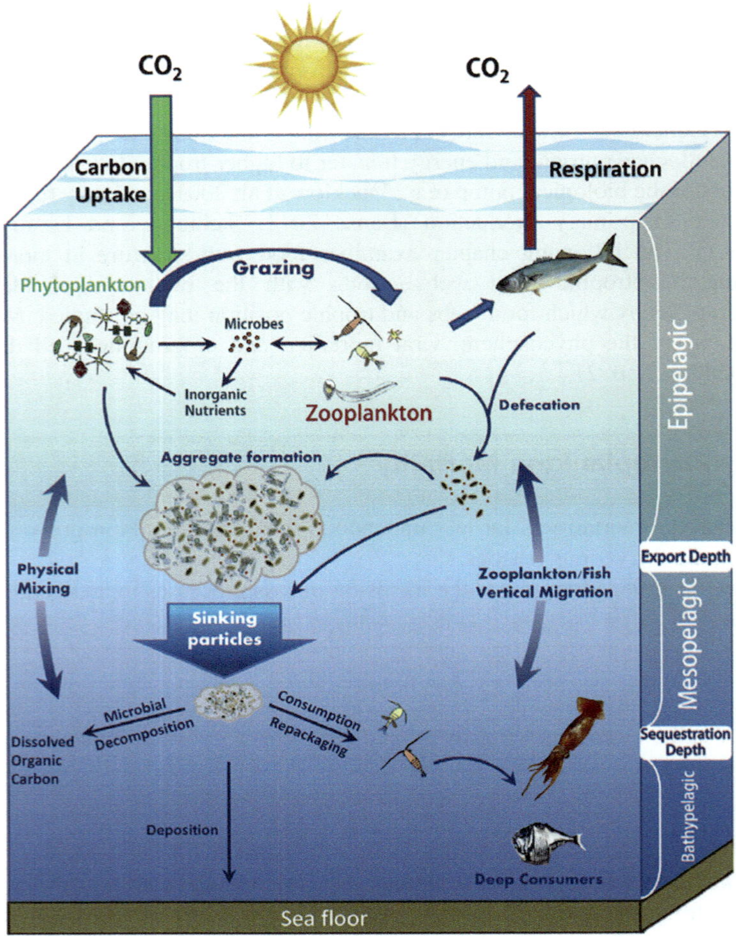

Fig. 19.1 Biological pump in the ocean. CO_2 is drawn into the ocean from the atmosphere, and dissolved CO_2 is processed into particulate organic carbon (POC) through photosynthesis by the phytoplankton in the euphotic layer of the epipelagic zone (0–200 m depth). Organic matter in epipelagic phytoplankton is processed by microbes, zooplankton, and their consumers into fecal pellets, organic aggregates ("marine snow"), and other forms of POC, which are exported to the mesopelagic and bathypelagic zones by sinking and vertical migration by zooplankton and fish. The export depth is sometimes identified as the bottom of the euphotic zone, or as the maximum mixed layer depth, both of which are variable, but usually lie at approximately 100 m depth. Dissolved organic carbon (DOC) is produced by microbial remineralization and can be mixed either upward or downward through the physical vertical mixing of water. In the mesopelagic zone (200–1000 m depth), microbial decomposition and zooplankton/fish consumption and repackaging of POC and DOC continue, with dramatic attenuation of export flux as organic compounds are remineralized into inorganic nutrients and other compounds. The sequestration depth is the depth below which carbon is stored in the ocean for 100 years or more. The sequestration depth has sometimes been designated as the bottom of the mesopelagic zone (Passow and Carlson 2012), but in other treatments as the maximum depth of winter mixing, which in the Atlantic can be between 100 and 800 m (Antia et al. 2001). Image reproduced from Turner (2015 Prog Oceanogr, doi.org10.1016/J.POCEAN.2014.08.005) with permission of Elsevier

temporally and regionally variable (Calbet 2008; Calbet et al. 2009; Campbell et al. 2009).

The mechanics of feeding vary with predator type, with an allocation of traits that may relate to finding, capturing, and ingesting particles (Kiørboe 2011). Zooplankton have to find their food in an environment where prey are ordinarily dilute. The solutions to acquire food when is it comparatively scant are few. Prey may be actively or passively hunted, or they may be concentrated. Some prey are engulfed, some particles are captured in feeding current or setae, and some are collected by the use of mucus or other sticky secretions. Some zooplankton are "ambush feeders," waiting motionless until prey pass; some are "feeding current" predators, capturing prey through feeding currents; and others are "cruise feeders," searching for prey as they move through the water (Almeda et al. 2018; Torres et al. 2023 and references therein; Fig. 19.2). All feeding types are found throughout the world, but feeding current types are the most common, followed by ambush predators (Torres et al. 2023; Fig. 19.3). Some zooplankton sense their prey from the chemical plumes the phytoplankton leave in their wake but very small phytoplankton do not leave such a trace. For these prey, some degree of contact may be necessary. Feeding mechanisms may depend on the size of the zooplankton.

Protozoa and small copepods live in a local low Reynold's number world, where water is more viscous. Reynold's number is the ratio of inertial to viscous forces and is quantified as

$$\mathrm{Re} = [\rho \upsilon \mathrm{L}]/\mu \qquad (19.3)$$

where υ is the relative velocity of a fluid across an object, L in the linear dimension of the object, ρ is the density, and μ is the viscosity of the fluid (Shapira 1961). Flow is laminar when Re is low and turbulent when Re is high. When Re is low, water appears thick—like honey—relative to the dimensions of a solid object. Water (and prey) stick to the appendages of the zooplankton, and for calanoid copepods, bristled appendages help to sieve particles from the water (Koehl and Strickler 1981). Larger copepods and other grazers may escape this constraint and function in a higher

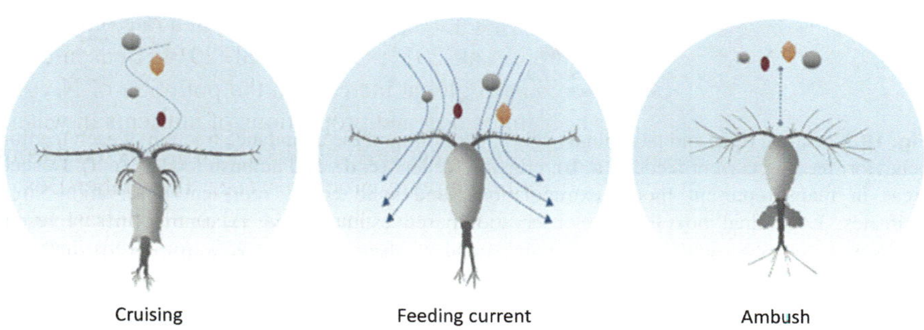

Fig. 19.2 Schematic of different copepod feeding types. Image reproduced from Torres et al. (2023 Env Sci Technol doi.org/10.1021/acs.est.2c04660) with permission of the American Chemical Society

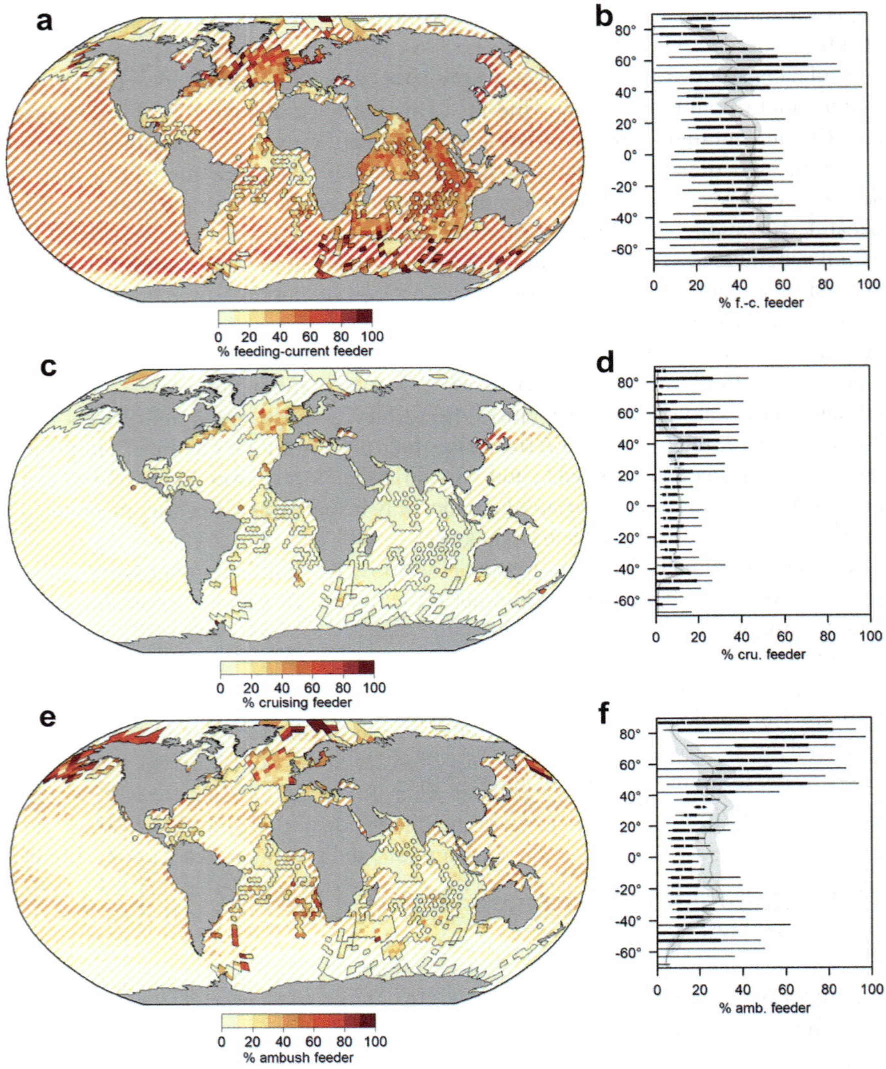

Fig. 19.3 Global (left) and latitudinal (right) distribution of the abundance fraction of each feeding behavior: feeding current feeders (**a**, **b**); cruising feeders (**c**, **d**), and ambush feeders (**e**, **f**). Dashed areas in maps represent model extrapolations, and solid colors represent observation-based estimates. Latitudinal boxplots are observation-based estimates, where central lines represent medians, boxes represent interquartile ranges, and whiskers represent 95% confidence intervals. Underlying polygons represent interquartile ranges of model-based estimates, and underlying gray line represents median of model-based estimates. The map has some limitations due to the lack of available field data for copepods smaller than 200 μm. Image reproduced from Torres et al. (2023 Env Sci Technol doi.org/10.1021/acs.est.2c04660) with permission of the American Chemical Society

Reynold's number world where water is more fluid. The former typically move by cilia, and when they locate a prey, they may ambush it. Larger copepods, such as some of the calanoids, create feeding current to pull in prey. Each feeding mechanism comes with its own trade-offs with respect to its own risk of predation and mating opportunities. The ratio of gain over risk varies with type of predator, their size, and local environment (Kiørboe 2011).

Assuming zooplankton encounter phytoplankton sufficiently to eat them, the typical perspective when considering grazing effects on phytoplankton is to assume that the zooplankton exert pressure based on the quantity of the phytoplankton. The rate of feeding has been characterized to generally follow a saturating response curve, although there are several parameterizations of this response (Holling 1959; Cushing 1978; Crowley and Martin 1989; Fig. 19.4). The response may be modeled linearly: a type I response, a hyperbolic relationship (type II), or sigmoidal response (type III). The type II response is described similarly to responses of enzyme kinetics or nutrient uptake rates (one of the "curves for all reasons"; Rao 2000; Chaps. 3 and 12).

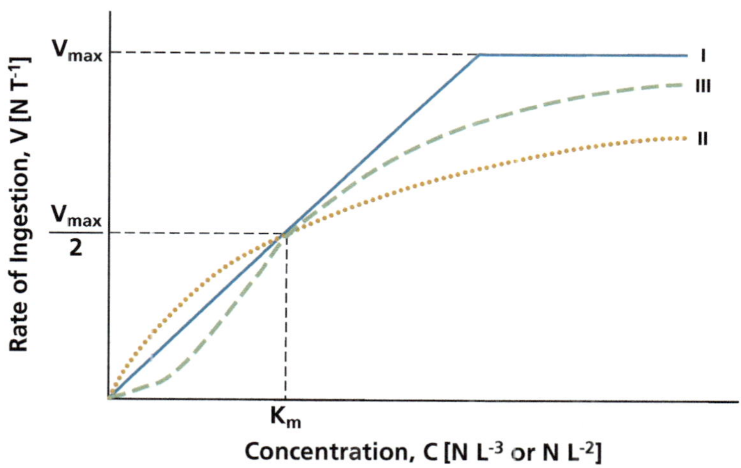

Fig. 19.4 Types of functional responses. Rate of ingestion is taken as example, and concentration of prey is used as a surrogate for encounter rate. The simplest response (type I) is rectilinear and indicates an abrupt shift from overall rate limitation by encounter rate to limitation by some subsequent step. The hyperbolic curve (type II) shows very gradual saturation of post-encounter processing. A sigmoidal response (type III) suggests some inefficiency or behavioral change that reduces encounter or its effectiveness at low concentrations and yet is saturated rapidly at high concentrations. Type I–III curves are roughly linear in concentration, respectively, right up to ($2K_m$, V_{max}), only well below K_m and only near K_m. For purposes of illustration, all three curves are forced through the point (K_m, $V_{max}/2$), where V_{max} is the maximal ingestion rate (reached asymptotically in curves of type II and III). Image redrawn based on Jumars (1993, Concepts in biological oceanography: an interdisciplinary primer, Oxford University Press)

$$V = V_{max} C/[K_m + C] \tag{19.4}$$

in which V in the ingestion rate, V_{max} is the maximal ingestion rate, K_m is the prey density at which ingestion rate at half maximal, and C is the concentration of prey. Parsons et al. (1967, 1969) suggested that there is a minimum food concentration, below which no feeding occurs, a modified type II response. That is, at low prey densities, the costs are too steep for the grazer to search for prey (and the prey may have a respite from their predators). In a type III response, there is an accelerating rate of grazing at low prey densities.

Predators cannot feed continuously and may cease feeding for any of a number of reasons. The prey may have saturated the predator's ability to process the food, or post-acquisition processes—related to digestion—may limit the ability of the predator to take in more food (Jumars 1993). Predators may also cease feeding depending on their nutritional state—i.e., what they ate and how long ago they ate it.

However, the relationship between predator and prey is much more complicated than these relationships imply and also much more complicated than the analogous phytoplankton-nutrient interactions. Predator–prey interactions encompass the suite of processes that include prey encounter, selection, capture, ingestion, digestion, and egestion (Flynn and Mitra 2016; Fig. 19.5). Zooplankton traits such as body size, feeding mode, and tolerance to phytoplankton metabolic composition, including their toxins, contribute to grazing pressure (e.g., Litchman et al. 2013; Kiørboe et al. 2018). Prey encounter depends on how the predator moves (on scales from instantaneous to diel and across different life stages), how turbulence may affect that motion and therefore encounter rates, and the energy needed to be expended under all types of conditions. Feeding is often discontinuous over time. These and other factors suggest that feeding relationships are variable and single response curves are not true representations of organisms in nature. The same case was made for single response curves for nutrient kinetics and photosynthetic-irradiance responses (Chaps. 8 and 12). Abrams (2022) reviewed some potential effects of other species on the function response of a particular predator to a particular prey and recent advances in functional responses.

19.2.1.2 Measuring Grazing Rates

Numerous methods have been used to measure rates of grazing. The simplest approaches involve estimating changes in phytoplankton (as cells or chlorophyll (chl) *a* or other cell proxy) over time in incubation experiments. Other approaches involve labeling specific food items with an isotopic tracer or other label. A variety of fluorescent dyes have been used to label prey. Also, plastic, bacterial, or phytoplankton-sized beads have been used to assess ingestion. These approaches each have their limitations.

Landry and Hassett (1982) introduced the dilution method for measuring rates of grazing in natural samples. This approach involves combining, using a gradient of proportions, unfiltered natural water (or filtered through a large sieve to remove the largest particles) with water that has been filtered to produce a grazer-free

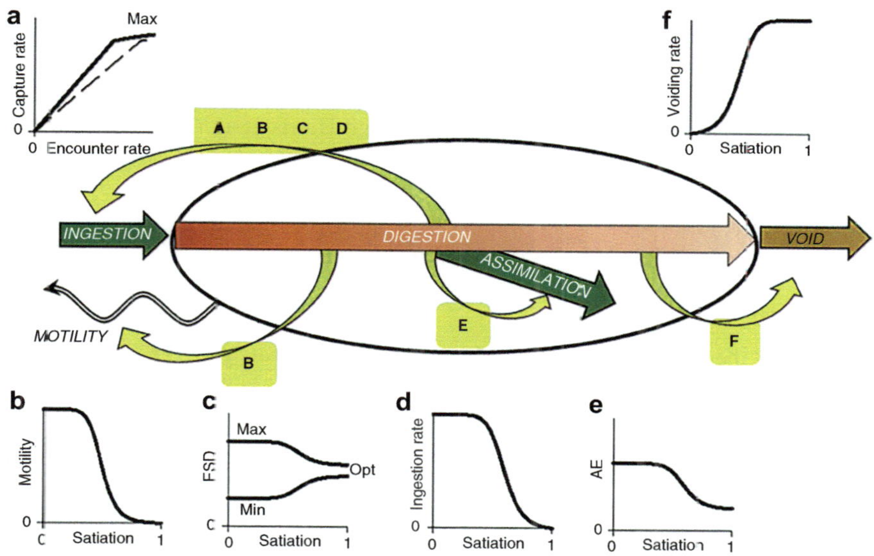

Fig. 19.5 Schematic showing range of interactions between prey (food) encounter, ingestion, digestion, assimilation, and voiding. Yellow curved arrows indicate feedbacks; the sinusoidal arrow indicates consumer motility. Plot **a** shows the relationship between encounter and capture, which is initially linear, but in the absence of any other feedbacks would be ultimately limited by handling at the point of capture or ingestion. The dashed line is for a prey type of lesser preference. Depending on preference, several prey items may be consumable over a period of time; consumption of any/all prey types contributes to satiation. Plot **b** shows satiation affecting motility, which in turn affects encounter rate **a**. Set against this feedback is the need for some consumers to move for reasons other than feeding, and motion of the prey relative to the consumer caused by their own motility and/or turbulence. Plot **c** shows how prey preference, indicated here by size selectivity, may be expected to tighten as satiation develops. Plot **d** shows how gut satiation itself also controls ingestion rate, effectively decreasing the maximum level (plateau value) of capture indicated in Plot **a**. Plot **e** shows assimilation efficiency declining with satiation, so that in the presence of abundant food digestion is less complete. Plot **f** shows how, especially in a peristaltic gut, satiation promotes voiding. This works with **e**, to counter **d** to enhance the flow of food. The relationship between prey abundance (per unit of space) and ingestion rate is an emergent property of all these interactions, coupled with other facets such as food quality and prior nutritional history of the consumer. Image reproduced from Flynn and Mitra (2016, Front Mar Sci doi.org/10.3389/fmars.2016.00165) under a creative commons license

community. By incubating samples over 24 h to encompass diel changes in feeding (using replicated treatments), the net growth rate versus dilution provides estimates of growth in the absence of grazing (intercept of the regression) and of the grazing mortality rate (slope of the regression). A modification of this approach was induced later (Landry et al. 1995) in which the dilution series was supplemented with an addition of the presumed limiting nutrient, based on the idea that mortality in the experiment should reflect grazing and not loss due to nutrient limitation. One needs to interpret such results with caution as over a period of a day, phytoplankton growth may occur due to the nutrient pulse, and the nutrient pulse may differentially simulate different fractions of the phytoplankton community.

In incubation experiments, the clearance rate (the volume of water cleared by the grazer) (F, μL predator^{-1} day^{-1}) and ingestion (I, cells predator^{-1} day^{-1}) rates are generally calculated according to Frost (1972) and Harris et al. (2005), such that

$$F = [\mathrm{V} \times \mathrm{g}]/n \tag{19.5}$$

and

$$I = F \times [C] \tag{19.6}$$

where V is the volume of water in the incubation vessel, g is the instantaneous feeding coefficient, n is the number of grazers, I and $[C]$ is the prey concentration averaged over the incubation period. It is this approach that has shown that microzooplankton herbivory is the major source of phytoplankton mortality in all marine regions (Sherr and Sherr 2002, 2009; Calbet and Landry 2004).

Using such an approach, another factor that should be accounted for, especially if on-deck incubation conditions differ from the light environment of the originally collected water, is photoacclimation (see Chaps. 6 and 8). If chl a is the proxy for phytoplankton change due to grazing, it must be recognized that pigment content may change due to changes in the light environment.

Menden-Deuer et al. (2021) have recently suggested an approach that may overcome the need for measuring grazing rates "one bottle at a time." Their proposed approach, specific for measuring rates of herbivorous protists, is based on optical changes that come with changing particle grazing due to the fact that grazers modify the abundance, size, and shape of their phytoplankton prey (Menden-Deuer and Kiørboe 2016). The concept is as follows. As particles are grazed, the contribution of unpigmented particles (e.g., detritus) may increase as that of the pigmented particles decreases. It may thus be possible to detect spectral changes using satellites by measuring the spectral slope of backscattering (Fig. 19.6). Such an approach will require considerable development and calibration, but holds promise for studying grazing—and all of its variability across space and time—in a more holistic or global context.

19.2.2 Grazing Effects on Phytoplankton Diversity

Through preferential grazing, zooplankton can alter the size structure or species composition of the phytoplankton (Frost 1980). Thus, according to Reynolds (1980), phytoplankton communities "...should be regarded not as the menu, but as the unserved portion of the meal," or as Lürling (2021) characterized, they are "the mostly inedible left-overs of the meal." There is an enormous variability in prey sizes that are selected by different types of grazers (Fig. 19.7; Flynn and Mitra 2016). Copepods selectively feed based on cell size, with both the smallest and largest cells

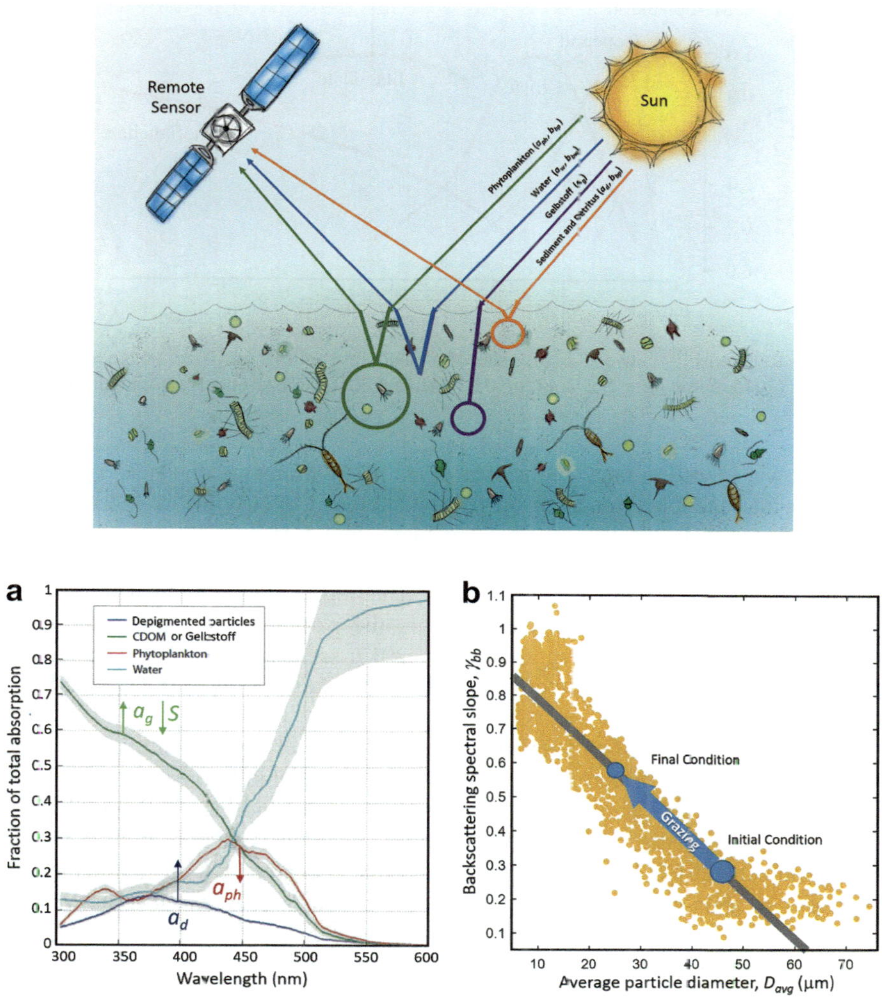

Fig. 19.6 Upper panel: Schematic of how common seawater constituents, including particulate and dissolved components, could both be generated and altered through the process of herbivorous grazing. Grazing is hypothesized to change spectral absorption and backscattering across UV and visible wavelengths. Lower panels: (**a**) As particles are grazed, absorption by gelbstoff (a_g) may increase and the spectral slope (S) may flatten compared to background gelbstoff subject to photobleaching. The contribution of depigmented particle absorption (a_d) may increase, and phytoplankton absorption (a_{ph}) may decrease as particles are grazed. The relative changes in these components result in spectral dips and troughs that may be observed in the remote sensing reflectance, particularly in sensors that can detect spectral changes in the UV-like PACE. Modified from Nelson and Siegel (2013). (**b**) Backscattering is expected to increase with increased detrital matter produced by grazing. In addition, we hypothesize that the spectral slope of backscattering g_{γ} will increase as the average particle size decreases with decreasing phytoplankton and increasing detrital particles. Modified from Slade and Boss (2015) showing trends in these parameters from field measurements. Images reproduced from Menden-Deuer et al. (2021, Front Mar Sci 8: doi.org/10.3389/fmars.2021.695938) under a creative commons license

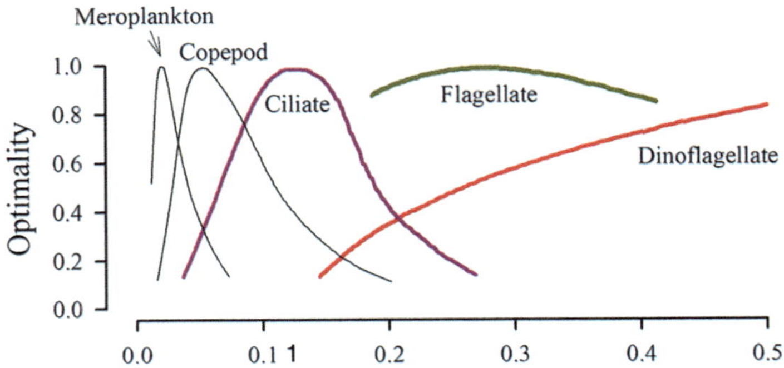

Fig. 19.7 Relationships between relative size of predator and optimal prey size. Image from Flynn and Mitra (2016, Front Mar Sci doi.org/10.3389/fmars.2016.00165) under a creative commons license

having lower ingestion rates (Gonçalves et al. 2014). Many copepods also prefer to graze on microzooplankton than on phytoplankton. Mixotrophic or heterotrophic dinoflagellates can take in prey that are many-fold larger than themselves (e.g., Menden-Deuer et al. 2005). Moreover, virtually all grazers have the ability to reject selected undesirable or unpalatable particles. The impact of grazing by heterotrophic protists can be so large that it can structure the particle size distribution of the plankton (Frost 1980; Steinberg and Landry 2017; see also Chap. 20). The success of any given prey is a function of how well zooplankton can feed preferentially on that prey.

Some grazers can feed on a range of prey, and a grazer can switch to an alternate food source if the preferred food becomes unavailable. Some predators are generalists, and others are specialists. In general, warmer and more eutrophic waters appear to favor smaller and more selective zooplankton (Hansson et al. 2007; de Senerpont et al. 2012). Switching may occur when a previously more abundant prey decreases in abundance. Accordingly, the less abundant prey can lose its refugee status.

Prey selection is also related to assimilation efficiency, which varies with both quantity and quality of food ingested (e.g., Tirelli and Mayzaud 2005). High ingestion rates may be inversely related to food quality: Grazers may eat more if the food quality deteriorates. The cost-benefit of prey switching varies as food quality in the ecosystem changes. Active prey switching may be an important driver of diversity and may help to alleviate an outcome of competitive exclusion, but there is much uncertainty in how these responses operate in situ (Vallina et al. 2014). Prey switching may also result in a different dietary composition, which may, in turn, affect egg production, egg hatching success, or naupliar survival (Vargas et al. 2006).

19.3 Living Among Enemies: Phytoplankton Defenses

Phytoplankton are "condemned to living with their enemies" (Lürling 2021, p. 237) and in order to survive must have defense strategies. Such strategies are an important force in phytoplankton diversity and selection (e.g., Smetacek 2001). The evolution of the diversity of traits in phytoplankton—and one of the explanations for the "paradox of the plankton" (Hutchinson 1961, Chap. 17)—is the interplay of competition for resources on the one hand and grazer avoidance on the other. Competition may favor small, fast-growing cells with high-nutrient uptake rates and low sinking losses, while grazing may favor large, or armored or structurally defended, slower-growing phytoplankton with higher sinking losses (Acevedo-Trejos et al. 2018; Lürling 2021). Ehrlich et al. (2020) demonstrated this trade-off for the phytoplankton community of Lake Constance, using two decades worth of data. They showed that seasonally increasing grazing pressure shifted the community to one with more defense traits, but at the cost of lower growth rates.

Strategies may be categorized in a variety of ways (Pančić and Kiørboe 2018; Lürling 2021) and may include such traits as morphological structures that may make ingestion difficult, movement or migrations to avoid predation, or production of allelopathic or toxic compounds that lead to predator avoidance (Fig. 19.8). Other strategies may include very high growth rates so that the phytoplankton outgrow their enemies. Making colonies may reduce zooplankton feeding rates. For diatoms, making thick silica walls may provide some protection against grazing or against digestion once ingested (as Hensen originally thought). The production of bioluminescent flashes in response to the presence of copepods by some dinoflagellates (Lindström et al. 2017) may also be a grazer defense strategy.

19.4 Grazing Effects on Nutrient Regeneration

Microzooplankton are the primary regenerators of nutrients in the sea (e.g., Fenchel and Harrison 1976; Caron and Goldman 1990). The contributions macrozooplankton and larger microzooplankton make to nutrient regeneration are complex: They can both enhance and reduce the flux of regenerated nutrients. On the one hand, they release nutrient directly (up to 40% of the nutrients from prey are excreted), but the form and proportion may depend on what they ate and how long ago they ate it (Glibert 1998). On the other hand, the net effect depends on the prey, as they may graze on phytoplankton, the primary consumers of nutrients, or protozoa, the primary regenerators of nutrients. Thus, whether macrozooplankton have a net positive impact on the regeneration of nutrients via their direct release or have a negative impact by reducing other regenerators will depend on the trophic interactions within and between the microbial loop and the microzooplankton. Such interactions were shown experimentally by Glibert et al. (1992) in which increasing numbers of copepods were added to natural coastal plankton assemblages. An increase in the rate of NH_4^+ regeneration was observed up to about 20 copepods L^{-1}, but when additional copepods were added, the rate of NH_4^+ regeneration

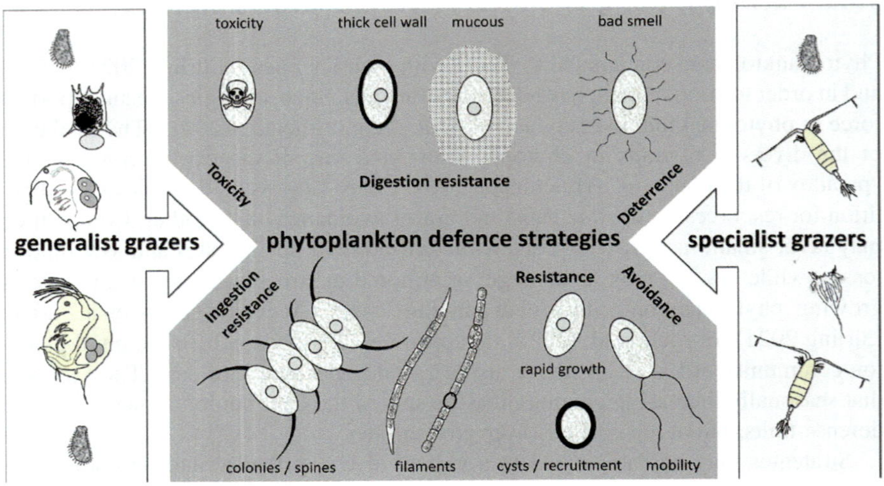

Fig. 19.8 Potential phytoplankton defenses against generalist and specialist zooplankton grazers. Defenses can be constitutive or inducible and include traits that reduce ingestion by zooplankton, reduce digestion, reduce the chance of encountering zooplankton, and even those that are directly detrimental to zooplankton. Image reproduced from Lürling (2021, Hydrobiol 848: 237–249. doi.org/10.1007/s10750-020-04370-3) under a creative commons license

actually decreased. That effect was hypothesized to be due to the impact of the copepods feeding on smaller regenerators. Miller et al. (1995), in similar experiments with summer estuarine plankton communities, also demonstrated that NH_4^+ regeneration rates were highest at densities of 10–20 copepods L^{-1}, with lower rates observed at copepod densities above and below that density.

Macrozooplankton, and larger microzooplankton, also tend to be "sloppy" grazers (Dagg 1974; Lampert 1978). This can result in the release of inorganic and organic substrates as cells are ruptured but not consumed entirely. They also release fecal pellets, which may decompose at different rates depending on local environmental factors, such as temperature. Such releases may, in turn, stimulate the uptake and growth of bacteria. Thus, through differential grazing and direct and indirect release of nutrients, macrozooplankton can alter the flux and composition of nutrients. This theme is continued in the following chapter, viewing the effects of nutrient regeneration through a stoichiometric lens.

19.5 Grazing Effects on Export from the Photic Zone

Zooplankton play an important role in export from the photic zone. Through ingestion and fecal pellet production and carcass production, as well as their daily migration in the water column, combined with greater migration during winter, phytoplankton production is actively exported to depth (Fig. 19.1). Some of this flux may be due to primary producers getting "caught" below a gradient such as the thermocline from which they cannot ascent (Falkowski et al. 2003).

High vertical fluxes from diatom blooms are fairly common, especially during spring blooms when mesozooplankton populations are still comparatively small (Smetacek 1985). Diatoms can sink fairly fast, and thus, they reach the sediment largely intact. It is therefore of no surprise that in the ocean, the export of particulate carbon (C) is related to the export of particulate silicate ($Si(OH)_4$), due to the "heavy" cell walls of diatoms (see Chap. 14). Diatoms can sink en masse if nutrient depletion in the euphotic zone sets in (Falkowski et al. 2003). However, cyanobacteria that may be more buoyant, or at least sink more slowly than diatoms, tend to be more degraded when they reach the sediment floor. This organic matter, whether fresh or degraded, may fuel the bacterial or other microbial communities deeper in the water column or in shallow sediments (Blomqvist and Heiskanen 2001). In general, diatoms have a lower C, nitrogen (N), and phosphorus (P) content compared to cyanobacteria (see Chap. 18). Accordingly, organic matter derived from cyanobacteria has been shown to support a greater micro-eukaryotic diversity than does diatom-derived organic matter.

Sinking or sedimentation of phytoplankton out of the euphotic zone can result in not only a loss of the phytoplankton, but also substantial nutrients. Such "export" production enriches the interior of the ocean with C, N, and P. The term "export production" encompasses all processes such as particle sinking, advection, or any process that transfers production out of the euphotic zone where it can support higher trophic levels. Export production may also occur when cells are aggregated into fecal pellets or other types of coagulated particles, such as transparent exopolymer particles (TEP), which can act as glue for phytoplankton, phytodetritus, and marine snow (Turner 2015 and references therein).

Export production is typically measured using sediment traps deployed at depth for some period of time, although there are many isotopic and other techniques and tools that have been—and are being—used to estimate sinking particles and their relation to productivity (see Chap. 2). Sediment traps have many associated biases, or, as stated by Antia (2005, p. 190), "Almost simultaneous with the enthusiasm in deploying traps in numerous environments came the sobering realization that they fell short of collecting the 'true' sinking flux. . ." Karl et al. (2012) showed, using a 13-year (1992–2004) sediment trap time series from the HOT Program at Station ALOHA off Hawaii that every summer (July 15–August 15), there is a large, rapid, and predictable pulse in particulate organic matter export to 4000 m depth. In the Sargasso Sea, Lomas et al. (2010) found that from 1996 to 2007 there was a >50% increase in POC export flux that coincided with a shift of the winter NAO index from consistently positive to neutral but variable and also correlated with increased primary production.

Using material collected from sediment traps can also yield clues to grazing rates. Welschmeyer and Lorenzen (1985) suggested that chl *a* is converted to phaeophorbide in the guts of grazing zooplankton, and thus, the rate of grazing can be back-calculated from the measuring phaeophorbide in sinking particles and making assumptions about the carbon:chl *a* ratio and the conversion efficiency of chl *a* to phaeophorbide.

Sedimentation, especially that of dinoflagellates, can be in the form of cysts. Cyst formation is generally considered an adaptation to adverse environmental conditions (e.g., Dale 1983), and massive encystment events may contribute to the collapse of blooms. Cysts may be calcified or non-calcified, and they may be temporary cysts or longer-term cysts, which can remain via for decades or longer. Cysts may provide a "seedbed" for new blooms, once conditions improve and they are resuspended in the water column. In fact, seedbeds of the dinoflagellate *Alexandrium fundyense* are used as an index of the potential for blooms to occur the following summer. Many dinoflagellate cysts have an endogenously controlled dormancy period, ranging from hours to months (Anderson 1980, 1998).

Resting cysts have a reduced metabolic rate, a resistant cell wall, and may have ample storage material (Fryxell 1983). Encystment may occur when environmental conditions, such as nutrients, temperatures, or light, become unfavorable. Encystment may also occur in response to grazer exudates, providing a survival mechanism (e.g., McQuoid and Hobson 1996; Rengefors et al. 1998).

19.6 Summary

The fate and manner of phytoplankton cell death are varied, with large implications for the fate and transfer of energy and nutrients through the food web, biogeochemical cycling, and export from the photic zone. Zooplankton grazing is only one process by which phytoplankton mortality occurs. Herbivorous protists are responsible for a far greater rate of removal of phytoplankton than microzooplankton, but microzooplankton have the ability, through their grazing, to structure size, composition, and diversity of phytoplankton. Effects depend on feeding strategies and on the defense strategies of the phytoplankton. By grazing, nutrient availability is also affected, as zooplankton not only release nutrients but also consume the consumers (the phytoplankton) and other microbial regenerators (microzooplankton). Export flux also is related to grazing activity, due to fecal matter production and coagulation of other particles. Concepts of how food webs are structured and the flow of energy and materials through food webs, including other pathways by which phytoplankton may die, are continued in the next chapter.

References

Abrams PA (2022) Food web functional responses. Front Ecol Evol 10. https://doi.org/10.3389/fevo.2022.984384

Acevedo-Trejos E, Marañón E, Merico A (2018) Phytoplankton size diversity and ecosystem function relationships across oceanic regions. Proc R Soc Lond B: Biol Sci 285:20180621

Almeda R, Van Gréve H, Kiørboe T (2018) Prey perception mechanism determines maximum clearance rates of planktonic copepods. Limnol Oceanogr 63:2695–2707. https://doi.org/10.1002/lno.10969

Anderson DM (1980) Effects of temperature conditioning on development and germination of *Gonyaulax tamarense* (Dinophyceae) hypnozygotes. J Phycol 16:166–172

Anderson DM (1998) Physiology and bloom dynamics of toxic *Alexandrium* species, with emphasis on life cycle transitions. In: Anderson DM, Cembella AD, Hallegraeff GM (eds) Physiological ecology of harmful algal blooms. NATO ASI series, vol G 41. Springer, Berlin, pp 30–47

Antia AN (2005) Solubilization of particles in sediment traps: revising the stoichiometry of mixed layer export. Biogeosciences 2:189–204

Antia A, Koeve W, Fischer G, Blanz T, Schultz-Bull D, Scholten J, Neur S et al (2001) Basin-wide particulate carbon flux in the Atlantic Ocean: regional export patterns and potential for atmospheric CO2 sequestration. Glob Biogeochem Cycles 15:845–862. https://doi.org/10.1029/000GB001376

Banse K (2013) Reflections about chance in my career, and on the top-down regulated world. Annu Rev Mar Sci 5(1):1–19. https://doi.org/10.1146/annurev-marine-121211-172359

Barber RT, Hilting AK (2002) History of the study of plankton productivity. In: Williams PJ, Thomas DN, Reynolds CS (eds) Phytoplankton productivity: carbon assimilation in marine and freshwater ecosystems. Blackwell Sci, Oxford, pp 16–43

Blomqvist S, Heiskanen A-S (2001) The challenge of sedimentation in the Baltic Sea. In: Wulff FV, Rahm LA, Larsson P (eds) A systems analysis of the Baltic Sea. Ecological studies Vol. 148. Springer, Berlin, pp 211–227

Calbet A (2001) Mesozooplankton grazing effect on primary production: a global comparative analysis in marine ecosystems. Limnol Oceanogr 46:1824–1830. https://doi.org/10.4319/lo.2001.46.7.1824

Calbet A (2008) The trophic role of microzooplankton in marine systems. ICES J Mar Sci 65:325–331. https://doi.org/10.1093/icesjms/fsn013

Calbet A, Landry MR (2004) Phytoplankton growth, microzooplankton grazing, and carbon cycling in marine systems. Limnol Oceanogr 49:51–57. https://doi.org/10.4319/lo.2004.49.1.0051

Calbet A, Atienza D, Henriksen CI, Saiz E, Adey TR (2009) Zooplankton grazing in the Atlantic Ocean: a latitudinal study. Deep Sea Res II 56:954–963. https://doi.org/10.1016/j.dsr2.2008.10.009

Campbell RG, Sherr EB, Ashjian CJ, Plourde S, Sherr BF, Hill V, Stockwell DA (2009) Mesozooplankton prey preference and grazing impact in the western Arctic Ocean. Deep Sea Res II 56:1274–1289. https://doi.org/10.1016/j.dsr2.2008.10.027

Caron DA, Goldman JC (1990) Protozoan nutrient regeneration. In: Capriulo GM (ed) Ecology of marine protozoa. Oxford University Press, New York, pp 283–306

Crowley PH, Martin EK (1989) Functional responses and interference within and between year classes of a dragonfly population. J N Am Benthol Soc 8:211–221. https://doi.org/10.2307/1467324

Cushing DH (1978) Upper trophic levels in upwelling areas. In: Boje R, Tomczak M (eds) Upwelling ecosystems. Springer, New York, pp 101–110

Dagg MJ (1974) Loss of prey contents during feeding by an aquatic predator. Ecology 55:903–906

Dale B (1983) Dinoflagellate resting cysts: 'benthic plankton'. In: Fryxell GA (ed) Survival strategies of the algae. Cambridge University Press, Cambridge, pp 69–136

de Senerpont DL, Elser JJ, Gsell AS, Huszar VLM, Ibelings BW, Jeppesen E, Kosten S et al (2012) Plankton dynamics under different climatic conditions in space and time. Freshw Biol 58:463–482. https://doi.org/10.1111/fwb.12053

Ducklow HW, Steinberg DK, Buesseler KO (2001) Upper ocean carbon export and the biological pump. Oceanography 14(4):50–58. https://doi.org/10.5670/oceanog.2001.16

Ehrlich E, Kath NJ, Gaedke U (2020) The shape of a defense–growth trade-off governs seasonal trait dynamics in natural phytoplankton. ISME J 14:1451–1462

Falkowski P, Laws E, Barber RT, Murray JW (2003) Phytoplankton and their role in primary, new, and export production. In: Fasham MJR (ed) Ocean biogeochemistry. Global change – the IGBP series. Springer, Berlin. https://doi.org/10.1007/978-3-642-55844-3_5

Fenchel T, Harrison P (1976) The significance of bacterial grazing and mineral cycling for the decomposition of particulate detritus. In: Anderson JM, MacFayden A (eds) The role of terrestrial and aquatic organisms in decomposition processes. Blackwell, Oxford, pp 285–299

Flynn KJ, Mitra A (2016) Why plankton modelers should reconsider using rectangular hyperbolic (Michaelis-Menten, Monod) descriptions of predator-prey interactions. Front Mar Sci 3:165. https://doi.org/10.3389/fmars.2016.00165

Franklin DJ, Brussaard CPD, Berges JA (2006) What is the role and nature of programmed cell death in phytoplankton ecology? Eur J Phycol 41:1–14. https://doi.org/10.1080/09670260500505433

Frost BW (1972) Effects of size and concentration of food particles on the feeding behavior of the marine planktonic copepod *Calanus pacificus*. Limnol Oceanogr 17:805–815

Frost BW (1980) Grazing. In: Morris I (ed) The physiological ecology of phytoplankton. University of California Press, Berkeley, pp 465–491

Fryxell GA (1983) Survival strategies of the algae. Cambridge University Press, New York, 144 pp

Glibert PM (1998) Interactions of top-down and bottom-up control in planktonic nitrogen cycling. Hydrobiologia 363:1–12

Glibert PM, Miller CA, Garside C, Roman MR, McManus GB (1992) NH_4^+ regeneration and grazing: interdependent processes in size-fractionated $^{15}NH_4^+$experiments. Mar Ecol Progr Ser 82:65–74

Gonçalves RJ, Van Someren GH, Couespel D, Kiørboe T (2014) Mechanisms of prey size selection in a suspension-feeding copepod, *Temora longicornis*. Mar Ecol Progr Ser 517:61–74. https://doi.org/10.3354/meps11039

Gran HH (1912) Pelagic plant life. In: Murray J, Hjort J (eds) The depths of the ocean. Macmillan and Co., London, pp 307–387

Hansson L-A, Gustafsson S, Rengefors K, Bomark L (2007) Cyanobacterial chemical warfare affects zooplankton community composition. Freshw Biol 52:1290–1301. https://doi.org/10.1111/j.1365-2427.2007.01765.x

Harris RP, Wiebe PH, Lenz J, Skjoldal HR, Huntley M (eds) (2005) ICES zooplankton methodology manual. Elsevier, Academic Press. 684 pp

Hensen V (1892) Einige ergebnisse der expedition. Ergebnisse der Plankton-Expedition der Humboldt-Stiftung, Band 1.A, pp 18–46

Holling CS (1959) The components of predation as revealed by a study of small mammal predation of the European pine sawfly. Can Entomol 91:293–320

Hutchinson GE (1961) The paradox of the plankton. Am Nat 95:137–114. https://doi.org/10.1086/282171

Jumars PA (1993) Concepts in biological oceanography: an interdisciplinary primer. Oxford University Press, Oxford, 348 pp

Karl DM, Church MJ, Dore JE, Letelier RM, Mahaffey C (2012) Predictable and efficient carbon sequestration in the North Pacific Ocean supported by symbiotic nitrogen fixation. Proc Natl Acad Sci USA 109:1842–1849

Kiørboe T (2011) How zooplankton feed: mechanisms, traits and trade-offs. Biol Rev 86:311–339

Kiørboe T, Visser A, Andersen KH (2018) A trait-based approach to ocean ecology. ICES J Mar Sci 75:1849–1863. https://doi.org/10.1093/icesjms/fsy090

Koehl MAR, Strickler JR (1981) Copepod feeding currents: food capture at low Reynolds numbers. Limnol Oceanogr 26:1062–1073

Lampert W (1978) Release of dissolved organic carbon by grazing zooplankton. Limnol Oceanogr 23:831–835

Landry MR, Hassett R (1982) Estimating the grazing impact of marine micro-zooplankton. Mar Biol 67:283–288. https://doi.org/10.1007/bf00397668

Landry MR, Kirshtein J, Constantinou J (1995) A refined dilution technique for measuring the community grazing impact of microzooplankton, with experimental tests in the central equatorial Pacific. Mar Ecol Prog Ser 120:53–63. https://doi.org/10.3354/meps120053

Lindström J, Grebner W, Rigby K, Selander E (2017) Effects of predator lipids on dinoflagellate defense mechanisms – increased bioluminescence capacity. Sci Rep 7:13104

Litchman E, Ohman MD, Kiørboe T (2013) Trait-based approaches to zooplankton communities. J Plankton Res 35:473–484. https://doi.org/10.1093/plankt/fbt019

Lomas MW, Steinberg DK, Dickey T, Carlson CA, Nelson NB, Condon RH, Bates NR (2010) Increased ocean carbon export in the Sargasso Sea linked to climate variability is countered by its enhanced mesopelagic attenuation. Biogeosciences 7:57–70

Lürling M (2021) Grazing resistance in phytoplankton. Hydrobiologia 848:237–249. https://doi.org/10.1007/s10750-020-04370-3

McQuoid MR, Hobson LA (1996) Diatom resting stages. J Phycol 32:889–902. https://doi.org/10.1111/j.0022-3646.1996.0089.x

Menden-Deuer S, Kiørboe T (2016) Small bugs with a big impact: linking plankton ecology with ecosystem processes. J Plankton Res 38:1036–1043. https://doi.org/10.1093/plankt/fbw049

Menden-Deuer S, Lessard E, Satterberg J, Grünbaum D (2005) Growth rates and starvation survival of three species of the pallium-feeding, thecate dinoflagellate genus *Protoperidinium*. Aquat Microb Ecol 41:145–152. https://doi.org/10.3354/ame041145

Menden-Deuer S, Homer SW, Dierssen H (2021) Promoting instrument development for new research avenues in ocean science: opening the black box of grazing. Front Mar Sci 8. https://doi.org/10.3389/fmars.2021.695938

Miller CA, Penry DL, Glibert PM (1995) The impact of trophic interaction on rates of nitrogen regeneration and grazing in Chesapeake Bay. Limnol Oceanogr 40:1005–1011

Mills EL (1989) Biological oceanography: an early history, 1870–1960. Cornell University Press, Ithaca, 378 pp

Nathansohn A (1908) Über die allgemeinen produktionsbedingunger im meere. Beiträge zur biologie des planktons, von H.H. Gran und Nathansohn. Internat Rev Hydrobiol 1:38–72

Nelson NB, Siegel DA (2013) The global distribution and dynamics of chromophoric dissolved organic matter. Annu Rev Mar Sci 5:447–476. https://doi.org/10.1146/annurev-marine-120710-100751

Pančić M, Kiørboe T (2018) Phytoplankton defense mechanisms: traits and trade-offs. Biol Rev 93:1269–1303

Parsons TR, Lebrasseur RJ, Fulton JD (1967) Some observations on the dependence of zooplankton grazing on the cell size and concentration of phytoplankton blooms. J Oceanogr Soc Jap 23:10–17

Parsons TR, Lebrasseur RJ, Fulton JD, Kennedy OD (1969) Production studies in the Strait of Georgia. Part 2. Secondary production under the Fraser River plume, February to May, 1967. J Exp Mar Biol Ecol 3:39–50

Passow U, Carlson CA (2012) The biological pump in a high CO_2 world. Mar Ecol Prog Ser 470:249–271. https://doi.org/10.3354/meps09985

Rao TR (2000) A curve for all reasons: the rectangular hyperbola in biology. Resonance 5:85–90

Rengefors K, Karlsson I, Hansson L-A (1998) Algal cyst dormancy: a temporal escape from herbivory. Proceed: Biol Sci 265:1353–1358

Reynolds CS (1980) Phytoplankton assemblages and their periodicity in stratifying lake systems. Ecogeography 3:141–159. https://doi.org/10.1111/j.1600-0587.1980.tb00721.x

Reynolds CS (1984) The ecology of freshwater phytoplankton. Cambridge University Press, Cambridge, 384 pp

Ryther JH (1969) Photosynthesis and fish production in the sea. Science 166:72–76

Shapira AH (1961) Shape and flow: the fluid dynamics of drag. Anchor Books/Doubleday, New York

Sherr EB, Sherr BF (2002) Significance of predation by protists in aquatic microbial food webs. Anton Leeuw Int J G Mol Microbiol 81:293–308

Sherr EB, Sherr BF (2009) Role of herbivorous protists in controlling initiation and development of marine phytoplankton blooms. Aquat Microb Ecol 57:253–262

Slade WH, Boss ES (2015) Spectral attenuation and backscattering as indicators of average particle size. Appl Opt 54:7264–7277. https://doi.org/10.1364/AO.54.007264
Smetacek V (1985) Role of sinking in diatom life-history cycles: ecological, evolutionary and geological significance. Mar Biol 84:239–251
Smetacek V (2001) A watery arms race. Nature 411:745
Steinberg DK, Landry MR (2017) Zooplankton and the ocean carbon cycle. Annu Rev Mar Sci 9: 413–444. https://doi.org/10.1146/annurev-marine-010814-015924
Tirelli V, Mayzaud P (2005) Relationship between functional response and gut transit time in the calanoid copepod *Acartia clausi*: role of food quantity and quality. J Plankton Res 27:557–568. https://doi.org/10.1093/plankt/fbi031
Torres R, Almeda R, Xu J, Hartmann N, Rist S, Brun P, Nielsen TG (2023) The behavior of planktonic copepods minimizes the entry of microplastics in marine food webs. Environ Sci Technol 57:179–189. https://doi.org/10.1021/acs.est.2c04660
Turner JT (2015) Zooplankton fecal pellets, marine snow, phytodetritus and the ocean's biological pump. Progr Oceanogr 130:205–248. https://doi.org/10.1016/j.pocean.2014.08.005
Vallina SM, Ward BA, Dutkiewicz S, Follows MJ (2014) Maximal feeding with active prey-switching: a kill-the-winner functional response and its effects on global diversity and biogeography. Prog Oceanogr 120:93–109. https://doi.org/10.1006/j.pocean.2013.08.001
Vargas CA, Escribano R, Poulet S (2006) Phytoplankton food quality determined time windows or successful zooplankton reproductive pulses. Ecology 87:2992–2999. https://doi.org/10.1890/0012-9658(2006)87[2992:pfqdtw]2.0.co;2
Welschmeyer NA, Lorenzen CJ (1985) Chlorophyll budgets: zooplankton grazing and phytoplankton growth in a temperate fjord and the Central Pacific gyres. Limnol Oceanogr 30:1–21
Wolff T, Peterson ME (1991) A brief biography of A.S.Ørsted, with noted on his travels in the West Indies and Central America and illustrations of collected polychaetes. Ophelia Supp 5:669–685

20 Food Webs, Cascades, Trophic Positions, and Stoichiometry

20.1 Introduction

G. Evelyn Hutchinson, the Yale limnologist most known for the "paradox of the plankton" (Hutchinson 1961, Chap. 17), was, together with his student of that time, Gordon Riley, undeterred in his novel mathematical approach to ecology (Chap. 1). He had the foresight to hire Raymond Lindeman as a postdoc, which was 1941 (Sterner 2014). Lindeman was a student of limnological ecology from Minnesota, broadly trained, and with the assistance of his wife, who was an algal specialist, captured the interest of his future employer by expressing that his interests were, "spectroscopic analysis in sediments, pollen analysis for chronology, microfossil analysis, analysis of chlorophyll, phosphorus and nitrogen to better understand the food cycle and population dynamics" (Sterner 2014, as paraphrased from a letter from Lindeman to Hutchison on November 11, 1940). Soon after Lindeman arrived at Yale (with a salary that tripled his annual income from $600 to $1800; Sterner 2014), he submitted his thesis work for publication—work based on a quantitative effort to understand the trophic interactions of Cedar Lake Bog in Minnesota. He used his data and understanding of the literature of the time to produce a food cycle diagram of the lake that captured energy flow artistically and elegantly (Fig. 20.1). At the center was ooze, the organic material derived from bacteria. Everything emanated from there, including the stocks and rates associated with each path. The concept of food web interactions was derived from the single lake he studied, but one which he had visited nearly 300 times and from which he and his wife had sifted each organism and (hand) analyzed all data. Lindeman also proposed that his data could be incorporated into mathematical formulae, but he himself felt "utterly incapable of applying such lofty principles of analysis to the aquatic ecosystem" and hence in interest in working with Hutchinson (quoted from Sterner 2014). Lindeman was guided in his thinking by many ecologists before him (e.g., Shelford 1918; Alsterberg 1924; Thienemann 1926; Strøm 1928; Rawson 1930) who examined food web linkages, but he was the "first to provide a quantitative accounting of all of these components in a single ecosystem" (Sterner 2014).

P. M. Glibert, *Phytoplankton Whispering: An Introduction to the Physiology and Ecology of Microalgae*, https://doi.org/10.1007/978-3-031-53897-1_20

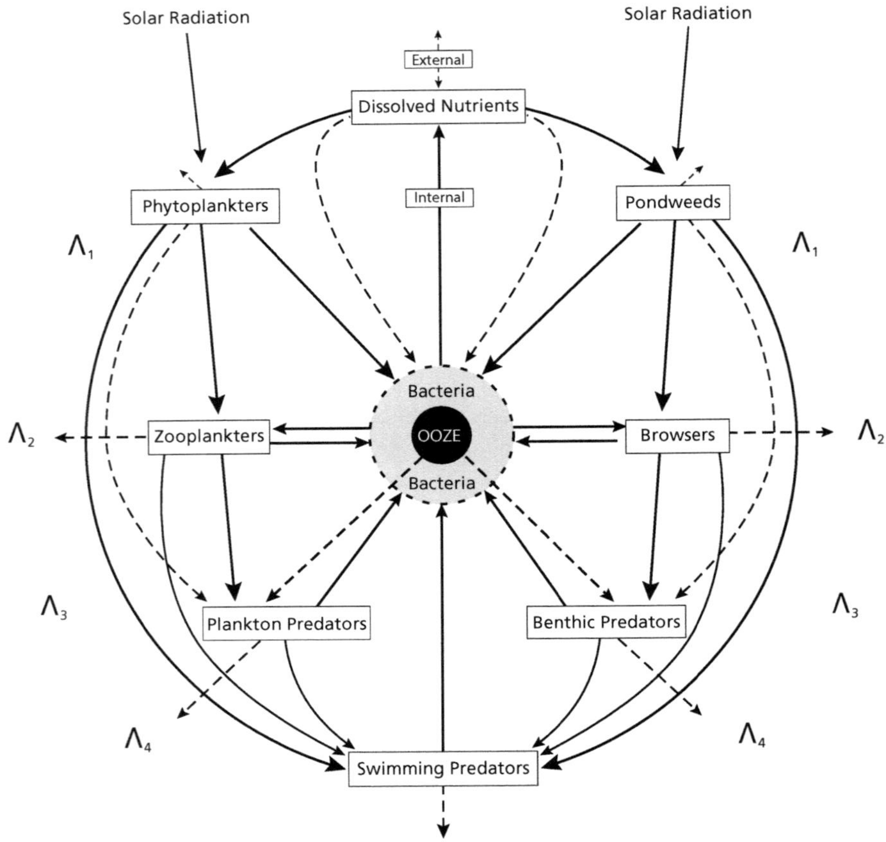

Fig. 20.1 Primary fundamentals of Cedar Lake Bog, the food cycle diagram Raymond Lindeman published. Image redrawn based on Lindeman (1942, Ecology 23:399–418)

Ahh, but that was not good enough for the reviewers of Lindeman's paper. His reviews included comments such as, "The chances are that the author's beliefs and imaginary lakes would be different entities if he had a background of observations of fifty or of a hundred of the 10,000 lakes he claimed by the State of Minnesota, instead only one, and that of a special type." And, "Limnology is not yet ready for generalizations of this kind. The basic background data for such a paper is far too fragmentary. If Dr. Lindeman could put this paper aside for ten years, then bring it out and see how it looks in the light of what we hope will be the added accumulation of limnological information, he might possibly congratulate himself that he deferred in publication." His paper was thus rejected by the editor of *Ecology*, on the basis that its conclusions were too far-reaching, that he studied only one lake, and that he should wait a decade or so to see if his ideas stood the test of time (quotes from Cook 1977, p. 23; Sterner 2014).

Raymond Lindeman could not put his paper aside for a decade. He was ill and died (at the age of 27) before his paper (one of just a few he ever published) was actually accepted (Lindeman 1942)—which was only made possible by rebuttal efforts of Hutchinson. The editor relented, acknowledging "time is a great sifter in these matters and it alone will judge the question" (quoted in Sobczak 2005). Time did judge well, but it took more than a decade for Lindeman's paper to be widely read. His reviewers (and the journal editor) would be amazed (embarrassed?) by the overwhelming acceptance of his theories of trophic dynamics now decades and decades later (and to have numerous awards and honors in his name; Sobczak 2005). His template for studying energy flow became central to the study of lake and marine ecosystems. We continue to be inspired by the science he created—and we can only imagine what he would have created if he had been given the gift of time.

Unifying principles have been presented throughout this book that allow a better understanding of phytoplankton traits and the forcing functions that drive phytoplankton diversity. Ecological stoichiometry links food webs, ecosystem metabolism, and biogeochemistry, influencing stocks and fluxes of key elements (Glibert et al. 2011, 2013; Welti et al. 2017). Investigating the interactions of temperature and nutrients by combining ecological stoichiometry (Sterner and Elser 2002) and metabolic theory of ecology (Brown et al. 2004; Moorthi et al. 2016) improves the understating of microbial and ecosystem ecology (Hall et al. 2010) on different levels of organization (individuals, populations, communities, food webs, ecosystem; see reviews by Cross et al. 2015; Vanni and McIntyre 2016). Diet-induced metabolic plasticity contributes to variation in metabolic allometry, at least at small scales of body size due to the greater respiratory response of smaller species to altered diets (Jeyasingh 2007). In all, the importance of stoichiometry requires that transformations of materials and nutrients be considered in trophic transfer, not just energy and collective effects of trophic interactions, food web structure, and biogeochemistry operate together, linked by biological and chemical diversity (Welti et al. 2017). This chapter examines our concepts of food web interactions and contemporary approaches for studying trophic interactions, including those associated with mixotrophs. Such an understanding is critically related to the functioning of ecosystems, their resilience and changes over time, and how we model food webs. The point is not just to add more complexity to "wiring diagrams," which is important in itself (as, e.g., the trophic network of a coastal lagoon by D'Alelio et al. 2022; Fig. 20.2), but also to understand control points and their sensitivity to change. The complexity of food webs, the approaches by which they can be teased apart, and how considerations of mixotrophy alter our perspectives of food web interactions are considered here.

20.2 Trophic Cascades

In its simplest form, the aquatic food chain was originally conceptualized based on understanding of terrestrial dynamics where the plants are the producers and the animals the consumers (Hairston et al. 1960). Thus, a "classic" food chain envisions

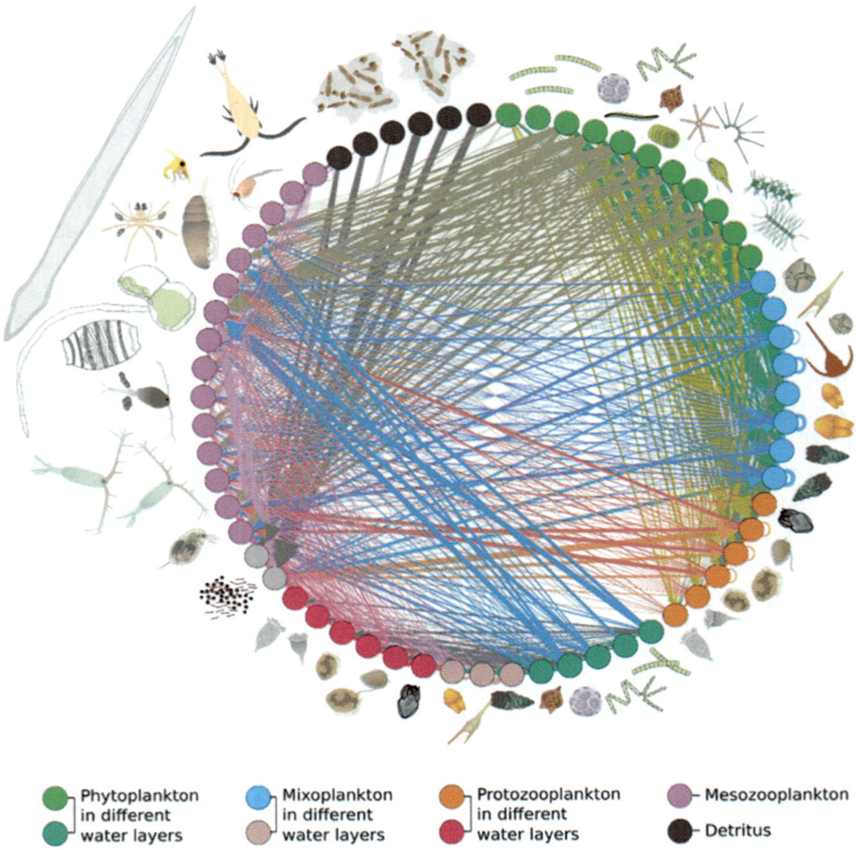

Fig. 20.2 Complexity of a plankton trophic network for a coastal lagoon. Arrows go from prey to predators. Image reproduced from D'Alelio et al. (2022, Water 14: doi: 10.3390/w14060974) under a creative commons license

photosynthesizing phytoplankton as the primary producers, the "herbivorous" zooplankton as the primary consumers, followed by the carnivorous zooplankton and fish as the secondary consumers, and so on.

Lindeman's concept of trophic dynamics is based on energy flow through the food web. Within this paradigm are embedded concepts such as growth and body size relationships, how they scale, and how nutrient availability or food availability affects consumption and growth. Within this concept also is the concept of inefficiency at each trophic level and associated losses from the food chain. By organizing organisms according to trophic level, and assuming an efficiency of transfer from one trophic level to the next of about 10%, early studies envisioned these relationships as a feeding or trophic pyramid (e.g., Elton 1927). Lindeman introduced the concept of energy transfer from trophic level to trophic level and linked all energy flow through an ecosystem to solar energy and the metabolism of individuals within each trophic level.

Another paradigm, developed shortly after Lindemen and notably associated with Lotka (1956) and earlier ideas of Volterra (1928), focuses on the impacts of predation and their impacts down the "chain." This paradigm emphasizes feedbacks and includes more attention to behaviors of predators and prey, life stage differences, and responses to change. Such a "top-down" perspective should lead to a *trophic cascade*, where grazing on one level relieves the next level down from grazing pressure and so on, resulting in alternating impacts across levels of a food web (Fig. 20.3, Pace et al. 1999). To be a cascade, effects must penetrate through more than one lower food web level. Time-dependent changes are implied. Instability or variability should be more common in a top-down controlled food web.

While the literature on trophic cascades and top-down food web control in lakes is quite large, there are comparatively few examples of trophic cascades in marine

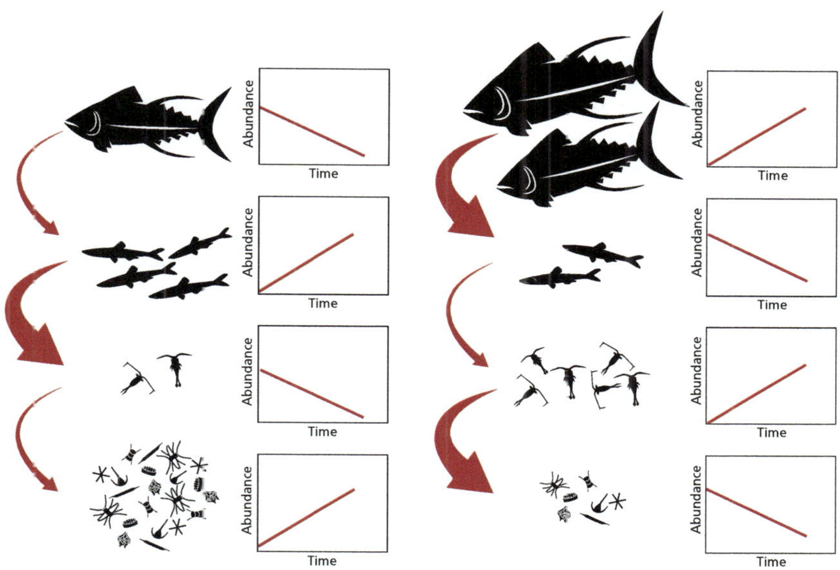

Fig. 20.3 Trophic cascades reflect the impact of predators at all levels of the food web. Via their grazing, they cause inverse relationship between abundances of different trophic levels. Changes in trophic cascades cause changes in biogeochemical cycling and can shift systems to alternate stable states. In the left-hand panels, the highest trophic level is comparatively less abundant than its abundance shown in the right-hand panels and thus would exert less grazing pressure on the next lower trophic level. As that level increases in abundance (as it is comparatively ungrazed), it increases in abundance and exerts more grazing pressure on the next trophic level (shown here as copepods), which decreases in abundance. As they decrease, their grazing pressure on phytoplankton decreases, allowing the phytoplankton to increase. The inverse pattern is shown in the right-hand panels, where the end result in this scenario is a decrease in the abundance of phytoplankton

pelagic systems. Not only the complexity of sampling all food web levels, but also tracking changes over time makes documenting such effects difficult. However, recent examples of cascades have been reported from the Black Sea and the Baltic Sea and do suggest strong top-down forcing (Pershing et al. 2015). Cascades are more common in lake systems as they are generally more spatially constrained than marine systems. Marine systems generally have high species diversity, which may also dilute the impact of change in the top predator; lakes typically have much lower biodiversity than oceans, as well as fewer trophic levels. Cascades thus appear to be more easily resolved in systems with low diversity, simple food webs, and restricted regions (Frank et al. 2005; Hessen and Kaartveft 2014).

Since these paradigms have come into common acceptance in aquatic ecology, complexities of trophic dynamics have become much better understood. Many different types of organisms cannot be assigned to a single trophic level—they can be an herbivore at one time and a carnivore another. They may change feeding modes at different life stages. Loss of energy between trophic levels can be highly variable and may be due to the export of organic matter from the system, nongrazing mortality, and respiration. Assimilation efficiency also depends on food quality, and stoichiometric balances and imbalances have effects that impact other trophic levels. Mixotrophs and the understanding of the microbial food web add further complexity. These complexities are explored below.

20.3 Beyond Food Quantity

Grazers may eat food based on food availability (Chap. 19), but their nutrition depends on the quality of that food. Trophic fitness parameters, such as rate of growth or fecundity, are related to the nutritional composition of the food on which they graze, and thus, ultimately food quality affects the transfer of energy and nutrients through the food web (e.g., De Troch et al. 2012). For those who consume phytoplankton, studies on food quality have mainly emphasized nutrient-rich biomolecules such as phosphorus (P)-rich lipids and nitrogen (N)-rich proteins or amino acids (Checkley 1980; Kleppel et al. 1998; Guisande et al. 2000; Arendt et al. 2005). Some diatom species, for instance, produce certain highly unsaturated fatty acids that are essential for zooplankton reproduction (reviewed by Kilham et al. 1997), while flagellates generally produce different fatty acids than diatoms (Olsen 1999). Different types of copepods preferentially graze on algae at different growth stages in order to obtain food of a specific fatty acid composition (De Troch et al. 2012).

Nutrient stoichiometry is another regulator of food quality. Stoichiometry affects various life stages of the predator differently (Moe et al. 2005). As a specific example, there is a greater need for carbon (C), N, and P for developing copepod juveniles, but at a later stage, while C is still needed for metabolism, more P must be allocated to eggs; thus, P-poor food can disproportionately affect egg production while not affecting adult survival (Færøvig and Hessen 2003; Laspoumaderes et al. 2010; Bentley et al. 2021).

Numerous studies of stoichiometric regulation of zooplankton in freshwater have been undertaken, leading to considerable insight into the efficiency of material and energy transfer as it relates to stoichiometric imbalances (Boersma and Elser 2006; Malzahn et al. 2007; Boersma et al. 2008; Shoo et al. 2009; Laspoumaderes et al. 2015; Bentley et al. 2021). Due to the typical role of P as a limiting nutrient in freshwater, most of these studies have focused on imbalances in C:P in prey. The work of Malzahn and Boersma (2012) convincingly demonstrated that *Acartia tonsa* reared on P-poor diets had lasting effects on copepod growth that were non-reversible even when the animals were subsequently exposed to more nutritious food. They also showed that the effect on growth was related to the duration of exposure to the low-quality food.

Stoichiometry of the prey can also affect other aspects of food quality, including toxin production, cell membrane thickness, and other chemical constituents that have been considered to turn good food "bad" (Mitra and Flynn 2005). Prey "chemistry," especially their toxin content, may be more important than cell size for copepods such as *Eudiaptomus gracilis* (DeMott and Moxter 1991). This was shown in experiments by Ger et al. (2018) in which different grazer types were mixed with toxic and nontoxic phytoplankton. By grazing on the nontoxic food preferentially, the system was "pushed" into a state where a toxic cyanobacterium dominated. The dominance of toxin-producing species can also result in a failure of normal predator-prey interactions, which, in turn, enhances the transfer of nutrients that sustain harmful algae at the expense of competing algal species (Iroigoen et al. 2005; Mitra and Flynn 2006).

Furthermore, Shoo et al. (2009) showed that poor-quality food for copepods has effects higher in the food web as the resulting nutritionally poor copepods themselves are grazed. The decline of copepod populations due to poor-quality food has far-reaching effects and may lead to the collapse of higher trophic levels that depend on these copepods as a food source (Malzahn et al. 2010; Winder et al. 2011; Hessen et al. 2013; Bentley et al. 2021). As evidence, Nobili et al. (2013) showed that changes in the quality and quantity of phytoplankton resulting from changing nutrient ratios in the North Pacific Subtropical Gyre have resulted in changes in zooplankton structure; as larger zooplankton declined, the biomass of smaller zooplankton increased, in turn affecting energy transfer and productivity at higher tropic levels. Beausgrand et al. (2003) showed that declines in the abundance of *Calanus* zooplankton in the North Sea have resulted in long-term declines in cod. More recently, the physiological responses of two contrasting estuarine copepods, *A. tonsa* and *Eurytemora carolleeae,* the former a broadcast spawner and the latter a brood spawner, fed a constant diatom diet based on C content, but which had a variable N:P content, resulted in differences in elemental composition (C, N, P) of tissue and eggs, as well as changes in the rates of grazing, excretion, egg production, and viability (Bentley et al. 2021).

While poor-quality food at the base of the food web has effects that may impact higher trophic levels, the opposite may also be true. That is, good-quality food can improve the food quality for higher trophic levels. Microbial prey are typically rich in C, N, and P, and therefore, if a consumer feeds to acquire one of those nutrients (if it is limiting, for example), it gains all elements. This can improve the food for the next consumer. Such trophic "upgrading" is highly species-specific at the level of both prey and predator (Klein Breteler et al. 1999, 2004; Tang and Taal 2005; Tang et al. 2018).

Food quality may also be altered on an episodic basis. A change in predator growth rate related to an altered diet depends on the extent to which the diet is sustained or switches and how frequently between low-quality food and high-quality food (Hood and Sterner 2010). From a stoichiometric perspective, detritus, high in C, may result in metabolic costs to consumers, including altered metabolic rate and growth rate (Hessen and Anderson 2008; Plath and Boersma 2001). Detritivores consume the least nutritionally balanced foods and thus have lower growth rates than their planktivorous or piscivorous counterparts (Sterner and Elser 2002).

20.4 Stoichiometry, Nutrient Regeneration, and Food Webs

Food webs structure nutrients. A grazer—any grazer—excretes, but how much and the nutrient composition of the excretion products vary with the food ingested and the grazer's metabolism. By excreting, egesting, or respiring what they do not need, an effective feedback develops with respect to the element stoichiometry of their resource or prey. Grazers may consume prey with an elemental composition similar to their own, but it is more likely that a difference between food resource N:P ratios and consumer N:P ratios can become established. Consumers excrete those nutrients ingested in excess (Vanni et al. 2002). Whether grazers are highly constrained or more flexible in their stoichiometry (see Chap. 18) has large consequences for nutrient regeneration, as those grazers that are highly constrained will, by necessity, regenerate the nutrients that are consumed but not needed to maintain their biomass. This creates wide variations in the stoichiometry of producers and consumers and presents further cellular challenges for producers to regulate both the limiting and saturating nutrient cellular levels.

Moreover, an animal may respond to the nutritional demand for one nutrient (e.g., C) and in so doing obtain an excess of another nutrient (e.g., P) requiring an investment of energy to release this excess. Alternatively, a consumer may need to release or dissipate excess C. This also appears to have major impacts on organismal fitness and, like the other stoichiometric concepts explained above, can affect ecological interactions at the ecosystem level and enhance stoichiometric regulation (Hessen and Anderson 2008). Peace et al. (2013) have suggested a number of possible mechanisms by which high food content may reduce growth: Reduction in feeding and reduction in C absorption sites should excess P hinder C absorption, or increase in respiration due to the costs of egestion, metabolism, or excretion.

Thus, not only can the metabolic processes of release by the consumers affect nutrient availability for phytoplankton, but changes in the external balance of nutrients can affect higher trophic levels by changing the relative elemental composition of the prey on which consumers depend (Glibert et al. 2013; Glibert 2015, 2017; Laspoumaderes et al. 2010; Bentley et al. 2021; Fig. 20.4). Ultimately, species

Fig. 20.4 Theoretical freshwater trophic cascades and their interactive effects with the proportion of N:P in dissolved nutrients and biomass of different trophic groups. Note that the N P of excretion should be the inverse of the N:P of the biomass of any given trophic group. The variable proportions of N:P in excretion products (consumer-driven stoichiometry) and new inputs of N or P are thought to result in these different stable state conditions

of grazers that can sequester the nutrient in least supply relative to their needs, while dissipating what they do not need, should become the dominant (and, in some cases, the keystone) species by outcompeting grazers that cannot effectively acquire what they need and/or dissipate what they do not.

There are numerous examples of these interactions. For example, calanoid copepods generally have a high biomass N:P ratio, ~20–35 by atoms, whereas freshwater *Daphnia* or cyclopoid copepods have N:P atomic ratios much closer to Redfield proportions (Walve and Larsson 1999; Sterner and Elser 2002; see Fig. 18.14). If a calanoid copepod, such as *Eurytemora*, eats phytoplankton with a lower N:P ratio, for example, a diatom, it will excrete proportionately more P and retain proportionately more N, thereby providing the phytoplankton a nutrient environment favorable to its needs. Conversely, a cyclopoid copepod, such as *Limnoithona*, which maintains a lower biomass N:P, would excrete more N relative to P and therefore could support a phytoplankton prey that has a more enriched N:P in its biomass (Fig. 20.5). Interestingly, in the San Francisco Bay Delta, such stoichiometric control can be inferred to have played a role in the shift of dominant copepod taxa when, over the period of several decades, the N:P ratio in the water column changed (Fig. 20.5; Glibert et al. 2013; Glibert 2015). The oscillation of the calanoid copepod *Eurytemora* with the cyclopoid copepod *Limnoithona* over time is a near mirror image of the oscillation in the N:P ratio in the water column, a pattern consistent with these grazers being responsive to changes in elemental stoichiometry and maintenance of altered dynamic equilibria on a long-term scale (Fig. 20.5). Other changes in community composition also have been shown to have paralleled

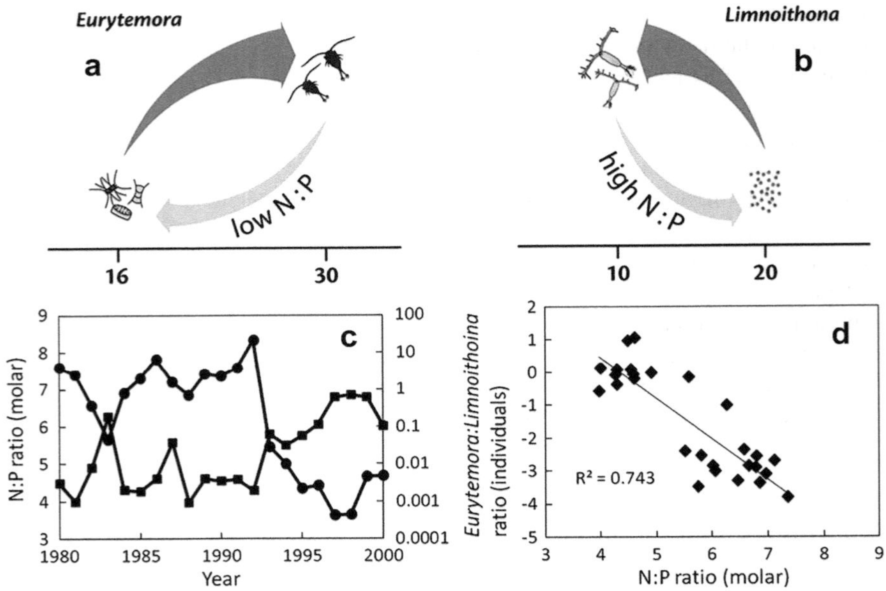

Fig. 20.5 Conceptual diagram of the ecological stoichiometric relationship between different phytoplankton and zooplankton genera. Panel (**a**) represents the relationships between diatoms and the copepod *Eurytemora affinis*; panel (**b**), the relationship between small flagellates and the cyclopoid copepod *Limnoithona tetraspina*. The x-axis represents the biomass N:P of the organisms. The wide arrow represents ingestion of the phototrophs by the grazer; the lighter arrow represents nutrient regeneration in the grazer's excretions. Note that the stoichiometry of the regenerated nutrients differs with zooplankton taxon and the N:P ratio of the food on which they graze. Panel (**c**) represents the comparison of the ratio of *Eurytemora/Limnoithona* and the ratio of dissolved inorganic nitrogen:phosphorus (DIN:DIP) for the period from 1980 to 2000 for Suisun Bay, San Francisco Bay Delta (circles—ratio of copepods; squares—nutrient ratios). Panel (**d**) is the correlation of the data from (**c**); the coefficient of determination was significant ($p < 0.05$). Images (**a**) and (**b**) reproduced from Glibert et al. (2011, Rev Fish Sci 19: doi: 10.1080/10641262.2011.611916) with permission of Taylor and Francis, and the composite image is reproduced from Glibert et al. (2013, J Mar Syst 125: doi: 10.1016/j.jmarsys.2012.10.004) with permission of Elsevier

changes in N:P ratio over the decadal scale in this system (Glibert et al. 2011). Clearly, changes in nutrient stoichiometry in this system, achieved through both external forces (altered land-based nutrient loads) and internal, organism-driven, assimilative, and dissimilative processes, are related to community compositional changes.

Studies from whole-lake experimentation also suggest that the N:P ratio is linked to alterations in zooplankton size, composition, and growth rate, as those animals with increased RNA allocation will grow at higher rates due to increased protein synthesis rates (Sterner and Elser 2002). Similar findings were reported from annual studies in the Baltic Sea (Walve and Larsson 1999). Hassett et al. (1997) compared the ecological stoichiometric constraints on zooplankton in 31 lakes and 21 marine systems, and their data were strongly suggestive of stoichiometric controls, especially P constraints in the lake systems.

Recently, the degree of homeostasis of the heterotrophic dinoflagellate *Noctiluca scintillans* was determined when fed different quality food (the diatom *Thalassiosira weissflogii* grown under nutrient-balanced, N-limited, and P-limited conditions). Their cellular elemental ratios were relatively homeostatic, resulting in their excretion of C, N, and P highly variable and dependent on the prey nutritional quality (Zhang et al. 2017).

20.5 Characterizing Trophic Transfer and Trophic Position Through Isotopes and Biochemistry

Food webs are descriptions of who eats what, or who eats whom. Multiple approaches have been used to develop data, concepts, and models about how different producers and consumers are connected. Lindeman enumerated all that he could. Others have inferred relationships based on food web structures, as well as gut contents. The latter approach may be useful for large predators, but often the guts are in varying states of digestibility and are unrecognizable. Moreover, teasing apart the guts of small protists is exceedingly difficult.

Stable isotopes of C and N have been widely used for analyzing trophic relationships (e.g., Peterson and Fry 1987; Fry 2006; Glibert et al. 2019). Such an approach provides a more time-integrated perspective of diet than do stomach contents, which are merely a snapshot in time. Stable isotopes in a predator reflect those of the food with some known (or assumed) level of fractionation. Fractionation occurs because isotopes of different atomic weights do not react with the same speed; discrimination occurs at each trophic step. Most commonly, ^{13}C and ^{15}N are applied in such studies, and the common axiom is that for C, "you are what you eat" (i.e., isotope ratios in consumers reflect those of their food), while for N, "you are what you eat +3‰" (i.e., consumers are typically enriched by 2–4‰ relative to their diet; Smith and Epstein 1970; DeNiro and Epstein 1978, 1981; Minagawa and Wada 1984; Fig. 20.6). Studies using natural abundance stable isotopes to make inferences about energy sources (^{13}C) and trophic position (^{15}N) have become commonplace and have been applied to a wide range of aquatic ecosystem types (Middelburg 2014 and references therein). While ^{13}C and ^{15}N have been the dominant stable isotope tracers, studies have also used ^{34}S to differentiate food sources in marine (Peterson and Fry 1987) and freshwater systems (Croisetière et al. 2009) and deuterium (H) isotopes to track aquatic-terrestrial linkages (Doucett et al. 2007; Cole et al. 2011; Solomon et al. 2011; Cole and Solomon 2012; Karlsson et al. 2012; Vander Zanden et al. 2016).

Stable isotope ratios of C have often used to distinguish between autochthonous and allochthonous sources of organic matter in freshwater systems (Cole et al. 2011; Cole and Solomon 2012), between pelagic and benthic primary producers (France 1995; Vadeboncoeur et al. 2003; Christianen et al. 2017), and between terrestrial- and marine-derived organic matter in estuaries (Salomons and Mook 1981; Peterson

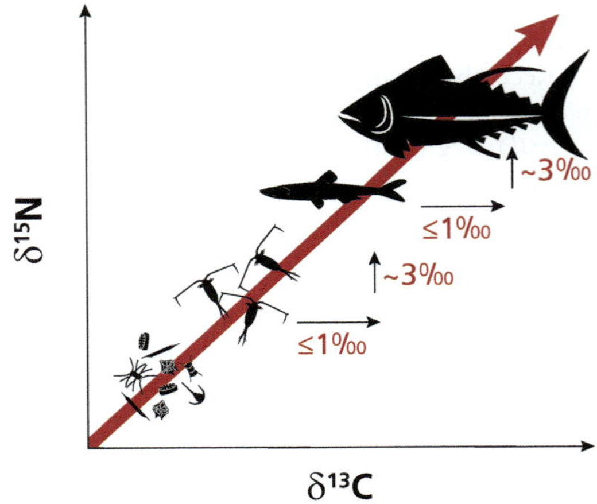

Fig. 20.6 Classic conceptual food relationships based on $\delta^{13}C$ and $\delta^{15}N$. For each trophic step, it is commonly assumed that $\delta^{13}C$ is enriched by 1‰ or less, while for $\delta^{15}N$, enrichment is on the order of 3‰

and Fry 1987; Cifuentes et al. 1988; Peterson et al. 1994; Middelburg and Herman 2007). Tank et al. (2011) applied stable isotopes to identify sources of dissolved organic matter (DOM) in lakes of the Mackenzie Delta (western Canadian Arctic) by coupling various measurements of DOM and dissolved organic C (DOC) with the $\delta^{13}C$ and $\delta^{15}N$ values of DOM and their precursor materials and found that distinct sources of DOM to adjacent lakes may contribute to differences in the fate and cycling of various C sources.

Stable isotopes are also frequently used to track anthropogenic N in aquatic systems (e.g., Owens 1987; Tucker et al. 1999; Costanzo et al. 2001; Lapointe et al. 2011; Loomer et al. 2014). In particular, increases in $\delta^{15}N$ (relative to a defined baseline or reference site) are often associated with contributions of sewage-derived N (Kendall 1998, reviewed by Glibert et al. 2019). The $\delta^{15}N$ of inorganic N derived from manure or sewage is often enriched (>10‰) due to isotopic fractionation that occurs at either the sewage treatment facility or downstream thereof. The $\delta^{15}N$ values of N in raw sewage are not necessarily enriched, but processes such as NH_3 volatilization and denitrification drive up the $\delta^{15}N$ values of the residual DIN during treatment and/or processing within the environment. This, in turn, imparts a ^{15}N -enriched signal to primary producers that take up the sewage-derived N (McCarthy et al. 1997; McCarthy and Valiela 1998).

Increasingly, however, it has been recognized that the common axioms of little to no fractionation with trophic transfer for C and approximately +3‰ for N are too simplistic (Glibert et al. 2019 and references therein). The trophic discrimination factor is variable, and applying a single value in a generic way can lead to misunderstanding (Lorrain et al. 2002; Phillips and Koch 2002; Fry 2006; Phillips 2012; Middelburg 2014). Variation in the N trophic discrimination factor depends on factors such as the organism, its form of excretion, diet, tissue type, and habitat type (e.g., Vander Zanden and Rasmussen 2001; Post 2002; Vanderklift and Ponsard 2003; Caut et al. 2009; Auerswald et al. 2010; Bunn et al. 2013). Moreover, the variability for herbivores is far greater than that for carnivores (Vander Zanden and Rasmussen 2001).

Food webs often deviate strongly from simple linear food chains and discrete trophic levels, and stable isotope applications have been used to resolve these complexities. For example, a multiyear stable isotope food web study of the California Current found separation of species into trophic groups based on $\delta^{13}C$ and $\delta^{15}N$ and that a wide diversity of prey was incorporated into upper-level predators (Madigan et al. 2012). Predators did not cluster together as expected if they were uniformly feeding on lower trophic-level prey. Bayesian mixing models revealed a high degree of omnivory and more complex trophic linkages, a finding that could have implications for ecosystem stability and resilience to perturbations.

With increasing recognition of the challenges and limitations of the bulk stable isotope approach for estimating trophic position, analyses of stable N isotope ratios of individual amino acids (AAs) have shown considerable consistency, leading to insight about trophic position from such analyses (McClelland and Montoya 2002; McClelland et al. 2003). The essence of this approach is that some AAs in heterotrophic eukaryotes exhibit high-trophic enrichment, while others show virtually no enrichment (McClelland and Montoya 2002). The difference between isotope-enriching (i.e., glutamic acid, glu) and non-enriching (phenylalanine, phe) AAs thus provides an indicator of consumer trophic position. The lack of trophic change in $\delta^{15}N$ in the non-enriching AAs is attributed to these AA either being assimilated directly into protein from food and/or that any catabolism does not incur N isotope discrimination during breakdown (reviewed by Glibert et al. 2019). Hence, the $\delta^{15}N$ of source AA is used to identify corresponding values for phytoplankton at the base of marine food webs and by extension their N sources. This approach cleverly provides a baseline ^{15}N correction internal to the organism of interest, thereby addressing the issue of accurately defining an external isotopic baseline signature in many ecosystems. Subsequent studies largely confirmed the trophic discrimination factors (TDFs) reported by McClelland and Montoya (2002). Chikaraishi et al. (2009) reported TDF_{AA} to be 7.6‰ from controlled feeding studies at the time and proposed the following equation to estimate consumer trophic position (TP) based on AAs (TP_{AA}):

$$TP_{AA} = \frac{\delta^{15}N_{Glu} - \delta^{15}N_{Phe} - \beta}{TDF_{AA}} + 1 \tag{20.1}$$

where $\delta^{15}N_{Glu}$ and $\delta^{15}N_{Phe}$ are ^{15}N for glutamic acid and phenylalanine, respectively, β is the isotopic difference between the two AAs in primary producers, and TDF_{AA} is the difference in TDF between the two AAs. While it appears that the AA-specific approach may provide higher precision and accuracy than traditional bulk methods for estimating consumer trophic position (Chikaraishi et al. 2009), a meta-analysis by McMahon and McCarthy (2016) highlighted substantial variability in TDF_{AA} values and suggested diet quality and mode of N excretion to be factors influencing this variation. A truly universal TDF_{AA} is unlikely to exist, but efforts to understand the patterns of variation are being undertaken to further refine this approach for elucidating food webs. McCarthy et al. (2013) identified two major groupings of AA in phytoplankton grown in culture that largely correlate with the source/trophic AA groups identified for heterotrophs. A non-fractionating group of AA (NF-AA) overlaps with the AA in the trophic group and has $\delta^{15}N$ values similar to Glu, which donates its amino-N to other AA through transamination. A fractionating group (F-AA) overlaps with the source group and shows relatively large deviations in $\delta^{15}N$ relative to Glu with Phe as an important example.

A related index that has been developed to understand complex trophic and nutritional pathways is that of the resynthesis index, ΣV (McCarthy et al. 2007). This index is useful for understanding trophic position by those organisms dependent on detrital or sedimented/degraded material. The principle is that microbial degradation of organic matter alters both the nutritional quality of this material and changes its $\delta^{15}N$ signal. Microbial heterotrophic AA resynthesis preferentially cleaves C–N bonds in selected AAS, thus altering the $\delta^{15}N$ signal of the AAs. More microbial resynthesis results in a higher ΣV index. Such an approach was recently used by Ledesma et al. (2020) to identify trophic position of two amphipods in the Baltic Sea, and they found the ΣV values varied between and within the species suggesting a broader feeding niche for one of their studied species and potentially differences in organic matter content or quality in the different sediment depths in which their species typically reside.

For microbial food webs, it is far more challenging to resolve the isotopic composition among bacteria, archaea, eukaryotic microbes, and detritus (dead organic matter), in particular for N, as there are few biomarkers for tracing N through the microbial domain (Boschker and Middelburg 2002). A biomarker for bacterial peptidoglycan, D-alanine, has been used to trace N uptake and flow through bacteria in coastal waters, seagrass beds, and sediments (Veuger et al. 2006, 2012; Eyre et al. 2016). Stable N isotope analyses of diaminopimelic acid have also been used to track bacterial N (Tobias et al. 2003a, b), although small yields can make this approach challenging. Assimilation of N by heterotrophic microbes feeding on detrital resources leads to elevated $\delta^{15}N$ (Caraco et al. 1998) with consequences for the isotope composition, hence trophic-level inference, of top consumers (Steffan et al. 2017). However, Guiterrez-Rodriguez et al. (2014) have shown that protozoan trophic steps do not necessarily lead to an increase in $\delta^{15}N$. Polar lipid-derived fatty acids (PLFAs) are another suitable biomarker, in this case for ^{13}C, for microbial food web studies; these fatty acids are present in relatively constant amounts in organisms of interest (Boschker and Middelburg 2002; Middelburg 2014). However, the specificity of PLFA (for identification of microorganisms) is rather low, and thus, bacteria and eukaryotes can be distinguished quite well from each other, but PLFA spectra of many phytoplankton groups overlap (Dijkman et al. 2009). This implies that resolving among phytoplankton groups should be done with utmost care (Van den Meersche et al. 2008; Dijkman et al. 2009). Work is rapidly advancing to unravel the extent of fractionation variation between major algal taxa and growth conditions. In all, the advancement in the ability to apply isotopic tracers to understand trophic position will be important in understanding climate forcing and anthropogenic activities and their influence on energy/material transfer, predation, and competition within food webs as well the ocean's biological CO_2 pump.

20.6 Microbial Food Webs and Inclusion of Mixoplankton

Food web interactions are even more complicated when one focuses on the microbes only. The microbial loop concept, first introduced by Pomeroy (1974), and more formally described by Azam et al. (1983), led to an understanding that the microbial components of marine food webs were consumers and decomposers. This conceptual change also recognized the importance of grazing of bacteria by protozooplankton (2–20 μm size range) wherein these grazers formed an important linkage between the marine microbial production and higher trophic levels. The microbial web, which encompasses not only bacteria, but also viruses, archaea, and heterotrophic protists, rather than unidirectionally decomposing materials, provides pathways by which dissolved organic matter (DOM), largely derived from phytoplankton exudates, is recycled and ultimately either passed to higher trophic levels or sequestered to deeper oceans in recalcitrant forms (Polimene et al. 2017). The heterotrophic protists, as consumers of bacteria, further contribute to the dissolved organic pool via excretion and to the particulate organic pool via production of fecal material. Thus, the protist predator-bacterial prey interactions, coupled with bacterial nutrient uptake and decomposition activities, modulate not only the amounts and rates of inorganic and organic material available for phytoplankton and bacterial growth, respectively, but also the transfer of energy. As methods have advanced, specifically those for determining microbial grazing (e.g., dilution experiments; see Chap. 19), it has been revealed that microzooplankton, rather than mesozooplankton, are responsible for most of the consumption of primary production. Thus, the acceptance of the role of bacteria evolved from simple decomposers, as a significant and actively cycling flux of DOM became recognized, and understanding of the different forms of DOM improved. Furthermore, it was recognized that this active flux could only be accounted for by an active and abundant bacterial community—a community that could be counted with the advancement of methods (reviewed by Williams and Ducklow 2019; see Chap. 2).

The microbial food web has been further advanced with the understanding of the role of mixoplankton (Chap. 15). Mixoplankton perform the dual role of being primary producers, photosynthesizing with either their own or acquired chloroplasts, as well as being primary consumers ingesting bacteria, phytoplankton, microzooplankton, and indeed other mixoplankton. These processes and interactions then impact physiological vital rates and quality of release of DOM and/or dissolved inorganic material (DIM) through digestion, excretion, and leakage. While mixoplankton are themselves consumers, due to their wide size range, they can also be prey (or predators) for many primary and secondary consumers (Jeong et al. 2010; Flynn et al. 2019).

The mixoplankton paradigm assumes mixoplankton dominate the base of microbial food webs and, via their activity, short circuit the flow of energy and materials to primary producers in ways that are fundamentally different than those envisioned in a microbial food web dependent on bacteria for this recycling (Glibert and Mitra 2022). Mixoplankton may selectively feed on prey that have a nutritional composition complementary to themselves, and this, in turn, can alter the elemental stoichiometry of the mixoplankton and of its release products (e.g., Lundgren et al. 2016; Lin et al. 2017; Zhang et al. 2017). When a macrograzer consumes a mixoplankter, this upgraded trophic benefit is further transferred (Traboni et al. 2020). If the grazer is itself a mixoplankter, as in the case of the multi-link mixoplankton consortia of a bacterivorous cryptophyte, a ciliate *Mesodinium*, and the dinoflagellate *Dinophysis*, one may expect trophic upgrading to be accomplished at each step. On the other hand, when release products are altered, this can also create both positive and negative feedbacks that, in turn, alter the composition of the primary producers, potentially generating those that are unpalatable (e.g., Mitra and Flynn 2006). Ultimately, how the resource stoichiometric proportions compare with those of the mixoplankton could have propagating effects through the microbial loop, with implications for food web structure and functioning and carbon sequestration (Mitra et al. 2014; Polimene et al. 2017; Fig. 20.7). Such advantages afforded to mixoplankton may outweigh their supposedly comparatively poor growth rates in otherwise non-limiting nutrient conditions (e.g., Burkholder et al. 2008). Similarly, in oceanic waters, mixoplanktonic activity may help to explain why oceanic "phytoplankton" have the ability to grow relatively rapidly even though inhabiting oligotrophic waters (Goldman et al. 1979); they may be opportunistic in taking up dissolved nutrients, and they may graze on

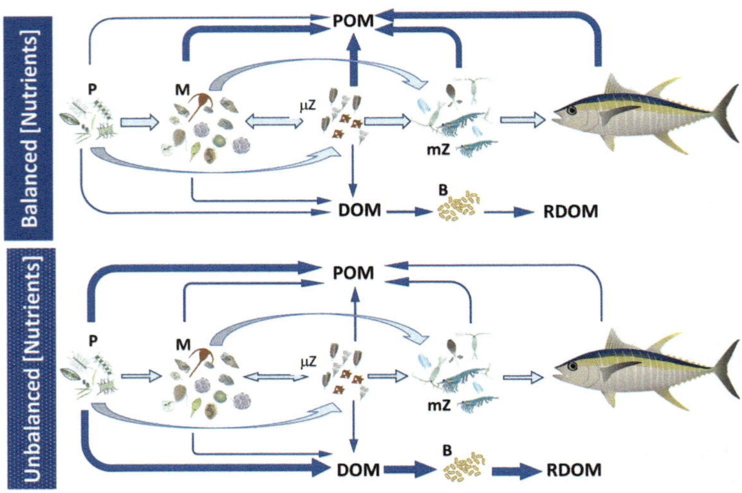

Fig. 20.7 Conceptual model demonstrating the impact of different environment nutrient concentrations on protist plankton stoichiometry and the biological and microbial carbon pumps (BP and MCP) under the mixoplankton paradigm. P is phytoplankton, M is mixoplankton, μZ is microzooplankton, mZ is mesozooplankton, B is bacteria, DOM is dissolved organic matter, and RDOM is refractory dissolved organic matter. Image reproduced from Glibert and Mitra (2022, Limnol Oceanogr 67: doi: 10.1002/lno.12018) under a creative commons license

bacteria or cyanobacteria or other prey to acquire nutrients (Zubkov and Tarran 2008; Hartmann et al. 2012; Mitra et al. 2014).

Much is yet to be explored—empirically and in models—with respect to the significance of mixoplankton in ocean biogeochemistry, C and nutrient cycling, and trophic dynamics. Modeling studies have shown the importance of considering phototrophy and phagotrophy in mixoplankton synergistically (Flynn and Mitra 2009). Indeed, models that attribute a portion of phytoplankton activity or zooplankton activity to mixotrophy cannot capture the implications for biogeochemical cycling and trophic dynamics (Mitra and Flynn 2010; see Chap. 29). In the absence of mixoplankton, the C fixation and production of dissolved organics in the open oceans could be severely underestimated with incorrect predictions associated with the microbial C pump (Mitra et al. 2014; Polimene et al. 2017). Experiments are required that explicitly explore the processes of photosynthesis, grazing, and nutrient status under varying environmental conditions—including conditions that are outside the traditionally considered "normal" range of temperature, CO_2, nutrients, pH, and allied factors.

20.7 Summary

Our understanding has evolved from Lindeman's original trophic dynamic concept, but the basic ideas of energy flow, trophic interactions, and cascading effects have stood the test of time. Understanding the flow of energy through food webs has increased with the recognition that elemental balance of food intake has ramifications beyond the immediate trophic level. Methods to quantify trophic position have greatly advanced with isotopic approaches. System complexity has becomes better understood. Our classical models of trophic interactions and energy transfer, while elegant, may need to be challenged as our understanding of food webs advances. Our understanding of biological complexity, multiple trophic interactions, mixoplankton that combine autotrophy and heterotrophy in a single organism, and nutrient regeneration has led to Lindeman's diagram being recast as tangled web of direct and indirect effects (Fig. 20.2). New approaches to assessing trophic position, along with advancing modeling approaches, will not only allow us better insight into plankton food webs but also their resilience to change, whether due to unforeseen regional and global changes or directed management actions.

References

Alsterberg G (1924) Die Nahrungszirkulation einiger Binnenseetypen. Arch Hydrobiol 15:291–338

Arendt K, Jónasdóttir S, Hansen P, Gartner S (2005) Effects of dietary fatty acids on the reproductive success of the calanoid copepod *Temora longicornis*. Mar Biol 146:513–530

Auerswald K, Wittmer MHOM, Zazzo A, Schaeufele R, Schnyder H (2010) Biases in the analysis of stable isotope discrimination in food webs. J Appl Ecol 47:936–941. https://doi.org/10.1111/j.1365-2664.2009.01764.x

Azam F, Fenchel T, Field JG, Gray JG, Meyer-Reil LA, Thingstad F (1983) The ecological role of water-column microbes in the sea. Mar Ecol Prog Ser 10:257–263. https://doi.org/10.3354/meps010257

Beausgrand G, Brander KM, Lindley JA, Souissi S, Reid PC (2003) Plankton effect on cod recruitment in the North Sea. Nature 426:661–664. https://doi.org/10.1038/nature02164

Bentley KM, Pierson JJ, Glibert PM (2021) Physiological responses of the copepods *Acartia tonsa* and *Eurytemora carolleeae* to changes in the nitrogen:phosphorus quality of their food. Nitrogen 2:62–85. https://doi.org/10.3390/nitrogen2010005

Boersma M, Elser JJ (2006) Too much of a good thing: on stoichiometrically balanced diets and maximal growth. Ecology 87:1325–1330

Boersma M, Aberle N, Hantzsche FM, Shoo KL, Wiltshire KH, Malzahn AM (2008) Nutritional limitation travels up the food chain. Int Rev Hydrobiol 93:479–488

Boschker HTS, Middelburg JJ (2002) Stable isotopes and biomarkers in microbial ecology. FEMS Microbiol Ecol 40:85–95. https://doi.org/10.1111/j.1574-6941.2002.tb00940.x

Brown JH, Gillooly JF, Allen AP, Savage VM, West GB (2004) Toward a metabolic theory of ecology. Ecology 85:1771–1789. https://doi.org/10.1890/03-9000

Bunn SE, Leigh C, Jardine TD (2013) Diet–tissue fractionation of δ15N by consumers from streams and rivers. Limnol Oceanogr 58:765–773. https://doi.org/10.4319/lo.2013.58.3.0765

Burkholder JM, Glibert PM, Skelton H (2008) Mixotrophy, a major mode of nutrition for harmful algal species in eutrophic waters. Harmful Algae 8:77–93

Caraco NF, Lampman GG, Cole JJ, Limburg KE, Pace ML, Fischer DT (1998) Microbial assimilation of DIN in a nitrogen rich estuary: implications for food quality and isotope studies. Mar Ecol Prog Ser 167:59–71

Caut S, Angulo E, Courchamp F (2009) Variation in discrimination factors ($\Delta^{15}N$ and $\Delta^{13}C$): the effect of diet isotopic values and applications for diet reconstruction. J Appl Ecol 46:443–453

Checkley DM (1980) Food limitation of egg production by a marine, planktonic copepod in the sea off southern California. Limnol Oceanogr 25:991–998

Chikaraishi Y, Ogawa NO, Kashiyama Y, Takano Y, Suga H, Tomitani A, Miyashita H et al (2009) Determination of aquatic food-web structure based on compound-specific nitrogen isotopic composition of amino acids. Limnol Oceanogr Meth 7:740–750

Christianen MJA, Middelburg JJ, Holthuijsen SJ, Jouta J, Compton TJ, van der Heide T, Piersma T et al (2017) Benthic primary producers are key to sustain the Wadden Sea food web - stable carbon isotope analysis at landscape scale. Ecology 98:1498–1512

Cifuentes LA, Sharp JH, Fogel ML (1988) Stable carbon and nitrogen isotope biogeochemistry in the Delaware estuary. Limnol Oceanogr 33:1102–1115

Cole JJ, Solomon CT (2012) Terrestrial support of zebra mussels and the Hudson River food web: a multi-isotope, Bayesian analysis. Limnol Oceanogr 57:1802–1815

Cole JJ, Carpenter SR, Kitchell JF, Pace ML, Solomon CT, Weidel B (2011) Strong evidence for terrestrial support of zooplankton in small lakes based on stable isotopes of carbon, nitrogen, and hydrogen. Proc Natl Acad Sci USA 108:1975–1980. https://doi.org/10.1073/pnas.1012807108

Cook RE (1977) Raymond Lindeman and the trophic-dynamic concept in ecology. Science 198: 22–26

Costanzo SD, O'Donohue MJ, Dennison WC, Loneragan NR, Thomas M (2001) A new approach for detecting and mapping sewage impacts. Mar Poll Bull 42:149–156

Croisetière L, Hare L, Tessier A, Cabana G (2009) Sulphur stable isotopes can distinguish trophic dependence on sediments and plankton in boreal lakes. Freshwat Biol 54:1006–1015. https://doi.org/10.1111/j.1365-2427.2008.02147.x

Cross WF, Hood JM, Benstead JP, Huryn AD, Nelson D (2015) Interactions between temperature and nutrients across levels of ecological organization. Glob Change Biol 21:1025–1040. https://doi.org/10.1111/gcb.12809

D'Alelio D, Russo L, del Gaizo G, Caputi L (2022) Plankton under pressure: how water conditions alter the phytoplankton-zooplankton link in coastal lagoons. Water 14:974. https://doi.org/10.3390/w14060974

De Troch M, Boeckx P, Cnuddle C, Van Gansbeke D, Vanreusel A, Vincx M, Caramujo MJ (2012) Bioconversion of fatty acids at the basis of marine food webs: insights from a compound-specific stable isotope analysis. Mar Ecol Prog Ser 465:53–67. https://doi.org/10.3354/meps09920

DeMott WR, Moxter F (1991) Foraging on cyanobacteria by copepods—responses to chemical defenses and resource abundance. Ecology 72:1820–1834

DeNiro MJ, Epstein S (1978) Influence of diet on the distribution of carbon isotopes in animals. Geochim Cosmochim Acta 42:495–506

DeNiro MJ, Epstein S (1981) Influence of diet on the distribution of nitrogen isotopes in animals. Geochim Cosmochim Acta 45:341–351

Dijkman NA, Boschker HTS, Middelburg JJ, Kromkamp JC (2009) Group-specific primary production based on stable-isotope labeling of phospholipid-derived fatty acids. Limnol Oceanogr Meth 7:612–625

Doucett RR, Marks JC, Blinn DW, Caron M, Hungate BA (2007) Measuring terrestrial subsidies to aquatic food webs using stable isotopes of hydrogen. Ecology 88:1587–1592. https://doi.org/10.1890/06-1184

Elton C (1927) Animal ecology. Macmillan, New York

Eyre BD, Oakes JM, Middelburg JJ (2016) Fate of microphytobenthos nitrogen in subtropical subtidal sediments: a ^{15}N pulse-chase study. Limnol Oceanogr 61:2108–2121

Færøvig PJ, Hessen DO (2003) Allocation strategies in crustacean stoichiometry: the potential role of phosphorus in the limitation of reproduction. Freshwat Biol 48:17821892. https://doi.org/10.1046/j.1365-2427.2003.01128.x

Flynn KJ, Mitra A (2009) Building the "perfect beast": modelling mixotrophic plankton. J Plankton Res 31:965–992

Flynn KJ, Mitra A, Anestis K, Anschütz AA, Calbet A, Ferreira GD, Gypens N et al (2019) Mixotrophic protists and a new paradigm for marine ecology: where does plankton research go now? J Plankton Res 41:375–391. https://doi.org/10.1093/plankt/fbz026

France R (1995) Carbon-13 enrichment in benthic compared to planktonic algae: foodweb implications. Mar Ecol Progr Ser 124:307–312

Frank KT, Petrie BD, Choi JS, Leggett (2005) Trophic cascades in a formerly cod-dominated ecosystem. Science 308:1621–1623. https://doi.org/10.1126/science.1113075

Fry B (2006) Stable isotope ecology. Springer, Dordrecht, p 308

Ger KA, Naus-Wiezer S, De Meester L, Lürling M (2018) Zooplankton grazing selectivity regulates herbivory and dominance of toxic phytoplankton over multiple prey generations. Limnol Oceanogr 64:1214–1227. https://doi.org/10.1002/lno.11108

Glibert PM (2015) More than propagule pressure: successful invading algae have physiological adaptations suitable to anthropogenically changing nutrient environments. Aquat Ecosyst Health Manag 18:334–341

Glibert PM (2017) Eutrophication, harmful algae and biodiversity - challenging paradigms in a world of complex nutrient changes. Mar Poll Bull 124:591–606. https://doi.org/10.1016/j.marpolbul.2017.04.027

Glibert PM, Mitra A (2022) From webs, loops, shunts, and pumps to microbial multitasking: evolving concepts of marine microbial ecology, the mixoplankton paradigm, and implications for a future ocean. Limnol Oceanogr 67(3):585–597. https://doi.org/10.1002/lno.12018

Glibert PM, Fullerton D, Burkholder JM, Cornwell JC, Kana TM (2011) Ecological stoichiometry, biogeochemical cycling, invasive species, and aquatic food webs: San Francisco estuary and comparative systems. Rev Fish Sci 19:358–417. https://doi.org/10.1080/10641262.2011.611916

Glibert PM, Kana TM, Brown K (2013) From limitation to excess: the consequences of substrate excess and stoichiometry for phytoplankton physiology, trophodynamics and biogeochemistry,

and the implications for modeling. J Mar Syst 125:14–28. https://doi.org/10.1016/j.jmarsys.2012.10.004
Glibert PM, Middelburg JJ, McClelland JW, Vander Zanden MJ (2019) Stable isotope tracers: enriching our perspectives and questions on sources, fates, rates, and pathways of major elements in aquatic systems. Limnol Oceanogr 64:950–981. https://doi.org/10.1002/lno.11087
Goldman JC, McCarthy JJ, Peavey DG (1979) Growth rate influence on the chemical composition of phytoplankton in oceanic waters. Nature 279:210–215
Guisande C, Riveiro I, Maneiro I (2000) Comparisons among the amino acid composition of females, eggs and food to determine the relative importance of food quantity and food quality to copepod reproduction. Mar Ecol Prog Ser 202:135–142
Guiterrez-Rodriguez A, Décima M, Popp BN, Landry MR (2014) Isotopic invisibility of protozoan trophic steps in marine food webs. Limnol Oceanogr 59:1590–1598. https://doi.org/10.4319/lo.2014.59.5.1590
Hairston NG, Smith FE, Slobodkin LB (1960) Community structure, population control, and competition. Am Nat 94:421–425
Hall EK, Singer GA, Kainz MJ, Lennon JT (2010) Evidence for a temperature acclimation mechanism in bacteria: an empirical test of a membrane-mediated trade-off. Funct Ecol 24: 898–908. https://doi.org/10.1111/j.1365-2435.2010.01707.x
Hartmann M, Grub C, Tarran GA, Martin AP, Burkill PH, Scanlan DJ, Zubkov MV (2012) Mixotrophic basis of Atlantic oligotrophic ecosystem. Proc Natl Acad Sci USA 109:5756–5760
Hassett RP, Cardinale B, Stabler LB, Elser JJ (1997) Ecological stoichiometry of N and P in pelagic ecosystems: comparison of lakes and oceans with emphasis on the zooplankton-phytoplankton interaction. Limnol Oceanogr 42:648–662
Hessen DO, Anderson TR (2008) Excess carbon in aquatic organ isms and ecosystems: physiological, ecological, and evolutionary implications. Limnol Oceanogr 53:1685–1696
Hessen DO, Kaartveft S (2014) Top-down cascades in lakes and oceans: different perspectives but same story? J Plankton Res 36:914–924. https://doi.org/10.1093/plankt/fbu040
Hessen DO, Elser JJ, Sterner RW, Urabe J (2013) Ecological stoichiometry: an elemental approach using basic principles. Limnol Oceanogr 58:2219–2236. https://doi.org/10.4319/lo.2013.58.6.2219
Hood JM, Sterner RW (2010) Diet mixing: do animals integrate growth or resources across temporal heterogeneity? Am Nat 176:651–663. https://doi.org/10.1086/656489
Hutchinson GE (1961) The paradox of the plankton. Am Nat 882:137–145
Iroigoen X, Flynn KJ, Harris RP (2005) Phytoplankton blooms: a 'loophole' in microzooplankton grazing impact? J Plankton Res 27:313–321
Jeong HJ, Yoo YD, Kim JS, Seong KA, Kang NS, Kim TH (2010) Growth, feeding and ecological roles of the mixotrophic and heterotrophic dinoflagellates in marine, planktonic food webs. Ocean Sci J 45:65–91. https://doi.org/10.1007/s12601-010-0007-2
Jeyasingh PD (2007) Plasticity in metabolic allometry: the role of dietary stoichiometry. Ecol Lett 10:282–289. https://doi.org/10.1111/j.1461-0248.2007.01023.x
Karlsson J, Berggren M, Ask J, Bystrom P, Jonsson A, Laudon H, Jansson M (2012) Terrestrial organic matter support of lake food webs: evidence from lake metabolism and stable hydrogen isotopes of consumers. Limnol Oceanogr 57:1042–1048
Kendall C (1998) Tracing nitrogen sources and cycling in catchments. In: Kendall C, McDonnell JJ (eds) Isotopic tracers in catchment hydrology. Elsevier, Amsterdam, pp 519–576
Kilham SS, Kreeger DA, Goulden CE, Lynn SG (1997) Effects of nutrient limitation on biochemical constituents of *Ankistrodesmus falcatus*. Freshwat Biol 38:591–596
Klein Breteler WCM, Schogt N, Baas M, Schouten S, Kraay GW (1999) Trophic upgrading of food quality by protozoans enhancing copepod growth: role of essential lipids. Mar Biol 135:191–198. https://doi.org/10.1007/s002270050616
Klein Breteler WCM, Koski M, Rampen (2004) Role of essential lipids in copepod nutritions: no evidence for trophic upgrading of food quality by a marine ciliate. Mar Ecol Prog Ser 274:199–208

Kleppel G, Buckart C, Houchin L (1998) Nutrition and the regulation of egg production in the calanoid copepod *Acartia tonsa*. Limnol Oceanogr 43:1000–1007

Lapointe BE, Thacker K, Hanson C, Getten L (2011) Sewage pollution in Negril, Jamaica: effects on nutrition and ecology of coral reef macroalgae. Chin J Oceanol Limnol 29:775–789

Laspoumaderes C, Modenutti B, Balseiro E (2010) Herbivory versus omnivory: linking homeostasis and elemental imbalance in copepod development. J Plankton Res 32:1573–1582

Laspoumaderes C, Modenutti B, Elser JJ, Balseiro E (2015) Does the stoichiometric carbon: phosphorus knife edge apply for predaceous copepods? Oecologia 178:557–569

Ledesma M, Gorokhova E, Holmstrand H, Garbaras A, Karlson AML (2020) Nitrogen isotope composition of amino acids reveals trophic positions in two sympatric amphipods. Ecol Evol 10: 10773–10784. https://doi.org/10.1002/ece3.6734

Lin C-H, Accoroni S, Glibert PM (2017) Mixotrophy in the dinoflagellate *Karlodinium veneficum* under variable nitrogen:phosphorus stoichiometry: feeding response and effects on larvae of the eastern oyster (*Crassostrea virginica*). Aquat Microb Ecol 79 101–114. https://doi.org/10.3354/ameo01823

Lindeman RE (1942) Trophic-dynamic aspect of ecology. Ecology 23:399–418

Loomer HA, Oakes KD, Schiff SL, Taylor WD, Servos MR (2014) Use of stable isotopes to trace municipal wastewater effluents into food webs within a highly developed river system. River Res Appl 31:1093–1100. https://doi.org/10.1002/rra.2826

Lorrain A, Paulet Y-M, Chauvaud L, Savoye N, Donval A, Saout C (2002) Differential δ^{13}C and δ^{15}N signatures among scallop tissues: implications for ecology and physiology. J Exp Mar Biol Ecol 275:47–61

Lotka AJ (1956) Elements of mathematical biology. Dover, New York

Lundgren V, Glibert PM, Granéli E, Vidyarathna NK, Fiori E, Ou L, Flynn KJ et al (2016) Metabolic and physiological changes in *Prymnesium parvum* when grown under, and grazing on, prey of variable nitrogen:phosphorus stoichiometry. Harmful Algae 55:1–12

Madigan DJ, Carlisle AB, Dewar H, Snodgrass OE, Litvin SY, Micheli F, Black BA (2012) Stable isotope analysis challenges wasp-waist food web assumptions in an upwelling pelagic ecosystem. Sci Rep 2:654. https://doi.org/10.1038/srep00654

Malzahn AM, Boersma M (2012) Effects of poor food quality on copepod growth are dose dependent and non-reversible. Oikos 121:1408–1416. https://doi.org/10.1111/j.1600-0706.2011.20186.x

Malzahn AM, Aberle-Malzahn N, Clemmesen C, Boersma M (2007) Nutrient limitation of primary producers affects planktivorous fish condition. Limnol Oceanogr 52:2062–2071

Malzahn AM, Hantzche F, Schoo KL, Boersma M, Aberle N (2010) Differential effects of nutrient-limited primary production on primary, secondary or tertiary consumers. Oecologia 162:35–48

McCarthy JW, Valiela I (1998) Linking nitrogen in estuarine producers to land-derived sources. Limnol Oceanogr 43:577–585

McCarthy JW, Valiela I, Michener RH (1997) Nitrogen-stable isotope signatures in estuarine food webs: a record of increasing urbanization in coastal watersheds. Limnol Oceanogr 42: 930–937

McCarthy MD, Benner R, Lee C, Fogel ML (2007) Amino acid nitrogen isotopic fractionation patterns as indicators of heterotrophy in plankton, particulate, and dissolved organic matter. Geochim Cosmochim Acta 71:4727–4744. https://doi.org/10.1016/j.gca.2007.06.061

McCarthy MD, Lehman J, Kudela R (2013) Compound-specific amino acid δ^{15}N patterns in marine algae: tracer potential for cyanobacterial vs. eukaryotic organic nitrogen sources in the ocean. Geochim Cosmochim Acta 103:104–120

McClelland JW, Montoya JP (2002) Trophic relationships and the nitrogen isotopic composition of amino acids in plankton. Ecology 83:2173–2180

McClelland JW, Holl CW, Montoya JP (2003) Relating low δ^{15}N values of zooplankton to nation in the tropical North Atlantic: insights provided by stable isotope ratios of amino acids. Deep Sea Res II 50:849–861

McMahon KW, McCarthy MD (2016) Embracing variability in amino acid $\delta^{15}N$ fractionation: mechanisms, implications, and applications for trophic ecology. Ecosphere 7:e01511. https://doi.org/10.1002/ecs2.1511

Middelburg JJ (2014) Stable isotopes dissect aquatic food webs from the top to the bottom. Biogeosciences 11:2357–2371. https://doi.org/10.5194/bg-11-2357-2014

Middelburg JJ, Herman PMJ (2007) Organic matter processing in tidal estuaries. Mar Chem 106: 127–147

Minagawa M, Wada E (1984) Stepwise enrichment of N-15 along food chains: further evidence and the relation between $\delta^{15}N$ and animal age. Geochim Cosmochim Acta 48:1135–1140. https://doi.org/10.1016/0016-7037(84)90204-7

Mitra A, Flynn KJ (2005) Predator–prey interactions: is 'ecological stoichiometry' sufficient when good food goes bad? J Plankton Res 27:393–399. https://doi.org/10.1093/plankt/fbi022

Mitra A, Flynn KJ (2006) Promotion of harmful algal blooms by zooplankton predatory activity. Biol Lett 2:194–197

Mitra A, Flynn KJ (2010) Modelling mixotrophy in harmful algal blooms: more or less the sum of the parts? J Mar Syst 83:158–169

Mitra A, Flynn KJ, Burkholder JM, Berge T, Calbet A, Raven JA, Granéli E et al (2014) The role of mixotrophic protists in the biological carbon pump. Biogeosciences 11:995–1005. https://doi.org/10.5194/bg-11-995-2014

Moe SJ, Stelzer RS, Forman MR, Harpole WS, Daufresne T, Yoshida T (2005) Recent advances in ecological stoichiometry: insights for population and community ecology. Oikos 109:29 –/39

Moorthi SD, Schmitt JA, Ryabov A, Tsakalakis I, Blasius B, Prelle L et al (2016) Unifying ecological stoichiometry and metabolic theory to predict production and trophic transfer in a marine planktonic food web. Philos Trans R Soc B Biol Sci 371:20150270. https://doi.org/10.1098/rstb.2015.0270

Nobili R, Robinson C, Buitenhuis E, Castellani C (2013) Food quality regulates the metabolism and reproduction of *Temora longicornis*. Biogeosci Discuss 10:3203–3239

Olsen Y (1999) Lipids and essential fatty acids in aquatic food webs: what can freshwater ecologists learn from mariculture? In: Arts MT, Wainman BC (eds) Lipids in freshwater ecosystems. Springer, New York, pp 161–202

Owens NJP (1987) Natural variations in nitrogen-15 in the marine environment. Adv Mar Biol 24: 389–451

Pace ML, Cole JJ, Carpenter SR, Kitchell JK (1999) Trophic cascades revealed in diverse ecosystems. Trends Ecol Evol 14:483–488

Peace A, Zhao Y, Loladze I, Elser JJ, Kuang Y (2013) A stoichiometric producer-grazer model incorporating the effects of excess food- nutrient content on consumer dynamics. Math Biosci 244:107–115

Pershing AJ, Mills KE, Record NR, Stamieszkin K, Wurtzell KV, Byron CJ, Fitzpatrick D et al (2015) Evaluating trophic cascades as drivers of regime shifts in different ocean ecosystems. Philos Trans R Soc B 370:20130265. https://doi.org/10.1098/rstb.2013.0265

Peterson BJ, Fry B (1987) Stable isotopes in ecosystem studies. Ann Rev Ecol Syst 18:293–320. https://doi.org/10.1146/annurev.ecolsys.18.1.293

Peterson BJ, Fry B, Hullar M, Saupe S, Wright R (1994) The distribution and stable isotopic composition of dissolved organic carbon in estuaries. Estuaries 18:111–121

Phillips DL (2012) Converting isotope values to diet composition: the use of mixing models. J Mammal 93:342–352. https://doi.org/10.1644/11-MAMM-S-158.1

Phillips DL, Koch PL (2002) Incorporating concentration dependence in stable isotope mixing models. Oecologia 130:114–125. https://doi.org/10.1007/s004420100786

Plath K, Boersma M (2001) Mineral limitation of zooplankton: stoichiometry constraints and optimal foraging. Ecology 82:1260–1269

Polimene L, Sailley S, Clark D, Mitra A, Allen JI (2017) Biological or microbial carbon pump? The role of phytoplankton stoichiometry in ocean carbon sequestration. J Plankton Res 39:180–186

Pomeroy LR (1974) The ocean's food web, a changing paradigm. BioSci 24:499–504. https://doi.org/10.2307/1296885

Post DM (2002) Using stable isotopes to estimate trophic position: models, methods, and assumptions. Ecology 83:703–718. https://doi.org/10.2307/3071875

Rawson DS (1930) The bottom fauna of Lake Simcoe and its role in the ecology of the lake. Ontario Fisheries Research Laboratory

Salomons W, Mook W (1981) Field observations of the isotopic composition of particulate organic carbon in the southern North Sea and adjacent estuaries. Mar Geol 41:11–20

Shelford VE (1918) Conditions of existence. In: Ward HB, Whipple GC (eds) Freshwater biology. Wiley, New York, pp 21–60

Shoo KL, Aberle N, Malzahn AM, Boersma M (2009) Does the nutrient stoichiometry of primary producers affect the secondary consumer *Pleurobrachia pileus*? Aquat Ecol 44:233–242

Smith BN, Epstein S (1970) Biogeochemistry of stable isotopes of hydrogen and carbon in salt marsh biota. Plant Physiol 46:738–742. https://doi.org/10.1104/pp.46.5.738

Sobczak WV (2005) Lindeman's trophic-dynamic aspect of ecology: "will you still need me when I'm 64?". Limnol Oceanogr Bull 14:53–57

Solomon CT, Carpenter SR, Clayton MK, Cole JJ, Coloso JJ, Pace ML, Vander Zanden MJ et al (2011) Terrestrial, benthic, and pelagic resource use in lakes: results from a three-isotope Bayesian mixing model. Ecology 92:1115–1125

Steffan SA, Chikaraishi Y, Dharampal PS, Pauli JN, Guédot C, Ohkouchi N (2017) Unpacking brown food-webs: animal trophic identity reflects rampant microbivory. Ecol Evol 7:3532–3541. https://doi.org/10.1002/ece3.2951

Sterner RW (2014) Raymond Laurel Lindeman and the trophic dynamic viewpoint. Limnol Oceanogr Bull 21:38–51. https://doi.org/10.1002/lob.201221237

Sterner RW, Elser JJ (2002) Ecological stoichiometry: the biology of elements from molecules to the biosphere. Princeton University Press, Princeton, NJ, 439 pp

Strøm KW (1928) Recent advances in limnology. Proc Linnean Soc London 2:96–110

Tang KW, Taal M (2005) Trophic modification of food quality by heterotrophic protists: species-specific effects on copepod egg production and egg hatching. J Exp Mar Biol Ecol 318:85–98

Tang Y, Yang X, Xu R, Zhang X, Liu Z, Zhang Y, Dumont HJ (2018) Heterotrophic microbes upgrade food value of a terrestrial carbon resource for *Daphnia magna*. Limnol Oceanogr 64: 474–482. https://doi.org/10.1002/lno.11052

Tank SE, Lesack LFW, Gareis JAL, Osborn CL, Hesslein RH (2011) Multiple tracers demonstrate distinct sources of dissolved organic matter to lakes of the Mackenzie Delta, western Canadian Arctic. Limnol Oceanogr 56:1297–1309

Thienemann A (1926) Der Nahrungskreislauf im Wasser. Zoologisher Anzeiger Supplement 2:29–79

Tobias C, Cieri M, Peterson BJ, Deegan LA, Vallino J, Hughes J (2003a) Processing watershed-derived nitrogen in a well-flushed New England estuary. Limnol Oceanogr 48:1766–1778

Tobias C, Giblin A, McClelland J, Tucker J, Peterson B (2003b) Sediment DIN fluxes and preferential recycling of benthic microalgal nitrogen in a shallow macrotidal estuary. Mar Ecol Prog Ser 257:25–36

Traboni C, Calbet A, Saiz E (2020) Effects of prey trophic mode on the gross-growth efficiency of marine copepods: the case of mixoplankton. Sci Rep 10:1–14

Tucker J, Sheats N, Giblin AE, Hopkinson CS, Montoya JP (1999) Using stable isotopes to trace sewage-derived material through Boston harbor and Massachusetts Bay. Mar Environ Res 48: 353–375

Vadeboncoeur Y, Jeppesen E, Vander Zanden MJ, Schierup H-H, Christoffersen K, Lodge DM (2003) From Greenland to green lakes: cultural eutrophication and the loss of benthic pathways in lakes. Limnol Oceanogr 48:1408–1418

Van den Meersche K, Soetaert K, Middelburg JJ (2008) A Bayesian compositional estimator for microbial taxonomy based on biomarkers. Limnol Oceanogr Meth 6:190–199

Vander Zanden HB, Soto DX, Bowen GJ, Hobson KA (2016) Expanding the isotopic toolbox: applications of hydrogen and oxygen stable isotope ratios to food web studies. Front Ecol Evol 4:20. https://doi.org/10.3389/fevo.2016.00020

Vander Zanden MJ, Rasmussen JB (2001) Variation in $\delta^{15}N$ and $\delta^{13}C$ trophic fractionation: implications for aquatic food web studies. Limnol Oceanogr 46:2061–2066

Vanderklift MA, Ponsard S (2003) Sources of variation in consumer-diet $\delta^{15}N$ enrichment: a meta-analysis. Oecologia 136:169–182. https://doi.org/10.1007/s00442-003-1270-z

Vanni MJ, McIntyre PB (2016) Predicting nutrient excretion of aquatic animals with metabolic ecology and ecological stoichiometry: a global synthesis. Ecology 97:3460–3471. https://doi.org/10.1002/ecy.1582

Vanni MJ, Flecker AS, Hood JM, Headworth JL (2002) Stoichiometry of nutrient recycling by vertebrates in a tropical stream: linking biodiversity and ecosystem function. Ecol Lett 5:285–293

Veuger B, van Oevelen D, Boscher HTS, Middelburg JJ (2006) Fate of peptidoglycan in an intertidal sediment: an in situ ^{13}C-labeling study. Limnol Oceanogr 51:1572–1580

Veuger B, van Oevelen D, Middelburg JJ (2012) Fate of microbial nitrogen, carbon, hydrolysable amino acids, monosaccharides, and fatty acids in sediment. Geochim Cosmochim Acta 83:217–233

Volterra V (1928) Variations and fluctuations of the number of individuals in animal species living together. ICES J Mar Sci 3:3–51. https://doi.org/10.1093/icesjms/3.1.3

Walve J, Larsson U (1999) Carbon, nitrogen and phosphorus stoichiometry of crustacean zooplankton in the Baltic Sea: implications for nutrient recycling. J Plankton Res 21:2309–2321

Welti N, Striebel M, Ulseth AJ, Cross WF, DeVilbiss S, Glibert PM, Guo L et al (2017) Bridging food webs, ecosystem metabolism, and biogeochemistry using ecological stoichiometry theory. Front Microbiol 8:1298. https://doi.org/10.3389/fmicb.2017.01298

Williams PJB, Ducklow H (2019) The microbial loop concept: a history, 1930–1974. J Mar Res 77 (Suppl):23–81

Winder M, Jassby AD, MacNally R (2011) Synergies between climate anomalies and hydrological modifications facilitate estuarine biotic invasions. Ecol Lett 14:749–757. https://doi.org/10.1111/j.1461-0248.2011.01635x

Zhang S, Liu H, Glibert PM, Guo C, Ke Y (2017) Elemental stoichiometry and nutrient excretion of *Noctiluca scintillans* in response to prey of different quality. Sci Rep 7:7622. https://doi.org/10.1038/s41598-017-05991-w

Zubkov MV, Tarran GA (2008) High bacterivory by the smallest phytoplankton in the North Atlantic Ocean. Nature 455:224–227

21 The Phycosphere-Living and Dying in Close Quarters

21.1 Introduction

Phytoplankton, even those in nutrient-deplete, oligotrophic waters, live in a crowded microenvironment. Their micro-neighborhood is teeming with other organisms, bacteria, viruses, and archaea. Some are helpful for their growth, some are not. The term "*phycosphere*," coined by Bell and Mitchell (1972), describes the region around algal cells "in which bacterial growth is stimulated by extracellular products of the that alga(e)." The term has since taken on the meaning of the diffusive boundary around algal cells in which other microbial interactions can occur via the organic compounds that are released by these cells. The "*microbiome*," the residents of this complex phycosphere, are diverse and complex. Their secrets have more recently been revealed through new tools and approaches, including metabarcoding and metagenomics and advanced chemical characterizations. Among those secrets is the identification of some of the *infochemicals* that exist within the phycosphere, those chemicals that carry cues, and which elicit various responses by the algal cell itself or its closely associated microbes. Community structure shapes the structure of these diverse microbes in ways that are only just beginning to be understood. Recall that it was not that long ago (see Chap. 1) when the very existence of bacteria in the oceans was questioned! This chapter examines the benefits and costs of living among a microbiome and the biogeochemical consequences of these networks of microbial interactions and their chemical signals, the "*plankton interactome*" (Lima-Mendez et al. 2015).

21.2 Benefits of Communal Living

The bioactive compounds produced by phytoplankton are many. Phytoplankton are leaky, and the organic carbon (C) compounds that exude, from simple C compounds to complex polysaccharides, become substrates for bacteria or may mediate

P. M. Glibert, *Phytoplankton Whispering: An Introduction to the Physiology and Ecology of Microalgae*, https://doi.org/10.1007/978-3-031-53897-1_21

allelopathic interactions with competing algae. Phytoplankton release many small hydrophilic compounds that may diffuse through the cell membrane, and they also produce many small and large molecules such as amino acids, monosaccharides, and proteins that may be translocated or actively transported to the extracellular space (e.g., Verner and Schatz 1988).

The very nature of these compounds can change depending on the type of phytoplankton and their physiological state. Small organic compounds, amino acids, sugar alcohols, and other highly labile compounds are most often produced by cells that are actively growing, while as cells begin to senesce, the compounds released may be more complex and of higher molecular weight (Seymour et al. 2017). Many of the small molecular weight compounds are chemical attractants for bacteria (e.g., Seymour et al. 2010). In turn, the ecological niches within the phycosphere for different bacteria and viruses may change over time. In other cases, specific alga taxa establish long-term relationships with specific taxa of bacteria, mostly Proteobacteria (e.g., Behringer et al. 2018).

Bacteria can become direct food for those phytoplankton having the ability to consume them (mixotrophs), and, if not directly eaten, bacteria can provide essential micronutrients and vitamins. Bacteria remineralize the organic compounds released by phytoplankton, in turn increasing inorganic nutrients such as NH_4^+. Concentration gradients of such nutrients may increase sufficiently to favor diffusion. Both bacteria and phytoplankton produce enzymes that sit on the cell surface, for example, alkaline phosphatase (see Chap. 13). Thus, when in close proximity, the breakdown of complex organic phosphorus (P) compounds may be of benefit to both the phytoplankton and the bacteria. Bacteria are also a source of vitamins for phytoplankton—something that the phytoplankton can obtain from the prokaryotes in exchange for the organic C they provide (e.g., Croft et al. 2005; Grant et al. 2014). For example, vitamin B_1, B_7, or B_{12} may be needed by specific taxa, but only available to them via their association with bacteria (e.g., Kazamia et al. 2012; Xie et al. 2013; Grant et al. 2014).

Dimethylsulfoniopropionate (DMSP), leaked by many types of phytoplankton, becomes a chemical attractant for bacteria (Seymour et al. 2010), in turn signaling quorum-sensing molecules in bacteria (Johnson et al. 2016). It also provides a substantial fraction of the C requirements of many bacteria (Kiene et al. 2000; Simó 2001). Nutrients, such as NH_4^+, are also cycled between bacteria and phytoplankton. Bacteria may transform sources of N that would otherwise not be available to phytoplankton. Iron (Fe) is another important micronutrient that exchanges between bacteria and phytoplankton. Bacteria can excrete small organic molecules with high affinity for Fe, becoming siderophores (see Chap. 16), facilitating Fe acquisition by the algal cell (e.g., Amin et al. 2007, 2012). In exchange, the bacteria benefit from the rich supply of organic C. Siderophores bind ferric ions with very high affinity, and they, in turn, provide a mechanism for the chelated Fe to be available for transport into phytoplankton cells via specific transporters. While some phytoplankton may have siderophores themselves (e.g., Kustka et al. 2007;

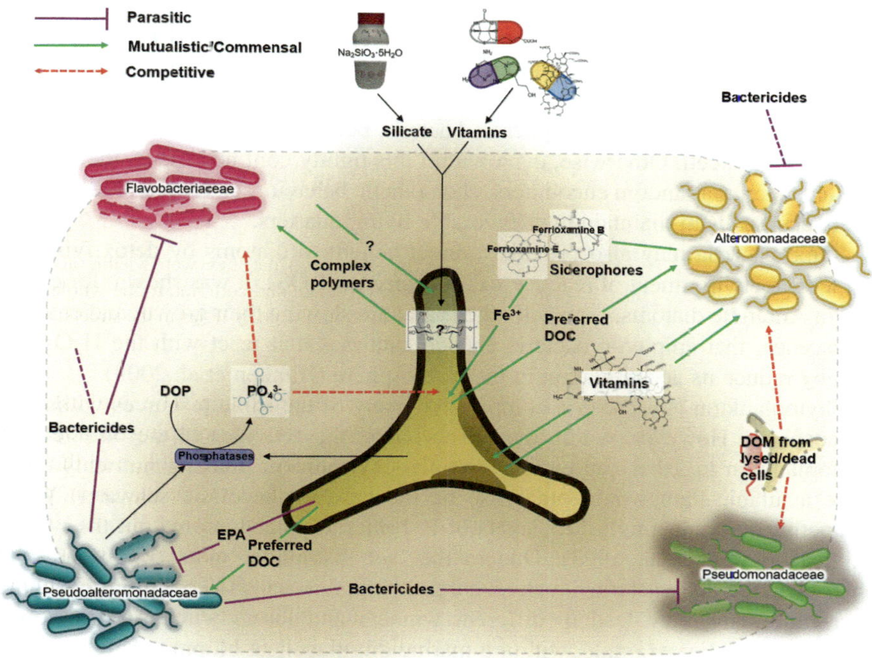

Fig. 21.1 Putative interactions between the diatom *Phaeodactylum tricornutum* and various bacterial families. Image reproduced from Moejes et al. (2017. Processes 5: 77. doi.org/10.3390/pr5040077) under a creative commons license

Sutak et al. 2020), other phytoplankton may make use of their microbial neighbor's such capability.

The microbiome of the diatom *Phaeodactylum tricornutum* has been well studied with regard to how different metabolites (Fe, vitamins, DOC, and PO_4^{3-}) are associated with different bacterial families (Moejes et al. 2017; Fig. 21.1). Metal acquisition is of particular relevance to the harmful diatom *Pseudo-nitzschia* spp., which produces domoic acid (DA, the neurotoxin—see Chap. 25). It appears that in the absence of bacteria, these diatoms fail to produce DA, or if they do, levels of production are comparatively low (e.g., Bates et al. 1995, 2004). The production of DA increases when cells are stressed for Fe or copper (Cu). DA binds Cu, but likely not Fe (due to its lower affinity). Interactions between Cu uptake, Fe affinity, and bacteria raise interesting questions (e.g., Guannel et al. 2011), including questions regarding the specific types of bacteria that may be favorable for this diatom when producing DA.

Diatoms, in particular, release copious amounts of organic material called transparent exopolymer particles (TEP), which are mostly acidic polysaccharides. Bacteria are attracted to these compounds and readily colonize them (Passow 2002; Amin et al. 2012). The attachment by bacteria appears to induce diatom cell aggregation. Bacteria, also, can release copious amounts of exopolysaccharides (EPS) in response to the presence of phytoplankton (Amin et al. 2012). These compounds are also involved in cell surface adhesion. Stickiness ensures that these cozy neighbors stay together. Within some phycospheres, bacteria become attached to the respective phytoplankton cell. Otherwise, even within this tightly confined space, the bacteria are dependent on random encounters, chemotactic behavior, or vertical transmission of attached cells to maintain their favorable living quarters.

Some bacteria may also provide a direct benefit to diatoms by detoxifying the immediate environment of reactive oxygen species (ROS). It was shown, for example, in Antarctic diatoms, especially those that are slowing their growth and entering senescence, that epiphytic bacteria encode catalases that react with the H_2O_2 and thereby reduce its accumulation in the phycosphere (Hünken et al. 2008).

Phytoplankton and bacteria compete for many of the same resources within the phycosphere. However, within the phycosphere, the activities of the bacteria and phytoplankton together can enrich the microenvironment, making nutrients much more available than would otherwise be the case in the dilute seawater. When competing for vanishingly low levels of P, bacteria typically come out the winner (e.g., Thingstad et al. 1993). Due to the fact that under most conditions both phytoplankton and bacteria are alive and growing, this leads to conditions changing over time. This may lead to different winners and losers within the microbial neighborhood. For example, within a marine dinoflagellate bloom, a succession of bacteria has been noted over time as the bloom developed and aged (e.g., Zhou et al. 2019).

21.3 Costs of Living in Close Quarters

Urban life, including microbial urban life, also comes with costs. There is competition for resources, in some cases with unfavorable outcomes for one or another, and there are defensive postures that develop to protect from urban neighbors that become too aggressive.

Allelopathic compounds that affect interactions between algal species are many (e.g., Fistarol et al. 2003; Ternon et al. 2018; Fernández-Herrera et al. 2021). Allelopathic chemicals from *Microcystis* have been shown to cause the collapse of *Peridinium gutanense* blooms (Vardi et al. 1999). Various diatom filtrates have been shown to cause DNA degradation and abnormal growth of the benthic dinoflagellate *Ostreopsis* cf. *ovata* (Pichierri et al. 2017). Studies by Wang et al. (2006) assessed the effects of *Alexandrium tamarense* on *Prorocentrum donghaiense*. When cultured together, *P. donghaiense* died when *A. tamarense* was provided at a density of one-fourth that of the *P. donghaiense* on a cell:cell basis, but both species survived when the cell density of *P. donghaiense* was increased tenfold (Wang et al. 2006). Allelopathic chemicals from algae can inflict harm on membrane integrity,

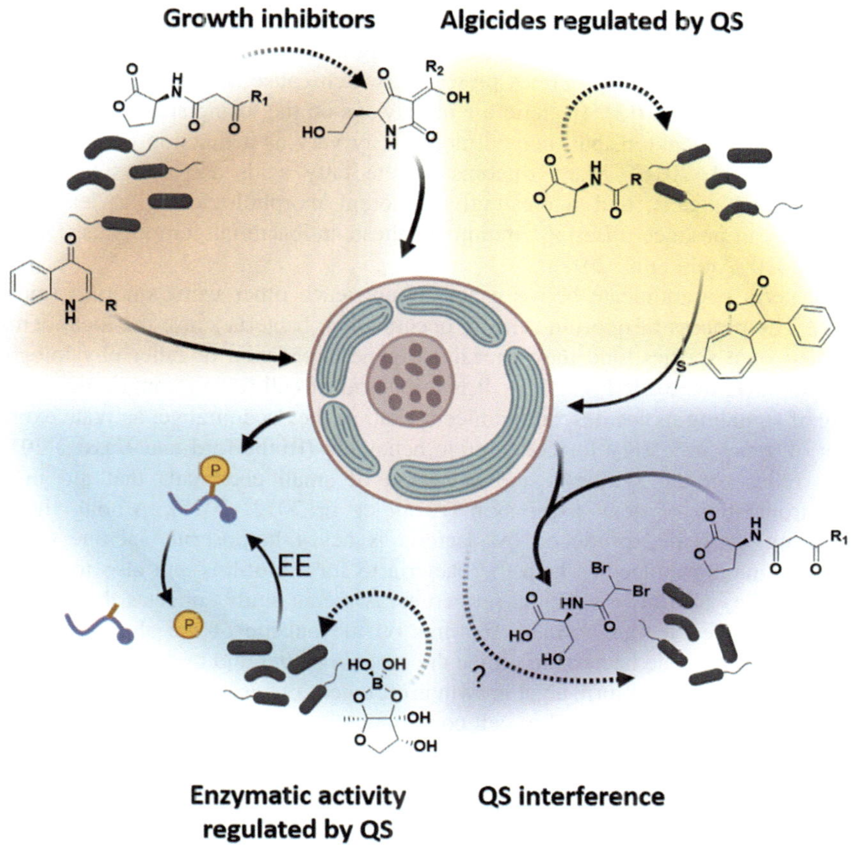

Fig. 21.2 Algae-bacteria interactions dictated by quorum-sensing signals in marine planktonic systems. These QS signals can be directly algicidal, as shown in the orange sector, such as alkylquinolones and tetramic acids that inhibit growth via photosynthesis inhibition. AHL-mediated quorum sensing (see text) regulates the production of other algicidal compounds (yellow sector). Quorum sensing regulates extracellular enzymatic activity in marine bacteria (blue sector), allowing these microbes to capitalize on encounters with nutrient hotspots, such as by liberating phosphates from organic molecules (blue curves with orange P) on marine particles and in phycospheres. Other ways by which microalgae interfere with quorum sensing are shown in the lilac sector. Dotted lines represent indirect actions of QS signals, while full lines represent direct effects. EE: extracellular enzymes. Image created with BioRender.com and reproduced from Dow (2021, Microorg 9: 1391. doi.org/10.3390/microorganisms9071391) under a creative commons license

photosynthetic activity or other growth functions (e.g., Ternon et al. 2018). Allelopathic chemicals can also induce changes in the activation of programmed cell death, as described further in Sect. 21.9.

Phytoplankton also have defense mechanisms for preventing the phycospheric bacteria from becoming too close, or too aggressive. In the case of the diatom *Chaetoceros didymus*, it secretes proteases, which cause cell lysis in the bacteria (Paul and Pohnert 2013). Bacteria are not always on the losing side. Some bacteria can kill phytoplankton, but their virulence may vary as a function of temperature (Mayers et al. 2016). Many diatoms secrete fatty acids and esters that act as antibacterial agents, and interestingly, different morphologies of at least some diatoms can produce differing amounts of these antibacterial compounds (Desbois et al. 2010; Amin et al. 2012).

Bacteria communicate between and among each other using quorum sensing (QS). The process of quorum sensing occurs when bacteria sense the local demise of their own species, and they coordinate gene expression or other physiological processes (Dow 2021; Fig. 21.2). It is a method of cell-to-cell communication by way of signaling molecules, autoinducers, that, as they accumulate, activate expression of genes necessary for cooperative behaviors (Rutherford and Bassler 2012). This relies on the synthesis and excretion of small chemicals that aid in the synchronization of gene expression (Amin et al. 2012, 2015). Among the QS signaling molecules produced by bacteria is acyl-L-homoserine lactone (AHL). These signaling molecules help the bacteria to form biofilms and also to degrade DMSP. They also have algicidal properties. A recent study suggests that once the signaling molecules are formed, biofilms on the outside of the algal cells form, thereby allowing the bacteria to utilize the organic matter and to degrade the DMSP (Zhang et al. 2022). In turn, algal growth is curtailed. Eukaryotes, also, engage in QS and other mechanisms of cell-to-cell communication. Both bacteria and eukaryotes also sense each other (Segev et al. 2016; Cirri and Pohnert 2019). Defensively, diatoms are able to disrupt the QS of motile bacteria or to modify QS signals using enzyme activity (Amin et al. 2012). Characterizing these compounds and the roles they play is an active area of investigation.

21.4 Living Close, but Not Too Close

The phycosphere exists as a diffusive boundary layer. It is not mixed with the surrounding fluid because—at these small scales—there is no turbulence. Accordingly, pure diffusion governs the flux of solutes (Amin et al. 2012, 2015). Strong gradients develop in NH_4^+ and dissolved organic matter (DOM) that may be favorable to bacteria (Fig. 21.3), but gradients also develop in allelopathic and other harmful compounds.

The region that may be enriched in these nutrients and chemical cues depends on the size of the phytoplankton cell and fluid motion (Amin et al. 2012; Seymour et al. 2017). For the tiniest of phytoplankton, such as *Prochlorococcus*, the phycosphere is accordingly tiny, but it increases many fold with larger cell-sized algae. A cell of 20 μm would have a sphere of influence of 90 μm, whereas a cell of 4 μm diameter

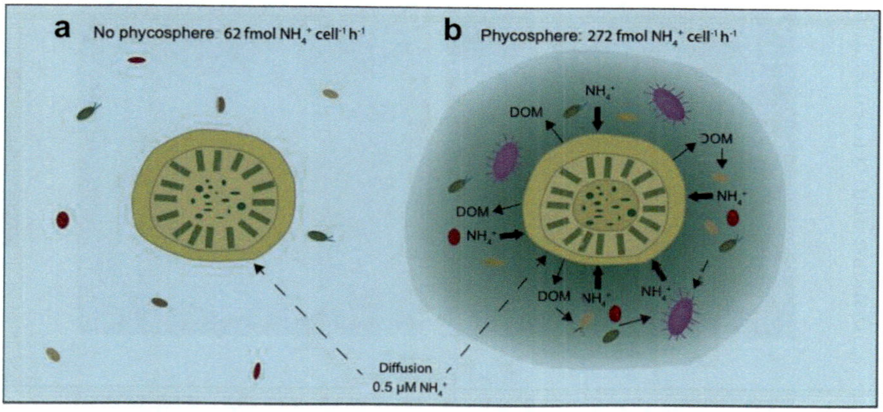

Fig. 21.3 Schematic image of ammonium (NH_4^+) assimilation via Panel (**a**) diffusion only and Panel (**b**) diffusion plus microbial interactions within the phycosphere. In Panel (**b**), bacterial recycling of dissolved organic matter (DOM) released by the diatom and protozoa grazing on bacteria together largely increases the ammonium availability, creating a gradient of elevated ammonium concentration around the diatom cell, decreasing with distance from the surface. Image reproduced from Olofsson et al. (2021, Water 13: 2301. doi.org/10.3390/w13162301) under a creative commons license

would have a sphere of influence of only about 18 μm. Fluid motion, sinking, or swimming motion can extend or distort this sphere of influence (Fig. 21.4).

21.5 Too Close: Viral Infection

According to Suttle (2005), if all the viruses of the planet, about 10^{31} of them, were lined up end to end, they would stretch past the nearest 60 galaxies. Viral infection may result in up to 60% of bacterial mortality and 30% of cyanobacterial mortality in coastal and estuarine systems (e.g., Proctor and Fuhrman 1990). Viruses are thought to account for about half of bacterial production in oligotrophic oceans (e.g., Boras et al. 2009), but the lysis rate of phytoplankton is highly variable. About 2–26% of primary production is thought to bypass trophic incorporation due to viral lysis; the viral shunt (Fuhrman 1999; Wilhelm and Suttle 1999). When viruses infect, they take over the biochemical machinery of the host cell, using it to replicate. Viral infection of phytoplankton hosts results in cell rupturing and therefore release of all the inorganic and organic contents—as well as many new viruses (Proctor and Fuhrman 1990; Suttle et al. 1991). Thus, in contrast to zooplankton that consume the cells they eat (with some sloppiness) and, after digestion, release inorganic and organic nutrients in relation to that which they consumed and used, in the case of viruses, the cells are lysed, releasing the entire contents at once. Since viruses infect hosts and kill them, they can be considered to exert top-down control. Generally, viruses are generally thought to be specific to specific hosts, but the actually specificity in nature is not known (Winter et al. 2010).

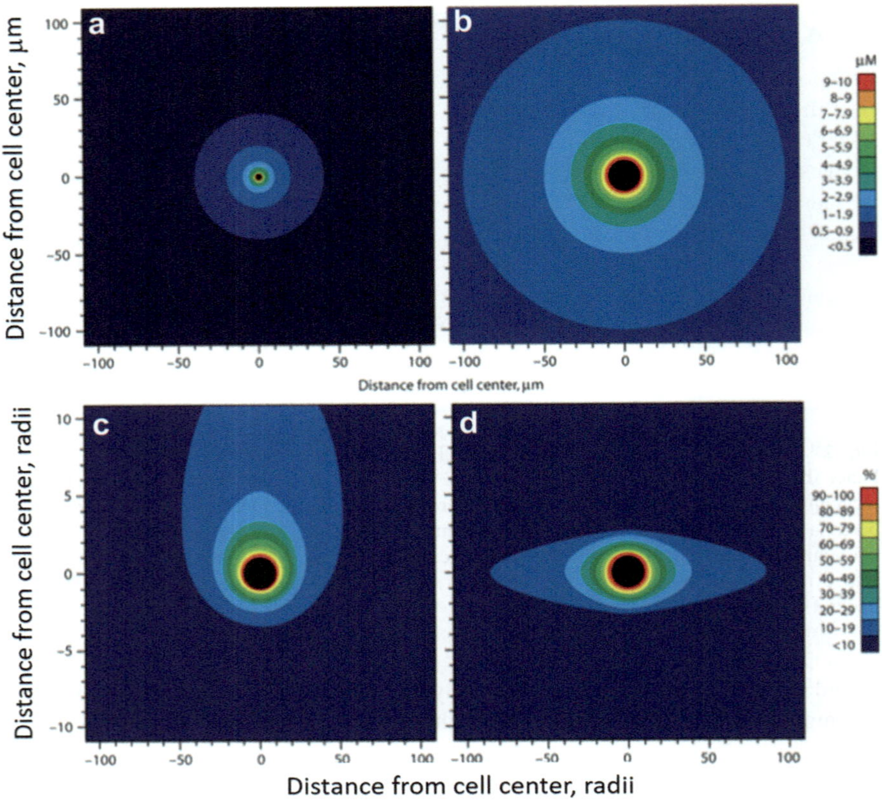

Fig. 21.4 Diffusive boundary layer around a spherical diatom cell (black circle) under different flow scenarios. Panels (**a** and **b**) Stationary cell (pure diffusion) with a 4 µm diameter (**a**) or with a 20 µm diameter (**b**). Panel (**c**) Sinking cell in the absence of flow. Panel (**d**) Cell in shear flow. Panels (**a**) and (**b**) demonstrate the effect of cell size on the concentration gradients of a compound excreted from the cell surface and consequently on the size of the diffusive boundary layer, and Panels (**c**) and (**d**) show the effect of sinking or swimming and shear flow on the shape of the diffusive boundary layer of a spherical cell. Panel (**d**) shows compression of the boundary layer in the y- and z-directions (with z being into the page) and stretch in the x-direction in response to uniaxial flow. Axes represent the distance away from the cell center either in mm (**a** and **b**) or in cell radius (**c** and **d**), and contours represent concentration ranges in µm (**a** and **b**) or in percentage of surface concentration (**c** and **d**). Image reproduced from Amin et al. (2012, Microb Mol Biol Rev 76: 667-684. doi.org/10.1128/MMBR.00007-12) with permission of the American Society for Microbiology

From the initial discovery of the prevalence of marine viruses, shown to outnumber bacteria by an order of magnitude (e.g., Bergh et al. 1989; Suttle 2007), followed by the identification of phages of specific taxa (e.g., Sullivan et al. 2003; Lindell et al. 2005), including phytoplankton-infecting viruses (e.g., Wilson et al. 2005; Derelle et al. 2008), and with the recognition of the production of viruses and their impact on the cycling of organic and inorganic material, the concept of the "viral shunt" was developed (Wilhelm and Suttle 1999; Jiao et al. 2010). Such a shunt diverts C from higher trophic levels to the DOM pool. Depending on the specific compounds released by viral lysis, the cycling or organic compounds can be modified, as some of these compounds can be readily used by bacteria, while other compounds may be more calcitrant (Weitz and Wilhelm 2012). Some evidence suggests that growth rates of bacteria may be faster in the presence of viruses because of the provision of rich organic material by the lysed hosts. For example, Zhao et al. (2019) showed that viral-induced DOM released by *Synechococcus* following infection was N-rich and sustained the surrounding bacteria. Others (e.g., Ma et al. 2018; Kuhlisch et al. 2021) have also shown that the DOM following viral infection of algal blooms may have a distinct chemical signature. The response by the host cell differs with infection by different viruses. By introducing auxiliary metabolic genes, biosynthetic pathways of metabolites may be changed (e.g., Graves et al. 1999; Lindell et al. 2005; Malitsky et al. 2016).

Viruses can be responsible for lysing harmful algal blooms (HABs) (e.g., Lawrence et al. 2002; Baudoux et al. 2006), and thus, the abundance of specific taxa can be modified through viral interactions. In the North Sea, blooms of the coccolithophore *Emiliania huxleyi* have been found to have a high rate infection rate toward the latter stages of the bloom (Bratbak et al. 1996). Thingstad (2000) has coined the concept "kill the winner" as the microbe that wins the competition for limiting resources becomes more abundant, but in so doing increases their contact with host-specific viruses. Viral lysis, then, reduces the abundance of the hosts.

In a recent study, Flynn et al. (2022) suggested that viruses orchestrate a "bust and busted" dynamic. In such a case, when a phytoplankton species manages to grow and achieve a degree of dominance, but then becomes infected by a virus, it "effectively sows the seeds of its own control" (Flynn et al. 2022 p. 991), because its infection will spread viruses that will prevent blooms going forward. This process alters succession as well, since the decline of the infected species alters competition with other species that interact for nutrients and grazer avoidance. Using a model configured with different combinations of nutrient loadings, phytoplankton cell sizes, and the possibility of cyst production, no clear patterns were established in the dynamics of the viral-host-predator-prey interactions. They were thus able to show that not only does virus infection play a role in terminating blooms, but it also modulates the growth of competitive species and their interactions. So, under a "boom and busted framework," viruses have the potential to "kill the winner" but can prevent the emergence of a potential winner over the long term. Such a concept adds yet another dimension to the understanding of the "paradox of the plankton" (Hutchinson 1961, Chap. 17): The asynchronous viral control of different members of the community contributes to different species co-existence.

21.6 Too Close: Parasite Infection

Infections by parasites also alter the metabolism of hosts, release of organic matter, and the surrounding microbiome. Of the parasitic infections within marine plankton, most are of the dinoflagellates Syndiniales (Lima-Mendez et al. 2015; Clarke et al.

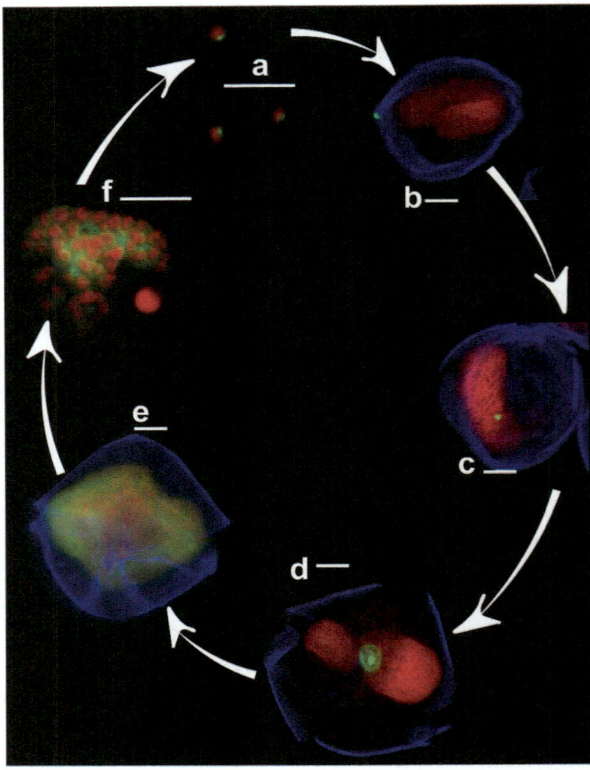

Fig. 21.5 Different stages of *Amoebophrya* sp. infecting *Alexandrium fundyense* in Salt Pond, Massachusetts, as revealed by fluorescent method applications. Green fluorescence shows an *Amoebophyra* genus-specific probe targeted to the parasite. Targeting on the parasite. Red and blue fluorescence marks the host nucleus and theca, respectively. (**a**) Free-living dinospores, (**b**) initiation of the infection, dinospore attached to the host theca, (**c**) early, (**d**) intermediate, and (**e**) later stages of infection (**f**) free-living "vermiform" stage. Scale bars = 10 mm. Image reproduced from Velo-Suárez et al. (2013, PLoS ONE 8: e81150. doi.org/10.1371/journal.pone.0081150) under a creative commons license

2019). These dinoflagellates are a diverse parasitic group, found from the ice edge to other marine environments. Parasites, such as viruses, can control algal blooms (Chambouvet et al. 2008).

A well-studied parasite is that of *Amoebophyra*, which does not kill its host (Fig. 21.5). Yet, its presence can induce a range of metabolic changes and infochemical production. The dinoflagellates, however, can also use exudates against the parasite (Long et al. 2021). Interestingly, the chemical cues given off by an infected cell are different and may alter their attraction to grazers. Some infected cells have been shown to have a stronger attraction to grazers (Park et al. 2021).

21.7 Infochemicals and Trophic Consequences

The microenvironment of phytoplankton, with its resident bacteria and viruses, as well as its potential parasites and grazers, has outsized impacts on the water column biogeochemistry. First, as discussed above, microbiomes and phycospheres are hotspots for nutrient cycling, which may directly impact the growth and success of different phytoplankton and bacterial taxa. As many HAB taxa are large enough to have a substantial phycosphere, the role of bacteria can be significant. As noted above, these associations with bacteria can affect toxins and other growth factors for the HAB phytoplankton. They may also be prey for those that are mixotrophic.

Many phytoplankton produce infochemicals that are grazer deterrents. Production of oxylipins (molecules derived from the oxidation of polyunsaturated fatty acids) by diatoms reduces the growth and grazing rate of microzooplankton and microzooplankton (Kâ et al. 2014; Ianora et al. 2015; Johnson et al. 2020). Diatom-produced polyunsaturated fatty acids may even trigger a microbial trophic cascade by reducing microzooplankton herbivory but enhancing herbivory on picoplankton (Franzè et al. 2018). Many algal toxins are also grazer deterrents.

There is also production by phytoplankton of some compounds that are now thought to grazer attractants. The metabolite DMSP and its breakdown product dimethyl sulfide (DMS) have been shown to be attractants for microzooplankton (Seymour et al. 2010; Shemi et al. 2021). Indeed, DMS plays a multitude of roles within microbial interactions, and in addition to its grazer effects, it may be an antiviral signal (Evans et al. 2006), and it may alter algal parasites and their life cycles (Garcés et al. 2013) and may also attract motile bacteria (Seymour et al. 2010).

21.8 Programmed Cell Death and the Black Queen Hypothesis

Some estimates, based on some field conditions, suggest that cell death—independent of grazing or viral lysis—may exceed 50% of cell growth (e.g., VanBoekel et al. 1992; Agustí et al. 1998). The process of "programmed cell death" (PCD) is fairly well understood in metazoan cells, but far less so in phytoplankton (Franklin et al. 2006; Bidle 2016). In essence, this "cell suicide" is mediated by a set of enzymes, caspases, that set in play a series of metabolic events that result in a distinctive dead cell morphology (Fig. 21.6). Although as a process PCD has been well studied, and the effects of stress are comparatively well understood, it is still not clear why only some cells carry out this process under certain conditions. It has been hypothesized that PCD is a "black queen"; that is, the traits are essential for community survival, but only some members bear this cost (Ndhlovu et al. 2021). The Black Queen Hypothesis relates a "public good" and a leaky function that are then exploited by other members of the plankton community.

Caspases are a family of proteases that cleave peptide chains and are responsible for morphological changes in cells undergoing PCD. They may be triggered by intercellular signaling or by environmental stress. Resulting dying or dead cells exhibit apoptosis, with symptomatic cell shrinkage, loss of membrane integrity, DNA fragmentation, induction of ROS, and other characteristics (Franklin et al. 2006). In other words, cell death may be triggered by any of a number of stressors. The induction of PCD and associated stress responses may be induced by numerous cell signaling molecules, including various types of ROS, hydrogen peroxide (H_2O_2), and superoxide (O_2^-), all of which alter the redox balance of the cell.

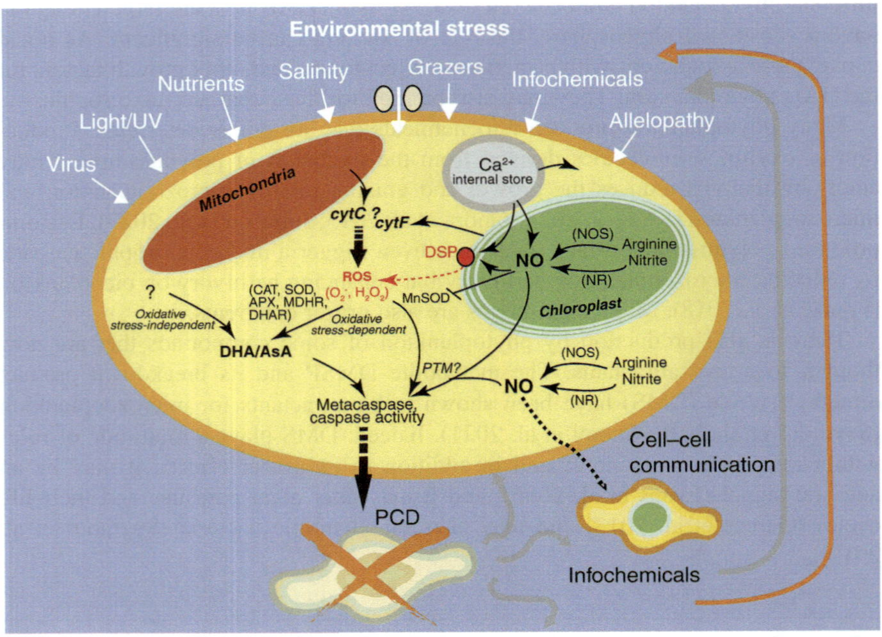

Fig. 21.6 Regulation of algal programmed cell death (PCD) and ROS- and NO-based stress and cell signaling pathways. Response pathways are depicted for a general algal cell. A variety of environmental stresses induce intracellular bursts of H_2O_2 and O^{2-}. This altered redox balance activates genes that help to deal with oxidative stress. In turn, this leads to various cellular changes that can elicit PCD via the expression of metacaspases and enhanced caspase activity. Further details are provided in Bidle (2016, Curr Biol 26: R594-R607. doi.org/10.1016/j.cub.2016.05.056). Image reproduced from Bidle (2016) with permission of Elsevier

Another signaling molecule that can induce PCD stress is nitric oxide (NO) (Vardi et al. 2008; see Fig. 12.8).

Evidence is emerging that at least some phytoplankton exhibit PCD. PCD-related genes may also have alternative physiological functions and appear to have a long evolutionary history (Bidle 2016). For example, Berman-Frank et al. (2004) suggested that PCD is a mechanism explaining colony formation in *Trichodesmium*—and the collapse of blooms after they form. PCD would enable small groups of cells to be released and dispersed from parent colonies after death or unfavorable conditions. The expression of PCD in *Trichodesmium* has been documented in cells that are under Fe limitation and high light (Bar-Zeev et al. 2013). These same stress conditions, Fe stress and high light, have also been associated with PCD responses, including caspase activity, in diatoms (Bidle 2016). Intraspecies communications that may occur upon cell stress may also send signals that result in PCD in neighboring cells. This has been shown in *Skeletonema marinoi*, for example, whereupon stress, sulfated sterols may accumulate that, in turn, may result in the production of ROS or NO in neighboring cells, thereby resulting in a crash of an entire population (Gallo et al. 2017).

In the freshwater colonial *Volvox*, cells die after a set number of divisions (Pommerville and Kochert 1981). The chlorophyte *Dunaliella tertiolecta* undergoes mass cell death when placed in darkness, with symptoms that can be characterized as PCD-like (Berges and Falkowski 1998). Many other marine and freshwater chlorophytes also induce PCD responses under a range of stresses, all with caspase involvement (Bidle 2016 and references therein). Caspase orthologues, including metacaspases, have also been observed in various phytoplankton. The regulatory process of cell death in *Skeletonema marinoi* has been studied by examining the expression of genes associated with oxidative stress and PCD (Wang et al. 2020). They found that genes associated with oxidative resistance and PCD, and activity of the caspase-like proteins, increased with N and P starvation.

Interactions between viral infections and PCD have been studied extensively in the coccolithophore *Emiliania huxleyi*. These phytoplankton are routinely infected by coccolithophorid-specific viruses, which alter the metabolism of the host in multiple ways, including triggering the production of ROS (H_2O_2) and NO, as well as elevated host caspase activity and metacaspase expression in a dose-dependent manner (e.g., Bidle et al. 2007; Bidle 2016, and references therein). Quorum sensing may also be important in viral infection of this alga. Whereas the QS molecule AHL may be harmful for algal growth, as described above (Sect. 21.3), a different QS molecule, 2-heptyl-4-quinolone (HHQ), has recently been shown to protect the coccolithophore *Emiliania huxleyi* from virus-induced mortality (Pollara et al. 2021; Harvey et al. 2023). The mechanism by which this occurs is complex. Viruses still infect the cell with this chemical present. However, the presence of HHQ appears to prevent the virus from exiting the cell and fewer viral particles are formed. When infected and in the presence of this QS molecule, PCD is suppressed, stalling mortality and further viral replication (Harvey et al. 2023).

21.9 Infochemicals and Biogeochemistry

As bacteria associations within phycospheres also affect the aggregation of particles, the very existence of these microenvironments affects the flux of C put of the euphotic zone. Viruses divert C via the viral shunt and may increase the export of C via settling of cell debris. Parasites further affect microbial grazers. Pathways of DMSP degradations in the phycosphere also impact the exchange of this compound with the atmosphere. The extent to which viral infection is controlled by QS molecules has important implications for C export and aggregate formation (Harvey et al. 2023).

With anthropogenic changes and climate change, the close-knit interaction among microbes will likely change. Species will change in timing and location, as temperatures and regional niches change. The introduction of new species will inevitability occur in new regions, altering phytoplankton, bacteria, viruses, and their grazers. New chemical releases, including novel infochemicals and/or toxins, will inevitably occur. It will remain a challenge to understand and quantify how these microscale interactions are occurring, how they are changing, and how they affect the transport of materials throughout the waters of the world.

21.10 Summary

Phytoplankton, bacteria, viruses, and other microbes live in close quarters, to the benefit—and detriment—of each other. The chemical ecology of these microenvironments is complex, involving the flux of both inorganic nutrients and highly complex organic compounds. Some of these compounds are of mutual benefit to the residents of the phycosphere, some are involved in signaling and sensing, and some affect toxins and their production or degradation. Viruses are ubiquitous and may exert powerful control on not only phytoplankton abundance but also on nutrient release. Also, if viral-host interactions are specific, competition and species succession may be altered. However, phytoplankton die independently of being eaten or infected. The metabolic and physiological stress responses and death activation pathways involve redox changes, caspases, or caspases-like proteases and a suite of cellular signaling processes. The complexities of these microenvironments are just beginning to be unraveled, but their importance in the broader ecology of phytoplankton can no longer be ignored. In all, the structuring of phytoplankton biodiversity is a dynamic balance between growth and loss, a concept that was introduced by Nathonsohn (1908; see Chap. 1), but one that is infinitely more complicated than ever could have been imagined at the time.

References

Agustí S, Satta MP, Mura MP, Benavent E (1998) Dissolved esterase activity as a trace of phytoplankton lysis: evidence of high phytoplankton lysis rates in the northwestern Mediterranean. Limnol Oceanogr 43:1836–1849

Amin SA, Küpper FC, Green DH, Harris WR, Carrano CJ (2007) Boron binding by a siderophore isolated from marine bacteria associated with the toxic dinoflagellate *Gymnodinium catenatum*. J Am Chem Soc 129:478–479

Amin SA, Parker MS, Armbrust EV (2012) Interactions between diatoms and bacteria. Microb Mol Biol Rev 76:667–684. https://doi.org/10.1128/MMBR.00007-12

Amin SA, Hmelo LR, van Tol HM, Durham BP, Carlson LT, Heal KR, Morales RL, Berthiame CT et al (2015) Interaction and signalling between a cosmopolitan phytoplankton and associated bacteria. Nature 522:98–101. https://doi.org/10.1038/nature14488

Bar-Zeev E, Avishay I, Bidle KD, Berman-Frank I (2013) Programmed cell death in the marine cyanobacterium *Trichodesmium* mediates carbon and nitrogen export. ISME J 7:2340–2348

Bates SS, Douglas DJ, Doucette GJ, Léger C (1995) Enhancement of domoic acid production by reintroducing bacteria to axenic cultures of the diatom *Pseudo-nitzschia multiseries*. Nat Toxins 3:428–435

Bates SS, Gaudet J, Kaczmarska I, Ehrman JM (2004) Interaction between bacteria and the domoic-acid-producing diatom *Pseudo-nitzschia multiseries* (Hasle) Hasle; can bacteria produce domoic acid autonomously? Harmful Algae 3:11–20

Baudoux A, Noordeloos A, Veldhuis M, Brussaard C (2006) Virally induced mortality of *Phaeocystis globosa* during two spring blooms in temperate coastal waters. Aquat Microb Ecol 44:207–217

Behringer G, Ochsenkühn MA, Fei C, Fanning J, Koester JA, Amin SA (2018) Bacterial communities of diatoms display strong conservation across strains and time. Front Microbiol 9:659. https://doi.org/10.3389/fmicb.2018.00659

Bell W, Mitchell R (1972) Chemotactic and growth responses of marine bacteria to algal extracellular products. Biol Bull 143:265–277

Berges JA, Falkowski PG (1998) Physiological stress and cell death in marine phytoplankton: induction of proteases in response to nitrogen or light limitation. Limnol Oceanogr 43:129–135

Bergh O, Borsheim KY, Bratbak G, Heldal M (1989) High abundance of viruses found in aquatic environments. Nature 340:467–468

Berman-Frank I, Bidle K, Haramaty L, Falkowski PG (2004) The demise of the marine cyanobacterium, *Trichodesmium* spp., via an autocatalyzed cell death pathway. Limnol Oceanogr 49: 997–1005

Bidle KD (2016) Programmed cell death in unicellular phytoplankton. Curr Biol 26:R594–R607. https://doi.org/10.1016/j.cub.2016.05.056

Bidle KD, Haramaty L, Barcelos-Ramos J, Falkowski PG (2007) Viral activation and recruitment of metacaspases in the unicellular coccolithophorid, *Emiliania huxleyi*. Proc Natl Acad Sci USA 104:6049–6054

Boras JA, Sala MM, Vázquez-Domínguez E, Weinbauer MG, Vaqué D (2009) Annual changes of bacterial mortality due to viruses and protists in an oligotrophic coastal environment (NW Mediterranean). Environ Microbiol 11:1181–1193. https://doi.org/10.1111/j.1462-2920.2008.01849.x

Bratbak G, Wilson W, Heldal M (1996) Viral control of *Emiliania huxleyi* blooms? J Mar Sys 9:75–81

Chambouvet A, Morin P, Marie D, Guillou L (2008) Control of toxic marine dinoflagellate blooms by serial parasitic killers. Science 322:1254–1257

Cirri E, Pohnert G (2019) Algae−bacteria interactions that balance the planktonic microbiome. New Phytol 223:100–106

Clarke LJ, Bestley S, Bissett A, Deagle BE (2019) A globally distributed Syndiniales parasite dominates the Southern Ocean micro-eukaryote community near the sea-ice edge. ISME J 13: 734–737. https://doi.org/10.1038/s41396-018-0306-7

Croft MT, Lawrence AD, Raux-Deery E, Warren MJ, Smith AG (2005) Algae acquire vitamin B_{12} through a symbiotic relationship with bacteria. Nature 438:90–93

Derelle E, Ferraz C, Escande M-L, Eychenié CR, Piganeau G, Dessevises Y et al (2008) Life-cycle and genome of OtV5, a large DNA virus of the pelagic marine unicellular green alga *Ostreococcus tauri*. PLoS One 3:e2250

Desbois AP, Walton M, Smith VJ (2010) Differential antibacterial activities of fusiform and oval morphotypes of *Phaeodactylum tricornutum* (Bacillariophyceae). J Mar Biol Assoc UK 90: 769–774

Dow L (2021) How do quorum-sensing signals mediate algae–bacteria interactions? Microorg 9: 1391. https://doi.org/10.3390/microorganisms9071391

Evans C, Malin G, Wilson WH, Liss PS (2006) Infectious titres of *Emiliania huxleyi* virus 86 are reduced by exposure to millimolar dimethyl sulfide and acrylic acid. Limnol Oceanogr 51: 2468–2471

Fernández-Herrera LJ, Band-Schmidt CJ, Zenteno-Savín T, Leyva-Valencia I, Hernández-Guerrero CJ, Muñoz-Ochoa M (2021) Cell death and metabolic stress in *Gymnodinium catenatum* induced by allelopathy. Toxins 13:506. https://doi.org/10.3390/toxins13070506

Fistarol GO, Legrand C, Granéli E (2003) Allelopathic effect of *Prymnesium parvum* on a natural plankton community. Mar Ecol Prog Ser 255:115–125

Flynn KJ, Mitra A, Wilson WH, Kimmance SA, Clark DR, Pelusi A, Polimene L (2022) 'Boom-and-busted' dynamics of phytoplankton–virus interactions explain the paradox of the plankton. New Phytol 234:990–1002. https://doi.org/10.1111/nph.18042

Franklin DJ, Brussaard CPD, Berges JA (2006) What is the role and nature of programmed cell death in phytoplankton ecology? Eur J Phycol 41:1–14. https://doi.org/10.1080/09670260500505433

Franzè G, Pierson JJ, Stoecker DK, Lavrentyev PJ (2018) Diatom-produced allelochemicals trigger trophic cascades in the planktonic food web. Limnol Oceanog 63:1093–1108

Fuhrman JA (1999) Marine viruses and their biogeochemical and ecological effects. Nature 399: 541–548

Gallo C, d'Ippolito G, Nuzzo G, Sardo A, Fontana A (2017) Autoinhibitory sterol sulfates mediate programmed cell death in a bloom-forming marine diatom. Nat Commun 8:1292

Garcés E, Alacid E, Reñé A, Petrou K, Simó R (2013) Host-released dimethylsulphide activates the dinoflagellate parasitoid *Parvilucifera sinerae*. ISME J 7:1065–1068

Grant MAA, Kazamia E, Cicuta P, Smith AG (2014) Direct exchange of vitamin B_{12} is demonstrated by modelling the growth dynamics of algal– bacterial cocultures. ISME J 8: 1418–1427

Graves MV, Burbank DE, Roth R, Heuser J, DeAngelis PL, Van Etteb JL (1999) Hyaluronan synthesis in virus PBCV-1-infected chlorella-like green algae. Virology 257:15–23

Guannel ML, Horner-Devine MC, Rocap G (2011) Bacterial community composition differs with species and toxigenicity of the diatom *Pseudo-nitzschia*. Aquat Microb Ecol 64:117–133

Harvey EL, Yang H, Castiblanco E, Coolahan M, Dallmeyer-Drennen G, Fukuda N, Greene E et al (2023) Quorum sensing signal disrupts viral infection dynamics in the coccolithophore *Emiliania huxleyi*. Aquat Microb Ecol 89:75–86. https://doi.org/10.3354/ame01998

Hünken M, Harder J, Kirst GO (2008) Epiphytic bacteria on the Antarctic ice diatom *Amphiprora kufferathii* Manguin cleave hydrogen peroxide produced during algal photosynthesis. Plant Biol 10:519–526

Hutchinson GE (1961) The paradox of the plankton. Am Nat 95:137–145

Ianora A, Bastianini M, Carotenuto CR, Roncalli V, Miralto A, Romano G (2015) Non-volatile oxylipins can render some diatom blooms more toxic for copepod reproduction. Harmful Algae 44:1–7

Jiao N, Herndl GJ, Hansell DA, Benner R, Kattner G, Wilhelm SW, Kirchman DL, Weinbauer MG, Luo T, Chen F et al (2010) Microbial production of recalcitrant dissolved organic matter: long-term carbon storage in the global ocean. Nat Rev Microbiol 8:593–599

Johnson MD, Edwards BR, Beaudoin DJ, Van Mooy BAS, Vardi A (2020) Nitric oxide mediates oxylipin production and grazing defense in diatoms. Environ Microbiol 22:629–645

Johnson WM, Kido Soule MC, Kujawinski EB (2016) Evidence for quorum sensing and differential metabolite production by a marine bacterium in response to DMSP. ISME J 10:2304–2316

Kâ S, Cartenuto Y, Romano G, Hwang J-S, Buttino I, Ianora A (2014) Impact of the diatom-derived polyunsaturated aldehyde 2-trans,4-trans decadienal on the feeding, survivorship and reproductive success of the calanoid copepod *Temora stylifera*. Mar Environ Res 93:31–37. https://doi.org/10.1016/j.marenvres.2013.07.009

Kazamia E, Czesnick H, Van Nguyen TT, Croft MT, Sherwood E, Sasso S, Hodson SJ et al (2012) Mutualistic interactions between vitamin B_{12}-dependent algae and heterotrophic bacteria exhibit regulation. Environ Microbiol 14:1466–1476. https://doi.org/10.1111/j.1462-2920.2012.02733.x

Kiene RP, Linn LJ, Bruton JA (2000) New and important roles for DMSP in marine microbial communities. J Sea Res 43:209–224

Kuhlisch C, Schleyer G, Shahaf N, Vincent F, Schatz D, Vardi A (2021) Viral infection of algal blooms leaves a unique metabolic footprint on the dissolved organic matter in the ocean. Sci Adv 7:eabf4680. https://doi.org/10.1126/sciadv.abf4680

Kustka AB, Allen AE, Morel FMM (2007) Sequence analysis and transcriptional regulation of iron acquisition genes in two marine diatoms. J Phycol 43:715–729

Lawrence JE, Chan AM, Suttle CA (2002) Viruses causing lysis of the toxic bloom-forming alga *Heterosigma akashiwo* (Raphidophyceae) are widespread in coastal sediments of British Columbia, Canada. Limnol Oceanogr 47:545–550

Lima-Mendez G, Faust K, Henry N, Decelle J, Colin S, Carcillo F, Chaffron S et al (2015) Determinants of community structure in the global plankton interactome. Science 348: 1262073. https://doi.org/10.1126/science.1262073

Lindell D, Jaffe JD, Johnson ZI, Church GM, Chisholm SW (2005) Photosynthesis genes in marine viruses yield proteins during host infection. Nature 438:86–89

Long M, Marie D, Szymczak J, Toullec J, Bigeard E, Sourisseau M, Le Gac M et al (2021) Dinophyceae can use exudates as weapons against the parasite *Amoebophrya* sp. (Syndiniales). ISME Comm 1:34

Ma X, Coleman ML, Waldbauer JR (2018) Distinct molecular signatures in dissolved organic matter produced by viral lysis of marine cyanobacteria. Environ Microbiol 20:3001–3011

Malitsky S, Ziv C, Rosenwasser S, Zheng S, Scatz D, Porat Z, Ben-Dor S et al (2016) Viral infection of the marine alga *Emiliania huxleyi* triggers lipidome remodeling and induces the production of highly saturated triacylglycerol. New Phytol 210:88–96. https://doi.org/10.1111/nph.13852

Mayers TJ, Bramucci AR, Yakimovich KM, Case RJ (2016) A bacterial pathogen displaying temperature-enhanced virulence of the microalga *Emiliania huxleyi*. Front Microbiol 7:892. https://doi.org/10.3389/fmicb.2016.00892

Moejes FW, Succurro A, Popa O, Maguire J, Ebenhöh O (2017) Dynamics of the bacterial community associated with *Phaeodactylum tricornutum* cultures. PRO 5:77. https://doi.org/10.3390/pr5040077

Nathonsohn A (1908) Über die allgemeinen Produktionsbedingungen in Meere. Beiträge zur Biologie des Planktons, von H.H. Gran and Nathansohn, Internationale Revue der gestamten. Hydrobiologie 1:38–72

Ndhlovu A, Durand PM, Ramsey G (2021) Programmed cell death as a black queen in microbial communities. Mol Ecol 30:1110–1119. https://doi.org/10.1111/mec.15757

Olofsson M, Power ME, Stahl DA, Vadeboncoeur Y, Brett MT (2021) Cryptic constituents: the paradox of high flux–low concentration components of aquatic ecosystems. Water 13:2301. https://doi.org/10.3390/w13162301

Park MG, Kim A, Jeon BS, Kim M (2021) Parasite-mediated increase in prey edibility in the predator-prey interaction of marine planktonic protists. Harmful Algae 103:101982. https://doi.org/10.1016/j.hal.2021.101982

Passow U (2002) Production of transparent exopolymer particles (TEP) by phyto- and bacterioplankton. Mar Ecol Prog Ser 236:1–12

Paul C, Pohnert G (2013) Induction of protease release of the resistant diatom *Chaetoceros didymus* in response to lytic enzymes from an algicidal bacterium. PLoS One 8:e57577. https://doi.org/10.1371/journal.pone.0057577

Pichierri S, Accoroni S, Pezzolesi L, Guerrini F, Romagnoli T, Pistocchi R, Totti C (2017) Allelopathic effects of diatom filtrates on the toxic benthic dinoflagellate *Ostreopsis* cf. *ovata*. Mar Environ Res 131:116–122. https://doi.org/10.1016/j.marenvres.2017.09.016

Pollara SB, Becker JW, Nunn BL, Boiteau R, Repeta D, Mudge MC, Downing G et al (2021) Bacterial quorum sensing signal arrests phytoplankton cell division and impacts virus-induced mortality. MSphere 6:e00009–e00021

Pommerville JC, Kochert GD (1981) Changes in somatic cell structure during senescence of *Volvox carteri*. Eur J Cell Biol 24:236–243

Proctor LM, Fuhrman JA (1990) Viral mortality of marine bacteria and cyanobacteria. Nature 343: 60–62

Rutherford ST, Bassler BL (2012) Bacterial quorum sensing: its role in virulence and possibilities for its control. Cold Spring Harb Perspect Med 2:a012427. https://doi.org/10.1101/cshperspect.a012427

Segev E, Wyche TP, Kim KH, Peterson J, Ellebrandt C, Vlamakis H, Barteneva N et al (2016) Dynamic metabolic exchange governs a marine algal-bacterial interaction. eLife 5:28. https://doi.org/10.7554/eLife.17473

Seymour JR, Simó R, Ahmed T, Stocker R (2010) Chemoattraction to dimethylsulfoniopropionate throughout the marine microbial food web. Science 329:342–345

Seymour JR, Amin SA, Raina J-B, Stocker R (2017) Zooming in on the phycosphere: the ecological interface for phytoplankton–bacteria relationships. Nat Microbiol 2:17065

Shemi A, Alcolombri U, Schatz D, Farstey V, Vincent F, Rotkopf R, Ben-Dor S (2021) Dimethyl sulfide mediates microbial predator-prey interactions between zooplankton and algae in the ocean. Nat Microbiol 6:1357–1366. https://doi.org/10.1038/s41564-021-00971-3

Simó R (2001) Production of atmospheric sulfur by oceanic plankton: biogeochemical, ecological and evolutionary links. Trends Ecol Evol 16:287–294

Sullivan MB, Waterbury JB, Chisholm SW (2003) Cyanophages infecting the oceanic cyanobacterium *Prochlorococcus*. Nature 424:1047–1051

Sutak R, Camadro J-M, Lesuisse E (2020) Iron uptake mechanisms in marine phytoplankton. Front Microbiol 11:566691. https://doi.org/10.3389/fmicb.2020.566691

Suttle CA (2005) Viruses in the sea. Nature 437:356–361

Suttle CA (2007) Marine viruses – major players in the global ecosystem. Nat Rev Microbiol 5: 801–812

Suttle CA, Chan AM, Cottrell MT (1991) Use of ultrafiltration to isolate viruses from seawater which are pathogens of marine phytoplankton. Appl Envir Microbiol 57:721–726

Ternon E, Pavaux AS, Marro S, Thomas OP, Lemée R (2018) Allelopathic interactions between the benthic toxic dinoflagellate *Ostreopsis* cf. *ovata* and a co-occurring diatom. Harmful Algae 75: 35–44. https://doi.org/10.1016/j.hal.2018.04.003

Thingstad TF (2000) Elements of a theory for the mechanisms controlling abundance, diversity, and biogeochemical role of lytic bacterial viruses in aquatic systems. Limnol Oceanogr 45:1320–1328

Thingstad TF, Skjoldal EF, Bohne RA (1993) Phosphorus cycling and algal-bacterial competition in Sandsfjord, western Norway. Mar Ecol Prog Ser 99:239–259

vanBoekel WHM, Hansen FC, Riegman R, Bak RPM (1992) Lysis-induced decline of a *Phaeocystis* spring bloom and coupling with the microbial foodweb. Mar Ecol Prog Ser 81: 269–276

Vardi A, Berman-Frank I, Rozenberg T, Hadas O, Kaplan A, Levine A (1999) Programmed cell death of the dinoflagellate *Peridinium gatunense* is mediated by CO_2 limitation and oxidative stress. Curr Biol 9:1061–1064. https://doi.org/10.1016/S0960-9822(99)80459-X

Vardi A, Bidle K, Kwityn C, Hirsch DJ, Thompson SM, Callow JA, Falkowski P, Bowler C (2008) A diatom gene regulating nitric oxide signaling and susceptibility to diatom-derived aldehydes. Curr Biol 18:1–5

Velo-Suárez L, Brosnahan M, Anderson DM, McGillicudy DJ (2013) A quantitative assessment of the role of the parasite *Amoebophrya* in the termination of *Alexandrium fundyense* blooms within a small coastal embayment. PLoS One 8:e81150. https://doi.org/10.1371/journal.pone.0081150

Verner K, Schatz G (1988) Protein translocation across membranes. Science 241:1307–1313

Wang H, Chen F, Mi T, Liu Q, Yu Z, Zhen Y (2020) Responses of marine diatom *Skeletonema marinoi* to nutrient deficiency: programmed cell death. Appl Environ Microbiol 86(3):e02460–e02419. https://doi.org/10.1128/AEM.02460-19

Wang Y, Zhiming Y, Song X, Zhang S (2006) Interactions between the bloom forming dinoflagellates *Prorocentrum donghaiense* and *Alexandrium tamarense* in laboratory cultures. J Sea Res 56:17–26

Weitz JS, Wilhelm SW (2012) Ocean viruses and their effects on microbial communities and biogeochemical cycles. F1000 Biol Rep 4:17. https://doi.org/10.3410/B4-17

Wilhelm SW, Suttle CA (1999) Viruses and nutrient cycles in the sea – viruses play critical roles in the structure and function of aquatic food webs. BioSci 49:781–788

Wilson WH, Schroeder DC, Allen MJ, Holden MTG, Parkhill J, Barrell BG, Churcher C et al (2005) Complete genome sequence and lytic phase transcription profile of a *Coccolithovirus*. Science 309:1090–1092. https://doi.org/10.1126/science.1113109

Winter C, Bouvier T, Weinbauer MG, Thingstad TF (2010) Trade-offs between competition and defense specialists among unicellular planktonic organisms: the "killing the winner" hypothesis revisited. Microbiol Mol Biol Rev 74:42–57. https://doi.org/10.1128/MMBR.00034-09

Xie B, Bishop S, Stessman D, Wright D, Spalding MH, Halverson LJ (2013) *Chlamydomonas reinhardtii* thermal tolerance enhancement mediated by a mutualistic interaction with vitamin B_{12}-producing bacteria. ISME J 7:1544–1555

Zhang Y, Zheng L, Wang S, Zhao Y, Xu X, Han B, Hu T (2022) Quorum sensing bacteria in the phycosphere of HAB microalgae and their ecological functions related to cross-kingdom interactions. Int J Environ Res Public Health 19:163. https://doi.org/10.3390/ijerph19010163

Zhao Z, Gonzior M, Schmitt-Kopplin P, Zhan Y, Zhang R, Jiao N, Chen F (2019) Microbial transformation of virus-induced dissolved organic matter from picocyanobacteria: coupling of bacterial diversity and DOM chemodiversity. ISME J 13:2551–2565

Zhou J, Chen G-F, Ying K-Z, Jun H, Song J-T, Cai Z-H (2019) Phycosphere microbial succession patterns and assembly mechanisms in a marine dinoflagellate bloom. App Envir Microbiol 85: e00349–e00319. https://doi.org/10.1128/AEM.00349-19

The Changing World

22

22.1 Introduction

The world is changing, rapidly. Due to climate change and eutrophication, many aquatic systems would be unrecognizable to Victor Hensen, Karl Brandt, H.H. Gran, Alexander Nathansohn, H.A. Harvey, Gordon Riley, or any of the investigators whose work was described in Chap. 1. We have upended many of the patterns and relationships that they studied: blooms are occurring in new and different places, species distributions are changing, and effects are felt from fresh to marine waters. The scale of the pace of many of these changes nearly defies comprehension. Responses in the plankton world to the many stressors are often complex, often nonlinear, interconnected and hard to predict, and even harder to control (Chap. 29). This chapter sets the stage of the magnitude of the global changes that are occurring and the reasons for such changes. The following chapters dive into the specifics of changes in temperature, dissolved oxygen, harmful algal blooms, ocean acidification, and their implications for phytoplankton in more detail.

22.2 Human Population

In the 1970s, it was hard to conceive that the human footprint would affect water quality in any substantial way. Buck Ketchum, a well-known oceanographer of the Woods Hole Oceanographic Institution of that era (Ketchum 1972, p. 157), summarized the then-current state of thinking as follows:

> Man's additions of plant nutrients, especially phosphate and nitrate, to the ocean is of considerable effect on the quality of the local coastal zone. These nutrients are found in domestic sewage and are also produced by runoff... In perspective of the world ocean, however, these additions are probably of little significance: it has been estimated that a world population of 6 x 10^9 injecting their metabolic wastes into the ocean would require over 10,000 years to double the fixed nitrogen compounds present.

P. M. Glibert, *Phytoplankton Whispering: An Introduction to the Physiology and Ecology of Microalgae*, https://doi.org/10.1007/978-3-031-53897-1_22

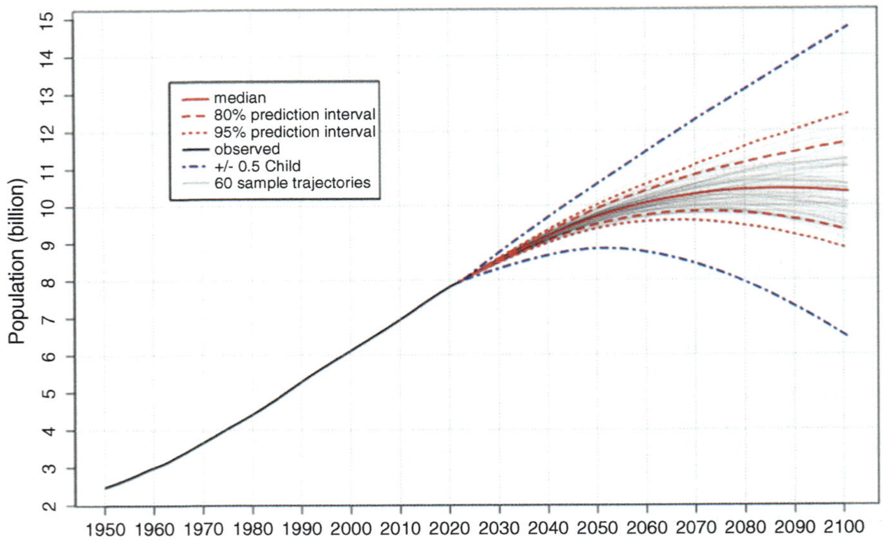

Fig. 22.1 Human population growth and projections of future total population. The population projections are based on the probabilistic projections of total fertility and life expectancy at birth. Image reproduced from the United Nations (https://population.un.org/wpp/Graphs/Probabilistic/POP/TOT/900) under a creative commons license

Ketchum could not envision a world with eight billion people, the current global population (Fig. 22.1, Glibert 2019). By the year 2050, human population will likely be nearly 10 billion, a 25% increase over current population estimates, with more than half of this increase expected to occur in Africa (http://www.un.org/en/development/desa/news/population/2015-report.html). Even though the actual population growth rate has begun to decline, or is expected to decline in the future, the impacts of an increasing world population will continue to be felt well into the future (Fig. 22.1).

It is expected that in the coming decades, half the population will reside in urban areas, and that megacities (>10 million people) will continue to be concentrated in China as well as other parts of Asia and Africa, most of which are along coastlines (United Nations 2014). These megacities are megacenters for anthropogenic nutrient pollution. Providing the space, food, and other resources for this enormous world population is increasingly a challenge to the sustainability of the Earth. Sadly, as the world continues to urbanize, sanitation has not kept pace. More than a quarter of the population, or 2.3 billion people, still live without basic sanitation (World Bank 2021). This includes 72% of the population in sub-Saharan Africa and 50% of the population in South Asia. The additions of nutrients that Ketchum referenced are hardly of little significance anymore.

22.3 Warming and CO_2

The world is warming, even faster than predictions of just a few years ago (Xu et al. 2018). Since 1980, the rate of warming per decade has doubled compared with the average rate of warming since 1880 (Lindsey and Dahlman 2023; Fig. 22.2). In late October 2018, the Intergovernmental Panel on Climate Change (IPCC) issued a special report on the impacts of global warming >1.5 °C above pre-industrial levels, which they concluded—with high confidence—is likely to occur between the years 2030 and 2052 (IPCC 2018). Reaffirming these projections, and extending the timeline further, the most recent Intergovernmental Panel on Climate Change (IPCC 2023) projected that global temperatures may rise by 4 °C during this century.

According to the US National Oceanic and Atmospheric Administration (NOAA), nine of the ten highest August surface temperatures (combined land and water) have occurred since 2009. In 2021, the June–August global surface temperature was the fourth highest for the 3-month season on record, while the 9-month period of January–August was the sixth warmest period on record (https://www.

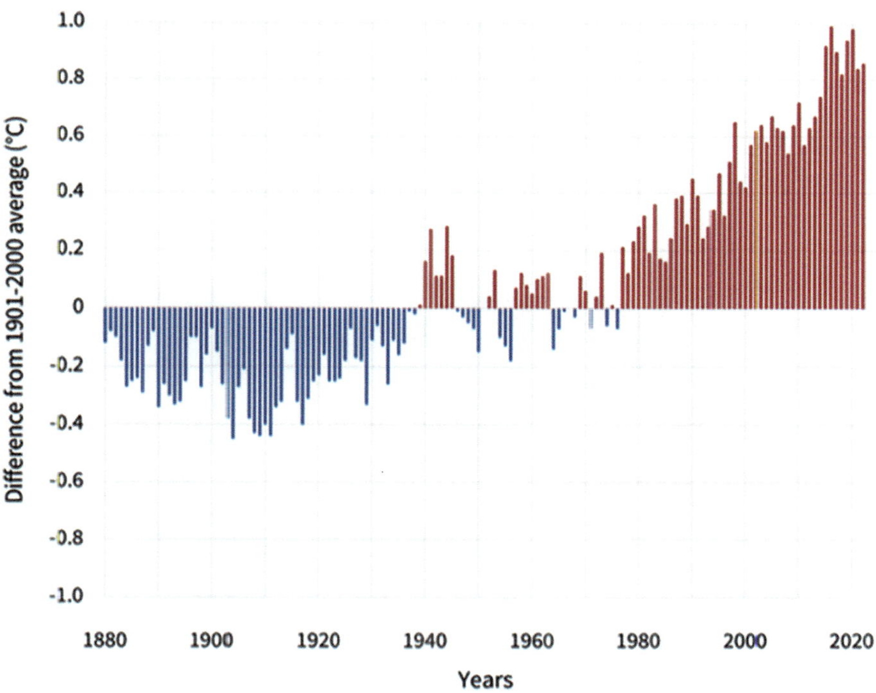

Fig. 22.2 Global, yearly surface temperature compared to the twentieth-century average from 1880 to 2022. Blue bars indicate cooler-than-average years; red bars show warmer-than-average years. NOAA http://Climate.gov graph, based on data from the National Centers for Environmental Information. Image reproduced from Lindsey and Dahlman (2023, https://www.climate.gov/media/15021), public access

ncdc.noaa.gov/sotc/global/202108). Each of the past 45 years have surpassed the twentieth-century average in temperature (Fig. 22.2; Richter 2022). The most current July 2023 had temperatures that have been "so off-the-charts hot with heat waves blistering three continents–North America, Europa and Asia" (Boresnstein 2023). The IPCC (2023) emphasized, with high confidence, that "every increment of global warming will intensify multiple and concurrent hazards." Thus, it is of no surprise that the intense recent heat has led to dangerous heat waves, wildfires, and other extreme events.

The oceans have a large heat storage capacity, and were it not for this, the atmosphere would be warming even faster than it is. That is, most of the thermal energy of the Earth due to global heating is stored in the oceans. Warming of the ocean surface waters (upper 700 m) represents about 60% of the total increase in stored heat in the climate system from 1971 to 2020. Yet, this heat content is not uniformly consistent; regions of the oceans that are warming comprise 56% of the ocean surface, whereas regions that are cooling only comprise 3% of the ocean surface (Johnson and Lyman 2020). The oceans experience heat waves, just as on land, with effects related to increased hypoxia, harmful algal blooms (HABs), and other ecosystem stresses. The most current ocean heat wave (as of this writing in 2023, Fig. 22.3), with temperatures in many parts of the world well above average, is due to not only to overall global warming, but also to the onset of El Niño.

Changes in water heat content not only affect organisms but also cause the oceans to expand, thereby altering sea level. Although hurricanes and tropical storms—by number—are not expected to increase in the warming world, their intensity is

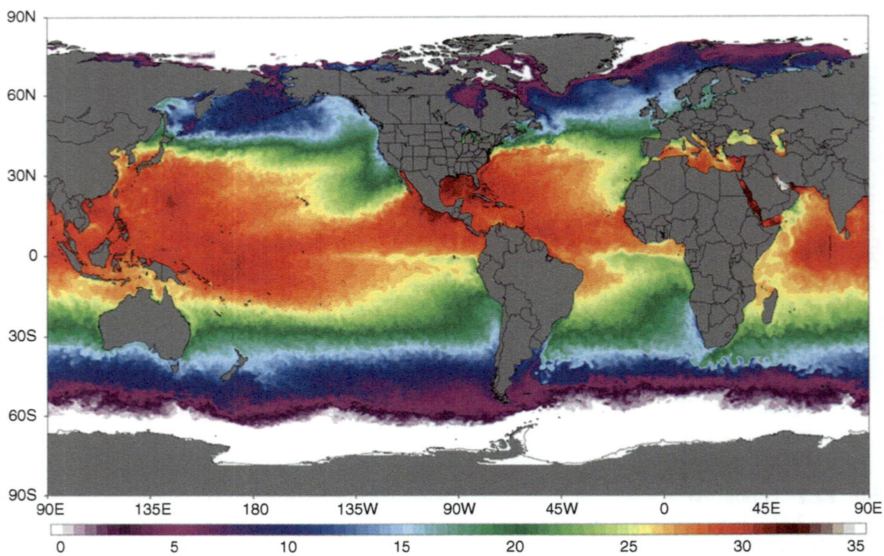

Fig. 22.3 Sea surface temperatures in July 2023. Image reproduced from http://ClimateReanalyzer.org/NOAA OISST v2.1 under a creative commons license

increasing. Warmer waters hold more moisture, and thus hurricanes and typhoons are becoming wetter, increasing property damage. Moreover, as urban areas expand, the number of people and property in vulnerable areas increases, the so-called "expanding bull's eye effect" (Ashley et al. 2014).

Temperatures are rising mainly due to increasing CO_2 concentrations in the atmosphere. The link between rising CO_2 and temperatures was suspected as early as the 1850s and possibly earlier (Tyndall 1861). In 1956, Dr. Gilbert Plass, a Canadian physicist, predicted that a doubling of CO_2 would warm the planet by 3.6 °C, that CO_2 levels in 2000 would be 30% higher than in 1900, and that the planet would be about 1 °C warmer in 2000 than in 1900 (Plass 1956). His theory—that humans were to blame for this increase—was summarized in the *New York Times* (Kaempffert 1956):

> Today more carbon dioxide is being generated by man's technological processes than by volcanoes, geysers and hot springs. Every century man is increasing the carbon dioxide content of the atmosphere by 30 percent—that is, at a rate of 1 1 °C in a century. It may be a chance coincidence that the average temperature of the world since 1900 has risen by this rate. But the possibility that man had a hand in the rise cannot be ignored.

Plass (1956, p. 315) predicted it would be centuries before this presented "a serious problem to future generations. Rather than centuries, it only took decades.

Global greenhouse emissions have increased 75% with a 25% increase from the 1990s to 2004 alone, primarily due to fossil fuel use. The USA, China, and Russia have emitted the most CO_2 since 1970, while African nations the least (Fig. 22.4). Other contributors to greenhouse gases are methane (CH_4) and nitrous oxide (N_2O).

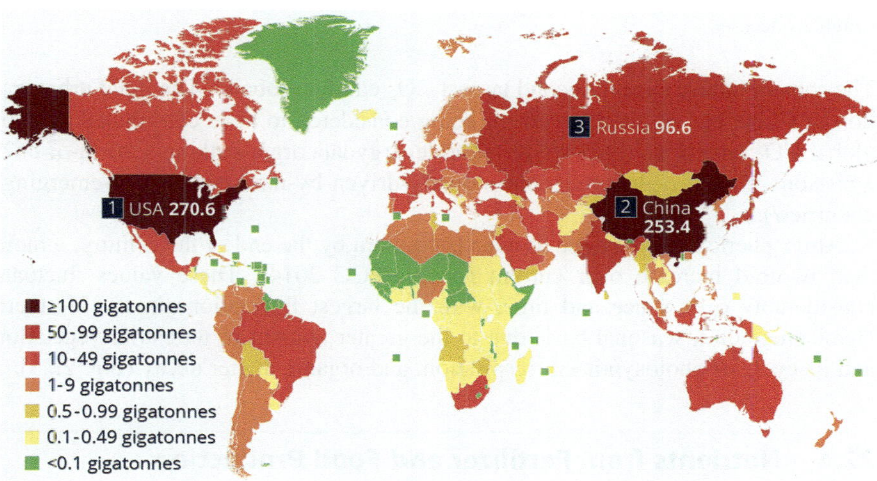

Fig. 22.4 The world's uneven CO_2 footprint: estimated cumulative CO_2 emissions by country between 1970 and 2021. Image reproduced from Richter (2022, https://www.statista.com/chart/28725/cumulative-co2-emissions-per-country-since-1970/) under a creative commons license

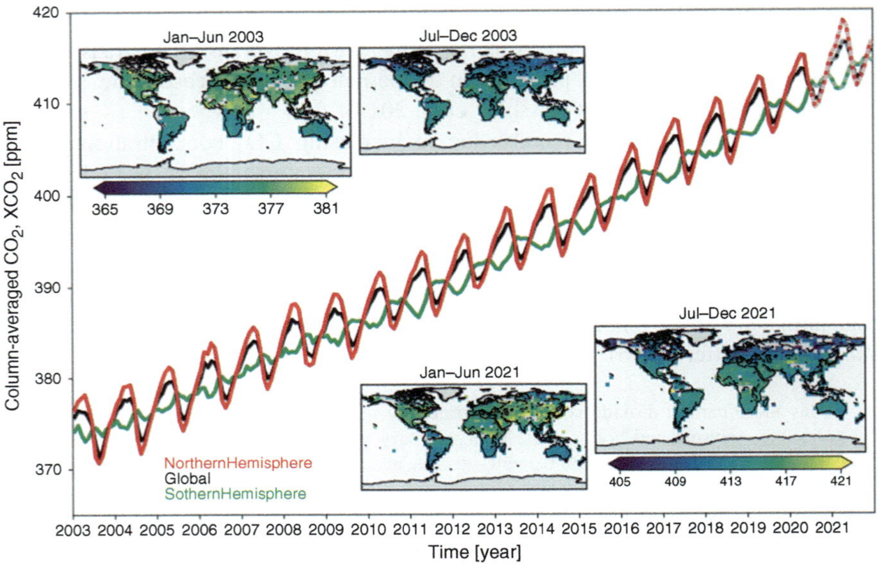

Fig. 22.5 Time series and charts of CO_2, in parts per million (ppm). Time series are shown for three latitude bands: global (black), Northern Hemisphere (red), and Southern Hemisphere (green). Data source: C3S/Obs4MIPs (v4.3) consolidated (2003–mid-2020) and CAMS preliminary near real-time data (mid-2020–2021) records. Note that the X in XCO_2 means that the values represent the concentrations averaged over the whole vertical atmospheric column. Credit: C3S/CAMS/ECMWF/University of Bremen/SRON. Inset maps show seasonal changes in 2003 and 2021. Image reproduced from Copernicus, the European Union's Earth Observation Programme (https://climate.copernicus.eu/climate-indicators/greenhouse-gas-concentrations) under a creative commons license

The cement industry is the second largest CO_2 emitter globally. The rapid urbanization of China has resulted in cement being considered to have contributed 25% of global CO_2 emissions (https://www.worldenergydata.org/a-striking-growth-of-co2-emissions-from-the-global-cement-industry-driven-by-new-facilities-in-emerging-countries/).

Atmospheric CO_2 may rise to over 1000 μatm by the end of the century, a more than twofold increase over current levels (IPCC 2014). These values fluctuate considerably over space and time, with the largest fluctuations in the Northern Hemisphere on a seasonal basis due to the greater amount of terrestrial vegetation and its cycle of photosynthesis, respiration, and organic matter decay (Fig. 22.5).

22.4 Nutrients from Fertilizer and Food Production

As human population grows, more food is required, but the very ability to grow more food sustains more people, in turn requiring more food production. Only about half the current global population could be sustained if there were no synthetic N-based fertilizer use (Ritchie et al. 2022). Food production is the largest source of nutrient pollution, and producing more food with low pollution remains one of the grand challenges (Sutton et al. 2013; Davidson et al. 2015). While Ketchum (1972) and others in the 1970s were concerned about the wastes from people, they were unconcerned about nutrient pollution from food production, as the agriculture industry as we now know it was in its infancy. Thus, in the 1970s, eutrophication, or the pollution from nitrogen (N) and phosphorus (P), was a problem largely localized to some freshwaters (e.g., Likens 1972; Ketchum 1972), and few reviews of phytoplankton ecology devoted more than token paragraphs to the issue.

"Eutrophication" is characterized by the excessive algal growth that occurs in response to the increase in human-derived nutrients. It has been formally defined in various ways (e.g., Nixon 1995; Richardson and Jørgensen 1996). Classically, eutrophication was described as the enrichment of a water body leading to enhanced organic production (e.g., Nixon 1995). This excessive enrichment of nutrients can lead to high biomass of phytoplankton that in turn can lead to reduction in oxygen when the biomass begins to decay. The biodiversity of phytoplankton communities may change not only due to the total load of nutrients but also due to their changing proportions (see Chap. 18). Some of this biomass may shade benthic producers causing their demise, and some of the algal biomass may also be harmful of toxic to other organisms or to human consumers. Much more will be discussed about hypoxia and harmful algal blooms in Chaps. 24 and 25. Fertilizers are but one factor contributing to changing nutrient loads and proportions, atmospheric deposition, aquaculture, and altered river flow through dam construction are other factors. The magnitude of these changes is introduced below.

22.4.1 Fertilizers

Farm operations changed in the 1960s and 1970s directly due to the invention of chemical nitrogen fertilizers, the method to chemically "fix" nitrogen gas (N_2) to ammonia (NH_3),

$$N_2 + 3H_2 \rightarrow 2NH_3 \tag{22.1}$$

This invention, from the German chemists Fritz Haber and Carl Bosch, known as the Haber-Bosch process, is considered one of the most important chemical reactions in the world (e.g., Smil 2001) and "the greatest single experiment in global geo-engineering ever made" (Sutton et al. 2013, p. 4). Prior to this invention, there were only two known pathways by which atmospheric N_2 was converted to NH_3: via lightening and via the biological N_2 fixation process that is carried out by some cyanobacteria and mutualistic bacteria.

The global manufacture of N fertilizers has increased from <10 million metric tons N year^{-1} in 1950 to >120 million metric tons year^{-1} in 2022, with the bulk of all chemical fertilizers having been produced since 1985 (Fig. 22.6; Howarth 2008; Glibert et al. 2014 and references therein, FAO 2017; Ritchie et al. 2022). The

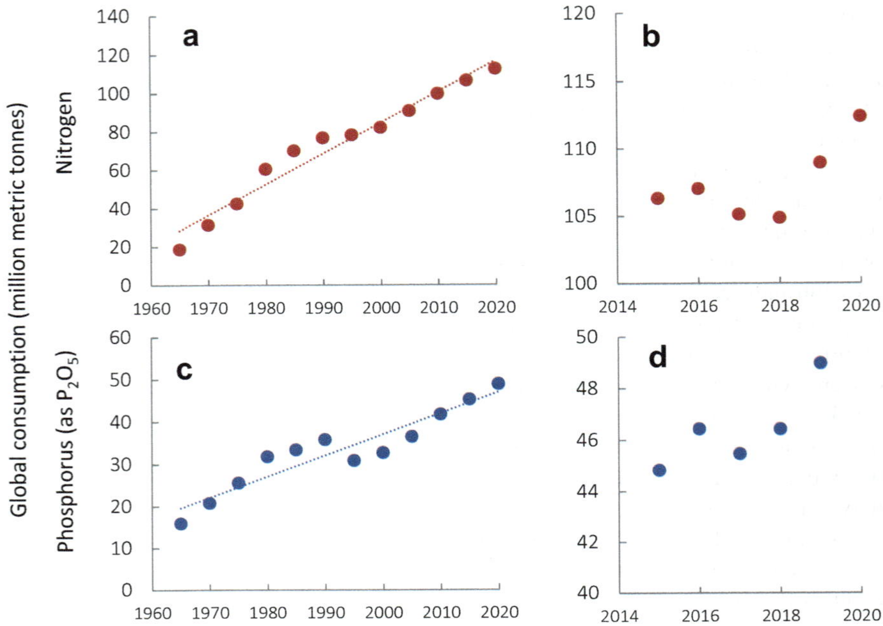

Fig. 22.6 Global consumption of world fertilizer, 1960–2020. Panels **a**, **b**: Nitrogen; Panels **c**, **d**: Phosphorus (as P_2O_5); Panel **a**, **c**: Trends by 5-year intervals; Panels **b**, **d**: Expanded view of trends from 2015 to 2020. Data obtained from https://www.statista.com/statistics/438967/fertilizer-consumption-globally-by-nutrient/

formulation of these fertilizers has also changed with time. The use of urea has been increasing globally (Glibert et al. 2006, 2014), which has important implications for the types of phytoplankton that may proliferate in waters receiving this form of runoff.

In contrast to the enormous expansion in the global use of chemical N fertilizers, the use of P-based fertilizers has shown a much smaller increase, at a rate only about a third that of N (Fig. 22.6c; Sutton et al. 2013; Glibert et al. 2014; Beusen et al. 2016; Bouwman et al. 2017), leading to global increase in the pollution of N vs P, and an alteration in the chemical composition and proportions of nutrients in waters with resulting impacts on phytoplankton composition. Unlike N, which is fixed from the atmosphere via natural and anthropogenic processes (i.e., the Haber-Bosch process), P must be mined from the Earth (see Chap. 13). Also in contrast to N, since the mid-1980s–1990s, the major industrialized nations began curtailing P use by removing it from laundry detergents and by upgrading sewage treatment plants to remove P (P is less expensive than N to remove; Litke 1999; van Drecht et al. 2009). Yet another important contrast is that P tends to adsorb to soils and sediments, creating a legacy reservoir of this nutrient that can accumulate and be released decades later. Such adsorption and desorption differs in fresh and salt water, one of the reasons for the typical limitation by P in freshwaters and by N in marine waters (see Chap. 13). Despite these differences, fertilizers applied as N and as P have an important similarity: of that which is applied as fertilizers to land for agricultural purposes, only 10–30% actually reaches human consumers (Galloway et al. 2002; Houlton et al. 2013), and more than half is lost to the environment in direct runoff and atmospheric volatilization or eventual deposition (Galloway et al. 2014).

22.4.2 Animal Production

As agriculture has grown and intensified, another trend in food production has occurred concurrently. As economies develop and people have more disposable income, their diets have changed. People consume more protein—as meat—when wealth increases and as the cost of meat production decreases. Meat production has demands for animal feed that, in turn, places demands on grain production and associated fertilizer use. To understand the magnitude of the impact of animal production for protein consumption, consider that if everyone in the world ate as much meat as Americans, "all inhabitable land on Earth would have to be used to raise livestock—and we'd still need more, an additional land mass as large as Africa and Australia combined" (Heath and Starr 2022, p. 110). China is catching up. China's meat production, in fact, rose 250% from 1986 to 2012, with another 30% increase as of 2020.

To accommodate the increasing demand for meat, animal production has become industrialized; animals are now largely grown in concentrated feeding animal operations (CAFOs; Fig. 22.7). Such operations are replacing small family farms throughout the world. In the USA, there are nearly 9 billion animals grown annually

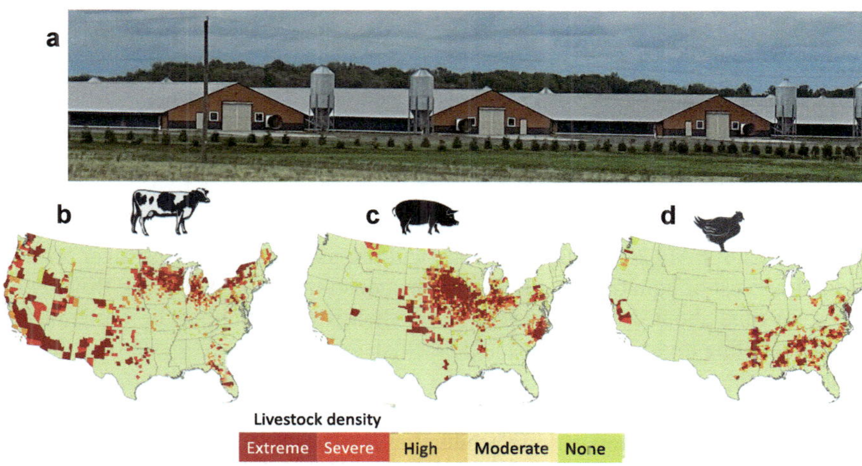

Fig. 22.7 Panel (**a**), Chicken farm (CAFO) on the eastern shore of the state of Maryland, USA. Photo by P. Glibert. Panel (**b–d**), Inventory for CAFOs in the USA in 2017 for dairy (Panel **b**), hogs (Panel **c**), and broiler chickens (Panel **d**). Inventory figures are from Food and Water Watch (2015, 2020). Panels (**b–d**) modified and reproduced from Glibert (2020, Biogeochem 150:139–180. https://doi.org/10.1007/s10533-020-00691-6) under a creative commons license

in such industrial farms, generating massive amounts of waste (Glibert 2020), and the global trend in manure generation from animal production shows a steady increase in the past six decades. As of 2015, the USA had over 20,000 CAFOs, a number that has substantially increased in the past several decades, but a number that likely underestimates the true value. The USA is the largest producer of beef and poultry and the third largest producer of pork in the world, which equates to a production of about 10 billion kg of beef and pork and 18 billion kg of chicken (USDA 2020), and this yields over 13,000 metric tons day (or five million metric tons per year) as N and over 5000 metric tons day^{-1} (or over 1.8 million metric tons per year) as P (Glibert 2020). In China, tens of thousands of CAFOs are estimated to produce 40 times more N pollution than other types of industries, as much of the manure produced in CAFOs is directly discharged into rivers (Ellis 2008). It has been further estimated that if unchecked, these dietary trends will, by 2050, be a major contributor to greenhouse gas emissions and global land clearing (Tilman and Clark 2014). There are two main sources of NH_3 from agriculture: it is emitted from fertilizer application, especially when those fertilizers are NH_4^+- and urea-based, and from management of manures. Of livestock waste, all animals contribute substantially to NH_3 emissions. Emissions of NH_3 are projected to increase and to become the dominant component of N emissions with rising temperatures (Fowler et al. 2013).

Enteric fermentation accounts for the largest fraction of methane (CH_4) from the agriculture sector. Emissions of CH_4 from manure management have grown sharply in the past 20 years as animal operations have intensified.

22.4.3 Aquaculture

In concert with changing diets – and in parallel with the increasing production of animals in CAFOs on land – aquaculture has been developing intensively. Aquaculture is simply the aquatic version of intensive animal production. This trend is also a function of declining natural fish stocks due to over-fishing. Based on data from the Food and Agriculture Organization on the United Nations (FAO), the annual growth rate in aquaculture was about 9% over the 1970–2000 period in both fresh and marine waters, with over 70% of this production in developing countries, mostly in Asia (Fig. 22.8, FAO 2016). By 2018, aquaculture production was just over 50% of the total harvest of aquatic animals for human consumption (FAO 2020; Boyd et al. 2022). Of this, about 75% of mariculture is shellfish production; the remainder is finfish, including high-value marine and brackish water species (e.g., salmon, bream) in intensive farming systems in cages and net pens (FAO 2016). In 2010, 60% of global crustacean and 98% molluscan production were in marine or brackish water culture. The volume of aquaculture production is expected to increase further in the coming decades (Delgado et al. 2003; Cork et al. 2005; Msangi and Batka 2015). Boyd et al. (2022) estimate that aquaculture production would need to increase from 82,087 Kt in 2018 to 129,000 Kt by 2050 to meet the demand of the greater population. However, aquaculture requires fishmeal and fish oil to feed the

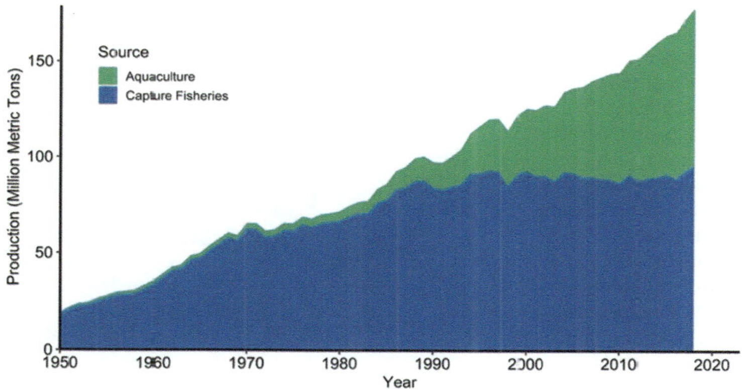

Fig. 22.8 Change in global production of aquaculture and capture fisheries. Image reproduced from Boyd et al. (2022, Food Sec 14: 805–827, https://doi.org/10.1007/s12571-021-01246-9) under a creative commons license

two-thirds of finfish and crustacean production that is feed-based. This may be unsustainable in the years to come.

Aquacultural finfish production takes place in a variety of systems in freshwater and marine environments with different species and production intensities. Using data for the period 1950–2010 (derived from the database FISHSTAT, FAO 2012a, b, 2016), based on annual production, per species, country, sea, and type of environment, a simple nutrient budget model was developed that describes the major flows of nutrients in various types of aquaculture systems (Fig. 22.9; Bouwman et al. 2011, 2013a). The model approach describes nutrients in feed inputs, feed conversion, nutrients in fish, and the outflow in the form of feces (solid forms based on apparent digestibility) and dissolved nutrients, as well as retention and recycling in pond and integrated aquaculture systems. Based on these models, global cultured production of finfish and crustacean contributed an estimated 1.7 million tons of N and 0.46 million tons of P to receiving waters during 2008 (Verdegem 2013). Within the relatively short period from 2000 to 2006, nutrient release from shellfish cultures increased by 2.5- to 3-fold, and much larger increases are predicted in nutrient contributions from shellfish cultures by 2050 (Bouwman et al. 2011). While globally these values are small compared to riverine input, on a regional basis they can be substantial (Bouwman et al. 2013b). Where marine aquaculture is confined to relatively small areas in coastal seas, primarily in Asia, the amount of nutrients locally released by aquaculture into these seas can be comparable to the nutrient input by rivers (Bouwman et al. 2013b).

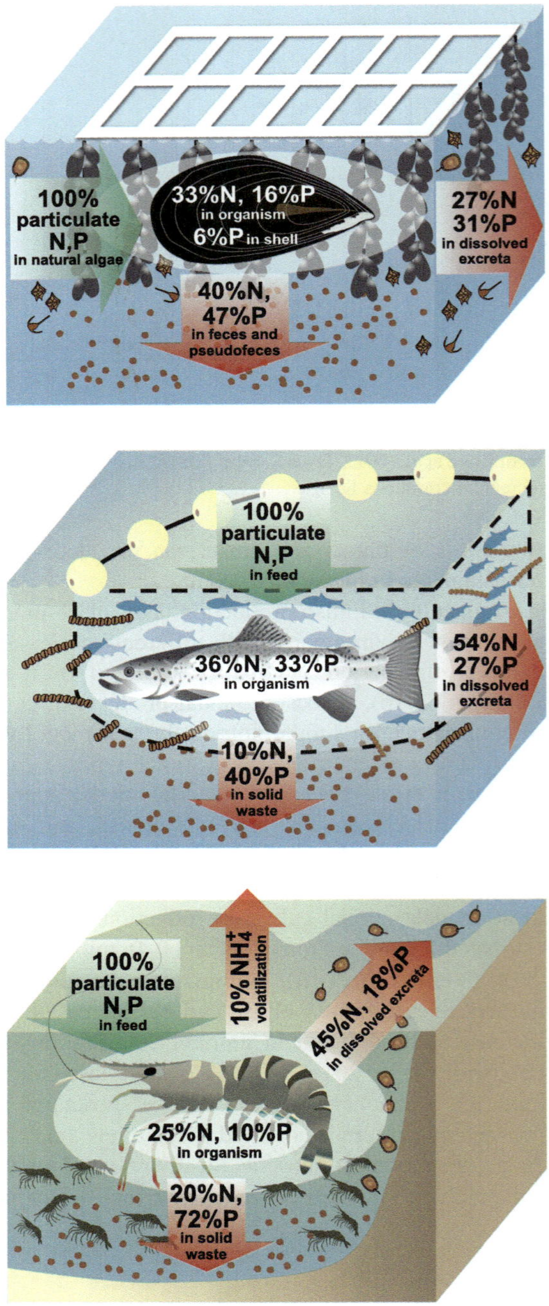

Fig. 22.9 Three types of mariculture and the associated N and P budgets in percent of the intake (food or feed intake is 100%): top panel, molluscan production in raft systems; middle, finfish production in cages; and bottom, crustacean production in ponds. The relative proportion of nutrients relative to exogenous or natural inputs are estimated based on the models outlined by Bowman et al. (2011, 2013a). The complement of the retention of nutrients in finfish (36% for N and 33% for P in salmon and rainbow trout), molluscan (33% for N and 22% for P), and crustacean production (25% for N and 10% for P) is released to the water columns. The proportion of nutrients in dissolved versus solid form depends on the digestibility of the food or feed. The larger the digestible fraction of a nutrient in the various food or feed ingredients, the larger the fraction in dissolved form. Image reproduced from Bouwman et al. (2013b Environ Res Letts 8(4): 044026. https://doi.org/10.1088/1748-9326/8/4/044026) under a creative commons license

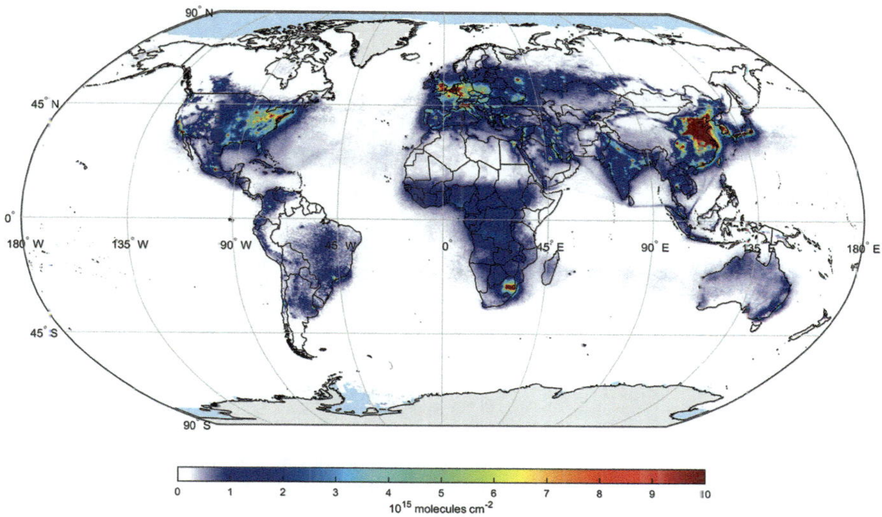

Fig. 22.10 Nitrogen emitted into the atmosphere by human activities. Global map of long-term average tropospheric NO_2 (given in units of × 10 15 molecules cm^{-2}). The average was computed over the entire time series from August 2002 to March 2012. Image reproduced from Schneider et al. (2015, Atmos Chem Phys 15: 1205–1220. https://doi.org/10.5194/acp-15-1205-2015) under a creative commons license

22.4.4 Atmospheric Deposition

In many regions, atmospheric deposition of N contributes significantly to pollution and is increasing (e.g., Jaworski et al. 1997; Howarth 2006; Galloway et al. 2008; Duce et al. 2008; Jia et al. 2016; Fig. 22.10). This N is derived from increasing NO_x emission from fossil fuel burning and from volatilization of animal manure and other land-based fertilizer applications. Atmospheric deposition of N is not only high, but increasing in many regions (e.g., Howarth 2006; Galloway et al. 2008; Duce et al. 2008) and in the downwind plumes of major animal operations and major cities (e.g., Whitall et al. 2003). Atmospheric deposition of N has reached >700 mg N m^2 $year^{-1}$ in many regions (Duce et al. 2008). Recent modeling has shown that there has been a threefold increase in soluble N deposition over land and twofold increase over the ocean due to human activities since 1850 and furthermore, while NOx emissions may decline by 20%, those of NH_3 are expected to increase by 22% by 2050 (Kanakidou et al. 2016). Atmospheric deposition of nutrients in the ocean is now ~20-fold greater than the Redfield ratio for N:P (Jickells 2006; Peñuelas et al. 2012), and these changes are also having large consequences for N:P stoichiometry in lakes (Elser et al. 2009). Moreover, emissions of NH_3 are further projected to increase and to become the dominant component of N emissions with rising temperatures (Fowler et al. 2013).

Fig. 22.11 Images of intensive agriculture (Panel **a**) and of housing and golf course developments (Panel **b**) that directly impact coastal or inland waters. Photos by P. Glibert

22.5 Enough!

Sustainability requires a balance between production, consumption, and decomposition. Such a balance is not evident in the intensity of agriculture or aquaculture in some areas. Nutrients exported from the land alter the nutrient balance of receiving waters – whether from human waste, animal waste from CAFOs (on land or in the sea), or fertilizer runoff from agricultural lands, or even from housing and golf course development close to shorelines (Fig. 22.11). It has been estimated that by 2050, human-related N and P sources will contribute up to 80% of river nutrient loads but, even as of the year 2000, the increased N and P loads had already been identified as a major driver of the loss of river biodiversity (Fig. 22.12; Burkholder and Glibert 2022).

22.6 Dam Those Rivers!

Modern water infrastructure has "replumbed the planet" (Boccaletti 2021). Dam and reservoir construction and other in-river consumptive uses, as well as channelization of flow, have greatly modified the timing, magnitude, form, and stoichiometry of nutrient delivery to the coast (see also Glibert et al. 2018). There are now 45,000 structures taller than 15 m that dam the waters of the world (Boccaletti 2021). As estimated by Beusen et al. (2016), retention of nutrients in lakes increased >60% between 1900 and 2000 due to reservoir construction and other anthropogenic flow alterations. Thousands of dams of a capacity of more than 1 MW have been constructed during a first construction boom in the 1950s and 1960s, and a second boom is now underway, mostly in developing economies (Grill et al. 2015, 2019; Zarfl et al. 2015; Fig. 22.13). Dam construction in many parts of the world

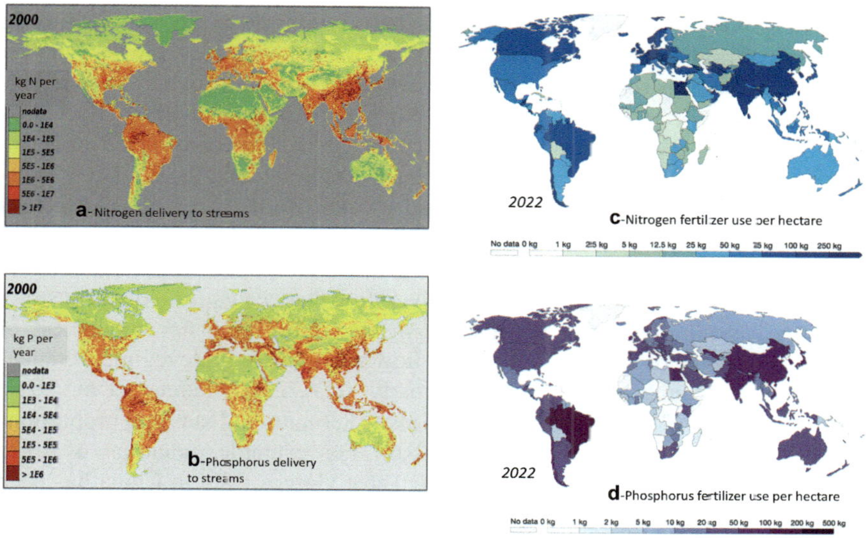

Fig. 22.12 Panels (**a**, **b**) Estimates of N and P export to streams for the year 2000 based on the Integrated Model to Assess Global Environment-Global Nutrient Model. Images reproduced from the Supplemental material of Beusen et al. (2016, Biogeoscience 13, 2441–2451, https://doi.org/10.5194/bg-13-2441-201) under a creative commons license. Panels (**c**, **d**) Comparison of N and P fertilizer use per hectare for 2022. Data from FAO as reported by Ritchie et al. (2022, https://ourworldindata.org/fertilizers) under a creative commons license

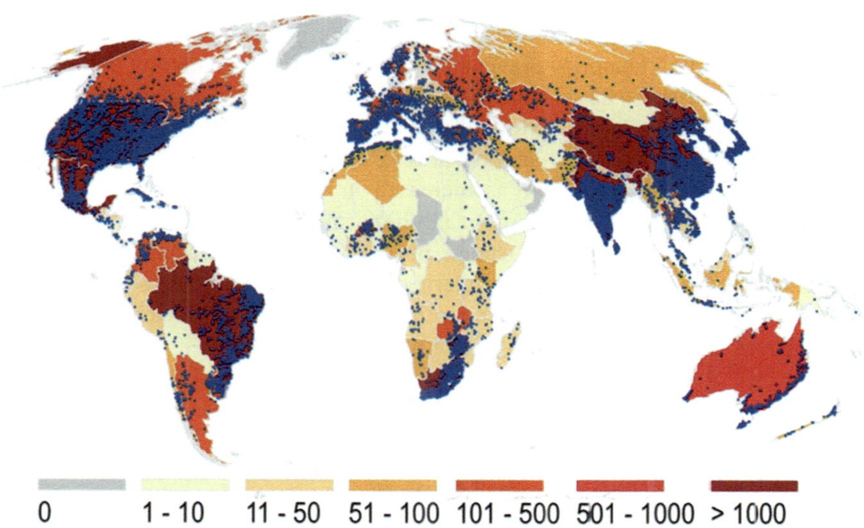

Fig. 22.13 Global distribution of dams by country. Image reproduced from Milligan et al. (2020, Scientific Data 7: 31. https://doi.org/10.1038/s41597-020-0362-5) under a creative commons license

contributes substantially to hydropower and also serves as flood protection, drinking water reservoirs, and other purposes. More than 3700 medium and large hydropower dams are under construction or planned worldwide that will double the currently installed capacity (Zarfl et al. 2015). By 2030, 93% of all rivers will be affected by dams (Grill et al. 2015). These dams are being, or will be, constructed in South America (Amazon and La Plata River basins) and in South and East Asia (Ganges-Brahmaputra and Yangtze basins) as well as in Africa (Zarfl et al. 2015). Such dams are expected to be developed in areas of the world where species richness is considered especially high, including the tropics and subtropics (Zarfl et al. 2015).

In addition to the reductions in overall river flow that occurs from dam construction, large river systems can become fragmented preventing free movement of organisms, and severe modification of river flow alters temperature regimes, dramatically reduces sediment transport (Vörösmarty et al. 2010; Lehner et al. 2011; Liermann et al. 2012). These modifications also alter nutrient loads and proportions in downstream waters (Beusen et al. 2016). For example, retention of silica (Si) upstream following construction of the Three Gorges Dam in China has been considered to be an important factor leading to altered nutrient proportions favoring HABs in East China Sea (e.g., Zhang et al. 2015). Altered retention further skews nutrient stoichiometry, as P is retained more efficiently than N during processing and transport in soils, groundwater, riparian zones and streams, rivers, lakes, and reservoirs (Beusen et al. 2016). For the major global dams, reactive P (total dissolved P + reactive particulate P) is more efficiently reduced (by 43%) than reactive Si (dissolved Si + reactive particulate Si) (by 21%) and total N (12%) (Maavara 2017; Maavara et al. 2014, 2015), thus increasing the N:P and N:Si ratios of riverine delivery to coastal areas.

Dams also alter stratification, because the density gradient of reservoir water—more inclined to solar heating—prevents the water column from mixing (Noori et al. 2021). Dam construction has also been shown to increase the rates of N_2 fixation due in part to associated warming (Moutinho et al. 2021). Also altered is the P cycle: under damming the rate of P retention in reservoirs is increased. Then, as waters warm, the release of sediment P is increased in the reservoirs, while P export downstream is decreased (Humborg et al. 2002; Wang et al. 2022). In all, dams impact nutrient cycling through a change in river connectivity, changes in nutrient cycling and nutrient export, as well as altering the physical retention of water and the consequence of such changes on other stressors, such as temperature changes. All of these changes have ultimate effects on the microbial community.

22.7 Browner and Darker

In addition to the increased use of organic N forms as fertilizer (e.g., Glibert et al. 2006, 2014) and as aquaculture sources, dissolved organic matter (DOM) loading is increasing with changes in precipitation, riverine discharge, sea level rise, altered land use, and the associated loss of many wetlands. The browning of waters has become an issue of great concern (Kritzberg et al. 2020). Images from Hurricane Ian, a powerful category 4 storm that struck Florida's southwest coast in September

Fig. 22.14 Panel (**a**) Image of export plumes from southwest Florida (USA) following Hurricane Ian in September 2022. The dark plumes are due to humic materials from decaying vegetation and other organic matter, such as sewage overflows. Note that the plumes extended 20–30 km from the coast. Image from the European Space Agency's Sentinel-2 mission on September 30, 2022, 8 days after the storm. Public domain. Panel (**b**) Differences in water color between surface waters due to the differences in concentration and composition of dissolved organic matter. The water in the glasses comes from natural waters in southern Sweden. Image reproduced from Kritzberg et al. (2020, Ambio 49: 375–390, https://doi.org/10.1007/s13280-019-01227-5) under a creative commons license

2022, illustrate how organic matter export from decaying vegetation and other land-based sources can impact coastal waters after an extreme event (Fig. 22.14). With coastal inundation due to sea level rise, coastal wetlands and mangrove swamps become stressed and migrate inland. In addition, stronger storms result in increased turbulence that increases scour near roots and trunks of nearshore vegetation, accelerating dieback. When this happens, organic matter and nutrient delivery is increased furthering erosion, leading to increasing nutrient delivery from land-based sources. Losses of marshes (including mangroves) accelerate nutrient supplies because the marshes are no longer available to take up nutrients and their decay and decomposition increases the supply of nutrients as organic matter decomposes (Charles and Dukes 2009; Glibert et al. 2020). Long-term (~10–30 year) data records have documented a significant recent increase in dissolved organic matter (DOM) concentrations in European (Hejzlar et al. 2003), Scandinavian (Forsberg and Petersen 1990; Andersson et al. 1991; Vuorenma et al. 2006), UK (Freeman et al. 2001; Worrall and Burt 2007), and North American rivers (Driscoll et al. 2003; Stoddard et al. 2003). This DOM includes dissolved humic substances that are naturally occurring, colored, high molecular weight organic acids. Increases in humic acids are resulting in the phenomenon of brownification of lakes and estuaries which is having profound effects on food webs and phytoplankton community composition. Brown lakes have shallower thermoclines which can lead to cold, potentially deoxygenated deep water (Solomon et al. 2015). Brown, humic-rich water may also alter pH, iron concentrations, and light penetration, all of which impact phytoplankton. These differences strongly impact metabolic rates, biogeochemical processes, and animal habitats (Solomon et al. 2015).

These brown, humic-rich waters have many consequences for phytoplankton (see also Chap. 29). Although classically considered to be biologically refractory in marine environments, humic acids have been shown to be bioavailable and used in growth of some algal species such as the toxic dinoflagellate *Karenia brevis* (Heil 2005; See et al. 2006). Summer may be a particularly important time for bioavailable DOM transport to estuaries, even though river discharge and DOM fluxes are low (e.g., Wiegner et al. 2006). Riverine DOM bioavailability is typically higher in the summer than during the winter (Stepanauskas et al. 1999, 2002), suggesting that its impact on the microbial community—including HABs—and contribution to eutrophication may be greater during this period. The largest fraction of the N in humic material is NH_4^+, loosely bound, and extractable using solid phase resins. NH_4^+ can be released from humic acids via photooxidation (Bushaw et al. 1996) and by ion exchange reactions, especially as salinities increase (as would occur with sea level rise and coastal inundation). Accordingly, humic acids have been hypothesized to act as a “shuttle” for NH_4^+ in estuaries, binding NH_4^+ at lower salinities and releasing it at higher salinities (Cochlan and Bronk 2001). Humic acids also decrease light penetration and can act as chelating agents that make Fe more available. Mixotrophic harmful algae would have an advantage in waters with such low light conditions, as they can augment energy and C inputs through feeding. Highly buoyant blooms of cyanobacteria would also be favored, both in low light and with higher concentrations of NH_4^+ and warming waters (Glibert et al. 2016).

22.8 Coastal Typology and the Susceptibility to Nutrient Enrichment

The magnitude and effect of land-based and atmospheric-based nutrient pollution on eutrophication and/or HAB proliferation depends not only on how much is exported but also how much is retained within the receiving waters. The fact that nutrient loads have generally increased is, in itself, not necessarily sufficient to cause eutrophication. The amount of loads matters, the forms of those loads matter, but so too does the physics of the receiving waters. From features such as eddies that concentrate plankton (Nishimoto and Washburn 2002) to small-scale turbulence and stratification, physical properties play important roles in how phytoplankton may aggregate. Coastal features such as banks, canyons, or islands have influences on upwelling and its local spatial effects (e.g., Pitcher et al. 2010). Coastal lagoons, a class of estuary characterized as shallow, highly enclosed, and typically quiescent in terms of wind, current, and wave energy develop different types of algal blooms in biomass and composition than their deeper, more dynamic counterparts, river-dominated estuaries (Madden et al. 2010; Glibert et al. 2010; see also Chap. 25).

Of the nutrient inputs that enter the coastal filter, an estimated 22% of the riverine N and 24% of the riverine P is retained (Laruelle 2009; Glibert et al. 2018). Estuaries in temperate regions of the Northern Hemisphere are particularly important as sinks for nutrients because of the prominence of relatively long residence time tidal systems. Fjords are particularly effective in retaining nutrients with residence times as long as decades, while small deltas have a low filtering capacity. Shallow, enclosed coastal lagoons can be subject to recurrent blooms, as for example, if a population of a harmful taxon becomes established. In tropical and subtropical regions, retention can be much less efficient and a large proportion of the nutrient inputs to the ocean bypasses the coastal filter and is discharged directly onto the shelves in large river plumes (McKee et al. 2004).

22.9 Summary

Recognizing the vast anthropogenic effects that rising CO_2, temperatures, nutrient pollution, greenhouse gas pollution, harvesting and the production of food (and its associated fertilizer use and animal confinement), and altered hydrodynamics for water consumption or electricity generation are having on the globe is fundamental to understanding how these changes affect ecological function and biodiversity or our aquatic systems. Nutrient pollution in retentive coastal and fresh waters has fundamentally different effects than nutrient inputs into systems with less retentive properties. Do we have effective clues for foreshadowing future states and the changing interactions and feedbacks between and among the major players—especially those at the base of the food web? Do we have the major conceptual foundational to understand changes that are occurring and to reliably predict what lies ahead? The next chapters dissect the individual stressors of temperature, hypoxia, harmful algae, and ocean acidification, and some of the systems experiencing extreme stresses are described in the final chapter of this unit.

References

Andersson T, Nilsson Å, Jansson M (1991) Coloured substances in Swedish rivers-temporal variation and regulating factors. In: Allard D, Borén H, Grimvall A (eds) Humic substances in the aquatic and terrestrial environment. Lecture notes in earth sciences, vol 33. Springer, Berlin, pp 243–253

Ashley WS, Strader S, Rosencrants T, Krmenec AJ (2014) Spatiotemporal changes in tornado hazard exposure: the case of the expanding bull's eye effect in Chicago, IL. WCAS 6:175–193

Beusen AF, Bouwman A, van Beek LPH, Mogollon J, Middelburg JJ (2016) Global riverine N and P transport to ocean increased during the 20th century despite increased retention along the aquatic continuum. Biogeosciences 13(8):2441–2451. https://doi.org/10.5194/bg-13-2441-2016

Boccaletti G (2021) Water, a biography. Pantheon Books, New York

Boresnstein S (2023) July is globe's hottest month in recorded history. PBS News Hour 27:2023. https://www.pbs.org/newshour/science/july-is-globes-hottest-month-in-recorded-history. Accessed 31 July 2023

Bouwman AF, Pawlowski M, Liu C, Beusen AWH, Shumway SE, Glibert PM, Overbeek CC (2011) Global hindcasts and future projections of coastal nitrogen and phosphorus loads due to shellfish and seaweed aquaculture. Rev Fish Sci 19:331–357. https://doi.org/10.1080/10641262.2011.603849

Bouwman AF, Beusen AHW, Overbeek CC, Bureau DP, Pawlowski M, Glibert PM (2013a) Hindcasts and future projects of global inland and coastal nitrogen and phosphorus loads due to finfish aquaculture. Rev Fish Sci 21:112–156

Bouwman AF, Beusen AHW, Glibert PM, Overbeek C, Pawlowski M, Herrera J, Mulsow S, Yu R, Zhou M (2013b) Mariculture: significant and expanding cause of coastal nutrient enrichment. Environ Res Lett 8(4):044026. https://doi.org/10.1088/1748-9326/8/4/044026

Bouwman AF, Beusen AHW, Lassaletta L, Van Apeldoorn D, Van Grinsven H, Zhang J (2017) Lessons from temporal and spatial patterns in global use of N and P fertilizer on cropland. Sci Rep 7:40366. https://doi.org/10.1038/srep40366

Boyd CE, McNevin AA, Davis RP (2022) The contribution of fisheries and aquaculture to the global protein supply. Food Sec 14:805–827. https://doi.org/10.1007/s12571-021-01246-9

Burkholder JM, Glibert PM (2022) Eutrophication and oligotrophication, Encyclopedia of biodiversity, vol 4. Elsevier. https://doi.org/10.1016/B978-0-12-384719-5.00047-2

Bushaw KL, Zepp RG, Tarr MA, Schulz-Jander D, Bourbonniere RA, Hodson RE, Miller WL et al (1996) Photochemical release of biologically available nitrogen from aquatic dissolved organic matter. Nature 381:404–407. https://doi.org/10.1038/381404a0

Charles H, Dukes JS (2009) Effects of warming and altered precipitation on plant and nutrient dynamics of a New England salt marsh. Ecol Appl 19:1758–1773. https://doi.org/10.1890/08-0172.1

Cochlan WP, Bronk DA (2001) Nitrogen uptake kinetics in the Ross Sea, Antarctica. Deep-Sea Res II Top Stud Oceanogr 48:4127–4153. https://doi.org/10.1016/S0967-0645(01)00083-2

Cork S, Peterson G, Petschel-Held G (2005) Four scenarios. In: Carpenter SR, Pingali PL, Bennett EM et al (eds) Ecosystems and human well-being, Scenarios, findings of the scenarios working Group of the Millennium Ecosystem Assessment, vol 2. Island Press, Washington, DC, pp 223–294

Davidson EA, Suddick EC, Rice CW, Prokopy LS (2015) More food, low pollution (MoFoLoPo): a grand challenge for the 21st century. J Environ Qual 44:305–311. https://doi.org/10.2134/jeq2015.02.0078

Delgado CL, Wada N, Rosegrant MW, Meijer S, Ahmed M (2003) Outlook for fish to 2020 meeting global demand. International Food Policy Research Institute, Washington, DC

Driscoll CT, Driscoll KM, Roy KM, Mitchell ML (2003) Chemical responses of lakes in the Adirondack region of New York to declines in acidic deposition. Environ Sci Technol 37:2036–2042

Duce RA, LaRoche J, Altieri K, Arrigo KR, Baker AR, Capone DG, Cornell S et al (2008) Impacts of atmospheric nitrogen on the open ocean. Science 320(5878):893–897

Ellis L (2008) A China environmental health project research brief: environmental health and China's concentrated animal feeding operations (CAFOs). www.wilsoncenter.org/sites/default/files/factory_farms_feb28.pdf

Elser JJ, Anderson T, Baron JS, Bergström A-K, Jansson M, Kyle M, Nydick KR, Steger L, Hessen DO (2009) Shifts in lake N:P stoichiometry and nutrient limitation driven by atmospheric nitrogen deposition. Science 326:835–837
FAO (2012a) Cultured aquatic species. FAO, Rome. http://fao.stat.fao.org
FAO (2012b) FAOSTAT database collections. FAO, Rome. http://fao.stat.fao.org
FAO (2016) Total fishery production 1950–2014. FishStatJ—Universal software for fishery statistical time series. Rome. http://www.fao.org/fishery/statistics/software/fishstatj/en
FAO (2017) World fertilizer trends and outlook to 2020. Summary report. http //www.fao.org/3/a-i6895e.pdf
FAO (2020) Fisheries and aquaculture statistics. Global aquaculture and fisheries production 1950–2018 (Fishstat). FAO Fisheries and Aquaculture Department, Rome
Food and Water Watch (2015) Factory Farm Nation 2015. www.foodandwaterwatch.org/sites/default/files/factory-farm-nation-report-may-2015.pdf
Food and Water Watch (2020) Factory Farm Nation 2020. www.foodandwaterwatch.org/sites/default/files/ib_2004_updfacfarmmaps-web2.pdf
Forsberg C, Petersen RC Jr (1990) A darkening of Swedish lakes due to increased humus inputs during the last 15 years. Verh Intern Vere Theor Ang Limnol 24:289–292
Fowler D, Coyle M, Skiba U, Sutton MA, Cape JN, Reis S, Sheppard LJ et al (2013) The global nitrogen cycle in the twenty-first century. Phil Trans Roy Soc B 368(1621):20130164. https://doi.org/10.1098/rstb.2013.0164
Freeman C, Evans CD, Monteith DT, Reynolds B, Fenner N (2001) Export of organic carbon from peat soils. Nature 412:785–786
Galloway JN, Cowling EB, Seitzinger SP, Socolow RH (2002) Reactive nitrogen: too much of a good thing? Ambio 31:60–63
Galloway JN, Townsend AR, Erisman JW, Bekunda M, Cai Z, Freney JR, Martinelli LA, Seitzinger SP, Sutton MA (2008) Transformation of the nitrogen cycle: recent trends, questions and potential solutions. Science 320:889–892
Galloway JN, Winiwarter W, Leip A, Leach AM, Bleeker A, Erisman JW (2014) Nitrogen footprints: past, present and future. Environ Res Lett 9(11):115003. https://doi.org/10.1088/1748-9326/9/11/115003
Glibert PM (2019) Phytoplankton in the aqueous ecological theater: changing conditions, biodiversity, and evolving ecological concepts. J Mar Res 77(S):83–137
Glibert PM (2020) From hogs to HABs: impacts of industrial farming in the US on nitrogen and phosphorus and greenhouse gas pollution. Biogeochem 150:1–42. https://doi.org/10.1007/s10533-020-00691-6
Glibert PM, Harrison JA, Heil CA, Seitzinger S (2006) Escalating worldwide use of urea—a global change contributing to coastal eutrophication. Biogeochem 77:441–463. https://doi.org/10.1007/s10533-005-3070-5
Glibert PM, Boyer J, Heil C, Madden C, Sturgis B, Wazniak C (2010) Blooms in lagoons: different from those of river-dominated estuaries. In: Kennish M, Paerl H (eds) Coastal lagoons: critical habitats of environmental change. Taylor and Francis, Boca Raton, FL, pp 91–114
Glibert PM, Manager R, Sobota DJ, Bouwman L (2014) The Haber-Bosch-harmful algal bloom (HB-HAB) link. Environ Res Lett 9:105001. https://doi.org/10.1088/1748-9326/9/10/105001
Glibert PM, Wilkerson FP, Dugdale RC, Raven JA, Dupont CL, Leavitt PR, Parker AE, Burkholder JM, Kana TM (2016) Pluses and minuses of ammonium and nitrate uptake and assimilation by phytoplankton and implications for productivity and community composition, with emphasis on nitrogen-enriched conditions. Limnol Oceanogr 61:165–197. https://doi.org/10.1002/lno.10203
Glibert PM, Beusen AHW, Harrison JA, Dürr HH, Bouwman AF, Laruelle GG (2018) Changing land-, sea-, and airscapes: sources of nutrient pollution affecting habitat suitability for harmful algae. In: Glibert PM, Berdalet E, Burford MA, Pitcher GC, Zhou M (eds) Global ecology and oceanography of harmful algal blooms. Ecol studies, vol 232. Springer, Cham. https://doi.org/10.1007/978-3-319-70069-4_4

Glibert PM, Beusen AHW, Bouwman AF, Burkholder JM, Flynn KJ, Heil CA, Li M et al (2020) Multifaceted climatic change and nutrient effects on harmful algae require multifaceted models. In: Botana LM, Louzao C, Vilariño N (eds) Climate change and marine and freshwater toxins, 2nd edn. DeGruyter Publishers, Berlin, pp 473–518. https://doi.org/10.1515/9783110625738-012

Grill G, Lehner B, Lumsdon AE, MacDonald GK, Zarfi C, Liermann CR (2015) An index-based framework for assessing patterns and trends in river fragmentation and flow regulation by global dams at multiple scales. Environ Res Lett 10(1):015001. https://doi.org/10.1088/1748-9326/10/1/015001

Grill G, Lehner B, Thieme M, Geenen B, Tickner D, Antonelli F, Babu S et al (2019) Mapping the world's free-flowing rivers. Nature 569:215–221

Heath C, Starr K (2022) Making numbers count. Avid Reader Press, New York, NY, p 110

Heil CA (2005) Influence of humic, fulvic and hydrophilic acids on the growth, photosynthesis and respiration of the dinoflagellate *Prorocentrum minimum* (Pavillard) Schiller. Harmful Algae 4: 603–618. https://doi.org/10.1016/j.hal.2004.08.010

Hejzlar J, Dubrovský M, Buchtele J, Ruzicka M (2003) The apparent and potential effects of climate change on the inferred concentration of dissolved organic matter in a temperate stream (the Malse River, South Bohemia). Sci Total Environ 310:143–152

Houlton BZ, Boyer E, Finzi A, Galloway J, Leach A, Liptzin D, Melillo J, Rosenstock TS, Sobota D, Townsend AR (2013) Intentional versus unintentional nitrogen use in the United States: trends, efficiency and implications. Biogeochem 114:11–23. https://doi.org/10.1007/s10533-012-9801-5

Howarth RW (2006) Atmospheric deposition and nitrogen pollution in coastal marine ecosystems. In: Visgilio G, Whitelaw DM (eds) Acid in the environment: lessons learned and future prospects. Springer, New York, pp 97–116

Howarth RW (2008) Coastal nitrogen pollution: a review of sources and trends globally and regionally. Harmful Algae 8:14–20

Humborg C, Blomqvist S, Avsan E, Bergensund Y, Smedberg E, Brink J, Mörth C-M (2002) Hydrological alterations with river damming in northern Sweden: implications for weathering and river biogeochemistry. Global Biogeochem Cycles 16(3):1039. https://doi.org/10.1029/2000gb001369

Intergovernmental Panel on Climate Change (IPCC) (2018) Summary for Policymakers. In: Masson-Delmotte V, Zhai P, Pörtner H-O, Roberts D, Skea J, Shukla PR, Pirani A et al. (eds) Global Warming of 1.5°C. An IPCC Special Report on the impacts of global warming of 1.5°C above pre-industrial levels and related global greenhouse gas emission pathways, in the context of strengthening the global response to the threat of climate change, sustainable development, and efforts to eradicate poverty. Cambridge University Press, Cambridge, UK and New York, NY, USA, pp 3–24. https://doi.org/10.1017/9781009157940.001

Intergovernmental Panel on Climate Change (IPCC) Core Writing Team (2014) Climate change 2014: Synthesis report. Contribution of Working Groups I, II and III to the Fifth Assessment Report of the Intergovernmental Panel on Climate Change. Pachauri RK, Meyer LA (eds), IPCC, Geneva, Switzerland

IPCC Core Writing Team (2023) Summary for policymakers. In: Lee H, Romero J (eds) Climate change 2023: synthesis report. A report of the intergovernmental panel on climate change. Contribution of working groups I, II and III to the sixth assessment report of the intergovernmental panel on climate change. IPCC, Geneva, Switzerland. 36 p

Jaworski NA, Howarth RW, Hetling LJ (1997) Atmospheric deposition of nitrogen oxides onto the landscape contributes to coastal eutrophication in the northeast United States. Environ Sci Technol 31:1995–2004

Jia Y, Yu G, Gao Y, He N, Wang Q, Jiao C, Zuo Y (2016) Global inorganic nitrogen dry deposition inferred from ground- and space-based measurements. Sci Rep 6:19810. https://doi.org/10.1038/srep19810

Jickells T (2006) The role of air-sea exchange in the marine nitrogen cycle. Biogeosci Discuss 3: 183–210

Johnson GC, Lyman JM (2020) Warming trends increasingly dominate global ocean. Nat Clim Chang 10:757–761. https://doi.org/10.1038/s41558-020-0822-0

Kaempffert W (1956) Science in review: warmer climate on earth may be due to more carbon dioxide in the air. New York Times, The Week in Review, p 191

Kanakidou M, Myriokefalitakis S, Daskalakis N, Fanourgakis G, Nenes A, Baker AR, Tsigaridis K, Mihalopoulos N (2016) Past, present, and future atmospheric nitrogen deposition. J Atmos Sci 73:2039–2047. https://doi.org/10.1175/JAS-D-15-0278.1

Ketchum BH (ed) (1972) Contamination and coastal pollution through waste disposal practices. The water's edge: critical problems of the coastal zone. The MIT Press, Cambridge, MA, pp 146–186

Kritzberg ES, Hasselquis EM, Škerlep M, Löfgren S, Olssen O, Stadmark J, Valinia S et al (2020) Browning of freshwaters: consequences to ecosystem services, underlying drivers, and potential mitigation measures. Ambio 49:375–390. https://doi.org/10.1007/s13280-019-01227-5

Laruelle GG (2009) Quantifying nutrient cycling and retention in coastal waster at the global scale. PhD dissertation, Geologica Ultraiectina. Mededelingen van de Faculteit Geowetenschappen Universiteit Utrecht No. 312, Utrecht University, Utrecht, 226 p

Lehner B, Liermann CR, Revenga C, Vörömsmarty C, Fekete B, Crouzet P, Döll P et al (2011) High-resolution mapping of the world's reservoirs and dams for sustainable river-flow management. Front Ecol Environ 9:494–502. https://doi.org/10.1890/100125

Liermann CR, Nilsson C, Robertson J, Ng RY (2012) Implications of dam obstruction for global freshwater fish diversity. Bioscience 62:539–548. https://doi.org/10.1525/bio.2012.62.6.5

Likens GE (ed) (1972) Eutrophication and aquatic ecosystems. Limnol Oceanogr Spec Symp 1.: 328 p. https://doi.org/10.1043/lo.1972.17.6.0965

Lindsey R, Dahlman L (2023) Climate change: global temperatures. https://www.climate.gov/news-features/understanding-climate/climate-change-global-temperature. Accessed 23 July 2023

Litke DW (1999) Review of phosphorus control measures in the United States and their effects on water quality. U.S. Geological Survey Water–Resources Investigations Report 99-4007. U.-S. Geological Survey, Denver, Colorado

Maavara T (2017) Perturbations to nutrient and carbon cycles by river damming. PhD thesis, University of Waterloo, 174 p

Maavara T, Dürr HH, Van Cappellen P (2014) Worldwide retention of nutrient silicon by river damming: from sparse data set to global estimate. Global Biogeochem Cycles 28:842–855

Maavara T, Parsons CT, Ridenour C, Stojanovic S, Dürr HH, Powley HR, Van Cappellen P (2015) Global phosphorus retention by river damming. Proc Natl Acad Sci 112:15603–15608. https://doi.org/10.1073/pnas.151179711

Madden CJ, Smith R, Dettman E, Kurtz J, Nelson W, Detenbeck N, Latimer J, Bricker S (2010) Estuarine typology development and application. In: Glibert PM, Madden CJ, Boynton W, Flemer D, Heil C, Sharp J (eds) Nutrients in estuaries: a summary report of the National Estuarine Experts Workgroup 2005–2007. EPA Office of Water, pp 29–44

McKee BA, Aller RC, Allison M, Bianchi TS, Kineke GC (2004) Transport and transformation of dissolved and particulate materials on continental margins influenced by major rivers: benthic boundary layer and seabed processes. Cont Shelf Res 24:899–926. https://doi.org/10.1016/j.csr.2004.02.009

Milligan M, van Soesbergen A, Sáenz L (2020) GOODD, a global dataset of more than 38,000 georeferenced dams. Sci Data 7:31. https://doi.org/10.1038/s41597-020-0362-5

Moutinho FHM, Marafao GA, Calijuri MD, Moreira MZ, Marcarelli AM, Cunha DGF (2021) Environmental factors and thresholds for nitrogen fixation by phytoplankton in tropical reservoirs. Int Rev Hydrobiol 106(1):5–17. https://doi.org/10.1002/iroh.202002057

Msangi S, Batka M (2015) Chapter 8: the rise of aquaculture. In: 2014-2015 Global food policy report. International Food Policy Research Institute. https://www.ifpri.org/sites/default/files/gfpr/2015/feature_3087.html

Nishimoto MM, Washburn L (2002) Patterns of coastal eddy circulation and abundance of pelagic juvenile fish in the Santa Barbara Channel, California, USA. Mar Ecol Prog Ser 241:183–199

Nixon SW (1995) Coastal marine eutrophication: a definition, social causes, and future concerns. Ophelia 41:199–219. https://doi.org/10.1080/00785236.1995.10422044

Noori R, Ansari E, Bhattarai R, Tang Q, Aradpour S, Maghrebi M, Haghighi AT et al (2021) Complex dynamics of water quality mixing in a warm mono-mictic reservoir. Sci Total Environ 777:146097. https://doi.org/10.1016/j.scitotenv.2021.146097

Peñuelas J, Sardans J, Rivas-Ubach A, Janssens IA (2012) The human-induced imbalance between C, N and P in earth's life system. Glob Chang Biol 18:3–6

Pitcher GC, Figueiras FG, Hickey BM, Moita MT (2010) The physical oceanography of upwelling systems and the development of harmful algal blooms. Prog Oceanogr 85:5–32. https://doi.org/10.1016/j.pocean.2010.02.002

Plass GN (1956) Carbon dioxide and the climate. Am Sci 44:302–316

Richardson K, Jørgensen BB (1996) Eutrophication: definition, history and effects. In: Jørgensen BB, Richardson K (eds) Eutrophication in coastal marine ecosystems, vol 52. Am Geophys Union. https://doi.org/10.1029/CE52p0001

Richter F (2022) The last 8 years have been the warmest on record. Statistica. www.statista.com/chart/28722/global-surface-temperature-anomalies/. Accessed 28 July 2023

Ritchie H, Roser M, Rosado P (2022) Fertilizers. https://ourworldindata.org/fertilizers. Accessed 15 July 2023

Schneider P, Lahoz WA, van der AR (2015) Recent satellite-based trends of tropospheric nitrogen dioxide over large urban conglomerations worldwide. Atmos Chem Phys 15:1205–1220. https://doi.org/10.5194/acp-15-1205-2015

See JH, Bronk DA, Lewitus AJ (2006) Uptake of Spartina-derived humic nitrogen by estuarine phytoplankton in nonaxenic and axenic culture. Limnol Oceanogr 51:2290–2299. https://doi.org/10.4319/lo.2006.51.5.2290

Smil V (2001) Enriching the earth: Fritz Haber, Carl Bosch, and the transformation of world food. The MIT Press, Cambridge, MA. 338 p

Solomon CT, Jones SE, Weidel BC, Buffam I, Fork ML, Karlsson J, Larsen S et al (2015) Ecosystem consequences of changing inputs of terrestrial dissolved organic matter to lakes: current knowledge and future challenges. Ecosystems 18:376–389. https://doi.org/10.1007/s10021-015-9848-y

Stepanauskas R, Leonardson L, Transvik LJ (1999) Bioavailability of wetland-derived DON to freshwater and marine bacterioplankton. Limnol Oceanogr 44:1477–1485

Stepanauskas R, Jørgensen NOG, Eigaard OR, Žvikas A, Transvik LJ, Leonardson L (2002) Summer inputs of riverine nutrients to the Baltic Sea: bioavailability and eutrophication relevance. Ecol Mongr 72:579–597. https://doi.org/10.2307/3100058

Stoddard JL, Kahl JS, Deviney FA, Dewalle DR (2003) Response of surface water chemistry to the Clean Air Act Amendments of 1990. US EPA, Research Triangle Park, NC, Rep. EPA/620/R-3/001

Sutton MA, Bleeker A, Howard CM, Bekunda M, Grizzetti B, de Vries W, van Grinsven HJM et al (2013) Our nutrient world: the challenge to produce more food and energy with less pollution. Centre for Ecology and Hydrology, Edinburgh, UK

Tilman D, Clark M (2014) Global diets link environmental sustainability and human health. Nature 515:518–522. https://doi.org/10.1038/nature13959

Tyndall J (1861) The Bakerian lecture: on the absorption and radiation of heat by gases and vapours, and on the physical connexion of radiation, absorption, and conduction. Phil Trans R Soc London 151:1–36

United Nations (2014) World urbanization prospects: 2014 revision, highlights (ST/ESA/SER.A/352). UN Department of Economic and Social Affairs, Population Division. https://esa.un.org/unpd/wup/publications/files/wup2014-highlights.pdf
US Department of Agriculture (USDA) (2020) Robust overseas demand for U.S. meats, led by pork and poultry, drove trade surplus in 2020. https://www.ers.usda.gov/data-products/chart-gallery/gallery/chart-detail/?chartId=101251. Accessed 10 July 2022
van Drecht G, Bouwman AF, Harrison J, Knoop JM (2009) Global nitrogen and phosphate in urban wastewater for the period 1970-2050. Global Biogeochem Cycles 23(4):GB0A03. https://doi.org/10.1029/2009GB003458
Verdegem MCJ (2013) Nutrient discharge from aquaculture operations in function of system design and production environment. Rev Aquac 5:158–171. https://doi.org/10.1111/raq.12011
Vörösmarty CJ, McIntyre PB, Gessner MO, Dudgeon D, Prusevich A, Green P, Glidden S, Bunn SE, Sullivan CA, Reidy Liermann C, Davies PM (2010) Global threats to human water security and river biodiversity. Nature 467:555–561. https://doi.org/10.1038/nature09440
Vuorenma J, Forsius M, Mannio J (2006) Increasing trends of total organic carbon concentrations in small forest lakes in Finland from 1987 to 2003. Sci Total Environ 365:47–65
Wang X, Chen Y, Yuan Q, Xing X, Hu B, Gan J, Zheng Y, Liu Y (2022) Effect of river damming on nutrient transport and transformation and its countermeasures. Front Mar Sci 9:1078216. https://doi.org/10.3389/fmars.2022.1078216
Whitall D, Hendrickson B, Paerl H (2003) Importance of atmospherically deposited nitrogen to the annual nitrogen budget of the Neuse River Estuary, NC. Envir Intern 29:393–399
Wiegner TN, Seitzinger SP, Glibert PM, Bronk DA (2006) Bioavailability of dissolved organic nitrogen and carbon from nine rivers in the eastern United States. Aquat Microb Ecol 43:277–287
World Bank (2021) Sanitation. www.worldbank.org/en/topic/sanitation. Accessed 6 Mar 2022
Worrall F, Burt TP (2007) Flux of dissolved organic carbon from U.K. rivers. Global Biogeochem Cycles 21:1–14
Xu Y, Ramanathan V, Victor DG (2018) Global warming will happen faster than we think. Nature 564:30–32. https://doi.org/10.1038/d41586-018-07586-5
Zarfl C, Lumsdon AE, Berlekamp J, Tydecks L, Tockner K (2015) A global boom in hydropower dam construction. Aquat Sci 77:161–170. https://doi.org/10.1007/s00027-014-0377-0
Zhang J, Wu Y, Zhang YY (2015) Plant nutrients and trace elements from the Changjiang watersheds and East China Sea Shelf. In: Zhang J (ed) Ecological continuum from the Changjiang (Yangtze River) watersheds to the East China Sea continental margin. Springer, pp 93–118

Additional Websites Accessed

https://climate.copernicus.eu/climate-indicators/greenhouse-gas-concentrations. Accessed January 2, 2023
https://www.climate.gov/media/15021. Accessed July 15, 2023
https://population.un.org/wpp/Graphs/Probabilistic/POP/TOT/900. Accessed July 15, 2023
https://www.statista.com/chart/28725/cumulative-co2-emissions-per-country-since-1970/. Accessed July 23, 2023
https://www.statista.com/statistics/438967/fertilizer-consumption-globally-by-nutrient. Accessed July 30, 2023
http://www.un.org/en/development/desa/news/population/2015-report.html. Accessed July 15, 2023
https://www.worldenergydata.org/a-striking-growth-of-co2-emissions-from-the-global-cement-industry-driven-by-new-facilities-in-emerging-countries/. Accessed January 2, 2023

Temperature and the Warming of Our Waters

23

23.1 Introduction

The name Dick Eppley (a renowned phytoplankton ecologist from Scripps Institution of Oceanography) is synonymous with phytoplankton-temperature relationships. His interest in temperature was motivated by a seeming lack of interest by the community in effects of temperature on phytoplankton. That opinion was developed by Steeman Nielsen (1960) and became widely adopted (see Chaps. 1 and 2 for an introduction to Steeman Nielsen and his development of ^{14}C as a tool for primary productivity measurements). So, in 1972, Eppley (1972, p. 1063) wrote a synthesis of temperature effects. He began as follows:

> Temperature does not seem to be important in the production of phytoplankton in the sea. For example, Steeman Nielsen (1960) has written, 'Recent investigations have shown... that the direct influence of temperature on organic production in the sea is fairly insignificant.' Other reviewers of photosynthesis in the sea likewise give little or no consideration to a role of temperature and Steeman Nielsen's statements find widespread endorsement in the published data on geographic and seasonal variation in marine phytoplankton photosynthesis.

He (Eppley 1972) then asked the questions, (1) why is temperature of so little importance? And (2) why would anyone write a review on temperature and phytoplankton growth in the oceans? Eppley, with his nonchalant style, intended to raise the profile of the importance of temperature.

The Intergovernmental Panel on Climate Change (IPCC 2023) projects—with a confidence of 50% or better—that global temperatures will rise by 4 °C during the twenty-first century, based on a range of modeling scenarios (see also Chaps. 22 and 28). Even with aggressive actions to mitigate warming and climate change, temperatures will continue to rise for decades and even into the next century. As of the time of writing of this chapter, global ocean temperatures have been well above average (Fig. 23.1). Indeed, summer water temperatures in 2023 in some waters of Florida were recorded to exceed 38°. Eppley's (or Steeman Nielsen's) initial

P. M. Glibert, *Phytoplankton Whispering: An Introduction to the Physiology and Ecology of Microalgae*, https://doi.org/10.1007/978-3-031-53897-1_23

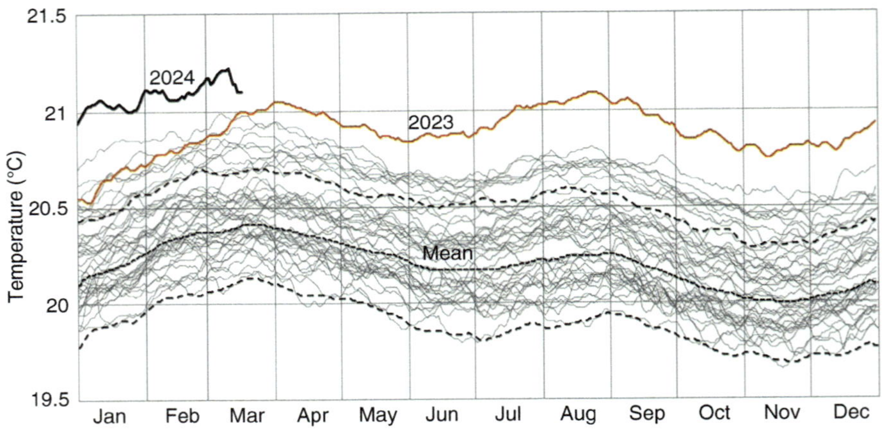

Fig. 23.1 Daily sea surface temperature for the world from 60S to 60N. The thick line showing highest temperatures is 2024; the orange line is 2023; the center dashed line is the mean from 1982 to 2011 and the other dashed lines represent ±2 σ. Faint gray lines represent individual years from 1988–2022. Image reproduced from https://ClimateReAnalyzer.org under a creative commons license

hypothesis that temperatures do not seem to be important in the production of phytoplankton in the sea seems almost unfathomable.

23.2 Eppley's Synthesis

Eppley summarized the then-available data on growth rates from approximately 130 species or clones derived from unicellular laboratory measurements on both fresh and marine taxa (Fig. 23.2). About half of these data had been previously aggregated by Hoogenhout and Amesz (1965). This synthesis revealed several patterns: (1) there is a gradual and exponential increase in growth rate (μ) with temperature up to about 40 °C, and (2) values of μ below 40 °C seem to fall within an envelope and a maximum expected temperature for any given temperature can be derived by the smooth curve of maximum values. This upper envelope relationship was described by the equation,

$$\mathrm{Log}_{10}\mu = 0.0275\mathrm{T} - 0.070 \tag{23.1}$$

Remarkably, this dataset includes examples of growth rates exceeding 10 doublings/day!

Eppley (1972) highlighted the temperature responses of several selected species which clearly illustrate the common pattern of a gradual increase in μ as temperatures rise to the optimum but an abrupt decline at supra-optimum

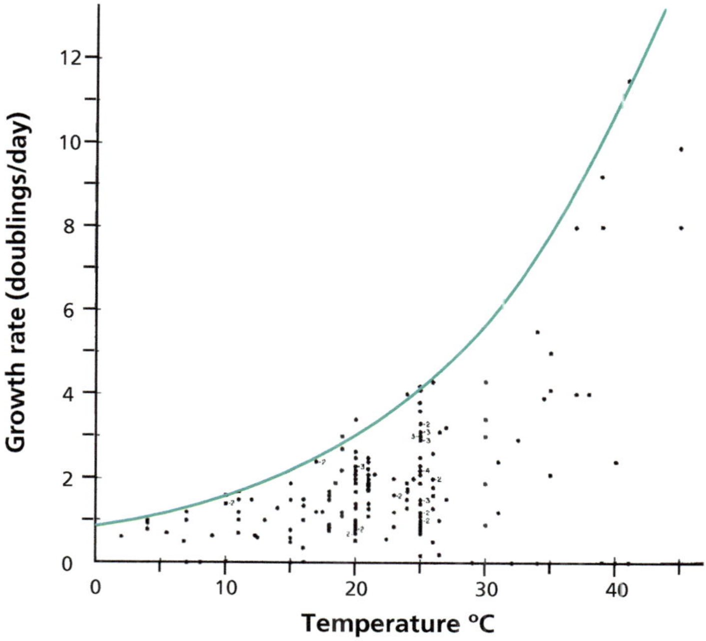

Fig. 23.2 Variation in the specific growth rate (m) of photoautotrophic algae with temperature. Data are all for laboratory cultures. Growth rate is expressed in doublings/day. Approximately 80 of the points are from the compilation of Hoogenhout and Ameaz (1965). That listing is restricted to maximum growth rates observed, largely in continuous light. The figure also includes additional data, mostly for cultures of marine phytoplankton, from the following sources: Lanskaya (1961), Eppley (1963), Castenholz (1964, 1969), Eppley and Sloan (1966), Swift and Taylor (1966), Thomas (1966), Paasche (1967,1968), Hulbert and Guillard (1968), Jørgensen (1968), Smayda (1969), Bunt and Lee (1970), Guillard and Myklestad (1970), Ignatiades and Smayda (1970), Polikarpov and Tokareva (1970). The latter papers include about 50 strains of marine phytoplankton. The line is the growth rate predicted by Eq. (23.1), i.e., the line of maximum expected μ. Small numbers by points indicate the number of values which fell on the point. Image redrawn based on Eppley (1972, Fish Bull 70: 1063–1085), public domain

temperatures (Fig. 23.3). He noted that growth in the thermal region close to—or above—the species optimum would be "risky" if the ambient temperature were to rise by a few degrees. How prescient! Another important observation in the Eppley (1972) synthesis was the recognition that temperature-response curves can be altered by other ambient conditions, as for example, salinity (Fig. 23.4). Both the shape of the thermal response curve and the interactions of temperature with other environmental factors are discussed in more detail below.

With even more insight well ahead of his time, Eppley also commented that the relationships he summarized represent natural phytoplankton growing under autotrophic conditions. He noted (p. 1066) that, "heterotrophic nutrition might lead to values of μ higher than predicted above, as appears to be the case when one compares doubling times of heterotrophic and photosynthetic bacteria or

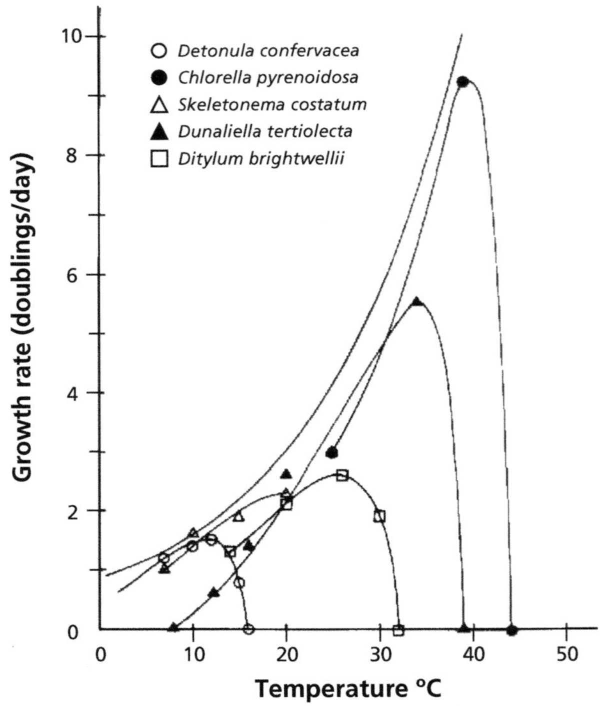

Fig. 23.3 Growth rate vs. temperature curves for five unicellular algae with different temperature optima: *Detonula confervacea* (Guillard and Ryther 1962), Smayda (1969), *Skeletonema costatum* (Jørgensen (1968), *Ditylum brightwellii* (Paasche 1968), *Dunaliella tertiolecta* (McLachlan 1960; Ukeles 1961; Eppley 1963; Eppley and Sloan 1966), and *Chlorella pyrenoidosa* (Sorokin and Krauss 1958, 1962). Image redrawn based on Eppley (1972, Fish Bull 70: 1063–1085), public domain

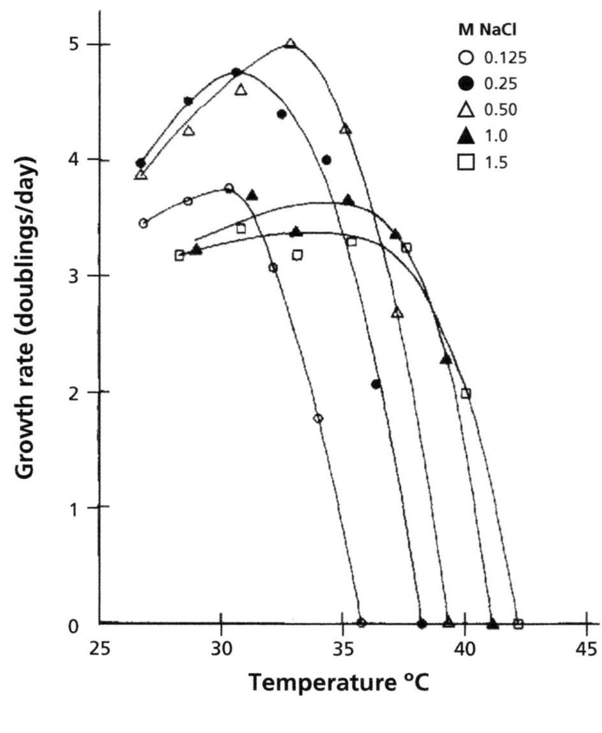

Fig. 23.4 Growth rate vs. temperature curves for *Dunaliella tertiolecta* measured in media containing different salt concentrations Image redrawn based on Eppley (1972, Fish Bull 70: 1063–1085), public domain

autotrophic vs. photoheterotrophic growth rates of the sewage alga *Chlamydomonas mundane*." He was right about that also.

Eppley related his laboratory synthesis to measurements in the field by analyzing available photosynthesis rates and rates of photosynthesis per unit carbon (C) (assimilation numbers). At that time, it was generally assumed that rates of photosynthesis and rates of growth would mirror each other. Moreover, photosynthesis rates at that time were generally derived from the C assimilation method of Steeman Nielsen (1952), which required long incubation times (a day). Eppley commented (1972, p. 1068), "What is needed is an instantaneous method not confounded by the complexities of long incubation either in situ, in enclosed vessels, or in shipboard cultures. Unfortunately, no such method is in view." Techniques such as those of PAM or FRRF fluorometry had not yet been invented (see Chap. 7).

Eppley concluded his analysis by stating (Eppley 1972, p. 1080), "given the current popularity of modeling, . . . neither the reader nor the author may be able to resist for long the temptation to combine Equation [23.1] with a realistic function for nutrient assimilation rate vs. ambient concentration, a function for the dependence of μ and assimilation number upon light, and a suitable function for describing effects of mixing, in line with critical depth theory, and to try it with his favorite set of oceanic data." Before describing some of the efforts that various investigators could not resist undertaking, several concepts and advances are first reviewed.

23.3 Thermal Optima and Temperature Adaptations: Formalizing the Thermal Response Curve

Broadly speaking, metabolism increases with increasing temperatures, but the response varies with species, specific reactions, or periods of adaptation. The value Q_{10} is defined as the ratio of reaction rates at two temperatures that differ by 10 °C. It is generally assumed that for many physiological processes, the Q_{10} doubles ($Q_{10} = 2$) for a 10 °C temperature rise. Eppley's relationship is described by a Q_{10} value of 1.88.

Another common temperature relationship is the Arrhenius equation (Arrhenius 1889):

$$k = \mathrm{A}e^{(-Ea/RT)} \tag{23.2}$$

where k is the rate, R is the gas constant, T is the temperature, A is the exponential factor, and Ea is the activation energy (Brown et al. 2004). This relationship is based on the concept that higher temperatures allow more molecules to reach an energy level greater than activation energy, allowing more chemical reactions to occur; cellular processes are thus driven faster.

The temperature response curve for a given species (or a given strain within a species) has a minimum (T_{min}), optimum (T_{opt}), and maximum (T_{max}). Thermal breadth is the total temperature range for growth between the T_{min} and T_{max}. The maximum growth rate (μ_{opt}) is the growth rate at T_{opt}. As Eppley noted, the curve is

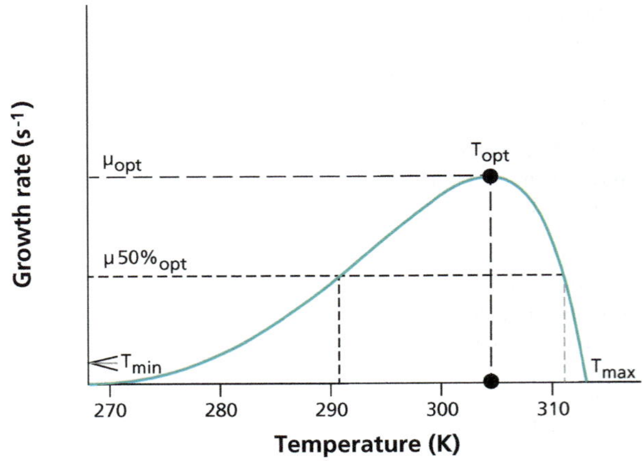

Fig. 23.5 Relationship of the thermal growth curve where the temperature (K) is plotted against the growth rate (s^{-1}). Cardinal temperatures (T_{min}, T_{opt}, and T_{max}) with their corresponding growth rates (μ_{opt} and $\mu_{50\%opt}$) are indicated. Image reproduced from Noll et al. (2020, Processes 8: 121. https://doi.org/10.3390/pr8010121) under a creative commons license

typically asymmetric, with a slower rise at temperatures $<T_{opt}$ and a sharper drop off at temperatures $>T_{opt}$ (Fig. 23.5). The steeper decline at high temperatures reflects the fact that proteins can begin to denature at temperatures well above their optimum. The shape of this curve has important implications: even small increases in temperature above T_{opt} lead to large declines in growth. There are various parameterizations of this relationship for microbial temperature-dependent growth rates and they have been well summarized in Noll et al. (2020).

23.4 Advances Since Eppley and Interactive Effects of Temperature and Other Factors

Since the publication of the "Eppley curve," there have been several efforts to update the relationship with new data. Brush et al. (2002) updated the curve and suggested the y-intercept of the "Eppley curve" should be increased by ~60% and also noted that the highest growth rates observed at the highest temperatures were associated with cultures grown under continuous light (suitable for understanding high latitude taxa, but not those from lower latitudes). Rose and Caron (2007) developed a comparable comparison for bacterivores and herbivores. Much higher growth rates at higher temperatures could be maintained by the bacterivores compared to the herbivores, and when comparable growth rates for the fastest growing bacterivores were compared with the phytoplankton from the "Eppley curve," rates were much faster. This is exactly what Eppley predicted. However, at low temperatures, growth rates of herbivores were lower than those of the corresponding phytoplankton, leading to the possibility of low grazing pressure on phytoplankton in cool or cold water, allowing phytoplankton a window of opportunity for growth during that time.

Various studies have examined the interactive effects of temperature and light availability or nutrient status, and just a few are highlighted here. Raven and Geider (1988) reviewed relationships between temperature and light. It is often expected that light-limited photosynthesis and growth will be less sensitive to temperature than light-saturated rates, due to limitation of photosynthesis by photon absorption at low irradiance. Compared to growth at high temperatures, at lower temperatures, the value of μ at a given photon flux density is higher; the ratio of temperature-independent light harvesting and temperature-dependent downstream photochemistry is lower than that for plants or algae growing at higher temperatures. Data available at the time of the Raven and Geider (1988) review revealed examples of the "expected" response, that is, for light-limited growth to be much less temperature sensitive than light-saturated growth (*Ceratium furca* and *Gonyaulax polyedra* (Meeson and Sweeney 1982), *Oscillatoria redekei* and *O. agardhii* (Foy 1983), and *Phaeodactylum tricornutum* (Fawley 1984)). On the other hand, Raven and Geider (1988) also reported on species for which the light-limited specific growth rate is temperature insensitive (*Microcystis aeruginosa*; Kruger and Eloff 1978) and where light effects on growth temperature are more complicated (*Leptocylindrus danicus*; Verity 1982).

Interactions between temperature and light availability have more recently been examined by Edwards et al. (2016), who synthesized data from 57 phytoplankton species. The Edwards et al. analysis showed that the value of the temperature optimum increased by about 4 °C from extreme light-limited conditions to irradiance levels of about 100 mmol photons m^{-2} s^{-1}, but then decreased again as irradiance levels continued to rise. In all, they found that light-limited growth, light-saturated growth, and optimal irradiance for growth are all temperature sensitive, but the maximum achievable growth rate increases with temperature for a given species under light saturation but not under light limitation. However, although individual species may show unimodal responses to temperature, in a natural community as temperatures change, species replace one another, and thus a community response may be insensitive to temperature even if individual species are sensitive.

Thomas et al. (2017) examined how a species optimum temperature varies with their nutrient status, testing whether optimum temperature decreases as nutrient limitation occurs, and they developed a model of population growth as a function of both factors. Given the highly nonlinear functions of growth with temperature and with nutrients, the interactive effects are not simple. By combining the relationships of specific growth rate as a function of temperature and of nutrients, they derived a joint function of both factors, the double-exponential model (Thomas et al. 2017):

$$\mu(T) = b_1 . . \exp(b_2 . T) - (d_0 + d_1 . \exp(d_2 . T) \qquad (23.3)$$

where $\mu(T)$ is the growth rate (day^{-1}), b_1 is the birth rate at 0 °C, b_2 is the exponential increase in birth rate (i.e., gross growth rate) with increasing temperature, d_0 is the temperature-independent mortality term, and d_1 and d_2 jointly describe the exponential increase in mortality rate with temperature. Thus, the first half of the equation describes the effect of temperature on birth rates, while the second half describes the mortality rates.

The Thomas et al. (2017) model showed that tolerance for high temperatures decreased under nutrient stress and the ability to take up nutrients decreases at both ends of the temperature spectrum. They also tested their model in the laboratory with the diatom *Thalassiosira pseudonana* under both phosphate (PO_4^{3-}) and nitrate (NO_3^-) limitation. The implication derived from their model, confirmed by the laboratory tests, is that species are more vulnerable to hot, low-nutrient conditions and that nutrient deprivation may exacerbate the effects of warming on phytoplankton communities. As warming also increases water column stratification and therefore increased nutrient limitation, temperature optima may decrease. The synergism of nutrient limitation and warming temperatures may have a greater effect than would be predicted if these factors were individually considered (Thomas et al. 2017). Warmer temperatures should lead to decreased elemental demands for N and P (Yvon-Durocher et al. 2015), due to the decreased demand for proteins and RNA that, in turn, may promote a decrease in cell size (Woods et al. 2003). On the other hand, warming may occur in conjunction with nutrient enrichment, as is the case with eutrophication in lakes, estuaries, and coastal regimes. Nutrient enrichment therefore may allow growth at temperatures that might be supra-optimal under nutrient limiting conditions.

Vidyarathna et al. (2023) studied the thermal niche of the dinoflagellate *Karlodinium veneficum* across different salinity and light levels. This dinoflagellate is common in estuarine waters and, based on literature, can grow across a wide range of salinities. It is common in Chesapeake Bay, for example. High light increased the thermal sensitivity of *K. veneficum* and lowered the thermal optima for growth (Fig. 23.6). Temperature increases above 28 °C were stressful and thus may suppress summer bloom formation in the Chesapeake Bay, especially under high light conditions. However, *K. veneficum* can inhabit the water column from the surface to several meters deep and may find habitats where light is optimum and therefore where temperature shifts are less stressful. In that case, bloom formation may prevail through the summer season. The high-salinity lower bay regions may provide over-summering habitats for *K. veneficum* during warm summer months. In a warming scenario, cells with mixotrophic feeding may also be able to overcome the negative direct effects of warm waters.

Siegel et al. (2023) examined the coexistence of species, in this case, two marine diatoms, under different temperature fluctuating regimes and under high and low N conditions. They found that N limitation reduces the magnitude of thermal variability over which stable species coexistence occurs and impacts the ability of the less competitive species to recover—exacerbating species displacement. Limitation of N diminished the competitiveness of the "winning" species and lengthened the time it would have otherwise taken to displace the "losing" species. Thermal variability apparently provided a refuge for the species that would otherwise be outcompeted if thermal conditions were stable, that is, fluctuation in temperature fostered ecological resistance. Their work suggests that community compositional may change in unpredictable ways as different abiotic variable fluctuate in conjunction with changes in temperature.

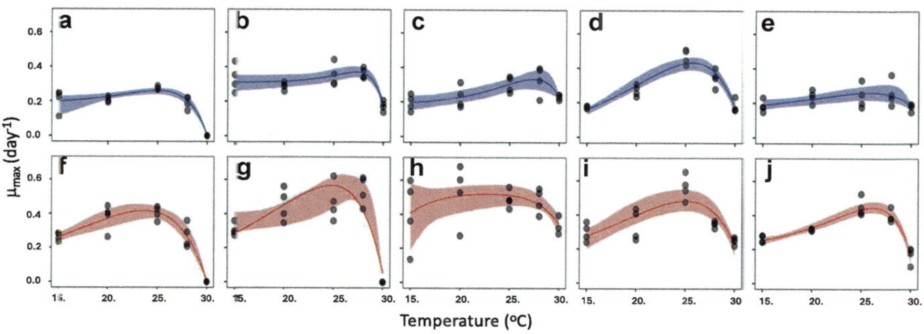

Fig. 23.6 Maximum growth rates (μ_{max}, day^{-1}) of *K. veneficum* across the temperature range of 15–30 °C at salinity 5 (Panels **a, f**), 10 (Panels **b, g**), 15 (Panels **c, h**), 20 (Panels **d, i**), and 30 (Panels **e, j**) at low light (upper Panels **a–e**) and at high light (lower Panel **f–j**) conditions. Each curve represents a thermal performance curve fit to the double-exponential model (Thomas et al. 2017, Eq. 23.3). The points represent μ_{max} of independent replicates at each experimental condition ($n = 4$ for each condition). The solid line represents the mean predictions, and the shaded bands represent the 95% confidence interval of model fits. Image reproduced from Vidyarathna et al. (2023, J Plankt Res: https://doi.org/10.1093/plankt/fbad019) with permission of Oxford University Press

23.5 Species and Functional Group Differences in Temperature Responses: Global Implications for Climate Change

Different phytoplankton species and functional groups have been shown to have different thermal responses, and numerous investigators have examined these differences and related them to regional or global distributions or their projected changes. Differences in temperature responses between taxa have been attributed to differences in protein structure and contents (especially enzymes), lipid composition of cell membranes, as well as other factors (Edwards et al. 2016).

Differences in the temperature responses of genotypes or species have often been related to their regional or global distributions. Chen (2015), based on literature compilations of available data, showed that T_{min}, T_{opt}, and T_{max} for marine phytoplankton decrease with latitude, but this was not the case for freshwater phytoplankton. Chen also showed, consistent with other studies, that T_{min} values for most cyanobacteria and dinoflagellates are higher than those of most diatoms; diatoms generally like it cool. Moreover, their thermal niches are typically narrower. Fast-growing taxa may tolerate a wider temperature range.

Thomas et al. (2012), expanding on the Eppley approach, examined growth responses to temperature from nearly 200 strains of 130 species reported in the literature and related them to their distribution. They found that strains exhibited a latitudinal gradient in T_{opt} and that T_{opt} was strongly related to mean annual temperature of the isolation location (Fig. 23.7). However, the width of the thermal niche was unrelated to temperature regimes. They projected an increased

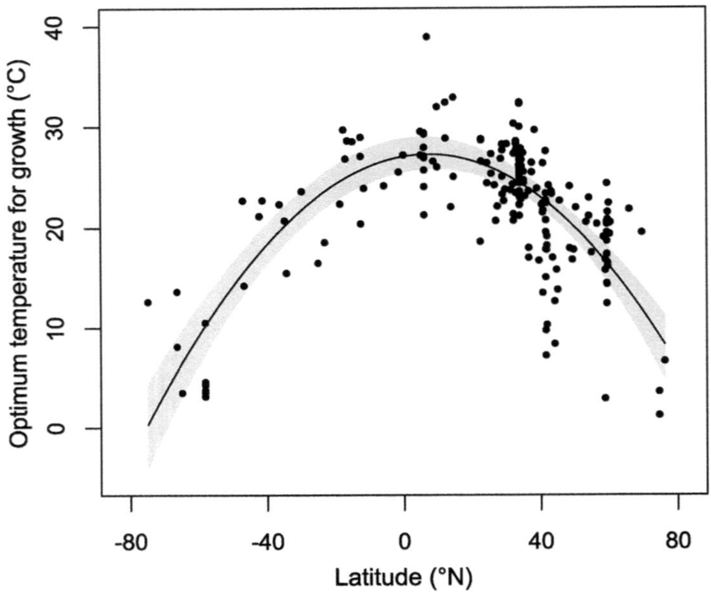

Fig. 23.7 Latitudinal gradient in the optimum temperature for growth of marine and estuarine phytoplankton strains ($n = 194$ strains, $R^2 = 0.55$, $P < 0.0001$). Each point represents the optimum temperature for growth of a single strain, estimated by fitting a thermal tolerance function to the data. The regression line (black) is shown, along with 95% confidence bands (gray). Image reproduced from Thomas et al. (2012, Science 338: 1085–1088. https://doi.org/10.1126/science.1224836) with permission of the American Association for the Advancement of Science (AAAS)

susceptibility of tropical strains to further warming and a loss of biodiversity in the tropics in the next hundred years.

Further building on Eppley's approach, and also compiling data from different functional groups, Anderson et al. (2021) reported different thermal responses for four major types: diatoms, dinoflagellates, cyanobacteria, and coccolithophores (Fig. 23.8). Their Q_{10} results differed somewhat from Eppley's, with cyanobacteria showing a Q_{10} of 2.13 and diatoms, dinoflagellates, and coccolithophores showing Q_{10} values of 1.55, 1.42, and 1.67, respectively (Anderson et al. 2021). Then, using sea surface temperature projections from an ensemble of models, they showed how different phytoplankton functional types might change in a future ocean. This exercise, like that of Thomas et al. (2012), showed the potential for a large reduction in tropical phytoplankton diversity over the next century, as well as poleward shifts in strain ranges, but a proportional growth increase at mid and high latitudes. The sharper decrease in tropical species is due their current existence being closer to their T_{opt} (Fig. 23.9).

A number of experimental and modeling efforts have converged on similar findings or trends with respect to cell size and the importance of mixotrophy in the

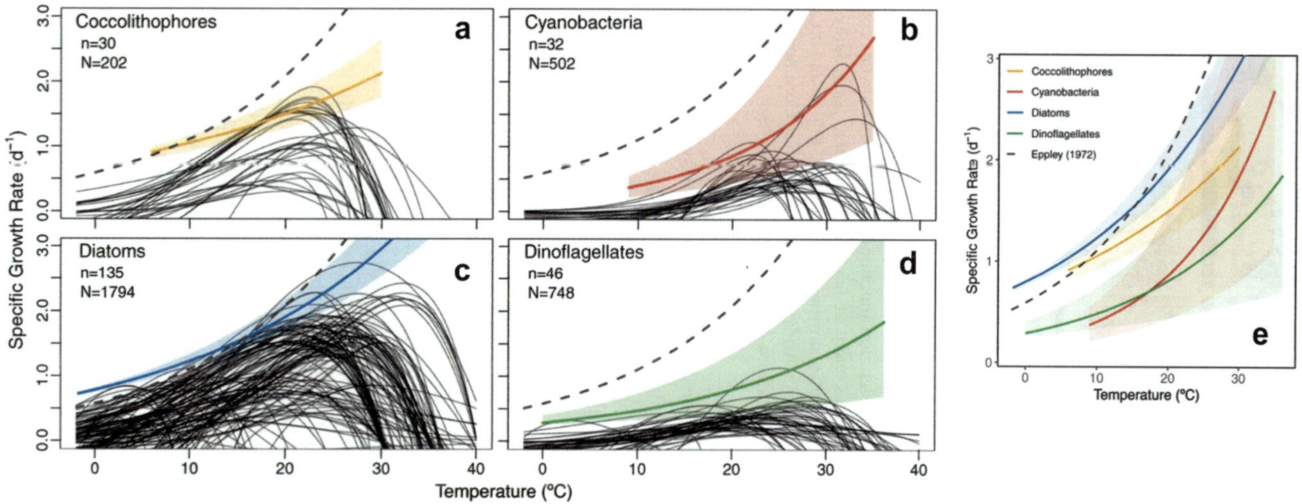

Fig. 23.8 Modeled exponential curves (colored lines) were fit to growth rates (N) and compared to the curve estimated by Eppley (gray dashed lines). Panels (**a**–**d**) The thermal responses for individual isolates for each functional group are shown in black. Modeled exponential curves (colored lines) were fit to measured growth rates (*N*) using a 99th quantile regression and compared to the Eppley curve, which assessed phytoplankton collectively (gray dashed line, same in each panel). Extent of modeled curves (colored lines) denotes limits of data. The 95% confidence intervals (shading) were determined using Markov chain marginal bootstrapping and are centered at the median. Thermal reaction norms (*n*) for each isolate characterized are shown in black. The cyanobacteria group does not include diazotrophic species. Panel (**e**) Functional group maximum growth rates (modeled μ_{max}) as a function of temperature fit using growth measurements according to Bissinger et al. (2008) and compared with that from Eppley. 95% confidence intervals (shading) were determined using Markov chain marginal bootstrapping. Dinoflagellates were characterized by the smallest performance changes overall, as evidenced by their relatively flatter reaction norms. Images reproduced from Anderson et al. (2021, Nat Commun 12: 6413, https://doi.org/10.1038/s41467-021-26651-8) under a creative commons license

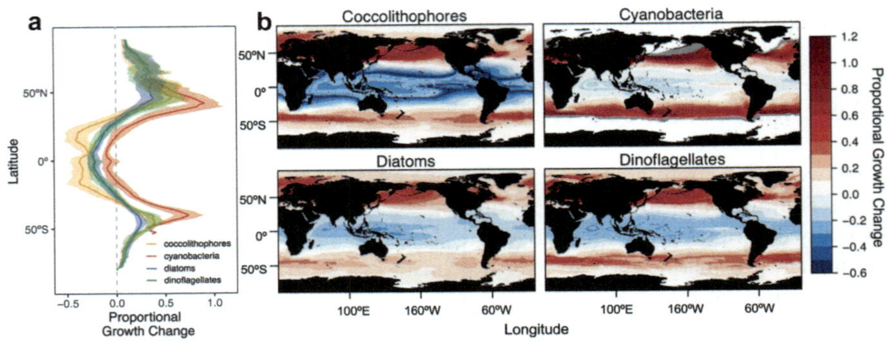

Fig. 23.9 Change in proportional growth predicted between historical (1950–1970) and future (2080–2100) temperature regimes under the RCP 8.5 climate scenario. Panel (**a**) Latitudinal averages in proportional growth change are depicted for each functional group (lines) with shading representing longitudinal variation from the mean (sd). Panel (**b**) Maps depict median proportional growth change from strains within each phytoplankton functional group. Strains were considered viable at a given location if their growth rates were at least 20% of their μ_{max}. Gray area at cyanobacteria boundary extent indicates growth occurring in the future, but absent in the past (i.e., a range expansion). Images reproduced from Anderson et al. (2021, Nat Commun 12: 6413, https://doi.org/10.1038/s41467-021-26651-8) under a creative commons license

future ocean. Due to the different relationships between photosynthesis and respiration, it can be hypothesized that phagotrophy (mixotrophy) would be favored at warmer temperatures compared to photoautotrophy (Wilken et al. 2013; Lin et al. 2018). Wilken et al. (2013) grew the mixotrophic chrysophyte *Ochromonas* sp. under autotrophic, heterotrophic, and mixotrophic conditions over a range of temperatures from 13 to 33 °C. Grazing rates on bacterial prey increased more strongly than did photosynthetic electron transport rates, but autotrophic growth reached a maximum at intermediate temperatures, leading to a decline in chlorophyll *a* content at the highest temperatures tested.

However, such effects also depend on the relative nutrient status of the cells. As shown by Lin et al. (2018) in mixotrophic modeling of a 10-day growth period, under a low N:P condition, the biomass (based on C) of the mixotrophic dinoflagellate *Karlodinium veneficum* was significantly higher in mixotrophic growth at warm temperatures compared to autotrophic growth for the same temperature (>25 °C), but such differences were reduced as N:P conditions changed in the model (Fig. 23.10). Under low N:P condition, temperature effects on mixotrophic growth rates of *K. veneficum* were also significant, with mixotrophic growth rates increasing faster than autotrophic growth rates at the highest temperature (i.e., 25 °C; Fig. 23.10a).

Franzè and Menden-Deuer (2020) examined temperature effects on mixotrophic (herbivorous) dinoflagellates. Growing three taxa over a temperature range from 0 to 22 °C, they observed that biomass-based growth rates were 10-fold higher than abundance-based growth rates at low temperatures due to the fact that cell biovolumes increased by 50%. Cold waters yield larger cells. While they observed

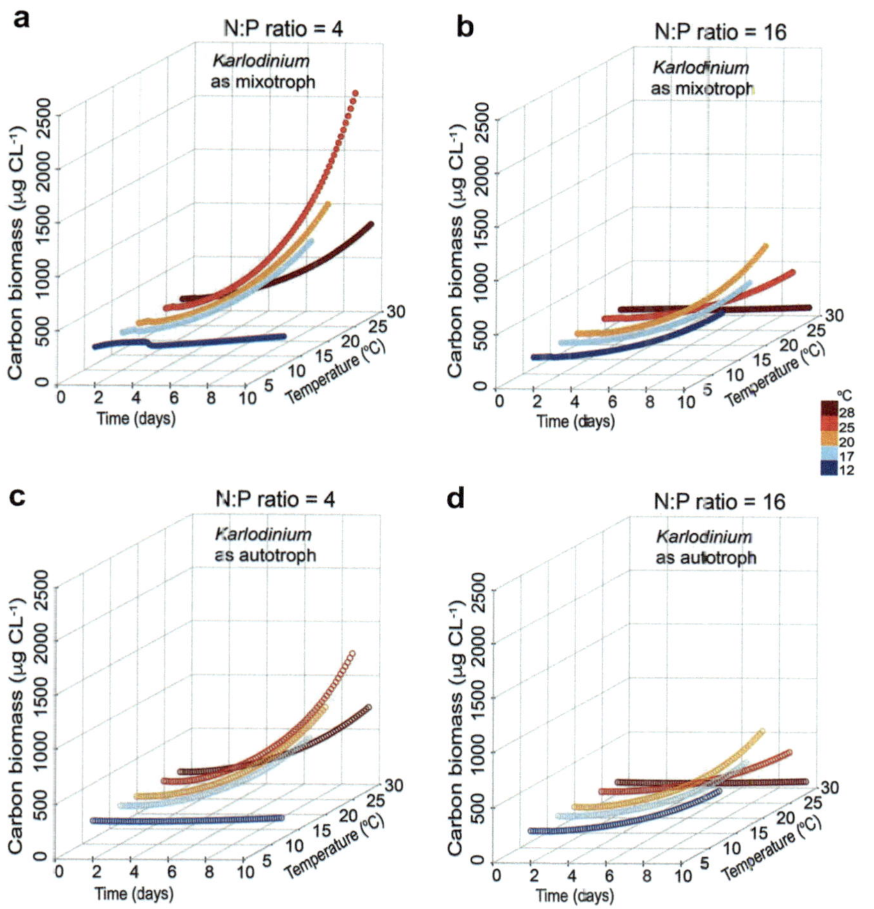

Fig. 23.10 Simulated changes in carbon biomass of *Karlodinium veneficum* in mixotrophic (Panels **a**, **b**) and autotrophic growth (Panels **c**, **d**) for cells simulated to grow under different temperatures and under low N:P conditions and at Redfield ratio N:P conditions. Image reproduced from Lin et al. (2018, Front Mar Sci 5:320. https://doi.org/10.3389/fmars.2018.00320) under a creative commons license

a linear increase in growth rates over the temperatures they tested, their growth study range may not have fully incorporated the full thermal niche. Their study design involved giving the dinoflagellates sufficient food during the course of their growth, and thus changes in cell size reflected temperature effects on both predator and prey uptake.

Warmer waters are thought to also favor smaller-sized cells because they may more efficiently harvest light and nutrients and maintain their position in the euphotic zone (Winder et al. 2009; Finkel et al. 2010; Zohary et al. 2020). Furthermore, if cells are more likely to be maintained higher in the water column, cells that are less susceptible to photoinhibition may be favored (Finkel et al. 2010). Smaller cells are afforded that advantage because they generally have larger cross-sectional area for photochemistry but faster repair of photo-damage (Key et al. 2010). This effect is synergistic with temperature. Cells increase light-harvesting pigment content as downstream photochemistry increases (e.g., with temperature) but this, in turn, increases the package effect (intracellular self-shading effects; Kirk 1994; Beardall et al. 2009). Small cells more easily overcome this effect. The lesser dependence on NO_3^- for cells higher in the water column would also be compounded by the decline in NO_3^- reductase (NR) with warmer temperatures (e.g., Lomas and Glibert 1999, 2000). Morán et al. (2010) illustrated clearly in the eastern and western North Atlantic Ocean that temperature alone explained the majority of the variance in the relationship of small cells to total phytoplankton.

Henson et al. (2021) developed a global model of the expected change in phytoplankton diversity at the end of this century, accounting for high emissions of greenhouse gases and associated changes in temperature and related biogeochemical cycling. Their models captured 2 prokaryote, 2 pico-prokaryote, 5 coccolithophore, 5 diazotroph, 11 diatom, 10 mixotrophic dinoflagellate, and 16 zooplankton types. They reported biomass declines in the tropics and temperate regions (and increases at high latitudes), alongside significant shifts in phytoplankton community composition. They too observed an increasing dominance of smaller phytoplankton cell types but relocation of some communities in terms of size classes and functional types (Fig. 23.11). Mixotrophic dinoflagellates increased in some regions and declined in others. Moreover, community structure became increasingly unstable, with a loss of ecological resilience.

23.6 Beyond Phytoplankton Growth

Natural communities may have weaker growth-temperature relationships than those determined in the laboratory. Natural communities have multiple strategies for adapting to different temperatures. For example, phytoplankton may adapt to low temperatures by modifying internal reserves to compensate for temperature-sensitive reactions (Raven and Geider 1988). Temperature affects more than growth rates. It affects motility, germination, pigment content, enzyme reactions, photosynthesis, and various other processes, changing the ability of cells to thrive in a particular habitat (e.g., Davis et al. 2009; Finkel et al. 2010; Fu et al. 2012; O'Neil et al. 2012; Boyd and Hutchins 2012; Wells et al. 2015; Glibert and Burford 2017; Sommer et al. 2015; Glibert and Burkholder 2018, among others).

Respiration rates generally increase with temperature faster than productivity rates. This implies that with warming, waters should become increasingly heterotrophic—as Eppley predicted. Regaudie-de-Gioux and Duarte (2012) showed that in waters warming more than 21 °C, this is indeed the case. This implies that if waters warm by 4 °C (IPCC 2023), the ratio of productivity to respiration should decline by 25%, shifting the oceans from C sinks to C sources. However, the relationship between the two rates may change with multigenerational adaptation,

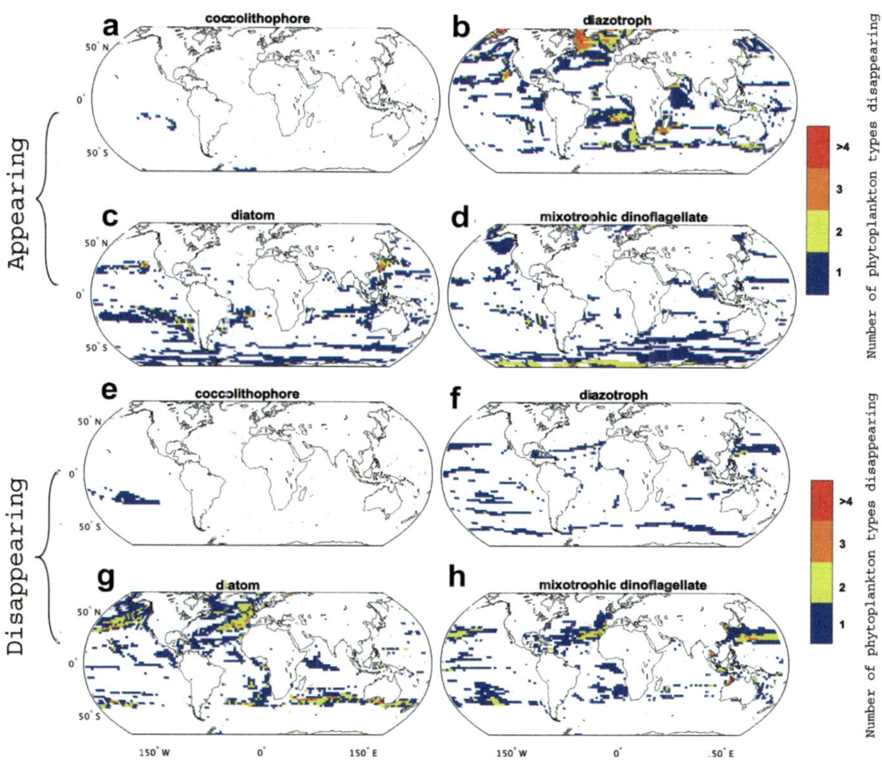

Fig. 23.11 The number of phytoplankton types appearing (Panels **a–d**) and disappearing (Panels **e–h**) between the baseline period (mean of 2005–2024) and end of century period (mean of 2081–2100) in each of six groups: coccolithophores (**a**, **e**), of which there are five types; diazotrophs (**b**, **f**), of which there are five types; diatoms (**c**, **g**), of which there are 11 types; mixotrophic dinoflagellates (**d**, **h**), of which there are 10 types. Prokaryotes and picoeukaryotes (of which there are two types of each) do not show any significant changes. Appearance (disappearance) is defined as a type contributing >0.1% (<0.1%) to total biomass. Image reproduced from Hensen et al. (Nat Comm 12: 5372. https://doi.org/10.1038/s41467-021-25699-w) under a creative commons license

as basal metabolism diminishes after adaptation (Padfield et al. 2016; Barton et al. 2020). Higher temperatures should also result in higher respiratory losses than comparable temperature increases of ingestion, reducing gross growth efficiency (Straile 1997; Li et al. 2011). The consequence of a higher response to temperature in respiration compared to photosynthesis should also result in a reduction in the C:N and C:P ratios of cells with warming.

The question of who will be favored depends not only on the individual species responses to temperature, but also how changes in rates of other members of the community are affected. Both growth and mortality rates are affected by temperature (Calbet and Landry 2004). Stimulation of grazer activity under warming conditions can result in complex interactions, including trophic cascades (Vidussi et al. 2011; Chap. 20). Cyanobacteria are well recognized to increase with warming; they generally have a high temperature optimum. Courboulès et al. (2022) observed, in a mesocosm conducted with Mediterranean coastal water, that warming of 3 °C increased cyanobacteria, but overall there was a reduction in phytoplankton biomass, but higher growth rates of bacteria occurred, as did rates of heterotrophic grazing. Bacterial responses to warming are as complex as those of phytoplankton: temperature does not consistently enhance rates, and in natural communities their rates of growth will also be affected by availability of dissolved organic C (DOC). Viruses may also increase their growth under warmer conditions, leading to a possibility of greater viral infection in a natural community under warming conditions. When temperatures exceed the optimum for phytoplankton, other stresses can set in. This can take the form of high membrane fluidity and enhanced degradation rates of proteins and enzymes, and if these effects become significant, they can both negatively affect primary production (Raven and Geider 1988; Staehr and Sand-Jensen 2006) and positively affect the rate of supply of organic matter for viruses. If viral lysis increases under warming, this could enhance this rate of supply (Danovaro et al. 2011). In all, the web of connected effects is extremely complicated.

Among the stressors associated with warmer waters is a change in density gradients associated with increased stratification (e.g., Dale et al. 2006; O'Neil et al. 2012). Higher temperatures promote increased water-column stability which has the dual effect of increased sinking rates for larger-sized taxa and decreased nutrient flux from deeper waters (e.g., Winder et al. 2009). At higher temperatures, water can become less viscous, favoring cyanobacterial species that can control their vertical position through internal buoyancy regulation (e.g., Walsby 1975; Visser et al. 2016 and references therein) and conversely selecting against diatoms that sink faster (e.g., O'Neil et al. 2012 and references therein). On larger scales, increasing temperatures may alter current patterns and physical water motion sufficient to alter species' ambits, promoting a southward-to-northward geographical shift (in the northern hemisphere) of tropical and subtropical species and a more global distribution of other species and a more circumglobal distribution of other species (e.g., Gilson et al. 2021; Benedetti et al. 2021).

These changes also alter the magnitude of the biological pump and the sequestration of C in the deep sea (Fig. 23.12). The data of Morán et al. (2010) that showed that picoplankton increased as temperatures warm, and the results of Courboulès et al. (2022) showing that bacterial production increased over phytoplankton, suggest a less efficient biological pump in a future, warmer ocean. Boyd (2015) further modeled both individual and cumulative effects of climate on the biological pump and showed that with warming more net primary production goes to the dissolved phase, decreasing export flux, but overall effects are confounded by increased phytoplankton growth under warmer conditions. Collectively these factors result in a substantial decrease in downward particulate organic C flux (POC) from the base of the euphotic zone at both high and low latitudes (Boyd 2015).

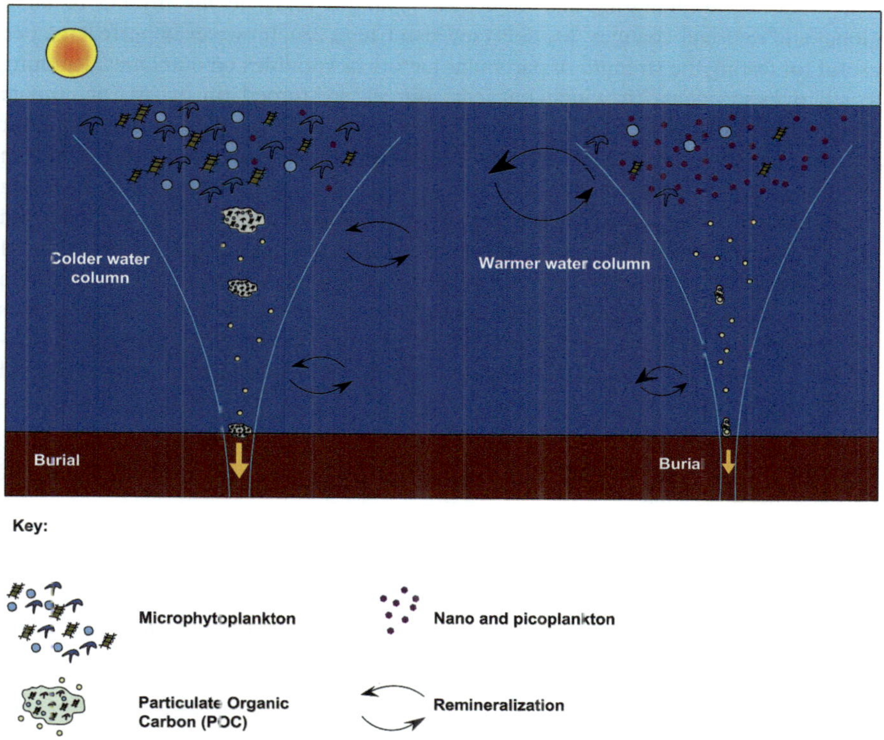

Fig. 23.12 A temperature-dependent biological pump. Conceptual diagram showing the effects of different water temperatures on organic carbon export and remineralization: more carbon is sequestered when temperature is colder compared to when is warmer. Image reproduced from Boscolo-Galazzo et al. (2018, Glob Planetary Change 170: 201–212. https://doi.org/10.1016/j.gloplacha.2018.08.017) under a creative commons license

23.7 Summary

Eppley had the foresight to synthesize temperature-response relationships at a time when responses to temperatures were commonly dismissed as unimportant. Clearly, temperature is important not only for growth rate, but many other metabolic reactions and morphology. With warming, evidence suggests a shift in cell size and a shift in the heterotrophic to autotrophic production proportion. Thermal niche varies with taxa, with strong implications for a future ocean.

Changes in temperatures and associated environmental changes are not occurring regionally along the smooth trajectory of increase indicated by long-term global average temperatures or oceanic pCO_2 changes (e.g., Neuer et al. 2017). The reality is that many climate effects are occurring with substantial gyrations at local scales. Simple, single-factor response relationships do not capture the complexities of how temperatures are changing, and how these changes relate to the myriad of other global and regional changes. Models (see also Chap. 28), however imperfect, can be useful for testing the strength of particular factors or variables on outcomes. Running in silico experiments in which nutrient sources are turned on or off, or climate variables altered, give clear clues as to the importance of such factors both for present and future conditions. Eppley recognized the importance of temperature in phytoplankton growth, and he anticipated the value of modeling and modeling "experiments." The rapidly changing global ocean makes our projections and predictions fraught with complexities and many uncertainties. Interactions of ocean chemistry, metabolic responses, to responses by different taxa, along with other climatic-related changes, means that our uncertainties in long-term responses are high. These complexities shine a bright light on the need for increasing understanding of the physiology and ecology of the interactions of the environmental factors that are changing with climate.

References

Anderson SI, Barton AD, Clayton S, Dutkiewicz S, Rynearson TA (2021) Marine phytoplankton functional types exhibit diverse responses to thermal change. Nat Commun 12:6413. https://doi.org/10.1038/s41467-021-26651-8

Arrhenius S (1889) Uber die reaktiongeschwindigkeit bei der inversion von rohrzucker durch säuren. Z für Phys Chem 4:226–248

Barton S, Jenkins J, Buckling A, Schaum CE, Smirnoff N, Raven JA, Yvon-Durocher G (2020) Evolutionary temperature compensation of carbon fixation in marine phytoplankton. Ecol Lett 23:722–733. https://doi.org/10.1111/ele.13469

Beardall J, Allen D, Bragg J, Finkel ZV, Quigg A, Rees TAV, Richardson A, Raven JA (2009) Allometry and stoichiometry of unicellular, colonial and multicellular phytoplankton. New Phytol 181:295–309. https://doi.org/10.1111/j.1469-8137.2008.02660

Benedetti F, Vogt M, Elizondo UH, Righetti D, Zimmermann NE, Gruber N (2021) Major restructuring of marine plankton assemblages under global warming. Nat Commun 12:5226. https://doi.org/10.1038/s41467-021-25385-x

Bissinger JE, Montagnes DJS, Sharples J, Atkinson D (2008) Predicting marine phytoplankton maximum growth rates from temperature: improving on the Eppley curve using quantile regression. Limnol Oceanogr 53:487–493

Boscolo-Galazzo F, Crichton KA, Barker S, Pearson PN (2018) Temperature dependencies of metabolic rates in the upper ocean: a positive feedback to global climate change? Glob Planetary Change 170:201–212. https://doi.org/10.1016/j.gloplacha.2018.08.017

Boyd PW (2015) Toward quantifying the response of the oceans' biological pump to climate change. Front Mar Sci 2:77. https://doi.org/10.3399/fmars.2015.00077

Boyd PW, Hutchins DA (2012) Understanding the responses of ocean biota to a complex matrix of cumulative anthropogenic change. Mar Ecol Prog Ser 470:125–135

Brown JH, Gillooly JF, Allen AP, Savage VM, West GB (2004) Towards a metabolic theory of ecology. Ecology 85:1771–1789

Brush MJ, Brawley JW, Nixon SW, Kremer JN (2002) Modeling phytoplankton production: problems with the Eppley curve and an empirical alternative. Mar Ecol Prog Ser 238:31–45

Bunt S, Lee CC (1970) Seasonal primary production in Antarctic Sea ice at McMurdo Sound in 1967. J Mar Res 28:304–320

Calbet A, Landry MR (2004) Phytoplankton growth, microzooplankton grazing, and carbon cycling in marine systems. Limnol Oceanogr 49:51–57. https://doi.org/10.4319/lo.2004.49.1.0051

Castenholz RW (1964) The effect of daylength and light intensity on the growth of littoral marine diatoms in culture. Physiol Plant 17:951–963

Castenholz RW (1969) Thermophilic blue-green algae and the thermal environment. Bact Rev 33:476–504

Chen B (2015) Patterns of thermal limits of phytoplankton. J Plankton Res 37:285–292. https://doi.org/10.1093/plankt/fbv009

Courboulès J, Mostajir B, Trombetta T, Mas S, Vidussi F (2022) Warming disadvantages phytoplankton and benefits bacteria during a spring bloom in the Mediterranean Thau Lagoon. Front Mar Sci 9. https://doi.org/10.3389/fmar.2022.878938

Dale B, Edwards M, Reid PC (2006) Climate change and harmful algal blooms. In: Granéli E, Turner JE (eds) Ecology of harmful algae. Springer, New York, NY, pp 367–378

Danovaro R, Corinaldesi C, Dell'Anno A, Fuhrman JA, Middelburg JJ, Noble RT, Suttle CA (2011) Marine viruses and global climate change. FEMS Microbiol Rev 35:993–1034. https://doi.org/10.1111/j.1574-6976.2010.00258.x

Davis TL, Berry DL, Boyer GL, Gobler CJ (2009) The effects of temperature and nutrients on the growth and dynamics of toxic and non-toxic strains of *Microcystis* during cyanobacteria blooms. Harmful Algae 8:715–725

Edwards KF, Thomas MK, Klausmeier CA, Litchman E (2016) Phytoplankton growth and the interaction of light and temperature: a synthesis at the species and community level. Limnol Oceanogr 61:1232–1244. https://doi.org/10.1002/lno.10282

Eppley RW (1963) Evaluation of certain marine algal flagellates for mass culture. Tech Doc Rep No. SAM- TDR-6:3–91. USAF School of Aerospace Medicine. Brooks Air Force Base, Tex

Eppley RW (1972) Temperature and phytoplankton growth in the sea. Fish Bull 70:1063–1085

Eppley RW, Sloan PR (1966) Growth rates of marine phytoplankton: correlation with light absorption by cell chlorophyll *a*. Physiol Plant 19:47–59

Fawley MW (1984) Effects of light intensity and temperature interactions on growth characteristics of *Phaeodactylum tricornutum* (Bacillariophyceae). J Phycol 20:67–72

Finkel ZV, Beardall J, Flynn KJ, Quigg A, Rees TAV, Raven JA (2010) Phytoplankton in a changing world: cell size and elemental stoichiometry. J Plankton Res 32:119–137

Foy RH (1983) Interaction of temperature and light on the growth rates of two planktonic *Oscillatoria* species under a short photoperiod regime. British Phycol J 18:267–273

Franzè G, Menden-Deuer S (2020) Common temperature-growth dependency and acclimation response in three herbivorous protists. Mar Ecol Prog Ser 634:1–13. https://doi.org/10.3354/meps13200

Fu FX, Tatters AO, Hutchins DA (2012) Global change and the future of harmful algal blooms in the ocean. Mar Ecol Prog Ser 470:207–233

Gilson AR, Smale DA, O'Connor N (2021) Ocean warming and species range shifts affect rates of ecosystem functioning by altering consumer–resource interactions. Ecology 102:e03341. https://doi.org/10.1002/ecy.3341

Glibert PM, Burford MA (2017) Globally changing nutrient loads and harmful algal blooms: recent advances, new paradigms, and continuing challenges. Oceanography 30(1):58–69. https://doi.org/10.5670/oceanog.2017.110

Glibert PM, Burkholder JM (2018) Causes of harmful algal blooms. In: Shumway S, Burkholder JM, Morton SL (eds) Harmful algal blooms: a compendium desk reference. Wiley Blackwell, Singapore, pp 1–38

Guillard RRL, Myklestad S (1970) Osmotic and ionic requirements of the marine centric diatom Cyclotella nana. Helgoländer wiss Meeresunters 20:104–110

Guillard RRL, Ryther JH (1962) Studies of marine planktonic diatoms. I. *Cyclotella nana* Hustedt and *Detonula confervacea* (Cleve) Gran. Can J Microbiol 8:229–239

Henson SA, Cael BB, Allen SR, Dutkiewicz S (2021) Future phytoplankton diversity in a changing climate. Nat Commun 12:5372. https://doi.org/10.1038/s41467-021-25699-w

Hoogenhout H, Amesz J (1965) Growth rates of photosynthetic microorganisms in laboratory cultures. Arch Mikrobiol 50:10–24

Hulbert EM, Guillard RRL (1968) The relationship of the distribution of the diatom *Skeletonema tropicum* to temperature. Ecology 49:337–339

Ignatiades L, Smayda TJ (1970) Autecological studies on the marine diatom *Rhizosolenia fragilissima* Bergon. I. The influence of light, temperature, and salinity. J Phycol 6:332–339

Intergovernmental Panel on Climate Change (IPCC) Core Writing Team (2023) Summary for policymakers. In: Lee H, Romero J (eds) Climate change 2023: synthesis report. A Report of the Intergovernmental Panel on Climate Change. Contribution of Working Groups I, II and III to the Sixth Assessment Report of the Intergovernmental Panel on Climate Change. IPCC, Geneva, Switzerland. 36 p

Jørgensen EG (1968) The adaptation of plankton algae. II. Aspects of the temperature adaptation of *Skeletonema costatum*. Physiol Plant 21:423–427

Key T, McCarthy A, Campbell DA, Six C, Roy S, Finkel ZV (2010) Cell size trade-offs govern light exploitation strategies in marine phytoplankton. Envir Microbiol 12:95–104

Kirk JTO (1994) Light and photosynthesis in aquatic ecosystems. Cambridge Univ Press, Cambridge, UK, p 401

Kruger GHJ, Eloff JN (1978) The effect of temperature on specific growth rate and activation energy of *Microcystis* and *Synechococcus* isolates relevant to the onset of natural blooms. J Limnol Soc South Afr 4:9–20

Lanskaya LA (1961) Fission rate of plankton algae of the Black Sea in cultures. In: Oppenheimer CH (ed) Marine microbiology. C. C. Thomas, Springfield, IL, pp 127–132

Li CC, Xu KD, Lei YL (2011) Growth and grazing responses to temperature and prey concentration of *Condylostoma spatiosum*, a large benthic ciliate, feeding on *Oxyrrhis marina*. Aquat Microb Ecol 64:97–104. https://doi.org/10.3354/ame01521

Lin C-H, Flynn KJ, Mitra A, Glibert PM (2018) Simulating effects of variable stoichiometry and temperature on Mixotrophy in the harmful dinoflagellate *Karlodinium veneficum*. Front Mar Sci 5:320. https://doi.org/10.3389/fmars.2018.00320

Lomas MW, Glibert PM (1999) Temperature regulation of nitrate uptake: a novel hypothesis about nitrate uptake and reduction in cool-water diatoms. Limnol Oceanogr 44:556–572

Lomas MW, Glibert PM (2000) Comparisons of nitrate uptake, storage and reduction in marine diatoms and flagellates. J Phycol 36:903–913

McLachlan J (1960) The culture of *Dunaliella tertiolecta* Butcher–a euryhaline organism. Can J Microbiol 6:367–379

Meeson BW, Sweeney BM (1982) Adaptation of *Ceratium furca* and *Gonyaulax polyedra* (Dinophyceae) to different temperatures and irradiances: growth rates and cell volumes. J Phycol 18:241–245

Morán XAG, Moran A, López-Urrutia A, Calvo-Díaz A, Li WKW (2010) Increasing importance of small phytoplankton in a warmer ocean. Glob Chang Biol 16:1137–1144. https://doi.org/10.1111/j.1365-2486.2009.01960.x

Neuer S, Benway HM, Bates N, Carlson CA, Church M, DeGrandpre M, Dunne J et al (2017) Monitoring ocean change in the 21st century. EOS 98. https://doi.org/10.1029/2017EO080045

Noll P, Lilge L, Hausmann R, Henkel M (2020) Modeling and exploiting microbial temperature response. Processes 8:121. https://doi.org/10.3390/pr8010121

O'Neil JM, Davis TW, Burford MA, Gobler CJ (2012) The rise of harmful cyanobacteria blooms: the potential roles of eutrophication and climate change. Harmful Algae 14:313–334

Paasche E (1967) Marine plankton algae grown with light- dark cycles. I. *Coccolith*us *huxleyi*. Physiol Plant 20:946–956

Paasche E (1968) Marine plankton algae grown with light- dark cycles. II. *Ditylum brightwellii* and *Nitzschia turgidula*. Physiol Plant 21:66–77. https://doi.org/10.1111/j.1399-3054.1968.tb07231.x

Padfield D, Yvon-Durocher G, Buckling A, Jennings S, Yvon-Durocher G, Hillebrand H (2016) Rapid evolution of metabolic traits explains thermal adaptation in phytoplankton. Ecol Lett 19:133–142. https://doi.org/10.1111/ele.12545

Polikarpov GG, Tokareva AV (1970) The cell cycle of the dinoflagellates *Peridinium trochoideum* (Stein) and *Gonyaulax polyedra* micro autoradiographic study. Hydrobiol J 6(5):54–55
Raven J, Geider R (1988) Temperature and algal growth. New Phytol 110:441–461. https://doi.org/10.1111/j.1469-8137.1988.tb00282.x
Regaudie-de-Gioux A, Duarte CM (2012) Temperature dependence of planktonic metabolism in the ocean. Glob Biogeochem Cycles 26:GB1015. https://doi.org/10.1029/2010GB003907
Rose JM, Caron DA (2007) Does low temperature constrain the growth rates of heterotrophic protists? Evidence and implications for algal blooms in cold waters. Limnol Oceanogr 52:886–895. https://doi.org/10.4319/lo.2007.52.2.0886
Siegel P, Baker KG, Low-Décarie E, Geider RJ (2023) Phytoplankton competition and resilience under fluctuating temperature. Ecol Evol 13:e9851. https://doi.org/10.1002/ece3.9851
Smayda TJ (1969) Experimental observations on the influence of temperature, light, and salinity on cell division of the marine diatom, *Detonula confervacea* (Cleve) Gran. J Phycol 5:150–157
Sommer U, Paul C, Moustaka-Gouni M (2015) Warming and ocean acidification effects on phytoplankton –from species shifts to size shifts within species in a mesocosm experiment. PLoS One 10:e0125239. https://doi.org/10.1371/journal.pone.0125239
Sorokin C, Krauss RW (1958) The effects of light intensity on the growth rates of green algae. Plant Physiol 33:109–113
Sorokin C, Krauss RW (1962) Effects of temperature and illuminance on *Chlorella* growth uncoupled from cell division. Plant Physiol 37:37–42
Staehr PA, Sand-Jensen K (2006) Seasonal changes in temperature and nutrient control of photosynthesis, respiration and growth of natural phytoplankton communities. Freshw Biol 51:249–262. https://doi.org/10.1111/j.1365-2427.2005.01490.x
Steeman Nielsen E (1952) The use of radioactive carbon (^{14}C) for measuring organic production in the sea. J Cons Int Explor Mer 18:117–140. https://doi.org/10.1093/icesjms/18.2.117
Steeman Nielsen E (1960) Productivity of the oceans. Ann Rev Plant Physiol 11:341–362
Straile D (1997) Gross growth efficiencies of protozoan and metazoan zooplankton and their dependence on food concentration, predator-prey weight ratio, and taxonomic group. Limnol Oceanogr 42:1375–1385. https://doi.org/10.4319/lo.1997.42.6.1375
Swift E 5th, Taylor WR (1966) The effect of pH on the division rate of the coccolithophorid *Cricosphaera elongata*. J Phycol 2:121–125
Thomas MK, Kremer CT, Klausmeier CA, Litchman E (2012) A global pattern of thermal adaptation in marine phytoplankton. Science 338:1085–1088. https://doi.org/10.1126/science.1224836
Thomas MK, Aranguren-Gassis M, Kremer CT, Gould MR, Anderson K, Klausmeier CA, Litchman E (2017) Temperature-nutrient interactions exacerbate sensitivity to warming in phytoplankton. Glob Change Biol 23:3269–3280. https://doi.org/10.1111/gcb.13641
Thomas WH (1966) Effects of temperature and illuminance on cell division rates of three species of tropical oceanic phytoplankton. J Phycol 2:17–22
Ukeles R (1961) The effect of temperature on the growth and survival of several marine algal species. Biol Bull (Woods Hole) 120:255–264
Verity PG (1982) Effects of temperature, irradiance and daylength on the marine diatom *Leptocylindrus danicus* Cleve. IV. Growth. J Exp Mar Biol Ecol 60:209–222
Vidussi F, Mostajir B, Fouilland E, Le Floc'H E, Nouguier J, Roques C, Got P et al (2011) Effects of experimental warming and increased ultraviolet B radiation on the Mediterranean plankton food web. Limnol Oceanogr 56:206–218. https://doi.org/10.4319/lo.2011.56.1.0206
Vidyarathna N, Ahn SH, Glibert PM (2023) Thermal niche of the dinoflagellate *Karlodinium veneficum* across different salinity and light levels. J Plankton Res. https://doi.org/10.1093/plankt/fbad019
Visser PM, Verspagen JMH, Sandrini G, Stal LJ, Mattijs HCP, Davis TW, Paerl HW, Huisman J (2016) How rising CO_2 and global warming may stimulate harmful cyanobacterial blooms. Harmful Algae 54:145–159
Walsby T (1975) Gas vesicles. Ann Rev Plant Phys 26:427–439

Wells ML, Trainer VL, Smayda TJ, Karlson BS, Trick CG, Kudela RM, Ishikawa A et al (2015) Harmful algal blooms and climate change: learning from the past and present to forecast the future. Harmful Algae 49:68–93

Wilken S, Huisman J, Naus-Wiezer S, Van Donk E (2013) Mixotrophic organisms become more heterotrophic with rising temperature. Ecol Lett 16:225–233. https://doi.org/10.1111/ele.12033

Winder M, Reuter JE, Schladow SG (2009) Lake warming favours small-sized plankton diatom species. Proc Roy Soc B 276:427–435

Woods HA, Makino W, Cotner JB, Hobbie SE, Harrison JF, Acharya K et al (2003) Temperature and the chemical composition of poikilothermic organisms. Funct Ecol 17:237–245

Yvon-Durocher G, Allen AP, Cellamare M, Dossena M, Gaston KJ, Leitao M, Montoya JM et al (2015) Five years of experimental warming increases the biodiversity and productivity of phytoplankton. PLoS Biol 13(12):e1002324

Zohary T, Flaim G, Sommer U (2020) Temperature and the size of freshwater phytoplankton. Hydrobiologia 848:143–155

Additional Websites Accessed

https://climatereanalyzer.org/clim/sst_daily/. Accessed March 17, 2024

Hypoxia and the Wheezing of Our Waters 24

24.1 Introduction and a Bit of History

Anaerobic for its first 2 billion years or so, the Earth was permanently disrupted when oxygen (O_2) evolved. Every living organism faced a new world; the rules of life changed, and organisms were challenged to adapt and evolve (Green 2007; Chap. 10). Now, the presence of O_2 is critical to the health of virtually all organisms in the world's lakes and oceans. And, we are losing it in aquatic systems, in large part by our own doing.

Before about 1870, most researchers thought that the deep oceans had no O_2—and no life. The *Challenger* expedition put an end to that thought, as the major oceans were shown to have O_2 even at considerable depths (Pilson 2013). Hypoxic and anoxic waters have existed throughout geological time, and hypoxic waters have naturally occurred in deep basins and O_2 minimum zones of the western boundary upwelling systems. However, increases in hypoxia in coastal and estuarine waters are comparatively recent. In freshwater lakes, hypoxia appears to have begun spreading before 1900, decades before its appearance led to alarm in coastal waters (Jenny et al. 2015). Local waters have long been the recipient of human wastes, fueling hypoxia.

In the 1800s, the River Thames in London was famous (infamous) for being a rotting mess, the repository of all things waste, from human sewage to industrial pollution. At that time, sewage was in the form of cesspits, which were cleaned from time to time by "nightmen" who would take the waste out of site (and into the river; Pomeroy 2021). The mess was bad, made worse by the heat wave of 1857, leading to a new moniker for the river condition, "The Great Stink." The river was hypoxic. Charles Dickens (1857) wrote in his novel *Little Dorrit*:

> Through the heart of the town a deadly sewer ebbed and flowed, in the place of a fine fresh river.

P. M. Glibert, *Phytoplankton Whispering: An Introduction to the Physiology and Ecology of Microalgae*, https://doi.org/10.1007/978-3-031-53897-1_24

And, to a friend he wrote (as reported by Flanders 2012):

> I can certify that the offensive smells, even in that short whiff, have been of a most head-and-stomach-distending nature.

The scientist Michael Faraday—of electrochemistry fame—described the condition in a letter to the newspaper in 1855, *The Times*:

> Near the bridges the feculence rolled up in clouds so dense that they were visible at the surface, even in water of this kind…The smell was very bad, and common to the whole of the water; it was the same as that which now comes up from the gully-holes in the streets; the whole river was for the time a real sewer.

A plan was devised, and sewage was diverted upstream. The biggest civil engineering project of that century was basically a series of pumping stations that moved the waste to less populated areas (Burns 2018). No more stink, and the river began to recover. But, within a hundred years, the river became dead, again, partly due to bombing of the pumping infrastructure in World War II, but also due to the rapidly expanding London population (who were responsible for more waste). By 1957, the river was declared "biologically dead" by the Natural History Museum of London. This prompted more action and sewage upgrades. It also led to the installation of river bubblers to increase the dissolved O_2 in order to maintain levels necessary to support fish populations (Edmonds-Brown 2022). Oxygen saturation recovered. A further improved sewer system is now being developed, expected to come on line in 2025. Such an improvement is sorely needed, as even in 2024, high levels of *E. coli* from sewage continue to pollute the river, even impacting competitive rowers on the river (Ott 2024). The River Thames is just one example of a river that has suffered due to sewage pollution. To this day, more than a quarter of the world's population still lacks adequate basic sanitation—and the waste has to go somewhere, typically in nearly waterways (World Bank 2021).

This chapter covers the processes that contribute to hypoxia, with a focus on eutrophication and takes a look at a few of the best studied hypoxic zones, and then examines in more detail the impacts of hypoxia on nutrient cycling and plankton. While the focus has often been on what organisms die as a result of hypoxia, the complexity of hypoxia on nutrient cycles and planktonic organisms is far more complicated.

24.2 Definitions of Hypoxia

Hypoxia is typically described as waters having <2 mg L^{-1} O_2 (equivalent to 1.4 ml L^{-1}). Some investigators use <3 mg L^{-1} as a criterion (Rabalais et al. 2010). Some use a biological standard, rather than a chemical concentration, a value that may vary by species (e.g., Slater et al. 2020). This value is the environmental oxygen (O_2) level below which an organism can no longer support a normal respiration rate and is often termed an organism's critical oxygen partial pressure (P_{crit}; Elliott et al. 2013; Roman et al. 2019 and references therein; Fig. 24.1).

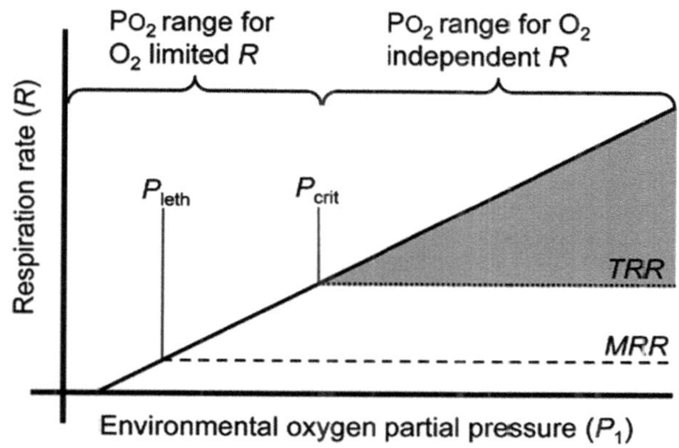

Fig. 24.1 Relationships between an organism's respiration rate and the environmental oxygen partial pressure. In this conceptual diagram, the solid line represents the linear relationship between environmental oxygen partial pressure and an organism's theoretical maximum respiration rate. Critical oxygen partial pressure (P_{crit}) is the environmental oxygen level below which an organism's theoretical maximum respiration rate is lower than its normal or target respiration rate (TRR, dotted line). Respiration rate is oxygen limited when environmental oxygen partial pressure is below this critical partial pressure. Respiration rate above the critical partial pressure is generally independent of environmental oxygen (indicated by gray shading). Lethal oxygen partial pressure (P_{leth}) is the environmental oxygen level below which an organism's theoretical maximum respiration rate is lower than its minimal survival (non-lethal) respiration rate (MRR, dashed line). Hypoxia-induced mortality will increase when environmental oxygen partial pressure is below this lethal oxygen partial pressure. Image reproduced from Elliott et al. (2013, PLoS ONE 85(5): e63987, https://doi.org/10.1371/journal.pone.0063987) under a creative commons license

Concentrations of O_2 below P_{crit} at which a minimum survivable respiration rate cannot be maintained are lethal and this value is termed P_{leth} (Fig. 24.1).

The definitions of hypoxia described above do not take into account the amount of time a system may be hypoxic. In some systems, waters may be hypoxic for extended periods—months or longer—while in other systems, such as shallow coastal lagoons, hypoxia may exist for only hours, as for example, when hypoxia develops on a day-night basis. Both the duration of hypoxia and associated ecosystem effects are related to system size (Fig. 24.2). Even if transient, the stress of not being able to breathe is experienced by many types of organisms when waters become hypoxic. Some fish are stressed by hypoxic conditions relatively quickly; other organisms, such as some invertebrate species, can survive severe hypoxia for days to weeks (Vaquer-Sunyer and Duarte 2008). Thus, the concept of hypoxic volume days (the cumulative period of time that a system is hypoxic seasonally or annually) is a useful metric for comparing systems or effects (e.g., Bever et al. 2013; Ni and Li 2023; Glibert and Li 2024).

When dissolved oxygen (DO) is low (hypoxic) or absent (anoxic), systems are often given the imperfect name, "dead zones." The term is a poor one because not all organisms die under such conditions. Organisms that can flee or swim away, often do so. Organisms that do not have the ability to escape may be stressed, with

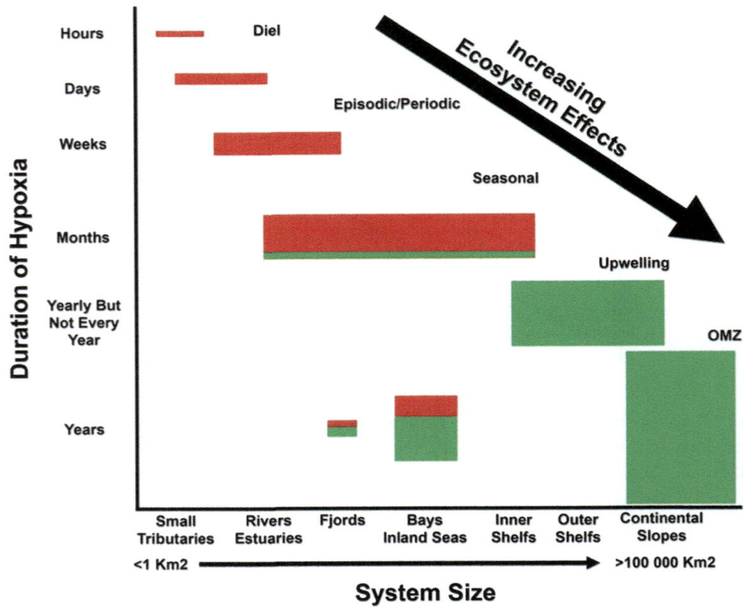

Fig. 24.2 Conceptual relationship between duration of hypoxia, system size, and ecosystem effects. The red and green colors represent the relative ratio of human versus natural influences, respectively. As human influences increase in shallower waters to the coastal ocean, there are more ecosystem effects. Upwelling areas and oxygen minimum zones (OMZs) are primarily natural, but human activities that indirectly affect global climate change are also affecting these areas. Image reproduced from Rabalais et al. (2010 Biogeosci 7: 585–619, https://www.biogeosciences.net/7/585/2010/) under a creative commons license

multiple metabolic impairments. On the other hand, some organisms—bacteria and some types of plankton—not only survive but can thrive.

24.3 Causes and Processes Contributing to Hypoxia

Sanitation—or lack thereof—only tells part of the story of the increasing deoxygenated waters of the world. Another part of the story is the timing of the invention for the industrial synthesis of nitrogen (N), the Haber-Bosch process (Chap. 22). Both advanced sanitation and the synthesis of ammonia are probably the two most important inventions that are responsible for the expansion of the human population. Sanitation is responsible for keeping water-borne illnesses from shortening lives, and the N fertilizer is responsible for expanding the food supply also lengthening lives (Chap. 23).

It is of no surprise that as production of ammonia fertilizer use increased, so did its runoff. No one noticed, initially. An interesting study in 1957 (Cook and Hulbert 1957) reported on N fertilizer application to sugar beets. After several manipulations,

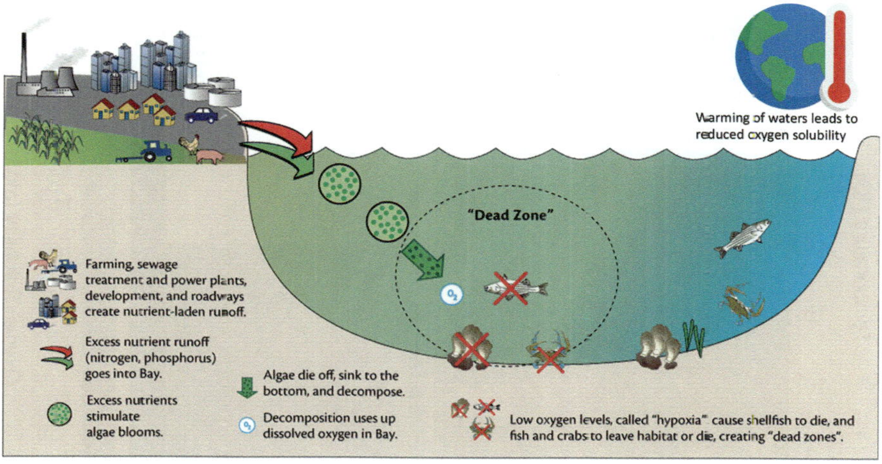

Fig. 24.3 Conceptual diagram illustrating how eutrophication contributes to hypoxia. Image modified from Boesch (2008, https://ian.umces.edu/publications/global-warming-and-the-free-state-comprehensive-assessment-of-climate-change-impacts-in-maryland/) under a creative commons license

they found that the amount of N applied that could be accounted for in the plant tissue ranged from 50 to 70%, but "the researchers did not discuss, for either case, where the rest of the fertilizer ended up" (Gorman 2013, p. 101). Moreover, compared to all the agrochemicals being applied (insecticides, pesticides), the lost fertilizer seemed harmless (Gorman 2013).

So, N fertilizer use increased, runoff increased, and so too did the hypoxic zones of the world's waters. Nutrients—whether from sewage inputs or runoff of fertilizers and other land-based nutrients—fuel primary production that, in turn, may be grazed or die when nutrient or other limitation sets in and which settles in the form of senescent cells, fecal pellets, or aggregates. Aerobic bacteria utilize this rich source of carbon (C), consuming O_2 (Fig. 24.3).

However, laying all the blame for deoxygenation on inadequate sanitation or the glut of anthropogenic nutrient pollution is not entirely correct. Organic matter loading unrelated to human pollution can also cause hypoxia. In East Africa, for example, organic matter loading from high populations of hippopotami has caused repeated occurrence of hypoxia in tropical rivers during flood inundation (Dutton et al. 2018). Dissolved organic matter (DOM) loading is also increasing with changes in precipitation, riverine discharge, sea level rise, altered land use, and the associated loss of many wetlands (see Chap. 22).

Besides nutrients that are fueling eutrophication, climate warming is another major contributor to the development of hypoxia (Glibert and Li 2024 and references therein). The solubility of O_2 decreases as waters warm. It decreases about 5% for every 2 °C of warming (Fig. 24.4). The solubility of O_2 also decreases with salinity. Thus O_2 solubility is higher in freshwater, compared with seawaters, and due to these factors alone, dissolved O_2 saturation decreases with as freshwaters discharge into

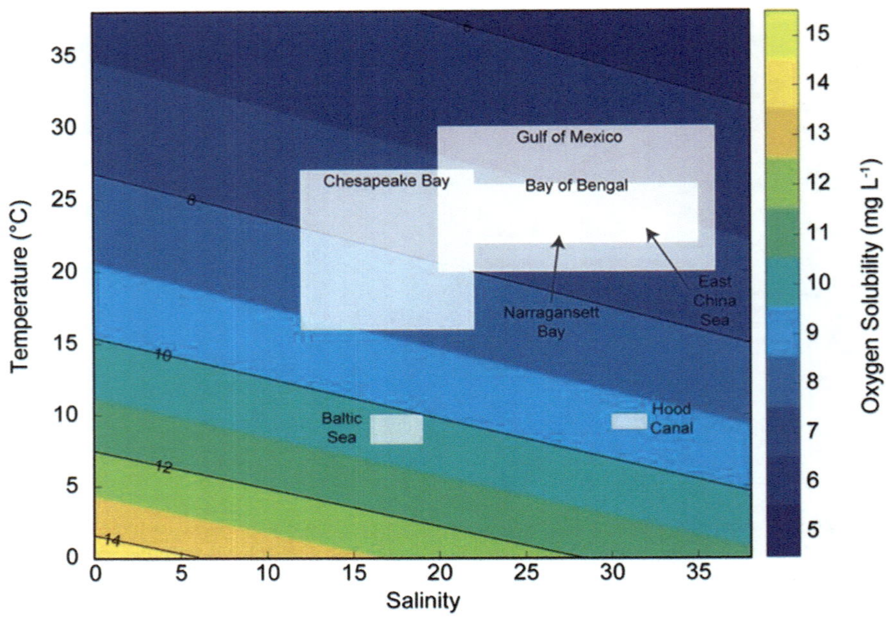

Fig. 24.4 Oxygen solubility (mg L^{-1}) is shown by the color scale at different salinity and temperature (°C) conditions, with boxes indicating the range of salinity and temperature during periods of seasonal deoxygenation for seven coastal and estuarine bodies of water around the world. Oxygen solubility calculated using the equations of Benson and Krause (1980, 1984), assuming surface water pressure. Image reproduced from Roman et al. (2019, Front Mar Sci 6: 139. https://doi.org/10.3389/fmars.2019.00139) under a creative commons license

coastal regions. Schmidtko et al. (2017) estimated that about 15% of the O_2 loss over the past decades is directly attributed to temperature-driven changes in solubility.

As waters warm, many organisms increase their metabolic rate, leading to a greater consumption of O_2. Furthermore, higher temperatures promote increased water column stability and a shallowing of the mixed layer and thus increased stratification. Long residence times and a high degree of stratification tend to isolate parcels of water without renewed O_2 supply but with abundance of organic matter for bacteria to decompose. With lower-salinity waters near surface and saltier waters at depth, lower O_2 waters remain at depth.

Open-coast upwelling systems are also subject to hypoxia and HAB events unrelated to eutrophication. Hypoxia is well documented to be associated with a number of eastern boundary currents (e.g., Diaz and Rosenberg 1995; Morales et al. 1999; Pitcher et al. 2021). In the open ocean, O_2 minimum zones (OMZs) are broadly distributed, and O_2 levels can reach zero, but the influence of these features affecting coastal shelf hypoxia is relatively recent. Wind-driven upwelling brings this low O_2, but nutrient-laden waters onto continental shelves, where these waters

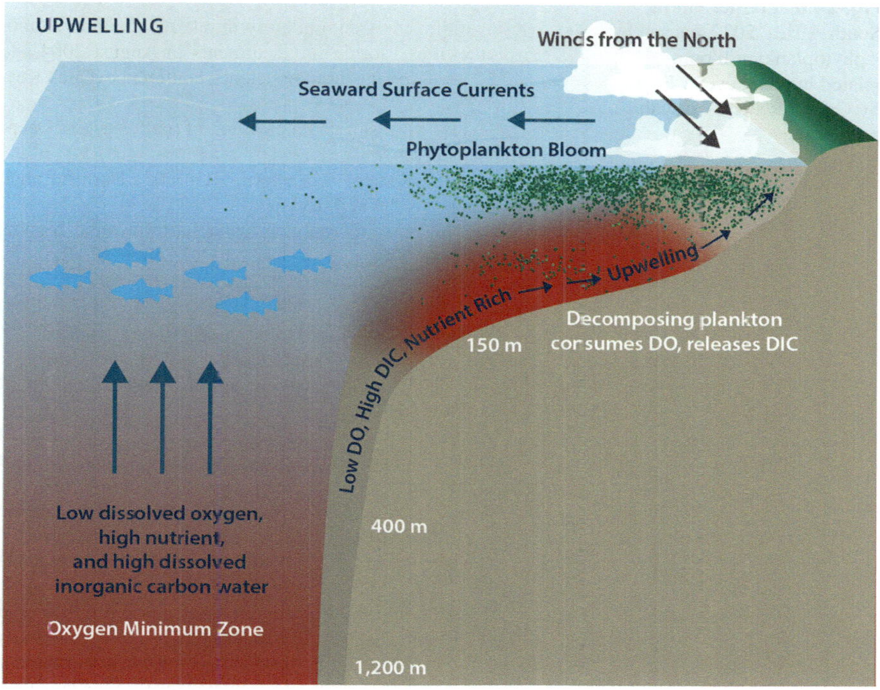

Fig. 24.5 Drivers of hypoxia and ocean acidification intensification in upwelling shelf systems. Equatorward winds drive the upwelling of low dissolved oxygen (DO), high nutrient, and high dissolved inorganic carbon (DIC) water from above the oxygen minimum zone. Cross-shelf gradients in productivity and bottom water residence times drive the strength of DO (DIC) decrease (increase) as water transits across a productive continental shelf. Variations in source water DO, DIC, and nutrient content, upwelling favorable winds, and shelf processes can all accentuate or dampen the severity of hypoxia and ocean acidifcation exposure. Image modified from Gewin (2010) by Moni Kovacs and reproduced from Chan et al. (2019 Oceanography 32(3):62–71, https://doi.org/10.5670/oceanog.2019.312) under a creative commons license

further warm, potentially stratify, and where local respiration losses of oxygen are enhanced (Chan et al. 2019; Glibert and Li 2024; Fig. 24.5).

24.4 Hypoxia and the Current "Dead Zone" Status

Mortality events often occur where water is poorly mixed or where habitat is restricted. Fish kills are commonly observed during hypoxic events. In 2022, a massive lobster kill was observed on near Cape Town, South Africa, following a phytoplankton-induced hypoxia event (Fig. 24.6). Longer-term impacts include loss of biodiversity and shifts in abundance and composition of benthic communities.

Fig. 24.6 Lobster kill in South Africa, 2022, caused by a phytoplankton bloom-related hypoxia event. Photo by Grant Pitcher used with permission

By 2008, there were about 400 dead zones in coastal oceans due to eutrophication (Diaz and Rosenberg 2008), a number that has since grown to over 700 (Fig. 24.7; Grégoire et al. 2021). The number of such dead zones in the ocean has doubled every decade from the 1960s. During the past 50 years, it is estimated that regions of low O_2 in the oceans have increased by 4.5 million km^2, and at a rate of loss of 1 gigaton of O_2 each year (Keeling and Garcia 2002; Breitburg et al. 2018). Thus, the size of oceanic O_2 minimum zones over the past five decades has increased by an area about the size of the European Union (Stramma et al. 2010). Furthermore, the global ocean data show an average decline in oceanic O_2 content of more than 2% since 1960, with a continued decline between 1 and 7% predicted by the year 2100 (Pitcher et al. 2021 and references therein). Thus, large regions of the oceans are deoxygenating even in the absence of large blooms (Fig. 24.8).

Hypoxia in freshwaters is also extensive (Blaszczak et al. 2022). An extensive global database of the hypoxia in rivers (including over 125,000 locations) recently showed that 12.6% of all locations exhibited at least one hypoxic measurement, and 35% of all sites had at least one measurement of O_2 below 5 mg L^{-1}. Within the USA, the proportion of sites with hypoxia was about the same as the global average, 12.6%, but Canada had 24% and Brazil 35% of sites exhibiting at least one hypoxic

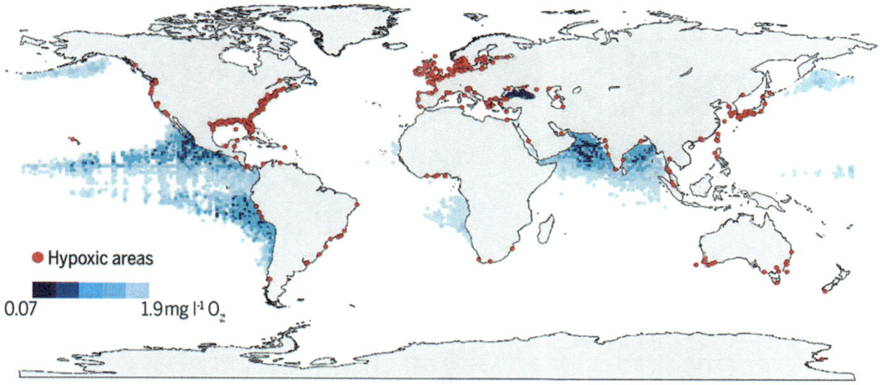

Fig. 24.7 Global distribution of low O_2 areas (<2 mg L^{-1}) in the coastal and global ocean. In the coastal areas, more than 500 sites have been inventories with low O_2 conditions in the past half century (red dots), while in the open ocean, the extent of low O_2 waters amounts to several million km^3 (blue dots refer to conditions at 300 m). Image reproduced from Breitburg et al. (2018 Science 59: 6371) with permission of Science

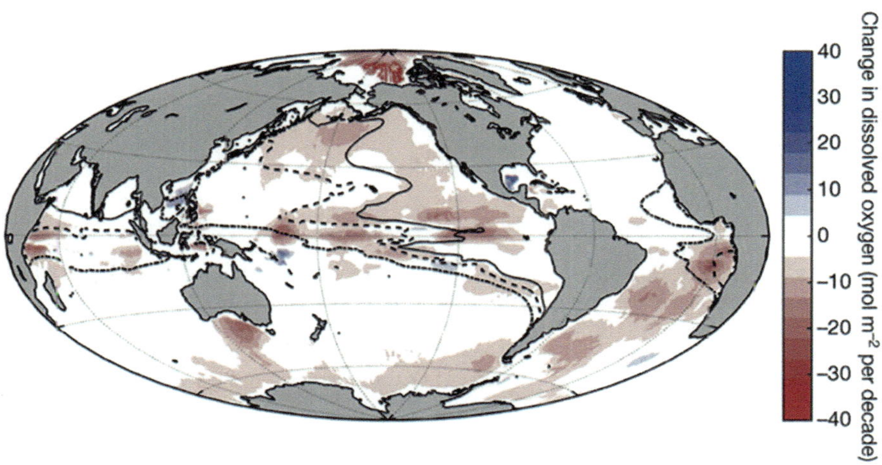

Fig. 24.8 Change in dissolved O_2 per decade since 1960 (color coded) from surface waters to the ocean bottom. Lines indicate boundaries of O_2 minimum zones; dashed-dotted lines indicate regions with less than 80 μmol kg^{-1} O_2 anywhere within the water column. Image reproduced from Schmidtko et al. (2017 Nature 542: 335–339) with permission of Springer Nature

measurement (Blaszczak et al. 2022; Fig. 24.9). Riverine hypoxic events that were prolonged, that is, lasting more than 1 week, were rare. Most hypoxic events were found to be in waters that were associated with wetlands or urban land cover, where sources of organic matter can be high.

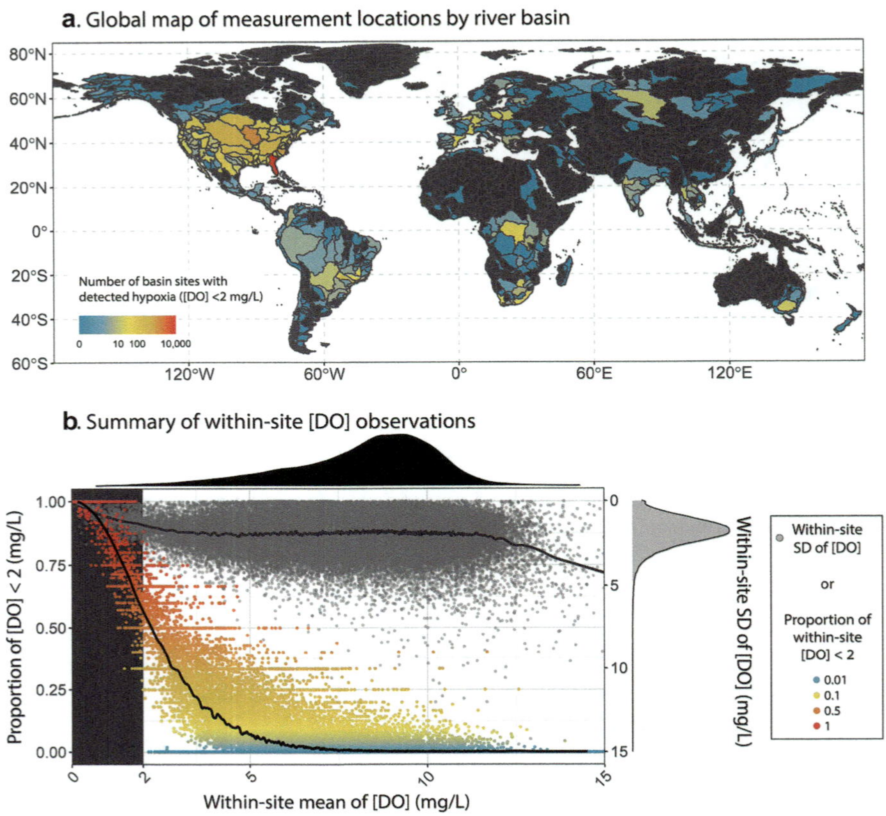

Fig. 24.9 The global distribution of riverine hypoxia. Panel (**a**) Global map of BasinATLAS level 4 subbasins (as defined in Linke et al. 2019) colored by the number of sites within each basin with DO observations < 2 mg DO L^{-1}. River basins with no DO observations are colored in gray (70% of all global basins). Panel (**b**) Within-site proportion of DO observations that are < 2 mg DO L^{-1} (left *y*-axis) and within-site standard deviation of DO concentration (right *y*-axis; gray points) for sites with at least three measurements ($n = 84{,}010$; 67% of all sites). Sixty-two sites with mean DO concentration >15 mg L^{-1} were excluded from this plot. Loess (locally estimated scatterplot smoothing) lines (black for the proportion of observations that are hypoxic; gray for the standard deviation of DO concentration) have a span of 1% of the dataset. The marginal distribution of within-site mean and standard deviation (SD) of DO concentration are shown at the top in black and the right side of the plot in gray, respectively. Image reproduced from Blaszczak et al. (2022 Limnol Oceanog Letts https://doi.org/10.5066/P99X6SIR) under a creative commons license

In all, there is no doubt that both hypoxia and HABs are pervasive in marine and freshwaters and they are increasing in frequency, geographic distribution, and duration in many, if not most, parts of the global ocean and freshwater systems.

24.5 Global "Dead Zones": Some Examples

24.5.1 Northern Gulf of Mexico

The northern Gulf of Mexico has one of the most well-recognized and well-studied "dead zones." This area is one of the largest human-caused areas of hypoxia globally and the largest in the coastal USA (Rabalais and Turner 2019). Since 1972 when hypoxia was first systematically documented, low bottom water O_2 levels in the coastal area westward along the Louisiana coast has seen persistent leading to severe hypoxia every year. The largest region of low O_2 reached 23,000 km^2 (Figs. 24.10 and 24.11).

The northern Gulf of Mexico receives waters from the Mississippi River watershed, the major agricultural region of the USA. The N sources in the runoff are many, representing the non-point source runoff from 41 states (Alexander et al. 2008; Rabalais et al. 2002, 2010; Rabalais and Turner 2019). Inputs of N to receiving waters have tripled since the 1970s (Rabalais et al. 2007), while phosphate (PO_4^{3-}) loads have doubled since about the 1950s, but dissolved silicate ($Si(OH)_4$) concentrations have roughly halved since the early 1900s (Justić et al. 1995; Turner and Rabalais 1991). Spring delivery of nutrients initiates a series of events that lead to summer hypoxia. Blooms begin to form on the continental shelf influenced by the Mississippi River, and with time, particulate matter is exported from the surface waters. This particulate matter is in the form of fecal pellets, individual cells,

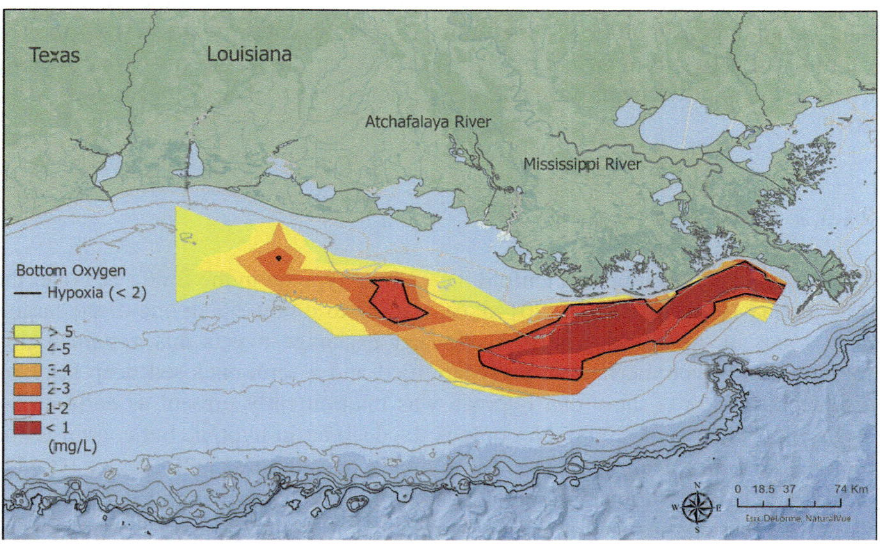

Fig. 24.10 Distribution of bottom water dissolved oxygen <2 mg L^{-1} along the Louisiana shelf for mid-summer 2022. Image reproduced from N. Rabalais, Louisiana Universities Marine Consortium/Louisiana State University, and R. E. Turner, Louisiana State University on-line at https://gulfhypoxia.net/research/shelfwide-cruises/

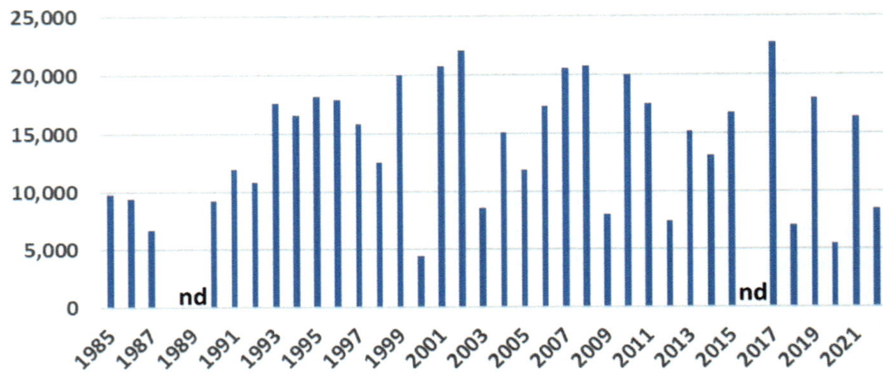

Fig. 24.11 Histogram of mid-summer bottom area of dissolved oxygen ≤2 mg L^{-1} on the continental shelf west of the Mississippi River delta since 1985. n.d. = no data. The only years not included are 1989 (lack of sufficient funding) and 2016 (lack of a suitable vessel). Image reproduced from Rabalais (2022 Shelfwide cruise, July 25–August 1, 2002, https://gulfhypoxia.net/research/shelfwide-cruise/?y=2022)

aggregates and other organic debris. By July, hypoxia is typically well established and is maintained for several months. This is also a time when water column stability is at its seasonal high. Strong correlations have been reported between Mississippi River flow and bottom O_2 deficits, with a time lag of 1 month (Justić et al. 1993).

Efforts to reduce N and P have been discussed for decades, but the magnitude of the agricultural basin of the watershed makes this a daunting problem to solve. The current goal is to reduce the hypoxic zone to 5000 km^2 or less over a 5-year running average, but improvements have not been demonstrable to date (Rabalais and Turner 2019).

24.5.2 Baltic Sea

There has been a documented tenfold increase in hypoxia in the Baltic Sea over the past century (Carstensen et al. 2014a, b), leading to it now being one of the largest dead zones in the world. The lack of water exchange makes this system highly susceptible to hypoxia. It is strongly stratified and a semi-enclosed deep sea, and sedimentary records show that hypoxia was intermittently present as early as the Holocene (Zillén et al. 2008). Not only is the deep basin hypoxic but coastal Baltic regions have increasingly experienced hypoxia over the last half century (Conley et al. 2011). River-borne nutrient loads from the intensive regional agriculture, wastewater discharges, and atmospheric deposition are the primary reasons for the expansion of hypoxia. Additionally, bottom waters have warmed substantially in the recent decades, leading to decreased O_2 solubility.

The Baltic's hypoxic zone, as with most anthropogenically induced hypoxic zones, is linked to its algal blooms as the source of the organic matter. The Baltic has been experiencing massive—toxic—cyanobacterial blooms for decades, and the extent of the hypoxic volume has increased over decades as well (Figs. 24.12 and 24.13). Various modeling simulations have all led to the same conclusion: little to no hypoxia would have developed in this system without anthropogenic nutrient loads

Fig. 24.12 Satellite image of phytoplankton swirling around the Swedish island of Gotland in the Baltic Sea, in July 2019. This image was captured by the Copernicus Sentinel-2 mission, processed by the European Space Agency and available through a creative commons license. Image accessed from: https://www.esa.int/ESA_Multimedia/Images/2019/12/Baltic_blooms

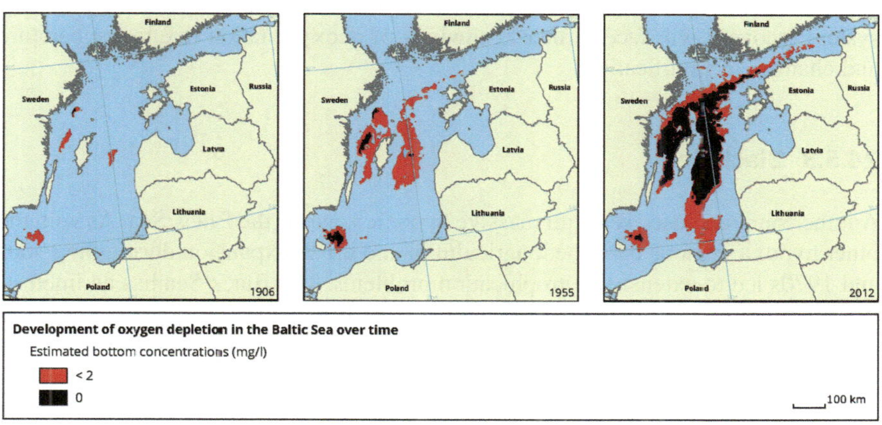

Fig. 24.13 Development of O_2 depletion in the bottom waters of the Baltic Sea over time. Red areas are those with dissolved $O_2 < 2$ mg L^{-1}; black areas are those with 0. The Swedish island of Gotland is visible in the middle of the image. Image reproduced from Carstensen et al. (2014a Proc Natl Acad Sci USA 111 5628–5633) under a creative commons license

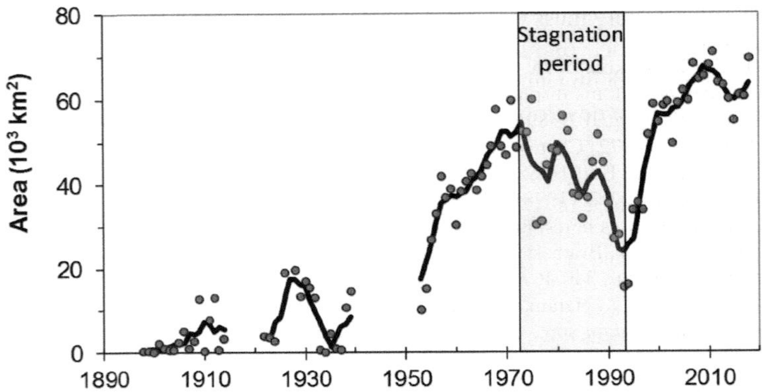

Fig. 24.14 Annual average areal extent of hypoxia Baltic Sea during the time period calculated from O_2 measurements. Image reproduced from Carstensen et al. (2014b Ambio 43: 26–36) under a creative commons license

and eutrophication (Pitcher et al. 2021 and references therein). Hypoxia in this system nearly disappeared in the mid-1990s but has since grown to an area >40,000 km^2 (Fig. 24.14; Carstensen et al. 2014a, b; Krapf et al. 2022).

While the Baltic's expansion in hypoxia is linked to the discharges in wastewater and agricultural fertilizer applications, this system provides a glimpse into the response of a system under nutrient reduction (Carstensen et al. 2006). The watershed of the Baltic has undergone some reductions in nutrient export due to management actions, but these have not reduced hypoxia due to the large nutrient pools that can become mobilized under hypoxic conditions. As discussed in more detail below, stored nutrients in sediments continue to be released and recycled, allowing hypoxia to continue even though the sources have been reduced. Moreover, with warming, system recovery will face further headwinds of deoxygenation due to temperature and changes in stratification.

24.5.3 Black Sea

Another semi-enclosed sea with massive hypoxic zones is the Black Sea. As with the other hypoxic regions described, agricultural and urban expansion during the 1960s and 1970s led to extensive eutrophication problems. The Black Sea has an interesting history of hypoxia/anoxia and historically has been naturally hypoxic, but this is in the open, deeper waters. Over 10,000 years ago, this body of water was an enclosed freshwater lake, but as seas rose roughly 8000 years ago, waters of the Mediterranean Sea entered it, resulting in permanent stratification and therefore loss of exchange of O_2 in the deep waters (Pitcher et al. 2021). This created a permanent anoxic region. The only current source of natural, oxygenated waters of the Black Sea is the Aegean Sea via the Bosphorus Straits, which is too narrow and too shallow to allow much replenishment of O_2 (Kirchman 2021). In spite of the fact that a condition of hypoxia/anoxia in the open Black Sea has long existed, the O_2 content of the water has been declining recently, with low O_2 waters climbing closer to the surface. This latter trend is likely due to warming and climate effects.

Hypoxia in the coastal regions along the Romanian and Ukranian coasts of the Black Sea is comparatively recent. As with other anthropogenically impacted waters, a massive increase in nutrient loads occurred with urbanization and agricultural development in the watershed. In the 1990s, with the decline in the Soviet Union, there was also a decline in agricultural and industrial activities in the region, leading to a decline—but not a cessation of—hypoxia in the coastal Black Sea. Remineralizaton of accumulated organic matter from eutrophication has delayed system recovery (Capet et al. 2013).

24.5.4 Chesapeake Bay

The Chesapeake Bay is the largest estuary in the USA, and its watershed includes both urban and agricultural lands, inhabited by >18,000,000 people. Hypoxia in the Chesapeake Bay is due to the excessive nutrient enrichment it receives from its large watershed (Fig. 24.15). Through analyses of long-term monitoring data (e.g., Kemp et al. 2005), as well as sediment cores (Cooper and Brush 1991, 1993), it is undeniable that the hypoxia this estuary experiences is a direct result of eutrophication.

After years of what seemed like a downward trend, hypoxia ticked backed up again in several of the most recent years, 2019 and 2021 (Fig. 24.16). A sense of progress toward nutrient pollution and hypoxia reduction may be biased by years of comparatively low precipitation (periods from 1997 to 2004, 2005 to 2018). The years with high hypoxia are generally associated with years that are unusually wet.

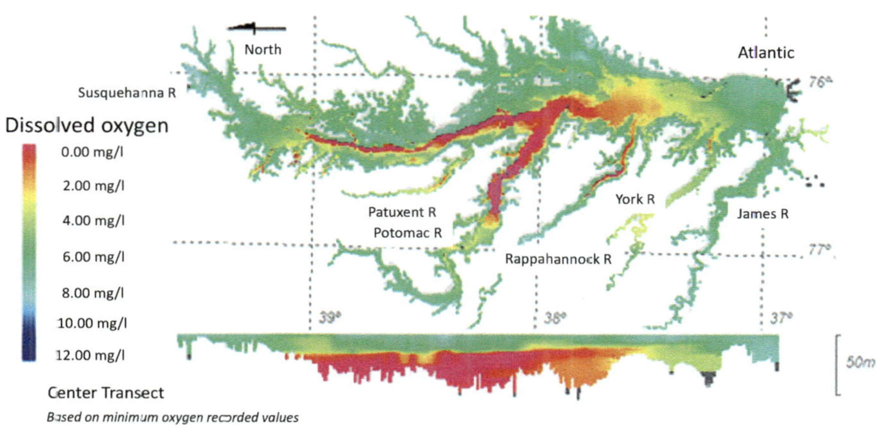

Fig. 24.15 Distribution of bottom water dissolved O_2 concentrations in Chesapeake Bay, USA, in August 2022. Image courtesy of http://eyesonthebay.dnr.maryland.gov

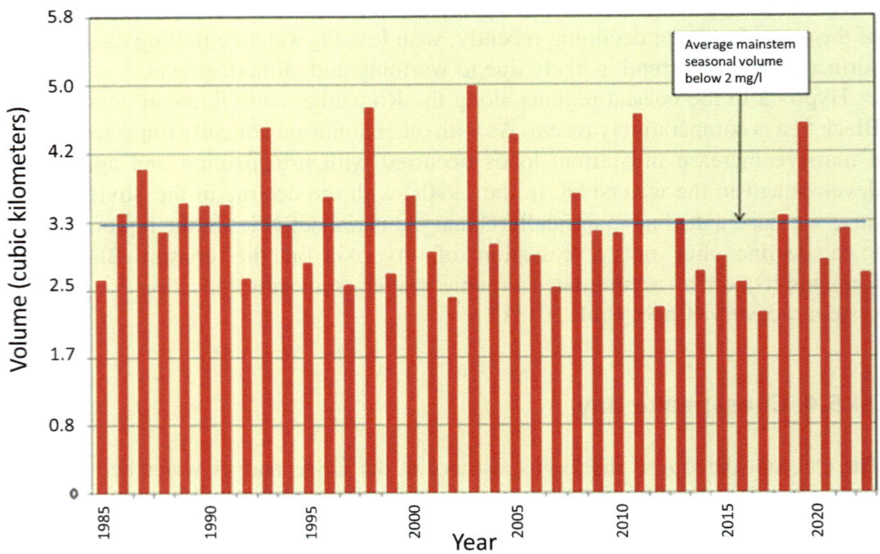

Fig. 24.16 Time series of dissolved O_2 volume for the mainstem of the Chesapeake Bay. Image reproduced from Maryland Department of Natural Resources, https://news.maryland.gov/dnr/2022/11/16/final-chesapeake-bay-hypoxia-report-for-2022/

Wet years not only bring enhanced runoff but can bring even higher nutrient loads that may have accumulated during the drier periods. There is some hopeful evidence, however, based on flow-corrected N loads and bottom water NH_4^+ levels that anoxia has been reduced somewhat (Li et al. 2016; Testa et al. 2017, 2018a, b).

The hypoxia of the Chesapeake Bay is complicated by many factors. Oysters have declined nearly 100-fold, due to a combination of overfishing, disease, and change in food quality from progressive eutrophication (Kemp et al. 2005). Thus, the natural filtering of the water has declined; algal blooms are larger than they would have been if grazed by this filter feeder. Although most of the concern over hypoxia in the Chesapeake Bay is focused on the low O_2 central channel, seiching of deoxygenated waters from winds can push low O_2 waters into shallower waters and tributaries (Breitburg 1992).

Tropical Storm Agnes in 1972 directly hit Chesapeake Bay and may have led to a system "reset," increasing its susceptibility to hypoxia. This hurricane brought record rainfall and a sediment load equivalent to the input of 30 average years (Schubel 1977). Peak hypoxic volume expanded 100% from ~5 km^3 during 1950–1969 pre-hurricane to ~10 km^3 during 1970–1989 in the post-Agnes years. Although this hypoxia expansion coincided with the gradual increase in the riverine nutrient loading and the climatic shift from the dry decade of the 1960s to the wet decade of the 1970s, the abrupt jump in the hypoxic volume around 1970 is striking (Ni and Li 2023). Post-storm surveys showed that Agnes overloaded the Chesapeake Bay with nutrients and organic matter that led to elevated plankton production for several years and decimated submerged aquatic plants (Loftus and Seliger 1976). Recent modeling analysis suggests that the hypoxic volume per unit nutrient load was 30% higher in the 1970s than in the 1960s, indicating a nonlinear system response to nutrient loading (Ni and Li 2023; Chap. 29).

The Chesapeake Bay community has been well aware of the impacts of nutrient loads on water quality, and efforts to reduce such loads have been underway since the 1980s, when the first multi-state Chesapeake Bay agreement was signed. Goals were set then to reduce loads by 40%, but for various reasons were unattainable, leading to new agreements and longer timelines for implementation. Nevertheless, substantial reductions have been taken. Recent modeling by Frankel et al. (2022) shows that if the N reductions since 1985 had not occurred, annual hypoxic volumes ($O_2 < 3$ mg L^{-1}) would have been ~50–120% greater during the average discharge years of 2016–2017 and ~20–50% greater during the wet years of 2018–2019. The effect was even greater for $O_2 < 1$ mg L^{-1}, where annual volumes would have been ~80–280% greater in 2016–2017 and ~30–100% greater in 2018–2019. Scavia et al. (2021) had previously estimated that without watershed management actions, average hypoxia would have been 7% higher compared to 1985, but if current management goals are met, long-term hypoxia could decrease by >30%. Moreover, they, as well as Ni et al. (2020), showed that warming in the Bay has offset the improvements in N reduction as well as effects of sea level rise. All of these studies underscore the ever-pressing need to reduce nutrients—and at least to meet management goals currently in place—but achieving such goals is extremely difficult against the headwinds of warming waters.

24.5.5 East China Sea

The East China Sea provides a vivid example of the effects of anthropogenic nutrient pollution and its effects on hypoxia/anoxia. This is the area formed by the convergence of the Changjiang (Yangtze) River plume and the Taiwan Warm Current (Zhou et al. 2008; Li et al. 2021). The Changjiang River is one of the largest in China. With over 440 million people in the watershed, and with massive expansion in agriculture and agricultural use of fertilizers over the decades, hypoxia—and associated algal blooms (Chap. 25)—began to develop in earnest in the 1980s (Zhu et al. 2011). Hypoxia typically develops late in summer, when waters are warm and stratified. Massive blooms—and massive hypoxic zones—up to 10,000–15,000 km^2 in areal extent, now occur regularly (Figs. 24.17 and 24.18; Zhou et al. 2008; Wang et al. 2017). These blooms occur offshore, beyond the region of high turbidity where phytoplankton is light limited (Li et al. 2021). Even though there is high transport of organic matter along with sediment from the turbid-rich Changjiang River, it is microbial respiration of the phytoplankton bloom production that is the major contributor to hypoxia (Wang et al. 2017).

Several factors suggest that hypoxia will exist well into the future in this region and that the zone of hypoxia may even be expanding northward (Zhu et al. 2011). Große et al. (2020) estimate that reductions of 50% in N loading will be required to

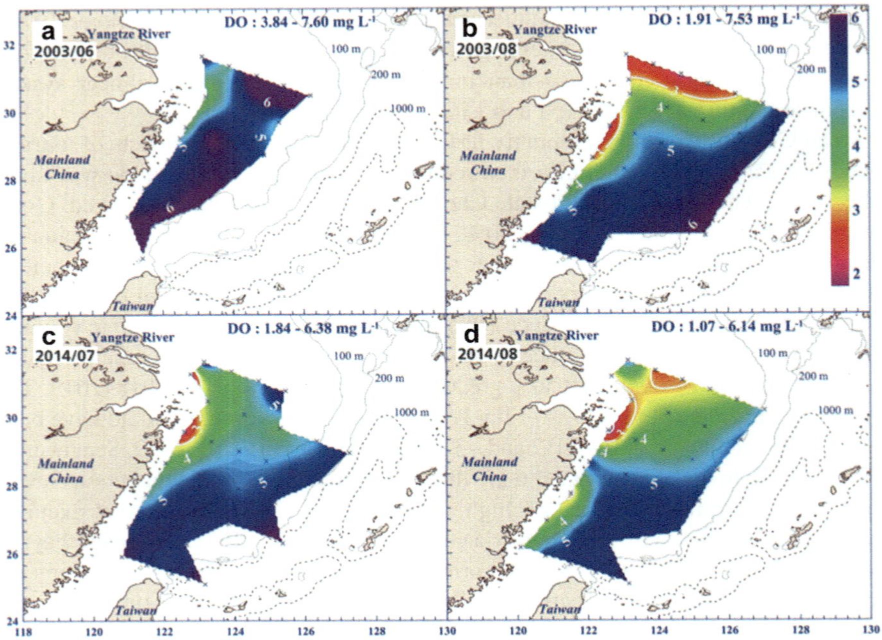

Fig. 24.17 Contours of concentration of dissolved O_2 (DO) in bottom water column or at 100 m if water column depth was >100 m in East China Sea in (Panel **a**) June and (Panel **b**) August 2003 and (Panel **c**) July and (Panel **d**) August 2014. Contour lines of DO = 3 mg L^{-1} are provided in white bold for reference. Note the wider area of hypoxia in August of both years compared to June or July. Image reproduced from Chen et al. (2022, Front Mar Sci 8: doi: 10.3389/fmars.2021.787808) under a creative commons license

Fig. 24.18 Hypoxic zone area in Changjiang Estuary. Data sources are summarized in Pitcher et al. (2021). Years with no data do not necessarily mean no hypoxia. Image redrawn based on Pitcher et al. (2021, Prog Oceanogr 197:102613. doi: 10.1016/j.pocean.2021.102613)

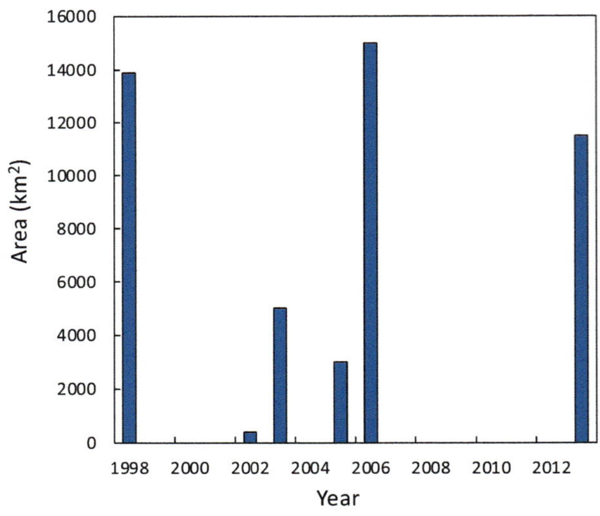

reduce the extent of hypoxia. While massive reductions in N have a high potential for mitigating blooms and hypoxia (Große et al. 2020), Wang et al. (2021) showed that due to the different sources of nutrients in China's coastal systems (agriculture, aquaculture, industry and wastewater), reductions in nutrient loads are and will continue to be exceedingly difficult, based on multiple scenarios. Moreover, as with virtually all the examples described, waters are warming accelerating deoxygenation (see also Chap. 27).

24.5.6 West Florida Shelf

Other mechanisms for hypoxia development have emerged in recent decades. Such mechanisms combine the changing climate with the adaptive strategy of many HABs to be not only consumers of dissolved inorganic nutrients but also consumers of organic particles, that is, to be mixotrophic, especially in warming waters (Chap. 15, Glibert and Li 2024). The convergence of climate change and evolving plankton community structure and productivity lays the foundation for O_2 depletion in nutrient-poor as well as nutrient-enriched coastal systems.

Hypoxia on the West Florida Shelf, the body of water along the Florida coast of the Gulf of Mexico, may result from this mechanism in some years. These waters have long been considered to be oligotrophic due to the general lack of strong upwelling that could bring up nutrient-rich and oxygen-poor deep water onto the shallow shelf (Vargo et al. 1987; Li et al. 2008). Furthermore, the small rivers draining onto the this shelf have low flows and do not routinely discharge substantial amounts of nutrients to the ocean except during tropical storms with associated heavy precipitation (Ager et al. 1991; Munson and Delfino 2007; Vargo et al. 2008). Blooms of the toxigenic dinoflagellate *Karenia brevis* are common, typically developing in late fall and terminating before the following summer. That is, they generally terminate before the hottest months of the years, when it is suspected that ambient temperatures exceed the thermal optimum for growth. However, in recent years, a number of blooms have been prolonged, lasting throughout the hot summer months. A recent retrospective analysis of water quality monitoring data between 2005 and 2019 revealed that hypoxia was present in 5 of the 16 years examined and occurred in years with large *K. brevis* summer blooms and was always found shoreward of the 50-m isobath (Turley et al. 2022). It has been suggested that hypoxia occurring in parallel to a *K. brevis* red tide is more likely to occur with warmer ocean temperatures and increased fluxes of organic nutrients and after hurricanes (Milbrandt et al. 2021). Enhanced heterotrophic metabolism may allow blooms to survive the heat of the summer, even if productivity is reduced due to stress on autotrophy. As *K. brevis* is mixotrophic, and grazes on picoplankton such as *Synechococcus*, it may be able to overcome the thermal stress-associated autotrophy, allowing it to survive the hottest months of the year. As heterotrophy increases, and as the solubility of O_2 decreases with warming waters, development of hypoxia becomes a more readily achievable end point, due to the fact that the euphotic waters have a lower starting point of O_2 saturation.

24.6 Impacts of Hypoxia on Nutrient Cycling

While hypoxia may be caused by nutrient enrichment, it, in turn, can alter nutrient regeneration and greenhouse gas fluxes through multiple mechanisms. The oxic zone in sediments may decrease (Cai and Sayles 1996), anerobic pathways of diagenesis increase (Middelburg and Levin 2009), and the abundance and activity of benthic organisms living on, or in, the sediment change (Levin et al. 2009; Middelburg and Levin 2009). Pathways of N cycling and their changes under low oxygen are complicated (Fig. 24.19). Nitrification requires molecular O_2 to convert NH_4^+ to NO_2^- and then to NO_3^-, and thus reductions in rates of nitrification are commonly observed (Kemp et al. 1990; Caffrey et al. 2003). Sulfide inhibits, and O_2 promotes this pathway (Joye and Anderson 2008). Nitrifiers may be able to recover from O_2 stress fairly quickly.

Effects on denitrification are more complicated. As a process that is normally inhibited by O_2, low dissolved O_2 and sulfide should promote denitrification. It does in many systems. McCarthy et al. (2015) reported that denitrification was indeed the dominant sink for N in northern Gulf of Mexico under hypoxic conditions. However, many factors affect rates of denitrification. Bioturbation can reduce the efficiency of denitrification (Owens and Cornwell 2020). During prolonged periods of hypoxia, due to reduced nitrification rates, denitrification rates can slow as they become

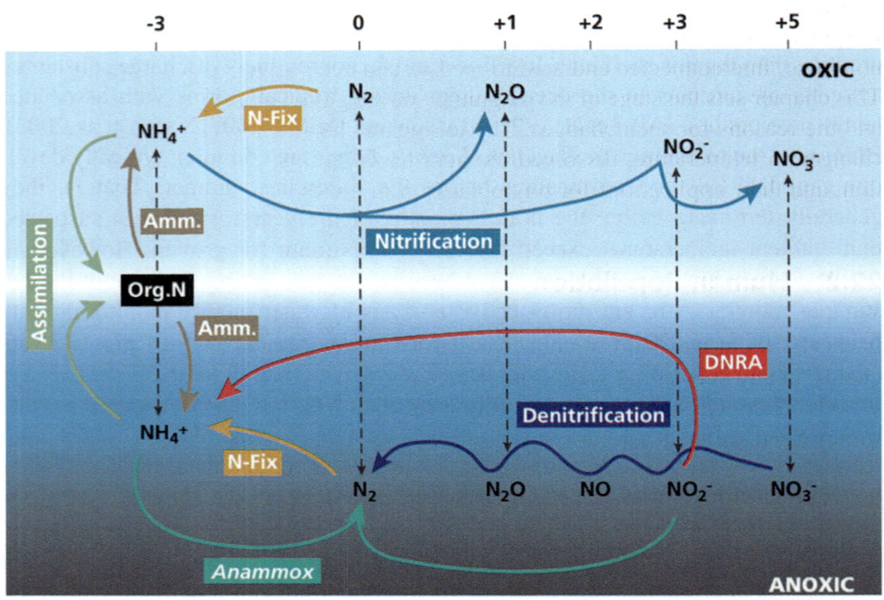

Fig. 24.19 A simplified representation of the marine nitrogen cycle and its microbial pathways. On top of the figure, numbers represent the oxidation state of nitrogen in the different compounds. N-fix = nitrogen fixation, Amm = ammonification, DNRA = dissimilatory nitrate reduction to ammonium. Image modified and redrawn based on Bonaglia (2015, Control factors of the marine nitriogen cycle: the role of meiofauna, macrofauna, oxygen and aggregates)

limited by availability of NO_3^-. That is, coupled nitrification-denitrification is reduced. Foster and Fulweiler (2019) suggested that with increasing hypoxia, nitrifiers may first produce more N_2O, but with increased hypoxia/anoxia, denitrification rates would be inhibited and the resulting lack of NO_3^-/ NO_2^- substrate would reduce N_2O (Foster and Fulweiler 2019). Hypoxic sediments generally exhibit higher rates of NH_4^+ efflux (Caffrey and Kemp 1990; Kemp et al. 1990; Conley et al. 2007), which may be due to decreased rates of nitrification or increased rates of dissimilatory NO_3^- reduction to NH_4^+ (DNRA, Chap. 11). These fluxes change as a function of duration of exposure, sediment type, and organic matter loading. Anammox generally decreases under hypoxic conditions and DNRA increases under hypoxia. While anammox, like denitrification, removes reactive N, DNRA does not, and NH_4^+ accumulates. With reduced nitrification, less NH_4^+ is available to be oxidized. When sediments are rich in organic matter and sulfidic, DNRA can dominate over denitrification.

For PO_4^{3-} flux, an enhanced release is expected under hypoxic conditions. Enhanced PO_4^{3-} flux has been well documented in Baltic Sea (e.g., Conley et al. 2002, 2007), Chesapeake Bay (Cowan and Boynton 1996), and Lake Erie, among many other systems. Conley et al. (2011) showed that interannual changes in PO_4^{3-} were positively correlated with the area of the Baltic Sea sediment that was hypoxic. The release of PO_4^{3-} can result in a positive feedback that leads to additional primary production and eutrophication (see also Chap. 13). This would be further enhanced with pH increases that accompany high biomass blooms. Rates of sediment water exchange vary as a function of O_2, but also salinity, and availability of terminal electron acceptors (Kemp et al. 2005; Cornwell et al. 2016).

24.7 Impacts of Hypoxia on Plankton

While impacts of hypoxia on various fish species, benthic and sessile organisms are quite well described (e.g., Diaz and Rosenberg 1995; Levin 2003; Wu 2009 and references therein), relatively few studies have examined the effects of hypoxia on many planktonic species groups. Of the plankton, zooplankton have often been better studied than phytoplankton with respect to effects of hypoxia (e.g., Marcus 2001; Roman et al. 2019 and references therein). Copepod survival, growth rates, egg production rates, and ingestion rates all decline as dissolved O_2 declines (Roman et al. 2019 and references therein). Vertical migration behaviors are also disrupted (Fig. 24.20). Effects of low O_2 on copepods also depend on the tolerance of species to low O_2. More hypoxia-tolerant species may thrive if they can use low O_2 waters as a refuge from predation by fish (that may be less tolerant) (e.g., Appeltans et al. 2003; Roman et al. 2019 and references therein). On the other hand, if zooplankton are not hypoxia-tolerant, they may aggregate at oxyclines where they are susceptible to predation (Taylor and Rand 2003; Roman et al. 2012). Dissolved O_2 also appears to affect the rate of feeding by copepods and their escape response (Elliott et al. 2013; Decker et al. 2004). Oxygen not only affects survival of adults but also affects egg production and hatching success. Egg viability depends on whether the copepods carry or release their eggs; hypoxia may favor brooders over spawners

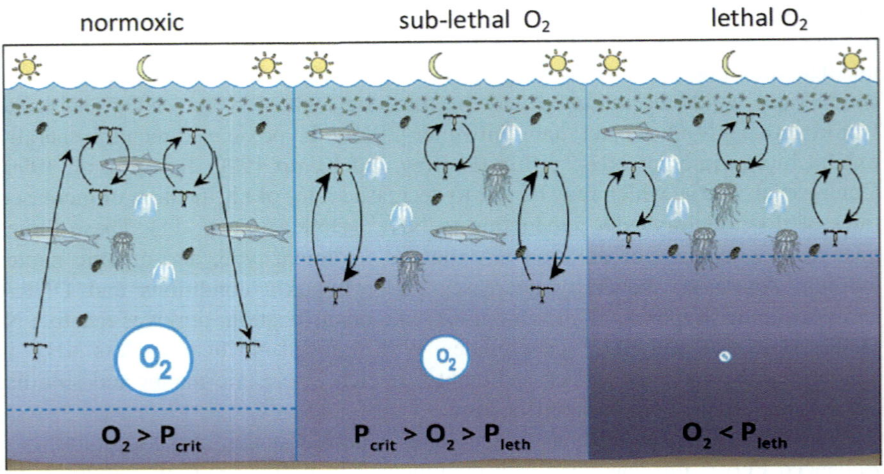

Fig. 24.20 Conceptual diagram of food web changes that may occur over diel cycles between normoxic conditions (left), sub-lethal deoxygenation conditions (middle), and lethal conditions (right). The size of the symbol for O_2 and the color indicates the relative amount of dissolved O_2 under each condition (with light blue indicating high O_2 concentrations and purple indicating lower O_2). Image reproduced from Roman et al. (2019, Front Mar Sci 6: doi: 10.3389/fmars.2019.00139) under a creative commons license

(Kimmel et al. 2012; Elliott et al. 2012). Some eggs may survive in sediment under anoxic waters and are able to hatch if waters are reaerated (Broman et al. 2017; Roman et al. 2019 and references therein). Smaller copepods seem to fare better with low dissolved O_2 than larger zooplankton, a finding documented in several laboratory experiments (Stadler and Marcus 1997; Roman et al. 1993). Pierson et al. (2017) observed that male *Acartia tonsa* copepods in Chesapeake Bay were more tolerant of low O_2 than females, as the later may have a higher O_2 demand due to egg production. Regardless, the trophic interactions are disrupted. If there is a community shift to smaller zooplankton, this alters the prey amount and quality for predators.

For an example of how hypoxia and increasing temperature collectively impact the phytoplankton community, the Arabian Sea presents an excellent example (see also Chap. 27). The permanent O_2 minimum zone has expanded due to climate change, leading to increased denitrification and changing nutrient stoichiometry (Goes et al. 2020). Beginning around 2000, the northern Arabian Sea shifted from a winter phytoplankton assemblage that had previously consisted mostly of diatoms to a contemporary phytoplankton assemblage dominated by the large dinoflagellate *Noctiluca scintillans* (up to 2000 μm in diameter (Gomes et al. 2014; and references therein; Goes and Gomes 2016; Goes et al. 2018)). These winter dinoflagellates are colorless mixotrophs, but blooms are green in color due to the presence of prasinophycean endosymbionts (*Pedinomonas noctilucae*). The mixotrophs rely upon the symbionts to fix carbon (C), and they also ingest a variety of prey. These blooms typically occur when the upwelled waters are hypoxic, as the endosymbiont of *Noctiluca* favors hypoxic conditions for its photosynthesis (Gomes et al. 2018). Due to the enhanced rates of denitrification, the previously prevalent diatoms may be experiencing NO_3^- stress, but *Noctiluca* has no requirement for NO_3^- due to its mixotrophic nutrition. Indeed, these dinoflagellates accumulate high intracellular concentrations of NH_4^+ when feeding; they also grow well in NH_4^+-enriched culture conditions. Moreover, they also prefer the calm, highly stratified waters of the winter upwelling conditions. When these blooms form, they are massive, and they result in further dissolved O_2 depletion and accumulations of elevated concentrations of NH_4^+. Jellyfish and salps have accumulated, as large zooplankton are unable to feed on this mixotroph. This enhances the cycling of organic C, sustaining the microbial loop (Naqvi et al. 2002; Naqvi 2021).

Stauffer et al. (2013) tracked the plankton assemblage of an embayment in southern California following a multi-week hypoxia and fish kill event. The proximate cause of hypoxia within the embayment was enhanced respiratory activity from a high concentration of fish. Phytoplankton became extremely stressed as hypoxia developed, as evidenced by depressed values of variable fluorescence, *Fv/Fm* (see Chap. 7 for description of *Fv/Fm*). Small bacterivorous scuticociliates became dominant when extremely low O_2 was present. A similar finding was reported by Fenchel et al. (1990) for hypoxic Danish fjords. It was suggested that the rise of this ciliate was due to the rise in bacteria from the decaying fish in the southern California embayment (Stauffer et al. 2013).

A quite different observation with regard to hypoxia effects has recently been reported by Sun et al. (2022). They found, based on a cruise in the Pearl River Estuary and a subsequent mesocosm experiment, that diatom production (*Thalassiosira weissflogii*) may have been enhanced, rather than depressed, in low O_2 waters. In their mesocosm experiment, they manipulated O_2 levels in conjunction with varying CO_2 levels. Measurements of photosynthetic yield (*Fv/Fm*) were higher in the low O_2 conditions regardless of CO_2 levels. They hypothesized that reductions in O_2 suppressed mitochondrial respiration and favored the carboxylase reaction over the oxygenase reaction of Rubisco (recall the dual affinities of this enzyme; Chap. 6). It also increased the efficiency of their C concentrating mechanisms (CCMs, Fig. 24.21; see Chaps. 6 and 26). Their experiments did document an increase in dinoflagellate and picoplankton abundance over the time of their experiment, but diatoms remained dominant. They further speculated that the enhancement in productivity could help to re-oxygenate low O_2 waters over time. Clearly, suppression of photosynthesis—or enhancement—depend on the degree of hypoxia, its duration, and the species present at the time of the event.

Each of these studies and examples has suggested a trend toward increased dinoflagellates, mixotrophs, or heterotrophs over diatoms (even though Sun et al. 2022 study showed diatom production enhanced, they too saw an increase in dinoflagellate abundance). Altered nutrient stoichiometry, a result of changing riverine loads as well as differential sediment flux effects due to hypoxia, underscores the difficulty in reversing hypoxia. Enhanced recycling of nutrients from hypoxia become a secondary nutrient problem, enhancing or prolonging

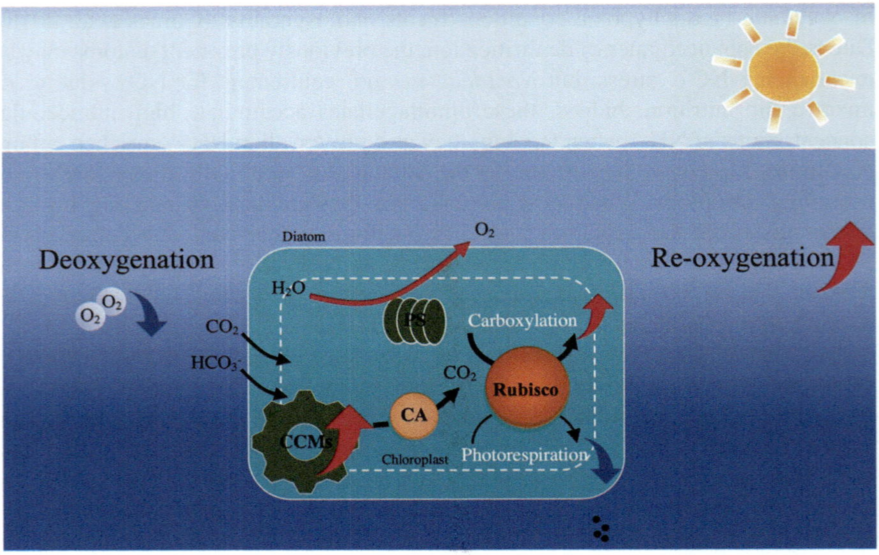

Fig. 24.21 Reduced O_2 levels enhance CO_2 concentrating mechanisms (CCMs) and photosynthetic performance with increased light energy use efficiency in diatoms and phytoplankton assemblages, promoting photosynthetic carbon fixation and O_2 evolution. The enhanced photosynthesis and reduced mitochondrial and photorespiration can substantially accelerate the re-oxygenation due to stimulated photosynthetic O_2 evolution. In natural environments, low O_2-enhanced production of phytoplankton biomass makes a more effective O_2 source that may help to counteract the negative effects of hypoxia on heterotrophs. Black, red, and blue arrows indicate directions, increase, and decrease, respectively. Image reproduced from Sun et al. (2022 Comm Biol 5: 54) under a creative commons license

plankton blooms (Kemp et al. 2005, 2009, Ma et al. 2006; Middelburg and Levin 2009).

24.8 Summary

Aquatic deoxygenation is occurring due to a complex suite of factors. In fresh and coastal waters, the input of nutrients fueling blooms and their subsequent decay is well correlated with hypoxia (Pitcher et al. 2021). Moreover, as waters warm, the solubility of oxygen decreases, in turn lessening the rate of re-oxygenation into deeper waters. However, regional and temporal variations are large, a function of not only differences in nutrient loadings but also physical and biogeochemical properties of individual systems. As continents heat up, coastal winds may be strengthened as the pressure differential between air over warm land and over cool waters increases. Unusually strong winds are often associated with coastal hypoxia due to upwelling.

The balance between respiration and system O_2 ventilation will ultimately determine whether hypoxia or anoxia becomes established and for how long. Climate change and variability is also altering the rate of deoxygenation in several ways. In addition to solubility, stratification is increasing, as is the frequency of precipitation events (bringing more runoff or delaying runoff events during droughts). Management actions are focused on nutrient reductions, which are required, but trajectories of recovery are complicated, and reduction measures are not easily implemented or not uniformly as successful as predicted. Hypoxia can promote more nutrient enrichment via a series of positive and nonlinear feedbacks. More nutrients lead to more blooms, more sinking and bacterial respiration, more efficient recycling of nutrients, inhibition of nitrification—but potentially increased denitrification, NH_4^+ efflux and DNRA and P release and solubility, further providing nutrients that support production in overlying waters, but under altered stoichiometric proportions. More work is needed to understand how phytoplankton and its microbial community change under hypoxic conditions. While algal blooms may lead to hypoxia, it is increasingly evident that hypoxia may also lead to HABs through these nutrient cycling mechanisms.

References

Ager L, Bass DG, Clewell AF, DeMort CL, Dixon LK, Epler JH, Estevez ED et al (1991) In: Livingston RJ (ed) The rivers of Florida, vol 83. Springer

Alexander RB, Smith RA, Schwarz GE, Boyer EW, Noaln JV, Brakebill JW (2008) Differences in phosphorus and nitrogen delivery to the Gulf of Mexico from the Mississippi River Basin. Environ Sci Tech 42:822–830. https://doi.org/10.1021/es0716103

Appeltans W, Hannouti A, van Damme S, Soetaert K, Vanthomme R, Tackx M (2003) Zooplankton in the Schelde estuary (Belgium/The Netherlands). The distribution of *Eurytemora affinis*: effect of oxygen? J Plankton Res 25:1441–1445. https://doi.org/10.1093/plankt/fbg101

Benson BB, Krause D Jr (1980) The concentration and isotope fractionation of gases dissolved in freshwater in equilibrium with the atmosphere. 1. Oxygen. Limnol Oceanogr 25:662–671

Benson BB, Krause D Jr (1984) The concentration and isotope fractionation of oxygen dissolved in freshwater and seawater in equilibrium with the atmosphere. Limnol Oceanogr 29:620–632

Bever AJ, Friedrichs MAM, Friedrichs CT, Scully ME, Lanerolle LWJ (2013) Combining observations and numerical model results to improve estimates of hypoxic volume within the Chesapeake Bay, USA. J Geophys Res Oceans 118:4924–4944. https://doi.org/10.1002/jgrc.20331

Blaszczak JR, Koening LE, Mejia FH, Gómez-Gener L, Dutton CL, Carter AM, Grimm NB et al (2022) Extent, patterns, and drivers of hypoxia in the world's streams and oceans. Limnol Oceanogr Letts. https://doi.org/10.5066/P99X6SIR

Boesch DF (ed) (2008) Global warming and the free state: comprehensive assessment of climate change impacts in Maryland. Report of the Scientific and Technical Working Group of the Maryland Commission on Climate Change. https://ian.umces.edu/publications/global-warming-and-the-free-state-comprehensive-assessment-of-climate-change-impacts-in-maryland/

Bonaglia S (2015) Control factors of the marine nitriogen cycle: the role of meiofauna, macrofauna, oxygen and aggregates. PhD Dissertation, Department of Geological Sciences, Stockhohm University

Breitburg DL (1992) Episodic hypoxia in Chesapeake Bay: interacting effects of recruitment, behavior, and physical disturbance. Ecological monographs 62:525–546. https://doi.org/10.2307/2937315

Breitburg DL, Levin LA, Oschlies A, Grégoire M, Chavez FP, Conley DJ, Garçon V et al (2018) Declining oxygen in the global ocean and coastal waters. Science 359:6371. https://doi.org/10.1126/science.aam7240

Broman E, Brusin M, Dopson M, Hylander S (2017) Oxygenation of anoxic sediment triggers hatching of copepod eggs. Proc R Soc 282:2015–2025. https://doi.org/10.1098/rspb.2015.2025

Burns J (2018) Too hot? In: 1858 a heatwave turned London into a stinking sewer. BBC News, 3 August 2018, www.bbc.com/news/education-45009749. Accessed 27 Nov 2022

Caffrey JM, Kemp WM (1990) Nitrogen cycling in sediments with estuarine populations of *Potamogeton perfoliatus* and *Zostera marina*. Mar Ecol Prog Ser 66:147–160. https://doi.org/10.3354/meps066147

Caffrey JM, Harrington NE, Solem I, Ward BB (2003) Biogeochemical processes in a small California estuary. 2. Nitrification activity, community structure and role in nitrogen budgets. Mar Ecol Prog Ser 248:27–40. https://doi.org/10.3354/meps248027

Cai W-J, Sayles FL (1996) Oxygen penetration depths and fluxes in marine sediments. Mar Chem 52(2):123–131. https://doi.org/10.1016/0304-4203(95)00081-X

Capet A, Beckers J-M, Grégoire M (2013) Drivers, mechanisms and long-term variability of seasonal hypoxia on the Black Sea northwestern shelf – is there any recovery after eutrophication? Biogeosciences 10(6):3943–3962

Carstensen J, Conley DJ, Andersen JH, Ærtebjerg G (2006) Coastal eutrophication and trend reversal: a Danish case study. Limnol Oceanogr 51:398–408. https://doi.org/10.4319/lo.2006.51.1_part_2.0398

Carstensen J, Anderson JH, Gustafsson BG, Conley DJ (2014a) Deoxygenation of the Baltic Sea during the last century. Proc Natl Acad Sci U S A 111:5628–5633. https://doi.org/10.1073/pnas.1323156111

Carstensen J, Conley DJ, Bonsdorff, Gustafsson BG, Hietanen S, Janas U, Voss M (2014b) Hypoxia in the Baltic Sea: biogeochemical cycles, benthic fauna, and management. Ambio 43:26–36

Chan F, Barth JA, Kroeker K, Lubchenco J, Menge BA (2019) The dynamics and impact of ocean acidification and hypoxia: insights from sustained investigations in the northern California current large marine ecosystem. Oceanography 32:62–71. https://doi.org/10.5670/oceanog.2019.312

Chen C-C, Ko DS, Gong G-C, Lien C-C, Chou W-C, Lee H-J, Shiah F-K, Huang Y-SW (2022) Reoxygenation of the hypoxia in the East China Sea: a ventilation opening for marine life. Front Mar Sci 8:787808. https://doi.org/10.3389/fmars.2021.787808

Conley DJ, Humborg C, Rahm L, Savchuk OP, Wulff F (2002) Hypoxia in the Baltic Sea and basin-scale changes in phosphorus biogeochemistry. Environ Sci Technol 36:5315–5320

Conley DJ, Carstensen J, Ærtebjerg G, Christensen PB, Dalsgaard T, Hansen JL, Josefson AB (2007) Long-term changes and impacts of hypoxia in Danish coastal waters. Ecol Appl 17(sp5):S165–S184. https://doi.org/10.1890/05-0766.1

Conley DJ, Carstensen J, Aigars J, Axe P, Bonsdorff E, Eremina T, Haahti B-M et al (2011) Hypoxia is increasing in the coastal zone of the Baltic Sea. Environ Sci Technol 45:6777–6783. https://doi.org/10.1021/es201212r

Cook RL, Hulbert WC (1957) Applying fertilizers. In: U.S. Departmetn of Agriculture, Soil, 1857 Yearbook of agriculture. Government Printing Office, Washington, DC, pp 216–228

Cooper SR, Brush GS (1991) Long-term history of Chesapeake Bay anoxia. Science 254:992–996

Cooper SR, Brush GS (1993) A 2,500-year history of anoxia and eutrophication in Chesapeake Bay. Estuaries 16:617–626

Cornwell JC, Owens MS, Boynton WR, Harris LA (2016) Sediment-water nitrogen exchange along the Potomac River estuarine salinity gradient. J Coast Res 32:776–787. https://doi.org/10.2112/JCOASTRES-D-15-00159.1

Cowan JLW, Boynton WR (1996) Sediment-water oxygen and nutrient exchanges along the longitudinal axis of Chesapeake Bay: seasonal patterns, controlling factors and ecological significance. Estuaries 19(3):562–580. https://doi.org/10.2307/1352518

Decker MB, Breitbur D, Purcell JE (2004) Effects of low dissolved oxygen on zooplankton predation by the ctenophore *Mnemiopsis leidyi*. Mar Ecol Prog Ser 280:163–172

Diaz RJ, Rosenberg R (1995) Spreading dead zones and consequences for marine ecosystems. Science 321:926–929

Diaz RJ, Rosenberg R (2008) Spreading dead zones and conseequences for marine ecosystems. Science 321:926–929. https://doi.org/10.1126/science.1156401

Dickens C (1857) Little Dorritt. Bradbury and Evans Publishers, London

Dutton CL, Subalusky AL, Hamilton SK, Rosi EJ, Post DM (2018) Organic matter loading by hippopotami causes subsidy overload resulting in downstream hypoxia and fish kills. Nat Commun 9:19851. https://doi.org/10.1038/s41467-018-04391-6

Edmonds-Brown V (2022) From 'biologically dead' to chart-toppingly clean: how the Thames made an extraordinary recovery over 60 years. The Conversation, April 21, 2022. https://theconversation.com/from-biologically-dead-to-chart-toppingly-clean-how-the-thames-made-an-extraordinary-recovery-over-60-years-180895. Accessed 9 May 2023

Elliott DT, Pierson JJ, Roman MR (2012) Relationship between environmental conditions and zooplankton community structure during summer hypoxia in the northern Gulf of Mexico. J Plankton Res 34:602–613. https://doi.org/10.1093/plankt/fbs029

Elliott DT, Pierson JJ, Roman MR (2013) Predicting the effects of coastal hypoxia on vital rates of the planktonic copepod *Acartia tonsa* Dana. PLoS One 8:e63987. https://doi.org/10.1371/journal.pone.0063987

Faraday M (1855) The state of the Thames. The Times, 9 July 1855, p 8

Fenchel T, Kristensen LD, Rasmussen L (1990) Water column aanoxia: vertical zonation of planktonic protozoa. Mar Ecol Prog Ser 62:1–10

Flanders J (2012) The Victorian city: everyday life in Dicken's London. Atlantic Books, London

Foster SQ, Fulweiler RW (2019) Estuarine sediment exhibit dynamic and variable biogeochemical responses to hypoxia. JGR Biogeosci 124:737–758. https://doi.org/10.1029/2018JG004663

Frankel LT, Friedrichs MAM, St-Laurent P, Bever AJ, Lipcius RM, Bhatt G, Shenk GW (2022) Nitrogen reduction have decreased hypoxia in the Chesapeake Bay: evidence from empirical and numerical modeling. Sci Total Environ 814:152722. https://doi.org/10.1016/j.scitotenv.2021.152722

Gewin V (2010) Oceanography: dead in the water. Nature 466(7308):812–814. https://doi.org/10.1038/466812a

Glibert PM, Li M (2024) Warming, wheezing, blooming waters: hypoxia and harmful algae. In: Baird D, Elliott M (eds) Treatise on estuarine and coastal science, vol 4, 2nd edn. Elsevier, Oxford, pp 401–444. https://doi.org/10.1016/B978-0-323-90798-9.00066-4

Goes JI, Gomes HR (2016) An ecosystem in transition: the emergence of mixotrophy in the Arabian Sea. In: Glibert PM, Kana TM (eds) Aquatic microbial ecology and biogeochemistry: a dual perspective. Springer, Cham, pp 155–170. https://doi.org/10.1007/978-3-319-30259-1_13

Goes JI, Gomes HR, Al-Hashimi K, Buranapratheprat A (2018) Ecological drivers of green Nocticula blooms in two monsoonal-driven ecosystems. In: Glibert PM, Berdalet E, Burford MA, Pitcher GC, Zhou MJ (eds) Global ecology and oceanography of harmful algal blooms. Springer, Cham, pp 327–336. https://doi.org/10.1007/978-3-319-70069-4_17

Goes JI, Tian H, Gomes HDR, Anderson OR, Al-Hashmi K, deRada S, Luo H et al (2020) Ecosystem state change in the Arabian Sea fueled by the recent loss of snow over the Himalayan-Tibetan Plateau region. Sci Rep 10(1):7422. https://doi.org/10.1038/s41598-020-64360-2

Gomes HR, Goes JI, Matondkar SGP, Buskey EJ, Basu S, Parab S, Thoppil P (2014) Massive outbreaks of *Noctiluca scintillans* blooms in the Arabian Sea due to spread of hypoxia. Nat Commun 5:4862. https://doi.org/10.1038/ncomms5862

Gomes HR, McKee K, Mile A, Thandapu S, Al-Hashmi K, Jiang X, Goes JI (2018) Influence of light availability and prey type on the growth and photo-physiological rates of the mixotroph *Noctiluca scintillans*. Front Mar Sci 5:374. https://doi.org/10.3389/fmars.2018.00374

Gorman HS (2013) The story of N: a social history of the nitrogen cycle and the challenge of sustainability. Rutgers University Press, New Brunswick, NJ. 241 p

Green BR (2007) Evolution of light-harvesting antennas in an oxygenated world. In: Falkowski PG, Knoll AH (eds) Evolution of primary producers in the sea. Elsevier, Amsterdam, pp 38–54

Grégoire M, Garçon V, Garcia H, Breitburg D, Isensee K, Oschlies A, Telszewski M et al (2021) A global ocean oxygen database and atlas for predicting deoxygenation and ocean health in the open and coastal ocean. Front Mar Sci 8. https://doi.org/10.3389/fmars.2021.724913

Große F, Fennel K, Zhang H, Laurent A (2020) Quantifying the contributions of riverine vs. oceanic nitrogen to hypoxia in the East China Sea. Biogeosciences 17:2701–2720. https://doi.org/10.5194/bg-17-2701-2020

Jenny JP, Francus P, Normandeau A, Lapointe F, Perga ME, Ojala A, Schimmelmann A, Zolitchka B (2015) Global spread of hypoxia in freshwater ecosystems during the last three centuries is caused by rising local human pressure. Glob Chang Biol 22:1481–1489

Joye SB, Anderson IC (2008) Nitrogen cycling in coastal sediments. In: Capone DG, Bronk DA, Mulholland MR, Carpenter EJ (eds) Nitrogen in the marine environment, 2nd edn. Academic Press, Amsterdam, pp 868–915

Justić D, Rabalais NN, Turner RE (1993) Effects of climatr change on hypoxia in coastal waters: a doubled CO_2 scenario for the northern Gulf of Mexico. Limnol Oceanogr 41:992–1003

Justić D, Rabalais NN, Turner RE (1995) Stoichiometric nutrient balance and origin of coastal eutrophication. Mar Pollut Bull 30:41–46. https://doi.org/10.1016/0025-326X(94)00105-I

Keeling RF, Garcia HE (2002) The change in oceanic O_2 inventory associated with recent global wamrmng. Proc Natl Acad Sci U S A 99:7848–7853

Kemp WM, Sampou P, Caffrey J, Mayer M, Henriksen K, Boynton W (1990) Ammonium recycling versus denitrification in Chesapeake Bay sediments. Limnol Oceanogr 35:1545–1563

Kemp WM, Boynton WR, Adolf JE, Boesch DF, Boicourt WC, Brush G, Cornwell JC, Fisher TR, Glibert PM, Hagy JD (2005) Eutrophication of Chesapeake Bay: historical trends and ecological interactions. Mar Ecol Prog Ser 303:1–29

Kemp WM, Testa JM, Conley DJ, Gilbert D, Hagy JD (2009) Temporal responses of coastal hypoxia to nutrient loading and physical controls. Biogeosciences 6:2985–3008

Kimmel DG, Boynton WR, Roman MR (2012) Long-term decline in the calanoid copepod *Acartia tonsa* in central Chesapeake Bay, USA: an indirect effect of eutrophication? Estuar Coast Shelf Sci 101:76–85. https://doi.org/10.1016/j.ecss.2012.02.019

Kirchman DL (2021) Dead zones: the loss of oxygen from rivers, lakes, seas, and the ocean. Oxford University Press, Oxford. 217 p

Krapf K, Naumann M, Dutheil C, Meier HEM (2022) Investigatin hypoxic and euxinic area changes based on various datasets from the Baltic Sea. Front Mar Sci:9. https://doi.org/10.3389/fmars.2022.823476

Levin LA (2003) Oxygen minimum zone benthos: adaptation and community response to hypoxia. In: Gibson RN, Atkinson RJA (eds) Oceanography and marine biology, an annual review, vol 41. CRC Press, London, pp 1–45

Levin LA, Ekau W, Gooday AJ, Jorissen F, Middelburg JJ, Naqvi SWA, Neira C et al (2009) Effects of natural and human-induced hypoxia on coastal benthos. Biogeosciences 6(10): 2063–2098. https://doi.org/10.5194/bg-6-2063-2009

Li M, Lee YJ, Testa JM, Li Y, Ni W, Kemp WM, Di Toro DM (2016) What drives interannual variability of hypoxia in Chesapeake Bay: climate forcing versus nutrient loading? Geophys Res Lett 43:2127–2134

Li QP, Hansell DA, Zhang JZ (2008) Underway monitoring of nanomolar nitrate plus nitrite and phosphate in oligotrophic seawater. Limnol Oceanogr Methods 6(7):319–326

Li W, Ge J, Ding P, Ma J, Glibert PM, Liu D (2021) Effects of dual fronts on the spatial pattern of phytoplankton assemblages off the Changjiang River estuary. Est Coasts 44:1408–1418. https://doi.org/10.1007/s12237-020-00893-z

Linke S, Lehner B, Dallaire CO, Ariwi J, Grill G, Anand M, Beames P (2019) Global hydro-environmental sub-basin and river reach characteristics at high spatial resolution. Sci Data 6: 283. https://doi.org/10.1038/s41597-019-0300-6

Loftus ME, Seliger HH (1976) A comparative study of primary production and standing crops of phytoplankton, in a portion of the upper Chesapeake Bay subsequent to tropical storm Agnes. In: Ruzecki EP, Schubel JR, Huggett RJ, Anderson AM, Wass ML, Marasco RJ, Lynch MP (eds) The effects of tropical storm Agnes on the Chesapeake estuarine system, CRC publication No. 54. The Johns Hopkins University Press, Baltimore, pp 509–521

Ma SF, Whereat EB, Luther GW (2006) Shift of algal community structure in dead end lagoons of the Delaware Inland Bays during seasonal anoxia. Aquat Microb Ecol 44:279–290

Marcus NH (2001) Zooplankton: responses to and consequences of hypoxia. In: Rabalais NN, Turner RE (eds) Coastal hypoxia: consequences for living resources and ecosystems. American Geophysical Union, Washington, DC. https://doi.org/10.1029/CE058p0049

McCarthy MJ, Newell SE, Carini SA, Gardner WS (2015) Denitrification dominates sediment nitrogen removal and is enhanced by bottom-water hypoxia in the northern Gulf of Mexico. Estuar Coasts 38:2279–2294. https://doi.org/10.1007/s12237-015-9964-0

Middelburg JJ, Levin LA (2009) Coastal hypoxia and sediment biogeochemistry. Biogeosciences 6(7):1273–1293. https://doi.org/10.5194/bg-6-1273-2009

Milbrandt EC, Martignette AJ, Thompson MA, Bartleson RD, Philips EJ, Badylak S, Nelson NG (2021) Geospatial distribution of hypoxia associated with a *Karenia brevis* bloom. Estuar Coastal Shelf Sci 259:107446. https://doi.org/10.1016/j.ecss.2021.107446

Morales CE, Hormazabal SE, Blanco JL (1999) Interannual variability in the mesoscale distribution of the depth of the upper boundary of the oxygen minimum layer off northern Chile (18–24S): implications for the pelagic system and biogeochemical cycling. J Mar Res 57:909–932

Munson AB, Delfino JJ (2007) Minimum wet-season flows and levels in southwest Florida rivers 1. J Am Water Resour Assoc 43(2):522–532

Naqvi SWA (2021) Deoxygenation in marginal seas of the Indian Ocean. Front Mar Sci 8:624322. https://doi.org/10.3389/fmars.2021.624322

Naqvi SWA, Sarma VVSS, Jayakumar DA (2002) Carbon cycling in the northern Arabian Sea during the northeast monsoon: significance of salps. Mar Ecol Prog Ser 226:35–44

Ni W, Li M (2023) What drove the nonlinear hypoxia response to nutrient loading in Chesapeake Bay during the 20th century? Sci Total Environ 861:160650. https://doi.org/10.1016/j.scitotenv.2022.160650

Ni W, Li M, Testa J (2020) Discerning effects of warming, sea level rise and nutrient management on long-term hypoxia trends in Chesapeake Bay. Sci Total Environ 737:139717. https://doi.org/10.1016/j.scitotenv.2020.139717

Ott H (2024) Oxford-Cambridge boat racers warned of “alarmingly high” E. coli levels in London’s sewage-infused Thames. CBS news, March 28, 2024. https://www.cbsnews.com/news/oxford-cambridge-boat-racers-warned-ofalarmingly-high-e-coli-levels-in-londons-sewage-infused-thames/

Owens MS, Cornwell JC (2020) Temporal enhancement of denitrification in bioirrigated estuarine sediments. Aquat Sci 82:1–12. https://doi.org/10.1007/s00027-020-00742-y

Pierson JJ, Slater WCL, Elliott D, Roman MR (2017) Synergistic effects of seasonal deoxygenation and temperature truncate copepod vertical migration and distribution. Mar Ecol Prog Ser 575: 57–68. https://doi.org/10.3354/meps12205

Pilson MEQ (2013) An introduction to the chemistry of the sea, 2nd edn. Cambridge Univ Press, New York. 524 p

Pitcher GC, Aguirre-Velarde A, Breitburg D, Cardich J, Carstensen J, Conley DC, Dewitte B et al (2021) System controls of coastal and open ocean oxygen depletion. Prog Oceanogr 197: 102613. https://doi.org/10.1016/j.pocean.2021.102613

Pomeroy R (2021) How the river Thames once became a cesspool in the name of public health. Real Clear Sci. January 18, 2021

Rabalais NN (2022) 2022 Shelfwide cruise, July 25–August 1, 2002. https://gulfhypoxia.net/research/shelfwide-cruise/?y=2022

Rabalais NN, Turner RE (2019) Gulf of Mexico hypoxia: past, present and future. Limnol Oceanogr Bull 28:117–124

Rabalais NN, Turner RE, Wiseman WJ Jr (2002) Hypoxia in the Gulf of Mexico, a.k.a. "The Dead Zone". Annu Rev Ecol Syst 33:235–263

Rabalais NN, Turner RE, Sen Gupta BK, Boesch DF, Chapman P, Murrell MC (2007) Hypoxia in the northern Gulf of Mexico: does the science support the plan to reduce, mitigate, and control hypoxia? Est Coasts 30:753–772

Rabalais NN, Díaz RJ, Levin LA, Turner RE, Gilbert D, Zhang J (2010) Dynamics and distribution of natural and human-caused coastal hypoxia. Biogeosciences 7:585–619. https://www.biogeosciences.net/7/585/2010/

Roman MR, Gauzens AL, Rhinehart WK, White JR (1993) Effects of low oxygen waters on Chesapeake Bay zooplankton. Limnol Oceanogr 38:1603–1614. https://doi.org/10.4319/lo.1993.38.8.1603

Roman MR, Pierson JJ, Kimmel DG, Boicourt WC, Zhang X (2012) Impacts of hypoxia on zooplankton spatial distributions in the northern Gulf of Mexico. Estuar Coasts 35:1261–1269. https://doi.org/10.1007/s12237-012-9531-x

Roman MR, Brandt SB, Houde ED, Pierson JJ (2019) Interactive effects of hypoxia and temperature on coastal pelagic zooplankton and fish. Front Mar Sci 6. https://doi.org/10.3389/fmars.2019.00139

Scavia D, Bertani I, Testa JM, Bever AJ, Blomquist JD, Friedrichs MAM, Linker LC et al (2021) Advancing estuarine ecological forecasts: seasonal hypoxia in Chesapeake Bay. Ecol Appl 31(6):e02384. https://doi.org/10.1002/eap.2384

Schmidtko S, Stranna L, Visbeck M (2017) Decline in global oceanic oxygen content during the past five decades. Nature 542:335–339. https://doi.org/10.1038/nature21399

Schubel JR (1977) The living Chesapeake. The Johns Hopkins University Press, Baltimore, MD. 112 p

Slater WL, Pierson JJ, Decker MB, Houde ED, Lozano C, Seuberling J (2020) Fewer copepods, fewer anchovies, and more jellyfish: how does hypoxia impact the Chesapeake Bay zooplankton community? Diversity 12:35. https://doi.org/10.3390/d12010035

Stadler LC, Marcus NH (1997) Zooplankton responses to hypoxia: behavioral patterns and survival of three species of calanoid copepods. Mar Biol 127:599–607

Stauffer BA, Schnetzer A, Gellene AG, Oberg C, Sukhatme GS, Caron DA (2013) Effects of an acute hypoxic event on microplankton community structure in a coastal harbor of southern California. Estuar Coasts 36:135–148. https://doi.org/10.1007/s12237-012-9551-6

Stramma L, Schmidtko S, Levin LA, Johnson GC (2010) Expanding oxygen-minimum zones in the tropcal oceans. Science 320:655–658. https://doi.org/10.1126/science.1153847

Sun JZ, Wang T, Huang R, Yi X, Zhang D, Beardall J, Hutchins DA, Liu X, Wang X, Deng Z, Li G, Gao G, Gao K (2022) Enhancement of diatom growth and phytoplankton productivity with reduced O_2 availability is moderated by rising CO_2. Commun Biol 5:54. https://doi.org/10.1038/s42003-022-03006-7

Taylor JC, Rand PS (2003) Spatial overlap and distribution of anchovies (*Anchoa* spp.) and copepods in a shallow stratified estuary. Acoust Fish Aquat Ecol 16:191–196. https://doi.org/10.1016/S0990-7440(03)00012-3

Testa JM, Clark JB, Dennison WC, Donovan CE, Fisher AW, Ni W, Parker M et al (2017) Ecological forecasting and the science of hypoxia in Chesapeake Bay. Bioscience 67:614–626. https://doi.org/10.1093/biosci/bix048

Testa JM, Kemp WM, Boynton WR (2018a) Season-specific trends and linkages of nitrogen and oxygen cycles in Chesapeake Bay. Limnol Oceanogr 63(5):2045–2064

Testa JM, Murphy RR, Brady DC, Kemp WM (2018b) Nutrient- and climate-induced shifts in the phenology of linked biogeochemical cycles in a temperate estuary. Front Mar Sci 5:114. https://doi.org/10.3389/fmars.2018.00114

Turley BD, Karnauskas M, Campbell MD, Hanisko DS, Kelble CR (2022) Relationships between blooms of *Karenia brevis* and hypoxia across the West Florida Shelf. Harmful Algae 114: 102223. https://doi.org/10.1016/j.hal.2022.102223

Turner RE, Rabalais NN (1991) Changes in Mississippi River water quality this century and implications for coastal food webs. Bioscience 41:140–147. https://doi.org/10.2307/1311453

Vaquer-Sunyer R, Duarte CM (2008) Thresholds of hypoxia for marine biodiversity. Proc Natl Acad Sci U S A 105:15452–15457

Vargo GA, Carder KL, Gregg W, Shanley E, Heil C, Steidinger KA, Haddad KD (1987) The potential contribution of primary production by red tides to the west Florida shelf ecosystem 1. Limnol Oceanogr 32:762–767

Vargo GA, Heil CA, Fanning KA, Dixon LK, Neely MB, Lester KA, Ault D et al (2008) Nutrient availability in support of *Karenia brevis* blooms on the central West Florida Shelf: what keeps *Karenia* blooming? Cont Shelf Res 28:73–98

Wang B, Chen J, Jin H, Li H, Huang D, Cai W-J (2017) Diatom bloom-derived bottom water hypoxia off the Changjiang estuary, with and without typhoon influence. Limnol Oceanogr 62: 1552–1569

Wang J, Bouwman AF, Liu X, Beusen AH, Van Dingenen R, Dentener F, Yao Y et al (2021) Harmful algal blooms in Chinese coastal waters will persist due to perturbed nutrient ratios. Environ Sci Tech Letts 8:276–284

World Bank (2021) Sanitation. https://wwww.worldbank.org/en/topic/sanitation. Accessed 13 Dec 2022

Wu RSS (2009) Effects of hypoxia on fish reproduction and development. Fish Physiol 27:79–141

Zhou MJ, Shen ZI, Yu RC (2008) Responses of a coastal phytoplankton community to increased nutrient input from the Changjiang (Yangtze) river. Cont Shelf Res 28:1483–1489

Zhu Z-Y, Zhang J, Wu Y, Zhang Y-Y, Lin J, Liu S-M (2011) Hypoxia off the Changjiang (Yangtze River) estuary; oxygen depletion and organic matter decomposition. Mar Chem 125:108–116

Zillén L, Conley DJ, Andrén T, Andrén E, Björck S (2008) Past occurrences of hypoxia in the Baltic Sea and the role of climate variability, environmental change and human impact. Earth Sci Rev 91:77–92

Additional Websites Accessed

https://eyesonthebay.dnr.maryland.gov. Accessed May 3, 2023

https://ian.umces.edu/media-library/hypoxia/. Accessed November 27, 2022

https://news.maryland.gov/dnr/2021/11/30/chesapeake-bay-hypoxia-report-year-end-2022/. Accessed June 23, 2023

https://www.esa.int/ESA_Multimedia/Images/2019/12/Baltic_blooms. Accessed October 3, 2022

Harmful Algae and the Blooming of Our Waters

25

25.1 Introduction

In 1525, Juan Sebastián Elcano was the first sailor to circumnavigate the globe, having seven ships in his fleet. During a stop in the Gulf of Guinea, the officers were served a large barracuda, after which they all suffered from diarrhea and lost consciousness (Urdaneta 1580, quoted in GEOHAB 2012).

In 1770, Captain James Cook, in his exploration of the coast of Australia, wrote, "There appeared in the water a kind of brown scum."

In 1793, a member of Captain George Vancouver's exploration team died and five crew members were sickened after eating contaminated mussels harvested in the uncharted coastline of what is now known as British Columbia.

In 1799, 100 members of a Russian expedition near Sitka, Alaska, died after consuming shellfish (Feder 1975).

In 1878, livestock in Australia were found to be poisoned after drinking from a brackish lake that had "thick scum like green oil paint," and their deaths were described as, "stupor and unconsciousness, falling and remaining quiet, as if asleep, unless touched when convulsions came on, with head and neck drawn back in rigid spasm, which subsides before death" (Francis 1878). The poison was later identified as nodularin toxins from cyanobacteria.

In 1899, a report was presented to the General Assembly of the state of Rhode Island, USA, detailing a bloom (a "plague") that caused major fish and shellfish deaths (Mead 1899; Fig. 25.1; Box 25.1).

All of these historical events, as well as the Biblical reference of waters turning to blood (Exodus—7:20,21), are now known to be associated with the natural phenomenon of blooms of toxic or harmful algae. The first report above was likely the first recorded incidence of a benthic HAB poisoning event (GEOHAB 2012). All of these observations show that harmful blooms are naturally occurring events or at least are events that have been occurring for a very long time. The fact that they have occurred throughout history does not negate the fact that their frequency and geographic extent is changing. Years ago, there was debate as to whether the increase in HABs that was

P. M. Glibert, *Phytoplankton Whispering: An Introduction to the Physiology and Ecology of Microalgae*, https://doi.org/10.1007/978-3-031-53897-1_25

State of Rhode Island and Providence Plantations.

TWENTY-NINTH ANNUAL REPORT

OF THE

COMMISSIONERS OF INLAND FISHERIES,

MADE TO THE

GENERAL ASSEMBLY

AT ITS

JANUARY SESSION, 1899.

PROVIDENCE, R. I.

E. L. FREEMAN & SONS, PRINTERS TO THE STATE.

1899.

Fig. 25.1 Cover of the Report to the Commissioners of the General Assembly of the State of Rhode Island with the description of the 1898 harmful algal bloom written by A.D. Mead

being documented was in fact real or whether increased monitoring or surveillance led to a perception of an increase in HAB. Now there is compelling evidence from many regions showing conclusively that increases in HAB proliferations are real, not sampling artifacts (Heisler et al. 2008; Hallegraeff et al. 2021).

This chapter reviews blooms and their harmful and toxic properties and effects on food webs across different habitats, why their distribution has changed and is changing, and several biomes in the ocean where HABs are commonly observed.

Box 25.1 Report of "The Investigation of the plague which destroyed multitudies of fish and crustatea during the fall of 1898" by A.D. Mead (1899)

During the last 2 months, the inhabitants of Rhode Island witnessed the following remarkable phenomenon. The water of a considerable portion of the bay became thick and red, emitting an odor almost intolerable to those living nearby. The situation became alarming when, on the 9th and 10th of September [1898], thousands of dead fish, crabs, and shrimps were found strewn along the shores or even piled up in windrows.

At the request of the Rhode Island Commission of Inland Fisheries, an investigation was made to determine the cause and extent of the unusual color of the water and of the great mortality of the fish. The results of this investigation are briefly as follows:

During the last of August, throughout September and a part of October, streaks of red or "chocolate" water were observed from near Quonset Point and Prudence Island, north to Providence, and, on the flood tide, up the Seekonk river, nearby to Pawtucket, a range of about 15 miles. In other parts of the bay, as far as could be learned, the phenomenon had not been observed.

On the 8th and 9th of September, the water became extremely red and thick in various localities from East Greenwich to Providence, and the peculiar

(continued)

Box 25.1 (continued)

behavior of the marine animals attracted much attention. Myriads of shrimps and blue crabs, and vast numbers of eels, menhaden, tautog, and flat-fish, came up to the surface and to the edge of the shore as though struggling to get out of the noxious water. Indeed, the shrimp and crabs were observed actually to climb out of the water upon stakes and buoys, and even upon the iron cylinders which support one of the bridges and which must have been very hot in the bright sun. In several instances, on these 2 days, hundreds of blue crabs were caught by a single individual in a few minutes' time, at the mouth of the Seekonk.

On the following day, September 10th, and for several days afterwards, hardly a live crab or shrimp could be found. Along the shores, however, in the same vicinity, cartloads of dead shrimp were piled up in windrows, and among them were strewn great numbers of crabs and fish of various kinds, especially menhaden and eels. This singular behavior and alarming mortality of marine animals was reported from nearly every station at which the red water occurred, and from no other station, which indicates that the two phenomena are related as cause and effect.

It was commonly believed that dye-stuffs or other refuse emptied into the rivers at the upper part of the bay gave the water its color and unpleasant odor, but microscopic examination showed that the water was swarming with minute organisms, species of *Peridinium*. The *Peridinium* is reddish brown in color and occurred in such excessive abundance that it gave to the water its peculiar color and odor, besides making it so opaque that one could hardly see a white shell 6 in. below the surface.

With regard to the systematic position of this organism, there is a difference of opinion. It is, in fact, ranked with the animals by some authors, and with the plants by others. … Besides a flagellum extending forward from the central groove, a very large flagellum lies in the equatorial sulcus and entirely encircles the body. No cilia could be demonstrated.

…

In the Seekonk River the shrimp and crabs gradually returned, and in about 3 weeks after the sudden mortality was nearly as numerous as before, though the water was at times distinctly colored. On the 23rd some shrimp, oysters, and small fish (*Fundulus*) were kept in the water where the *Peridinium* were the thickest and suffered no apparent injury. In consideration of these facts, it has been doubted whether the *Peridinium* was the immediate cause of the peculiar behavior and death of the fish which occurred on the 9th and 10th of September, especially as the weather had been phenomenally hot for several weeks previous to those dates. I believe, however, that the *Peridinium* was the cause of the trouble, and not the hot weather or manufacturers' waste for the following reasons:

(continued)

Box 25.1 (continued)

On the 2 or 3 days in which the mortality took place, the water was extremely red.

The hot weather was followed by a cold wave a day or two before the mortality commenced.

The phenomenon occurred in Greenwich Bay and off Nayatt, many miles from any considerable source of contamination.

Finally, the phenomena in question were noticed by very many persons throughout the whole range of the red water, while in the neighboring portions of the bay, for example, in the Warren River and Bristol harbor, where the temperature of the water is quite as high as in the red-water districts, no *Peridinium* and no mortality or unusual behavior of the marine animals was reported, though the regions were carefully canvassed.

25.2 What Is a Bloom? What Is a HAB?

Defining a bloom is not straightforward. It is even more difficult to define a HAB. The term HAB actually has no specific definition; it is an operational term only (Glibert et al. 2018; Glibert and Li 2024). In fact, the imprecise nature of the term is purposeful, representing a concept that embraces many types of species, many types of harm, and many types of blooms. Some HAB taxa are technically not algae at all (a theme that has been repeated throughout this volume with regard to the use of the term "algae"). For example, the cyanobacterial HABs, or CyanoHABs, most common in freshwaters, are technically not "algae." Also, many dinoflagellate HABs are mixoplankton and do not depend (in part or in entirety) on photosynthesis for their nutrition; they can obtain nutrition through grazing and particle ingestion (Chap. 15). The term "HAB" is also used to refer to blooms of some nontoxic micro- or macroalgae (seaweeds) that can grow out of control and cause major ecological impacts such as the displacement of indigenous species, habitat alteration, or oxygen depletion. The term HAB can also refer to those benthic species that may be toxic or harmful.

With the frequency of blooms beginning to gain attention, Ted Smayda (1997, p. 1135; who was introduced in Chap. 17) stated that, "What constitutes a bloom...has regional, seasonal, and species-specific aspects; it is not simply a biomass issue...The salient criterion to use in defining whether a "harmful" species is in bloom and the distinctive feature of such blooms lie not in the level of abundance, but whether its occurrence has harmful consequences." Since the publication of that paper, biomass criteria for a few HAB species have been defined, but more generally HABs continue to be defined in terms of the extent to which they cause harmful events (fish kills), toxic events (shellfish poisoning), ecosystem disruption (nutritional and/or prey-size mismatches, such as picocyanobacterial blooms), or large biomass events (hypoxia or anoxia). A few regions/states or

Fig. 25.2 Many HABs are produced by the accumulation of vividly colored cells on surface water. These high biomass blooms can cause hypoxia, can contribute to toxicity of fish and shellfish, kill fish directly or indirectly, and can cause other environmental problems. Panels (**a**, **b**) a *Karenia brevis* bloom and its associated fish kill in Florida, and Panel (**c**) a beach sign warning of the same *K. brevis* red tide; Panel (**d**) a *Microcystis* bloom in a freshwater lake, and Panel (**e**), a HAB bloom off San Diego, California. Photos (**a**–**d**) by P. Glibert, photo (**e**) by Melvil, reproduced from https://en.wikipedia.org/wiki/Harmful_algal_bloom#/media/File:La-Jolla-Red-Tide.780.jpg, under a creative commons license

countries have defined blooms in legal frameworks, requiring specific fisheries to impose testing or closures should a specific cell concentration be reached in the water column. Mostly the term is applied when there is evidence of harm; however an investigator or management agency chooses to define it (Glibert and Burkholder 2018).

In all cases, the HAB species must be present regardless of their biomass relative to other species in the assemblage. The HAB organism does not need to be dominant or in high abundance to elicit harmful effects, although high biomass HABs are the most easily recognized (Fig. 25.2). On the other hand, HAB taxa can be present without causing harm, and not all HAB taxa are harmful if introduced to new environments. In a later paper on the global expansion of HABs, Smayda (2002, p. 292) wrote:

> Anthropogenic seedings are not, in themselves, bloom stimulation events; they are only the first phase of a multi-phase process. A newly vectored, non-indigenous species is initially pioneering: it must either find an open niche or displace a niche occupant as its first step towards successful accommodation within the community... Until colonization is achieved, alien species introduced into water masses that have been modified by cultural nutrient enrichment, water mass conditioning by aquaculture, or climatological disturbances, will not bloom. Successful colonization alone is not decisive, it usually must be accompanied at some point, or coincide with habitat disturbance – a pre-condition for many HAB occurrences.

25.3 How Are HABs Harmful?

HABs can be harmful in several fundamental ways. Many (although not all) HABs produce toxins, some of which are among the most potent toxins known. Toxins vary by HAB taxa, and they can also vary within a species, being present under some conditions and not present under other conditions (Glibert et al. 2018). Even within a class of toxins, multiple congeners may occur. HAB toxins may be cytotoxic, genotoxic, mutagenic, pathogenic, or immunosuppressive (Burkholder et al. 2018).

The major marine HAB toxins that have human or marine mammal health effects include saxitoxin, the cause of paralytic shellfish poisoning (PSP); brevetoxin, the cause of neurotoxic shellfish poisoning (NSP); okadaic acid, the cause of diarrhetic shellfish poisoning (DSP); domoic acid, the cause of amnesic shellfish poisoning (ASP); ciguatoxins, the cause of ciguatera fish poisoning (CGP); and cyanotoxins, which can cause skin rashes and irritation and with substantial exposure, cancers (Table 25.1). Other toxins, such as azaspiracid, the cause of azaspiracid shellfish poisoning (AZP), and karlotoxins, derived from *Karlodinium* sp., among others, are becoming better understood, especially with regard to their ecosystem effects, which are less well recognized. The major toxins and their effects are next reviewed. Other emerging toxins such as the spirolides, yessotoxins, pectenotoxins, and palytoxins have been summarized by the US National HAB Office (https://hab.whoi.edu/impacts/impacts-human-health/emerging-toxins/).

Table 25.1 Major poisonings associated with harmful algae, the vectors, and human health symptoms. Reproduced and modified from Glibert et al. (2018) with permission from Springer Nature

Syndrome or illness	Major vector	Symptoms
Paralytic shellfish poisoning (PSP)	Saxitoxin from *Alexandrium* spp. and other species in shellfish	Numbness around lips and mouth Respiratory paralysis Death
Neurotoxic shellfish poisoning (NSP)	Brevetoxin from *Karenia* spp. in shellfish, aerosolized toxins	Nausea, diarrhea, respiratory distress Eye irritation
Diarrhetic shellfish poisoning (DSP)	Okadaic acid from *Dinophysis* spp. in shellfish	Diarrhea, vomiting, cramping
Amnesic shellfish poisoning (ASP)	Domoic acid from *Pseudo-nitzschia* sp. in shellfish	Short-term memory loss; vomiting, cramps
Ciguatoxin poisoning	Ciguatoxin from *Gambierdiscus* in fish	Nausea, vomiting, and neurologic symptoms including reversal of sensation of hot and cold
Cyanotoxin poisoning	Microcystins and other toxins from cyanobacteria in water	Skin irritation; respiratory irritation; tumor-promotion; liver cancer, failure

25.3.1 Saxitoxins

Saxitoxins were among the first HAB toxins to be identified. They occur in both dinoflagellates and in some cyanobacteria, and all are low molecular weight, heat-stable, water-soluble hydrophilic tetrahydropurine neurotoxins (Steidinger et al. 2008; Fig. 25.3). Among the dinoflagellates, the three most common saxitoxin-producing genera are *Alexandrium*, *Gymnodinium*, and *Pyrodinium*, although different species may produce different congeners (Landsberg 2002; Broadwater et al. 2018). Some cyanobacteria also produce saxitoxins. Over 20 congeners have been identified, with varying toxin potencies for mammals (Landsberg 2002). When people consume shellfish that have accumulated these toxins, they may experience tingling around the lips and mouth, numbness of extremities, and loss of motor control. The action of the toxin is similar to that of brevetoxin (see Sect. 25.3.2), in that these toxins bond to sodium channels blocking the passage of nerve impulses. With a large enough dose of these toxins, respiratory paralysis—and death—may occur (Cusick and Sayler 2013; Broadwater et al. 2018). It is this toxin that killed the early explorers of Alaska mentioned in the introduction to this chapter. Due to the water-soluble nature of the toxins, they are readily absorbed across the gastrointestinal tract, and symptom onset can be very rapid after ingestion (Baden et al. 1995; Steidinger et al. 2008). Marine mammals and sea birds are also affected by these toxins. These toxins have been implicated in the mass mortality of humpback whales and in reproductive failures of right whales. Chronic exposure by whales may occur when they ingest copepods that have been feeding on *Alexandrium*, and this appears to be related to decreased feeding, and a reduction in calving of the exposed whales (Durbin et al. 2002; Steidinger et al. 2008). Bottlenose dolphins are also suspected to be impacted by saxitoxin, and a mass mortality event in Florida was ascribed to this toxin, as it co-occurred with an outbreak of another saxitoxin-producing dinoflagellate, *Pyrodinium bahamense* var. *bahamense* (Fire and Van Dolah 2012; Broadwater et al. 2018). A recent massive whale mortality event in Chile in 2015, including at least 340 primarily sei whales, has been linked to

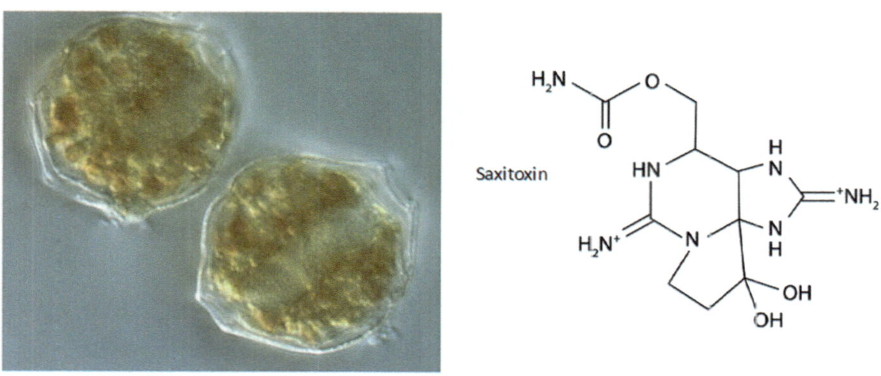

Fig. 25.3 *Alexandrium tamarense* and the structure of its toxin, saxitoxin. Photo by Nancy Lewis, National Research Council of Canada, reproduced under a creative commons license. Chemical structure after Graham et al. (2016, Algae. LJLM)

PSP toxicity from feeding near shore (Häussermann et al. 2017). It is important to note that these marine mortality events are often difficult to directly relate to toxin exposure—even when toxins are found in the stomach or fecal matter, as co-occurring factors and overall health status may increase susceptibility to toxin effects or other diseases. Birds are also affected by saxitoxin poisonings (Landsberg et al. 2007; Shumway et al. 2003; Steidinger et al. 2008).

25.3.2 Brevetoxins

Brevetoxins are primarily produced by *Karenia brevis*, found virtually exclusively in the Gulf of Mexico, although these toxins may be produced by a few other taxa (Fig. 25.4). Brevetoxin has a ladder-like polycyclic ether structure, which has a high affinity to voltage-gated sodium channels, causing the channel to remain open when it normally would be closed, thus causing inappropriate neuronal transmission (Ramsdell and Zabka 2008; Broadwater et al. 2018). To date, as many as 14 different brevetoxins have been identified from *K. brevis* cultures, with varying potencies (Baden et al. 2005; Steidinger et al. 2008). Human exposure to brevetoxins occurs via two main pathways. First, it may be consumed when people eat shellfish. Ingestion of brevetoxins causes nausea, lip tingling sensations, loss of muscle control, and in extreme cases, seizures (Poli et al. 2000; Broadwater et al. 2018).

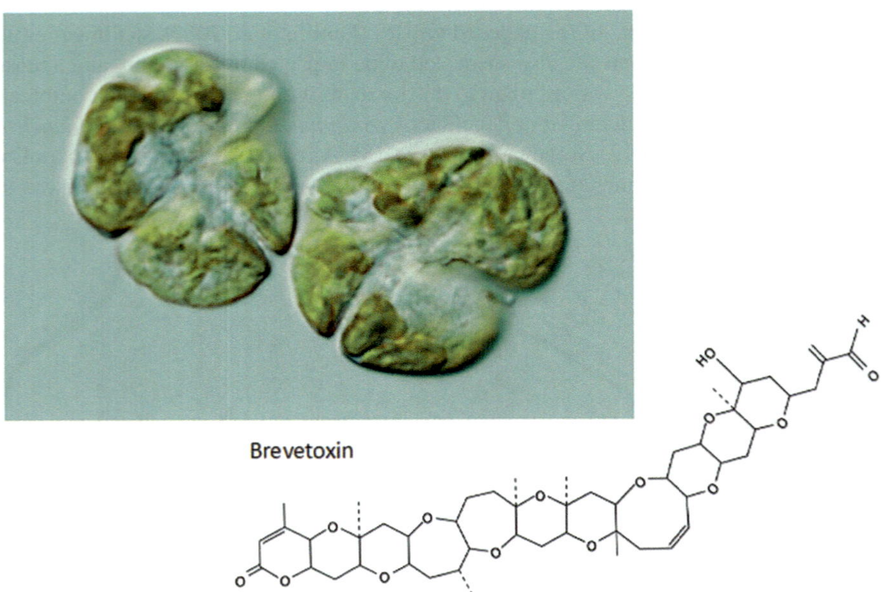

Fig. 25.4 *Karenia brevis* and the structure of its toxin, brevetoxin. Photo by Florida Fish and Wildlife Research Institute (https://www.flickr.com/photos/myfwc/5807660937), reproduced under a creative commons license. Chemical structure after Graham et al. (2016, Algae. LJLM)

To protect human health, the State of Florida historically set 5000 cells L^{-1} of *Karenia brevis* as the threshold for shellfish closures. Repeated negative cell counts must be documented before shellfish beds are legally allowed to reopen for commercial harvest. Marine mammals are exposed to brevetoxins when they consume food that have incorporated the toxins. They may show signs of disorientation, seizures, and other effects (Bossert 2011) and during the most severe blooms, hundreds of manatees and bottlenose dolphins may be poisoned (Fire et al. 2007, 2020; Landsberg et al. 2009; Twiner et al. 2012; Broadwater et al. 2018). Manatee deaths may even occur weeks after the *Karenia* cells have diminished in the plankton, as the manatees may ingest seagrasses with attached filter feeding organisms that themselves may be contaminated with brevetoxins (Flewelling et al. 2005; Broadwater et al. 2018).

Toxin exposure may also occur via inhalation when brevetoxins are carried by sea spray. Sea spray can become laced with toxins when cells are lysed by breaking waves or wind action, and the brevetoxins are released, becoming attached to the salt particles of the aerosols. Then, they are breathed in by those near to the shore. When brevetoxins are inhaled, people can experience coughing, gagging, or burning sensation in the upper respiratory track—symptoms that can last for days following exposure (Backer et al. 2003; Steidinger et al. 2008; Kirkpatrick et al. 2011). These symptoms are induced at brevetoxin concentrations several orders of magnitude lower than those that induce illness from ingestion. These symptoms are especially problematic for those who are prone to asthma or other respiratory complications.

25.3.3 Okadaic Acid

Okadaic acid and its derivatives are primarily produced by the very different dinoflagellates *Dinophysis* and *Prorocentrum* sp. (Fig. 25.5). These *Dinophysis* toxins are polyether compounds that can cause vomiting, nausea, and diarrhea

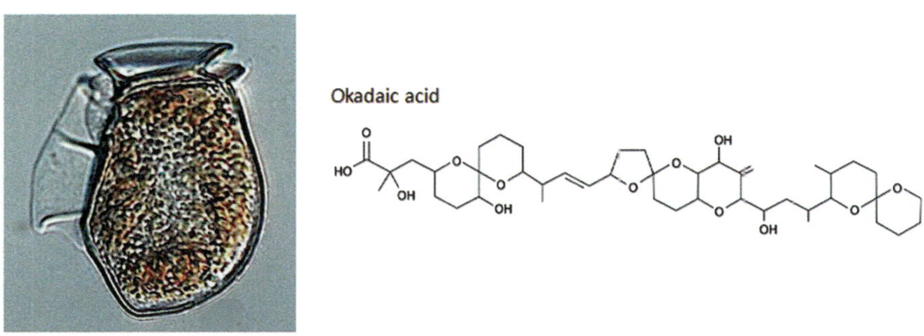

Fig. 25.5 *Dinophysis* and the structure of its toxin, okadaic acid. Photo by Alexandra Kraberg and reproduced from Wikipedia (https://en.wikipedia.org/wiki/Dinophysis_acuta#/media/File:Dinophysis_acuta.jpg) under a creative commons license. Chemical structure after Graham et al. (2016, Algae. LJLM)

lasting up to several days when shellfish is consumed (Fire and Van Dolah 2012; Broadwater et al. 2018). Symptoms are usually relatively short-lived, and not life-threatening, although there is some evidence that okadaic acid may be a powerful tumor promoter if exposure is long term (Backer et al. 2005). The mode of toxin action is via the inhibition of phosphatase enzymes that help to regulate many important cell processes such as the cell cycle and transcription of oncogenes (Dounay and Forsyth 2002). These same phosphatases are involved in intracellular signaling mechanisms that can be altered in tumor formation, and correlations between okadaic acid exposure and tumor-promoting effects on rats and on turtles have been demonstrated (e.g., Fujiki et al. 1988; Landsberg et al. 1999; Steidinger et al. 2008). Given the broad distribution of these dinoflagellates globally, these effects are well known and present threats to shellfish-growing regions or aquaculture sites. Shellfish are well monitored in most regions, but closures based on cell counts differ regionally and with specific species of *Dinophysis*. Of the *Prorocentrum* species that produce this toxin, *P. lima* is most common. This is a benthic species that can attach to both living and dead substrates, making it particularly problematic in aquaculture region (Van Dolah 2000; Steidinger et al. 2008). A number of other toxins may co-occur with okadaic acid. These include pectenotoxins, made by the same organisms, and yessotoxins, made by different dinoflagellates, but also cyclic polyether compounds.

25.3.4 Domoic Acid

Domoic acid is an unusual marine HAB toxin, as it is not derived from a dinoflagellate, but instead from several species of the diatom of the genus *Pseudo-nitzschia* (Fig. 25.6). It was first identified in eastern Canada in 1987 when several hundred

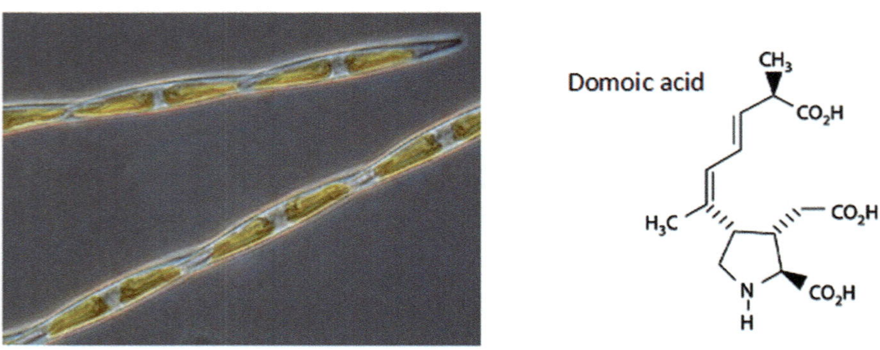

Fig. 25.6 *Pseudo-nitzschia* and its toxin, domoic acid. Photo by Rozalind Jester and reproduced from Wikipedia under a creative commons license (https://en.wikipedia.org/wiki/Pseudo-nitzschia#/media/File:Pseudonitzschia2.jpg). Chemical structure after Graham et al. (2016, Algae. LJLM)

people became ill, and several died, all after consuming mussels. Of those affected, most had abdominal cramps and experienced vomiting, diarrhea, and loss of short-term memory, but some required intensive care because of seizures and disorientation (Bates et al. 1989; Trainer et al. 2008). Because of the memory loss issues, the syndrome was termed amnesic shellfish poisoning. Before its identification in *Pseudo-nitzschia* was confirmed, it had been known to be produced in red macroalgae, *Chondria armata*, which in Japan was noted to have insecticidal properties (Trainer et al. 2008). Domoic acid is a tricarboxylic amino acid, similar to the neurotransmitter glutamine, and therefore competes for receptor sites in nerve cells (Trainer et al. 2008; Broadwater et al. 2018; Fig. 25.6). In the brain, it concentrates in regions of the hippocampus that are responsible for learning and memory (Van Dolah et al. 2003). *Pseudo-nitzschia* species are now recognized to be widely distributed around the world, mostly in coastal regions associated with high nutrient concentrations (Trainer et al. 2003, Trainer et al. 2008; Parsons et al. 2013).

Poisoning events of marine mammals by domoic acid have been well documented and are being detected in more locations. The first major event conclusively documenting a domoic acid marine mortality event involved over 400 California sea lions in the late 1990s, although substantial evidence suggests that similar events involving sea lions and fur seals occurred decades prior (Scholin et al. 2000; Broadwater et al. 2018). The animals experienced disorientation and seizures, among other symptoms—indicative of neurological dysfunction (Gulland 2000; Broadwater et al. 2018)—and domoic acid was detected in anchovies, one of their foods, as well as in the feces, concurrent with a *Pseudo-nitzschia* bloom. Sea lion pups also had a decreased survival rate. In addition to acute exposure, it is now known that these marine mammals may experience chronic domoic acid epileptic disease that can follow sublethal exposures over a period of time (Ramsdell and Gulland 2014; Broadwater et al. 2018). These animals may subsequently become less efficient predators, as spatial memory is relied upon for forage (Cook et al. 2016; Broadwater et al. 2018). Using preserved samples, it was also shown that domoic acid caused disoriented and agitated birds in the 1960s in California (Box 25.2, Fig. 25.7). Domoic acid poisonings of marine mammals, from sea lions to otters to bottlenose dolphin and right whales, have now been documented off the coast of Peru, in the Gulf of Mexico, and elsewhere. Blooms of the associated toxic diatoms appear to occur more frequently on eastern continental boundaries associated with upwelling (Trainer et al. 2008).

Fig. 25.7 Theater release poster for Alfred Hitchcock's "The Birds." Image reproduced from Wikipedia (https://upload.wikimedia.org/wikipedia/commons/c/c0/The_Birds_original_poster.jpg), public domain

Box 25.2 "The Birds"

Alfred Hitchcock produced the famous thriller *The Birds* based on a real event in the 1960s when "crazed seabirds pelted the shores of North Monterey Bay, California, regurgitating anchovies" (from a California newspaper, cited in Bargu et al. 2012). When another such event in the same area occurred decades later, involving disoriented pelicans, the cause was identified: domoic acid poisoning from a *Pseudo-nitzschia* bloom. Archived samples of zooplankton from the 1960s were subsequently retrieved and their gut contents identified (Bargu et al. 2012). These toxic diatoms constituted nearly 80% of the gut flora, leading to the conclusion that domoic acid poisoning was the likely cause of the crazed Sooty shearwater birds that inspired Alfred Hitchcock (Fig. 25.7).

25.3.5 Ciguatoxins

One of the more complex, but under-reported and under-appreciated, toxins is ciguatoxin, a polyether toxin from the dinoflagellate genus *Gambierdiscus* (Fig. 25.8). The toxin is similar to brevetoxin in structure but more potent and the associated symptoms often last much longer (Fire and Van Dolah 2012; Broadwater et al. 2018). It affects thousands of people worldwide annually – a number that was estimated in the 1970s to be about 5 times more than the reported incidences of PSP and brevetoxin poisoning combined (Steidinger and Baden 1985). It is the leading cause of non-bacterial food poisoning in the USA and the most frequent seafood-related illness worldwide (Arena et al. 2004; Bienfang et al. 2008). Different from the other major marine dinoflagellate toxins which are acquired after eating shellfish, ciguatoxins are associated with consumption of reef fish, such as barracuda or grouper, that themselves have consumed the dinoflagellate by way of consumption of herbivorous fish. Thus, the processes of bioaccumulation and biomagnification

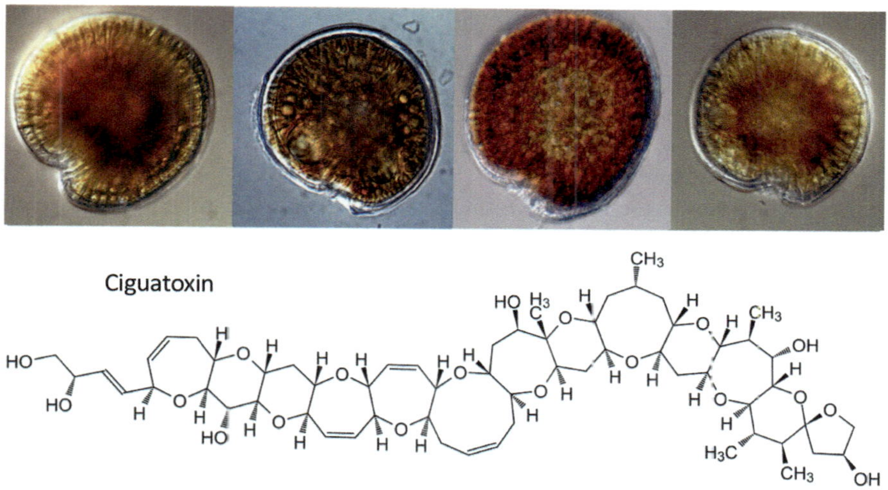

Fig. 25.8 *Gambierdiscus* and its toxin, ciguatoxin. Upper panels represent four *Gambierdiscus* species, *G. caribaeus, G.carolinianus, G. carpenteri*, and *G. belizeanus*. Image from M. Vandersea, NOAA, under a creative commons license. Chemical structure after Graham et al. (2016, Algae. LJLM)

come into play before the toxic fish are eaten by humans (Bienfang et al. 2008). Although Captain Cook, in the quote at the beginning of this chapter, recognized floating scums of what is likely to have been cyanobacterium *Trichodesmium*, it is thought that his men also suffered ciguatera poisoning during their exploration of what is now New Caledonia (Helfrich 1964; Bienfang et al. 2008). Also, soldiers stationed in the Pacific during World War II frequently experienced ciguatera poisoning (Hokama and Yoshikawa-Ebesu 2001; Bienfang et al. 2008).

Ciguatoxin is a polar, lipid-soluble, highly oxygenated polyether compound (Fig. 25.8), and there are numerous congeners (Bienfang et al. 2008). Symptoms of ciguatera poisoning develop within 12–24 h after eating contaminated fish and involve gastrointestinal, neurological, and cardiovascular symptoms (Bienfang et al. 2008). The neurological symptoms include numbness and tingling of hands and feet, dizziness, and most interestingly, the reversal of sensations of hot and cold—symptoms that can last for weeks to even years depending on the dosage (Arena et al. 2004). Also interesting is the general pattern that gastrointestinal symptoms (diarrhea, cramping, vomiting) are more commonly in the Caribbean, while neurological symptoms appear to be more common in the Pacific (Nicholson and Lewis 2006; Bienfang et al. 2008). This may be due to different common strains of *Gambierdiscus* or different toxin composition between the locales. Repeated exposures to ciguatoxin appear to exacerbate the symptoms, as does exercise, ingestion of alcohol, and high protein foods. Toxicity of the *Gambierdiscus* cells appears to increase with temperature and light, being highest at mid to late exponential phase. In addition to effects on humans, ciguatoxins have effects on marine mammals, including the critically endangered Hawaiian monk seals (Broadwater et al. 2018).

Other benthic species are also toxic and common throughout the subtropics. *Ostreopsis* is common throughout the south Pacific, and more recently in waters of Europe, including the Mediterranean. These cells live on benthic rocks and in close contact with macroalgae and can form intense blooms as described further below (GEOHAB 2012).

25.3.6 Cyanotoxins

Cyanobacteria, due to their immense diversity, have been well studied with regard to their diversity of secondary natural products. Although their toxins are the topic of consideration herein, their natural products are used in pharmaceuticals, food, cosmetic, and agricultural industries. Cyanotoxins have caused acute intoxication events in animal and humans and have contaminated drinking water supplies. Some cyanotoxins are carcinogenic (Buratti et al. 2017).

There are three major groups of cyanobacterial toxins, the cyclic peptides (e.g., microcystins, nodularins, and anabaenopeptins), the alkaloids (e.g., anatoxin-a, saxitoxins), and the polyketide/polycyclic derivatives (e.g., cylindrospermopsins; Fig. 25.9). Most of the toxic cyanobacteria are freshwater in origin, although knowledge about marine cyanotoxins is rapidly advancing. Some species make

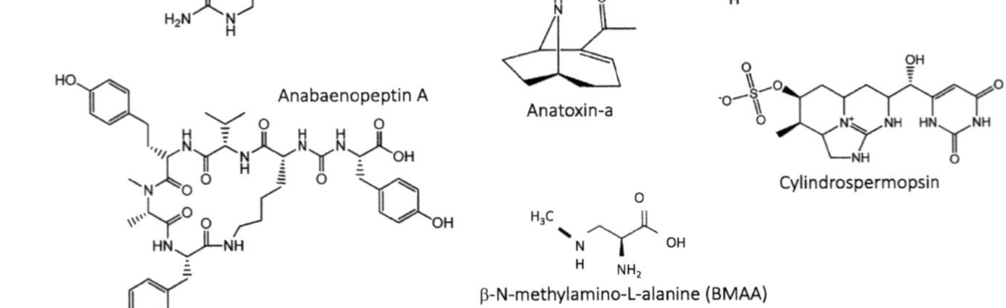

Fig. 25.9 Various cyanotoxins. Image modified from Roy-Lachapelle et al. (2019, Anal Meth 11: 5289–5300) under a creative commons license. BMAA structure reproduced from Wikipedia (https://en.wikipedia.org/wiki/Beta-Methylamino-L-alanine#/media/File:3-Methylamino-L-ala nine.svg), public domain

more than one toxin and different toxins or variants may be made by the same species when growing in different locales. For example, *Cylindrospermopsis raciborskii* appears to produce a neurotoxin when it is found in South American waters, a cytotoxin when it is found in North American or Australasian waters, and yet a third type of toxin when it is found in Europe (Stewart and Falconer 2008).

Among the cyclic peptide, the microcystins are the most well recognized, and among those, microcystin-LR is of the most concern (Vichi et al. 2016). The microcystins, nodularins, and anabaenopeptins are all families of compounds, with environmental conditions one of the important factors in regulating which variant is produced at any given time. Microcystins are most commonly known to be derived from cyanobacteria within the *Microcystis* genus, but they can also be made by species within the *Planktothrix*, *Anabaena, Nostoc*, and *Anabaenopsis* and *Snowella* genera (Sivonen and Jones 1999; Stewart and Falconer 2008). *Microcystis* is now known to bloom in every continent except Antarctica (Zurawell et al. 2005) and is associated with increasing eutrophication. A number of early human poisoning cases are well documented. One of the most well-known comes from the poisoning of over 130 patients at a Brazilian dialysis treatment facility and the death of nearly half of these people. The origin of the toxin came from a drinking water reservoir, and although the Brazilian clinic had ion exchange and activated carbon filters in its purification system, such treatment removed the cells but not the toxin. Microcystin was found in the liver and blood of the patients, leading to the conclusion that the injuries and deaths were caused by liver failure (Carmichael et al. 2001; Jochimsen et al. 1998; Stewart and Falconer 2008). The action of microcystins and nodularins in the liver cells is the protein phosphatase enzymes which normally regulate the amino acid constituents of particular proteins. When phosphate removal from the enzymes is inhibited by the action of the toxin, the phosphate builds up, in turn altering the structure of the liver and acting as a tumor promoter (Stewart and Falconer 2008). The toxicity of nodularins is similar to that of microcystin-LR, and more than seven variants of nodularins have been identified. The anabaenopeptins are quite diverse, with the majority of cyanobacteria able to biosynthesize anabaenopeptins belonging to genera such as *Anabaena*, *Microcystis*, *Nodularia*, *Oscillatoria*, and *Planktothrix*. Except for *Microcystis*, those genera are filamentous cyanobacteria belonging to the order Nostocales and Oscillatoriales. The variants of these toxins varies environmentally, likely a function of the different amino acids available to the cells. They have also been found to be inhibitory toward phosphatases and proteases (Monteiro et al. 2021).

Microcystins affect people by other routes in addition to ingestion of contaminated drinking water. They may also be consumed when seafood, including shellfish, fish, and even duck, is eaten. There is some contradiction in the literature as to the effect of cooking procedures in microcystin content. Zhang et al. (2010) found that microcystin levels in mussels increased substantially when the animals were cooked, while Guzman-Guillen et al. (2011) did not find such an effect.

The alkaloid toxins are themselves also quite diverse in structure and action, but all are neurotoxins. The smallest chemical alkaloid toxin is anatoxin-a (Fig. 25.9), first isolated from *Anabaena flos-aquae*. The structure is similar to that of cocaine, blocks synaptic transmission, and is stable in water for up to several weeks (Sivonen and Jones 1999; Stewart and Falconer 2008). It is produced by the genera *Anabaena*, *Phormidium*, *Raphidiopsis*, *Aphanizomenon*, and *Oscillatoria* (both planktonic and benthic strains). Saxitoxins, the same tricyclic alkaloids produced by some marine dinoflagellates that cause PSP (Fig. 25.3), are produced by a variety of cyanobacteria, including *Aphanizomenon flos-aquae*, *Anabaena circinalis*, *Cylindrospermopsis raciborskii*, and *Planktothrix*, among others (Stewart and Falconer 2008). These toxins produced by cyanobacteria have the same action as those produced by dinoflagellates: they block the movement of sodium ions through sodium channels in motor nerves and can cause acute respiratory paralysis (Pomati et al. 2006; Stewart and Falconer 2008).

The third group of alkaloid toxins are the cylindrospermopsins: polyketide alkaloids with a tricyclic guanido group with a sulfate and a methyl group attached (Stewart and Falconer 2008; Fig. 25.9). The molecule has high water solubility (Ohtani et al. 1992; Stewart and Falconer 2008). These toxins, like many of the other cyanotoxins, were identified from human poisoning events and traced to water reservoirs, from which the organisms were identified and tested for toxicity. Cylindrospermopsins are protein synthesis inhibitors, which tend to cause death through progressive organ failure, such as the liver, and thus they contrast in the mode of action with the neurotoxins (Stewart and Falconer 2008). As water-soluble toxins, they are not removed with the removal of cells from the water, and any chemical control—such as copper sulfate treatment—may increase the toxicity by breaking the cells.

There are a number of additional toxins identified in fresh and marine cyanobacteria about which knowledge is rapidly advancing (Rastogi et al. 2015; Fig. 25.9). Aplysiatoxin and lyngbyatoxin are potent tumor promotors that have been reported from cyanobacteria such as *Lyngbya majuscula* (note: this species has underdone taxonomic changes and several new genera have been elevated, now named *Dapis*), *Schizothrix calcicola*, *Trichodesmium erythraeum*, and *Oscillatoria nigroviridis* (Rastogi et al. 2015). These toxins may also induce acute inflammatory responses on the skin—the so-called seaweed dermatitis. The bioactive peptide β-*N*-methylamino-L-alanine (BMAA; Fig. 25.9) has also received considerable attention recently as it is suggested to be the causative agent for a number of neurological diseases (Lobner et al. 2007). The association of BMAA with neurodegenerative diseases was first made in the 1950s in several South Pacific islands where prevalence of disease was 2 orders of magnitude higher than that in continental US populations (Kurland and Mulder 1954). The source is thought to be exposure to cycad seed, which synthesizes the BMAA by symbiotic *Nostoc* and from which flour was eaten by the local populations. Work remains to be done to understand the full relationship between BMAA exposure and neurodegenerative diseases and cyanobacteria. The cyanotoxins antillatoxin, kalkitoxin, and jamaicamide are also neurotoxic compounds (Aráoz et al. 2010).

25.3.7 Indirect Harm

Many types of harmful effects occur from HABs that are not directly toxic. Many HABs are harmful simply through their high biomass accumulation that may lead to environmental damage, including hypoxia, anoxia, and shading of submerged vegetation, each of which, in turn, can lead to a multitude of negative environmental consequences. These are the blooms that cause visible water discoloration (Fig. 25.2). As discussed in Chap. 24, the global occurrences of such "dead zones" are expanding (Diaz and Rosenberg 2008; Breitburg et al. 2018), and nutrient regeneration from such events can promote (or sustain) the growth of more HABs.

Some HABs have physical structures, such as spines, that can lodge in fish gills and can cause irritation and eventual suffocation. Some HABs have more subtle effects on the ecosystem or effects that are more difficult to quantify in the wild. Blooms that are known to disrupt ecosystem function but which may or may not necessarily be toxic are considered ecosystem disruptive algal blooms (EDABs; Sunda et al. 2006). Sunda et al. (2006) note that these EDAB species are often unpalatable and thereby reduce grazing. Blooms of picoplankton, such as the pelagophyte *Aureococcus anophagefferens* (brown tide) and the CyanoHAB *Synechococcus* sp. that can be sustained for long periods of time once established, can have multiple negative ecosystem effects which, in turn, may help to sustain these blooms. The positive feedbacks of reduced grazing and/or bottom shading contribute to the availability of nutrients for these blooms (Sunda et al. 2006). In Laguna Madre, Texas, which experienced a nearly decade-long bloom of the EDAB species *Aureoumbra lagunensis*, the density of protozoan grazers was found to be greatly reduced during blooms, and it was suggested that a thick polysaccharide layer around the cells may make it difficult for the protozoa to feed (Buskey and Stockwell 1993; Buskey et al. 2001). Allelopathic chemicals may also play an important role in maintaining EDAB species (Sunda et al. 2006; Granéli et al. 2008).

In addition to ecological and human health impacts of HABs, the economic impacts can also be very large (Hoagland and Scatasta 2006). However, most countries have neither conducted economic analyses of HABs nor collected data that can be used to generate reliable quantitative estimates of net economic losses and impacts (Trainer et al. 2020a). The overall economic impact of HABs is difficult to determine, but the costs range from the direct expenses of public health and medical care for those affected to losses in commercial and recreational fishing and in tourism-related activities from water quality deterioration. Single HAB fish kill events in Korea and shellfish kill events in China and Japan have been estimated to have cost from \$1 to 300 million in lost fish or shellfish in aquaculture settings (GEOHAB 2010; Trainer and Yoshida 2014). Loss of revenue from shellfish harvesting area closures also takes enormous economic toll on communities affected. For example, in 2003, it was estimated that more than \$6 million in revenue from sport fishing and tourism due to fish kills were lost in Texas due to a bloom of a small flagellate, *Prymnesium parvum*, in inland rivers and reservoirs. Annual damages from HABs on Korean aquaculture facilities have been estimated to be in the range of \$4 million dollars and over \$16 million on the recreational fishery (Jin et al. 2020). Recent blooms in Florida may have cost over a billion dollars when loss of coastal tourism is considered (Alvarez et al. 2024). Yet another economic cost in some regions is the effect of HABs on seawater reverse osmosis (SWRO) desalination plants (e.g., Villacorte et al. 2015). In this case, the main concern is biofouling of the RO membranes, which can cause plants to shut down, resulting in a loss of an important utility service in regions where this is a major source of freshwater. In addition, where recurrent HABs are documented, another economic cost of HABs is that of sustained environmental monitoring to protect human health or industries such as aquaculture.

25.4 Food Web Effects

Food web effects of HABs are loosely defined as the effects on organisms from one or more trophic levels. There is increasing evidence that algal toxins have a vast array of effects, from direct lethality to indirect effects modifying recruitment, biodiversity, and trophic interactions (e.g., Burkholder et al. 2018). Effects are characterized differently depending on study conditions. Complicating the quantification of bloom effects is the presence and extent of lag effects, indirect effects or long-term cascading, or accumulating effects (Burkholder et al. 2018).

Many of the acute effects of HAB toxins were described above. Lethal or acute impacts are the easiest to quantify; lingering or accumulating effects and their variability under different environmental conditions much more difficult. Some toxins, such as brevetoxins, may be depurated over the course of days to weeks; thus impacts on their grazers may be felt differentially for a long period of time. As an example, acute effects of brevetoxin (from *Karenia brevis*) on fish have been reported to include altered swimming behavior (swimming in a spiral, twisting behavior, loss of equilibrium), defecation and regurgitation, paralysis of the pectoral fin, curvature of the caudal fin, and convulsions (Landsberg 2002). Sublethal effects are also difficult to characterize—especially in the field, where organisms may exhibit reduced feeding, altered rates of reproduction, or other effects that may make them increasingly susceptible to other sources of mortality—such as other diseases or grazing. Allelopathic effects may also alter ecosystem function. As an example, effects were pronounced when exudate of a natural bloom of *K. brevis* (uncharacterized allelochemicals) were tested on the co-occurring species *Amphora* sp., *Asterionellopsis glacialis* (diatoms), and *Prorocentrum minimum*. These three taxa, as well as *Akashiwo* c.f. *sanguinea*, were also growth-inhibited when exposed to exudate from a toxic culture of *K. brevis*. The exudates suppressed photosynthetic efficiency and damaged cell membranes (Prince et al. 2008; Burkholder et al. 2018).

Early life stages of marine organism may be especially susceptible to effects of HABs, but such effects are difficult to quantify. However, much has been learned from laboratory exposure experiments, Experiments conducted on embryos and larvae of freshly spawned oysters, *Crassostrea virginica* and *C. ariakensis*, exposed to the toxic dinoflagellate *Karlodinium veneficum* showed increased mortality relative to similar embryos and larvae exposed to a control diet (Glibert et al. 2007; Stoecker et al. 2008; Pease et al. 2021; Fig. 25.10). These experiments also

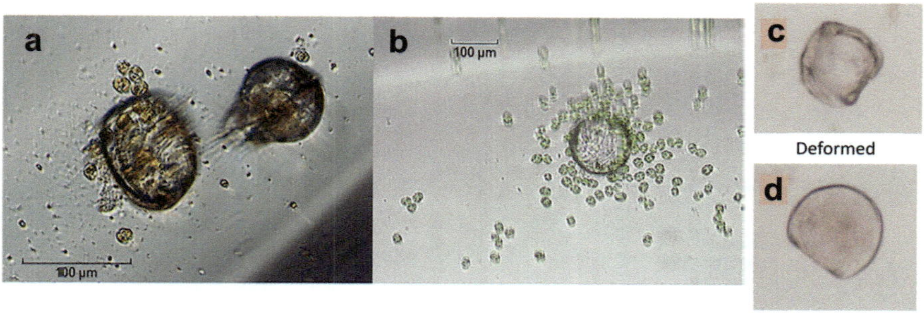

Fig. 25.10 Oyster larvae and effects of HABs . Panel (**a**) *Karlodinium veneficum* inside of a larval oyster shell with a live oyster larva swimming in the background; Panel (**b**) *Karlodinium veneficum* swarming around an empty larval oyster shell; Panel (**c**) a deformed larva after exposure to *K. veneficum*, and Panel (**d**) a normal oyster larva. Images (**a**) and (**b**) reproduced from Pease et al. (2021, Harmful Algae 101: 101965. https://doi.org/10.1016/j.hal.2020.101965) with permission from Elsevier, and (**c**) and (**d**) from Glibert et al. (2007, J Shellfish Res 26: 919–925)

demonstrated important behavioral changes in pediveliger larvae when exposed to *K. veneficum* in their diet, leading them to stop swimming and to sink. Even if such effects did not immediately result in mortality, any change in behavior may influence larval dispersal, reduce feeding and growth, and likely increase the susceptibility of larvae to predation. Related results have been documented for the HAB dinoflagellate *Prorocentrum minimum*. Larvae had poorer survival and lower settling success with only *P. minimum* in the diet, but survival improved when the proportion of *P. minimum* in their diet was reduced (Luckenbach et al. 1993; Wikfors and Smolowitz 1995). In Australia, the related HAB species, *P. rhathymum*, has been associated with mortalities of spat of the Japanese or Pacific oyster, *Crassostrea gigas* (Pearce et al. 2005). Moreover, developmental rates of *C. virginica* and the northern quahog, *Mercenaria mercenaria*, larvae were protracted when exposed to *Karenia brevis* densities of 1000 cells ml^{-1}, while at 5000 cells ml^{-1}, survival was reduced to 37%, 26%, and 19% for the bay scallop *Argopecten irradians*, *M. mercenaria*, and *C. virginica*, respectively (Leverone et al. 2006). Larvae of the same oyster (*C. virginica*) have also been shown to die in the presence of the dinoflagellates *Alexandrium catenella* and *Dinophysis acuminata*. The latter dinoflagellates produce pectenotoxin which casues larval mortality upon direct exposure (Pease et al. 2022). As another example, bay scallops showed decreased feeding, reduced fecundity and larval shell growth rates, and recruitment failure when exposed to the pelagophyte *Aureococcus anophagefferens* (Bricelj et al. 1987; Gallagher et al. 1989).

One of the more interesting examples of food web effects that illustrate the trophic transfer of algal toxins is the case of the eagle-killer cyanobacterium. For years it was observed that eagles were displaying neurological impairments and deaths in several southeast US states, but the causative organism remained elusive. The eagles lost their ability to fly, became paralyzed, and displayed tremors and

Fig. 25.11 Schematic of the neurotoxin accumulation in eagles from the cyanobacterium *Aetokthonos hydrillicola* that grows on invasive hydrilla. Hydrilla accumulates bromine which is a component of the herbicides often used to eradicate these invasive plants and other pollutants. Coots and other organisms feed on hydrilla or the community of organisms that live on hydrilla, in turn becoming intoxicated with the toxin of *A. hydrillicola.* Then, eagles may feed on the coots and they become intoxicated with the algal toxin. Symbols for the herbicide sprayer and the eagle are from VectorStock used under an expanded license. Image based on Breinlinger et al. (2021, Science 371:eaax9050. doi: 10.1126/science.aax9050)

seizures. Their natural prey, American coots, showed many of the same symptoms. They had brain lesions that looked like parts of their brains were eaten away. While algal toxins were suspected, it was years before they were identified. The cause was eventually found to be an epiphytic cyanobacteria that grows on invasive hydrilla (*Hydrilla verticillata*) (Wilde et al. 2005; Breinlinger et al. 2021; Fig. 25.11). One of the keys to unraveling these relationships was the discovery that hydrilla hyper-accumulates bromide (an ion of bromine), which is an essential element of the toxin. Bromine enters the environment from human activities such as coal-fired power plants and from various herbicides—including those used for hydrilla control. The cyanobacterium was ultimately named *Aetokthonos hydrillicola* (Greek for eagle killer that lives on hydrilla).

While the eagle-killer story may be sensational, blooms of the toxic *Microcystis aeruginosa* have been known to be lethal to various zooplankton, fish, waterfowl, and mammals, including domestic animals that have orally ingested the toxins (Burkholder et al. 2018; Fig. 25.12). The routes by which microcystins can affect organisms throughout the food web are both direct and indirect. Bioaccumulation can occur, but it is also a fact that the microcystin content within *Microcystis* varies substantially by strain and by environmental conditions. These toxins can be transferred to benthic habitats by way of feces. Effects tend to be higher at higher temperatures (Zurawell et al. 2005). However, at extreme high temperatures, >35 °C, the microcystin cell quota has been shown to decrease (Bui et al. 2018).

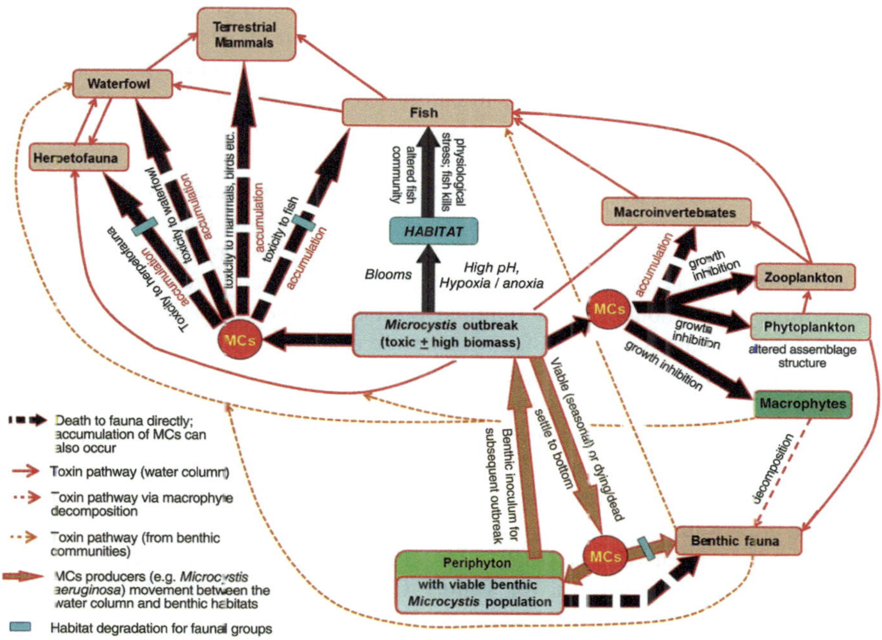

Fig. 25.12 A conceptual model of microcystin (MC) pathways in aquatic ecosystems. These toxins are known to bioaccumulate in various fauna (e.g., Ibelings et al. 2005; Xie et al. 2005), and they are contained within cyanobacteria cells but can also be released into the surrounding medium. Not represented in this diagram are other cyanotoxins and other bioactive substances from *M. aeruginosa*, and toxins and bioactive substances from other sources that may be present. Also not shown, other than for macrophytes, are pathways for MC distribution via dead/dying flora and fauna that. Sink out of the water column to benthic habitats. Image reproduced from Burkholder et al. (2018, Harmful algal blooms: a compendium desk reference) with permission from Wiley

25.5 HABs and Their Global Distribution

25.5.1 Global Patterns and Drivers

HABs are now found worldwide, but different types of HABs are common in specific regions. HABs are increasing in frequency, magnitude, and geographic extent and virtually every coast and major water body in the USA has had major HABs in the past few years (Glibert et al. 2018; Glibert 2020; Hallegraeff et al. 2021; Glibert and Li 2024). A recent synthesis of globally reported HAB data suggests that while there is no uniform increase in all parts of the world, there are distinct areas and species that are increasing (Hallegraeff et al. 2021). One of the problems with characterizing global changes in HABs is that the global databases are imperfect at best, with varying countries reporting consistently and others reporting fewer than the number of actual HAB or toxic events. Nevertheless, there are distinct patterns that have emerged. Overall, the types of HABs and their related seafood poisoning syndromes clearly vary across the globe (Fig. 25.13). Some regions are more prone to seafood toxin events, others to mortality events, and others to high biomass events. Over time, most regions have showed increases in reported events (Hallegraeff et al. 2021).

Taking a closer look at several specific trends in the global analysis of Hallegraeff et al. (2021), clear increases in *Dinophysis*, *Pseudo-nitzschia*, and *Alexandrium* can be seen (Fig. 25.14). Of note is the increasing trend in parallel with increasing observations; that is, more monitoring will inevitably detect more events, but at the same time, more monitoring is required because of the increasing frequency of events (an apt analogy is Covid for which more cases necessitated more testing).

Recent global trends in freshwater blooms were established using 2.91 million Landsat images from 1982 to 2019, covering nearly 250,000 freshwater lakes, or just over half of the world's lake area (Hou et al. 2022). Nearly 10% of these lakes experienced detectable blooms (Figs. 25.15 and 25.16) and the decadal change in bloom occurrence is apparent, especially since about the year 2000.

Many freshwater HAB events are increasing, especially in Asia and Africa, where fertilizer use is extensive (see Chap. 22). An extreme example of freshwater bloom expansion is that of Lake Tai (or Taihu), China, where blooms of the toxigenic cyanobacterium (CyanoHAB) *Microcystis* have increased in duration from ~1 month per year to nearly 10 months per year over the past 15 years (Duan et al. 2009), concomitant with increasing fertilizer use in the watershed and other nutrient sources (Glibert et al. 2014; Fig. 25.17).

The distribution of specific toxins from freshwater cyanobacterial proliferations have also been characterized globally (Wood et al. 2020). In the USA, over roughly the past decade, toxic *Microcystis* blooms, easily visible from satellite imagery, have expanded to cover the entirety of Lake St. Clair (Michigan and Ontario, Canada) and much of Lake Erie (Michalak et al. 2013). In August 2014, the city of Toledo, Ohio, issued a "Do not drink or boil" advisory to about 500,000 people after microcystins in the city's finished drinking water were measured at up to 2.5 $\mu g\ L^{-1}$ (Fitzsimmons 2014). *Microcystis* blooms have become common features in Florida's major river systems, the St. Johns and Caloosahatchee; in Lake Okeechobee (the tenth largest lake in the USA); and in the freshwater tidal St. Lucie Estuary, where huge outbreaks have been sustained seasonally every year over the past decade (Neuhaus 2016). Blooms in coastal lagoons of Florida in recent years have been described as "guacamole thick."

The global distributions of benthic HABs, both marine and freshwater, have also been quantified recently. In the marine realm, as ciguatera poisoning is associated with the consumption of reef fish, it most commonly occurs in the tropical oceans of the world (Fig. 25.18).

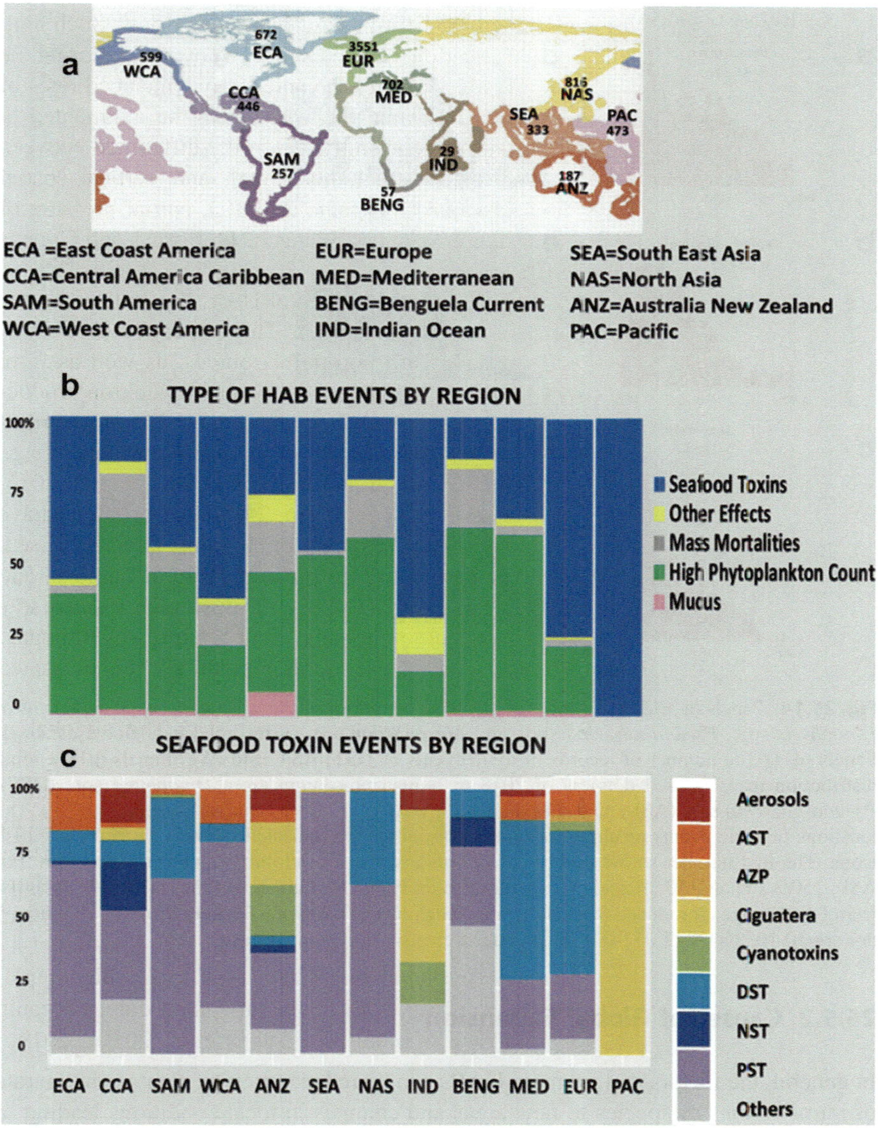

Fig. 25.13 Panel (**a**) Twelve different geographic regions, with total number of recorded events indicated; Panel (**b**) Different types of harmful algal phenomena; and (**c**) Different types of seafood toxin syndromes. Paralytic shellfish toxins were dominant in East Coast America (ECA), South America (SAM), West Coast America (WCA), South East Asia (SEA), and North East Asia (NAS), diarrhetic shellfish toxins were prevalent in the Mediterranean (MED) and Europe (EUR), and ciguatera was predominant in the Indian Ocean (IND) and tropical Pacific (PAC). Australia/New Zealand (ANZ) and Central America/Caribbean (CCA) displayed mixtures of events. Image reproduced from Hallegraeff et al. (2021, Commun Earth Environ 2: 117. https://doi.org/10.1038/s43247-021-00178-8) under a creative commons license

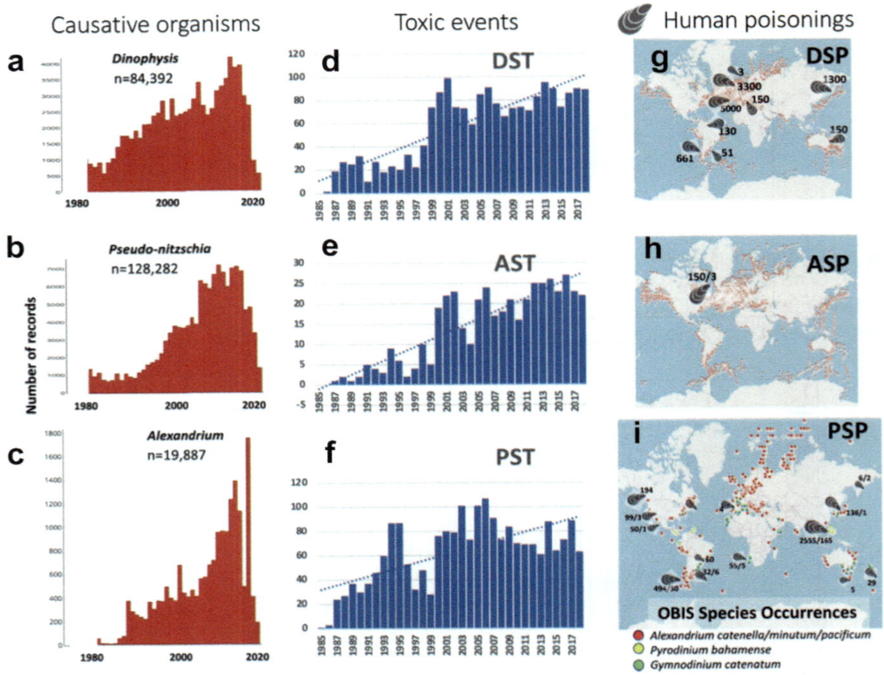

Fig. 25.14 Panels (**a**–**c**) Total number of global observations of causative microalgal organisms of *Dinophysis* spp., *Pseudo-nitzschia* spp., and *Alexandrium* spp. as recorded in a global database; Panels (**d**–**f**) The number of records of toxic events of DST, AST, and PST; Panels (**g**–**i**) Global distribution maps (as red dots) as of 2018 of human poisonings from *Dinophysis* spp. (DSP), *Pseudo-nitzschia* spp. (ASP), and *Alexandrium, Pyrodinium, Gymnodinium catenatum* (PSP). The locations of toxic events resulting in human poisonings are indicated by the size of the shellfish icons. The first number shows number of poisonings, the second number indicates fatalities. For ASP, 150/3 signals 150 clinical cases with three fatalities. No human fatalities have ever occurred from DSP. Image reproduced from Hallegraeff et al. (2021, Commun Earth Environ 2: 117. https://doi.org/10.1038/s43247-021-00178-8) under a creative commons license

25.5.2 Causes of Global Expansion

In general, the factors that promote HABs can be reduced to two: changes in the rate of introductions of species to new areas and changes in local conditions leading to conditions more conducive to the growth of individual species (Glibert 2020). The latter may be due to anthropogenic changes, such as changes in nutrient loads resulting from expanding human population and associated nutrient pollution from agriculture and animal operations, alterations due to human changes in fishing pressure or aquaculture development, or large-scale changes in flow from major water diversion projects (see Chap. 22). However, they may also be due to interactions between trophic and biogeochemical changes that occur once new species become established or to changes in abiotic parameters or physical

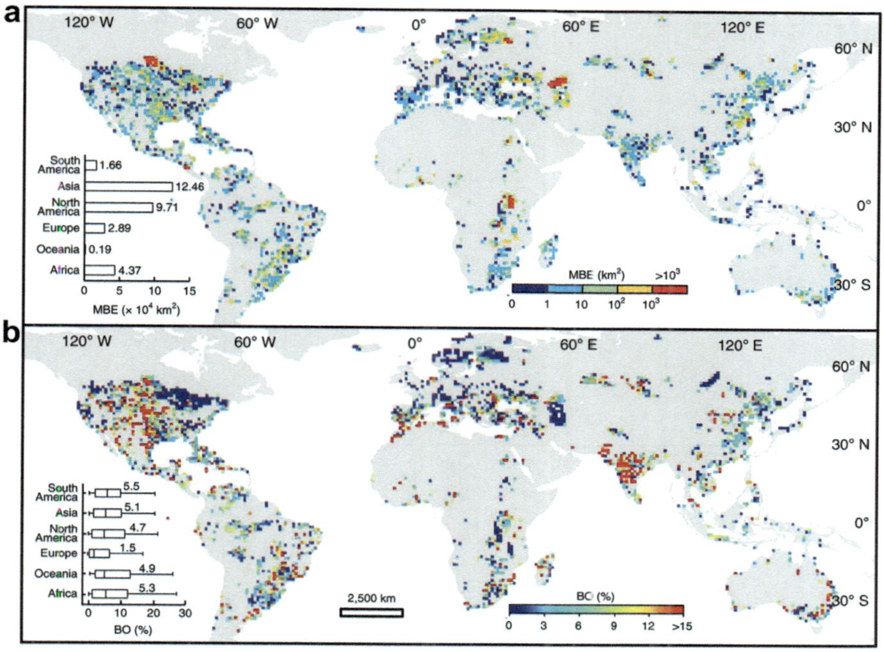

Fig. 25.15 Maximum bloom extent (MBE; Panel **a**) and percent bloom occurrence (BO; Panel **b**) aggregated into 1° × 1° grid cells. Continental statistics (summarized MBE values, box plots for BO) are presented in each panel. In the box plots, the left and right bounds of the boxes are the first and third quartiles, respectively, the bar in the middle shows the median, and the whiskers show the minimum and maximum values. Image reproduced from Hou et al. (2022 Nat Geosci 15: 130–134. https://doi.org/10.1038/s41561-021-00887-x) with permission from Springer Nature

dynamics, such as temperature and stratification, that are caused by climate-related changes (e.g., Sunda et al. 2006; Glibert et al. 2011; Glibert 2015; Wells et al. 2016). The complex set of adaptive strategies that different species have will lead to those species being more or less successful in contrasting environmental conditions (e.g., Margalef 1978; Collos 1986; Glibert and Burkholder 2011; Glibert 2015, 2016). The growth of some species themselves can alter the biological and biogeochemical environment, in some cases changing the environment favorably for further growth, or for growth of other harmful species. Of course, no amount of pressure from an altered rate of species introductions will ensure success of that species in a new environment unless conditions are suitable for its growth (e.g., Smayda 2002; Glibert 2015). As stated by Wells et al. (2016, p. 69) in their review of HABs and climate change, for HABs to be successful depends on the "species 'getting there'...'being there' as indigenous species …, and 'staying there' …." The same is true for nutrients and related environmental conditions. They must "get there," often from anthropogenic sources, they must "be there," and they must "stay there,"

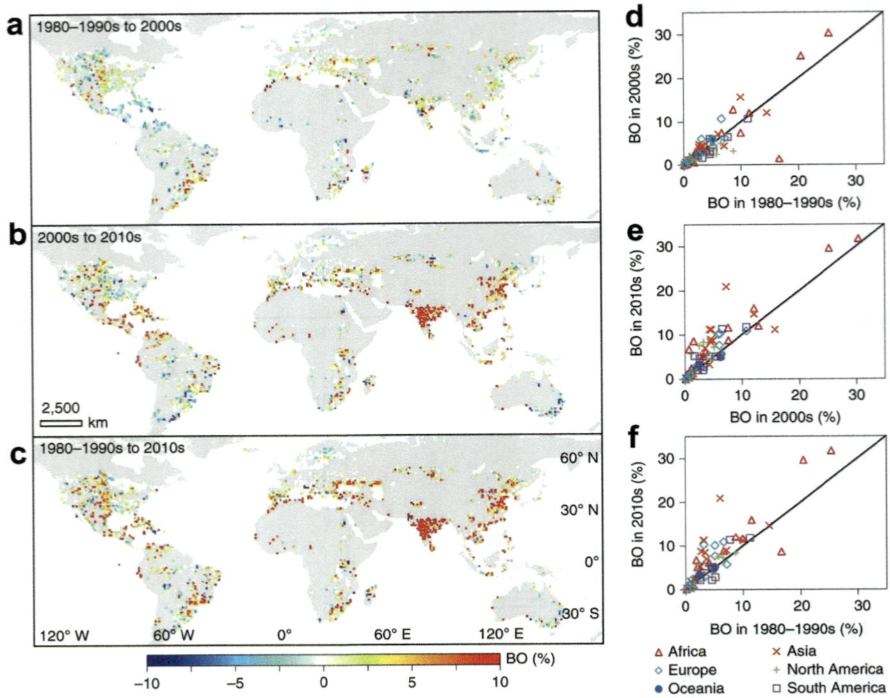

Fig. 25.16 Panels (**a**–**c**) Global maps present changes aggregated into 1° × 1° grid cells for different temporal comparisons: 1980–1990s to 2000s (Panel A); 2000s to 2010s (Panel **b**); and 1980–1990s to 2010s (**c**). Panels (**d**–**f**) Country-level changes in bloom occurrence (BO; $n = 65$) that correspond to the temporal periods in **a** (Panel **d**), **b** (Panel **e**), and **c** (Panel **f**). Image reproduced from Hou et al. (2022 Nat Geosci 15: 130–134. https://doi.org/10.1038/s41561-021-00887-x) with permission from Springer Nature

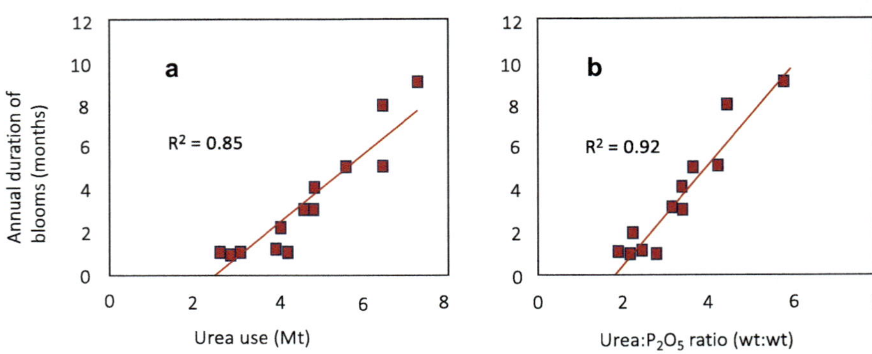

Fig. 25.17 Panel (**a**) Annual duration of *Microcystis* blooms in Lake Taihu, China, in relation to urea use in the watershed. Panel (**b**) The urea: P (as P_2O_5) ratio of fertilizer use in the watershed. Image reproduced from Glibert et al. (2014, Environ Res Lett 9: 105001. https://doi.org/10.1088/1748-9326/9/10/105001) under a creative commons license

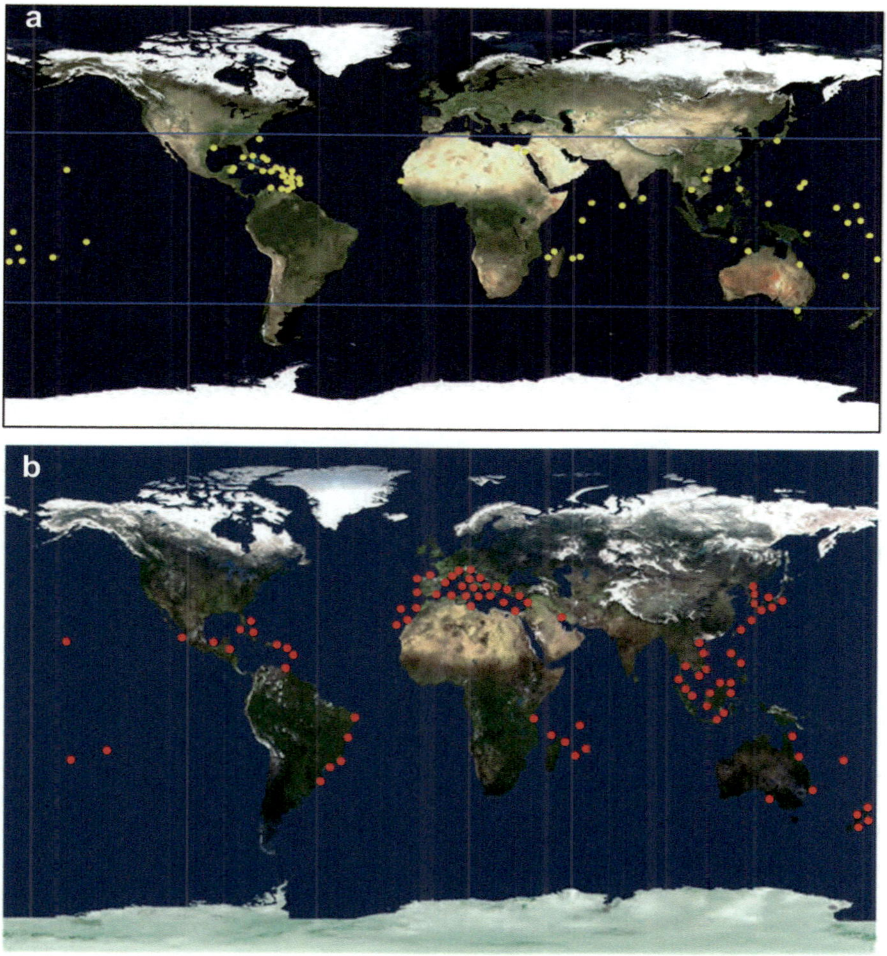

Fig. 25.18 Panel (**a**) Ciguatera fish poisoning incidences from data updated from Tester et al. (2010). Longitude lines (light blue) are marked at 35° North and 35° South. Panel (**b**) Geographical distribution of the genus *Ostreopsis*, redrawn and updated from Rhodes (2010) by A. Zingone. Images reproduced from GEOHAB (2012, IOC of UNESCO and SCOR) with permission from the Intergovernmental Oceanographic Commission

often through physical dynamics, changes in trophodynamics and biogeochemical processing, or climate-induced changes (Glibert and Burkholder 2018).

The relationship between nutrient export and HAB occurrences was shown globally for the HAB species *Prorocentrum cordatum* (reported as *P. minimum*), which has long been associated with eutrophication. Using spatially explicit global maps of nutrient export by form, and documented occurrences of *P. cordatum*, clear associations could be seen (Glibert et al. 2008; Fig. 25.19). In interpreting these

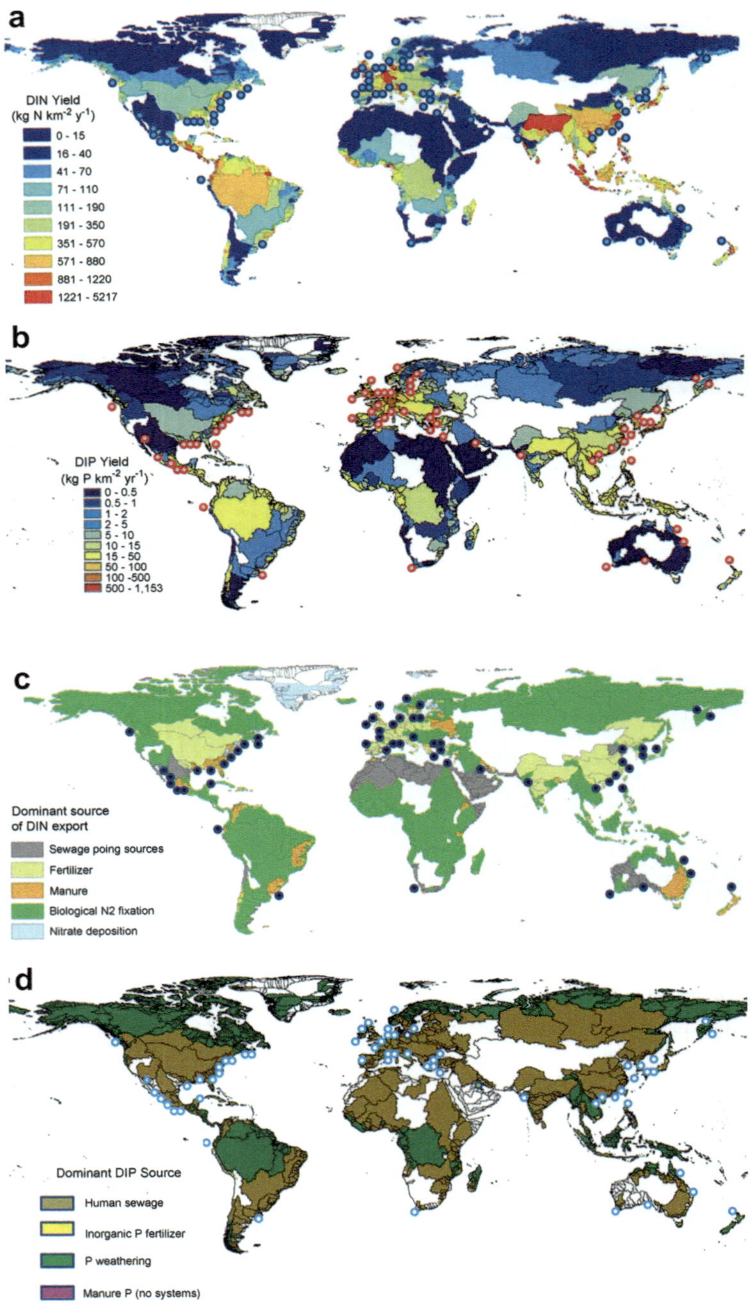

Fig. 25.19 Panel (**a**) Model of DIN yield by watershed (kg N km^{-2} year^{-1}) by exoreic basins (based on models of Dumont et al. 2005) and the global distribution of *Prorocentrum cordatum* indicated by circles (based on review of Heil et al. 2005). Panel (**b**) as for (**a**), except for DIP; Panel (**c**) as for (**a**) except that DIN is shown by the dominant source (based on Dumont et al. 2005), and Panel (**d**) as for (**c**) except for DIP (based on Harrison et al. 2005). Composite images reproduced from Glibert et al. (2008, Harmful Algae 8: 33–38) with permission from Elsevier

maps, it should be considered that nutrient yields do not necessarily reflect the nutrient that the cells may "see" at any particular point in time. Nutrient yields are annual averages, whereas HABs frequently are ephemeral events, and thus, there may be a temporal mismatch between nutrient loads and HAB events. In general, however, this HAB taxa is associated with regions of high dissolved inorganic nitrogen (DIN) and phosphorus (DIP) exports that are strongly influenced by anthropogenic sources (such as fertilizers and manures for DIN). Blooms of this species are also linked to regions with relatively high anthropogenic contributions to dissolved organic N and P export. Comparing the distribution of *P. cordatum* with different sources of N and P (as distinguished by spatially explicit models), it is apparent that this HAB species proliferates where fertilizers and manures dominate the export of N and where human sources dominate the export of P (Fig. 25.19a, b). Observations of *P. cordatum* are rare where biological N_2 fixation and P weathering are dominant, such as throughout Indonesia, or where human sewage is the dominant source of N, as in northern Africa.

One of the major factors of concern for HAB expansion is the expansion of aquaculture throughout many parts of the world (Bouwman et al. 2013a, b; Fig. 25.20). World aquaculture has grown significantly over the past several decades, with a growth that has outpaced capture fisheries (Boyd et al. 2022, see also Chap. 22). Growth has been driven largely by activities in China, and collectively Asian aquaculture activities produce almost 90% of marine aquaculture products worldwide. Aquaculture has several direct effects on HAB growth and distribution. Aquaculture products are often shipped worldwide, and harmful algae can be carried with these products. Similarly, seed stock and feed are shipped worldwide, creating opportunities for HAB introductions (Glibert and Burkholder 2018).

More importantly, aquaculture produces considerable organic waste from undigested feed (depending on the aquaculture product, not applicable in shellfish aquaculture systems) and from excretion products that can alter the amounts, forms, and proportions of nutrients. Modeling by Bouwman et al. (2011, 2013a, b) demonstrated how nutrient composition may change in the presence of molluscan production in raft systems, finfish production in cage systems, and crustacean production in ponds. The nutrients not retained in finfish (36% for N and 33% for P in salmon and rainbow trout), molluscan (33% for N and 22% for P), and crustacean production (25% for N and 10% for P) are released to the water column (see Chap. 22, Fig. 22.9). The proportion of nutrients in dissolved versus solid form depends on the digestibility of the food or feed. The larger the digestible fraction of a nutrient in the various food or feed ingredients, the larger the fraction in dissolved form. Bouwman et al. (2013b) projected that nutrients from aquaculture will increase up to sixfold by 2050 with exceedance of the nutrient assimilative capacity in parts of the world where aquaculture activities are already intensive. Accordingly, it is not surprising that in five regions of the world where aquaculture is expanding, the meta-analysis of Hallegraeff et al. (2021) revealed an expansion in HABs (Fig. 25.21). This trend, however, may be magnified by the need for improved monitoring in aquaculture regions.

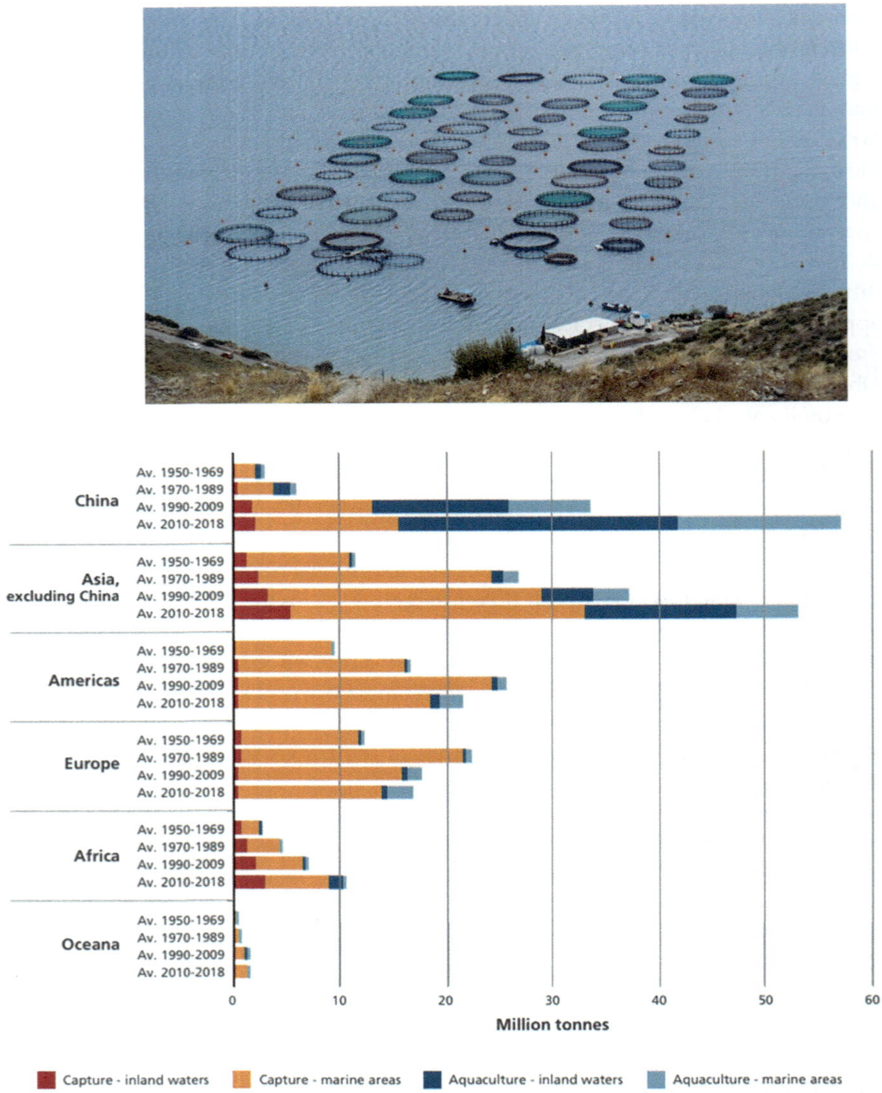

Fig. 25.20 Upper panel, A fish farm off Euboea, Greece. Photo by Jebulon and reproduced from Wikipedia https://en.wikipedia.org/wiki/Fish_farming#/media/File:Fish_farm_Amarynthos_Euboea_Greece.jpg) under a creative commons license. Lower panel, Regional contribution to world fisheries and aquaculture production. Image reproduced from FAO (2020, https://doi.org/10.4060/ca9229en) under a creative commons license

One of the areas of the world where aquaculture has impacted HABs, and in turn where HABs have impacted aquaculture, is Chile. The aquaculture industry has grown rapidly since the 1970s, due to the production of salmon. Chile has become

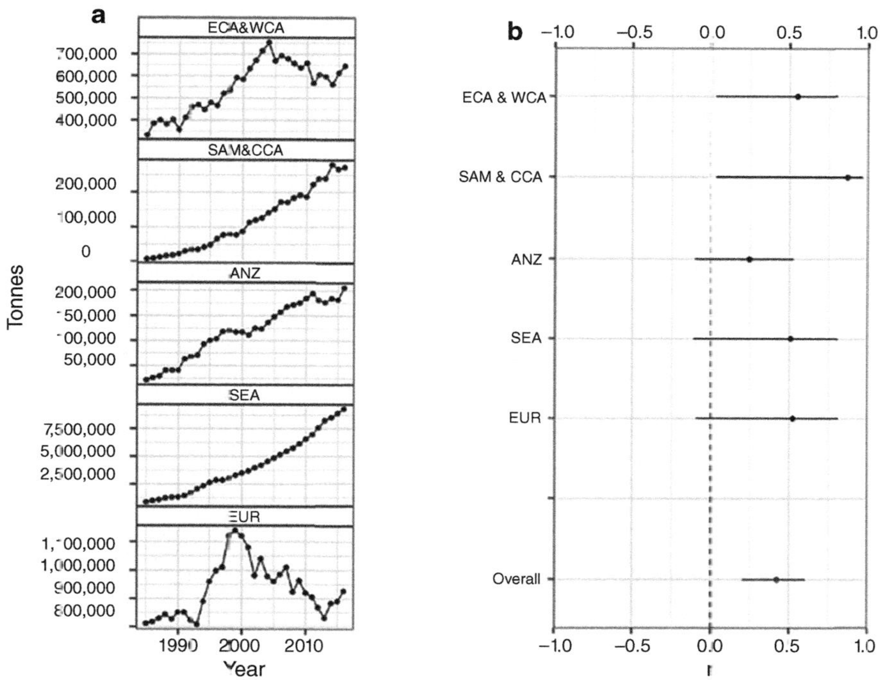

Fig. 25.21 Panel (**a**) Changes in tonnage of aquaculture production of fish, molluscs, crustaceans, and aquatic plants from 1985 to 2018 in five regions (ECA + WCA—East Coast America and West Coast America; SAM + CCA—South America and Central American and Caribbean; ANZ—Australia and new Zealand; SEA—South East Asia; EUR—Europe) and (**b**) Meta-analysis of global HAB events over time vs. aquaculture. The overall number of events over time was significantly correlated with aquaculture production (bottom). Weighted mean correlations (filled circles) are shown with 99% confidence limits (bars) in Panel (**b**). Image reproduced from Hallegraeff et al. (2021, Commun Earth Environ 2: 117. https://doi.org/10.1038/s43247-021-00178-8) under a creative commons license

the second largest salmon-producing region in the world, with Norway being the first (Avendaño-Herrera 2018). In Chile, the early life stages of the fish are grown in the central to southern regions, in land-based farms that rely on streamwater. These farms release feces and unconsumed food, but the environmental impacts of such releases are still being studied and debated. An important recent study has shown that these organic nutrients increase bacteria downstream and decrease benthic algae, turning the systems into more heterotrophic conditions (Kamjunke et al. 2017). Other studies have also suggested environmental degradation below these freshwater farms (Figueroa et al. 2017).

The fattening stages of farmed salmon occurs mostly in the Patagonian fjords (Quiñones et al. 2019). This region is recognized as one of high biodiversity and ecological importance (Fernández et al. 2000; Hucke-Gaete et al. 2006;

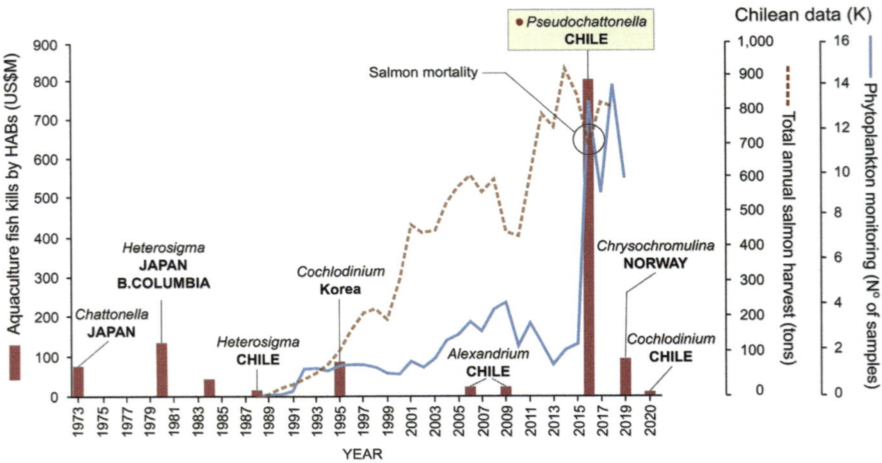

Fig. 25.22 Historical (1973–2020) worldwide aquaculture fish kills by fish-killing algae (FKA) blooms (red bars), total annual harvest (tons) since the beginning of the Chilean salmon farming in the 1980s (orange dotted line), and total annual number of water samples for phytoplankton analysis (blue continuous line). Image reproduced from Mardones et al. (2021, Sci Tot Environ 766: 144383. https://doi.org/10.1016/j.scitotenv.2020.144383) with the permission of Elsevier

Häussermann and Försterra 2009). Eutrophication of this area began to be recognized in the 1970s mainly due to enrichment by feces and excretion and undigested food from these operations. It has been estimated that the annual nutrient discharges from 154 salmon farms operating in this region of Chile in 2010 is equivalent to 12,300 t N and 1600 t P (Niklitschek et al. (2013).

The HABs associated with these fish farms in Chilean fjords are many. Of major concerns are the dinoflagellate *Alexandrium catenella* and the raphidophyte *Pseudochattonella* cf. *verruculosa*, both of which can kill or harm the farmed salmon (Quiñones et al. 2019 and references therein). In 2016, *Pseudochattonella verruculosa* was responsible for the worst damage from a HAB in Chile's history and possibly the largest farmed fish mortality ever recorded with losses estimated at 100,000 metric tons of Atlantic salmon (*Salmo salar*), coho salmon (*Oncorhynchus kisutch*), and rainbow trout (*Oncorhynchus mykiss*) and an export loss of US$800 million (Trainer et al. 2020b; Mardones et al. 2021; Fig. 25.22).

25.6 Marine Biomes and HABs

In addition to regions influenced by eutrophication, there are a number of specific biomes of the oceans that have characteristics HABs (see also Glibert et al. 2018). Some of these are described below.

25.6.1 Stratification and Thin Layer Systems

The persistence and strength of stratification appears to one of the important factors favoring the selection of certain types of HABs. Among the stressors associated with warmer waters is a change in density gradients and increased stratification (e.g., Dale et al. 2006; O'Neil et al. 2012). Higher temperatures promote increased water column stability which has the dual effect of increased sinking rates for larger-sized taxa and decreased nutrient flux from deeper waters (e.g., Winder et al. 2009). Increased thermal stratification favors known bloom-forming, toxigenic cyanobacterial species that control their vertical position through internal buoyancy regulation or that are diazotrophic (e.g., Walsby 1975; O'Neil et al. 2012; Visser et al. 2016 and references therein).

In particular, thin layers, those distinct stratification features that may range from millimeters to several meters in the vertical, and up to kilometers in the horizontal dimension, and which can persist for periods of days to weeks, can develop distinct chemical and biological signatures and distinct HABs (Raine et al. 2018). Thin layer development and maintenance requires interactions between physical, chemical, and biological processes (Bertalet et al. 2017; Fig. 25.23). Different convergence mechanisms can favor a persistent and stable layer structure, leading to what has been described as a "magic carpet" of HABs (Bjørsen and Nielsen 1991). Some thin layers are short-lived, some are sustained for longer periods of time. The species that dominate in such layers are those that may be susceptible to photoinhibition if exposed to high light in surface waters, those which can access the increased nutrient concentrations found in the layer, those that may be mixotrophic and find increased food in the layer, or those avoiding increased surface temperatures. Thin layers may provide the dual advantage of light from above and nutrients from below. Off the French coast, a thin layer of dinoflagellates, including the mixotroph *Dinophysis* cf. *acuminata*, has been observed in the region of the thermocline (Gentien et al. 2005), The same has been reported for *Dinophysis norvegica* in the Baltic Sea, where

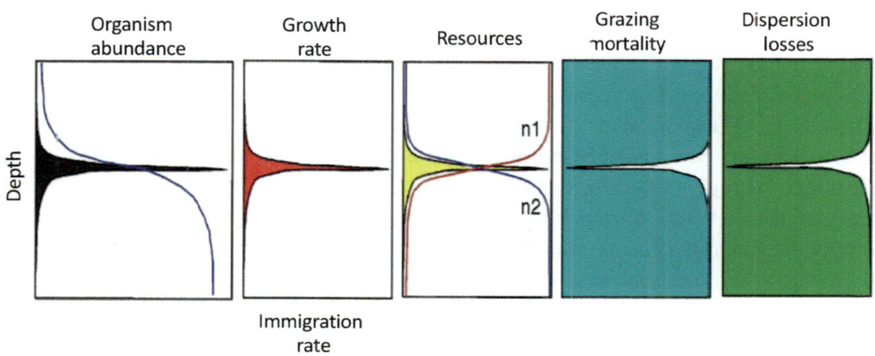

Fig. 25.23 Conceptual diagram of thin layers and their controlling mechanisms. Image based on P.L. Donaghay and reproduced from Berdalet et al. (2017, Oceanography 30(1): 46–57. https://doi.org/10.5670/oceanog.2017.109) under a creative commons license

a 1–2 m thick layer up to 80,000 cells L^{-1} is often situated between 20 and 25 m depth where light is <1% of surface irradiance (Gisselson et al. 2002). Gentien et al. (1995) noted that some thin layers have up to 100% dominance of dinoflagellates. McManus et al. (2003) showed that bacterial production rates can increase in thin layers, potentially yielding increased prey availability for mixoplankton taxa.

It is important to recognize that some cyanobacteria in freshwaters also can develop in deep layers. For example, a deep layer of the cyanobacteria *Planktothrix isothrix*, with a low cyanotoxin cell quota, and *P. rubescens* at lower cell abundances, but with a higher cyanotoxin cell quota was recently reported for a Canadian lake (Erratt et al. 2022). The chl concentration within the thin layer was more than fivefold higher than that at the near surface.

25.6.2 Coastal Embayments

Shallow coastal embayments, and their deeper counterparts, fjords, can be favorable habitats for HABs for a number of reasons. They are surrounded by land, so they are impacted by runoff and other land-sea interactions (Roy et al. 2018). Furthermore, as enclosed or semi-enclosed systems, they may have long residence times and strong benthic coupling and associated nutrient fluxes. Shallow, retentive bays are also favored habitats of those HABs that are cyst-forming. Stable, deep cyst beds may serve as long-term reservoirs of cells, the important inoculum for recurrent blooms.

An important consideration in the determination of what type of species, including HAB species, that may proliferate in estuaries and coastal embayments may depend on the physical typology of the system. Coastal lagoons differ substantially from river-dominated estuaries in the type of high-biomass phytoplankton blooms that generally proliferate, the seasonal timing of the blooms, the forms of nutrients that support these blooms, and the influence of hydrology on blooms and their nutrient dynamics (Glibert et al. 2010). Coastal lagoons represent a class of estuary characterized as shallow, highly enclosed, and typically quiescent in terms of wind, current, and wave energy relative to their deeper, more dynamic counterparts, river-dominated estuaries. Coastal lagoons also tend to have high surface to volume ratios and can have high wetland to water ratios as well (Bricker et al. 2007). Coastal lagoons may support more frequent summer-winter blooms of picoplankton (<3 μm in size), EDAB taxa, whereas riverine estuaries are characterized by high-biomass spring diatom blooms. These picoplankton blooms are also more likely to be sustained on regenerated forms of N, as NH_4^+, urea, or dissolved organic substrates, compared to riverine spring blooms which are often supported by seasonal NO_3^- inputs. Regenerated forms of nutrients are sustained in lagoons due to the generally high surface:volume ratios of these systems and long residence times, both of which lead to a strong coupling between the benthos and the plankton and high benthic nutrient fluxes. The picoplankton genera which dominate have physiological characteristics that make them well-suited to compete effectively for these substrates but must have sustained nutrient sources and limited grazing losses to maintain bloom biomass for prolonged periods. Blooms in river-dominated estuaries, by contrast, tend to be highly seasonal and dominated by larger-sized phytoplankton (i.e., >10 μm), typically diatoms or dinoflagellates.

Nutrient composition within coastal lagoons tends to have a nutrient composition that differs significantly from that of river-dominated systems. Given their reduced riverine influx, lagoonal systems do not generally have elevated concentrations of NO_3^- that are frequently observed in river-dominated estuaries, particularly in late winter-spring. Whereas concentrations of NO_3^- in many riverine systems can exceed many tens of μM (e.g., Chesapeake Bay, Kemp et al. 2005; Neuse River Estuary, Christian et al. 1991; Burkholder et al. 2006), lagoonal systems more typically have higher relative concentrations of NH_4^+ than NO_3^- (Burkholder et al. 2006). In fact, some lagoons can accumulate very significant levels of NH_4^+, >30 μM (Boyer et al. 1999; Burkholder et al. 2006; Glibert et al. 2007, 2010). In addition, organic forms of N and P tend to dominate the nutrient pool in lagoons compared to their respective inorganic nutrient forms (Boyer et al. 1999, 2006; Glibert et al. 2007). The shallow nature of these systems also means that nutrient regeneration from benthic fauna would be more important to the overlying algal populations than would be the case for deeper estuaries. Microbially mediated dissimilatory NO_3^- reduction to NH_4^+ (DNRA), a process by which NO_3^- is converted to NH_4^+ in the presence of organic matter (see Chap. 11), has also been shown to be significant in Florida Bay, where measured rates have ranged from 11 to 170 μmol m^{-2} h^{-1} (Gardner and McCarthy 2005).

25.6.3 Upwelling Systems

Eastern boundary coastal upwelling systems provide another common habitat for some types of HABs (Pitcher et al. 2018; Glibert and Li 2024). There is a considerable diversity of HAB dinoflagellates found in upwelling systems, but the upwelling systems themselves provide a diversity of habitats. During active upwelling—especially when diatoms are considered the most common bloom-formers—dinoflagellates may be found offshore, at fronts of outer shelf currents (Pitcher et al. 2018). In these unique sub-habitats of upwelling systems, these cells may find refuge, serve as seed banks and inoculum for future blooms, and may thrive. Thin layers within upwelling systems may be another unique sub-habitat, with consequences for cross-shelf and along-shelf transport of cells.

Upwelling typically brings N in the form of NO_3^-, the form of N that normally would favor diatoms in the typically cooler waters associated with upwelling. However, cyst-forming toxic species, such as *Alexandrium catenella*, *Gymnodinium catenatum*, *Lingulodinium polyedra*, and *Protoceratium reticulatum* (see Pitcher et al. 2018) suggest that a combination of environmental condition and life strategy play a role in determining which species will be successful in such environments. Cycles of mixing and stratification may be important in regulating the microenvironment leading to spatial and temporal differences in phytoplankton functional groups, and individual taxa, that dominate at any given time. Upwelling favorable winds serve as an initial source of *Pseudo-nitzschia* in Juan de Fuca and Heceta

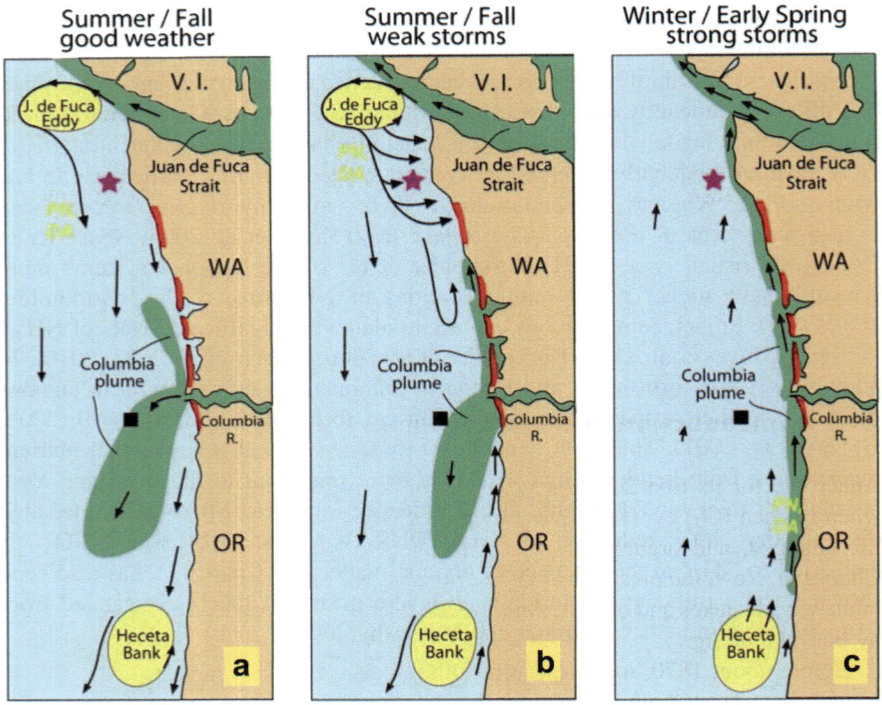

Fig. 25.24 A conceptual model of the transport of toxic *Pseudo-nitzschia* cells from the Juan de Fuca eddy and Heceta Bank during summer and autumn under (Panel **a**) prevailing upwelling favorable winds and (Panel **b**) during weak reversals; and (Panel **c**) during late winter and early spring. Arrows depict surface currents, clamming beaches are marked in red, and freshwater plumes in green. Image reproduced from Hickey et al. (2013 Harmful Algae 25: 1–14) with permission from Elsevier

Bank, northwest US coast, but down-welling favorable wind reversals bring these cells on shore (Fig. 25.24).

25.6.4 Benthic Systems

Benthic dinoflagellates are generally under-studied but are now being investigated in new locations because of the rising awareness of their potential health, economic, and ecological impacts at a global scale (Berdalet et al. 2017; Berdalet and Tester 2018). Benthic HABs, both marine and freshwater, are being reported with increasing frequency. Marine benthic HABs (Fig. 25.25), which have generally been most common in the tropics, appear to be showing a range expansion to some temperate areas. Environmental factors, both abiotic and biotic, are known to influence several aspects of benthic HABs, from their proliferation, toxin production, and accumulation to their effects on environment and human health. Growth, distribution, and

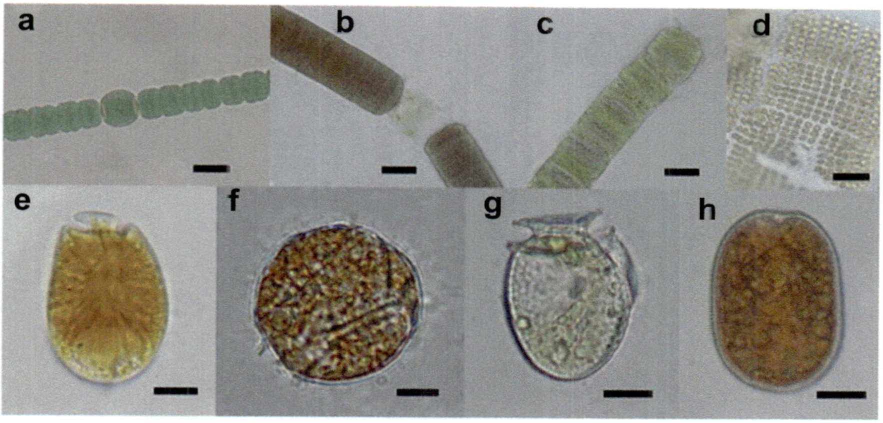

Fig. 25.25 Light microscopy photomicrographs of some distinctive potentially toxic microalgal species found in Paranaguá Bay (southern Brazil): (Panel **a**) *Anabaena* sp.; (Panels **b**, **c**) *Lyngbya* cf. *aestuarii*; (Panel **b**) vegetative trichome; (Panel **c**) Reproduction of short trichome, showing the formation of hormogonia by cell division; (Panel **d**) *Merismopedia elegans*; (Panel **e**) *Amphidinium operculatum*; (Panel **f**) *Coolia malayensis*; (Panel **g**) *Dinophysis* cf. *acuminata*; (Panel **h**) *Prorocentrum* cf. *lima*. Scale bars; (Panel **a–c**): 5 μm; (Panel **d–h**): 10 μm. Image reproduced from Moreira-González et al. (2019, Ocean Coastal Res 68: e20276. https://doi.org/10.1590/S2675-28242020068276) under a creative commons license

abundance of these toxic HABs are largely temperature driven and expected to shift in response to climate induced changes as ocean temperatures rise (Kibler et al. 2015; Tester et al. 2010).

There has long been uncertainty regarding the nutritional status of benthic HAB taxa and relationship to other environmental parameters. Calm conditions are a prerequisite for blooms, and only when this condition exists do temperature and nutrient start to have a decisive effect. There are some conflicting results regarding the relationships between *Ostreopsis* bloom dynamics and both water temperature and nutrient concentrations. These benthic taxa are also mixotrophic. In a study by Accoroni et al. (2015) on *Ostreopsis* in the northern Adriatic Sea, it was found that in the bloom onset period PO_4^{3-} concentrations were significantly higher than in either the bloom maintenance phase or the pre-bloom conditions, resulting in low N:P ratios (24.5 ± 7.6). Both Vanucci et al. (2012) and Pezzolesi et al. (2014) found, in experimental studies of different Adriatic *O.* cf. *ovata* strains, that the depletion of P was proportionately more rapid than that of N, highlighting the strong P demand of this dinoflagellate. Interestingly, after bloom onset, blooms were maintained in a condition of elevated N:P ratios, well above Redfield proportions, a condition normally taken to be indicative of P limitation. However, PO_4^{3-} was never depleted from the water column, and thus true P limitation likely never developed physiologically. However, net growth rates were slower than during the initial phase of blooms development. Accoroni et al. (2015) developed a conceptual model of bloom sequence and nutrients. They suggested that blooms may be initiated at low N:P levels (even less than Redfield ratio, but not in severely N limiting conditions), often

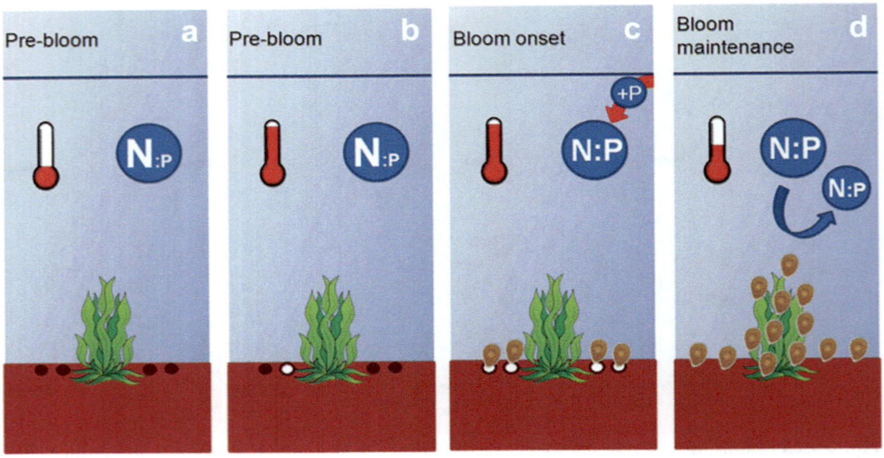

Fig. 25.26 Conceptual diagram of the effect of N:P ratio and water temperature on *Ostreopsis* cf. *ovata* bloom. Panel (**a**) represents the typical pre-bloom condition, showing low temperature and very high N:P ratio. Panel (**b**) represents the pre-bloom condition just before the bloom onset: although the temperature (>25 °C) is optimal for cyst germination, the high N:P ratio does not allow the start of the bloom. Bloom onset (Panel **c**) occurs at T > 25 °C and at N:P ratio (around 24) balanced by an injection of P-rich water, allowing both cyst germination and cell proliferation. Panel (**d**) shows the optimal environmental conditions for the bloom maintenance: temperature is not limiting, as bloom peak occurs at T around 20 °C, and the slight increase of N:P ratio accompanied by a general decrease of N and P could be attributable to *O.* cf. *ovata* growth. Image reproduced from Accoroni et al. (2015, Harmful Algae 45: 14–25) with permission from Elsevier

stimulated by a "flush" of nutrients or organic materials that may allow a latent resident population to increase growth rate. After that growth rate increase, bloom biomass is able to increase often accompanied by N:P values greater than Redfield. At this point, these blooms are able to sustain high biomass levels, not necessarily by high growth rates. Thus, while high growth rates (stimulated by an injection of P-rich water) may allow blooms to initiate, adaptive physiology would enable the maintenance of blooms at less than maximal growth rates and at non-optimal N:P ratios (Fig. 25.26).

25.7 Summary

HABs cause a wide range of harms in fresh and marine waters and are expanding in many parts of the world's waters. The species are diverse, the toxins are diverse, the understanding of what constitutes harm is diverse, but the consequences are very real. Eutrophication and climate change are contributing to the expansion of global HAB events. The proliferation of aquaculture, especially fish culture in both freshwater and marine waters, is also a factor in their regional expansion. New toxins, new vectors, and new trophic impacts are continuing to be described. The following chapters place HABs in the context of the multiple changes in the global environment, and how blooms may be predicted and perhaps controlled.

References

Accoroni S, Glibert PM, Pichierri T, Romagnoli T, Marini M, Totti C (2015) A conceptual model of annual *Ostreopsis* cf. *ovata* blooms in the northern Adriatic Sea based on the synergistic effects of hydrodynamics, temperature, and the N:P ratio of the water column. Harmful Algae 45:14–25

Alvarez S, Brown CE, Diaz MG, O'Leary H, Solis D (2024) Non-linear impacts of harmful algal blooms on the coastal tourism economy. J Environ Manag 351 119811. https://doi.org/10.1016/j.envman.2023.119811

Aráoz R, Molg J, Tandeau de Marsac N (2010) Neurotoxic cyanobacterial toxins. Toxicon 56:813–823. https://doi.org/10.1016/j.toxicon.2009.07.036

Arena P, Levin B, Fleming LE, Friedman MA, Blythe DG (2004) A pilot study of the cognitive and psychological correlates of chronic ciguatera poisoning. Harmful Algae 3:50–60

Avencaño-Herrera R (2018) Proper antibiotics use in the Chilean salmon industry: policy and technology bottlenecks. Aquaculture 495:803–805

Backer LC, Fleming LE, Rowan A, Cheng YS, Benson J, Pierce RH, Zaias J et al (2003) Recreational exposure to aerosolized brevetoxins during Florida red tide events. Harmful Algae 2:19–28

Backer LC, Schurz-Rogers H, Fleming LE, Kirkpatrick B, Benson J (2005) Marine phycotoxins in seafood. In: Dabrowski WM, Sikorski ZE (eds) Toxins in food. CRC Press, Boca Raton, FL, pp 155–189

Baden DG, Fleming LE, Bean JA (1995) Marine toxins. In: DeWolf FA (ed) Handbook of clinical neurology: intoxications of the nervous system, part H, natural toxins and drugs. Elsevier, Amsterdam, pp 141–175

Baden DG, Bourdelais AJ, Jacocks H, Michelliza S, Naar J (2005) Natural and derivative brevetoxins: historical background, multiplicity and effects. Environ Health Perspect 113: 621–625

Bargu S, Silver MW, Ohman MD, Benitez-Nelson CR, Garrison DL (2012) Mystery behind Hitchcock's birds. Nat Geosci 5:2–3

Bates SS, Bird CJ, de Freitas ASW, Foxall R, Gilgan M, Hanic LA, Johnson GR et al (1989) Pennate diatom *Nitzschia pungens* as the primary source of domoic acid, a toxin in shellfish from eastern Prince Edward Island, Canada. Can J Fish Aquat Sci 46:1203–1215

Berdalet E, Tester PA (2018) Key questions and recent research advances on harmful algal blooms in benthic systems. In: Glibert PM, Berdalet E, Burford M, Pitcher G, Zhou M (eds) Global ecology and oceanography of harmful algal blooms (GEOHAB). Springer, Cham, pp 261–286. https://doi.org/10.1007/978-3-319-70069-4-13

Berdalet E, Montresor M, Roy S, Yamazaki H, Cembella A, Raine R (2017) Harmful algal blooms in fjords, coastal embayments, and stratified systems, recent progress and future research. Oceanography 30(1):46–57. https://doi.org/10.5670/oceanog.2017.109

Bienfang PK, Parsons ML, Bidigare RR, Laws EA, Moeller PDR (2008) Ciguatera fish poisoning: a synopsis from ecology to toxicity. In: Walsh PJ, Smith SL, Fleming LE, Solo-Gabriele HM, Gerwick WH (eds) Oceans and human health—risks and remedies from the seas. Elsevier, Amsterdam, pp 257–270

Bjørsen PK, Nielsen TG (1991) Decimeter scale heterogeneity in plankton during a pycnocline bloom of *Gyrodinium aureolum*. Mar Eco Progr Ser 73:263–267

Bossert GD (2011) Marine mammals as sentinel species for oceans and human health. Vet Pathol 48:676–690. https://doi.org/10.1177/0300985810388525

Bouwman AF, Pawlowski M, Liu C, Beusen AHW, Shumway SE, Glibert PM, Overbeek CC (2011) Global hindcasts and future projections of coastal nitrogen and phosphorus loads due to shellfish and seaweed aquaculture. Rev Fish Sci 19:331–357

Bouwman AF, Beusen AHW, Overbeek CC, Bureau DP, Pawlowski M, Glibert PM (2013a) Hindcasts and future projections of global inland and coastal nitrogen and phosphorus loads due to finfish aquaculture. Rev Fish Sci 21:112–158

Bouwman AF, Beusen AHW, Glibert PM, Overbeek C, Pawlowski M, Herrera J, Mulsow S et al (2013b) Mariculture: significant and expanding cause of coastal nutrient enrichment. Environ Res Lett 8.: 044026, 5 p. https://doi.org/10.1088/1748-9326/8/4/044026

Boyd CE, McNevin AA, Davis RP (2022) The contribution of fisheries and aquaculture to the global protein supply. Food Sec 14:805–827. https://doi.org/10.1007/s12571-021-01246-9

Boyer JN, Fourqurean JW, Jones RD (1999) Seasonal and long-term trends in the water quality of Florida Bay (1989–1997). Estuaries 22:417–430

Boyer JN, Dailey SK, Gibson PJ, Rogers MT, Mir-Gonzalez D (2006) The role of dissolved organic matter bioavailability in promoting phytoplankton blooms in Florida Bay. Hydrobiologia 561: 71–85

Breinlinger S, Phillips TJ, Haram BN, Mareš J, Martínez Yerena JA, Hrouzek P, Sobotka R et al (2021) Hunting the eagle killer: a cyanobacterial neurotoxin causes vacuolar myelinopathy. Science 371:eaax9050. https://doi.org/10.1126/science.aax9050

Breitburg D, Levin LA, Oschlies AA, Grégoire M, Chavez FP, Conley DJ, Garçon V et al (2018) Declining oxygen in the global ocean and coastal waters. Science 359:eaam7240. https://doi.org/10.1126/science.aam7240

Bricelj VM, Epp J, Malouf RE (1987) Intraspecific variation in reproductive and somatic growth cycles of bay scallops *Argopecten irradians*. Mar Ecol Prog Ser 35:123–137

Bricker SB, Longstaff B, Dennison W, Jones A, Boicourt K, Wicks C, Woerner J (2007) Effects of nutrient enrichment in the nation's estuaries: a decade of change. NOAA Coastal Ocean Program Decision Analysis Series No. 26. National Center for Coastal Ocean Science, Silver Spring, Maryland

Broadwater MH, Van Dolah FM, Fire SE (2018) Vulnerabilities of marine mammals to harmful algal blooms. In: Shumway SE, Burkholder JM, Morton SL (eds) Harmful algal blooms: a compendium desk reference. Wiley Blackwell, Singapore, pp 191–222

Bui T, Dao T-S, Vo T-G, Lürling M (2018) Warming affects growth rates and microcystin production in tropical bloom-forming *Microcystis* strains. Toxins 10:123. https://doi.org/10.3390/toxins10030123

Buratti FM, Manganelli M, Vichi S, Stefanelli M, Scardala S, Testai E, Funari E (2017) Cyanotoxins: producing organisms, occurrence, toxicity, mechanism of action and human health toxicological risk evaluation. ArchToxicol 91:1049–1130. https://doi.org/10.1007/s00204-016-1913-6

Burkholder JM, Dickey DA, Kinder C, Reed RE, Mallin MA, Melia G, McIver MR et al (2006) Comprehensive trend analysis of nutrients and related variables in a large eutrophic estuary: a decadal study of anthropogenic and climatic influences. Limnol Oceanogr 51:463–487

Burkholder JM, Shumway SE, Glibert PM (2018) Food web and ecosystem impacts of harmful algae. In: Shumway SE, Burkholder JM, Morton SL (eds) Harmful algal blooms: a compendium desk reference. Wiley Blackwell, Singapore, pp 243–336

Buskey EJ, Stockwell DA (1993) Effects of a persistent 'browntide' on zooplankton populations in the Laguna Madre of South Texas. In: Smayda TJ, Shimizu Y (eds) Toxic phytoplankton blooms in the sea, Proceedings of the fifth international conference toxic marine phytoplankton. Elsevier, Amsterdam, pp 659–666

Buskey EJ, Liu H, Collumb C, Bersano JGF (2001) The decline and recovery of a persistent Texas brown tide algal bloom in the Laguna Madre (Texas, USA). Estuaries 24:337–346

Carmichael WW, Azevedo SM, An JS, Molica RJR, Jochimsen EM, Lau S, Rinehart KL et al (2001) Human fatalities from cyanobacteria: chemical and biological evidence for cyanotoxins. Environ Health Perspect 109:663–668

Christian RR, Boyer JN, Stanley DW (1991) Multi-year distribution patterns of nutrients within the Neuse River Estuary, North Carolina. Mar Ecol Prog Ser 71:259–274

Collos Y (1986) Time-lag algal growth dynamics: biological constraints on primary production in aquatic environments. Mar Ecol Prog Ser 33:193–206

Cook PF, Reichmuth C, Rouse A, Dennison S, VanBonn B, Gulland F (2016) Natural exposure to domoic acid causes behavioral perseveration in wild sea lions: neural underpinnings and diagnostic application. Neurotoxicol Teratol 57:95–105

Cusick KD, Sayler GS (2013) An overview on the marine neurotoxin, saxitoxin: genetics, molecular targets, methods of detection and ecological functions. Mar Drugs 11:991–1018

Dale B, Edwards M, Reid PC (2006) Climate change and harmful algal blooms. In: Granéli E, Turner J (eds) Ecology of harmful algae. Springer, Heidelberg, pp 367–378

Diaz RJ, Rosenberg R (2008) Spreading dead zones and consequences for marine ecosystems. Science 321:926–929

Dounay AB, Forsyth CJ (2002) Okadaic acid: the archetypal serine/threonine protein phosphatase inhibitor. Curr Med Chem 9:1939–1980

Duan HR, Ma X, Xu F, Kong F, Zhang S, Kong W, Hao J, Shang L (2009) Two decade reconstruction of algal blooms in China's Lake Taihu. Environ Sci Technol 43:3524–3528

Dumont E, Harrison JA, Kroeze C, Bakker E-J, Seitzinger S (2005) Global distribution and sources of dissolved inorganic nitrogen export to the coastal zone: results from a spatially explicit, global model. Glob Biogeochem Cycles 19:GB4S02

Durbin E, Teegarden G, Campbell R, Cembella A, Baumgartner MF, Mate BR (2002) North Atlantic right whales, *Eubalaena glacialis,* exposed to paralytic shellfish poisoning (PSP) toxins via a zooplankton vector, *Calanus finmarchicus*. Harmful Algae 1:243–251

Erratt K, Creed IF, Freeman EC, Trick CG, Westrick J, Birbeck JA, Watson C, Zastepa A (2022) Deep cyanobacteria layers: an overlooked aspect of managing risks of cyanobacteria. Environ Sci Technol 56:17902–17912

Feder B (1975) When the red tide strikes. The New York Times, August 24, 1975. https://www.nytimes.com/1975/08/24/archives/when-the-red-tide-strikes-when-the-red-tide-strikes.html

Fernández M, Jaramillo E, Marquet PA, Moreno CA, Navarrete SA, Ojeda FP, Valdovinos CR, Vasquez JA (2000) Diversity, dynamics and biogeography of Chilean benthic nearshore ecosystems: an overview and guidelines for conservation. Rev Chilena de Hist Nat 73:797–830

Figueroa D, Aguayo C, Valdivia P, Encina-Montoya F, Nimptsch J, Esse C (2017) Evaluación y análisis de los posibles parámetros ambientales a ser incorporados en las normas de emisión y/o de calidad de aguas fluviales y lacustres, destinados a centros de cultivo ubicados en tierra. Informe Final. Subsecretaría de Pesca y Acuicultura FIP N°2015-05

Fire SE, Van Dolah FM (2012) Marine biotoxins: emergence of harmful algal blooms as health threats to marine wildlife. In: Aguirre AA, Ostfeld R, Daszak P (eds) New directions in conservation medicine. Oxford University Press, NewYork, pp 374–389

Fire SE, Fauquier D, Flewelling LJ, Henry M, Naar J, Pierce R, Wells RS (2007) Brevetoxin exposure in bottle nose dolphins (*Tursiops truncatus*) associated with *Karenia brevis* blooms in Sarasota Bay, Florida. Mar Biol 152:827–834

Fire SE, Browning JA, Durken WN, Stoken MK (2020) Comparison of during-bloom and inter-bloom brevetoxin and saxitoxin concentrations in Indian River Lagoon bottlenose dolphins, 2002–2011. Aq Tox 218:105371

Fitzsimmons EG (2014) Tap water ban for Toledo residents. The New York Times, August 3, 2014. Accessed 3 Oct 2016

Flewelling LJ, Naar JP, Abbott JP, Baden DG, Barros NB, Bossart GD, Bottein MY et al (2005) Brevetoxicosis: red tides and marine mammal mortalities. Nature 435:755–756

Food and Agriculture Organization on the United Nations (FAO) (2020) The state of world fisheries and aquaculture 2020. Sustainability in action. Rome. https://doi.org/10.4060/ca9229en

Francis G (1878) Poisonous Australian lake. Nature 18:11–12

Fujiki H, Suganuma M, Suguri H, Yoshizawa S, Takagi K, Uda N, Wakamatsu K et al (1988) Diarrhetic shellfish toxin, dinophysistoxin 1, is a potent tumor promoter on mouse skin. Japan J Cancer Res 79:1089–1093

Gallagher SM, Stoecker DK, Bricelj VM (1989) Effects of the brown tide alga on growth, feeding physiology and locomotory behavior of scallop larvae (*Argopecten irradians*). In: Cosper EM, Bricelj VM, Carpenter EJ (eds) Novel phytoplankton blooms. Springer, Berlin, pp 511–541

Gardner WS, McCarthy MJ (2005) Denitrification versus dissimilatory nitrate reduction to ammonium (DNRA) or nitrite (DNRN) in hypersaline Florida Bay sediments in August 2004 and January 2005. In: 2005 Florida bay and adjacent marine systems science conference. Duck Key, FL, pp 85–86

Gentien P, Lunven M, Lehaître M, Duvent JL (1995) In situ depth profiling of particle sizes. Deep-Sea Res 42:1297–1312

Gentien P, Donaghay PL, Yamazki H, Raine R, Reguera B, Osborn TR (2005) Harmful algal blooms in stratified environments. Oceanography 18:172–183

GEOHAB (2010) GEOHAB Asia, global ecology and oceanography of harmful algal blooms in Asia: a regional comparative programme. In: Furuya K, Glibert PM, Zhou M, Raine R (eds) IOC and SCOR, Baltimore and Paris. 69 p

GEOHAB (2012) Global ecology and oceanography of harmful algal blooms, GEOHAB Core research project: HABs in benthic systems. In: Berdalet E, Tester P, Zingone A (eds) IOC of UNESCO and SCOR, Paris. 64 p

Gisselson L-A, Carlsson P, Granéli E, Pallon J (2002) *Dinophysis* blooms in the deep euphotic of the Baltic Sea: do they grow in the dark? Harmful Algae 2:401–418

Glibert PM (2015) More than propagule pressure: successful invading algae have physiological adaptations suitable to anthropogenically changing nutrient environments. Aquat Ecosyst Health Manage 18:334–341

Glibert PM (2016) Margalef revisited: a new phytoplankton mandala incorporating twelve dimensions including nutritional physiology. Harmful Algae 55:25–30

Glibert PM (2020) Harmful algae at the complex nexus of eutrophication and climate change. Harmful Algae 91:101583. https://doi.org/10.1016/j.hal.2019.03.001

Glibert PM, Burkholder JM (2011) Eutrophication and HABs: strategies for nutrient uptake and growth outside the Redfield comfort zone. Chinese J Oceanol Limnol 29:724–738

Glibert PM, Burkholder JM (2018) Causes of harmful algal blooms. In: Shumway S, Burkholder JM, Morton SL (eds) Harmful algal blooms: A compendium desk reference. Wiley Blackwell, Singapore, pp 1–38

Glibert PM, Li M (2024) Warming, wheezing, blooming waters: hypoxia and harmful algae. In: Baird D (ed) Treatise on estuarine and coastal science, 2nd edn. Elsevier

Glibert PM, Alexander J, Meritt DM, North EW, Stoecker DK (2007) Harmful algae pose additional challenges for oyster restoration: impacts of the harmful algae *Karlodinium veneficum* and *Prorocentrum minimum* on early life stages of the oysters *Crassostrea virginica* and *Crassostrea ariakensis*. J Shellfish Res 26:919–925

Glibert PM, Mayorga E, Seitzinger SP (2008) *Prorocentrum minimum* tracks anthropogenic nitrogen and phosphorus inputs on a global basis: application of spatially explicit nutrient export models. Harmful Algae 8:33–38

Glibert PM, Boyer J, Heil C, Madden C, Sturgis B, Wazniak C (2010) Blooms in lagoons: different from those of river-dominated estuaries. In: Kennish M, Paerl H (eds) Coastal lagoons: critical habitats of environmental change. CRC Press, Taylor and Francis Group, Boca Raton, FL, pp 91–114

Glibert PM, Fullerton D, Burkholder JM, Cornwell JC, Kana TM (2011) Ecological stoichiometry, biogeochemical cycling, invasive species and aquatic food webs: San Francisco estuary and comparative systems. Rev Fish Sci 19:358–417

Glibert PM, Manager R, Sobota DJ, Bouwman L (2014) The Haber-Bosch- harmful algal bloom (HB-HAB) link. Environ Res Lett 9:105001. https://doi.org/10.1088/1748-9326/9/10/105001

Glibert PM, Berdalet E, Burford M, Pitcher G, Zhou M (2018) Harmful algal blooms and the importance of understanding their ecology and oceanography. In: Glibert PM, Berdalet E, Burford M, Pitcher G, Zhou M (eds) Global ecology and oceanography of harmful algal blooms (GEOHAB). Springer, Cham, pp 9–25. https://doi.org/10.1007/978-3-319-70069-4_2

Graham LE, Graham JM, Wilcox LW, Cook ME (2016) Algae, 3rd edn. LJLM Press. 689 p

Granéli E, Weberg M, Salomon PS (2008) Harmful algal blooms of allelopathic microalgal species: the role of eutrophication. Harmful Algae 8:94–102

Gulland F (2000) Domoic acid toxicity in California sea lions (*Zalophus californianus*) stranded along the central California coast, May–October,1998. NOAA Technical Memorandum. U.-S. Department of Commerce, Washington

Guzman-Guillen R, Prieto AI, Moreno I, Eugenia Soria M, Camean AM (2011) Effects of thermal treatments during cooking, microwave oven and boiling, on the unconjugated microcystin concentration in muscle of fish (*Oreochromis niloticus*). Food Chem Toxicol 49:2060–2067. https://doi.org/10.1016/j.fct.2011.05.018

Hallegraeff GM, Anderson DM, Belin C, Bottein M-YD, Bresnan E, Chinain M, Enevoldsen H et al (2021) Perceived global increase in algal blooms is attributable to intensified monitoring and emerging bloom impacts. Commun Earth Environ 2:117. https://doi.org/10.1038/s43247-021-00178-8

Harrison JA, Seitzinger SP, Caraco N, Bouwman AF, Beusen AHW, Vörösmarty C (2005) Dissolved inorganic phosphorus export to the coastal zone: results from a new, spatially explicit, global model (NEWS-SRP). Glob Biogeochem Cycles 19:GB4S03

Häussermann V, Försterra G (2009) Marine benthic fauna of Chilean Patagonia. Illustrated identification guide. Nature in Focus, Chile. 1000 p

Häussermann V, Gutstein CS, Bedington M, Cassis D, Olavarria C et al (2017) Largest baleen whale mass mortality during strong El Nino event is likely related to harmful toxic algal bloom. PeerJ 5:e3123. https://doi.org/10.7717/peerj.3123

Heisler J, Glibert PM, Burkholder JM, Anderson DA, Cochlan WP, Dennison WC, Dortch Q et al (2008) Eutrophication and harmful algal blooms: a scientific consensus. Harmful Algae 8:3–13

Helfrich P (1964) Fish poisoning in Hawaii. Hawaii Med J 22:361–372

Hickey BM, Trainer VL, Kosro PM, Adams NG, Connolly TP, Kachel NB, Geier SL (2013) A springtime source of toxic *Pseudo-nitzschia* cells on razor clam beaches in the Pacific Northwest. Harmful Algae 25:1–14

Hoagland P, Scatasta S (2006) The economic effects of harmful algal blooms. In: Granéli E, Turner J (eds) Ecology of harmful algae. Springer, Heidelberg, pp 391–402. https://doi.org/10.1007/978-3-540-32210-8_30

Hokama Y, Yoshikawa-Ebesu JSM (2001) Ciguatera fish poisoning: a foodbourne disease. J Toxicol–Toxins Rev 20:85–139

Hou X, Feng L, Dai Y, Hu C, Gibson L, Tang J, Lee Z et al (2022) Global mapping reveals increase in lacustrine algal blooms over the past decade. Nat Geosci 15:130–134. https://doi.org/10.1038/s41561-021-00887-x

Hucke-Gaete R, Viddi FA, Bello ME (2006) Marine conservation in southern Chile: the importance of the Chiloe-Corcovado area for blue whales, biological diversity and sustainability.Biological diversity and sustainable development. Imprenta America. Valdivia, Chile. http://www.biblioteca.uach.cl/biblioteca_virtual/libros/2006/333.956HUC.pdf

Ibelings BW, Bruning K, de Jonge J, Wolfstein K, Pires LMD, Postma J, Burger T (2005) Distribution of microcystins in a lake food web: no evidence for biomagnification. Microb Ecol 49:487–500

Jin D, Moore S, Holland D, Anderson L, Lim W-A, Kim D-H, Jardine S et al (2020) Gulf of Maine: PSP (*Alexandrium catenella*). In: Trainer VL (ed) GlobalHAB, evaluating, reducing and mitigating the cost of harmful algal blooms: a compendium of case studies. PICES Sci. Rep. No. 59, p 21

Jochimsen EM, Carmichael WW, An JS, Cardo DM, Cookson ST, Holmes CEM, Antunes MBC et al (1998) Liver failure and death after exposure to microcystins at a hemodialysis center in Brazil. N Engl J Med 338:873–878

Kamjunke N, Nimptsch J, Harir M, Herzsprung P, Schmitt-Kopplin P, Neu TR et al (2017) Land-based salmon aquacultures change the quality and bacterial degradation of riverine dissolved organic matter. Sci Rep 7:43739. https://doi.org/10.1038/srep43739

Kemp WM, Boynton WR, Adolf JE, Boesch DF, Boicourt WC, Brush G, Cornwell JC et al (2005) Eutrophication in Chesapeake Bay: historical trends and ecological interactions. Mar Ecol Prog Ser 303:1–29

Kibler SR, Tester PA, Kunkel KE, Moore SK, Litaker RW (2015) Effects of ocean warming on growth and distribution of dinoflagellates associated with ciguatera fish poisoning in the Caribbean. Ecol Model 316:194–210. https://doi.org/10.1016/J.ECOLMODEL.2015.08.020

Kirkpatrick BA, Fleming LE, Bean JA, Nierenberg K, Backer LC, Cheng YS, Pierce R et al (2011) Aerosolized red tide toxins (brevetoxins) and asthma: continued health effects after 1 hour beach exposure. Harmful Algae 10:138–143

Kurland LT, Mulder DW (1954) Epidemiologic investigations of amyotrophic lateral sclerosis. I. Preliminary report on geographic distribution and special reference to the Mariana Islands, including clinical and pathologic observations. Neurology 4:438–448

Landsberg JH (2002) The effects of harmful algal blooms on aquatic organisms. Rev Fish Sci 10: 113–390

Landsberg JH, Balazs GH, Steidinger KA, Baden DG, Work TM, Russell DJ (1999) The potential role of natural tumor promoters in marine turtle fibropapillomatosis. J Aquat Anim Health 11: 199–210

Landsberg JH, Vargo GA, Flewelling LJ, Wiley F (2007) Algal biotoxins. In: Thomas NJ, Hunter DB, Atkinson CT (eds) Infectious diseases of wild birds. Blackwell, Ames, IA, pp 431–455

Landsberg JH, Flewelling LJ, Naar J (2009) *Karenia brevis* red tides ,brevetoxins in the food web, and impacts on natural resources: decadal advancements. Harmful Algae 8:598–607

Leverone JR, Blake NJ, Pearce RH, Shumway SE (2006) Effects of the dinoflagellate Karenia brevis on larval development in three species of bivalve mollusc from Florida. Toxicon 48:75–84. https://doi.org/10.1016/j.toxicon.2006.04.012

Lobner D, Piana PMT, Salous AK, Peoples RW (2007) β-N-methylamino-L-alanine enhances neurotoxicity through multiple mechanisms. Neurobiol Dis 25:360–366. https://doi.org/10.1016/j.nbd.2006.10.002

Luckenbach MW, Sellner KG, Shumway SE, Greene K (1993) Effects of two bloom forming dinoflagellates, *Prorocentrum minimum* and *Gyrodinium uncatenum,* on the growth and survival of the eastern oyster, *Crassostrea virginica*. J Shell Res 12:411–415

Mardones JI, Paredes J, Godoy M, Suarez R, Norambuena L, Vargas V, Fuenzalida G et al (2021) Disentangling the environmental processes responsible for the world's largest farmed fish-killing harmful algal bloom: Chile, 2016. Sci Total Environ 766:144383. https://doi.org/10.1016/j.scitotenv.2020.144383

Margalef R (1978) Life-forms of phytoplankton as survival alternatives in an unstable environment. Oceanol Acta 1:493–509

McManus MA, Alldredge AL, Barnard AH, Boss E, Case JF, Cowles TJ, Donaghay PL et al (2003) Characteristics, distribution and persistence of thin laers over a 48 hour period. Mar Ecol Prog Ser 261:1–19

Mead AD (1899) The investigation of the plague which destroyed multitudes of fish and crustacea during the fall of 1989. Twenty-ninth annual report of the Commissions of Inland fisheries made to the General Assembly of the State of Rhode Island. E.L. Freeman & Sons, State Printers, Providence, R.I. pp 31–34

Michalak AM, Anderson EJ, Beletsky D, Boland S, Bosch NS, Bridgeman TB, Chaffin JD et al (2013) Record-setting algal bloom in Lake Erie caused by agricultural and meteorological trends consistent with expected future conditions. Proc Natl Acad Sci U S A 110:6448–6452

Monteiro PR, do Amaral SC, Siqueira AS, Xavier LP, Santos AV (2021) Anabaenopeptins: what we know so far. Toxins 13:522. https://doi.org/10.3390/toxins13080522

Moreira-González AR, Brustolin MC, Mafra LL Jr (2019) Composition and abundance of benthic microalgae from the estuarine complex of Paranaguá Bay (southern Brazil) with special emphasis on toxic species. Ocean Coastal Res 68:e20276. https://doi.org/10.1590/S2675-28242020068276

Neuhaus L (2016) Miles of algae and a multitude of hazards. The New York Times July 18, 2016. Accessed 28 Sept 2016

Nicholson GM, Lewis RJ (2006) Ciguatoxins: cyclic polyether modulators of voltage-gated ion channel function. Mar Drugs 4:82–118

Niklitschek EJ, Soto D, Lafon A, Molinet C, Toledo P (2013) Southward expansion of the Chilean salmon industry in the Patagonian Fjords: main environmental challenges. Rev Aquacul 4:1–24

Ohtani I, Moore RE, Runnegar MTC (1992) Cylindrospermopsin: a potent hepatotoxin from the blue-green alga *Cynlindrospermopsis raciborskii*. J Am Chem Soc 114:7941–7942. https://doi.org/10.1021/ja00046a067

O'Neil JM, Davis TW, Burford MA, Gobler CJ (2012) The rise of harmful cyanobacteria blooms: the potential roles of eutrophication and climate change. Harmful Algae 14:313–334

Parsons ML, Dortch Q, Doucette GJ (2013) An assessment of *Pseudo-nitzschia* population dynamics and domoic acid production in coastal Louisiana. Harmful Algae 30:65–77

Pearce I, Handlinger JH, Hallegraeff GH (2005) Histopathology in Pacific oyster (*Crassostrea gigas*) spat caused by the dinoflagellate *Prorocentrum rhathymum*. Harmful Algae 4:61–74

Pease SKD, Reece KS, O'Brien J, Hobbs PLM, Smith JL (2021) Oyster hatchery breakthrough of two HABs and potential effects on larval eastern oysters (*Crassostrea virginica*). Harmful Algae 101:101965. https://doi.org/10.1016/j.hal.2020.101965

Pease SKD, Brosnahan ML, Sanderson MP, Smith JL (2022) Effects of two toxin-producing harmful algaee, *ALexandrium carenella* and *Dinophysis acuminata* (Dinophceae), on activity and mortality of larval shellfish. Toxins 14:335. https://doi.org/10.3390/toxins14050335

Pezzolesi L, Pistocchi R, Fratangeli F, Dell'Aversano C, Dello Iacovo E, Tartaglione L (2014) Growth dynamics in relation to the production of the main cellular components in the toxic dinoflagellate *Ostreopsis cf. ovata*. Harmful Algae 36:1–10

Pitcher GC, Figueiras FG, Kudela RM, Moita T, Reguera B, Ruiz-Villareal M (2018) Key questions and recent advances on harmful algal blooms in eastern boundary upwelling systems. In: Glibert PM, Berdalet E, Burford M, Pitcher G, Zhou M (eds) Global ecology and oceanography of harmful algal blooms (GEOHAB). Springer, Cham, pp 205–227. https://doi.org/10.1007/978-3-319-70069-4_11

Poli MA, Musser SM, Dickey RW, Eilers PP, Hall S (2000) Neurotoxic shellfish poisoning and brevetoxin metabolites: a case study from Florida. Toxicon 38:981–993

Pomati F, Kellmann R, Cavalieri R, Burns BP, Neilan BA (2006) Comparative gene expression of PSP-toxin producing and non-toxic *Anabaena circinalis* strains. Environ Int 32:743–748

Prince EK, Myers TL, Naar J, Kubanek J (2008) Competing phytoplankton undermines allelopathy of a bloom-forming dinoflagellate. Proc Biol Sci 275:2733–2741

Quiñones RA, Fuentes M, Montes RM, Soto D, León-Muñoz J (2019) Environmental issues in Chilean salmon farming; a review. Rev Aquacul 11:375–402. https://doi.org/10.1111/raq.12337

Raine R, Berdalet E, Yamazaki H, Jenkinson I, Reguera B (2018) Key questions and recent research advances on harmful algal blooms in stratified systems. In: Glibert PM, Berdalet E, Burford M, Pitcher G, Zhou M (eds) Global ecology and oceanography of harmful algal blooms (GEOHAB). Springer, Cham, pp 165–186

Ramsdell JS, Gulland FM (2014) Domoic acid epileptic disease. Mar Drugs 12:1185–1207

Ramsdell JS, Zabka TS (2008) Inutero domoic acid toxicity: a fetal basis to adult disease in the California Sea lion (*Zalophus californianus*). Mar Drugs 6:262–290

Rastogi RP, Madamwar D, Incharoen-sakdi A (2015) Bloom dynamics of cyanobacteria and their toxins: environmental health impacts and mitigations strategies. Front Microbiol. https://doi.org/10.3389/fmicb.2015.01254

Rhodes L (2010) A review of the ecological factors driving biogeography of Gambierdiscus species. GEOHAB open science meeting: HABs in benthic systems, Honolulu, Hawai'i, USA, 21–23 June, 2010

Roy S, Montresor M, Cembella A (2018) Key questions and recent research advances on harmful algal blooms in fjords and coastal embayments. In: Glibert PM, Berdalet E, Burford M, Pitcher G, Zhou M (eds) Global ecology and oceanography of harmful algal blooms (GEOHAB). Springer, Cham

Roy-Lachapelle A, Duy SV, Munoz G, Dinh QT, Bahl E, Simon DF, Sauvé S (2019) Analysis of multiclass cyanotoxins (microcystins, anabaenopeptins, cylindrospermopsin and anatoxins) in lake waters using on-line SPE liquid chromatography high-resolution Orbitrap mass spectrometry. Anal Methods 11:5289–5300

Scholin CA, Gulland F, Doucette GJ, Benson S, Busman M, Chavez FP, Cordaro J et al (2000) Mortality of sea lions along the central California coast linked to a toxic diatom bloom. Nature 403:80–84
Shumway SE, Allen SM, Boersma PD (2003) Marine birds and harmful algal blooms: sporadic victims or under-reported events? Harmful Algae 2:1–17
Sivonen K, Jones G (1999) Cyanobacterial toxins. In: Chorus I, Bartram J (eds) Toxic cyanobacteria in water: a guide to their public health consequences, monitoring and management. E&FN Spon on London, pp 41–111. www.int/water_sanitation_health/resourcesquality/toxcyanchap3.pdf
Smayda TJ (1997) What is a bloom? A commentary. Limnol Oceanogr 42:1132–1136
Smayda TJ (2002) Turbulence, watermass stratification and harmful algal blooms: an alternative view and frontal zones as "pelagic seed banks." Harmful Algae 1:95–112
Steidinger KA, Baden DG (1985) Toxic marine dinoflagellates. In: Spector DL (ed) Dinoflagellates. Academic Press, Orlando, FL, pp 210–261
Steidinger KA, Landsberg JH, Flewelling LJ, Kirkpatrick BA (2008) Toxic dinoflagellates. In: Walsh PJ, Smith SL, Fleming LE, Solo-Gabriele HM, Gerwick WH (eds) Oceans and human health–risks and remedies from the seas. Elsevier, Amsterdam, pp 239–256
Stewart I, Falconer IR (2008) Cyanobacteria and cyanobacteria toxins. In: Walsh PJ, Smith SL, Fleming LE, Solo-Gabriele HM, Gerwick WH (eds) Oceans and human health–risks and remedies from the seas. Elsevier, Amsterdam, pp 271–296
Stoecker DK, Adolf JE, Place AR, Glibert PM, Meritt D (2008) Effects of the dinoflagellates *Karlodinium veneficum* and *Prorocentrum minimum* on early life history stages of the eastern oyster, *Crassostrea virginica*. Mar Biol 154:81–90
Sunda WG, Granéli E, Gobler CJ (2006) Positive feedback and the development and persistence of ecosystem disruptive algal blooms. J Phycol 42:963–974
Tester PA, Feldman RL, Nau AW, Kibler SR, LitakerRW (2010) Ciguatera fish poisoning and sea surface temperatures in the Caribbean Sea and the West Indies. Toxicon 56: 698–710 https://doi.org/10.1016/J.TOXICON.2010.02.026
Trainer, VL, Yoshida T (eds) (2014) Proceedings of the workshop on economic impacts of harmful algal blooms on fisheries and aquaculture. PICES Scientific Report No. 47, 85 p
Trainer VL, Eberhart B-TL, Wekell JC, Adams NG, Hanson L, Cox F, Dowell J (2003) Paralytic shellfish toxins in Puget Sound, Washington state. J Shellfish Res 22:213–223
Trainer VL, Hickey BM, Bates SS (2008) Toxic diatoms. In: Walsh PJ, Smith SL, Fleming LE, Solo-Gabriele HM, Gerwick WH (eds) Oceans and human health–risks and remedies from the seas. Elsevier, Amsterdam, pp 219–237
Trainer VL, Davidson K, Wakita K, Berdalet E, Suddleson M, Myre G, Trethewey D (2020a) Introduction. In: Trainer VL (ed) GlobalHAB, evaluating, reducing and mitigating the cost of harmful algal blooms: a compendium of case studies. PICES Scientific Report No. 59, pp 1–4
Trainer VL, Moore SK, Hallegraeff G, Kudela RM, Clement A, Mardones JI, Cochlan WP (2020b) Pelagic harmful algal blooms and climate change: lessons from nature's experiments with extremes. Harmful Algae 91:101591. https://doi.org/10.1016/j.hal.2019.03.009
Twiner MJ, Flewelling LJ, Fire SE, Bowen-Stevens SR, Gaydos JK, Johnson CK, Landsberg JH et al (2012) Comparative analysis of three brevetoxin-associated bottlenose dolphin (*Tursiops truncatus*) mortality events in the Florida panhandle region (USA). PLoS One 7:e42974
Urdaneta A (1580) Relación de los sucesos de la armada del comendador Loaisa a las Islas de la Especiería o Molucas en 1525 y sucesos acaecidos en ellas hasta el 1536. Real Biblioteca II/1465
Van Dolah FM (2000) Marine algal toxins: origins, health effects, and their increased occurrence. Environ Health Perspect 108:133–141
Van Dolah FM, Doucette GJ, Gulland FMD, Rowles TL, Bossart GD (2003) Impacts of algal toxins on marine mammals. In: Vos JG, Bossart GD, Fournier M, O'Shea TJ (eds) Toxicology of marine mammals. Taylor and Francis, NewYork, pp 247–269
Vanucci S, Pezzolesi L, Pistocchi R, Ciminiello P, Dell'Aversano C, Dello Iacovo E, Fattorusso E et al (2012) Nitrogen and phosphorus limitation effects on cell growth, biovolume, and toxin production in *Ostreopsis* cf. *ovata*. Harmful Algae 15:78–90

Vichi S, Buratti FM, Testai E (2016) Microcystins: toxicological profile. In: Gopalakrishnakone P, Haddad V Jr, Tubaro A, Kim E, Kem WR (eds) Marine and freshwater toxins. Springer, Dordrecht, pp 219–238

Villacorte LO, Tabatabai SAA, Anderson DA, Amy G, Schippers JC, Kennedy MD (2015) Seawater reverse osmosis desalinization and (harmful) algal blooms. Desalinization 360:61–80

Visser PM, Verspagen JMH, Sandrini G, Stal LJ, Mattijs HCP, Davis TW, HW, Huisman J (2016) How rising CO_2 and global warming may stimulate harmful cyanobacterial blooms. Harmful Algae 54: 145–159

Walsby T (1975) Gas vesicles. Annu Rev Plant Physiol 26:427–439

Wells ML, Trainer VL, Smayda TJ, Karlson BSO, Trick CG, Kudela RM, Ishikawa A et al (2016) Harmful algal blooms and climate change: learning from the past and present to forecast the future. Harmful Algae 49:68–93

Wikfors GH, Smolowitz RM (1995) Experimental and histological studies of four life-history stages of the eastern oyster, *Crassostrea virginica*, exposed to a cultured strain of the dinoflagellate *Prorocentrum minimum*. Biol Bull 188:313–328

Wilde SB, Murphy TM, Hope CP, Habrun SK, Kempton J, Birrebkott A, Wiley F et al (2005) Avian vacuolar myelinopathy linked to exotic aquatic plants and a novel cyanobacterial species. Environ Toxicol 20:348–353. https://doi.org/10.1002/tox.20111

Winder M, Reuter JE, Schladow SG (2009) Lake warming favours small-sized plankton diatom species. Proc R Soc B 276:427–435

Wood SA, Kelly LT, Bouma-Gregson K, Humbert J-F, Laughinghouse HD IV, Lazorchak J et al (2020) Toxic benthic freshwater cyanobacterial proliferations: challenges and solutions for enhancing knowledge and improving monitoring and mitigation. Freshwat Biol 65:1824–1842

Xie LQ, Xie P, Guo LG, Li L, Miyabara Y, Park H-D (2005) Organ distribution and bioaccumulation of microcystins in freshwater fish at different trophic levels from the eutrophic Lake Chaohu, China. Environ Toxicol 20:293–300

Zhang D, Chen J, Xie P (2010) Effects of temperature on the stability of microcystins in muscle of fish and its consequences for food safety. Bull Environ Contam Toxicol 84:202–207. https://doi.org/10.1007/s00128-009-9910-6

Zurawell RW, Chen H, Burke JM, Prepas EE (2005) Hepatotoxic cyanobacteria: a review of the biological importance of microcystins in freshwater environments. J Toxicol Environ Health Part B 8:1–37. https://doi.org/10.1080/10937400590889412

Additional Websites Accessed

https://en.wikipedia.org/wiki/Beta-Methylamino-L-alanine#/media/File:3-Methylamino-L-alanine.svg. Accessed August 14, 2023

https://en.wikipedia.org/wiki/Dinophysis_acuta#/media/File:Dinophysis_acuta.jpg). Accessed August 14, 2023

https://en.wikipedia.org/wiki/Fish_farming#/media/File:Fish_farm_Amarynthos_Euboea_Greece.jpg. Accessed August 14, 2023

https://en.wikipedia.org/wiki/Harmful_algal_bloom#/media/File:La-Jolla-Red-Tide.780.jpg. Accessed August 14, 2023

https://en.wikipedia.org/wiki/Pseudo-nitzschia#/media/File:Pseudonitzschia2.jpg). Accessed August 14, 2023

https://hab.whoi.edu/impacts/impacts-human-health/emerging-toxins/. Accessed August 2, 2023

https://upload.wikimedia.org/wikipedia/commons/c/c0/The_Birds_original_poster.jpg. Accessed August 14, 2023

https://www.flickr.com/photos/myfwc/5807660937. Accessed August 14, 2023

Ocean Acidification and the Souring of Our Waters 26

26.1 Introduction

Atmospheric CO_2 has increased about 100 ppm (~30%) since the late 1950s, when David Keeling of the Scripps Institution of Oceanography began measuring it. The year-over-year increase shows no signs of slowing. The year 2013 was when atmospheric concentrations exceeded 400 ppm for the first time (in at least a million years). Time series from the Atlantic and Pacific all show the same trends: increasing CO_2 levels and associated decreasing pH (Fig. 26.1). There is seasonality in these values, due to the influence of plant photosynthesis, which is weighted toward northern hemisphere productivity, drawing down the concentration to a greater extent in summer.

Global oceans have absorbed about 30% of these carbon (C) emissions and 90% of anthropogenic heat since preindustrial times (Gruber et al. 2019). CO_2 concentrations in the surface waters of the ocean are in equilibrium with atmospheric CO_2 concentrations. The absorption of this increasing amount of CO_2 by the oceans was initially considered a favorable flux, thought to be a mechanism whereby the heating of the earth would be minimized even with heat trapped under the growing blanket of atmospheric CO_2. It was this line of thinking that led John Martin and others to propose fertilizing the oceans with iron (Fe), so that a phytoplankton bloom could absorb even more CO_2 (Chap. 16). Then the ocean started responding. It was only in 2003 that the term Ocean Acidification (OA) came into common use, when the first models of effects of increasing atmospheric CO_2 on ocean chemistry were published (Caldeira and Wickett 2003). Caldeira and Wickett (2003) used one of the scenarios from the Intergovernmental Panel on Climate Change (IPCC) to estimate CO_2 emissions through 2100, then used a logistic function to take the estimate further into the future. With an assumed unabated CO_2 trajectory, and with their understanding of ocean chemistry, these authors concluded that the expected future changes in ocean pH are greater than those experienced in the past 300 million years. Indeed, pH values in the oceans have changed significantly in the past decades (Fig. 26.1). To date, this decline has been modest, ~0.1 pH units (Fig. 26.1). Although the drop in pH

P. M. Glibert, *Phytoplankton Whispering: An Introduction to the Physiology and Ecology of Microalgae*, https://doi.org/10.1007/978-3-031-53897-1_26

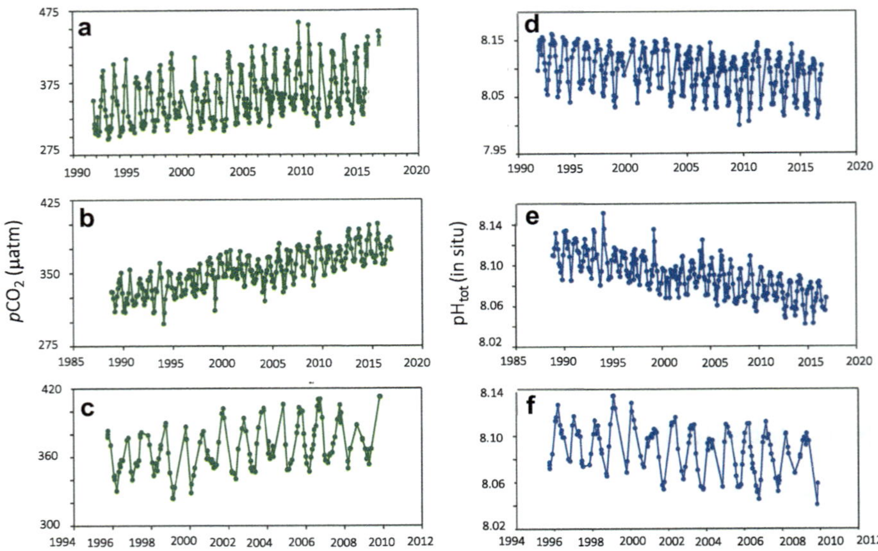

Fig. 26.1 Time series of in situ pCO_2 (Panels **a-c**) and pH_T (Panels **d–f**) for three time series stations, including the Bermuda Atlantic Time Series (BATS, Panels **a** and **d**), the Hawaii Ocean Time Series (HOTS, Panels **b** and **e**) and the European Station for Time Series in the Canary Islands (ESTOC, Panels **c** and **f**). Data sources: BATS data: http://batsftp.bios.edu/BATS/bottle/; Hot data: University of Hawaii (http://hahana.soest.hawaii.edu/hot/products/HOT_surface_CO2; ESTOC data: https://www.nodc.noaa.gov/archive/arc0051/0100064/1.1/data/0-data/). Image reproduced from Tilbrook et al. (2019, Front Mar Sci 6: doi.org/10.3389/fmars.2019.00337) under a creative commons license

does not currently seem large, it represents a 30% increase in seawater [H+]. The effects of OA on productivity and ocean chemistry are large (Fig. 26.2).

This chapter first explores the effects of increasing CO_2 on ocean chemistry. Then, the complex effects of rising CO_2 and decreasing pH on phytoplankton photosynthesis, N_2 fixation, calcification, algal toxicity, and community interactions are explored. The following chapter examines the responses of phytoplankton to the intersection of the multiple stressors of hypoxia, warming, HABs and C chemistry changes that are now co-occurring in several rapidly changing regions of the oceans, coastal waters, and estuaries.

26.2 Ocean Carbon Chemistry

Several reactions occur when atmospheric CO_2 comes in contact with oceanic water. First, the gas becomes dissolved (aqueous, aq). The aqueous CO_2 reacts with water forming carbonic acid, which rapidly dissociates to bicarbonate and H^+. Bicarbonate

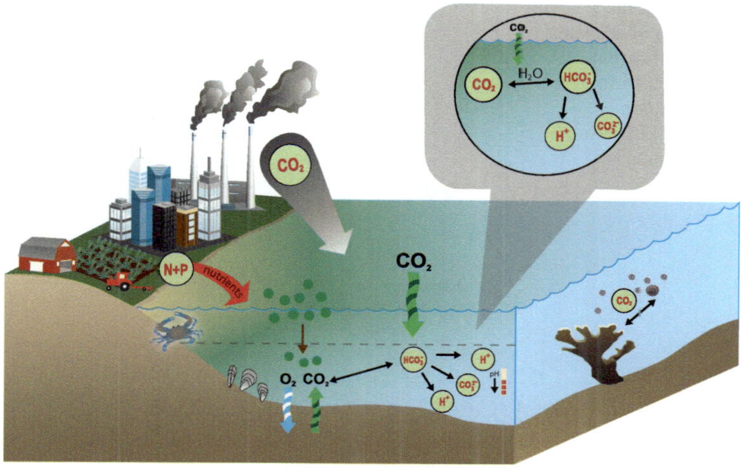

Fig. 26.2 Conceptual diagram of the impact of acidification can have on waters that may be subject also to nutrient enrichment. Image from the Integration and Application Network, University of Maryland Center for Environmental Science (https://ian.umces edu/media-library/atmospheric-and-eutrophic-acidification/), reproduced under a creative commons license

dissociates to carbonate, also producing H^+ (Fig. 26.2). With more H^+, pH goes down, and the water acidifies. Expressed formally, these relationships are:

$$CO_{2gas} \leftrightarrow CO_{2aq} \tag{26.1}$$

$$CO_{2aq} + H_2O \leftrightarrow H_2CO_3 \tag{26.2}$$

$$H_2CO_3 \leftrightarrow HCO_3{}^- + H^+ \tag{26.3}$$

$$HCO_3{}^- \leftrightarrow CO_3{}^{2-} + H^+ \tag{26.4}$$

As more CO_2 is dissolved in seawater, the balance between the carbonate species changes (Fig. 26.3), with CO_{2aq} increasing and $CO_3{}^{2-}$ decreasing (Barker and Ridgewell 2012).

$$CO_{2aq} + H_2O + CO_3{}^{2-} \leftrightarrow 2HCO_3{}^- \tag{26.5}$$

Dissolved CO_2 in water can also react with carbonate minerals ($CaCO_3$). This mineral is fundamental for many marine organisms. Its reaction with CO_{2aq} also produces carbonate,

$$CaCO_3 + CO_{2aq} + H_2O \leftrightarrow Ca^{2+} + 2\,HCO_3{}^- \tag{26.6}$$

The aragonite saturation state, Ω, is used as the common measure of the relationship between $CO_3{}^{2-}$ in seawater and its value at saturation,

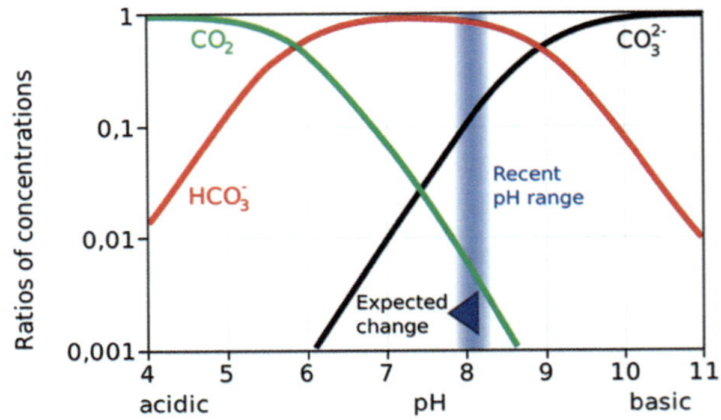

Fig. 26.3 Change in carbonate system of seawater from ocean acidification. Image reproduced from Wikipedia (https://en.wikipedia.org/wiki/Bjerrum_plot), public domain

$$\Omega = \left(\left[Ca^{2+}\right]_{SW} \times \left[CO_3{}^{2-}\right]\right) / [CaCO_3] \qquad (26.7)$$

or

$$\Omega = \left[CO_3{}^{2-}\right]_{seawater} / \left[CO_3{}^{2-}\right]_{saturation} \qquad (26.8)$$

At a value of $\Omega = 1$, the seawater is in equilibrium with respect to aragonite. An aragonite saturation state (Ω) greater than 1.0 indicates supersaturation (indicating aragonite precipitation), while values less than 1.0 indicate undersaturation which favors dissolution. The value of Ω will decrease with OA as $[CO_3{}^{2-}]_{seawater}$ decreases. This change has varying effects on organisms that calcify. It is the waters that are undersaturated with respect to carbonate ions that are corrosive or detrimental to organisms with calcified shells or exoskeletons. These corrosive effects are the most commonly considered impacts of OA. For phytoplankton, this impact would have significant consequences for coccolithophores, which possess a $CaCO_3$ shell. However, there are also effects on photosynthesizing phytoplankton, but these depend on the type of phytoplankton and their photosynthesis physiology. Characterizing and understanding these effects also depends on the methods used to alter seawater chemistry. These methods and effects on photosynthesis are next reviewed.

26.3 Methods Considerations: Altering Seawater Carbon Chemistry Is Not Easy

Studies on the effects of OA require some degree of manipulation of seawater C chemistry. There are multiple ways to do this, and each alters the C chemistry in a somewhat different way. Typically, studies of elevated atmospheric CO_2 are studied either by changing pCO_2 or changing pH. There are advantages and disadvantages of each approach.

Seawater pCO_2 can be altered by bubbling with the gas. Cultures can be bubbled directly. A variation on this theme is to mix CO_2-saturated water with the growth media in known proportions. Using both approaches, an investigator can achieve the desired pCO_2 concentration. It is of note that if one is conducting culture experiments where sterile media is required, the process of autoclaving degasses the media, so a re-equilibration must be done or taken into account when new gases are added (LaRoche et al. 2011). Another approach to changing C chemistry is to add a strong acid or base, such as HCl or NaOH to the growth media. This approach is comparatively easy to implement, but it modifies total alkalinity without changing total inorganic C concentration.

Adding CO_2 (by whichever delivery mechanism is chosen) may be the more "realistic" approach to changing C chemistry than adding acid, but there are also drawbacks to this approach. One drawback is that aeration may enhance surface coagulation of organic matter (Engel et al. 2004). This has consequences for microbial activity. Furthermore, some phytoplankton, especially dinoflagellates or filamentous cyanobacteria, do not grow well under conditions of continuous bubbling due to the turbulence created in the growth chambers (Shi et al. 2009). Moreover, in any culture experiment involving the effects of C chemistry, one of the major difficulties in maintaining a target CO_2 or pH level is due to the biology itself associated with photosynthetic and respiratory fluxes of CO_2. As biomass builds up, CO_2 and O_2 can change significantly in the culture, including on a diel basis. Where possible, optically thin cultures are desirable, which limits the magnitude of photosynthesis and respiration. With respect to mesocosm experiments (see Chap. 3) that allow a more complex community to be assessed, CO_2 control is exceedingly difficult. In field trials, remotely operated vehicles (ROVs) have been used to release liquid CO_2 in situ (e.g., Brewer et al. 2001).

26.4 OA and Phytoplankton Photosynthesis

Photosynthesis requires CO_2, so an increase in CO_2 should, in principle, be beneficial for phytoplankton. The fixation of CO_2 occurs via the carboxylase reaction of the enzyme Rubisco (ribulose-1,5-bisphosphatecarboxylase oxygenase, Chap. 6). Rubisco is an enigmatic enzyme in that it has dual catalytic reactions—it reacts with CO_2 (carboxylase activity) but also with O_2 (oxygenase activity). The challenges for algae in acquiring sufficient CO_2 for growth are several-fold. First, not only does Rubisco have a comparatively low affinity for CO_2 relative to O_2, but the concentrations of O_2 in the cell are much higher than those of CO_2. Also, the diffusion of CO_2 in water is 10,000 slower than the diffusion of CO_2 in air (Moroney and Ynalvez 2007). Third, with fluctuating levels of inorganic C and pH in water, the availability of CO_2 and HCO_3^- changes. Thus, it is an "abominably perplexing puzzle" (Tcherkez et al. 2006, p. 7246) as to how this enzyme can be responsible for biomass production.

To overcome these challenges, algae have evolved mechanisms to concentrate CO_2 at the site of Rubisco. These mechanisms are generally referred to as "carbon concentrating mechanisms" or CCMs. By concentrating CO_2 in the vicinity of Rubisco, the problem of the dual affinity of Rubisco is lessened. Algae have evolved multiple mechanisms to concentrate CO_2 at the site of carboxylation involving transporters and subcellular processes that speed up the equilibrium of the carbonate system. Earlier evolved algal groups (e.g., cyanobacteria and dinoflagellates) appear to possess more active CCMs and also have a higher degree of CCM plasticity (Van de Waal et al. 2019). The advantages of CCMs to the cell also include energy savings relative to photorespiratory metabolism and a reduced cell quota of Rubisco per unit biomass (Kupriyanova et al. 2023). In all, algae have been so successful at these adaptations that it is now estimated that more than 80% of marine primary productivity is facilitated by some form of CCM (Raven and Beardall 2016; Raven et al. 2017; Griffiths et al. 2017). Freshwater red algae, chrysophyte and synurophyte algae are more likely to lack CCMs (Giordano et al. 2005).

Common to the CCMs of all algae is the possession of carbonic anhydrase (CA) enzymes (both internal and external) that are zinc-containing enzymes responsible for speeding up the pH-dependent equilibrium of the $CO_2 - H_2CO_3^-$ reaction which is otherwise quite slow (seconds time scale) without a catalyzer. The CA enzyme speeds up the reaction ~$10^5 - 10^6$ times. The CA enzyme does not increase the CO_2 concentration above the pH-dependent equilibrium, but it increases the CO_2 concentration above that which it would be when CO_2 is actively removed by photosynthesis. It creates an internal HCO_3^- pool. One can think of CA as facilitating a supply chain so that CO_2 availability is not limiting photosynthetic production, whereas without CA, the CO_2 supply can be a limiting step in production. The manner in which algae utilize CA and transporters varies among taxa.

26.4.1 Cyanobacteria

Cyanobacteria have well-developed and well-characterized CCMs (Giordano et al. 2005; Mackey et al. 2015; Kupriyanova et al. 2023). Cyanobacteria have up to two CO_2 and three HCO_3^- uptake transporters. Once in the cytoplasm, CO_2 is converted to HCO_3^- via activity of CA. This occurs because CO_2 would otherwise efflux from the cell via diffusion. Then, the HCO_3^- is subsequently concentrated into carboxysome structures (Fig. 26.4). These are small protein bodies and it is there that much of the cell's Rubisco and CA enzymes are concentrated. Different types of carboxysomes have been identified but they serve the same purpose. The activity of CA ensures that CO_2 at the site of Rubisco remains high during carboxylase activity. Such a CCM system has the potential of concentrating CO_2 in the vicinity of Rubisco about 1000-fold, thereby inhibiting reaction with O_2 (Sui et al. 2020).

The transport of HCO_3^- and CO_2 across the cell membrane is under pH control. As pH drops (increasing pCO_2), the efflux rate of CO_2 increases, necessitating more energy to be expended by the cell to transport and maintain internal inorganic C. Increasing acidity increases the energetic cost of the CCM (Mangan et al. 2016).

Oceanic taxa have fewer CO_2 and HCO_3^- uptake transporters than estuarine taxa (Price et al. 2008). For picoplankton such as *Prochlorococcus*, which has a streamlined genome, the CCM is simple, with one or two HCO_3^- pumps and a carboxysome but it is very efficient (Hopkinson et al. 2014). Even among the picoplankton, *Synechococcus* and *Prochlorococcus,* there are substantial differences

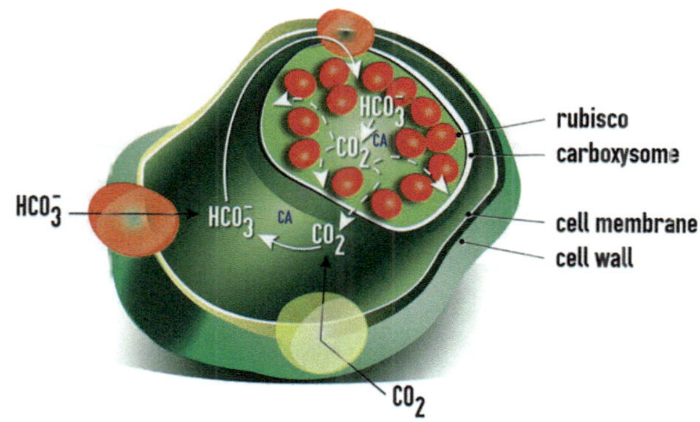

Fig. 26.4 The cyanobacterial carbon concentrating mechanism (CCM). Carbon acquisition occurs via active uptake of HCO_3^- and CO_2 via transport proteins on the cell surface (an exception is *Prochlorococcus*, which appears to lack CO_2 transporters). In the cytoplasm, inorganic C is maintained as HCO_3^- via the thylakoid NDH complex, which acts as a unidirectional carbonic anhydrase (CA). This step is thought to prevent diffusive loss of C due to the lower permeability of the cell membrane for HCO_3^- than for CO_2. The HCO_3^- enters a proteinaceous microcompartment, the carboxysome, via diffusion or possibly transport. The carboxysome contains most of the cell's rubisco. Inside the carboxysome, HCO_3^- is converted into CO_2 via CA at a rate sufficient to saturate rubisco and minimize photorespiration. The CO_2 is fixed by Rubisco into organic molecules as it passively diffuses through the carboxysome. Any CO_2 lost via leakage can be recovered in the cytoplasm and re-enter the concentration pathway. Image reproduced from Mackey et al. (2015, Oceanography 28: 74–91. doi.org/10.5670/oceanog.2105.33) under a creative commons license, as modified from Price et al. (2008, J Exp Bot 59:1441–1461. doi.org/10.1093/jxb/erm112)

in the functioning of the CCM. In one study, when CO_2 levels were increased to 750 ppm, no change in the photosynthesis of *Prochlorococcus* was observed, but the response by *Synechococcus* was pronounced. With its slightly larger size, *Synechococcus* and its higher phycobilin and chlorophyll *a* (chl *a*) content, higher light harvesting efficiency (α) and higher photosynthetic rates were observed when higher pCO_2 conditions were provided, with effects enhanced with warming (Fu et al. 2007). Moreover, in a field study, Lomas et al. (2012) found that the ratio of *Synechococcus* to *Prochlorococcus* increased with increasing pCO_2, suggesting a competitive edge under such conditions.

With regard to those cyanobacteria that are N_2-fixing, there are further complexities with changing pCO_2. Those complexities are described in Sect. 26.5.

26.4.2 Diatoms

Among diatoms, the CCM capacity varies. Many diatoms have well-developed CCMs (Raven and Beardall 2003; Armbrust et al. 2004; Raven et al. 2008; Mackey et al. 2015), and large diatoms have greater CCM capacity than do small diatoms, which helps overcome the disadvantage of their lower surface/volume ratio and associated additional diffusional constraints (Wu et al. 2014). Diatoms have

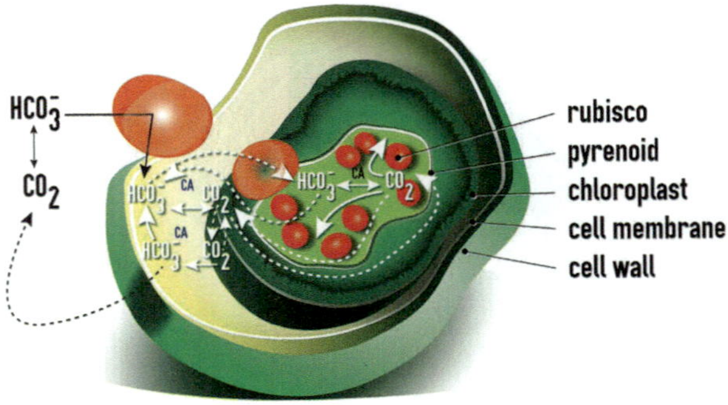

Fig. 26.5 The diatom CCM. Diatom cells acquire C through the active transport of HCO_3^- and passive diffusion of CO_2^-. All of the cell's membranes are highly permeable to CO_2, which diffuses between the compartments. The main driver of the CCM is active transport of HCO_3^- from the cytoplasm into the chloroplast. The HCO_3^- in the chloroplast then diffuses into the pyrenoid, a proteinaceous microcompartment enclosing Rubisco. Carbonic anhydrase (CA) in the pyrenoid generates CO_2 from HCO_3^-, saturating Rubisco and minimizing photorespiration. CO_2 lost through diffusion at any step in the concentration process can be recovered through conversion back into HCO_3^- by CA in the cytoplasm and stroma. Image reproduced from Mackey et al. (2015, Oceanography 28: 74–91. doi.org/10.5670/oceanog.2105.33) under a creative commons license, as modified from Hopkinson et al. (2011, Proc Natl Acad Sci USA 108:3830–3837)

transporters at the cell surface for transport of HCO_3^-, but CO_2 more freely diffuses into (and out of) the cells (Burkhardt et al. 2001). The cytoplasmic CA in diatoms may also continually maintain an inward CO_2 diffusive flux by facilitating the conversion of CO_2 to HCO_3^- (Matsuda et al. 2017; Fig. 26.5). The CCM transports HCO_3^- to the chloroplast, and from there into the pyrenoid, a substructure of the chloroplast in which Rubisco is housed.

Some diatoms use a "C4 pathway," which involves the fixation of HCO_3^- as opposed to CO_2. In the "C4 pathway," the enzyme phosphoenolpyruvate carboxylase (PEPCase) fixes HCO_3^- into the C4 sugar oxaloacetate (Reinfelder 2011). The fixed C is then decarboxylated near the site of Rubisco (Fig. 26.6, Mackey et al. 2015). In the fully sequenced *Thalassiosira pseudonana* and *Phaeodactylum tricornutum*, all genes required for a C4 pathway have been found to be present (Armbrust et al. 2004; Bowler et al. 2008). In a study by Kroth et al. (2008), it was shown that in C4 diatoms, the processes of decarboxylation of the C4 sugar, oxaloacetate, and malate, as well as carboxylation by Rubisco, occur in the mitochondria and plastid, respectively. Oxaloacetate is shuttled by way of malate and cleaved by a decarboxylating enzyme in the vicinity of Rubisco (Sage 2004). The conversion of CO_2 to HCO_3^- and back again requires energy; this has been suggested to also act as one of a number of energy-dissipating processes of diatoms (see also Chaps. 6 and 12).

Interestingly, the activity of PEPCase changes with exposure to different forms of N, and is generally higher under NO_3^- availability compared with NH_4^+ availability (Guo et al. 2007a, b). Therefore, under a constant inorganic C supply, the affinity

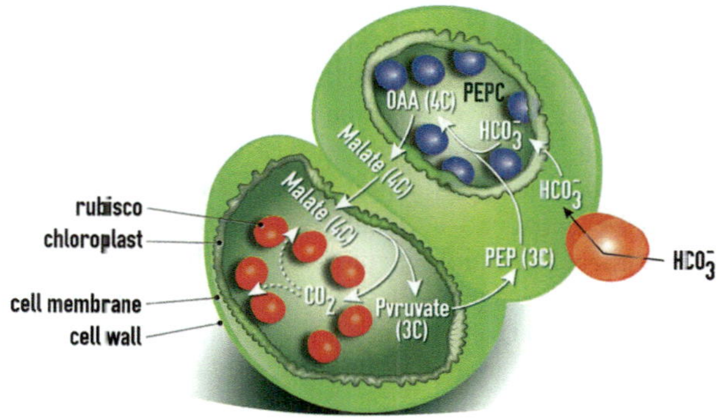

Fig. 26.6 The C4 photosynthetic pathway. In C4 carbon fixation, CO_2 is incorporated with phosphoenolpyruvate (PEP) into the 4-C molecule oxaloacetate by the enzyme phosphoenol-pyruvate carboxylase (PEPC), which may be located in the chloroplast or the cytoplasm. PEPC has a higher affinity for CO_2 than Rubisco and does not interact with O_2, so the efficiency of C fixation is high in this step. The C is transported as malate between cells to the other chloroplast, where decarboxylation liberates CO_2 in the vicinity of Rubisco, thereby saturating the enzyme and preventing photorespiration. Pyruvate is also generated in that step and is converted back into PEP by ATP-dependent phosphorylation (not shown). Carbon fixation then proceeds via the Calvin Cycle as in the C3 pathway. Due to the additional C fixation step, the C4 pathway requires an extra 12 molecules of ATP in addition to the 18 molecules required in C3 photosynthesis. Image reproduced from Mackey et al. (2015, Oceanography 28: 74–91. doi.org/10.5670/oceanog.2105.33) under a creative commons license

for CO_2 increases when NO_3^- is the growth substrate, resulting in higher C assimilation per unit N when a CCM is operating (Raven 1991; Raven et al. 2005; Glibert et al. 2016). Inversely, when NH_4^+ is the predominant growth substrate, the affinity of PEPCase for CO_2 is less, resulting in lower CO_2 at the site of Rubisco, and photorespiration may increase accordingly. The balance between NH_4^+ and NO_3^- also alters the redox balance and therefore growth conditions other than CO_2 availability may contribute to some of the variable effects of CO_2 change on diatom productivity.

Diatoms possess one of the most efficient forms of Rubisco for carboxylation over oxygenation, Form ID (Badger et al. 1998; Tortell 2000; Young et al. 2016). Thus, photorespiratory losses in diatoms are generally less than in other phytoplankton taxa (Rech et al. 2008). A modeling study on CCMs in diatoms (Zhang and Luo 2022) suggested that anthropogenically produced CO_2 would reduce the HCO_3^- requirement of diatoms—but the effect would be greatest for diatoms that are medium sized, ~7 μm diameter. This idea is in contrast with earlier studies suggesting larger diatoms would be favored (Wu et al. 2014). The Zhang and Luo model accounted for CO_2 diffusion and HCO_3^- transport as well as optimal allocation of energy between HCO_3^- acquisition and synthesis. Their model showed that with higher CO_2 there is higher permeability across the cell membrane, but this diminishes with higher cell C demand determined by higher rates of cell growth potential. These authors thus suggest that larger cells would not be favored in a future ocean (and export of such cells would not be stimulated).

Numerous studies focused on diatoms have shown they are able to adjust the CCM to adapt to variable CO_2 levels (Hoppe et al. 2018; Wolf et al. 2018; Karlusich et al. 2021). Under rising CO_2 conditions, it has been hypothesized that CCMs in diatoms would be downregulated, which would save intracellular energy, and productivity would be enhanced accordingly (Taylor et al. 2012; Biswas et al. 2017). Although some stimulation in diatoms has been observed in laboratory experiments with favorable light and nutrients (Wu et al. 2014), according to Hopkinson et al. (2011) and Goldman et al. (2017), in diatoms, a doubling of CO_2 would only reduce CCMs by about 20%. This would decrease energy expended on CO_2 fixation by about 3–6%, in turn, increasing primary productivity by only a few percent. Thus, they conclude that the overall effects of OA on growth, photosynthesis, and respiration of diatoms should be small over the coming century change in *p*CO2. In a rising CO_2 world, diatoms may also be disfavored in the competition with those cells that lack CCMs (e.g., Riesbesell et al. 2000; Fu et al. 2012; King et al. 2015).

Even though the presence of CCMs in diatoms in laboratory conditions has been well documented, until recently there has been some uncertainty regarding the functioning of CCMs in natural communities of diatoms. Using the metagenomic and metatranscriptomic data from the Tara Oceans project, the diversity of CCMs was investigated by searching for sequences that coded for the enzymes associated with diatom CCMs. These enzymes included CA, CbbX (Calvin-Benson-Bassham protein complex), PEPC (phosphoenolpyruvate carboxylase), PEPCK (phosphoenolpyruvate carboxykinase), and ME (malic enzyme) in the eukaryote catalog of the expedition (Carradec et al. 2018; Pierella Karlusich et al. 2021). These sequences were collected at all sites in all oceans, which covered a wide range in CO_2 availability (Fig. 26.7). With 203 metagenomes and 224 metatranscriptomes from 66 stations in the database, they were able to document global hotspots of CCMs by size fraction and gain insights into relationships with temperature, CO_2, and nutrient availability.

26.4.3 Dinoflagellates

Dinoflagellates also vary in their CCMs. Dinoflagellates have different forms of Rubisco, resulting in a progression of selectivity for CO_2 over O_2 This impacts the type of CCM they need. The peridinin-containing dinoflagellates possess Form II Rubisco that is less favorable for carboxylase activity than Forms 1A and 1B (cyanobacteria and Chl *b* containing algae and higher plants), and Form 1D (diatoms and haptophytes; Whitney and Andrews 1998; Tcherkez et al. 2006). Those dinoflagellates having tertiary endosymbiotic plastids or kleptoplastids (see Chap. 15) also have Rubisco Forms 1D or 1B (Raven et al. 2017, 2020). Accordingly, their need for a CCM is less. Common dinoflagellates with Form 1D are *Karenia brevis*, *Karenia mikimotoi*, and *Karlodinium veneficum* (Raven et al. 2020).

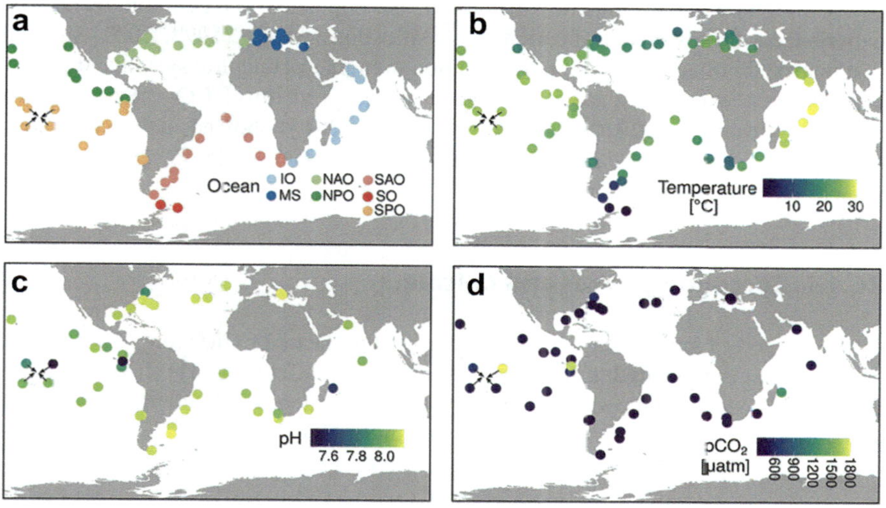

Fig. 26.7 Tara Ocean study. Panel (**a**) Ocean regions samples; Panel (**b**) Temperature; Panel (**c**) Panel (**d**) pCO_2. The sampling covers almost all main ecogeographic locations. The ocean regions included: IO, Indian Ocean; MS, Mediterranean Sea; NAO, North Atlantic Ocean; NPO, North Pacific Ocean; SAO, South Atlantic Ocean; SO, Southern Ocean; SPO, South Pacific Ocean. Image reproduced from Pierella Karlusich et al. (2021, Front Plant Sci 12:657821. doi.org/10.3389/fpls.2021.657821) under a creative commons license

Phagotrophic dinoflagellates also have a source of CO_2 from digestion of particles, but this is not a replacement for the need for a CCM (Raven et al. 2020).

Those dinoflagellates with Form II Rubisco appear to have a C3 pathway of C assimilation (Raven et al. 2020). The CCM of many dinoflagellates is not compartmentalized into a pyrenoid; only about half of the dinoflagellates studied to date have been found to have pyrenoids. For those without, Rubisco is dispersed in the stroma (Raven et al. 2020 and references therein). There is no convincing evidence of active CO_2 transport, and although mechanisms of inorganic C transport are not well understood, accumulation of inorganic CO_2 in the pyrenoid, or more generally in the stroma, must involve active transport of HCO_3^- at the chloroplast envelope with CA activity in the cytosol and chloroplast (Raven et al. 2020). Some dinoflagellates have transporters for HCO_3^- uptake, but this does not appear to be the case for all dinoflagellates (Dason et al. 2004; Mackey et al. 2015; Raven et al. 2020). More information is needed as to the location of intracellular CA and intracellular C transport in dinoflagellates (Raven et al. 2020). The PEPCase enzyme plays an important role in the CCMs of dinoflagellates.

Bloom-forming dinoflagellates appear to benefit from altered pCO_2, but once again, relationships are complex. Eberlein et al. (2016) showed that with elevated $pCO2$, CCMs in two dinoflagellates were downregulated and energy was reallocated to N acquisition, leading to an increased quota of N-rich compounds such as toxins and chl *a*. In blooms, as described in more detail in Chap. 27, conditions of elevated pH develop due to CO_2 drawdown, and thus the opposite of OA conditions may be experienced, that is, alkalinification (or basification). Hinga (1992, 2002) showed that marine dinoflagellates are more common in higher pH natural systems; this may be consistent with their capacity to photosynthesize at lower $CO_2 : HCO_3^-$ than diatoms (Raven et al. 2020). Van de Waal et al. (2013) showed, for a calcareous dinoflagellate, *Thoracosphaera heimii*, that elevated pCO_2 conditions reduced both growth rate and calcification.

26.4.4 Coccolithophores and Calcifiers

It was formerly thought that coccolithophores either lacked CCMs (e.g., Nimer et al. 1992) or that they tended to be of low efficiency (Reinfelder 2011). It was thought that CCMs were unnecessary because one consequence of calcification is the generation of H^+ which can drive the formation of CO_2 in the cell via the reaction,

$$2HCO_3^- + Ca^+ \rightarrow CaCO_3^- + CO_2 + H_2O \tag{26.9}$$

Thus, the calcification process was thought to act as a form of CCM (e.g., Reinfelder 2011).

Recently, new insights into coccolithophore CCMs have come to light, and evidence of highly active CCMs in coccolithophores *Emiliania huxleyi* and *Gephryoocapsa oceanica* has emerged (Bach et al. 2013; Larsen and Beardall 2023). Even so, these processes within these organisms have been difficult to unravel. Bach et al. (2013) showed that while photosynthesis and calcification are linked in these organisms, the CCMs are not driven by calcification. Their experimental design allowed them to separate the effects of CO_2 and HCO_3^- and pH on growth rates and on photosynthesis and calcification. They were able to show that coccolithophores appear to rely mainly on dissolved CO_2 for photosynthesis, but HCO_3^- for calcification (Bach et al. 2013). Bach et al. (2013) showed that at low DIC levels (which would be relevant when they form dense blooms), calcification rates dropped, leading to an upregulation of CCMs under such conditions. Thus, increasing CO_2 and OA should enhance photosynthesis over calcification. Indeed, many coccolithophores have been shown to reduce their rate of calcification at elevated CO_2 and may even produce degraded or aberrant cells (Mackey et al. 2015 and references therein). The effects of OA on calcification are discussed in more detail in Sect. 26.6.

26.4.5 Raphidophytes and Haptophytes

Far less is known about the existence of, or regulation of, CCMs in raphidophytes. The raphidophyte *Heterosigma akashiwo* has been shown to depend solely on diffusive CO_2 uptake (Nimer et al. 1997; Fu et al. 2008b). Accordingly, it should benefit from higher pCO_2. The haptophyte *Phaeocystis globosa* has an effective

CCM which appears to be saturated at present-day CO_2 levels. Thus, no change in growth rate with more CO_2 has been reported (Rost et al. 2003).

26.4.6 Chlorophytes

The model chlorophyte, *Chlamydomonas reinhardtii*, has been well characterized with respect to CCMs, and it is now well documented that across the entire green algal lineage, CCMs operate (Goudet et al. 2020; Kupriyanova et al. 2023). All chlorophytes have C3 biochemistry. The CCM of these cells relies on the pH gradient across the chloroplast thylakoid membrane that is established under light conditions. The chloroplast stroma has a pH of ~8, while the thylakoid lumen has a pH of ~4–5. This difference shifts the equilibrium of HCO_3^- and CO_2 (Fig. 26.3). Accordingly, HCO_3^- predominates in the chloroplast stroma while CO_2 predominates in the thylakoid (Moroney and Ynalvez 2007). There is thus a need for CA in the thylakoid to convert the HCO_3^- to CO_2. Then, the CO_2 must be made available to Rubisco before it can be converted back to HCO_3^- in the higher pH environment of the chloroplast stroma. A pyrenoid separates the CA in the stroma from the Rubisco. The Rubisco contained within is in a matric of knotted thylakoid tubules which are thought to be the mechanism of delivery of the CO_2 (Mukherjee et al. 2019; Goudet et al. 2020 and references therein). The CCM is inducible upon transfer to lower CO_2 levels. A recent conceptual model of CCMs in *C. reinhardtii* shows how the pathways of inorganic C change under conditions ranging from high CO_2 to low and very low CO_2. That model suggests that at high CO_2 the pyrenoid structures are weakened and Rubisco is found throughout the stroma (Kupriyanova et al. 2023).

26.5 OA Effects on N_2-Fixing Cyanobacteria

One of the important questions with respect to OA is the effect on N_2 fixation rates and how they may change in a future ocean. This question is far from resolved, however, as the results to date are confounded by different laboratory approaches for changing OA (changing $p$$CO_2$ or pH), and whether or not other limiting factors were imposed.

The energy for N_2 fixation competes with the energy required for photosynthesis and C fixation (Luo et al. 2019). When N_2-fixation rates are high, C fixation rates are generally not (e.g., Berman-Frank et al. 2001). One of the most abundant N_2-fixer in the oceans, *Trichodesmium*, has been the subject of most studies on the effects of OA and N_2 fixation, and the findings are rather mixed. Multiple studies have shown elevated N_2 fixation rates under elevated CO_2 conditions (e.g., Hutchins et al. 2007; Barcelos e Ramos et al. 2007; Levitan et al. 2010; Kranz et al. 2009; Hutchins and Fu 2017). Such increases have been suggested to be due to the lower energy expenditure on CCMs as opposed to increased photosynthetic rates. This is based on observations that neither O_2 evolution nor electron transport rates increased under the altered conditions (Levitan et al. 2010; Kranz et al. 2009). Similarly,

Crocosphaera displayed increased N_2 fixation rates under elevated CO_2 conditions, as well as increased growth and cellular pigment content (Fu et al. 2008a).

Other results have not been so consistent. The effects of OA on N_2 fixation are clearly dependent on the Fe conditions of the cells. Enzymes associated with N_2-fixation and photosynthesis are Fe rich (Chap. 16). Shi et al. (2010, 2012) have shown that under OA and Fe limitation, N_2 fixation is repressed in *Trichodesmium.* They suggested this was due to the fact that OA changes the binding characteristics of Fe ligands, resulting in a reduced bioavailability of this important element.

Hong et al. (2017) reported somewhat controversial results (see Hutchins et al. 2017) that an increase in pCO_2 and a decrease in pH may have independent effects on physiology, including N_2 fixation of *Trichodesmium* when Fe is limiting. By adjusting CO_2 and pH independently, they documented that growth increased under increased pCO_2, as did integrated daily rates of N_2 fixation—but the same effects were not seen with reduced pH (Fig. 26.8). At low pH, nitrogenase requires more energy and more enzymes and weakens intracellular pH buffering. Their study also showed that the inhibitory effects of pH can overwhelm the stimulatory effects of increased pCO_2. Thus, they observed that nitrogenase (as NifH abundance) increased under conditions of reduced pH, but they did not see this in the daily N_2 fixation rate. Overall, they projected that there could be a decline in N_2 fixation rates in a future ocean. They further explored the regulation of key proteins under different experimental conditions and, based on those data, proposed a conceptual model of the effects of acidification on transporters and on electron transport (Fig. 26.8). In all, they suggested that acidification with Fe limitation results in a reallocation of Fe among proteins, compensating for the loss of N_2-fixation efficiency, in turn affecting electron flow and therefore productivity.

Luo et al. (2019) also projected a decrease in N_2 fixation rates under future conditions. They used literature results in a model framework and showed that with a pH decrease of 0.2 units, CCMs of *Trichodesmium* could be downregulated, only marginally increasing growth rates, but decreasing nitrogenase activity substantially. Using this approach, they modeled the potential decrease in *Trichodesmium* N_2-fixation from the 1990s to 2090 and found that N_2 fixation would decrease by ~27%, with a higher decrease percentage where Fe is naturally limiting (Fig. 26.9).

Zhang et al. (2022) explored the interaction of OA and PO_4^{3-} limitation on N_2 fixation rates of *Trichodesmium.* Using PO_4^{3-}-limited chemostat-grown cultures, they showed that the negative effects of OA were amplified and that together these factors would cause a substantial decrease in global *Trichodesmium* N_2 fixation by 2100. Interestingly, the Zhang study reported that this synergistic effect was due to an accumulation of polyphosphate in the cells, potentially requiring the cells to reallocate energy reserves.

It is of note that studies of the freshwater cyanobacterium *Nodularia spumigena* from the Baltic Sea also showed decreases in N_2 fixation rates as pCO_2 increased (Czerny et al. 2009). These authors suggested that this was due to the deleterious effects of lower pH on the transport of N from the heterocysts to the vegetative cells. Resolving relationships between pCO_2 and pH changes and other limiting factors and environmental changes, how they interact and the synergistic or antagonistic

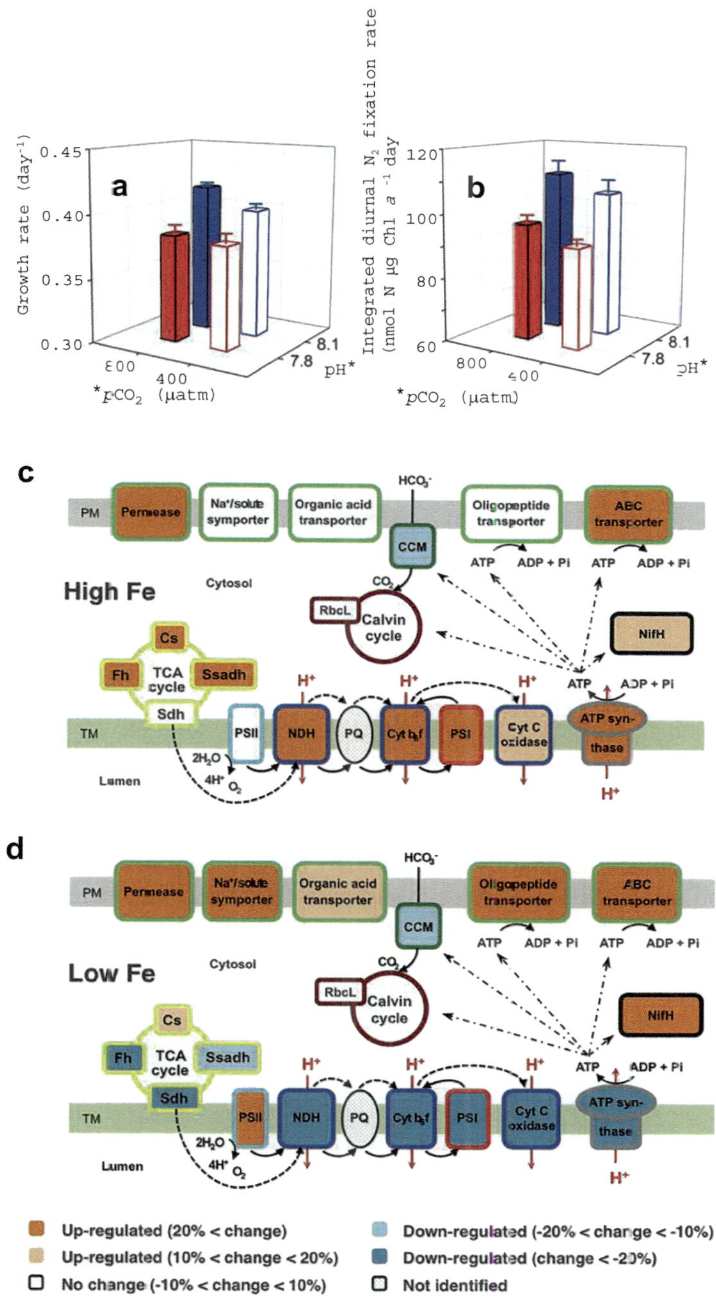

Fig. 26.8 Panels (**a**, and **b**) Separate effects of increasing pCO_2 and decreasing pH on *Trichodesmium erythraeum*. Panel (**a**) Specific growth rates. Panel (**b**) Integrated diurnal

Fig. 26.9 The projected *Trichodesmium* N_2 fixation potential. Panel (**a**) Results [mmol N (mol C)$^{-1}$ d^{-1}] in the 1990s and Panel (**b**) relative change from 1990s to 2090s. Image reproduced from Luo et al. (2019, Nat Comm 10: 1521. doi.org/10.1038/s41467-019-09554-7) under a creative commons license

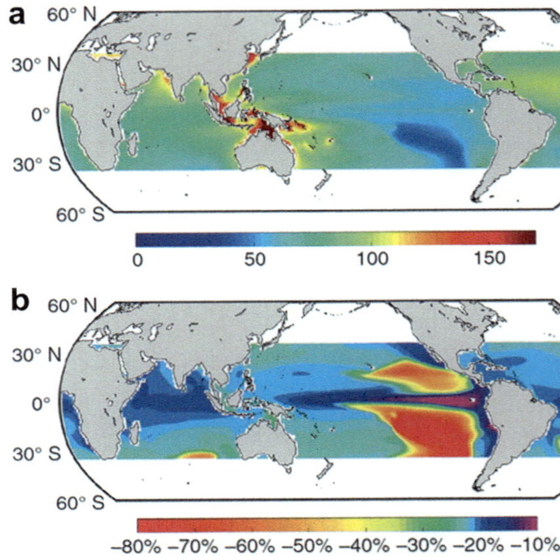

effects they cause, open up many questions and are the subject of much current research (Chap. 27, Glibert et al. 2022; Lin 2023).

26.6 OA and Calcification

Ocean acidification is largely thought to reduce calcification rates for those phytoplankton that calcify (Gao et al. 1993; Riebesell et al. 2000). As CO_2 in water increases, concentrations of HCO_3^- and H^+ increase, and the Ω decreases. Under current conditions, ocean waters are generally above the saturation point for calcite, so organisms favorably precipitate this material. However, with increased CO_2, this is not the case, and waters may become undersaturated with respect to Ω by the year 2100 (Orr et al. 2005). This should increase the availability of inorganic C for photosynthesis (Beardall et al. 2009; Raven and Beardall 2014). However, due to the thinning of shells, acidification can augment the impact of UV on calcifiers, leading to increased photoinhibition of photosynthesis (Gao and Zheng 2010). Thus,

Fig. 26.8 (continued) N_2-fixation rates, where pCO_2 and pH were varied independently. Blue and red denote pH 8.1 and 7.8, respectively; solid bars denote 800 μatm CO_2, and open bars denote 400 μatm CO_2. Error bars represent the SD of biological replicates ($n = 4$). Panels (**c**, **d**) Schematic representations of the functions of various proteins and of their change in abundance in response to acidified conditions at high and low Fe concentrations. Images reproduced from Hong et al. (2017 Science 356, 527–531. doi.org/10.1126/science.aal2981) with permission from the American Association for the Advancement of Science the response of calcifiers to OA is complicated (Beardall and Raven 2013). Ziveri et al. (2023) showed that in the North Pacific, a large proportion of pelagic $CaCO_3$, most of which is due to coccolithophores, is remineralized within the photic zone, in terms impacting the estimate of export flux of calcifying organisms.

There has been considerable genomic and transcriptomic work on the dominant calcifier *Emiliania huxleyi* and its responses to environmental perturbations (reviewed by Taylor and Brownlee 2016). Results are confounded by the use of varying methods by different investigators, various periods of acclimation and strain-specific responses. For example, while Riebesell et al. (2000) and many others have shown a decrease in calcification with pCO_2 increases, Lohbeck et al. (2012) showed higher rates of calcification in *E. huxleyi* after growing it for more than 500 generations. Thus, the extent to which these calcifiers may be advantaged or disadvantaged in a future ocean remains unclear.

The Tara Oceans expeditions again provide some supporting evidence for major effects on calcifiers. Using material collected on these expeditions, CT scans were performed on foraminifera samples from the equatorial Pacific and were compared to scans performed on historically collected samples from the Challenger expedition (1872–1876) and held in the Natural History Museum of London (Fox et al. 2020). All recent samples had up to 76% thinner shells than the historic samples (Fig. 26.10). The effect is clear, even if all the mechanisms of interaction are not.

26.7 OA and HAB Toxicity

The potential for increases in toxicity of HABs with increasing pCO_2 is also of concern (Boyd and Hutchins, 2012; Gobler et al. 2017). Some patterns and effects on key species or genera are emerging. For example, various studies have suggested an increase in toxicity of different *Alexandrium* species. An overall trend of increasing toxicity with increasing pCO_2 has been reported for numerous strains of cultured *Alexandrium ostenfeldii* (Kremp et al. 2012) and for *A. minutum* (Lian et al. 2022). The latter study showed that both intracellular and extracellular toxins were significantly enhanced at high (800 and 1200 ppm) pCO_2 levels. In the toxigenic diatoms *Pseudo-nitzschia multiseries* and *P. fraudulenta,* concentrations of their toxin, domoic acid, have been shown to increase at high CO_2/low pH in some studies (e.g., Sun et al. 2011; Tatters et al. 2012). Changes in the growth and toxicity of another toxigenic dinoflagellate, *Karenia brevis*, have also been shown to differ when exposed to different pCO_2 levels (e.g., Errera et al. 2014; Bercel and Kranz 2019). The latter study found an increasing trend in brevetoxin with enhanced pCO_2 levels.

As discussed previously with regard to OA and N_2 fixation, culturing approaches can also affect the extent to which relationships are observed between elevated pCO_2 and toxicity in some harmful algal species (Van de Waal et al. 2011). The synthesis of at least some toxins is light dependent, as for example, karlotoxin production by *Karlodinium veneficum* and saxitoxin production by *Alexandrium catenalla* (Proctor et al. 1975; Adolf et al. 2008). Thus, light exposure during experimentation may alter

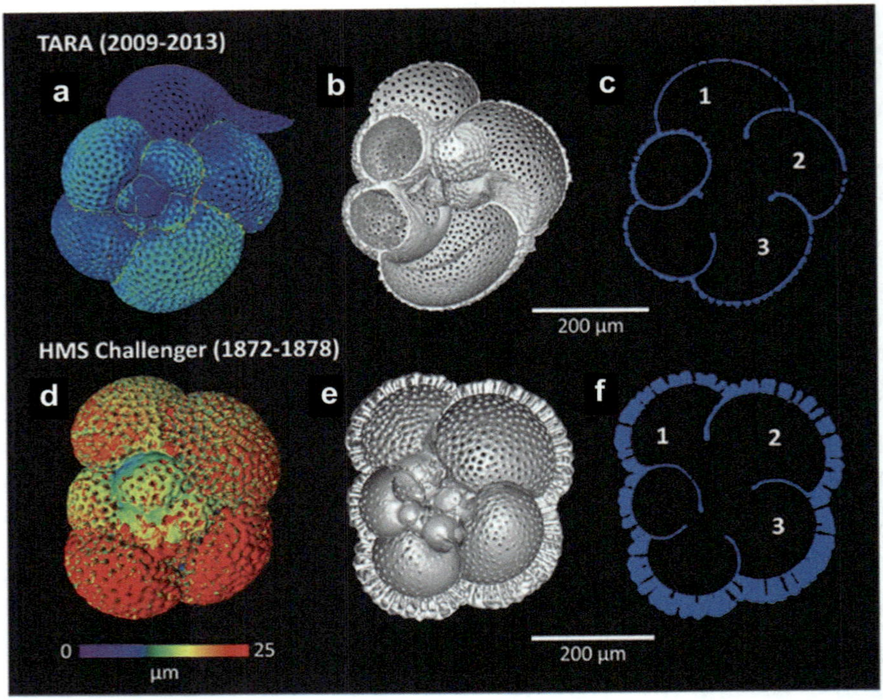

Fig. 26.10 Nano-CT representative reconstructions and measurements for *Neogloboquadrina dutertrei* from Tara (Panels **a**–**c**), and Challenger (Panels **d**–**f**). Panels (**a**) and (**d**)—color map illustrating shell thickness, warm colors representing areas of relatively thicker test wall; Panels (**b**) and (**e**)—cross section through rendered foraminifera test; and Panels (**c**) and (**f**)—slice through foraminifera test. Image reproduced from Fox et al. (2020, Sci Rep 10: 1620. doi.org/10.1038/s41598-020-58501-w) under a creative commons license

toxin production. Nutrient availability can also affect the response. Fu et al. (2010) showed that toxin production by *Karlodinium veneficum* was stimulated with an increase in CO_2 under P-limited conditions, but no change under nutrient replete conditions. On the other hand, Müller et al. (2019) reported that both the growth and ichthyotoxicity of *Karlodinium veneficum* decreased when pH was reduced from 8.03 to 7.51.

A recent meta-analysis of then-available studies on OA and HABs underscores the potential for increased proliferation and toxicity of HABs in a world with elevated pCO_2 and one that is simultaneously warmed (Brandenburg et al. 2019). The Brandenburg et al. (2019) study showed that there was a significant overall increase in the growth of dinoflagellate HABs by 20% under future CO_2 conditions but there were no consistently reported changes with respect to toxicity. The latter finding is, at least in part, due to the limited amount of data on toxicity changes under changing pCO_2 for many species. Gobler et al. (2017) modeled growth rate and bloom duration for species common to the North Atlantic and North Pacific oceans.

They found warming temperatures facilitated the amplification of HAB events. Thus, even without firm trends with regard to future CO_2 conditions and toxicity, the growth and duration of blooms at least several HAB species will likely continue to increase.

26.8 OA and Plankton Community Responses

The complexity of individual species responds to multiple factors such as OA and nutrient availability, but the interaction between and among species is even more complicated. To test the interactions of changing pH conditions and nutrient availability simulating eutrophic conditions, Flynn et al. (2015) explored the interaction of three phytoplankton taxa. They studied the weakly calcifying strain of the haptophyte (*Emiliania huxleyi*), the motile cryptophyte (*Rhodomonas* sp.), and a silicifying diatom (*Thalassiosira weissflogii*), maintaining cultures in present-day pH conditions (8.2), more acidic conditions expected in the future (7.6), and conditions that were more basic, representative of a dense bloom (8.8). Cultures were also compared when the pH was allowed to drift as growth progressed, versus when initial conditions were maintained by adding acid or base. Growth rates of *T. weissflogii* and *E. huxleyi* were almost halved in the alkaline cultures where pH was allowed to drift further upward during growth, whereas their growth rates were similar or slightly enhanced in systems that started the pH drift from more acidic conditions (Fig. 26.11). There was little difference between *Rhodomonas* grown in drift systems of different initial pH, at least during the nutrient replete phase. These results were coupled with models revealing potential windows of opportunity for each competing species under different conditions. Growth of *E. huxleyi* was favored under extant pH, with more acidic and basic conditions being unfavorable. *Rhodomonas*, which had the lowest maximum growth rate, had its highest competitive scope at elevated pH but only when nutrient replete; these are the conditions likely to characterize eutrophic areas during blooms. In eutrophic systems, acidification begins at a higher pH, while with acidification, basification begins at a lower pH. These different pH changes alter the scope for competitive interactions, community competition and succession, nutrient stoichiometry, and biogeochemical cycling in complex ways.

Other investigators have explored interactions in more realistic mesocosm studies (see also Chap. 3). In Sweden, Taucher et al. (2017) used 10 mesocosms that extended 19 m into the water column, encasing ~50 m^3 of water each, and used treatments representing end-of-century conditions and controls. Their additions of CO_2 to the treatments occurred at various intervals during the study. They found that copepod biomass increased by ~40% but the size class shifted, becoming dominated by smaller size classes. Large copepods nearly disappeared. They speculated that this was most likely due to bottom-up effects, rather than direct effects on the copepods per se. The CO_2 treatments resulted in higher phytoplankton productivity, with the most prominent effect on the picoeukaryotes. This may have been due to increased grazing by ciliates.

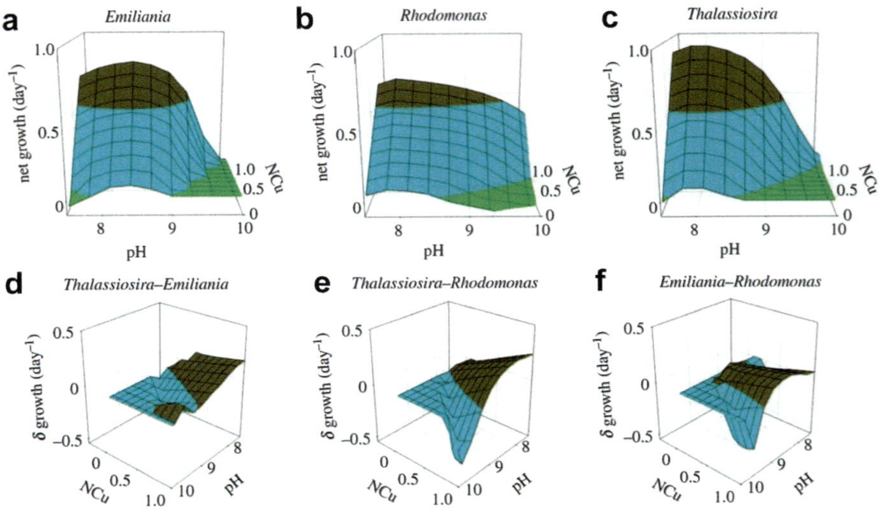

Fig. 26.11 Emergent modeled relationships between growth rate, pH, and nutrient status (NCu; where 0 is nutrient-starved and 1 is replete) for the prymnesiophyte *Emiliania huxleyi* (Panel **a**), the cryptophyte *Rhodomonas* sp. (Panel **b**) and the diatom *Thalassiosira weissflogii* (Panel **c**) and differences in net growth potential between pairs of algae (Panels **d**–**f**), with pH and nutrient status. Darkest zones (brown) indicate zones with growth rates >0.5 day^{-1}. In each plot, the light zones (blue) indicate where the second named species would outgrow the first named species; a value of d growth = zero indicates where neither species exhibited positive net growth. Images reproduced from Flynn et al. (2015, Proc R Soc B 282: 20142604. doi.org/10.1098/rspb.2014.2604) under a creative commons license

Hyun et al. (2020) studied the effects of present day (400 ppm CO_2), acidification (~900 ppm CO_2), and combined effects of acidification and warming of ~3 °C using 2400 L mesocosms in the southern coast of Korea. They found that picoplankton and small, chain-forming diatoms may benefit under future conditions. Large diatoms did not do well. The abundance of heterotrophic dinoflagellates also increased. Spisla et al. (2021), using large mesocosms, exposed to extreme high pCO_2 conditions for >50 days, reported lower chlorophyll *a*, decreased primary production and enhanced export flux of particulate material and pronounced top-down control relative to controls.

In a recent, large-scale study conducted in mesocosms across the world, Taucher et al. (2022) documented community changes and their elemental composition. They found striking differences in the elemental composition of the diatom debris at the bottom of the mesocosms. The ratio of silica (Si):N of the sinking material increased by nearly 20% when CO_2 conditions representative of 2100 were imposed. They suggested that diatoms, which have a protective organic coating around the Si cell wall, dissolved more slowly than the N-containing compounds in the cells. Thus, Si was exported to deeper waters at a rate faster than the export of N. They used a global database of sediment trap values in conjunction with earth system model simulations

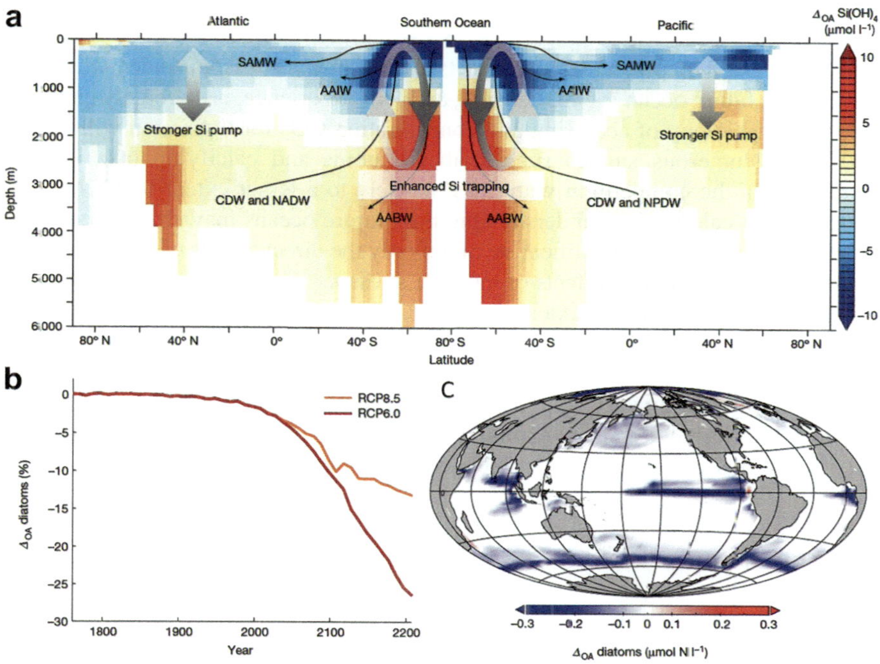

Fig. 26.12 Global impacts of slower silica dissolution under OA. Panel (**a**) Differences in the global vertical distribution of Si(OH)$_4$ (zonally averaged for the Atlantic and Pacific Oceans), including a depiction of the OA-driven downward transport of Si and its trapping in the deep ocean. AABW, Antarctic bottom water; AAIW, Antarctic intermediate water; CDW, circumpolar deep water; NADW, North Atlantic deep water; NPDW, North Pacific deep water; SAMW, sub-Antarctic mode water. Panel (**b**) OA effect on global diatom biomass under RCP8.5 and RCP6.0 emission scenarios of the IPCC. Panel (**c**) Spatial distribution of the difference in diatom biomass. Results show the net effect of the OA-driven slowdown of silica dissolution (Δ_{OA}), given as the difference to the standard model configuration (i.e., including impacts of climate change, but excluding OA effects on Si dissolution). Panels (**a**) and (**c**) show results from year 2200 of the RCP8.5 simulation. From Taucher et al. (2022 Nature 605: 696–700. doi.org/10.1038/s41586-022-04637-0) under a creative commons license

to suggest that the net effect will be a decline in diatoms by 13–26% by 2200. The change is projected to be highest in the Southern Ocean where diatom productivity is currently high (Fig. 26.12). The global effect stems from the trapping of the marine Si inventory in the deep ocean. Thus, based on these findings, the negative effect of OA on diatoms should be greater than any positive growth effect due to warming. Collectively, these findings underscore the complexity of OA interactions and add further evidence to the projection that diatoms will see a decline in the future ocean and vertical fluxes to the deep ocean will be altered.

26.9 Summary

The complexity of projecting the effects of the many stressors under which our present and future ocean is experiencing, and will experience, is extremely challenging. The physiology of acquiring and concentrating CO_2 for C fixation will change with OA. Numerous studies suggest that diatoms and calcifying taxa may be disfavored in the competition with cells that rely to a lesser extent on CCMs, and which do not calcify. Yet, understanding how future oceans may change requires a greater recognition and quantification of not only the direct effects of single stressors but the direct and indirect effects of multiple stressors. The effects of OA on rates of N_2 fixation depend on the extent to which other stressors, such as Fe or PO_4^{3-} limitation, may be simultaneously imposed. Toxicity of many HABs will change with effects of both acidification and warming. These interactions, and the added complexity of C chemistry changes in systems undergoing rapid change, from the Arctic to the coastal ocean, are explored in more detail in the following chapter. The need for further research on effect interactions is great.

References

Adolf JE, Bachvaroff TR, Place AR (2008) Environmental modulation of karlotoxin levels in strains of the cosmopolitan dinoflagellate, *Karlodinium veneficum* (Dinophyceae). J Phycol 45: 176–192

Armbrust EV, Berges JA, Bowler C, Green BR, Martinez D, Putnam NH, Zhou S et al (2004) The genome of the diatom *Thalassiosira pseudonana*: ecology, evolution, and metabolism. Science 306:79–86

Bach LT, Mackinder LCM, Schulz KG, Wheeler G, Schroeder DC, Brownlee C, Riebesell U (2013) Dissecting the impact of CO_2 and pH on the mechanisms of photosynthesis and calcification in the coccolithophore *Emiliania huxleyi*. New Phytol 199:121–134

Badger MR, Andrews TJ, Whitney SM, Ludwig M, Yellowlees DC (1998) The diversity and co-evolution of rubisco, plastids, pyrenoids and chloroplast-based CO_2-concentrating mechanisms in algae. Can J Bot 76:1052–1071. https://doi.org/10.1139/b98-074

Barcelos e Ramos J, Biswas H, Schulz KG, LaRoche J, Riebesell U (2007) Effect of rising atmospheric carbon dioxide on the marine nitrogen fixer *Trichodesmium*. Glob Biogeochem Cycles 21:GB2028

Barker S, Ridgewell A (2012) Ocean acidification. Nat Educ Knowledge 3(10):21

Beardall J, Raven JA (2013) Calcification and ocean acidification: new insights from the coccolithophore *Emiliania huxleyi*. New Phytol 199:1–3

Beardall J, Stojkovic S, Larsen S (2009) Living in a high CO_2 world: impacts of global climate change on marine phytoplankton. Plant Ecol Divers 2:191–205. https://doi.org/10.1080/17550870903271363

Bercel TL, Kranz SA (2019) Insights into carbon acquisition and photosynthesis in *Karenia brevis* under a range of CO_2 concentrations. Prog Oceanogr 172:65–76. https://doi.org/10.1016/j.pocean.2019.01.011

Berman-Frank I, Lundgren P, Chen YB, Kupper H, Kolber Z, Bergman B, Falkowski P (2001) Segregation of nitrogen fixation and oxygenic photosynthesis in the marine cyanobacterium *Trichodesmium*. Science 294:1534–1537. https://doi.org/10.1126/science.1064082

Biswas H, Shaik AUR, Bandyopadhyay D, Chowdhury N (2017) CO_2 induced growth response in a diatom dominated phytoplankton community from SW Bay of Bengal coastal water. Estuar Coast Shelf Sci 198:29–42. https://doi.org/10.1016/j.ecss.2017.07.022

Bowler C, Allen AE, Badger JH, Grimwood J, Jabbari K, Kuo A, Maheswari U et al (2008) The *Phaeodactylum* genome reveals the evolutionary history of diatom genomes. Nature 456:239–244

Boyd PW, Hutchins DA (2012) Understanding the responses of ocean biota to a complex matrix of cumulative anthropogenic change. Mar Ecol Prog Ser 470:125–135. https://doi.org/10.3354/meps10121

Brandenburg KM, Velthuis M, Van de Waal DB (2019) Meta-analysis reveals enhanced growth of marine harmful algae from temperate regions with warming and elevated CO_2 levels. Glob Change Biol 25:2607–2618. https://doi.org/10.1111/gcb.14678

Brewer PG, Peltzer ET, Rehder G (2001) Experiments on the deep ocean disposal of fossil fuel CO_2. EOS Trans Am Geophys Union 82:U32B-05

Burkhardt S, Amoroso G, Riebesell U, Sultemeyer D (2001) CO_2 and HCO_3^- uptake in marine diatoms acclimated to different CO_2 concentrations. Limnol Oceanogr 46:1378–1391. https://doi.org/10.4319/lo.2001.46.6.1378

Caldeira K, Wickett ME (2003) Anthropogenic carbon and ocean pH. Nature 425:365. https://doi.org/10.1038/425365a

Carradec Q, Pelletier E, Da Silva C, Alberti A, Seeleuthner Y, Blanc-Mathieu R et al (2018) A global ocean atlas of eukaryotic genes. Nat Commun 9:373

Czerny J, Barcelos e Ramos J, Riebesell U (2009) Influence of elevated CO_2 concentrations on cell division and nitrogen fixation rates in the bloom-forming cyanobacterium *Nodularia spumigena*. Biogeosciences 6:1865–1875. https://doi.org/10.5194/bg-6-1865-2009

Dason JS, Huertas IE, Colman B (2004) Source of inorganic carbon for photosynthesis in two marine dinoflagellates. J Phycol 40:285–292

Eberlein T, Van de Waal DB, Brandenburg KM, John U, Voss M, Achterberg EP, Rost B (2016) Interactive effects of ocean acidification and nitrogen limitation on two bloom-forming dinoflagellate species. Limnol Oceanogr 443:127–140. https://doi.org/10.3354/meps11568

Engel A, Delille B, Jacquet S, Riebesell U, Rochelle-Newall E, Terbrüggen A, Zondervan I (2004) Transparent exopolymer particles and dissolved organic carbon production by *Emiliania huxleyi* exposed to different CO_2 concentrations: a mesocosm experiment. Aquat Microb Ecol 34:93–104

Errera RM, Yvon-Lewis S, Kessler JD, Campbell L (2014) Reponses of the dinoflagellate *Karenia brevis* to climate change: pCO_2 and sea surface temperatures. Harmful Algae 37:110–116. https://doi.org/10.1016/j.hal.2014.05.012

Flynn KJ, Clark DR, Mitra A, Fabian H, Hansen PJ, Glibert PM, Wheeler GL et al (2015) Ocean acidification with (de)eutrophication will alter future phytoplankton growth and succession. Proc R Soc B 282:20142604. https://doi.org/10.1098/rspb.2014.2604

Fox L, Stukins S, Hill T, Miller CG (2020) Quantifying the effect of anthropogenic climate change on calcifying plankton. Sci Rep 10:1620. https://doi.org/10.1038/s41598-020-58501-w

Fu F-X, Warner ME, Zhang Y, Feng Y, Hutchins DA (2007) Effects of increased temperature and CO_2 on photosynthesis, growth, and elemental ratios in marine *Synechococcus* and *Prochlorococcus* (cyanobacteria). J Phycol 43:485–496. https://doi.org/10.1111/j.1529-8817.2007.00355.x

Fu F-X, Mulholland MR, Garcia N, Beck A, Bernhardt PW, Warner ME et al (2008a) Interactions between changing pCO_2, N_2 fixation, and Fe limitation in the marine unicellular cyanobacterium *Crocosphaera*. Limnol Oceanogr 53:2472–2484. https://doi.org/10.4319/lo.2008.53.6.2472

Fu F-X, Zhang Y, Warner ME, Feng Y, Sun J, Hutchins DA (2008b) A comparison of future increased CO_2 and temperature effects on sympatric *Heterosigma akashiwo* and *Prorocentrum minimum*. Harmful Algae 7:76–90. https://doi.org/10.1016/j.hal.2007.05.006

Fu F-X, Place AR, Garcia NS, Hutchins DA (2010) CO_2 and phosphate availability control of toxicity of the harmful bloom dinoflagellate *Karlodinium venefium*. Aquat Microb Ecol 59:55–65

Fu M, Wang Z, Pu X, Xu Z, Zhu M (2012) Changes in nutrient concentrations and N:P:Si ratios and their possible impacts on the Huanghai Sea ecosystem. Acta Oceanol Sinica 31:101–112

Gao K, Zheng Y (2010) Combined effects of ocean acidification and solar UV radiation on photosynthesis, growth, pigmentation and calcification of the coralline alga *Corallina sessilis*(-Rhodophyta). Glob Change Biol 16:2388–2398. https://doi.org/10.1111/j.1365-2486.2009.02113.x

Gao K, Aruga Y, Asada K, Ishihara T, Akano T, Kiyohara M (1993) Calcification in the articulated coralline alga Corallina pilulifera, with special reference to the effect of elevated CO_2concentration. Mar Biol 117:129–132. https://doi.org/10.1007/bf00346434

Giordano M, Beardall J, Raven JA (2005) CO_2 concentrating mechansism in algae: mechanisms, environmental modulation, and evolution. Annu Rev Plant Biol 56:99–131. https://doi.org/10.1146/annurev.arplant.56.032604.144052

Glibert PM, Wilkerson FP, Dugdale RC, Raven JA, Dupont C, Leavitt PR, Parker AE et al (2016) Pluses and minuses of ammonium and nitrate uptake and assimilation by phytoplankton and implications for productivity and community composition, with emphasis on nitrogen-enriched conditions. Limnol Oceanogr 61:165–197. https://doi.org/10.1002/lno.10203

Glibert PM, Cai W-J, Hall E, Li M, Main K, Rose K, Testa J, Vidyarathna N (2022) Stressing over the complexities of multiple stressors in marine and estuarine systems. Ocean-Land-Atmos Res 2022:9787258. https://doi.org/10.34133/2022/9787258

Gobler CJ, Doherty OM, Hattenrath-Lehmann T, Griffith AW, Kang Y, Litaker RW (2017) Ocean warming since 1982 has expanded the niche of toxic algal blooms in the North Atlantic and North Pacific oceans. Proc Natl Acad Sci USA 114(19):4975–4980. https://doi.org/10.1073/pnas.1619575114

Goldman JAL, Bender L, Morel FMM (2017) The effects of pH and pCO_2 on photosynthesis and respiration in the diatom *Thalassiosira weissflogii*. Photosynth Res 132:83–93. https://doi.org/10.1007/s11120-016-0330-2

Goudet MMM, Orr DJ, Melkonian M, Müller KH, Meyer MT, Carmo-Silva E, Griffiths H (2020) Rubisco and carbon-concentrating mechanism co-evolution across chlorophyte and streptophyte green algae. New Phytol 227:810–823. https://doi.org/10.1111/nph.16577

Griffiths H, Meyer MT, Rickaby REM (2017) Overcoming adversity through diversity: aquatic carbon concentrating mechanisms. J Exp Bot 68:3689–3695. https://doi.org/10.1093/jxb/erx278

Gruber N, Clement D, Carter BR, Feely RA, van Heuven S, Hoppema M, Key RM et al (2019) The oceanic sink for anthropogenic CO_2 from 1994 to 2007. Science 363:1193–1199

Guo S-W, Zhou Y, Gao YX, Li Y, Shen QR (2007a) New insights into the nitrogen form effect on photosynthesis and photorespiration. Pedosphere 17:601–610

Guo S-W, Zhou Y, Shen QR, Zhang F (2007b) Effect of ammonium and nitrate nutrition on some physiological processes in higher plants- growth, photosynthesis, photorespiration, and water relations. Plant Biol 9:21–29

Hinga KR (1992) Co-occurrence of dinoflagellate blooms and high pH in marine enclosures. Mar Ecol Prog Ser 86:181–187

Hinga KR (2002) Effects of pH on coastal marine phytoplankton. Mar Ecol Prog Ser 228:281–300

Hong H, Shen R, Zhang F, Wen Z, Chang S, Lin W, Kranz SA et al (2017) The complex effects of ocean acidification on the prominent N2-fixing cyanobacterium *Trichodesmium*. Science 356:527–531. https://doi.org/10.1126/science.aal2981

Hopkinson BM, Dupont CL, Allen AE, Morel FMM (2011) Efficiency of the CO_2-concentrating mechanism of diatoms. Proc Natl Acad Sci USA 108:3830–3837

Hopkinson BM, Young JN, Tansik AL, Binder BB (2014) The minimal CO_2-concentrating mechanism of *Prochlorococcus* spp. MED4 is effective and efficient. Plant Physiol 166:2205–2217

Hoppe CJM, Wolf KK, Schuback N, Tortell PD, Rost B (2018) Compensation of ocean acidification effects in arctic phytoplankton assemblages. Nat Clim Chang 8:529–533. https://doi.org/10.1038/s41558-018-0142-9

Hutchins DA, Fu F-X (2017) Microorganisms and ocean global change. Nat Microbiol 2:17508. https://doi.org/10.1038/nmicrobiol.2017.58

Hutchins DA, Fu F-X, Zhang Y, Warner ME, Feng Y, Portune K et al (2007) CO_2 control of *Trichodesmium* N_2 fixation, photosynthesis, growth rates, and elemental ratios: implications for past, present, and future ocean biogeochemistry. Limnol Oceanogr 52:1293–1304. https://doi.org/10.4319/lo.2007.52.4.1293

Hutchins DA, Fu F-X, Walworth NG, Lee MD, Saito MA, Webb EA (2017) Comment on "The complex effects of ocean acidification on the prominent N_2-fixing cyanobacterium *Trichodesmium*". Science 357:527–531. https://doi.org/10.1126/science.aao0067

Hyun B, Kim J-M, Jang P-G, Jang M-C, Choi K-H, Lee K, Yang EJ et al (2020) The effects of ocean acidification and warming on growth of a natural community of coastal phytoplankton. J Mar Sci Eng 8:821. https://doi.org/10.3390/jmse8100821

Karlusich JJP, Bowler C, Biswas H (2021) Carbon dioxide concentration mechanisms in natural populations of marine diatoms: insights from *Tara* oceans. Front Plant Sci 12:657821. https://doi.org/10.3389/fpls.2021.657821

King AL, Jenkins BD, Wallace JR, Liu Y, Wikfors GH, Milke LM, Meseck SL (2015) Effects of CO_2 on growth rate, C:N:P, and fatty acid composition of seven marine phytoplankton species. Mar Ecol Prog Ser 537:59–69

Kranz SA, Sultemeyer D, Richter K-U, Rost B (2009) Carbon acquisition in Trichodesmium: the effect of pCO_2 and diurnal changes. Limnol Oceanogr 54:548–559. https://doi.org/10.4319/lo.2009.54.2.0548

Kremp A, Godhe A, Egardt J, Dupont S, Suikkanen S, Casabianca S et al (2012) Intraspecific variability in the response of bloom-forming marine microalgae to changed climate conditions. Ecol Evol 2:1195–1207. https://doi.org/10.1002/ece3.245

Kroth PG, Chiovitti A, Gruber A, Martin-Jezequel V, Mock T, Parker MS et al (2008) A model for carbohydrate metabolism in the diatom *Phaeodactylum tricornutum* deduced from comparative whole genome analysis. PLoS One 3:e1426. https://doi.org/10.1371/journal.pone.0001426

Kupriyanova EV, Pronina NA, Los DA (2023) Adapting from low to high: an update to CO_2-concentrating mechanisms of cyanobacteria and microalgae. Plants 12:1569. https://doi.org/10.3390/plants12071569

LaRoche J, Rost B, Engel A (2011) Bioassay, batch culture and chemostat experimentation. In: Reibesell U, Fabry VJ, Hansson L, Gattuso J-P (eds) EUR24872–guide to best practices for ocean acidification research and data reporting. Publication Office for the European Union, Luxembourg, p 260. https://doi.org/10.2777/66906

Larsen SH, Beardall J (2023) Measurement of DIC acquisition and evidence for a CO_2concentrating mechanism in *Gephyrocapsa oceanica* (Isochrysidales, Coccolithophyceae). Phycologia 62:29–35. https://doi.org/10.1080/00318884.2022.2136445

Levitan O, Sudhaus S, LaRoche J, Berman-Frank I (2010) The influence of pCO_2 and temperature on gene expression of carbon and nitrogen pathways in *Trichodesmium* IMS101. PLoS One 5: e15104

Lian Z, Li F, He X, Chen J, Yu R-C (2022) Rising CO_2 will increase toxicity of marine dinoflagellate *Alexandrium minutum*. J Hazard Mater 431:128627

Lin S (2023) Phosphate limitation and ocean acidification co-shape phytoplankton physiology and community structure. Nat Commun 14:2699. https://doi.org/10.1038/s41467-023-38381-0

Lohbeck KT, Riebesell U, Reusch TBH (2012) Adaptive evolution of a key phytoplankton species to ocean acidification. Nat Geosci 5:346–351

Lomas MW, Hopkinson BM, Losh JL, Ryan DE, Shi DL, Xu Y, Morel FMM (2012) Effect of ocean acidification on cyanobacteria in the subtropical North Atlantic. Aquat Microb Ecol 66: 211–222. https://doi.org/10.3343/ame01576

Luo YW, Shi D, Kranz SA, Hopkinson BM, Hong H, Shen R, Zhang F (2019) Reduced nitrogenase efficiency dominates response of the globally important nitrogen fixer *Trichodesmium* to ocean acidification. Nat Commun 10:1521. https://doi.org/10.1038/s41467-019-09554-7

Mackey K, Morris JJ, Morel FMM, Kranz SA (2015) Response of photosynthesis to ocean acidification. Oceanography 28:74–91. https://doi.org/10.5670/oceanog.2105.33

Mangan NM, Flamholz A, Hood RD, Savage DF (2016) pH determines the energetic efficiency of the cyanobacterial CO_2 concentrating mechanism. Proc Natl Acad Sci USA 113:E5354–E5362. https://doi.org/10.1073/pnas.1525145113

Matsuda Y, Hopkinson BM, Nakajima K, Dupont CL, Tsuji Y (2017) Mechanisms of carbon dioxide acquisition and CO_2 sensing in marine diatoms: a gateway to carbon metabolism. Philos Trans R Soc Lond B 372:20160403. https://doi.org/10.1098/rstb.2016.0403

Moroney JV, Ynalvez RA (2007) Proposed carbon dioxide concentrating mechanism in *Clamydomonas reinhardtii*. Eukaryot Cell 6:1251–1259. https://doi.org/10.1128/EC.00064-07

Mukherjee A, Lau CS, Walker CE, Rai AK, Prejean CI, Yates G, Emrich-Mills T et al (2019) Thylakoid localized bestrophin-like proteins are essential for the CO_2 concentrating mechanism in *Chlamydomonas reinhardtii*. Proc Natl Acad Sci USA 116:16915–16920

Müller MN, Durantes-Aranda JJ, Seger A, Botana MT, Brandini FP, Hallegraeff GM (2019) Icthyotoxicity of the dinoflagellate *Karlodinium veneficum* in response to changes in seawater pH. Front Mar Sci 6:82. https://doi.org/10.3389/fmars.2019.00082

Nimer NA, Dixon GK, Merret MJ (1992) Utilization of inorganic carbon by the coccolithophorid *Emiliania huxleyi* (Lohman) Kamptner. New Phytol 120:153–158

Nimer NA, Iglesias-Rodriguez MD, Merrett MJ (1997) Bicarbonate utilization by marine phytoplankton species. J Phycol 33:625–631. https://doi.org/10.1111/j.0022-3646.1997.00625.x

Orr JC, Fabry VJ, Aumont O, Bopp L, Doney SC, Feely RA, Gnanadesikan A et al (2005) Anthropogenic Ocean acidification over the twenty-first century and its impact on calcifying organisms. Nature 437:681–686

Price GD, Badger MR, Woodger FJ, Long BM (2008) Advances in understanding the cyanobacterial CO_2-concentraing-mechanism (CCM): functional components, ci transporters, diversity, genetic regulation and prospects for engineering into plants. J Exp Bot 59:1441–1461. https://doi.org/10.1093/jxb/erm112

Proctor NH, Chan SL, Trevor AJ (1975) Production of saxitoxin by cultures of *Gonyaulax catenella*. Toxicon 13:1–9. https://doi.org/10.1016/0041-0101(75)90152-X

Raven JA (1991) Physiology of inorganic carbon acquisition and implications for resource use efficiency by marine phytoplankton: relation to increased CO_2 and temperature. Plant Cell Environ 14:779–794. https://doi.org/10.1111/j.1365-3040.1991.tb01442.x

Raven JA, Beardall J (2003) Carbon acquisition mechanisms of algae: carbon dioxide diffusion and carbon dioxide concentrating mechanisms. In: Larkum AWD, Douglas SE, Raven JA (eds) Photosynthesis in algae, 1st edn. Kluwer Academic Publ, pp 205–224

Raven JA, Beardall J (2014) CO_2 concentrating mechanisms and environmental change. Aquat Bot 118:24–37

Raven JA, Beardall J (2016) The ins and outs of CO_2. J Exp Bot 67:1–13

Raven JA, Ball LA, Beardall J, Giordan M, Maberly SC (2005) Algae lacking carbon-concentrating mechanisms. Can J Bot 83:879–890. https://doi.org/10.1139/b05-074

Raven JA, Cockell CS, De La Rocha CL (2008) The evolution of inorganic carbon concentrating mechanisms in photosynthesis. Philos Trans R Soc Lond B 363:2641–2650

Raven JA, Beardall J, Sánchez-Baracaldo P (2017) The possible evolution, and future, of CO_2-concentrating mechanisms. J Exp Bot 68:3701–3716

Raven JA, Suggett DJ, Giordano M (2020) Inorganic carbon concentrating mechanisms in free-living and symbiotic dinoflagellates and chromerids. J Phycol 56:1377–1397. https://doi.org/10.1111/jpy.13050

Rech M, Morant-Manceau A, Tremblin G (2008) Carbon fixation and carbonic anhydrase activity in Haslea ostrearia (Bacillariophyceae) in relation to growth irradiance. Photosynthetica 46:56–62. https://doi.org/10.1007/s11099-008-0011-2

Reinfelder JR (2011) Carbon concentrating mechanisms in eukaryotic marine phytoplankton. Annu Rev Mar Sci 3:291–315

Riebesell U, Zondervan I, Rost B, Tortell PD, Zeebe RE, Morel FM (2000) Reduced calcification of marine plankton in response to increased atmospheric CO_2. Nature 407:364–367. https://doi.org/10.1038/35030078

Rost B, Riebesell U, Burkhardt S, Sültemeyer D (2003) Carbon acquisition of bloom-forming marine phytoplankton. Limnol Oceanogr 48:55–67. https://doi.org/10.4319/lo.2003.48.1.0055

Sage RF (2004) The evolution of C_4 photosynthesis. New Phytol 161:341–370

Shi D, Xu Y, Morel FMM (2009) Effects of the pH/pCO_2 control method on medium chemistry and phytoplankton growth. Biogeosciences 6:1199–1207

Shi D, Xu Y, Hopkinson BM, Morel FMM (2010) Effect of ocean acidification on iron availability to marine phytoplankton. Science 327:676–679. https://doi.org/10.1126/science.1183517

Shi D, Kranz A, Kim JM, Morel FMM (2012) Ocean acidification slows nitrogen fixation and growth in the dominant diazotroph *Trichodesmium* under low-iron conditions. Proc Natl Acad Sci USA 109:E3094–E3100

Spisla C, Taucher J, Bach LT, Haunost M, Boxhammer T, King AL, Jenkins BD et al (2021) Extreme levels of ocean acidification restructure the plankton community and biogeochemistry of a temperate coastal ecosystem: a Mesocosm study. Front Mar Sci 7:611157. https://doi.org/10.3389/fmars.2020.611157

Sui N, Huang F, Liu L-N (2020) Photosynthesis in phytoplankton: insights from the newly discovered biological inorganic carbon pumps. Mol Plant 13:949–951

Sun J, Hutchins DA, Feng YY, Seubert EL, Caron DA, Fu F-X (2011) Effects of changing pCO_2 and phosphate availability on domoic acid production and physiology of the marine harmful bloom diatom *Pseudo-nitzschia multiseries*. Limnol Oceanogr 56:829–840. https://doi.org/10.4319/lo.2011.56.3.0829

Tatters AO, Fu F-X, Hutchins DA (2012) High CO_2 and silicate limitation synergistically increase the toxicity of Pseudo-nitzschia fraudulenta. PLoS One 7:e32116. https://doi.org/10.1371/jounral.pone.0032116

Taucher J, Haunost M, Boxhammer T, Bach LT, Algueró-Muñiz M, Riebesell U (2017) Influence of ocean acidification on plankton community structure during a winter-to-summer succession: an imaging approach indicates that copepods can benefit from elevated CO_2 via indirect food web effects. PLoS One 12(2):e0169737. https://doi.org/10.1371/journal.pone.0169737

Taucher J, Bach LT, Prowe AEF, Boxhammer T, Kvale K, Riebesell U (2022) Enhanced silica export in a future ocean triggers global diatom decline. Nature 605:696–700. https://doi.org/10.1038/s41586-022-04687-0

Taylor AR, Brownlee C (2016) Calcification. In: Borowitzka AM, Beardall J, Raven JA (eds) The physiology of microalgae. Springer International Publ, Cham, Switzerland, pp 301–318

Taylor AR, Brownlee C, Wheeler GL (2012) Proton channels in algae: reasons to be excited. Trends Plant Sci 17:675–684. https://doi.org/10.1016/j.tplants.2012.06.009

Tcherkez GGB, Farquhar GD, Andrews TJ (2006) Despite slow catalysis and confused substrate specificity, all ribulose bisphosphate carboxylases may be nearly perfectly optimized. Proc Natl Acad Sci USA 103:7246–7251

Tilbrook B, Jewett EB, DeGrandpre MD, Hernandez-Ayon JM, Feely RA, Gledhill DK, Hansson L et al (2019) An enhanced ocean acidification observing network: from people to technology to data synthesis and information exchange. Front Mar Sci 6:337. https://doi.org/10.3389/fmars.2019.00337

Tortell PD (2000) Evolutionary and ecological perspectives on carbon acquisition in phytoplankton. Limnol Oceanogr 45:744–750. https://doi.org/10.4319/lo.2000.45.3.0744

Van de Waal DB, Verspagen JMH, Finke JF, Vournazou V, Immers AK, Kardinaal WEA et al (2011) Reversal in competitive dominance of a toxic versus non-toxic cyanobacterium in response to rising CO_2. ISME J 5:1438–1450

Van de Waal DB, John U, Ziveri P, Resichart GJ, Hons M, Sluijs A, Rost B (2013) Ocean acidification reduces growth and calcification in a marine dinoflagellate. PLoS One 8:e95987

Van de Waal DB, Brandenburg KM, Keuskamp J, Trimborn S, Rokitta S, Kranz S, Rost B (2019) Highest plasticity of carbon concentrating mechanisms in earliest evolved phytoplankton. Limnol Oceanogr Lett 4:37–43. https://doi.org/10.1002/lol2.10102

Whitney SM, Andrews TJ (1998) The CO_2/O_2 specificity of single-subunit ribulose-bisphosphate carboxylase from the dinoflagellate, *Amphidinium carterae*. Australian J Plant Phys 25:131–138

Wolf KK, Hoppe CJ, Rost B (2018) Resilience by diversity: large intraspecific differences in climate change responses of an Arctic diatom. Limnol Oceanogr 63:397–411. https://doi.org/10.1002/lno.10639

Wu Y, Campbell DA, Irwin AJ, Suggett DJ, Finkel ZV (2014) Ocean acidification enhances the growth rate of larger diatoms. Limnol Oceanogr 59:1027–1034. https://doi.org/10.4319/lo.2014.59.3.1027

Young JN, Heureux AM, Sharwood RE, Rickaby RE, Morel FMM, Whitney SM (2016) Large variation in the rubisco kinetics of diatoms reveals diversity among their carbon-concentrating mechanisms. J Exp Bot 67:3445–3456. https://doi.org/10.1093/jxb/erw163

Zhang F, Wen Z, Wang S, Tang W, Luo Y-W, Kranz SA, Hong H, Shi D (2022) Phosphate limitation intensifies negative effects of ocean acidification on globally important nitrogen fixing cyanobacterium. Nat Commun 13:6730

Zhang Q, Luo Y-W (2022) A competitive advantage of middle-sized diatoms from increasing seawater CO_2. Front Microbiol 13:838629. https://doi.org/10.3389/fmicb.2022.838629

Ziveri P, Gray WR, Anglada-Ortiz G, Manno C, Greland M, Incarbona A, Rae JWB et al (2023) Pelagic calcium carbonate production and shallow dissolution in the North Pacific Ocean. Nat Commun 14:805. https://doi.org/10.1038/s41467-023-36177-w

Additional Websites Accessed

http://batsftp.bios.edu/BATS/bottle/. Accessed 2 Aug 2023

https://en.wikipedia.org/wiki/Bjerrum_plot. Accessed 2 Aug 2023

http://hahana.soest.hawaii.edu/hot/products/HOT_surface_CO2. Accessed 2 Aug 2023

https://ian.umces.edu/media-library/atmospheric-and-eutrophic-acidification/. Accessed 2 Aug 2023

https://www.nodc.noaa.gov/archive/arc0051/0100064/1.1/data/0-data/. Accessed 2 Aug 2023

27 Happening Now: Systems Experiencing Extreme Change and Multiple Stressors

27.1 Introduction

James J. (Jim) McCarthy (Fig. 27.1), an acclaimed phytoplankton ecologist, Nobel Prize Winner as co-Chair of the Intergovernmental Panel on Climate Change from 1997 to 2001, and recipient of many accolades, was a leading voice on the rapid changes of the Earth and ocean systems. In 2009, he wrote (McCarthy 2009), "A profound lesson from the past few decades of scientific discovery across the Earth and life sciences is that the weight of the human footprint on essential life-supporting services of the Earth system has grown dramatically since the time of Darwin." Although many of his writings on the topic of climate change are now more than a decade old, they are just as relevant today.

Jim described how profoundly and deeply he was impacted by the contrast of his visits to the Arctic nearly two decades apart (https://climatechange.environment.harvard.edu/james-mccarthy),

> The first time I was in the Arctic, the high Arctic, was late 1980s and I got up to about 80°N to Svalbard and I was just struck by the extraordinary beauty, the contrast of the clear sky, the brilliant ice, the glaciers, wildlife of spectacular form. It was enthralling in a way that brought together for me a lot of my interest in the ocean with natural history and climate. Eleven years later, on my second visit to the North Pole, this was in 2000, again on a Russian ice breaker, we were all expecting to see an Arctic Ocean like the one I had seen in [the 1980s], but as we moved north, we never encountered the really thick ice that we had expected. As we got to the North Pole, we found not just thin ice but actually a large area of open water...What we now know is that the amount of ice covering the Arctic Ocean in summer right now is about 20% of what was in 1980. It is not 20% less in area. It is less in area and in thickness. The combination of the two is what gives you the 20%...The Arctic is warming faster than most other regions on Earth...The ocean is interacting with the atmosphere in a way that it hasn't in the past...As the Arctic continues to lose ice and that's what all the trends show, we will see effects on climate effects on ocean currents that will affect people living in heavily populated areas of North America, Europe, and Asia.

P. M. Glibert, *Phytoplankton Whispering: An Introduction to the Physiology and Ecology of Microalgae*, https://doi.org/10.1007/978-3-031-53897-1_27

Fig. 27.1 James J. McCarthy. Photo by Christopher Michel. Photo reproduced from the Union of Concerned Scientists under a creative commons license

This chapter examines how the stressors of the world's oceans, temperature, hypoxia, HABs, and ocean acidification (OA) (Fig. 27.2; see also Chaps. 23, 24, 25, and 26), combine to alter phytoplankton composition and dynamics. Several case studies are described to illustrate not only how stressors interact, but also how the complexity of stressors may differ for ocean and estuarine ecosystems.

27.2 Multiple Stressors

Multiple stressor interactions include direct ecological effects, indirect ecological effects, and those that occur between stressors. These stressors may affect systems additively and synergistically but may also counteract each other. One recent meta-analysis (Crain et al. 2008) found that simple additive interactions occur in only 26% of studies, with synergistic (amplified effects) and antagonistic (dampened effects) interactions occurring in 36 and 38% of studies, respectively. Multiple stressors impact all aspects of food webs (Arrigo et al. 2020; Glibert et al. 2022). Some stressors can be a consequence of other stressors (e.g., HABs can result from nutrient pollution and can lead to hypoxia, and changes in CO_2 and pH can result from altered rates of productivity, including from HABs), but they can also be stressors in their own right (Fig. 27.3). While direct ecological effects may follow dose-response characteristics (i.e., are correlative), indirect ecological effects are more complex often resulting in responses that are different from those that would occur if operated singly. Interactions between stressors occur when one stressor affects the timing or intensity of another stressor (e.g., warming increases HABs or may change the timing of blooms). Indirect ecological responses and interactions among stressors in higher trophic levels are especially complex because of differing sensitivities and exposures among life stages within a species and stressors can also alter species interactions such as competition and predation that cascade through the food web.

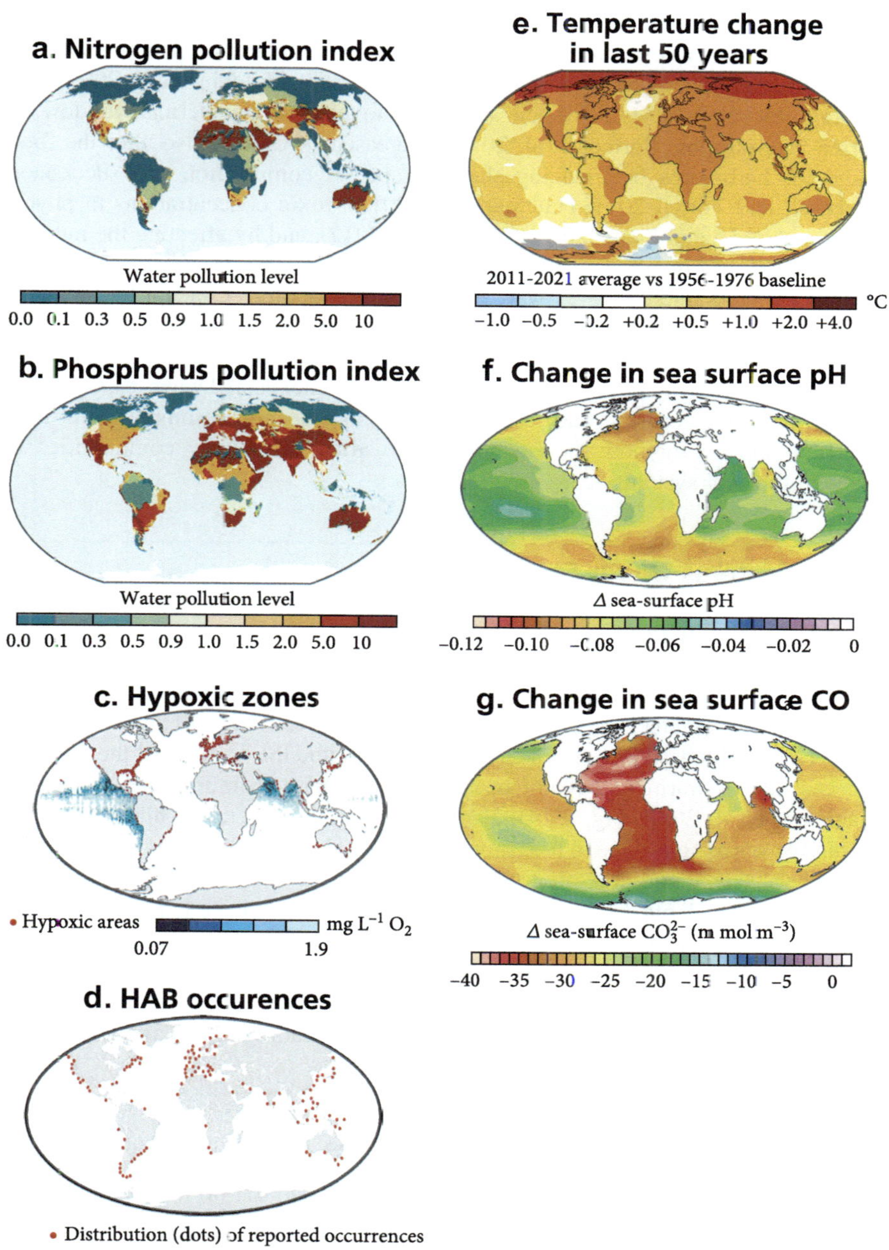

Fig. 27.2 Global multiple stressors. Global maps showing: Panels (**a** and **b**) index of nitrogen and phosphorus pollution per river basin, where values >1 indicate that the waste assimilation capacity of the river basin is insufficient and nutrient pollution results; Panel (**c**): hypoxic zones (<2 mg L^{-1} O_2, red dots) and oxygen minimum zones at 300 m (blue shaded regions); Panel (**d**): distribution of reported occurrences of HABs that produce paralytic shellfish poisoning; Panel (**e**): average air temperature change in last 50 years (2011–2021 average versus 1956–1976 baseline); Panel (**f**):

Changes in nutrient availability, form and proportion, food availability, food quality, community structure, and habitat availability among other factors all affect food web responses in complex ways (Feng et al. 2009; Hutchins et al. 2009). For example, elevated pCO_2 has been suggested to narrow the thermal window of marine heterotrophs (Pörtner et al. 2011). Increased pCO_2 can also alter the food quality of herbivores by inducing shifts in species composition and succession (Reibesell and Tortell 2011; Hutchins et al. 2013), toxin concentrations in phytoplankton (Bercel and Kranz 2019; Kremp et al. 2012), and by affecting the nutrient stoichiometry of organic matter. An important consideration is the ability (or lack thereof) of some microbes—as well as larger heterotrophs—to adapt to new CO_2 (or other) conditions with multigenerational exposures (e.g., Collins 2011; Schaum and Collins 2014). Beyond the intuitive and direct responses, eutrophication and CO_2 enrichment alter our perspective of OA in some unexpected ways, especially in nutrient-impacted coastal and estuarine systems. Several examples are next presented illustrating how these stressors are affecting aquatic communities in different types of systems.

27.3 Systems Under Stress

27.3.1 Arctic Ocean

The Arctic is warming at an alarming rate, 3–4 times faster than the global average (AMAP 2021; Rantanen et al. 2022), a phenomenon referred to as Arctic Amplification (AA). Although there is much uncertainty around the estimate of the AA, this extreme rate of warming has not been fully incorporated into predictive models (Rantanen et al. 2022; Fig. 27.4). Even more alarming is that while the annual average of warming is three to four times the global average, the period of October to December has warmed at a rate five times the global average (Fig. 27.4; Rantanen et al. 2022). Accelerated warming and increased irradiance have led to reduced ice cover in winter and greater losses in spring and summer (Perovich et al. 2007; Ardyna and Arrigo 2020). Such extreme rates of warming are due to the reduction of sea-ice cover and changes in atmospheric circulation. With a less reflective surface (less ice), there is more absorption of solar radiation, leading to this accelerated rate

Fig. 27.2 (continued) estimated change in seawater pH caused by oceanic uptake of anthropogenically created CO_2 between the 1700s and the 1990s; Panel (**g**): estimated change in annual mean sea surface CO_3^{2-} between the 1700s and the 1990s. Panels (**a**) and (**b**) from Mekonnen and Hoekstra (2015, 2017); Panel (**c**) modified from Breitburg et al. (2018); Panel (**d**) is modified from https://hab.whoi.edu/maps/regions-world-distribution; Panel (**e**) is modified from https://data.giss.nasa.gov/gistemp/maps/index_v4.html; Panel (**f**) is modified from https://www.epa.gov/climatechange/indicators; and Panel (**g**) is modified from https://commons.wikimedia.org/wiki/File:WOA05_GLODAP_del_co3_AYool. All images are reproduced under creative commons licenses, and the composite is reproduced from Glibert et al. (2022 Ocean-Land-Atmos. Res, 9787258, doi.org/10.34133/2022/9787258) also under a creative commons license

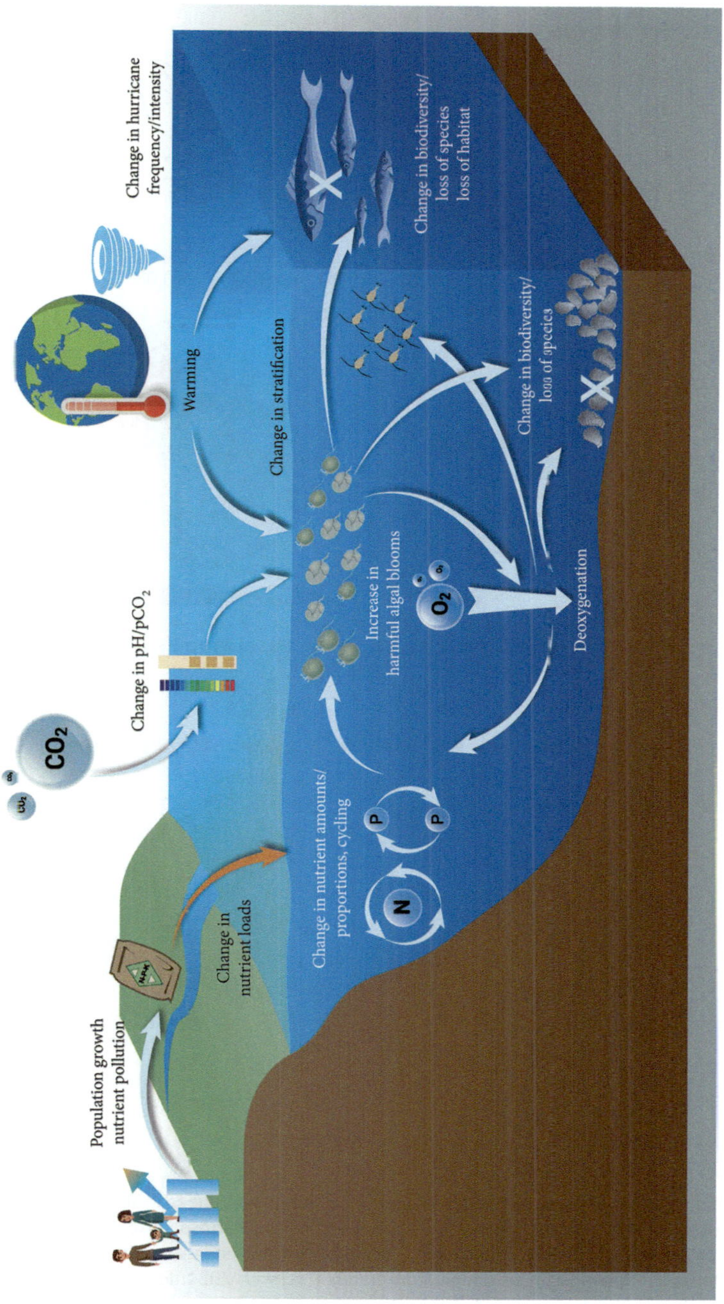

Fig. 27.3 Conceptual diagram of the complexity of interactions of multiple stressors in aquatic systems. Symbols courtesy of the Integration and Application Network, University of Maryland Center for Environmental Science. Image reproduced from Glibert et al. (2022, Ocean-Land-Atmos. Res, 9787258, doi.org/10.34133/2022/9787258) under a creative commons license

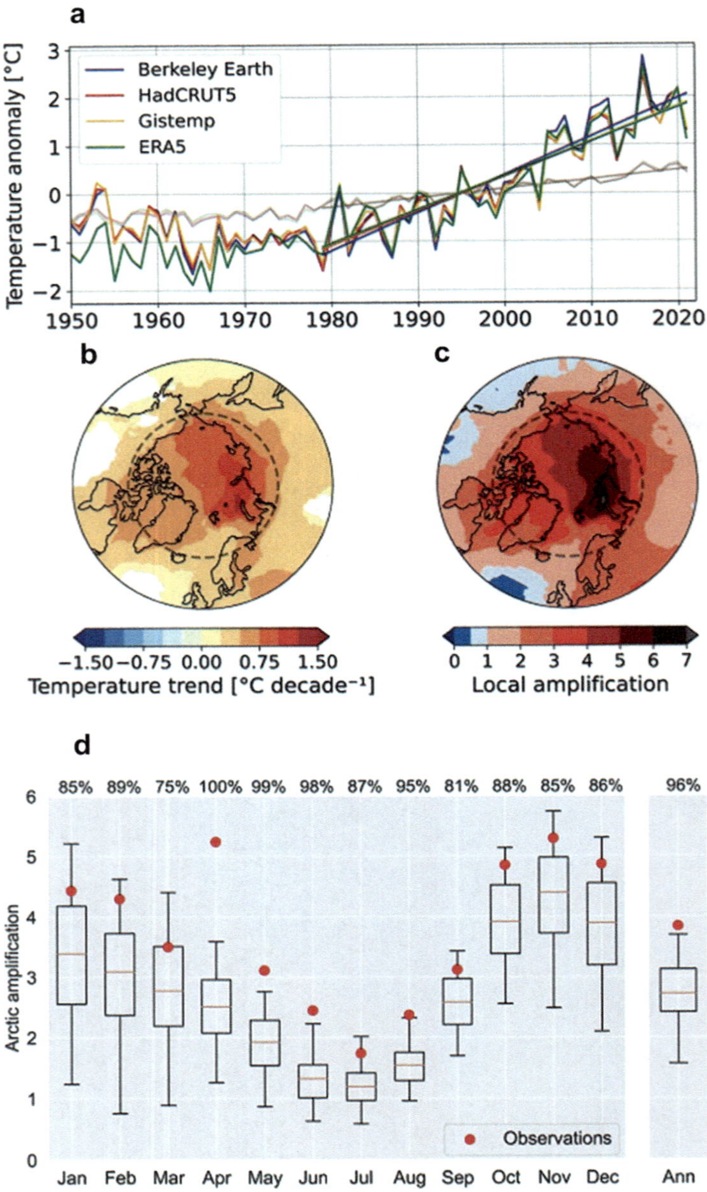

Fig. 27.4 Panel (**a**) Annual mean temperature anomalies in the Arctic (66.5°–90°N) (dark colors) and globally (light colors) during 1950–2021 derived from the various observational datasets. Temperature anomalies have been calculated relative to the standard 30-year period of 1981–2010. Shown are also the linear temperature trends for 1979–2021. Panel (**b**) Annual mean temperature trends for the period 1979–2021, derived from the average of the observational datasets. Areas without a statistically significant change are masked out. Panel (**c**) Local amplification ratio calculated for the period 1979–2021, derived from the average of the observational datasets. The dashed line in Panels (**b**) and (**c**) depicts the Arctic Circle (66.5°N latitude).

of warming. Sea ice is retreating earlier and advancing later in the year. More clouds warm the lower atmosphere. Accordingly, the change in open water area is also not spatially uniform (Fig. 27.4). These increases in open water are further leading to increased cloud cover, increased precipitation (as rain rather than snow), decreased pH, and rising seas (Meredith et al. 2019).

These changes have led to increased phytoplankton production in the extended ice-free periods. It was only a few years ago that it was estimated that production had increased by 20% from 1998 to 2009 (Arrigo and Dijken 2015), but more recent estimates suggest that this increase is closer to 60% since the late 1990s (Ardyna and Arrigo 2020; Lewis et al. (2020). These changes are directly related to the open water area and length of the open water season (Pabi et al. 2008; Arrigo and Dijken 2015).

Historically Arctic waters were considered to be limited by nitrogen (N), but increased coastal erosion and rivers are delivering more of this critical limiting nutrient (Tank et al. 2012). As conceptualized by Juranek (2022), due to changes in winds and stratification, the duration of upwelling has increased in the past decades, further enhancing productivity. However, there remains some uncertainty how—or the extent to which—changing nutrient concentrations are fueling enhanced primary productivity (Juranek 2022). There is considerable regional variation in nutrient delivery, and local factors, including stratification, wind patterns and terrestrial influence, and coastal erosion are important in certain regions (Fig. 27.5). Woodgate et al. (2012) showed that the freshwater flow from the Bering Sea to the Arctic Ocean increased by 50% between 2001 and 2011. Increased freshwater and coastal erosion may support up to one-third of Arctic primary production on an annual basis (Terhaar et al. 2021).

The carbon (C) chemistry of the Arctic is also changing rapidly. Increasing pCO_2 has been observed in a number of studies (e.g., Cai et al. 2010). The western Arctic is acidifying faster than any other ocean, with a further change of ~0.45 pH units expected over the coming century (Steinacher et al. 2009). This is a function of melting ice. The ice melt dilutes the alkalinity of the seawater, reducing buffering capacity and the ability to resist acidification (Rysgaard et al. 2011). These waters have traditionally been considered to be C-rich and N-poor (McClelland et al. 2012). The expansion of ice-free regions and enhanced primary productivity yields more uptake of CO_2. The increased primary production leads to increased C export and its respiration, preventing the accumulated respiratory CO_2 from outgassing at the

Fig. 27.4 (continued) Panel (**d**) annual Arctic Amplification by month. The red circles indicate the observed Arctic Amplification (AA), as derived from the average of the four observational datasets. The orange lines indicate the medians of the current generation climate model (the Coupled Model Intercomparison Project phase 6, CMIP6, model), boxes show the first and third quartiles, and whiskers extend to the 5–95th percentiles of the realizations. The numbers in the upper row give the percent rank of the observed AA in the CMIP6 ensemble distribution. The Arctic is defined as the area north of 66.5°N. Image reproduced from Rantanen et al. (2022 Comm Earth Environ 3168) under a creative commons license

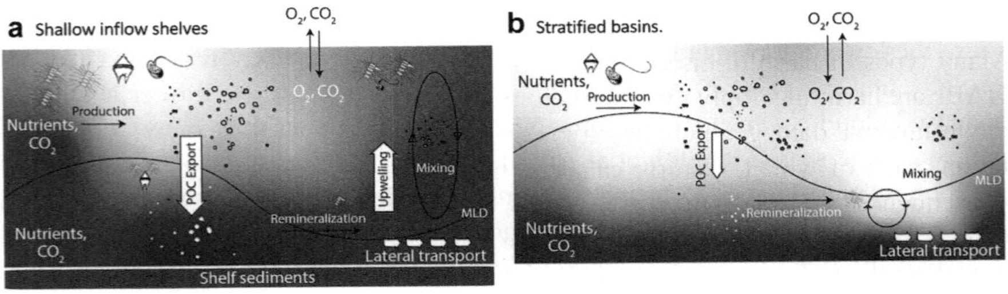

Fig. 27.5 Conceptual diagram of nutrient replenishment mechanisms on shallow inflow shelves (Panel **a**), and stratified deep basins (Panel **b**). Green shading indicates relative nutrient availability. On inflow shelves, advective nutrient inputs from lower latitudes and strong replenishment due to storm-induced and winter mixing support high rates of primary production (PP) and organic matter export. Stratified interior shelves and basins have weaker winter nutrient replenishment and storm-induced mixing, with consequent lower PP rates. MLD = mixed layer depth. Image reproduced from Juranek (2022, Oceanogr, doi.org/10.5670/oceanogr.2022.120) under a creative commons license

surface. At depth, this leads to seasonal undersaturation of calcium carbonate (e.g., Bates and Mathis 2009).

All of these changes are contributing to a changed food web in the Arctic. Poleward advection is leading to the introduction of species including *Synechococcus* (Paulsen et al. 2016). Acidification and increased pCO_2 is favoring picoeukaryotes (including *Micromonas polaris*), disfavoring coccolithophores, and having variable effects on diatoms (reviewed by Ardyna and Arrigo 2020). These picoeukaryotes may also be osmotrophic or mixotrophic (see Chap. 15), and the projection is that this mode of nutrition will be favored in a future Arctic compared to pure autotrophy (e.g., diatoms) as sea ice declines and waters continue to warm (Stoecker and Lavrentyev 2018). Further stressing diatoms is a decrease in silicate from inflowing Atlantic waters. Recent poleward expansion of the coccolithophore *Emiliania huxleyi* has been attributed to substantial changes in the northward flow of North Atlantic Waters to the Arctic (Oziel et al. 2020). Collectively these changes are leading to a likelihood of more HABs and for major changes in food webs and biogeochemical cycles to occur.

27.3.2 Arabian Sea/Indian Ocean

The Arabian Sea has historically had one of the largest and most extreme oxygen (O_2) minimum zones, with hypoxia extending throughout most of the depth range of 150–1250 m (Fig. 27.6; Lachkar et al. 2021, 2023; see also Chap. 24). Accordingly, this region has also historically been a region of high rates of denitrification (Codispoti et al. 2001; e.g., see Chap. 11). Even with this historical condition of low O_2, the depletion of O_2 within the Arabian Sea has accelerated in recent decades, with summertime hypoxia now more frequent and extensive (Al-Yamani and Naqvi 2019).

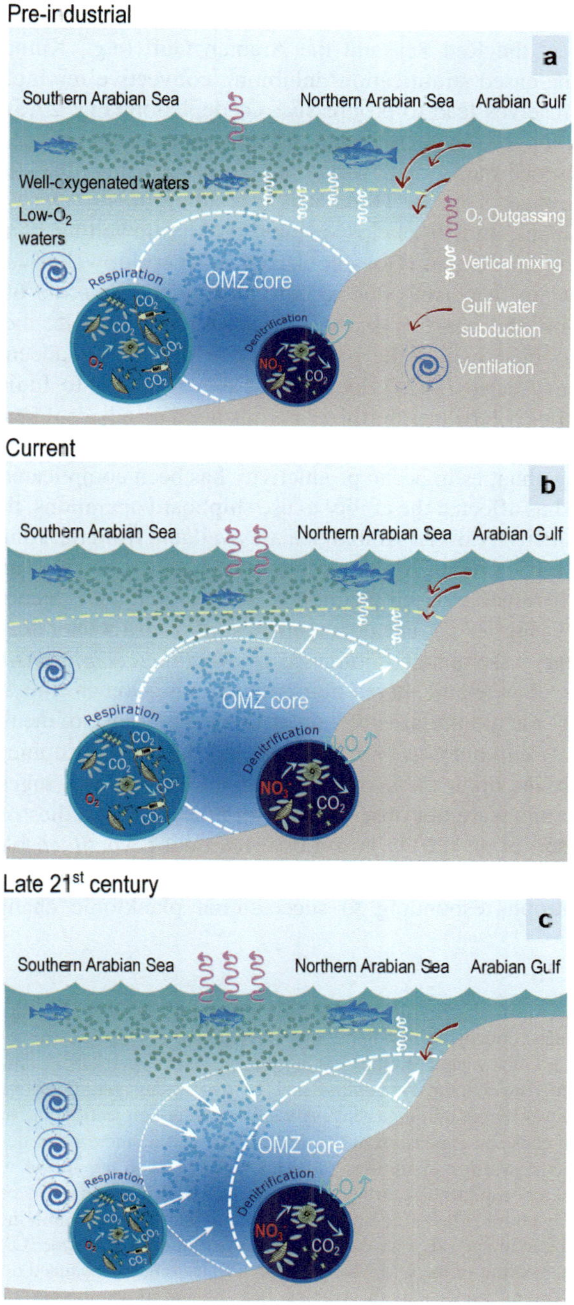

Fig. 27.6 Summary of recent and potential future changes in Arabian Sea OMZ. This schematic summarizes the recent and main potential future changes in the Arabian Sea OMZ as well as their drivers and implications. Panel (**a**) Pre-industrial conditions: cool conditions favor strong O_2 solubility in the seawater and weak outgassing. Well-oxygenated waters ($O_2 > 100$ mmol m^{-3})

Warming has also accelerated in the Arabian Sea since the 1990s, particularly in its marginal seas, the Red Sea and the Arabian Gulf (e.g., Kumar et al. 2009). Warming has increased stratification, inhibiting convective mixing. Warming surface waters themselves lead to progressive O_2 depletion (Fig. 27.6). However, the complexity of change is even greater.

Stronger summer monsoonal winds are now occurring than in past decades. Receding snow cover from the Himalayas is resulting in stronger winds across the Indian subcontinent, and this helps to intensify the upwelling of nutrients off of Somalia, Yemen, and Oman (Goes et al. 2005). These upwelled nutrients supplement those that are increasing due to eutrophication. The scale of the change in eutrophication off the coast of India is massive, for example, the population of Mumbai has doubled in just the past decade with little advancement in sewage treatment. Swaney et al. (2014) documented that N inputs to Indian waters have excessive runoff of N fertilizer; fertilizer is consumed on adjacent lands at among the highest rates in the world having increased by 50% between 2000 and 2011.

Documenting changes in ocean productivity has been complicated due to the rise in piracy which has affected the ability to use shipboard operations, though historical data on the productivity of the Arabia Sea are available from the Joint Global Ocean Flux Study (JGOFS) cruises of the early to mid-1990s (Gomes et al. 2008). Ocean color has been invaluable in unraveling the changes that have occurred since (Goes and Gomes 2016; Fig. 27.7). The documented shift from a winter diatom assemblage to a contemporary assemblage dominated by *Noctiluca scintillans* (Chap. 24) is a result of climate change and altered nutrient loadings (Gomes et al. 2014; Goes and Gomes 2016). These blooms are vivid green due the presence of the the prasinophyte endosymbiont (*Pedinomonas noctilucae*) in the heterotrophic dinoflagellate (Fig. 27.8). It relies upon the symbionts to fix C, but it also ingests a variety of prey. These blooms were not observed during the JGOFS studies of the 1990s. As noted by Harrison et al. (2011) in a global review of *Noctiluca* blooms and their expansion, "*Noctiluca* may well be a coastal or offshore manifestation of eutrophication, a mixotroph responding to successional planktonic changes in nutrient

Fig. 27.6 (continued) occupy a thick upper layer overlaying the OMZ. Strong winter convection and subduction of Gulf waters in the north ventilate the upper OMZ, thus maintaining its denitrifying suboxic core contained at depth. Panel (**b**) Present-day conditions: weaker O_2 solubility in the seawater causes moderate outgassing. A weaker winter convection and subduction of Gulf waters in the north cause an expansion of the upper OMZ and an increase of denitrification. Panel (**c**) Future conditions: warmer conditions further lower O_2 solubility in the water and enhance outgassing. Moreover, enhanced stratification limits the vertical penetration of well-oxygenated waters, contained in a thinner layer (compressing potential habitat). Reduction in winter convection and Gulf water subduction causes a weakening of the ventilation of the upper OMZ in the northern Arabian Sea and a shoaling of the OMZ core, potentially leading to enhanced denitrification there. However, in the central and southern Arabian Sea, enhanced ventilation as well as reduced productivity and respiration lead to important oxygenation at depth, causing a substantial shrinking of the OMZ core. Image reproduced from Lachkar et al. (2023 Front Mar Sci doi.org/10.3389/fmars.2023.1122043) under a creative commons license

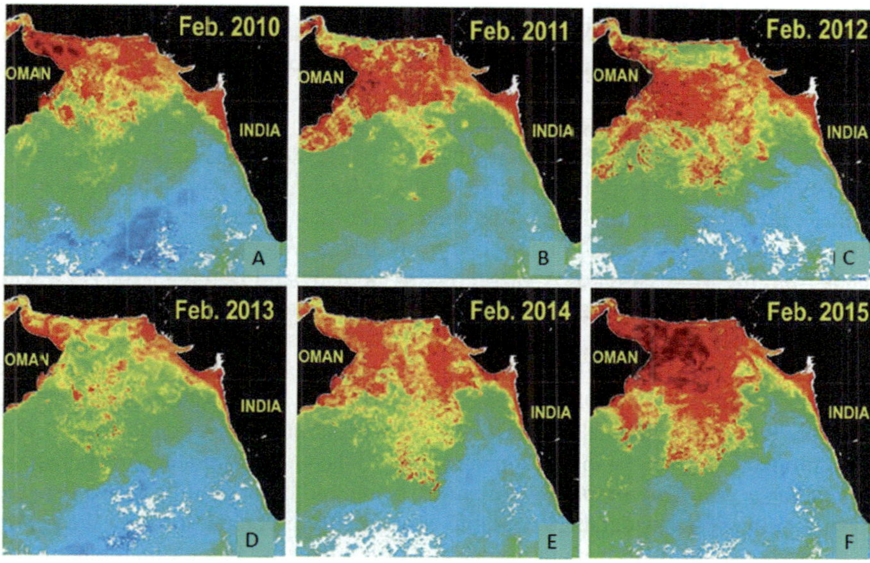

Fig. 27.7 NASA MODIS-Aqua monthly composite images of chlorophyll *a* in the Arabian Sea showing the spatial expanse of *Noctiluca* blooms in February of 2010, 2011, 2012, 2013, 2014, and 2015

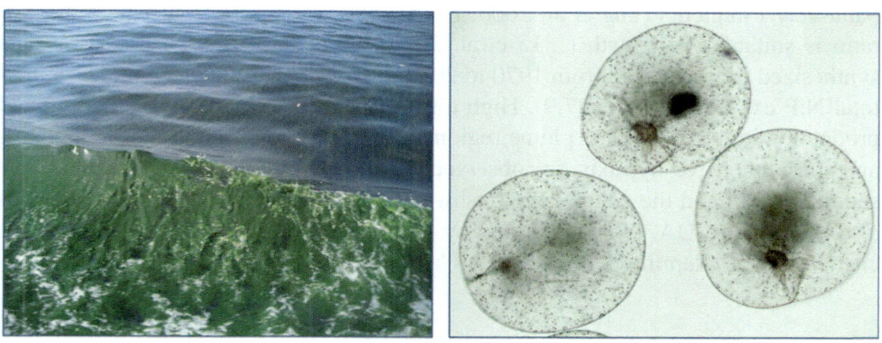

Fig. 27.8 Green *Noctiluca* bloom and *Noctiluca* cells with endosymbionts. Image reproduced from Harrison et al. (2011, Chin J Oceanol Limnol 29: 807–831) with permission of Springer

availability." Moreover, even though *Nocticula* blooms have dominated the plankton, other dinoflagellate blooms have expanded. An example is the development of massive *Margalefidinium* (reported as *Cochlodinium*) *polykrikoides* blooms that have become increasingly observed in the Arabian Gulf and Sea of Oman. For example, in 2008–2009 such a bloom expanded along the western side of the Arabian Gulf ultimately covering about 14,000 km^2 (Richlen et al. 2010; Al-Azri et al. 2014).

27.3.3 East China Sea

Among the most spectacular examples of the association between nutrient loading, hypoxia, and HABs is the East China Sea in the region formed by the convergence of the Yangtze (Changjiang) River plume and the Taiwan Warm Current (Zhou et al. 2008; Glibert et al. 2018a, 2022; Li et al. 2021; Fig. 27.9). This region is warming at a rate several times the global average, with the largest increases in temperature in winter (Sasaki and Umeda 2021). The watershed of the Changjiang area delivers an annual dissolved inorganic N (DIN) load from coastal waters that is higher than DIN loads from the Mississippi or Amazon rivers (Goolsby and Battaglin 2001; Li et al. 2009), a consequence of the 440 million people and intensive agriculture.

Hypoxia began to be reported from the late 1950s, and became a substantial problem beginning in the 1980s (Zhu et al. 2011; Große et al. 2020) as blooms were expanding (Zhou et al. 2008; Li et al. 2009). These hypoxic zones result from the large phytoplankton blooms that have also been expanding. Though diatoms are still the dominant phytoplankton group, blooms of the dinoflagellates *Prorocentrum donghaiense* and *Karenia mikimotoi* are common and have been linked to fish and shrimp kills with considerable economic consequences. The rapid rise in the export of N and phosphorus (P) from the expansion of fertilizer use in the Changjiang River watershed has been strongly associated with the expansion of these blooms (Zhou et al. 2008; Glibert et al. 2018a). However, there is considerable complexity in how nutrient export leads to blooms. Blooms form when the N-rich Changjiang River plume mixes with upwelled water from the Taiwan Warm Current, which is comparatively P-replete (Tang et al. 2000; Fang 2004), forming a front where the N:P ratio is suitable for growth (J. Li et al. 2009; W. Li et al. 2021). Based on data synthesized for the period from 1970 to 2010, blooms occur when the molar ratio of total N:P exceeds 30 (Fig. 27.9). High total suspended sediment reduces light and production in the immediate plume region, and highest chlorophyll (chl) *a* amounts and most intensive hypoxia are observed in the transitional regions between the sediment front and the edge of the freshwater plume (Li et al. 2021). The region is also undergoing OA, as documented by the decadal differences in pH and these changes are accelerating with hypoxia (Cai et al. 2011; Xiong et al. 2020).

27.3.4 Estuaries: Chesapeake Bay and Comparative Systems

Estuaries are particularly complex with regard to the effects of multiple stresses. Due to the fact that these systems, where there are substantial interactions with watershed inputs, the adjacent ocean, and sediments, the temporal and spatial fluctuations in pH, O_2, temperature, and primary production can be extensive, leading to extremes and high variability (Lauchlan and Nagelkerken 2019; Glibert et al. 2022). Compared with the open ocean, coastal and large estuarine systems are experiencing an accelerating pace of acidification as organic matter respiration contributes to dissolved organic C production, in addition to atmospheric CO_2 (Cai et al. 2011, 2017, 2021). On the other hand, reduced CO_2 availability is driven by high primary

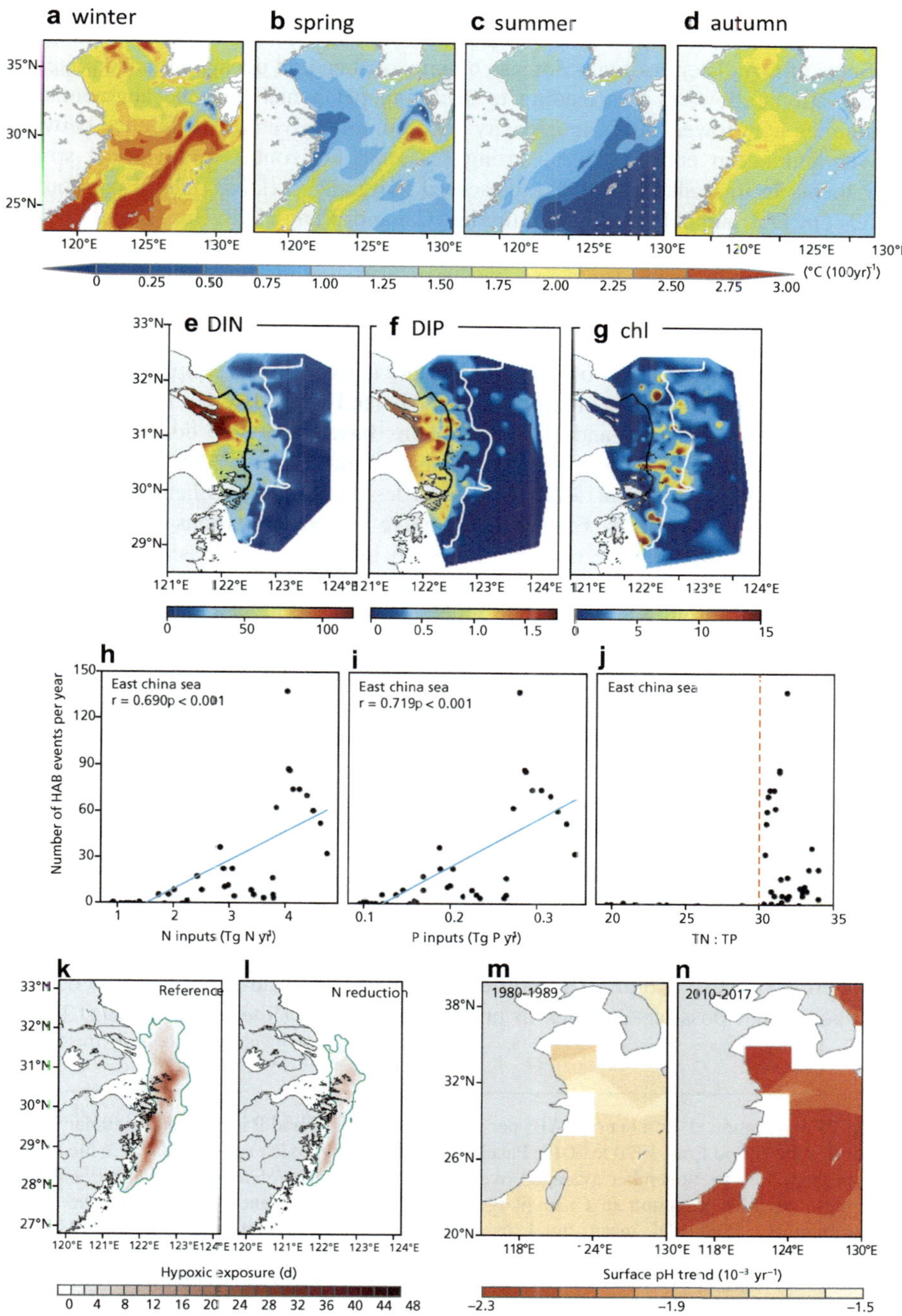

Fig. 27.9 Composite of multiple stressors in the East China Sea. Panels (**a**–**d**) Modeled linear sea surface temperature trends by season; Panels (**e**, **f**, and **g**) Dissolved inorganic N and P (μM) and chlorophyll *a* (μg L^{-1}) based on multiyear observational data from 1988 to 2016; Panels (**h**, **i**, and **j**)

production—as occurs during the development of high-biomass blooms. Such high biomass can lead to elevated pH (basification) due to CO_2 drawdown, countering the effects of OA. Then, as high-biomass blooms decline and decay, enhanced rates of respiration follow, in turn leading to declining pH, which may be augmented in estuaries with lower buffering capacity. Accordingly, given the difference between river and ocean end members, strong seasonal, horizontal, and vertical spatial gradients in total alkalinity (TA), dissolved inorganic C (DIC), partial pressure of CO_2 (pCO_2), and pH develop in estuaries (Cai and Wang 1998; Borges and Gypens 2010; Carstensen and Duarte 2019).

Chesapeake Bay, the largest estuary in the continental USA, illustrates why the problem of multiple stressors in estuaries is so complex. This estuary receives large nutrient loads from sewage effluent and from non-point sources, as well as from atmospheric deposition, all of which contribute to excessive phytoplankton production and hypoxia (Kemp et al. 2005; Li et al. 2016; Fig. 27.10a, b). The problems of eutrophication, hypoxia, and HABs problems have been well documented (e.g., Glibert et al. 2001. Kemp et al. 2005; Marshall and Egerton 2009; Li et al. 2015; Ni et al. 2020; Wolny et al. 2020; Figs. 27.10c–e; Chap. 24). Other problems have compounded these effects. For example, native oyster populations have collapsed, and shifts in fisheries have occurred due to fishing pressure, loss of habitats, and declining water quality (e.g., Kemp et al. 2005). The loss of oysters has resulted in a loss of natural filtration of the water, which in the nineteenth century was estimated to be at a rate of a few days, whereas with current oyster populations, it is estimated to take several 100 days to filter the same volume of water.

The Chesapeake Bay also has a very complex C chemistry. The oligohaline heterotrophic upper bay contrasts with the mesohaline autotrophic mid-bay that experiences most of the hypoxia (Testa and Kemp 2014), and with a polyhaline lower bay with a balanced net metabolism (Kemp et al. 1997). Each of these bay regions has a distinct C chemistry regime. In the upper bay, where pH in near-surface waters has historically been low due to the low pH of the incoming river water, there has been a long-term basification trend, influenced by changing freshwater input and increased total alkalinity from the Susquehanna River (Fig. 27.10f; Kaushal et al. 2013). By contrast, in the lower bay, which historically has had a higher pH, modeling suggests a multi-decadal decrease in pH and acidification due to oceanic influence and possibly changes in primary production (Shen et al. 2019a, 2019b;

←

Fig. 27.9 (continued) number of HABs per year in relation to N and P inputs and the molar TN:TP ratio for the period from 1970 to 2010; Panels (**k** and **l**) The hypoxic exposure (days) in the bottom layer from July to November averaged over 2008–2013 for the reference simulation, and for a scenario of 50% reduction in Changjiang River N loads; and Panels (**m** and **n**) The spatial distributions of decadal mean trend in ocean surface pH for the years 1980–1989 and 2010–2017. Panels (**a–d**) are from Sasaki and Umeda (2021), (**e–g**) from Li et al. (2009), (**h–j**) from Wang et al. (2021), (**k** and **l**) from Große et al. (2020), and (**m** and **n**) from Zhang and Wang (2019), all reproduced under creative commons licenses. Compositive image reproduced from Glibert et al. (2022, Ocean-Land-Atmos. Res: 9787258, doi.org/10.34133/2022/9787258) also under a creative commons license

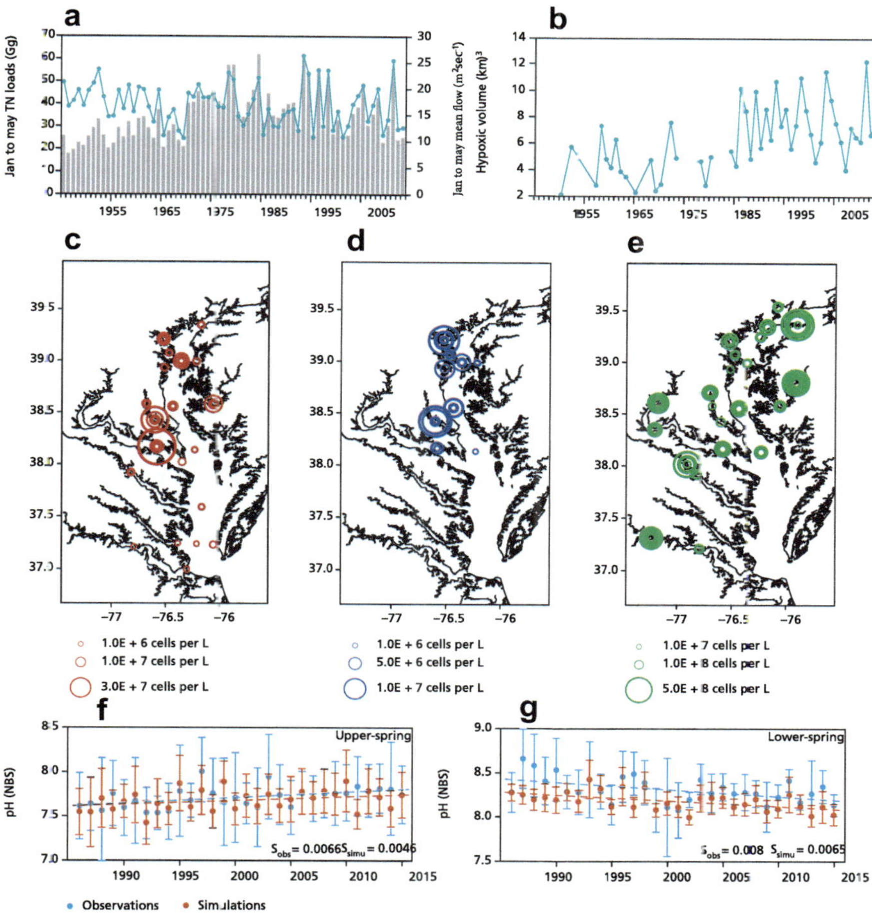

Fig. 27.10 Composite of multiple stressors in Chesapeake Bay. Panels (**a** and **b**) Time series of (**a**) winter-spring (January–May) flow (blue dots) and total nitrogen loading (green bars) from the Susquehanna River and Panel (**b**): summer (July) hypoxic volume (defined as the DO concentration below 2 mg/L) in the Chesapeake Bay. The hypoxic volume increased twofold between 1950s and 1980s, corresponding to a similar increase in nutrient loading. Although nutrient loading started to level off from 1990s, the hypoxic volume continued to increase and exhibited large interannual variations. Panels (**c–e**) Examples of HABs in the Chesapeake Bay based on long-term data. Panel (**c**) represents *Prorocentrum minimum*, panel (**d**) represents *Karlodinium veneficum* and panel (**e**) represents cyanobacterial HABs. Panels (**f** and **g**) The pH trends with time for the upper and lower Chesapeake Bay in spring. Panels (**a** and **b**) reproduced from Li et al. (2016), Panels (**c–e**) from Li et al. (2015), and panels (**f** and **g**) from Shen et al. (2020). Panels (**a–e**) reprinted with permission of Elsevier; Panels (**f** and **g**) reproduced under a creative commons license. Composite image reproduced from Glibert et al. (2022, Ocean-Land-Atmos. Res: 9787258, doi.org/10.34133/2022/9787258) under a creative commons license

Fig. 27.10g). The counterbalance between OA and river basification (Shen et al. 2020) results in the most variability in pH and carbonate chemistry in the autotrophic mid-bay (Brodeur et al. 2019), with strong pH fluctuations driven by phytoplankton photosynthesis.

Bay blooms can lead to, or can accelerate, an elevation in pH values, particularly in the upper bay. Indeed, Fan et al. (2003) reported pH values up to 9.5 for a bloom of *Prorocentrum cordatum* (reported as *P. minimum)* in Chesapeake Bay. Collectively, these changes in pH have important ramifications for HABs and their competitors or their prey (for mixoplanktonic species). From a biogeochemical perspective, a bloom-induced rise in pH has various consequences, especially in shallow waters, such as lakes or estuaries. As pH increases, the fundamental physical–chemical relationships related to P adsorption–desorption change (Chap. 13). Enhancement of sediment P release under elevated water column pH conditions has been observed in eutrophic lakes (e.g., Andersen 1974; Drake and Heaney 1987) and tidal freshwater/oligohaline estuaries (Seitzinger 1991; Gao et al. 2012; see Chap. 13). Consequently, the very existence of high-biomass blooms in waters with a legacy of P enrichment effectively creates a self-sustaining flux of P that can prolong bloom events.

Not all estuaries experience the same types of stresses as the Chesapeake. Coastal topology, retention, and latitude all play a role (Glibert et al. 2018b, 2022). Dürr et al. (2011) defined four major types of near-shore coasts, including small deltas, tidal systems, lagoons, and fjords based on hydrological, lithological, and morphological criteria. These systems represent the estuarine filter, where riverine waters mix with those of the coastal ocean and a complex interplay of physical and biogeochemical processes takes place. Compared to Chesapeake Bay, the nearby Delaware Bay has no hypoxia and limited acidification and HABs due to consistent vertical mixing and shorter residence times. Shallow coastal lagoons, such as the Maryland Coastal Bays and Delaware Inland Bays, may have spatially limited, but localized, diel cycling hypoxia and acidification as their biogeochemical interactions with the benthos become more important, leading to high-frequency changes in CO_2, pH, and nutrient fluxes (Ganju et al. 2020). Retentive zones are also important for HABs, especially those that form cysts, as the cysts accumulate and can lead to blooms whenever favorable conditions occur. Shallow, enclosed coastal lagoons are thus subject to recurrent blooms once a population of HAB becomes established. Fjords are particularly effective in retaining nutrients with characteristic residence times as long as decades, while small deltas have a low filtering capacity. In tropical and sub-tropical regions, retention is much less efficient and a large proportion of the nutrient inputs to the ocean bypass the coastal filter by discharging directly onto the shelves of large rivers. Large rivers largely bypass the near-shore filter, while karstic and arheic coasts act as inactive filters (McKee et al. 2004).

27.4 Challenges of Adding Complexity and Predicting Multiple Stressor Effects

In order to represent the physical and biogeochemical aspects of coastal ecosystems sufficiently well to simultaneously model HABs, deoxygenation, acidification, warming, and their interactions, complex biogeochemical models are required. Although modeling is covered in more detail in Chap. 28, it is worth highlighting how complex such models can become when considering multiple stressors. The key outcome of any such model is the ability to dynamically predict key stressor variable concentrations (e.g., nutrients form and proportion, C chemistry variables, and dissolved O_2) and their interactions with biological and biogeochemical processes.

The first consideration for models is that stressors need to be represented so that their possible effects on each other can be simulated (Glibert et al. 2022). Some of these linkages are relatively straightforward and commonly implemented, for example, the simultaneous impact of phytoplankton respiration on dissolved inorganic C production, dissolved O_2 consumption, total alkalinity balance, and thus its impact on deoxygenation and acidification. More advanced models include linkages such as a formal representation of calcification as it occurs in the water column and sediments, or parameterization of the production of calcium carbonate as related to primary production, or descriptions of the fate or dissolution of calcium carbonate and its effect as a buffer against acidification (e.g., Su et al. 2020). Having all of the stressors within a hydrodynamic–biogeochemistry–biological coupled model is extremely complex even if one just considers the primary producers.

27.5 Summary

The stressors associated with anthropogenic inputs and climate changes are complex and interactive. From the Arctic to the Arabian Sea to the Chinese coast and Chesapeake Bay, warming, nutrient pollution, hypoxia, and HABs are increasing, coincident with changes in CO_2 and pH. No longer is it sufficient to study the effects of a single stressor on primary producers if the goal is to predict how systems will change in the coming years. Stressors are interactive, often reacting in nonlinear ways, amplified and propagated through food web interactions. The needs to understand dose–response effects are great, but so too is the need to expand consideration of interaction of effects. New approaches for studying the collective effects are needed, as are new models that incorporate the complexities of these effects. Such models are next considered.

References

Al-Azri A, Piontkovski SA, Al-Hashmi KA, Goes JI, Gomes HR, Glibert PM (2014) Mesoscale and nutrient conditions associated with the massive 2008 *Cochlodinium polykrikoides* bloom in the sea of Oman/Arabian gulf. Estuar Coasts 37:325–338

Al-Yamani F, Naqvi SWA (2019) Chemical oceanography of the Arabian Sea. Deep Sea Res 161: 72–80. https://doi.org/10.1016/j.dsr2.2018.10.003

AMAP (Arctic Monitoring and Assessment Programme). (2021). Arctic climate change update 2021: key trends and impacts. Summary for policy-makers. https://www.amap.no/

Andersen JM (1974) Nitrogen and phosphorus budgets and the role of sediments in six shallow Danish lakes. Arch Hydrobiol 74:528–550

Ardyna M, Arrigo KR (2020) Phytoplankton dynamics in a changing Arctic Ocean. Nat Clim Chang 10(10):892–903. https://doi.org/10.1038/s41558-020-0905-y

Arrigo KR, van Dijken GL (2015) Continued increases in Arctic Ocean primary production. Prog Oceanogr 136:60–70. https://doi.org/10.1016/j.pocean.2015.05.002

Arrigo KR, van Dijken GL, Cameron MA, van der Grient J, Wedding LM, Hazen L, Leape J et al (2020) Synergistic interactions among growing stressors increase risk to an Arctic ecosystem. Nat Commun 11:6255. https://doi.org/10.1038/s41467-020-19899-z

Bates NR, Mathis JT (2009) The Arctic Ocean marine carbon cycle: evaluation of air-sea CO_2 exchanges, ocean acidification impacts and potential feedbacks. Biogeosciences 6(11): 2433–2459

Bercel TL, Kranz S (2019) Insights into carbon acquisition and photosynthesis in *Karenia brevis* under a range of CO_2 concentrations. Prog Oceanogr 172:65–76. https://doi.org/10.1016/j.pocean.2019.01.011

Borges AV, Gypens N (2010) Carbonate chemistry in the coastal zone responds more strongly to eutrophication than to ocean acidification. Limnol Oceanogr 55:346–353. https://doi.org/10.4319/lo.2010.55.1.0346

Breitburg D, Levin LA, Oschlies A, Grégoire M, Chavez FP, Conley DJ, Garçon V et al (2018) Declining oxygen in the global ocean and coastal waters. Science 359(6371):eaam7240. https://doi.org/10.1126/science.aam7240

Brodeur JR, Chen B, Su J, Xu Y-Y, Hussain N, Scaboo KM, Zhang Y et al (2019) Chesapeake Bay inorganic carbon: spatial distribution and seasonal variability. Front Mar Sci 6:99. https://doi.org/10.3389/fmars.2019.00099

Cai W-J, Wang Y (1998) The chemistry, fluxes, and sources of carbon dioxide in the estuarine waters of the Satilla and Altamaha Rivers, Georgia. Limnol Oceanogr 43:657–668. https://doi.org/10.4319/lo.1998.43.4.0657

Cai W-J, Chen L, Chen B, Gao Z, Lee SH, Chen J, Pierrot D et al (2010) Decrease in the CO_2 uptake capacity in an ice-free Arctic Ocean basin. Science 329(5991):556–559. https://doi.org/10.1126/science.1189338

Cai W-J, Hu X, Huang W-J, Murrell MC, Lehrter JC, Lohrenz SE, Chou W-C et al (2011) Acidification of subsurface coastal waters enhanced by eutrophication. Nat Geosci 4(11): 766–770. https://doi.org/10.1038/ngeo1297

Cai W-J, Huang W-J, Luther GW, Pierrot D, Li M, Testa J, Xue M et al (2017) Redox reactions and weak buffering capacity lead to acidification in the Chesapeake Bay. Nat Commun 8:1–12. https://doi.org/10.1038/s41467-017-00417-7

Cai W-J, Feely RA, Testa JM, Li M, Evans W, Alin SR, Xu Y-Y et al (2021) Natural and anthropogenic drivers of acidification in large estuaries. Annu Rev Mar Sci 13:23–55. https://doi.org/10.1146/annurev-marine-010419-011004

Carstensen J, Duarte CM (2019) Drivers of pH variability in coastal ecosystems. Environ Sci Technol 53:4020–4029. https://doi.org/10.1021/acs.est.8b03655

Codispoti L, Brandes JA, Christensen J, Devol A, Naqvi S, Paerl HW et al (2001) The oceanic fixed nitrogen and nitrous oxide budgets: moving targets as we enter the Anthropocene? Sci Mar 65: 85–105. https://doi.org/10.3989/scimar.2001.65s285

Collins S (2011) Competition limits adaptation and productivity in a photosynthetic alga at elevated CO_2. Proc R Soc B 278:247–255. https://doi.org/10.1098/rspb.2010.1173

Crain CM, Kroeker K, Halpern BS (2008) Interactive and cumulative effects of multiple human stressors in marine systems. Ecol Lett 11:1304–1315. https://doi.org/10.1111/j.1461-0248.2008.01253x

Drake JC, Heaney SI (1987) Occurrence of phosphorus and its potential remobilization in the littoral sediments of a productive English lake. Freshw Biol 17:513–523

Dürr HH, Laruelle GG, Van Kempen C, Slomp CP, Meybeck M, Middelkoop H (2011) Worldwide typology of near-shore coastal systems: defining the estuarine filter of river inputs to the ocean. Estuar Coasts 34:441–458. https://doi.org/10.1007/s12237-12011-19381-y

Fan C, Glibert PM, Burkholder JM (2003) Characterization of the affinity for nitrogen, uptake kinetics, and environmental relationships for *Prorocentrum minimum* in natural blooms and laboratory cultures. Harmful Algae 2:283–299

Fang TH (2004) Phosphorus speciation and budget of the East China Sea. Cont Shelf Res 24:1285–1299. https://doi.org/10.1016/j.csr.2004.04.003

Feng Y, Hare CE, Leblanc K, Rose J, Zhang Y, DiTullio GR, Lee PA et al (2009) The effects of increased pCO_2 and temperature on the North Atlantic spring bloom. I. The phytoplankton community and biogeochemical response. Mar Ecol Prog Ser 388:13–25. https://doi.org/10.3354/meps08133

Ganju NK, Testa JM, Suttles SE, Aretxabaleta AL (2020) Spatiotemporal variability of light attenuation and net ecosystem metabolism in a back-barrier estuary. Ocean Sci 16:593–614. https://doi.org/10.5194/os-16-593-2020

Gao Y, Cornwell JC, Stoecker DK, Owens MS (2012) Effects of cyanobacterial-driven pH increases on sediment nutrient fluxes and coupled nitrification-denitrification in a shallow fresh water estuary. Biogeosciences 9:2697–2710. https://doi.org/10.5194/bg-9-2697-2012

Glibert PM, Magnien R, Lomas MW, Alexander J, Fan C, Haramoto E, Trice M, Kana TM (2001) Harmful algal blooms in the Chesapeake and coastal bays of Maryland, USA: comparisons of 1997, 1998, and 1999 events. Estuaries 24:875–883. https://doi.org/10.2307/1353178

Glibert PM, Al-Azri A, Allen JI, Bouwman AF, Beusen AHW, Burford MA, Harrison PJ, Zhou MJ (2018a) Key questions and recent research advances on harmful algal blooms in relation to nutrients and eutrophication. In: Glibert PM, Berdalet E, Burford M, Pitcher G, Zhou M (eds) Global ecology and oceanography of harmful algal blooms (GEOHAB). Springer Cham, Switzerland, pp 229–259. https://doi.org/10.1007/978-3-319-70069-4_12

Glibert PM, Beusen AHW, Harrison JA, Dürr HH, Bouwman A, Laruelle GG (2018b) Changing land, sea- and airscapes: sources of nutrient pollution affecting habitat suitability for harmful algae. In: Glibert PM, Berdalet E, Burford M, Pitcher G, Zhou M (eds) Global ecology and oceanography of harmful algal blooms (GEOHAB). Springer Cham, Switzerland, pp 53–76. https://doi.org/10.1007/978-3-319-70069-4_4

Glibert PM, Cai W-J, Hall E, Li M, Main K, Rose K, Testa J, Vidyarathna N (2022) Stressing over the complexities of multiple stressors in marine and estuarine systems. Ocean-Land-Atmos Res 27:9787258. https://doi.org/10.34133/2022/9787258

Goes JI, Gomes HR (2016) An ecosystem in transition: the emergence of mixotrophy in the Arabian Sea. In: Glibert PM, Kana TM (eds) Aquatic microbial ecology and biogeochemistry: a dual perspective. Springer International Publ, Switzerland, pp 155–170

Goes JI, Prasad TG, Gomes HR, Fasullo JT (2005) Warming of the Eurasian landmass is making the Arabian Sea more productive. Science 308:545–547

Gomes HR, Goes JI, Matondkar SGP, Parab SG, Al-Azri ARN, Thoppil PG (2008) Blooms of *Nocticula miliaris* in the Arabian Sea—an in situ and satellite study. Deep Sea Res 55:751–765

Gomes HR, Goes JI, Matondkar SGP, Buskey EJ, Basu S, Parab S, Thoppil P (2014) Massive outbreaks of *Nocticula scintillans* blooms in the Arabian Sea due to spread of hypoxia. Nat Commun 5:4862. https://doi.org/10.1038/ncomms5862

Goolsby DA, Battaglin WA (2001) Long-term changes in concentrations and flux of nitrogen in the Mississippi River basin, USA. Hydrol Proc 15:1209–1226. https://doi.org/10.1002/hyp.210

Große F, Fennel K, Zhang H, Laurent A (2020) Quantifying the contributions of riverine vs. oceanic nitrogen to hypoxia in the East China Sea. Biogeosciences 17:2701–2714. https://doi.org/10.5194/bg-17-2701-2020

Harrison PJ, Furuya K, Glibert PM, Xu J, Liu HB, Yin K, Lee JHW et al (2011) Geographical distribution of red and green *Noctiluca scintillans*. Chin J Oceanol Limnol 29:807–831

Hutchins DA, Mulholland MR, Fu F (2009) Nutrient cycles and marine microbes in a CO_2-enriched ocean. Oceanography 22:128–145. https://doi.org/10.5670/oceanog.2009.103

Hutchins DA, Fu FX, Webb EA, Walworth N, Tagliabue A (2013) Taxon-specific response of marine nitrogen-fixers to elevated carbon dioxide concentrations. Nat Geosci 6:790–785. https://doi.org/10.1038/ngeo1858

Juranek L (2022) Changing biogeochemistry of the Arctic Ocean: surface nutrient and CO_2 cycling in a warming, melting north. Oceanography 35(3/4):144–155. https://doi.org/10.5670/oceanog.2022.120

Kaushal SS, Likens GE, Utz RM, Pace ML, Grese M, Yepsin M (2013) Increased alkalinization in the eastern U.S. Environ Sci Technol 47:10302–10311. https://doi.org/10.1021/es401046s

Kemp WM, Smith EM, Marvin-DiPasquale M, Boynton WR (1997) Organic carbon balance and net ecosystem metabolism in Chesapeake Bay. Mar Ecol Prog Ser 150:229–248. https://doi.org/10.3354/meps150229

Kemp WM, Boynton WR, Adolf JE, Boesch DF, Boicourt WC, Brush G, Cornwell JC et al (2005) Eutrophication in Chesapeake Bay: historical trends and ecological interactions. Mar Ecol Prog Ser 303:1–29. https://doi.org/10.3354/meps303001

Kremp A, Godhe A, Egardt J, Dupont S, Suikkanen S, Casablanca S, Penna A (2012) Strain variability in the response of bloom-forming marine microalgae to changed climate conditions. Ecol Evol 2:1195–1207. https://doi.org/10.1002/ece3.245

Kumar SP, Roshin RP, Narvekar J, Kumar PD, Vivekanandan E (2009) Response of the Arabian Sea to global warming and associated regional climate shift. Mar Environ Res 68:217–222

Lachkar Z, Mehari M, Al Azhar M, Lévy M, Smith S (2021) Fast local warming is the main driver of recent deoxygenation in the northern Arabian Sea. Biogeosci: article 18(20):14069. https://doi.org/10.5194/bg-18-5831-5849

Lachlar Z, Lévy M, Hailegeorgis D, Vallivattathillam P (2023) Differences in recent and future trends in the Arabian Sea oxygen minimum zone: processes and uncertainties. Front Mar Sci 10: 1122043. https://doi.org/10.3389/fmars.2023.1122043

Lauchlan SS, Nagelkerken I (2019) Species range shifts along multistressor mosaics in estuarine environments under future climate. Fish Fisheries 21:32–46. https://doi.org/10.1111/faf.12412

Lewis KM, van Dijken GL, Arrigo KR (2020) Changes in phytoplankton concentration now drive increased Arctic Ocean primary production. Science 369:198–202. https://doi.org/10.1126/science.aay8380

Li J, Glibert PM, Zhou M, Lu S, Lu D (2009) Relationships between nitrogen and phosphorus forms and ratios and the development of dinoflagellate blooms in the East China Sea. Mar Ecol Prog Ser 383:11–26. https://doi.org/10.3354/meps07975

Li J, Glibert PM, Gao Y (2015) Temporal and spatial changes in Chesapeake Bay water quality and relationships to *Prorocentrum minimum, Karlodinium veneficum*, and CyanoHAB events, 1991–2008. Harmful Algae 42:1–14. https://doi.org/10.1016/j.hal.2014.11.003

Li M, Lee YJ, Testa JM, Li Y, Ni W, Kemp WM, Di Toro DM (2016) What drives interannual variability of hypoxia in Chesapeake Bay: climate forcing versus nutrient loading? Geophys Res Lett 43:2127–2134. https://doi.org/10.1002/2015GL067334

Li W, Ge J, Ding P, Ma J, Glibert PM, Liu D (2021) Effects of dual fronts on the spatial pattern of phytoplankton assemblages off the Changjiang River estuary. Estuar Coasts 44:1408–1418. https://doi.org/10.1007/s12237-020-00893-z

Marshall HG, Egerton TA (2009) Phytoplankton blooms: their occurrence and composition within Virginia's tidal tributaries. Va J Sci 60(3):149–164. https://digitalcommons.odu.edu/biology_fac_pubs/113. Accessed 4 Jan 2022

McCarthy JJ (2009) Reflections on: our planet and its life, origins, and futures. Science 326:1646–1655. https://doi.org/10.1126/science.1184937

McClelland JW, Holmes RM, Dunton KH, Macdonald RW (2012) The Arctic Ocean estuary. Estuar Coasts 35(2):353–368. https://doi.org/10.1007/s12237-010-9357-3

McKee BA, Aller RC, Allison M, Bianchi TS, Kineke GC (2004) Transport and transformation of dissolved and particulate materials on continental margins influenced by major rivers: benthic boundary layer and seabed processes. Cont Shelf Res 24:899–926
Mekonnen MM, Hoekstra AY (2015) Global gray water footprint and water pollution levels related to anthropogenic nitrogen loads to fresh water. Environ Sci Technol 49:12860–12868. https://doi.org/10.1021/acs.est.5b03191
Mekonnen MM, Hoekstra AY (2017) Global anthropogenic phosphorus loads to freshwater and associated grey water footprints and water pollution levels: a high-resolution global study. Water Resour Res 54:345–358. https://doi.org/10.1002/2017WR020448
Meredith M, Sommerkorn M, Cassotta S, Derksen C, Ekaykin A, Hollowed A, Kofinas G et al (2019) Polar regions. In: Pörtner HO, Roberts DC, Masson-Delmotte V, Zhai P, Tignor M, Poloczanska E, Mintenbeck K et al (eds) IPCC special report on the ocean and cryosphere in a changing climate. Cambridge University Press, Cambridge, UK and New York, NY, USA, pp 203–320. https://doi.org/10.1017/9781009157964.005
Ni W, Li M, Testa JM (2020) Discerning effects of warming, sea level rise and nutrient management on long-term hypoxia trends in Chesapeake Bay. Sci Total Environ 737:139717. https://doi.org/10.1016/j.scitotenv.2020.139717
Oziel L, Baudena A, Ardyna M, Massicotte P, Randelhoff A, Sallée J-B, Ingvaldsen RB, Devred E, Babin M (2020) Faster Atlantic currents drive poleward expansion of temperate phytoplankton in the Arctic Ocean. Nat Commun 11:1705. https://doi.org/10.1038/s41467-020-15485-5
Pabi S, van Dijken GL, Arrigo KR (2008) Primary production in the Arctic Ocean, 1998-2006. J Geophys Res Oceans 113(C8):C08005. https://doi.org/10.1029/2007JC004578
Paulsen ML, Doré H, Garczarek L, Seuthe L, Müller O, Sandaa R-A, Bratbak G, Larsen A (2016) *Synechococcus* in the Atlantic gateway to the Arctic Ocean. Front Mar Sci 3:191. https://doi.org/10.3389/fmars.2016.00191
Perovich DK, Light B, Eichen H, Jones KF, Runciman K, Nghiem SV (2007) Increasing solar heating of the Arctic Ocean and adjacent seas, 1979–2005: attribution and role in the ice-albedo feedback. Geophys Res Lett 34:L19505. https://doi.org/10.1029/2007GL031480
Pörtner HO, Gutowska M, Ishimatsu A, Lucassen M, Melzner F, Seibel B (2011) Effects of ocean acidification on nektonic organisms. In: Gattuso JP, Hansson L (eds) Ocean acidification. Oxford University Press, Oxford, pp 154–175
Rantanen M, Karpechko AY, Lipponen A, Nordling K, Hyvärinen O, Ruosteenoja K, Vihma T, Laaksonen A (2022) The Arctic has warmed nearly four times faster than the globe since 1979. Commun Earth Environ 3:168. https://doi.org/10.1038/s43247-022-00498-3
Richlen ML, Morton SL, Jamali EA, Rajan A, Anderson DM (2010) The catastrophic 2008–2009 red tide in the Arabian gulf region, with observations on the identification and phylogeny of the fish-killing dinoflagellate *Cochlodinium polykrikoides*. Harmful Algae 9:163–172. https://doi.org/10.1016/j.hal.2009.08.013
Riebesell U, Tortell PD (2011) Effects of ocean acidification on pelagic organisms and ecosystems. In: Gattuso JP, Hansson L (eds) Ocean acidification. Oxford University Press, Oxford, pp 99–121
Rysgaard S, Bendtsen J, Delille B, Dieckmann GS, Glud RN, Kennedy H, Mortensen J et al (2011) Sea ice contribution to the air-sea CO_2 exchange in the Arctic and southern oceans. Tellus Ser B Chem Phys Meteorol 63(5):823–830. https://doi.org/10.1111/j.1600-0889.2011.00571.x
Sasaki YN, Umeda C (2021) Rapid warming of sea surface temperature long the Kuroshio and China coast in the East China Sea during the twentieth century. J Clim 34(12):4803–4815. https://doi.org/10.1175/JCLI-D-20-0421.1
Schaum CE, Collins S (2014) Plasticity predicts evolution in a marine alga. Proc Royal Soc B 281: 20141486. https://doi.org/10.1098/rspb.2014.1486
Seitzinger SP (1991) The effect of pH on the release of phosphorus from Potomac estuary sediments: implications for blue-green algal blooms. Estuar Coast Shelf Sci 33:409–418

Shen C, Testa J, Ni W, Cai W-J, Li M, Kemp WM (2019a) Ecosystem metabolism and carbon balance in Chesapeake Bay: a 30-year analysis using a coupled hydrodynamic-biogeochemical model. J Geophys Res Oceans 124(8):6141–6153. https://doi.org/10.1029/2019JC015296

Shen C, Testa JM, Li M, Cai W-J, Waldbusser GG, Ni W, Kemp WM et al (2019b) Controls on carbonate system dynamics in a coastal plain estuary: a modelling study. J Geophys Res Biogeosci 124:61–78. https://doi.org/10.1029/2018JG004802

Shen C, Testa JM, Li M, Cai W-J (2020) Understanding anthropogenic impacts on pH and aragonite saturation in Chesapeake Bay: insights from a 30-year model study. J Geophys Res Biogeosci 125(7):e2019JG005620. https://doi.org/10.1029/2019JG005620

Steinacher M, Joos F, Frölicher TL, Plattner G-K, Doney SC (2009) Imminent Ocean acidification in the Arctic projected with the NCAR global coupled carbon cycle-climate model. Biogeosciences 6:515–533. https://doi.org/10.5194/bg-6-515-2009

Stoecker DK, Lavrentyev PJ (2018) Mixotrophic plankton in the polar seas: a pan-Arctic review. Front Mar Sci 5:292. https://doi.org/10.3389/fmars.2018.00292

Su J, Cai W-J, Brodeur J, Chen B, Hussain N, Yao Y, Ni C, Testa JM, Li M, Xie X (2020) Chesapeake Bay acidification buffered by spatially decoupled carbonate mineral cycling. Nat Geosci 13:441–444

Swaney DP, Hong B, Paneer Selvam A, Howarth RW, Ramesh R, Purvaja R (2014) Net anthropogenic nitrogen inputs and nitrogen fluxes from Indian watersheds: an initial assessment. J Mar Syst 141:45–58. https://doi.org/10.1016/j.jmarsys.2014.09.004

Tang TY, Tai JH, Yang YJ (2000) The flow pattern north of Taiwan and the migration of the Kuroshio. Cont Shelf Res 20:349–371. https://doi.org/10.1016/S0278-4343(99)00076-X

Tank SE, Manizza M, Holmes RM, McClelland JW, Peterson BJ (2012) The processing and impact of dissolved riverine nitrogen in the Arctic Ocean. Estuar Coasts 35:401–415

Terhaar J, Lauerwald R, Regnier P, Gruber N, Bopp L (2021) Around one third of current Arctic Ocean primary production sustained by rivers and coastal erosion. Nat Commun 12:169. https://doi.org/10.1038/s41467-020-20470-z

Testa JM, Kemp WM (2014) Spatial and temporal patterns of winter-spring oxygen depletion in Chesapeake Bay bottom water. Estuar Coasts 37:1432–1448. https://doi.org/10.1007/s12237-014-9775-8

Wang J, Bouwman AF, Liu X, Beusen AHW, Van Dingenen R, Detener F, Yao Y et al (2021) Harmful algal blooms in Chinese coastal waters will persist due to perturbed nutrient ratios. Environ Sci Technol Lett 8:276–284. https://doi.org/10.1021/acs.estlett.1c00012

Wolny JL, Tomlinson MC, Schollaert Uz S, Egerton TA, McKay JR, Meredith A, Reece KS et al (2020) Current and future remote sensing of harmful algal blooms in the Chesapeake Bay to support the shellfish industry. Front Mar Sci 7:337. https://doi.org/10.3389/fmars.2020.00337

Woodgate RA, Weingartner TJ, Lindsay R (2012) Observed increases in Bering Strait oceanic fluxes from the Pacific to the Arctic from 2011 to 2011 and their impacts on the Arctic Ocean water column. Geophy Res Lett 39:L24603. https://doi.org/10.1029/2012GL054092

Xiong T-Q, Wei Q-S, Zhai W-D, Li C-L, Wang S-Y, Zhang Y-X, Liu S-J, Yu S-Q (2020) Comparing subsurface seasonal deoxygenation and acidification in the Yellow Sea and Northern East China Sea along the north-to-south latitude gradient. Front Mar Sci 7:686. https://doi.org/10.3389/fmars.2020.00686

Zhang H, Wang K (2019) Simulated CO_2-induced ocean acidification for ocean in the East China Sea: historical conditions since preindustrial time and future scenarios. Sci Rep 9: 18559. https://doi.org/10.1038/s41586-019-0911-2

Zhou MJ, Shen ZL, Yu RC (2008) Responses of a coastal phytoplankton community to increased nutrient input from the Changjiang (Yangtze) river. Cont Shelf Res 28:1483–1489. https://doi.org/10.1016/j.csr.2007.02.009

Zhu Z-Y, Zhang J, Wu Y, Zhang Y-Y, Lin J, Liu S-M (2011) Hypoxia off the Changjiang (Yangtze River) estuary: oxygen depletion and organic matter decomposition. Mar Chem 125:108–116. https://doi.org/10.1016/j.marchem.2011.03.005

Additional Websites Accessed

https://climatechange.environment.harvard.edu/james-mccarthy. Accessed 13 Nov 2022
https://commons.wikimedia.org/wiki/File:WOA05_GLODAP_del_co3_AYool.png. Accessed 13 Feb 2022
http://data.giss.nasa.gov/gistemp/maps/index_v4.html. Accessed 13 Feb 2022
http://www.epa.gov/climatechange/indicators. Accessed 13 Feb 2022
https://hab.whoi.edu/maps/regions-world-distribution/. Accessed 13 Feb 2022

28 Modeling Our Past, Present, and Future

28.1 Introduction

Modeling has not always been embraced by biologists and ecologists. As a pioneer of applying mathematics to biology in the nineteenth century, Victor Hensen (Chap. 1) found that his efforts in that regard were "repugnant to many German biologists" (Mills 1989, p. 37). The mathematical approaches of H.W. Harvey in the Plymouth Laboratory, and later G. Evelyn Hutchinson, at Yale were equally rebuffed. However, quantitative approaches persisted. Taking up the mantle of the importance of mathematical approaches was Thomas Park, an ecologist of the 1930s. In his writings, he argued that ecology needed a "thorough housecleaning followed by a well-planned program of refurbishing" and that "ultimately ecology must cease to be a descriptive, qualitative science and become quantitative in character" (Park 1939, p. 332, 336 quoted in Mills 1989, p. 262).

Around the time of Park, the mid-1930s, Gordon Riley, a limnologist, showed—based on regression analyses—that the amount of chlorophyll *a* (chl *a*) in his study ponds could be related to the amount of organic matter, and the amount that was contributed by phytoplankton could be calculated (summarized by Mills 1989). Riley's work later shifted to the marine realm, where his ideas and his application of multiple regression analysis were not always embraced. Henry Bigelow of the Woods Hole Oceanographic Institution bluntly said, "Anybody who thinks he can predict more than 10% of plankton variability is a damn fool, but good luck" (Riley 1979, pp. 23–24 quoted in Mills 1989, p. 273). Bigelow went on to explain predictions were so difficult because of "our general ignorance of the inter-relationships in the very complex chain of events in the sea that govern the comparative success of failure of its inhabitants in the struggle for life . . .[but] nothing in the sea falls haphazard; if we cannot predict it, it is because we do not know the cause or how the cause works" (Bigelow 1931, p. 206, quoted in Mills 1989, p. 273). Over a period of several years, and many cruises to Georges Bank in the Gulf of Maine, which included many light–dark incubation experiments of productivity, combined with physical and chemical measurements, Riley ultimately

P. M. Glibert, *Phytoplankton Whispering: An Introduction to the Physiology and Ecology of Microalgae*, https://doi.org/10.1007/978-3-031-53897-1_28

succeeded in establishing "a technique of mathematically analyzing and modeling plankton dynamics that was quite unlike anything attempted before in biological oceanography" (Mills 1989, p. 285). Riley's resulting model expressed phytoplankton productivity as a function of light intensity, water transparency, nutrients, depth of the mixed layer, surface temperature, and abundance of grazers (Riley 1946). With all its limitations and flaws, he ushered in a new way of thinking. However, it was years before many biologists could undertake (by long hand) the complex mathematics he introduced.

John Steele, in the 1950s, a Scottish ecologist (and decades later, Director of the Woods Hole Oceanographic Institution), inspired by Riley's approach, developed a model of the North Sea that connected plankton production and fishing. His models both underscored what research needed to be done to advance understanding of plankton production, physics, and fisheries and showed, "the fact that these bases [simple observable variables] are sufficient for adequate description of the main features of the changing plant life in the sea implies that the factors controlling this life must be simple in character and few in number" (Steele, 1959, p. 148 quoted in Mills, 1989, p. 315).

The idea that application of models was, at one time, a new and controversial subject may seem foreign to marine ecologists today. Models are now not only accepted as tools, but are expected to be in the toolbox, to advance hypothesis testing and prediction. From Hensen to Riley and Steele, to current day oceanographers, it has always been clear that the challenge in model constructs is balancing realism with complexity. Not wanting to be the "damn fool," who can only "predict 10% of plankton variability" (as Henry Bigelow once warned, Mills 1989, p. 273), the temptation is to make models more and more computationally complex. As depictions of nature, models should be as accurate as possible, but only as complex as necessary (Flynn and McGillicuddy 2018). The more complex a model (whether Riley doing his multiple regression analyses by hand or current analysts using super computers), the more the demands for both operator time and data acquisition, and thus any decrease in computational steps that do not compromise the model performance is necessary. Choice of model frameworks, of course, also depends on the needs for the model application (Glibert et al. 2020). Managers may need near real-time predictions to protect human health, to understand potential for seafood toxin risk, or to protect water supplies. Such models require high frequency data and operational models must be run regardless of the calendar or personal schedule of the operator. If computational time is intensive, the cost of frequently running such models can be prohibitive. Longer term predictions, such as end-of-century climate predictions, may need to be complex to account for many complex relationships, but serve a different purpose. They give researchers and managers insights into how a system functions or will function with future nutrient loading or climate change, for example.

This chapter can only begin to scratch the surface of modeling approaches. The focus here is on those types of models that have application for understanding plankton dynamics and especially for predicting and understanding harmful algal blooms (HABs). Various model applications and outputs have been presented throughout various chapters of this volume, but a more systematic approach to modeling and applications is described here. After a review of some general kinds of models, some of the complexity of modeling HABs is reviewed. A few examples of the kinds of models that are in use operationally and for more strategic analysis of future conditions are illustrated, focusing on models that have been applied in predicting hypoxia and HABs in the eutrophic Chesapeake Bay.

28.2 General Types of Models

28.2.1 Statistical Models

Riley used statistical models. Using his extensive measurements of such variables as temperature, phosphate (PO_4^{3-}) and nitrate (NO_3^-) concentrations, and phytoplankton and zooplankton abundance, he was able to develop relationships and therefore predict phytoplankton concentration at a given temperature, for example. Statistical models have since seen a wide and varied suite of applications to ecology (Legendre and Legendre 1998). Explanatory models, such as Riley's, seek to establish the strength of a response (such as phytoplankton biomass) to a given variable (or set of variables). Predictive models seek statistical relationships that can be applied, with some degree of explanatory power, to yield a response with some as-yet-unmeasured variable or variables. Linear regression is the oldest and most commonly used approach to explain a response as a function of a measured variable. Multiple linear regression uses several explanatory variables to predict the response. While past performance is no guarantee of future results, there are many examples of multi-decadal observations of HAB responses to known environmental changes (Ralston and Moore 2020). As but one example, Anderson et al. (2009) used a regression model of satellite ocean color and sea surface temperature to predict *Pseudo-nitzschia* blooms in Santa Barbara Channel, California.

Numerous statistical advances have been made since the time of Riley, among which is the application of generalized linear models (GLMs). In GLMs, the predictor variables are combined into a single function which is then related to the response. This approach allows variables to have non-normal distribution, something not permitted in ordinary linear regression models. Thus, in GLM applications, while the response variable is related by a linear function to a number of explanatory variables, the explanatory variables themselves may have any number of probability distributions, such as normal, binomial, Poisson, negative binomial, or others (Guisan et al. 2002) Accordingly, fitting a GLM is not unlike fitting a multiple regression, but GLMs can be applied where linear regression fails to make adequate predictions. As examples in HAB applications, Anderson et al. (2010) used such a modeling approach to predict toxigenic *Pseudo-nitzschia* blooms in Chesapeake Bay, including such variables as time of year, location, temperature, salinity, light, nutrients, and freshwater flow, and Singh et al. (2014) used a similar wide range of variables to predict *Dinophysis* abundance in coastal Arabian Sea.

Generalized additive models (GAMs) are similar to GLMs, but instead of doing a linear regression, additive regressions are used, using backfitting as appropriate, resulting in a more flexible model. The functions may be non-parametric. The underlying assumption is that functions are additive and the component variables are smooth (Guisan et al. 2002). Thus, GAMs can deal with highly nonlinear and non-monotonic relationships. Noisy data becomes easier to handle, and multiple predictors can be used, each having its distinct functional form. As an example in HAB applications, Díaz et al. (2016) used GAMs to predict *Dinophysis acuta* blooms in an upwelling-influenced ria in Galicia.

28.2.2 Process (Trait-Based) Models

Process, trait-based, or mechanistic models take a different approach than statistical models. In this case, the state variable is described by a mechanistic transfer function. Traits are those biotic and abiotic factors that affect fitness (Martini et al. 2021; Litchman 2022; see Chap. 18). Among ecosystem functions, resource acquisition and growth rates are key traits that are most frequently parameterized in process models. The previous chapters in this book have described many of these transfer functions: growth rate as a function of nutrients, as a function of light, and as a function of temperature, as examples (Fig. 28.1). Many of these relationships follow the "curve for all reasons" (Rao 2000; see Chap. 3), a saturating hyperbola; growth as a function of temperature is the obvious exception.

Simple process models are based on predicting responses at a given time or space given known variables. Individual box models, or zero-dimensional models, are

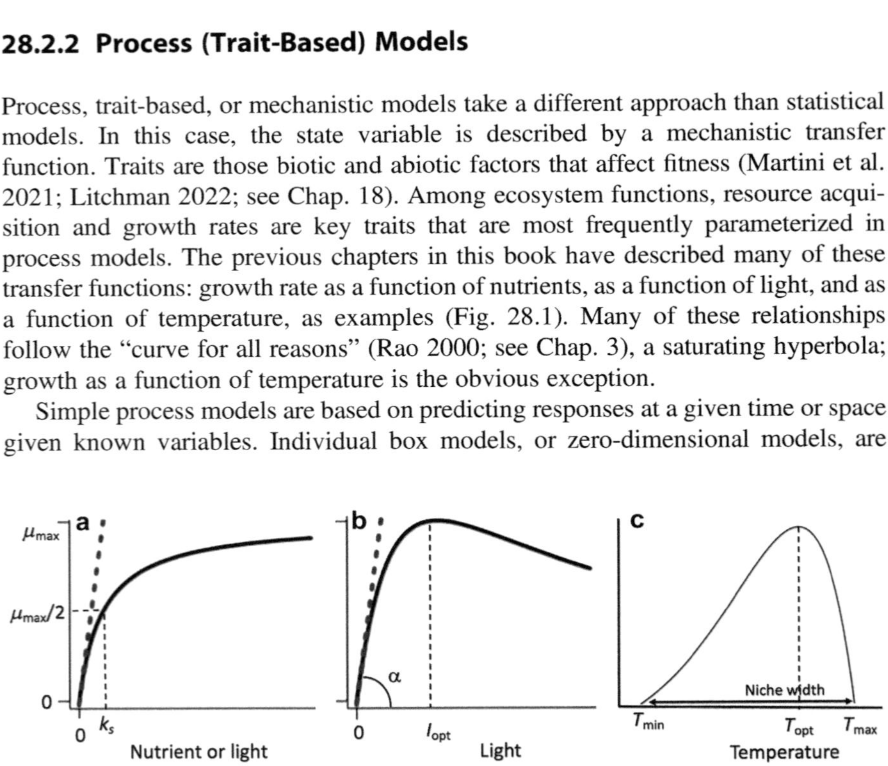

Fig. 28.1 Recap of key functional traits of phytoplankton (see previous chapters of this volume). Panel (**a**) Typical relationship between growth rate and resource (nutrient or light) level and the associated traits, the maximum growth rate μ_{max}, the growth affinity (initial slope of the growth-resource curve), and the half-saturation constant for growth k_s. Panel (**b**) Light-dependent growth curve with photoinhibition; the traits are the maximum growth rate, the light affinity α (slope), and the optimum irradiance I_{opt}, above which photoinhibition occurs. Panel (**c**) Typical thermal performance curve (TPC) (growth dependence on temperature) and the associated traits, the optimum temperature for growth T_{opt}, the minimum and maximum temperatures where growth is non-negative, T_{min} and T_{max}, respectively, and the thermal niche width. Image reproduced from Litchman (2022, Limnol Oceanogr Letts 8: 229–246. 10.1002/lol2.10294) under a creative commons license

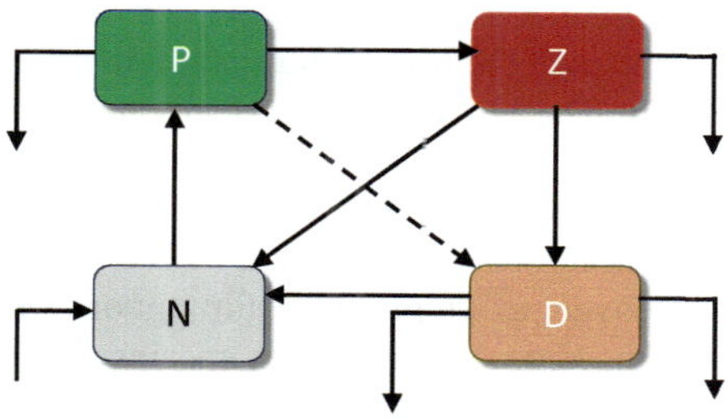

Fig. 28.2 Basic configuration of an N-P-Z-D model

simple process models well used in aquatic ecology. One such construct is the nutrient–phytoplankton–zooplankton (NPZ) model, in which only three state variables–nutrients (most often nitrogen (N)), phytoplankton, and zooplankton—are modeled, using equations that relate phytoplankton production to light and nutrients, zooplankton grazing, and losses due to excretion, or death (Fasham et al. 1990; Franks 2002). Simple NPZ models have been widely used, exploring questions of how a system responds if changes are made in the transfer functions, or if different parameters are used. Underscoring what was noted above, as depictions of nature, models should be as accurate as possible, but only as complex as necessary. Thus, NPZ models provide a simple model construct that may be suitable in a wide range of applications.

Adding complexity means adding more "boxes" and more transfer functions. Some NPZ models have added complexity, such as the addition of boxes representing detritus, D, or bacteria, B (Fig. 28.2). Basic NPZ models do not give species-specific information which may be needed in interpreting a HAB event. That is, phytoplankton are all included in a single box. Boxes may be added to represent different functional groups. Phytoplankton groups have been partitioned as "small" vs "large" or "diatom" vs "non-diatom" or at most a few phytoplankton functional types in these types of models (Arora et al. 2013). Then, the responses of the different functional types are each parameterized with their individual transfer functions.

Adding different nutrients other than N also adds complexity. In typical eutrophic conditions, C:N:P may be required, but some systems may also require inclusion of silica (Si), and yet others may require inclusion of iron (Fe). Multi-element descriptions in such stoichiometric models, as opposed to Redfield constructs (fixed C:N:P ratio; see Chap. 18), provide a basis for the development of mechanistic models that contain functional response descriptors with recognized physiological bases.

Parameterizing process models is challenging in a number of ways. First, the rates of growth as a function of individual abiotic factors are poorly known for many species, laboratory data may be available only for a limited number of species or strains or conditions, and varying interactive effects of different factors are rarely considered (Ralston and Moore 2020; Glibert et al. 2020, 2022). Such interactions may be additive, synergistic, or even antagonistic (Glibert et al. 2022; Chap. 27). Nutrient uptake processes, for example, are a function not only of factors such as temperature, salinity, and pH but also of the nutritional or growth state of the organism at the time of nutrient delivery (Flynn et al. 1999; Glibert and Burkholder 2006). The range of nutritional capability of many HAB species also adds to the complexity of adequately parameterizing nutrient acquisition. Our advancing knowledge of the complexity of nutrient acquisition, including mixotrophy (see Chap. 15 and discussed in more detail below), is only recently being included in models. Moreover, as emphasized throughout this volume, the seeming definitiveness of response curves for a given species at a given point in time under a given suite of conditions does not reflect the ability of cells to acclimate or adapt, to regulate as conditions change.

A number of approaches have been developed in recent years to capture the complexity of different species, functional types, and their traits, adaptations, and diversity (Table 28.1; reviewed by Acevedo-Trejos et al. 2022). As mentioned above, it is possible to add "boxes" to represent different plankton functional types (PFT), but adding diversity introduces more parameters and increases the computational cost. Basin-scale and global-scale models have often used this approach (e.g., Follows et al. 2007; Follows and Dutkiewitz 2011). Individual-based models (IBM) use algorithms to apply a set of rules to individuals or populations, allowing

Table 28.1 Characteristics of different approaches for modeling plankton diversity. Table modified from Acevedo-Trejos et al. (2022, Front Mar Sci 9: 10.3389/fmar.2022.975414) under a creative commons license

Model type	Strengths, advantages, and disadvantages
Statistical model (STM)	Describes patterns but difficult to define mechanisms Various types of data applicable, including DNA/RNA sequencing data Analysis of various types of diversity data
Individual-based model (IBM)	Individual interactions where complex patterns emerge from simple rules Captures population and community heterogeneity and stochastic processes Computationally expensive
Acclimation model (ACC)	Describes acclimation based on time-integrated performance Assumes instantaneous shifts in trait values at the individual level (e.g., intracellular stoichiometry) Can describe selection processes at the population or community level
Ecological network analysis (ENA)	Synthetic indicators of ecosystem status Quantification of resilience and stability Detection of early signs of stress
Plankton functional type model (PFT)	Applicable across tropic levels Modular and easy to hybridize with other approaches
Adaptive trait-based model (ATM)	Describes a population or community by the mean variance of a trait Assumes an environment-mediated trade-off dictates the dynamics of the trait Computationally efficient

properties to emerge from traits and interactions. Hellweger and Bucci (2009) used such an approach to investigate global plankton biogeography. The limitation of these models, similarly, is their computational cost. Acclimation models (ACC; see also Chap. 12) approach the question of the ability of an organism to adapt to a changing environment. Flexibility in nutrient response curves, for example, or photosynthetic responses have been modeled (e.g., Smith et al. 2011; Bonachela et al. 2013). However, adding flexibility and acclimation responses adds complexity, but such models have been used in conjunction with other models to represent diversity (Acevedo-Trejos et al. 2022). Similarly, adaptive trait models (ATM) allow responses to "evolve" in response to environmental changes, but the difference is that the adaptive traits are key outputs rather than inputs into models (e.g., Klausmeier et al. 2020). Ecological network analysis (ENA) depicts flow of energy and materials through different trophic components. Such an approach simplifies details on individual species but explicitly represents the species that comprise the entire food web (or parts of interest thereof; Glibert et al. 2022). Each of these different modeling approaches resolves individual organisms and the plankton community differently and each has their relative advantages (Table 28.1; Fig. 28.3; Acevedo-Trejos et al. 2022). Moreover, different modeling types are better suited to incorporate different types of data. Each of these modeling types has applications related to questions relevant to societal needs, from eutrophication to climate change to ecosystem management (Fig. 28.4; Acevedo-Trejos et al. 2022).

Modeling an individual species, for example a HAB species, against the background of a plankton community is complicated. For parameterizing a HAB species in an ecological context, knowledge of the transfer functions is needed not only for the HAB taxa of interest but also for the non-HAB taxa. A detailed description of

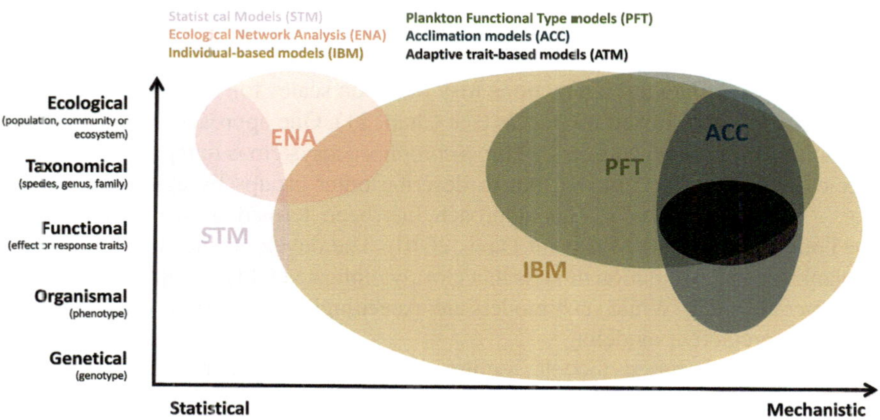

Fig. 28.3 Six modeling approaches commonly used to capture various dimensions of plankton biodiversity arranged along an axis spanning from those models that require explicit descriptions of ecological or biological processes (mechanistic) to those methods that do not (statistical). Image reproduced from Acevedo-Trejos et al. (2022, Front Mar Sci 9: 10.3389/fmar.2022.975414) under a creative commons license

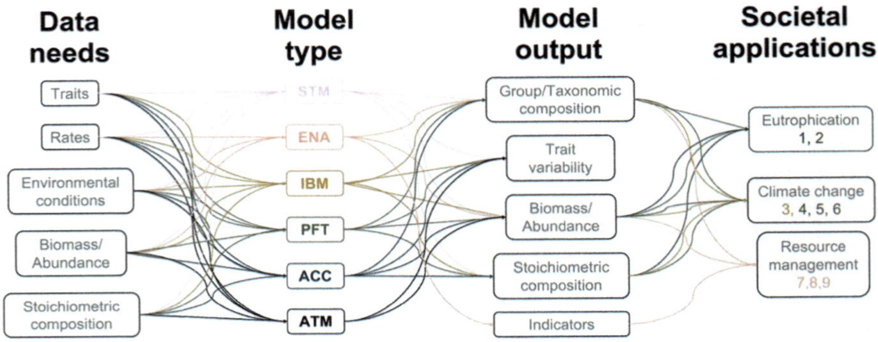

Fig. 28.4 Summary of the main data requirements, outputs, and common societal applications for different modeling approaches described in text. Some entries can be both data needs and model outputs. The numbers in the boxes of societal applications refer to examples of each approach: (1) Kerimoglu et al. (2018) (ACC), (2) Lancelot et al. (2007) (PFT), (3) Toseland et al. (2013) (IBM), (4) Kwiatkowski et al. (2018) (PFT), (5) Kwiatkowski et al. (2019) (PFT), (6) Henson et al. (2021) (PFT), (7) de la Vega et al. (2018) (ENA), (8) Meddeb et al. (2018) (ENA), (9) Saint-Béat et al. (2020) (ENA). Image reproduced from Acevedo-Trejos et al. (2022, Front Mar Sci 9: 10.3389/fmar.2022.975414) under a creative commons license

non-HAB species may be warranted, for example, for descriptions of prey growth dynamics in support of mixotrophic HABs, of specific predators, or of indicator species that may not even have a close trophic association with the HAB species. The role of non-HAB species and processes not only includes competitors and grazers, and can extend to the benthos, as activity of these organisms can directly affect the plankton community and they alter nutrient availability and cycling, among many other processes and interactions (Glibert et al. 2010). Microbial association, including interactions with bacteria and viruses, is also an area about which knowledge is rapidly advancing, and the interactions between nutrient availability and bacterial/viral interactions may occur on scales for which very little data are available for inclusion in models (see Chap. 21). One approach to modeling the physiology of a harmful species within a complex ecosystem is to represent the HAB species with a detailed model, but to describe other groups by aggregate models (e.g., "other flagellates"). This approach has been termed a rhomboid strategy (DeYoung et al. 2004; Mitra and Davis 2010). The development of 'digital twins' of plankton, i.e., simulation models that closely replicate reality, is rapidly advancing (Flynn et al. 2022). While such models are exceedingly complex, they hold promise for future predictive modeling.

While process-based models require a considerable amount of complicated data—and are more complex to develop and to run than statistical models, there are many applications of such models operating with success, and their applications are described in more detail below. Before turning to those examples, however, additional model types and model considerations are next addressed.

28.2.3 Habitat Models

One approach that has been applied in applying complex ecosystem models to HAB predictions (and predictions of species more generally) is to examine habitat suitability, as opposed to predicting biomass or growth rates of specific taxa directly. Habitat models take the output of other models, including process models, to define a suite of environmental conditions (e.g., temperature, salinity, nutrient forms, and availability) that intersect with the growth environment niche of a particular species (Glibert et al. 2022). The product of habitat models is not biomass or growth, but rather a growth "window of opportunity" in space or time. In one such effort, Glibert et al. (2014) developed a suite of model projections of the effect of climate change on HAB distribution in several regions of the globe, NW European Shelf-Baltic Sea system, NE Asia, and SE Asia. The projections (end of century, 2090–2100) were based on one scenario of the Intergovernmental Panel on Climate Change (IPCC), applied in a coupled physical-biogeochemical model (Holt et al. 2009). Habitat suitability was based on literature reviews (Heil et al. 2005; Vargo et al. 2008; Glibert et al. 2012; Brand et al. 2012), and a suite of physical and chemical (nutrient) parameters previously associated with positive growth and/or accumulated biomass of the selected HAB genera were defined. The model output showed the potential for the effects of nutrient loading and climate change in terms of projected change in area and fraction of time annually conducive to the specific HAB taxa studied.

The model output from Glibert et al.'s (2014) exercise was also instructive in that it showed that bloom expansion (or contraction) not only depends on the change in abiotic conditions supportive of their growth, but also on the temporal or spatial match of expansion of these parameters. In order for blooms to occur, all criteria for growth must be met at the same time and location. Any scenario that alters one set of matching parameters in time or space without altering the other criteria likely leads to a "mismatch" of conditions and a lesser likelihood for blooms to develop. The temporal mismatch in suitability of conditions for growth should not be viewed as evidence for limited potential for expansion. Rather, such a difference points to the possibility of expansion should there be a change in the timing of any one of the parameters. Climate forcing may alter the timing of nutrient loads relative to seasonal warming which may, in turn, alter the alignment of suitable conditions.

Brown et al. (2013) developed a habitat model of various HABs as well as pathogens in Chesapeake Bay using probabilities derived from multivariate models of multiple environmental factors (salinity, temperature, chl *a*, nutrients, etc.). This model was used in an ecological prediction mode for several years.

28.2.4 Machine Learning Models

Due to their powerful nonlinear modeling capability, machine learning methods are beginning to be used to predict HAB events. Neural networks (NN) use layers of nodes to connect observations, each weighted differently. The networks are trained with a data set; then predictions are made using new input data. Artificial neural networks (ANNs) have been used for more than 20 years to model freshwater cyanobacterial blooms. For example, in the late 1990s, ANNs were used to forecast the occurrence of cyanobacterial blooms and their associations with various environmental factors in the Murray River, Australia (Maier et al. 1998). An ANN model was used to predict algal blooms in Hong Kong coastal waters (Lee et al. 2003) and to predict outbreaks of the dinoflagellate *Dinophysis acuminata* in southern Spain (Velo-Suarez and Gutierrez-Estrada 2007). More recently, a NN approach was used to predict the presence/absence and abundance of the dinoflagellate *Karlodinium* and the diatom *Pseudo-nitzschia* in Alfacs Bay in the northwest Mediterranean Sea (Guallar et al. 2016). Another machine learning approach is support vector machine (SVM) models, which use a machine learning algorithm that seeks the hyperplane that best separates two labeled classes from each other; i.e., it classifies data. Such an approach has been used to predict blooms in freshwater reservoirs (Xie et al. 2012, Garcia Nieto et al. 2013).

Machine learning approaches have also been used in multiple applications in predicting HABs in the Gulf of Mexico, but with different objectives. Liu and Weisberg (2012) used such approaches to demonstrate the role of deep-ocean forcing on the western Florida Shelf in the major bloom occurrences. Weisberg et al. (2014) reported that the position of the Loop Current can affect blooms. When the Loop Current is in its southern position, it creates an upwelling of deep nutrients and fosters a diatom bloom that may outcompete any nascent blooms of the toxigenic dinoflagellate, *Karenia brevis*. Hill et al. (2019) used satellite remote sensing of chl *a* from 2003 to 2018, as well as sea surface temperature and bathymetry, as inputs to a convolutional NN to detect the presence of *K. brevis* blooms on western Florida Shelf, achieving a maximum detection accuracy of 91%. More recently, Li et al. (2021) applied four types of machine learning approaches (support vector machine, relevance vector machine, Naïve Bayes classifier, and ANN to predict *K. brevis* blooms and found that while all models captured the long-term time series of these blooms, the relevance vector machine model performed best.

28.3 Adding More Complexity

No matter what type of model approach is used, predicting plankton diversity and variability remains challenging in an era of rapid environmental change. The challenges can be found on multiple fronts: getting the environmental conditions right, getting the processing of energy and nutrients right (the transfer functions), and having the right kind of input data. To gain flexibility to represent all environmental factors and responses and interactions, models may be built as independent entities and then coupled as needs or questions demand (Glibert et al. 2022; Fig. 28.5). In a coupled model system, the outputs of one model become inputs to other modules. For example, when land use models are coupled to HAB or food web models, integrated predictions from river and land inputs through physics to the biological community can be attained. Climate is another obvious driver of phytoplankton change, and downscaling of climate models to regional and local conditions may be

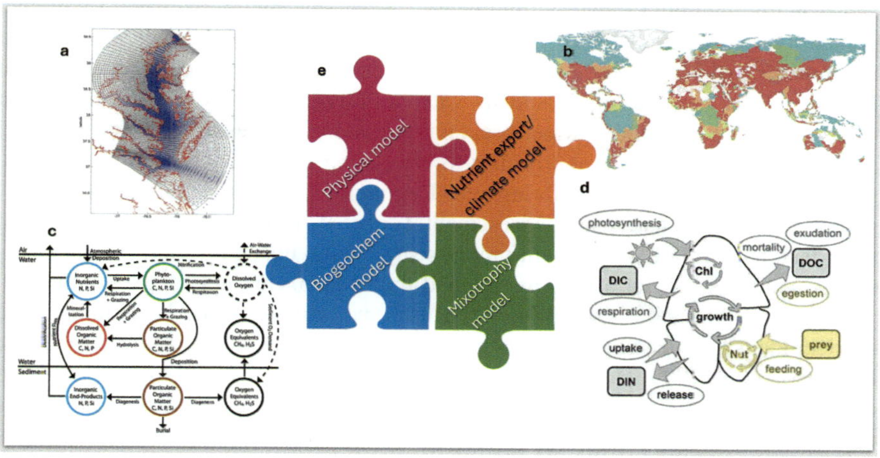

Fig. 28.5 Modular, coupled models that link physical models (Panel **a**), land-based changes in nutrient loads and/or climate change models (Panel **b**), biogeochemical models (Panel **c**) and complex biological models (Panel **d**) are required for modeling dynamic systems such as estuaries. The outputs of nutrient loading models or climate change models become inputs for biogeochemical models (appropriately downscaled as necessary). Sub-models, such as mixotrophy models (Panel **d**) may be applied as necessary. Like pieces of a puzzle (Panel **e**), sub-models may be coupled or added as needed. Panels **a**, **c** reproduced from Testa et al. (2014 J. Mar. Syst. 139:139–158), panel **b** from Beusen and Bouwman (2022 Front Water doi.org/10.3389/frwa.2022.893585), and panel **d** from Leles et al. (2021, Prog Oceanogr, https://doi.org/10.1016/j.pocean.2020.102481). Images **a**, **b**, **c** reproduced with permission of Elsevier and image **b** reproduced under a creative commons license

required to obtain regionally relevant C chemistry input factors. Modular, coupled, multi-trophic-level models may provide the flexibility to explore additive, amplified, propagated, or even reduced or antagonistic effects that can emerge from the complex interactions of multiple stressors (Glibert et al. 2022).

28.3.1 Environmental Complexity: Land-Based Nutrient Loads

Many models, especially those related to eutrophication, require input terms related to nutrient loading. These models themselves are complex and subject to many assumptions and complex parameterizations. Spatially explicit global maps of dissolved river N and P export and dominant sources have been derived using various models, including the Global Nutrient Export from WaterSheds (Global NEWS) models (Seitzinger et al. 2005; Dumont et al. 2005; Harrison et al. 2005, 2010), and models that have advanced since then, such as the Integrated Model to Assess the Global Environment–Global Nutrient Model (IMAGE-GNM; Beusen et al. 2015, 2016; Vilmin et al. 2018, 2020; Chap. 22). The nutrient source terms that are considered in both the Global NEWS and IMAGE-GNM models include natural sources, such as N_2 fixation and P weathering, and anthropogenic sources (non-point inputs from fertilizer by crop type, N_2 fixation by crops, atmospheric N deposition, and manure by animal species; point sources from sewage, as estimated by human population and treatment level) (Seitzinger et al. 2005, 2010). These models also account for hydrological and physical factors including water runoff, precipitation intensity, land use and slope, and in-water removal processes such as dams and reservoirs and consumptive water use, and ultimately, they estimate nutrient export at the river mouths. Other approaches to estimating nutrient export on regional to global scales are the Spatially References Regressions on Watersheds (SPARROW) model and the Soil and Water Assessment Tool (SWAT) model that use statistical relationships to relate water quality monitoring data to upstream sources and watershed characteristics that affect the fate and transport of nutrients (e.g., Smith et al. 1997). The SPARROW model has been used extensively in the USA to estimate nutrient loads to receiving waters such as the Gulf of Mexico (Alexander et al. 2008; Robertson et al. 2009; Hoos and McMahon 2009). Collectively, these models have documented large differences between regions and over time globally, as well as differences in nutrient export by element, but also require considerable input data as well as assumptions.

28.3.2 Physical and Biogeochemical Modeling

There are well-established approaches for configuring dynamic, physical models of aquatic systems. The hydrodynamic models commonly used in coastal and estuarine systems are the Regional Ocean Modeling System (ROMS) (Haidvogel et al. 2008) and the Finite Volume Coastal Ocean Model (FVCOM) (Chen et al. 2003; Curchitser et al. 2013), but many other modeling systems are available. One of the primary differences in these systems is flexibility in the shape and size of the grid used to represent the bathymetry and area of the system of interest. The more flexible the grid, the better it may resolve abrupt changes in bathymetry (and thus to simulate the true depth distribution of the system) and the fine-scale representation of shorelines and freshwater–marine transitions.

The diversity of biogeochemical models is even greater. Some biogeochemical models share common features (Friedrichs et al. 2007; Fennel et al. 2019) in that they may represent a single phytoplankton group and are limited to tracing 1–2 key constituents (e.g., N, P). Other models include more complex nutrient pools, including organic and inorganic forms of N, P, Si, and C (Isleib et al. 2007; Zhang and Li 2010) and simulate multiple phytoplankton groups. For example, the ERSEM (European Regional Seas Ecosystem) model (Blackford et al. 2004) has four plankton functional types for phytoplankton (picoplankton, flagellates, diatoms, and dinoflagellates), three for zooplankton (microzooplankton, mesozooplankton, and heterotrophic dinoflagellates), and one for bacteria. ERSEM also parameterizes phytoplankton and bacteria with variable C:N:P stoichiometry. Some biogeochemical models fully represent sediment biogeochemistry through multiple-layer modules (e.g., Shen et al. 2019) or through simple parameterization (Fennel et al. 2016). Most of these models represent dissolved oxygen (O_2), given the important feature of deoxygenation in coastal ecosystems (Chap. 24), and many modern biogeochemical models represent the carbonate system (Chap. 26), but with a

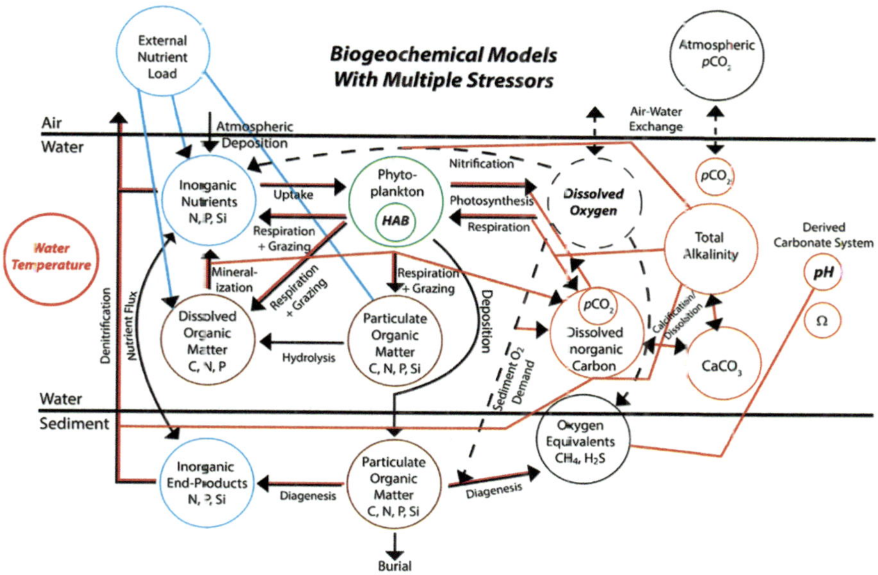

Fig. 28.6 Schematic diagram of state variables, biogeochemical reactions, and their interactions. The complexity of the diagram underscores the tight connectivity of the major biogeochemical processes as they impact nutrients, phytoplankton, dissolved oxygen, and both organic and inorganic carbon dynamics, as well as the ubiquitous role of temperature in modulating these processes. Image by J. Testa, published in modified form by Glibert et al. (2022, Ocean-Land-Atmos. Res. 2022: 9787258. 10.34133/2022/9787258.) and reproduced under a creative commons license

range of formulations, where some include calcification and dissolution and anaerobic metabolism, while others may just represent the impacts of aerobic metabolism on dissolved inorganic C and total alkalinity (Fig. 28.6).

28.3.3 Climate Change and Changing Carbon Chemistry

Modeling climate effects on biological or biogeochemical processes requires, on the one hand, high-resolution climate models that allow realistic simulations of not only temperature changes and CO_2 changes, but also extreme heat and precipitation events, and, on the other hand, knowledge of how the C chemistry regime alters organismal dynamics (Glibert et al. 2020). With warming of the ocean waters and the atmosphere, storms are becoming stronger and wetter. Diffenbaugh et al. (2018), for example, using both historical observations and a large climate model ensemble, quantified the influence of global warming on extreme heat and precipitation events around the world and found that climate forcing increased the probability of wettest 5-day period by 41% globally. Warmer waters are the accelerant for hurricanes or typhoons, as warmer atmospheres hold more moisture. Stronger and wetter storms have significant implications for nutrient delivery to coastal systems, via both upwelling and coastal runoff. As atmospheric CO_2 rises, oceanic carbonate chemistry is changing, resulting in acidification over time in many waters (see Chap. 26). Capturing these changes globally is difficult enough. Downscaling these models regionally adds even more complexity.

The complexity of the interactions and competition between phytoplankton species that may occur as pCO_2 or pH change has been studied in the laboratory and clearly demonstrated in models (e.g., Flynn et al. 2015; see also Chap. 26). The implications for phytoplankton composition and diversity are many.

In a rising CO_2 world, one could argue that diatoms may be disfavored in competition with those cells that either lack C concentrating mechanisms (CCMs) or have less efficient CCMs (Fu et al. 2012; Glibert et al. 2022 and references therein; see also Chap. 26). Smaller cells, as well as cells with mixotrophic capability, are also likely to be favored (see Chap. 23). As pH declines, there is also some evidence that some cyanobacteria can increase growth rates and therefore outcompete eukaryotic algae under such conditions (O'Neil et al. 2012). Decreasing pH may also alter harmful algal characteristics, including swimming behavior via alteration in the transmembrane pH gradients that provide energy for movement (Williams 1988). Faster swimming speeds and downward velocities (Kim and Spivack 2013) and dispersive swimming behavior (Hauser et al. 1978) have resulted from lower pH, potentially altering encounter rates with nutrients and predators. Changes in HAB toxicity have also been linked to changing pH and pCO_2 levels (e.g., Fu et al. 2010; Hattenrath-Lehmann et al. 2015; Pang et al. 2017; Clark and Flynn 2000; Errera et al. 2014; Bercel and Kranz, 2019). These effects may be species specific or even clone specific or may change with nutrient availability, increasing the challenge of incorporating such effects in models.

28.3.4 Physiological Complexity: Osmo- and Mixotrophy

One of the most challenging aspects of modeling many harmful species is their flexibility in nutritional modes. Many—if not most—HAB species are now recognized to be osmo-mixotrophs (see Chap. 15). Linking these nutritional modes to plankton dynamics is complicated on several levels.

Many freshwaters are becoming more and more enriched with dissolved organic matter (DOM); they are becoming "browner" (see Chap. 22). This is having profound effects on phytoplankton and food web dynamics. Recent work has shown that the brownification of freshwaters synergistically interacts with warming temperatures to enhance cyanobacterial blooms and their toxin (Urrutia-Cordero et al. 2016). These investigators used a factorial experiment with different temperatures and humic enrichment levels, and found that temperature effects on *Microcystis* and its toxin levels were amplified in the brownification treatment, especially when combined with warming. Either more toxins were produced per cell or more toxin strains were selected for the combined humic enrichment + higher temperature treatments. They supplemented these findings modeling of lake data, which revealed more toxic blooms under conditions of warm temperatures and high

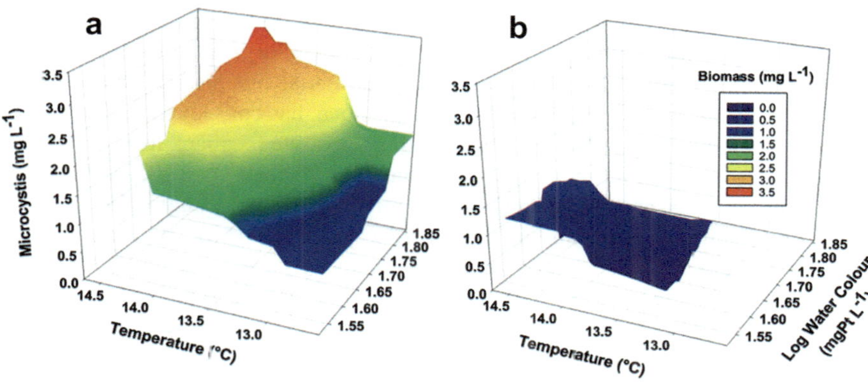

Fig. 28.7 The effects of climatic warming and brownification on *Microcystis* in Lake Ringsjön, a shallow eutrophic lake in Sweden. The response of *Microcystis* biomass (mg L^{-1}) to temperature (°C) and water color (used as a proxy of brownification; mg Pt L^{-1}). Panel (**a**): before management (1996–2004; n = 9) and Panel (**b**): after management (2005–2014; n = 10) through biomanipulation. The surface 3D plots were produced through interpolation of the raw data (monthly means April–October from each year) using inverse square weighting. After biomanipulation was performed in the lake. *Microcystis* biomass was not related to variations in either temperature or water color. Image reproduced from Urritia-Cordero et al. (2016, Sci Rep 6: 29542. 10.1038/srep29542) under a creative commons license

humic acid loads. A parallel model application showed that food web manipulation of the lake decreased these impacts, suggesting that food web interactions and related nutrient cycling effects are highly intertwined with these effects, but such results perhaps offer potential options for managers to keep blooms under control (Fig. 28.7). Yet, comparatively few models are addressing these changes, especially considering HABs. Among the modeling frameworks that are addressing marsh production and accretion and changing DOM inputs to coastal and freshwaters is the Delft3D modeling system, including one of its modules, the Delft3D-WAQ (water quality) model (Van Gils et al. 1993; Los et al. 2004). It should be noted that this model system characterizes changing DOM loads but does not currently include any detailed biological responses.

Modeling the growth of organisms that are mixoplankton requires a new level of understanding and the development, testing, and deployment of new types of models. The comparatively straightforward transfer functions of growth rate as a function of nutrients, or photosynthetic rate as a function of light, become much more complicated when mixotrophy is considered. As previously described (Chap. 15), mixoplanktonic activity fundamentally changes our understanding of planktonic food webs, enabling primary producers to acquire nutrients directly from ingestion of prey such as bacteria and algal competitors and even from their predators. Mixotrophy in these plankton is not simply additive or substitutional with photosynthesis; rather, it is synergistic (Mitra and Flynn 2010). Thus, through mixotrophy, there is provision of additional nutrients (N, P, Fe) from feeding to support primary production, together with a contribution of C to supplement photosynthesis under conditions of light limitation (including night). Moreover, the ecophysiology (nutritional quality) of the component organisms present in the ecosystem affects not only their own growth potential but also the activities of others. The cellular composition of algae, for instance, has consequences for grazers and their emergent properties, with potential for a positive feedback loop generating ungrazable primary producers (Mitra and Flynn 2006; Glibert et al. 2010). Predator–prey dynamics thus change in light of the mixoplankton paradigm (Flynn et al. 2019). In some instances, toxins are used in the capture of prey and thus are directly related to the mixotrophic capacity of the organisms (e.g., Granéli and Flynn 2006; Stoecker et al. 2006; Adolf et al. 2007). Consider also the complexity of the specialist non-constitutive mixoplankton, such as *Dinophysis*, which depend on a series of specific trophic interactions to provide the plastids for their growth (Hansen et al. 2019).

Expanding from a classic NPZ model to a mixotrophic model is not straightforward. In a classic NPZ type of model, in which the only nutrient described is N (i.e., as N, but without further differentiation of N form), one could make the assumption that feeding to obtain N by mixotrophs is already essentially covered through the description of the activity of microzooplankton in the "Z" box. However, as noted above, that is not sufficient because a mixotrophic consumer has the potential to have assimilation and growth efficiencies approaching 100%, coupled with phototrophy, while within a non-mixotrophic protist member of the "microzooplankton," efficiencies will be ~50%. Inclusion of mixoplankton within an NPZ model results in a substantial difference in planktonic trophic dynamics (Mitra et al. 2014b, Anschütz and Flynn 2020). Mitra et al. (2014b) extended an NPZ model in several ways. First, they added a B box, for bacteria, as bacteria are not only an important source of nutrients through their remineralization, but they are a source of food also, expressing the new mixotrophic paradigm in a new model structure. To the flow diagram, another new box is also added: an M box, into which all those protists that are mixotrophic through a combination of phototrophy and phagotrophy are placed. The zooplankton (Z) box was also further fractionated. It is immediately clear that the complexity of interactions increased substantially with related consequences for computational complexity (Fig. 28.8).

A critical issue in modeling mixotrophs is controlling the growth rate as a function of the supply of energy and nutrients presented by different modes of nutrition. The classic way of describing both phytoplankton and zooplankton growth rates in models depicts the maximum growth rate (μ_{max}) as a reflection of the maximum nutrient acquisition rate (Fasham et al. 1990). Thus, in models of N-limited phytoplankton, μ_{max} is matched to the maximum rate of N assimilation. For zooplankton, μ_{max} is a function of the maximum grazing rate (G_{max}) and constants defining assimilation efficiency (AE) and respiration (Mitra et al. 2014a). However, nutrient transport in phytoplankton may exceed that required to support μ_{max} by an order of magnitude (e.g., McCarthy and Goldman 1979, Goldman and Glibert 1983; Chap. 8). Likewise, and setting aside the fact that AE is not a constant, but rather declines with ingestion rate and food quality, the value of G_{max} greatly exceeds the grazing rate needed to match μ_{max} (Mitra et al. 2014a). For mixotrophs,

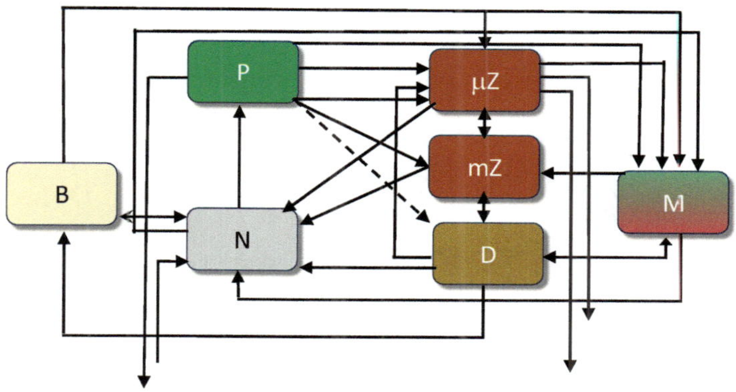

Fig. 28.8 Configuration of an NPZ model in which a bacteria box (B)and a mixotroph (M) box are added, and in which the zooplankton (Z) box is partitioned into micro (μZ) and macro (mZ) zooplankton. Compare the complexity of these interactions with those depicted in the simple NPZD model in Fig. 28.2

the mismatch between such simplifying modeling approaches and reality is incompatible because of the interactions between phototrophy and phagotrophy. Those interactions are more complex because of the multi-nutrient basis of the metabolic processes involved.

Conceptually, a mechanistic model of mixotrophy has to consider multiple points of interaction (Flynn and Mitra 2009). The first, full model construction of mixotrophy, the *Perfect Beast* model, is multi-nutrient (C-N-P), and describes the interactions between photosynthesis, inorganic nutrient acquisition, and prey acquisition (Flynn and Mitra 2009, Fig. 28.9; see Chap. 15). Using the *Perfect Beast* model of Flynn and Mitra (2009), and applying experimental data from the mixoplankter *Karlodinium veneficum* and its prey, *Rhodomonas*, grown individually under varying temperature, under autotrophic conditions, and in mixed batch cultures in which N:P stoichiometry of both predator and prey was varied, the dynamic of the predator–prey interactions was demonstrated. Biomass of *K. veneficum* was highest when it was parameterized as a mixoplankter, and when it consumed prey under the highest N:P conditions considered (Lin et al. 2018a, b). The modeled scenarios under all N:P conditions showed large differences between mixoplanktonic and autotrophic growth, but these differences also varied with temperature (see Chap. 23). When inorganic nutrients were in balanced proportions, lower biomass of the mixoplankton was attained at all temperatures in the simulations, suggesting that natural systems might be more resilient against *K. veneficum* development in warming temperatures if nutrients were available in balanced proportions. Gonzalez et al. (2022) developed a different model of adaptive dynamics of mixotrophic plankton in order to explore impacts of rising temperatures. Indeed, they found that mixotrophy rose as temperatures rose, but if

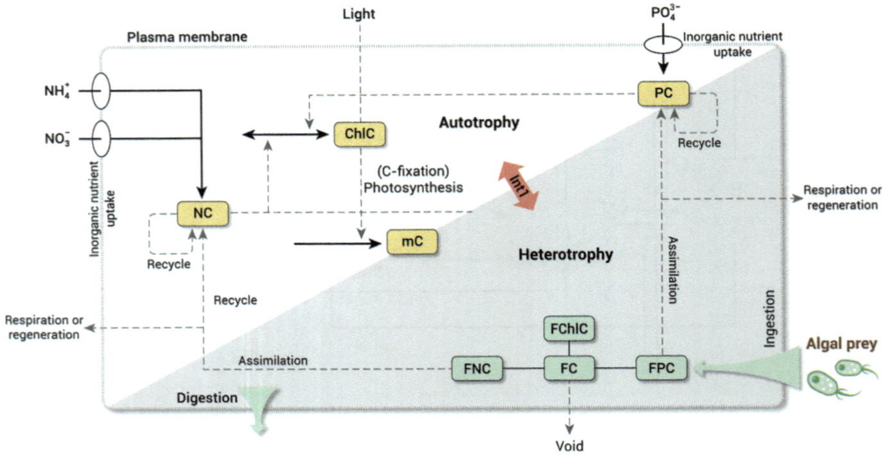

Fig. 28.9 Schematic of the structure of the mixotrophy model, showing major flows in and out of state variables (solid arrows and boxes) from the external parameters (NO_3^-, PO_4^{3-}, and light), and the major feedback processes (dashed arrows) (based on Flynn and Mitra 2009). Autotrophic growth uses inorganic nutrients and light via the photosystems of the mixotroph (phototrophy; white part). A proportion of activity leading to growth is required to support synthesis of those photosystems. Predation brings algal prey into the food vacuole within the confines of the mixotroph cell (heterotrophy; gray part). Interactions between phototrophic and heterotrophic nutrition (Int1) influence the growth of the mixotrophy. The state variables (yellow boxes) that describe carbon (C), nitrogen (N), phosphorus (P), and chlorophyll (Chl) associated with core mixotroph biomass are mC (C-biomass of the mixotroph), ChlC (chlorophyll C quota), NC (cellular NC quota), and PC (cellular PC quota), while the same constituents (green boxes) associated with the content of food vacuole are FC (food vacuole C content relative to mC), FChlC (food vacuole Chl content relative to mC), FNC (food vacuole N content relative to mC), and FPC (food vacuole P content relative to mC). Image reproduced from Lin et al. (2018a, Front Mar Sci 320, 10.3389/fmars.2018.00320) as revised by Glibert et al. (2022, Ocean-Land-Atmos. Res. 2022: 9787258, doi.org/10.34133/2022/9787258), both under creative commons licenses

prey abundance was reduced (due to intensive grazing), greater reliance on photosynthesis emerged.

The *Perfect Beast* mixoplankton model (Flynn and Mitra 2009) has been configured in the ERSEM structure to represent all mixoplankton functional groups allometrically and functionally and applied to long-term data from the Western English Channel (Leles et al. 2018). That effort demonstrated that the simulation of many C fluxes, including trophic transfer, is dependent on how and which mixotrophic functional groups are represented in the model (Leles et al. 2021). The *Perfect Beast* model has also been parameterized for the mixotroph dinoflagellate *Karlodinium veneficum* and incorporated into a 3D-coupled hydrodynamic (ROMS)-biogeochemical model of eutrophic Chesapeake Bay (M. Li et al. 2022a; see Sect. 28.4).

28.3.5 Data Consistency, Complexity

Models depend on parameters that are not always easily measured or available and knowledge of fluxes, dynamics, or physiological variable is often not adequately captured in models. Measuring rates often takes intensive experimentation, especially if one considers changes as a function of changing environmental or physiological conditions. Concepts of microbial diversity, mismatches between model structure and data, and the linking of biogeochemical processes to gene expression are difficult enough, even without the computational difficulties of increasingly complicated models with enormously complex data (Coles and Hood 2016); Glibert et al. 2020). In many cases, data on a relevant process may be obtained using multiple methods, but such techniques may not be inter-comparable or may have changed over time, challenging how best to calibrate models. One such example is the measurement of productivity, which may be by use of ^{14}C, ^{18}O, or variable fluorescence or other techniques. Whereas ^{14}C measurements determine the rate of incorporation of C, ^{18}O experiments measure the water splitting reaction and non-cyclic electron transport; variable fluorescence instruments (e.g., FRRF, PAM) measure photochemical efficiency of the photosystem II pathway; these rates are not equivalent (Suggett et al. 2009, see Chap. 7). Such a distinction may not be readily apparent to those incorporating such data in models. The long-term data from *Karenia brevis* blooms in Florida highlight another challenge for modeling long-term trends: that is, shifting protocols. In the early years of monitoring *K. brevis* blooms, for example, the only reliable method available for measuring the brevetoxins of the harmful alga was the mouse bioassay. Over time, advanced analytical approaches, including molecular assays and kits, have been developed and applied for toxin analysis (e.g., Twiner et al. 2012). In addition to sparing mice, these new methods have improved detection limits and have yielded insights into potentially different forms of toxin. Yet, each advancement in methodology may complicate time series analyses, model parameterization, and application.

In contrast, the rapid pace of advancement in tools and techniques, while improving detection capability, is creating other challenges for data incorporation into models (Stachelek and Madden 2015, see Chap. 2). The revolution in imaging and probe techniques, miniaturization, and in situ continuous data streams are, on the one hand, helping to advance our knowledge on new organisms, pathways, and their variation across species and ecosystems, in space and time, while at the same time creating new challenges in bioinformatics and data interpretation (Coles and Hood 2016; Glibert et al. 2018, 2022). New instrumentation and methodologies are inundating researchers and modelers with a "data deluge" (Baraniuk 2011). The advancement in instrumentation (sequencing, mass spectrometry), bioinformatic algorithms (for analyzing genomic and proteomic data), and public databases of DNA sequences, peptide profiles, and metabolites continues at a rapid pace that will make 'omics approaches increasingly applied in both organismal and toxin detection and on a range of scales to understand HABs, their cell mechanisms, evolution, and interactions (Hennon and Dyhrman 2020, Fig. 28.10). Such efforts are only in their infancy in terms of models.

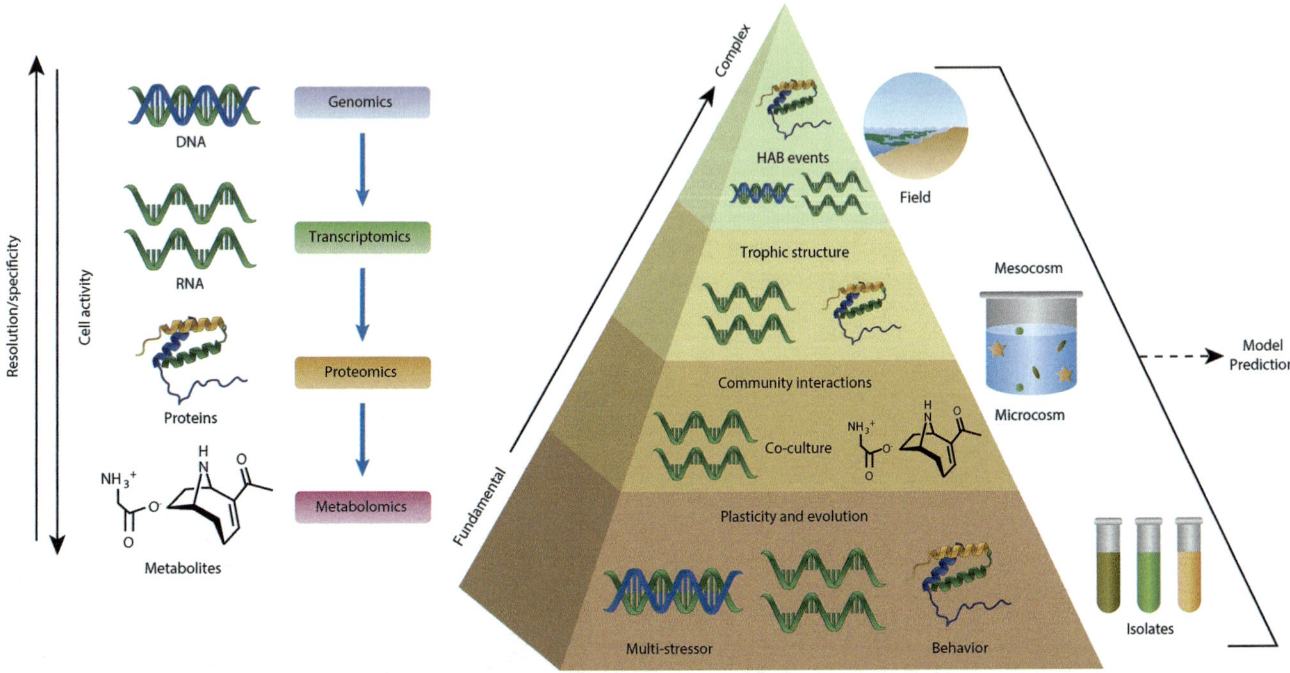

Fig. 28.10 Hierarchy of complexity in climate change experiments on HAB species. Conceptual diagram of experimental and field approaches for measuring impacts of climate change and the role of omics in disentangling HAB species responses. Image reproduced from Hennon and Dyhrman (2020, Harmful Algae 91:101587, doi.org/10.1016/j.hal.2019.03.005) with permission of Elsevier

28.4 Applying Coupled Models to Dynamic Estuaries: The Chesapeake Bay Example

Estuaries are particularly complex with regard to multiple stressors, in turn affecting the propensity for hypoxia and HABs. As previously described, temporal fluctuations in O_2, temperature, and primary production in estuaries can be extensive leading to extremes and high variability (Lauchlan and Nagelkerken 2019); Chap. 27). In the Chesapeake Bay (Fig. 28.11), for example, excessive nutrient loading from the watershed stimulates primary production in surface euphotic waters in spring, leading to unassimilated organic matter settling to deeper waters and the development of hypoxia by summer (Kemp et al. 2005). The Chesapeake Bay is an exceptionally well-studied, eutrophic estuary, but one with many challenges in terms of modeling (Glibert and Li 2024).

Climate change has played, and is expected to play, an important role in driving changes in Chesapeake Bay, and climate changes have interacted, and will continue to interact, with other factors in complex ways. Wang et al. (2017) investigated the individual and combined effects of warming and sea level rise on hypoxia by 2050. Summer anoxic volume was estimated to increase by 1.4% due to warming, but sea level rise resulted in a 12% reduction in the anoxic volume. In addition to warming and sea level rise, Irby et al. (2018) found that warming reduced O_2 solubility year

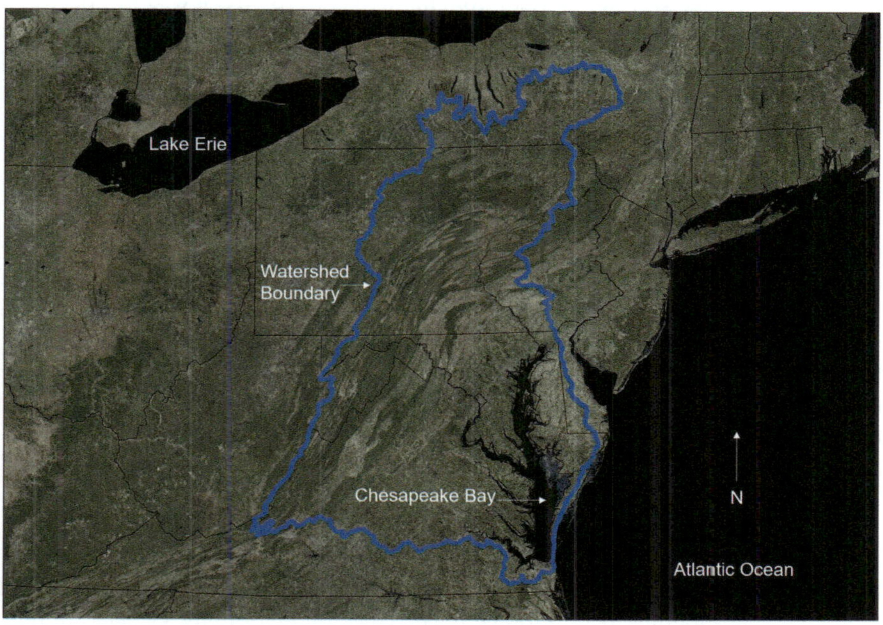

Fig. 28.11 The mid-Atlantic coast of the USA, showing the Chesapeake Bay and its watershed. Image credit: P. Haggerty, U.S. Geological Survey

around, but that changes in precipitation and river flow fueled increased primary production, and sea level rise increased bottom-water O_2 but decreased O_2 at mid-depths. Ni et al. (2020), using climate downscaling projections for dissolved O_2, found that the hypoxic and anoxic volumes would increase by 10–30% between the late twentieth and mid-twenty-first century. To further understand what drove the long-term changes in the Chesapeake Bay over the past 30 years, Ni et al. (2020) used a coupled hydrodynamic-biogeochemical model to conduct hindcast simulations between 1985 and 2016. They removed seasonal and interannual variations, and then observed that dissolved O_2 concentration in all regions of the Chesapeake had a statistically significant declining trend: ~0.1 mg L^{-1} per decade. They also showed that warming was the dominant driver of the long-term O_2 decline, overwhelming the effects of sea level rise and modest O_2 increases associated with nutrient reduction. In a following study, Ni and Li (2023) further analyzed the O_2 budget in a fixed control volume of the bottom water most susceptible to hypoxia. Hypoxia reduced nitrification and denitrification processes in the sediment, increasing nutrient recycling by 15% and water column respiration by 12%. These further analyses by Ni and Li (2023) demonstrated a feedback mechanism for driving the nonlinear hypoxia response to nutrient loading. Their results suggest that recent warming has prevented O_2 improvements in Chesapeake Bay expected from nutrient input reductions and support the expectation that continued warming will serve to counter future nutrient management actions.

The Chesapeake Bay also suffers from numerous HABs (reviewed in Chap. 27), of which *Prorocentrum cordatum* (previously reported as *P. minimum*) and *Karlodinium veneficum* (previously reported as *K. micrum*) have been well studied and modeled. Using a coupled hydrodynamic–biogeochemical model, a habitat model for the Chesapeake Bay dinoflagellates *P. cordatum* and *K. veneficum* was developed to show how these taxa may change in the future. A set of physical and chemical parameters which produced favorable habitat conditions for growth were first identified (Li et al. 2020); then the outcomes of the model, for the time and space over which these habitat conditions were met, were compared. Regional climate downscaling, using three different climate scenarios, was further used to estimate habitat change by mid-century (Fig. 28.12). Model output showed that in both the spring and fall, *P. cordatum* blooms shift away from the hot summer; the spring bloom shifts earlier while the fall bloom shifts later in the year. For *K. veneficum*, the results were similar, with summer blooms expected to be fewer. Higher summer temperature also moves some days in July and August out of the preferred temperature range for *K. veneficum* blooms and extends the bloom season earlier to May and later to October.

The *Perfect Beast* model (Flynn and Mitra 2009) was also configured for *K. veneficum* in Chesapeake Bay (M. Li et al. 2022a). The model was applied in hindcast mode to simulate spatial and seasonal distributions of this mixotroph and the factors contributing to its variable dependence on autotrophy vs heterotrophy. The model showed that autotrophic growth dominated in the spring and heterotrophic growth dominated in summer. The ingestion rate of the cryptophyte prey of

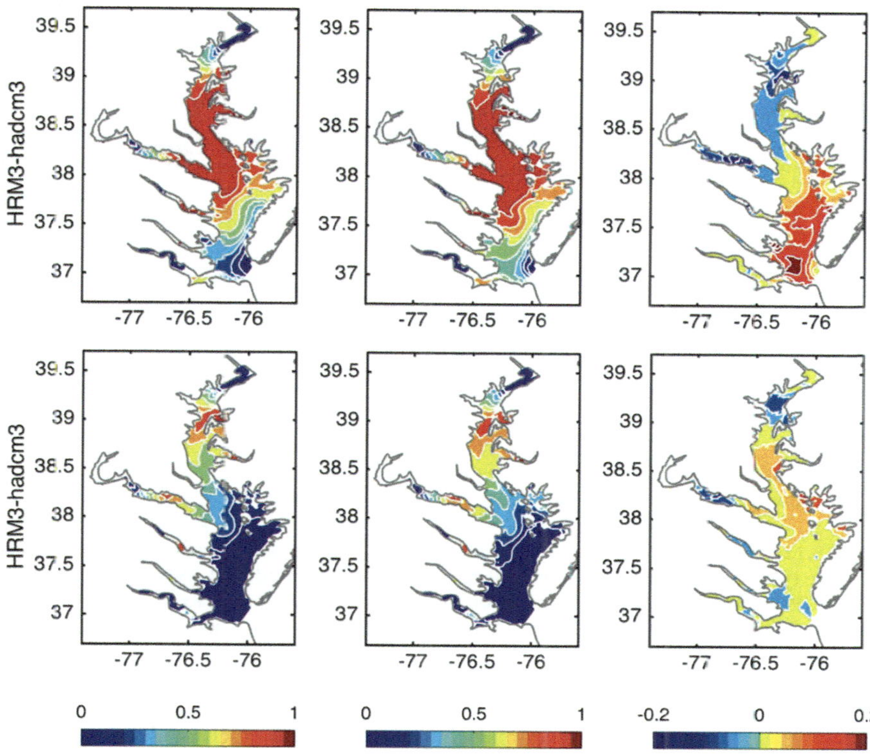

Fig. 28.12 Changes in the favorable habitat areas for *Prorocentrum cordatum* (upper panels) and for *Karlodinium veneficum* (lower panels) for Chesapeake Bay, showing the results of one scenario of regional climate change. The left-hand panels show current conditions, the middle panels show the projected future conditions, and the right-hand panels show the difference. Images reproduced from Li et al. (2020, Sci Total Envir 744:140947, 10.1016/j.scitotenv.2020.140947) with permission of Elsevier

K. veneficum increased with prey density and also increased when the N:P ratio was ~33 (i.e., when there was inorganic P deficiency) with the onset of summer, affirming that this species appears to switch to mixotrophic feeding when inorganic nutrient conditions develop, coincident with available prey (Fig. 28.13).

Using the mixotroph model of *K. veneficum*, scenarios were then tested to determine effects of nutrient reduction and/or further nutrient loading (M. Li et al. 2022a). The model showed that a 40% reduction in the riverine N loading—a target set by Chesapeake Bay managers—does not reduce the bloom size of the mixotroph in any substantive way (Fig. 28.14e), but has some measurable impacts on its autotrophic prey (Fig. 28.14f). A 40% increase in the riverine N loading also had small impacts, with small increases in the cell density of *K. veneficum* and cryptophytes (Figs. 28.14g, h). These scenarios thus underscored the complexity

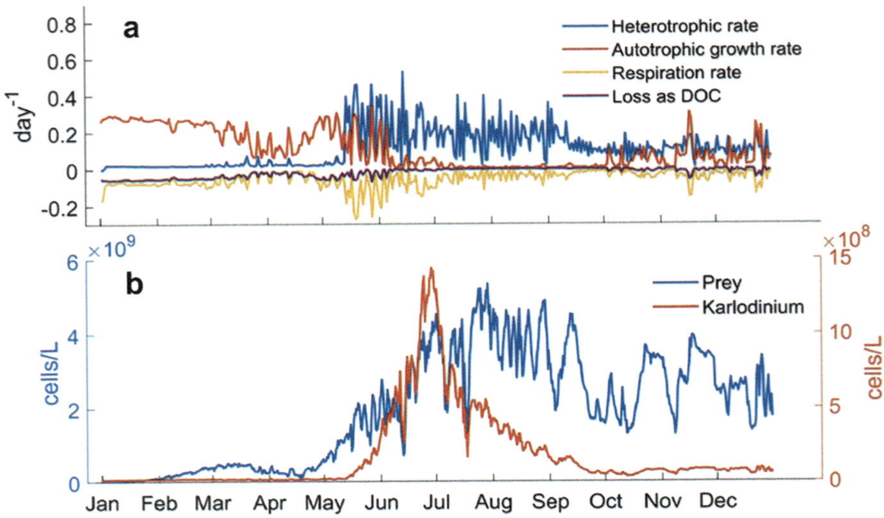

Fig. 28.13 Time series of autotrophic growth rate (red), heterotrophic growth rate (blue), respiration rate (yellow), and void rate (purple) (Panel **a**), *K. veneficum* (red) and cryptophyte (blue) cell density (Panel **b**) at one of the mid-bay stations in Chesapeake Bay. Images reproduced from M. Li et al. (2022a, Harmful Algae 113: 102203, 10.106/j.hal.2022.102203) under a creative commons license

of, as well as importance of, incorporating mixotrophy in such models: organisms such as *K. veneficum* with mixotrophic feeding may shift their growing season in a warmer world, and are likely to be maintained, even with nutrient reductions, if prey are available.

The same coupled hydrodynamic-biogeochemical model for Chesapeake Bay was also used to assess carbonate changes for the twenty-first century. It was shown that Chesapeake Bay may experience a near doubling of CO_2 uptake through increased primary production, a decrease in pH of up to 0.3, and a 90% increase in acidic volume (Li et al. 2023). Those model projections also suggested that hypoxia and acidification will be initiated earlier in the year in future climate conditions (Fig. 28.15). These projections could be ameliorated with a 40% reduction in riverine nutrients, which, in turn, would reduce phytoplankton uptake of inorganic C.

Further modeling that accounts for changes in pH on the growth of *P. cordatum* in the Chesapeake suggests relatively modest effects. Peak concentrations of this species are projected to increase by 2.9% in 2050 and 6.2% in 2100 as pH decreases by 0.2 or 0.4, respectively (R. Li et al. 2022b). Furthermore, while the toxicity of *P. cordatum* in Chesapeake Bay has not been documented, and the toxicity of this species is a topic of debate (Heil et al. 2005), reduced pH could increase the toxicity of some other harmful algae (e.g., Sun et al. 2011; Tatters et al. 2012).

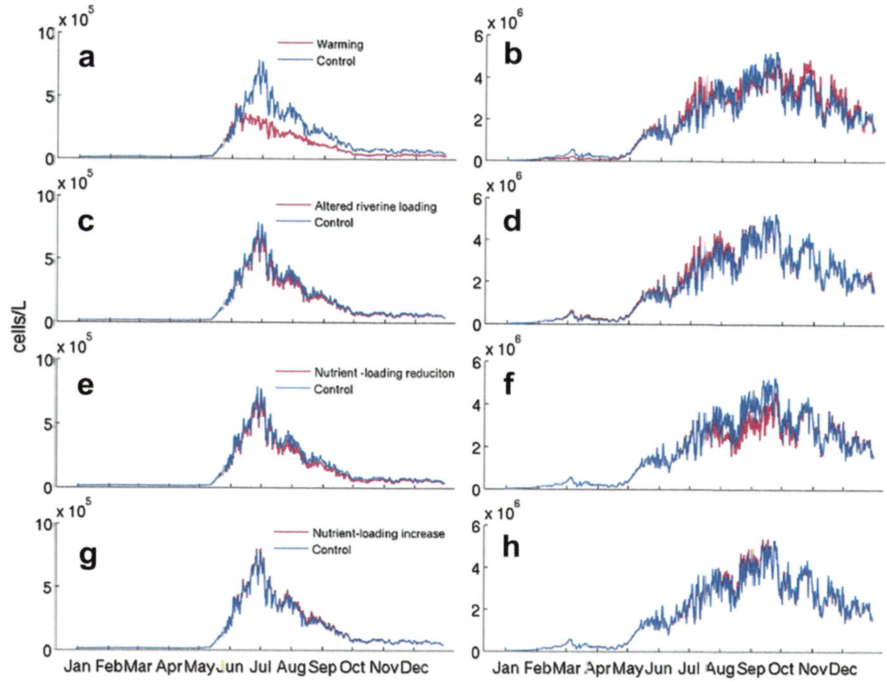

Fig. 28.14 Time series of *K. veneficum* (left) and cryptophytes (right) at station CB 3.3C obtained from 4 groups of numerical experiments examining the potential effects of climate change and nutrient management: warming (Panel **b**, **b**), altered riverine loading (Panel **c**, **d**), nutrient-loading reduction (Panel **e**, **f**), and increase (Panels **g**, **h**). Images reproduced from M. Li et al. (2022a, Harmful Algae 113:102203, 10.1016/j.hal.2022.102203) under a creative commons license

In all, these modeling efforts have shown that the seasonality of hypoxia and at least some HABs may change in the future in the dynamic Chesapeake Bay in complex ways. Warming has, and will continue to, counter any O_2 improvements that would be expected from nutrient reduction efforts, and mixotrophic capabilities of a common HAB also counteract positive effects on HAB occurrence due to nutrient reduction as the nutrient reductions effects only have modest impacts on the HAB prey. Thus, reducing the extent of hypoxia and HABs in Chesapeake Bay in years to come will remain exceptionally difficult even with the best efforts to reduce nutrient loads. Such findings are not unique to this one estuary; there are many lessons for temperate estuaries and coastal systems experiencing similar stresses. These findings are not, however, reason to forego nutrient reduction efforts. The pace of change as waters warm only make these efforts all the more important.

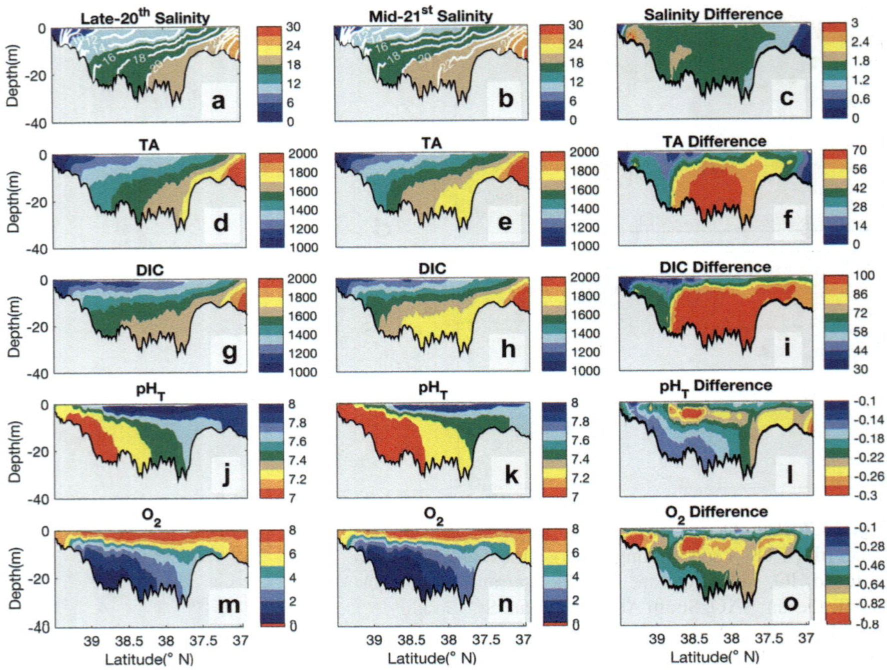

Fig. 28.15 Along-channel distributions of summer mean (June to August) salinity (psu, Panels **a**–**c**), TA (total alkalinity, μmol kg^{-1}, Panel **d**–**f**), DIC (dissolved inorganic carbon, μmol kg^{-1}, Panels **g**–**i**), pH_T (Panels **j**–**l**), and O_2 (mg L^{-1}, Panels **m**–**o**) in the late-twentieth century, mid-twenty-first century, and difference between the two periods. Images reproduced from Li et al. (2023, Comm Earth Envir 4: 68. 10.1038/s43247-023-00733-5) under a creative commons license

28.5 Forecasting and Prediction

The previous example illustrated prediction decades in the future and the challenges associated. These exercises are useful in understanding how climate may alter habitat, and in silico experiments on nutrient changes illustrated how important different factors may interact in altering future habitat. However, managers more often need near real-time predictions on the scale of days to weeks, not decades or centuries. Highly complex models are not suitable for this application, and repeating a theme of this chapter, models should be as accurate as possible, but only as complex as necessary. In the Gulf of Mexico, where HABs can cause respiratory illness, and where blooms are patchy, weekly and monthly maps of statewide *K. brevis* abundance (Fig. 28.16) are posted by the Florida Fish and Wildlife Research Institute, and 3–4 day predictions are now made available for different regions of the coast, using satellite remote sensing of chl *a*, in situ sampling of *Karenia brevis*, and wind buoy data (Stumpf et al. 2003, 2009). In Lake Erie, an

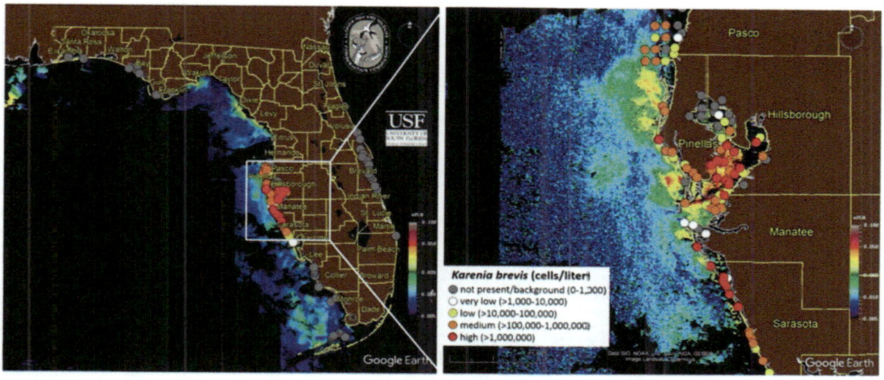

Fig. 28.16 Example of the monitoring maps provided for the west coast of Florida, this one representing mid-July 2021. The colored dots represent individual cell counts collected and analyzed by the Florida Fish and Wildlife Research Institute. The dots are overlain on ocean color satellite images. Image provided by the Integrated Red Tide Information System (IRIS) at https://optics.marine.usf.edu/projects/iris.html. The hotter colors show higher levels of chl *a*

operational forecast model provides the extent and 5-day outlooks of where *Microcystis* blooms may travel and at what concentrations, allowing alerts or management actions if necessary (Fig. 28.17). Similar forecasting efforts have developed for the Gulf of Maine and for central California, as well as throughout Europe, yielding 3–5 day outlooks of HAB location, intensity, and toxicity (Franks 2018, Flynn and McGillicuddy 2018, and references therein). In all cases, the transition from the development of a model used for research to one used operationally requires dedication of personnel and committed funding from agencies.

28.6 Summary

Capturing the effects of the many stressors on our present and future aquatic systems is extremely challenging. Models are advancing rapidly, but there are many challenges. The plasticity of nutritional pathways, as well as the plasticity of food web and other interactions, creates immense challenges for model constructs. New data streams and new approaches and methods give new opportunities for validating models, but also create their own challenges. New model approaches, combined with open-source availability of many models, provide the opportunity for advancing computational capabilities. With many factors changing in today's environment, including temperature, deoxygenation, carbonate chemistry, and nutrient pollution, different types of models are necessary to explore the complexity of interactions. Ultimately forecast models must be robust but must be simple enough to be operational and affordable to managers. Ensembles of models and integrated ecosystem models that couple the atmosphere, land, and coastal ocean are required to enable the

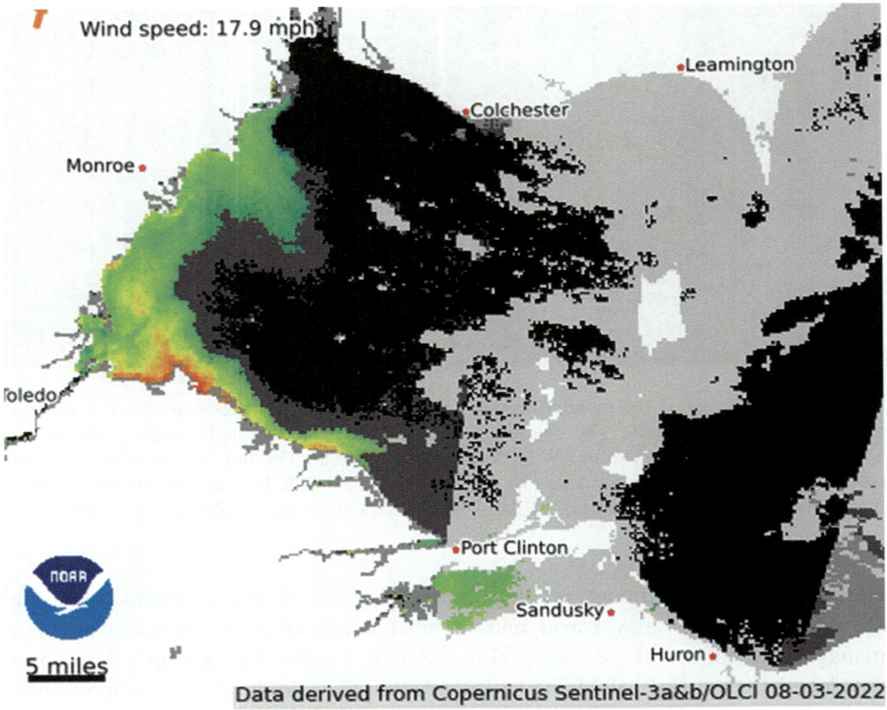

Fig. 28.17 Observed *Microcystis* cyanobacteria bloom biomass in western Lake Erie (August 3, 2023). The images were derived from Copernicus Sentinel-3 satellite data from the European Organisation for the Exploitation of Meteorological Satellites (EUMETSAT) and were processed by NOAA, National Centers for Coastal Ocean Science. Image from https://coastalscience.noaa.gov/science-areas/habs/hab-forecasts/lake-erie/satellite-imagery/ Open access

quantitative estimation from air-shed to the ocean and to investigate ecosystem response to climate changes and to further explore the changes in HABs that are to be expected in the future as eutrophication and climate impacts increase.

References

Acevedo-Trejos E, Cadier M, Chakraborty S, Chen B, Cheung S, Grigoratou M, Guill C et al (2022) Modelling approaches for capturing plankton diversity (MODIV), their societal applications and data needs. Front Mar Sci 9:975414. https://doi.org/10.3389/fmar.2022.975414

Adolf JE, Krupatkina D, Bachvaroff TR, Place AR (2007) Karlotoxin mediates grazing of *Oxyrrhis marina* on strains of *Karlodinium veneficum*. Harmful Algae 6:400–412. https://doi.org/10.1016/j.hal.2006.12.003

Alexander RB, Smith RA, Schwarz GE, Boyer EW, Nolan JV, Brakebill JW (2008) Differences in phosphorus and nitrogen delivery to the Gulf of Mexico from the Mississippi River basin. Environ Sci Technol 42(3):822–830

Anderson CR, Siegel DA, Kudela RM, Brzezinski MA (2009) Empirical models of toxigenic *pseudo-nitzschia* blooms: potential use as a remote detection tool in the Santa Barbara Channel. Harmful Algae 8:478–492. https://doi.org/10.1016/j.hal.2008.10.005

Anderson CR, Sapiano M, Prasad MB, Long W, Tango P, Brown C, Murtugudde R (2010) Predicting potentially toxigenic *pseudo-nitzschia* blooms in the Chesapeake Bay. J Mar Syst 83:127–140. https://doi.org/10.1016/j.jmarsys.2010.04.003

Anschütz A-A, Flynn KJ (2020) Niche separation between different functional types of mixoplankton: results from NPZ-style N-based model simulations. Mar Biol 167:3. https://doi.org/10.1007/s00227-019-3612-3

Arora VK, Boer GJ, Friedlingstein P, Eby M, Jones CD, Christian JR, Bonan G et al (2013) Carbon–concentration and carbon–climate feedbacks in CMIP5 earth system models. J Clim 26(15):5289–5314. https://doi.org/10.1175/JCLI-D-12-00494.1

Baraniuk RG (2011) More is less: signal processing and the data deluge. Science 331:717–719

Bercel TL, Kranz S (2019) Insights into carbon acquisition and photosynthesis in *Karenia brevis* under a range of CO_2 concentrations. Prog Oceanogr 172:65–76. https://doi.org/10.1016/j.pocean.2019.01.011

Beusen AHW, Van Beek LPH, Bouwman AF, Mogollon JM, Middelburg JJ (2015) Coupling global models for hydrology and nutrient loading to simulate nitrogen and phosphorus retention in surface water–description of IMAGE–GNM and analysis of performance. Geosci Model Dev 8:4045–4067. https://doi.org/10.5194/gmd-8-4045-2015

Beusen AHW, Bouwman AF, van Beek LPH, Mogollon J, Middelburg JJ (2016) Global riverine N and P transport to ocean increased during the 20th century despite increased retention along the aquatic continuum. Biogeosciences 13:2441–2451. https://doi.org/10.5194/bg-13-2441-2016

Beusen AHW, Bouwman AF (2022) Future projections of river nutrient export to the global coastal ocean show persisting nitrogen and phosphorus distortion. Front Water 4: https://doi.org/10.3389/frwa.2022.893585

Bigelow HB (1931) Oceanography, its scope, problems and economic importance. Houghton Mifflin, Boston, p 263

Blackford JC, Allen JI, Gilbert FJ (2004) Ecosystem dynamics at six contrasting sites: a generic modelling study. J Mar Syst 52:191–215. https://doi.org/10.1016/j.jmarsys.2004.02.004

Bonachela JA, Allison SD, Martiny AC, Levin SA (2013) A model for variable phytoplankton stoichiometry based on cell protein regulation. Biogeosciences 10:4341–4356. https://doi.org/10.5194/bg-10-4341-2013

Brand LE, Campbell L, Bresnan E (2012) *Karenia*: the biology and ecology of a toxic genus. Harmful Algae 14:156–178

Brown CW, Hood RR, Long W, Jacobs J, Ramers DL, Wazniak C, Wiggert ID et al (2013) Ecological forecasting in Chesapeake Bay: using a mechanistic-empirical modeling approach. J Mar Syst 125:113–125

Chen C, Liu H, Beardsley RC (2003) An unstructured grid, finite-volume, three-dimensional, primitive equations ocean model: application to coastal ocean and estuaries. J Atmos Ocean Technol 20:159–186. https://doi.org/10.1175/1520-0426(2003)020<0159:AUGFVT>2.0.CO;2

Clark R, Fynn KJ (2000) The relationship between the dissolved inorganic carbon concentration and growth rate in marine phytoplankton. Proc R Soc Lond B Biol Sci 267:953–959

Coles VJ, Hood RR (2016) Approaches and challenges for linking marine biogeochemical models with the "omics" revolution. In: Glibert PM, Kana TM (eds) Aquatic microbial ecology and biogeochemistry: a dual perspective. Springer, Geneva, Switzerland, pp 45–66

Curchitser EN, Batchelder HP, Haidvogel DB, Fiechter J, Runge J (2013) Advances in physical, biological, and coupled ocean models during the US GLOBEC program. Oceanography 26:52–67. https://doi.org/10.5670/oceanog.2013.75

de la Vega C, Schückel U, Horn S, Kröncke I, Asmus R, Asmus H (2018) How to include ecological network analysis results in management? A case study of three tidal basins of the

Wadden Sea, south-eastern north Sea. Ocean Coast Manag 163:401–416. https://doi.org/10.1016/j.ocecoaman.2018.07.019

deYoung B, Heath M, Werner F, Chai F, Megrey B, Monfray P (2004) Challenges of modeling ocean basin ecosystems. Science 304:1463–1466

Díaz PA, Ruiz-Villareal M, Pazos Y, Moita T, Reguera B (2016) Climate variability and *Dinophysis acuta* blooms in an upwelling system. Harmful Algae 53:145–159

Diffenbaugh NS, Singh D, Mankin JS (2018) Unprecedented climate events: historical changes, aspirational targets, and national commitments. Sci Adv 4:eaaoo3354. https://doi.org/10.1126/sciadv.aao3354

Dumont E, Harrison JA, Kroeze C, Bakker EJ, Seitzinger SP (2005) Global distribution and sources of dissolved inorganic nitrogen export to the coastal zone: results from a spatially explicit, global model. Glob Biogeochem Cycles 19:GB4S02

Errera RM, Yvon-Lewis S, Kessler JD, Campbell L (2014) Responses of the dinoflagellate *Karenia brevis* to climate change: pCO_2 and sea surface temperatures. Harmful Algae 37:110–116. https://doi.org/10.1016/j.hal.2014.05.012

Fasham MJR, Ducklow H, McKelvie SM (1990) A nitrogen-based model of plankton dynamics in the oceanic mixed layer. J Mar Res 48:591–639. https://doi.org/10.1357/002224090784984678

Fennel K, Laurent A, Hetland R, Justić D, Ko DS, Lehrter J et al (2016) Effects of model physics on hypoxia simulations for the northern Gulf of Mexico: a model inter-comparison. J Geophys Res Oceans 121:5731–5750. https://doi.org/10.1002/2015JC011577

Fennel K, Gehlen M, Brasseur P, Brown CW, Ciavatta S, Cossarini G, Crise A et al (2019) Advancing marine biogeochemical and ecosystem reanalyses and forecasts as tools for monitoring and managing ecosystem health. Front Mar Sci 6:89. https://doi.org/10.3389/fmars.2019.00089

Flynn KJ, McGillicuddy DJ Jr (2018) Modeling marine harmful algal blooms: current status and future prospects. In: Shumway SE, Burkholder JM, Morton SL (eds) Harmful algal blooms and their management: a compendium desk reference. John Wiley & Sons, Ltd., Hoboken, NJ, pp 115–134

Flynn KJ, Mitra A (2009) Building the "perfect beast": modelling mixotrophic plankton. J Plankton Res 31:965–992. https://doi.org/10.1093/plankt/fbp044

Flynn KJ, Page S, Wood G, Hipkin CR (1999) Variations in the maximum transport rates for ammonium and nitrate in the prymnesiophyte *Emiliania huxleyi* and the raphidophyte *Heterosigma carterae*. J Plankton Res 21:355–371

Flynn KJ, Clark DR, Mitra A, Fabian H, Hansen PJ, Glibert PM, Wheeler GL et al (2015) Ocean acidification with (de)eutrophication will alter future phytoplankton growth and succession. Proc R Soc B 282:20142604. https://doi.org/10.1098/rspb.2014.2604

Flynn KJ, Mitra A, Anestic K, Anschutz A, Calbet A, Ferreira GD, Gypens N et al (2019) Mixotrophic protists and a new paradigm for marine ecology: where does plankton research go now? J Plankton Res 41:375–391. https://doi.org/10.1093/plankt/fbz026

Flynn KJ, Torres R, Irogoien X, Blackford JC (2022) Plankton digital twins–a new research tool. J Plank Res 44:805. https://doi.org/10.1093/plankt/fbac042

Follows MJ, Dutkiewicz S (2011) Modeling diverse communities of marine microbes. Annu Rev Mar Sci 3:427–451. https://doi.org/10.1146/annurev-marine-120709-142848

Follows MJ, Dutkiewicz S, Grant S, Chisholm SW (2007) Emergent biogeography of microbial communities in a model ocean. Science 315:1843–1846. https://doi.org/10.1126/science.1138544

Franks PJS (2002) NPZ models of plankton dynamics: their construction, coupling to physics, and application. J Oceanogr 58:379–387. https://doi.org/10.1023/A:1015874028196

Franks PJS (2018) Recent advances in modeling of harmful algal blooms. In: Glibert PM, Berdalet E, Burford M, Pitcher G, Zhou MJ (eds) Global ecology and oceanography of harmful algal blooms. Springer, Cham, Switzerland, pp 359–380

Friedrichs MA, Dusenberry JA, Anderson LA, Armstrong RA, Chai F, Christian JR, Doney SC et al (2007) Assessment of skill and portability in regional marine biogeochemical models: role of multiple planktonic groups. J Geophys Res Oceans 112(C8). https://doi.org/10.1029/2006JC003852

Fu F-X, Place AR, Garcia NS, Hutchins DA (2010) CO_2 and phosphate availability control of toxicity of the harmful bloom dinoflagellate *Karlodinium veneficum*. Aquat Microb Ecol 59:55–65

Fu F-X, Tatters AO, Hutchins DA (2012) Global change and the future of harmful algal blooms in the ocean. Mar Ecol Prog Ser 470:207–233

Garcia Nieto PJ, Fernández JRRA, de Cos Juez FJ, Sánchez-Lasheras F, Muñiz CD (2013) Hybrid modelling based on support vector regression with genetic algorithms in forecasting the cyanotoxins presence in the Trasona reservoir (Northern Spain). Environ Res 122:1–10. https://doi.org/10.1016/j.envres.2013.01.001

Glibert PM, Burkholder JM (2006) The complex relationships between increasing fertilization of the earth, coastal eutrophication and proliferation of harmful algal blooms. In: Granéli E, Turner J (eds) Ecology of harmful algae. Springer Verlag, Heidelberg, pp 341–354

Glibert PM, Li M (2024) Warming, wheezing, blooming waters: hypoxia and harmful algae. In: Baird D (ed) Treatise on estuarine and coastal science, 2nd edn. Elsevier

Glibert PM, Allen JI, Bouwman AF, Brown CW, Flynn KJ, Lewitus AJ (2010) Modeling of HABs and eutrophication: status, advances, challenges. J Mar Syst 83:262–275

Glibert PM, Burkholder JM, Kana TM (2012) Recent advances in understanding of relationships between nutrient availability, forms and stoichiometry and the biogeographical distribution, ecophysiology, and food web effects of pelagic and benthic *Prorocentrum* spp. Harmful Algae 14:231–259

Glibert PM, Allen JI, Artioli Y, Beusen A, Bouwman AF, Harle J, Holmes R, Holt J (2014) Vulnerability of coastal ecosystems to changes in harmful algal bloom distribution in response to climate change: projections based on model analysis. Glob Change Biol 20:3845–3858. https://doi.org/10.1111/gcb.12662

Glibert PM, Pitcher GC, Bernard S, Li M (2018) Advancements and continuing challenges of emerging technologies and tools for detecting harmful algal blooms, their antecedent conditions and toxins, and applications in predictive models. In: Glibert PM, Berdalet E, Burford M, Pitcher GC, Zhou MJ (eds) Global ecology and oceanography of harmful algal blooms. Springer, Cham, Switzerland, pp 339–358

Glibert PM, Beusen AHW, Bouwman AF, Burkholder JM, Flynn KJ, Heil CA, Li M et al (2020) Multifaceted climatic change and nutrient effects on harmful algae require multifaceted models. In: Botana LM, Louzao C, Vilariño N (eds) Climate change and marine and freshwater toxins, 2nd edn. de Gruyter Publishers, Berlin, pp 473–518. https://doi.org/10.1515/9783110625738-012

Glibert PM, Cai WJ, Hall E, Li M, Main K, Rose K, Testa J, Vidyarathna N (2022) Stressing over the complexities of multiple stressors in marine and estuarine systems. Ocean-Land-Atmos Res 2022:9787258. https://doi.org/10.34133/2022/9787258

Goldman JC, Glibert PM (1983) Kinetics of inorganic nitrogen uptake. In: Carpenter EJ, Capone DG (eds) Nitrogen in the marine environment. Academic Press, pp 233–274

Gonzalez LM, Proulx SR, Moeller HV (2022) Modeling the metabolic evolution of mixotrophic phytoplankton in response to rising ocean surface temperatures. BMC Ecol Evol 22:136

Granéli E, Flynn K (2006) Chemical and physical factors influencing toxin content. In: Granéli E, Turner J (eds) The ecology of harmful algae. Springer-Verlag, Heidelberg, pp 229–241

Guallar C, Delgado M, Diogène J, Fernándeez-Tegedor M (2016) Artificial neural network approach to population dynamics of harmful algal blooms in Alfacs Bay (NW Mediterranean): case studies of *Karlodinium* and *pseudo-nitzschia*. Ecol Model 338:37–50. https://doi.org/10.1016/j.ecolmodel.2016.07.009

Guisan A, Edwards TC Jr, Hastie T (2002) Generalized linear and generalized additive models in studies of species distributions: setting the scene. Ecol Model 157:89–100

Haidvogel DB, Arango H, Budgell WP, Cornuelle BD, Curchitser E, Di Lorenzo E, Fennel K et al (2008) Ocean forecasting in terrain-following coordinates: formulation and skill assessment of the Regional Ocean modeling system. J Comput Phys 227:3595–3624. https://doi.org/10.1016/j.jcp.2007.06.016

Hansen PJ, Anderson R, Stoecker DK, Decele J, Altenburger A, Blossom HE, Drumm K, Mitra A, Flynn KJ (2019) Mixotrophy among freshwater and marine protists. In: Schmidt TM (ed) Encyclopedia of microbiology, 4th edn. Academic Press, pp 199–210. https://doi.org/10.1016/B978-0-12-809633-8.20685-7

Harrison JA, Seitzinger SP, Bouwman AF, Caraco NF, Beusen AHW, Vörösmarty CJ (2005) Dissolved inorganic phosphorus export to the coastal zone: results from a spatially explicit, global model. Glob Biogeochem Cycles 19:1–15. https://doi.org/10.1029/2004GB002357

Harrison JA, Bouwman AF, Mayorga E, Seitzinger SP (2010) Magnitudes and sources of dissolved inorganic phosphorus inputs to surface fresh waters and the coastal zone: a new global model. Glob Biogeochem Cycles 24:GB1003. https://doi.org/10.1029/2009GB003590

Hattenrath-Lehmann TK, Smith JL, Wallace RB, Merlo L, Koch F, Mittelsdorf H, Goleski JA et al (2015) The effects of elevated CO_2 on the growth and toxicity of field populations and cultures of the saxitoxin-producing dinoflagellate *Alexandrium fundyense*. Limnol Oceanogr 60:197–214. https://doi.org/10.1002/lno.10012

Hauser DCR, Petrylak D, Singer G, Levandowsky M (1978) Calcium-dependent sensory motor response of a marine dinoflagellate to CO_2. Nature 273:230–231. https://doi.org/10.1038/273230a0

Heil CA, Glibert PM, Fan C (2005) *Prorocentrum minimum* (Pavillard) schiller –a review of a harmful algal bloom species of growing worldwide importance. Harmful Algae 4:449–470

Hellweger FL, Bucci V (2009) A bunch of tiny individuals-individual-based modeling for microbes. Ecol Model 220:8–22. https://doi.org/10.1016/j.ecolmodel.2008.09.004

Hennon G, Dyhrman S (2020) Progress and promise of omics for predicting the impacts of climate change on harmful algal blooms. Harmful Algae 91:101587. https://doi.org/10.1016/j.hal.2019.03.005

Henson SA, Cael B, Allen SR, Dutkiewicz S (2021) Future phytoplankton diversity in a changing climate. Nat Commun 12:1–8. https://doi.org/10.1038/s41467-021-25699-w

Hill PR, Kumar A, Temini M, Bull DR (2019) HABNet: machine learning, remote sensing based detection and prediction of harmful algal blooms. IEEE J Sel Topics Appl Earth Observ Rem Sens arXiv:1912.02305

Holt J, Harle J, Proctor R, Michel S, Ashworth M, Batstone C, Allen JI et al (2009) Modelling the global coastal ocean. Philos Trans Roy Soc A: Math Phys Eng Sci 367:939–951

Hoos AB, McMahon G (2009) Spatial analysis of instream nitrogen loads and factors controlling nitrogen delivery to stream sin the United States using spatially references regression on watershed attributes (SPARROW) and regional classification frameworks. Hydrol Process 23: 2275–2294

Irby ID, Friedrichs MAM, Da F, Hinson KE (2018) The competing impacts of climate change and nutrient reductions on dissolved oxygen in Chesapeake Bay. Biogeosciences 15(9):2649–2668. https://doi.org/10.5194/bg-15-2649-2018

Isleib RR, Fitzpatrick JJ, Mueller J (2007) The development of a nitrogen control plan for a highly urbanized tidal embayment. Proc Water Environ Fed 5:296–320. https://doi.org/10.2175/193864707786619152

Kemp WM, Boynton WR, Adolf JE, Boesch DF, Boicourt WC, Brush G, Cornwell JC et al (2005) Eutrophication of Chesapeake Bay: historical trends and ecological interactions. Mar Ecol Prog Ser 303:1–29

Kerimoglu O, Große F, Kreus M, Lenhart H-J, van Beusekom JE (2018) A model-based projection of historical state of a coastal ecosystem: relevance of phytoplankton stoichiometry. Sci Total Environ 639:1311–1323. https://doi.org/10.1016/j.scitotenv.2018.05.215

Kim H, Spivack AJ, Menden-Deuer S (2013) pH alters the swimming behaviors of the raphidophyte *Heterosigma akashiwo:* implications for bloom formation in an acidified ocean. Harmful Algae 26:1–11. https://doi.org/10.1016/j.hal.2013.03.004

Klausmeier CA, Kremer CT, Koffel T (2020) Trait-based ecological eco-evolutionary theory. In: McCann KS, Gellner G (eds) Theoretical ecology: concepts and applications. Oxford University Press, Oxford, UK

Kwiatkowski L, Aumont O, Bopp L, Ciais P (2018) The impact of variable phytoplankton stoichiometry on projections of primary production, food quality, and carbon uptake in the global ocean. Glob Biogeochem Cycles 32:516–528. https://doi.org/10.1002/2017GB005799

Kwiatkowski L, Aumont O, Bopp L (2019) Consistent trophic amplification of marine biomass declines under climate change. Glob Change Biol 25:218–229. https://doi.org/10.1111/gcb.14468

Lancelot C, Gypens N, Billen G, Garnier J, Roubeix V (2007) Testing an integrated river–ocean mathematical tool for linking marine eutrophication to land use: the phaeocystis-dominated Belgian coastal zone (southern north sea) over the past 50 years. J Mar Syst 64(1–4):216–228. https://doi.org/10.1016/j.jmarsys.2006.03.010

Lauchlan SS, Nagelkerken I (2019) Species range shifts along multistressor mosaics in estuarine environments under future climate. Fish Fish 21:32–46. https://doi.org/10.1111/faf.12412

Lee JHW, Huang Y, Dickman M, Jayawardena AW (2003) Neural network modelling of coastal algal blooms. Ecol Model 159:179–201

Legendre P, Legendre L (1998) Numerical ecology, 2nd edn. Elsevier, Amsterdam

Leles SG, Polimene L, Bruggeman J et al (2018) Modelling mixotrophic functional diversity and implications for ecosystem function. J Plankton Res 40:627–642

Leles SG, Bruggeman J, Polimene L, Blackford J, Flynn KJ, Mitra A (2021) Differences in physiology explain succession of mixoplankton functional types and affect carbon fluxes in temperate seas. Prog Oceangr 190:102481. https://doi.org/10.1016/j.pocean.2020.102481

Li M, Ni W, Zhang F, Glibert PM, Lin C-H (2020) Climate-induced interannual variability and projected change of two harmful algal bloom taxa in Chesapeake Bay. USA Sci Total Environ 744:140947. https://doi.org/10.1016/j.scitotenv.2020.140947

Li M, Chen Y, Zhang F, Song Y, Glibert PM, Stoecker DK (2022a) A three-dimensional mixotrophic model of *Karlodinium veneficum* blooms in a eutrophic estuary: seasonal and spatial dynamics and effects of nutrient ratios, prey concentration and temperature. Harmful Algae 113:102203. https://doi.org/10.1016/j.hal.2022.102203

Li M, Guo Y, Cai W-J, Testa JM, Shen C, Li R, Su J (2023) Projected increase in carbon dioxide drawdown and acidification in large estuaries under climate change. Commun Earth Environ 4: 68. https://doi.org/10.1038/s43247-023-00733-5

Li MF, Glibert PM, Lyubchich V (2021) Machine learning classification algorithms for predicting *Karenia brevis* blooms on the West Florida shelf. J Mar Sci Eng 9. https://doi.org/10.3390/jmse9090999

Li R, Li M, Glibert PM (2022b) Coupled carbonate chemistry - harmful algae bloom models for studying effects of ocean acidification on *Prorocentrum minimum* blooms in a eutrophic estuary. Front Mar Sci 9:889233. https://doi.org/10.3389/fmars.2022.889233

Lin C-H, Flynn KJ, Glibert PM, Mitra A (2018a) Simulating effects of variable nutrient stoichiometry and temperature on mixotrophy in the harmful dinoflagellate *Karlodinium veneficum*. Front Mar Sci 5. https://doi.org/10.3389/fmars.2018.00320

Lin C-H, Lyubchich V, Glibert PM (2018b) Time series models of decadal trends in the harmful algal species *Karlodinium veneficum* in Chesapeake Bay. Harmful Algae 73:110–118

Litchman E (2022) Understanding and predicting harmful algal blooms in a changing climate: a trait-based framework. Limnol Oceanogr Lett 8:229–246. https://doi.org/10.1002/lol2.10294

Liu Y, Weisberg RH (2012) Seasonal variability on the West Florida shelf. Prog Oceanogr 104:80–98. https://doi.org/10.1016/j.pocean.2012.06.001

Los FJ, Tatman S, Minns AW (2004) Flyland—a future airport in the North Sea? An integrated modelling approach for marine ecology. In: Liong SY, Phoon KK, Babovic V (eds) 6th intern Conf on Hydroinformatics. World Scientific, Publishing Co. Pte. Ltd, Singapore. isbn:981-238-787-0

Maier HR, Dandy GC, Burch MD (1998) Use of artificial neural networks for modelling cyanobacteria anabaena spp. in the river Murray, South Australia. Ecol Model 105:257–272

Martini S, Larras F, Boyé A, Faure E, Aberle N, Archambault P, Bacouillard L et al (2021) Functional trait-based approaches as a common framework for aquatic ecologists. Limnol Oceanogr 66:965–994. https://doi.org/10.1002/lno.11655

McCarthy JJ, Goldman JC (1979) Nitrogenous nutrition of marine phytoplankton in nutrient-depleted waters. Science 203:670–672. https://doi.org/10.1126/science.203.4381.670

Meddeb M, Grami B, Chaalali A, Haraldsson M, Niquil N, Pringault O et al (2018) Plankton food-web functioning in anthropogenically impacted coastal waters (SW Mediterranean Sea): an ecological network analysis. Prog Oceanogr 162:66–82. https://doi.org/10.1016/j.pocean.2018.02.013

Mills EL (1989) Biological oceanography: an early history, 1870–1960. Cornell University Press, Ithaca and London, p 378

Mitra A, Davis K (2010) Defining the "to" in end-to-end models. Prog Oceanogr 84:39–42

Mitra A, Flynn KJ (2006) Promotion of harmful algal blooms by zooplankton predatory activity. Biol Lett 2:194–197

Mitra A, Flynn KJ (2010) Modelling mixotrophy in harmful algal blooms: more or less the sum of the parts? J Mar Syst 83:158–169

Mitra A, Castellani C, Gentleman WC, Jónasdóttir SH, Flynn KJ, Bode A, Halsband C et al (2014a) Bridging the gap between marine biogeochemical and fisheries sciences; configuring the zooplankton link. Prog Oceanogr 129:176–199

Mitra A, Flynn KJ, Burkholder JM, Berge T, Calbet A, Raven JA, Granéli E, Glibert PM et al (2014b) The role of mixotrophic protists in the biological carbon pump. Biogeosciences 11: 995–1005. https://doi.org/10.5194/bg-11-995-2014

Ni W, Li M (2023) What drove the nonlinear hypoxia response to nutrient loading in Chesapeake Bay during the 20th century? Sci Total Environ 861:160650. https://doi.org/10.1016/j.scitotenv.2022.160650

Ni W, Li M, Testa J (2020) Discerning effects of warming, sea level rise and nutrient management on long-term hypoxia trends in Chesapeake Bay. Sci Tot Environ 737:139717. https://doi.org/10.1016/j.scitotenv.2020.139717

O'Neil JM, Davis TW, Burford MA, Gobler CJ (2012) The rise of harmful cyanobacteria blooms: the potential roles of eutrophication and climate change. Harmful Algae 14:313–334. https://doi.org/10.1016/j.hal.2011.10.027

Pang M, Xu J, Qu P, Mao X, Wu Z, Xin M, Sun P et al (2017) Effect of CO_2 on growth and toxicity of *Alexandrium tamarense* from the East China Sea, a major producer of paralytic shellfish toxins. Harmful Algae 68:240–247. https://doi.org/10.1016/j.hal.2017.08.008

Park T (1939) Ecology looks homeward. Quart Rev Biol 14:3332–3336

Ralston DK, Moore SK (2020) Modeling harmful algal blooms in a changing climate. Harmful Algae 91:101729. https://doi.org/10.1016/j.hal.2019.101729

Rao TR (2000) A curve for all reasons: the rectangular hyperbola in biology. Resonance 5:85–90

Riley GA (1946) Factors controlling phytoplankton populations on Georges Bank. J Mar Res 6:54–73

Riley GA (1979) Summation. In Second informal workshop on the oceanography of the Gulf of Maine and adjacent seas, May 14–17, 1979. Halifax, N.S., Dalhousie Univ

Robertson DM, Schwarz GE, Saad DA, Alexander RB (2009) Incorporating uncertainty into the ranking of SPARROW model nutrient yields from Mississippi/Atchafalaya River basin watersheds. J Am Water Resour Assoc 45:534–549

Saint-Béat B, Fath BD, Aubry C, Colombet J, Dinasquet J, Fortier L et al (2020) Contrasting pelagic ecosystem functioning in eastern and western Baffin bay revealed by trophic network modeling. Elementa: Sci Anthropocene 8:1. https://doi.org/10.1525/elementa.397

Seitzinger SP, Harrison JA, Dumont E, Beusen AHW, Bouwman AF (2005) Sources and delivery of carbon, nitrogen and phosphorous to the coastal zone: an overview of global nutrient export from watersheds (NEWS) models and their application. Glob Biogeochem Cycles 19:GB4S01. https://doi.org/10.1029/2005GB002606

Seitzinger SP, Mayorga E, Bouwman AF, Kroeze C, Beusen AHW, Billen G, Van Drecht G et al (2010) Global river nutrient export: a scenario analysis of past and future trends. Global Biogeochem Cycles 24. https://doi.org/10.1029/2009GB003587

Shen C, Testa JM, Li M, Cai W-J, Waldbusser GG, Ni W, Kemp WM et al (2019) Controls on carbonate system dynamics in a coastal plain estuary: a modelling study. J Geophys Res 124: 7478. https://doi.org/10.1029/2018JG004802

Singh A, Hårding K, Reddy HRV, Godhe A (2014) An assessment of *Dinophysis* blooms in the coastal Arabian Sea. Harmful Algae 34:29–35. https://doi.org/10.1016/j.hal.2014.02.006

Smith RA, Schwartz GE, Alexander RB (1997) Regional interpretation of water-quality monitoring data. Water Resour Res 33:2781–2798

Smith SL, Pahlow M, Merico A, Wirtz KW (2011) Optimality-based modeling of planktonic organisms. Limnol Oceanogr 56:2080–2094. https://doi.org/10.4319/lo.2011.56.6.2080

Stachelek J, Madden CJ (2015) Application of inverse path distance weighting for high-density spatial mapping of coastal water quality patterns. Int J Geogr Inf Sci 29:1240–1250. https://doi.org/10.1080/13658816.2015.1018833

Steele JH (1959) The quantitative ecology of marine phytoplankton. Biol Rev 34:129–158

Stoecker DK, Tillmann U, Granéli E (2006) Phagotrophy in harmful algae. In: Granéli E, Turner J (eds) The ecology of harmful algae. Springer-Verlag, Heidelberg, pp 177–187

Stumpf RP, Holderied K, Sinclair M (2003) Determination of water depth with high-resolution satellite imagery over variable bottom types. Limnol Oceanogr 48:547–556. https://doi.org/10.4319/lo.2003.48.1_part_2.0547

Stumpf RP, Tomlinson MC, Calkins JA, Kirkpatrick B, Fisher K, Nierenberg K, Currier R, Wynne TT (2009) Skill assessment for an operational algal bloom forecast system. J Mar Syst 76:151–161. https://doi.org/10.1016/j.jmarsys.2008.05.016

Suggett D, Moore CM, Hickman A, Geider RJ (2009) Interpretation of fast repetition rate (FRR) fluorescence: signatures of phytoplankton community structure versus physiological state. Mar Ecol Prog Ser 376:1–9. https://doi.org/10.3354/meps07830

Sun J, Hutchins DA, Feng YY, Seubert EL, Caron DA, Fu FX (2011) Effects of changing pCO_2 and phosphate availability on domoic acid production and physiology of the marine harmful bloom diatom. *Pseudo-nitzschia multiseries*. Limnol Oceanogr 56:829–840. https://doi.org/10.4319/lo.2011.56.3.0829

Tatters AO, Fu FX, Hutchins DA (2012) High CO_2 and silicate limitation synergistically increase the toxicity of *Pseudo-nitzschia fraudulenta*. PLoS One 7:e32116. https://doi.org/10.1371/journal.pone.0032116

Testa JM, Li Y, Lee YJ, Li M, Brady DC, DiToro DM, Kemp WM, Fitzpatrick JJ (2014) Quantifying the effects of nutrient loading on dissolved O2 cycling and hypoxia in Chesapeake bay using a coupled hydrodynamic-biogeochemical model. J Mar Sys 139:139–158

Toseland A, Daines SJ, Clark JR, Kirkham A, Strauss J, Uhlig C et al (2013) The impact of temperature on marine phytoplankton resource allocation and metabolism. Nat Clim Chang 3:1–6. https://doi.org/10.1038/nclimate1989

Twiner MJ, Flewelling LJ, Fire SE, Bowen-Stevens SR, Gaydos JK, Johnson CK, Landsberg JH et al (2012) Comparative analysis of three brevetoxin-associated bottlenose dolphin (*Tursiops truncatus*) mortality events in the Florida panhandle region (USA). PLoS One 7(8):e42974. https://doi.org/10.1371/journal.pone.0042974

Urrutia-Cordero P, Ekvall MK, Hansson L-A (2016) Local food web management increases resilience and buffers against global change effects on freshwaters. Sci Rep 6:29542. https://doi.org/10.1038/srep29542

Van Gils JAG, Ouboter MRL, De Rooij MN (1993) Modelling of water and sediment quality in the Scheldt estuary, Netherland. J Aquat Ecol 27:257–265. https://doi.org/10.1007/BF02334789

Vargo GA, Heil CA, Fanning KA et al (2008) Nutrient availability in support of *Karenia brevis* blooms in the central West Florida shelf: what keeps *Karenia* blooming? Cont Shelf Res 28:73–98

Velo Suárez L, Gutiérrez-Estrada JC (2007) Artificial neural network approaches to one-step weekly prediction of *Dinophysis acuminata* blooms in Huelva (Western Andalucía, Spain). Harmful Algae 6:363–371

Vilmin L, Mogollón JM, Beusen AHW, Bouwman AF (2018) Forms and subannual variability of nitrogen and phosphorus loading to global river networks over the 20th century. Glob Planet Change 163:67–85

Vilmin L, Mogollón JM, Beusen AHW, van Hoek WJ, Liu X, Middelburg JJ, Bouwman AF (2020) Modelling process-based biogeochemical dynamics in surface freshwaters with the IMAGE-DGNM framework. J Adv Model Earth Syst 12(11):e2019MS001796

Wang P, Linker LC, Wang H, Bhatt G (2017) Assessing water quality of the Chesapeake Bay by the impact of sea level rise and warming. IOP Conf Ser Earth Environ Sci 82:012001. https://doi.org/10.1088/1755-1315/82/1/012001

Weisberg RH, Zheng L, Liu Y, Lembke C, Lenes JM, Walsh JJ (2014) Why no red tide was observed on the West Florida continental shelf in 2010. Harmful Algae 38:119–136. https://doi.org/10.1016/j.hal.2014.04.010

Williams RJP (1988) Proton circuits in biological energy interconversions. Ann Rev Biophys Biophys Chem 17:71–97

Xie Z, Lou I, Ung WK, Mok KM (2012) Freshwater algal bloom prediction by support vector machine in Macau storage reservoirs. Math Probl Eng 2012:397473. https://doi.org/10.1155/2012/397473

Zhang H, Li S (2010) Effects of physical and biochemical processes on the dissolved oxygen budget for the Pearl River estuary during summer. J Mar Syst 79:65–88

Additional Websites Accessed

https://coastalscience.noaa.gov/science-areas/habs/hab-forecasts/lake-erie/satellite-imagery/. Accessed 4 Aug 2023

https://optics.marine.usf.edu/projects/iris.html. Accessed 3 Aug 2023

https://ourworldindata.org. Accessed 31 Jul 2023

Reversing Course: Ecosystem Restoration, Geoengineering, and Algal Bloom Mitigation

29

29.1 Introduction

There is no question that aquatic ecosystems around the world have been degraded. From the previous chapters, it should be obvious that we have enriched the waters of the world with nutrients, and in turn, they are experiencing more hypoxia and more HABs, and the ecosystem services provided are being diminished (Chaps. 24–25). Waters are warming at an alarming rate (Chap. 23). They are also becoming acidified as more CO_2 is pumped into the atmosphere and dissolved in the warming waters (Chap. 26). Enough! Can anything be done? What can be done? What is being done? On what time and space scales are mitigation measures applicable? This chapter examines why it is so difficult to reverse course and restore ecosystems to their unpolluted past. This chapter also examines some of the mitigation measures that are being explored on local to national to global scales to reduce algal blooms in real time. This chapter cannot address all actions being used, but the actions described provide an overview of how ecosystem restoration and geoengineering is advancing, and how such efforts are working, or when and why some approaches may prove to be problematic.

29.2 Return to Neverland or Humpty Dumpty's Great Fall

Carlos Duarte, a well recognized aquatic scientist, and his colleagues, suggested that the expectation that ecosystems could be restored to their idealized, unpolluted, past by reducing human pressures is "as likely as the existence of *Neverland*" (Duarte et al. 2009, p. 29), the fictional island in the children's novel *Peter and Wendy* (commonly referred to as *Peter Pan*; Barrie 1911). The primary theme of the novel was the conflict between innocence of childhood and responsibility of adulthood, and the sense of loss as systems change. Peter never grows up, but Wendy and the other characters of the story do. Over time, it became harder and harder for Peter to remember Neverland the way it was. Lamothe et al. (2019) also invoked children's

P. M. Glibert, *Phytoplankton Whispering: An Introduction to the Physiology and Ecology of Microalgae*, https://doi.org/10.1007/978-3-031-53897-1_29

literature to describe the difficulty of reversing ecological damage. Those authors suggested that at least in some cases, the path was more analogous to the fate of Humpty Dumpty: ecological disturbance and damage may be so severe that recovery cannot occur. The pieces cannot be put back together.

Systems change. Baselines shift. System responses have time lags and/or may display alternate state stables, hysteresis, and legacy effects. Accordingly, the responses of systems as they are restored, illustrated here using nutrient reductions as examples, may take a range of trajectories (Duarte et al. 2009; Fig. 29.1). While it is possible for some systems to revert to their original state following the trajectory of change in reverse (Fig. 29.1a), more often than not the system does not show signs

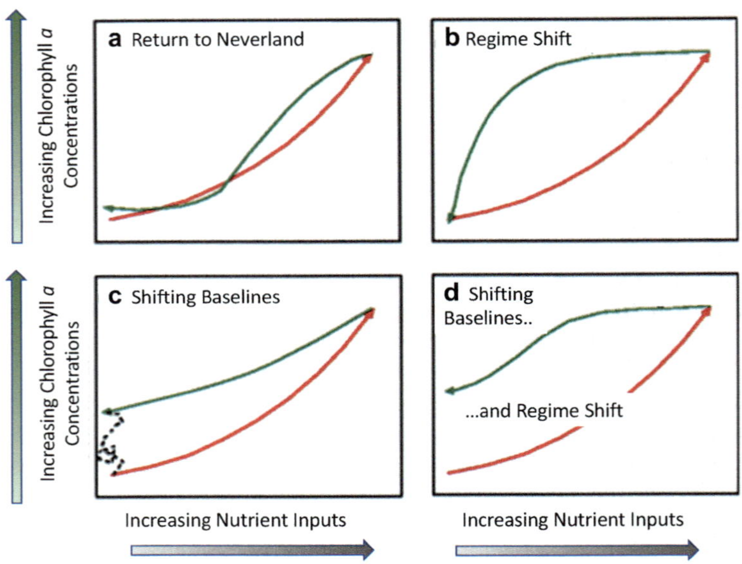

Fig. 29.1 Idealized trajectories of chlorophyll *a* concentrations as an indicator of ecosystem status, and nutrient inputs to coastal ecosystems under increasing (*red line*) and decreasing (*green line*) nutrient inputs under different response scenarios: (Panel **a**) "Return to Neverland" scenario implying a direct reversible relationship between chlorophyll a concentrations and nutrient inputs; (Panel **b**) a trajectory resulting from a "Regime Shift" in ecosystem status in response to nutrient inputs. This trajectory results in an apparent time lag, or hysteresis effect, in the response to reducing nutrient inputs; (Panel **c**) "Shifting Baselines" scenario, where changes in forcing factors other than nutrients (e.g., climate, food web structure) forces a trajectory for the ecosystem independent of that forced by nutrients, depicted by the *dotted line*, preventing the ecosystem to return to the "reference condition" after reducing nutrient inputs; and (Panel **d**) a trajectory displaying "Regime Shift and Shifting Baselines" combined. Image reproduced from Duarte et al. (2009, Estuar Coasts 32:29–36. doi.org/10.1007/s12237-008-9111-2) under a creative commons license

of reversal until substantial reductions have been made (Fig. 29.1b), and some systems never return to their original baseline (Fig. 29.1c, d).

Hypoxia in the Chesapeake Bay perfectly illustrates this concept of shifting baselines. The hypoxic volume and duration in the Chesapeake Bay experienced a rapid expansion between the 1950s and 1980s, associated with large increases in N loading (Officer et al. 1984; Hagy et al. 2004; Kemp et al. 2009). Indeed, there has been a near doubling of the hypoxic volume per unit spring nitrogen (N) load in the recent decades (Hagy et al. 2004; Kemp et al. 2005; Fig. 29.2). Kemp et al. (2005)

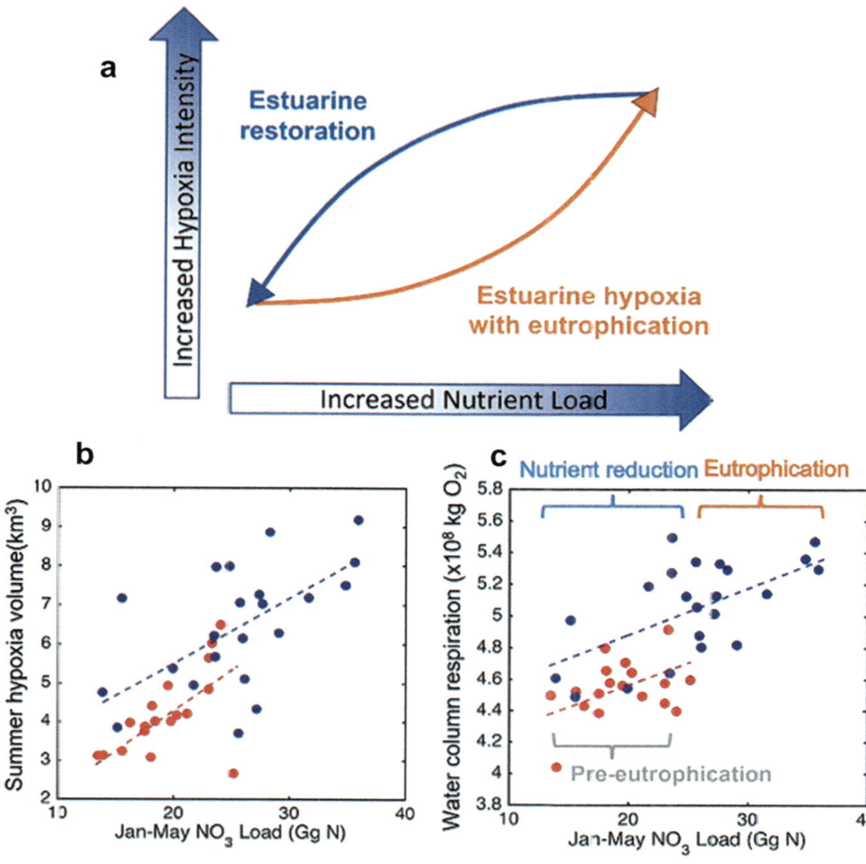

Fig. 29.2 (Panel **a**) Schematic of a nonlinear hysteretic response of hypoxia to changes in nutrient load. (Panel **b**) Scatter plots between Jan–May Susquehanna River nitrate load and summer average hypoxic volume. (Panel **c**) Scatter plots between Jan–May Susquehanna River nitrate load and water column respiration. The *blue* and *red* dots in B, C represent the data from 1950–1967 and 1968–1989, respectively. The *dashed blue* and *red* lines in (**b**, **c**) are the linear regression fits to 1950–1967 and 1968–1989, respectively. Image reproduced from Ni and Li (2023, Sci Total Environ 861:160650. https://doi.org/10.1016/j.scitotenv.2022.160650) under a creative commons license

suggested that loss of benthic plant biomass and increased efficiency of N and phosphorus (P) recycling may be a contributing factor to this increase in hypoxia. Analyses by Testa and Kemp (2012) supported this view. Ni and Li (2023) further analyzed the O_2 budget in a fixed control volume of the bottom water most susceptible to hypoxia and showed an abrupt (~50%) jump around 1968. They documented that this was caused by an upward shift in the relationship between the water column respiration and winter-spring nutrient load (Fig. 29.2). Hypoxia altered the rates of N cycling in the sediment, amplifying nutrient recycling by 15% and water column respiration by 12%. Thus, these analyses have all supported the notion of a feedback mechanism for driving the nonlinear hypoxia response to nutrient loading.

An alternate model of change and its potential reversal is that of the alternate stable state theory. This theory was developed to describe the general state of shallow, productive lakes (Scheffer et al. 1993, 2003). It conceptualizes that a system will develop a stable state condition and will remain stable until an environmental change or disturbance occurs. Such shifts can be abrupt (e.g., Tátrai et al. 2009). This change alters the positive reinforcing feedbacks of homeostasis and shifts the system to a new stable state: hysteresis overcomes homeostasis (Scheffer et al. 1993; Scheffer and Carpenter 2003). The typical stable states in such shallow lakes are pelagic-phytoplankton-dominated systems and littoral-macrophyte-dominated systems (Scheffer et al. 2003; Mieczan 2010). Recent interest in stable state theory has occurred because of efforts to restore macrophyte dominance in systems that have become dominated by phytoplankton due to eutrophication (e.g., Bachmann et al. 1999).

Regime shifts represent yet another way to conceptualize alternate stable states. Regime shifts have often been illustrated by marble-in-cup diagrams (e.g., Willems 1970; Scheffer et al. 1993, 2003; Amemiya et al. 2007; Fig. 29.3). The likelihood of an ecosystem (marble) being moved to a new state (new cup) is a function of system resilience. When the dynamic structure of the ecosystems is changed, restoration of

Fig. 29.3 Ball-and-cup heuristic as illustrated for the Chesapeake Bay. The first ball represents the pre-eutrophication state in the 1970s. Following Hurricane Agnes and subsequent eutrophication, the current stable state is degraded relative to the pre-1970s condition. Restoration efforts aim to restore the Bay to its earlier condition but face a steep uphill climb. Image adapted from Gunderson (2000) by Goldman (2004, Chesapeake Quarterly 3:6)

the original stable state is much more difficult, if even possible. Again, Chesapeake Bay illustrates this trajectory. The pre-1970s condition was disrupted with Hurricane Agnes in 1972 (recall that Ni and Li (2023) showed an abrupt change around 1968). The Bay's current stable state is degraded and restoration efforts are an uphill climb (Goldman 2004; Fig. 29.3).

29.3 Searching for Solutions

With the myriad of changes in the oceans, there is no question that a portfolio of solutions is required to solve today's problems. While individual lakes or estuaries may be restored, the oceans will not return to their pristine state. What actions can and should be taken to reduce eutrophication? Acidification? Algal blooms?

In Chap. 16, the idea of adding a tanker of iron (Fe) to reverse global warming was described. New ideas are emerging for accelerating ocean processes that remove and store CO_2 away from the atmosphere. The US National Academy of Sciences recently released a report on ocean-based CO_2 removal strategies (NAS 2022). Ocean-based CO_2 removal approaches range from developing methods for deep sea storage, enhancing ocean alkalinity via various approaches, and increasing micro- and macro-algal growth. The NAS (2022) report highlighted that adherence to a code of conduct (such as that detailed in Box 29.1) should be made a condition of government or private funding for ocean fertilization or CO_2 reduction mitigation research.

Crucial questions remain unresolved. Whether it is adding Fe to stimulate primary productivity, or alkaline agents to reduce acidification, or chemical or biological agents to reduce or kill algal blooms, or even adding nutrients to enhance their growth, there are scientific, legal, societal, and ethical issues that must be considered. Developing approaches that are effective while minimizing risk is necessary.

Research on ocean fertilization, CO_2 sequestration, and algal mitigation needs a scaled approach. For Fe enrichment experiments, the idea initially was generated in bottle "bioassay" experiments, and over time, the size, amount of enrichment, and area covered increased as new experiments were conducted (Chap. 16). With algal control measures, the USA—in its recent funding program—has adopted a tiered system of bottle assays, mesocosm experiments, and finally field pilot tests. At each stage, the research can be halted if efficacy has not been demonstrated (to both researchers and funders), or if other environmental or other impacts outweigh benefits.

Geoengineering research and research on ecosystem restoration are advancing on seemingly different tracks, but they are two sides of the same coin. Both themes are aimed at reserving course and returning the oceans (or planet) to a state more representative of a previous time, a time less polluted, a time when the oceans were cooler or less acidified, a time with fewer HABs. At times these approaches may lead to a collision course. For example, proposals to enrich water with urea to promote algal blooms to enhance fisheries and ultimately increase sequestration through the biological pump, or similarly proposals for artificial upwelling to achieve the same goals, are subjecting systems to accelerated eutrophication (Glibert et al. 2008). Even if efforts for enhancing nutrients to enhance sequestration are proposed and conducted at sites away from sensitive shoreline and inland waters, their goals are ultimately at odds with efforts to reduce nutrients and the associated effects of eutrophication. As shown for the Fe-enrichment experiments, there is a potential for creation of toxic or HABs (Chap. 16). Much has been learned from nutrient enrichment experiments with different nutrient forms and proportions to understand that phytoplankton community responses, and therefore responses of the biological pump, are highly variable.

The next sections describe some of the contentious debates regarding nutrient reduction strategies to control eutrophication, i.e., ecosystem restoration efforts, and some of the approaches that are advancing for controlling algal blooms, i.e., mitigation efforts. The former are long-term strategies for ecosystem change, and the latter are short-term solutions to mitigate immediate algal blooms. Science is advancing and lessons have been learned on both fronts. An overarching lesson is that solutions, either for restoration or for mitigation, are not simple, responses are not linear. Understanding the full suite of processes and factors that underlie restoration or mitigation at all scales is a grand ecological, social, and economic challenge. The approaches mentioned here are but a small sample of the restoration and mitigation approaches being tested or advocated.

Box 29.1 The Oxford Principles of Geoengineering Research

The Oxford Principles were developed by academic researchers at the University of Oxford, University College London, and the University of Cardiff in 2009. Later that year, the UK House of Commons Science and Technology Select Committee on the Regulation of Geoengineering endorsed the principles as "provid[ing] a basis to begin the discussion of . . . the regulation of geoengineering" and recommended that they be further developed (UK House of Commons Science and Technology Committee 2010). Support for the principles was also expressed by geoengineering researchers at the International Conference on Climate Intervention Technologies in 2010. The principles are as follows:

Principle 1: Geoengineering to be regulated as a public good. While the involvement of private actors in the delivery of a geoengineering technique should not be prohibited, and may indeed be encouraged to ensure that deployment of a suitable technique can be effected in a timely and efficient manner, regulation of such techniques should be undertaken in the public interest by the appropriate bodies at the state and/or international levels.

Principle 2: Public participation in geoengineering decision-making. Wherever possible, those conducting geoengineering research should be required to notify, consult, and ideally obtain the prior informed consent of those affected by the research activities...

(continued)

Box 29.1 (continued)

Principle 3: Disclosure of geoengineering research and open publication of results. There should be complete disclosure of research plans and open publication of results in order to facilitate better understanding of the risks and to reassure the public as to the integrity of the process ...

Principle 4: Independent assessment of impacts. An assessment of the impacts of geoengineering research should be conducted by a body independent of those undertaking the research; where techniques are likely to have transboundary impact, such assessment should be carried out through the appropriate regional and/or international bodies. Assessments should address both the environmental and socioeconomic impacts of research, including mitigating the risks of lock-in to particular technologies or vested interests.

Principle 5: Governance before deployment. Any decisions with respect to deployment should only be taken with robust governance structures already in place, using existing rules and institutions wherever possible.

29.4 Controlling the Inputs: Linear Thinking Regarding Nutrient Reductions in a Nonlinear World Doesn't Work

It is well understood that nutrient pollution leads to eutrophication. This has led to the common "straight line" thinking that removing the limiting nutrient will reverse the problem. Returning to Neverland is not always a straight line and instead may be a difficult and arduous path (Fig. 29.1). Most often, especially in freshwater systems, the single nutrient most recommended for control is P. The origin of this notion is often attributed to the body of research on experimental lakes in Canada in the 1970s. Additions of P resulted in intensive algal blooms, while additions of N did not (Schindler et al. 1971; see Chap. 3). This strategy of reducing the limiting nutrient, especially P, is often argued by those in the freshwater community (e.g., Schindler et al. 2008, 2016). This straight-line thinking has led to the common conclusion, as articulated by Wang and Wang (2009, p. 1449), "To promote growth of living organisms, every essential element should be provided... To mitigate eutrophication, only one element needs to be controlled. Because [P] is relatively easy to control, it should be focused upon to mitigate eutrophication." Straight-line thinking focuses only on the trajectory illustrated in Fig. 29.1a.

With nutrient reduction, such a straight-line perspective fails to recognize that reductions in supply of only the limiting nutrient will alter nutrient stoichiometry and this, in turn, can have large consequences on nutrient cycling, biogeochemistry, and algal biodiversity (Glibert et al. 2011, 2015, 2017; see Chap. 18). Even relatively small changes in nutrient supply have been shown to have large consequences on many important properties of the ecosystems (Nielsen 2003; Glibert et al. 2011, 2015). Modifying the N:P ratio of a system can alter the phytoplankton community

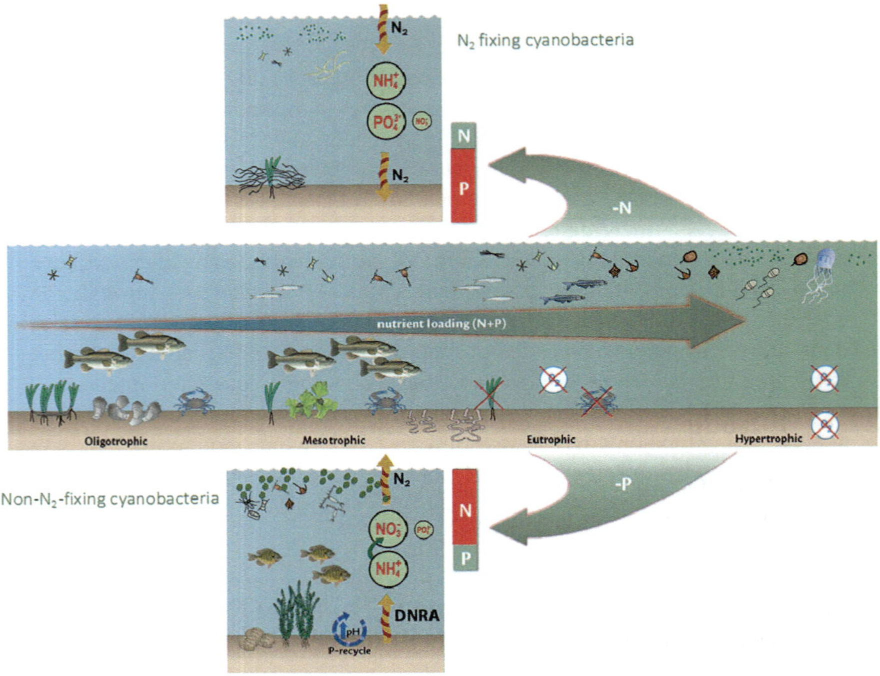

Fig. 29.4 (Middle Panel) Conceptual schematic showing the progression of an estuarine system toward eutrophication with the increased loading of both nitrogen and phosphorus. The panel above illustrates how a system may change with increased P and decreased N. Increased N_2 fixation and cyanobacteria may be outcomes depending on other ambient conditions. The panel below illustrates the conceptual change in the system with increased N or decreased P. Biogeochemistry, in turn, may be altered, leading to increased sediment recycling of P. This condition may also lead to non-N_2 fixing cyanobacteria. Both strategies alter the environmental suitability for certain types of species. Image modified from Glibert (2015, Aquat Ecosyst Health Manag 18:334–341)

composition and associated biogeochemistry (Fig. 29.4). If P is reduced without concomitant N reductions, chlorophyll *a* (chl *a*) may indeed decline, and light penetration and benthic productivity may increase. However, this may open the possibility for invasive species and for major changes in the biogeochemistry of the sediment. Alternatively, if N only is reduced, there may be an increase in N_2-fixing cyanobacteria and a shift in phytoplankton dominance. Both of these trajectories argue for a balanced approach.

Those who promote a P-only reduction strategy have argued that many of the N_2-fixing cyanobacteria that thrive in a low N environment are toxic (e.g., Schindler et al. 2008, 2016; Fig. 29.4 upper panel). This may occur in some systems, no doubt. However, their argument can be countered by the observation of increasing abundance of the most common freshwater cyanobacterium, *Microcystis*, that is not

N_2-fixing (Glibert et al. 2014). The occurrence of *Microcystis* is increasing in many freshwaters around the world in direct proportion to increasing N loads. The clearest example of this is Lake Tai (Taihu) in China where *Microcystis* blooms are now a year-round phenomenon and the increasing frequency of their occurrence over the past three decades is directly proportional to the increasing input of N fertilizer and the increasing N:P of nutrient loads (Glibert et al. 2014; Chap. 25). In the Philippines, increased *Microcystis* has been associated with high N loading but not P loading (Baldia et al. 2007), and in the Huron River, Michigan, USA, *Microcystis* has been associated with molar water-column N:P ratios ranging from 40 to 80 (Lehman 2007). Similar relationships were reported for a field survey of Hirosawa-no-ike Pond, Kyoto, Japan (Ha et al. 2009). Many other examples of increasing *Microcystis* blooms are also supportive of the importance of increasing N loads (e.g., Gobler et al. 2007; Moisander et al. 2009, Davis et al. 2010; Glibert et al. 2011 and references therein, Glibert 2017). Thus, reducing P only does not solve this problem and may, in fact, contribute to it.

The resource ratio theory (Tilman 1977, 1982, 1985; Smayda 1990, 1997; see Chap. 18) predicts that as ratios of different essential elements change, the community structure will change due to competition between algae with different optimal nutrient ratios. Diatoms, being generally considered a faster growing, "*r*" selected group, and with their comparatively low optimal N:P ratio (Hillebrand et al. 2013), would be expected to be outcompeted by species with higher optimal N:P (e.g., chlorophytes, cyanobacteria, dinoflagellates) if N:P in the environment increases. Optimal nutrient ratios give clues as to changes in community assembly as nutrient composition changes.

Many cyanobacteria and marine dinoflagellate HABs (among other HAB taxa) are also more toxic when N is in stoichiometric excess over P. Thus, contrary to dogma that control of P only is required to avoid promotion of toxic cyanobacteria under N-limiting conditions, it must be recognized that the toxicity of many HABs also increases at the opposite end of the stoichiometric spectrum, when N availability is high relative to P. The evidence for toxin production increasing under non-stoichiometrically balanced conditions is considerable. Most notably, excess N over P availability has been related to microcystin production under controlled chemostat conditions and in natural populations (Van de Waal et al. 2009; Harris et al. 2016). In Daechung Reservoir, Korea, *Microcystis* toxicity was related not only to an increase in N in the water, but also to cellular N content (Oh et al. 2001). In the dinoflagellate *Alexandrium tamarense*, saxitoxin production increased by three- to fourfold under high N:P conditions (Boyer et al. 1987; Guisande et al. 2002, reviewed by Granéli 2005; Granéli and Flynn 2006; Glibert and Burkholder 2011; Fig. 29.5). Similarly, toxin production by the dinoflagellates *Gymnodinium catenatum*, *Alexandrium excavatum*, and the diatom *Pseudo-nitzschia multiseries* also increased with increasing N:P (Granéli and Flynn 2006). Many toxins are rich in N and accordingly N-rich toxins can accumulate in excess under P limitation (e.g., Granéli and Flynn 2006; Van de Waal et al. 2014). These examples do not negate the

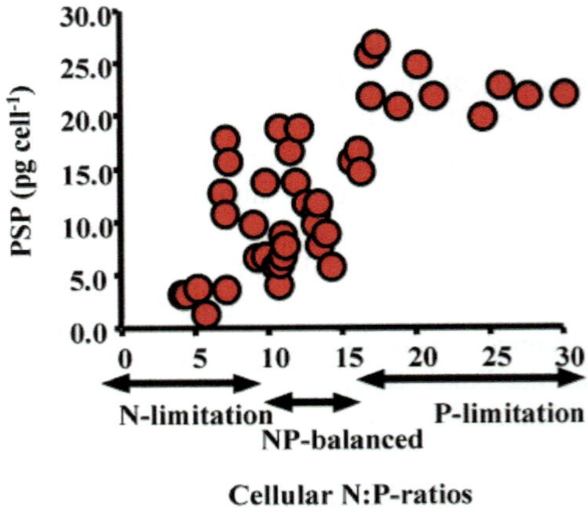

Fig. 29.5 Relation between toxicity and nitrogen to phosphorus ratio for the paralytic shellfish toxin-producing species *Alexandrium tamarense*. Image reproduced from Granéli (2005, www.ut.ee/~olli/eutr/), open access

fact that some taxa are indeed toxic when N availability is low relative to stoichiometric conditions. Nutrient imbalanced conditions, whether from excess or limiting nutrients relative to balanced conditions are the conditions most likely for toxic species to proliferate or for taxa to become more toxic (e.g., Glibert and Burkholder 2011; Van de Waal et al. 2009).

Given the current rates of export of N to aquatic system (and the fact that N:P in nutrient export is increasing globally), there is much that can be done to reduce N without driving systems into extreme N limitation. With stoichiometric ratios many-fold higher than Redfield proportions (see Chap. 11), even moderate N reduction measures may be beneficial without driving systems into extreme N limitation. Finally, it must be recognized that regulation of P without controlling N can also displace the effects of eutrophication in space or time; in estuaries, for example, the excess N not taken up at the point of discharge may promote blooms in place or it may be displaced downstream where it may eventually intercept adequate levels of P from other sources and other blooms can occur there (Glibert et al. 2011). Such effects have been documented for the Neuse River Estuary, North Carolina, USA (Paerl et al. 2004), the mesohaline reach of Chesapeake Bay (Fisher et al. 1992), among many other regions.

The need for dual nutrient control has been increasingly recognized (Boesch 2019; Conley et al. 2009; Glibert and Burford 2017; Paerl 2009; Wurch et al. 2019; Wurtsbaugh et al. 2019). Agencies throughout Europe and the USA are now either enacting or considering dual nutrient reductions. Not only are the stoichiometric arguments being accepted, but so too is the understanding that limiting nutrients vary from fresh to marine systems and may vary within systems spatially or temporally. Biogeochemistry changes when salinity changes. Also, most systems are poised to be co-limited by major nutrients. Bioassay experiments have revealed that biomass typically responds to additions of both N and P (Paerl et al. 2016). A meta-analysis undertaken by Elser et al. (2007) of 485 bioassays from coastal waters reported that both phytoplankton and periphyton were more limited by N than by P, but like the freshwater studies, additions of N + P usually acted synergistically. Another meta-analysis by Downing et al. (1999) also found that phytoplankton in estuaries, harbors, bays, and land-locked seas were more frequently stimulated by N than by P.

McCrackin et al. (2017) undertook a meta-analysis of 89 eutrophication recovery efforts. This study included control efforts that employed dual management of N and P (64%), P alone (33%), and only 3% on N alone (McCrackin et al. 2017). Interestingly, water-column N concentrations responded more quickly to N reductions than they responded to P reductions because denitrification removes N, whereas internal recycling of legacy P from the sediments can slow recovery, despite external reductions (Søndergaard et al. 2003; Vahtera et al. 2007).

Nevertheless, arguments remain about how to advance improved water quality across many systems and how to do so at a cost that society will accept. Understanding the dual roles of N and P and the importance of their co-reductions in reducing eutrophication does not make implementation of reductions equally easy. Nor does it make development of numeric N and P criteria of different types of waters easy. The high cost of nutrient reduction, the inequitable distribution of responsibility and cost, remain almost intractable challenges. The Dutch N crisis (Box 29.2) highlights how the need for reduction in N leads to major economic issues, societal anger, and political ramifications.

Box 29.2 The Dutch Nitrogen Crisis: How Dairy You Dutch Our Nitrogen!

The Dutch are farmers. They are the predominant agricultural exporter in Europe. Accordingly, they use a massive amount of N. The Dutch also have large, industrialized animal production facilities, and thus the Netherlands is a major producer of NH_3 from animal manures and wastes. The gross amount of N added to farmland in the Netherlands is three times the European average, and while it has come down in the past two decades, it still represents the second highest per-hectare load in Europe (Fig. 29.6). A series of court cases has brought the N situation in the Netherlands to a crisis. These high loads of N are negatively impacting nature preserves within the European Union, and the resulting solution (and court action) required a combination of voluntary and mandatory reductions—to reduce N emissions by 50% by 2030. To accomplish this, the agricultural sector needs to be reduced. The government proposed forced buyouts of farms, as many as 10,000 farms, and massive reductions in livestock, but this has resulted in anger, protests, and the rise of a new rural populist movement which swept to power in the most recent elections. Time will tell whether N reductions will, in fact, occur in the Netherlands.

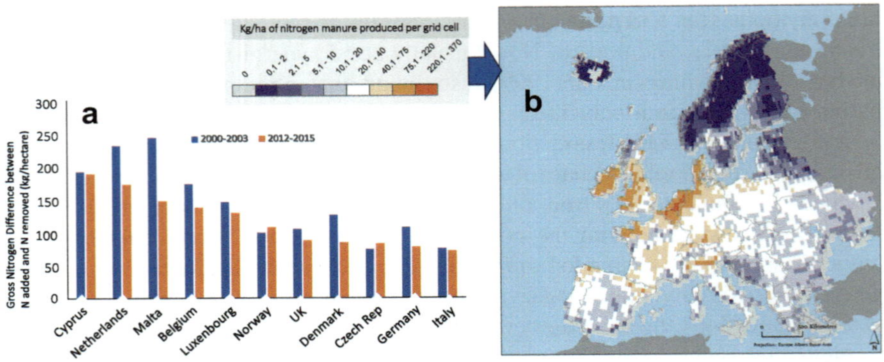

Fig. 29.6 (Panel **a**) Gross nitrogen difference between N added and N removed for the top N-producing countries in Europe for the time periods indicated. (Panel **b**) Amount of N in manure produced within 0.5 degree grid cells (kg/hectare). Panel (**a**) plotted from data from the European Environment Agency. Image (**b**) reproduced from Potter and Ranakutty (2010, https://sedac.ciesin.columbia.edu/data/set/ferman-v1-nitrogen-in-manure-production/maps) under a creative commons license

29.5 Controlling the Effects: Reducing Algal Blooms

When algal blooms impact systems and result in toxic impacts, hypoxia, or even economic impacts on communities, the goal of managers may be to reduce the bloom or its impacts—quickly, which may or may not involve a longer-term goal of ecosystem restoration. Restoration goals are important in the long term, but these measures do not solve immediate problems related to blooms. Control measures are those that focus on removing or killing the bloom organism and/or their toxins. Mitigation measures are those that involve minimizing effects with or without directly controlling the causative organism. Thus, mitigation involves also monitoring, harvesting bans, forecast models, and public education in addition to directed actions (Sengco 2009).

When public health is impacted, such as when toxins are detected in drinking water supplies or when blooms are intoxicating commercial or recreational shellfish, or when public use areas are impacted by fish kills, control measures are necessary. There are physical, chemical, and biological control methods, each of which has advantages but also disadvantages. Some methods are more targeted for individual species; some are more general-acting. These methods are reviewed below, recognizing that this overview cannot cover all treatments that are in use or under study as potential mitigation or control measures.

29.5.1 Physical Control Measures

Physical control measures are those that physically remove the offending algae or move water in such a way that dilution or other physical properties are altered, making the environment less conducive to growth of the algae. For blooms that form visible and large surface accumulations, various methods have been developed and applied to "skim" the cells from the surface. This method, of vacuuming the surface waters, is much like mowing the lawn. Material may be removed, but it likely grows back. This approach works for macroalgae, is suitable for small areas, but was also used for massive bloom removal off the Chinese coast in 2008 (Fig. 29.7). At that time, a bloom of the macroalgal species *Ulva* (also reported as *Enteromorpha prolifera* or sea lettuce) occurred at the venue of the Olympic Games sailing competition, almost blanketing the water with filamentous scum (Hu et al. 2010). One of the features of this species and its blooms is that it tends to float, making detection from remote sensing (satellite imaging) feasible. Approximately one million tons of algae were removed from the waters before the sailing regatta, requiring about 15,000 members of the Army to assist in the removal effort (Furuya et al. 2010). This bloom has re-occurred annually since, linked to the increased aquaculture production of the seaweed *Porphyra yezoensis* (or nori). Culturing nori without *U. prolifera* "contaminants" is difficult. When the nori are harvested, *U. prolifera* detaches and begins to float away. When the winds and currents are favorable, *U. prolifera* is transported into the Yellow Sea, where it can continue to access nutrients (from eutrophication of the Chinese coast) and grow (Liu et al. 2013). Physical control measures are routinely used, along with other mitigation measures, to remove these blooms (Xia et al. 2022).

Other physical measures involve mechanisms to increase turbulence, decrease stratification, or decrease residence time. Aerators, bubblers, and mixers (e.g., Hudnell 2010) are often employed in ponds and small lakes. Altering flushing times may be achieved if reservoir water can be released or if other opportunities

Fig. 29.7 Environmental impacts of the *Ulva prolifera* green tides as seen from the situation in the Qingdao coasts. Image from Y. Zhang et al. (2019a Nat Sci Rev. 6: 825–838. https://doi.org/10.1093/nsr/nwz026) under a creative commons license

exist to manipulate flow via engineering. However, increasing flushing rates may release bloom water to new regions, thereby spreading, rather than solving, the problem. This issue arises in Florida when water is released from Lake Okeechobee, water that is often laden with blooms of *Microcystis* and associated nutrients. These releases can precipitate intense blooms in the adjacent estuaries (e.g., Phlips et al. 2012).

One of the promising and effective physical mechanisms for controlling some types of blooms is the application of clay. Clay applications have been shown to be effective in both freshwater blooms (*Microcystis* and other cyanobacteria) and numerous dinoflagellates (Park et al. 2013; Sellner and Rensel 2018; Anderson et al. 2017). There is now a wealth of data on the most effective types of clay, treatment of clay (Rensel and Anderson 2004; Sengco and Anderson 2004; Yu et al. 2017), and application rates and protocols. Most effective are clays that have been modified to change the surface properties of the clay particles from negative to positive. This increases the flocculation efficiency and the rate of settling (Yu et al. 2017). Most of the experience with clay applications comes from South Korea and China, where clay applications are routinely used to protect fish cages and pens when blooms occur (Fig. 29.8). The mechanism of action is via attraction of the algae to the clay particles, then the clay-algal particles sink, thereby removing the offending algae from the surface waters. Some cells die due to lack of sunlight or due to the fact that they become buried with falling clay, but others remain viable or may encyst, thereby remaining a seed population for a future bloom. Although widely used in Asia, the use of clay as a control measure in the USA is mainly in the testing stage, as permitting for wide-scale use has been limited or unavailable for rapid response application.

Several other approaches, including UV exposure, ozonation, and ultrasound, all are effective in killing algae, but generally not practical in field deployments (Sellner

Fig. 29.8 Application of clay to bloom water in South Korea. Photo by P. Glibert

and Rensel 2018). An additional concern with these methods is that while the cells may be killed, the toxins may remain, or other organisms may also be impacted.

29.5.2 Chemical Control Measures

There are numerous chemicals that have been used to control algal blooms, and chemical control methods date back decades. One of first attempts to use a chemical control agent against the toxic dinoflagellate *Karenia brevis* took place in the 1950s, when copper sulfate ($CuSO_4 \cdot 5H_2O$) was spread along a 50 km stretch of the Florida coast (Rounsefell and Evans 1958). Quoting from the report and a subsequent addendum on this large-scale application, "The copper very quickly reduced . . .the red tide organisms, from several million to practically none per liter relieving the area of the respiratory irritation caused by the airborne toxin. . .In 2 out of the 5 areas the organisms rose again to concentrations lethal to fish in 10 to 14 days after dusting. . . .After mature consideration of the copper sulfate experiment (by an expert review panel), ... the combination of excessive cost, short duration of control, and possibility of harm to other marine life render application of the method inadvisable." However, $CuSO_4$ is still used in small ponds and lakes for control of cyanobacteria (reviewed by Sellner and Rensel 2018). Like many other approaches that kill or lyse cells, there can be a release of intracellular toxins into the water (Lam et al. 1995).

Another chemical treatment that has been employed is NH_4Cl with $CuSO_4$ or NH_4Cl with phosphoric acid (reviewed by Sellner and Rensel 2018). Needless to say, one of the problems with these approaches is the enrichment of the water with NH_3/NH_4^+ which can serve as an N source for the phytoplankton. Hydrogen peroxide (H_2O_2) has also been used with varying success. There have been successful demonstrations of use of peroxide against *Microcystis* and other algae (Lürling et al. 2014), but detrimental effects on zooplankton have also been shown in some cases. Another chemical that has been used is potassium permanganate. It has been tried experimentally in aquaculture ponds in Maryland, USA (Deeds et al. 2004), to control *Karlodinium veneficum*, but there are many concerns with its use in the natural environment, as it is an effective oxidizer of all organic matter and biota.

Other chemicals have been used, especially in freshwater, to remove cyanobacteria in drinking water. These compounds coagulate the cells, helping to remove them—and their intracellular toxins—from reservoir supplies. Coagulation, the key step in conventional drinking water treatment for pollutant removal, has been verified as an effective approach to remove intracellular toxins with intact cells (Xu et al. 2016).

One chemical agent that has been fairly extensively studied, especially in fresh waters, is the P adsorbent, Phoslock®. When added to a system and settled on the sediments, it serves as a barrier to P release, thereby reducing this necessary nutrient for further growth of cyanobacteria (e.g., Meis et al. 2012; Bishop and Richardson 2018; Ding et al. 2018; Li et al. 2021). Phoslock® has rapid adsorption kinetics but repeated applications may be necessary to effectively sustain the inhibition of algal growth. As with some other chemical methods, however, increase release of NH_4^+ may be another negative outcome of the use of Phoslock®, with its potential to further fuel blooms.

Effective reductions of blooms have been documented with applications of barley straw (*Hordeum vulgare*). In both the USA and the UK, where either bales of straw or loose straw have been added to lakes, there have been subsequent successful reductions in *Microcystis* (e.g., Newman and Barrett 1993; Brownlee et al. 2003; Sellner et al. 2015). Straw treatments take time for effects to be observed, and the time scale is on the order of days to weeks depending on the amount of straw added and the size of the bloom or impacted water body. The straw decomposes, yielding polypenolic compounds that result in inhibition of autotrophic growth via damage to cell membranes (Xiao et al. 2013; Huang et al. 2015). Effectiveness may be less in controlling brackish or marine HABs (Terlizzi et al. 2002).

29.5.3 Biological Control Measures

In recent years, there has been considerable effort in understanding and developing potential methods for application of, algicidal bacteria. Algicidal bacteria are those that inhibit the growth of, or kill, algae. Algicidal bacteria operate either directly or

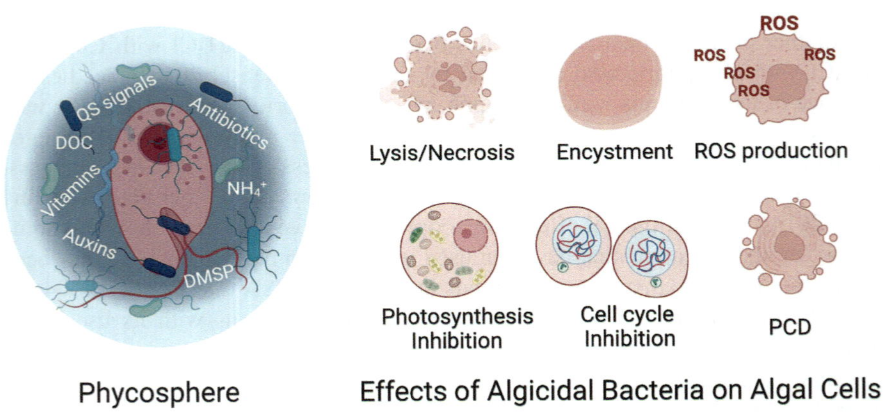

Fig. 29.9 (Left) Bacteria-algae interactions occur in the phycosphere where chemicals are exchanged. Algicidal bacteria may exert control over algal cell growth and viability through direct physical contact or indirectly through production of algicidal compounds. (Right) Algal response to algicidal bacteria varies. Algicidal interactions may cause lysis, or necrosis, in which the algal cell is lysed through chemical or physical means. Algal cells may also form temporary cysts as a defense mechanism. Chemical compounds produced by algicidal bacteria may induce production of reactive oxygen species (ROS), impact photobiology, inhibit cell cycle progression, and/or induce programmed cell death (PCD), or apoptosis, in phytoplankton. DOC, dissolved organic carbon; QS, quorum sensing; DMSP, dimethylsulfoniopropionate. Image created using BioRender.com and reproduced from Coyne et al. (2022, Front Microbiol 13:871177. https://doi.org/10.3389/fmicb.2022.871177) under a creative commons license

indirectly (Coyne et al. 2022; Fig. 29.9). Many unique algicidal bacteria have been identified against specific HABs. As examples, Shi et al. (2022) identified a novel bacterial against *Karenia mikimotoi* and C.C. Zhang et al. (2019b) identified a novel bacterium against *Microcystis aeruginosa.* A high proportion of algicidal bacteria are of the phylum *Proteobacteria* (Coyne et al. 2022). The bacteria may antagonize the alga cell via direct contact, or from a compound that may accumulate in the phycosphere (see Chap. 21). By killing the algal cell, the bacteria can gain access to the dissolved organic material (DOM) that is generated when the cell ruptures (e.g., Wang et al. 2019). These interactions can be quite complicated. When one algal cell dies, another species may become dominant, so the community may become restructured (Tilney et al. 2014). The community of bacteria is also complicated, with interactions between bacteria with a phycosphere leading to dominants that may be favorable or unfavorable. As an example, the dinoflagellate *Karenia brevis*, if maintained without its natural bacteria, loses resistance to an algicidal bacterium (Roth et al. 2008a).

Effects of the algicidal bacteria on algal cells may take a number of forms. Cells may be lysed. Algal cells may also encyst (Roth et al. 2008b). And, due to effects on metabolism and/or photochemistry, the algicidal compounds may induce production of reactive oxygen species (ROS), furthering the likelihood of cell lysis (Tan et al. 2016, reviewed by Coyne et al. 2022). Some algicidal bacteria have been shown to induce programmed cell death (PCD) responses (see also Chap. 21). Pokrzywinski et al. (2017) identified several PCD indicators, including caspase-like activity in dinoflagellates after exposure to an algicidal bacterium.

Some algicidal bacteria are effective in controlling algae due to their production of the compound prodigiosin, a family of red pigments. These compounds have shown to be effective against *Margalefidinium polykrikoides* (reported as *Cochlodinium*), *Phaeocystis globosa*, and *Microcystis aeruginosa* (M.-J. Kim et al. 2008a; Yang et al. 2013; Zhang et al. 2016). Yang et al. (2017) recently showed that prodigiosin compounds produce reactions similar to PCD; that is, they result in the production of ROS, which induces lipid peroxidation, damages the membrane system, and destroys the function of the photosystem (Fig. 29.10).

There is emerging evidence of important algicidal interactions with bacteria that are important in the field. For those interactions that require direct contact, much is being learned about the phycosphere and its role in concentrating bacteria abundance and accordingly, interactions between them (Seymour et al. 2017; Chap. 21). Late-stage blooms are also increasingly documented to have associations with algicidal bacteria (e.g., Imai et al. 2001; D. Kim et al. 2008b; Zhang et al. 2021; Shi et al. 2018). On the other hand, some algicidal bacteria isolated in the laboratory do not seem to show the same effects in the field, where concentrations are lower, and the nutrient regime also differs (e.g., Amaro et al. 2005).

Field trials of algicidal bacteria are fraught with difficulties. Not only are high concentrations of the target bacteria required, but dispersal mechanisms are difficult, and overall acceptance of introduction of a killing bacteria and potential secondary effects are of major concern. One way to do this is to immobilize the bacteria into porous matrices to prevent inadvertent dispersion (e.g., Mehrotra et al. 2021,

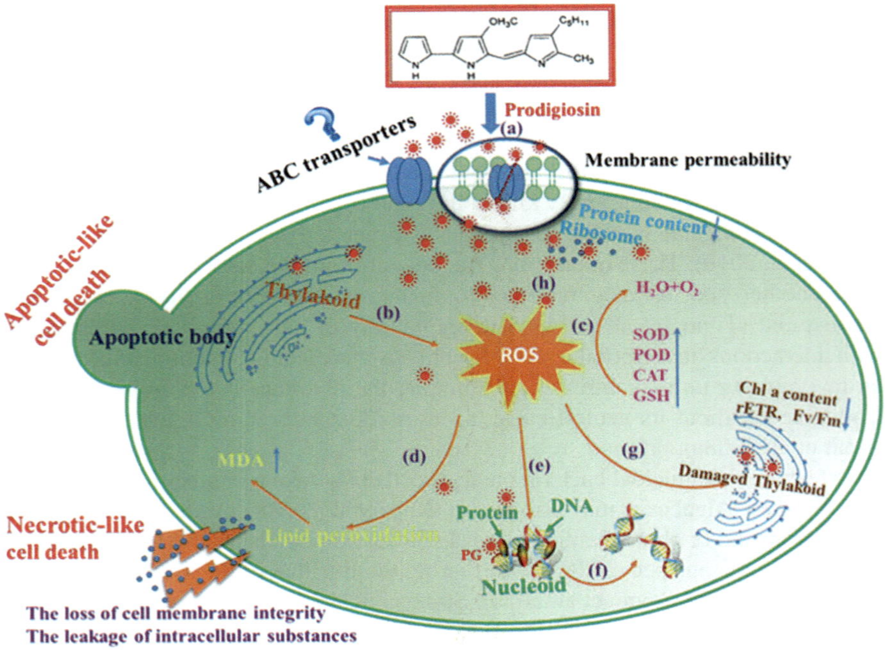

Fig. 29.10 A schematic diagram depicting the mechanism of algal cell death under prodigiosin (PG) stress. (**a**) PG may be transferred across the cell membrane by ABC transporters; (**b**) PG likely stimulates photosynthesis to generate excess ROS; (**c**) the antioxidant system is activated to scavenge ROS; (**d**) the excess ROS could lead to lipid peroxidation, and the cellular membrane permeability and integrity are damaged, which likely results in necrotic-like cell death; (**e**) PG may bind to nucleotides, especially purine (ribo)nucleosides, and the excess ROS could cause DNA damage; (**f**) the damaged DNA may be partially repaired by the nucleotide excision repair system; (**g**) the excess ROS could disrupt the thylakoid membrane and inhibit photosynthesis as the treatment time increases; (**h**) PG may target the ribosomes, which could interfere with protein synthesis. Image reproduced from Yang et al. (2017, Sci Rep 7:7750. https://doi: 10.1038/s41598-017-08132-5) under a creative commons license

reviewed by Coyne et al. 2022). Another important concern, similar to other mitigation measures described above, is the potential for toxins to be released to the environment upon cell rupture (Ndlela et al. 2018).

An interesting association between algicidal bacteria, seagrass, and several harmful taxa have been noted. Seagrass are sites with large concentrations of bacteria, especially in the complex epiphytic communities and biofilms that develop (Sakami et al. 2017).

29.6 Summary and a Comment on Issues of Scope and Scale

The techniques to control and mitigate HABs are numerous and effective—but with major caveats. Many chemical control methods are expensive and have side effects that are detrimental to other aspects of the ecosystem. Biological control agents also have issues associated with dispersal and societal acceptance of introducing bacterial control agents. Nutrient control is the ultimate control strategy, but single nutrient control measures may shift a system into a new stable state, potentially including increased toxicity, due to changing stoichiometry, or may displace effects of eutrophication in space. Remediation and restoration efforts take time. Systems do not snap back; they may take years to return to Neverland, if ever.

As the above description of methods has shown, there are many approaches for preventing, controlling, and mitigating algal blooms. Many such methods work—and have worked—for protecting fish cages and netpens for aquaculture and show great promise in ponds and small areas. Yet, virtually all of these approaches have many impracticalities when large blooms arise. Cost is one. Cost is complicated, because it is not always obvious who should bear the burden of the costs, or the burden of implementation. An agency or municipality? A commercial enterprise (e.g., an aquaculture farm)? Are monies and equipment available to enact a measure in a timely way? Another complicating fact is that regardless of the treatment application, whether chemical or biological, it is not possible to disperse control treatments over a wide area. Not only are the chemical or other agents expensive, but the infrastructure for broad scale dispersion is significant. Permitting is another issue. Permits take considerable time to obtain, usually not on the time scale of an immediate need. Some blooms cross borders, from local borders, to state borders, and even country borders. Are permitting requirements the same as a bloom moves through different locations? Does the application violate local, national, or international laws or resolutions? Some approaches are proprietary, and only available only through specific commercial enterprises. Across all methods is the concern of secondary effects. How does the public view the control measure and its potential negative effects relative to the scale of the bloom? Who makes all of the decisions? The science is advancing, but mechanisms to ensure the best science (and not pseudoscience) is applied may not always be in place.

References

Amaro AM, Fuentes MS, Ogalde SR, Venegas JA, Suárez-Isla BA (2005) Identification and characterization of potentially algal-lytic marine bacteria strongly associated with the toxic dinoflagellate *Alexandrium catenella*. J Eukaryot Microbiol 52:191–200

Amemiya T, Enomoto T, Rossberg AG, Tamamoto T, Inamori Y, Itoh K (2007) Stability and dynamical behavior in a lake-model and implications for regime shifts in real lakes. Ecol Model 206:54–62

Anderson CR, Sellner KG, Anderson DM (2017) Bloom prevention and control. In: Anderson D, Boerlage SFE, Dixon MB (eds) Harmful algal blooms (HABs) and desalination: a guide to impacts, monitoring, and management. UNESCO, Paris, France, pp 205–222

Bachmann RW, Hoyer MV, Canfield DE Jr (1999) The restoration of Lake Apopka in relation to alternative stable states. Hydrobiologia 394:219–232

Baldia SF, Evangelista AD, Aralar EV, Santiago AE (2007) Nitrogen and phosphorus utilization in the cyanobacterium *Microcystis aeruginosa* isolated from Laguna de Bay, Philippines. J Appl Phycol 19: 607–613

Barrie JM (1911) Peter and Wendy. Charles Scribner's Sons, p 267.

Bishop WM, Richardson RJ (2018) Influence of Phoslock® on legacy phosphorus, nutrient ratios, and algal assemblage composition in hypereutrophic water resources. Environ Sci Pollut Res Int 25(5):4544–4557. https://doi.org/10.1007/s11356-017-0832-2

Boesch DF (2019) Barriers and bridges in abating coastal eutrophication. Front Mar Sci 6:123. https://www.frontiersin.org/article/10.3389/fmars.2019.00123

Boyer GL, Sullivan JJ, Andersen RJ, Harrison PJ, Taylor FRJ (1987) Effects of nutrient limitation on toxic production and composition in the marine dinoflagellate *Protogonyaulax tamarensis*. Mar Biol 96: 123–128

Brownlee EF, Sellner KG, Sellner SG (2003) The effects of barley straw (*Hordeum vulgare*) on freshwater and brackish phytoplankton and cyanobacteria. J Appl Phycol 15:525–531

Conley DJ, Pearl HW, Howarth RW, Boesch DF, Seitzinger SP, Havens KE, Lancelot C, Likens GE (2009) Controlling eutrophication: nitrogen and phosphorus. Science 323:1014–1015. https://doi.org/10.1126/science.1167755

Coyne KJ, Wang Y, Johnson G (2022) Algicidal bacteria: a review of current knowledge and applications to control harmful algal blooms. Front Microbiol 13:871177. https://doi.org/10.3389/fmicb.2022.871177

Davis TW, Harke MJ, Marcoval MA, Goleski J, Orano-Dawson C, Berry DL, Gobler CJ (2010) Effects of nitrogenous compounds and phosphorus on the growth of toxic and non-toxic strains of *Microcystis* during cyanobacterial blooms. Aquat Microb Ecol 61: 149–162

Deeds JR, Mazzaccaro AP, Terlizzi DE, Place AR (2004) Treatment options for the control of an ichthyotoxic dinoflagellate in an estuarine aquaculture facility: a case study. In hall S, Anderson D, Kleindinst J, Zhu M, Zou Y (eds), Proceedings of the second international conference on harmful algae management and mitigation, Asia-Pacific economic cooperation, APEC Publ. #204-MR-04.2, Singapore, pp 177–181

Ding S, Sun Q, Chen X, Liu Q, Wang D, Lin J, Zhang C, Tsang DC (2018) Synergistic adsorption of phosphorus by iron in lanthanum modified bentonite (Phoslock®): new insight into sediment phosphorus immobilization. Water Res 134:32–43

Downing JA, Osenberg CW, Sarnelle O (1999) Meta-analysis of marine nutrient-enrichment experiments: variation in the magnitude of nutrient limitation. Ecology 80:1157–1167. https://doi.org/10.1890/0012-9658(1999)080[1157:maomne]2.0.co;2

Duarte CM, Conley DJ, Carstensen J, Sánchez-Camacho M (2009) Return to Neverland: shifting baselines affect eutrophication restoration targets. Estuar Coasts 32:29–36. https://doi.org/10.1007/s12237-008-9111-2

Elser JJ, Bracken MES, Cleland EE, Gruner DS, Harpole WS, Hillebrand H Ngai JT et al. (2007). Global analysis of nitrogen and phosphorus limitation of primary producers in freshwater, marine and terrestrial ecosystems. Ecol Lett, 10: 1135–1142. https://doi.org/10.1111/j.1461-0248.2007.01113.x

Fisher TR, Peele ER, Ammerman JW, Harding LW Jr (1992) Nutrient limitation of phytoplankton in Chesapeake Bay. Mar Ecol Prog Ser 82:51–63

Furuya K, Glibert PM, Zhou RR (2010) GEOHAB Asia. Global ecology and oceanography of harmful algal blooms in Asia: a regional comparative programme. IOC and SCOR, Baltimore and Paris, p 69

Glibert PM (2015) More than propagule pressure: successful invading algae have physiological adaptations suitable to anthropogenically changing nutrient environments. Aquat Ecosyst Health Manag 18:334–341

Glibert PM (2017) Eutrophication, harmful algae and biodiversity—challenging paradigms in a world of complex nutrient changes. Mar Pollut Bull 124:591–606. https://doi.org/10.1016/j.marpolbul.2017.04.027

Glibert PM, Burford MA (2017) Globally changing nutrient loads and harmful algal blooms: recent advances, new paradigms, and continuing challenges. Oceanography 30:58–69. https://doi.org/10.5670/oceanog.2017.110

Glibert PM, Burkholder JM (2011) Eutrophication and HABs: strategies for nutrient uptake and growth outside the Redfield comfort zone. Chin J Oceanol Limnol 29:724–738

Glibert PM, Azanza R, Burford M, Furuya K, Abal E, Al-Azri A, Al-Yamani F et al (2008) Ocean urea fertilization for carbon credits poses high ecological risks. Mar Pollut Bull 56:1049–1056

Glibert PM, Fullerton D, Burkholder JM, Cornwell JC, Kana TM (2011) Ecological stoichiometry, biogeochemical cycling, invasive species, and aquatic food webs: San Francisco estuary and comparative systems. Rev Fish Sci 19(4):358–417. https://doi.org/10.1080/10641262.2011.611916

Glibert PM, Maranger R, Sobota DJ, Bouwman L (2014) The Haber Bosch–harmful algal bloom (HB–HAB) link. Environ Res Lett 9(10):13. https://doi.org/10.1088/1748-9326/9/10/105001

Gobler CJ, Davis TW, Coyne KJ, Boyer GL (2007) The interactive influences of nutrient loading and zooplankton grazing on the growth and toxicity of cyanobacteria blooms in a eutrophic lake. Harmful Algae 6:119–133

Goldman E (2004) Recovering resilience: can restoration bring back the Bay's buffers? Chesapeake Quarterly 3:6

Granéli E (2005) Harmful algal blooms. In: Wassmann P, Olli K (eds), Drainage basin inputs and eutrophication: an integrated approach. University of Tromso, Norway, pp. 99–112. www.ut.ee/~olli/eutr/

Granéli E, Flynn K (2006) Chemical and physical factors influencing toxin content. In: Granéli E, Turner J (eds) Ecology of harmful algae. Springer, Heidelberg, pp 229–241

Guisande C, Frangópulos M, Maneiro I, Vergara AR, Riveiro I (2002) Ecological advantages of toxin production by the dinoflagellates *Alexandrium minutum* under phosphorus limitation. Mar Ecol Prog Ser 225: 169–176

Gunderson LH (2000) Ecological resilience-in theory and application. Ann Rev Ecol System 31: 425–439

Ha JH, Hidaka T, Tsuno H (2009) Quantification of toxic *Microcystis* and evaluation of its dominance ratio in blooms using real-time PCR. Environ Sci Technol 43: 812–818

Hagy JD, Boynton WR, Keefe CW, Wood KV (2004) Hypoxia in Chesapeake Bay, 1950-2001: long-term change in relation to nutrient loading and river flow. Estuaries 27:634–658

Harris TD, Smith VH, Graham JL, Van de Waal DB, Tedesco LP, Clercin N (2016) Combined effects of nitrogen and phosphorus and nitrate to ammonia ratios on cyanobacterial metabolite concentrations in eutrophic midwestern USA reservoirs. Inland Waters 6:199–210

Hillebrand H, Steinert G, Boersma M, Malzahn A, Meunier C, Plum C, Ptacnik R (2013) Goldman revisited: faster-growing phytoplankton has lower N:P and lower stoichiometric flexibility. Limnol Oceanogr 58: 2076–2088. https://doi.org/10.4319/lo.2013.58.3.2076

Hu CM, Li DQ, Chen CS, Ge JZ, Muller-Karger FE, Liu JP, Yu F, He MX (2010) On the recurrent *Ulva prolifera* blooms in the Yellow Sea and East China Sea. J Geophys Res 115. https://doi.org/10.1029/2009jc005561

Huang H, Xiao X, Ghadouani A, Wu J, Nie Z, Peng C, Xu X, Shi J (2015) Effects of natural flavonoids on photosynthetic activity and cell integrity in *Microcystis aeruginosa*. Toxins 7:66–80

Hudnell HK (2010) The state of US freshwater harmful algal blooms assessments, policy and legislation. Toxicon 55:1024–1034. https://doi.org/10.1016/j toxicon.2009.07.021

Imai I, Sunahara T, Nishikawa T, Hori Y, Kondo R, Hiroishi S (2001) Fluctuations of the red tide flagellate *Chattonella* spp. (Raphidophyceae) and the algicidal bacterium *Cytophaga* sp. in the Seto Inland Sea. Japan Mar Biol 138:1043–1049

Kemp WM, Boynton WR, Adolf JE, Boesch DF, Boicourt WC, Brush G, Cornwell JC, Fisher TR et al (2005) Eutrophication in Chesapeake Bay: historical trends and ecological interactions. Mar Ecol Prog Ser 303:1–29

Kemp WM, Testa JM, Conley DJ, Gilbert D, Hagy JD (2009) Temporal responses of coastal hypoxia to nutrient loading and physical controls. Biogeosciences 6:2985–3008

Kim D, Kim JF, Yim JH, Kwon S-K, Lee CH, Lee HK (2008a) Red to red-the marine bacterium *Hahella chejuensis* and its product prodigiosin for mitigation of harmful algal blooms. J Microbiol Biotechnol 18:1621–1629

Kim M-J, Jeong S-Y, Lee, S-J (2008b) Isolation, identification, and algicidal activity of marine bacteria against *Cochlodinium polykrikoides*. J Appl Phycol 20:1069–1078

Lam AK-Y, Prepas EE, Spink D, Hrudey SE (1995) Chemical control of hepatotoxic phytoplankton blooms: implications for human health. Water Res 29:1845–1854

Lamothe KA, Somers KM, Jackson DA (2019) Linking the ball-and-cup analogy and ordination trajectories to describe ecosystem stability, resistance, and resilience. Ecosphere 10:e02629. https://doi.org/10.1002/ecs2.2629

Lehman E (2007) Seasonal occurrence and toxicity of *Microcystis* in impoundments of the Huron River, Michigan, USA. Water Res 41: 795–802

Li J, Sellner K, Place A, Cornwell J, Gao Y (2021) Mitigation of CyanoHABs using Phoslock® to reduce water column phosphorus and nutrient release from sediment. Int J Environ Res Public Health 18:13360. https://doi.org/10.3390/ijerph182413360

Liu DY, Keesing JK, He PM, Wang ZL, Shi YJ, Wang WJ (2013) The world's largest macroalgal bloom in the Yellow Sea, China: formation and implications. Estuar Coast Shelf Sci 129:2–10. https://doi.org/10.1016/j.ecss.2013.05.021

Lürling M, Meng D, Faassen EJ (2014) Effects of hydrogen peroxide and ultrasound on biomass reduction and toxin release in the cyanobacterium *Microcystis aeruginosa*. Toxins 6: 3260–3280

McCrackin ML, Jones HP, Jones PC, Moreno-Mateos D (2017) Recovery of lakes and coastal marine ecosystems from eutrophication: a global meta-analysis. Limnol Oceanogr 62:507–518. https://doi.org/10.1002/lno.10441

Mehrotra T, Dev S, Banerjee A, Chatterjee A, Singh R, Aggarwal S (2021) Use of immobilized bacteria for environmental bioremediation: a review. J Environ Chem Eng 9:10590. https://doi.org/10.1016/j.jece.2021.105920

Meis S, Spears BM, Maberly SC, O'Malle MB, Perkins RG (2012) Sediment amendment with Phoslock® in Clatto reservoir (Dundee, UK): investigating changes in sediment elemental composition and phosphorus fractionation. J Environ Manag 93:185–193

Mieczan T (2010) Periphytic ciliates in three shallow lakes in eastern Poland: a comparative study between a phytoplankton-dominated lake, a phytoplankton-macrophyte lake and a macrophyte-dominated lake. Zool Stud 49:589–600

Moisander PH, Ochiai M, Lincoff A (2009) Nutrient limitation of *Microcystis aeruginosa* in northern California Klamath River reservoirs. Harmful Algae 8: 889–897

National Academies of Sciences, Engineering, and Medicine (NAS) (2022) A research strategy for ocean-based carbon dioxide removal and sequestration. The National Academies Press, Washington, DC. https://doi.org/10.17226/26278

Ndlela LL, Oberholster PJ, Van Wyk JH, Cheng PH (2018) Bacteria as biological control agents of freshwater cyanobacteria: is it feasible beyond the laboratory? Appl Microbiol Biotechnol 102: 9911–9923

Newman JR, Barrett PRF (1993) Control of *Microcystis aeruginosa* by decomposing barley straw. J Aquat Plant Manag 31: 203–206

Ni W, Li M (2023) What drove the nonlinear hypoxia response to nutrient loading in Chesapeake Bay during the 20th century? Sci Total Environ 861:160650. https://doi.org/10.1016/j.scitotenv.2022.160650

Nielsen K (2003) Nutrient loading and consumers: agents of change in open-coast macrophyte assemblages. Proc Nat Acad Sci 100:7660–7665. https://doi.org/10.1073/pnas.0932534100

Oh H-M, Lee SJ, Kim J-H, Kim H-S, Yoon B-D (2001) Seasonal variation and indirect monitoring of microcystin concentrations in Daechung Reservoir, Korea. Appl Envir Microbiol 67:1484–1489

Officer C, Biggs R, Taft J, Cronin L, Tyler M, Boynton WR (1984) Chesapeake Bay anoxia: origin, development, and significance. Science 223:22–27

Paerl HW (2009) Controlling eutrophication along the freshwater-marine continuum: dual nutrient (N and P) reductions are essential. Estuar Coasts 32:593–601. https://doi.org/10.1007/s12237-009-9158-8

Paerl HW, Valdes LM, Joyner AR, Pehler MF, Lebo ME (2004) Solving problems resulting from solutions: evolution of a dual nutrient management strategy for the eutrophying Neuse River estuary, North Carolina. Environ Sci Technol 38:3068–3073. https://doi.org/10.1021/es0352350

Paerl HW, Scott JT, McCarthy MJ, Newell SE, Gardner WS, Havens KE, Hoffman DK et al (2016) It takes two to tango: when and where dual nutrient (N & P) reductions are needed to protect lakes and downstream ecosystems. Environ Sci Technol 50:10805–10813. https://doi.org/10.1021/acs.est.6b02575

Park TG, Lim WA, Park YT, Lee CK, Jeong HJ (2013) Economic impact, management and mitigation of red tides in Korea. Harmful Algae 30S:S131–S143

Phlips EJ, Badylak S, Hart J, Haunert D, Lockwood J, Manley H, O'Donnell K et al (2012) Climatic influences on autochthonous and allochthonous phytoplankton blooms in a subtropical estuary, St. Lucie estuary, Florida, USA. Estuar Coasts 35:335–352

Pokrzywinski KL, Tilney CL, Modla S, Caplan JL, Ross J, Warner ME, Coyne KJ (2017) Effects of the bacterial algicide IRI-160AA on cellular morphology of harmful dinoflagellates. Harmful Algae 62:127–135

Potter P, Ranakutty N (2010) Nitrogen in manure production, V1, (1994–2001) https://sedacciesincolumbiaedu/data/set/ferman-v1-nitrogen-in-manure-production/maps. Accessed 15 Aug 2023

Rensel JE, Anderson DM (2004) Effects of phosphatic clay dispersal to control harmful algal blooms in Puget Sound, Washington. In Steidinger KA, Landsberg JH, Tomas CR, Vargo GA (eds) Proceedings of the Xth international conference on harmful algae, St. Petersburg Beach, Florida. Florida Fish and Wildlife Conservation Commission, Florida Institute of Oceanography, and intergovernmental oceanographic commission of UNESCO, Paris, pp 522–524

Roth PB, Mikulski CM, Doucette GJ (2008a) Influence of microbial interactions on the susceptibility of *Karenia* spp. to algicidal bacteria. Aquat Microb Ecol 50:251–259

Roth PB, Twiner MJ, Mikulski CM, Barnhorst AB, Doucette GJ (2008b) Comparative analysis of two algicidal bacteria active against the red tide dinoflagellate *Karenia brevis*. Harmful Algae 7:682–691

Rounsefell GA, Evans JE (1958) Large-scale experimental test of copper sulfate as a control for the Florida red tide. United States Department of the Interior, Fish and Wildlife Service, Special Scientific Report-Fisheries No. 270

Sakami T, Sakamoto S, Takagi S, Inaba N, Imai I (2017) Distribution of three algicidal *Alteromonas* sp. strains in seagrass beds and surrounding areas in the Seto Inland Sea, Japan Fish Sci 83:113–121

Scheffer M, Carpenter SR (2003) Catastrophic regime shifts in ecosystems: linking theory to observation. Trends Ecol Evol 18:648–656

Scheffer M, Hosper H, Meijer ML, Moss B, Jeppesen E (1993) Alternative equilibria in shallow lakes. Trends Ecol Evol 8:260–262

Scheffer M, Szabo S, Gragnani A, van Nes EH, Rinaldi S, Kautsky N, Norberg J et al (2003) Floating plant dominance as a stable state. Proc Natl Acad Sci USA 100:4040–4045

Schindler DW, Armstrong FAJ, Holmgren SK, Brunskill GJ (1971) Eutrophication of Lake 227, Experimental Lakes area, northwestern Ontario, by addition of phosphate and nitrate. J Fish Res Board Can 28:1763–1782. https://doi.org/10.1139/f71-261

Schindler DW, Hecky RE, Findlay DL (2008) Eutrophication of lakes cannot be controlled by reducing nitrogen input: results of a 37-year whole-ecosystem experiment. Proc Natl Acad Sci USA 105:11254–11258. https://doi.org/10.1073/pnas.0805108105

Schindler DW, Carpenter SR, Chapra SC, Hecky RE, Orihel DM (2016) Reducing phosphorus to curb lake eutrophication is a success. Environ Sci Technol 50:8923–8929. https://doi.org/10.1021/acs.est.6b02204

Sellner KG, Rensel J (2018) A review of prevention, control, and mitigation options for harmful algal blooms. In: Shumway SE, Burkholder JM, Morton SL (eds) Harmful algal blooms: a compendium desk reference. Wiley-Blackwell, Hoboken, pp 433–492

Sellner KG, Place AR, Williams E, Gao Y, Van Dolah E, Paolisso M, Bowers HA, Roche S (2015) Hydraulics and barley straw (*Hordeum vulgare*) as effective treatment options for a cyanotoxin-impacted lake. In MacKenzie AL (ed) Marine and freshwater harmful algae: proceedings of the 16th international conference on harmful algae, Wellington, New Zealand, 27th–31st October 2014. Cawthron institute and International Society for the Study of harmful algae (ISSHA), Nelson, pp 218–221

Sengco MR (2009) Prevention and control of *Karenia brevis* blooms. Harmful Algae 8: 623–628

Sengco MR, Anderson DM (2004) Controlling harmful algal blooms through clay flocculation. J Eukaryot Microbiol 51:169–172

Seymour JR, Amin SA, Raina J-B, Stocker R (2017) Zooming in on the phycosphere: the ecological interface for phytoplankton-bacteria relationships. Nat Microbiol 2:17065. https://doi.org/10.1038/nmicrobiol.2017.65

Shi X, Liu L, Li Y, Xiao Y, Ding G, Lin S, Chen J (2018) Isolation of an algicidal bacterium and its effects against the harmful-algal-bloom dinoflagellate *Prorocentrum donghaiense* (Dinophyceae). Harmful Algae 80:72–79

Shi X, Zou Y, Yheng W, Liu L, Xie Y, Ma R, Chen J (2022) A novel algicidal bacterium and its effects against the toxic dinoflagellate *Karenia mikimotoi* (Dinophyceae) Microbiol Spectrum 10: https://doi.org/10.1128/spectrum.00429-22

Smayda TJ (1990) Novel and nuisance phytoplankton blooms in the sea: evidence for a global epidemic. In: Granéli E, Sundstrom B, Edler L, Anderson DM (eds) Toxic marine phytoplankton. Elsevier, New York, pp 29–40

Smayda TJ (1997) Harmful phytoplankton blooms: their ecophysiology and general relevance. Limnol Oceanogr 42:1137–1153

Søndergaard M, Jensen JP, Jeppesen E (2003) Role of sediment and internal loading of phosphorus in shallow lakes. Hydrobiologia 506–509:135–145. https://doi.org/10.1023/B:HYDR.0000008611.12704.dd

Tan S, Hu XL, Yin PH, Zhao L (2016) Photosynthetic inhibition and oxidative stress to the toxic *Phaeocystis globosa* caused by a diketopiperazine isolated from products of algicidal bacterium metabolism. J Microbiol 54: 364–375. https://doi.org/10.1007/s12275-016-6012-0

Terlizzi, DE, Ferrier MD, Armbrester EA, Anlauf KA (2002) Inhibition of dinoflagellate growth by extracts of barley straw (*Hordeum vulgare*). J Appl Phycol 14:275–280

Testa JM, Kemp WM (2012) Hypoxia-induced shifts in nitrogen and phosphorus cycling in Chesapeake Bay. Limnol Oceanogr 57:835–850. https://doi.org/10.4319/lo.2012.57.3.0835

Tilman D (1977) Resource competition between plankton algae: an experimental and theoretical approach. Ecology 58:338–348

Tilman D (1982) Resource competition and community structure in population biology. Princeton University Press, Princeton, p 296

Tilman D (1985) The resource-ratio hypothesis of plant succession. Am Nat 125:827–852

Tilney CL, Pokrzywinski KL, Coyne KJ, Warner ME (2014) Effects of a bacterial algicide, IRI-160AA, on dinoflagellates and the microbial community in microcosm experiments. Harmful Algae 39:210–222. https://doi.org/10.1016/j.hal.2014.08.001

Tátrai I, Boros G, Á, György I, Mátyás K, Korponai J, Pomogyi P, Havasi M, Kucserka T (2009) Abrupt shift from clear to turbid state in a shallow eutrophic, biomanipulated lake. Hydrobiol 620:149–161

UK House of Commons Science and Technology Committee (2010) Science and Technology Committee-fifth report: The regulation of geoengineering. https://publications.parliament.uk/pa/cm200910/cmselect/cmsctech/221/22102.htm. Accessed 23 Aug 2023

Van de Waal DB, Verspagen JM, Lürling M, Van Donk E, Visser PM, Huisman J (2009) The ecological stoichiometry of toxins produced by harmful cyanobacteria: an experimental test of the carbon-nutrient balance hypothesis. Ecol Lett 12:1326–1335

Van de Waal DB, Smith VH, Declerck SA, Stam ECM, Elser JJ (2014) Stoichiometric regulation of phytoplankton toxins. Ecol Lett 17:736–742. https://doi.org/10.1111/ele.12280

Vahtera, E, Conley DJ, Gustafsson BG, Kuoa H, Pitkänen H, Savchuk OP, Tamminen T et al. 2007. Internal ecosystem feedbacks enhance nitrogen-fixing cyanobacteria blooms and complicate management in the Baltic Sea. Ambio 36:186–194. https://doi.org/10.1579/0044-7447(2007)36[186,IEFENC]2.0.CO;2

Wang H, Wang H (2009) Mitigation of lake eutrophication: loosen nitrogen control and focus on phosphorus abatement. Prog Nat Sci 19:1445–1451

Wang YY, Li HX, Fan QL, Wei JJ, Wang XY, Jiang XX et al. (2019) Impacts of identified bacterium *Ensifer adhaerens* on *Microcystis aeruginosa* and subsequent microcystin release. Water Air Soil Pollut 230:1–17. https://doi.org/10.1007/978-3-319-23534-9_1

Willems JL (1970) Stability theory of dynamical systems. Wiley, New York

Wurch LL, Alexander H, Frischkorn KR, Haley ST, Gobler CJ, Dyhrman ST (2019) Transcriptional shifts highlight the role of nutrients in harmful brown tide dynamics. Front Microbiol 10: 1–11. https://doi.org/10.3389/fmicb.2019.00136

Wurtsbaugh WA, Paerl HW, Dodds WK (2019) Nutrients, eutrophication and harmful algal blooms along the freshwater to marine continuum. WIREs Water 6:e1373. https://doi.org/10.1002/wat2.1373

Xia Z, Yuan H, Liu J, Sun Y, Tong Y, Zhao S, Xia J et al (2022) A review of physical, chemical, and biological green tide prevention methods in the Southern Yellow Sea. Mar Pollut Bull 180: 113772. https://doi.org/10.1016/j.marpolbul.2022.113772

Xiao X, Huang H, Ge Z, Rounge TB, Shi J, Xu X, Li R, Chen Y (2013) A pair of chiral flavonolignans as novel anti-cyanobacterial allelochemicals derived from barley straw (*Hordeum vulgare*): characterization and comparison of their anti-cyanobacterial activities. Environ Microbiol 16:1238–1251

Xu H, Pei H, Xiao H, Jin Y, Li X, Hu W, Ma C et al. (2016) Behaviors of *Microcystis aeruginosa* cells during floc storage in drinking water treatment process. Sci Rep 6: 34943 https://doi.org/10.1038/srep34943

Yang F, Wei HY, Li XQ, Li YH, Li XB, Yin LH, Pu YP (2013) Isolation and characterization of an algicidal bacterium indigenous to Lake Taihu with a red pigment able to lyse *Microcystis aeruginosa*. Biomed Environ Sci 26:148–154

Yang K, Chen Q, Zhang D, Zhang H, Lei X, Chen Z, Li Y, et al. (2017) The algicidal mechanism of prodigiosin from *Hahella* sp. KA22 against *Microcystis aeruginosa*. Sci Rep 7:7750. https://doi.org/10.1038/s41598-017-08132-5

Yu Z, Song X, Cao X, Liu Y (2017) Mitigation of harmful algal blooms using modified clays: theory, mechanisms, and applications. Harmful Algae 69:48–64

Zhang CC, Massey IY, Liu Y, Huang FY, Gao RH, Ding M (2019b) Identification and characterization of a novel indigenous algicidal bacterium *Chryseobacterium species* against *Microcystis aeruginosa*. J Toxicol Environ Health A 82:845–853. https://doi.org/10.1080/15287394.2019.1660466

Zhang H, Peng Y, Zhang S, Cai G, Li Y, Yang X, Yang K et al. (2016) Algicidal effects of prodigiosin on the harmful algae *Phaeocystis globosa*. Front Microbiol 7:602

Zhang Y, He P, Li H, Li G, Liu J, Jiao F, Zhang J et al. (2019a) *Ulva prolifera* green-tide outbreaks and their environmental impact in the Yellow Sea, China. Natl Sci Rev 6:825–838. https://doi.org/10.1093/nsr/nwz026

Zhang YL, Chen D, Cai J, Zhang N, Li F, Li CL et al. (2021). Complete genome sequence analysis of *Brevibacillus laterosporus* bl-zj reflects its potential algicidal response. Curr Microbiol 78:1409–1417. https://doi.org/10.1007/s00284-021-02378-z

30 Postscript: Changing of the Guard

30.1 Another Look Back

This book has highlighted the contributions of many men who have contributed to plankton ecology over the decades (and centuries). It was only upon completing this effort that I realized just how many men had been highlighted (some in flattering tones, some not so much). Through my career I have been highly aware that this field has been male-dominated. In my early cruises and workshops, I was the only woman. I am old enough to have been prohibited in my early career from participating in cruises due to age-old taboos against women at sea. I am now honored to have participated in cruises with women-only scientific parties and also to have a daughter who earned her mariner's license and has spent many months at sea.

To bookend the introduction of this volume which highlighted men such as Hensen, Nathansohn, Brandt, Steeman Nielsen, Lindeman, Redfield, and many others, it is worth bringing to light some of the early pioneering women in the field and to highlight why there are comparatively fewer women. The taboo against women at sea has run deep. Perhaps best summarized by Henry Stommel of the Woods Hole Oceanographic Institution in the period of the 1950s (Eike 2021), "Part of the issue is an emotional one. For many men going to sea represents a temporary reversion to boyhood, and they don't want their glorious chance to get away from it all endangered. . ." Or, perhaps, it can best be summarized by one of the port captains at that time (quoted in Eike 2021), "There are many thoughts against the employment of women on vessels, ranging from superstition, which may be overcome by gradual adjustment to practical reasons. It would be in the interest of science to let women go to sea, but with the present ships I am against it. On one-day trips or even overnight the lack of privacy wouldn't present too serious a problem, but to throw women in with men on two to three-week cruises would be unwise. . .."

Early pioneering women had guts. As Bonatti and Crane (2012) wrote, based on the historical research of Goodrich (1890) and Ridley (2010), one of the women who dared break the taboo against women at sea was Jeanne Baret, a botanist who sailed

P. M. Glibert, *Phytoplankton Whispering: An Introduction to the Physiology and Ecology of Microalgae*, https://doi.org/10.1007/978-3-031-53897-1_30

on the French ships *La Boudeuse* and *l'Étoile*, which circumnavigated the globe in a 1676–1679 expedition led by Louis Antoine de Bougainville. She disguised herself as a man for the first year of the voyage, only to be discovered later and ultimately allowed to stay with the voyage because her captain thought she was not attractive! The rule against women at sea was formalized by the Scripps Institution of Oceanography in 1949 (Bonatti and Crane 2012), crystalizing the then-policy of "discourag[ing] them as much as possible" (Day 1999; E. Shor, Scripps Institution of Oceanography historian, pers. comm., 2010, cited in Bonatti and Crane (2012). For some institutions, women were allowed if their husbands could go with them.

Mary Sears, a most accomplished marine biologist from the 1940s to the 1970s, head of the Oceanographic Unit of the Navy's Hydrography office and for many years editor of *Deep Sea Research*, was not allowed on the Woods Hole Oceanographic Institution vessels (Cullen 2005, cited in Bonatti and Crane 2012). She was allowed on Peruvian fishing vessels, where she studied plankton and El Niño conditions. The alleged reason, again summarized by Bonatti and Crane (2012), "for not letting her and other WHOI women oceanographers go to sea was, of course, that there were 'no proper facilities' for a lady." According to Cullen (2005), Sears commented, "But there were pails" and added that, "though there were no proper facilities on the Peruvian vessels, what made the difference was that there were proper gentlemen aboard!"

A scandal erupted in 1956 at the Woods Hole Oceanographic Institution when a woman named Roberta Eike (see Eike 2021), a graduate student in marine geology, stowed away in the bilge of the R/V Caryn (Fig. 30.1). When discovered, she was confined to the captain's quarters (who had to find other quarters) and the ship returned to port immediately. She was promptly dismissed from Woods Hole with her fellowship cancelled, thus ending her career (Bonatti and Crane 2012).

Just as women were barred from going to sea at that time, women were also barred from research in remote regions such as Antarctica. Downes and Lancaster (2020) uncovered some of these reasons—or as they note, flimsy excuses. Based on a letter sent to a woman who applied in the 1960s to be part of the British Antarctic Survey, the response to her reeked of his bias (Jones 2012, quoted in Downes and Lancaster 2020), "Women wouldn't like it in Antarctica as there are no shops and no hairdresser." Echoing the sentiment of H. Stommel above, a Commander of US Antarctic operations said, "…the presence of women would wreck the illusion of the frontiersman—the illusion of [a man] being a hero" [quotation from Commander of US Antarctic operations] (Chipman 1986, p. 87, quoted in Downes and Lancaster 2020).

One of the early pioneers of freshwater ecology was Ruth Patrick. No book on plankton ecology can be complete without her mention. She was a leading voice in the first Clean Water Act. She developed theories of biodiversity and plankton structure in river systems. Ultimately receiving the national's highest honor in scientific achievement, the National Medal of Science, and for whom a National Academy of Sciences limnology center is named, as is a major award of the Association of the Sciences of Limnology and Oceanography (ASLO), she had to endure the humiliation of being offered a job at the Academy of Natural Sciences in

July 20, 1956 Four Sections — Thirty-two Pages Price Ten Cents

Girl Scientist Sails As Stowaway

Protests Ban On Women Aboard Research Craft

A petite and very earnest Radcliffe graduate student who thinks oceanography discriminates against women, challenged regulations at Woods Hole Oceanographic institution this week by stowing away on the research vessel, Caryn.

The stowaway was Miss Rogerta Eike, 23. Miss Eike is a marine biologist and she is serious about it. At Woods Hole she found herself in the position of a geologist who was never allowed on field trips. The frequent research cruises of the oceanographic vessels from Woods Hole are for men only. It was all right for Miss Eike to dabble in the laboratory. She could not go to sea. In her opinion this was more than an irksome restriction. She argued and pleaded in vain.

Monday afternoon the ketch, Caryn, put to sea for a five-day cruise over the continental shelf. Chief scientist was Dr. George E. Clarke, under whom Miss Eike had been studying marine crustacea. It seemed probable this trip of the Caryn would provide a chance to examine the so-called "scattering layer", a submarine phenomenon of particular interest to all who, like Miss Eike, are interested in tiny crustaceans. Had Miss Eike been a man she might logically turn back without accomplishing anything would be expensive. Pending orders from the institution, Chief Scientist Clarke advised Captain Munns to hold seaward. They sailed about 200 miles southeast of Woods Hole. Dr. Clarke supervised some plankton tows and they made measurements of submarine illumination.

Confined To Quarters

On Tuesday orders were radioed from Port Captain John F. Pike in Woods Hole. The stowaway was to be confined to quarters. The captain's cabin became the "brig", dispossessing him. The captain had to seek other quarters. Miss Eike emerged from duress to dine with crew and scientists. There were 16 on the Caryn, six at each table sitting in the cramped wardroom. Tuesday morning was overcast, but the sun broke through. Caryn pitched and Miss Eike became seasick. She swallowed a pill and took a short walk on deck and felt better. On Wednesday Caryn turned homeward.

Because she returned earlier than expected there was not the usual crowd of wives and children on the dock to greet Caryn's return. Institution employes who saw her enter the harbor and some Boy Scouts who scampered over from their nearby meeting

Fig. 30.1 Portion of the front page of the *Falmouth Enterprise*, July 20, 1956, with an account of the stowaway story of Rɔberta Eike. Image reproduced from Boratti and Crane (2012, Oceanography 25: 32–39. doi.org/.0.5670/oceanogr.2012.103) under a creative commons license

1934 only to be told she would not be paid (Zauzmer 2013). She was eventually paid—after 7 years—and ultimately rose to become the academy's board of trustee's first female chair. She worked at that esteemed institution every day until she turned 97. She focused on diatoms and their importance, how they changed with water quality, as well as understanding the degrading health the bodies of water, a dominant voice in the early work on water pollution. Her life's work lives on.

30.2 Progress and a Look Forward

There has been progress in diversity, but not enough.

Progress in representation of women has greatly improved over the past three decades, and women now represent approximately 50% of doctoral degrees in oceanography in the US (O'Connell 2014; Orcutt and Cetinic 2014; Wilson 2019; Lima and Rheuban 2021). It took 30-plus years, but on my last cruise, the science team was only women.

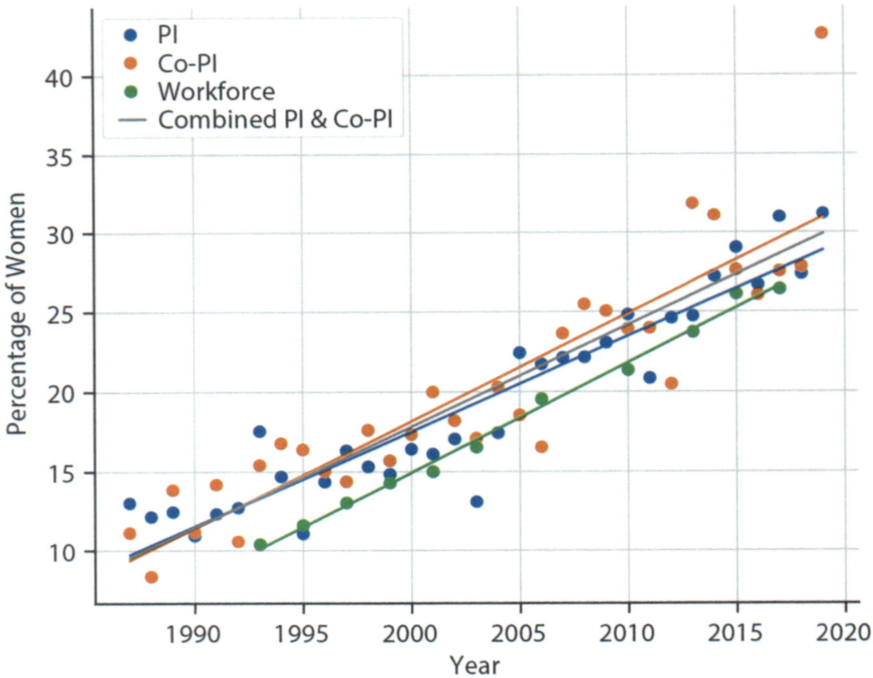

Fig. 30.2 Annual mean percentages of women principal investigators (PIs) and co-PIs in National Science Foundation Division of Ocean Sciences (NSF-OCE) awards, and of women in the academic Earth, atmospheric, and ocean sciences workforce. The lines represent the linear regression fits for each category (PI, co-PI, and academic workforce). The *gray* line represents the linear regression fit for the combined PI and co-PI data. Table 1 shows the parameter values for the different linear regression fits. The academic workforce data were obtained from NSF's Survey of Doctoral Recipients, a biennial survey conducted through the National Center for Science and Engineering Statistics (NCSES) to provide demographic information about individuals holding doctoral degrees in science, engineering, or health fields from US academic institutions (SDR; NSF 2021). Image reproduced from Lima and Rheuban (2021, Oceanography 35(1):68–75. doi.org/10.5670/oceanog.2021.401) under a creative commons license

A recent analysis of how women are represented in the portfolio of NSF awards in oceanography shows that there continues to be significant progress, but differences remain (Lima and Rheuban 2021). From 1990 to 2019, the percentage of women as principal or co-principal investigators increased from <15% to >25% (Fig. 30.2) and about the same number of women now serve as chief scientists on cruises (Orcutt and Cetinic 2014). Institutions such as Woods Hole Oceanographic Institution finally now hire women as crew on scientific vessels as well.

Within these gains, there are disparities within subdisciplines (Fig. 30.3). Clustering award statistics by themes or broad topics, Lima and Rheuban (2021) show that highest mean net increases in women's success with NSF as principal or co-principal investigators include workshop science international (8) and student

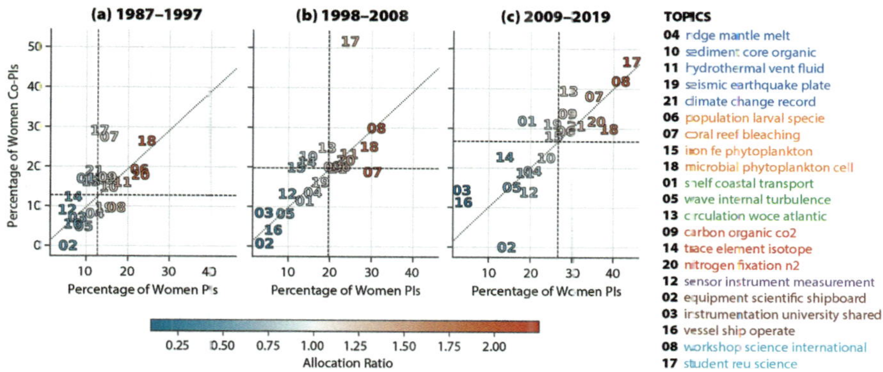

Fig. 30.3 Scatter plot of the percentages of women PIs vs co-PIs for the 21 extracted research topics in three different 11-year periods: (**a**) 1987–1997, (**b**) 1998–2008, and (**c**) 2009–2019. The *dotted diagonal line* represents the 1:1 ratio. The vertical and horizontal *dashed lines* represent the mean percentage of women in the academic workforce for the time period computed from the linear trend fitted to the combined PI and co-PI data for all awards. The topics in the upper right quadrant of each panel have higher women's participation than the academic workforce for both PIs and co-PIs, while those in the lower left quadrant have lower participation than the workforce for both PIs and co-PIs. For the topics in the lower right quadrant, women's participation as PIs is higher than in the workforce but participation as co-PIs is lower than in the workforce. Conversely, for topics in the upper left quadrant, women's participation as co-PIs is higher than in the workforce but participation as PIs is lower than in the workforce. The *marker* color represents the allocation ratio. An allocation ratio greater than one means that women PIs assign a higher proportion of their awards to that topic than men do, and vice versa, and provides an indicator of the relative proportion of women to men PIs in that particular topic. The topic legends are grouped and colored according to the NSF-OCE program. *Blue* represents topics in marine geology and geophysics and the Ocean Drilling Program, *orange* represents topics in biological oceanography, *green* represents topics in physical oceanography, *red* represents topics in chemical oceanography, *purple* represents topics associated with instrumentation development (Oceanographic Technology and Interdisciplinary Coordination and Major Research Instrumentation program), *brown* represents topics related to ship instrumentation and operations (ocean instrument and shipboard scientific support equipment), and *cyan* represents topics in education and human resources. Image reproduced from Lima and Rheuban (2021, Oceanography 35(1):68–75. doi.org/10.5670/oceanog.2021.401) under a creative commons license

undergraduate research opportunities (REU) (17). Also included in the topics in the highest cluster are coral reef bleaching (07), climate change (21), carbon organic CO_2 (09), and nitrogen fixation (20). In contrast, the topics with the smallest net gains are equipment scientific shipboard (2), vessel ship operations (16), instrumentation university shared (3), and hydrothermal vent fluid (11). The topics that showed a net decrease in the percentage of women PIs between 1987–1997 and 2009–2019 include vessel ship operate (16) and instrumentation university shared (3). There is hope that the next book to be written on phytoplankton physiological ecology will highlight the increasing contribution of women. There is no lack of women whose contributions could play a prominent role (see *Oceanography* volume 27(4) for example).

Progress has also been made in the acceptance of dual career couples. The academic collaboration of Raymond Lindeman and his wife was recounted in Chap. 20. For years, women were not allowed to work in the field if they were married, but if they were married, they were allowed to work subserviently to their husbands. As noted by Slack (1996), "Collaborative marriages were common in botany even in the first half of the nineteenth century in England. However, many of these partnerships were highly imbalanced, as was the case, for example, of the collaborative marriage of Elizabeth G. Knight and Nathaniel Lord Britton. He, a professor of Botany at Columbia by 1891, and she a Hunter College graduate. Their collaboration was one in which he had the sole paid position, and she conducted her research unpaid. Nevertheless, she went on to publish over 300 articles and reviews, edited several journals and even supervised at least one graduate student." Theirs was a "personal and botanical partnership," but their story highlights the anti-nepotism regulations, sexism, and bias that have pervaded academia for decades. Glibert and Kana (2016), experiencing many of these biases as a dual career couple (albeit not to the same extent), complied a synthesis of many contemporary spousal collaborations in microbial ecology and biogeochemistry. As recently reported in an article in *Nature* highlighting dual career couples, and based on a National Science Foundation survey, over 25% of "married people with doctorates had a spouse working in science or engineering" (Smith 2014). Fortunately, pathways for dual career couples are improving, with many institutions recognizing that spousal collaborative teams often bring a unique and special strength to a collaboration. It is now increasingly accepted that dual career couples are a fact of scientific life. Whether they work closely on a daily basis, or whether they share only a similar passion for their work but maintain distinct disciplines, couples share "in the joys and the sorrows of their life-companions, but they also have a part in their thoughts, their studies, their labors, their achievements" (Mozens 1913, cited in Pycior et al. 1996).

Where progress is lacking is broader diversity. Biologists in the geosciences are among the least diverse of the natural sciences (Bernard and Cooperdick 2018), and among the geosciences, diversity in the aquatic sciences is especially poor. Representation of people of color and with disabilities has not increased in any substantial way in the past decades (Fig. 30.4). A recent survey of graduate students in the ocean sciences showed that $<9\%$ identified as coming from a group that is considered part of a minority community in the USA (Black, Hispanic/Latinx, Native Alaskan, Native American, Pacific Islander) (Garza 2021). The statistics on the percentage of women from underrepresented minorities with PhDs in the geosciences is even more grim (Fig. 30.5). Inclusivity does not stop with sex or skin color. Inclusivity must also include those with disabilities (Solomon 2021; NSF 2021). It is a common misconception that the geosciences, "are only for those who are physically fit and able to engage in rigorous fieldwork activities (Locke 2005; Sexton et al. 2014; Atchison and Libarkin 2016). This ableist perspective places undue social discrimination on [students with disabilities]" (Carabajal et al. 2017 and references therein). This perspective sounds so much like the bias against women at sea decades ago. Change has to happen.

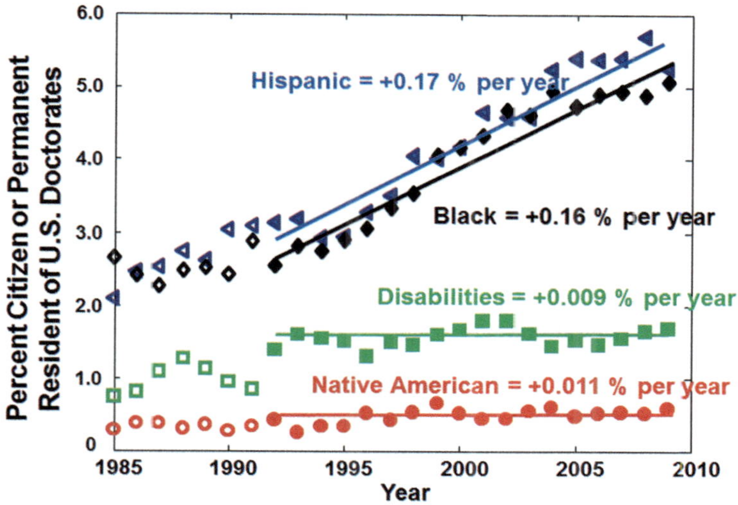

Fig. 30.4 Percent change in STEMM doctorates earned by the respective groups shown through 2006. Data compiled by University of Delaware Science and Engineering Leadership Initiative, based on Bureau of Labor Statistics

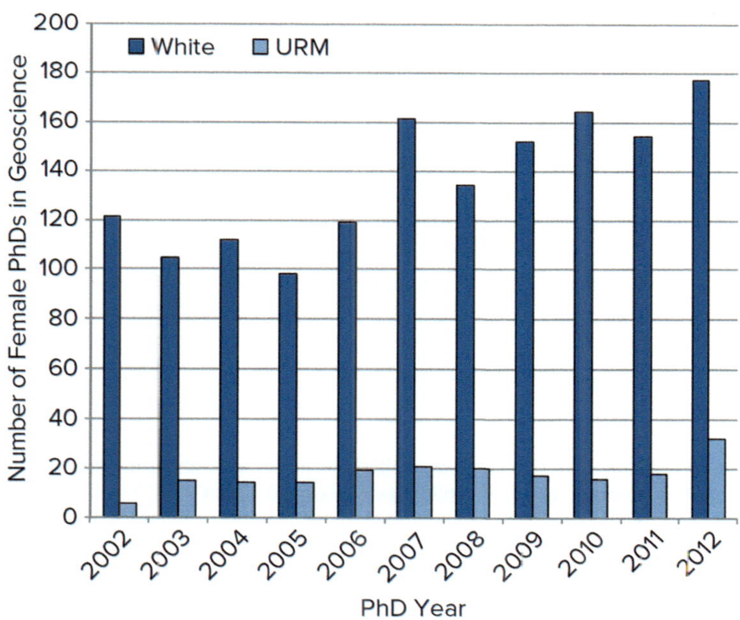

Fig. 30.5 Numbers of white and under-represented minority (URM) females (US citizens and permanent residents) receiving PhDs in Earth and Ocean Sciences. Image reproduced from O'Connell (2014, Oceanography 27(4):15–22. doi.org/10.5670/oceanog.2014.108) under a creative commons license

The presence of diverse perspectives (including gender and ethnic diversity and disability) can lead to more innovative approaches, methods, and ideas in tackling today's challenging issues. Science is not a zero-sum game. Strong evidence has emerged that more diverse research teams produce higher-impact research and result in more effective and efficient problem-solving in real-world contexts. Given the problems of the world's waters, from eutrophication, hypoxia, HABs, and altered carbon chemistry, the great minds of the world are needed. In addition, the perspectives of local communities must be heard. Often, they are the ones most vulnerable. Vulnerable communities—worldwide—impacted by rising temperatures, hurricanes, deteriorating water quality may be the least able to afford to implement mitigation measures. They have local knowledge and a real sense of the fragility of their ecosystems. They are—or should be—a critical part of any decision process (Whitty 2023). Their communities are being disrupted, their way of life threatened, and their knowledge and values need to be recognized. Many more perspectives, innovations, and ways of knowing are needed to co-create robust solutions that address complex water issues that affect everyone on Earth.

We all need to help make this happen. Individuals, institutions, journals, and societies all have a role to play. Deliberate communications and engaged communities can help in making change (e.g., Behl et al. 2021). I hope the following words from Cory Garza (2021) will not be necessary in the next volume of this or related books.

> A homogeneous community of ocean scientists that lacks representation of diverse groups limits the likelihood that the urgent challenges associated with the Anthropocene will be met, while also decreasing the likelihood that findings will be accepted by an increasingly diverse public. Now is the time for the ocean sciences to engage a new generation of diverse individuals who can bring new perspectives and approaches for dealing with the emerging needs of twenty-first century ocean science.

The only change necessary in Garza's words is that it is time for all aquatic scientists—not just ocean scientists—to engage a new generation of diverse individuals. I look forward to seeing the contributions they will make to phytoplankton ecology and beyond.

References

Atchison CL, Libarkin JC (2016) Professionally held perceptions about the accessibility of the geosciences. Geosphere 12:1154–1165. https://doi.org/10.1130/GES01264.1

Behl M, Cooper S, Garza C, Kolesar SE, Legg S, Lewis JC, White L, Jones B (2021) Changing the culture of coastal, ocean, and marine sciences: strategies for individual and collective actions. Oceanography 27(3):53–60

Bernard RE, Cooperdick EHG (2018) No progress on diversity in 40 years. Nat Geosci 11:292–295. https://doi.org/10.1038/s41561-018-0116-6

Bonatti E, Crane K (2012) Oceanography and women: early challenges. Oceanography 25:32–39. https://doi.org/10.5670/oceanogr.2012.103

Carabajal IG, Marshall AM, Atchison CL (2017) A synthesis of instructional strategies on geoscience education literature that address barriers to inclusion for students with disabilities. J Geosci Educ 65:531–541. https://doi.org/10.5408/16-211.1

Chipman E (1986) Women on the ice: a history of women in the far south. Melbourne University Press, Melbourne

Cullen V (2005) Down to the sea for science: 75 years of ocean research, education and exploration at the Woods Hole Oceanographic Institution. Woods Hole Oceanographic Institution, Woods Hole, p184

Day D (1999) Overview of the history of women at Scripps institution of oceanography. UCSD Women's Center Panel, 5 pp. Available at: http://scilibucsdedu/sio/hist/day_overview_women_siopdf. Accessed 4 Sep 2022

Downes BJ, Lancaster J (2020) Celebrating women conducting research in freshwater ecology. . .and how the citation game is damaging them. Mar Freshw Res 71:139–155

Eike R (2021) Box 1. Woman and oceanography. Oceanography 25(4):36

Garza C (2021) Diversifying the ocean sciences: thoughts on the challenge ahead. Oceanography 34(2):184–185. https://doi.org/10.5670/oceanog.2021.204

Glibert PM, Kana TM (eds) (2016) Aquatic microbial ecology and biogeochemistry: a dual perspective. Springer, Cham, 300

Goodrich FB (1890) The history of the sea. Edgewood, Hubbard Bros, p 682

Jones B (2012) Women won't like working in Antarctica as there are no shops and hairdressers. In The Telegraph, 20 May 2012. Available at www.telegraph.co.uk/news/earth/environment/9260864/Women-wont-like-working-in-Antarctica-as-there-are-no-shops-and-hairdressers.html. Accessed 13 March 2023

Lima ID, Rheuban JE (2021) Gender differences in NSF Ocean sciences awards. Oceanography 35(1):68–75. https://doi.org/10.5670/oceanog.2021.401

Locke, S.M. 2005. The status of persons with disabilities in the geosciences. White Paper. New Mexico State University, Las Cruces, NM

Mozens HJ (1913) Women in science. University of Notre Dame Press, Cambridge, MA, p 412

National Science Foundation (NSF) and National Center for Science and Engineering Statistics (NCSES) (2021) Diversity and STEM: women, minorities, and persons with disabilities. National Science Foundation, Alexandria VA. https://ncses.nsf.gov/pubs/nsf23315/

O'Connell S (2014) Women of the academy and the sea: 2000–2014. Oceanography 27(4):15–22. https://doi.org/10.5670/oceanog.2014.108

Orcutt B, Cetinic I (2014) Women in oceanography: continuing challenges. Oceanography 27(4): 5–13. https://doi.org/10.5670/oceanog.2014.106

Pycior HM, Slack NG, Abir-Am PG (eds) (1996) Creative couples in the sciences. Rutgers University Press, New Brunswick, p 369

Ridley G (2010) The discovery of Jeanne Baret: a story of science, the high seas, and the first woman to circumnavigate the globe. Crown Publishers, New York, p 304

Sexton JM, Connell SO, Banning J (2014) Characteristics and culture of geoscience departments as interpreted from their website photographs. J Women Minorities Sci Eng 20(3):257–278. 10.1615/JWomenMinorScienEng.2014009751

Slack NG (1996) Botanical and ecological couples: a continuum of relationships. In: Pycior HM, Slack NG, Abir-Am PG (eds) Creative couples in the sciences. Rutgers University Press, New Brunswick, pp 235–253

Smith K (2014) Love in the lab: close collaborations. Nature 510:460

Solomon C (2021) An often overlooked element of diversity: disability. Limnol Oceanogr Bull 30: 20–21. https://doi.org/10.1002/lob.10425

Whitty TS (2023) Oceans and communities. In: Obaidullah F (ed) The ocean and us. Springer, Cham, pp 257–266

Wilson C (2019) U.S. female geoscience degree rates in atmospheric science, geography, geoscience and ocean science, 2000–2012. American Geosciences Institute, https://www.americangeosciences.org/geoscience-currents/us-female-geoscience-degree-rates-atmospheric-science-geography-geoscience-and

Zauzmer J (2013) Ruth Patrick, ecology pioneer, dies at 105. The Washington Post, 23 Sep 2013

GPSR Compliance

The European Union's (EU) General Product Safety Regulation (GPSR) is a set of rules that requires consumer products to be safe and our obligations to ensure this.

If you have any concerns about our products, you can contact us on ProductSafety@springernature.com

In case Publisher is established outside the EU, the EU authorized representative is:

Springer Nature Customer Service Center GmbH
Europaplatz 3
69115 Heidelberg, Germany

Batch number: 10732023

Printed by Printforce, the Netherlands